AF332941

Proceedings of the 2005 Marseille Singularity School and Conference

SINGULARITY THEORY

Dedicated to

Jean–Paul Brasselet
on his 60th birthday

Proceedings of the 2005 Marseille Singularity School and Conference

SINGULARITY THEORY

Dedicated to

Jean-Paul Brasselet
on his 60th birthday

CIRM, Marseille, France   24 January – 25 February 2005

Editors

Denis Chéniot
Nicolas Dutertre
Claudio Murolo
David Trotman
University of Provence, France

Anne Pichon
University of Méditerranée, France

World Scientific

NEW JERSEY · LONDON · SINGAPORE · BEIJING · SHANGHAI · HONG KONG · TAIPEI · CHENNAI

Published by

World Scientific Publishing Co. Pte. Ltd.

5 Toh Tuck Link, Singapore 596224

USA office: 27 Warren Street, Suite 401-402, Hackensack, NJ 07601

UK office: 57 Shelton Street, Covent Garden, London WC2H 9HE

British Library Cataloguing-in-Publication Data
A catalogue record for this book is available from the British Library.

First published 2007
Reprinted 2008

SINGULARITY THEORY
Dedicated to Jean-Paul Brasselet on His 60th Birthday
Proceedings of the Singularity School of Conference

ISBN-13 978-981-270-410-8
ISBN-10 981-270-410-8

Printed in Singapore by Mainland Press Pte Ltd

From left to right:
Denis Chéniot, Claudio Murolo, Anne Pichon, David Trotman and Nicolas Dutertre

Introduction

The Singularity School and Conference was organized from 24 January to 25 February 2005, in the CIRM, Centre International de Rencontres Mathématiques de Luminy, Marseille, France, as a Special School of the Formation Permanente of the French CNRS. Organizers of the five weeks were successively Anne Pichon, Jean-Paul Brasselet, David Trotman, Nicolas Dutertre and Claudio Murolo, Denis Chéniot. Around 200 mathematicians from 31 countries participated in this very successful event.

The five weeks were organized in the following way : Elementary and advanced courses were given during the first week by high level specialists of the subject : Jean-Paul Brasselet, *Characteristic classes for Singular Varieties*, Herwig Hauser, *Basic techniques for resolution of singularities*, Anatoly Libgober, *Topology of the complements to hypersurfaces in projective space*, David Trotman, *Stratifications of subanalytic and semialgebraic sets*, Vladimir Zakalyukin, *Lagrangian and Legendrian singularities*. The second week was devoted to general lectures on singularity theory. During the third week users of singularity theory gave lectures on applications of the theory to various domains: robot manipulation, visualization techniques, medical imaging, nerve cells, resonance tongues, gravitational lensing, etc... During the fourth week young researchers explained their recent results and during the last week lectures were concentrated on geometry and topology in singularity theory.

The purpose of the event was to cover recent developments in the area and to introduce young researchers to singularities in geometry and topology.

The book is a collection of courses, surveys and new results in singularity theory and its applications presented during the five weeks. It provides a focus for senior researchers and postgraduate students, summarizing the current main ideas in singularity theory and suggesting future research directions in the area.

The main interesting features of the papers presented for the Proceedings of the Singularity School and Conference are: the elementary and advanced courses in the first part of the book, and the papers on applications of singularity theory in the second part, which is an original point in comparison with classical proceedings on the subject. The third part contains surveys providing development and evolution of the most important themes of the theory from the beginning up to now, and research papers on important recent results in the field. They represent the current problems under investigation in the area and the present state of the art of the subject.

All papers published in the present book have been refereed.

The volume is complementary to the volume published by World Scientific Publishing for the School and Workshop on Singularity theory held in Trieste, 15 August - 3 September 2005.

The event was made possible with the help of the CNRS, Centre National de la Recherche Scientifique, especially the Formation Permanente of CNRS and local institutions of the CNRS: FRUMAM, IML and LATP, the CIRM, the French Ministry of Research and Technology, the French Ministry of Foreign Affairs, the Science Faculty of the Université de la Méditerranée, the University of Provence, the Town of Marseille, France, the Conseil Général des Bouches-du-Rhône, France. The publication of the present volume was made possible with the help of the Conseil Régional of Provence-Alpes-Côte d'Azur. The organizers thank particularly those participants who funded themselves partially or totally .

It is a great pleasure to thank all the speakers and the participants whose presence was the real success of the School and Conference. We are also very grateful to the staff of the IML and the CIRM for their help in organizing the event. Finally, our thanks go to the editorial staff of World Scientific Publishers for their patience and efficient help with publication process of the volume.

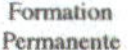

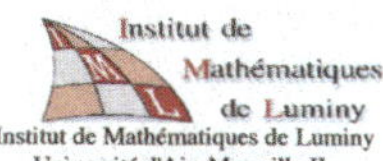
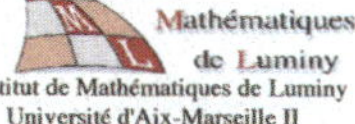

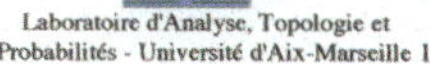

| Formation Permanente | Fédération de Recherche des Unités de Mathématiques de Marseille | Institut de Mathématiques de Luminy Université d'Aix-Marseille II | Laboratoire d'Analyse, Topologie et Probabilités - Université d'Aix-Marseille I | CIRM |

5 Semaines de Singularités ❦ 5 Weeks of Singularities
CIRM - Luminy - Marseille
24/01/2005 - 25/02/2005

24-28 January	31 January-4 February	7-11 February	14-18 February	21-25 February
Introductory Courses on Singularity Theory	*General Conference on Singularities*	*Applications of Singularities*	*Young Singularists*	*Geometry and Topology of Singularities*

Centre International de Rencontres Mathématiques

Organizing Committee

Jean-Paul Brasselet *IML - Marseille*
Denis Chéniot *LATP - Marseille*
Nicolas Dutertre *LATP - Marseille*
Victor Goryunov *Univ. of Liverpool*
Lê Dung Tráng *ICTP - Trieste*
Françoise Michel *Univ. of Toulouse*
Claudio Murolo *LATP - Marseille*
Anne Pichon *IML - Marseille*
Dirk Siersma *Univ. of Utrecht*
David Trotman *LATP - Marseille*

Scientific Committee

Maria Emilia Alonso *Madrid*
Georges Comte *Nice*
Jim Damon *North Carolina*
Peter Giblin *Liverpool*
Helmut Hamm *Münster*
Stanisław Janeczko *Warsaw*
Ignacio Luengo *Madrid*
Alejandro Melle *Madrid*
Donal O'Shea *Mount Holyoke*
Mutsuo Oka *Tokyo*
Claudio Procesi *Rome*
Bernard Teissier *Paris*
Terry Wall *Liverpool*
Claude Weber *Geneva*
Yosef Yomdin *Weizmann Inst*

PRE-INSCRIPTION / PRE-REGISTRATION FORM : **http://frumam.cnrs-mrs.fr/**
INFORMATIONS : France BODIN, F.R.U.M.A.M. - FR CNRS 2291, Parc Scientifique de Luminy - Case 909
13288 - Marseille Cedex 9, FRANCE - Tel./Fax : 00 33 (0)4 91 26 95 62
E-mail : **bodin@frumam.cnrs-mrs.fr**

Jean-Paul Brasselet

This volume is dedicated to our friend and colleague Jean-Paul Brasselet, who reached the round age of 60 in March 2005. We recall here some of his achievements, in particular his influence on research in singularities in Marseille.

Until 1991 Jean-Paul was at the University of Lille 1, directing the research laboratory from 1985 to 1991 and was instrumental in the creation and construction of a fine new library for the mathematics department. He was one of Henri Cartan's last research students, defending his Thèse d'Etat in 1977 (directed jointly by Cartan and Shih Weishu), and was also strongly influenced by Marie-Hélène Schwartz, his colleague at Lille, who had been working on extending the theory of characteristic classes to singular analytic spaces since the early 50's (and has continued to do so almost to this day - at the age of 87, she published a 216 page research monograph in the year 2000 on Chern classes for analytic spaces, and is thus an inspiration to us all !). The theory of characteristic classes for singular spaces became the dominant theme in Jean-Paul's research. His lecture notes in the present volume provide an introduction to the theory and bear witness to the longevity of his efforts and thus to the richness of the theory.

An unusual feature of his latter activity in Lille was a collaboration with the French National Railways, the SNCF, applying differential geometry to the design of TGV tracks between Paris and Lille. Before leaving for Marseille to become Director of the CIRM, and of the IML, he organised a major singularities conference in Lille with over 230 participants and traditional giant figures, and edited the proceedings in a volume published as volume 201 of the London Mathematical Society Lecture Notes.

He organised another huge singularities conference in Marseille in 1993 (around 150 participants) as well as collaborating in running regular Marseille-Nice meetings, up to 6 a year, from 1991 to 2003. Three doctoral students accompanied Jean-Paul to Marseille - Belkacem Bendiffalah, Michal Kwiecinski and Jianji Zhou. The three made entertaining speeches at the celebration dinner in Marseille in honour of Jean-Paul. The 90's saw a regular expansion of activity in singularity theory in Marseille: Jean-Paul and his students joining David Trotman with his students Laurent Noirel, Stéphane Simon, and then Claudio Murolo, Georges Comte, Dwi Juniati and Guillaume Valette. Denis Chéniot arrived from Nice in 1993 and soon acquired a student in Christophe Eyral, while in 1994 Lê Dũng Tráng came from Paris with his students including Jawad Snoussi and Meral Tosun, and later Caroline Ausina and Eric Akeke. Hélène Maugendre was appointed Maître de Conférences in 1996, Anne Pichon in 1998, Claudio

Murolo in 2000 and Nicolas Dutertre in 2002. In the creation of a stimulating research environment for this group of mostly young researchers to develop and thrive, Jean-Paul played a leading role by organising many international meetings and by bringing his collaborators to Marseille for extended periods (Paolo Aluffi, Gottfried Barthel, Lev Birbrair, Karl-Heinz Fieseler, Ludger Kaup, André Legrand, Ursula Ludwig, Markus Pflaum, Jörg Schürmann, José Seade, Tatsuo Suwa and Nicolae Teleman).

Recently Jean-Paul was the French organiser of French-Japanese singularities meetings in Sapporo (1998 and 2004), and Luminy (2002), and the 8th São Carlos Singularities Workshop (in Luminy !) in 2004. He has also spent much time in Brussels as a European expert. All this activity meant he had the competence and the confidence to propose that the six of us plan the 5 Weeks of Singularity Theory which took place in January and February 2005. That the whole event went so smoothly and on such a large scale was due in great part to his experience and know-how.

All Jean-Paul's colleagues know his sociability and good humour, also his efficiency in realising ambitious projects. Let this book be a mark of our respect and appreciation and an encouragement for future researchers in singularity theory.

Denis Chéniot
Nicolas Dutertre
Claudio Murolo
Anne Pichon
David Trotman

Marseille, October 12th, 2006

CONTENTS

PART I

Five Courses on Singularity Theory

The Schwartz Classes
of complex analytic singular Varieties

Jean-Paul BRASSELET

Directeur de recherche CNRS
Institut de Mathématiques de Luminy - Marseille - France
E-mail: jpb@iml.univ-mrs.fr

One provides a detailed construction of the Schwartz classes. They are characteristic classes associated to complex analytic singular varieties. In a first step, one gives the construction of Schwartz classes by obstruction theory. Then one relates these classes to Mather's and MacPherson's ones. The third part is devoted to the computation of examples. The last section deals with polar varieties and definitions of characteristic classes via polar varieties. These are old and new results, partly obtained jointly with M.-H. Schwartz, with G. Gonzalez-Sprinberg, with G. Barthel, K.-H. Fieseler and L. Kaup and with P. Aluffi.

Keywords: Singular Varieties, characteristic classes, Chern classes, polar varieties

1. Introduction

The Euler-Poincaré characteristic has been the first characteristic class to be introduced. For a triangulated (possibly singular) compact variety X without boundary, it has been defined, as

$$\chi(X) = \sum (-1)^i n_i \,,$$

where n_i is the number of i-dimensional simplices of the triangulation of X. It is also equal to $\sum (-1)^i b_i$ where b_i is the i-th Betti number, rank of $H_i(X)$. The Poincaré-Hopf theorem says that, if X is a manifold and v a continuous vector field with a finite number of isolated singularities a_k with indices $I(v, a_k)$, then

$$\chi(X) = \sum I(v, a_k) \,.$$

This means that the Euler-Poincaré characteristic is a measure of the obstruction to the construction of a non-zero vector field tangent to X, i.e. of a non-zero section of the tangent bundle.

3

The first definition of highest characteristic classes has been given in terms of obstruction of linearly independent sections of the tangent bundle. In a parallel way, Todd defined the so-called polar varieties and shown that some linear combinations of them are invariant. In fact, it appears that they coincide with Chern classes of complex analytic manifolds (see §8).

In years 1960, Hirzebruch shown that characteristic classes can be characterized by a system of axioms. During several years, the attractiveness of the axiomatic properties of Chern classes caused the viewpoints of obstruction theory and polar varieties to be somewhat forgotten. It is interesting to see that these viewpoints came back on the scene with the question of defining characteristic classes for singular varieties (see Teissier [Te]).

There are various definitions of characteristic classes for singular varieties. In the real case, there is a combinatorial definition, which simplifies the problem. In the complex case, the situation is more complicated (and certainly more interesting !), due to the fact that there is no combinatorial definition of Chern classes. Thinking of the obstruction theory point of view, one has to find a substitute to the tangent bundle. In fact there are various candidates to substitute the tangent bundle and to each of them corresponds a different definition of Chern class for singular varieties.

If X is a singular complex analytic variety, equipped with a Whitney stratification and embedded in a smooth complex analytic manifold M, one may consider the union of tangent bundles to the strata, that is a subspace E of the tangent bundle to M. The space E is not a bundle but it generalizes the notion of tangent bundle in the following sense: A section of E over X is a section v of $TM|_X$ such that in each point $x \in X$, then $v(x)$ belongs to the tangent space of the stratum containing x. To consider E as a substitute to the tangent bundle of X and to use obstruction theory is the M.-H. Schwartz point of view (1965, [Sc1]).

Another possible substitute for the tangent bundle is to consider, in each point $x \in X$, the space of all possible limits of tangent vector spaces $T_{x_i}(X_{\mathrm{reg}})$ where x_i is a sequence of points in the regular part X_{reg} of X converging to x. That point of view leads to the notion of Mather class, which is an ingredient in the MacPherson definition, in the case of algebraic complex varieties (1974, [MP]) . The other main ingredient for these classes is the notion of Euler local obstruction.

Finally, when there exists a normal bundle N to X in M, for example in the case of local complete intersections, one can consider the virtual bundle $TM|_X \setminus N$ as a substitute to the tangent bundle of X. That point of view is the one of Fulton (1980, [Fu]).

There are relations between the classes obtained by the previous constructions. First of all, the Schwartz and MacPherson classes coincide, via Alexander duality (1979, [BS]). The relation between Mather classes on the one side and Schwartz-MacPherson classes on the other side follows form the MacPherson's definition itself: His construction uses Mather classes, taking into account the local complexity of the singular locus along Whitney strata. This is the role of the local Euler obstruction.

A natural question arose to compare the Schwartz-MacPherson and the Fulton-Johnson classes. A result of Suwa [Su] shows that in the case of isolated singularities, the difference between these classes is given by the sum of the Milnor numbers in the singular points. It was natural to call Milnor classes the difference arising in the general case (see [Yo]). This difference has been described by several authors by different methods: P. Aluffi, J.P. Brasselet-D. Lehmann-J. Seade-T. Suwa, A. Parusiński-P. Pragacz and S. Yokura. We will not discuss of these classes in the present course. Interested reader can refer to [Br2], [Br5].

We provide in §8 explicit examples and computations of Schwartz-MacPherson classes in the case of Thom spaces associated to Segre and Veronese embeddings and iterated cones. Results of this section and other examples (for instance the case of toric varieties) have been obtained in join works with G. Barthel, K.-H. Fieseler and L. Kaup (see [BBF], [BFK]).

The definition of polar varieties is classical in the smooth case. In §10, we give the construction of Chern classes using polar varieties in the context of manifolds. The notion of polar varieties has been extended in the singular case by Lê D. T. and B. Teissier [LT] and R. Piene [Pi1,Pi2] gave formula for Mather classes in terms of polar varieties, for singular varieties. The result of P. Aluffi and J.-P. Brasselet relates other definitions of classes to polar varieties, in a particular case. A general formula has been conjectured in [Br3].

Two books related to the subject will appear soon, where the reader will find a complete view on the subject ([Br5], [BSS2])

The very nice survey on Characteristic Classes of Singular Spaces, by Jörg Schürmann and Shoji Yokura [SY], in this volume provides, for the students and researchers, an useful and interesting complement to this course.

The author thanks the editors and the referee for useful comments and remarks.

2. The substitutes of the tangent bundle

A complex analytic manifold M admits a (complex) tangent bundle TM. In the case of a complex analytic singular variety X, there is no longer tangent bundle. The different notions of Chern classes, in the singular setting, correspond to different notions of substitute to the tangent bundle. There are (at least) three ways to define such a substitute in the case of a singular variety X embedded in a manifold M:

1. let us consider the union E of tangent spaces to the strata of a stratification of X and consider the sections of TM whose images are in E. This is the method used by M.-H. Schwartz, showing that it is not possible to proceed to obstruction theory, using any section, but that one has to use vector fields and frames obtained by radial extension.

2. let us consider in each point $x \in X$, the set of all possible limits of tangent spaces $T_{x_i}(X_{\mathrm{reg}})$ to sequences of points $x_i \in X_{\mathrm{reg}}$ converging to x. That is the Nash transformation and the Nash bundle on it.

3. let us consider the virtual bundle. That is the method used by Fulton. If X is smooth, one has the exact sequence

$$0 \to TX \to TM|_X \to N_X M \to 0$$

where $N_X M$ is the normal bundle of X in M. In the case of a singular variety such that the normal bundle $N_X M$ exists (for instance hypersurfaces or local complete intersections), one can define the virtual bundle (in the Grothendieck group $KU(X)$) as

$$\tau_X = TM|_X - N_X M.$$

A common setting of the different notions is that they define characteristic classes in homology. One has to notice that, in the case of singular spaces, there are no longer characteristic classes in cohomology.

3. Obstruction theory - The smooth case

Let M be a complex manifold of (complex) dimension m, endowed with a hermitian metric. The tangent bundle to M, denoted by TM, is a complex vector bundle of rank m, whose fiber in a point x of M is the tangent vector space to M in x, denoted by $T_x(M)$ and is isomorphic to $\mathbb{C}^m$. The vector bundle TM is locally trivial, i.e. there is a covering of M by open subsets $\{U_i\}$ such that the restriction of TM to U_i is isomorphic to $U_i \times \mathbb{C}^m$.

The Poincaré-Hopf Theorem says that $\chi(M)$ is a measure of the obstruction for the construction of a vector field tangent to the manifold M.

In the same way, the objective of the obstruction theory is, for r fixed, $1 \leq r \leq \dim_{\mathbb{C}} M$, to define a measure of the obstruction to the construction of r linearly independent vector fields tangent to the manifold, i.e. to the construction of r sections of TM linearly independent (over $\mathbb{C}$) in each point.

Definition 3.1. An *r-field* on a subset A of M is a set $v^{(r)} = \{v_1, \ldots, v_r\}$ of r continuous vector fields defined on A. A singular point of $v^{(r)}$ is a point where the vectors (v_i) fail to be linearly independent. A non-singular r-field is also called an *r-frame*.

One observes that considering r-frames which are linearly independent r-fields or orthogonal r-fields provides the same construction and result (see [St]). It will be sometimes easier to consider orthogonal r-fields.

The r-frames are sections of the fibre bundle $T^r(M)$, with basis M, associated to TM and whose fiber in the point x of M is the set of r-frames of $T_x(M)$. This bundle is no longer a vector bundle. The "typical" fiber of $T^r(M)$ is the set of all r-frames of $\mathbb{C}^m$, called the Stiefel manifold and denoted by $V_{m,r}(\mathbb{C})$. These manifolds, in particular their homotopy groups, have been studied by Stiefel and by Whitney (see Steenrod [St]).

Let us consider the following situation: (D) is a cell decomposition of M sufficiently small so that every cell d lies in an open subset U over which $T^r(M)$ is trivial. We notice that trivialization open sets for $T^r(M)$ are the same that those of $T(M)$.

Let us consider the following question:

Let us suppose that one has a section $v^{(r)}$ of $T^r(M)$ on the boundary ∂d of the k-dimensional cell d. Is it possible to extend this section in the interior of d ? If the answer is no, what is the obstruction for such an extension ?

The section $v^{(r)}$, defined on the boundary of d, provides a map

$$\partial d \xrightarrow{v^{(r)}} T^r(M)|_U \cong U \times V_{m,r}(\mathbb{C}) \xrightarrow{pr_2} V_{m,r}(\mathbb{C}) \tag{1}$$

where pr_2 is the second projection. One obtains a map

$$\mathbb{S}^{k-1} \cong \partial d \xrightarrow{pr_2 \circ v^{(r)}} V_{m,r}(\mathbb{C})$$

hence an element of $\pi_{k-1}(V_{m,r}(\mathbb{C}))$ denoted by $[\gamma(v^{(r)}, d)]$.

Let us suppose that $[\gamma(v^{(r)}, d)] = 0$, then, by classical homotopy theory, the map $\mathbb{S}^{k-1} \to V_{m,r}(\mathbb{C})$ defined on the boundary $\mathbb{S}^{k-1}$ of the ball $\mathbb{B}^k$ can be extended inside the ball.

$$\partial d \cong \mathbb{S}^{k-1} \longrightarrow V_{m,r}(\mathbb{C})$$
$$\cap \qquad ? \nearrow$$
$$d \cong \mathbb{B}^k$$

In another words, if $[\gamma(v^{(r)}, d)] = 0$, then the map $\partial d \to V_{m,r}(\mathbb{C})$ can be extended inside d. This means that there is no obstruction to the extension of the section $v^{(r)}$ inside d. This happens, in particular, in the case $\pi_{k-1}(V_{m,r}(\mathbb{C})) = 0$.

In order to answer to the previous question, we need to know the homotopy groups of $V_{m,r}(\mathbb{C})$. They are equal to (see [St]):

$$\pi_{k-1}(V_{m,r}(\mathbb{C})) = \begin{cases} 0 & \text{for } k < 2(m - r + 1) \\ \mathbb{Z} & \text{for } k = 2(m - r + 1) \end{cases} \tag{2}$$

Let us denote $2p = 2(m - r + 1)$. A generator of the first non-zero homotopy group $\pi_{2p-1}(V_{m,r}(\mathbb{C}))$ can be described in the following way. Let us fix a $(r - 1)$-frame in $\mathbb{C}^m$. It defines a $(r - 1)$-subspace of $\mathbb{C}^m$ whose complementary is a complex space $\mathbb{C}^p$. The unit sphere in $\mathbb{C}^p$, denoted by $\mathbb{S}^{2p-1}$, is oriented with orientation induced by the natural one of $\mathbb{C}^p$. Let us consider, for every point w of the sphere, the r-frame consisting of the fixed $(r - 1)$-frame and the vector w, one obtains an element of $V_{m,r}(\mathbb{C})$. The induced map from the oriented sphere $\mathbb{S}^{2p-1}$ to $V_{m,r}(\mathbb{C})$ defines a generator of $\pi_{2p-1}(V_{m,r}(\mathbb{C}))$.

One obtains:

Proposition 3.1. *Let $v^{(r)}$ be a r-frame defined on the boundary ∂d of the k-cell d.*

(i) If $k < 2(m - r + 1)$, one has $[\gamma(v^{(r)}, d)] = 0$, then one can extend the r-frame, already defined on ∂d, inside d without singularity.

(ii) If $k = 2(m - r + 1)$, the r-frame defines an integer $[\gamma(v^{(r)}, d)]$ that we denote by $I(v^{(r)}, \hat{d})$. That index which measures the obstruction to the extension of $v^{(r)}$ inside d.

The dimension $2p = 2(m - r + 1)$ is called the obstruction dimension for the construction of an r-frame tangent to M.

If $v^{(r)}$ is a r-frame defined on the boundary ∂d of the $2p$-cell d, there are many ways to extend $v^{(r)}$ inside d with an isolated singularity. We will proceed in the following way: let us consider the $(r - 1)$-frame $v^{(r-1)} = (v_1, \ldots, v_{r-1})$ corresponding to the first $(r - 1)$ vectors of $v^{(r)}$. It defines a section of $T^{r-1}(M)$ over ∂d. The obstruction dimension for the extension of

$v^{(r-1)}$ is $2(m-(r-1)+1) = 2p+2$. That means that one can extend $v^{(r-1)}$ inside d without singularity. The extension defines a $(r-1)$-sub-bundle of $TM|_d$, whose complementary is a sub-bundle Q of (complex) rank p. The last vector v_r of $v^{(r)}$ defines a section of Q over ∂d. The obstruction dimension for the extension of v_r as a section of Q over d is $2(p-1+1) = 2p$, that is the dimension of d. That means that one can extend v_r inside d as a section of Q with an isolated singularity at the barycenter $\hat{d}$. The index of the r-frame $v^{(r)} = (v^{(r-1)}, v_r)$ at the singular point $\hat{d}$ is defined as $[\gamma(v^{(r)}, d)]$, we denote it by $I(v^{(r)}, \hat{d})$. One observes that this corresponds to the classical definitions of the index of a r-frame at a singular point.

The Chern classes can be defined now in the following way: One choose arbitrary r-frames on the 0-cells and one extends them without singularity, i.e. as a section of $T^r M$, on the 1-cells. By (i) of Proposition 3.1 one can extend that section by induction process on higher dimensional cells till we reach the obstruction dimension $2p$. For each $2p$-cell d, the section $v^{(r)}$ being defined on the boundary, provides an index $I(v^{(r)}, \hat{d})$. The generators of $\pi_{2p-1}(V_{m,r}(\mathbb{C}))$ being consistent (see [St]), one define a cochain

$$\gamma \in C^{2p}(M; \pi_{2p-1}(V_{m,r}(\mathbb{C}))), \qquad \text{such that } \gamma(d) = I(v^{(r)}, \hat{d}),$$

for each $2p$-cell d, and then extend by linearity. This cochain is actually a cocycle, called the obstruction cocycle.

Proposition 3.2. *The cohomology class of the obstruction cocycle γ does not depend on the various choices involved in its definition.*

Definition 3.2. The p-th (cohomology) *Chern class* of M,

$$c^p(M) \in H^{2p}(M; \mathbb{Z})$$

is the class of the obstruction cocycle.

By Poincaré duality isomorphism, cap-product by the fundamental class $[M]$ of M

$$H^{2p}(M; \mathbb{Z}) \longrightarrow H_{2(r-1)}(M; \mathbb{Z})$$

the image of $c^p(M)$ in $H_{2(r-1)}(M)$ is the $(r-1)$-st homology Chern class of M, denoted by $c_{r-1}(M)$. It is represented by the cycle

$$\sum_{\dim s = 2(r-1)} I(v^{(r)}, \hat{d}) \, s. \tag{3}$$

where s is the simplex in the simplicial triangulation (K) of which d is dual. The barycenter $\hat{d}$ is the intersection point of s and d, that is the barycenter of s, as well.

In particular, the evaluation of $c^m(M)$ on the fundamental class $[M]$ of M yields the Euler-Poincaré characteristic.

4. The Schwartz classes

The first definition of Chern class for singular varieties has been given in 1965 by M.-H. Schwartz in two "Notes aux CRAS" [Sc1].

In order to define characteristic classes of singular varieties, it is necessary to know the local structure of the singular variety. That is given by the structure of stratified space and by suitable definition of triangulation on the variety.

4.1. *Stratifications, triangulations and cell decompositions*

In the following, M will be a complex analytic manifold equipped with a semi-analytic stratification $\{V_\alpha\}$, i.e. a partition into analytic manifolds V_α, called the strata such that, for each stratum V_α, the closure $\overline{V}_\alpha$ and the boundary $\dot{V}_\alpha = \overline{V}_\alpha \setminus V_\alpha$ are semi-analytic sets, union of strata. We denote by $X \subset M$ a complex analytic compact subset stratified by $\{V_\alpha\}$.

As we know, on a singular variety, there is no more tangent space in the singular points. One way to find a substitute for the tangent bundle is to stratify the singular variety into submanifolds. One can proceed to the following construction: If X is a singular complex analytic variety, equipped with a stratification and embedded in a smooth complex analytic manifold M one can consider the union of tangent bundles to the strata. That is a subspace E of the tangent bundle to M. The space E is not a bundle but it generalizes the notion of tangent bundle in the following sense: A section of E over X is a section v of $TM|_X$ such that in each point $x \in X$, then $v(x)$ belongs to the tangent space of the stratum containing x. Such a section is called a *stratified vector field* over X:

Definition 4.1. A *stratified vector field* v on a part A of X is a (continuous) section of the tangent bundle TM defined on A and such that, for every $x \in A$, one has $v(x) \in T(V_{\alpha(x)})$ where $V_{\alpha(x)}$ is the stratum containing x.

To consider E as the substitute to the tangent bundle of X and to use obstruction theory is the M.-H. Schwartz point of view (1965, [Sc1]), in the case of analytic varieties.

When one considers stratifications of singular varieties, it is natural to ask for conditions with which the strata glue together. The so-called Whitney conditions [Wh] are the one which allow to proceed to the construction of radial extension vector fields. According to a result of Whitney, every analytic complex variety can be equipped with a Whitney stratification.

Definition 4.2. One says that the Whitney conditions are satisfied for the stratification $\{V_\alpha\}$ of X if, for any pair of strata (V_α, V_β) such that V_α is in the closure of V_β, one has:

a) if (x_n) is a sequence of points in V_β with limit $y \in V_\alpha$ and if the sequence of tangent spaces $T_{x_n}(V_\beta)$ admits a limit T (in the suitable Grassmanian space) when n goes to $+\infty$, then T contains $T_y(V_\alpha)$.

b) if (x_n) is a sequence of points in V_β with limit $y \in V_\alpha$ and if (y_n) is a sequence of points in V_α with limit y, such that the sequence of tangent spaces $T_{x_n}(V_\beta)$ admits a limit T for n going to $+\infty$ and such that the sequence of directions $\overline{x_n\, y_n}$ admits a limit λ when n goes to $+\infty$, then λ lies in T.

Let $X \subset M$ be a singular n-dimensional complex analytic variety embedded in a complex m-dimensional manifold. Let us consider a Whitney stratification $\{V_\alpha\}$ of M such that X is a union of strata and let us denote by (K) a triangulation of M compatible with the stratification, i.e. each open simplex is contained in a stratum.

The first nice observation of M.-H. Schwartz concerns the triangulations:

We denote by (K') a barycentric subdivision of (K) and by (D) the associated dual cell decomposition. Each cell in (D) is transverse to the strata. This implies that if d is a cell of (real) dimension k and V_α is a stratum of (complex) dimension n_α, then $d \cap V_\alpha$ is a cell whose (real) dimension is

$$\dim(d \cap V_\alpha) = k - 2(m - n_\alpha).$$

Consequence: This means that if d is a cell whose dimension is the dimension of obstruction to the construction of an r-frame tangent to M, i.e. $2p = 2(m - r + 1)$, then $d \cap V_\alpha$ is a cell whose dimension is exactly the dimension of obstruction to the construction of an r-frame tangent to the stratum V_α, i.e. $2(n_\alpha - r + 1)$.

We will use two important properties of the dual cells:

The (D)-cells which meet X are duals of (K)-simplices lying in X. Union of such cells is a neighbourhood $N(X)$ around X. That is not a fibre bundle on X but one has the following construction:

Dual cells are union of simplices of the barycentric subdivision (K'). For each (closed) simplex τ in (K') such that $\tau \cap X \neq \emptyset$, then one calls $\tau_X = \{a_0, a_1, \ldots, a_i\}$ the set of vertices in τ which are in X and $\tau'_X =$

12

$\{a_{i+1}, \ldots, a_k\}$ the set of vertices in τ which are not in X. Let us call $N_\varepsilon(\tau)$ the set of points in τ such that $\sum_{j=0}^{i} \lambda_j \leq \varepsilon$ and we use the following notation:

$$N_\varepsilon(X) = \bigcup_{\tau \subset N(X)} N_\varepsilon(\tau). \tag{4}$$

That is a "tube" around X and there is a retraction of $N_\varepsilon(X)$ on X along "rays": two points x and x' in τ belong to the same ray if their barycentric coordinates corresponding to vertices in τ_X are proportional on the one hand and their barycentric coordinates corresponding to vertices in τ'_X are proportional on the other hand (see Figure 1, Page 30 in [Sc3]).

The second nice construction of M.-H. Schwartz is the construction of radial extension of vector fields that we explicit below.

4.2. *Radial extension process - the local case*

One gives a description of the local radial extension process. This will be used for the global process in the next section.

Let us consider $X, M, \{V_\alpha\}, (K), (D)$ as before. Let $V_\alpha \subset X$ be a stratum, with complex dimension n_α, let a be the barycenter of a $2(r-1)$-simplex s of the triangulation (K), lying in V_α. One denotes by $d = d(s)$ the dual cell. Then $d_\alpha = d \cap V_\alpha$ is a $2(n_\alpha - r + 1)$-cell in V_α. Note that in general, $d \cap X$ is not a cell.

Let $v^{(r)}$ a r-frame defined in d_α with an isolated singularity at a. One will construct inside d the parallel extension of $v^{(r)}$ and a particular vector field, the transversal vector field:

4.2.1. *Parallel Extension*

Provided the simplices of (K) are sufficiently small, the cell d can be identified with $d_\alpha \times \mathbb{D}^{k_\alpha}$ where $\mathbb{D}^{k_\alpha}$ is a disk which is transverse to V_α and whose dimension is $k_\alpha = 2(m - n_\alpha)$. For a precise identification, one works by induction on the dimension of cells in $d_\alpha = d \cap V_\alpha$: the 0-cells in the boundary of d_α are barycenters of $2n$-simplices s_i^{2n} containing s in their boundary, the dual cells $d(s_i^{2n})$ are homeomorphic to $\mathbb{D}^{k_\alpha}$. By induction one extends the identification on cells of the boundary, then on d_α itself.

Let us consider the parallel extension $\hat{v}^{(r)}$ of $v^{(r)}$ in d along the fibers $\mathbb{D}^{k_\alpha}$. Let V_β be a stratum such that $a \in \overline{V_\beta}$. At a point $x \in d \cap V_\beta$ the parallel extension $\hat{v}^{(r)}(x)$ is not necessarily tangent to V_β. However, the Whitney condition (a) guarantees that if d is sufficiently small, then the angle between $T_a(V_\alpha)$ and $T_x(V_\beta)$ is small. That implies that the orthogonal projection of $\hat{v}^{(r)}(x)$ on $T_x(V_\beta)$ does not vanish. Of course, considering for each stratum the projection of the parallel extension on the tangent space to the stratum at the given point does not provide a continuous frame. In order to obtain a continuous frame, one has to consider a slight modification of the construction, in the neighbourhood of the strata, which is easy to understand, but complicated to describe into details. The good extension will be $\hat{v}^{(r)}(x)$ away from V_β and continuously going to the projection of $\hat{v}^{(r)}(x)$ on $T_x(V_\beta)$ when approaching V_β, using a suitable partition of unity. That construction is correctly and entirely described in M.-H. Schwartz book [Sc2]. In fact, one has to work simultaneously for all strata V_β such that $a \in \overline{V_\beta}$, that complicates a detailed construction.

In conclusion, the Whitney (a) condition implies that one can proceed to the construction of a stratified r-frame, denoted by $\hat{v}^{(r)}(x) = \{\hat{v}_1, \dots, \hat{v}_r\}$ which is a "parallel extension" of the given frame on V_α, in the cell d, identified to a tube around d_α.

One observes that the singular locus of $\hat{v}^{(r)}$ corresponds to a k_α-dimensional disk which is transversal to d_α at the point $\hat{d} = a$.

4.2.2. *Transversal vector field*

Let us consider the transversal vector field $g(x)$, which is the gradient of the square of the function distance to V_α, for an appropriate Riemannian metric. The vector field $g(x)$ is not necessarily tangent to the strata V_β such that $a \in \overline{V_\beta}$. However, the Whitney condition (b) guarantees that in d, which is identified to a sufficiently small "tube" around d_α and for $x \in d \cap V_\beta$, the angle between $g(x)$ and $T_x(V_\beta)$ is small. That means that the orthogonal projection of $g(x)$ on $T_x(V_\beta)$ does not vanish. In the same way than for the parallel extension, considering for each stratum the projection of $g(x)$ on the tangent space to the stratum at the given point does not provide a continuous vector field. In order to obtain a continuous vector field, one has to consider a similar modification of the construction. The good vector field will be $g(x)$ away from V_β and continuously going to the projection of $g(x)$ on $T_x(V_\beta)$ when approaching V_β. That construction is also completely described in M.-H. Schwartz book [Sc2], and one has to work simultaneously for all strata V_β such that $a \in \overline{V_\beta}$.

Let us call horizontal part of the boundary of the tube $d \cong d_\alpha \times \mathbb{D}^{k_\alpha}$, the part of the boundary corresponding to $d_\alpha \times \partial \mathbb{D}^{k_\alpha} = d_\alpha \times \mathbb{S}^{k_\alpha - 1}$ by the previous identification. The vector field g is pointing outward d along the

14

horizontal part of the boundary.

In conclusion, one obtains a stratified "transversal" vector field still denoted by g which vanishes along V_α, which is growing with the distance to V_α and which is pointing outward d along the horizontal part of the boundary of the "tube" d provided that the tube is sufficiently small.

4.2.3. *Local radial extension*

Definition 4.3. [Sc3] [BS] Let s be a simplex in V_α and let d be the dual cell of s. Let $v^{(r)} = \{v_1, \ldots, v_r\}$ be an r-frame defined in $d_\alpha = d \cap V_\alpha$, possibly with an isolated singularity at the barycenter of d_α, the (local) radial extension of $v^{(r)}$ is the r-frame $\tilde{v}^{(r)}$ defined in the cell d as the parallel extension $\hat{v}^{(r)}$ of $v^{(r)}$ to which one adds the transversal vector field on the last coordinate, i.e.

$$\tilde{v}^{(r)} = (\tilde{v}^{(r-1)}, \tilde{v}_r) = \{\hat{v}_1, \ldots, \hat{v}_{r-1}, \hat{v}_r + g(x)\}.$$

Proposition 4.1. *[Sc3] [BS] Let $v^{(r)} = \{v_1, \ldots, v_r\}$ be an r-frame defined in $d_\alpha = d \cap V_\alpha$, with an isolated singularity at the barycenter a of d_α, then the (local) radial extension of $v^{(r)}$ is defined inside the cell d and it has an isolated singularity at a. The index of $\tilde{v}^{(r)}$ at a, computed in the cell d as a section of $T^r M$, is the same than the index of $v^{(r)}$ at a, computed in the cell $d_\alpha = d \cap V_\alpha$ as a section of $T^r V_\alpha$. We write*

$$I(\tilde{v}^{(r)}, a; d) = I(v^{(r)}, a; d \cap V_\alpha).$$

That property is the main property of the radial extension, that is precisely the property which allows to construct the obstruction classes for singular varieties.

4.3. *Chern classes for singular varieties*

In that section, one proceeds to the construction of a "global" radial extension of an r-frame and one shows the following Theorem (see also [BS]):

Theorem 4.1. *[Sc1], [Sc3] One can construct, on the cells d of the $2p$-skeleton $(D)^{2p}$ which intersect X, a stratified r-frame $v^{(r)} = (v^{(r-1)}, v_r)$ called radial extension frame, whose singularities satisfy the following properties:*

(i) $v^{(r)}$ has only isolated singular points, which are zeroes of the last vector v_r. On $(D)^{2p-1}$, the r-frame $v^{(r)}$ has no singular point. On $(D)^{2p}$ the $(r-1)$-frame $v^{(r-1)}$ has no singular point.

(ii) Let $a \in V_\alpha \cap (D)^{2p}$ be a singular point of $v^{(r)}$ in the n_α-dimensional stratum V_α. If $n_\alpha > r - 1$, the index of $v^{(r)}$ at a, denoted by $I(v^{(r)}, a)$, is the same as the index at a of the restriction of $v^{(r)}$ to $V_\alpha \cap (D)^{2p}$ considered as an r-frame tangent to V_α. If $n_\alpha = r - 1$, then $I(v^{(r)}, a) = +1$.

(iii) Inside a 2p-cell d which meets several strata, the only singularities of $v^{(r)}$ lie in the lowest dimensional one (in fact located at the barycenter of d).

(iv) The last vector v_r of the r-frame is pointing outward (particular) regular neighbourhoods U of X in M. The r-frame $v^{(r)}$ has no singularity on the boundary ∂U.

Proof. The "global" construction of the radial extension frame is as follows:

One consider on M a Whitney stratification compatible with X.

The proof will go by induction on the dimension of the strata. We will show that, for each stratum V_α, the theorem is true for $X = \overline{V}_\alpha$.

a) Let us denote by n_0 the lowest dimension of strata in X such that $n_0 \geq r - 1$. The strata whose dimension is less than n_0 do not contribute to the corresponding Chern class. The reason is that such a stratum does not meet the 2p-skeleton $(D)^{2p}$.

a) Let us prove the theorem for the n_0-dimensional stratum, denoted by V_{α_0}. We distinguish the cases $n_0 = r - 1$ and $n_0 > r - 1$:

a1) If $n_0 = r - 1$, let d be a 2p-cell which intersects V_{α_0}. The intersection is a point: the barycenter $\hat{s}$ of the $2(r - 1)$-simplex s of which d is dual. At each such point $\hat{s}$, let us fix a $(r - 1)$- frame $v^{(r-1)}(\hat{s})$ tangent to V_{α_0}. One can extend $v^{(r-1)}(\hat{s})$ on d by the local parallel extension process (4.2.1) as a stratified $(r - 1)$-frame $\hat{v}^{(r-1)}$ on such a cell d. Using the transversal vector field g constructed in 4.2.2, one obtains a stratified r-frame $\tilde{v}^{(r)}(x) = (\hat{v}^{(r-1)}(x), g(x))$ on each d. One has

$$I(\tilde{v}^{(r)}, \hat{s}; d) = +1$$

a2) If $n_0 > r - 1$, on the one hand, one observes that if a 2p-cell d intersects V_{α_0}, then the intersection is a $2(n_0 - (r - 1))$-cell which is dual, in V_{α_0} of a $2(r - 1)$-simplex s. On the other hand, the (D)-cells with dimension less than $2p$ do not meet the strata which lie in $\overline{V}_{\alpha_0} \setminus V_{\alpha_0}$.

Let us call (D_{α_0}) the set of cells in V_{α_0}, intersections $d_{\alpha_0} = d \cap V_{\alpha_0}$ for all (D)-cells d whose dimension k satisfies $2(m - n_0) \leq k \leq 2p = 2(m - (r - 1))$. If $\dim d = k$, then $\dim d_{\alpha_0} = k - 2(m - n_0)$. So, if $d_{\alpha_0} \in (D_{\alpha_0})$, one has $0 \leq \dim d_{\alpha_0} \leq 2(n_0 - (r - 1))$. One observes that the obstruction dimension for the construction of an r-frame tangent to V_{α_0} is $2(n_0 - (r - 1))$.

One chooses an r-frame $v^{(r)}$ in each 0-dimensional cell d_{α_0}, one can extend these frames as a section of $T^r V_{\alpha_0}$ on higher dimensional cells d_{α_0} by classical obstruction theory, by induction on dimensions of cells till we reach the obstruction dimension $2(n_0 - (r-1))$. One obtain an r-frame $v^{(r)}$ with isolated singularities at the barycenters $\hat{d}$ of the $2(n_0 - (r-1))$-dimensional cells d_{α_0}, with index $I(v^{(r)}, \hat{d}; d_{\alpha_0})$.

The r-frame can be extended by the local extension process 4.2.3 as an r-frame $\tilde{v}^{(r)}$ on the $2p$-cells d such that $d \cap V_{\alpha_0} = d_{\alpha_0}$, with an isolated singularity at $\hat{d}$ in each cell d and such that

$$I(v^{(r)}, \hat{d}; d_{\alpha_0}) = I(\tilde{v}^{(r)}; \hat{d}, d).$$

That proves the theorem for the lowest dimensional strata V_{α_0}, of dimension bigger than $2(r-1)$. In both cases a1) and a2), the neighbourhood U is the set of cells in (D) which intersect V_{α_0}.

b) Let us now consider a stratum V_γ with (complex) dimension n_γ and let us suppose that the theorem has been proved for all strata of dimension lower than (and equal to) n_γ, i.e. that the theorem is true for $X = \overline{V}_\gamma$. One denotes by $N_\varepsilon(\overline{V}_\gamma)$ the neighbourhood defined in (4). Let us denote by $v^{(r)}$ an r-frame satisfying conditions of the Theorem for $X = \overline{V}_\gamma$ and for $U = N_\varepsilon(\overline{V}_\gamma)$.

Let us call V_δ the next stratum, i.e. the one whose dimension n_δ is strictly bigger than n_γ and such that there is no other strata whose dimension is between n_δ and n_γ. One has to show that the theorem is true for $X = \overline{V}_\delta$.

The r-frame $v^{(r)}$ is defined on the $2p$-skeleton of $N_\varepsilon(\overline{V}_\gamma)$ with singularities situated in V_γ. That means that $v^{(r)}$ is defined on the skeleton of dimension $2(n_\delta - r + 1)$ of $U_\delta = V_\delta \cap N_\varepsilon(\overline{V}_\gamma)$. Moreover, the last vector of $v^{(r)}$ is pointing inward V_δ along $\partial U_\delta = \overline{U}_\delta \setminus U_\delta$. By classical obstruction theory, one can extend $v^{(r)}$ inside V_δ on the $2p_\delta = 2(n_\delta - r + 1)$-skeleton and such that:

- $v^{(r)}$ has only isolated singularities, which are zeroes of the last vector v_r,
- on $(D)^{2p-1}$, the r-frame $v^{(r)}$ has no singular point,
- on $(D)^{2p}$ the $(r-1)$-frame $v^{(r-1)}$ has no singular point.

One can extend, by the local extension process 4.2.3, the obtained r-frame $v_\delta^{(r)}$ on the $2(n_\delta - r + 1)$-cells of (D_δ) as a stratified r-frame $\tilde{v}_\delta^{(r)}$ on the $2p$-cells in $(D)^{2p}$ which meet $V_\delta \setminus \Omega_\delta$. One has:

$$I(\tilde{v}_\delta^{(r)}; \hat{d}, d) = I(v_\delta^{(r)}, \hat{d}; d \cap V_\delta)$$

for each $2p$-cell d which meet $V_\delta \setminus \Omega_\delta$.

One obtains an r-frame $v^{(r)}$ defined in $(D)^{2p} \cap U(V_\gamma)$ and an r-frame $\widetilde{v}_\delta^{(r)}$ defined in $(D)^{2p} \cap (V_\delta \setminus \Omega_\delta)$. The problem is that, while these two frames agree on V_δ, i.e. on $(D)^{2p} \cap \partial U(V_\gamma) \cap V_\delta$, they do not agree a priori on the common part $(D)^{2p} \cap \partial U(V_\gamma) \cap U'(V_\delta)$ where $U'(V_\delta)$ is the union of the (D)-cells which meet $V_\delta \setminus \Omega_\delta$.

The solution is rather technical. One has to work with two systems of neighbourhood on the following way:

The previous construction can be performed for all $0 < \varepsilon \leq 1$. Let us suppose that the construction has been performed for say, $N_\varepsilon(\overline{V}_\gamma)$ and $N_1(\overline{V}_\gamma)$, the r-frame on the first neighbourhood being restriction of the r-frame on the second. Then one obtains :

- an r-frame inside $N_\varepsilon(\overline{V}_\gamma)$, we will not modify it,
- two r-frames inside $\left(N_1(\overline{V}_\gamma) \setminus N_\varepsilon(\overline{V}_\gamma)\right) \cap U(V_\gamma)$, the first one being $v^{(r)}$ defined in $N_1(\overline{V}_\gamma)$, the second one is $\widetilde{v}_\delta^{(r)}$. In fact, they coincide on V_γ.

Let us call λ the radius of $N_1(\overline{V}_\gamma)$ and

$$\lambda' = \frac{1}{1-\varepsilon}\lambda - \frac{\varepsilon}{1-\varepsilon}$$

In $\left(N_1(\overline{V}_\gamma) \setminus N_\varepsilon(\overline{V}_\gamma)\right) \cap U(V_\gamma)$, one considers the r-frame

$$\lambda'\widetilde{v}_\delta^{(r)} + (1 - \lambda')v^{(r)}$$

and one obtains a continuous stratified r-frame in

$$N_\varepsilon(\overline{V}_\delta) = N_\varepsilon(\overline{V}_\gamma) \cup U'_\varepsilon(V_\delta)$$

which solves the problem. As we said, to verify all conditions is rather tedious. That is completely performed in M.-H. Schwartz's work. $\quad\square$

4.4. *Obstruction cocycles and classes*

Let us denote by $N = N(X)$ the tubular neighborhood of X in M consisting of the (D)-cells which intersect X. The dual cell of a (K)-simplex s in X is denoted by $d = d(s)$ and their common barycenter by $\hat{s} = d(s) \cap s$. Let us denote by d^* the elementary (D)-cochain whose value is 1 at d and 0 at all other cells. We can define a $2p$-dimensional (D)-cochain in $C^{2p}(N, \partial N)$ by:

$$\sum_{\substack{d(s) \in N \\ \dim d(s) = 2p}} I(v^{(r)}, \hat{s})\, d^*.$$

This cochain actually is a cocycle whose class $c^p(X)$ lies in

$$H^{2p}(N, \partial N) \cong H^{2p}(N, N \setminus X) \cong H^{2p}(M, M \setminus X),$$

where the first isomorphism is given by retraction along the rays of N and the second by excision (by $M \setminus N$).

Definition 4.4. [Sc1], [Sc3] The p-th Schwartz class of X is the class

$$c^p(X) \in H^{2p}(M, M \setminus X).$$

The Schwartz class is independent of the choices : stratification, triangulation, dual cell decomposition. The direct proof is rather tedious (see [Sc2]). The fact that the Schwartz class is dual of the MacPherson one helps a lot for such a proof.

5. Euler local obstruction

5.1. *Nash transformation*

Let M be an analytic manifold, of complex dimension m. Let X be an subanalytic complex variety in M of complex dimension n, equipped with a Whitney stratification. Let us denote by $\Sigma = X_{\mathrm{sing}}$ the singular part of X and by $X_{\mathrm{reg}} = X \setminus \Sigma$ the regular part.

The Grassmanian manifold of complex n-planes in $\mathbb{C}^m$ is denoted by $G(n, m)$. Let us consider the Grassmann bundle of n (complex) planes in TM, denoted by G. The fibre G_x over $x \in M$ is the set of n-planes in $T_x(M)$, it is isomorphic to $G(n, m)$. An element of G is denoted by (x, P) where $x \in M$ and $P \in G_x$.

On the regular part of X, one can define the Gauss map $\sigma : X_{\mathrm{reg}} \longrightarrow G$ by

$$\sigma(x) = (x, T_x(X_{\mathrm{reg}})).$$

Definition 5.1. The Nash transformation $\widetilde{X}$ is defined as the closure of the image of σ in G. It is equipped with a natural analytic projection $\nu : \widetilde{X} \longrightarrow X$.

$$
\begin{array}{ccc}
G & \qquad \widetilde{X} = \overline{\mathrm{Im}\sigma} \hookrightarrow G & \\
{}^{\nearrow \sigma} \downarrow & \nu \downarrow \qquad \downarrow & \qquad (5) \\
X_{\mathrm{reg}} \hookrightarrow M & X \qquad \hookrightarrow M &
\end{array}
$$

In general, $\widetilde{X}$ is not smooth, nevertheless, it is an analytic variety and the restriction $\nu : \widetilde{X} \to X$ of the bundle projection $G \to M$ is analytic.

Let us denote by Θ the tautological bundle over G. The fiber Θ_P of the tautological bundle Θ over G, in a point $(x, P) \in G$, is the set of the vectors v of the n-plane P.

$$\Theta_P = \{v(x) \in T_x M : v(x) \in P, \quad x = \nu(P)\}$$

Let us define $\widetilde{\Theta} = \Theta|_{\widetilde{X}}$, then $\widetilde{\Theta}|_{\widetilde{X}_{\text{reg}}} = T(X_{\text{reg}})$ where $\widetilde{X}_{\text{reg}} = \nu^{-1}(X_{\text{reg}}) \cong X_{\text{reg}}$ and

$$\widetilde{\Theta} = \Theta \times_G \widetilde{X} = \{(v(x), \tilde{x}) \in \Theta \times \widetilde{X} : v(x) \in \tilde{x}\}$$

$\tilde{x} \in \widetilde{X}$ is an n-complex plane in $T_x(M)$ and $x = \nu(\tilde{x})$.

One has a diagram:

$$
\begin{array}{ccc}
\widetilde{\Theta} & \hookrightarrow & \Theta \\
\downarrow & & \downarrow \\
\widetilde{X} & \hookrightarrow & G \\
\nu \downarrow & & \downarrow \\
X & \hookrightarrow & M
\end{array}
\tag{6}
$$

We will denote by $\widetilde{\Theta}^r$ the bundle of "r-frames" associated to $\widetilde{\Theta}$, i.e. the bundle whose fiber in a point (x, P) of $\widetilde{X}$ is the set of r linearly independent vectors in P.

The following lemma is fundamental for the understanding of the geometrical definition of the local Euler obstruction.

Lemma 5.1. *([BS], Proposition 9.1) A stratified vector field v on a subset $A \subset X$ admits a canonical lifting $\tilde{v}$ on $\nu^{-1}(A)$ as a section of $\widetilde{\Theta}$.*

One defines the map $\nu_* : \widetilde{\Theta} \to TM|_X$ by $\nu_*(v(x), \tilde{x}) = v(\nu(\tilde{x})) = v(x)$. One has a commutative diagram:

$$
\begin{array}{ccc}
\widetilde{\Theta} & \xrightarrow{\nu_*} & TM|_X \\
\tilde{v} \uparrow\downarrow & & v \uparrow\downarrow \\
\widetilde{X} & \xrightarrow{\nu} & X
\end{array}
$$

Let us recall that a *radial* vector field v in a neighbourhood of the point $a \in X$ is a stratified vector field so that there exists $\varepsilon_0 > 0$ such that for all ε, $0 < \varepsilon < \varepsilon_0$, the vector $v(x)$ is pointing outward the ball $B_\varepsilon = B_\varepsilon(a)$ over the boundary $S_\varepsilon = S_\varepsilon(a) = \partial B_\varepsilon$. By the Bertini-Sard theorem, S_ε is transverse to the strata V_α if ε is small enough, so the definition takes sense.

Theorem 5.1. *Theorem-Definition [BS] Let v be a radial vector field over $X \cap S_\varepsilon$ and $\tilde{v}$ the lifting of v over $\nu^{-1}(X \cap S_\varepsilon)$. The local Euler obstruction*

$\mathrm{Eu}_a(X)$ *is the obstruction to extend $\tilde{v}$ as a nowhere zero section of $\tilde{\Theta}$ over* $\nu^{-1}(X \cap B_\varepsilon)$, *evaluated on the orientation class* $\mathcal{O}_{\nu^{-1}(B_\varepsilon), \nu^{-1}(S_\varepsilon)}$:

$$Eu_a(X) = Obs(\tilde{v}, \tilde{\Theta}, \nu^{-1}(X \cap B_\varepsilon)).$$

Theorem 5.2. *([BS], Théorème 11.1) (Proportionality Theorem). Let $v^{(r)}$ be a stratified r-frame on the $2p$-cell $d = d(s)$ with an isolated singularity with index $I(v^{(r)}, \hat{s})$ at the barycenter $\{\hat{s}\} = d \cap s$. Let us denote by $\tilde{v}^{(r)}$ the lifting of $v^{(r)}$ on $\nu^{-1}(\partial d \cap X)$. The obstruction to the extension of $\tilde{v}^{(r)}$ as a section of $\tilde{\Theta}^r$ on $\nu^{-1}(d \cap X)$ is equal to:*

$$Obs(\tilde{v}^{(r)}, \tilde{\Theta}^r, \nu^{-1}(d \cap X)) = Eu_{\hat{s}}(X) \cdot I(v^{(r)}, \hat{s}).$$

6. MacPherson and Mather classes

Let us recall firstly some basic definitions. In this section, one considers the category of complex algebraic varieties.

A *constructible set* in a variety X is a subset obtained by finitely many unions, intersections and complements of subvarieties. A *constructible function* $\alpha : X \to \mathbb{Z}$ is a function such that $\alpha^{-1}(n)$ is a constructible set for all n. The constructible functions on X form a group denoted by $\mathbb{F}(X)$. If $A \subset X$ is a subvariety, we denote by $\mathbf{1}_A$ the characteristic function whose value is 1 over A and 0 elsewhere.

If X is triangulable, α is a constructible function if and only if there is a triangulation (K) of X such that α is constant on the interior of each simplex of (K). Such a triangulation of X is called α-adapted.

The correspondence $\mathbb{F} : X \to \mathbb{F}(X)$ defines a contravariant functor when considering the usual pull-back $f^* : \mathbb{F}(Y) \to \mathbb{F}(X)$ for a morphism $f : X \to Y$. One interesting fact is that it can be made a covariant functor when considering the pushforward f_* defined on characteristic functions by:

$$f_*(\mathbf{1}_A)(y) = \chi(f^{-1}(y) \cap A), \qquad y \in Y$$

for a morphism $f : X \to Y$, and linearly extended to elements of $\mathbb{F}(X)$. The following result was conjectured by Deligne and Grothendieck in 1969:

Theorem 6.1. *[MP] Let $\mathbb{F}$ be the covariant functor of constructible functions and let $H_*(\ ; \mathbb{Z})$ be the usual covariant $\mathbb{Z}$-homology functor. Then there exists a unique natural transformation*

$$c_* : \mathbb{F} \to H_*(\ ; \mathbb{Z})$$

satisfying $c_*(\mathbf{1}_X) = c^*(X) \cap [X]$ if X is a manifold.

The theorem means that for every algebraic complex variety, one has a functor $c_* : \mathbb{F}(X) \to H_*(X; \mathbb{Z})$ satisfying the following properties:

(1) $c_*(\alpha + \beta) = c_*(\alpha) + c_*(\beta)$ for α and β in $\mathbb{F}(X)$,
(2) $c_*(f_*\alpha) = f_*(c_*(\alpha))$ for $f : X \to Y$ morphism of algebraic varieties and $\alpha \in \mathbb{F}(Y)$,
(3) $c_*(\mathbf{1}_X) = c^*(X) \cap [X]$ if X is a manifold.

6.1. *Mather classes*

The first approach to the proof of the Deligne-Grothendieck's conjecture is given by the construction of Mather classes. Let $X \subset M$ a possibly singular algebraic complex variety embedded in a smooth one. Let us define the Nash transformation $\widetilde{X}$ of X, as in section 5.1 and the Nash bundle $\widetilde{\Theta}$ on $\widetilde{X}$.

Definition 6.1. The Mather class of X is defined by:

$$c^M(X) = \nu_*(c^*(\widetilde{\Theta}) \cap [\widetilde{X}])$$

where $c^*(\widetilde{\Theta})$ denotes the usual (total) Chern class of the bundle $\widetilde{\Theta}$ in $H^*(\widetilde{X})$ and the cap-product with $[\widetilde{X}]$ is the Poincaré duality homomorphism (in general not an isomorphism, see [Br1]).

The Mather classes do not satisfy the Deligne-Grothendieck's conjecture. One has to take into account the complexity of the singular variety along the strata. That is the role of the local Euler obstruction, used in MacPherson's construction.

6.2. *MacPherson classes*

The MacPherson's construction uses both the constructions of Mather classes and local Euler obstruction.

For a Whitney stratification $\{V_\alpha\}$ of X, we have the following lemma:

Lemma 6.1. *[MP] There are integers n_α such that, for every point $x \in X$, one has:*

$$\sum_\alpha n_\alpha \mathrm{Eu}_x(\overline{V_\alpha}) = 1.$$

Definition 6.2. [MP] The MacPherson class of X is defined by

$$c_*(X) = c_*(\mathbf{1}_X) = \sum_\alpha n_\alpha\, i_* c_*^M(\overline{V_\alpha})$$

where i denotes the inclusion $\overline{V_\alpha} \hookrightarrow X$.

Theorem 6.2. *([BS], see also [AB2]) The MacPherson class is image of the Schwartz class by the Alexander duality isomorphism [Br1]*

$$H^{2(m-r+1)}(M, M \setminus X) \xrightarrow{\cong} H_{2(r-1)}(X).$$

One calls Schwartz-MacPherson class the class $c_*(X)$ in $H_*(X)$.

Corollary 6.1. *The Schwartz-MacPherson class $c_{r-1}(X)$ is represented by the cycle:*

$$\sum_{\substack{s \subset X \\ \dim s = 2(r-1)}} I(v^{(r)}, \hat{s})\ s$$

Corollary 6.2. *[BS] The Chern-Mather class $c^M_{r-1}(X)$ is represented by the cycle:*

$$\sum_{\substack{s \subset X \\ \dim s = 2(r-1)}} Eu_{\hat{s}}(X) I(v^{(r)}, \hat{s})\ s$$

7. Schwartz-MacPherson classes of Thom spaces associated to embeddings

In this section and as a matter of example, we compute the Schwartz-MacPherson classes of the Thom spaces associated to Segre and Veronese embeddings. Results of this section have been obtained with Gerard Gonzalez-Sprinberg following ideas of Jean-Louis Verdier and Mark Goresky and with Gottfried Barthel, Karl-Heinz Fieseler and Ludger Kaup.

7.1. *The projective cone*

Let us consider an n-dimensional projective variety Y in $\mathbb{P}^m = \mathbb{PC}^m$ and let us denote by H_Y the restriction of the hyperplane bundle of $\mathbb{P}^m$ to Y. We denote by Q the completed projective space of the total space of H_Y, i.e. $Q = \mathbb{P}(H_Y \oplus 1_Y)$ where 1_Y is the trivial bundle of complex rank 1 on Y. The canonical projection $p \colon Q \to Y$ admits two sections, zero and infinite, with images $Y_{(0)}$ and $Y_{(\infty)}$. The Thom space, i.e. the projective cone $X = KY$ is obtained as a quotient of Q by contraction of $Y_{(\infty)}$ in a point $\{s\}$. It is the Thom space associated to the bundle H_Y, with basis Y.

Let us consider $p \colon Q \to Y$ as a sphere bundle with fiber $\mathbb{S}^2$, subbundle of a bundle $\bar{p} \colon \bar{Q} \to Y$ with fiber the ball $\mathbb{B}^3$. We denote by $\theta_{\bar{Q}} \in H^3(\bar{Q}, Q)$ the associated Thom class; one has a Gysin exact sequence

$$\ldots \to H_{j+1}(Y) \to H_{j-2}(Y) \xrightarrow{\gamma} H_j(Q) \xrightarrow{p_j} H_j(Y) \to \ldots;$$

in which the Gysin map γ is the composition

$$H_{j-2}(Y) \xrightarrow{(\bar{p}_{j-2})^{-1}} H_{j-2}(\bar{Q}) \xrightarrow{(\cap\theta_{\bar{Q}})^{-1}} H_{j+1}(\bar{Q},Q) \xrightarrow{\partial} H_j(Q)$$

and can be explicited in the following way: If ζ is a cycle in the class $[\zeta] \in H_{j-2}(Y)$, then $\gamma([\zeta])$ is the class of the cycle $p^{-1}(\zeta)$ in $H_j(Q)$.

Let us denote by π the canonical projection $\pi \colon Q \to KY$.

Proposition 7.1. *The Chern classes of Q and Y are related by the formula*

$$c_*(Q) = (1 + \eta_0 + \eta_\infty) \cap \gamma(c_*(Y)), \tag{7}$$

where $\eta_j := c^1\big(\mathcal{O}(Y_{(j)})\big) \in H^2(Q)$ for $j = 0$, ∞, and $\cap$ denotes the usual cap-product.

Proof. The vertical tangent bundle T_v of $p \colon Q \to Y$ is defined by the exact sequence:

$$0 \to T_v \to TQ \to p^*TY \to 0.$$

One has, in $H^*(Q)$

$$c^*(Q) = c^*(T_v) \cup c^*\big(p^*(TY)\big). \tag{8}$$

The sheaf of sections of the bundle T_v is the sheaf canonically associated to the divisor $Y_{(0)} + Y_{(\infty)}$, it is denoted by $\mathcal{O}_Q(Y_{(0)} + Y_{(\infty)})$. By Poincaré isomorphism in Y, the divisor $[Y_{(j)}] \in H_{2n}(Q)$ is identified to the class $\eta_j \in H^2(Q)$. The Chern class of T_v is

$$c^*(T_v) = 1 + \eta_0 + \eta_\infty.$$

By definition of the Gysin map γ, one has a commutative diagram

$$
\begin{array}{ccc}
H^i(Y) & \xrightarrow{\cap[Y]} & H_{2n-i}(Y) \\
\downarrow{\scriptstyle p^i} & & \downarrow{\scriptstyle \gamma} \\
H^i(Q) & \xrightarrow{\cap[Q]} & H_{2n+2-i}(Q)
\end{array}
\tag{9}
$$

and by Poincaré duality

$$c^*(p^*(TY)) \cap [Q] = p^*\big(c^*(TY)\big) \cap [Q] = \gamma(c^*(TY) \cap [Y]) = \gamma(c_*(Y)). \tag{10}$$

Using formulae (8) and (10), one obtains the formula (7):

$$c_*(Q) = (1 + \eta_0 + \eta_\infty) \cap \gamma(c_*(Y)) \qquad \square$$

24

7.2. *Schwartz-MacPherson classes of the projective cone*

Definition 7.1. We call homological projective cone and we denote by K the composition $K = \pi_*\gamma : H_{j-2}(Y) \to H_j(KY)$ for $j \geq 2$. For $j = 0$, i.e. for $H_{-2}(Y) = 0$, we let $K(0) := [a] \in H_0(KY)$ where $\{a\}$ is the vertex of the projective cone KY.

Let us remark that, for $j \geq 2$, K is an homomorphism.

Theorem 7.1. *Let $Y \subset \mathbb{P}^m$ be a projective variety and $\imath : Y \hookrightarrow KY$ the canonical inclusion into the projective cone KY on Y with vertex $\{a\}$. Let us denote also by $K : H_*(Y) \to H_{*+2}(KY)$ the homological projective cone, one has*

$$c_j(KY) = \imath_* c_j(Y) + K c_{j-1}(Y), \tag{11}$$

where $Kc_{-1}(Y)$ denotes the class $[a] \in H_0(KY)$

Proof. Let $\mathbf{1}_Q$ be the constructible function which is the characteristic function of Q, one has

$$\pi_*(\mathbf{1}_Q)(x) = \begin{cases} \chi(Y), & \text{if } x = a \\ 1, & \text{elsewhere,} \end{cases}$$

i.e.

$$\pi_*(\mathbf{1}_Q) = \mathbf{1}_{KY} + (\chi(Y) - 1)\mathbf{1}_{\{a\}}.$$

On the one hand, one has

$$\pi_* c_*(\mathbf{1}_Q) = c_*(\pi_*(\mathbf{1}_Q)),$$

one obtains

$$\pi_* c_*(Q) = c_*(KY) + (\chi(Y) - 1)\,[a]. \tag{12}$$

On the other hand, from the formula (7) one obtains:

$$\pi_* c_*(Q) = \pi_*\gamma(c_{*-1}(Y)) + \pi_*(\eta_0 \cap \gamma(c_*(Y))) + \pi_*(\eta_\infty \cap \gamma(c_*(Y))). \tag{13}$$

Let $\iota_0 : Y \hookrightarrow Q$ and $\iota_\infty : Y \hookrightarrow Q$ be the inclusions of Y as zero and infinite sections of Q respectively. By definition of γ, one has for every cycle ζ in Y and for $j = 0$ or ∞

$$\eta_j \cap \gamma([\zeta]) = (\iota_j)_*([\zeta])$$

then

$$\pi_*(\eta_j \cap \gamma(c_*(Y))) = \pi_* \iota_{j*} c_*(\mathbf{1}_Y) = \pi_* c_*(\mathbf{1}_{Y_{(j)}}) = c_* \pi_*(\mathbf{1}_{Y_{(j)}}).$$

Let us denote by $\iota = \pi \circ \iota_0 : Y \hookrightarrow KY$ the natural inclusion of Y in KY, one has

$$\pi_*(\mathbf{1}_{Y_{(0)}}) = \mathbf{1}_{\iota(Y)} \quad \text{and} \quad \pi_*(\mathbf{1}_{Y_{(\infty)}}) = \chi(Y)\mathbf{1}_{\{a\}}.$$

One obtains

$$\pi_*(\eta_0 \cap \gamma(c_*(Y))) = c_*(\mathbf{1}_{\iota(Y)}) = \iota_* c_*(Y),$$

and

$$\pi_*(\eta_\infty \cap \gamma(c_*(Y))) = \chi(Y)c_*(\mathbf{1}_{\{a\}}) = \chi(Y)[a],$$

where $[a]$ is the class of the vertex a in $H_0(KY)$. The comparizon of the formulae (12) and (13) gives:

$$c_*(KY) = \iota_* c_*(Y) + \pi_* \gamma c_{*-1}(Y) + [a],$$

and the Theorem 7.1. $\qquad\square$

7.3. *Case of the Segre and Veronese embeddings*

The previous construction associates canonically a Thom space $X = KY$ to the embedding of a smooth variety Y in $\mathbb{P}^m$. As examples, let us consider the image of the Segre embedding $\mathbb{P}^1 \times \mathbb{P}^1 \hookrightarrow \mathbb{P}^3$, defined in homogeneous coordinates by

$$(x_0 : x_1) \times (y_0 : y_1) \mapsto (x_0 y_0 : x_0 y_1 : x_1 y_0 : x_1 y_1),$$

and the image of the Veronese embedding $\mathbb{P}^2 \hookrightarrow \mathbb{P}^5$ defined by

$$(x_0 : x_1 : x_2) \mapsto (x_0^2 : x_0 x_1 : x_0 x_2 : x_1^2 : x_1 x_2 : x_2^2).$$

Chern classes and intersection homology of these exemples have been computed in [BG1] and, in a more systematic way in [BFK] . In the case of the Segre embedding, let d_1 and d_2 two fixed lines belonging each to a system of generatrices of the quadric $Y = \mathbb{P}^1 \times \mathbb{P}^1$. Let us denote by ω the canonical generator of $H^2(\mathbb{P}^1)$, one has $c^*(\mathbb{P}^1) = 1 + 2\omega$ and

$$c_*(Y) = c_*(\mathbb{P}^1 \times \mathbb{P}^1) = ([Y] + 2[d_1]) * ([Y] + 2[d_2]) = [Y] + 2([d_1] + [d_2]) + 4[y]$$

where y is a point in Y and where $*$ denotes the intersection of cycles or homology classes. One has

$$K(c_*(Y)) = [KY] + 2([Kd_1] + [Kd_2]) + 4[Ky].$$

Let us denote by $\sim$ the homology relation of cycles. In KY, one has ([BG1], 3):

$$Y \sim Kd_1 + Kd_2, \quad d_1 \sim d_2 \sim Ky, \quad y \sim a,$$

and, by Theorem 7.1

$$c_*(KY) = \underbrace{[KY]}_{H_6(KY)} + \underbrace{3([Kd_1] + [Kd_2])}_{H_4(KY)} + \underbrace{8[Ky]}_{H_2(KY)} + \underbrace{5[a]}_{H_0(KY)} \, ,$$

which is the result of [BG1].

In the case of the Veronese embedding, let d be a projective line in $Y = \mathbb{P}^2$, one has: $c^*(\mathbb{P}^2) = 1 + 3\omega + 3\omega^2$ where ω is the canonical generator of $H^2(\mathbb{P}_2)$, and is dual, by Poincaré isomorphism of the class $[d] \in H_2(\mathbb{P}_2)$. One has, by Poincaré duality

$$c_*(Y) = [Y] + 3[d] + 3[y]$$

where y is a point in Y. One has

$$K(c_*(Y)) = [KY] + 3[Kd] + 3[Ky]$$

such that, in KY, ([BG1], 3.b), $Y \sim 2Kd$, $d \sim 2Ka$ and $y \sim a$. One has

$$c_*(KY) = \underbrace{[KY]}_{H_6(KY)} + \underbrace{5[Kd]}_{H_4(KY)} + \underbrace{9[Ky]}_{H_2(KY)} + \underbrace{4[a]}_{H_0(KY)}$$

8. Polar varieties

A very nice and interesting historical introduction and complete bibliography for relation of characteristic classes with polar variieties can be found in the Teissier's paper [Te]. We use (and abuse of) it.

History of polar varieties began with Todd, in 1936. The basic idea is to consider what Todd calls the "Polar Loci" of a projective variety $X \subset \mathbb{P}^n$. It turns out that certain formal linear combinations of the intersections of general polar loci of X with general linear sections (of various dimensions) of X are invariants of X, i.e. do not depend upon the projective embedding of X and the choices of polar loci and linear sections.

More precisely, given a non singular $d - 1$-dimensional variety X in $\mathbb{P}^{N-1}$, for a linear subspace $L \subset \mathbb{P}^{N-1}$ of dimension $N - d + k - 2$, i.e. of codimension $d - k + 1$, let us set

$$P_k(X; L) = \{x \in X \,|\, \dim(T_{X,x} \cap L) \geq k - 1\}.$$

this is the polar variety of X associated to L. If L is general, it is either empty or the (pure) codimension in X is k.

Todd shows that the following formal linear combinations of varieties

$$X_k = \sum_{j=0}^{k} (-1)^j \binom{d-k+j+1}{j} P_{k-j}(X;L) \cap H_j$$

where H_j is a linear subspace of codimension j, are independent of all the choices made and of the embedding of X in a projective space, provided that the L's and the H_j's have been chosen general enough.

The linear combination is at first sight a rather awkward object to deal with. The idea is that X_k represents a generalized variety of codimension k in X, also any numerical character $e(Y)$ associated to algebraic varieties Y and which is additive in the sense that $e(Y_1 \cup Y_2) = e(Y_1) + e(Y_2)$ whenever Y_1 and Y_2 have the same dimension, can be extended by linearity to such a generalized variety. Given a partition $i_1, \ldots, i_k$ of $d - 1$, the intersection numbers

$$(X_{i_1} * \ldots * X_{i_k})$$

are well defined since the intersection of the corresponding varieties is zero dimensional. Here each X_i is assumed to be a general representative obtained by taking general and independant linear spaces. The intersection numbers depend only upon the structure of X as an algebraic variety.

Todd considered an equivalence relation between varieties, called rational equivalence. One of the main results of Todd is that the numbers $X_{i_1} * \ldots * X_{i_k}$ depend only upon X, that they are independent invariants and the arithmetic genus of X is a function of them.

The topological Euler-Poincaré characteristic of X can be computed to show the equality

$$\chi(X) = \deg X_d = \sum_{j=0}^{d} (j+1)(P_{d-j}(X).H_j)$$

where $(a.b)$ denotes the intersection number. In this case since we intersect with a linear space of complementary dimension, it is the degree of the projective variety $P_{d-j}(X)$.

After Nakano, Hirzebruch, Serre, Gamkrelidze, the invariants X_k of Todd (or rather their cohomology classes) coincide with the Chern classes of the tangent bundle of X.

9. Chern classes via polar varieties (smooth case)

The construction of Chern classes using Schubert varieties was already present in Chern's original paper. This construction was emphasized by

Gamkrelidze in [Ga1] and [Ga2].

The Schubert cell decomposition of the Grassmann manifold $G = G(n, m)$ of n-planes in $\mathbb{C}^m$ has been described by Ehresmann [Eh] and has been used by Chern to give an alternative definition of his characteristic classes. Let $(\mathcal{D})$

$$\{0\} = D_m \subset D_{m-1} \subset \cdots \subset D_1 \subset D_0 = \mathbb{C}^m \tag{14}$$

be a flag in $\mathbb{C}^m$, with $\operatorname{codim}_\mathbb{C} D_j = j$.

For each integer k, with $0 \le k \le n$, the k-th Schubert variety associated to (14), defined by

$$M_k(\mathcal{D}) = \{T \in G(n, m) : \dim(T \cap D_{n-k+1}) \ge k\}$$

is an algebraic subvariety of $G(n, m)$ of pure codimension k. The inequality condition is equivalent to saying that T and D_{n-k+1} do not span $\mathbb{C}^m$.

Let θ^n be the universal bundle over $G(n, m)$. The cycle $(-1)^k M_k(\mathcal{D})$ represents the image, under the Poincaré duality isomorphism, of the Chern class $c^k(\theta^n) \in H^{2k}(G(n, m))$. If V is an n-dimensional complex analytic manifold and $f : V \to G(n, m)$ is the classifying map for TV, i.e. such that $TV \cong f^*(\theta^n)$, then the cohomological Chern classes of V are $c^k(V) = c^k(TV) = f^*(c^k(\theta^n))$ (see [MS]).

Let us now consider the projective situation. We denote by $\tilde{G}(n, m)$ the Grassmann manifold of n-dimensional linear subspaces in $\mathbb{P}^m$. We fix a flag of projective linear subspaces $(\mathcal{L})$

$$L_m \subset L_{m-1} \subset \cdots \subset L_1 \subset L_0 = \mathbb{P}^m \tag{15}$$

where $\operatorname{codim}_\mathbb{C} L_j = j$. The k-th Schubert variety associated to $\mathcal{L}$ is defined by

$$M_k(\mathcal{L}) = \{\tilde{T} \in \tilde{G}(n, m) : \dim(\tilde{T} \cap L_{n-k+2}) \ge k - 1\}$$

Let us remark that we always have $\dim(\tilde{T} \cap L_{n-k+2}) \ge k - 2$. The Schubert variety $M_k(\mathcal{L})$ has codimension k in $\tilde{G}(n, m)$.

Let us denote $N = nm = \dim_\mathbb{C} \tilde{G}(n, m)$ and fix $0 \le \alpha \le m$. The Schubert variety

$$M_k^{N-\alpha} = \{(x, \tilde{T}) : x \in L_{\alpha-k}, \ x \in \tilde{T}, \ \dim(\tilde{T} \cap L_{n-k+2}) \ge k - 2\}$$

$$= L_{\alpha-k} \cap M_k(\mathcal{L}) \tag{16}$$

is the intersection of $M_k(\mathcal{L})$ with a general $(\alpha - k)$-codimensional plane and it has codimension α in $\tilde{G}(n, m)$. The (homological) Chern classes of $\tilde{G}(n, m)$ are

$$c_{N-\alpha}(\tilde{G}(n, m)) = \sum_{k=0}^{\alpha}(-1)^k \binom{n - \alpha + 1}{n - k + 1} M_k^{N-\alpha}. \tag{17}$$

Let us now consider the case of an n-projective manifold $V \subset \mathbb{P}^m$.

The k-th polar variety is defined by

$$P_k = \{x \in V : \dim(T_x(V) \cap L_{n-k+2}) \geq k - 1\},$$

where $T_x(V)$ is the projective tangent space to V at x. For L_{n-k+2} sufficiently general, the codimension of P_k in V is equal to k. Also, the class $[P_k]$ of P_k modulo rational equivalence in the Chow group $A_{n-k}(V)$ does not depend on L_{n-k+2} for L_{n-k+2} sufficiently general. This class is called the k-th polar class of V.

Let $\gamma : V \to \tilde{G}(n, m)$ be the Gauss map, i.e. the map defined by

$$\gamma(x) = T_x(V) \subset \mathbb{P}^m.$$

Then

$$P_k = \gamma^{-1}(M^k(\mathcal{L})).$$

The relation between Chern classes and Todd invariant has been described by Nakano [Na], Gamkrelidze [Ga1], [Ga2] and indirectly by Hirzebruch and Serre.

If $\mathcal{L} = \mathcal{O}_{\mathbb{P}^m}(1)|_V$, then one obtains the Todd formula (compare with (3)):

$$c_{n-\alpha}(V) = \sum_{k=0}^{\alpha}(-1)^k \binom{n - \alpha + 1}{n - k + 1} c^1(\mathcal{L})^{\alpha-k} \cap [P_k] \tag{18}$$

where the cap-product with $c^1(\mathcal{L})^{\alpha-k}$ is equivalent to the intersection with a general $(\alpha - k)$-codimensional plane.

10. Mather classes via polar varieties

The Mather classes have be defined in §6.1. One can provide an alternative definition, by using polar varieties. Let us firstly consider the situation of an affine variety $X^n \subset \mathbb{C}^m$. For a general flag $\mathcal{D}$ as in (14), one define (see diagramme (9))

$$\begin{array}{ccccc} \tilde{X} & \hookrightarrow & G(n,m) \times \mathbb{C}^m & \xrightarrow{\pi_1} & G(n,m) \\ {\scriptstyle \sigma \nearrow \downarrow \nu} & & \downarrow \pi_2 & & \\ X_{\mathrm{reg}} & \hookrightarrow & X & \hookrightarrow & \mathbb{C}^m \end{array}$$

and we denote by $\tilde{\gamma} = \pi_1|_{\tilde{X}} : \tilde{X} \to G(n,m)$ the Gauss map.

Let us define the following analytic subspace of X [LT]:

$$N_k(\mathcal{D}) = \nu \circ \tilde{\gamma}^{-1}(M_k(\mathcal{D})) = \overline{\nu(\tilde{\gamma}^{-1}(M_k(\mathcal{D})) \cap \sigma(X_{\mathrm{reg}}))}.$$

We will say that the flag $\mathcal{D}$ is good if it is sufficiently general, i.e. if $\tilde{\gamma}$ is transverse to the strata

$$M_{k,i}(\mathcal{D}) = \{W \in G(n,m) : \mathrm{codim}(W + D_{n-k+i-1}) = k+1\}$$

of $M_k(\mathcal{D})$. In that case, the cycle $N_k(\mathcal{D})$ is well defined and independent of the choice of the good flag, it is called the polar variety (Lê - Teissier).

If the flag $\mathcal{D}$ is good, and still in the affine situation, let $\pi : X \to \mathbb{C}^{n-k+1}$ be the restriction to X of a linear projection with kernel D_{n-k+1}, then $N_k(\mathcal{D})$ is the closure (in X) of the critical locus of the restriction of π to X_{reg} [LT].

Let us consider now the projective case, the Mather class can also be defined using polar varieties [LT]. Let us denote by $X^n \subset \mathbb{P}^m$ a projective variety, we define the k-th polar variety P_k as the closure of

$$\{x \in X_{\mathrm{reg}} : \dim(T_x(X_{\mathrm{reg}}) \cap L_{n-k+2}) \geq k-1\}.$$

Then one has [Pi2]:

$$c_{n-s}^M(X) = \sum_{k=0}^{s} (-1)^k \binom{n-s+1}{n-k+1} c^1(\mathcal{L})^{s-k} \cap [P_k],$$

where $\mathcal{L} = \mathcal{O}_{\mathbb{P}^m}(1)|_X$.

That provides an expression of Schwartz-MacPherson classes in terms of polar varieties. In particular cases, the Fulton and Milnor classes can also be expressed in terms of polar varieties (see [AB1]).

References

AB1. P. Aluffi and J.P. Brasselet, *Interpolation of Characteristic Classes of Singular Hypersurfaces.* Advances in Maths. 180 N2, (2003), 692-704.

AB2. P. Aluffi and J.P. Brasselet, *Une nouvelle preuve de la coïncidence des classes définies par M.-H. Schwartz et par R. MacPherson.* Preprint MPI, 2006 - 84.

BBF. G. Barthel , J.P. Brasselet et K.-H. Fieseler, *Classes de Chern des variétés toriques singulières*, C.R.A.S. t. 315, Srie I, p. 187-192, 1992.

Br1. J.P. Brasselet, *Définition combinatoire des homomorphismes de Poincaré, Alexander et Thom pour une pseudo-variété*, Astérisque n 82-83, 1981.

Br2. J.P. Brasselet, *Characteristic classes and Singular Varieties*, Vietnam Journal of Mathematics, 33 (2005) 1-16.

Br3. J.P. Brasselet, *Milnor classes via polar varieties,* Contemporary Mathematics, 266 (2000), 181 - 187.

Br4. J.P. Brasselet, *Poincaré-Hopf Theorems on Singular Varieties*, to appear in the Proceedings of the Trieste school and Worshop 2005, World Scientific.

Br5. J.P. Brasselet, *Characteristic classes*, book in preparation.

BFK. J.P. Brasselet, K.-H. Fieseler et L. Kaup, *Classes caractéristiques pour les cônes projectifs et homologie d'intersection*, Comment. Math. Helvetici 65 (1990) 581-602.

BG1. J.P. Brasselet et G. Gonzalez-Springberg. *Espaces de Thom et contre-exemples de J.L. Verdier et M. Goresky*, Bol. Soc. Brazil. Mat. **17** (1986), no 2, 23–50.

BG2. J.P. Brasselet et G. Gonzalez-Springberg. *Sur l'homologie d'intersection et les classes de Chern des variétés singulières* with an appendix of Jean-Louis Verdier: *Un calcul triste*, Travaux en cours n° 23, 5 – 11 Hermann (1987).

BS. J.P. Brasselet, M.-H. Schwartz: *Sur les classes de Chern d'un ensemble analytique complexe*, Astérisque 82-83 (1981), 93-147.

BSS1. J.-P. Brasselet, J. Seade, T. Suwa *An explicit cycle representing the Fulton-Johnson class I*, to appear in Séminaires et Congrès, SMF, Franco-japanese congress, Marseille, 2002.

BSS2. J.P. Brasselet, J. Seade, T. Suwa *Indices of Vector fields and characteristic Classes of singular Varieties*, book in preparation.

Eh. C. Ehresmann, *Sur la topologie de certains espaces homogènes,* Annals of Math., **35**, No 2 (1934).

Fu. W. Fulton, *Intersection Theory*, Springer-Verlag, (1984).

Ga1. P.B. Gamkrelidze, *Computation of the Chern cycles of algebraic manifolds* (in Russian) Doklady Akad. Nauk.,**90**, No 5 (1953), 719–722.

Ga2. P.B. Gamkrelidze, *Chern's cycles of complex algebraic manifolds* (in Russian) Izv. Akad. Nauk. SSSR, Math. Ser. **20** (1956), 685–706.

LT. Lê D. T. et B. Teissier. *Variétés polaires locales et classes de Chern des variétés singulières*, Ann. of Math 114 1981, 457-491.

MP. R.MacPherson, *Chern classes for singular algebraic varieties,* Ann. of Math. **100**, no 2 (1974), 423-432.

MS. J. Milnor and J. Stasheff, *Characteristic Classes,* Princeton University Press (1974).

Na. S. Nakano. *Tangential vector bundle and Todd canonical systems on an algebraic variety,* Mem. Coll. Sci. Univ. Kyoto Ser. A. Math. **29** (1955) 145–149.

Pi1. R. Piene, *Polar classes of singular varieties,* Ann. Sc. E.N.S. **11**, (1978),

247-276.

Pi2. R. Piene, *Cycles polaires et classes de Chern pour les variétés projectives singulières,* Séminaire Ecole Polytechnique, Paris, 1977-78 and Travaux en cours **37**, Hermann Paris (1988), 7-34.

SY. J. Schürmann and S. Yokura, *A Survey of Characteristic Classes of Singular Spaces,* in this volume.

Sc1. M.-H.Schwartz, *Classes caractéristiques définies par une stratification d'une variété analytique complexe,* CRAS **260**, (1965), 3262-3264 et 3535-3537.

Sc2. M.-H. Schwartz: *Champs radiaux sur une stratification analytique,* Travaux en cours, 39 (1991), Hermann, Paris.

Sc3. M.-H. Schwartz, *Classes obstructrices des ensembles analytiques* 2001.

St. N. Steenrod, *The Topology of Fibre Bundles,* Princeton Univ. Press (1951).

Su. T. Suwa, *Classes de Chern des intersections complètes locales,* C.R.Acad.Sci. Paris, **324**, (1996), 67-70.

Te. B. Teissier, *Quelques points de l'histoire des variétés polaires, de Poncelet à nos jours.* Séminaire d'Analyse, 1987–1988, Clermont-Ferrand, Exp. No. 4, Univ. Clermont-Ferrand II, 1990.

Wh. H. Whitney *Tangents to an analytic variety,* Ann of Math 81, 496 – 549 (1965).

Yo. S. Yokura*On a Milnor class,* Preprint 1997.

DESINGULARIZATION OF IDEALS AND VARIETIES

HERWIG HAUSER

University of Innsbruck, Austria

Singular mobiles were introduced by Encinas and Hauser in order to conceptualize the information which is necessary to prove strong resolution of singularities in characteristic zero. It turns out that after Hironaka's Annals paper from 1964 essentially all proofs rely – either implicitly or explicitly – on the data collected in a mobile, often with only small technical variations. The present text explains why mobiles are the appropriate resolution datum and how they are used to build up the induction argument of the proof.

Keywords: resolution, blowups, singularities.

CLASS 1: Examples etc.

The Cylinder $X_1 : x^2 + y^2 = 1$ in $\mathbb{A}^3$ contracts under $(x, y, z) \to (xz, yz, z)$ to the Cone $X_2 : x^2 + y^2 = z^2$. The linear change $(x, y, z) \to (x, 2y, z + y)$ transforms this equation into $X_2 : x^2 + (y - z)z = 0$. This Cone contracts under $(x, y, z) \to (xy, y, yz)$ to the Calypso $X_3 : x^2 + y^2 z = z^2$. From there, we get via $(x, y, z) \to (xz, y, z)$ the Calyx of equation $X_4 : x^2 + y^2 z^3 = z^4$. In this way, the Calyx is represented as the image of a smooth scheme under a rational map. We have parametrized a singular surface by a regular one (see fig. 1-4).

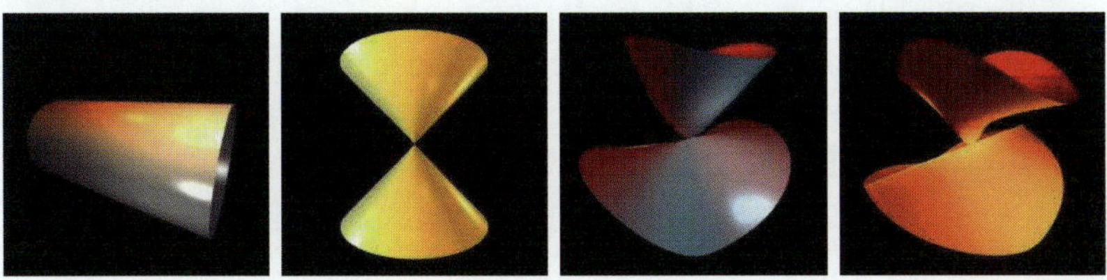

Figures 1-4: Resolution of Calyx by three successive blowups.

Start now with Calyx. We propose in this course to describe methods which allow to reconstruct from its equation the above or some other paramatrization.

Setting: X reduced singular scheme of finite type over a field K, mostly affine, $X = \operatorname{Spec} A$, with A a nilpotent-free finitely generated K-algebra. Choose a presentation $K[x_1, \ldots, x_n] \to A = K[x_1, \ldots, x_n]/I$ for some ideal I of $K[x] = K[x_1, \ldots, x_n]$. This corresponds to an embedding $X \subset \mathbb{A}^n = \mathbb{A}^n_K$ with $X = V(I)$. We may also choose generators $I = (g_1, \ldots, g_k)$. The singular locus $\operatorname{Sing} X$ of X is a closed reduced subscheme of X.

Example: The Spitz of equation $(z^3 - x^2 - y^2)^3 = x^2 y^2 z^3$ in $\mathbb{A}^3$ (fig. 7). The singular locus consists of two cusps (one in xz-plane, one in yz-plane) with the same tangent at 0. Isomorphic to the cartesian product of plain cusp with itself.

Resolution of singularities: Surjective morphism $\tilde{X} \to X$ with $\tilde{X}$ regular. Also: Desingularization, parametrization, projection, shadow.

Embedded resolution: Given X in a regular W, a proper birational morphism $\Pi : \tilde{W} \to W$ and a regular $\tilde{X} \subset \tilde{W}$ which maps under Π onto X and is transversal to exceptional divisor $E = \pi^{-1} Z$, where $Z \subset W$ is the locus above which π is not an isomorphism (usually: $Z = \operatorname{Sing} X$).

Strong resolution of $X \subset W$: Embedded resolution $\pi : \tilde{X} \to X$ induced by $\Pi : \tilde{W} \to W$ such that:

- π isomorphism outside $\operatorname{Sing} X$ (*economy*);
- π independent of embedding $X \subset W$ (*excision*);
- π commutes with smooth morphisms (*equivariance*), in particular with open immersions, localization, completion, with taking cartesian product with regular scheme, field extensions, group actions on X lift to action on $\tilde{X}$;
- π is composition of blowups in regular centers (*explicitness*);
- centers of blowup are the top locus of a local upper semicontinuous invariant (*effectiveness*).

Exercises: (1) Prove that the maps given at the very beginning yield indeed a resolution of the Calyx. Show that all properties of an embedded resolution are fulfilled. Determine the centers of blowup as well as all exceptional components.

(2) Find for the Kolibri of equation $x^2 = y^2 z^2 + z^3$ a resolution (fig. 5). Determine first the geometry and the singular locus. Try as first centers both the origin and the singular locus.

(3) Show that the map $\mathbb{A}^2 \to \mathbb{A}^3$ given by $(s,t) \to (st, s, t^2)$ parametrizes the Whitney-umbrella $X : x^2 = y^2 z$ (fig. 6). Is it a resolution? Check if all required properties hold.

(4) Show that the blowup of the Whitney-umbrella with center the origin yields a surface which has one cone-like isolated singularity and at another point the singularity of the Whitney-umbrella (fig. 6'). Conclude from this that the singularities need not improve if the centers are too small.

(5) Determine all finite symmetries of the Spitz (fig. 7). Then show that it is isomorphic to the cartesian product of the cusp $x^2 = y^3$ in the plane $\mathbb{A}^2$ with itself. Find other embeddings of this product into $\mathbb{A}^3$.

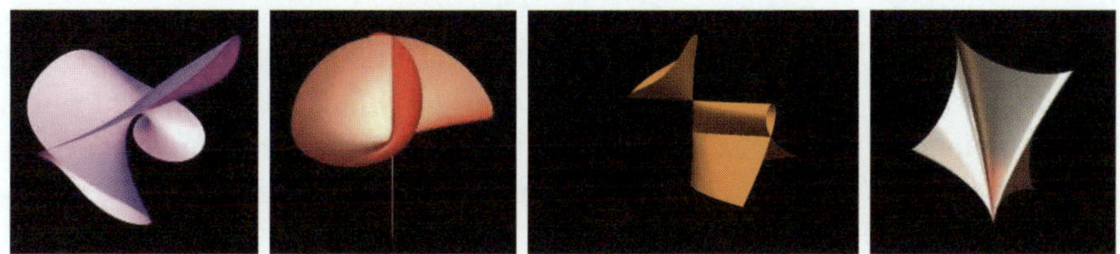

Figures 5, 6, 6' and 7.

CLASS 2: Blowups

For convenience, we restrict to blowups of affine space $\mathbb{A}^n$ whose centers are regular closed subschemes Z of $\mathbb{A}^n$. All constructions extend naturally to arbitrary regular ambient schemes and centers therein.

The center Z is defined in $\mathbb{A}^n$ by an ideal I_Z of $K[x_1, \ldots, x_n]$, for which we may choose generators $g_1, \ldots, g_k \in K[x]$. Consider then the map

$$\gamma : \mathbb{A}^n \setminus Z \to \mathbb{P}^{k-1} : a \to (g_1(a) : \ldots : g_k(a))$$

where $(u_1 : \ldots : u_k)$ denote projective coordinates in $\mathbb{P}^{k-1}$. The graph Γ of γ lives in $(\mathbb{A}^n \setminus Z) \times \mathbb{P}^{k-1}$. We define $\tilde{\mathbb{A}}^n$, the blowup of $\mathbb{A}^n$ in Z, as the Zariski closure of this graph

$$\tilde{\mathbb{A}}^n = \overline{\Gamma} \subset \mathbb{A}^n \times \mathbb{P}^{k-1}.$$

It comes with a natural projection $\Pi : \tilde{\mathbb{A}}^n \to \mathbb{A}^n$, the blowup map, induced from the projection $\mathbb{A}^n \times \mathbb{P}^{k-1} \to \mathbb{A}^n$ on the first n components (cf. fig. 8) . Different choices of the generators of I_Z yield isomorphic blowups. The preimage $Y' = \Pi^{-1}(Z)$ is a hypersurface in $\tilde{W}$ called the exceptional divisor. Letting $(u_1 : \ldots : u_k)$ denote projective coordinates in $\mathbb{P}^{k-1}$, the equations of $\tilde{\mathbb{A}}^n$ in $\mathbb{A}^n \times \mathbb{P}^{k-1}$ are

$$u_i g_j(g) - u_j g_i(x) = 0 \ \text{ for all } \ i \ \text{ and } \ j.$$

We may cover projective space $\mathbb{P}^{k-1}$ by k affine charts isomorphic to $\mathbb{A}^{k-1}$ and given by $u_j \neq 0$ for $j = 1, \ldots, k$. This, in turn, yields a covering of $\tilde{\mathbb{A}}^n$ by k affine charts isomorphic to $\mathbb{A}^n$, so that the chart expressions of Π can be read off as polynomial maps from $\mathbb{A}^n$ to $\mathbb{A}^n$. It will always be this description we use to carry out computations and proofs.

If Z is a coordinate subspace, defined by, say, $x_j, j \in J$, for some subset J of $\{1, \ldots, n\}$ (this can always be achieved locally after passing to completions), the chart expression in the j-th chart is

$$\Pi_j : \mathbb{A}^n \to \mathbb{A}^n : x_i \to x_i \ \text{ for } \ i \notin J \setminus j,$$
$$x_i \to x_i x_j \ \text{ for } \ i \in J \setminus j.$$

There are several other ways to define blowups, e.g. by a universal property or as the Proj of the Rees algebra associated to the ideal of Z. See the

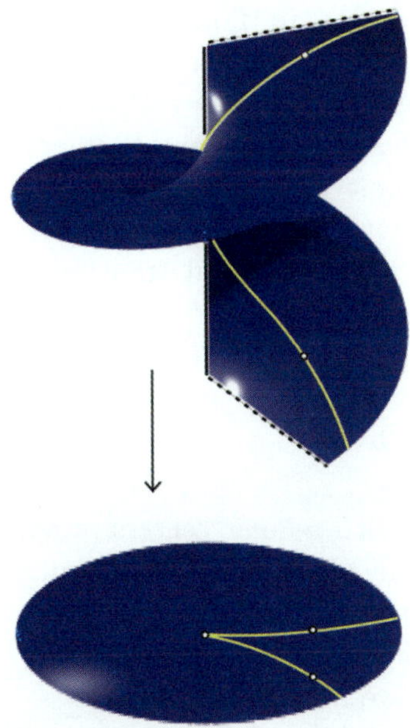

Figure 8: The blowup of $\mathbb{A}^2$ in the origin.

lectures [65] or the book of Eisenbud and Harris [54] for more details.

Properties of blowups:

- They are proper birational maps.
- They induce an isomorphism over the complement of the center.
- Blowups commute with localization, completion, restriction to (open or closed) subschemes containing the center (make precise what is meant here – you will have to take the strict transform of the subscheme).
- If $W_1 = W \times L$ with L regular, the blowup of W_1 in $Z_1 = Z \times L$ is the cartesian product of the blowup of W in Z with the identity on L.
- Compositions of blowups are again blowups. There is a procedure by G. Bodnár to determine an appropriate center whose blowup yields the composition; it is defined by a non-reduced ideal.
- Local blowups $(\tilde{W}, a') \to (W, a)$ (specify what shall mean "local")

admit coordinates in W at a which make the map monomial. (What happens if a' moves along Y', how must the coordinates change?)

- If Z is regular and transversal to $X \subset W$ regular or normal crossings (in the sense of the exercises below) then the total transform X^* is a scheme with at most normal crossings. If X is regular, the strict transform X^s is again regular (and transversal to Y').
- The ideal I^* of the total transform $X^* = \Pi^{-1}(X)$ of X in $\tilde{W}$ factors into $I^* = I^o_{Y'} \cdot I^\Upsilon$ for a certain ideal I^Υ in $\tilde{W}$ (the weak transform of I) where o denotes the order of X along Z in W (see the next section). Here, o is the maximal power with which $I_{Y'}$ can be factored from I^*.

Remarks. Blowups with regular centers provide a simple algebraic modification of regular schemes W and their singular subschemes X, being just a monomial substitution of the variables. Heuristically speaking, the blowup reveals the shape of X along Z up to the first available order of the Taylor expansion. By this we mean the following:

Example. Consider the line $L : y = 0$ in $\mathbb{A}^2$ and the tangent k-th order parabola $P : y = x^k$. Both meet at the origin with multiplicity k (i.e., the intersection is a k-fold zero). Blowing up the origin, the y-chart with map $(x, y) \to (xy, x)$ is the relevant one (explain why). There, the strict transforms of L and P have equations $y = 0$ and $y = x^{k-1}$, so the order of tangency has decreased by 1.

This shows that blowups in regular centers are a very rough device to resolve singularities. They take into account only a small portion of the geometry of X. There exist other modifications, for instance the Nash modification or normalization, which are somewhat more sophisticated procedures. However, they lack some of the basic algebraic features blowups have and which make them so useful.

Exercises: (1) Show that if Z is a regular hypersurface, the blowup map Π is an isomorphism.

(2) Show that for regular centers Z in $\mathbb{A}^n$, the blowup $\tilde{\mathbb{A}}^n$ is again regular and of dimension n.

(3) Determine explicitly the covering of $\tilde{\mathbb{A}}^n$ by affine charts and the corresponding chart expressions of the blowup map. Then express it in terms of the respective coordinate rings as a certain ring extension.

(4) Show that if W is a cartesian product $W_1 \times Z$ and a a point in Z, the blowup $\tilde{W} \to W$ of W in Z is the cartesian product of the blowup $\tilde{W}_1 \to W_1$ of W_1 in $Z_1 = \{a\}$ with the identity on Z. Then make the explicit local computations of the blowup of a circle in $\mathbb{A}^3$.

(5) Two (or several) schemes are called transversal at a point a if the product of their ideals is a monomial ideal (locally at a, with respect to suitable formal coordinates). Take three regular surfaces in $\mathbb{A}^3$ so that each two meet transversally. Show that all three need not meet transversally. What happens if you require in addition that all possible intersections of two of the schemes meet transversally?

(6) Consider the blowups $\tilde{W}$ and $\tilde{U}$ of $W = U = \mathbb{A}^3$ in the two centers Z and Z_1 of ideals (xy, z), respectively $(xy, z)(x, z)(y, z)$. What do you observe? Then apply a second blowup (with center a point of your choice) to $\tilde{W}$ and show that the composition equals the blowup $\tilde{U} \to U$ (provided that you have chosen the correct point on $\tilde{W}$).

(7) Define and compute the strict transform of a plane vector field under the blowup of $\mathbb{A}^2$ in a point. Do you always get a vector field on $\tilde{\mathbb{A}}^2$?

(8) Blow up the Fanfare $x^2 + y^2 = z^3$ in $\mathbb{A}^3$ once with center the origin and once with center the z-axis. Compute the orders of the respective strict transforms.

(9) Show that the blowup of $\mathbb{A}^2$ with center the non-reduced origin of ideal $(x, y^k)(x, y^{k-1}) \cdots (x, y)$ gives a regular scheme $\tilde{\mathbb{A}}^2$ and separates the two components of $x(x - y^k) = 0$. Interpret this blowup as a composition of blowups in regular (reduced) centers.

(10) What are the total and strict transforms of a regular hypersurface X in W if the center equals X?

CLASS 3: Transforms

Throughout, $\pi : W' \to W$ denotes the blowup of a scheme W in a regular center Z with exceptional divisor $Y' = \pi^{-1}(Z) \subset W'$. We shall describe various ways how to lift schemes and ideals in W to schemes and ideals in W'. Again, we shall stick to an affine scheme $W = \mathbb{A}^n$ with coordinate ring $\mathbb{K}[x] = \mathbb{K}[x_1, \ldots, x_n]$ and even work locally at a chosen point a of W – taken to be the origin of $\mathbb{A}^n$, so that we may argue in the formal power series ring $\mathbb{K}[[x]] = \mathbb{K}[[x_1, \ldots, x_n]]$. By a' we shall always denote a point in Y' mapping under π to a. Choosing the local coordinates $x_1, \ldots, x_n$ suitably at a we may assume that a' is the origin of one of the affine charts on $W' = \tilde{\mathbb{A}}^n$ and the respective chart expression of the blowup map is given by an algebra-homomorphism $\varphi : \mathbb{K}[x] \to \mathbb{K}[x]$ sending x_i to either x_i or $x_i x_j$ as specified earlier (and certainly proven by you in the exercises).

Let X be closed in W and given by the ideal J of $K[x]$ (X need not be reduced, but we assume that X is rare in W, i.e., not equal to one or several components of W). The total transform X^* of X is the pullback $\pi^{-1}(X)$ of X in W' under π. Thus, locally at a', its ideal equals $J^* = \varphi(J) = (f \circ \pi, f \in J)$. If $Z \subset X$ then X^* contains Y' as a component (because $X \subsetneq W$ locally at all points). As Y' is a hypersurface, we get a factorization $J^* = M' \cdot I'$, where M' is a suitable power of the principal ideal $I_{Y'}$ defining Y' in W', say $M' = I_{Y'}^o$ for some $o > 0$. The maximal power o of $I_{Y'}$ which can be factored from J^* is given by the behavior of X along Z. More precisely:

The order $\mathrm{ord}_Z X$ of X along Z is defined as the maximal integer k so that $J \subset I_Z^k$. In particular, if $Z = \{a\}$ is a (reduced) point and X is a hypersurface $f = 0$, say $J = (f)$, then $\mathrm{ord}_a X$ is just the order of vanishing of f at a, i.e., the order of the Taylor expansion of f at a. If X is not a hypersurface, the order equals the minimum of the orders at a of a generator system of the defining ideal of X. Of course, it depends only on the stalk of J at a, and the order remains the same when passing to completions. Note that the order depends on the embedding of X in W at a. If X is not minimally embedded locally at a, (i.e., the dimension of W at a is not minimal among all local embeddings of X at a in a regular ambient scheme) the order of X at a is 1. In this case, the order is not significant for describing the complexity of the singularity of X at a.

For $c \in \mathbb{N}$, we let $\mathrm{top}(X, c)$ be the locus of points where the order of X in W is at least c. By the upper semicontinuity of the order, the top locus is a closed (reduced) subscheme. We let $\mathrm{top}(X)$ be the locus of points where the order of X in W is maximal. Of course, we can also define $\mathrm{top}(X)$ locally at a point a, as the local subscheme where the order of X equals $\mathrm{ord}_a X$.

With these definitions we get the factorization of the total transform $J^* = M' \cdot I'$, where $M' = I_{Y'}^o$, for $o = \mathrm{ord}_Z X = \mathrm{ord}_Z I$. This order is the maximal power with which $I_{Y'}$ can be factored from J^*. We call $X^\curlyvee$ and $J^\curlyvee = I_{Y'}^{-o} \cdot J^*$ the weak transform of X and J under the blowup $\pi : W' \to W$. If X is a hypersurface, it coincides with the strict transform X^s of X.

One of the basic facts for allowing resolution in the spirit of Hironaka is the following: *If the center Z is contained in the top locus of X, the order of the weak transform $X^\curlyvee$ at points of Y' is less or equal the order of X along Z,*

$$\mathrm{ord}_{a'} X^\curlyvee \le \mathrm{ord}_a X.$$

This holds also for the strict transform (as a consequence of the inequality), and for the Hilbert-Samuel function of X at points a, requiring that it is constant along Z and taking a natural ordering among all Hilbert-Samuel functions (see Bennett's paper [22] or [80], [71]).

Properties: As blowups did, passing to the weak transform commutes with restriction to open subschemes, localization and completion. Also, if X and Z are invariant under a group action, the group action lifts to $X^\curlyvee$. There are three algebraic properties of weak transforms which we will use repeatedly.

If P and Q are ideals in W, we have $(P \cdot Q)^\curlyvee = P^\curlyvee \cdot Q^\curlyvee$. However, $(P + Q)^\curlyvee \ne P^\curlyvee + Q^\curlyvee$ in general, it suffices to take two principal ideals of different order along Z. If $\mathrm{ord}_Z P = \mathrm{ord}_Z Q$, the the weak transform is distributive, say $(P + Q)^\curlyvee = P^\curlyvee + Q^\curlyvee$. There is a nice trick to achieve this equality also in case $p = \mathrm{ord}_Z P \ne \mathrm{ord}_Z Q = q$. Replace $P + Q$ by the weighted sum $P^q + Q^p$ and get $(P^q + Q^p)^\curlyvee = (P^q)^\curlyvee + (Q^p)^\curlyvee$. As we have $\mathrm{ord}_a P^q = q \cdot \mathrm{ord}_a P$, we do not lose information on the order when passing to powers of ideals.

The third commutation property of weak transforms is with respect to coefficient ideals. These play a decisive role in the induction on the dimension as they allow to pass to ideals in less variables. Their definition is somewhat cumbersome. Let $W = \mathbb{A}^n$ with local coordinates $(x_n, \ldots, x_1)$ at a. For simplicity, we take $a = 0$. Let V be the hypersurface in W defined by $x_n = 0$. Let I be an ideal in W at a of order $o = \mathrm{ord}_a I$. The coefficient ideal of I at a in V is defined as the ideal in V generated by certain powers of the coefficients of the elements of I when expanding these with respect to x_n. More precisely, write $f = \sum_{i \ge 0} a_{i,f}(x')x_n^i$ for $f \in I$ and with $x' = (x_{n-1}, \ldots, x_1)$. Then

$$\operatorname{coeff}_V(I) = ((a_{i,f},\ f \in I)^{\frac{o}{o-i}},\ i < o).$$

For example, if $f(x) = x_n^o + g(x')$ has no mixed terms, we get $\operatorname{coeff}_V(f) = (g)$. You will object that, in general, the exponents $\frac{o}{o-i}$ are rational numbers. This can be remedied by taking instead as exponents $\frac{o!}{o-i}$, producing for $f(x) = x_n^o + g(x')$ the ideal $(g^{(o-1)!})$. Taking factorials loadens the notation without improving the understanding, so we will allow rational exponents and leave it to the reader to define the correct equivalence relation on rational powers of ideals in order to circumvent any traps. As the order of ideals is just multiplied with a constant when passing to powers of ideal, there is no harm in having rational exponents (once you got used to it).

Let $Z \subset \operatorname{top}(I)$ be the center of the blowup $\pi : W' \to W$, with weak transform $I^\curlyvee$. Let $a \in Z$ be a point, $V \subset W$ a local hypersurface of W at a (i.e., defined in a neighborhood of a), let $a' \in Y'$ be a point in Y' so that a' lies above a and in the strict ($=$ weak) transform V' of V. We already know that $\operatorname{ord}_{a'} I^\curlyvee \leq \operatorname{ord}_a I$. If $\operatorname{ord}_{a'} I^\curlyvee < \operatorname{ord}_a I$ we are happy because something has improved. If equality holds (we then say that a' is an equiconstant point for I), we have at least the following commutativity relation

The coefficient ideal of $I^\curlyvee$ at a' in V' is the transform of the coefficient ideal of I at a in V,

$$\operatorname{coeff}_{V'} I^\curlyvee = (\operatorname{coeff}_V I)^!.$$

This equality does not hold if the order has dropped, and it neither holds if we take on the right hand side the weak transform $(\operatorname{coeff}_V I)^\curlyvee$ of $\operatorname{coeff}_V I$. Instead, we have to take a new transform, the so called controlled transform. Let $c = \operatorname{ord}_Z I$ and define $(\operatorname{coeff}_V I)^! = I_{Y' \cap V'}^{-c} \cdot (\operatorname{coeff}_V I)^*$ (the number c is called the control). This is not hard to prove after passing to local coordinates, using that we always have $\operatorname{ord}_a \operatorname{coeff}_V I \geq \operatorname{ord}_a I$. The magic formula with the controlled transform of the coefficient ideal allows to compare ideals in smaller dimension precisely in the case where the order of the original ideal I could not tell us that the singularities improved under blowup. This output recompenses by far the lack of elegance we had to accept in the definition of coefficient ideals.

Be careful: The coefficient ideal nor its order are intrinsic objects. We will have to make an effort to extract coordinate independent information from them.

Exercises: (1) Show that $\mathrm{ord}_Z X = \min_{a \in Z} \mathrm{ord}_a X$ and that $\mathrm{ord}_a X$ defines an upper semicontinuous function on W. Look up in Hironaka's or Bennett's Annals papers [77], [22] why it does not increase under localization. (This holds also for the Hilbert-Samuel function of X at a.)

(2) Compute for several schemes X in W the order of X along a subscheme Z of W. Then determine for each X the stratification of X by the strata of constant order (with respect to points).

(3) If X is not locally minimally embedded in W at a point $a \in X$, the order of X at a equals 1.

(4) Try to find (natural) equations for the top locus $\mathrm{top}(X, c)$, first in characteristic 0 (easy), then in arbitrary characteristic. In the first case, show that $\mathrm{top}(X)$ lies locally in a regular hypersurface of W.

(5) In $J^* = M' \cdot I'$ the order $o = \mathrm{ord}_Z X = \mathrm{ord}_Z I$ is the maximal power with which $I_{Y'}$ can be factored from J^*.

(6) Let X be a subscheme of W of codimension at least 2, with strict and weak transforms X^s and $X^\curlyvee$ under the blowup $\pi : W' \to W$. Figure out in three examples which components of $X^\curlyvee$ do not show up in X^s.

(7) (Mandatory) Show that the order of an ideal I in W does not increase when passing to its weak transform, provided the center is included in $\mathrm{top}(I)$. Hint: You may work locally in the completion, rectifying thus the center to a coordinate subspace, and then choose coordinates for which the local blowup $(W', a') \to (W, a)$ is monomial. (In exceptional cases you are allowed to consult [70] to convince you that it would have been easy.)

(8) Determine in three examples the equiconstant points of an ideal I under blowup, i.e., the points $a' \in Y'$ where the order of $I^\curlyvee$ has remained constant.

(9) Compute the coefficient ideals of $f = x^3 + yz^2$ and $f = x(y^7 - z^8)$ at 0 with respect to the three coordinate hypersurfaces. Compare the respective orders of the resulting ideals.

(10) Compute the coefficient ideals of the polynomials of (9) after blowing up the origin and compare them with the controlled transforms of the coefficient ideals below. Then prove the commutativity of the passage to coefficient ideals with blowups at equiconstant points.

44

CLASS 4: Construction of Mobiles

In this section we wish to guide you towards the correct definition of mobiles. They shall be intrinsic, globally defined objects at a certain stage of the resolution process containing all information we need in order to define the local resolution invariant and to choose the center of the next blowup. In the last section we shall give the precise definition of mobiles and show how they transform under the blowup with the chosen center. This, in turn, will be used to compute the local invariant after blowup and to show that it decreases at each point of the new exceptional component.

So let us start with an ideal sheaf $\mathcal{J}$ on our regular ambient scheme W. We choose a point $a \in W$ and let J denote the stalk of $\mathcal{J}$ at a. Taking an affine neighborhood of a in W we may simply assume that J is an ideal of polynomials in n variables with coefficients in the ground field $\mathbb{K}$.

Mobiles control two features of the resolution process: The factorization of ideals into a monomial and a singular part (the exceptional and the not yet resolved portion of the ideal), and the transversality of the chosen centers with the respective exceptional locus. The first task is accomplished by the combinatorial handicap D of the mobile, the second by the transversal handicap E. It is appropriate to introduce them in separate sections.

The combinatorial handicap

We have already seen that after blowup powers of the exceptional components will be factored from the total transform of the ideals, so in order to keep things systematic (which is not very original but helpful) we write $J = M \cdot I$ with $M = 1$ the trivial ideal (the whole local ring) and $I = J$.

Here is a nice idea: We proceed as we would know what the center of the first blowup is (you remember: often uniqueness is proven before existence, because then you already know how your object has to look like when you try to construct it). So let Z be a closed regular subscheme of W with induced blowup $\pi : W' \to W$ of W along Z and exceptional divisor $Y' \subset W'$. We denote by $I_{Y'} = I_{W'}(Y')$ the ideal defining Y' in W'. We let $J' = J^*$ be the total transform (inverse image) of J under π. Its order will have increased, so that's not a good number to look at. Much more interesting is the weak transform, and, to keep things straight, we denote it by $I' = I^{\curlyvee}$. Thus $J' = M' \cdot I'$ with $M' = I_{Y'}^{\operatorname{ord}_z I}$ a normal crossings divisor (even regular for we have blown up only once). Locally, M' is just a power of a variable (the variable defining Y' in W').

Set now $o = \mathrm{ord}_Z I$, let a be a point in Z (outside of Z nothing will happen since π is an isomorphism there), and let a' be any point in Y' above a, say $\pi(a') = a$. Set $o' = \mathrm{ord}_{a'} I'$. The next thing to do is to compare o' with o. Here we remember the key inequality from earlier sections: If the order of I along Z is constant, in particular if $Z \subset \mathrm{top}(I)$, and hence $o' = \mathrm{ord}_a I$ for all $a \in Z$, we have $o' \le o$, because I' is the weak transform of I. In view of this pleasant event ("the order does not increase") we immediately agree to allow only centers inside $\mathrm{top}(I)$. "Ah", you respond, "maybe we even have $o' < o$ for all a' above a." Then we would be done. – Sorry, this is too optimistic, the equlity $o' = o$ may occur and the points where this happens form a closed subscheme of Y' (but prove that $o' < o$ if $n = 1$ and $o > 0$). These are the equiconstant points of I in Y'.

Before confronting this situation, we do some book-keeping. We will call D_n and D'_n the (non-reduced) divisors defined by M and M' in W and W' (of course, $D_n = \emptyset$). They are globally defined and tell us how to factorize the ideals J and J'. That is information we will need later on. As we shall soon perform the descent in dimension, we write $J_n = M_n \cdot I_n$ for $J = M \cdot I$ and similarly $J'_n = J'$.

So what shall we do at an equiconstant point a' above a where the order of I'_n has remained constant? Now, generically along Y', the order drops. Only a few points admit constant order. We suspect that at an equiconstant point a' the ideal I' must have a special shape. Possibly we can profit of it. Let us therefore observe what happens in two variables, say plane curves. It is immediately seen that $x^p + y^q$ with $q \ge p$ has weak transform $x^p + y^{q-p}$ (in the relevant chart). If $q < 2p$, the order drops and we are done. If $q \ge 2p$, the order remains constant. However, the degree of the monomial y^q has dropped to $q - p$.

This strongly suggests to associate to I_n an ideal in one variable less and to look at its order. This is done via coefficient ideals. Choose locally at a a regular hypersurface V. You harshly protest because we agreed to choose never an object ad hoc, everything has to be natural. I respond that we are not interested in V, neither in the resulting coefficient ideal, but only in its order. It suffices to make this order independent of the choice of V. There are two options: either the minimum of all possible orders, over all choices of V, or the maximum.

You will have to convince yourself that the minimum is not significant, it just equals o. Therefore we take V so that the order of the coefficient ideal is maximized. Such V's are called hypersurfaces of weak maximal contact with I_n at a. They exist, and only in case that I is bold regular, i.e., a power of a variable, the resulting order is infinite. In this case we redefine the coefficient ideal to be the trivial ideal 1.

We stop briefly for book-keeping, setting $W_{n-1} = V$ and $J_{n-1} = \operatorname{coeff}_{W_{n-1}} I_n$. The letter J is taken instead of I because, as we saw in class II, coefficient ideals do not pass to the weak transform under blowup (the letter I is reserved for ideals which pass to weak transforms). For accurateness, we factorize $J_{n-1} = M_{n-1} \cdot I_{n-1}$ with $M_{n-1} = 1$ and $I_{n-1} = J_{n-1}$, and set $o_{n-1} = \operatorname{ord}_a I_{n-1}$.

Let's go to W' at a' and $J'_n = M'_n \cdot I'_n$. Denote by J'_{n-1} the coefficient ideal of I'_n with respect to a local hypersurface W'_{n-1} which maximizes its order. The curve case suggests that J'_{n-1} has something to do with J_{n-1}. We are now curious to explore this connection between the coefficient ideals of I_n and I'_n (we have seen portion of it in class II). Remember that J'_{n-1} equals the controlled transform of J_{n-1} if W'_{n-1} is the strict transform of W_{n-1} (in particular, a' must be included in W'_{n-1}).

At this point, where things seem to become more and more involved, there pop up a few very favorable coincidences. They will make everything work marvellously – provided we are in characteristic 0. Such lucky strokes are rare in mathematics, and I see no substantial reason why they occur precisely here and now. Once Abhyankar and Hironaka discovered them in the fifties (stories tell that the latter was visiting the former and insisted for four days until he had completely clarified the former's vision of using Tschirnhaus' transformation for resolution purposes), the rest was only technique (as other stories tell).

In positive characteristic these coincidences do not occur – and nobody has found a working substitute for them. At least for the arguments and constructions to follow, the characteristic p case is much less accessible, if at all.

Stroke 1: If W_{n-1} maximizes the order of $J_{n-1} = \operatorname{coeff}_{W_{n-1}} I_n$ at a (everything is local), it contains locally the top locus $\operatorname{top}(I_n)$ of I_n. False in characteristic $p > 0$, see [104], [103] or [67].

Stroke 2: There is a simple procedure to construct such hypersurfaces of weak maximal contact (not all), via osculating hypersurfaces, see [55] or [70]. Look up the definition there or see the exercises. This construction appears in various forms in most of the resolution papers. Hypersurfaces of weak maximal contact can also be constructed (by different means) in positive characteristic, but do not enjoy the same nice properties.

Stroke 3: If W_{n-1} maximizes the order of $J_{n-1} = \operatorname{coeff}_{W_{n-1}} I_n$ at a, its strict transform W'_{n-1} contains all equiconstant points of I_n in Y'. First observed by Zariski. Proof: Computation in local coordinates. Also ok in positive characteristic.

Stroke 4: If W_{n-1} is osculating for I_n, in particular, maximizes the order of $J_{n-1} = \operatorname{coeff}_{W_{n-1}} I_n$ at a, and if the order has remained constant, $o'_n = o_n$ at a', its strict transform W'_{n-1} is osculating for I'_n, in particular, maximizes the order of $J'_{n-1} = \operatorname{coeff}_{W'_{n-1}} I'_n$ at a'. Proof: Computation in local coordinates. False in positive characteristic, see [66].

With this gambling things become easy. Fix $a \in Z$ and $a' \in Y'$ above a. Choose W_{n-1} osculating at a for I_n, let J_{n-1} be the corresponding coefficient ideal of I_n in W_{n-1}. Then, at each equiconstant point a' above a, $W'_{n-1} = W^s_{n-1}$ is osculating for the weak transform $I'_n = I^Y_n$. This ideal has as coefficient ideal J'_{n-1} the controlled transform $(J_{n-1})^! = I_{Y'_{n-1}}^{-\operatorname{ord}_a I_n} \cdot J^*_{n-1}$ of J_{n-1} in W'_{n-1}, where $I_{Y'_{n-1}} = I_{W'_{n-1}}(Y' \cap W'_{n-1})$ denotes the ideal defining in W'_{n-1} the exceptional divisor $Y' \cap W'_{n-1}$ of the blowup $W'_{n-1} \to W_{n-1}$. Recall here that, locally at a, Z is contained in W_{n-1}, so that $Y' \cap W'_{n-1}$ is regular.

In particular, we may factorize $J'_{n-1} = M'_{n-1} \cdot I'_{n-1}$ with $I'_{n-1} = I^Y_{n-1}$ the weak transform of I_{n-1} and M'_{n-1} a normal crossings divisor in W'_{n-1} supported by the exceptional component $Y' \cap W'_{n-1}$. Hence the divisor $D'_{n-1} = (\operatorname{ord}_a I_{n-1} - \operatorname{ord}_a I_n) \cdot Y'$ of W' has normal crossings at a' and defines the principal monomial ideal $I_{W'_{n-1}}(D'_{n-1} \cap W'_{n-1}) = M'_{n-1}$ locally at a'.

This looks a little bit complicated. And indeed, it is complicated, especially, if you are not yet used to this type of constructions. But always keep in mind the corresponding commutative diagram, with vertical arrows the blowups in W_n and W_{n-1}, and horizontal arrows the descent in dimension. If you draw it for yourself on a sheet of paper things will clarify immediately (after having done one explicit computation for, say, a surface singularity). And you will realize that, again, everything is absolutely systematic.

Let us collect our data at the point a and at the equiconstant point a' above a:

$$J_n = M_n \cdot I_n \text{ in } W_n,$$

$$J'_n = M'_n \cdot I'_n \text{ in } W'_n,$$

$$J_{n-1} = M_{n-1} \cdot I_{n-1} \text{ in } W_{n-1},$$

$$J'_{n-1} = M'_{n-1} \cdot I'_{n-1} \text{ in } W'_{n-1}, \text{ and}$$

$$I'_n \text{ and } I'_{n-1} \text{ are the weak transforms of } I_n \text{ and } I_{n-1}.$$

Moreover,

$$D'_n = \operatorname{ord}_a I_n \cdot Y' \text{ and } D'_{n-1} = (\operatorname{ord}_a I_{n-1} - \operatorname{ord}_a I_n) \cdot Y'.$$

By the way, what are these data at a point a' where the order of I'_n has dropped? Either we refuse to define them, since our induction on the order already works, or, as we shall do, we choose any (new) osculating hypersurface W'_{n-1} for I'_n at a', set $J'_{n-1} = \operatorname{coeff}_{W'_{n-1}}(I'_n)$ with trivial factorization $J'_{n-1} = 1 \cdot I'_{n-1}$ (no other factorization need hold). Of course, I'_{n-1} is no longer the weak transform of I_{n-1}, so that its order may be quite arbitrary, but we don't care, since – lexicographically – the pair $(o'_n, o'_{n-1}) < (o_n, o_{n-1})$ has dropped at a'. You may notice that D'_{n-1}, though globally defined on W', is only a stratified divisor, since the multiplicity of Y' depends on the point a'. Specify what are the strata along which D'_{n-1} is coherent?

At this point, you may wish to see a concrete example. Here it is: Let $J = J_2$ be the principal ideal in $W = W_2 = \mathbb{A}^2$ generated by $f = x^p + y^q$ with $0 < p \le q$. We place ourselves at the origin $a = 0$ of $\mathbb{A}^2$, which is the only singular point of the plane curve X defined by f. As no blowup has occured so far, $J_2 = M_2 \cdot I_2$ with $M_2 = 1$ and $I_2 = J_2$. The order $o_2 = \operatorname{ord}_a I_2$ equals 2. In characteristic 0, the hypersurface W_1 defined by $x = 0$ in W_2 maximizes the order of the coefficient ideal $J_1 = \operatorname{coeff}_{W_1}(I_2) = (y^q)$. (If the characteristic equals p, this is not true if q is a multiple of p.) We get $J_1 = M_1 \cdot I_1$ with $M_1 = 1$ and $I_1 = J_1$. Clearly, $o_1 = \operatorname{ord}_a I_1 = q \ge p$. The invariant is the pair $(o_2, o_1) = (\operatorname{ord}_a I_2, \operatorname{ord}_a I_1)$ and attains at $a = 0$ its maximal value (p, q). This will therefore be our first center of blowup, $Z = \{0\}$ in $\mathbb{A}^2$ with blowup $\pi : W'_2 \to W_2$ and exceptional divisor $Y' \subset W'_2$. Let a' be a point of Y'. If a' is the origin of the x-chart, the order of $I'_2 = I^Y_2 = (1 + x^q p^q)$ has dropped to 0, so that there $J'_2 = I^p_{Y'} \cdot 1$ and W'_1 can be chosen arbitrarily, with $J'_1 = 1$ (by definition of the coefficient ideal of the trivial ideal). Hence $M'_1 = I'_1 = 1$ and the orders are $(o'_2, o'_1) = (0, 0) < (o_2, o_1) = (p, q)$. The same phenomenon occurs at all point a' of Y' outside the origin of the y-chart.

So let us look at this origin. It is the most interesting point. There, the order o'_2 of $I'_2 =$ is $q - p$ if $q < 2p$ and p if $q \ge 2p$. In the first case, the order has dropped, $I'_2 = (y^{q-p} + x^p)$ and our local hypersurface W'_1 will now be chosen as $y = 0$ with coefficient ideal $J_1' = (x^p)$. The factorization is $J_1' = M_1' \cdot I_1'$ with $M_1' = 1$ and $I_1' = J_1'$. You see that I_1' is not the weak transform of I_1', which does not matter because o_2' has dropped so that the pair of orders $(o_2', o_1') = (q - p, p)$ has dropped lexicographically.

We are left with the case $q \geq 2p$. The order of $I_2' = (x^p + y^{q-p})$ at a' (the origin of the y-chart) has remained constant equal to p. Therefore we will really need the descent in dimension here. The local hypersurface W_1' can be chosen equal to the strict transform W_1^s of W_1. It has equation $x = 0$ in this chart. The coefficient ideal J_1' is generated by y^{q-p} and factorizes into $J_1' = M_1' \cdot I_1'$ with $I_1' = 1$ the weak transform of $I_1 = (y^q)$ under the blowup of W_1 in $Z = \{0\}$. Hence $M_1' = J_1' = (y^{q-p})$. As for the orders at a', we get $(o_2', o_1') = (p, 0)$ which is lexicogaphically smaller than $(o_2, o_1) = (p, q)$. Our induction is thus completed at all points a' of Y'.

To make things more explicit, we write down the two combinatorial handicaps before and after blowup. In $W = W_2$ we have $D = (D_2, D_1) = (\emptyset, \emptyset)$ everywhere. If $q < 2p$, the combinatorial handicap D' in $W' = W_2'$ equals everywhere $(D_2', D_1') = (p \cdot Y', \emptyset)$. If $q \geq 2p$, we stratify W_2' into $S = W_2' \setminus \{0_{y-\text{chart}}\}$ and $T = \{0_{y-\text{chart}}\}$. At all points of S we have $D_2' = p \cdot Y'$ and $D_1' = \emptyset$. In contrast, at the origin of the y-chart we have $D_2' = p \cdot Y'$ and $D_1' = (q - p) \cdot Y'$, so that indeed $M_2' = (y^p)$ in W_2' and $M_1' = (y^{q-p})$ in W_1'. Notice here that W_1' defined by $x = 0$ is transversal to Y'.

Exercises: (1) Prove that $o' < o$ if $n = 1$ and $o > 0$, for $o = \text{ord}_a I$ and $o' = \text{ord}_{a'} I'$, taking $I' = I^Y$ the weak transform. Hint: Determine first the center Z.

(2) Figure out why the four lucky strokes hold in characteristic 0? Look up the counterexamples in positive characteristic.

(3) Show that in characteristic 0, the local top locus of an ideal is contained in a regular hypersurface whose weak transform contains all equiconstant points (this hypersurface will be defined by a suitable derivative of the generators of the ideal). Then look up the example of Narasimhan in positive characteristic (see [70]).

(4) Assume that $q \geq 3p$ and compute the combinatorial handicap for the plane curve $x^p + y^q = 0$ after the second blowup. What would happen in characteristic p for $q = 3p$?

(5) Resolve the Whitney-umbrella $x^2 + yz^2 = 0$ by taking as center the top

50

locus of the triple of orders (o_3, o_2, o_1). If you got tired of the computations, write a program which computes all data.

(6) Do the same for the surface $x^2 + y^3 + z^4 = 0$. Then find out why we really need the combinatorial handicap D at all stages of the resolution process and what its transformation rule is.

The transversal handicap

Assume that we are at a certain stage of our resolution process and wish to make the next blowup. In this section we address the question how to ensure that our chosen center is transversal to the already existing exceptional components which were produced by the preceding blowups. Recall that this transversality is necessary to get after the blowup a new exceptional locus having again normal crossings.

Let us denote W our present ambient scheme, $\mathcal{J}$ the ideal sheaf we wish to resolve, a a point of W and J the stalk of $\mathcal{J}$ at a. Let F be the exceptional locus in W produced by the prior blowups. By induction on the number of blowups we may assume that F has normal crossings. As transversality of two schemes is a local property compatible with completion, we may stick to a neighborhood of a in W and pass, if necessary, to the completion of the local rings. Thus we may suppose that $W = \mathbb{A}^n$ and that J is a polynomial ideal.

In order to know how to factorize J and the subsequent local coefficient ideals at a into a product of a principal monomial ideal and a remaining factor, we have introduced and constructed in the last section the combinatorial handicap D in W. It consists of normal crossings divisors $D_n, \ldots, D_1$ in W so that $J_i = M_i \cdot I_i$ for all $n \geq i \geq 1$, where $M_i = I_{W_i}(D_i \cap W_i)$ are the ideals associated to a local flag of regular schemes $W = W_n \supset W_{n-1} \supset \cdots \supset W_1$ at a.

Neglecting transversality problems, the center of blowup would be, locally at a, the scheme W_{d-1} with d minimal so that $I_d \neq 1$ (then $W_{d-1} = \text{top}(I_d)$ is just the support of I_d). Despite the fact that the flag $W_n \supset \cdots \supset W_1$ is not intrinsic (there are many possible choices), we saw that the so defined center does not depend on these choices and gives a global closed and regular subscheme of W. Let us call it the virtual center Z^{virt}. Virtual, because, in practice, the actual center $Z = Z^{actu}$ of the next blowup will mostly be different from Z^{virt} (it will be contained in Z^{virt}), precisely for transversality reasons with the exceptional locus F.

So let us investigate the precise constellation of Z^{virt} and F. Again, the question is local. We may assume that the point a lies in the intersection of both, otherwise Z^{virt} and F are trivially transversal at a. At an intersection point a, several things may happen. Recall here that we consider two schemes to be transversal at a if the product of their ideals in W defines a normal crossings scheme (i.e., if the ideal generated by the product in the completion of the local ring of W at a can be generated by monomials).

If Z^{virt} is contained in all components of F passing through a (the intersection of these components is just the local top locus of F at a), it is certainly transversal to F. If it is not contained, it may be transversal to some components of F and not transversal to others. In this case, we will have to choose a smaller center Z inside Z^{virt}. But which one? Taking simply for Z the intersection of Z^{virt} with all components of F to which it fails to be transversal does not work because this intersection will in general be singular scheme-theoretically.

Sticking to our philosophy from earlier sections, we proceed again upside down and assume that we already know how to choose the actual center Z^{actu} transversal to F. This is not a bad idea, but once in a while we will have to stop waving hands and to start making *Nägel mit Köpfen*. In any case, let's see what happens.

If $Z^{actu} = Z^{virt}$, everything is fine, our invariant introduced (vaguely) in the section on the combinatorial handicap will drop (this will be explicited more carefully in the last section) and (vertical) induction applies. By transversality of Z with F, the new exceptional divisor F' in W' will have again normal crossings. Fine!

So let us look at the case $Z^{actu} \subsetneq Z^{virt}$. Something surprising is happening (in retrospection, it won't be such a surprise): The resolution invariant remains constant, the situation seems not to improve. Why is this the case? The clue is the upper semicontinuity of the invariant: By construction, it is constant along Z^{virt} and attains its maximal value there (Z^{virt} is the top locus of the invariant). Along the open subscheme $U = Z^{virt} \setminus Z^{actu}$ of Z^{virt} the blowup is a local isomorphism, so at points of the strict transform U^s the invariant will remain constant. By upper semicontinuity, it has the same value on the closure $\overline{U}$ in W'. As this closure meets the new exceptional component $Y' = \pi^{-1}(Z^{actu})$, there will be points of Y' where the invariant has not dropped. We are stuck.

We suggest that you digest briefly this last paragraph by taking for X in $\mathbb{A}^3$ the cartesian product of the plane cusp $z^3 = y^2$ with the x-axis and for F the the cartesian product of the parabola $y = x^2$ with the z-axis. The virtual center Z^{virt} will be the x-axis (make sure that no other choice makes sense), which is tangent to F and hence not allowed as center. Instead, we have to take $Z^{actu} = \{0\}$ the origin, and the transform X' of X in $W' = \widetilde{\mathbb{A}^3}$ looks quite the same.

Doing mathematics is – aside genuine *Geistesblitze* – a *Wechselspiel* of computing examples; observing; pointing out obstructions; finding the reasons for the obstructions; observing again; trying to isolate the obstruction so as to see clearly its *Ursprung*; computing once more, etc. Looking carefully at phenomena and complicated configurations is one of the most delicate jobs for mathematicians. Often we just do not see what is there, and of what we could profit of. And only afterwards the solution to the problem seems so natural, so evident. If we had just seen it earlier.

In view of these "profound" philosophical and pedagogical contemplations, we look once again at our situation.

By transversality, we are forced to choose a center smaller than the one we would like to take and which would make the invariant drop. Being too small, the invariant remains the same (at least at some points of the new exceptional component) and our induction breaks down. The invariant is not able to detect any improvement of the singularities.

At this point we will ask ourselves why we blow up at all if it does not help to advance the induction. We could as well do nothing and resignate. This question is precisely the correct one, so we repeat it: Why blowing up at all if the virtual center is not transversal to the exceptional locus.

The question contains, at least in this case, also the answer. We blow up because we wish to improve our resolution problem, which consists in making an ideal a monomial ideal. But our non-transversality problem we encounter on the way is precisely of the same nature as our original problem: An ideal (in this case the product of the ideal of Z^{virt} and of F) is not a monomial ideal.

After all this much-talking-and-little-saying it should have become clear what to do: We interpret the non-transversality problem as a separate resolution problem and try to resolve it first in order to be able afterwards – once it is solved – to choose indeed the virtual center as the actual center. Therefore our present blowup with center $Z^{actu} \subsetneq Z^{virt}$ has the intention to help to make Z^{virt} transversal to F. This is the true purpose of the blowup, and obviously the invariant associated to the ideal J won't recognize that. Looking back at the example from before, we see that the blowup does improve the transversality problem, after the blowup the virtual center is again the x-axis, but the transform F' of F is now transversal to the x-axis.

This is encouraging and we immediately start to build up the data for our secondary (= transversality) resolution problem. The approach indeed works, though it burdens considerably the whole setting and constructions. Just imagine that along the solution of the subproblem new exceptional components will pop up, and while solving the transversality issue we may confront another transversality problem, which we have to solve first before we are allowed to attack the original one. And so on. This is technically (very) frightening.

There is an elegant solution to this annoying superposition of subordinate resolution problems suggested by Villamayor in [121]. In each step of the descent in dimension via local flags $W_n \supset \ldots \supset W_1$ take care in advance of the transversality problem by modifying the ideals I_i so that the resulting center is already contained in all exceptional components to which the virtual center may not be transversal. Just multiply I_i by the ideal Q_i of dangerous components, i.e., those to which the next local hypersurface W_{i-1} may not be transversal. Then the top locus of $K_i = I_i \cdot Q_i$ is contained in W_{i-1} locally at a.

The dangerous exceptional components are collected in the transversal handicap $E = (E_n, \ldots, E_1)$. Here, E_i is the normal crossings divisor formed by those exceptional components to which W_{i-1} may not be transversal. Even though W_{i-1} is not intrinsic, E_i will not depend on any choices and will obey a precise law of transformation under blowup. We will specify this law in the next section.

Meanwhile, let us see the impact of the construction. First, the components $o_i = \mathrm{ord}_a I_i$ of the invariant will be replaced by pairs (o_i, q_i) where $q_i = \mathrm{ord}_a Q_i$ measures the advance of the transversality problem in dimension $i - 1$ (the shift by 1 has notational reasons). If the components $(o_n, q_n, o_{n-1}, \ldots, o_i)$ have remained constant under blowup, the transformation law for E says that Q_i passes to its weak transform $Q_i^\curlyvee$. As the center will lie in $\mathrm{top}(K_i) = \mathrm{top}(I_i) \cap \mathrm{top}(Q_i)$ (here, the top loci have to be considered locally at a), the order of Q_i won't increase. This immediately implies the fabulous inequality

$$(o'_n, q'_n, o'_{n-1}, \ldots, o'_i, q'_i) \leq (o_n, q_n, o_{n-1}, \ldots, o_i, q_i),$$

where the two vectors are compared lexicographically. And by exhaustion of the dimensions, when looking at the whole new invariant $(o_n, q_n, \ldots, o_1, q_1)$,

54

it must have decreased.

There are some technical details which still have to be filled in. For instance, the local hypersurface W_{i-1} at a will be chosen to be osculating for I_i (and not, as one may think, for K_i). Also, one has to take care for establishing the necessary inclusions of the various top loci, for expliciting the transformation laws for all the ideals J_i, M_i, I_i, Q_i, K_i, and for ensuring that the resulting center is indeed transversal to the current exceptional locus F. All this can be done. Due to the systematic approach, it is even not as breathtaking as one might expect. We will see portion of it in the next section.

The determination of the dangerous exceptional components has a computational drawback. For each i, we have to take all possibly non-transversal components of F with respect to W_{i-1}, and many of these could already be transversal, but we just don't see it, because our invariant is unable to check it out. This inconvenience increases considerably the complexity of the algorithm. However, concerning the theoretical part of the construction of the resolution, it is quite useful because it follows in each dimension the same pattern and uses only information prescribed by the local invariant. Thus it is automatically intrinsic (i.e., independent of the local choices of hypersurfaces, hence global), and allows a systematic treatment via inductions on the dimensions.

If you look up the paper [55] you will realize that the hardest part is to become familiar with all the constructions and definitions collected in the section *Concepts*. The purpose of these lecture notes and [70] and [65] is precisely to motivate these constructions and to give you some feeling for them. But then, the actual proofs are rather short and almost routine. See the sections *Transversality* or *Top loci* in [55].

Exercises. (1) Two regular subschemes U and V of W meet transversally (in the sense defined above), if and only if their intersection $U \cap V$ is a regular scheme. Does this hold also for three regular subschemes, taking all pairwise intersections? (You may remember an earlier exercise.)

(2) Let F be a normal crossings scheme. Show that all possible intersections of components of F meet transversally. Does the converse hold?

(3) If, locally at a point a in W, a regular scheme Z is contained in all components of a normal crossings divisor F passing through a, then Z is transversal to F at a.

(4) Assume that a regular scheme Z meets all intersections of the components of a normal crossings scheme F in W transversally. Determine the cases when Z meets F transversally and when not.

(5) In the situation of exercise (4), consider the blowup $W' \to W$ of W with center Z, and let F^* be the total transform of F in W'. Is F^* again a normal crossings scheme in W'?

(6) Start at zero, i.e., with empty exceptional locus, and blow up once $W = \mathbb{A}^3$ at 0. Figure out whether in W' there can already occur a transversality problem, and if yes, determine the dimensions where it becomes virulent. Then indicate the transversal handicap $E' = (E'_3, E'_2, E'_1)$ in W'. Hint: E' will again consist of stratified divisors, the strata being given by the values of the invariant along the new exceptional component Y'.

(7) If you have done all the exercises up to now you are allowed to take a break. Otherwise return to the last exercise you did not do and give it a new try.

CLASS 5: Resolution of Mobiles

Here is now the precise definition of mobiles. A *singular mobile* on a regular n-dimensional ambient scheme W is a quadruple $\mathcal{M} = (\mathcal{J}, c, D, E)$ where $\mathcal{J}$ is a coherent ideal sheaf on W (one could also allow $\mathcal{J}$ to live on a regular, locally closed subscheme V of W, cf. [55]), c is a positive integer, the control, and $D = (D_n, \ldots, D_1)$ and $E = (E_n, \ldots, E_1)$ are strings of stratified normal crossings divisors D_i and E_i on W. Stratified means that there is a finite stratification of W by locally closed subschemes such that each D_i and E_i is coherent along the strata.

We call D, respectively E, the *combinatorial* and *transversal handicap* of $\mathcal{M}$. The divisors D_i are in general not reduced; they carry a small additional information, their *label*, which allows to order the components of D_i, but which shall not bother us here (for details, see [55]). The divisors E_i are reduced, have no components in common, and their union $|E|$ will equal the exceptional locus in W at the current stage of the resolution process.

You should think of a mobile as follows (cf. the last chapter): The ideal $\mathcal{J}$ is the ideal defining the singular scheme X in W we wish to resolve. It passes under blowup to its controlled transform $\mathcal{J}^! = I(Y')^{-c} \cdot J^*$ with respect to c. At the beginning, the handicaps are trivial, $D_i = E_i = \emptyset$. Under blowup, they obey a precise law of transformation, which we shall describe later on. This will allow to associate to any mobile $\mathcal{M}$ in W and blowup $W' \to W$ the transformed mobile $\mathcal{M}'$ in W'. We say that the mobile $\mathcal{M}$ is resolved, if the order of $\mathcal{J}$ at all points of W is less than c. Notice here that for $c = 2$ and $\mathcal{J}$ a principal ideal, this signifies that $\mathcal{J}$ defines a regular scheme. However, as the order is not so significant for non-hypersurfaces (order 1 at a point just means that the scheme is locally not minimally embedded in the ambient scheme), it is more convenient to take the control $c = 1$, in which case the mobile is resolved if $\mathcal{J}$ is the structure sheaf of W. This, in turn, signifies that the scheme we started with has as total transform a normal crossings divisor.

Mobiles are not as complicated as one might think. They are globally defined objects which do not depend on any ad hoc or local choices. The delicate part is to associate to them a local invariant and to define the transformation law. These two things are strongly related to each other. In the course of their definition we will have to consider objects which are not intrinsic and only locally defined. But we don't care as long as the final output is intrinsic.

For a mobile $\mathcal{M}$ and a point a in W, the local invariant $i_a(\mathcal{M})$ of $\mathcal{M}$ at a will be a vector of integers, and these integers are the orders of certain ideals defined locally at a. Thus we have perfect control on them under blowup as long as the ideal in question passes to its weak transform. Namely, in this case, the order of the ideal won't increase. As we have already seen in earlier sections, the respective ideals will indeed pass to their weak transforms provided that the earlier components of our invariant have remained constant. This suggests to consider $i_a(\mathcal{M})$ with respect to the lexicographic ordering.

Let us now see the details. We shall associate to $\mathcal{M}$ and a ideals $J_n, \ldots, J_1$, $I_n, \ldots, I_1$ and $K_n, \ldots, K_1$ defined in local flags $W_n \supset \ldots \supset W_1$ at a. The W_i are regular hypersurfaces in W_{i+1} defined in a neighborhood of a, where $W_n = W$ is the ambient scheme. There will be a certain rule how to choose them, but in any case they are not unique nor intrinsic. The ideals J_i, I_i and K_i are defined in W_i, and are neither intrinsic. We denote them by roman letters, because we think of them as the stalks at a of ideal sheaves. Our invariant is then simply the vector

$$i_a(\mathcal{M}) = (o_n, k_n, o_{n-1}, \ldots, o_1, k_1)$$

where $o_i = \mathrm{ord}_a I_i$ and $k_i = \mathrm{ord}_a K_i$. Again, this is not too complicated. The motivation for doing so was given in the last chapter. The point is that the components o_i and k_i do *not depend* on our choice of the flag $W_n \supset \ldots \supset W_1$ and of the ideals J_i, I_i and K_i (which, of course, are subject to certain conditions). So it is justified to call $i_a(\mathcal{M})$ an invariant of the mobile $\mathcal{M}$ at a.

We cheat here a little bit, because in reality, $i_a(\mathcal{M})$ has some more components, the combinatorial components m_i which are squeezed in between k_i and o_{i-1}. But all of them are zero except one, and this non-zero component is only used in a very special case in which the mobile is already almost resolved (the so called monomial or combinatorial case, see below and [55]). We do not wish to discuss it in these notes.

We now describe the rules which relate all the local ideals between each other and with the mobile. The problem here is that everything is motivated only a posteriori when you see how the rules make the induction argument work. So we ask you a little patience.

The relation between J_i and I_i is simple, and prescribed by the i-th component D_i of the combinatorial handicap. We have $J_i = M_i \cdot I_i$ where $M_i = I_{W_i}(D_i \cap W_i)$ denotes the ideal defining $D_i \cap W_i$ in W_i. By the law of transformation for D_i and the restrictions on the choice of W_i, both will intersect transversally so that the factor $I_{W_i}(D_i \cap W_i)$ is indeed a principal monomial ideal. It is the exceptional portion we wish to factor from J_i, and I_i is the interesting part of J_i which is not yet resolved. Observe here that J_n is just the stalk of $\mathcal{J}$ at a, and that at the beginning when all D_i are still empty the factorizations trivially exist. After some blowups, it will have to be proven that the factorizations exist, but this will follow directly from the definition of the D_i. Actually, the transformation law for D_i is precisely chosen so as to allow the factorization of J_i and moreover so that the factor I_i is the weak transform of the respective factor before blowup. It also shows that the component $o_i = \mathrm{ord}_a I_i$ captures interesting information, namely how far J_i is from being a principal monomial ideal. The ideal K_i equals, up to a small technical detail which we omit, the product of I_i with the transversality ideal $Q_i = I_{W_i}(E_i \cap W_i)$ of the mobile $\mathcal{M}$ in dimension i. Its order k_i (or, equivalently, the difference $q_i = k_i - o_i$) measures how far E_i and W_i are from being separated at a. In any case, and this is the important thing, the local top locus of Q_i at a will be contained in all components of E_i which pass through a. This ensures that also the center of blowup will be contained in these components. Recall here that E_i collects the dangerous components, i.e., those to which otherwise the chosen virtual center may fail to be transversal.

To repeat: o_i tells us how far we are with the resolution of J_i, and k_i how far we are with our transversality problem.

We are left to indicate how we choose the local flag $W_n \supset \ldots \supset W_1$ and how the ideals in different dimensions relate. As for the flag, W_i is a local hypersurface of W_{i+1} at a which is chosen so as to maximize the order at a of the coefficient ideal $\mathrm{coeff}_{W_i} K_{i+1}$ of K_{i+1} in W_i. There are several ways how to construct such hypersurfaces, and in characteristic 0 these constructions are particularly nice and behave well. But what is clear and crucial is that the order of $\mathrm{coeff}_{W_i} K_{i+1}$ does not depend on the choice of W_i. We then impose our last correlation rule among the various local ideals. It is $J_i = \mathrm{coeff}_{W_i} K_{i+1}$. Again, there is a slight technical complication which we only sketch. It occurs when I_i is already *bold regular*, i.e., generated by a power of one variable. In this case, the coefficient ideal would be 0, which is unpleasant for notational regards. Therefore one then sets $J_i = 1$.

With these settings, it can be shown that the resulting invariant $i_a(\mathcal{M})$ is well defined, upper semicontinuous and has all the properties required. In particular, its top locus Z is regular and transversal to the exceptional locus. So Z can be chosen as the center of the next blowup.

It is time that you perform the construction of the local invariant in a concrete example. Only then you will get a feeling for it. Take a principal ideal

$\mathcal{J}$ in three variables where you are still able to compute the coefficient ideal by hand. You start with trivial handicaps. You get an invariant, and you let Z be its top locus. Then blow up the ambient three-space in this center and consider the transformed mobile above, with new local invariants.

Ah, we have not defined the transform of mobiles yet. Right! Here is the transformation law. Let a' be a point of W' above $a \in Z$. We only define the transformed mobile $\mathcal{M}'$ locally at a', and leave it as exercise to show that this also makes sense globally. And we assume that the center Z is the top locus of $i_a(\mathcal{M})$ in W. This ensures that Z lies in all top loci of the ideals I_i and K_i.

We already said that $\mathcal{J}$ passes to its controlled transform $\mathcal{J}' = \mathcal{J}^!$. The control c' remains the same $c' = c$, except if the order of $\mathcal{J}'$ has dropped everywhere below c, in which case we are done. The formulas for the combinatorial and transversal handicaps depend on the behaviour of the invariant under blowup. The definition is recursive and a bit involved. So please sharpen your pencil.

We set $D'_n = D^*_n + (o_n - c) \cdot Y'$ so that $J'_n = I_{W'_n}(D'_n \cap W'_n) \cdot I'_n$ with $I'_n = I^\curlyvee_n$ the weak transform of I_n. We thus dispose of $o'_n = \mathrm{ord}_{a'} I'_n$. If $o'_n < o_n$ we set $E'_n = \emptyset$, if $o'_n = o_n$ we set $E'_n = E^*_n$ (pullback). Now assume that we have already defined $D'_n, E'_n, \ldots, D'_{i+1}, E'_{i+1}$ in W'. We thus dispose of the truncated invariant

$$(o'_n, k'_n, \ldots, o'_{i+1}, k'_{i+1})$$

at a'. If $(o'_n, k'_n, \ldots, o'_{i+1}, k'_{i+1}) <_{lex} (o_n, k_n, \ldots, o_{i+1}, k_{i+1})$ we set $D'_i = \emptyset$, if $(o'_n, k'_n, \ldots, o'_{i+1}, k'_{i+1}) = (o_n, k_n, \ldots, o_{i+1}, k_{i+1})$ we set $D'_i = D^*_i + (o_i - k_{i+1}) \cdot Y'$. We have thus defined also the component o'_i of our invariant. If $(o'_n, k'_n, \ldots, k'_{i+1}, o'_i) <_{lex} (o_n, k_n, \ldots, k_{i+1}, o_i)$ we set $E'_i = \emptyset$, if $(o'_n, k'_n, \ldots, k'_{i+1}, o'_i) = (o_n, k_n, \ldots, k_{i+1}, o_i)$ we set $E'_i = E^*_i$. So the definition of D'_i and E'_i depends on whether the earlier components of the invariant have dropped or not.

The transformation formulas look complicated, but they are precisely chosen so that the ideals J'_i, I'_i, Q'_i and K'_i satisfy the same rules as their sisters below. This is a computation in local coordinates which is not too difficult. Moreover, whenever $(o'_n, k'_n, \ldots, o'_{i+1}, k'_{i+1})$ has not dropped lexicographically, the ideal I'_i is the weak transform of I_i and hence $o'_i \leq o_i$. Similarly, whenever $(o'_n, k'_n, \ldots, k'_{i+1}, o'_i)$ has not dropped lexicographically, the ideal K'_i is the weak transform of K_i and hence $k'_i \leq k_i$. This shows that the invariant never increases.

To show that it actually decreases, we have to distinguish two circumstances. We place ourselves at the point a. Let d be the smallest index so that $o_d > 0$. We have seen earlier that the center Z then equals W_{d-1}. In case that the ideal K_d is bold regular (i.e., a power of a variable) and the truncated invariant $(o'_n, k'_n, \ldots, k'_{d+1}, o'_d)$ has not dropped, the transform K'_d equals the weak transform $K_d^{\curlyvee} = 1$ and hence $k'_d = 0 < k_d$. Note here that in this case $J_{d-1} = 1$ by definition and the further components of the invariant are all zero.

The second case is when K_d is not bold regular and hence $J_{d-1} \neq 1$. By the choice of d we have $I_{d-1} = 1$, so that $J_{d-1} = I_{W_{d-1}}(D_{d-1} \cap W_{d-1})$ is a principal monomial ideal. This is the monomial or combinatorial case, in which the hidden components m_i of the invariant come into play. To give you a feeling, just think of the polynomial $z^4 + x^a y^b$ with $a + b \geq 4$. You should have no problems in figuring out how to choose the center Z (according to the values of a and b) so that after finitely many blowups the order has dropped below 4. And in the general case, with J_{d-1} a principal monomial ideal the choice of the center and the reasoning are quite the same.

It looks strange, but we are finished – modulo some breadcrumbs. We have defined the transform $\mathcal{M}'$ in W' of our mobile $\mathcal{M}$ and given some hints why the local ideals J'_i, I'_i, Q'_i and K'_i exist again and satisfy the required relations. Actually, the members W'_i of the local flag $W'_n, \ldots, W'_1$ at a' coincide with the strict = weak transform of W_i if the truncated invariant $(o'_n, k'_n, \ldots, k'_{d+1}, o'_{d+1})$ has not dropped at a', the remaining members have to be chosen from scratch (which does not matter since the later components of the invariant are irrelevant).

The author of these lines is well aware that the above indications cannot please a critical reader – there is too much hand waving and too little substance, say proof. But precisely this shortcome may motivate you to look at the complete argument as given in the paper [55], and you will realize that there is not so much to add. The constructions are the same (including one or the other additional detail) and they are so systematic that (all) the proofs are really short. No one takes more than half a page, or at most one page.

So how to conclude these notes? One question is whether there is really a need for the non-expert to understand the proof of resolution of singularities in characteristic zero, aside curiosity. There are two answers: First, Hironaka's proposal for the inductive argument – remember that the above is nothing but a conceptualization of the original proof (with the help of the techniques developed by the successors of Hironaka) – is a paradigm of mathematical organisation. While reading these notes you should have observed that the clue to everything is the systematic definition of mobiles and their transforms, the rest are almost routine verifications.

Secondly, the problem of resolution is still wide open in positive characteristic and in the arithmetic case. Either somebody invents a completely new approach for these cases (which should not be discarded) or we succeed to understand the characteristic zero proof so much better that we get an idea how to tackle the other cases. Along the lines of Hironaka's proof, when translated to positive characteristic, funny things tend to happen. The invariant may increase, but only in quite special cases which can be pinned down explicitly. And if it increases, the increase is very small, namely at most one (at least in the relevant examples). So you immediately think that if it increases only by one, maybe in the next blowup it drops by two and we have won again. This is almost the case, but only almost. If you are curious to know what type of phenomena may happen, you may look at the article [66]. See you then!

62

References

1. S. Abhyankar. Three dimensional embedded uniformization in characteristic p. Lectures at Purdue University, Notes by M.F. Huang 1968.
2. S. Abhyankar. Local uniformization on algebraic surfaces over ground fields of characteristic $p \neq 0$. *Ann. of Math. (2)*, 63:491–526, 1956.
3. S. Abhyankar. On the valuations centered in a local domain. *Amer. J. Math.*, 78:321–348, 1956.
4. S. Abhyankar. Ramification theoretic methods in algebraic geometry. *Princeton Univ. Press*, 1959.
5. S. Abhyankar. Current status of the resolution problem. In *Summer Institute on Algebraic Geometry.* Proc. Amer. Math. Soc., 1964.
6. S. Abhyankar. Resolution of singularities of arithmetical surfaces. In *Arithmetical Algebraic Geometry (Proc. Conf. Purdue Univ., 1963)*, pages 111–152. Harper & Row, New York, 1965.
7. S. Abhyankar. An algorithm on polynomials in one indeterminate with coefficients in a two dimensional regular local domain. *Ann. Mat. Pura Appl. (4)*, 71:25–59, 1966.
8. S. Abhyankar. *Resolution of singularities of embedded algebraic surfaces.* Pure and Applied Mathematics, Vol. 24. Academic Press, New York, 1966.
9. S. Abhyankar. Three dimensional embedded uniformization on characteristic p. Lectures at Purdue University, Notes by M.F. Huang, 1968.
10. S. Abhyankar. Resolution of singularities of algebraic surfaces. In *Algebraic Geometry (Internat. Colloq., Tata Inst. Fund. Res., Bombay, 1968)*, pages 1–11. Oxford Univ. Press, London, 1969.
11. S. Abhyankar. Desingularization of plane curves. In *Singularities, Part 1 (Arcata, Calif., 1981)*, volume 40 of *Proc. Sympos. Pure Math.*, pages 1–45. Amer. Math. Soc., Providence, RI, 1983.
12. S. Abhyankar. Good points of a hypersurface. *Adv. in Math.*, 68(2):87–256, 1988.
13. S. Abhyankar. *Algebraic geometry for scientists and engineers*, volume 35 of *Mathematical Surveys and Monographs*. American Mathematical Society, Providence, RI, 1990.
14. S. Abhyankar and T. T. Moh. Newton-Puiseux expansion and generalized Tschirnhausen transformation. I, II. *J. Reine Angew. Math.*, 260:47–83; ibid. 261 (1973), 29–54, 1973.
15. D. Abramovich and A. J. de Jong. Smoothness, semistability, and toroidal geometry. *J. Algebraic Geom.*, 6(4):789–801, 1997.
16. J. M. Aroca, H. Hironaka, and J. L. Vicente. *The theory of the maximal contact.* Instituto "Jorge Juan" de Matemáticas, Consejo Superior de Investigaciones Científicas, Madrid, 1975. Memorias de Matemática del Instituto "Jorge Juan", No. 29. [Mathematical Memoirs of the "Jorge Juan" Institute, No. 29].
17. J. M. Aroca, H. Hironaka, and J. L. Vicente. *Desingularization theorems*, volume 30 of *Memorias de Matemática del Instituto "Jorge Juan" [Mathematical Memoirs of the Jorge Juan Institute]*. Consejo Superior de Investigaciones Científicas, Madrid, 1977.

18. M. Artin. Lipman's proof of resolution of singularities. In J.H. Silverman G. Cornell, editor, *Arithmetic Geometry*. Springer, 1986.

19. M. Artin. Lipman's proof of resolution of singularities for surfaces. In *Arithmetic geometry (Storrs, Conn., 1984)*, pages 267–287. Springer, New York, 1986.

20. C. Ban and L. J. McEwan. Canonical resolution of a quasi-ordinary surface singularity. *Canad. J. Math.*, 52(6):1149–1163, 2000.

21. MacEwan L. Ban, C. Canonical resolution of a quasi-ordinary surface singularity. Preprint Ohio State University 1998.

22. B. M. Bennett. On the characteristic functions of a local ring. *Ann. of Math. (2)*, 91:25–87, 1970.

23. P. Berthelot. Altérations de variétés algébriques (d'après A. J. de Jong). *Astérisque*, (241):Exp. No. 815, 5, 273–311, 1997. Séminaire Bourbaki, Vol. 1995/96.

24. E. Bierstone and P. D. Milman. Semianalytic and subanalytic sets. *Inst. Hautes Études Sci. Publ. Math.*, (67):5–42, 1988.

25. E. Bierstone and P. D. Milman. Uniformization of analytic spaces. *J. Amer. Math. Soc.*, 2(4):801–836, 1989.

26. E. Bierstone and P. D. Milman. A simple constructive proof of canonical resolution of singularities. In *Effective methods in algebraic geometry (Castiglioncello, 1990)*, volume 94 of *Progr. Math.*, pages 11–30. Birkhäuser Boston, Boston, MA, 1991.

27. E. Bierstone and P. D. Milman. Canonical desingularization in characteristic zero by blowing up the maximum strata of a local invariant. *Invent. Math.*, 128(2):207–302, 1997.

28. E. Bierstone and P. D. Milman. Resolution of singularities. In *Several complex variables (Berkeley, CA, 1995–1996)*, volume 37 of *Math. Sci. Res. Inst. Publ.*, pages 43–78. Cambridge Univ. Press, Cambridge, 1999.

29. E. Bierstone and P. D. Milman. Desingularization algorithms. I. Role of exceptional divisors. *Mosc. Math. J.*, 3(3):751–805, 1197, 2003. Dedicated to Vladimir Igorevich Arnold on the occasion of his 65th birthday.

30. G. Bodnár. Computation of blowing up centers. *J. Pure Appl. Algebra*, 179(3):221–233, 2003.

31. G. Bodnár and J. Schicho. Automated resolution of singularities for hypersurfaces. *J. Symbolic Comput.*, 30(4):401–428, 2000.

32. G. Bodnár and J. Schicho. A computer program for the resolution of singularities. In *Resolution of singularities (Obergurgl, 1997)*, volume 181 of *Progr. Math.*, pages 231–238. Birkhäuser, Basel, 2000.

33. G. Bodnár and J. Schicho. Two computational techniques for singularity resolution. *J. Symbolic Comput.*, 32(1-2):39–54, 2001. Computer algebra and mechanized reasoning (St. Andrews, 2000).

34. F. A. Bogomolov and T. G. Pantev. Weak Hironaka theorem. *Math. Res. Lett.*, 3(3):299–307, 1996.

35. M. Brandenberg. *Aufblasungen affiner Varietäten*. PhD thesis, 1992. Thesis Zürich.

36. A. Bravo and O. Villamayor. A strengthening of resolution of singularities

in characteristic zero. *Proc. London Math. Soc. (3)*, 86(2):327–357, 2003.

37. A. M. Bravo, S. Encinas, and O. Villamayor. A simplified proof of desingularization and applications. *Rev. Mat. Iberoamericana*, 21(2):349–458, 2005.

38. E. Brieskorn. Über die Auflösung gewisser Singularitäten von holomorphen Abbildungen. *Math. Ann.*, 166:76–102, 1966.

39. E. Brieskorn. Die Auflösung der rationalen Singularitäten holomorpher Abbildungen. *Math. Ann.*, 178:255–270, 1968.

40. E. Brieskorn and H. Knörrer. *Ebene algebraische Kurven*. Birkhäuser Verlag, Basel, 1981.

41. F. Cano and R. Piedra. Characteristic polygon of surface singularities. In *Géométrie algébrique et applications, II (La Rábida, 1984)*, volume 23 of *Travaux en Cours*, pages 27–42. Hermann, Paris, 1987.

42. V. Cossart. Desingularization of embedded excellent surfaces. *Tôhoku Math. J. (2)*, 33(1):25–33, 1981.

43. V. Cossart. Polyèdre caractéristique et éclatements combinatoires. *Rev. Mat. Iberoamericana*, 5(1-2):67–95, 1989.

44. V. Cossart. Contact maximal en caractéristique positive et petite multiplicité. *Duke Math. J.*, 63(1):57–64, 1991.

45. V. Cossart. Uniformisation et désingularisation des surfaces d'après Zariski. In *Resolution of singularities (Obergurgl, 1997)*, volume 181 of *Progr. Math.*, pages 239–258. Birkhäuser, Basel, 2000.

46. V. Cossart, J. Giraud, and U. Orbanz. *Resolution of surface singularities*, volume 1101 of *Lecture Notes in Mathematics*. Springer-Verlag, Berlin, 1984. With an appendix by H. Hironaka.

47. S. D. Cutkosky. When are the blow ups of two ideals the same scheme? *Comm. Algebra*, 19(4):1119–1124, 1991.

48. S. D. Cutkosky. Local factorization of birational maps. *Adv. Math.*, 132(2):167–315, 1997.

49. S. D. Cutkosky. Local monomialization and factorization of morphisms. *Astérisque*, (260):vi+143, 1999.

50. S. D. Cutkosky. *Monomialization of morphisms from 3-folds to surfaces*, volume 1786 of *Lecture Notes in Mathematics*. Springer-Verlag, Berlin, 2002.

51. S. D. Cutkosky. Resolution of morphisms. In *Valuation theory and its applications, Vol. I (Saskatoon, SK, 1999)*, volume 32 of *Fields Inst. Commun.*, pages 117–140. Amer. Math. Soc., Providence, RI, 2002.

52. S. D. Cutkosky. Generically finite morphisms and simultaneous resolution of singularities. In *Commutative algebra (Grenoble/Lyon, 2001)*, volume 331 of *Contemp. Math.*, pages 75–99. Amer. Math. Soc., Providence, RI, 2003.

53. S. D. Cutkosky. *Resolution of singularities*, volume 63 of *Graduate Studies in Mathematics*. American Mathematical Society, Providence, RI, 2004.

54. D. Eisenbud and J. Harris. *The geometry of schemes*, volume 197 of *Graduate Texts in Mathematics*. Springer-Verlag, New York, 2000.

55. S. Encinas and H. Hauser. Strong resolution of singularities in characteristic zero. *Comment. Math. Helv.*, 77(4):821–845, 2002.

56. S. Encinas and O. Villamayor. Good points and constructive resolution of singularities. *Acta Math.*, 181(1):109–158, 1998.

57. S. Encinas and O. Villamayor. A course on constructive desingularization and equivariance. In *Resolution of singularities (Obergurgl, 1997)*, volume 181 of *Progr. Math.*, pages 147–227. Birkhäuser, Basel, 2000.

58. S. Encinas and O. Villamayor. A new proof of desingularization over fields of characteristic zero. In *Proceedings of the International Conference on Algebraic Geometry and Singularities (Spanish) (Sevilla, 2001)*, volume 19, pages 339–353, 2003.

59. J. Giraud. *Desingularization in low dimension*, volume Resolution of surface singularities of *Lecture Notes in Math.* Springer 1984, 51-78.

60. J. Giraud. *Étude locale des singularités*. U.E.R. Mathématique, Université Paris XI, Orsay, 1972. Cours de 3ème cycle, 1971–1972, Publications Mathématiques d'Orsay, No. 26.

61. J. Giraud. Sur la théorie du contact maximal. *Math. Z.*, 137:285–310, 1974.

62. J. Giraud. Contact maximal en caractéristique positive. *Ann. Sci. École Norm. Sup. (4)*, 8(2):201–234, 1975.

63. R. Goldin and B. Teissier. Resolving singularities of plane analytic branches with one toric morphism. In *Resolution of singularities (Obergurgl, 1997)*, volume 181 of *Progr. Math.*, pages 315–340. Birkhäuser, Basel, 2000.

64. A. Grothendieck. Travaux de Heisouké Hironaka sur la résolution des singularités. In *Actes du Congrès International des Mathématiciens (Nice, 1970)*, *Tome 1*, pages 7–9. Gauthier-Villars, Paris, 1971.

65. H. Hauser. *Seven short stories on blowups and resolution.* Proceedings of the Gökova Geometry Topology Conference 2005, pp. 1-48. International Press 2006.

66. H. Hauser. Why Hironaka's resolution fails in positive characteristic. preprint on www.hh.hauser.cc.

67. H. Hauser. Seventeen obstacles for resolution of singularities. In *Singularities (Oberwolfach, 1996)*, volume 162 of *Progr. Math.*, pages 289–313. Birkhäuser, Basel, 1998.

68. H. Hauser. Excellent surfaces and their taut resolution. In *Resolution of singularities (Obergurgl, 1997)*, volume 181 of *Progr. Math.*, pages 341–373. Birkhäuser, Basel, 2000.

69. H. Hauser. Resolution of singularities 1860–1999. In *Resolution of singularities (Obergurgl, 1997)*, volume 181 of *Progr. Math.*, pages 5–36. Birkhäuser, Basel, 2000.

70. H. Hauser. The Hironaka theorem on resolution of singularities (or: A proof we always wanted to understand). *Bull. Amer. Math. Soc. (N.S.)*, 40(3):323–403 (electronic), 2003.

71. H. Hauser. Three power series techniques. *Proc. London Math. Soc. (3)*, 89(1):1–24, 2004.

72. H. Hauser, J. Lipman, F. Oort, and A. Quirós, editors. *Resolution of singularities*, volume 181 of *Progress in Mathematics*. Birkhäuser Verlag, Basel, 2000. A research textbook in tribute to Oscar Zariski, Papers from the Working Week held in Obergurgl, September 7–14, 1997.

73. H. Hauser and G. Regensburger. Explizite Auflösung von ebenen Kurvensingularitäten in beliebiger Charakteristik. *Enseign. Math. (2)*, 50(3-4):305–

66

353, 2004.

74. M. Herrmann, S. Ikeda, and U. Orbanz. *Equimultiplicity and blowing up*. Springer-Verlag, Berlin, 1988. An algebraic study, With an appendix by B. Moonen.

75. H. Hironaka. Desingularization of excellent surfaces. Notes by B. Bennett at the Conference on Algebraic Geometry, Bowdoin 1967. Reprinted in: Cossart, V., Giraud, J., Orbanz, U.: Resolution of surface singularities. Lecture Notes in Mathematics 1101, Springer 1984.

76. H. Hironaka. Preface to Zariski's collected papers i, 307-312, mit press 1979.

77. H. Hironaka. Resolution of singularities of an algebraic variety over a field of characteristic zero. I, II. *Ann. of Math. (2) 79 (1964), 109-203; ibid. (2)*, 79:205-326, 1964.

78. H. Hironaka. Characteristic polyhedra of singularities. *J. Math. Kyoto Univ.*, 7:251-293, 1967.

79. H. Hironaka. Additive groups associated with points of a projective space. *Ann. of Math. (2)*, 92:327-334, 1970.

80. H. Hironaka. Certain numerical characters of singularities. *J. Math. Kyoto Univ.*, 10:151-187, 1970.

81. H. Hironaka. Desingularization of complex-analytic varieties. In *Actes du Congrès International des Mathématiciens (Nice, 1970), Tome 2*, pages 627-631. Gauthier-Villars, Paris, 1971.

82. H. Hironaka. Schemes, etc. In *Algebraic geometry, Oslo 1970 (Proc. Fifth Nordic Summer School in Math.)*, pages 291-313. Wolters-Noordhoff, Groningen, 1972.

83. H. Hironaka. Bimeromorphic smoothing of a complex-analytic space. *Acta Math. Vietnam.*, 2(2):103-168, 1977.

84. H. Hironaka. Idealistic exponents of singularity. In *Algebraic geometry (J. J. Sylvester Sympos., Johns Hopkins Univ., Baltimore, Md., 1976)*, pages 52-125. Johns Hopkins Univ. Press, Baltimore, Md., 1977.

85. H. Hironaka. Stratification and flatness. In *Real and complex singularities (Proc. Ninth Nordic Summer School/NAVF Sympos. Math., Oslo, 1976)*, pages 199-265. Sijthoff and Noordhoff, Alphen aan den Rijn, 1977.

86. H. Hironaka. *On the presentations of resolution data*, volume Algebraic Analysis, Geometry and Number Theory 1988. 1989.

87. H. Jung. Darstellung der Funktionen eines algebraischen Körpers zweier unabhängiger Veränderlicher x,y in der Umgebung einer Stelle x=a, y=b. *J. Reine Angew. Math.*, 133:289-314, 1908.

88. K. Kiyek and J. L. Vicente. *Resolution of curve and surface singularities*, volume 4 of *Algebras and Applications*. Kluwer Academic Publishers, Dordrecht, 2004.

89. J. Kollár. Lecutres on resolution of singularities. Preprint Princeton 2005.

90. E. Kunz. Algebraische Geometrie IV. Vorlesung Regensburg.

91. D. T. Le and M. Oka. On resolution complexity of plane curves. *Kodai Math. J.*, 18:1-36, 1995.

92. M. Lejeune and B. Teissier. Contribution à l'étude des singularités du point de vue du polygone de Newton. Thèse d'Etat, Paris 1973.

93. J. Lipman. Rational singularities, with applications to algebraic surfaces and unique factorization. *Inst. Hautes Études Sci. Publ. Math.*, (36):195–279, 1969.

94. J. Lipman. Introduction to resolution of singularities. In *Algebraic geometry (Proc. Sympos. Pure Math., Vol. 29, Humboldt State Univ., Arcata, Calif., 1974)*, pages 187–230. Amer. Math. Soc., Providence, R.I., 1975.

95. J. Lipman. Desingularization of two-dimensional schemes. *Ann. Math. (2)*, 107(1):151–207, 1978.

96. J. Lipman. On complete ideals in regular local rings. In *Algebraic geometry and commutative algebra, Vol. I*, pages 203–231. Kinokuniya, Tokyo, 1988.

97. J. Lipman. Topological invariants of quasi-ordinary singularities. *Mem. Amer. Math. Soc.*, 74(388):1–107, 1988.

98. T. T. Moh. On a stability theorem for local uniformization in characteristic *p*. *Publ. Res. Inst. Math. Sci.*, 23(6):965–973, 1987.

99. T. T. Moh. Quasi-canonical uniformization of hypersurface singularities of characteristic zero. *Comm. Algebra*, 20(11):3207–3249, 1992.

100. T. T. Moh. On a Newton polygon approach to the uniformization of singularities of characteristic *p*. In *Algebraic geometry and singularities (La Rábida, 1991)*, volume 134 of *Progr. Math.*, pages 49–93. Birkhäuser, Basel, 1996.

101. H. T. Muhly and O. Zariski. The Resolution of Singularities of an Algebraic Curve. *Amer. J. Math.*, 61(1):107–114, 1939.

102. S. B. Mulay. Equimultiplicity and hyperplanarity. *Proc. Amer. Math. Soc.*, 89(3):407–413, 1983.

103. R. Narasimhan. Hyperplanarity of the equimultiple locus. *Proc. Amer. Math. Soc.*, 87(3):403–408, 1983.

104. R. Narasimhan. Monomial equimultiple curves in positive characteristic. *Proc. Amer. Math. Soc.*, 89(3):402–406, 1983.

105. T. Oda. Hironaka's additive group scheme. In *Number theory, algebraic geometry and commutative algebra, in honor of Yasuo Akizuki*, pages 181–219. Kinokuniya, Tokyo, 1973.

106. T. Oda. Hironaka group schemes and resolution of singularities. In *Algebraic geometry (Tokyo/Kyoto, 1982)*, volume 1016 of *Lecture Notes in Math.*, pages 295–312. Springer, Berlin, 1983.

107. T. Oda. Hironaka's additive group scheme. II. *Publ. Res. Inst. Math. Sci.*, 19(3):1163–1179, 1983.

108. T. Oda. Infinitely very near singular points. In *Complex analytic singularities*, volume 8 of *Adv. Stud. Pure Math.*, pages 363–404. North-Holland, Amsterdam, 1987.

109. M. Oka. Geometry of plane curves via toroidal resolution. In *Algebraic geometry and singularities (La Rábida, 1991)*, volume 134 of *Progr. Math.*, pages 95–121. Birkhäuser, Basel, 1996.

110. U. Orbanz. *Embedded resolution of algebraic surfaces after Abhyankar (characteristic 0)*, volume Resolution of surface singularities of *Lecture Notes in Math.* Springer 1984.

111. J. Pfeifle. Aufblasung monomialer Ideale. Master's thesis, Diplomarbeit

Innsbruck, 1996.

112. C. Rodriguez Sanchez. *Good points and local resolution of threefold singularities*. PhD thesis, Thesis, Univ. Leon, 1998.

113. J. Rosenberg. Blowing up nonreduced toric subschemes of $\mathbb{A}^n$. Preprint 1998.

114. B. Singh. Effect of a permissible blowing-up on the local Hilbert functions. *Invent. Math.*, 26:201–212, 1974.

115. B. Singh. Formal invariance of local characteristic functions. In *Seminar D. Eisenbud/B. Singh/W. Vogel, Vol. 1*, volume 29 of *Teubner-Texte zur Math.*, pages 44–59. Teubner, Leipzig, 1980.

116. M. Spivakovsky. A counterexample to Hironaka's "hard" polyhedra game. *Publ. Res. Inst. Math. Sci.*, 18(3):1009–1012, 1982.

117. M. Spivakovsky. A solution to Hironaka's polyhedra game. In *Arithmetic and geometry, Vol. II*, volume 36 of *Progr. Math.*, pages 419–432. Birkhäuser Boston, Boston, MA, 1983.

118. M. Spivakovsky. A counterexample to the theorem of Beppo Levi in three dimensions. *Invent. Math.*, 96(1):181–183, 1989.

119. B. Teissier. Valuations, deformations, and toric geometry. In *Valuation theory and its applications, Vol. II (Saskatoon, SK, 1999)*, volume 33 of *Fields Inst. Commun.*, pages 361–459. Amer. Math. Soc., Providence, RI, 2003.

120. W. V. Vasconcelos. *Arithmetic of blowup algebras*, volume 195 of *London Mathematical Society Lecture Note Series*. Cambridge University Press, Cambridge, 1994.

121. O. Villamayor. Constructiveness of Hironaka's resolution. *Ann. Sci. École Norm. Sup. (4)*, 22(1):1–32, 1989.

122. O. Villamayor. Patching local uniformizations. *Ann. Scient. Ec. Norm. Sup. Paris*, 25:629–677, 1992.

123. O. Villamayor. An introduction to the algorithm of resolution. In L. Narváez A. Campillo, editor, *Algebraic Geometry and Singularities*. Birkhaeuser, 1996. Proc. Conf. on Singularities La Rábida.

124. O. Villamayor. On equiresolution and a question of Zariski. *Acta Math.*, 185:123–159, 2000.

125. R. J. Walker. Reduction of the singularities of an algebraic surface. *Ann. of Math. (2)*, 36(2):336–365, 1935.

126. R. J. Walker. *Algebraic curves*. Dover Publications Inc., New York, 1962.

127. B. Youssin. Newton polyhedra without coordinates. *Mem. Amer. Math. Soc.*, 87(433):i–vi, 1–74, 1990.

128. O. Zariski. Polynomial Ideals Defined by Infinitely Near Base Points. *Amer. J. Math.*, 60(1):151–204, 1938.

129. O. Zariski. The reduction of the singularities of an algebraic surface. *Ann. of Math. (2)*, 40:639–689, 1939.

130. O. Zariski. Local uniformization on algebraic varieties. *Ann. of Math. (2)*, 41:852–896, 1940.

131. O. Zariski. Normal varieties and birational correspondences. *Bull. Amer. Math. Soc.*, 48:402–413, 1942.

132. O. Zariski. A simplified proof for the resolution of singularities of an alge-

braic surface. *Ann. of Math. (2)*, 43:583–593, 1942.

133. O. Zariski. The compactness of the Riemann manifold of an abstract field of algebraic functions. *Bull. Amer. Math. Soc.*, 50:683–691, 1944.

134. O. Zariski. Reduction of the singularities of algebraic three dimensional varieties. *Ann. of Math. (2)*, 45:472–542, 1944.

135. O. Zariski. The concept of a simple point of an abstract algebraic variety. *Trans. Amer. Math. Soc.*, 62:1–52, 1947.

136. O. Zariski. Exceptional singularities of an algebroid surface and their reduction. *Atti Accad. Naz. Lincei Rend. Cl. Sci. Fis. Mat. Natur. (8)*, 43:135–146, 1967.

137. O. Zariski. *Collected papers. Vol. I: Foundations of algebraic geometry and resolution of singularities*. The MIT Press, Cambridge, Mass.-London, 1972. Edited by H. Hironaka and D. Mumford, Mathematicians of Our Time, Vol. 2.

138. O. Zariski. A new proof of the total embedded resolution theorem for algebraic surfaces (based on the theory of quasi-ordinary singularities). *Amer. J. Math.*, 100(2):411–442, 1978.

139. O. Zariski. *Collected papers. Vol. IV*, volume 16 of *Mathematicians of Our Time*. MIT Press, Cambridge, Mass., 1979. Equisingularity on algebraic varieties, Edited and with an introduction by J. Lipman and B. Teissier.

140. O. Zariski. *Algebraic surfaces*. Classics in Mathematics. Springer-Verlag, Berlin, 1995. With appendices by S. S. Abhyankar, J. Lipman and D. Mumford, Preface to the appendices by Mumford, Reprint of the second (1971) edition.

Lectures on topology of complements and fundamental groups

A. Libgober

Department of Mathematics
University of Illinois at Chicago
851 S.Morgan Str. Chicago, Illinois, 60607
e-mail: libgober@math.uic.edu

This is an introduction to the topology of the complement to plane curves and hypersurfaces in projective space. It is based on lectures given in Lumini in February and in ICTP (Trieste) in August of 2005. We discuss key problems concerning the families of singular curves, the one variable Alexander polynomials and the orders of the homotopy groups of the complements to hypersurfaces with isolated singularities. We also discuss multivariable generalizations of these invariants and the Hodge theory of infinite abelian covers used in calculations of multivariable invariants. A historical overview is included as the opening section.

1. Introduction

The study of the topology of plane algebraic curves is an old subject. In fact, its problems come up naturally after the very first definitions in a basic course on algebraic curves. And yet, the answers obtained so far are often elusive or incomplete. If C is an algebraic curve in a complex projective plane $\mathbf{P}^2$, what is the fundamental group of $\mathbf{P}^2 - C$? Which properties of C affect the complexity of this group? For which group G does there exist C such that G is the fundamental group of the complement to C? When are two curves isotopic in an appropriate sense, so that complements stay unchanged during such isotopies? What are the invariants of such isotopies? These are obvious questions, and much is known about them, but complete or even satisfactory answers are still out of reach. Below I want to describe some recent developments, and I hope that this can serve as an introduction to these ideas and methods.

Perhaps the real beginning of this subject should be credited to Enriques, though some important work on the construction of interesting singular curves and the numerology (i.e., calculations of the number of singular points of a given type etc.) started much earlier. For example in the early 19th century, Plücker discovered important formulas relating the degree, number of nodes, and cusps of a curve to similar invariants of the dual curve. From Newton to Puiseaux and beyond the methods were developed for analyzing singular points of plane curves, and already Newton had classified the types of singular cubics. Lefschetz ([40]) used Plücker's work to obtain the first non trivial information on how many nodes and cusps a plane curve of a given degree can have (a problem which still remains largely unresolved).

At the end of the 19th century, undoubtedly influenced by Picard's and Severi's works on the topology of complex surfaces, Enriques initiated a program to extend Riemann's and Hurwitz's results on multivalued functions, or in a more modern terminology, covering spaces of Riemann surfaces, to higher dimensions (cf. [24]). According to Riemann, a multivalued function in one variable, (e.g. $w = \sqrt{z}$ or more generally a solution to the equation $w^d + a_1(z)w^{d-1} + \cdots + a_d(z)$, where $a_i(z)$ are single valued holomorphic functions of $z \in \mathbf{C}$), is specified by the following data: first, the collection of its ramification points $B \subset \mathbf{C} \subset \mathbf{P}^1$; second, the number n of values of the multivalued function; and finally, the monodromy representation $\pi_1(\mathbf{P}^1 - B) \to \Sigma_n$ of the fundamental group into the symmetric group on n letters. What makes Riemann's approach very effective for the description of multivalued functions is the fact that the fundamental group in question is always a free group since the ramification locus is just a collection of points in $\mathbf{P}^1$. Therefore the whole multivalued function is specified by the ramification locus B and the assignment of arbitrary permutations $\sigma_1, ..., \sigma_{\mathrm{Card}(B)}$ in the symmetric group Σ_n to the generators of $\pi_1(\mathbf{P}^1 - B)$ with the only restriction that $\sigma_1 \cdots \sigma_{\mathrm{Card}(B)} = id.$

It was realized by Enriques (and others; a rather complete account of the work before the mid 1930's is given by Zariski in his seminal book [85]) that a similar description of multivalued functions of several variables is still valid, but also that in higher dimensions such a result is much less efficient since $\sigma_1, ..., \sigma_{\mathrm{Card}(B)}$ must satisfy additional relations. For example, any algebraic curve in $\mathbf{P}^2$ can be a branching curve of a multivalued function, but one cannot assign arbitrary elements of Σ_n to generators of $\pi_1(\mathbf{P}^2 - B)$ since this group is almost never free. Rather, the permutations should satisfy certain compatibility conditions (one should note that the concept of the fundamental group did not completely crystallized at the time of the work of Enriques, and therefore his statements are much less straightforward than those presented here). Enriques described these conditions very explicitly. In modern terms, his description amounts to the calculation of the quotient of the fundamental group by the intersection of subgroups of finite index in terms of geometric generators (those discussed in section 3; note that it is still unknown if this intersection is trivial, i.e. the fundamental group is residually finite; cf. section 2.2). For example, if the branching curve has degree d and is *non-singular*, the fundamental group of the complement is cyclic of order d. At the same time, the number of geometric generators is d (cf. section 2.3) which therefore must satisfy several relations. In particular, in non-singular case, one can assign to a geometric generator only a permutation of order dividing d in Σ_n and the assignment to the rest of the generators is determined by the latter.

O. Zariski (after arriving in the US and visiting Princeton where Lefschetz and Alexander were working at the time) understood that the fundamental group of the complement is the central object in this theory and introduced many ideas that were new at the time, even in the context of similar problems in the knot theory. He showed how subtle the questions on the fundamental groups can be: not only that the fundamental group depends on the degree of the branching curve, as is the case for multivalued functions in one variable, but even knowing the number of nodes and cusps is not sufficient. He proved that a curve with 6 cusps can have as its fundamental group the cyclic group $\mathbf{Z}_6$ or the free product $\mathbf{Z}_2 * \mathbf{Z}_3 = PSL_2(\mathbf{Z})$. He also showed that such sextics can be distinguished by a geometric condition: in the first case, the cusps must be in general position, i.e. not to belong to a curve of degree 2, while in the second, they must belong to a conic. Zariski also used many technical ideas that were just appearing at the time in topology, e.g. studying the homology of cyclic covers (which in knot theory can be traced to Alexander and Reidemeister [2], [70]). The systematic study of the branched coverings using the theory of adjoints (cf. section 5.2) allowed him to relate the homology of branched covers to the superabundances of linear systems defined by the cusps (cf. [82]). He found a close relationship between the fundamental groups of the complements and braid groups by considering the duals for rational nodal and elliptic nodal curves. One of the tools was his celebrated theorem on fundamental groups of hyperplane sections extending Lefschetz homological results. In the context of branched coverings, Zariski even obtained expressions close to the Alexander polynomial (cf. [83]), as was noticed by D. Mumford (cf. [85]). This was the basis of Mumford's question about the role of

Alexander polynomial in algebraic geometry (*). *

After 1937 Zariski abruptly changed the scope of his interests and turned to the ambitious project of reconstructing algebraic geometry on the firm foundations of commutative algebra. Some of his students, however, continue to develop this subject, (cf. [78], [41]); much later, but in a similar spirit, M. Oka ([68]) generalized Zariski's calculation of the fundamental group of the complement of a sextic with six cusps on a conic by proving that for the curve C given by the equation $(x^p + y^p)^q + (y^q + z^q)^p = 0$, $gcd(p,q) = 1$ one has $\pi_1(\mathbf{P}^2 - C) = \mathbf{Z}_p * \mathbf{Z}_q$. The study of the topology continued mostly in the works of O. Chisini and his students ([12]) who initiated the use of braids for the study of fundamental groups and covering spaces. Abhyankar (cf. [1]), who studied with Zariski at Harvard in the 50's, was investigating fundamental groups, and in particular obtained important results on the fundamental groups of the complements, but the main focus was the algebraization of the fundamental groups.

One of the driving problems in the study of the fundamental groups in the 60's and 70's was the question of commutativity of fundamental groups of the complements of curves having nodes as the only singularities. Severi ([76]) outlined an argument which, as was realized later, happend to be incomplete. It was based on an assertion that the variety of plane curves of fixed degree with a fixed number of nodes is irreducible. Zariski repeated Severi's argument in [85] but did return to this issue much later (cf. [87]). Severi's statement eventually was confirmed by J. Harris ([30]). A direct algebraic proof of commutativity was found by W. Fulton (cf [29]) using Abhyankar's work and, shortly after that, a topological argument was given by P. Deligne.(cf. [16]) . A little later, M. Nori (cf. [67]) clarified these results further by obtaining conditions for the commutativity of the fundamental group of the complement of curves on arbitrary surfaces, in this respect continuing the work of Abhyankar (cf. [67])

In the 70's the problems about fundamental groups of complements were mentioned infrequently. Mumford, in the already quoted appendix to [85], also raised the problem of investigating the quotient G'/G'' for the fundamental groups of the complements. In the introduction to volume III of the collected papers by Zariski, containing the papers on the topology of the complements, Artin and Mazur, after discussing Zariksi's study of cyclic multiple planes, note:

**These questions were answered later in the author's papers [42] [43] and further extended in [49], [53] [57] (see references to other related works in these papers).

"Also, as far as the editors are aware, there has been no further progress in the delicate study of cyclic multiple planes for general d. There are many tantalizing questions here-there are even a number of less delicate topological issues to sort out. For example, for irreducible plane curve C with arbitrary singularities can one give some reasonable sufficient conditions for regularity of H_d in terms of zeros of "local Alexander polynomials"- that is, the Alexander polynomials of the knots associated with singularities of C?"

The answers to these questions were obtained in the author's papers [42] and [43]. If $G = \pi_1(\mathbf{C}^2 - C)$ one has $G/G' = H_1(\mathbf{C}^2 - C) = \mathbf{Z}^r$ where r is the number of irreducible components (cf. 2.2.1). In the case when C is irreducible, one has the exact sequence:

$$0 \to G'/G'' \to G/G'' \to \mathbf{Z} \to 0$$

This sequence defines the action of $\mathbf{Z}$ on G'/G''. This action coincides with the action on H_1 induced by the action of the group $\mathbf{Z}$ of covering transformations on the universal cyclic cover after Hurewicz identification of G'/G'' with the first homology group of the latter. The advantage of replacing a projective curve by an affine one is that in the affine situation one has an infinite tower of covering spaces, while in the projective case the degree of the cover must divide the degree of the curve. On the other hand, if the line at infinity is transversal to a projective curve, the group of the affine curve is just a central extension of the projective one (in the non-transversal case the relation is more subtle). The interpretation of Alexander polynomials in terms of infinite cyclic covers was discussed in the context of knot theory by J. Milnor in [63]. It is shown in [42] that $G'/G'' \otimes \mathbf{Q}$, as a module over the group ring of $\mathbf{Z}$ i.e. the ring $\mathbf{Q}[t, t^{-1}]$, is a torsion module and hence the order $\Delta_C(t)$ of $G'/G'' \otimes \mathbf{Q}$ is well defined (up to a unit of $\mathbf{Q}[t, t^{-1}]$). This is a *global* invariant of the curve in $\mathbf{C}^2$. On the other hand, with each singular point of C one associates link, i.e. the intersection of C with the boundary of a small ball about this singular point. As a result one obtains a set of local Alexander polynomials Δ_P corresponding to all singularities P of the curve C (as was suggested by Artin and Masur). However, one need another important ingredient: in [42] the author introduced the Alexander polynomial at infinity Δ_∞ which is the Alexander polynomial of the intersection of C with a ball in $\mathbf{C}^2$ of a sufficiently large radius. This is the link of C at infinity. The answer to the question of Artin and Mazur in the above quote is given by the following divisibility theorem from [42] for the Alexander polynomials associated with the curve:

$$\Delta(C) \mid \Pi_{P \in \mathrm{Sing}(C)} \Delta_P(C)$$

$$\Delta(C) \mid \Delta_\infty(C) \tag{1}$$

and the theorem expressing the homology of cyclic covers in terms of Alexander polynomials (cf. [42] and theorem 2.7 below). For example, for sextic curves with six cusps, which Zariski was considering in [81], the Alexander polynomial $\Delta_C(t)$ is equal to $t^2 - t + 1$ or 1 depending on whether the six cusps are on conic or not. Note that both divide $(t^2 - t + 1)^6$ and $(t^6 - 1)^4(t - 1)$ (which are the product of local Alexander polynomials and the Alexander polynomial at infinity respectively). The divisibility relation in [42] yields certain information on the structure of the fundamental groups. For example, $G'/G'' \otimes \mathbf{Q}$ is trivial if Δ_∞ and $\Pi_{P \in \mathrm{Sing}(C)} \Delta_P(C)$ are relatively prime. In particular, if the only singularities are cusps, then $G'/G'' \otimes \mathbf{Q} = 0$, unless the degree of C is divisible by 6. The regularity condition, which was conjectured by Artin-Masur is the following: the cyclic multiple plane H_d or degree d is regular (i.e. the irregularity $q = \dim H^1(\mathcal{O}) = 0$) if none of the roots of the local Alexander polynomials is a root of unity of degree $\deg C$ or degree d (cf. [42]).

The work [42] is topological and many of the results were extended to the differential category (cf. [44]). The issue of the dependence of the Alexander polynomial on the position of singularities, was dealt with in [43]. Zariski's results were generalized as follows. As in [82], the irregularity of the cyclic multiple planes was obtained in terms of superabundances of certain linear systems associated with the collections of singular points of the curve. However, for singularities that are more complicated than cusps, the systems are specified by more subtle geometric conditions: the local equations for the elements of the linear systems responsible for the irregularity of the cyclic branched covers must belong to certain ideals called in [43] the *ideals of quasiadjunction*. Later, these ideals appeared in many other contexts and became known as *multiplier ideals* (cf. [39]). Other important numerical invariants of plane curve singularities that were introduced in [43] were later identified in [59] with the part of the spectrum introduced in the 70's by Arnold and Steenbrink (cf [71]). The work [25] also related the irregularity of multiple planes to the position of singularities. The ideas of [25] rely on vanishing theorems which later led to a much better understanding of those (cf. [26]): a key development in algebraic geometry in the 90's.

In the early 80's, about the time when the work on Alexander polynomials described above appeared, there was another important development in the study of plane singular curves. B. Moishezon initiated a program for describing the topology of algebraic surfaces in terms of branching curves in $\mathbf{P}^2$. Branching curves of generic projections form a subclass in the class of curves having nodes and cusps as the only singularities. If one starts with a projective surface, considers a pluricanonial embedding using a fixed multiplicity of the canonical class, and then uses a generic projection, the branching curve in $\mathbf{P}^2$ becomes an invariant of the deformation type of the surface (the fact that one does not need the monodromy representation into the symmetric group was conjectured by Chisini and subsequently proven in [38]). Moishezon's first calculations deal with the branching curves of generic projections of non-singular surfaces in $\mathbf{P}^3$. If the degree d of a surface is 3, one obtains as the branching curve Zariski's sextic given by the equation $f_2^3 + f_3^2 = 0$. For surfaces of arbitrary degree Moishezon obtains, as the fundamental groups of the complements of the branching curves, the quotients of Artin's braid groups by the centers (which for $d = 3$ gives $PSL_2(\mathbf{Z})$). Moishezon's important idea was that the primary invariant is not the fundamental group but rather the braid monodromy which implicitly is present in van Kampen's method of calculation of the fundamental group (Moishezon was unaware of Chisini's work [12] until he completed [64]). In this vein, the author showed that the braid monodromy defines not just the fundamental group but also the homotopy type (cf. [46], and further works by M. Teicher cf. [77]). Later Moishezon continued this work jointly with M. Teicher. Methods of braid monodromy recently found applications in symplectic geometry (cf. [4]). More recently Teicher and her students have continued the systematic study of the braid monodromy and the fundamental groups of the complements of the branching curves of generic projections and arrangements of lines and quadrics.

In the late 80's work started on a generalization of the theory of complements of singular curves to higher dimensions. The case of hypersurfaces with isolated singularities it turns out is remarkably similar to the case of curves. In [49], the author showed that for $n > 1$ the role of Alexander polynomial is played by the order of the homotopy group $\pi_n(\mathbf{C}^{n+1} - V) \otimes \mathbf{Q}$ considered as a module over $\pi_1(\mathbf{C}^{n+1} - V) = \mathbf{Z}$. The point is that this homotopy group can be canonically identified with the homology $H_n(\widetilde{\mathbf{C}^{n+1} - V}, \mathbf{Z})$ of the infinite cyclic cover of the complement. The divisibility relations (1) extend to the orders of the homotopy groups, and examples of hypersurfaces with non-trivial homotopy appear as a natural generalization of Zariski's sextics. For example $\pi_2(\mathbf{C}^3 - V) \neq 0$ for V given in $\mathbf{P}^3$ by the equation: $f_{21}^2 + f_{14}^3 + f_6^7 = 0$ (f_n is the generic form of degree n in four variables). Analytic theory developed by the author in [43] was also extended to higher dimensions in [52]. In it, the mixed Hodge structure on homotopy groups was introduced and one of its Hodge components was related to the super-abundance of linear systems defined by singularities of the hypersurface.

In the 90's the first results on a multivariable generalization of the Alexander invariants were obtained (cf. [48]). The theory of multivariable Alexander polynomials of links, due to R. Fox, depends on a very special feature of the link groups: the first Fitting ideal of the Alexander module is "almost" principal. The fundamental groups of the complements of re-ducible algebraic curves in $\mathbf{C}^2$ are similar to the link groups in the sense that both have surjections onto $\mathbf{Z}^r (r > 1)$. However for algebraic curves the first Fitting ideal of the Alexander module is far from being princi-pal. As a result one cannot define a multivariable Alexander polynomial in a meaningful way. The puzzle of the existence of multivariable invariants of algebraic curves was resolved in author's paper [48] by introducing the *characteristic varieties* (cf. chapter 4 below) which are the zero sets of Fit-ting ideals of the Alexander modules. In the case of one-variable Alexander polynomials no information is lost by replacing the Alexander polynomial by its set of zeros (at least for curves in $\mathbf{P}^2$ for which the Alexander module is semisimple) but for reducible curves the zero sets provide a non-trivial and very interesting invariant.

Applications followed shortly. In [33] the characteristic varieties were re-lated to the cohomology of local systems. They also appeared in the study of polynomial periodicity of Betti numbers of branched covering spaces (cf. [32]). For the curves for which all components have degree 1, i.e. ar-rangements of lines, the components of the characteristic varieties were related to the cohomology algebra of the complement (cf. [13]). The calcu-lation of the homology of abelian covers constructed by Hirzebruch, which have universal covers biholomorphic to the ball, did fall naturally in the gen-eral scheme that was valid for arbitrary arrangements and covers (cf. [53]). An analytic (rather than topological) theory was developed in [53] and characteristic varieties were expressed in terms of superabundances of the linear systems. Essential in this calculation were the results in [6], on the structure of the jumping loci for the cohomology of local systems. They represent an extension to quasiprojective varieties of the results of Green-Lazarsfeld, Beauville, Catanese, Simpson, Deligne and others which assert that the jumping loci for the cohomology of local systems are cosets of certain subgroups of the group of characters of the fundamental group.

During the late 90's, the study of the topology of plane algebraic curves became a much more active area of research. Many new examples of Zariski pairs due to E.Artal-Bartolo and collaborators and independently to M.Oka showed how common the phenomenon is of curves having different equisingular isotopy type with the same local data. Many new calculations were carried out of the fundamental groups of the complements by M.Teicher's school which finally led to a general conjecture on the structure of the fundamental groups of the branching curves of generic projections (cf. [77]). Interactions with the combinatorics of arrangements were important and led to at least a conjectural description of the characteristic varieties and much stronger vanishing for the cohomology of local systems than were available earlier (cf. [54], [53], [13]). Connections with symplectic topology should be noted (cf. [4]). There was further progress in the study of the complements in higher dimensions using generalizations of the Zariski-van Kampen's theorem (cf. [49], [11], [27], [79]). Nevertheless, despite tremendous progress, since the first works by Enriques, Zariski, van Kampen and Chisini, many problems still remain open and a complete understanding of the topology of the complements of curves and hypersurfaces still remains elusive (*) †.

In the text below we outlined some of the problems whose resolutions may clarify substantially the situation. The exposition is very elementary in the beginning, describing a motivation for the study of the following sections. In the later parts a reader will need more and more to rely on material covered in standard courses in algebraic geometry. Moreover, some familiarity with the mixed Hodge theory is needed in the last sections. The textbook [18] is a very good reference for the background material and also for other related issued omitted below. Most of the material has appeared already in the literature some time ago but some results appear to be new.

I want to thank J.P. Brasselet, D. Cheniot, J. Damon, M. Oka, A. Pichon D. Trotman, N. Dutertre and C. Murolo who organized the conferences in Lumini and Trieste for the opportunity to present this beautiful area of mathematics. I also want to thank L. Kauffman for a discussion of the history of polynomial invariants in knot theory.

Finally I want to dedicate this paper to J.P. Brasselet on the occasion of his 60th birthday.

---

† *cf. discussion of some open problems in A.Libgober, Problems in topology of the complements to plane singular curves, Proc. of School on Singularities, Trieste, 2005. J. Damon and M. Oka Editors.

2. Fundamental groups of the complement

2.1. *Problem of classification up to isotopy*

2.1.1. *Stratification of the discriminant*

Classically, many problems in the topology of plane curves and hypersurfaces were rooted attempts of some kind of classification (cf. [81]). We shall start by discussing what kind classification of curves or hypersurfaces one may expect.

Hypersurfaces of a fixed degree d in $\mathbf{P}^n$ are parameterized by $\mathbf{P}^{\binom{n+d}{d}-1}$. The parametrization is given by assigning to a defining equation the collection of coefficients of its monomials (in some fixed order). The discriminant $Disc(n,d)$ is the hypersurface in $\mathbf{P}^{\binom{n+d}{d}-1}$ consisting of the points corresponding to singular hypersurfaces. $Disc(n,d)$ has singularities in codimension one. An interesting problem is to understand the stratification of the discriminant hypersurfaces $Disc(n,d)$. By this we mean to describe the singular locus of the discriminant hypersurface (having codimension two in $\mathbf{P}^{\binom{n+d}{d}-1}$), then the singular locus of singular locus (having in $\mathbf{P}^{\binom{n+d}{d}-1}$ the codimension 3) and so on. More precisely, we consider the universal hypersurface of degree d i.e. $\mathcal{V} \subset \mathbf{P}^n \times \mathbf{P}^{\binom{n+d}{d}-1}$ consisting of pairs (P,V) such that $P \in V$. $Disc(n,d)$ is the image of the critical set of the projection on the second factor and its preimage in $\mathcal{V}$ is the universal singular hypersurface. The critical set of the projection on the second factor is the singular set $Sing(n,d)$ of the universal singular hypersurface. The restriction on $Sing(n,d)$ of the second projection is a surjective map onto $Disc(n,d)$. Moreover, this map is one to one outside of a codimension one algebraic subset $Sing_2(n,d)$ of $Sing(n.d)$ containing as a dense subset the singular points of hypersurfaces admitting more than one singularity. Then we consider the critical set $Sing(Sing(n,d))$ of the restriction of the projection on the non singular part of $Sing(n,d)$. As $Sing_2(n,d)$ it also has codimension one in $Sing(n,d)$ and the codimension one stratum of $Disc(n,d)$ is the union of the image of $Sing_2(n,d)$ and $Sing(Sing(n,d))$ in $Disc(n,d)$ and so on. With such a definition, Thom's isotopy theorem yields that the hypersurfaces belonging to each stratum are equisingular so the strata represent equisingular families of hypersurfaces. Note that the subset in $\mathbf{P}^{\binom{n+d}{d}-1}$ parameterizing equisingular hypersurfaces is singular in general (cf. [80]).

The case $n = 1$ is already very interesting and non trivial. The discriminant consist of homogeneous polynomials $\prod_i(\alpha_i u - \beta_i v)$ in two variables u, v having multiple roots, i.e. factors such that (α_i, β_i) and (α_j, β_j) satisfy

$det\begin{vmatrix} \alpha_i & \beta_i \\ \alpha_i & \beta_i \end{vmatrix} = 0$. The strata correspond to partitions of d, i.e. the conjugacy classes of the symmetric group Σ_d. A lot is known about the geometry of these strata, for example the degrees of their closures as well as other algebro-geometric information. The cases of discriminants with $n > 1$ are much more complicated. Many pieces of information are known. For example, in the case $n = 2$ the degrees of the strata corresponding to rational nodal curves have the interpretation as Gromov-Witten invariants of a projective plane and as such satisfy beautiful recurrence relations (cf. [37]). Indeed, the dimension of this stratum is $3d - 1$ where d is the degree of the curves (i.e. $\frac{(d+1)(d+2)}{2} - \frac{(d-1)(d-2)}{2}$) so the degree of the corresponding stratum is the number of nodal curves of degree d passing through generic $3d - 1$-points. The degrees of strata of nodal curves are subject to a conjecture of Göttsche discussed, for example, in [35].

2.1.2. *Classification of quadrics, cubics definition of local type*

Another class of discriminants which is well understood consists of the cases with $d = 2$. Each stratum corresponds to the quadrics of a fixed rank. In particular each stratum is a determinantal variety.

Classification of plane cubics goes back to Newton. Codimension one stratum consists of cubic curves with one node. It has the degree equal to 12. There are two strata having codimension 2. One consists of curves with one cusp and another formed by the reducible curves having two components: a nonsingular quadric and a non tangent to it line. The rest of the strata correspond to reducible curves and each is determined by strata of curves of lower degree and the mutual position. The strata of codimension three are: unions of a non singular quadric and a tangent line (in the closure of both strata of codimension 2) and the union of three lines in general position. Note that each of these strata is described by the local type of singularities: the number of nodes, cusps, tacnodes etc. A definition of the local type is the following:

Definition 2.1. Two reduced curves C and C' (of the same degree) have the same local type if there are one to one correspondences between their irreducible components and singular points of C and C' satisfying the following. The incidence relation between the singularities and components is preserved and for each pair of corresponding singularities P and P' there are neighborhoods $B_{P,\epsilon}$ and $B'_{P',\epsilon'}$ of P and P' respectively and homeomorphisms $\phi_P : B_\epsilon \to B'_{\epsilon'}$ such that $\phi_P(C \cap B_\epsilon) = C' \cap B'_{\epsilon'}$.

Two possibly non reduced curves have the same local type if:
(a) corresponding reduced curves have the same local type
(b) there are one to one correspondences between the components and singular points such that corresponding components have the same degrees and multiplicities.

2.1.3. *Examples with disconnected strata*

The classification of strata of curves of degree 4 provides the first example when the local type of singularities (in the sense of the first part of the definition 2.1) yields the strata with several connected components. The quartics with three nodes have two types: firstly the irreducible ones and quartics which are the unions of a non singular cubic and a generic line. The strata are distinguished by a global property.

For each degree there are finitely many irreducible families of plane curves having the same local type.

Problem 2.1. *Find discrete invariants of families of curves having the same local type.*

This problem is similar to the problem of classification of knots in S^3. Thom's isotopy theorem implies that the curves (or hypersurfaces) in a *connected* equisingular family are isotopic and hence have diffeomorphic complements. The main tool in the study of knots is the fundamental group of the complement which is one of the reasons suggesting to look at $\pi_1(\mathbf{P}^2 - C)$ or also into $\pi_k(\mathbf{P}^n - V)$ with $k > 1$ in the case of hypersurfaces of higher dimensions.

2.2. *Fundamental groups of the complements*

The classification problem of the strata of the discriminant brings in the fundamental group of the complements as a potentially important invariant but there are many other reasons for looking at the fundamental groups. One is that the fundamental groups of the complements of hypersurfaces control the covers of projective space and any projective algebraic variety having the dimension n is a branched covering space of $\mathbf{P}^n$.

Linear representations of the fundamental groups appear as the monodromy representations of differential equations and the correspondence between the monodromy groups (i.e. the quotients of the fundamental group) and differential equations is the subject of the Riemann-Hilbert problem. For example monodromy representation of KZ equation yields an interesting representation of the pure braid group closely related to the discriminant $Disc(1, d)$.

Each of these "applications" lead to concrete questions about the fundamental groups. For example, the use of π_1 for the study of covering spaces suggests the following. In the above presentation of algebraic varieties as the cover of $\mathbf{P}^n$ the degree of the cover is always finite. So the coverings are determined already by the quotient of the intersection of all subgroups of finite index. A natural problem (already mentioned in the introduction) is the following: does this intersection contain only the identity or in other words is the fundamental group of the complement to an algebraic hypersurface *residually finite*. Alternatively this can be stated as follows: is the map $\pi_1(\mathbf{P}^n - V) \to \pi_1^{alg}(\mathbf{P}^n - V)$ into the algebraic fundamental group injective. Note that the fundamental group of an algebraic variety does not have to be residually finite (D. Toledo). In general the problem of finding the properties of the fundamental groups of the complements or characterizing the algebraic structure of these groups is one of the central and the most difficult problems in algebraic geometry.

2.2.1. *Homology of the complements*

An easily available information about the fundamental groups $\pi_1(\mathbf{P}^{n+1}-V)$ comes from calculation of the homology $H_1(\mathbf{P}^{n+1}-V)$ which, by Hurewicz theorem, is the quotient of the fundamental group by its commutator. Here is the answer:

Proposition 2.2. *Let V be the union of irreducible components $V_1, ..., V_r$ having the degrees $d_1, ..., d_r$. Then $H_1(\mathbf{P}^{n+1} - V, \mathbf{Z}) = \mathbf{Z}^r/(d_1, ..., d_r)$.*

For example, if $g.c.d.(d_1, ..., d_r) = 1$ then the homology group is torsion free. This is the case when one of components has the degree equal to 1 or in other words for the complements to hypersurfaces in $\mathbf{C}^{n+1}$.

2.2.2. *Examples of calculations of the fundamental groups*

In the last twenty years quite a few calculations of the fundamental groups were made. For example, as was mentioned in the introduction, Moishezon-Teicher calculated the fundamental groups of the complements to the branching curves of generic projections of many algebraic surfaces (cf. [77]). Oka calculated the fundamental groups of the complements to many curves having low degree, in particular to various classes of curves of degree 6 (cf. [69]). Many calculations were carried out by Artal-Carmona-Cogolludo (cf. [3]). Some of the techniques for these calculations I will discuss in the next chapter, but here I want to explain some short and elegant calculations made by Zariski 80 years back.

Proposition 2.3. *Let $\hat{C}_d$ be a curve dual to a rational nodal curve C_d having the degree d (the degree of $\hat{C}_d$ is equal to $2(d-1)$, it has $3d-6$ cusps and $2(d-2)(d-3)$ nodes). The group $\pi_1(\mathbf{P}^2 - \hat{C}_d)$ is isomorphic to the braid group of sphere on d strings. In particular the fundamental group of the complement to the quartic with 3 cusps is a non abelian group having order 12.*

Indeed, C_d is a generic projection on $\mathbf{P}^2$ of a rational normal curve C in $\mathbf{P}^d$ and the dual to C_d curve is a section of the hypersurface $\hat{C}$ in $\mathbf{P}^d$ dual to C by a generic plane H. The complement to this hypersurface $\hat{C}$ consists of hyperplanes in $\mathbf{P}^d$ intersecting C transversally i.e. at d distinct points which can be chosen arbitrary. In fact, the hyperplanes transversal to C are in one to one correspondence with the d-tuples of disctinct points on C. Hence the space of based loops in this complement is identified with the space of braids of $\mathbf{P}^1(\mathbf{C}) = S^2$. Finally the isomorphism $\pi_1(\mathbf{P}^2 - C_d) = B_d(S^2)$ follows from Lefschetz hyperplane section theorem applied to the embedding of the complement in H into the complement in $\mathbf{P}^d$. In the case $d = 3$, the pure braid group of sphere can be identified with $\pi_1(PGL_1(\mathbf{C})) = \mathbf{Z}_2$ and hence one has the exact sequence: $1 \to \mathbf{Z}_2 \to B(S^2) \to S_3 \to 1$.

2.2.3. *Alexander invariants of the fundamental groups*

Since the problem of characterization and understanding the fundamental group is very complicated it is reasonable to try to rather understand some invariants of the fundamental groups. An accessible and interesting invariant is the Alexander invariant of a group.

Let G be an arbitrary group together with a surjective homomorphism $\phi : G \to \mathbf{Z}^r$. Let $\operatorname{Ker}\phi = K$ and let $K' = [K, K]$ be the commutator. If ϕ is the abelianization $G \to G/G'$ then $K = G', K' = G'' = [G', G']$. We have:

$$0 \to K/K' \to G/K' \to \mathbf{Z}^r \to 0 \tag{2}$$

In particular K/K' receives the action of $\mathbf{Z}^r$ and hence K/K' becomes the module over the group ring of $\mathbf{Z}^r$. This module is called the Alexander invariant of the pair (G, ϕ). In the case when ϕ is the abelianization one obtains an invariant depending on the group G only. It is denoted below as $A(G,\phi)$ or if ϕ is the abelianization as $A(G)$.

This definition can be interpreted geometrically. If X_G is a CW-complex having G as its fundamental group then the homomorphism ϕ defines the covering space $\tilde{X}_{G,\phi}$. One has $\pi_1(\tilde{X}_{G,\phi}) = K$ and $K/K' = H_1(\tilde{X}_{G,\phi}, \mathbf{Z})$. The action of $\mathbf{Z}$ corresponds to the action of the group $\mathbf{Z}$ of deck transformations on $\tilde{X}_{G,\phi}$.

For perfect groups, i.e. such that $G = G'$, this invariant is trivial (since $r = 0$ is the only possibility), but since the fundamental groups of the complements in $\mathbf{P}^{n+1}$ are perfect only if the hypersurface is the hyperplane (cf. 2.2) for them the Alexander invariant is always interesting.

There is an algorithmic procedure for calculation of the Alexander modules due to R.Fox ("Fox calculus") (cf. [28]).

Let G be a finitely generated, finitely presented group i.e. one has a surjective map $\Phi : F_s \to G$ of the free group F_s on s generators, $x_1, ..., x_s$ with the kernel being the normal closure of a finite set of elements given by the words $R_1, ..., R_N$ in F_s. Consider the maps of the group rings: $\frac{\partial}{\partial x_j}$: $\mathbf{Z}[F_s] \to \mathbf{Z}[F_s]$ uniquely specified by the conditions:

$$\frac{\partial(uv)}{\partial x_j} = \frac{\partial u}{\partial x_j} a(v) + u \frac{\partial u}{\partial x_j}; \qquad \frac{\partial x_i}{\partial x_j} = \delta_{i,j} \qquad (3)$$

where $a : \mathbf{Z}[F_s] \to \mathbf{Z}$ is the augmentation surjection. Using operators $\frac{\partial}{\partial x_j}$ one can define the map of free $\mathbf{Z}[\mathbf{Z}_r]$-modules given by the Jacobi matrix:

$$(\phi_* \circ \Phi_* \frac{\partial R_i}{\partial x_j}) : \mathbf{Z}[\mathbf{Z}_r]^N \to \mathbf{Z}[\mathbf{Z}_r]^r \qquad (4)$$

which entries are obtained by applying the homomorphisms $\Phi_* : \mathbf{Z}[F_s] \to \mathbf{Z}[G]$ and $\phi_* : \mathbf{Z}[G] \to \mathbf{Z}[\mathbf{Z}^r]$ of group rings induced by Φ and ϕ respectively. The geometric meaning of this map is the following. With a presentation Φ one can associate the 2-complex X_G with single 0-cell, r 1-cells forming wedge $S^1 \vee ... \vee S^1$ of circles corresponding to the generators of G and N 2-cells attached so that the boundary of each is represented by the word R_i $(i = 1, ..., N)$ in $S^1 \vee ... \vee S^1$. The covering space $\tilde{X}_{G,\phi}$ corresponding to the homomorphism of the fundamental group has a canonical cell structure given by the preimages of cells in the above cell decomposition of X: each cell in X_G is replaced by cells of the same dimension corresponding to the elements of the covering group. Hence we obtain the isomorphisms $C_2(\tilde{X}_{G,\phi}) = \mathbf{Z}[\mathbf{Z}^r]^N$ and $C_1(\tilde{X}_{G,\phi}) = \mathbf{Z}[\mathbf{Z}^r]^s$. Moreover, after this identification, the boundary operator $\partial_2 : C_2(\tilde{X}_{G,\phi}) \to C_1(\tilde{X}_{G,\phi})$ becomes identified with the operator given by (4). Since $H_0(\tilde{X}_{G,\phi}, \mathbf{Z}) = \mathbf{Z}$ and $C_0(\tilde{X}_{G,\phi}, \mathbf{Z}) = \mathbf{Z}[\mathbf{Z}]$ we have the isomorphism $\mathrm{Im}\partial_1 = \mathrm{Ker}C_0(\tilde{X}_{G,\phi}, \mathbf{Z}) \to \mathbf{Z} = I_{\mathbf{Z}[\mathbf{Z}^r]}$ where $I_{\mathbf{Z}[\mathbf{Z}^r]}$ is the augmentation ideal of the group ring. Hence, (4) determines the presentation of the module very closely related to $H_1(\tilde{X}_{G,\phi})$. More precisely, if $\hat{H}(X_{G,\phi})$ is the module having presentation (4) then we have:

$$0 \to H_1(X_{G,\phi}) \to \hat{H}(X_{G,\phi}) \to I_{\mathbf{Z}[\mathbf{Z}^r]} \to 0 \tag{5}$$

For example for (the affine portions of) the curves in proposition 2.3, the Alexander module $A(G)$ coinciding with $H_1(X_{G,\phi})$ can be calculated as follows (in these examples $\phi : \pi_1 \to H_1 = \mathbf{Z}$ is the canonical homomorphism):

$$A(\pi_1(\mathbf{P}^2 - C_d)) = \mathbf{Z}[t, t^{-1}]/(t^2 - t + 1) \ (d = 4), A(\pi_1(\mathbf{P}^2 - C_d)) = 0 \ (d \geq 5) \tag{6}$$

For the links of algebraic singularities, which all belong to the class of iterated torus link, the Alexander polynomial, *i.e. the order of $A(G) \otimes \mathbf{Q}$ as a $\mathbf{Q}[t, t^{-1}]$-module*, can be found using the data of iterations and the values of Alexander polynomial of for the torus knot: for the link of singularity $x^p = y^q \ \ g.c.d.(p, q) = 1$ one has the following:

$$\Delta(t) = \frac{(t^{pq} - 1)(t - 1)}{(t^p - 1)(t^q - 1)} \tag{7}$$

Another way to calculate the Alexander polynomial is to use the A'Campo formula for the zeta-function of the monodromy in terms of a resolution of the singularity (cf. [23]):

$$\zeta(t) = \Pi(1 - t^{m_i})^{\chi(E_i^o)} \tag{8}$$

Here E_i are the exceptional curves of a resolution, E_i^o is set of points in E_i which are non-singular of the exceptional divisor, m_i is the order along E_i of the pullback of the equation of the singularity and χ denotes the topological Euler characteristic. The $\zeta(t)$ determines the Alexander polynomial of a curve singularity via: $\zeta(t) = \frac{(t-1)}{\Delta(t)}$.

2.2.4. *Alexander polynomials of plane algebraic curves: divisibility theorems*

There are two types of general results concerning the Alexander invariants of the fundamental groups $\pi_1(\mathbf{C}^2 - C)$. The Alexander polynomials of plane algebraic curves are restricted by the degree of the curve, by the local type of singularities and by position of the curve relative to the line at infinity. These restrictions sometimes yield triviality of the Alexander polynomial. On the other hand, the Alexander polynomial is completely determined by the local type of the singularities of the curve and the superabundances of certain linear systems given by the data depending on the singularities.

We shall start, with discussion of the first group of results. Let C be a projective curve and L be the line at infinity. One has the linking number homomorphism: $\mathrm{lk} : \mathbf{C}^2 - C \to \mathbf{Z}$ associating to a loop γ in $\mathbf{C}^2 - C$ the (oriented) number of intersection points of C and an immersed disk with the boundary γ. In the case when C is irreducible the homomorphism $H_1(\mathbf{C}^2 - C) \to \mathbf{Z}$ is the abelianization and was already used above. In general, lk defines the Alexander module and the Alexander polynomial $\Delta_C(t)$ (we shall omit mentioning the linking homomorphism used in its definition).

With each singular point $P \in C \in \mathbf{P}^2$ we associate the local Alexander polynomial which is the Alexander polynomial of the link defined as follows. In the case when P does not belong to the line at infinity L, the link is the intersection of C with a sufficiently small ball about P (so that the link type is independent of the radius). If this link has several components (i.e. P has several branches) the Alexander polynomial again is calculated using the homomorphism given by the total linking number (in S^3). In the case when $P \in L$, i.e. the curve has singularities at infinity, the local Alexander polynomial is defined as above but P considered as the singular point of $P \in C \cup L$. Note that, as follows from the definitions, the local Alexander polynomials can be calculated as the characteristic polynomials of the monodromy operators (cf. [23], [61] for examples and algorithms).

On the other hand, one can define the Alexander polynomial at infinity $\Delta_{\infty,C}$ as the Alexander polynomial of the link which is the intersection of C with the boundary of a sufficiently small tubular neighborhood of L in $\mathbf{P}^2$ (this boundary is the sphere of a sufficiently large radius in $\mathbf{C}^2 = \mathbf{P}^2 - L$). For example, is C is a union of d lines passing through a point in $\mathbf{P}^2$ outside of L then the link at infinity is the Hopf link with d components and hence its Alexander polynomial is:

$$\Delta_{\infty,C} = (t^d - 1)^{d-2}(t - 1) \tag{9}$$

The same equality holds for a curve which is transversal to the line at infinity since there is a deformation of such a curve to a union of d lines as above, such that transversality holds for all curves appearing during the deformation.

With these definitions we have the following:

Theorem 2.4. *([42])*

$$\Delta_C(t) \mid \Pi_{P \in \mathrm{Sing} C} \Delta_P(t)$$

$$\Delta_C(t) \mid \Delta_{\infty, C}(t)$$

Consider, for example an irreducible curve in $\mathbf{P}^2$ having ordinary cusps (i.e. having $x^2 = y^3$ as the local equation) and nodes (local equation: $x^2 = y^2$) as the only singularities. Then, as follows from (7), the local Alexander polynomial for each singularity is $t^2 - t + 1$ (cusp) or $t - 1$ (node). Moreover, it is not hard to show that the multiplicity of the factor $(t - 1)$ is $r - 1$ where r is the number of irreducible component of C (cf. [42]). Hence we obtain:

Corollary 2.5. *Let C be an irreducible curve in $\mathbf{P}^2$ having cusps and nodes as the only singularities. Then:*

$$\Delta_C(t) = (t^2 - t + 1)^s$$

for some integer $s \geq 0$.

Combining this corollary, the divisibility and the formula (9) we obtain:

Corollary 2.6. *Let C be an irreducible curve in $\mathbf{P}^2$ having cusps and nodes as the only singularities. Then $\Delta_C(t) = 1$ unless d is divisible by 6.*

We leave as an exercise for a reader to work out that $pq \nmid d$ is a sufficient condition for triviality of the Alexander polynomial for an irreducible curve of degree D with singularities locally given by $x^p = y^q$.

Since the curves discussed in Proposition 2.3 (and also the branching curves of generic projections of non-singular surfaces in $\mathbf{P}^3$ cf. [64]) have the degree $d(d - 1)$ it follows that the Alexander polynomial is trivial if $d \equiv 2 (\mathrm{mod} 3)$ which explains with no calculation the triviality part of equation (6) (at least after tensoring with $\mathbf{Q}$). Many additional examples of calculations of the Alexander polynomials can be found in [69]. Note, finally that it is also beneficial to consider the Alexander polynomials over finite fields $\mathbf{F}_p$, rather than over $\mathbf{Q}$ i.e. $H_1(X_{G,\phi}, \mathbf{F}_p)$ (cf. [44]).

2.2.5. *Alexander polynomials of plane algebraic curves: position of singularities*

Now we shall discuss the dependence of the Alexander polynomial on the positions of singularities of the curve. To this end we consider the invariants of plane curve singularities (introduced in [43]) which are the collections of rational numbers, $\kappa_1^P, ..., \kappa_{n(P)}^P$, called the constants of quasiadjunction and corresponding to each point P in the set $\mathrm{Sing}C \subset \mathbf{P}^2$ of singular points of C. Moreover, to each $\kappa \in \mathbf{Q}$, which is a constant of quasiadjunction of a point $P \in \mathrm{Sing}C$, and each $Q \in \mathrm{Sing}C$, we associate the ideal $\mathcal{J}_\kappa \subset \mathcal{O}_Q$ in the local ring of $Q \in \mathbf{P}^2$. (P and Q may be distinct). This data, consisting of constants of quasiadjunction and the ideals in the local rings of singular points, determines the global Alexander polynomial $\Delta_C(t)$ completely (cf. [43] and theorem (2.10) below).

The idea of calculation is based on the relation between the Alexander polynomial and the homology of cyclic covers on one side and the classical method of adjoints for description of the holomorphic forms on hypersurfaces in projective space (cf. [85]).

The relationship between the Alexander polynomial and the homology of cyclic branched covering is the following:

Theorem 2.7. *Let $f(x,y) = 0$ be the equation of a curve $C \in \mathbf{C}^2$. Let $\tilde{V}_n$ be a desingularization of a compactification of the surface $z^n = f(x,y)$ in $\mathbf{C}^3$. If $A(\pi_1(\mathbf{C}^2 - C)) \otimes \mathbf{Q} = \oplus \mathbf{Q}[t, t^{-1}]/(\delta_i(t))$ is the cyclic decomposition of the Alexander module of C (i.e. $\Delta_C(t) = \Pi_i \delta_i(t)$) then $\mathrm{rk}H_1(V_n, \mathbf{Q})$ is equal to the sum over i of the numbers of common roots of $t^n - 1$ and $\delta_i(t)$. If the line at infinity is transversal to C then the Alexander module is semisimple and the dimension of the ω_n-eigenspace of a generator of the Galois group $\mathbf{Z}_n$ acting on $H_1(V_n, \mathbf{C})$ (ω_n is a root of unity of degree n) is equal to the multiplicity of ω_n as a root of the Alexander polynomial.*

Note that the first Betti number of a non-singular projective algebraic surface is a birational invariant and hence the first Betti number of a resolution of a compactification is a well defined invariant of an affine surface $z^n = f(x,y)$. Therefore it is also an invariant of affine curve C. Similar to 2.7 result is valid for branched covering of S^3 branched over a link: the idea of using covering spaces to derive invariants of knots goes back to Alexander and Reidemister (cf [2], [70], [83], [85]). A consequence of this theorem is that the homology of cyclic covers, in the case when line at infinity is transversal to C, determine the Alexander polynomial. Another consequence is periodicity of the homology of cyclic covers. In the abelian case the growth of the homology is polynomial periodic (cf. [32]).

The calculation of the homology of cyclic covers using theory of adjoints was carried out in [82] (the case when C has cusps and nodes), [41] (the case when C has singularities of the form $x^k = y^k$ or $x^k = y^{k+1}$) and, much later, for the curves with arbitrary singularities, in [43]. The proofs for a generalization to situation including hypersurfaces having arbitrary dimension is given in [50]. In fact all these proofs yields the irregularity $q = \dim H^1(\tilde{V}_n, \mathcal{O}_{\tilde{V}_n}) = \dim H^0(\tilde{V}_n, \Omega^1_{\tilde{V}_n}) = \frac{1}{2}\dim H^1(\tilde{V}_n, \mathbf{C})$ (and in [50] the Hodge number $h^{n,0}$ for cyclic coverings of $\mathbf{P}^{n+1}$).

For details of the using this method we shall refer to [50] and section 5.2, but here we shall only remark that the adjoint ideal of a germ $(W, P) \in (\mathbf{C}^3, P)$ of isolated singularity at P consists of germs in $\mathcal{O}_P$ which restriction to W belongs to $\Phi_*(\Omega^2_{\tilde{W}})$ where $\Phi : \tilde{W} \to W$ is a resolution of singularities of W. If W is given by the equation $F = 0$, then the 2-forms on $W - P$ are residues of 3-forms $\frac{\psi(x,y,z)\,dx \wedge dy \wedge dz}{F(x,y,z)}$ on $\mathbf{C}^3$ having pole of order one along W i.e. the restrictions of 2-forms:

$$\frac{\psi(x,y,z)dx \wedge dy}{\frac{\partial F(x,y,z)}{\partial z}} \tag{10}$$

on $W - P$.

On the other hand the 2-forms on a resolution can be described as the 2-forms on $W - P$ which can be extended over the exceptional locus of Φ. Hence a germ $\psi(x,y,z)$ is in the adjoint ideal of W if the pull back of the form (10) on resolution $\tilde{W}$ extends over the exceptional set. Such interpretation of 2-forms on resolutions allows to relate the dimensions of space 1-forms on $\tilde{V}_n$ (which is isomorphic to $H^1(\Omega^2_{\tilde{V}_n})$) to H^1 of certain sheaf of ideals on $\mathbf{P}^2$ which we are going to describe.

Let $\phi(x,y)$ be a germ of a holomorphic function. Let us consider the function $\Xi_\phi(n)$ which assigns to a n the minimal k such that $z^k\phi(x,y)$ belongs to the adjoint ideal of the singularity $z^n = f(x,y)$.

Lemma 2.8. *There exist $\kappa_\phi \in \mathbf{Q}$ (also depending on singularity $f(x,y)$) such that:*

$$\Xi_\psi(n) = [\kappa_\phi n]$$

([..] denotes the integer part).

The adjoint ideal of a function $F(x,y,z)$, which is generic for its Newton polytope, can be described as follows: a monomial $x^\alpha y^\beta z^\gamma$ is in the adjoint ideal of $F(x,y,z)$ if and only if the point $(\alpha+1, \beta+1, \gamma+1)$ is inside the Newton polytope of $F(x,y,z)$ (cf. [62]). Hence if $f(x,y) = x^a + y^b$ and $\phi(x,y) = x^i y^j$ then $z^k x^i y^j$ is in the adjoint ideal of $z^n = f(x,y) = x^a + y^b$ if and only if $(i+1)bn + (j+1)an + (k+1)ab > abn$ or $k + 1 > n(1 - (i +$

$1)\frac{1}{a} - (j+1)\frac{1}{b})$. Therefore:

$$\Xi_{x^i y^j}(n) = \max([n(1 - (i+1)\frac{1}{a} - (j+1)\frac{1}{b})], 0) \tag{11}$$

This construction can be used to associate to a constant $\kappa \in \mathbf{Q}$ the following ideal in the local ring of the singular point of germ $f(x,y)$:

Definition 2.9. Let $\kappa \in \mathbf{Q}$. The corresponding ideal of quasiadjunction is defined as follows:

$$J_\kappa = \{\phi(x,y) | \kappa > \kappa_\phi\}$$

For example if $f(x,y) = x^2 + y^3$ then:

$$\kappa_{x^i y^j}(n) = \begin{cases} [\frac{n}{6}], & (i,j) = 0 \\ 0, & i+j \geq 1 \end{cases} \tag{12}$$

and hence there is only one constant of quasiadjunction $\kappa = \frac{1}{6}$. Moreover, $J_{\frac{1}{6}}$ is generated by monomials such that $i+j \geq 1$ i.e. is the maximal ideal.

Loeser and Vaquié (cf. [59]) showed that the constants of quasiadjuction are precisely the elements of Arnold-Steenbrink spectrum of singularity $f(x,y)$ which are inside the interval $(0,1)$. In particular $exp(2\pi i \kappa_\phi)$ are the eigenvalues of the monodromy of $f(x,y) = 0$ and hence are the roots of the Alexander polynomial of the link of $f(x,y)$. After introduction of multiplier ideals it was soon realized that the ideals of quasiadjunction are closely related to multiplier ideals (cf. section 5.4 below). J.Kollar noticed the connection between the log-canonical threshold and the constants of quasiadjunction (cf. [36] and section 5.4).

Using the ideals J_κ in the local rings of points in $\mathbf{P}^2$, which are the singular points of a curve $C \in \mathbf{P}^2$, one defines the ideal sheaf

$$\mathcal{J}_\kappa = \mathrm{Ker}\mathcal{O}_{\mathbf{P}^2} \to \oplus_{P \in \mathrm{Sing}C} \mathcal{O}_P / J_{\kappa,P} \tag{13}$$

where $J_{\kappa,P}$ is the ideal corresponding to the singularity of C at P and the constant κ. Using this we can calculate the Alexander polynomial as follows:

Theorem 2.10. *Let C be a curve in $\mathbf{P}^2$ having degree d and let $\kappa_1, ..., \kappa_N$ be the collection of all constants of quasiadjunction of all singular points of C. Then the Alexander polynomial $\Delta_C(t)$ is given by:*

$$\Pi_{i, d\kappa_i \in \mathbf{z}}[(t - exp(2\pi\sqrt{-1}\kappa_i))(t - exp(2\pi\sqrt{-1}\kappa_i))]^{\dim H^1(\mathbf{P}^2, \mathcal{J}_{\kappa_i}(d-3-d\kappa_i))}$$

Note that the exponent can be written as follows:

$$\dim H^1(\mathbf{P}^2, \mathcal{J}_{\kappa_i}(d-3-d\kappa_i)) = \dim H^0(\mathbf{P}^2, \mathcal{J}_{\kappa_i}(d-3-d\kappa_i)) - \chi(\mathcal{J}_{\kappa_i}) \tag{14}$$

(since $H^2(\mathbf{P}^2, \mathcal{J}_{\kappa_i}(d-3-d\kappa)) = 0$). In other words the exponent is the difference between the actual and "expected" dimensions of the linear system of curves of degree $d-3-d\kappa_i$ which local equations belong to the ideals of quasiadjunction corresponding to the constant κ. Therefore, (14) is what is classically called the superabundance of this linear system.

As an example, let us consider the sextics with six ordinary cusps. Since only one type of singularities is present and (12) shows that there is only one constant quasiadjunction $\kappa = \frac{1}{6}$, the Alexander polynomial has the form

$$[(t - exp(\frac{2\pi i}{6}))(t - exp(-\frac{2\pi i}{6}))]^s = (t^2 - t + 1)^s$$

Now the linear system in question consists on the curves having degree $6 - 3 - \frac{6}{6} = 2$ with local equations belonging to the maximal ideals of the singular points. Since the dimension of the space of quadrics is 6, the expected dimension of our linear system is 0 and if a quadric containing all six cusps does exist then the actual dimension is 1 (one can show that this is the maximal possible value). Hence $s = 1$ and the Alexander polynomial is $t^2 - t + 1$.

For a sextic $\hat{C}_3$ with nine cusps, which is dual to a non singular cubic, one has $\dim H^0(\mathbf{P}^2, \mathcal{J}_{\kappa_i}(d-3-d\kappa_i)) = 0$ and $\chi(\mathcal{J}_{\kappa_i}) = -3$. Therefore

$$\Delta_{\hat{C}_3}(t) = (t^2 - t + 1)^3 \tag{15}$$

For the curve from [78] given by the equation: $f_{3n}^2 + f_{2n}^3 = 0$ where f_n is a generic form of degree n, which has only ordinary cusps at $6n^2$ points forming a complete intersection of curves of degrees $2n$ and $3n$ the exponent of $t^2 - t + 1$ in theorem 2.10 is the superabundance of the curve of degree $6n - 3 - \frac{6n}{6}$ passing through this complete intersection. By a theorem of Cayley-Bacharach this superabundance is 1 and hence the Alexander polynomial is $t^2 - t + 1$.

2.3. *Commutative fundamental groups*

2.3.1. *Commutativity in terms of local type of singularities. Nori's theorem*

Historically, much of the work on the fundamental groups of the complements, was focused on the cases when the fundamental group is abelian.

In this case Prop. 2.2 yields the complete calculation of π_1. For example, as was pointed out in the introduction, F.Severi was claiming that the fundamental group of the complement to a curve having nodes as the only singularities is abelian. More precisely he claimed the irreducibility of the stratum of nodal curves (this was proven much later by J.Harris in [30]). The irreducibility of this stratum yields that each nodal curve can be degenerated into a union of lines in general position and for such union (these days called a generic arrangement of lines) a direct calculation shows that the fundamental group of the complement is free abelian. More generally than in the case of nodal curves, one expects, vaguely speaking, that if a curve has not too many singularities or if the singularities are sufficiently mild then the fundamental groups of the complement will be abelian. A precise result in this direction follow from a theorem of M.Nori:

Theorem 2.11. *Let D and E be a pair of curves on a non singular surface X. Assume that D has nodes as the only singularities, that D and E intersects transversally and that for an irreducible component C of D one has $C^2 > 2r(C)$ where $r(C)$ is the number of nodes on C. Then $N = \mathrm{Ker}\,\pi_1(X - D - E) \to \pi_1(X - E)$ is abelian.*

For plane curves one obtains the following which extends the earlier commutativity results of S.Abhyankar.

Theorem 2.12. *For a germ ϕ of a curve singularity in $\mathbf{C}^2$ let us define the invariant $e(\phi)$ as follows. Let $\Phi : S \to \mathbf{C}^2$ be a resolution of the singularity of ϕ and $\Phi^*(\phi) = F + G$ where F is the proper transform of $\phi = 0$, G is the exceptional set and F and G_{red} meet transversally. Let $e(\phi) = G(G + 2F)$ and let, for a curve C on a non singular projective surface X, $F(C)$ be the sum over all singularities of C of the invariants $e(\phi)$. If $C^2 > F(C)$ then the extension $\pi_1(X - C) \to \pi_1(X)$ is central.*

Proof. Apply Nori's theorem 2.11 to the curves C' and E on a surface X' such that $C' \subset X'$ is the proper transform of C in an embedded resolution $\Phi : X' \to X$ of singularities of $C \subset X$ and E is the exceptional set. Then, if $C' + G$ is the total Φ-transform of C, we have $C^2 = (C' + G)^2 = C'^2 + 2(C', G) + G^2 = C'^2 + F(C)$. Hence the assumed inequality translates into $C'^2 > 0$. Now Nori's theorem yields the conclusion. Note that for a node we have $G = 2E$ where E is the exceptional line and $C' = L_1 + L_2$. Hence $G^2 + 2(G, C') = 4E^2 + 2 \cdot 2E(L_1 + L_2) = 4$. For a cusp $F(\phi) = 6$. In particular on a simply-connected surface the fundamental group of the complement to a curve with δ nodes and κ cusps is abelian if $C^2 > 6\kappa + 4\delta$.

The following question is still open:

Question 2.13. Let N be a normal subgroup of $\pi_1(X)$ generated by the images of the fundamental groups of non singular models of components. Does N has a finite index in $\pi_1(X)$

If so, then the fundamental group of a surface, containing a rational curve with positive self-intersection, must be finite.

2.3.2. *On a proof of Nori's theorem*

Let us consider a special case when $E = \emptyset$, and C is an irreducible nonsingular curve on X. Let U be a tubular neighborhood of C. Then $U - C \to C$ is a circle fibration and the fiber δ is the element of $\pi_1(U - C)$ belonging to the center of the latter group. Since in this case the assumption of the theorem is $C^2 > 0$, the theorem of Nakai and Moishezon (cf. [31]) yields that C is ample. Hence a small deformation D of nC, which we may assume belongs to U, is very ample and also smooth. By Zariski-Lefschetz theorem $\pi_1(D - C) \to \pi_1(X - C)$ is surjective and hence $\pi_1(U - C) \to \pi_1(X - C)$ is surjective as well. Therefore the image if the class of γ in $\pi_1(X - C)$ belongs to its center. On the other hand, any element in $N = \mathrm{Ker}\pi_1(X - D) \to \pi_1(X)$ is product of elements conjugated to γ. Indeed, take such element δ and consider 2-disk Δ which it bounds in X. We can assume that $\Delta \cap C$ consists of finitely many transversal intersections. Therefore $\delta = \Pi\delta_i$ where $\delta_i = \alpha_i\gamma_i\alpha^{-1}$ with γ_i being a fiber of $U - C \to C$ and α_i is a path going from the base point to a point on the boundary of U. In particular δ_i is conjugate to γ in $\pi_1(X - C)$ and hence is equal to γ. Hence δ is a power of γ i.e. N is cyclic.

Crucial in the proof of Nori's theorem in the case of nodal C is the following Nori's weak Lefschetz theorem which is very interesting by itself.

Theorem 2.14. *Let $i : H \to U$ be an embedding of a connected compact complex analytic subspace (possibly non reduced) into a connected complex manifold U in which H is defined by a locally principal sheaf of ideals. Assume that $\mathcal{O}_U(H)|H$ is ample and that $\dim U > 2$. Let $q : U \to X$ be a holomorphic local isomorphism with the target being a smooth projective variety and $h = q \circ i$. Let R be an arbitrary Zariski closed subset and $G = \mathrm{Im}\pi_1(U - q^{-1}(R)) \to \pi_1(X - R)$. Then G is a subgroup of finite index.*

2.4. *Higher homotopy groups*

Other natural invariants of the homotopy type of the complement are the higher homotopy groups. However for curves, the higher homotopy groups, unlike the fundamental groups, it seems, do not have an algebro-geometric significance. Moreover, in most cases the higher homotopy groups, considered as abelian groups are infinitely generated. A more useful way to consider them is by using the action of π_1 on π_k i.e. consider π_k as a module over π_1. But, unless π_1 is abelian, understanding modules over π_1 involves a subtle non commutative algebra. For curves however, as will be explained in the next section, the homotopy type of the complement is determined by another invariant of the pair $(\mathbf{P}^2, C)$ called the *braid monodromy*. On the other hand for hypersurfaces in $\mathbf{P}^{n+1}$ with $n > 1$ the homotopy groups in dimensions up to n have interesting algebro-geometric meaning which we shall proceed to discuss.

2.4.1. *Action of the fundamental group on higher homotopy groups*

Let us start with the example which shows why the homotopy groups of simplest topological spaces are infinitely generated.

Example 2.15. Let us consider π_2 of the wedge $S^1 \vee S^2$. Clearly $\pi_1(S^1 \vee S^2) = \mathbf{Z}$. On the other hand $\pi_2(S^1 \vee S^2)$ can be identified with π_2 of the universal cover of $S^1 \vee S^2$. Viewing the universal covering map of the circle as the the quotient of $\mathbf{R}$ by the subgroup of integers makes it natural to view the universal cover of $S^1 \vee S^2$ as the real line with S^2's attached at the integer points. Hence the universal cover has H_2, and by Hurewicz theorem also π_2, isomorphic to $\mathbf{Z}^\infty$. On the other hand, since the deck transformation of the universal cover acts transitively on S^2's attached to $\mathbf{R}$, both H_2 and π_2 are cyclic modules over the group of deck transformations i.e $\pi_2(\widetilde{S^1 \vee S^2}) = \mathbf{Z}[t, t^{-1}]$ ($\widetilde{S^1 \vee S^2}$ denotes the universal cover).

In general, the homotopy groups can be given the structure of a module over the fundamental group using the Whitehead product: $\pi_n \times \pi_m \to \pi_{n+m-1}$ with $n = 1$. In the cases when $\pi_i(X) = 0$ for $2 \le i \le n-1$, denoting by $\tilde{X}$ the universal cover, we have $\pi_n(X) = \pi_n(\tilde{X}) = H_n(\tilde{X})$ and the action of $\pi_1(X)$ is just the action of the deck transformations on the homology.

Such X come up naturally:

Theorem 2.16. *Let V be a hypersurface in $\mathbf{P}^{n+1}$ having only isolated singularities. Let H be a generic hyperplane. Then $\pi_1(\mathbf{P}^{n+1} - V \cap H) = \mathbf{Z}$*

and $\pi_i(\mathbf{P}^{n+1}-V\cap H) = 0$ for $2 \le i \le n-1$. Moreover, $\pi_n(\mathbf{P}^{n+1}-V\cap H)\otimes\mathbf{Q}$ is a $\mathbf{Q}[t, t^{-1}]$-torsion module.

More generally, the Lefschetz hyperplane section theorem yields that the conclusion of the theorem holds for arbitrary hypersurfaces in $\mathbf{P}^N$ for which the singular locus has codimension $n + 1$. To see this (and also the first part of theorem 2.16) recall it:

Theorem 2.17. *(Lefschetz hyperplane section theorem)*
(a) Let X be a projective subvariety having dimension n and let L be a codimension d linear subspace such that X is a local complete intersection outside of L. Then

$$\pi_i(X \cap L) \to \pi_i(X)$$

is isomorphism for $0 \le i < n - d$ and surjective for $i = n - d$.
(b)Let X be a quasiprojective. The conclusion of (a) take place for generic L.

Vanishing statement in theorem 2.16 follows from this and calculation of the homotopy groups of the complement to non-singular hypersurfaces.

Recently, L.Maxim ([60]) showed that the homology of infinite cyclic covers of the complement to an affine hypersurface, generic relative to the hyperplane at infinity, are torsion modules in all dimensions except the top one (cf. also [22]).

2.4.2. *Orders of the homotopy groups*

It follows from the theorem 2.16 and the classification of modules over PIDs that

$$\pi_n(\mathbf{P}^{n+1} - V \cap H) \otimes \mathbf{Q} = \oplus\mathbf{Q}[t, t^{-1}]/\Delta_i(t)$$

for some polynomials $\Delta_i(t)$ defines up to a unit in $\mathbf{Q}[t, t^{-1}]$. We call $\Delta(t) = \Pi_i\Delta_i(t)$ the order of the group π_n. Though $\Delta(t)$ cannot be calculated in terms of a local data of singularities there is the following divisibility relation, generalizing the divisibility relation for the Alexander polynomials:

Theorem 2.18. *(Divisibility theorem I) The order of $\pi_n(\mathbf{C}^{n+1}-V)$ divides the product of characteristic polynomials of the monodromy operators of singularities of V:*

$$\Delta(\mathbf{C}^{n+1} - V)|\Pi_{P_i \in Sing(V)}\Delta_{P_i}(t)$$

Note that as it stated, one should assume that V it transversal to the hyperplane at infinity. However one can define correction factors corresponding to the singularities at infinity so that, after multiplication by these correction factors the right side in 2.18, the divisibility relation holds.

Theorem 2.19. *(Divisibility theorem II) Let V be a hypersurface transversal to the hyperplane at infinity H_∞. Let S_∞ be the boundary of a small tubular neighborhood of H_∞ and let $L_\infty = V \cap S_\infty$. Then the homology of the infinite cyclic cover of $S_\infty - L_\infty$ is a torsion $\mathbf{C}[t, t^{-1}]$-module and Δ_∞ and $\Delta(\mathbf{C}^{n+1} - V)|\Delta_\infty$.*

(see [49] for a statement in the case with a weaker than transversality to H_∞ assumption).

3. Homotopy groups via pencils

3.1. *Van Kampen theorem and braid monodromy*

Now let us consider how one can actualy calculate the fundamental group of a complement in the case of curves and how to calculate the first non trivial homotopy group of the complement in the case of hypersurfaces. In this section we shall deal with the curves (cf. also [27] where the case of possibly singular quasiprojective varieties is discussed).

Let C be a curve on a projective surface X for which we want to describe $\pi_1(X - C)$. Consider a line bundle $\mathcal{L}$ on X such that $\dim H^0(X, \mathcal{L})) \geq 2$ and select a 2-dimensional linear system $\mathbf{L} \subseteq H^0(X, \mathcal{L})$. Let B be the base locus of $\mathbf{L}$ (it contains at most $c_1(\mathcal{L})^2$ points). We shall assume for simplicity that $B \cap C = \emptyset$. The classical case is $X = \mathbf{P}^2$, $\mathcal{L} = \mathcal{O}(1)$ and $\mathbf{L} \subset H^0(\mathbf{P}^2, \mathcal{O}(1))$ consists of sections with the zerosets containing a fixed point. We have a regular map onto the projectivization of $\mathbf{L}$:

$$p : X - B \to \mathbf{P}(\mathbf{L}) = \mathbf{P}^1 \tag{16}$$

with generic fiber $L_{t_0} - L_{t_0} \cap B, t_0 \in \mathbf{P}^1$ being non singular by Bertini's (or Sard's) theorem. Though generic element of $\mathbf{L}$ may be singular at points of B, we shall make additional assumtion that L_{t_0} is non singular at any $p \in B \cap C$.

The curve L_{t_0} is ample and hence $\pi_1(L_{t_0} - L_{t_0} \cap C) \to \pi_1(X - C)$ is surjective by Lefschetz theorem. We want to describe the kernel of this map. Let $Sing \subset \mathbf{P}^1$ be the (finite) subset of points $t_1, ..., t_N$ corresponding to singular members of the pencil. Each fiber of the pencil (16) is a punctured curve (which, if L_{t_0} is non singular at the points of B, has genus $g(L_{t_0}) =$

98

$\frac{c_1(\mathcal{L})(K_X + c_1(\mathcal{L}))}{2} + 1$). Using pencil $\mathbf{P(L)}$, we also will form the *affine portion* of X as $X - L_{t_\infty}$ where $t_\infty \in \mathbf{P(L)}$ is a generic point. We shall see below that $\pi_1(X - C)$ has presentation obtained from $\pi_1(X - L_{t_\infty} - C \cap L_{t_\infty})$ by adding one additional relation (cf. theorem 3.2).

For each d one can define the braid group $B_d(L_{t_0} - B)$ which is the group of isotopy classes of orientation preserving diffeomorphisms of L_{t_0} which are constant in a neighborhood of B in L_{t_0}. In the case $L_{t_0} - B \cap L_{t_0} = \mathbf{C}$ one obtains the classical Artin's braid group with generators $\sigma_i, i = 1, ..., d - 1$ and relations

$$\sigma_i \sigma_j = \sigma_j \sigma_i \; |i - j| \geq 2, \sigma_i \sigma_{i+1} \sigma_i = \sigma_{i+1} \sigma_i \sigma_{i+1} \quad i = 1, .., d - 2 \qquad (17)$$

(for presentations of braid groups similar to $B_d(L_{t_0} - B)$ by generators and relations and extending this one, see [74]).

We want to construct the homomorphism $\pi_1(\mathbf{P}^1 - Sing, t_0) \to B_d(L_{t_0} - B)$ called the *braid monodromy* which will yield a system of generators of Ker $\pi_1(L_{t_0} - L_{t_0} \cap C) \to \pi_1(X - C)$. Let $\mathbf{C} = \mathbf{P}^1 - t_\infty$. We shall start by defining "good" systems of generators of $\pi_1(\mathbf{C} - Sing, t_0) = \pi_1(\mathbf{P}^1 - t_\infty - Sing, t_0)$, and then assign the braids in $B_d(L_{t_0} - B)$ to them.

Definition 3.1. Let $Sing = \{t_1, ..., t_N\}$. A system of generators $\gamma_i \in \pi_1(\mathbf{C} - \bigcup_i t_i, t_0)$ is called *good* if each of the loops $\gamma_i : S^1 \to \mathbf{C} - \bigcup_i t_i$ extends to a map of the disk $D^2 \to \mathbf{C}$ with non-intersecting images for distinct i's.

One way to construct a good system of generators is the follwoing. Select a system of small disks Δ_i about each point $t_i \; i = 1, ..., N$, and choose a system of N non-intersecting paths δ_i connecting the base point t_0 with a point of $\partial \Delta_i$. Then $\gamma_i = \delta^{-1} \circ \partial \Delta_i \circ \delta_i$ is a good system of generators (with, say, the counterclockwise orientation of $\partial \Delta_i$). We shall need also good systems of generators of the fundamental groups of the complements to a finite set of N points on a *compact* Riemann surface having genus $g \geq 0$ which will be used in the statement of the theorem 3.2. Those are the systems of generators $\gamma'_1, ..., \gamma'_{2g}$, consisting of the images $2g$ sides of a $4g$-gon for some presentation of the surface as a $4g$-gon with identified sides and a good system of generators $\gamma_1, ..., \gamma_N$ of the complement to N points in this $4g$-gon in the above sense. We have the only relation

$$R: \; \Pi \gamma_1 \cdot ... \cdot \gamma_N = \Pi[\gamma'_i, \gamma'_{i+1}] \qquad (18)$$

In the case $g = 0$ this relation becomes $\Pi \gamma_1 \cdot ... \cdot \gamma_N = 1$.

Now let us define the braid monodromy corresponding to an element $\gamma \in \pi_1(\mathbf{P}^1 - \text{Sing})$. Let $\gamma \in \mathbf{P}^1 - \text{Sing}$ be the image of an embedding of S^1 taking the base point to t_0. We can view γ as the image of the map $\iota : I \to \mathbf{P}^1 - Sing$ (I is the unit interval) such that $\iota(0) = \iota(1) = t_0$. Then $(X - B - C) \times_{\mathbf{P}^1 - Sing} I$ is a locally trivial fibration over I and hence is a trivial fibration. This means that there is a map $\Phi : L_{t_0} - L_{t_0} \cap C \times I \to X - B$ such that $\Phi(t)|_{L_{t_0} - L_{t_0} \cap C \times t}$ is a homeomorphism onto $L_t - L_t \cap C$. Note that though Φ is not unique any two choices are isotopic via isotopies commuting with projections on I. Hence we obtain the map $\Phi(1) : L_{t_0} - L_{t_0} \cap C \to L_{t_0} - L_{t_0} \cap C$ and the isotopy class of this map is well defined. We can assume that this map keeps B fixed. One checks immediately that dependence on ι yields homotopic maps $\Phi(1)$ and a homotopy of γ extends to a homotopy of $\Phi(1)$ (but B may not be possible to preserve). Hence we obtain the braid monodromy homomorphism:

$$\pi_1(\mathbf{P}^1 - Sing) \to \pi_0(Diff(L_{t_0})) = B_d(L_{t_0}) \tag{19}$$

where $d = (C, L_t)$ and the last group is the braid group of Riemann surface L_{t_0}.

There is a useful way to encode algebraically the homomorphism (19) using the choice of a good system of generators of $\pi_1(\mathbf{P}^1 - Sing)$. Recall that we fixed a fiber L_{t_∞} of the pencil which we call the fiber at infinity. We can select monodromy transformations all fixing a neighborhood of B for all γ_i i.e. we obtain ordered system of braids: $\beta(\gamma_i) = \Phi_{\gamma_i}(1) \in B_d(L_{t_0} - L_{t_0} \cap C - B)$ with the order given by the order of the good systems of generators. The latter is given by the counterclockwise ordering of loops about the point t_0. Moreover, the product is a fixed word in $B_d(L_{t_0})$ independend of C. For example we obtain in the case of curves in $\mathbf{C}^2$:

$$\Pi \beta(\gamma_i) = \Delta^2 \tag{20}$$

where Δ^2 is the generators of the center of the Artin's braid group B_d (cf. [47]).

We have the following calculation in terms of the braid monodromy originated by Zariski-van Kampen:

Theorem 3.2. *Let* $b \in \partial T(B) \cap L_{t_0}$ *where* $T(B)$ *is a neighborhood of* B *in* X *and let* α_j *be a good system of generators of* $\pi_1(L_{t_0} - L_{t_0} \cap C, b)$. *Then*

$$\pi_1(X - C - L_{t_\infty}) = \pi_1(L_{t_0} - L_{t_0} - B, b)/(\beta(\gamma_i)(\alpha_i)\alpha_i^{-1})$$

(quotient by the normal subgroup generated by specified elements). The group $\pi_1(X - C)$ *can be obtained by adding to the above the relation* R *defined in (18).*

In the case of plane curves we have just the homomorphism into Artin's braid group which by itself is an interesting invariant of plane curves containing more information than the fundamental group. For example the braid monodromy determines the homotopy type of the complement $\mathbf{C}^2 - C$ (cf. [46]). Many calculations are done for curves C which are the branching curves of generic projections of surfaces (cf. [77]). Recently braid monodromy found applications in symplectic geometry (cf. [4]).

3.2. *Homotopy groups via pencils*

Now let V be a hypersurface in $\mathbf{C}^{n+1}$ transversal to the hyperplane at infinity and having only isolated singularities. We want to describe a calculation of the first non trivial homotopy group $\pi_n(\mathbf{C}^{n+1} - V)$ (recall that the lower homotopy groups were described already in the theorem 2.16) in terms of pencils generalizing the Zariski-van Kampen procedure described above in the theorem 3.2.

We start with a high dimensional analog of the braid group and a linear representation generalizing the Burau representation of the braid group. In higher dimensions we have several candidates for such a generalization.

Let us consider a sphere S^{2n-1} in $\mathbf{C}^n$ of a sufficiently large radius. Let $\partial_\infty V = V \cap S^{2n-1}$ and let $Emb(V, \mathbf{C}^n)$ be the space of submanifolds of $\mathbf{C}^n$ with the following property: each is diffeomorphic to V and, moreover, is isotopic to the chosen embedding of V. In addition we require that for any $V' \in Emb(V, \mathbf{C}^n)$ one has $V'(V) \cup S^{2n-1} = \partial_\infty V$ i.e. the link at infinity is fixed. We shall use the topology with the basis consisting of sets $U(V, \epsilon)$ of submanifolds $V' \subset \mathbf{C}^n$ which belong to the tubular neighborhood of V having radius ϵ and which are isotopic to V.

We want to contsruct a linear representation:

$$\pi_1(Emb(V, \mathbf{C}^n)) \to Aut\pi_n(\mathbf{C}^n - V) \tag{21}$$

After a choice of a basis in the $\pi_1(\mathbf{C}^n - V)$-module $\pi_n(\mathbf{C}^n - V)$ this homomorphism becomes the homomorphism into $GL_r(\mathbf{Z}[t, t^{-1}])$ where r is the rank of $\tilde{H}_n(\mathbf{C}^n - V, \mathbf{Z})$ (the reduced homology of the complement). The group $\pi_1(Emb(V, \mathbf{C}^n))$ for $n = 1$ is the Artin's braid group B_d where $d = \deg V$ which makes $\pi_1(Emb(V, \mathbf{C}^n))$ a high dimensional analog of B_d. The map (21) is given in terms of the representation of another group which also is a candidate for the high-dimensional braid group.

Let $Diff(\mathbf{C}^n, S^{2n-1})$ be the group of diffeomorphisms of $\mathbf{C}^n$ acting as the identity outside S^{2n-1}. This group can be identified with the group $Diff(S^{2n}, D^{2n})$ of the diffeomorphisms of the sphere fixing a disk. Let

$Diff(\mathbf{C}^n, V)$ be the subgroup of $Diff(\mathbf{C}^n, S^{2n-1})$ of the diffeomorphisms which take V into itself.

The group $Diff(\mathbf{C}^n, S^{2n-1})$ acts transitively on $Emb(V, \mathbf{C}^n)$ with the stabilizer $Diff(\mathbf{C}^n, V)$. Therefore we have the following exact sequence:

$$\pi_1(Diff(S^{2n+2}, D^{2n+2})) \to \pi_1(Emb(\mathbf{C}^n, V)) \to \pi_0(Diff(\mathbf{C}^n, V)) \quad (22)$$

$$\to \pi_0(Diff(S^{2n+2}, D^{2n+2})) \to$$

Any element in $Diff(\mathbf{C}^n, V)$ induces the self map of $\mathbf{C}^n - V$ and also the self map of the universal (in the case $n = 1$ universal cyclic) cover of this space. Hence it induces an automorphism of $H_n(\widetilde{\mathbf{C}^n - V}, \mathbf{Z}) = \pi_n(\mathbf{C}^n - V)$, $n > 1$. This gives the representation:

$$\pi_0(Diff(\mathbf{C}^n, V)) \to Aut\pi_n(\mathbf{C}^n - V) \quad (23)$$

The composition of the boundary homomorphism in (22) with the map (23) results in the representation (21). The groups $\pi_1(Emb(\mathbf{C}^n, V))$ and $\pi_0(Diff(\mathbf{C}^n, V))$ both are high-dimensional analogs of the braid groups but their algebraic study was not carried out so far. However some resembling high-dimensional analogs of the mapping class groups were studied in [34].

In the case $n = 1$, as already was mentioned, V is just a collection of points in $\mathbf{C}$, $\pi_1(Emb(\mathbf{C}, V)) = \pi_0(Diff(\mathbf{C}, V))$ is the Artin's braid group, and this construction gives the homomorphism of the braid group into $AutH_1(\widetilde{\mathbf{C} - V}, \mathbf{Z})$ which, after a choice of the basis in $H_1(\widetilde{\mathbf{C} - V})$ corresponding to the choice of the generators of the braid group, gives the reduced Burau representation. In higher dimensions the isomorphism $\pi_1(Emb(V, \mathbf{C}^n)) = \pi_0(Diff(\mathbf{C}^n, V))$ fails.

An interesting unsolved problem is the following:

Problem 3.1. *Calculate the groups* $\pi_1(Emb(V, \mathbf{C}^n))$ *and* $\pi_0(Diff(\mathbf{C}^n, V))$

Now we can define the relevant monodromy operators which correspond to the loops in the parameter space of a linear pencil of hyperplane sections. These operators are the high dimensional analogs of the braid monodromy in the case of curves.

Recall that by our assumptions the projective closure of V is a hypersurface in $\mathbf{P}^{n+1}$ having only isolated singularities. Let $H \subset \mathbf{P}^{n+1}$ be the hyperplane at infinity (which is transversal to V). Let L_t, $t \in \mathbf{C}$, be a pencil of hyperplanes in $\mathbf{C}^{n+1}$ the projective closure of which has a hyperplane M belonging to H as its base locus. We shall assume that the pencil is sufficiently generic so that M is transversal to V. Let $t_1, ..., t_N$ denote the collection of those $t \in \mathbf{C}$ for which the hypersurface $V \cap L_t$ is singular. We shall also assume that for any i the singularity of $V \cap L_{t_i}$ is outside of H.

The pencil L_t over $\mathbf{C} - \bigcup_i t_i$ defines a locally trivial fibration τ of $\mathbf{C}^{n+1} - V$ with a non-singular hypersurface in $\mathbf{C}^n$ transversal to the hyperplane at infinity as a fiber. The restriction of this fibration on the complement to a sufficiently large ball in $\mathbf{C}^{n+1}$ (i.e. a sufficiently small tubular neighbourhood in $\mathbf{P}^{n+1}$ of H) is trivial, as follows from the assumptions on the singularities at infinity. Let $\gamma : [0,1] \to \mathbf{C} - \bigcup_i t_i$ $(i = 1, ..., N)$ be a loop with the base point t_0. A choice of a trivialization of the pull back of the fibration τ on $[0,1]$ using γ, defines a loop e_γ in $Emb(L_{t_0}, V \cap L_{t_0})$. Different trivializations produce homotopic loops in this space.

Definition 3.3. The monodromy operator corresponding to γ is the element in
$Aut(\pi_n(L_{t_0} - L_{t_0} \cap V))$ corresponding in (21) to e_γ.

Next we will need to associate the following homomorphism, called the *degeneration operator* to the data consisting of a singular fiber L_{t_i} and a loop γ with the base point t_0 in the parameter space of the pencil where γ bounds a disk Δ_{t_i} not containing other singular points of the pencil :

$$\pi_{n-1}(L_{t_i} - L_{t_i} \cap V) \to \pi_n(L_{t_0} - L_{t_0} \cap V)/Im(\Gamma - I). \qquad (24)$$

Here Γ is the monodromy operator corresponding to γ.

To construct (24) let us note that the π_1-module on the right in (24) is isomorphic to the homology $H_n(\tau^{-1}(\partial \Delta_{t_i}), \mathbf{Z})$ of the infinite cyclic cover of the restriction of the fibration τ on the boundary of Δ_{t_i}. This follows immediately from the Wang sequence of a fibration over a circle and the vanishing of the homotopy of $L_{t_0} - L_{t_0} \cap V$ in dimensions below n. Let B_i be a polydisk in $\mathbf{C}^{n+1}$ such that $B_i = \Delta_i \times B$ for a certain polydisk B in L_{t_0}. Then $\tau^{-1}(\Delta_i) - B_i$ is a trivial fibration over Δ_i with the infinite cyclic cover $L_{t_i} - L_{t_i} \cap V$ as a fiber. In particular, one obtains the map:

$$\pi_{n-1}(L_{t_0} - L_{t_0} \cap V) = H_{n-1}(L_{t_0} - L_{t_0} \cap V, \mathbf{Z}) \to H_n(\tau^{-1}(\Delta_i) - B_i, \mathbf{Z}) =$$

$$H_{n-1}(L_{t_0} - L_{t_0} \cap V, \mathbf{Z}) \oplus H_n(L_{t_0} - L_{t_0} \cap V, \mathbf{Z}) \qquad (25)$$

Definition 3.4. The degeneration operator is the map (24) given by composition of the map (25) with the map

$$H_n(\tau^{-1}(\Delta_i) - B_i, \mathbf{Z}) \to H_n(\tau^{-1}(\Delta_i), \mathbf{Z}) = \pi_n(L_{t_0} - L_{t_0} \cap V)$$

induced by inclusion.

The following is the high-dimensional analog of the van Kampen theorem (cf. [49]).

Theorem 3.5. *Let V be a hypersurface in $\mathbf{P}^{n+1}$ having only isolated singularities and transversal to the hyperplane H at infinity. Consider a pencil of hyperplanes in $\mathbf{P}^{n+1}$ the base locus M of which belongs to H and is transversal in H to $V \cap H$. Let $\mathbf{C}_t^n$ ($t \in \Omega$) be the pencil of hyperplanes in $\mathbf{C}^{n+1} = \mathbf{P}^{n+1} - H$ defined by L_t (where $\Omega = \mathbf{C}$ is the set parameterizing all elements of the pencil L_t excluding H). Denote by $t_1, ..., t_N$ the collection of those t for which $V \cap L_t$ has a singularity. We shall assume that the pencil was chosen so that $L_t \cap H$ has at most one singular point outside of H. Let t_0 be different from either of t_i ($i = 1, ..., N$). Let γ_i ($i = 1, ..., N$) be a good collection, in the sense described in definition (3.1), of paths in Ω based in t_0 and forming a basis of $\pi_1(\Omega - \bigcup_i t_i, t_0)$. Denote by $\Gamma_i \in Aut(\pi_n(\mathbf{C}_t^n - V \cap \mathbf{C}_t^n))$ be the monodromy automorphism corresponding to γ_i. Let finally $\sigma_i : \pi_{n-1}(\mathbf{C}_{t_i}^n - V \cap \mathbf{C}_{t_i}^n) \to \pi_n(\mathbf{C}_{t_0}^n - V \cap \mathbf{C}_{t_0}^n)^{\Gamma_i}$ be the degeneration operator of the homotopy group of a special element of the pencil into the corresponding quotient of covariants constructed above. Then*

$$\pi_n(\mathbf{C}^{n+1} - V \cap \mathbf{C}^{n+1}) = \pi_n(\mathbf{C}^n - V \cap \mathbf{C}^n)/(Im(\Gamma_1 - I), Im\sigma_1, ..., Im(\Gamma_N - I), \sigma_N)$$

We refer for a proof to the paper [49]. There is another way to describe this homotopy group replacing the degeneration operator by variation operators on the homotopy groups to which we shall turn now.

3.3. *Variation operators*

Variation operators classically defined in homology (or cohomology). The idea of defining homotopy variation operators comes from the fact that the homotopy groups on question are the homology groups of covering spaces. A description of the homotopy groups using variation operators was carried out in [11].

We shall continue to use the notations from the previous section but in addition let us select $e \in M - M \cap V$ which we shall use as the base point for the homotopy groups. The homotopy variation operator is a certain homomorphism of $\mathbf{Z}[t, t^{-1}]$-modules:

$$\mathcal{V}_i: \ \pi_n(L_{t_0} - L_{t_0} \cap V, M - M \cap V, e) \longrightarrow \pi_n(L_{t_0} - L_{t_0} \cap V, e) \qquad (26)$$

associated with each γ_i for $1 \leq i \leq N$.

104

As in definition (3.4) of degeneration operators we shall go to the d fold cover and use the homological variation operators. Let $W \subset \mathbf{P}^{n+2}$ be the d-fold cyclic branched over V cover of $\mathbf{P}^{n+1}$, $j : V \to W$ the embedding and $\mathcal{L}_t$ be the pull back of the pencil L_t on W. $\mathcal{L}_{t_0} \cap W$ is the d-fold cover of L_{t_0} branched over $V \cap L_{t_0}$. We shall consider the homological variation operators:

$$V_i: \ H_n(\mathcal{L}_{t_0} \cap (W - j(V)), \mathcal{M} \cap (W - j(V))) \longrightarrow H_n(\mathcal{L}_{t_0} \cap (W - j(V))) \quad (27)$$

associated, for $1 \le i \le N$, with the homotopy classes γ_i.

The definition and the properties of operators V_i are discussed in [9]. For a relative n-cycle Ξ on $\mathcal{L}_{t_0} \cap (W - j(V))$ with boundary in $\mathcal{M} \cap (W - j(V))$, one defines ($[\cdot]$ denotes the class of a cycle):

$$V_i([\Xi]_{(\mathcal{L}_{t_0} \cap (W - j(V)), \mathcal{M} \cap (W - j(V)))}) = [H_{i*}(\Xi) - \Xi]_{\mathcal{L}_{t_0} \cap (W - j(V))} \quad (28)$$

where H_i is the geometric monodromy corresponding to γ_i. H_i leaves the points of $\mathcal{M} \cap (W - j(V))$ fixed. Therefore, the chain $H_{i*}(\Xi) - \Xi$ is actually an absolute cycle and the correspondence $\Xi \mapsto H_{i*}(\Xi) - \Xi$ induces a homomorphism V_i at the homology level ([9, Lemmas 4.6 and 4.8]). This homomorphism depends only on the homotopy class γ_i ([9, Lemma 4.8]).

Now, if $n \ge 2$ then for $1 \le i \le N$, using the isomorphism α_{t_0} and the homomorphism $\bar{\alpha}_{t_0}$, V_i yields the homotopical variation operator $\mathcal{V}_i$ by requiring that the following diagram will be commutative:

$$
\begin{array}{ccc}
H_n(\mathcal{L}_{t_0} \cap (W - j(V)), \mathcal{M} \cap (W - j(V))) & \xrightarrow{V_i} & H_n(\mathcal{L}_{t_0} \cap (W - j(V))) \\
\uparrow \bar{\alpha}_{t_0} & & \uparrow \alpha_{t_0} \\
\pi_n(L_{t_0} - L_{t_0} \cap V, M - M \cap V, e) & \xrightarrow{\mathcal{V}_i} & \pi_n(L_{t_0} - L_{t_0} \cap V, e).
\end{array}
$$
$$(29)$$

As V_i depends only on the homotopy class γ_i so do the operators $\mathcal{V}_i$.

With these definitions we have the following (cf. [11]):

Theorem 3.6. *Let V be a hypersurface in $\mathbf{P}^{n+1}$ with $n \ge 2$ having only isolated singularities. Consider a pencil $(L_t)_{t \in \mathbf{P}^1}$ of hyperplanes in $\mathbf{P}^{n+1}$ with the base locus M transversal to V. Denote by $t_1, \ldots, t_N$ the collection of those t for which $L_t \cap V$ has singularities. Let t_0 be different from either of $t_1, \ldots, t_N$. Let γ_i be a good collection of paths in $\mathbf{P}^1$ based in t_0. Let $e \in M - M \cap V$ be a base point. Let $\mathcal{V}_i$ be the variation operator corresponding to γ_i. Then the inclusion induces an isomorphism:*

$$\pi_n(\mathbf{P}^{n+1} - V, e) \longleftarrow \pi_n(L_{t_0} - L_{t_0} \cap V, e) \Big/ \sum_{i=1}^{N} \mathcal{V}_i \quad (30)$$

There is an affine version of this theorem equivalent to this one since $\pi_n(\mathbf{P}^{n+1} - V) = \pi_n(\mathbf{C}^{n+1} - V)$ in the case when H is transversal to V.

Recently, Cheniot and Eyral proposed definition of homotopy variation operator in general showed that the map as in the above theorem is surjective (cf. [10]; see also [79] for another discussion of variation operators).

4. Local multivariable Alexander invariants: topological theory

Now we want to develop an abelian version of the cyclic theory presented so far. Our goal at this point, as in the link theory, is to study abelian covers, though what will follow deviates from the link-theoretical point of view at several points. The most important one is that the Alexander type invariants are not polynomials: the substitute for the orders of the modules over PID which were discussed before are the subvarieties of commutative algebraic groups called the characteristic varieties.

4.1. *Characteristic varieties of groups*

4.1.1. *Definitions*

Let us start with a classical construction of commutative algebra. Let R be a Noetherian commutative ring with a unit and let M be a finitely generated R-module. Let the homomorphism $\Phi : R^m \to R^n$ be such that $M = \mathrm{Coker}\Phi$. The k-th Fitting ideal of M is the ideal $\mathcal{F}_k(M)$ generated by $(n - k + 1) \times (n - k + 1)$ minors of the matrix of Φ. $\mathcal{F}_k(M)$ depends only on M rather than on Φ. The k-th characteristic variety M is the reduced sub-scheme of $\mathrm{Spec}R$ defined by $\mathcal{F}_k(M)$.

If $R = \mathbf{C}[H]$ where H is an abelian group then $\mathrm{Spec}R$ is a torus having the dimension equal to the rank of H. If H is free then after a choice of generators of H, R can be identified with the ring of Laurent polynomials and $\mathrm{Spec}R = (\mathbf{C}^*)^{\mathrm{rk}H}$ is a complex torus. In particular each k-th characteristic variety of an R-module is a subvariety $V_k(M)$ of $(\mathbf{C}^*)^{\mathrm{rk}H}$. If H has a torsion then the number of connection components of $\mathrm{Spec}\mathbf{C}[H]$ is the order of the torsion and the connected component of the unit can be identified with $\mathrm{Spec}\mathbf{C}[H/Tor(H)]$.

A more functorial description can be given as follows (cf. [8]):

$$V_k(M) = \mathrm{Supp}_{red}(\wedge^k M) = \mathrm{Supp}_{red}(R/F_k(M)) \tag{31}$$

We shall apply this construction to the modules $A(G, \phi)$ defined in section 2.2.3. for pairs (G, ϕ) where $\phi : G \to \mathbf{Z}^r$. Prime examples which we shall consider are the following:

Example 4.1.
(i) Links in S^3. In this case $H_1(S^3 - L, \mathbf{Z}) = \mathbf{Z}^r$ where L is such a link and r is the number of its components.
(ii) Algebraic curves in $\mathbf{C}^2$ having r irreducible components (cf. section 2.2.1)

We shall denote the corresponding characteristic varieties as $V_k(G, \phi)$ omitting ϕ when no confusion is possible.

Definition 4.2. (cf. [53]) The *depth* of a component V of a characteristic variety $V_k(G)$ is the integer $i = \max\{j | V \subset V_j(G)\}$.

In the case $r = 1$ and G is one of the groups as above, $V_1(G)$ is the zero set of the Alexander polynomial and $V_k(G)$ is determined by the zero sets of elementary divisors of the Alexander module. Vice versa, the zero sets of Fitting ideals determine the zero sets of elementary divisors of a module over PID. Since the orders of $\mathbf{Q}[t, t^{-1}]$-modules in a cyclic decomposition are determined up to a unit of the ring of Laurent polynomials the depth of each root of the Alexander polynomial, given in terms of V_k's, determines the Alexander module completely.

If $\mathrm{codim} V_1(G, \phi)$ in $\mathrm{Spec} \mathbf{C}[G/G']$ is equal to one then the information carried by V_1 is equivalent to the multivariable Alexander polynomial up to the exponent of each factor (this is the case when ϕ is the abelianization of a link group). On the other hand if the codimension is bigger than one then for the pair (G, ϕ) the Alexander polynomial is not defined (or is trivial depending on convention) but $V_1(G)$ can be very interesting.

Now, as the first example, let us calculate the characteristic varieties of a free group. If $G = F_r$ is a free group on r-generators then $G'/G'' = H_1(\widetilde{\bigvee_r S^1}, \mathbf{Z})$, where $\widetilde{\bigvee_r S^1}$ is the universal abelian cover of the wedge of r circles. It fits into the exact sequence:

$$0 \to H_1(\widetilde{\bigvee_r S^1}, \mathbf{C}) \to \mathbf{C}[\mathbf{Z}^r]^r \to I \to 0$$

with I denoting the augmentation ideal of the group ring of $\mathbf{Z}^r$: $I = \mathrm{Ker} \mathbf{C}[\mathbf{Z}^r] \to \mathbf{C}$ where the homomorphism sends each generator to $1 \in \mathbf{C}$. Indeed, as an universal abelian cover of $\widetilde{\bigvee_r S^1}$ one can take the subset of $\mathbf{R}^r$ of points having at least $r - 1$ integer coordinates with the action of $\mathbf{Z}^r$ given by translations; unit vectors of the standard basis provide identification of 1-chains with $\mathbf{C}[\mathbf{Z}^r]^r$ while the module of 0-chains is identified with

$\mathbf{C}[\mathbf{Z}^r]$. The boundary map sends each generator $e_i, i = 1, ..., r$ of $\mathbf{C}[\mathbf{Z}^r]^r$ to $(t_i - 1) \in \mathbf{C}[\mathbf{Z}^r]$. This is the map which also appears in the Koszul complex (cf. [75]) in which we put $R = \mathbf{C}[\mathbf{Z}]$:

$$\wedge^i R^r \longrightarrow \wedge^{i-1} R^r \longrightarrow \cdots \to R^r \to R \tag{32}$$

where $\partial_i(e_{j_1} \wedge ... \wedge e_{j_i}) = \sum (-1)^k (t_k - 1) e_{j_1} \wedge ... \wedge \hat{e}_{j_k} \wedge ... \wedge e_{j_i}$. Since $(t_1 - 1, ..., t_r - 1)$ is a system of parameters the complex (32) is exact. Therefore

$$H_1(\widetilde{\bigvee_r S^1}, \mathbf{C}) = \mathrm{Coker}\Lambda^{\binom{r}{3}} \mathbf{C}[\mathbf{Z}^r]^r \to \Lambda^{\binom{r}{2}} \mathbf{C}[\mathbf{Z}^r]^r \tag{33}$$

in the Koszul resolution corresponding to the $(t_1 - 1), ..., (t_r - 1)$. This implies that $V_i(F_r) = \mathbf{C}^{*r}$ for $0 < i \leq r - 1$ and $V_i(F_r) = (1, ..., 1)$ for $r \leq i \leq \binom{r}{2}$ i.e. $\mathbf{C}^{*r}$ is component having depth $r - 1$, and $1 \in \mathbf{C}^{*r}$ has depth $\binom{r}{2}$.

For arbitrary group G, as was pointed out in earlier sections, the Fox calculus provides presentation for the extension of the homology of universal abelian cover by the augmentation ideal of the group ring of the covering group. This is sufficient to determine the characteristic varieties outside of the identity character.

4.1.2. *Unbranched covering*

The homology of a cyclic unbranched covering X_n of a CW-complex X with $\pi_1(X) = G$ corresponding to the homomorphism $G \overset{\phi}{\to} \mathbf{Z} \to \mathbf{Z}/n\mathbf{Z}$ can be found using Milnor's exact sequence (cf. [63]) i.e. the homology sequence corresponding to the exact sequence of chain complexes

$$0 \to C_*(\tilde{X}) \overset{t^n - 1}{\longrightarrow} C_*(\tilde{X}) \to C_*(X_n) \to 0 \tag{34}$$

The induced homology sequence:

$$\to H_1(\tilde{X}, \mathbf{C}) \overset{t^n - 1}{\longrightarrow} H_1(\tilde{X}, \mathbf{C}) \to H_1(X_n, \mathbf{C}) \to \mathbf{C} \to 0 \tag{35}$$

shows that $\mathrm{rk} H_1(X_n, \mathbf{C}) = \mathrm{rk} \mathrm{Coker}(t^n - 1)|_{H_1(\tilde{X}, \mathbf{C})} + 1$. In abelian case, to order to find the homology of the covering $X_{n_1, ..., n_r}$ corresponding to the homomorphism $G \overset{\phi}{\to} \mathbf{Z}^r \to \oplus_{i=1}^{i=r} \mathbf{Z}/n_i \mathbf{Z}$ the Milnor's sequence (35) should be replaced by the five term exact sequence corresponding to the spectral sequence of the covering group $H = \mathrm{Ker} \mathbf{Z}^r \to \oplus_i \mathbf{Z}/n_i \mathbf{Z}$ acting on the covering space $\tilde{X}$ corresponding to the homomorphism ϕ:

$$H_p(\mathbf{Z}^r, H_q(\tilde{X}, \mathbf{C})) \Rightarrow H_{p+q}(X_{n_1, ..., n_r}, \mathbf{C}) \tag{36}$$

108

This exact sequence is

$$H_2(X_{n_1,\ldots,n_r}, \mathbf{C}) \to H_2(H, \mathbf{C}) \to H_1(\tilde{X})_H \to H_1(X_{n_1,\ldots,n_r}, \mathbf{C}) \to H_1(H, \mathbf{C}) \to 0$$
$$(37)$$

where for a H-module M, $M_H = M/I(H)M$ detotes the module of covariants $(I(H)$, as above, is the augmentation ideal of the group ring of H). This yields the following formula for the first Betti number of abelian covers:

Proposition 4.3. *([48]) Let $X_{n_1,\ldots,n_r}$ be the finite unbranched abelian cover of a CW-complex X as above which is the quotient of the infinite abelian cover corresponding to $\phi : G \to \mathbf{Z}^r$. Let $V_i(G, \phi)$ be the characteristic varieties of (G, ϕ). For $P \in \mathbf{C}^{*r}$ let $f(P, G, \phi) = \{max \quad i | P \in V_i(G, \phi)\}$. Then*

$$\mathrm{rk}H_1(X_{n_1,\ldots,n_r}) = r + \Sigma_{\omega_i^{n_i}=1, (\omega_{n_1},\ldots,\omega_{n_r})\neq(1,\ldots,1)} f((\omega_{n_1}, .., \omega_{n_r}), (G, \phi))$$

4.1.3. *Homology of local systems*

Homology of rank one local systems also can be described in terms of the characteristic varieties. Such a local system is a homomorphism $\chi : G \to \mathbf{C}^*$ i.e. a character of the fundamental group. There is a natural identification of $\mathrm{Spec}\mathbf{C}[G/G']$ and $\mathrm{Char}G$. Moreover, $\mathrm{Spec}\mathbf{C}[G/\mathrm{Ker}\phi]$ can be identified with the subgroup of $\mathrm{Char}G$ of characters which can be factored through ϕ. Let $\tilde{X}_{G/G'}$ denotes the infinite cover corresponding to the subgroup G'. Recall, that the homology $H_1(X, \chi)$ of a local system χ, where $\chi \in \mathrm{Char}\pi_1(X)$, is defined as the homology of the chain complex:

$$\to C_i(\tilde{X}_{G/G'}) \otimes_{\mathbf{C}[G/G']} \mathbf{C}_\chi \to C_i(\tilde{X}_{G/G'}) \otimes_{\mathbf{C}[G/G']} \mathbf{C}_\chi \qquad (38)$$

where the chains $C_i(\tilde{X}_{G/G'})$ of the universal abelian covers are given the structure of $\mathbf{C}[G/G']$-module and $\mathbf{C}_\chi$ is $\mathbf{C}$ endowed with the module structure given by the character χ.

One has the following:

Proposition 4.4. *(cf. [33], [53]) If $\chi \neq 1$ then*

$$H_1(X, \chi) = H_1(\tilde{X}_{G/G'}, \mathbf{C}) \otimes_{\mathbf{C}[H_1(X,\mathbf{Z})]} \mathbf{C}_\chi$$

In particular, $\chi \in \mathrm{Char}G, \chi \neq 1$ belongs to $V_k(G)$ if and only if $H_1(X, \chi) \geq k$.

4.2. *Links of plane curves and multivariable Alexander polynomial*

For a link in S^3 with r components the characteristic varieties form a class of affine subvarieties of the torus without many apparent specific properties. An interesting problem is the following:

Problem 4.1. *Which sequences of subvarieties can occur as $V_i(G)$ where $G = \pi_1(S^3 - L)$ for some link in S^3.*

For a multivariable Alexander polynomial one has:

$$\Delta(t_1^{-1}, ..., t_r^{-1}) = \Delta(t_1, ..., t_r) \tag{39}$$

(up to a unit of the ring $\mathbf{Z}[\mathbf{Z}^r]$ i.e. a factor $\pm t_1^{a_1}...t_r^{a_r}$ where $a_i \in \mathbf{Z}$) which provides one of conditions asked for in the realization problem 4.1

The characteristic varieties of links of algebraic singularities are very special. Let us call a translated subgroup of $\mathbf{C}^{*r}$ a coset of a subgroup isomorphic to $\mathbf{C}^{*s}$ $(s < r)$. Such a "subgroup" is called *translated by an element of a finite order* if this coset has a finite order in $\mathbf{C}^{*r}/\mathbf{C}^{*s}$. Using the fact that links of algebraic singularities are iterated torus links one can prove the following:

Proposition 4.5. *(cf. [55]) The characteristic varieties of algebraic links are unions of translated subgroups.*

For example the link of singularity $x^r - y^r = 0$ has the Alexander polynomial $t_1 \cdot ... \cdot t_r = 1$. The Alexander polynomial of $(x^2 - y^3)(x^3 - y^2) = 0$ is $(t_1^2 t_2^3 - 1)(t_1^2 t_2^3 - 1) = 0$ (cf. [55]).

4.3. *Links of isolated non normal crossings*

Disjoint non intersecting spheres of dimension greater than one and having codimension 2 in an ambient sphere never can form a link of an algebraic singularity. Therefore, the high dimensional link theory does not play in the singularity theory the role similar to the role played by the link theory in S^3. There is nevertheless a *local abelian* analog of the local cyclic theory of the links of high dimensional algebraic singularities. It appears when one looks at the isolated non normal crossings (cf. [56], [21])

Definition 4.6. (cf. [56]) Let $D_1, .., D_k$ be divisors on a complex manifold X and $P \in D_1 \cap ... \cap D_k$. These divisors have a normal crossing at P if there exist in a neighborhood U_P of P in X together with a system of complex analytic local coordinates $(z_1, ..., z_{\dim X})$ in U_P such that D_i in U_P is given by the equations $z_i = 0$. $D_1, ..., D_k$ have an isolated non normal crossing at P if there exist a ball B_ϵ in X centered at P having sufficiently small radius ϵ such that for any $Q \neq P$ in B_ϵ the divisors D_i containing Q form at Q a divisor with normal crossings.

In particular each D_i has at most isolated singularity at P. A more general case, when the ambient space X is allowed to have a singularity at P is considered in [21].

The theory we shall describe here is invariant under analytic changes of variables so we can assume that $X = \mathbf{C}^{n+1}$. The starting point is the following vanishing result:

Theorem 4.7. *(cf. [56]) Let $X = \bigcup_{i=1}^{r} D_i \subset \mathbf{C}^{n+1}$ be a union of r irreducible germs of hypersurfaces with normal crossings outside of the origin. If $n \geq 2$, then*

$$\pi_1(\partial B_\epsilon - \partial B_\epsilon \cap X) = \mathbf{Z}^r \quad \text{and} \quad \pi_k(\partial B_\epsilon - \partial B_\epsilon \cap X) = 0 \quad \text{for} \ 2 \leq k < n.$$

In the case when $r = 1$ this result follows from Milnor's fibration theorem and connectivity of Milnor fibers (cf. [61]). In fact, the universal cyclic cover of the complement to a link of isolated hypersurface singularity D is homotopy equivalent to the Milnor fiber M_D. In particular $\pi_n(\partial B_\epsilon - D \cap \partial B_\epsilon) = H_n(M_D, \mathbf{Z})$. For a general isolated non-normal crossing (INNC) the main invariant is $\pi_n(\partial B_\epsilon - \cup D_i \cap \partial B_\epsilon)$. This, as usual, is the module over $\mathbf{Z}[\pi_1(\partial B_\epsilon - \cup D_i \cap \partial B_\epsilon] = \mathbf{Z}[t_1, t_1^{-1}, ..., t_r, t_r^{-1}]$. We shall call it the *homotopy module of INNC*. In the case $r = 1$ this module structure is equivalent to the $\mathbf{Z}[\mathbf{Z}]$-module structure on an abelian group endowed with an automorphism where the abelian group is the middle homology of the Milnor fiber and the automorphism is the monodromy operator. Notice that in the case of normal crossing (i.e. when the "singularity" is absent), the universal abelian cover of $\partial B_\epsilon - \cup D_i \cap \partial B_\epsilon$ is contractible and all homotopy groups are trivial.

Definition 4.8. (cf. [56]) k-th characteristic variety $V_k(X)$ of an isolated non-normal crossing $X = \cup_{i=1}^{i=r} D_i$ is the subset in $\mathrm{Spec} \mathbf{C}[\pi_1(\partial B_\epsilon - \partial B_\epsilon \cap (\bigcup_{1 \leq i \leq r} D_i))]$ formed by the zeros of the k-th Fitting ideal of $\pi_n(\partial B_\epsilon - \cup D_i \cap \partial B_\epsilon)$

Let us consider an example of a non normal crossing. The simplest non trivial case is when all components are given by linear equations i.e when

INNC is given in $\mathbf{C}^{n+1}$ by the equation $l_1 \cdot ... \cdot l_r = 0$, where l_i are *generic* linear forms. This INNC is a cone over a generic arrangement of hyperplanes in $\mathbf{P}^n$. Since the complement to a generic arrangement of r hyperplanes in $\mathbf{P}^n$ has the homotopy type of an n-skeleton of the product of $r-1$-copies of the circle S^1 it is enough to calculate the module structure on the π_n of such skeleton. It can be done as follows (cf. [56]).

We shall use $\pi_n(\partial B_\epsilon - D) = H_n(\widetilde{Sk_n((S^1)^{r-1})}, \mathbf{Z})$, where $\widetilde{Sk_n((S^1)^{r-1})}$ is the universal cover of the n-skeleton, and will calculate the homology. Notice that in the minimal cell decomposition of $S^{1^{r-1}}$ one has $\binom{r-1}{i}$ cells of dimension i. The universal cover of the above skeleton is obtained by removing the $\mathbf{Z}^{r-1}$ orbits of all open faces having the dimension greater than n in the unit cube in $\mathbf{R}^{r-1}$. The chain complex of the universal cover of $(S^1)^{r-1}$ can be identified with the Koszul complex of the group ring of $\mathbf{Z}^{r-1} = \mathbf{Z}^r/(1,...,1)$ (so that the generators of $\mathbf{Z}^r$ correspond to the standard generators of $H_1(\partial B_\epsilon - D)$). The system of parameters of this Koszul complex is $(t_1 - 1, .., t_r - 1)$. Hence $H_n(\widetilde{\partial B_\epsilon - D}, \mathbf{Z}) = Ker \Lambda^n R \to \Lambda^{n-1} R$ where $R = \mathbf{Z}[t_1, .., t_r, t_1^{-1}, ..., t_r^{-1}]/(t_1 \cdot ... \cdot t_r - 1)$. As a result, one has the following presentation:

$$\Lambda^{n+1}([\mathbf{Z}[t_1, t_1^{-1}, ..., t_r, t_r^{-1}]/(t_1..., t_r - 1)]^r) \to$$

$$\Lambda^n([\mathbf{Z}[t_1, t_1^{-1}, ..., t_r, t_r^{-1}]/(t_1..., t_r - 1)]^r) \to \qquad (40)$$

$$\pi_n(\mathbf{C}^{n+1} - \bigcup D_i) \to 0$$

In particular, the support of the π_n is the subgroup $t_1 \cdot ... \cdot t_r = 1$.

The relation between the characteristic varieties, the unbranched covering spaces and the local systems described in the case of links in S^3 extends to this high dimensional situation as well. We have the following:

Proposition 4.9. *(cf. [56]) (a) For each $P \in \mathrm{Spec}\mathbf{C}[\pi_1(\partial B_\epsilon - \partial B_\epsilon \cap X)]$ let*

$$f(P, X) = \{\max k \mid P \in V_k(X)\}$$

Let $U_{m_1,...,m_r}$ be unbranched cover of $\partial B_\epsilon - \partial B_\epsilon \cap (\bigcup_{1 \le i \le r} D_i)$ corresponding to the homomorphism $\pi_1(\partial B_\epsilon - \partial B_\epsilon \cap (\bigcup_{1 \le i \le r} D_i)) = \mathbf{Z}^r \to \oplus_{1 \le i \le r} \mathbf{Z}/m_i\mathbf{Z}$. Then

$$\mathrm{rk} H_p(U_{m_1,...,m_r}, \mathbf{C}) = \Lambda^p(\mathbf{Z}^r) \text{ for } p \le n - 1,$$

$$\mathrm{rk} H_n(U_{m_1,...,m_r}, \mathbf{C}) = \sum_{(...,\omega_j,...), \omega_j^{m_j} = 1} f((...,\omega_j,...), \bigcup_{1 \le i \le r} D_i)$$

(b) If $1 \neq \chi \in \mathrm{Char}\,\pi_1(\partial B_\epsilon - \partial B_\epsilon \cap (\bigcup_{1 \leq i \leq r} D_i)) = \mathrm{Spec}\,\mathbf{C}[\partial B_\epsilon - \partial B_\epsilon \cap (\bigcup_{1 \leq i \leq r} D_i)]$ is a character of the fundamental group then

$$H_i(\partial B_\epsilon - \partial B_\epsilon \cap (\bigcup_{1 \leq i \leq r} D_i), \chi) = 0 \quad 1 \leq i \leq n-1$$

$$H_n(\partial B_\epsilon - \partial B_\epsilon \cap (\bigcup_{1 \leq i \leq r} D_i), \chi) = \pi_n(\partial B_\epsilon - \partial B_\epsilon \cap (\bigcup_{1 \leq i \leq r} D_i)) \otimes_{\mathbf{Z}} \mathbf{C} \otimes_{\mathbf{C}[H_1(X,\mathbf{Z})]} \mathbf{C}_\chi$$

Milnor theory [61] is applicable to INNC as to any hypersurface and one can relate relate Milnor's invariants to the characteristic varieties discussed here. We have the following:

Proposition 4.10. *(cf. [56]) The homology of the Milnor fiber M_D of an INNC singularity D is given by:*

$$H_p(M_D, \mathbf{Z}) = \Lambda^p(\mathbf{Z}^{r-1}) \quad \text{for } 1 \leq p < n$$

The action of the monodromy on this homology is trivial. The multiplicity of $\omega \neq 1$ as a root of the characteristic polynomial $\Delta_n(D, t)$ of the Milnor's monodromy on $H_n(M_D, \mathbf{C})$ is equal to:

$$m_\omega = f((...,\omega,...), D) = \max\{i | (\omega, ..., \omega) \in V_i(D) \subset \mathrm{Spec}\,\mathbf{C}[\pi_1(\partial B_\epsilon - \partial B_\epsilon \cap D)]\}$$

In the case of INNC the unbranched covering admits a natural compactification which provides model for the abelian branched covering of the sphere S^{2n+1} with the link of INNC as the branching locus. The branching cover itself is a link of an isolated complete intersection singularity.

If the local equations of the locally irreducible components $D_1, ..., D_r$ are $f_1 = 0, ..., f_r = 0$, then we can use the link of the singularity:

$$z_1^{m_1} = f_1(x_1, ..., x_{n+1}), ..., z_1^{m_1} = f_1(x_1, ..., x_{n+1}) \tag{41}$$

as a model of the abelian branched cover.

The link of singularity (41) is a $(n-1)$-connected manifold having the dimension equal to $2n+1$ since it is a link of ICIS. We shall express the homology of this link in terms of the characteristic varieties of the homotopy modules associated to INNCs formed by various components of $D = \bigcup D_i$.

Proposition 4.11. *([56]) Let $V_{m_1,...,m_r}$ be the link of singularity (41) which is the branched cover of ∂B_ϵ having $\partial B_\epsilon \cap (\bigcup_{1 \leq i \leq r} D_i)$ as its branching locus. Let $G = \oplus_{1 \leq i \leq r} \mathbf{Z}/m_i\mathbf{Z}$ be the Galois group of this cover. For each $\chi \in \mathrm{Char}\,G$ let $I_\chi = \{i | 1 \leq i \leq r, \chi(\mathbf{Z}_{m_i}) \neq 1\}$ where $\mathbf{Z}_{m_i}$ is* *the i-th summand of G. Any χ can also be considered as a character of $\pi_1(\partial B_\epsilon - \partial B_\epsilon \cap (\bigcup_{i \in I_\chi} D_i))$ in which case it will be called I_χ-reduced and denoted χ_{red}. Let V_χ be the branched cover of ∂B_ϵ branched over $\partial B_\epsilon \cap (\bigcup_{i \in I_\chi} D_i)$ and having $\mathrm{Im}\chi = G/\mathrm{Ker}\chi$ as its Galois group. Then*

$$\pi_p(V_{m_1,\ldots,m_r}) = 0 \text{ for } 1 \le p \le n - 1,$$

$$\mathrm{rk}H_n(V_{m_1,\ldots,m_r}, \mathbf{C}) = \sum_{\chi \in \mathrm{Char}} f(\chi_{red}, \bigcup_{i \in I_\chi} D_i).$$

This proposition shows that there is a close relationship between the homology of the tower of abelian covers and the characteristic varieties (at least in local case). In fact the homology of covers in this tower can be used to describe the characteristic varieties. We shall use this point in the following section for the calculation of the homology of infinite abalian covers in terms algebro-geometric data such as resolution of singularities and the ideals of quasiadjunction.

5. Hodge decomposition of local Alexander invariants

5.1. *Zeros of Fitting ideals and Hodge numbers in cyclic case*

Our goal in this section is to study the structure of characteristic varieties in the local case i.e. to describe in terms of resolutions of singularities the Alexander invariants of the germs of reducible plane curves and INNCs. This will give an algebro-geometric description of these invariants. The global counterparts of the local invairants from this section will be considered in the section 6. All structures introduced in this section will be used there for the calculation of corresponding global invariants.

We shall start by considering the cyclic i.e. the one variable case and review the relationship between the Hodge structure on the cohomology of Milnor fiber (recall that it is homotopy equivalent to the infinite cyclic cover) and the Alexander invairants for a link of an isolated singularity. Calculation of the Alexander *polynomial* does not require the mixed Hodge theory and is a special case of the A'Campo's formula for the zeta function of the monodromy in terms of a resolution ([5]). But calculation of the zeros of higher Fitting ideals does depend on the data of MHS (we refer to [15] or [18] for the formalism of the latter).

Indeed, if D has only one component with isolated singularity, the order of $\pi_n(S^{2n+1} - D \cap S^{2n+1}) \otimes \mathbf{Q}$ is equivalent to the zeta function of the

114

monodromy. Hence, if E_i are the components of the exceptional set of a resolution π, $\pi^*(D)$ has the multiplicity m_i along a component E_i of the exceptional locus and the euler characteristic of the set of points in E_i nonsingular in the union of $\bigcup E_i$ and the proper preimage of D is $\chi(E_i^\circ)$ then (cf. [5]):

$$\Delta_n((S^{2n+1} - D \cap S^{2n+1})^{(-1)^n}(1-t) = \Pi(1 - t^{m(E_i)})^{-\chi(E_i^\circ)} \qquad (42)$$

For the rest of this section we focus mainly on the case of curves i.e. assume that $\dim D = 1$. The cohomology (*) group $H^{1\ddagger}$ of the Milnor fiber of a plane curve singularity supports the mixed Hodge structure with weights 0,1 and 2, with the identification

$$N : W_2/W_1 \to W_0 \qquad (43)$$

given by the logarithm of an appropriate power of the monodromy(cf. [72]). Recall that this means that one has canonically defined (weight) filtration $H^1 = W_2 \supseteq W_1 \supseteq W_0 \supseteq 0$ such that for each quotient $W_n/W_{n-1} = \oplus_{p+q=n}H^{p,q}$. In fact there is a strong relation between these groups $H^{p,q}$: they all come from increasing Hodge filtration. Moreover, if T is the monodromy operator on $H^1(M,\mathbf{C})$ and $T = T_sT_u$ is the factorization into semisimple and unipotent part and if $N = log(T_u) = \sum_{i \geq 1}(-1)^{i-1}\frac{(T_u-I)^i}{i}$ then N induces the isomorphism in (43).

All Hodge groups are invariant under the action of the semisimple part T_s of the monodromy. Let $h_\zeta^{p,q}$ (cf. [72]) be the dimension of the eigenspace of this semisimple part acting on the space $H^{p,q}$. The numbers $h_\zeta^{p,q}$ determine the Jordan form of the monodromy as follows. The size of the Jordan blocks of the monodromy does not exceed 2 and the number of blocks corresponding to an eigenvalue ζ of size 1×1 (resp. 2×2) is equal to $h_\zeta^{1,0} + h_\zeta^{0,1}$ (resp. $h_\zeta^{0,0}$). As a consequence, the generators of the Fitting ideals have the form:

$$\Delta_i = \prod_{(\zeta)}(t - \zeta)^{a_{\zeta,i}}$$

where

$\ddagger *$ we shall work with the cohomology as is more common in Hodge theory though one has the dual structures on homology. One of the differences is the presence of the negative weights in MHS in homology. See [65] where the author works with MHS on homotopy groups (also discussion below of the homotopy groups) having negative weights and where natural dual theory with positive weights is not available

$$a_{\zeta,i} = \begin{cases} h_\zeta^{1,0} + h_\zeta^{0,1} + 2h_\zeta^{0,0} - 2(i-1) & \text{if } 1 \le i \le h_\zeta^{0,0} \\ h_\zeta^{1,0} + h_\zeta^{0,1} - (i-1-h_\zeta^{0,0}) & \text{if } h_\zeta^{0,0} < i \le h_\zeta^{0,0} + h_\zeta^{1,0} + h_\zeta^{0,1} \\ 0 & \text{if } i > h_\zeta^{0,0} + h_\zeta^{1,0} + h_\zeta^{0,1} \end{cases}$$

In particular, Δ_i can be calculated algebraically in terms of a resolution of the singularity since all Hodge numbers $h_\zeta^{p,q}$ can be found in terms of a resolution (cf. [72]). A calculation of the Hodge numbers $h_\zeta^{p,q}$ is equivalent to the identifying the following subsets in the set of zeros of the characteristic polynomial of the monodromy operator:

$$\mathcal{H}_{p,q,k} = \{\zeta | h_\zeta^{p,q} \ge k\} \tag{44}$$

Arnold-Steenbrink spectrum [72] is also equivalent to this data. Our goal for the remaining sections of this lecture will be to describe the partition of (the unitary part of) the zero sets of Fitting ideal i.e. the characteristic varieties of plane curve singularities (and also INNC) into sets having the same definittion (44). We will call the partition which we shall obtain *the Hodge decomposition of characteristic varieties.*

In the abelian case the multivariable Alexander polynomial, and hence $V_1(G)$, can be found from a resolution of the singularity by a formula similar to the formula of A'Campo (cf. [23]):

Theorem 5.1. *Let $f_i = 0, i = 1, ...r$ be the equations of branches of a reducible curve C. Let $\pi : \tilde{\mathbf{C}}^2 \to \mathbf{C}^2$ be a resolution of singularities and $E_j, j = 1, ..., N$ be the exceptional components. Let $m_{i,j} = \mathrm{ord}_{E_j}\pi^*(f_i)$ and let E_j° be a Zariski open subset in E_j consisting of points which are non singular on the reduced total preimage of C. Then the Alexander polynomial of the link of singularity of C is:*

$$\Pi_j(1 - t_1^{m_{1,j}} \cdot ... \cdot t_r^{m_{r,j}})^{-\chi(E_i^\circ)}$$

For example for the Hopf link with r-components we obtain $(1 - t_1 \cdot ... \cdot t_r)^{r-2}$.

We shall use the description of the cohomology of branched coverings from the last lecture to calculate higher $V_i(G)$. The plan is as follows. As already was mentioned, the information about the characteristic varieties is closely related to the information about the cohomology of branched abelian covers (cf. Prop. 4.11). Those are the links of singularities of complete intersection and hence have the canonical mixed Hodge structure. We shall calculated the eigenspaces of the deck transformations acting on the Hodge spaces of the cohomology of branched coverings in terms of the algebraic data of the singularity i.e. some associated ideals generalizing the ideals of quasiadjunction defined earlier in these lectures (cf. 2.9). This will give us the calculation of the sets (44) and hence the characteristic varieties. The MHS on the cohomology of a link, or equivalently the appropriate local cohomology (cf. [71]), can be described in terms of differential forms on a resolution of the singularity so we shall need an explicit description of the these forms. We shall obtain it using the theory of adjoint hypersurfaces.

5.2. *Theory of adjoints*

The classical theory of adjoints (which already was briefly mentioned in section 2.2.5 and which was used there to calculated one variable Alexander polynomials) gives a presentation of the geometric genus of a resolution of singularities of a hypersurface or a complete intersections in $\mathbf{P}^N$ in terms of the degree (or the multidegree in the case of a complete intersections) and the local data from the singularities.

A starting point maybe the observation that the genus of a non-singular plane curve C of degree d (i.e. $g = h^{1,0} = \dim H^0(\Omega_C)$) is equal to $\frac{(d-1)(d-2)}{2}$ which also is equal to the dimension of the space of plane curves of degree $d - 3$ (i.e. $\dim H^0(\mathbf{P}^2, \mathcal{O}(d - 3))$). If C is not smooth but has, say δ, nodes as the only singularities then the genus of desingularization is equal to $\frac{(d-1)(d-2)}{2} - \delta$ which also is the dimension of the space of plane curves of degree $d-3$ passing through the nodes. In the case when C has singularities more complicated than nodes one can associate with each singular point P the ideal in the local ring $\mathcal{O}_P$ (adjoint ideal) such that the genus of desingularization is the dimension of the space of curves of degree $d - 3$ passing through the singularities of C and which local equations at each $P \in \text{Sing}C \subset \mathbf{P}^2$ belong to the adjoint ideals in the corresponding local rings $\mathcal{O}_P$.

Explicitly, these ideals can be described as follows. Let $X \subset \mathbf{P}^{n+1}$ be a hypersurface and let $f_* : \tilde{X} \to X$ be a resolution of singularities. Let $\mathcal{A} = f_*(\Omega_{\tilde{X}}^n)(-d + n + 2)$ and $\mathcal{A}' = \pi^{-1}(\mathcal{A})$ where $\pi : \mathcal{O}_{\mathbf{P}^{n+1}} \to \mathcal{O}_X$ is the restriction map. Then $\mathcal{A}'$ is called the sheaf of adjoint ideals. It follows easily that $H^0(\mathcal{A}'(d - n - 2)) = h^{n,0}(\tilde{X})$.

The sheaf $f_*(\Omega_{\tilde{X}}^n)$ is a subsheaf of $i_*(\Omega_{X-\text{Sing}X}^n)$ where $i : X - \text{Sing}X \to X$ is the embedding. One has the residue map which fits into the exact sequence:

$$0 \to \Omega_{\mathbf{P}^{n+1}}^{n+1} \to \Omega_{\mathbf{P}^{n+1}}^{n+1}(X) \overset{\text{Res}}{\to} i_*(\Omega_{X-\text{Sing}X}^n) \to 0 \tag{45}$$

This residue map sends a form $\omega = \frac{f(z_1,\dots,z_n)dz_1 \wedge \dots \wedge dz_n}{F}$, defined in a chart with coordinates $z_1, ..., z_n$ and having the pole of order one along X (given by in this chart by the equation $F = 0$) to $(-1)^{j-1} \frac{f\, dz_1 \wedge ... \hat{dz_j} \wedge dz_n}{F_{z_j}}|_{X - \mathrm{Sing} X}$ (the restriction is independent of $j, 1 \leq j \leq n$). ¿From this point of view the stalk of the sheaf $\mathcal{A}'$ at $P \in \mathbf{P}^{n+1}$ consists of $f \in \mathcal{O}_P$ such that $Res(\omega)$ extends to a holomorphic form on some resolution $\tilde{X}$.

In the case when a subvariety $X \subset \mathbf{P}^{n+r}$ is a complete intersection of hypersurfaces D_i given by the equations $F_i = 0, i = 1, ..., r$ the stalk of the sheaf of adjoint ideals at a singular point can be described using the fact that a holomorphic n-form can be obtained as a residue of a $(n + r)$-form on $\mathbf{P}^{n+r} - \bigcup_i D_i$ having poles of order one along each hypersurface $F_i = 0$ and that

$$Res \frac{dw_1 \wedge ... \wedge dw_{n+r}}{F_1 \cdot ... \cdot F_r} = \frac{... \wedge \hat{dw}_{i_1} ... \wedge \hat{dw}_{i_r} ...}{\frac{\partial(F_1, ..., F_r)}{\partial(w_{i_1}, ..., w_{i_r})}}|_X \qquad (46)$$

where $\frac{\partial(F_1, ..., F_r)}{\partial(w_{i_1}, ..., w_{i_r})}$ is the Jacobian of partial derivatives of the system $(F_1, ..., F_r)$ relative to variables $(w_{i_1}, ..., w_{i_r})$ (it is easy to check that the restriction up to sign is independent of collection of variables $(w_{i_1}, ..., w_{i_r})$).

This construction describes the differential forms on a resolution in terms of the linear systems of hypersurfaces in $\mathbf{P}^{n+r}$ given by ideal sheaves on the latter. The cohomology and the Hodge structure on the link of a complete intersection singularity, as already was mentioned, can be described in terms of differential forms (cf [72]) and this description can be used to calculate the cohomology of the link in terms of certain ideals in the local ring of singular point. We shall review the construction of MHS on various cohomology associated with singular points of complex spaces and in the next section we shall discuss the connections with the ideals of quasiadjunction.

The cohomology of the link L of an isolated singularity x of a complex space X $(\dim X = n)$ can be given a Mixed Hodge structure, for example using canonical identification $H^k(L) = H^*_{\{x\}}(X)$ with the local cohomology (*) §. The mixed Hodge structure on the latter was described in [71]. The Hodge numbers: $h^{kpq}(L) = \dim Gr_F^p Gr_{p+q}^W H^k(L)$ have the following symmetry properties:

$$h^{kpq} = h^{2n-k-1, n-p, n-q} \qquad (47)$$

---

§* Recall that if Y is a subset in a topological space X and $\mathcal{F}$ is a sheaf on X then $H_Y^i(X)$ is the right derived functor of the functor $\Gamma_Y(X, \mathcal{F})$ of sections of $\mathcal{F}$ supported on Y. It fits into long exact sequence: $... \to H_Y^i(X, \mathcal{F}) \to H^i(X, \mathcal{F}) \to H^i(X - Y, \mathcal{F}) \to H_Y^{i+1}(X, \mathcal{F}) \to ...$ (cf [31])

If E is the exceptional divisor for a resolution, then for $k < n$ one has

$$h^{kpq}(L) = h^{kpq}(E) \quad if \; p + q < k$$

$$h^{kpq}(L) = h^{kpq}(E) - h^{2n-k,n-p,n-q}(E) \quad if \; p + q = k \tag{48}$$

$$h^{kpq}(L) = 0 \quad if \; p + q > k$$

The local cohomology $H_E^*(\tilde{X})$ ([71]) where $\tilde{X}$ is a resolution of X support the canonical mixed Hodge structure. We shall consider in more detail the case $dim_{\mathbb{C}}\tilde{X} = 2$ which we shall use to describe the characteristic varieties of germs of plane curves. We have

$$H_E^*(\tilde{X}) = Hom(H^{4-*}(E), \mathbf{Q}(-2)) \tag{49}$$

where $\mathbf{Q}(-2)$ is the Tate Hodge structure of type $(2,2)$. Since the Hodge and weight filtrations on $H^1(E)$ have the form:

$$H^1(E) = W_1 \supset W_0 \supset 0, \quad H^1(E) = F^0 \supset F^1 \supset F^2 = 0$$

on $H_E^3(\tilde{X})$ we have:

$$H_E^3(\tilde{X}) = W_4 \supset W_3 \supset W_2 = 0, H_E^3(\tilde{X}) = F^1 \supset F^2 \supset F^3 = 0$$

Moreover

$$F^1 H^1(L) = F^1 H^1(E) = F^2 H_E^3(\tilde{X}) \tag{50}$$

In order to relate this mixed Hodge structure to the differential forms one can use the following complex:

$$0 \to A_E^2(\tilde{X}) \to A_E^3(\tilde{X}) \to 0 \tag{51}$$

where

$$A_E^2(\tilde{X}) = \Omega_{\tilde{X}}^1(log \; E)/\Omega_{\tilde{X}}^1, A_E^3(\tilde{X}) = \Omega_{\tilde{X}}^2(log \; E)/\Omega_{\tilde{X}}^2$$

with filtrations given by

$$F^2 A_E^p(\tilde{X}) = 0 \; for \; p < 3, F^2 A_E^p(\tilde{X}) = A_E^p(\tilde{X}) \; for \; p \geq 3$$

$$W_3 A_E^3(\tilde{X}) = W_1 \Omega_{\tilde{X}}^2(log \; E)/\Omega_{\tilde{X}}^2$$

Since $H^3(E) = 0$, the relations (48) and (49) yield that the complex (51) completely determines h^{1pq} (and hence all Hodge numbers h^{kpq} by (47)).

Putting all this together we obtain the following isomorphism:

$$H^0(\Omega_{\tilde{X}}^2(log\ E))/H^0(\Omega_{\tilde{X}}^2) = F^1 H^1(L) \tag{52}$$

Our next goal will be to apply this to the links of the complete intersection singularities which are the abelian covers branched over links of plane curve singularities. Such a link (41), in the case of curves with r components, has the following equations in $\mathbf{C}^{r+2}$:

$$z_1^{m_1} = f_1(x,y), \quad \cdots \quad z_1^{m_1} = f_r(x,y) \tag{53}$$

We want to calculate the eigenspaces corresponding to the characters of the Galois group acting on the Hodge spaces $H^{p,q,k}$ of singularity (53) by interpreting the left hand side of (52) in terms of ideals in the local ring of the singularity $f_1 \cdots f_r = 0$ in $\mathbf{C}^2$.

5.3. *Ideal of quasiadjunction and log-quasiadjunction*

We shall start with the following multivariable generalization of the ideals of quasiadjunction introduced in section 2.2.5.

Definition 5.2. (cf. [51] [53]) *An ideal of quasiadjunction* of type $(j_1,...,j_r|\ m_1,...,m_r)$ is the ideal in the local ring of a plane curve singularity $O \in C \subset \mathbf{C}^2$ consisting of the germs ϕ such that the 2-form:

$$\omega_\phi = \frac{\phi(x,y)z_1^{j_1} \cdots z_r^{j_r}\,dx \wedge dy}{z_1^{m_1-1} \cdots z_r^{m_r-1}},$$

extends to a holomorphic form on a resolution of the singularity of the abelian cover of a ball B of type $(m_1,...,m_r)$, i.e. a resolution of (53) (we suppress dependence of ω_ϕ on $j_1,..j_r, m_1,...m_r$). In other words, this ideal consists of germs such that $\phi z_1^{j_1} \cdots z_r^{j_r}$ belongs to the adjoint ideal of the singularity (53). In particular the condition on ϕ is independent of a resolution.

Note that ω_ϕ in 5.2 is the residue of the form $\frac{\phi z_1^{j_1} \cdots z_r^{j_r}\,dx \wedge dy}{(z_1^{m_1}-f_1(x,y))\cdots(z_r^{m_r}-f_r(x,y))}$ (cf. (46)) We always shall assume that $0 \le j_1 < m_1,..,0 \le j_r < m_r$. Also, notice that forms ω_ϕ in definition (5.2) are exactly the forms on the abelian cover (53) which are the eigenforms corresponding to the character of the Galois group taking value $exp(\frac{2\pi i(j_i-m_i+1)}{m_i})$ on the automorphism of the surface (53) induced by multiplication of the i-th coordinate by $exp(\frac{2\pi i}{m_i})$.

120

An ideal of log-quasiadjunction (resp. *an ideal of weight one log-quasiadjunction*) of type $(j_1, .., j_r | m_1, .., m_r)$ is the ideal in the same local ring consisting of germs ϕ such that ω_ϕ extends to a log-form (resp. weight one log-form) on a resolution of the singularity of the same abelian cover. Recall (cf. [15]) that a holomorphic 2-form is weight one log-form if it is a combination of forms having poles of order at most one on each component of the exceptional divisor and not having poles of order one on a pair of intersecting components. These ideals are also independent of a resolution (cf. [55]).

One can show (cf. [53]) that an ideal of quasiadjunction $\mathcal{A}(j_1, ..., j_r | m_1, ..., m_r)$ is determined by the vector (i.e. depends only on the collection of ratios):

$$(\frac{j_1 + 1}{m_1}, ..., \frac{j_r + 1}{m_r}). \tag{54}$$

This is also the case for the ideals of log-quasiadjunction and weight one log-quasiadjunction. Indeed, these ideals can be described in terms of resolutions as follows. For a given embedded resolution $\pi : V \to \mathbf{C}^2$ of the germ $f_1 \cdots f_r = 0$ with the exceptional curves $E_1, .., E_k, ..., E_s$ let $a_{k,i}$ (resp. c_k, resp. $e_k(\phi)$) be the multiplicity of the pull back on V of f_i $(i = 1, .., r)$ (resp. $dx \wedge dy$, resp. ϕ) along E_k. Then ϕ belongs to the ideal of quasiadjunction of type $(j_1, .., j_r | m_1, .., m_r)$ if and only if for any k

$$a_{k,1}\frac{j_1 + 1}{m_1} + ... + a_{k,r}\frac{j_r + 1}{m_r} > a_{k,1} + ... + a_{k,r} - e_k(\phi) - c_k - 1 \tag{55}$$

(cf. [53]). Similar calculation shows that a germ ϕ belongs to the ideal of log-quasiadjunction corresponding to $(j_1, .., j_r | m_1, .., m_r)$ if and only if the inequality

$$a_{k,1}\frac{j_1 + 1}{m_1} + ... + a_{k,r}\frac{j_r + 1}{m_r} \geq a_{k,1} + ... + a_{k,r} - e_k(\phi) - c_k - 1 \tag{56}$$

is satisfied for any k. In addition, a germ ϕ belongs to the ideal of weight one log-quasiadjunction if and only if this germ is a linear combination of germs ϕ satisfying inequality (56) for any collection of k's such that corresponding components do not intersect and satisfying the inequality (55) for k outside of this collection. We shall denote the ideal of quasiadjunction (resp. log-quasiadjunction, resp. weight one log-quasiadjunction) corresponding to $(j_1, .., j_r | m_1, .., m_r)$ as $\mathcal{A}(j_1, .., j_r | m_1, .., m_r)$ (resp. $\mathcal{A}''(j_1, .., j_r | m_1, .., m_r)$, resp. $\mathcal{A}'(j_1, .., j_r | m_1, .., m_r)$). Note the inclusions:

$$\mathcal{A}(j_1, .., j_r | m_1, .., m_r) \subseteq \mathcal{A}'(j_1, .., j_r | m_1, .., m_r) \subseteq \mathcal{A}''(j_1, .., j_r | m_1, .., m_r)$$

Both (55) and (56) follow from the following calculation (cf. [53] section 2 for complete details). One can use the normalization of the fiber product $\tilde{V}_{m_1,...,m_r} = V \times_{\mathbf{C}^2} V_{m_1,..,m_r}$ as a resolution of singularity (53) in the category of manifolds with quotient singularities (cf. [58]). We have:

$$
\begin{array}{ccc}
\tilde{V}_{m_1,..,m_r} & \xrightarrow{\tilde{p}} & V \\
\tilde{\pi} \downarrow & & \pi \downarrow \\
V_{m_1,..,m_r} & \xrightarrow{p} & \mathbf{C}^2
\end{array}
\tag{57}
$$

The preimage of the exceptional divisor of $V \to \mathbf{C}^2$ in $\tilde{V}_{m_1,...,m_r}$ forms a divisor with normal crossings (cf. [72]), though the preimage of each component is reducible in general. In this case the irreducible components above each exceptional curve do not intersect. If the Galois group G of $\tilde{p}$ is abelian (as we always assume here) and, in particular, is the quotient of $H_1(B - C \cap B, \mathbf{Z})$, then the Galois group of $\tilde{p}^{-1}(E_i) \to E_i$ is $G/(\gamma_i)$ where for an exceptional curve E_k, γ_k is the image in the Galois group of the homology class of the boundary of a small disk transversal to E_k in V. The components of $\tilde{p}^{-1}(E_i)$ correspond to the elements of $G/(\gamma_i, ...\gamma_l..)$ where l runs through indices of all exceptional curves intersecting E_i, while $\tilde{p}_i$ restricted on each component has $(\gamma_i, ...\gamma_l..)/(\gamma_i)$ as the Galois group. The points $\tilde{p}^{-1}(E_i \cap E_j)$ correspond to the elements of $G/(\gamma_i, \gamma_j)$ and the points of $\tilde{p}^{-1}(E_i \cap E_j)$ belonging to a fixed component correspond to cosets in $(\gamma_i, ...\gamma_l..)/(\gamma_i, \gamma_j)$. The order of the vanishing of ω_ϕ on $\tilde{V}_{m_1,...,m_r}$ along E_k is equal to:

$$
\Sigma_{i=1}^{i=r}(j_i - m_i + 1)\frac{m_1 \cdots \hat{m}_i \cdots m_r \cdot a_{k,i}}{g_{k,1} \cdots g_{k,r} s_k} + \frac{m_1 \cdots m_r \cdot ord_{E_k}(\pi^*(\phi))}{g_{k,1} \cdots g_{k,r} \cdot s_k} + \tag{58}
$$

$$
+ \frac{c_k \cdot m_1 \cdots m_r}{g_{k,1} \cdots g_{k,r} \cdot s_k} + \frac{m_1 \cdots m_r}{g_{k,1} \cdots g_{k,r} \cdot s_k} - 1
$$

where $g_{k,i} = g.c.d.(m_i, a_{k,i})$ and $s_k = g.c.d.(..., \frac{m_i}{g_{k,i}}, ...)$.

A consequence of (58) is that ω_ϕ has an order of pole equal to one (resp. zero) along the component E_k of the above resolution if and only if for such ϕ one has equality in (56) (resp. (55) is satisfied).

Proposition 5.3. *(cf. [55]) 1. Let $\mathcal{A}''$ be an ideal of log-quasiadjunction. There is a unique polytope $\mathcal{P}(\mathcal{A}'')$ such that a vector $(\frac{j_1+1}{m_1}, ..., \frac{j_r+1}{m_r}) \in \mathcal{P}(\mathcal{A}'')$ if and only if the ideal $\mathcal{A}''(j_1, .., j_r|m_1, .., m_r)$ contains $\mathcal{A}''$ ¶.*

¶(*) i.e. a subset in $\mathbf{R}^r$ given by a set of linear inequalities $L_s \geq k_s$. We say that an affine hyperplane in $\mathbf{R}^r$ supports a codimension one face of a polytope if the intersection

2. The set of vectors (54) for which $\mathcal{A}(j_1,..,j_r|m_1,..,m_r) \neq \mathcal{A}''(j_1,..,j_r|m_1,..,m_r)$ is a dense subset in the boundary of the polytope having as its closure a union of faces of such a polytope. The closure of the set of vectors (54) for which $\mathcal{A}'(j_1,..,j_r|m_1,..,m_r) \neq \mathcal{A}''(j_1,..,j_r|m_1,..,m_r)$ is also a union of certain faces of such a polytope.

3. The ideal $\mathcal{A}(j_1,..,j_r|m_1,...,m_r)$ (resp. $\mathcal{A}'(j_1,..,j_r|m_1,...,m_r)$ and $\mathcal{A}''(j_1,..,j_r|m_1,...,m_r)$) is independent of the array $(j_1,..,j_r|m_1,..,m_r)$ as long as the vector (54) varies within the interior of the same face of quasi-adjunction.

We shall call the above faces *the faces of quasiadjunction* (resp. *weight one faces of quasiadjunction*). $\mathcal{A}_\Sigma$ will denote $\mathcal{A}(j_1,...,j_r|m_1,...,m_r)$ with corresponding vector (54) belonging to the interior of a face of quasiadjunction Σ (similarly for $\mathcal{A}'_\Sigma$ and $\mathcal{A}''_\Sigma$).

In the case $r = 1$ and when $f(x,y)$ is weighted homogeneous one can use the description of the adjoint ideals given by M.Merle and B.Tessier (cf. [62] and section 2.2.5). The polytopes of quasiadjunction are in $\mathbf{R}$ and hence are just constants. They are the constants of quasiadjunction introduced in [43]. As we mentioned already, it was shown in [59] that they are the elements of Arnold-Steenbrink spectrum which belong to the interval (0,1).

The polytopes of quasiadjunctions are subsets of a unit cube $\mathcal{U}$ with the coordinates corresponding to the components of the link. We shall view it also as the fundamental domain for the Galois group $H^1(S^3 - L, \mathbf{Z})$ of the universal abelian cover $H^1(S^3 - L, \mathbf{R})$ of the group $H^1(S^3 - L, \mathbf{R}/\mathbf{Z})$ of the unitary characters of $H_1(S^3 - L, \mathbf{Z})$ (i.e. the maximal compact subgroup of $Char(H_1(S^3 - L, \mathbf{Z})) = H^1(S^3 - L, \mathbf{C}^*))$. $exp : \mathcal{U} \to Char(H_1(S^3 - L, \mathbf{Z}))$ will denote the restriction of $H^1(S^3 - L, \mathbf{R}) \to H^1(S^3 - L, \mathbf{R}/\mathbf{Z})$ on $\mathcal{U}$.

For any sub-link $\tilde{L}$ of L, i.e. a link formed by components of L, we have surjection $\pi_1(S^3 - L) \to \pi_1(S^3 - \tilde{L})$ induced by inclusion. Hence $Char H_1(S^3 - \tilde{L}, \mathbf{Z})$ is a sub-torus of $Char H_1(S^3 - L, \mathbf{Z}))$ (in coordinates in the latter torus corresponding to the components of L it is given by equations of the form $t_\alpha = 1$ where subscripts correspond to components of L absent in $\tilde{L}$). Moreover, since the homology of the universal abelian cover $H_1(\widetilde{S^3 - L})$ surjects onto $H_1(\widetilde{S^3 - \tilde{L}})$, it follows that $V_i(S^3 - \tilde{L})$ belongs to

---

of this hyperplane with the boundary of the polytope has dimension $r - 1$. A face of a polytope is the intersection of a supporting face of the polytope with the boundary. A codimension one face of a polytope in $\mathbf{R}^r$ is a polytope of dimension $r - 1$. By induction one obtains faces of arbitrary codimension for original polytope (for $r = 3$ those are called edges and vertices). The boundary of the polytope is the union of its faces.

a component of $V_i(S^3 - L)$ (cf. [53]). We shall call a character of $\pi_1(S^3 - L)$ (or a connected component of $V_i(S^3 - L)$) *essential* if it does not belong to a subtorus $Char H^1(S^3 - \tilde{L})$ for any sublink $\tilde{L}$ of L.

Let $L_{m_1,..,m_r}$ be the link of singularity (53) or equivalently the cover of S^3 branched over the link L and having a quotient $H_{m_1,..,m_r} = \mathbf{Z}/m_1\mathbf{Z} \oplus ... \oplus \mathbf{Z}/m_1\mathbf{Z}$ of $H_1(S^3 - L, \mathbf{Z})$ as its Galois group. We shall view $Char H_{m_1,..,m_r}$ as a subgroup of $Char H_1(S^3 - L, \mathbf{Z})$. The group $H_{m_1,..,m_r}$ acting on $H^1(L_{m_1,..,m_r})$ preserves both Hodge and weight filtrations.

Theorem 5.4. *(cf. [55]) An essential character* $\chi \in Char(H_1(S^3 - L, \mathbf{Z}))$ *is a character of the representation of* $H_{m_1,..,m_r}$ *acting on* $F^1(H^1(L_{m_1,..,m_r}))$ *if and only if it factors through the Galois group* $H_{m_1,..,m_r}$ *and belongs to the image of a face of quasiadjunction under the exponential map.*

The multiplicity of χ *in this representation of the Galois group is equal to* $dim A_\Sigma'' / A_\Sigma$ *where* A_Σ'' *(resp.* A_Σ*) is the ideal of log-quasiadjunction (resp. ideal of quasiadjunction) corresponding to a vector (54) belonging to the face of quasiadjunction* Σ.

A character χ *is a character of the representation of the Galois group of the cover on* $W_0(H^1(L_{m_1,..,m_r}))$ *if and only if it factors through the Galois group* $H_{m_1,..,m_r}$ *and it belongs to the image under the exponential map of a weight one face of quasiadjunction.*

5.4. *Multiplier ideals and log-canonical thresholds*

The ideals and polytopes of quasiadjunction are closely related to recently studies multiplier ideals (cf. [66], [39]) and log-canonical thresholds.

For a $\mathbf{Q}$-divisor D on a non singular manifold X its multiplier ideal $\mathcal{J}(D)$ (cf. ([66]) can be defined as follows. Let $f : Y \to X$ be an embedded resolution of D and $f^*(D) = -E$. Then $\mathcal{J}(D) = f_*(\mathcal{O}_Y(K_Y - f^*(K_X) - \lfloor E \rfloor))$ where $\lfloor E \rfloor$ is round-down of a $\mathbf{Q}$-divisor. In this terminology one can define the ideals of quasiadjunction as follows. For an array $(\gamma_1, .., \gamma_r), (\gamma_i \in \mathbf{Q})$ let $D_{\gamma_1,..,\gamma_r}$ be given by equation $f_1^{\gamma_1} \cdots f_r^{\gamma_r}$. Then $\mathcal{J}(D_{\gamma_1,..,\gamma_r}) = A(j_1, .., j_r | m_1, ..., m_r)$ where $\gamma_i = 1 - \frac{j_i+1}{m_i}$ for $i = 1, .., r$. This follows immediately from (55).

To describe the relation with the log-canonical thresholds, recall ([36]) that a pair (X, D) where X is normal and D is a $\mathbf{R}$-divisor such that $K_X + D$ is $\mathbf{R}$-Cartier is called log-canonical at $x \in X$ if for any birational morphism

124

$f : Y \to X$, with Y normal, in the decomposition

$$K_Y = f^*(K_X + D) + \sum_E a(E, X, D)E \qquad (59)$$

for each irreducible E having center at x one has $a(E, X, D) \geq -1$. This coefficient is called *discrepancy* of divisor D on X along E.

Proposition 5.5. *(cf. [55]) The local ring $\mathcal{O}_O$ of a singularity $f_1 \cdots f_r = 0$ at the origin O of $\mathbf{C}^2$ considered as the ideal in itself is an ideal of log-quasiadjunction. Let $\mathcal{P}$ be the corresponding polytope of log-quasiadjunction. Let D_i be the divisor in $\mathbf{C}^2$ with the local equation $f_i = 0$ near the origin. Then for $\{(\gamma_1, ..., \gamma_r)\} \in \mathbf{R}^r$ the divisor $\gamma_1 D_1 + ... + \gamma_r D_r$ is log-canonical at $(0, 0) \in \mathbf{C}^2$ if and only if $(1 - \gamma_1, .., 1 - \gamma_r)$ belongs to the polytope $\mathcal{P}$.*

To see why this is the case, let us consider the polytope given by inequalities (55) in which one puts $e_k(\mathcal{A}'') = 0$, i.e.

$$a_{k,1}x_1 + ... + a_{k,r}x_r \geq a_{k,1} + ... + a_{k_r} - c_k - 1 \qquad (60)$$

Let $(j_1, .., j_r | m_1, .., m_r)$ be such that the corresponding vector (54) belongs to the boundary of this polytope. Then $1 \in \mathcal{A}''(j_1, .., j_r | m_1, ..., m_r)$ and hence the ideal
$\mathcal{A}''(j_1, .., j_r | m_1, .., m_r)$ is the local ring of the origin (i.e is not proper).

If $\pi : V \to \mathbf{C}^2$ is an embedded resolution then the discrepancy of $f_1^{\gamma_1} \cdot \cdots f_r^{\gamma_r}$ along E_k is:

$$c_k - (a_{k,1}\gamma_1 + ... + a_{k,r}\gamma_r)$$

i.e. the discrepancy along each E_k is not less than -1 if and only if $(1 - \gamma_1, ..., 1 - \gamma_r)$ satisfies (60).

As an example to this proposition one can consider the ordinary cusp $x^2 + y^3$ the log-canonical threshold is $\frac{5}{6}$ and the constant of quasiadjunction is $\frac{1}{6}$ (cf. section 2.2.5).

The polytopes of quasiadjunction are "pieces" of the zeros of the (multivariable) Alexander polynomail and in this sense are analogs of the spectrum.

Problem 5.1. *Find a generalization of the semicontinuity of spectrum of a singularity*

For some results in this direction cf. [55].

5.5. *Hodge decomposition for INNCs*

The calculation from the last section can be partially extended to INNC. Namely we extend the calculation which involve the forms of top degree and hence will obtain at least a part of the components of characteristic varieties.

We shall start with the definition:

Definition 5.6. (cf. [55]) Let $f_i = 0$ be the equation of divisor D_i and let $\pi : \tilde{\mathbf{C}}^{n+1} \to \mathbf{C}^{n+1}$ be a resolution of the singularities of $\bigcup D_i$ (i.e. the proper preimage of the latter is a normal crossings divisor). Let $\mathcal{V}_{m_1,...,m_r}$ be the singularity (41) having $V_{m_1,...,m_r}$ as its link. Let $\tilde{V}$ be a normalization of $\tilde{\mathbf{C}}^{n+1} \times_{\mathbf{C}^{n+1}} V_{m_1,..,m_r}$ (cf. (41)) The ideal of quasiadjunction of type $(j_1,...,j_r|m_1,...,m_r)$ is the ideal $\mathcal{A}(j_1,...,j_r|m_1,...,m_r)$ of germs $\phi \in \mathcal{O}_{0,\mathbf{C}^{n+1}}$ such that the $(n+1)$-form:

$$\omega_\phi = \frac{\phi z_1^{j_1} \cdot ... \cdot z_r^{j_r} \, dx_1 \wedge ... \wedge dx_{n+1}}{z_1^{m_1-1} \cdot ... \cdot z_r^{m_r-1}} \tag{61}$$

on the non singular locus of $V_{m_1,...,m_r}$ after the pull back on $\tilde{V}$ extends over the exceptional set.

The l-th ideal of log-quasiadjunction $\mathcal{A}_l(\log E)(j_1,...,j_r|m_1,...,m_r)$ is the ideal of $\phi \in \mathcal{O}_{0,\mathbf{C}^{n+1}}$ such that the the pull back of the corresponding form ω_ϕ on $\tilde{V}$ is log-form on $(\tilde{V}, E)$ having weight at most l.

We have the following:

Proposition 5.7. *(cf. [55], [56]) There exist a collection of subsets $\mathcal{P}_\kappa, (\kappa \in \mathcal{K})$ in the unit cube*

$$\mathcal{U} = \{(x_1,...,x_r)|0 \le x_i \le 1\}$$

in $\mathbf{R}^r$ and a collection of affine hyperplanes $l_i(x_1,...,x_r) = \alpha_i$ such that each $\mathcal{P}_\kappa$ is the boundary of the polytope consisting of solutions to the system of inequalities:

$$l_i(x_1,...,x_r) \ge \alpha_i$$

and such that

$$\left(\frac{j_1+1}{m_1},...,\frac{j_r+1}{m_r}\right) \in \mathcal{U} \tag{62}$$

belongs to $\mathcal{P}_\kappa$ if and only if

$$\dim \mathcal{A}(\log E)(j_1,...,j_r|m_1,...,m_r)/\mathcal{A}(j_1,...,j_r|m_1,...,m_r) \ge 1 \tag{63}$$

Moreover

$$\dim \mathcal{A}_l(\log E)(j_1, ..., j_r | m_1, ..., m_r)/\mathcal{A}_{l-1}(j_1, ..., j_r | m_1, ..., m_r) \geq k \qquad (64)$$

if only if (54) belongs to a collection of certain faces $\mathcal{P}_{\kappa,\iota}^{k,l}(\iota \in \mathcal{I}^{k,l})$ of polytopes $\mathcal{P}_\kappa$.

Now the exponents of the polytopes of quasiadjunction land in the characteristic varieties. More precisely we have:

Theorem 5.8. *(cf. [56]) A character of $\pi_1(\partial B_\epsilon - \partial B_\epsilon \cap (\bigcup_{1 \leq i \leq r} D_i))$ acting on*
$W_l(F^n H^n(V_{m_1,...,m_r}))$ *via the action of the Galois group has the eigenspace of dimension at least k if and only if it has the form:*

$$(exp 2\pi\sqrt{-1}a_1, ..., exp 2\pi\sqrt{-1}a_r)$$

where $(a_1, ..., a_r)$ belongs to one of the faces $\mathcal{P}_{\kappa,\iota}^{k,l}$ of a polytope $\mathcal{P}_\kappa$ of quasiadjunction of $\bigcup D_i$. In particular, the Zariski closures of exponents of polytopes of quasiadjunction are components of characteristic varieties. These components are the translated subgroups by points of finite order

We conjecture that all components are the translated subgroups by points of finite order.

Conjecture 5.9. *Characteristic variety is a union of translated subtori of $\mathrm{Spec}\mathbf{C}[\pi_1(\partial B_\epsilon - \partial B_\epsilon \cap X)]$ with each translations given by a point of finite order.*

An interesting problem is to calculate them in terms of resolution.

6. Homotopy groups of the complements to hypersurfaces in projective space and linear systems determined by singularities

In this section we want to discuss the characteristic varieties associated with hypersurfaces which are divisors with isolated non normal crossings in a projective space. This can be viewed as a global version of the results from the previous lecture. An interesting case occurs already when all hypersurfaces have degree one i.e. the case of arrangements of hyperplanes. The advantage of the case of INNC is that one does not have the problems associated with complexity of the fundamental group since the fundamental groups for such arrangements are abelian (unless we are dealing with an arrangement of lines). The theory of INNC arrangements is still highly non trivial and is far from being well understood. Note that a more general case of divisors with normal crossings in general projective manifolds (rather then in $\mathbf{P}^{n+1}$) is considered in [57]. The main results and conjectures of this section show how the local characteristic varieties plus certain linear systems associated with the points of non normal crossings determine the global characteristic varieties. This generalizes the results on the Alexander polynomial discussed earlier in section 2.2.5.

6.1. *Homotopy groups of the complements to INNC*

Let us consider the a divisor D in projective space $\mathbf{P}^{n+1}$ which has as its singularities normal crossings with only finitely many exceptions i.e. has isolated non normal crossings. This situation includes as its special cases:

a) Arbitrary reduced curves in $\mathbf{P}^2$

b) Hypersurfaces in $\mathbf{P}^{n+1}$ with isolated singularities and hypersurfaces in $\mathbf{C}^{n+1}$ with isolated singularities and transversal to the hyperplanes at infinity.

c) Arrangements of hyperplanes in $\mathbf{P}^{n+1}$ such that each intersection of hyperplanes having codimension $k \neq n+1$ belongs to exactly k hyperplanes of the arrangement.

The starting point is the the following vanishing of the homotopy groups generalizing already discussed result from [49]:

Theorem 6.1. *(cf. [57]) Let $X = \mathbf{P}^{n+1}$ and D be a divisor having finitely many non-normal crossings. Assume that one of the components has degree 1. Then $\pi_i(\mathbf{P}^{n+1} - D) = 0$ for $2 \leq i \leq n - 1$. If all intersections are the normal crossings, then $\pi_n(\mathbf{P}^{n+1} - D) = 0$ and hence $\mathbf{P}^{n+1} - D$ is homotopy equivalent to the wedge of the $n + 1$-skeleton of the torus $(S^1)^k$ and several copies of S^{n+1}.*

One also has a similar vanishing for the homology of local systems.

Theorem 6.2. *(cf. [57]) Let $\chi \in \mathrm{Char}\pi_1(\mathbf{P}^{n+1} - D)$ be a character of the fundamental group different from the identity and let $\mathbf{C}_\chi$ be the one dimensional space vector space considered as the $\mathbf{C}[\pi_1(\mathbf{P}^{n+1} - D)]$ module via the character χ. Then, viewing χ as a local system we have:*

$$H_i(\mathbf{P}^{n+1} - D, \chi) = 0 \ (i \leq n - 1)$$

$$H_n(\mathbf{P}^{n+1} - D, \chi) = \pi_n(\mathbf{P}^{n+1} - D) \otimes_{\mathbf{C}[\pi_1(\mathbf{P}^{n+1} - D)]} \mathbf{C}_\chi$$

The main problem for INNC hence is to understand the first non trivial homotopy group $\pi_n(\mathbf{P}^{n+1} - D)$. Similarly to the local case the starting point is the following:

Definition 6.3. (cf. [57]) The k-th characteristic variety $V_k(\pi_n(\mathbf{P}^{n+1}-D))$ of the homotopy group $\pi_n(\mathbf{P}^{n+1}-D)$ is the zero set of the k-th Fitting ideal of $\pi_n(\mathbf{P}^{n+1}-D)$, i.e. the zero set of minors of order $(n-k+1) \times (n-k+1)$ of Φ in a presentation

$$\Phi : \mathbf{C}[\pi_1(\mathbf{P}^{n+1} - D)]^m \to \mathbf{C}[\pi_1(\mathbf{P}^{n+1} - D)]^l \to \pi_n(X) \to 0$$

of $\pi_1(\mathbf{P}^{n+1} - D)$ module $\pi_n(\mathbf{P}^{n+1} - D)$ via generators and relations. Alternatively (cf. theorem 6.2) outside of $\chi = 1$, $V_k(\pi_n(\mathbf{P}^{n+1} - D))$ is the set of characters $\chi \in \text{Char}[\pi_1(\mathbf{P}^{n+1} - D)]$ such that $\dim H_n(\mathbf{P}^{n+1} - D, \chi) \geq k$.

6.2. *Jumping loci on quasiprojective varieties*

A remarkable fact is that the characteristic varieties of the complements have a very simple structure (unlike in the similar situations outside of algebraic geometry). We did see this already in the case of links of curve singularities and in the case of charactersstic variety V_1 for local INNCs. The local systems on *a non singular projective variety* correspond to holomorphic bundles which are topologically trivial. The jumping loci for the cohomology of such bundles are unions of translates of abelian subvarieties of the Picard variety. These results, having long history, are due to Catanese, Beauville, Green-Lazarsfeld, Simpson and Deligne. We shall use the following quasi-projective version dealing with the cohomology of local systems which will allow us eventually to describe the characteristic varieties.

Theorem 6.4. *([6]) Let $\hat{X}$ be a projective manifold such that $H^1(\hat{X}, \mathbf{C}) = 0$. Let $\hat{D}$ be a divisor with normal crossings. Then there exists a finite number of unitary characters $\rho_j \in \text{Char}\pi_1(\hat{X} - \hat{D})$ and holomorphic maps $f_j : \hat{X} - \hat{D} \to T_j$ into complex tori T_j such that the set $\Sigma^k(\hat{X} - \hat{D}) = \{\rho \in \text{Char}\pi_1(\hat{X} - \hat{D}) | \dim H^k(\hat{X} - \hat{D}, \rho) \geq 1\}$ coincides with $\bigcup \rho_j f_i^* H^1(T_j, \mathbf{C}^*)$. In particular, Σ^k is a union of translated by unitary characters subgroups of $\text{Char}\pi_1(\hat{X} - \hat{D})$.*

Hence we also obtain:

Corollary 6.5. *The characteristic variety $V_k(\pi_n(\mathbf{P}^{n+1} - D))$ is a union of translated subgroups S_j of the group $\text{Char}\pi_1(\mathbf{P}^{n+1} - D)$ by unitary charac-*

ters ρ_j:

$$V_k(\pi_n(\mathbf{P}^{n+1} - D)) = \bigcup \rho_j S_j$$

In the case $k = 1$ the components of characteristic variety having a positive dimension correspond to the maps onto hyperbolic curves. This has many applications for example to calculations of characteristic varieties (cf. [53]), estimating the order of the group of automorphisms of the complements (cf. [7]), classification of arrangements of lines (cf. [54]) among others but we won't discuss them here.

6.3. *The Hodge numbers of abelian covers of projective spaces and linear systems*

Let, as before, $D = \cup_{i=0,..r} D_i$ be a divisor in $\mathbf{P}^{n+1}$. We shall assume, to simplify the exposition, that one of components, say D_0 has degree equal to one and that there are D has non non normal crossings on D_0. Let $\pi_1(\mathbf{P}^{n+1} - D) \to \oplus \mathbf{Z}/m_i\mathbf{Z}$ be a surjective homomorphism and let $X_\mathbf{m}$ ($\mathbf{m} = (..., m_i, ...)$) be a normalization of a compactification of unbranched cover of $\mathbf{P}^{n+1} - D$ corresponding to this homomorphism. Let $f : X_\mathbf{m} \to \mathbf{P}^{n+1}$ be the corresponding projection.

Our goal is to calculate the Hodge number $h^{n,0}(X_\mathbf{m})$. Starting from D, we shall define global polytopes of quasiadjucntion so that with each face δ of the polytope is associated the ideal sheaf $\mathcal{J}_\delta$. The above Hodge number is equal to the number of lattice points is δ counted with the weight given by the dimension of linear system of hypersurface of degree given by δ and with local conditions given by the ideal sheaf $\mathcal{J}_\delta$.

Let is consider the unit cube $\mathcal{U} = \{(x_1, ..., x_r) \in \mathbf{R}^r | 0 \le x_i \le 1\}$ coordinate of which correspond to irreducible components $D_1, ..., D_r$ of the divisor D. We view $\mathbf{R}^r$ as the universal cover of the group $(S^1)^r$ of unitary characters of $\pi_1(\mathbf{P}^{n+1} - D)$ and $\mathcal{U}$ as the fundamental domain for the action of the covering group on the cover. With each point P in $\mathbf{P}^{n+1}$ where D has a non-normal crossing the definition 5.7 associates a polytope $\mathcal{P}_\kappa$ in the unit cube in $\mathbf{R}^s$ with coordinates corresponding to the components of D. Since one has the canonical projection $\pi : \mathbf{R}^r \to \mathbf{R}^s$, forgetting the coordinates corresponding to D_i's not containing P, each $\mathcal{P}_\kappa$ defines the polytope $\pi^{-1}(\mathcal{P}_\kappa)$ in $\mathcal{U}$ which we shall denote by the same letter. This defines a finite collection of polytopes $\mathcal{P}_{\kappa,P} \subset \mathcal{U}$.

Definition 6.6. Consider the equivalence relation on points in $\mathcal{U}$ calling two points equivalent if the collections of polytopes $\mathcal{P}_{\kappa,P}$ containing these

130

two points are identical. The equivalence class is called the global polytope
of quasiadjucntion.

A global face of quasiadjunction is a face of a global polytope of quasiadjunction.

Let $\mathcal{S}_\delta$ be the set of non normal crossings P of D for which there exist the
polytopes $\mathcal{P}_{\kappa,P}$ contaning δ.

The ideal sheaf corresponding to δ is a sheaf $\mathcal{J}_\delta \subset \mathcal{O}_{\mathbf{P}^{n+1}}$ such that
$\mathcal{O}_{\mathbf{P}^{n+1}}/\mathcal{J}_\delta$ is supported at $\mathcal{S}_\delta$ and such that the stalk at P is the ideal
which is the intersection of local ideals of quasiadjunction corresponding to
local polytopes containing δ.

Clearly such an equivalence class is a polytope i.e. consists of points
satisfying a system of linear inequalities. Also the collections S_δ of non
normal crossings are defined entierly by the local data of D. The Hodge
number $h^{n,0}(X_{\mathbf{m}})$ depends on additional piece of information.

Theorem 6.7. *(cf. [53], [57]) Let D as above and let d_i be the degree of the
irreducible component D_i. For each $\chi \in \mathrm{Char} \oplus \mathbf{Z}/m_i\mathbf{Z} \subset \mathrm{Char}\pi_1(\mathbf{P}^{n+1}-D)$
let $\delta(\chi)$ be the global face of quasiadjunction containing $\frac{1}{2\pi\sqrt{-1}}\log(\chi) \in \mathcal{U}$.
Let l be such that the hyperplane $d_1 x_1 + ... + d_r x_r = l \ (l \in \mathbf{Z})$ contains δ.
Then*

$$h^{n,0}(X_{\mathbf{m}}) = \sum_\delta \sum_{\chi \in \delta} \dim H^1(\mathbf{P}^{n+1}, \mathcal{J}_\delta(l - n - 1))$$

A proof in the cyclic case and in the case of curves which generalizes
Zariski's approach ([82]) is given in [50] and [53] and the case of INNC is
similar. Alternatively, one can also use the approach in [25].

Example 6.8. For an irreducible curve of degree d with nodes and the ordinary cusps as the only singularities the global polytope of quasiadjunction
coincides with the local one of the cusp. The only face of quasiadjunction
is $x = \frac{1}{6}$. The contributing hyperplane is given by $dx = \frac{d}{6}$ and its level is $\frac{d}{6}$.
The sheaf of quasiadjunction corresponding to this face of quasiadjunction
is the ideal sheaf having stalks different from the local ring only at the
points of $\mathbf{P}^2$ where the curve has cusps and the stalks at those points are
the maximal ideals of the corresponding local rings.

For characters not on the global faces of quasiadjunction one still can
define the ideal sheaves looking at the polytopes containing the lifts of the
characters $\frac{1}{2\pi\sqrt{-1}}\log(\chi) \in \mathcal{U}$ into universal cover of the torus of unitary

character and also the integer l such that $d_1 x_1 + ... + d_r x_r = l$ contains the lift. However the corresponding group $H^1(\mathbf{P}^{n+1}, \mathcal{J}_\delta(l-n-1))$ is vanishing. For plane curves with nodes and cusps one obtains the following classical result (for the most part already discussed earlier).

Corollary 6.9. *(Zariski's theorem) Let C be a plane curve of degree d having nodes and cusps as the only singularities. Let $\mathcal{J}$ be the subsheaf of the sheaf of regular functions whose sections belong to the maximal ideals at the points in $\mathbf{P}^2$ which are the cusps of C. If $k > \frac{5d}{6}$ then*

$$H^1(\mathbf{P}^2, \mathcal{J}(k-3)) = 0$$

If $6|d$ then $H^1(\mathbf{P}^2, \mathcal{J}(\frac{5d}{6} - 3)) = h^{1,0}(X_d)$ is equal to the irregularity of a a resolution of singularities of a d-fold cyclic cover of $\mathbf{P}^2$ branched over C.

6.4. *Mixed Hodge structure on homotopy groups*

The theorem 6.2 suggests an additional structure on the characteristic varieties coming from the mixed Hodge structure on the cohomology of local systems. This is an analog of discussed earlier in local case the Hodge decomposition of characterstic varities. The MHS on the cohomology of local systems can be understood by interpreting the cohomology of local systems having finite order as the eigenspaces of the Galois group acting on the abelian covers as follows. A word of warning however: MHSs on the cohomology of local systems are coming with two F-filtrations which do not have to be conjugate i.e. in terminology of [6] are $\mathbf{C}$-mixed Hodge structures. In particular $h^{p,q} \neq h^{q,p}$ in general.

Theorem 6.10. *Let G be a finite group and $g : \pi_1(X) \to G$ be a surjection. Let $\chi \in \mathrm{Char}\pi_1(X)$ be a character which is the pull back of a character of G. Assume that $\pi_i(X) = 0$ for $2 \leq i \leq n-1$. Finally let X_G be the unbranched cover of X corresponding to G. Then the eigenspace $H^n(X_G)_\chi$ is isomorphic to the homology of $H^n(\mathbf{C}_\chi)$ of the local system $\mathbf{C}_\chi$ corresponding to χ. In particualar, the cohomology classes in $H^n(\mathbf{C}_\chi)$ acquire the Hodge type.*

If X is quasiprojective and non singular, so is X_G and hence $H^n(X_G)$ admits the mixed Hodge structure with the weights $n, ..., 2n$.

Definition 6.11. (cf. [57]) Let $\mathbf{P}^{n+1} - D$ be a complement to an INNC in $\mathbf{P}^{n+1}$. For a local system χ of finite order let $h_\chi^{p,q,n}$ be the dimension of the space of cohomology classes in $H^n(\mathbf{C}_\chi)$ having the Hodge type (p,q). The following subset of $V_k(\pi_n(\mathbf{P}^{n+1} - D))$:

$$\mathcal{P}_k^{p,q,n} = \{\chi | h_\chi^{p,q,n} \geq k\}$$

is called the component of the characteristic variety of type (p, q, n)

One has $\mathcal{P}_k^{p,q,n} \neq \emptyset$ only if $n \le p+q \le 2n$ and $\bigcup_{p,q,n} \mathcal{P}^l = V_1(\pi_1(\mathbf{P}^{n+1} - D))$.

6.5. *A relation between the Hodge numbers of branched and unbranched abelian covers*

We want to use the theorem 6.7 to detect some components of characteristic varieties of the homotopy groups. Here is a relation between branched and unbranched covers which we shall need since the theorem 6.7 works in compact case.

Theorem 6.12. *(cf. [57]) Let* $\chi \in Char(\pi_1(\mathbf{P}^{n+1} - D))$ *be a character of a finite quotient* G *of* $\pi_1(\mathbf{P}^{n+1} - D)$. *Let* $\bar{U}_G$ *be a* G-*equivariant non-singular compactification of* U_G *and let* $H^{p,q}(\bar{U}_G)_\chi$ *be the* χ-*eigenspace of* G *acting on* $H^{p,q}(\bar{U}_G)$. *Then*

$$h^{n,0,n}(\mathbf{C}_\chi) = h^{n,0}(U_G)_\chi = h^{n,0}(\bar{U}_G)_\chi$$

6.6. *Main theorem and Open Problems*

Combining the relationship between the cohomology of local systems and the cohomology of unbranched covers, the relation between the cohomology of branched and unbranched covers in the theorem 6.12 and the calculation of the Hodge numbers of branched covers in theorem 6.7 we obtain the following, extending results on Alexander polynomials and the case of reducible curves in [53]:

Theorem 6.13. *(cf. [57]) Let* $D \subset \mathbf{P}^{n+1}$ *be a union of hypersurfaces* $D_0, D_1, ..., D_r$ *of degrees* $1, d_1, ..., d_r$ *respectively, which is a divisor with isolated non-normal crossings. Let* $\mathcal{F}$ *be a face of global polytope of quasi-adjunction, i.e. a face of an intersection of polytopes of quasi-adjunction corresponding to a collection* S *of non-normal crossings of* D. *Let* $d_1 x_1 + ... + d_r x_r = l$ *be a hyperplane containing the face of quasi-adjunction* $\mathcal{F}$. *If* $H^1(\mathcal{A}_\mathcal{F} \otimes \mathcal{O}(l - 3)) = k$, *then the Zariski closure of* $exp(\mathcal{F}) \subset Char H_1(\mathbf{P}^{n+1} - D)$ *belongs to a component of* $V_k(\pi_n(\mathbf{P}^{n+1} - D))$.

There is a generalization to INNC divisors on arbitrary projective simply-connected varieties. I refer to [57] for conjectures. Here is a short list of the open problems in the case of divisors in $\mathbf{P}^{n+1}$.

Problem 6.1.
Are there components of characteristic variety $V_k(\pi_1(\mathbf{P}^{n+1} - D))$ which are not Zariski closures of $\mathcal{P}_{n,0,n}$?

Problem 6.2. *Find methods for detecting the sets $\mathcal{P}_{p,q,n}$ with $(p,q) \neq (n,0)$*

A difficulty here is that one cannot work with arbitrary compactifications since the Hodge numbers $h^{p,q}$ are not birational invariants. It would be good to have techniques which will allow to work directly with the complement and avoiding to some extent the compactification.

Problem 6.3. *Generalize the main theorem to projective algebraic varieties and beyond the cases when $\mathcal{O}(D_i) = \mathcal{L}^{m_i}$.*

See discussion of this in [57]

Problem 6.4. *Find additional interesting examples beyond the one described in [57].*

Problem 6.5. *Solve realization problem for characteristic varieties i.e. describe how many components and what are their dimensions depending on the numerical data of the divisor D on X.*

References

1. S.Abhynakar, Appendix to Ch. VIII in [85]
2. J.W. Alexander, Topological invariants of knots and links, Transactions AMS, 30 (1928), p.275-306.
3. E.Artal Bartolo, R.Carmona, J.I.Cogolludo. Braid monodromy and topology of plane curves. Duke Math. J. 118 (2003), no. 2, 261–278.
4. D. Auroux, S. K. Donaldson, L. Katzarkov Singular Lefschetz pencils, math.AG/0410332.
5. A'Campo, N. La fonction zeta d'une monodromie. Comment. Math. Helv. 50 (1975), 233–248.
6. D.Arapura, Geometry of cohomology of support loci for local systems I, Alg. Geom., vol. 6, p. 563, (1997).
7. T.Bandman, Libgober, A. Counting rational maps onto surfaces and fundamental groups. Internat. J. Math. 15 (2004), no. 7, 673–690.
8. D.A.Buchsbaum and D.Eisenbud, What annihilates a module?, J.Algebra, 47 (1977),231-243.
9. D. Chéniot, Vanishing cycles in a pencil of hyperplane sections of a nonsingular quasi-projective variety. Proc. London Math. Soc. (3) 72 (1996), 515–544.

10. D. Chéniot, C.Eyral, Homotopical variations and Zariski-van Kampen theorems. math.AG/0204176.
11. D Chéniot, D.; Libgober, A. Zariski-van Kampen theorem for higher-homotopy groups. J. Inst. Math. Jussieu 2 (2003), no. 4, 495–527.
12. O. Chisini, Courbes de diramation des planes multiples et tresses algebrique, Deuxieme Colloque de Géométrie Algebrique (Liege, 1952), Masson, Paris, 1952.
13. D.Cohen, A.Suciu, Characteristic varieties of arrangements. Math. Proc. Cambridge Philos. Soc. 127 (1999), no. 1, 33–53.
14. A.Degtyarev, A divisibility theorem for the Alexander polynomial of a plane algebraic curve. (Russian) Zap. Nauchn. Sem. S.-Peterburg. Otdel. Mat. Inst. Steklov. (POMI) 280 (2001), Geom. i Topol. 7, 146–156, 300; translation in J. Math. Sci. (N. Y.) 119 (2004), no. 2, 205–210
15. P.Deligne, Theorie de Hodge. II, Publ. Math. IHES 40, 5-58 (1971).
16. P.Deligne, Le groupe fondamental du complement d'une courbe plane n'ayant que des points doubles ordinaires est abelien (d'apres W. Fulton). Bourbaki Seminar, Vol. 1979/80, pp. 1–10, Lecture Notes in Math., 842, Springer, Berlin-New York, 1981.
17. A.Dimca, Betti numbers of hypersurfaces and defects of linear systems. Duke Math. J. 60 (1990), no. 1, 285–298.
18. A.Dimca, Singularities and topology of hypersurfaces. Universitext. Springer-Verlag, New York, 1992.
19. A.Dimca, S.Papadima, Hypersurface complements, Milnor fibers and higher homotopy groups of arrangements. Ann. of Math. (2) 158 (2003), no. 2, 473–507
20. A.Dimca,A.Nemethi, Hypersurface complements, Alexander modules and monodromy. Real and complex singularities, 19–43, Contemp. Math., 354, Amer. Math. Soc., Providence, RI, 2004.
21. A.Dimca, A.Libgober, Local topology of reducible divisors, math.AG/0303215.
22. A.Dimca, A.Libgober, Regular Functions Transversal at Infinity, math.AG/0504128.
23. D.Eisenbud, W.Neumann, Three dimensional link theory and invariants of plane curve singularities, Annals of mathematical studies 110, Princeton, N.J. 1985.
24. F.Enriques, Sulla construzione delle funzioni algebriche di due variabili possendenti una data curve diramzione, Annali di Matematica pure ed applicata, Ser. 4, Vol. 1, 1923, p.185-198.
25. H.Esnault, Fibre de Milnor d'un cne sur une courbe plane singulir. Invent. Math. 68 (1982), no. 3, 477–496.
26. H.Esnault, E.Viehweg, Lectures on vanishing theorems. DMV Seminar, 20. Birkhauser Verlag, Basel, 1992.
27. C.Eyral, Fundamental groups of singular quasi-projective varieties. Topology 43, (2004), p. 749-764.
28. R.Fox, Quick trip through knot theory. Topology of 3-manifolds and related topics (Proc. The Univ. of Georgia Institute, 1961) pp. 120–167 Prentice-Hall,

Englewood Cliffs, N.J. 55.20

29. W.Fulton, On the fundamental group of the complement of a node curve. Ann. of Math. (2) 111 (1980), no. 2, 407–409.

30. J.Harris, On the Severi problem. Invent. Math. 84 (1986), no. 3, 445–461.

31. R.Hartshorne, Algebraic Geometry, Springer, 1977.

32. E,Hironaka, Polynomial periodicity for Betti numbers of covering surfaces. Invent. Math. 108 (1992), no. 2, 289–321. 14E20 (14J99)

33. E.Hironaka Alexander stratifications of character varieties. Ann. Inst. Fourier (Grenoble) 47 (1997), no. 2, 555–583.

34. Krylov, N.A. On the Jacobi group and the mapping class group of $S^3 \times S^3$. Trans. Amer. Math. Soc. 355 (2003), no. 1, 99–117.

35. S.Kleiman, R.Piene, Node polynomials for families: methods and applications. Math. Nachr. 271 (2004), 69–90.

36. J.Kollar, Singularities of pairs, Algebraic Geometry, Santa Cruz 1995, Proc. Symp. Pure Math. vol.62, part 1, AMS 1997.

37. Kontsevich, M.; Manin, Yu. Gromov-Witten classes, quantum cohomology, and enumerative geometry. Comm. Math. Phys. 164 (1994), no. 3, 525–562.

38. Vik. Kulikov, On Chisini's conjecture. (Russian) Izv. Ross. Akad. Nauk Ser. Mat. 63 (1999), no. 6, 83–116;

39. R.Lazarsfeld, Positivity in algebraic geometry. I,II. Ergebnisse der Mathematik und ihrer Grenzgebiete. 3. Folge. A Series of Modern Surveys in Mathematics, 48. Springer-Verlag, Berlin, 2004.

40. S.Lefschetz, On the existence of loci with given singularities. Trans. Amer. Math. Soc. 14 (1913), no. 1, 23-41.

41. M.Lehr. Regular linear systems of curves with singularities of a given curve as base points, Amer. J. Math. vol. 54, p. 471-488 (1932).

42. A.Libgober, Alexander polynomial of plane algebraic curves and cyclic multiple planes. Duke Math. J. 49 (1982), no. 4, 833–851.

43. A.Libgober, Alexander invariants of plane algebraic curves, Proc. Symp. Pure Math. 1983. Amer. Math. Soc. vol. 40. 1983.

44. A.Libgober, Alexander modules of plane algebraic curves. Low-dimensional topology (San Francisco, Calif., 1981), 231–247, Contemp. Math., 20, Amer. Math. Soc., Providence, R.I., 1983.

45. Libgober, A. Homotopy groups of the complements to singular hypersurfaces. Bull. Amer. Math. Soc. (N.S.) 13 (1985), no. 1, 49

46. A.Libgober, On the homotopy type of the complement to plane algebraic curves. J. Reine Angew. Math. 367 (1986), 103–114.

47. A.Libgober, Fundamental groups of the complements to plane singular curves. Algebraic geometry, Bowdoin, 1985 (Brunswick, Maine, 1985), 29–45, Proc. Sympos. Pure Math., 46, Part 2,

48. A.Libgober, On homology of finite abelian coverings. Topology and application, 43, 1992, p.157-166.

49. A.Libgober, Homotopy groups of the complements to singular hypersurfaces, II. Ann. Math. (2) 139 (1994), 117–144.

50. A.Libgober, Position of singularities of hypersurfaces and the topology of their complements. Algebraic geometry, 5. J. Math. Sci. 82 (1996), no. 1,

3194–3210.

51. A.Libgober, Abelian covers of projective plane, London Math. Soc. Lecture Notes Series, vol. 263. Singularity Theory, W.Bruce and D.Mond editors, p. 281-290. 1999.

52. A.Libgober, Position of singularities of hypersurfaces and the topology of their complements. Algebraic geometry, 5. J. Math. Sci. 82 (1996), no. 1, 3194–3210.

53. A.Libgober, Characteristic varieties of algebraic curves. Applications of algebraic geometry to coding theory, physics and computation (Eilat, 2001), 215–254, NATO Sci. Ser. II Math. Phys. Chem., 36, Kluwer Acad. Publ., Dordrecht, 2001.

54. A.Libgober, S.Yuzvinski, Cohomology of the Orlik-Solomon algebras and local systems. Compositio Math. 121 (2000), no. 3, 337–361.

55. A.Libgober, Hodge decomposition of Alexander invariants. Manuscripta Math. 107 (2002), no. 2, 251–269.

56. A.Libgober, Isolated non-normal crossings. Real and complex singularities, 145–160, Contemp. Math., 354, Amer. Math. Soc., Providence, RI, 2004.

57. A.Libgober, Homotopy groups of complements to ample divisors math.AG/0404341.

58. J.Lipman, Introduction to resolution of singularities. Algebraic geometry (Proc. Sympos. Pure Math., Vol. 29, Humboldt State Univ., Arcata, Calif., 1974), pp. 187–230. Amer. Math. Soc., Providence, R.I., 1975.

59. F.Loeser, Vaquié M. The Alexander polynomial of a projective plane curve. Topology 29 (1990), no. 2, 163–173.

60. L.Maxim, Intersection Homology and Alexander Modules of Hypersurface Complements. math.AT/0409412

61. J.Milnor, Singular points of complex hypersurfaces. Annals of Mathematics Studies, No. 61 Princeton University Press, Princeton, N.J.; University of Tokyo Press, Tokyo 1968.

62. M.Merle, B.Tessier, In Seminar on the Singularities of Surfaces Centre de Mathematiques de l'Ecole Polytechnique, Palaiseau, 1976–1977. Edited by M. Demazure, H. Pinkham and B. Teissier. Lecture Notes in Mathematics, 777.

63. J.Milnor, Infinite cyclic coverings. 1968 Conference on the Topology of Manifolds (Michigan State Univ., E. Lansing, Mich., 1967) pp. 115–133 Prindle, Weber & Schmidt, Boston, Mass.

64. B.Moishezon, Stable branch curves and braid monodromies. Algebraic geometry (Chicago, Ill., 1980), pp. 107–192, Lecture Notes in Math., 862, Springer, Berlin-New York, 1981.

65. J.Morgan, The algebraic topology of smooth algebraic varieties. Inst. Hautes Etudes Sci. Publ. Math. No. 48 (1978), 137–204;

66. A.M. Nadel, Multiplier ideal sheaves and Kähler- Einstein metrics of positive scalar curvature, Ann. Math, 132, (1990), 549-596.

67. M.Nori, Zariski's conjecture and related problems. Ann. Sci. Ecole Norm. Sup. (4) 16 (1983), no. 2, 305–344.

68. M.Oka, Some plane curves whose complements have non-abelian fundamental

groups. Math. Ann. 218 (1975), no. 1, 55–65.

69. M.Oka, Geometry of cuspidal sextics and their dual curves. Singularities—Sapporo 1998, 245–277, Adv. Stud. Pure Math., 29, Kinokuniya, Tokyo, 2000.

70. K.Reidemeister, Knottenthorie, Chelsea Publishing Company, NY 1948, Copyright 1932, Springer, Berlin.

71. Steenbrink, J. H. M. Mixed Hodge structures associated with isolated singularities. Singularities, Part 2 (Arcata, Calif., 1981), 513–536, Proc. Sympos. Pure Math., 40, Amer. Math. Soc., Providence, RI, 1983. (

72. J.Steenbrink Mixed Hodge structure on the vanishing cohomology. Real and complex singularities (Proc. Ninth Nordic Summer School/NAVF Sympos. Math., Oslo, 1976), pp. 525–563.

73. C.Sabbah, Modules d'Alexander et $\mathcal{D}$-modules, Duke Math. Journal, vo. 60, no.3, 1990, p.729-814.

74. G.P. Scott, Braid groups and the group of homeomorphisms of a surface, Proc. Cambridge Philos. Soc. 68 (1970) 605–617.

75. J.P. Serre, Local Algebra, Springer Monographs in Mathematics. Springer-Verlag, Berlin, 2000.

76. F.Severi, Vorlesungen über Algebraische Geometrie, Leipzig, 1921.

77. M.Teicher, Braid groups, algebraic surfaces and fundamental groups of complements of branch curves. Algebraic geometry—Santa Cruz 1995, 127–150, Proc. Sympos. Pure Math., 62, Part 1, Amer. Math. Soc., Providence, RI, 1997.

78. W.S. Turpin, On the fundamental group of a certain class of plane curves, Amer. J. Math 96, 59 (1937).

79. M.Tibar, Homotopy variation and nongeneric pencils, math.AG/0207108

80. Wahl, J. Deformations of planes curves with nodes and cusps. Amer. J. Math. 96 (1974), 529–577.

81. O.Zariski, On the problem of existence of alebraic functions of two variables possesing a given branch curve, Amer. J. of Math. vol. LI, 2, 1929

82. O. Zariski, Oscar On the irregularity of cyclic multiple planes. Ann. of Math. (2) 32 (1931), no. 3, 485–511.

83. O.Zariski On linear connection index of the algebraic surfaces $z^n = f(x,y)$. Proc. National Acad. Sci. USA. vol.15 (1929).

84. O.Zariski, A theorem on the Poincare group of an algebraic hypersurface. Ann. of Math. (2) 38 (1937), no. 1, 131–141.

85. O.Zariski, Algebraic surfaces, Second Edition, Springer Verlag, 1971.

86. O.Zariski, Collected papers. Vol. III. Topology of curves and surfaces, and special topics in the theory of algebraic varieties. Edited and with an introduction by M. Artin and B. Mazur. Mathematicians of Our Time. The MIT Press, Cambridge, Mass.-

87. O.Zariski, On the problem of irreducibility of the algebraic system of irreducible plane curves of a given order and having a given number of nodes. Arithmetic and geometry, Vol. II, 465–481, Progr. Math., 36, Birkhauser Boston, Boston, MA, 1983.

LECTURES ON REAL STRATIFICATION THEORY

David TROTMAN

LATP-UMR 6632
University of Provence
Centre de Mathématique et Informatique,
39 rue Joliot-Curie,
13453 Marseille Cedex 13, France
E-mail: trotman@cmi.univ-mrs.fr

1. Stratifications

What is a stratification ?

The idea is to decompose a singular space into smooth manifolds with some control on how these manifolds fit together.

In 1957 Whitney [W1] showed that every algebraic variety $V = f^{-1}(0)$, where $f : \mathbf{R}^n \to \mathbf{R}^p$ has polynomial coordinates, can be partitioned into finitely many connected smooth submanifolds of $\mathbf{R}^n$. This he called a *manifold complex*. Such a partition is obtained by showing that the singular part of V is again algebraic and of dimension strictly less that that of V. One obtains thus a filtration of V by algebraic subvarieties,

$$V \supset SingV \supset Sing(SingV) \supset \ldots$$

Thom proposed that a partition should exist for which transversality to strata of a map $g : \mathbf{R}^m \to \mathbf{R}^n$ is an open condition on maps in $C^\infty(\mathbf{R}^m, \mathbf{R}^n)$, and that there should be some "local triviality" in a neighbourhood of each stratum.

As a result Whitney refined his definition in 2 papers [W2], [W3] which appeared in 1965, concerning stratifications of real and complex analytic varieties. Thom then developed a theory of C^∞ stratified sets, described in detail in his 1969 paper entitled *"Ensembles et morphismes stratifiés"* [Th2].

I will now describe what has become the accepted notion of Whitney stratification (due to Thom and Whitney).

Definition 1.1. (C^k stratification). Let Z be a closed subset of a differentiable manifold M of class C^k. A C^k *stratification* of Z is a filtration by closed subsets

$$Z = Z_d \supset Z_{d-1} \supseteq \cdots \supseteq Z_1 \supseteq Z_0$$

such that each difference $Z_i - Z_{i-1}$ is a differentiable submanifold of M of class C^k and dimension i, or is empty. Each connected component of $Z_i - Z_{i-1}$ is called a *stratum* of dimension i. Thus Z is a disjoint union of the strata, denoted $\{X_\alpha\}_{\alpha \in A}$.

Example 1.1. The filtration of a realisation of a simplicial complex defined by skeleta, where the strata are the open simplices.

We would like the stratification to "look the same" at different points on the same stratum. This turns out to be possible if "looking the same" is interpreted as "having neighbourhoods which are homeomorphic". Various equisingularity conditions have been introduced ensuring this. An obvious necessary condition is as follows:

Definition 1.2. A stratification $Z = \bigcup_{\alpha \in A} X_\alpha$ satisfies the *frontier condition* if $\forall (\alpha, \beta) \in A \times A$ such that $X_\alpha \cap \overline{X_\beta} \neq \emptyset$, one has $X_\alpha \subseteq \overline{X_\beta}$. As the strata are disjoint this means that $X_\alpha = X_\beta$ or that $X_\alpha \subset \overline{X_\beta} \setminus X_\beta$.

One says that the stratification is *locally finite* if the number of strata is locally finite.

2. Whitney's conditions (a) and (b)

The most successful of the different regularity conditions proposed so as to provide adequate "equisingularity" are the conditions (a) and (b) of Whitney ([W2], [W3]).

Definition 2.1. Take two adjacent strata X and Y, i.e. two C^1 submanifolds of M such that $Y \subset \overline{X} \setminus X$. The pair (X, Y) is said to satisfy Whitney's condition (a) at $\in Y$, or to be (a)-regular at y if : $\forall$ sequences $\{x_i\} \in X$ with limit y such that, in a local chart at y, $\{T_{x_i} X\}$ tends to τ in the grassmannian G_{dimX}^{dimM}, one has $T_y Y \subseteq \tau$.

The pair (X, Y) is said to satisfy Whitney's condition (b) at $y \in Y$, or to be (b)-regular at y if : $\forall$ sequences $\{x_i\} \in X$ and $\{y_i\} \in Y$ with limit y such that, in a local chart at y, $\{T_{x_i} X\}$ tends to τ and the lines $\overline{x_i y_i}$ tend to λ, one has $\lambda \in \tau$.

When $Z = \bigcup_{\alpha \in A} X_\alpha$ is a locally finite stratification such that all pairs of adjacent strata satisfy the frontier condition and are (b)-regular at all points, we say we have a *Whitney stratification* of Z.

Remark 2.1. It will be a nontrivial consequence of the theory that the frontier condition is automatically satisfied by pairs of adjacent strata of a locally finite (b)-regular stratification.

Definition 2.2. Let $\pi : T_Y \to Y$ be a C^1 tubular neighbourhood of Y in M. A pair of adjacent strata (X, Y) is said to be (b^π)-regular if for all sequences $\{x_i\}$ in X such that x_i tends to y and the lines $\overline{x_i \pi(x_i)}$ tend to λ and the tangent planes $T_{x_i} X$ tend to τ, then $\lambda \in \tau$.

Exercises.
1. $(b) \Rightarrow (a)$.
2. $(b) \Leftrightarrow (b^\pi) \quad \forall \pi$.
3. $(a) + (b^\pi)$ for some $\pi \Leftrightarrow (b)$.
4. If (X, Y) is (b)-regular at $y \in Y$, then $dim Y < dim X$.

The following standard example due to Whitney shows that (a) does not imply (b).

Example 2.1. Let $Z = Z_2 = \{y^2 = t^2 x^2 + x^3\} \subset \mathbf{R}^3$. Set $Z_1 = \{(0, 0, t) | t \in \mathbf{R}\}$ and $Z_0 = \emptyset$. Then $Z_2 \supset Z_1 \supset Z_0 = \emptyset$ is a filtration defining a stratification with 4 strata of dimension 2 and one stratum of dimension 1. The strata are defined as follows : $X_1 = (Z_2 - Z_1) \cap \{t > 0\} \cap \{x < 0\}, X_2 = (Z_2 - Z_1) \cap \{t < 0\} \cap \{x < 0\}, X_3 = (Z_2 - Z_1) \cap \{y < 0\} \cap \{x > 0\}, X_4 = (Z_2 - Z_1) \cap \{y > 0\} \cap \{x > 0\}, Y = Z_1$. You can check that the pairs of strata (X_3, Y) and (X_4, Y) are (b)-regular, and in fact they are each C^∞ manifolds with boundary, while (X_1, Y) and (X_2, Y) are not (b)-regular at $(0, 0, 0)$, although they are (a)-regular. Note that the frontier property does not hold for (X_1, Y) and (X_2, Y). It is possible to unite X_1 and X_2 into one connected stratum by turning Y into a circle, so that the frontier condition would hold. But (b) will still fail.

Next we give an example showing that (b^π) does not imply (a).

Example 2.2. (Koike and Kucharz [Tr2]). Let $Z = \{x^3 - 3xy^5 + ty^6 = 0\} \subset \mathbf{R}^3$, with filtration $Z_2 \supset Z_1 = (Ot) \supset Z_0 = \emptyset$.

Theorem 2.1. *(Whitney 1965 [W2], [W2]). Every analytic variety (in $\mathbf{R}^n$ or $\mathbf{C}^n$) admits a Whitney stratification whose strata are analytic (hence C^∞) manifolds.*

Hironaka proved that the same is true of every subanalytic set (in particular every semialgebraic set). His proof uses resolution of singularities. A more elementary proof is due to Denkowska, Wachta and Stasica [DWS], [DS].

One can ask why one should study Whitney's condition (a), as it is strictly weaker than condition (b). One reason is that it is both simple to understand and easy to check. A second reason is that it is both necessary and sufficient for transversality to the strata of a stratification to be an *open* condition, as we shall see in the next theorem.

Definition 2.3. We say that a map $f : N \to M$ between C^1 manifolds is *transverse* to a stratification of a closed set $Z \subset M$, if $\forall x \in N$ such that $f(x) \in Z$, then

$$(df)_x T_x N + T_{f(x)} X = T_{f(x)} M$$

where X is the stratum containing $f(x)$.

Theorem 2.2. *(Trotman 1979 [Tr1]). A locally finite stratification of a closed subset Z of a C^1 manifold M is (a)-regular if and only if for every C^1 manifold N, $\{f \in C^1(N, M) | f$ is transverse to the strata of $Z\}$ is an open set in the Whitney C^1 topology.*

Condition (a) for (X, Y) says that the distance between the tangent space to X at x and the tangent space to Y at y tends to zero as x tends to y. Kuo and Verdier studied what happens when the rate of vanishing of this distance is $O(|x - \pi_Y(x)|)$.

Definition 2.4. Two adjacent strata (X, Y) are (w)-regular at $y_0 \in Y$, or satisfy the Kuo-Verdier condition (w), if there exists a constant $C > 0$ and there exists a neighbourhood U of y_0 in M such that

$$d(T_y Y, T_x X) < C||x - y||$$

$\forall x \in U \cap X, y \in U \cap Y$. Here, for vector subspaces V and W of an inner product space E,

$$d(V, W) = sup\{inf\{sin\theta(v, w) | w \in W^*\} | v \in V^*\}$$

where $\theta(v, w)$ is the angle between v and w.

Note that $d(V, W) = 0 \Leftrightarrow V \subset W$, and that $d(V, W) = 1 \Leftrightarrow \exists v \in V^*, v \perp W$.

Proposition 2.1. *(Kuo [Ve]) For subanalytic X and Y, $(w) \Rightarrow (b)$.*

So (w)-regularity is a stronger regularity condition than (b). It turns out to be generic too, as the following theorem shows.

Theorem 2.3. *(Verdier 1976 [Ve]) Every subanalytic set admits a locally finite (w)-regular stratification. This is also true for definable sets in arbitrary o-minimal structures (Loi 1998).*

For complex analytic strata, $(b) \Leftrightarrow (w)$ ([HeM], [Te2]). Real algebraic examples showing that (b) does not imply (w) are common because (b) is a C^1 invariant while (w) is not.

Example 2.3. (Brodersen-Trotman [BT]) Let $Z = \{y^4 = t^4 x + x^3\} \subset \mathbf{R}^3$. Then the stratification of Z defined by $Z = Z_2 \supset Z_1 = (Ot)$ is (b)-regular but not (w)-regular. Z is actually the graph of the C^1 function $f(x, t) = (t^4 x + x^3)^{1/4}$.

As we want our stratifications to "look the same" at different points of a given stratum one might hope that there is a C^1 diffeomorphism mapping neighbourhoods of a point y_1 on Y to neighbourhoods of another point y_2 on Y. This is not true in general.

Example 2.4. (Whitney [W2]). Let $Z = \{(x, y, t) | xy(x - y)(x - ty), t \neq 1\} \subset \mathbf{R}^3$, stratified by $Z = Z_2 \supset Z_1 = (Ot)$. This is a family of 4 lines parametrised by t. The stratification is both (b)-regular and (w)-regular, but there is no C^1 diffeomorphism mapping Z_{t_1} to Z_{t_2} where $Z_t = Z \cap \mathbf{R}^2 \times \{t\}$, because of the crossratio obstruction. (A linear isomorphism of the plane preserving 3 lines preserves also any 4th line.)

In the next sections we will discuss the Thom-Mather isotopy theorem ensuring local topological triviality and more recent work of Mostowski and Parusinski giving generic local bilipschitz triviality of analytic varieties and subanalytic sets.

3. Transversality and stratified isotopy

3.1. *Transversal intersection of stratifications*

Suppose Z and Z' are two closed stratified sets of a manifold M. Denote the set of strata by Σ and Σ' respectively. We can stratify $Z \cap Z'$ by $\Sigma \cap \Sigma' = \{X \cap X' | X \in \Sigma, X' \in \Sigma'\}$ if Σ and Σ' are transverse, i.e. $\forall X \in \Sigma, \forall X' \in \Sigma'$, X and X' are transverse.

Theorem 3.1. *If (Z, Σ) and (Z', Σ') are Whitney (b)-regular (resp. (a)-regular, resp. (w)-regular), and have transverse intersections in M, then $(Z \cap Z', \Sigma \cap \Sigma')$ is (b)-regular (resp. (a)-regular, resp. (w)-regular).*

This can often be useful. The case of (b)-regularity was treated in the book by Gibson, Wirthmuller, du Plessis and Looijenga (1976 [GWPL]). A more general theorem of this kind was proved by Orro and Trotman in 2002 [OT], including (w)-regularity.

Products: If Z and Z' are Whitney stratified then so is $Z \times Z'$.

Triangulation: It is known that all Whitney stratified sets are triangulable (Goresky [G1], Johnson, Shiota).

Open question [G2]: Does a Whitney stratified set (Z, Σ) have a triangulation whose open simplexes are the strata of a Whitney stratification refining Σ ?

For semialgebraic sets, this has recently been proved in 2005 by Shiota [Sh2]. Moreover he obtains a semialgebraic triangulation.

I will now make more precise what is known about local topological triviality.

Theorem 3.2. *(Thom-Mather [Th2], [M]). Let (Z, Σ) be a Whitney stratified subset of a C^2 manifold M. Then for each stratum $Y \in \Sigma$ and each point $y_0 \in Y$ there is a neighbourhood U of y_0 in M, a stratified set $L \subset S^{k-1}$ and a homeomorphism*

$$h : (U, U \cap Z, U \cap Y) \to (U \cap Y) \times (B^k, c(L), y_0)$$

such that $p_1 \circ h = \pi_Y$, where $c(L)$ is the cone on the link L with vertex y_0, $k = codim Y$, B^k is the k-ball, and π_Y is the projection onto $U \cap Y$ of a tubular neighbourhood.

This theorem applies without any hypothesis of analyticity or subanalyticity. The proof of Mather [M] uses the notion of controlled vector field, and the homeomorphism is obtained by integrating such controlled vector fields.

Definition 3.1. A (stratified) vector field v on a stratified set (Z, Σ) is defined by a collection of vector fields $\{v_X | X \in \Sigma\}$. It is *controlled* when $(\pi_Y)_* v_X(x) = v_Y(\pi_Y(x))$ and $(\rho_Y)_* v_X(x) = 0$ on a tubular neighbourhood T_Y of Y, where T_Y is part of a set of compatible tubular neighbourhoods called control data.

See Mather's notes [M] for details of the theory of controlled vector fields. It was not until around 1996 that a complete proof was published that these stratified controlled vector fields could be assumed to be *continuous* (Shiota-du Plessis-Bekka [P], [Sh1]) : given a vector field v_Y on a stratum Y of a Whitney stratified set, or indeed a Bekka stratified set, there actually exists a continuous controlled stratified vector field $\{v_X\}$ on M extending v_Y. This result has been used (for example) by Hamm [Ham] to simplify some of the fundamental results in stratified Morse theory [GM], and by S. Simon to prove a stratified version of the Poincaré-Hopf theorem [Si].

One can characterise (w)-regularity using stratified vector fields as follows.

Proposition 3.1. *(Brodersen-Trotman). A stratification is (w)-regular $\Leftrightarrow$ every vector field on a stratum Y extends to a rugose stratified vector field in a neighbourhood of Y.*

Definition 3.2. A stratified vector field is called *rugose* near y_0 when there exists a neighbourhood U of y_0 and a constant $C > 0$, such that $\forall x \in U \cap X, \forall y \in U \cap Y$,

$$\| v(x) - v(y) \| \leq C \| x - y \| .$$

This resembles an asymmetric Lipschitz condition, and poses the question of whether the extension of a Lipschitz vector field can be chosen to be Lipschitz.

4. Lipschitz stratifications

Mostowski in 1985 [Mo] introduced certain conditions (L) on a stratification, strengthening (w), which imply the possibility of extending Lipschitz vector fields and are (almost) characterised by the existence of Lipschitz extensions [Pa].

Here are the definitions, which are necessarily somewhat complicated.

Definition 4.1. (cf. Mostowski [Mo]). Let $Z = Z_d \supset \cdots \supset Z_\ell \neq \emptyset$ be a closed stratified set in $\mathbf{R}^n$. Write $\overset{\circ}{Z}_j = Z_j - Z_{j-1}$.

Let $\gamma > 1$ be a fixed constant. A *chain* for a point $q \in \overset{\circ}{Z}_j$ is a strictly decreasing sequence of indices $j = j_1, j_2, \ldots, j_r = \ell$ such that each $j_s (s \geq 2)$ is the greatest integer less than j_{s-1} for which

$$dist(q, Z_{j_s-1}) \geq 2\gamma^2 dist(q, Z_{j_s}).$$

146

For each $j_s, 1 \le s \le r$, choose $q_{j_s} \in \overset{\circ}{Z}_{j_s}$ such that $q_{j_1} = q$ and $|q - q_{j_s}| \le \gamma dist(q, Z_{j_s})$.

If there is no confusion we call $\{q_{j_s}\}_{s=1}^r$ a chain of q.

For $q \in \overset{\circ}{Z}_j$, let $P_q : \mathbf{R}^n \to T_q(\overset{\circ}{Z}_j)$ be the orthogonal projection to the tangent space and let $P_q^\perp = I - P_q$ be the orthogonal projection to he normal space $(T_q(\overset{\circ}{Z}_j)^\perp$.

Definition 4.2. (cf. [Mo] and [Pa]) A stratification $\Sigma = \{Z_j\}_{j=\ell}^d$ of Z is said to be a *Lipschitz stratification*, or to satisfy the (L)-conditions, if for some constant $C > 0$ and for every chain $\{q = q_{j_1}, \ldots, q_{j_r}\}$ with $q \in \overset{\circ}{Z}_{j_1}$ and each $k, 2 \le k \le r$,

$$| P_q^\perp P_{q_{j_2}} \cdots P_{q_{j_k}} | \le C \, | \, q - q_{j_2} \, | \, /d_{j_k-1}(q) \qquad (L1)$$

and for each $q' \in \overset{\circ}{Z}_{j_1}$ such that $| \, q - q' \, | \le (1/2\gamma) \, d_{j_1-1}(q)$,

$$| (P_q - P_{q'}) P_{q_{j_2}} \cdots P_{q_{j_k}} | \le C \, | \, q - q' \, | \, /d_{j_k-1}(q) \qquad (L2)$$

and

$$| P_q - P_{q'} | \le C \, | \, q - q' \, | \, /d_{j_1-1}(q) \qquad (L3).$$

Here $dist(-, Z_{\ell-1}) \equiv 1$, by convention.

It is not hard to show that for a given Lipschitz stratification $\exists \, C > 0$ such that $\forall x \in \overset{\circ}{Z}_j, \forall y \in \overset{\circ}{Z}_k, k < j$ then

$$|P_x^\perp P_y| \le \frac{C|x - y|}{dist(y, Z_{k-1})},$$

so that because $|P_x^\perp P_y| = d(T_y \overset{\circ}{Z}_k, \overset{\circ}{Z}_j)$, then (w)-regularity follows, with a precise estimation for the constant (which can tend to infinity as y approaches Z_{k-1}).

Theorem 4.1. *(Parusinski 1994). Every subanalytic set admits a Lipschitz stratification. Moreover such Lipschitz stratifications are locally bilipschitz trivial.*

It is not true that definable sets in arbitrary o-minimal structures admit Lipschitz stratifications.

Example 4.1. (Parusinski). Let $X(t)$ be the union of the x-axis and the graph $y = x^t (x > 0)$ in $\mathbf{R}^3 = (x, y, t)$. Then the Lipschitz types of $X(t)$ are distinct for all $t > 1$.

Question 4.1. : Do definable sets in polynomially bounded o-minimal structures admit Lipschitz stratifications ?

It is clear that the (L)-conditions are much more of a constraint than is (w).

Example 4.2. (Mostowski). In $\mathbf{C}^4$ or $\mathbf{R}^4$ let $Z = \{y = z = 0\} \cup \{y = x^3, z = tx\}$. Then (w) holds along the t-axis, but (L) fails.

Example 4.3. (Koike-Juniati). In $\mathbf{R}^3$ let $Z = \{y^2 = t^2 x^2 + x^3, x \geq 0\}$ stratified by $Z = Z_2 \supset Z_1 = \langle Ot \rangle$. It is easy to check that (w) holds for this semialgebraic example, while $(L2)$ fails : let $q = q_{j_1} = q_2 = (t^2, \sqrt{2t^3}, t), q' = (t^2, -\sqrt{2t^3}, t), q_{j_2} = q_1 = (0, 0, t)$, as $t \to 0$. See [JTV].

In his 1974 Arcata lectures Teissier gave criteria for a good equisingularity condition E on a stratification of a complex analytic set; E-regularity should:

1) be as strong as possible;

2) be generic, i.e. every complex analytic set should possess an E-regular stratification;

3) imply local topological triviality along strata;

4) imply equimultiplicity;

5) be preserved after intersection with generic linear spaces containing a given stratum, locally linearised $(E \Rightarrow E^*)$;

6) have a Zariski equisingularity property.

Criteria 2) to 6) hold for Whitney (b)-regularity (see Teissier 1982), which turns out to be equivalent to (w) in the complex case as noted above. Criterion 5) is an essential part of the proof, via the equimultiplicity of polar varieties. (Recall that (b) does not imply (w) for real algebraic varieties.)

Definition 4.3. (E^*)-regularity.

Let M be a C^2 manifold. Let Y be a C^2 submanifold of M and let $y \in Y$. Let X be a C^2 submanifold of M such that $y \in \overline{X}$ and $Y \cap X = \emptyset$. Let (E) denote an equisingularity condition (e.g. $(b), (w), (L)$). Then (X, Y) is said to be (E_{codk})-regular at $y (0 \leq k \leq codY)$ if there exists an open dense subset U^k of the grassmannian of codimension k subspaces of $T_y M$ containing $T_y Y$, such that if W is a C^2 submanifold of M with $Y \subset W$

near y, and $T_y W \in U^k$, then W is transverse to X near y, and $(X \cap W, Y)$ is (E)-regular at y.

One says finally that (X, Y) is (E^*)-regular at y if (X, Y) is (E_{codk})-regular for all $k, 0 \le k < codY$.

Theorem 4.2. *(Navarro Aznar-Trotman). For subanalytic stratifications, $(w) \Rightarrow (w^*)$, and if $dimY = 1, (b) \Rightarrow (b^*)$.*

Question 4.2. : Does $(b) \Rightarrow (b^*)$ for all subanalytic stratifications ?

Theorem 4.3. *([Te2], [HeM]) For complex analytic stratifications, $(b) \Rightarrow (b^*)$.*

Theorem 4.4. *([JTV]) For subanalytic stratifications, $(L) \Rightarrow (L^*)$.*

We conclude that the (L)-regularity of Mostowski is possibly the best equisingularity condition. However it has disadvantages:

1) it is not generic for definable sets over non polynomially bounded o-minimal structures (while (b) and (w) are generic, as proved by Tà Lê Loi),

2) it has a long and complicated definition which is hard to work with (while (b) and (w) have simple definitions).

5. Definable trivialisations

We have seen that Whitney (b)-regularity ensures local topological triviality. Mostowski and Parusinski proved that a (L)-regular stratification is locally bilipschitz trivial. It is natural to ask if such trivialisations can be chosen to be definable. Or generally, if Z is a semialgebraic set is there some stratification of Z which is locally semialgebraically trivial ? This was proved by Hardt in 1980 [Har]; his method was very recently improved by G. Valette [Va1] who obtained semialgebraic bilipschitz triviality.

Theorem 5.1. *(Hardt). Semialgebraic sets admit locally semialgebraically trivial stratifications.*

Theorem 5.2. *(Valette). Semialgebraic sets admit locally semialgebraically bilipschitz trivial stratifications.*

There are also subanalytic versions of these results. For semialgebraic (b)-regular stratifications Coste and Shiota [CS] proved a semialgebraic isotopy theorem using real spectrum methods. See the book of Shiota [Sh1] for further details and references.

6. Bekka's (c)-regularity

It can be important to be more precise as to when a stratification is locally topologically trivial, for example when classifying topologically or studying topological stability (cf. work of the Liverpool School by Bruce, Giblin, Gibson, Wall, Looijenga, Wirthmuller and the book of du Plessis and Wall [PW]). Then one needs the weakest regularity condition on a stratification which will ensure local topological triviality.

Definition 6.1. (K. Bekka [B]).A stratified set (Z, Σ) in a manifold M is (c)-regular if for every stratum Y of Σ there exists an open neighbourhood U_Y of Y in M and a C^1 function $\rho_Y : U_Y \to [0, \infty)$ such that $\rho_Y{}^{-1}(0) = Y$ and the restriction $\rho_Y|_{U_Y \cap Star(Y)}$ is a Thom map, where $Star(Y) = \bigcup\{X \in \Sigma | X \geq Y\}$, i.e. $\forall X \in Star(Y)$, with $\rho_{XY} = \rho_Y|_X$ and $x \in X$,

$$lim_{x \to y} T_x(\rho_{XY}{}^{-1}(\rho_Y(x))) \supseteq T_y Y \quad \forall y \in Y.$$

Note that $\rho_Y : U_Y \to [0, \infty)$ is defined globally on a neighbourhood of Y. So this is not a local condition.

Theorem 6.1. *(Bekka [B]). (c)-regular stratifications are locally topologically trivial along strata.*

The proof is by proving the existence of an abstract stratified structure of Mather which allows the use of Mather's theory of controlled stratified vector fields [M]. If one only requires constance of homological/cohomological data then one can weaken (c) even further - see the book of Schurmann [Sch].

We saw how (w) and (L) are characterised by the existence of appropriate lifts of vector fields. Here is the corresponding result for (c)-regularity.

Theorem 6.2. *(du Plessis-Bekka [P]) A stratification is (c)-regular $\Leftrightarrow$ every C^1 vector field on a stratum Y admits a continuous controlled stratified extension to a neighbourhood of Y.*

This means that there exists a family of vector fields $\{v_X | X \in Star(Y)\}$ such that $v = \bigcup v_X$ is continuous (in TM), while being controlled as defined above.

How do (c) and (b) compare ?

I proved in my thesis (see [Tr0]) that (b) over a stratum Y is equivalent to the property that for every C^1 tubular neighbourhood T_Y of Y the restriction to neighbouring strata of the associated map (π_Y, ρ_Y) is a submersion, where $\pi_Y : T_Y \to Y$ is the canonical retraction and $\rho_Y : T_Y \to [0, 1)$ the canonical distance function .

In comparison, (c) says that there exists some $C^1 \rho$ (not necessarily associated to a tubular neighbourhood; ρ can be degenerate, e.g. weighted homogeneous, or even flat on Y) such that for every C^1 tubular neighbourhood T_Y of Y the restriction to neighbouring strata of the map (π_Y, ρ) is a submersion [B].

One can prove that (b) implies (c) while the converse is false.

7. Condition (t^k)

We return to the first example of Whitney, $Z = \{y^2 = t^2 x^2 + x^3\}$. Slice the surface by a plane S transverse to the t-axis at 0. Then the topological type of the germ at 0 of the intersection $Z \cap S$ is constant, i.e. independent of S. Remember that Whitney $a)$ holds. Thom noticed this and mentioned it to Kuo, who proved the following theorem [K].

Theorem 7.1. *(Kuo 1978). If (X, Y) is (a)-regular at $y \in Y$ then (h^∞) holds, i.e. the germs at y of intersections $S \cap X$, where S is a C^∞ submanifold transverse to Y at $y \in S \cap Y$ and $dim S + dim Y = dim M$, are homeomorphic.*

It later turned out that one can replace (h^∞) by (h^1), meaning one considers all C^1 transversals S, and weaken (a) to (t^1), defined as follows.

Definition 7.1. A pair of strata (X, Y) is (t^k)-regular at $y \in Y$ if for every C^k submanifold S transverse to Y at $y \in Y \cap S$, there is a neighbourhood U of y such that S is transverse to X on $U \cap X$ $(1 \le k \le \infty)$.

Theorem 7.2. *(Trotman 1985 [Tr3]). (t^1) is equivalent to (h^1).*

Theorem 7.3. *(Trotman-Wilson 1999 [TW]). For subanalytic strata, (t^k) is equivalent to the finiteness of the number of topological types of germs at y of $S \cap X$ for S a C^k transversal to Y $(k \ge 1)$.*

The proofs developed with Kuo and Wilson use the "Grassmann blowup" introduced by Kuo and myself [KT]. Let

$$E^{n,d} = \{(L, x) | x \in L\} \subset G^{n,d} \times \mathbf{R}^n$$

for $d < n$, with projection to $G^{n,d}$, denote the canonical d-plane bundle. Let $\beta = \beta_{n,d}$ denote projection to $\mathbf{R}^n$. When $d = 1$ this is the usual blowup of $\mathbf{R}^n$ with centre 0.

Suppose $X, Y \subset \mathbf{R}^n$ and $0 \in Y$ with $d = codim Y$.

Let $\tilde{X} = \beta^{-1}(X)$ and let $\tilde{Y} = \{(L, 0)|L \text{ is transverse to } Y \text{ at } 0\}$. the following striking theorem results from work by Kuo and myself [KT], completed by work with Wilson [TW].

Theorem 7.4. (X, Y) *is* (t^k)-*regular at* $0 \in Y$ *if and only if* $(\tilde{X}, \tilde{Y})$ *is* (t^{k-1})-*regular at every point of* Y $(k \geq 1)$.

Explanation when $k = 1$: here (t^0) is equated with (w), the Kuo-Verdier condition. So in particular, (w)-regularity is the first in a sequence of (t^k)-regularity conditions !

Now we can see how to prove that (t^1) implies (h^1) by using the Verdier isotopy theorem ([Ve]) for (w)-regular stratifications in the Grassmann blowup, although this was not the orginal proof.

The (t^k)- conditions were used to characterise jet sufficiency by Trotman and Wilson, generalising theorems of Bochnak, Kuo, Lu and others, and realising part of the early programme of Thom (1964). See [TW] for details. Very recent work with Gaffney and Wilson [GTW] develops an algebraic approach to the (t^k)- conditions, using integral closure of modules.

To illustrate the difference between (t^2) and (t^1), and the previous theorem, look at the Koike-Kucharz example (1979) given by $Z = \{x^3 - 3xy^5 + ty^6 = 0\} \subset \mathbf{R}^3$ stratified as usual by (X, Y) with Y the t-axis and X its complement $Z - Y$. Then (X, Y) is (t^2) but not (t^1) at 0. It is easy to check that there are 2 topological types of germs at 0 of intersections $S \cap X$ where S is a C^2 submanifold transverse to Y at 0. However the number of topological types of such germs for S of class C^1 is infinite, even uncountable.

It is easy to construct similar examples showing (t^k) does not imply (t^{k-1}).

8. Density and normal cones

I mentioned Hironaka's theorem that complex analytic Whitney stratifications are equimultiple along strata.

What is a real version of this statement ?

Define the multiplicity $m(V, p)$ at a point p of a complex analytic variety V to be the number of points near p in the intersection of V with a generic plane L missing p of complementary dimension to that of V. This positive integer is equal to the Lelong number, or density $\theta(V, p)$ of V at p defined as the limit as ϵ tends to 0 of the quotient $\frac{vol(V \cap B_\epsilon(p))}{vol(P \cap B_\epsilon(p))}$.

Kurdyka and Raby showed that the density is well-defined for subanalytic sets, as a positive real number. It is thus natural to conjecture (I did so in 1988) that the density of a subanalytic set is continuous along strata of a Whitney stratification, as a generalisation of Hironaka's theorem. This was partially proved by Georges Comte in his thesis (1998) for subanalytic Verdier (w)-regular stratifications [C], or more generally for subanalytic (b^*)-regular stratifications. The general conjecture was proved for subanalytic (b)-regular stratifications by Guillaume Valette in 2003 [Va2]. Valette also showed that the density is a lipschitz function along strata of a subanalytic (w)-regular stratification.

In the paper [Hi] about equimultiplicity, Hironaka also proved results about the *normal cones* of analytic Whitney stratifications.

Definition 8.1. Suppose Z is a stratified subset of $\mathbf{R}^n$ and let Y be a stratum. Let π_Y be the projection of a tubular neighbourhood of Y and let $\mu(v) = \frac{v}{\|v\|}$. The normal cone is defined to be:

$$C_Y Z = \overline{\{(x, \mu(x\pi_Y(x)))|x \in Z - Y\}}|_Y \subset \mathbf{R}^n \times S^{n-1}.$$

Let $p : C_Y Z \to Y$ be the canonical projection.

Theorem 8.1. *A (b)-regular subanalytic stratification of a subanalytic set is*

(npf) normally pseudo-flat, i.e. p is an open map, and

(n) for each stratum Y and each point y of Y, the fibre $(C_Y Z)_y$ of the normal cone at y is equal to the tangent cone $C_y(Z_y)$ at y to the special fibre $\pi_Y - 1(y)$.

The proof is by integration of vector fields (cf. [Hi], also [OT]).

Example 8.1. The result is not true for definable sets in non-polynomially bounded o-minimal structures. For an example one can take Z in $\mathbf{R}^3$ to be the graph of the function $f : [0, \infty) \times \mathbf{R} \to \mathbf{R}$ defined by

$$z = f(x, y) = x - \frac{x}{ln(x)} ln(y + (x^2 + y^2)^{\frac{1}{2}}).$$

Stratify Z by $Z_1 = \{0y\} \subset Z$. One checks easily that $(C_Y Z)_0$ is an arc, while $C_0(Z_0)$ is a point so that the criterion (n) above fails. Moreover the example is not normally pseudoflat, nor (b^*)-regular, but it is Whitney (b)-regular.

Example 8.2. In [OT] real algebraic (a)-regular examples are given showing that (n) does not imply (npf) and conversely. First let $(0z) = Z_1 \subset Z = \{x(x^2 + y^2)z^2 - (x^2 + y^2)^2 + xy^2 = 0\}$. Then (a) and (n) hold but (npf) fails. Finally look yet again at $\{y^2 = t^2x^2 + x^3\}$, stratified by the t-axis and its complement. Although (n) fails, because $(C_Y Z)_0$ consists of 2 points while $C_0(Z_0)$ consists of 1 point, it is normally pseudoflat.

References

B.   K. Bekka, C-régularité et trivialité topologique, *Singularity theory and its applications, Warwick 1989, Part I*, Lecture Notes in Math. 1462, Springer, Berlin, 1991, 42-62.

BT.   H. Brodersen, D. Trotman, Whitney (b)-regularity is weaker than Kuo's ratio test for real algebraic stratifications, *Mathematica Scandinavia* 45 (1979), 27-34.

C.   G. Comte, Équisingularité réelle : nombres de Lelong et images polaires, *Ann. Sci. École Norm. Sup.* (4) 33 (2000), no. 6, 757–788.

CS.   M. Coste, M. Shiota, *Thom's first isotopy lemma: a semialgebraic version, with uniform bound*, Real analytic and algebraic geometry (Trento, 1992), 83–101, de Gruyter, Berlin, 1995.

DS.   Z. Denkowska, J. Stasica, *Ensembles sousanalytiques à la polonaise*, manuscript, 1985.

DWS.   Z. Denkowska, K. Wachta, J. Stasica, Stratification des ensembles sous-analytiques avec les propriétés (A) et (B) de Whitney, *Univ. Iagel. Acta Math.* 25 (1985), 183–188.

GTW.   T. Gaffney, D. Trotman, L. Wilson, *Equisingularity of sections, (t^r) condition, and the integral closure of modules*, preprint, 2005.

GWPL.   C. G. Gibson, K. Wirthmüller, A. A. du Plessis and E. J. N. Looijenga, *Topological stability of smooth mappings*, Lecture Notes in Math. 552, Springer-Verlag, 1976.

G1.   M. Goresky, Triangulation of stratified objects, *Proc. Amer. Math. Soc.* 72 (1978), no. 1, 193–200.

G2.   M. Goresky, Whitney stratified chains and cochains, *Trans. Amer. Math. Soc.* 267 (1981), no. 1, 175–196.

GM.   M. Goresky, R. MacPherson, *Stratified Morse theory*, Ergebnisse der Mathematik und ihrer Grenzgebiete (3), 14. Springer-Verlag, Berlin, 1988.

Ham.   H. Hamm, On stratified Morse theory, *Topology* 38 (1999), no. 2, 427–438.

Har.   R. Hardt, Semi-algebraic local-triviality in semi-algebraic mappings, *Amer. J. Math.* 102 (1980), no. 2, 291–302.

HeM.   J.-P. Henry, M. Merle, Limites de normales, conditions de Whitney et éclatement d'Hironaka, *Proceedings of Symposia in Pure Mathematics, Volume 40, Arcata 1981–Singularities, Part 2*, American Mathematical Society, Providence, Rhode Island, 1983, 575-584.

Hi.   H. Hironaka, Normal cones in analytic Whitney stratifications, *Inst. Hautes Études Sci. Publ. Math.* 36 1969 127–138.

JTV.   D. Juniati, D. Trotman, G. Valette, Lipschitz stratifications and generic, *Journal of the London Mathematical Society*, (2) 68 (2003), no. 1, 133–147.

K.   T.-C. Kuo, On Thom-Whitney stratification theory, *Math. Ann.* 234 (1978), no. 2, 97–107.

KT.   T.-C. Kuo, D. Trotman, On (w) and (t^s)-regular stratifications, *Inventiones Mathematicae* 92, 1988, 633–643.

Lo.   Ta Lê Loi, Verdier and strict Thom stratifications in o-minimal structures, *Illinois J. Math.* 42 (1998), no. 2, 347–356.

M.   J. Mather, *Notes on topological stability*, Mimeographed notes, Harvard University, 1970.

Mo.   T. Mostowski, Lipschitz equisingularity, *Dissertationes Math. (Rozprawy Mat.)* 243 (1985), 46 pp.

NT.   V. Navarro Aznar, D. Trotman, Whitney regularity and generic wings, *Annales de l'Institut Fourier, Grenoble*, 31, 1981, 87–111.

OT.   P. Orro, D. Trotman, Cône normal et régularités de Kuo-Verdier, *Bulletin de la Société Mathématique de France*, 130 (2002), 71–85.

Pa.   A. Parusinski, Lipschitz stratification of subanalytic sets, *Ann. Sci. cole Norm. Sup.* (4) 27 (1994), no. 6, 661–696.

P.   A. du Plessis, Continuous controlled vector fields, *Singularity theory (Liverpool, 1996, edited by J. W. Bruce and D. M. Q. Mond), London Math. Soc. Lecture Notes* **263**, Cambridge Univ. Press, Cambridge, (1999), 189-197.

PW.   A. A. du Plessis, C. T. C. Wall, *The Geometry of Topological Stability*, Oxford University Press, Oxford, 1995. Oxford University Press, 1995.

Sch.   J. Schurmann, *Topology of singular spaces and constructible sheaves*, Mathematics Institute of the Polish Academy of Sciences. Mathematical Monographs (New Series), 63. Birkhuser Verlag, Basel, 2003.

Sh1.   M. Shiota, *Geometry of Subanalytic and Semialgebraic Sets*, Birkhaüser, Boston, 1997.

Sh2.   M. Shiota, Whitney triangulations of semialgebraic sets. *Ann. Polon. Math.* 87 (2005), 237–246.

Si.   S. Simon, Champs totalement radiaux sur une structure de Thom-Mather, *Ann. Inst. Fourier (Grenoble)* 45 (1995), no. 5, 1423–1447.

Te1.   B. Teissier, Introduction to equisingularity problems, *Algebraic geometry (Proc. Sympos. Pure Math.*, Vol. 29, Humboldt State Univ., Arcata, Calif., 1974), pp. 593–632. Amer. Math. Soc., Providence, R.I., 1975.

Te2.   B. Teissier, Variétés polaires. II. Multiplicités polaires, sections planes, et conditions de Whitney, *Algebraic geometry (La Ràbida, 1981)*, 314–491, Lecture Notes in Math., 961, Springer, Berlin, 1982.

Th1.   R. Thom, Local topological properties of differentiable mappings, *Differential Analysis, Bombay Colloq.*, 1964, pp. 191–202

Th2.   R. Thom, Ensembles et morphismes stratifiés, *Bull. Amer. Math. Soc.* 70, 1969, pp. 240–284.

Tr0.   D. Trotman, Geometric versions of Whitney regularity for smooth stratifications, *Ann. Sci. École Norm. Sup.* (4) 12 (1979), 453–463.

Tr1.   D. Trotman, Stability of transversality to a stratification implies Whitney (a)-regularity, *Inventiones Mathematicae* 50, 1979, 273–277.

Tr2.   D. Trotman, Comparing regularity conditions on stratifications, *Proceedings of Symposia in Pure Mathematics, Volume 40, Arcata 1981–Singularities, Part 2*, American Mathematical Society, Providence, Rhode Island, 1983, 575–586.

Tr3.   D. Trotman, Transverse transversals and homeomorphic transversals, *Topology* 24 (1985), no. 1, 25–39.

TW.   D. Trotman, L. Wilson, Stratifications and finite determinacy, *Proceedings of the London Mathematical Society*, (3) 78, 1999, no. 2, 334–368.

Va1.   G. Valette, Lipschitz triangulations, *Illinois J. of Math.*, 49, no. 3 (2005), 953–979

Va2.   G. Valette, *Volume, density and Whitney conditions*, preprint.
Ve.   J.-L. Verdier, Stratifications de Whitney et théorème de Bertini-Sard, *Inventiones Math.* 36 (1976), 295-312.
W1.   H. Whitney, Elementary structure of real algebraic varieties, *Ann. of Math.* (2) 66 (1957), 545–556.
W2.   H. Whitney, Local properties of analytic varieties, *Differential and Combinatorial Topology,* Princeton Univ. Press, (1965), 205–244.
W3.   H. Whitney, Tangents to an analytic variety, *Annals of Math.* (2) 81 (1965), 496–549.

LAGRANGIAN AND LEGENDRIAN SINGULARITIES

V.V.GORYUNOV and V.M.ZAKALYUKIN

Department of Mathematical Sciences
University of Liverpool
Liverpool L69 3BX, UK
e-mail: goryunov@liv.ac.uk
Department of Mechanics and Mathematics
Moscow State University
Leninskie gory, 1
119992 Moscow,Russia
e-mail: vzakal@liv.ac.uk

These are notes of the introductory courses we lectured in Trieste in 2003 and Luminy in 2004. The lectures contain basic notions and fundamental theorems of the local theory of singularities of wave fronts and caustics with some recent applications to geometry.

Keywords: Symplectic and contact geometry, Lagrangian and Legendrian submanifolds and singularities, generating families

R.Thom and V.Arnold noticed that the singularities that can be visualized in many physical models are of special nature.

This was the starting point of the theory of Lagrangian and Legendrian mappings developed by V.I.Arnold and his school some thirty years ago. Since then the significance of Lagrangian and Legendrian submanifolds of symplectic and respectively contact spaces has been recognised throughout all mathematics, from algebraic geometry to differential equations, optimisation problems and physics.

Alternatively these singularities are called singularities of caustics and wave fronts.

Suppose, for example, that a disturbance (such as a shock wave, light, an epidemic or a flame) is propagating in a medium from a given submanifold (called *initial wave front*). To determine where the disturbance will be at time t (according to the Huygens principle) we must lay a segment of length t along every normal to the initial front. The resulting variety is called an *equidistant* or a wave front.

Along with wave fronts, ray systems may also be used to describe propagation of disturbances. For example, we can consider the family of all normals to the initial front. This family has the envelope, which is called *caustic* – "burning" in Greek – since the light concentrates at it. A caustic is clearly visible on the inner surface of a cup put in the sunshine. A rainbow in the sky is the caustic of a system of rays which have passed through drops of water with total internal reflection.

Generic caustics in three-dimensional space have only standard singularities. Besides regular surfaces, cuspidal edges and their generic (transversal) intersections, these are: the swallowtail, the 'pyramid' (or 'elliptic umbilic') and the 'purse' (or 'hyperbolic umbilic'). They are a part of R.Thom's famous list of simple catastrophes. It is not so difficult to see that the singularities of a propagating wave front slide along the caustic and trace it out.

Symplectic space is essentially the phase space (space of positions and momenta) of classical mechanics, inheriting a rich set of important properties.

It turns out that caustics and wave fronts are the loci of critical values of special non-generic mappings of manifolds of equal dimensions or mappings from n to $n + 1$ dimensional manifolds. The general definition of these mapping was given by V.Arnold via the projections of Lagrangian and Legendre submanifolds embedded into symplectic and contact spaces.

These construction describes many special classes of mappings: Gauss mapping, gradient mappping, etc.

In fact, Lagrangian or Legendre mapping is determined by a single family of fuctions. This crucial fact makes the theory transparent and constructive.

In particular, stable wave fronts and caustics are discriminants and bifurcation diagrams of singularities of functions. That is why their generic low dimensional singularities are governed by famous simple Weyl groups.

Recently new areas in the theory of integrable systems in mathematical physics (Frobenuous structures, D-modules) yield new field of applications of Lagrangian and Legendre singularities.

In these lecture notes, we do not touch the fascinating results in symplectic and contact topology, a young branch of mathematics which answers questions on global behaviour of Lagrangian and Legendrian submanifolds. An interested reader may be addressed to the paper [[4]]. Our lectures is an introduction to the original local theory, with an accent on applications in geometry. We hope that they will inspire the reader to do more extensive reading. Items on our bibliography list [[1–3]]. may be rather useful for this.

1. Symplectic and contact geometry

1.1. *Symplectic geometry*

A <u>symplectic form</u> ω on a manifold M is a closed 2-form, non-degenerate as a skew-symmetric bilinear form on the tangent space at each point. So $d\omega = 0$ and ω^n is a volume form, $\dim M = 2n$. Manifold M equipped with a symplectic form is called <u>symplectic</u>. It is necessarily even-dimensional.

If the form is exact, $\omega = d\lambda$, the manifold M is called *exact symplectic*.

Examples.

1. The basic model of a symplectic space is the vector space $K = \mathbf{R}^{2n} = \{q_1, \ldots, q_n, p_1, \ldots, p_n\}$ with the form

$$\lambda = pdq = \sum_{i=1}^{n} p_i dq_i, \qquad \omega = d\lambda = dp \wedge dq.$$

In these coordinates the form ω is constant. The corresponding bilinear form on the tangent space at a point is given by the matrix

$$J = \begin{pmatrix} 0 & -I_n \\ I_n & 0 \end{pmatrix}$$

Any non-degenerate skew-symmetric bilinear form on a linear space, has a <u>Darboux</u> basis in which the form has this matrix.

2. $M = T^*N$. $\quad \lambda = pdq$ - Take for λ the *Liouville form* defined in a coordinate free way as

$$\lambda(\alpha) = \pi(\alpha)(\rho_*(\alpha)),$$

where

$$\alpha \in T(T^*N), \quad \pi : T(T^*N) \to T^*N \quad \text{and} \quad \rho : T^*N \to N.$$

The manifold M, $d\lambda$ is exact symplectic. For local coodinates $q_1, \ldots, q_n$ on N, the dual coordinates $p_1, \ldots, p_n$ are the coefficients of the decomposition of a covector into a linear combination of the differentials dq_i:

$$\lambda = \sum_{i=1}^{n} p_i dq_i.$$

3. On a Kähler manifold M, the imaginary part of its Hermitian structure $\omega(\alpha, \beta) = Im(\alpha, \beta)$ is a skew-symmetric 2-form which is closed.

4. Product of two symplectic manifolds. Given two symplectic manifolds (M_i, ω_i), $i = 1, 2$, their product $M_1 \times M_2$ equipped with the 2-form $(\pi_1)_* \omega_1 - (\pi_2)_* \omega_2$, where the π_i are the projections to the corresponding factors, is a symplectic manifold.

A diffeomorphism $\varphi : M_1 \to M_2$ which sends the symplectic structure ω_2 on M_2 to the symplectic structure ω_1 on M_1,

$$\varphi^* \omega_2 = \omega_1 \,,$$

is called a <u>symplectomorphism</u> between (M_1, ω_1) and (M_2, ω_2). When the (M_i, ω_i) are the same, a symplectomorphism preserves the symplectic structure. In particular, it preserves the volume form ω^n.

Symplectic group.
For $K = (\mathbf{R}^{2n}, dp \wedge dq)$ of our first example, the group $Sp(2n)$ of <u>linear</u> symplectomorphisms is isomorphic to the group of matrices S such that

$$S^{-1} = -J S^t J \,.$$

Here t is for transpose. The characteristic polynomial of such an S is reciprocal: if α is an eigenvalue, then α^{-1} also is. The Jordan structures for α and α^{-1} are the same.

Introduce an auxiliary scalar product $(\cdot, \cdot)$ on K, with the matrix I_{2n} in our Darboux basis. Then

$$\omega(a, b) = (a, \widetilde{J}b) \,,$$

where $\widetilde{J}$ is the operator on K with the matrix J. Setting $q = \mathrm{Re}\, z$ and $p = \mathrm{Im}\, z$ makes K a complex Hermitian space, with the multiplication by $i = \sqrt{-1}$ being the application of $\widetilde{J}$. The Hermitian structure is

$$(a, b) + i\omega(a, b) \,.$$

From this,

$$Gl(n, \mathbf{C}) \bigcap O(2n) = Gl(n, \mathbf{C}) \bigcap Sp(2n) = O(2n) \bigcap Sp(2n) = U(n) \,.$$

Remark. The image of the unit sphere $S_1^{2n-1} : q^2 + p^2 = 1$ under a linear symplectomorphism can belong to a cylinder $q_1^2 + p_1^2 \leq r$ only if $r \geq 1$.

The non-linear analog of this result is rather non-trivial: $S_1^{2n-1} \in T^* \mathbf{R}^n$ (in the standard Euclidean structure) cannot be symplectically embedded into the cylinder $\{q_1^2 + p_1^2 < 1\} \times T^*\mathbf{R}^{n-1}$. This is Gromov's theorem on symplectic camel.

Thus, for $n > 1$, symplectomorphisms form a thin subset in the set of diffeomorphisms preserving the volume ω^n.

The dimension k of a linear subspace $L^k \subset K$ and the rank r of the restriction of the bilinear form ω on it are the complete set of $Sp(2n)$-invariants of L.

Define the skew-orthogonal complement $L^{\angle}$ of L as

$$L^{\angle} = \{v \in K | \omega(v, u) = 0 \quad \forall u \in L\}.$$

So $\dim L^{\angle} = 2n - k$. The kernel subspace of the restriction of ω to L is $L \cap L^{\angle}$. Its dimension is $k - r$.

A subspace is called <u>isotropic</u> if $L \subset L^{\angle}$ (hence $\dim L \leq n$).
Any line is isotropic.

A subspace is called <u>co-isotropic</u> if $L^{\angle} \subset L$ (hence $\dim L \geq n$).
Any hyperplane H is co-isotropic. The line $H^{\angle}$ is called the <u>characteristic direction</u> on H.

A subspace is called <u>Lagrangian</u> if $L^{\angle} = L$ (hence $\dim L = n$).

Lemma.*Each Lagrangian subspace $L \subset K$ has a regular projection to at least one of the 2^n co-ordinate Lagrangian planes (p_I, q_J), along the complementary Lagrangian plane (p_J, q_I). Here $I \bigcup J = \{1, \ldots, n\}$ and $I \bigcap J = \emptyset$.*

Proof. Let L_q be the intersection of L with the q-space and $\dim L_q = k$. Assume $k > 0$, otherwise L projects regularly onto the p-space. The plane L_q has a regular projection onto some q_I-plane (along q_J) with $|I| = k$. If L does not project regularly to the p_J-plane (along (q, p_I)) then L contains a vector $v \in (q, p_I)$ with a non-trivial p_I-component. Due to this non-triviality, the intersection of the skew-orthogonal complement $v^{\angle}$ with the q-space has a $(k-1)$-dimensional projection to q_I (along q_J) and so does not contain L_q. This contradicts to L being Lagrangian.

A Lagrangian subspace L which projects regularly onto the q-plane is the graph of a self-adjoint operator S from the q-space to the p-space with its matrix symmetric in the Darboux basis.

162

Splitting $K = L_1 \oplus L_2$ with the summands Lagrangian is called a polarisation. Any two polarisations are symplectomorphic.

The Lagrangian Grassmanian $Gr_L(2n)$ is diffeomorphic to $U(n)/O(n)$. Its fundamental group is $\mathbf{Z}$.

The Grassmanian $Gr_k(2n)$ of isotropic k-spaces is isomorphic to $U(n)/(O(k) + U(n-k))$.

Even in a non-linear setting a symplectic structure has no local invariants (unlike a Riemannian structure) according to the classical

Darboux Theorem. *Any two symplectic manifolds of the same dimension are locally symplectomorphic.*

Proof. We use the homotopy method. Let ω_t, $t \in [0,1]$, be a family of germs of symplectic forms on a manifold coinciding at the distinguished point A. We are looking for a family $\{g_t\}$ of diffeomorphisms such that $g_t^* \omega_t = \omega_0$ for all t. Differentiate this by t:

$$\mathcal{L}_{v_t} \omega_t = -\gamma_t$$

where $\gamma_t = \partial \omega_t / \partial t$ is a known closed 2-form and $\mathcal{L}_{v_t}$ is the Lie derivative along the vector field to find. Since $\mathcal{L}_v = i_v d + d i_v$, we get

$$d i_{v_t} \omega_t = -\gamma_t .$$

Choose a 1-form α_t vanishing at A and such that $d\alpha_t = -\gamma_t$. Due to the non-degeneracy of ω_t, the equation $i_{v_t} \omega_t = \omega(\cdot, v_t) = \alpha_t$ has a unique solution v_t vanishing at A. $\qquad\qquad\square$

Weinstein's Theorem. *A submanifold of a symplectic manifold is defined, up to a symplectomorphism of its neighbourhood, by the restriction of the symplectic form to the tangent vectors to the ambient manifold at the points of the submanifold.*

In a similar local setting, the inner geometry of a submanifold defines its outer geometry:

Givental's Theorem. *A germ of a submanifold in a symplectic manifold is defined, up to a symplectomorphism, by the restriction of the symplectic* *structure to the tangent bundle of the submanifold.*

Proof of Givental's Theorem. It is sufficient to prove that if the restrictions of two symplectic forms, ω_0 and ω_1, to the tangent bundle of a submanifold $G \subset M$ at point A coincide, then there exits a local diffeomorphism of M fixing G point-wise and sending one form to the other. We may assume that the forms coincide on $T_A M$.

We again use the homotopy method, aiming to find a family of diffeomorphism-germs g_t, $t \in [0,1]$, such that

$$g_t|_G = id_G\,, \quad g_0 = id_M\,, \quad g_t^*(\omega_t) = \omega_0 \quad (*) \qquad \text{where} \quad \omega_t = \omega_0 + (\omega_1 - \omega_0)t\,.$$

Differentiating $(*)$ by t, we again get

$$\mathcal{L}_{v_t}(\omega_t) = d(i_{v_t}\omega_t) = \omega_0 - \omega_1$$

where v_t is the vector field of the flow g_t. Using the "relative Poincare lemma", it is possible to find a 1-form α so that $d\alpha = \omega_0 - \omega_1$ and α vanishes on G. Then the required vector field v_t exists since ω_t is non-degenerate. $\qquad\qquad\square$

Darboux theorem is a particular case of Givental's theorem: take a point as a submanifold.

If at each point x of a <u>submanifold</u> L of a symplectic manifold M the subspace $T_x L$ is Lagrangian in the symplectic space $T_x M$, then L is called <u>Lagrangian</u>.

Examples.
1. In T^*N, the following are Lagrangian submanifolds: the zero section of the bundle, fibres of the bundle, graph of the differential of a function on N.

2. The graph of a symplectomorphism is a Lagrangian submanifold of the product space (it has regular projections onto the factors). An arbitrary Lagrangian submanifold of the product space defines a so-called Lagrangian relation which, in a sense, is a multivalued generalization of a symplectomorphism.

Weinstein's theorem implies that a tubular neighbourhood of a Lagrangian submanifold L in any symplectic space is symplectomorphic to a tubular neighbourhood of the zero section in T^*N.

A <u>fibration</u> with Lagrangian fibres is called <u>Lagrangian</u>.

Locally all Lagrangian fibrations are symplectomorphic (the proof is similar to that of the Darboux theorem).

A cotangent bundle is a Lagrangian fibration.

Let $\psi : L \to T^*N$ be a Lagrangian embedding and $\rho : T^*N \to N$ the fibration. The product $\rho \circ \psi : L \to N$ is called a <u>Lagrangian mapping</u>. It critical values

$$\Sigma_L = \{q \in N | \exists p : (p,q) \in L, \ \operatorname{rank} d(\rho \circ \psi)|_{(p,q)} < n\}$$

form the <u>caustic</u> of the Lagrangian mapping. The equivalence of Lagrangian mappings is that up to fibre-preserving symplectomorphisms of the ambient symplectic space. Caustics of equivalent Lagrangian mappings are diffeomorphic.

Hamiltonian vector fields.

Given a real function $h : M \to \mathbf{R}$ on a symplectic manifold, define a <u>Hamiltonian vector field</u> v_h on M by the formula

$$\omega(\cdot, v_h) = dh.$$

This field is tangent to the level hypersurfaces $H_c = h^{-1}(c)$:

$$\forall a \in H_c \quad dh(T_a H_c) = 0 \qquad \Longrightarrow \qquad T_a H_c = v_h^{\measuredangle}, \quad \text{but} \quad v_h \in v_h^{\measuredangle}.$$

The directions of v_h on the level hypersurfaces H_c of h are the <u>characteristic directions</u> of the tangent spaces of the hypersurfaces.

Associating v_h to h, we obtain a Lie algebra structure on the space of functions:

$$[v_h, v_f] = v_{\{h,f\}} \qquad \text{where} \quad \{h, f\} = v_h(f),$$

the latter being the Poisson bracket of the Hamiltonians h and f.

A Hamiltonian flow (even if h depends on time) consists of symplectomorphisms. Locally (or in $\mathbf{R}^{2n}$), any time-dependent family of symplectomorphisms that starts from the identity is a phase flow of a time-dependent Hamiltonian. However, for example, on a torus $\mathbf{R}^2/\mathbf{Z}^2$ (the quotient of the plane by an integer lattice) the family of constant velocity displacements are symplectomorphisms but they cannot be Hamiltonian since a Hamiltonian function on a torus must have critical points.

Given a time-dependent Hamiltonian $\widetilde{h} = \widetilde{h}(t, p, q)$, consider the extended space $M \times T^*\mathbf{R}$ with auxiliary co-ordinates (s, t) and the form $pdq - sdt$. An auxiliary (extended) Hamiltonian $\widehat{h} = -s + \widetilde{h}$ determines a flow in the extended space generated by the vector field

$$\dot{p} = -\frac{\partial \widehat{h}}{\partial q}, \quad \dot{q} = \frac{\partial \widehat{h}}{\partial p}, \quad \dot{t} = -\frac{\partial \widehat{h}}{\partial s} = 1, \quad \dot{s} = \frac{\partial \widehat{h}}{\partial t}.$$

The restrictions of this flow to the $t = const$ sections are essentially the flow mappings of $\widetilde{h}$.

The integral of the extended form over a closed chain in $M \times \{t_o\}$ is preserved by the $\widehat{h}$-Hamiltonian flow. Hypersurfaces $-s + \widetilde{h} = const$ are invariant. When $\widetilde{h}$ is autonomous, the form pdq is also a relative integral invariant.

A (transversal) intersection of a Lagrangian submanifold $L \subset M$ with a Hamiltonian level set $H_c = h^{-1}(c)$ is an isotropic submanifold L_c. All Hamiltonian trajectories emanating from L_c form a Lagrangian submanifold $exp_H(L_c) \subset M$. The space Ξ_{H_c} of the Hamiltonian trajectories on H_c inherits, at least locally, an induced symplectic structure. The image of the projection of $exp_H(L_c)$ to Ξ_{H_c} is a Lagrangian submanifold there. This is a particular case of a symplectic reduction which will be discussed later.

Example. The set of all oriented straight lines in $\mathbf{R}_q^n$ is T^*S^{n-1} as a space of characteristics of the Hamiltonian $h = p^2$ on its level $p^2 = 1$ in $K = \mathbf{R}^{2n}$.

1.2. *Contact geometry*

An odd-dimensional manifold M^{2n+1} equipped with a maximally non-integrable distribution of hyperplanes (contact elements) in the tangent spaces of its points is called a <u>contact manifold</u>.

The maximal non-integrability means that if locally the distribution is determined by zeros of a 1-form α on M then $\alpha \wedge (d\alpha)^n \neq 0$ (cf. the Frobenius condition $\alpha \wedge d\alpha = 0$ of complete integrability).

Examples.

1. A projectivised cotangent bundle PT^*N^{n+1} with the projectivisation of the Liouville form $\alpha = pdq$ is a contact manifold. This is also called the space of contact elements on N. The spherisation of PT^*N^{n+1} is a 2-fold covering of PT^*N^{n+1} and its points are co-oriented contact elements.

2. The space J^1N of 1-jets of functions on N^n is another standard model of contact space. (Two functions have the same m-jet at a point x if their Taylor polynomials of degree k at x coincide). The space of all 1-jets at all points of N has local coordinates $q \in N$, $p = df(q)$ which are the partial derivatives of a function at q, and $z = f(q)$. The contact form is $pdq - dz$.

<u>Contactomorphisms</u> are diffeomorphisms preserving the distribution of contact elements.

Contact Darboux theorem. *All equidimensional contact manifolds are locally contactomorphic.*

An analog of Givental's theorem also holds.

Symplectisation.
Let $\widetilde{M}^{2n+2}$ be the space of all linear forms vanishing on contact elements of M. The space $\widetilde{M}^{2n+2}$ is a "line" bundle over M (fibres do not contain the zero forms). Let $\widetilde{\pi} : \widetilde{M} \to M$ be the projection. On $\widetilde{M}$, the symplectic structure (which is homogeneous of degree 1 with respect to fibres) is the differential of the canonical 1-form $\widetilde{\alpha}$ on $\widetilde{M}$ defined as

$$\widetilde{\alpha}(\xi) = p(\widetilde{\pi}_*\xi), \qquad \xi \in T_p\widetilde{M}.$$

A contactomorphism F of M lifts to a symplectomorphism of $\widetilde{M}$:

$$\widetilde{F}(p) := (F^*_{F(x)})^{-1}p.$$

This commutes with the multiplication by constants in the fibres and preserves $\widetilde{\alpha}$. The symplectisation of contact vector fields (= infinitesimal contactomorphisms) yields Hamiltonian vector fields with homogeneous (of degree 1) Hamiltonian functions $h(rx) = rh(x)$.

Assume the contact structure on M is defined by zeros of a fixed 1-form β. Then M has a natural embedding $x \mapsto \beta_x$ into $\widetilde{M}$.

Using the local model $J^1\mathbf{R}^n$, $\beta = pdq - dz$, of a contact space we get the following formulas for components of the contact vector field with a homogeneous Hamiltonian function $K(x) = h(\beta_x)$ (notice that $K = \beta(X)$ where X is the corresponding contact vector field):

$$\dot{z} = pK_p - K, \quad \dot{p} = -K_q - pK_z, \quad \dot{q} = K_p.$$

where the subscripts mean the partial derivations.

Various homogeneous analogs of symplectic properties hold in contact geometry (the analogy is similar to that between affine and projective geometries).

In particular, a hypersurface (transversal to the contact distribution) in a contact space inherits a field of characteristics.

Contactisation.
To an exact symplectic space M^{2n} associate $\widehat{M} = \mathbf{R} \times M$ with an extra co-ordinate z and take the 1-form $\alpha = \lambda - dz$. This gives a contact space.

Here the vector field $\chi = -\frac{\partial}{\partial z}$ satisfies $i_\chi \alpha = 1$ and $i_\chi d\alpha = 0$. Such a field is called a <u>Reeb</u> vector field. Its direction is uniquely defined by a contact structure. It is transversal to the contact distribution. Locally, projection along χ produces a symplectic manifold.

A <u>Legendrian submanifold</u> $\widehat{L}$ of M^{2n+1} is an n-dimensional integral submanifold of the contact distribution. This dimension is maximal possible for integral submanifolds due to maximal non-integrability of the contact distribution.

Examples.
1. To a Lagrangian $L \subset T^*M$ associate $\widehat{L} \subset J^1 M$:

$$\widehat{L} = \left\{ (z, p, q) \mid z = \int pdq, \ (p, q) \in L \right\}.$$

Here the integral is taken along a path on L joining a distinguished point on L with the point (p, q). Such an $\widehat{L}$ is Legendrian.

2. The set of all covectors annihilating tangent spaces to a given submanifold (or variety) $W_0 \subset N$ form a Legendrian submanifold (variety) in PT^*N.

3. If the intersection I of a Legendrian submanifold $\widehat{L}$ with a hypersurface Γ in a contact space is transversal, then I is transversal to the characteristic vector field on Γ. The set of characteristics emanating from I form a Legendrian submanifold.

A <u>Legendrian fibration</u> of a contact space is a fibration with Legendrian fibres. For example, $PT^*N \to N$ and $J^1N \to J^0N$ are Legendrian. Any two Legendrian fibrations of the same dimension are locally contactomorphic.

The projection of an embedded Legendrian submanifold $\widehat{L}$ to the base of a Legendrian fibration is called a <u>Legendrian mapping</u>. Its image is called the <u>wave front</u> of $\widehat{L}$.

Examples.

1. Embed a Legendrian submanifold $\widehat{L}$ into J^1N. Its projection $W(\widehat{L})$ to J^0N, which is the wave front, is a graph of a multivalued action function $\int p\,dq + c$ (again we integrate along paths on the Lagrangian submanifold $L = \pi_1(\widehat{L})$, where $\pi_1 : J^1N \to T^*N$ is the projection dropping the z coordinate). If $q \in N$ is not in the caustic Σ_L of L, then over q the wave front $W(\widehat{L})$ is a collection of smooth sheets.

If at two distinct points $(p', q), (p'', q) \in L$ with a non-caustical value q, the values z of the action function are equal, then at (z, q) the wave front is a transversal intersection of graphs of two regular functions on N.

The images under the projection $(z, q) \mapsto q$ of the singular and transversal self-intersection loci of $W(\widehat{L})$ are respectively the caustic Σ_L and so-called <u>Maxwell</u> (conflict) <u>set</u>.

2. To a function $f = f(q), q \in \mathbf{R}^n$, associate its Legendrian lifting $\widehat{L} = j^1(f)$ (also called the 1-jet extension of f) to $J^1\mathbf{R}^n$. Project $\widehat{L}$ along the fibres parallel to the q-space of another Legendrian fibration

$$\pi_1^{\wedge}(z, p, q) \mapsto (z - pq, p)$$

of the same contact structure $p\,dq - dz = -q\,dp - d(z - pq)$. The image $\pi_1^{\wedge}(\widehat{L})$ is called the <u>Legendre transform</u> of the function f. It has singularities if f is not convex.

This is an affine version of the projective duality (which is also related to Legendrian mappings). The space PT^*P^n (P^n is the projective space) is isomorphic to the projectivised cotangent bundle $PT^*P^{n\wedge}$ of the dual space $P^{n\wedge}$. Elements of both are pairs consisting of a point and a hyperplane, containing the point. The natural contact structures coincide. The set of all hyperplanes in P^n tangent to a submanifold $S \subset P^n$ is the front of the dual projection of the Legendrian lifting of S.

Wave front propagation.

Fix a submanifold $W_0 \subset N$. It defines the (homogeneous) Lagrangian submanifold $L_0 \subset T^*N$ formed by all covectors annihilating tangent spaces to W_0.

Consider now a Hamiltonian function $h : T^*N \to \mathbf{R}$. Let I be the intersection of L_0 with a fixed level hypersurface $H = h^{-1}(c)$. Consider the Lagrangian submanifold $L = exp_H(I) \subset H$ which consists of all the characteristics emanating from I. It is invariant under the flow of H.

The intersections of the Legendrian lifting $\widehat{L}$ of L into J^1N ($z = \int p\,dq$) with co-ordinate hypersurfaces $z = const$ project to Legendrian submanifolds (varieties) $\widehat{L}_z \subset PT^*N$. In fact, the form $p\,dq$ vanishes on each tangent vector to $\widehat{L}_z$. In general, the dimension of $\widehat{L}_z$ is $n-1$.

The wave front of $\widehat{L}$ in J^0N is called the <u>big wave front</u>. It is swept out by the family of fronts W_z of the $\widehat{L}_z$ shifted to the corresponding levels of the z-co-ordinate. Notice that, up to a constant, the value of z at a point over a point (p,q) is equal to $z = \int p\frac{\partial h}{\partial p}dt$ along a segment of the Hamiltonian trajectory going from the initial I to (p,q).

When h is homogeneous of degree k with respect to p in each fibre, then $z_t = kct$. Let $I_t \subset L$ be the image of I under the flow transformation g_t for time t. The projectivised I_t are Legendrian in PT^*N. The family of their fronts in N is $\{W_{kct}\}$. So the W_t are momentary wave fronts propagating from the initial W_0. Their singular loci sweep out the caustic Σ_L.

The case of a time-depending Hamiltonian $h = h(t,p,q)$ reduces to the above by considering the extended phase space $J^1(N \times \mathbf{R})$, $\alpha = p\,dq - r\,dt - dz$. The image of the initial Legendrian subvariety $\widehat{L}_0 \subset J^1(N \times \{0\})$ under g_t is a Legendrian $L_t \subset J^1(N \times \{t\})$.

When z can be written locally as a regular function in q, t it satisfies the Hamilton-Jacobi equation $-\frac{\partial z}{\partial t} + h(t, \frac{\partial z}{\partial q}, q) = 0$.

2. Generating families

2.1. *Lagrangian case*

Consider a co-isotropic submanifold $C^{n+k} \subset M^{2n}$. The skew-orthogonal complements $T_c^{\angle}C$, $c \in C$, of tangent spaces to C define an integrable distribution on C. Indeed, take two regular functions whose common zero level set contains C. At each point $c \in C$, the vectors of their Hamiltonian fields belong to $T_c^{\angle}C$. So the corresponding flows commute. Trajectories of all such fields emanating from $c \in C$ form a smooth submanifold I_c integral for the distribution.

By Givental's theorem, any co-isotropic submanifold is locally symplectomorphic to a co-ordinate subspace $p_I = 0$, $I = \{1, \ldots, n - k\}$, in $K = \mathbf{R}^{2n}$. The fibres are the sets $q_J = const$.

Proposition. Let L^n and C^{n+k} be respectively Lagrangian and co-isotropic submanifolds of a symplectic manifold M^{2n}. Assume L meets C transversally at a point a. Then the intersection $X_0 = L \bigcap C$ is transversal to the isotropic fibres I_c near a.

The proof is immediate. If $T_a X_0$ contains a vector $v \in T_a I_c$, then v is skew-orthogonal to $T_a L$ and also to $T_a C$, that is to any vector in $T_a M$. Hence $v = 0$.

Isotropic fibres define the fibration $\xi : C \to B$ over a certain manifold B of dimension $2k$ (defined at least locally). We can say that B is the manifold of isotropic <u>fibres</u>.

It has a well-defined induced symplectic structure ω_B. Given any two vectors u, v tangent to B at a point b take their liftings, that is vectors $\widetilde{u}, \widetilde{v}$ tangent to C at some point of $\xi^{-1}(b)$ such that their projections to B are u and v. The value $\omega(\widetilde{u}, \widetilde{v})$ depends only on the vectors u, v. For any other choice of liftings the result will be the same. This value is taken for the value of the two-form ω_B on B.

Thus, the base B gets a symplectic structure which is called a <u>symplectic reduction</u> of the co-isotropic submanifold C.

Example. Consider a Lagrangian section L of the (trivial) Lagrangian fibration $T^*(\mathbf{R}^k \times \mathbf{R}^n)$. The submanifold L is the graph of the differential of a function $f = f(x, q)$, $x \in \mathbf{R}^k$, $q \in \mathbf{R}^n$. The dual coordinates y, p are given on L by $y = \frac{\partial f}{\partial x}$, $p = \frac{\partial f}{\partial q}$. Therefore, the intersection $\widetilde{L}$ of L with the co-isotropic subspace $y = 0$ is given by the equations $\frac{\partial f}{\partial x} = 0$. The intersection is transversal iff the rank of the matrix of the derivatives of these equations, with respect to x and q, is k. If so, the symplectic reduction of $\widetilde{L}$ is a Lagrangian submanifold L_r in $T^* \mathbf{R}^n$ (it may not be a section of $T^* \mathbf{R}^n \to \mathbf{R}^n$).

This example leads to the following definition of a generating function (the idea is due to Hörmander).

Definition. A <u>generating family</u> of the Lagrangian mapping of a submanifold $L \subset T^* N$ is a function $F : E \to \mathbf{R}$ defined on a vector bundle E over N such that

$$L = \left\{ \ (p, q) \ \mid \ \exists x : \frac{\partial F(x, q)}{\partial x} = 0, \quad p = \frac{\partial F(x, q)}{\partial q} \ \right\}.$$

Here $q \in N$, and x is in the fibre over q. We also assume that the following Morse condition is satisfied:

$$0 \text{ is a regular value of the mapping } \quad (x, q) \mapsto \frac{\partial F}{\partial x}.$$

The latter guarantees L being a smooth manifold.

Remark. The points of the intersection of L with the zero section of $T^* N$ are in one-to-one correspondence with the critical points of the function F. In symplectic topology, when interested in such points, it is desirable to avoid a possibility of having no critical points at all (as it may happen on a non-compact manifold E).

Therefore, dealing with global generating families defining Lagrangian submanifolds globally, generating families with good behaviour at infinity should be considered.

A generating family F is said to be <u>quadratic at infinity</u> (QI) if it coincides with a fibre-wise quadratic non-degenerate form $Q(x, q)$ outside a compact.

On the topological properties of such families and on their rôle in symplectic topology see the papers by C.Viterbo, for example [4].

Existence and uniqueness (up to a certain equivalence relation) of QI generating families for Lagrangian submanifolds which are Hamiltonian isotopic to the zero section in $T^* N$ of a compact N was proved by Viterbo, Laundeback and Sikorav in the 80s:

Given any two QI generating families for L, there is a unique integer m and a real ℓ such that $H^k(F_b, F_a) = H^{k-m}(F_{b-\ell}, F_{a-\ell})$ for any pair of

$a < b$. Here F_a is the inverse image under F of the ray $\{t \le a\}$.

However, we shall need a local result which is older and easier.

Existence.

Any germ L of a Lagrangian submanifold in $T^*\mathbf{R}^n$ has a regular projection to some (p_J, q_I) co-ordinate space. In this case there exists a function $f = f(p_J, q_I)$ (defined up to a constant) such that

$$L = \left\{\ (p,q)\ \ \Big|\ \ q_J = -\frac{\partial f}{\partial p_J}, \quad p_I = \frac{\partial f}{\partial q_I}\ \right\}.$$

Then the family $F_J = xq_J + f(x, q_I)$, $x \in \mathbf{R}^{|J|}$, is generating for L. If $|J|$ is minimal possible, then $\mathrm{Hess}_{xx} F_J = \mathrm{Hess}_{p_J p_J} f$ vanishes at the distinguished point.

Uniqueness.

Two family-germs $F_i(x, q)$, $x \in \mathbf{R}^k$, $q \in \mathbf{R}^n$, $i = 1, 2$, at the origin are called <u>R-equivalent</u> if there exists a diffeomorphism $\mathcal{T} : (x, q) \mapsto (X(x, q), q)$ (i.e. preserving the fibration $\mathbf{R}^k \times \mathbf{R}^n \to \mathbf{R}^n$) such that $F_2 = F_1 \circ \mathcal{T}$.

The family $\Phi(x, y, q) = F(x, q) \pm y_1^2 \pm \ldots, \pm y_m^2$ is called a <u>stabilisation</u> of F.

Two family-germs are called <u>stably R-equivalent</u> if they are R-equivalent to appropriate stabilisations of the same family (in a lower number of variables).

Lemma. *Up to addition of a constant, any two generating families of the same germ L of a Lagrangian submanifold are stably $\mathcal{R}$-equivalent.*

Proof. Morse Lemma with parameters implies that any function-germ $F(x, q)$ (with zero value at the origin which is taken as the distinguished point) is stably $\mathcal{R}_0$-equivalent to $\widetilde{F}(y, q) \pm z^2$ where $x = (y, z)$ and the matrix $\mathrm{Hess}_{yy}\widetilde{F}|_0$ vanishes. Clearly $\widetilde{F}(y, q)$ is a generating family for L if we assume that $F(x, q)$ is.

Since the matrix $\partial^2 \widetilde{F}/\partial y^2$ vanishes at the origin, the Morse condition for $\widetilde{F}$ implies that there exists a subset J of indices such that the minor

$\partial^2 \widetilde{F}/\partial y \partial q_J$ is not zero at the origin. Hence the mapping

$$\Theta : (y, q) \mapsto (p_J, q) = (\partial \widetilde{F}/\partial q_J, q)$$

is a local diffeomorphism. The family $G = \widetilde{F} \circ \Theta^{-1}$, $G = G(p_J, q)$, is also a generating family for L.

The variety $\partial \widetilde{F}/\partial y = 0$ in the domain of Θ is mapped to the Lagrangian submanifold L in the (p, q)-space by setting $p = \partial \widetilde{F}/\partial q$ and forgetting y. Therefore, the variety $X = \{\partial G/\partial p_J = 0\}$ in the (p_J, q)-space is the image of L under its (regular) projection $(p, q) \mapsto (p_J, q)$.

Compare now G and the standard generating family F_J defined above (with p_J in the role of x). We may assume their values at the origin coinciding. Then the difference $G - F_J$ has vanishing 1-jet along X. Since X is a regular submanifold, $G - F_J$ is in the square of the ideal $\mathcal{I}$ generated by the equations of X, that is by $\partial F_J/\partial p_J$.

The homotopy method applied to the family $A_t = F_J + t(G - F_J)$, $0 \leq t \leq 1$, shows that G and F_J are $\mathcal{R}_0$-equivalent. Indeed, it is clear that the homological equation

$$-\frac{\partial A_t}{\partial t} = F_J - G = \frac{\partial A_t}{\partial p_J} \dot{p}_J$$

has a smooth solution $\dot{p}_J$ since $F_J - G \in \mathcal{I}^2$ while the $\partial A_t/\partial p_J$ generate $\mathcal{I}$ for any fixed t. $\qquad\square$

2.2. Legendrian case

Definition. A <u>generating family</u> of the Legendrian mapping $\pi|_L$ of a Legendrian submanifold $L \subset PT^*(N)$ is a function $F : E \to \mathbf{R}$ defined on a vector bundle E over N such that

$$L = \left\{ \ (p, q) \ \mid \ \exists x : \ F(x, q) = 0, \ \ \frac{\partial F(x, q)}{\partial x} = 0, \ \ p = \frac{\partial F(x, q)}{\partial q} \ \right\},$$

where $q \in N$ and x is in the fibre over q, provided that the following Morse condition is satisfied:

$$0 \text{ is a regular value of the mapping } (x, q) \mapsto \{F, \frac{\partial F}{\partial x}\}.$$

Definition. Two function family-germs $F_i(x, q)$, $i = 1, 2$, are called <u>V-equivalent</u> if there exists a fibre-preserving diffeomorphism $\Theta : (x, q) \mapsto (X(x, q), q)$ and a function $\Psi(x, q)$ not vanishing at the distinguished point such that $F_2 \circ \Theta = \Psi F_1$.

Two function families are called stably V-equivalent if they are stabilisations of a pair of V-equivalent functions (may be in a lower number of variables x).

Theorem. *Any germ $\pi|_L$ of a Legendrian mapping has a generating family. All generating families of a fixed germ are stably V-equivalent.*

Proof. For an n-dimensional N, we use the local model $\pi_0 : J^1 N' \to J^0 N'$, $N' = \mathbf{R}^{n-1}$, for the Legendrian fibration.

Consider the projection $\pi_1 : J^1 N' \to T^* N'$ restricted to L. Its image is a Lagrangian germ $L_0 \subset T^* N$. If $F(x,q)$ is a generating family for L_0, then $F(x,q) - z$ considered as a family of functions in x with parameters $(q,z) \in J^0 N' = N$ is a generating family for L and vice versa. Now the theorem follows from the Lagrangian result and an obvious property: multiplication of a Legendrian generating family by a function-germ not vanishing at the distinguished point gives a generating family. After multiplication by an appropriate function Ψ, a generating family (satisfying the regularity condition) takes the form $F(x,q) - z$ where (q,z) are local coordinates in N. $\qquad\square$

Remarks.

A symplectomorphism φ preserving the bundle structure of the standard Lagrangian fibration $\pi : T^* \mathbf{R}^n \to \mathbf{R}^n$, $(q,p) \mapsto q$ has a very simple form

$$\varphi : (q,p) \mapsto \left(Q(q), DQ^{-1*}(q)(p + df(q)) \right) ,$$

where $DQ^{-1*}(q)$ is the dual of the derivative of the inverse mapping of the base of the fibration, $Q \circ \pi = \pi \circ \varphi$, and f is a function on the base.

To see this, it is sufficient to write in the coordinates the equation $\varphi_* \lambda - \lambda = df$.

The above formula shows that fibres of any Lagrangian fibration posses a well-defined affine structure.

Consequently, a contactomorphism ψ of the standard Legendrian fibration $PT^* \mathbf{R}^n \to \mathbf{R}^n$ acts by projective transformations in the fibres:

$$\psi : (q,p) \mapsto \left(Q(q), DQ^{-1*}(q)p \right) .$$

Hence, there is a well-defined projective structure on the fibres of any Legendrian fibration.

We also see that Lagrangian equivalences act on generating families as

R-equivalences $(x, q) \mapsto (X(x, q), Q(q))$ and additions of function in parameters q.

Legendrian equivalences act on Legendrian generating families just as R-equivalences.

2.3. *Examples of generating families*

The importance of the constructions introduced above for various applications is illustrated by the following examples.

1. Consider a Hamiltonian $h : T^*\mathbf{R}^n \to \mathbf{R}$ which is homogeneous of degree k with respect to the impulses p: $h(\tau p, q) = \tau^k h(p, q)$, $\tau \in \mathbf{R}$.

An initial submanifold $W_0 \subset \mathbf{R}^n$ (initial wave front) defines an exact isotropic $I \subset H_c = h^{-1}(c)$. Assume I is a manifold transversal to v_h. Put $c = 1$.

The exact Lagrangian flow-invariant submanifold $L = exp_h(I)$ is a cylinder over I with local coordinates $\alpha \in I$ and time t from a real segment (on which the flow is defined).

Assume that in a domain $U \subset T^*R^n \times \mathbf{R}$ the restriction to L of the phase flow g_t of v_h is given by the mapping $(\alpha, t) \mapsto (Q(\alpha, t), P(\alpha, t))$ with $\frac{\partial P}{\partial \alpha, t} \neq 0$. Then the following holds.

Proposition. a) The family $F = P(\alpha, t)(q - Q(\alpha, t)) + kt$ of functions in α, t with parameters $q \in \mathbf{R}^n$ is a generating family of L in the domain U.

b) For any fixed t, the family $\widetilde{F}_t = P(\alpha, t)(q - Q(\alpha, t))$ is a Legendrian generating family of the momentary wave front W_t.

The proof is an immediate verification of the Hörmander definition using the fact that value of the form pdq on each vector tangent to $g_t(I)$ vanishes and on the vector v_h it is equal to $p\frac{\partial h}{\partial p} = kh = k$.

2. Let $\varphi : T^*\mathbf{R}^n \to T^*\mathbf{R}^n$, $(q, p) \mapsto (Q, P)$ be a symplectomorphism close to the identity. Thus the system of equations $q' = Q(q, p)$ is solvable for q. Write its solution as $q = \widetilde{q}(q', p)$.

Assume the Lagrangian mapping of a Lagrangian submanifold L has a generating family $F(x, q)$. Then the following family G of functions in x, q, p with parameters q' is a generating family of $\varphi(L)$:

$$G(x, p, q; q') = F(x, \widetilde{q}) + p(\widetilde{q} - q) + S(p, q').$$

Here $S(q', p)$ is the "generating function" in the sense of Hamiltonian mechanics of the canonical transformation φ, that is

$$dS = PdQ - pdq .$$

Notice that, if φ coincides with the identity mapping outside a compact, then G is a quadratic form at infinity with respect to the variables (q, p).

The expression $p(\widetilde{q} - q) + S(p, q')$ from the formula above is the generating family of the symplectomorphism φ.

3. Represent a symplectomorphism φ of $T^*\mathbf{R}^n$ into itself homotopic to the identity as a product of a sequence of symplectomorphisms each of which is close to the identity. Iterating the previous construction, we obtain a generating family of $\varphi(L)$ as a sum of the initial generating family with the generating families of each of these transformations. The number of the variables becomes very large, $\dim(x) + 2mn$, where m is the number of the iterations. Namely, consider a partition of the time interval $[0, T]$ into m small segments $[t_i, t_{i+1}]$, $i = 0, \ldots, m-1$. Let $\varphi = \varphi_m \circ \varphi_{m-1} \circ \ldots \varphi_1$ where $\varphi_i : (Q_i, P_i) \mapsto (Q_{i+1}, P_{i+1})$ is the flow map on the interval $[t_i, t_{i+1}]$. Then the generating family is

$$G(x, Q, P, q) = F(x, Q_0) + \sum_{i=0}^{m-1} \left(P_i(U_i(Q_{i+1}, P_i) - Q_i) + S_i(P_i, Q_{i+1}) \right) ,$$

where: $Q = Q_0, \ldots, Q_{m-1}, q = Q_m, Q_i \in \mathbf{R}^n, \quad q \in \mathbf{R}^n, S_i$ is a generating function of φ_i, and $U_i(Q_{i+1}, P_i)$ are the solutions of the system of equations $Q_{i+1} = Q_{i+1}(Q_i, P_1)$ defined by φ_i.

One can show that if φ is a flow map for time $t = 1$ of a Hamiltonian function which is convex with respect to the impulses then the generating family G is also convex with respect to the P_i and these variables can be removed by the stabilisation procedure. This provides a generating family of $\varphi(L)$ depending just on x, Q, q for the image of Lagrangian submanifold L admitting a generating family itself. Usually these variables x, Q, q are taken from a compact domain. In this case, the generating family has nice properties. For example, the family attains minimal and maximal values on the fibre over point q. This means, that for any point q from the image of the projection of $\varphi(L)$ among all projections of the Hamilton vector field trajectories emanating at $t = 0$ from L and coming to $\pi^{-1}(q)$ at time $t = 1$ there are some which provide global minimal value of the action function.

In particular, this implies that going from the initial point along a generic geodesic the first point where the geodesic segment fails to be minimal is a a conflict point. At generic (regular) point of the caustic the generating family does not have minimum value.

3. Applications

3.1. *Singularities of wave fronts and caustics*

Famous results of Arnold and Thom relating stable singularities of low-dimensional wave fronts to the discriminants of the Weyl groups are based on relation between caustics and wave fronts and discriminants and bifurcation diagrams of families of functions depending on parameters.

Caustics

Singularities of Lagrangian projections are essentially the singularities of their generating families treated as families of functions depending on parameters. In particular, the caustic $\Sigma(L)$ of Lagrangian submanifold L projection coincides with the stratum of the bifurcation diagram of the generating family $f(x, q)$ which is the collection of parameter q values such that the restriction $f(\cdot, q)$ has a non-Morse critical point.

Stability of Lagrangian projection with respect to symplectomorphisms preserving the fibration structure corresponds to the versality of the generating family with respect to the R_+- equivalence group (diffeomorphisms of the source space and additions of the function with constants).

In the space of germs of functions in k variables there are only finitely many orbits of codimension $k + n$ with $n \leq 5$ of the R_+-equivalence group [[3]]. Those are the orbits of simple A, D, E singularity classes. This fact implies the following

Theorem.*Let $E(L)$ be the space of Lagrangian embeddings of a compact manifold L (of the dimension $n \leq 5$) into a Lagrangian fibration space, equipped with C^∞ topology. Then a Lagrangian projection $\pi \circ i$ for an embedding i from an open and dense subset of $E(L)$ at any point is equivalent to a Lagrangian projection determined by the germ at the origin of some of the following standard versal deformations of simple singularities of functions*

178

with $m \leq n+1$:

$$A_m : \quad F = \pm x^{m+1} + q_1 x^{m-1} + \cdots + q_{m-1}x;$$
$$D_m : \quad F = x_1^2 x_2 \pm x_2^{m-1} + q_1 x_2^{m-2} + \cdots + q_{m-2}x_2 + q_{m-1}x_1;$$
$$E_6 : \quad F = x_1^3 \pm x_2^4 + q_1 x_1 x_2^2 + q_2 x_1 x_2 + q_3 x_2^2 + q_4 x_1 + q_5 x_2.$$

Remarks.

1. In particular, the caustics of generic local Lagrangian projections to 3-space are diffeomorphic of the caustics of A_2(smooth surface), A_3 (cuspidal ridge), A_4 (swallowtail), $D_4^{\pm}$ (purse or pyramid). Germs of generic caustics can have several several components of these types which are mutually transversal.

2. Starting from $n = 6$ some R_+ orbits have continuous invariants (moduli). Therefore, respective Lagrangian projections have invariants which are functional moduli (invariants depending on parameters. However even in this cases generating families provide some useful information of topological structure of caustics.

3. For the dimensions $n \geq 3$ the list of generic singularities of Lagrangian projections differs from the list of singularities of arbitrary mappings of spaces of equal dimensions. Lagrangian mapping are special. However, they arise in many physical and geometrical problems. For example, Gauss map is Lagrangian. Envelope of geodesics emanating from an initial point on a Riemannian manifold is the caustic of so-called exponential Lagrangian mapping. The intensity of light at caustic points of a family of optical rays increases. The asymptotics of the intensity given by the oscillation integral was studied by A.Varchenko, P.Pham and others. It is related to the spectrum and mixed Hodge structure of respective function singularity [[1]].

4. Some specific applied problems involve non-generic Lagrangian singularities. They can be symmetric, or even determined by a non-smooth Lagrangian varieties projections. The study of the correponding generating families require special singularity theory techniques (equivariant mappings, non-isolated singularities, etc.). A recent example of a caustic related to a generating family with non-isolated singularities is given by the exponential mapping on subriemannian 3-space with contact distribution [[5]].

Wave fronts

The wave front of a local Legendrian projection is the discriminant $D(F)$ of its generating family $F(x, q)$ that is the set of parameters q such that the zero level set of $F(cdot, q)$ contains a critical point.

Legendre equivalent Legendre projections have V-equivalent generating families and diffeomorphic wave fronts. Under some mild conditions, the converse holds also [[6]].

A Legendre submanifold germ L embedded into a Legendre fibration is called regular if the regular points of the Legendre projection are dense in L and the projection is proper.

Proposition. *If the front of a Legendre submanifold germ $\widetilde{L}$ coincides with the front of the regular germ L, then $\widetilde{L}$ coincides with L.*

The proof follows from the fact that near the regular point (p, q) the Legendre submanifold $L \subset PT^*M$ coincides with the set of contact elements p annihilating tangent vectors to wavefront $\pi(L)$. The entire Legendre submanifold is the closure of its regular points.

So, regular Legendrian submanifolds having diffeomorphic wavefronts are Legendre equivalent and the respective generating families are V-equivalent. Hence the classification of the generic singularities of Legendre projections is essentially the classification (up to diffeomorphisms) of generic singularities of wave fronts. Notice that in the Lagrangian case the similar statement is false.

Theorem. *Let $\widehat{E}(L)$ be the space of Lagendrian embeddings of a compact manifold L (of the dimension $n \leq 5$) into a Lagrangian fibration space, equipped with C^∞ topology. Then a Legendrian projection $\pi \circ i$ for an embedding i from an open and dense subset of $\widehat{E}(L)$ at any point is equivalent to a Legendrian projection determined by the germ at the origin of some of the following standard V-versal deformations of simple singularities of functions with $m \leq n + 1$:*

$$A_m : \quad F = \pm x^{m+1} + q_1 x^{m-1} + \cdots + q_{m-1}x + q_m;$$
$$D_m : \quad F = x_1^2 x_2 \pm x_2^{m-1} + q_1 x_2^{m-2} + \cdots + q_{m-2}x_2 + q_{m-1}x_1 + q_m;$$
$$E_6 : \quad F = x_1^3 \pm x_2^4 + q_1 x_1 x_2^2 + q_2 x_1 x_2 + q_3 x_2^2 + q_4 x_1 + q_5 x_2 + q_m.$$

Generic wave fronts germs in 3-space are diffeomorphic either to swallowtails (A_4) or to collection of mutually transversal smooth surfaces (A_2) and cuspidal ridges (A_3).

3.2. *Metamorphosis of wave front*

Consider a germ at the origin of the R-versal deformation depending on parameters q

$$F(x, q) = f(x) + \sum_{i=1}^{n} q_i \varphi_i(x) \tag{1}$$

of the polynomial $f(x)$ having at the origin a critical point of multiplicity $\mu \leq n$.

Assume the germs φ_i, $i = 1, \ldots, \mu$ form a basis of local gradient factor algebra

$$Q_f = C^\infty(x)/C^\infty(x)\left\{\frac{\partial f}{\partial x}\right\},$$

of the algebra $C^\infty(x)$ of germs at the origin of smooth functions in x. Assume that $\varphi_{]m+1}, \ldots, \varphi_n$ are equal to zero.

Proposition.(see[[6]]) *The real-analytic vector fields which are tangent to the wave front of the Legendrian projection germ determined by the generating family F form a free module over the ring of germs of functions in q with n generators.*

In other words the wave fronts of R-versal families of functions are Saito's free divisors. A distinguished system of generators $V_1, \ldots, V_n$ are easy to describe. For any φ_j, $j = 1, \ldots, \mu$ consider the decomposition

$$-F(x, q)\varphi_j(x) = \sum_{i=1}^{\mu} e_{i,j}\varphi_i(x)\mathrm{mod}(C^\infty(x, q)\{\frac{\partial F(x, q)}{\partial x}\}).$$

Then vector fields

$$V_j = \sum_{i=1}^{\mu} e_{ij}(q)\frac{\partial}{\partial q_i} \quad j = 1, \ldots, \mu \text{ and } V_j = \frac{\partial}{\partial q_j}, \quad \text{for } j = \mu + 1, \ldots, n$$

form the basis of tangent (logarithmic) vector fields. An easy proof of this is bases just on the wave front property mentioned above. An one-parameter group of diffeomotphisms mapping the wave front to itself correspond to a family of V-equivalences of the family F with itself. The infinitesimal version of the latter condition provides the decomposition being the linear combination of the decompositions for V_j.

Let L_t be a family of Legendre submanifolds germs in PT^*M smoothly depending on $t \in \mathbf{R}$. We can choose generating families $F(x, q, t)$ of L_t also smoothly depending on t. Then $F(x, q, t)$ considered as a family of functions in x with parameters q, t is a generating family for a (big) Legendre submanifold in $PT^*(M \times R)$.

Hence, to reduce the family of Legendrian projections L_t to a normal form we can reduce the big Legendre projection to a normal form, and then using diffeomorphisms preserving the big front normalize the fibration of the extended configuration space q, t by level hypersurfaces of the function t.

Let the big generating family be R–stable (this holds generically in small dimensions) and be equivalent to the family (1). Take a distinguished generator φ_μ from the annihilator of the maximal ideal of the algebra Q_f. Applying logarithmic vector fields to a generic function t on the parameter space, the normal forms of a generic function are either q_j, for $j > \mu$ or

$$\pm q_* + \sum_{j=\mu+1}^{n} \pm q_j^2.$$

These formulas describe singularities of moving wave front transformations. In variational problems the family of wave fronts are given by the level sets of the action function. The distance function $f(x, q)$ between point x from certain initial variety X_0 and a point q in the ambient space determines a family of equidistants of X_0 whose generic metamorphosis are also described by these normal forms.

3.3. *Affine generating families*

An example of wave front propagation different from the Riemannian distance function is provided by the generating families related to the systems of chords described in [[7]].

Let M, a_0 and N, b_0 be two germs at points a_0 and b_0 of smooth hypersurfaces in an affine space $\mathbf{R}^n$. Let $\mathbf{r}_i : U_i^{n-1} \to \mathbf{R}^n \qquad i = 1, 2$ be local regular parametrizations of M and N, where U_i are viscinities of the origin in $\mathbf{R}^{n-1}$ with local coordinates u and v respectively, $\mathbf{r}_1(0) = a_0$, $\mathbf{r}_2(0) = b_0$.

A $\underline{\text{parallel pair}}$ is a pair of points $a \in M$, $b \in N$, $a \neq b$ such that the hyperplane $T_a M$ which is tangent to M at a is parallel to the tangent hyperplane $T_b N$.

Suppose the distinguished pair a_0, b_0 is a parallel one. A *chord* is the straight line $l(a, b)$ passing through a parallel pair: $l(a, b) = \{q \in \mathbf{R}^n \mid q = \lambda a + \mu b, \ \lambda \in \mathbf{R}, \ \mu \in \mathbf{R}, \ \lambda + \mu = 1\}$.

An <u>affine</u> (λ, μ)–<u>equidistant</u> E_λ of the couple (M, N) is the set of all $q \in \mathbf{R}^n$ such that $q = \lambda a + \mu b$ for given $\lambda \in \mathbf{R}$, $\mu \in \mathbf{R}$, $\lambda + \mu = 1$ and all parallel pairs a, b (close to a_0, b_0).

The *extended* affine space is the space $\mathbf{R}_e^{n+1} = \mathbf{R} \times \mathbf{R}^n$ with baricentric cooordinate $\lambda \in \mathbf{R}$, $\mu \in \mathbf{R}$, $\lambda + \mu = 1$ on the first factor (called <u>affine time</u>).

Denote by $pr : w = (\lambda, q) \mapsto q$ the projection of $\mathbf{R}_e^{n+1}$ to the second factor.

An <u>affine extended wave front</u> $W(M, N)$ of the couple (M, N) is the union of all affine equidistants each embedded into its own slice of the extended affine space: $W(M, N) = \{(\lambda, E_\lambda)\} \subset \mathbf{R}_e^{n+1}$.

The <u>bifurcation set</u> $\mathrm{Bif}(M, N)$ of a family of affine equidistants (or of the family of chords) of the couple M, N is the image under pr of the locus of the critical points of the restriction $pr_r = pr \,|_{W(M,N)}$ A point is <u>critical</u> if pr_r at this point fails to be a regular projection of a smooth submanifold.

In general $\mathrm{Bif}(M, N)$ consists of two components: the <u>caustic</u> Σ being the projection of singular locus of extended wave front $W(M, N)$ and the <u>envelope</u> Δ being the (closure of) the image under pr_r of the set of regular points of $W(M, N)$ which are the critical points of the projection pr restricted to the regular part of $W(M, N)$.

The caustic consists of the singular points of momentary equidistants E_λ while the envelope is the envelope of family of regular parts of momentary equidistants.

On the other hand the affine wave front is swept out by the liftings to $\mathbf{R}_e^{n+1}$ of chords. Each of them has regular projection to configuration space $\mathbf{R}^n$. Hence the bifurcation set $B(M, N)$ is essentially the envelope of the family of chords.

A germ of a family $F(x, w)$ of functions in $x \in \mathbf{R}^k$ with parameters $w = (t, q) \in \mathbf{R}_e^{n+1}$ where $t \in \mathbf{R}$ and $q \in \mathbf{R}^n$ determines the following collection of varieties:

The <u>fiberwise critical set</u> is the set $\mathbf{C}_F \subset \mathbf{R}^k \times \mathbf{R} \times \mathbf{R}^n$ of the solutions

(x, w) of so-callled <u>Legendre equations</u>:

$$F(x, w) = 0, \qquad \frac{\partial F}{\partial x} = 0.$$

The wave front (discriminant) is $W(F) = \{(t, q) \mid \exists x : (x, w) \in \mathbf{C}_F\}$.

The intersections of (big) wave front with $t = \text{const}$ subspaces are called *momentary* wave fronts $W_t(F)$.

The bifurcations set $\text{Bif}(F)$ is the image under the projection pr : $(t, q) \mapsto q$ of the points of $W(F)$ where the restriction $pr\,|_{W(F)}$ fails to be a regular projection of a smooth submanifold. Projections of singular points of $W(F)$ form the <u>caustic</u> $\Sigma(F)$, and singular projections of regular points of $W(F)$ determines the <u>envelope</u> or <u>criminant</u> $\Delta(F)$.

Family F is generating family of a Legendre subvariety $\widehat{L}(F) \subset PT^*(\mathbf{R}^{n+1})$ which is smooth provided that the Legendre equations are locally regular, i.e standard Morse conditions are fulfilled [[1]].

Two germs of families F_i $i = 1, 2$ are called *space-time-contact–* equivalent ("v" - for short) if there exist a non-zero function $\phi(x, t, q)$ and a diffeomorphism $\theta : \mathbf{R}^k \times \mathbf{R}^{n+1} \to \mathbf{R}^k \times \mathbf{R}^{n+1}$, of the form

$$\theta : (x, t, q) \mapsto (X(x, t, q), T(t, q), Q(q))$$

such that $\phi F_1 = F_2 \circ \theta$.

The sum of the family $F(x, t, q)$ with a non-degenerate quadratic form in extra variables $y_1, \ldots, y_m$ is called a <u>stabilization</u> of F. Two germs of families are *v-stable* equivalent if they are *v*-equivalent to stabilizations of one and the same family in fewer variables.

The bifurcation diagrams of *v*-stable equivalent families are diffeomorphic. Theory of singularities of functions with respect to this equivalence group see in [[8,9]].

The critical points of the projection $pr\,|_{\mathbf{C}_F}$ satisfy the equation:

$$\det \begin{vmatrix} \dfrac{\partial F}{\partial x} & \dfrac{\partial F}{\partial t} \\[2ex] \dfrac{\partial^2 F}{\partial x^2} & \dfrac{\partial^2 F}{\partial t \partial x} \end{vmatrix} = 0.$$

Since the first k entries $\frac{\partial F}{\partial x}$ of the first row vanish, the determinant factorises. Hence the bifurcation diagram $B(F)$ splits into two components. One of them (which is the *criminant* $\Delta(F)$) is the image of the projection $(x, t, q) \mapsto q$ of the subvariety $\mathbf{C}_d \subset \mathbf{C}$ determined by the equation $\frac{\partial F}{\partial t} = 0$. The other one (which is the *caustic* $\Sigma(F)$) is the image of the projection $(x, t, q) \mapsto q$ of the subvariety $\mathbf{C}_s \subset \mathbf{C}$ determined by the equation $det\left(\frac{\partial^2 F}{\partial x^2}\right) = 0$.

The following version of Hyugens principle holds: the criminant (envelope) coincides with the wave front of F, considered as a family in variables x and t with parameters q only.

Definition. An <u>affine generating family</u> $\mathcal{F}$ of a pair M, N is a family of functions in $u, v, p \in U_1 \times U_2 \times ((\mathbf{R}^n)^\wedge \backslash \{0\},\ 0$ with parameters $\lambda, q \in \mathbf{R} \times \mathbf{R}^n$ of the form

$$\mathcal{F}(u, v, p) = \lambda < \mathbf{r}_1(u) - q, p > + \mu < \mathbf{r}_2(v) - q, p > .$$

Here $\lambda, \mu = 1 - \lambda$ are baricentric cooordinates on $\mathbf{R}$, and $<,>$ is the standard pairing of vectors from $\mathbf{R}^n$ and covectors p from the dual space $(\mathbf{R}^n)^\wedge$.

Proposition. *The germ at a point $q_0 = \lambda_0 a_0 + \mu_0 b_0$ of affine equidistants generated by a pair (M, a_0), (N, b_0) coincides with the family of momentary wave fronts generated by the germ $\mathcal{F}$ at the point $x = 0$, $y = 0$, $[p] = [dr_1|_{a_0}] = [dr_2|_{b_0}]$. The wave front $W_\mathcal{F}$ coincides with the affine extended wavefront $W(M, N)$. Bifurcation diagram* $\mathrm{Bif}(\mathcal{F})$ *coincides with the set* $B(M, N)$.

The classification of germs of functions $f(x, t)$ with zero one-jet with respect to stable v-equivalence (without parameters) starts with the orbits $[\,[8,9]]\ (x \in \mathbf{R})$:

$$B_k: \quad \pm x^2 + t^k; \qquad C_k: \quad x^k + tz; \quad k = 2, 3, 4 \qquad F_4: x^3 + t^2.$$

The complement to them has codimension 4. Their miniversal deformations in parameters $q \in \mathbf{R}^3$ are as follows:

$$B_k: \quad \pm x^2 + t^k + q_{k-1}t^{k-2} + \cdots + q_1;$$

$$C_k: \quad x^k + xt + q_{k-1}x^{k-2} + \cdots + q_3 x^2 + q_2 t + q_1;$$

$$F_4 : x^3 + t^2 + q_3 xt + q_1 x + q_2 t + q_1.$$

Introduce a non-generic singularity class (related to D.Mond classification of mappings from plane to space):

$$\tilde{C}_4 : F = q_1 + t(q_2 + t + x_1 q_3 + x_1^3 + x_2^2).$$

The following results were proven in [[7]].

Theorem.(Transversal case) *If $n \leq 5$ and the intial chord (a, b) is not parallel to $T_a M$ then there is an open dense subset of the space of germs of hypersurfaces M and N such that at any point the criminant is void and caustic is diffeomorphic to that of some of simple singularities A_m, D_m, E_m provided that $m \leq n + 1$.*

Theorem. (Tangential case) *If $n = 3$ and the intial chord (a, b) is parallel to $T_a M$ then there is an open dense subset of the space of germs of surfaces M and N such that the criminant coinsides with the ruled surface swept by bitangent chords, and the bifurcation set $B(M, N)$ germs at any point of a bitangent chord is diffeomorphic to the bifurcation diagram of some of simple classes $B_k, C_k, \ k = 2, 3, 4, \ F_4$ or of the exceptional class $\tilde{C}_4$.*

References

1. V. I. Arnold, *Singularities of caustics and wave fronts*, Kluwer Academic Publ., Dordrecht-Boston-London, 1990.
2. V. I. Arnold, V. V. Goryunov, O. V. Lyashko and V. A. Vassiliev, *Singularities II. Classification and Applications*, Encyclopaedia of Mathematical Sciences, vol.39, Dynamical Systems VIII, Springer Verlag, Berlin a.o., 1993.
3. V. I. Arnold, S. M. Gusein-Zade and A. N. Varchenko, *Singularities of Differentiable maps. Vol. I*, Monographs in Mathematics **82**, Birkhäuser, Boston, 1985.
4. C. Viterbo, *Generating functions, symplectic geometry and applications*, Proc. Intern. Congr. Math., Zurich, 1994.
5. A. A. Agrachev, G. Charlot, J. P. Gauthier, V. M. Zakalyukin, *On subriemannian caustics and wave fronts for contact distributions in three-space*, Journal of Dynamical and Control Systems, **6** (2000) n.3 365-395
6. V. M. Zakalyukin, *Reconstructions of fronts and caustics depending on parameters, versality of mappings*, Journal of Soviet Mathematics **27** (1984) 2785-2811.
7. P. J. Giblin, V. M. Zakalyukin, *Singularities of centre symmetry sets*, Proc. London Math. Soc. **90** 2005 n.3 132-166.

186

8. V. V. Goryunov, *Projections of generic surfaces with boundary*, Theory of singularities and its applications (ed. V.I.Arnold), Advances in Soviet Mathematics 1, AMS, Providence, RI, 1990 157-200.
9. V. M. Zakalyukin, *Envelopes of families of wave fronts and control theory*, Proc. Steklov Inst. Math. **209** (1995) 133-142

PART II

Applications of Singularity Theory

SINGULARITIES OF ROBOT MANIPULATORS

P. S. Donelan

School of Mathematics, Statistics and Computer Science,
Victoria University of Wellington,
PO Box 600, Wellington, New Zealand
E-mail: peter.donelan@vuw.ac.nz

Engineers have for some time known that singularities play a significant role in the design and control of robot manipulators. Singularities of the kinematic mapping, which determines the position of the end–effector in terms of the manipulator's joint variables, may impede control algorithms, lead to large joint velocities, forces and torques and reduce instantaneous mobility. However they can also enable fine control, and the singularities exhibited by trajectories of the points in the end–effector can be used to mechanical advantage.

A number of attempts have been made to understand kinematic singularities and, more specifically, singularities of robot manipulators, using aspects of the singularity theory of smooth maps. In this survey, we describe the mathematical framework for manipulator kinematics and some of the key results concerning singularities. A transversality theorem of Gibson and Hobbs asserts that, generically, kinematic mappings give rise to trajectories that display only singularity types up to a given codimension. However this result does not take into account the specific geometry of manipulator motions or, *a fortiori*, to a given class of manipulator. An alternative approach, using screw systems, provides more detailed information but also shows that practical manipulators may exhibit high codimension singularities in a stable way. This exemplifies the difficulties of tailoring singularity theory's emphasis on the generic with the specialized designs that play a key role in engineering.

Keywords: Singularity, Robot Manipulator, Screw System

1. Robot Manipulators

The International Federation for the Promotion of Mechanism and Machine Science [1] has defined a *mechanism* to be "a system of bodies designed to convert motions of, and forces on, one or several bodies into constrained motions of, and forces on, other bodies"; a *robot* is a "mechanical system under automatic control that performs operations such as handling and automation", while a *manipulator* is a "device for gripping and the controlled movement of objects". More comprehensively, the International Organisation for Standardisation (ISO) [2] defines a *manipulator* to be "a machine, the mechanism of which usually consists of a series of segments, jointed or sliding relative to one another, for the purpose of grasping and/or moving objects (pieces or tools) usually in several degrees of freedom. It may be controlled by an operator, a programmable electronic controller, or any logic system (for example cam device, wired, etc.)". Certainly these definitions are necessarily imprecise, but we can construct succinct mathematical descriptions that capture the important aspects of robot kinematics and dynamics. These must capture the relationship between inputs and outputs so, in their simplest form, consist of a function, the *kinematic mapping*, between a manifold of inputs and the manipulator's configuration space.

Typically, the component bodies or segments of a robot manipulator are rigid, or at least can be treated as such for the purposes of kinematic analysis. Each can be furnished with an orthonormal coordinate frame. One component is usually designated as the base and assigned a fixed coordinate frame, though, for some problems, it may be preferable to regard all components as mobile relative to an ambient coordinate frame. Of primary interest in most problems is the motion of the *end-effector*—the grasping component, or component to which the operative tool is attached. Its *pose* (position and orientation) relative to the base can be described by a Euclidean isometry mapping its coordinate frame to that of the base.

The components of a manipulator are connected by joints of various kinds: *revolute* (R), slider or *prismatic* (P), screw or *helical* (H), ball or *spherical* (S), *planar* (E). This is not an exhaustive list but these and other joints all arise from the contact of surfaces (*lower kinematic pairs*), or curves and points (*higher kinematic pairs*) in the components. The classic book of Hunt [3] provides an engineering perspective. Those joints which are subject to an input (e.g. via a servo-motor) are termed *actuators*, while others are *passive* joints. The R and P–joints are simplest to engineer and their kinematic analysis is generally more straightforward. Nearly all practical manipulators use these as actuators. Some of the other joints can be synthesized by means of combinations of R and P joints. A universal (U) joint is a combination of 2 R–joints with intersecting axes while a spherical joint can be synthesized by 3 R–joints and both occur frequently as passive joints.

The global architecture of a manipulator (or any rigid–body mechanism) may be partially encoded combinatorially by means of a graph whose vertices are the components and edges denote a joint between components or, dually, with vertices the joints and edges the components (see, for example, Ref. 4). Additional information is required to fully specify the robot. This information is partly topological—the nature of each joint, including its number of degrees of freedom (*dofs*)—and partly geometric, namely the *design parameters* that specify the size and relative placement of the components and joints.

A mechanism whose graph is a path is called a *kinematic chain* and a manipulator with this architecture is called *serial*. Related to serial manipulators are those whose graph is a tree, a class that includes most robot hands having fingers or other gripping mechanisms. In such manipulators, all joints are actuated.

Parallel mechanisms are those in which the end-effector, usually called the *platform* in this context, is connected to the base by two or more independent kinematic chains [5]. Typically, only some of the joints in each chain are actuated and if this number is 1, the mechanism is called *fully parallel*, otherwise *hybrid parallel*. In particular, the graphs of parallel mechanisms contain cycles which impose equational constraints on the joint variables.

An important question regarding parallel mechanisms is to determine their mobility. This can be determined in the generic case from the graph using the *Grübler–Kutzbach mobility formula* [3] for a spatial mechanism with n components, g joints, the ith joint having f_i dof, $i = 1, \ldots, g$:

$$\mathfrak{M} = 6(n - g - 1) + \sum_{i=1}^{g} f_i. \tag{1}$$

The *mobility* $\mathfrak{M}$ represents the number of inner degrees of freedom of the mechanism as a whole. An extension of this formula that takes into account symmetries has recently been derived [6]. For planar and spherical mechanisms, for example, one needs to replace 6 in (1) by 3, the dimension of the relevant isometry subgroup. The mobility also tells us the number of joints that need to be actuated. Note however that we provide a more precise definition of mobility in the next section.

2. Geometry of Manipulator Kinematics

We will assume that our manipulators consist of rigid bodies connected by standard joints, operating in 3–dimensional Euclidean space. The *Euclidean group* of isometries, $SE(3)$ is isomorphic, via choice of coordinates in the moving and fixed spaces, to the semi-direct product of the proper rotation group $SO(3)$ and the additive group of translations, $\mathbb{R}^3$. In this form the elements of $SE(3)$ are pairs $(A, \mathbf{a}) \in SO(3) \ltimes \mathbb{R}^3$, A a 3×3 orthogonal matrix with determinant $+1$ and $\mathbf{a}$ a 3–vector. The group product, representing composition of isometries, is

$$(A_2, \mathbf{a}_2) \cdot (A_1, \mathbf{a}_1) = (A_2 A_1, A_2 \mathbf{a}_1 + \mathbf{a}_2).$$

The group acts on $\mathbb{R}^3$ by

$$(A, \mathbf{a}).\mathbf{x} = A\mathbf{x} + \mathbf{a}$$

and we can regard this is a map from the end–effector coordinates to base coordinates, describing the *pose* of the end–effector.

The Euclidean group is a 6–dimensional Lie group. Pure translations form a 3–dimensional normal subgroup $\mathbb{R}^3$, and the set of rotations about *any* point of $\mathbb{R}^3$ is also a 3–dimensional subgroup isomorphic to $SO(3)$. Two–dimensional Euclidean and spherical motion can both be regarded as special cases since the relevant isometry groups, $SE(2)$ and $SO(3)$ are (3–dimensional) subgroups of $SE(3)$. The connected Lie subgroups of $SE(3)$ have been classified by Hervé [7] (and in a different setting by Beckers *et al* [8]) and play an important role in mechanism theory.

Note that $SE(3)$ is also a linear algebraic group that may, for example, by defined as a real algebraic variety in $\mathbb{R}^{12}$ (representing 9 entries in a 3×3 matrix A and a 3–vector $\mathbf{a}$) by means of the equations $A^t A = AA^t = I$, $A = \mathrm{adj}\, A$. Lazard [9] introduced this approach to analyze poses of the Gough–Stewart platform described below.

Elements of $SE(3)$ may be represented or parametrized in various forms [10–12]. For example, the rotation component can be described by

- Euler angles (ϕ, θ, ψ), representing successive rotations about specific axes, so that

$$A = \begin{pmatrix} \cos\psi & -\sin\psi & 0 \\ \sin\psi & \cos\psi & 0 \\ 0 & 0 & 1 \end{pmatrix} \begin{pmatrix} 1 & 0 & 0 \\ 0 & \cos\theta & -\sin\theta \\ 0 & \sin\theta & \cos\theta \end{pmatrix} \begin{pmatrix} \cos\phi & -\sin\phi & 0 \\ \sin\phi & \cos\phi & 0 \\ 0 & 0 & 1 \end{pmatrix}$$

  This can also be written as a product of exponentials: $\exp(\psi Z)\exp(\theta X)\exp(\phi Z)$, where X and Z are elements of the Lie algebra $\mathfrak{so}(3)$.

- Rodrigues parameters, writing $A = (I - B)^{-1}(I + B)$, B skew-symmetric.

- Unit quaternions $q = c_0 + c_1 i + c_2 j + c_3 k \in \mathbb{H}$, which form a double cover of $SO(3)$ and act on $\mathbb{R}^3$, embedded in $\mathbb{H}$ as the pure imaginary quaternions, by $x \mapsto qx\bar{q}$ (where $\bar{q}$ denotes the quaternionic conjugate formed by negating the imaginary part of q).

The last of these extends to representation of elements of $SE(3)$ by Clifford's dual quaternions, $q + \epsilon p$, where q, p are unit quaternions and $\epsilon^2 = 0$.

An important observation is that the relative displacements permitted by the various types of joints are represented, for a choice of coordinates in each of the connected components, by subgroups of $SE(3)$, which are embedded analytic submanifolds whose dimensions are, by definition, the number of *degrees of freedom* of the joints. In particular, the motions defined by R, P and H–joints are its 1–parameter subgroups.

The terminology for describing the kinematics of manipulators is not standardized. The following definitions are adapted from those used by the Parallel Mechanisms Information Centre [5].

Definition 2.1. Given two rigid bodies B_1 and B_2, equipped with choices of orthonormal coordinates and connected by a joint J, the **joint space** of J is the submanifold of $SE(3)$ corresponding to the set of possible displacements of B_2 relative to B_1.

The **joint space** $P_{\mathcal{R}}$ of a robot manipulator $\mathcal{R}$ is the product of the joint spaces of all its joints, its **articular space** $Q_{\mathcal{R}}$ is the product of the joint spaces of its actuated joints. Its **configuration space** is the subset $M_{\mathcal{R}} = f^{-1}(c) \subseteq P_{\mathcal{R}}$ of attainable values for the joint variables, where $f : P_{\mathcal{R}} \to \mathbb{R}^k$ is a function determining the constraints.

The **kinematic mapping** for $\mathcal{R}$, with an identified end–effector or platform B carrying an orthonormal coordinate frame, is the function $\lambda : M_{\mathcal{R}} \to SE(3)$ which to each attainable set of joint variables assigns the pose of B. The **workspace** $W_{\mathcal{R}} \subseteq SE(3)$ of $\mathcal{R}$ is the image of the kinematic mapping. In the case that $M_{\mathcal{R}}$ is a manifold, its dimension is the **mobility** of $\mathcal{R}$.

Given a point $\mathbf{w}$ in the end–effector, the **evaluation map** of $\mathbf{w}$ is the smooth function

$$e_w : SE(3) \to \mathbb{R}^3, \qquad (A, \mathbf{a}) \mapsto A\mathbf{w} + \mathbf{a}$$

The **trajectory** of $\mathbf{w}$ under the kinematic mapping λ is the smooth function

$$\tau_w = e_w \circ \lambda : M_{\mathcal{R}} \to \mathbb{R}^3.$$

Note that if one treats $SE(3)$ as an algebraic group then, for manipulators with R, P and S–joints, $M_{\mathcal{R}}$ is a subvariety, even in the singular cases, so the mobility can be defined via the dimension of that variety.

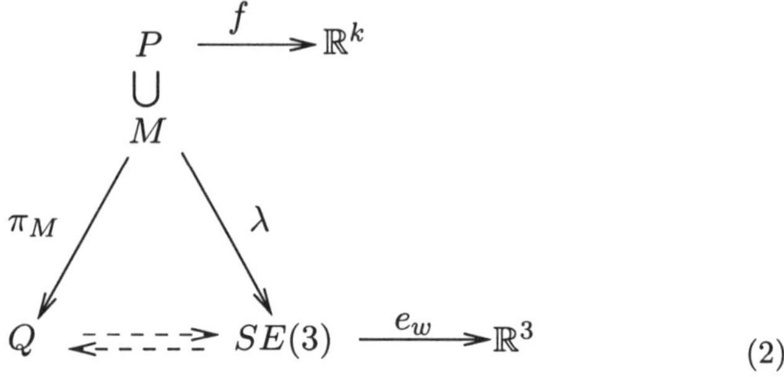

$$\tag{2}$$

The relation between the joint, articular and configuration spaces (the subscript $\mathcal{R}$ has been suppressed) is illustrated in (2), where $\pi_M : M \to Q$ denotes the restriction to M of the projection from the joint space to the articular space. The broken arrows between Q and $SE(3)$ indicate the two fundamental relationships that underlie the control and use of manipulators. The arrow $Q \to SE(3)$ denotes the *forward* (or *direct*) *kinematics* of the manipulator. If it is possible to find such a function then it determines the configuration of the end–effector for a given set of joint variables. The arrow $SE(3) \to Q$ is the *inverse kinematics* of the manipulator, which determines the joint variables required for a given pose of the end–effector. Where distinct points in M give rise to the same pose, the mechanism configurations are sometimes referred to as *postures*. The presence of singularities in either or both of λ and π_M obstructs the existence of global inverse or forward kinematics.

Topologically, the joint space for an R–joint is a circle S^1 and for P and H–joints an embedded real line $\mathbb{R}$. The joint space of an S–joint is an embedded $SO(3)$. In practice, there may be physical limitations to a manipulator, restricting the effective joint space of each joint to some subset, say an interval. However to simplify the mathematical analysis we will assume the joint space to be the entire subgroup.

Hence, the joint and articular spaces of a robot manipulator are also smooth manifolds. However whether the configuration space is a manifold will depend on whether the constraint function f is submersive along $f^{-1}(c)$, and that may depend on the design parameters.

3. Three Classes of Manipulator

3.1. *Serial manipulators*

For a serial manipulator, every joint is actuated so, ignoring any engineering limitations on joints, the joint space, articular space and configuration space coincide. That means the left-hand side of (2) collapses and the forward kinematics is simply the kinematic mapping. In all standard industrial manipulators the joints are either R or P so the configuration space M is a product of a generalized torus $T^r = S^1 \times \ldots \times S^1$ (r copies) and a Euclidean space $\mathbb{R}^p$. We will see shortly that the kinematic mapping is analytic, so in order for the workspace to have non-empty interior (the end–effector has maximum freedom of translation and orientation) we require the mobility $m = r + p \geq 6$. If equality holds the manipulator is called *non-redundant*, while if $m > 6$, it is called *redundant*.

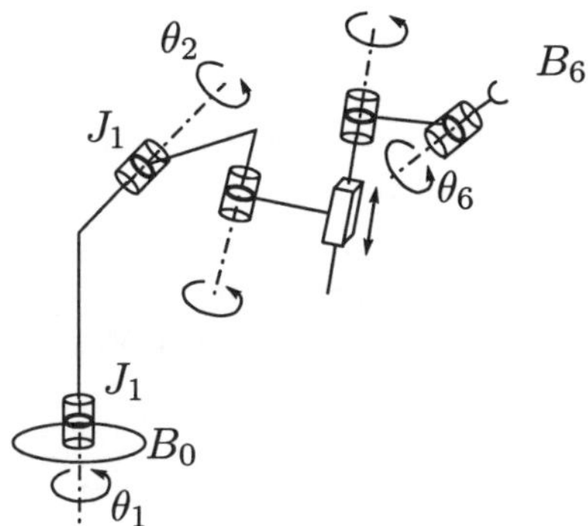

Fig. 1.   Serial manipulator

Associate a coordinate frame to each component and choose a 'home' configuration. Label the components in order from the base $B_0, \ldots, B_m$ and joints $J_1, \ldots, J_m$ so that joint J_i connects B_{i-1} and B_i. In the home configuration the coordinates of these components are related by an element $U_i \in SE(3)$. Displacements arising from J_i form a one–parameter subgroup which can therefore be written in the form $\exp(\theta_i Y_i)$ for some Y_i, $i = 1, \ldots, m$ in the Lie algebra $\mathfrak{se}(3)$ (see Section 6). In particular, as was shown by Brockett [13], the kinematic mapping (= forward kinematics) for the end–effector has the form:

$$\lambda(\theta_1, \ldots, \theta_m) = U_1 \exp(\theta_1 Y_1) \, U_2 \exp(\theta_2 Y_2) \, \ldots \, U_m \exp(\theta_m Y_m). \quad (3)$$

Judicious choice of coordinates enables one to express the matrices Y_i

in a standard form and the isometries U_i in terms of a small number of parameters defined by the manipulator geometry (Denavit–Hartenberg parameters—see, for example, Refs. 11,14,15). Alternatively, by choosing coordinate frames that coincide in the home configuration, $U_2 = \cdots = U_m = 1$ (where we use 1 to represent the identity in the group) the kinematic mapping can be expressed as a product of exponentials.

In practice, many industrial serial manipulators are *wrist–partitioned.* Orientation of the end–effector is achieved by means of a 3R spherical wrist—that is, the 3 axes of rotation intersect at a point, the *wrist centre.* Motion of the wrist relative to its centre can be represented, for example, using Euler angles. Location of the wrist centre is achieved by means of a 3–joint arm, often referred to as a *regional* manipulator having $p = 0, 1, 2$ or 3 P–joints and $r = 3 - p$ R–joints. This partitioning simplifies singularity analysis as discussed in Section 4.

3.2. *Planar 4–bar mechanisms*

The study of the kinematics of mechanisms, dating back at least to Watt's parallel motion [16] used for converting linear to rotary motion in steam engines, can be seen as a forerunner of robot kinematics. A simple but informative example of a parallel mechanism is provided by the planar 4–bar (classically referred to as the 3–bar) mechanism [17,18], of which Watt's motion is an example. The mechanism (Figure 2) consists of 4 components linked in a closed quadrilateral $ABCD$ by revolute joints. Regard the base AD as fixed. AB is referred to as the *input bar*, BC as the *coupler bar* and CD as the *output bar*. The design parameters are the lengths of the 4 bars d_i, $i = 0, 1, 2, 3$. The planar Grübler–Kutzbach formula confirms that the mobility of this mechanism is generically

$$\mathfrak{M} = 3(4 - 4 - 1) + 4 = 1.$$

The joint space is a 3–dimensional torus, parametrized by the angles (α, β, γ) and the constraint equations, arising from the closure of the quadrilateral, are:

$$d_1 \cos \alpha + d_2 \cos \beta + d_3 \cos \gamma = d_0$$
$$d_1 \sin \alpha + d_2 \sin \beta + d_3 \sin \gamma = 0 \tag{4}$$

Gibson and Newstead [18] showed that the configuration space is indeed a 1–dimensional manifold (diffeomorphic to either S^1 or $S^1 \times \{0, 1\}$) so long as $\max\{d_i : i = 0, 1, 2, 3\}$ is less than the sum of the other 3 sides and the

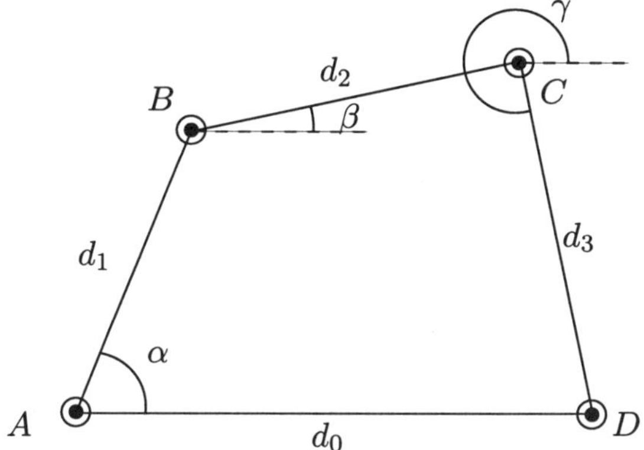

Fig. 2.   Planar 4–bar mechanism

Grashof condition

$$d_0 \pm d_1 \pm d_2 \pm d_3 \neq 0$$

is satisfied. This can be readily derived from the constraint equations (4) via the Submersion Theorem (see, for example Ref. 19).

In classical industrial applications, the actuated joint is at A and the output bar is the end–effector. The restriction to $M_{\mathcal{R}}$ of the projection of the joint space $P_{\mathcal{R}}$ onto the joint space of each joint is either onto, in which case the joint is called a *crank*, or not, when it is a *rocker*. A crank–rocker (referring to the behaviour of the input and output bars) enables conversion of rotary motion into rectilinear motion. However, from the perspective of parallel mechanisms, it makes more sense to take the coupler bar BC as the end–effector (platform). The kinematic mapping in terms of $(\alpha, \beta, \gamma) \in M_{\mathcal{R}}$ is given, in one form, by

$$\lambda(\alpha, \beta, \gamma) = \left(\begin{pmatrix} \cos\beta & -\sin\beta \\ \sin\beta & \cos\beta \end{pmatrix}, \begin{pmatrix} a\cos\alpha \\ a\sin\alpha \end{pmatrix} \right) \in SE(2) \cong SO(2) \ltimes \mathbb{R}^2,$$

but recall that the joint variables are constrained by (4).

3.3. *The Gough–Stewart Platform*

Gough devised this famous parallel mechanism as a tyre–testing rig for Dunlop Tyres in the 1950s [20]. A similar design was later proposed by Stewart for use as the platform for a flight simulator [21]. MacCallion [22] first considered its use as a workspace manipulator. The vertices of an equilateral triangle in the base are connected pairwise to those of a similar triangle in the platform by articulated struts as in Figure 3. The 6 struts are joined to the base by universal joints and to the platform by spherical joints and each has a single prismatic actuator enabling the strut length to be altered (Figure 3). The platform is therefore a 6-UPS mechanism. This architecture is also called an *octahedral hexapod*. In greatest generality, the 6 struts may connect arbitrary points in the base and platform. Between these extremes are numerous architectures embodying different degrees of symmetry. In fact, the original platform properly has planar hexagonal base and platform, each with triangular symmetry. A brief history of such mechanisms can be found in Bonev [23] and a recent review of the theory is in Ref. 24.

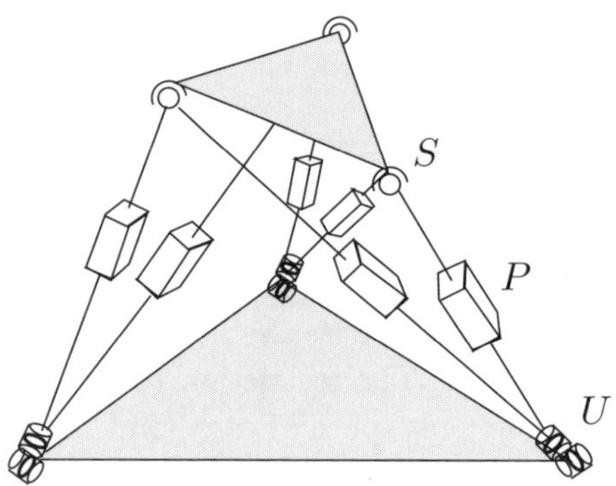

Fig. 3.   Gough–Stewart platform

The Grübler–Kutzbach formula predicts that such structures generically have mobility 6 (there are 14 links: 6×2 in the struts, plus base and platform; $6 \times 3 = 18$ joints having $6 \times 6 = 36$ degrees of freedom). Realizing this result in terms of Definition 2.1 is a little harder. It is simpler to replace the configuration space M by $\tilde{M}$, defined to be the subset of $(\mathbb{R}^3)^3 \times \mathbb{R}^6 \cong \mathbb{R}^{15}$ representing the 3 sets of coordinates for platform vertices together with the variable strut lengths, subject to 3 independent equations fixing the distances between the vertices and 6 equations relating the coordinates to the strut lengths in terms of the base joint coordinates. The Submersion Theorem gives mobility 6. In fact, it is clear that $\tilde{M}$ is diffeomorphic to $SE(3)$ since there is a correspondence between platform poses and feasible coordinates for the joints. The relation between M and $\tilde{M}$ is less clear since it is possible that a family of joint coordinates all give rise to the same pose.

(This occurs, for example of the UPS struts are replaced by SPS struts, in which case the struts are free to rotate about their axes in any pose.)

If we allow the replacement of M by $\tilde{M}$ in (2), then the right-hand side of the triangle collapses and the inverse kinematics are well defined. The forward kinematics of the Gough–Stewart platform has attracted interest amongst both kinematicians and mathematicians. Methods from topology and algebraic geometry have been used to show that for a given set of actuator variables there may be up to 16 poses for the platform in the architecture described here (40 for the general architecture). This number depends on both the actuator variables and the design parameters and a full understanding of the way in which the number changes presents a highly technical challenge in singularity theory and topology. (For more detail, see Refs. 9,46, for example.)

4. Instantaneous Kinematics and Singularities

While singularities have non-local implications for the control and use of manipulators, they arise as local or instantaneous phenomena from the rank deficiency of a derivative. The diagram (2) illustrates that this may occur in a number of ways in relation to manipulators. For serial manipulators, it is the singularities of the kinematic mapping/forward kinematics and trajectories that are of interest, whereas for fully parallel manipulators it is those of the constraint function defining the configuration space and of the projection onto the articular space (inverse kinematics). The distinction between the classes of mechanisms in respect of their singularities was first recognized by Gosselin and Angeles [25] and subsequently refined by Zlatanov *et al* [26,27]. Simaan and Shoham [28] have used their ideas to analyze singularities of hybrid serial/in–parallel mechanisms.

The importance of singularities from an engineering perspective arises for several reasons:

(1) *Loss of freedom.* The derivative of the kinematic mapping or forward kinematics represents the conversion of joint velocities into generalized end-effector velocities, i.e. linear and angular velocities. This linear transformation is generally referred to as the *manipulator Jacobian* in the robotics literature. A drop in rank reduces the dimension of the image, representing a loss of instantaneous motion for the end-effector of one or more degrees. The proper setting for describing this is the theory of screw systems, discussed below in Section 6, which is used in many papers in the robotics literature, for example Refs. 29–31.

200

(2) *Workspace.* When a manipulator is at a boundary point of its workspace, the manipulator is necessarily at a singular point of its kinematic mapping, though the converse is not the case. Interior components of the singular set separate regions with different numbers or topological types of inverse kinematics. These are usually associated with a change of posture in some component of the manipulator. Therefore knowledge of the manipulator singularities provides valuable information about its workspace [32].

(3) *Loss of control.* A variety of control systems is used for manipulators. Rate control systems require the end–effector to traverse a path at a fixed rate and therefore determine the required joint velocities by means of the inverse of the derivative of the (known) forward kinematics. Near a singularity, this matrix is ill-conditioned and either the control algorithm fails or the joint velocities and accelerations may become unsustainably great. Conversely, force control algorithms, well-adapted for parallel manipulators, may result in intolerable joint forces or torques near singularities of the projection onto the joint space.

(4) *Mechanical advantage.* Near a singular configuration, large movement of joint variables may result in small motion of the end–effector. Therefore there is mechanical advantage that may be realised as a load-bearing capacity (interesting examples in human activities are presented by Kieffer and Lenarčič [33]) or as fine control of the end–effector (an example of a telescope-focussing device is given by Carretero *et al* [34,35]). Another aspect of this is in the design of mechanisms possessing trajectories with specific singularity characteristics. In traditional 1–dof mechanisms (such as the planar 4–bar) a cusp singularity provides 'dwell'— the trajectory is close to stationary for a period of time allowing some step in a production process to be performed [36]. A higher–dimensional example is the use of a corank 3 singularity by the remote centre compliance device [37–39] (see also Section 7).

The following theorems are central examples illustrating the necessity of singularity analysis for serial and parallel manipulators. The following result of Gottlieb [40], also discussed in Ref. 41, is for the forward kinematics of serial manipulators.

Theorem 4.1. *For any serial manipulator with configuration space $M = T^p \times \mathbb{R}^{n-p}$ with $n \geq 6$, the kinematic mapping $\lambda : M \to SE(3)$ possesses singularities.*

The proof is based on the observation that if there were no singularities then λ would be a submersion, giving rise to a fibration of $\mathbb{R}^n$ (the universal covering space for M) over $SO(3)$. But this is ruled out on topological grounds. Moreover, it is not possible globally to avoid singularities by introducing redundancy in the manipulator. That is, for $n > 6$ there is no continuous function $\theta : SE(3) \to M$ such that $\lambda(\theta(X)) = X$.

There has been extensive analysis of the actual singularity configurations for industrial robot manipulators. Wang and Waldron [31] analyzed the general 6–dof serial manipulator using screw theory (see section 6) and showed that the *singularity field* (set of singular configurations) is independent of the joint variables θ_1 and θ_6. Litvin *et al* [30,42] explicitly examined 6R manipulators and showed that there were three sets of singularities relating to different configurations, subsequently dubbed wrist, elbow and shoulder singularities. Stanišić and Engelberth [43] looked at wrist-partitioned manipulators, where the Jacobian determinant factors into one component for the wrist and one for the arm subassembly. They showed that there are surfaces in $\mathbb{R}^3$, dependent on the manipulator's configuration, such that when the wrist centre lies on them then the manipulator is in a singular configuration. The surfaces themselves are determined by the associated screw system of the arm subassembly in its current configuration, described in further detail in Sections 6 and 7.

Tchoń and Muszynski [44,45] sought to characterize serial manipulator singularities by finding normal forms with respect to $\mathcal{A}$–equivalence (right–left equivalence). They showed that among corank 1 singularities (i.e. away from intersections of the singular surfaces), the elbow and shoulder singularities are folds and hence stable, but the wrist singularities have infinite $\mathcal{A}$–codimension ('differential degree' in their terminology).

In relation to fully parallel manipulators of the Gough–Stewart kind, Merlet [46] showed the following:

Theorem 4.2. *The inverse kinematics of a 6-UPS parallel mechanism are singular if and only if the lines spanned by the 6 struts are linearly dependent.*

This follows because the rows of the inverse Jacobian can be shown to be the Plücker line coordinates of the struts. Again this is closely linked to the theory of screw systems.

It is worth noting here that for parallel manipulators, the full joint space is important for singularity analysis, since there may be configurations for which platform motion is possible because of passive joint velocities. This was originally observed by di Gregorio and Parenti-Castelli [47] and has been the subject of further exploration, under the terminology *constraint singularities* by Zlatonov *et al* [48].

5. Genericity and Transversality Theorems

5.1. *One-genericity*

A general theory for kinematic mappings λ can be set up by considering spaces of smooth (or analytic) mappings $C^\infty(M, SE(3))$ where M is the configuration space of a manipulator or, locally, germs of such mappings. The Whitney Immersion Theorem (see, for example, Ref. 19) assures us that for manipulators with up to 3 dof, there is an open and dense set of mappings that are immersions (and if M is compact, then this is true for 1–1 immersions). In practice, we are frequently interested in cases where $\dim M \geq 6$, and for dimension 4 and upwards singularities will occur stably.

Several authors have sought to identify generic properties for kinematic mappings—that is, properties possessed by a suitably large subset, say open and dense, or at least residual. A starting point is usually to identify relevant submanifolds of a jet bundle $J^k(M, SE(3))$ (or multi-jet bundle) and require transversality to these for the associated jet extension. Then the Thom Transversality Theorem [19] guarantees genericity.

Pai and Leu [49] adopt this approach, using the stratification of the 1–jet bundle by corank, that is $\Sigma^r = \{\sigma \in J^1(M, SE(3)) : \operatorname{corank} \sigma = r\}$. A kinematic mapping transverse to this stratification is called *1–generic*. They analyze the standard architectures for serial manipulators from this perspective, distinguishing the orientation singularities of the wrist from the translational singularities of the regional manipulator, by composing the kinematic mapping with projection onto its components. In the case of translations, this projection is coordinate dependent and amounts to analyzing the trajectory of the wrist centre. The singular point set in the joint space is always invariant under rotation about the first axis. Hence, in the generic case where $\Sigma^r \lambda = (j^1 \lambda)^{-1}(\Sigma^r)$ can only be non-empty for $r = 0, 1$, the singular point set can be identified with a union of circles in the 2–torus corresponding to θ_2, θ_3 in equation (3). This led Burdick [50] to propose a classification of 3R regional manipulators with generic kinematic mapping based on the homotopy class(es) of the components of $\Sigma^1 \lambda$.

Subsequent work by Wenger *et al* [51–53] resolved some conjectures of Burdick concerning 3R manipulators. In particular, they showed that the presence of a cusp singularity, that is $\Sigma^{1,1}$ in the Thom–Boardman classification, is a necessary and sufficient condition for the existence of a path in the configuration space realizing a change of posture *without* encountering a singularity. This is an interesting result in the area of singularity avoidance, which is concerned with the topology of the singularity field.

5.2. *Trajectory singularities and a transversality theorem*

A deeper approach was pursued by Gibson *et al.* They sought to examine the relation between the kinematic mapping $\lambda = (A, \mathbf{a})$ and its family of trajectories

$$\tau_\lambda = e_w \circ \lambda : M \times \mathbb{R}^p \to \mathbb{R}^p, \quad (x, w) \mapsto A(x)\mathbf{w} + \mathbf{a}(x)$$

for planar ($p = 2$) and spatial ($p = 3$) kinematics. The key result is the following theorem of Gibson and Hobbs [54]. For integers $k, r \geq 1$, there is a multijet extension

$$_r j_1^k \tau_\lambda : M^{(r)} \times \mathbb{R}^p \to {}_r J^k(M, \mathbb{R}^p)$$

where $_1$ means take jets w.r.t. first component only. This map assigns the k–jets of the p–parameter family of trajectories to a set of r distinct configurations in M.

Theorem 5.1. *Given a finite stratification S of the multijet bundle $_r J^k(M, \mathbb{R}^p)$, the set of $\lambda \in C^\infty(M, SE(p))$ such that $_r j_1^k \tau_\lambda$ is transverse to S is residual.*

The original proof followed that of Wall [55] for singularities of projections of generic immersions; a simpler proof, using the fact that the evaluation map is a submersion, follows from a theorem of Montaldi [56] on composite maps. Note that this, in some sense, subsumes Pai and Burdick's approach for regional manipulators, since that concerns the wrist–centre trajectory, by taking $r = k = 1$ and S the corank stratification. The theorem underpins a programme, described in Ref. 57, for exploring singularities of trajectories.

- Classify $\mathcal{A}$-types of multigerm singularities up to relevant codimension. Subject to amalgamating orbits with moduli, this typically provides an $\mathcal{A}$–invariant finite stratification of the multijet bundles.
- Transversality imposes constraints on codimension of strata that can be encountered. Gibson and Hobbs [54] show that the requirement for a non-stable multigerm with $\mathcal{A}$–modality m to occur transversely is that its $\mathcal{A}_e$–codimension is $\leq \dim M + m$.

- Versal unfoldings of these singularity types give local models for the bifurcation sets in $\mathbb{R}^p$.

In a sequence of papers [54,58–65], Gibson and co-workers filled out details of this programme for planar and spatial motions up to 3–dof, in the process generating new lists of (multi-)singularities of maps between spaces of dimension up to 3. It was shown [58] that the stable singularity type of the kinematic mapping germ itself imposes restrictions on the singularity type of its trajectory germs. However the $\mathcal{A}$–classification of map-germs becomes computationally harder as the number of degrees of freedom increase. For example, there are more than 50 classes up to $\mathcal{A}_e$–codimension 3 for 3–dof spatial motions [64].

5.3. *Problems with genericity*

There is a fundamental difficulty with the genericity approach. The idea of genericity is to identify properties of mappings that are typical among all such mappings. The space $C^\infty(M, SE(3))$ is infinite-dimensional. However in robotics one is almost always concerned with a specific class of manipulators defined by a finite number of design parameters. It is by no means sure that this finite–dimensional space will lie in such a way that a given property that is generic in $C^\infty(M, SE(3))$ will remain so on restriction to the subset. One can use the Elementary Transversality Theorem [19] in this setting but that is likely to require explicit calculation of transversality, stratum by stratum.

For serial manipulators, the design parameters can be chosen to be Denavit–Hartenberg parameters. Pai and Leu showed explicitly that within this parameter space, all the standard regional manipulator architectures (i.e. with a combination of 3 R or P joints) and orientation (wrist) architectures have an open and dense set—the complement of an analytic function in the parameters—of 1–generic mappings.

However even in one of the simplest cases involving parallelism, coupler curves of planar 4–bar mechanisms, it has not yet been established that the restricted version of the Gibson–Hobbs Transversality Theorem holds, the remaining obstruction being the monogerm stratum of A_4 singularities (ramphoid cusps) [66].

A second drawback is that these approaches do not explicitly take into account the structure of $SE(3)$ or its tangent spaces. In a singular configuration we are interested not only in the dimension of the image of the derivative but also in how it lies with respect to that structure.

Finally, it is important to note that from an engineering perspective it is often advantageous to use *special*, that is non-generic, architectures in order to achieve desirable motion characteristics. One example is the emphasis on wrist-partitioned serial manipulators with only R and P joints. Another example is the interest in so-called *over-constrained* structures, where the mobility exceeds that predicted by the Grübler–Kutzbach formula. A well-known example is the Bennett mechanism [67,68], a closed 4–link kinematic chain with 1 dof, though the formula predicts that spatial closed chains with up to 6 links should be rigid. The study of singularities is relevant to over-constrained mechanisms because their motion can be regarded as a motion contained entirely within the subset of singular configurations of an associated open chain or serial mechanism (realised by unlinking one of the joints). This approach to closed mechanisms is exploited, for example, by Lerbet and Hao [69].

6. Screw Systems

Theorems 4.1 and 4.2 indicate the importance of the tangent spaces to the Lie group $SE(3)$ as either the range or domain of the relevant derivatives. The essential structure is that of the Lie algebra $\mathfrak{se}(3)$, which can be variously identified as the tangent space to $SE(3)$ at the identity, the space of one-parameter subgroups of $SE(3)$ or the space of Killing vector fields on $\mathbb{R}^3$ (see, for example, Ref. 70). It inherits from the group structure described in Section 1 the structure of a semi-direct product of the Lie algebras $\mathfrak{so}(3)$ of the rotation group and $\mathfrak{t}(3)$ of the translation group. Thus, elements may be represented by a pair $(B, \mathbf{v}) \in \mathfrak{so}(3) \ltimes \mathfrak{t}(3)$ where B is a skew-symmetric 3×3 matrix and $\mathbf{v}$ a 3-vector.

If B has the form

$$\begin{pmatrix} 0 & -u_3 & u_2 \\ u_3 & 0 & -u_1 \\ -u_2 & u_1 & 0 \end{pmatrix}$$

then it can be identified with the vector $\mathbf{u} = (u_1, u_2, u_3)^t$ that (if non-zero) spans its kernel. Thus elements of $\mathfrak{se}(3)$ can be represented by 6–vectors $(\mathbf{u}, \mathbf{v})$. The orbits of the associated Killing vector fields are illustrated in Figure 4.

Following ideas in Refs. 71,72, the following local equivalence was defined [73], where it is assumed that coordinates are chosen so that at the configuration $x \in M$, $\lambda(x) = 1$ (the group identity).

Definition 6.1. Two kinematic mapping germs $\lambda_i : M, x \to SE(n), 1,$

206

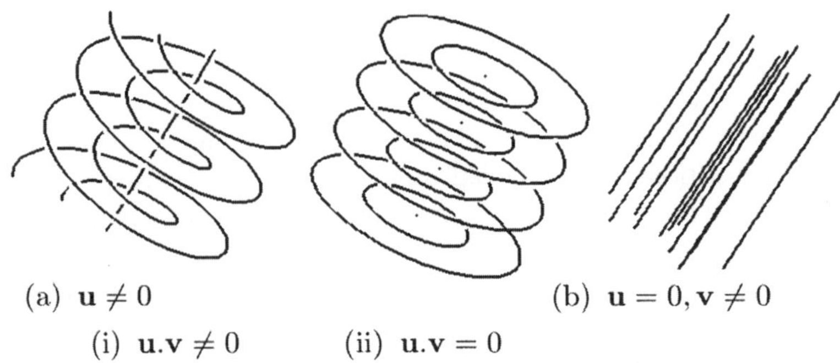

(a) $\mathbf{u} \neq 0$ (b) $\mathbf{u} = 0, \mathbf{v} \neq 0$

(i) $\mathbf{u}.\mathbf{v} \neq 0$ (ii) $\mathbf{u}.\mathbf{v} = 0$

Fig. 4. Infinitesimal motions of $\mathfrak{se}(3)$

$i = 1, 2$ are $\mathcal{I}$-**equivalent** if there exists a germ of a diffeomorphism $\phi :$ $M, x \to M, x$ and an element $g \in SE(n)$ such that

$$\lambda_2 = g^{-1}.(\lambda_1 \circ \phi).g.$$

First-order invariants for $\mathcal{I}$–equivalence arise from the adjoint action of $SE(3)$ on $\mathfrak{se}(3)$. The Lie algebra has non-trivial radical (maximal solvable ideal) $1 \ltimes \mathfrak{t}(3)$ so is not semisimple. In particular its Killing form is degenerate and the polynomial invariant theory for reductive algebras does not apply. Donelan and Gibson [73] determined generators for the ring of invariant polynomials for the adjoint action of $SE(n)$ and in particular:

Theorem 6.1. *The ring of invariant polynomials of the adjoint action of $SE(3)$ on $\mathfrak{se}(3)$ is generated by the Killing form $\langle \mathbf{u}, \mathbf{u} \rangle$ and the Klein form $\langle \mathbf{u}, \mathbf{v} \rangle$.*

This is a special case of the theorem of Panyushev [74] for semi-direct products $\mathfrak{g} \ltimes V$ with $\mathfrak{g}$ reductive.

The effect of ϕ in Definition 6.1 is that we are only interested in the subspace $T_x \lambda(M) \subseteq \mathfrak{se}(3)$, which can be regarded as an element of the Grassmannian of subspaces of dimension $k = \operatorname{rank} T_x \lambda$. Elements of the projective space $P\mathfrak{se}(3)$ are called *screws*. Since the invariants in Theorem 6.1 are both quadratic, their ratio $h = \langle \mathbf{u}, \mathbf{v} \rangle / \langle \mathbf{u}, \mathbf{u} \rangle$ is a projective invariant, called the *pitch* of the screw, and indeed it measures the displacement parallel to the axis during one revolution, as in Figure 4 (a)(i). In the case $\mathbf{u} = 0$, set $h = \infty$.

Since the adjoint action is linear it induces an action on the Grassmannians. The theory of screws and screw systems was first developed extensively by Ball [75] and later revived by Hunt [3]. Hunt put forward a classification scheme based on geometric and engineering intuition. He noted that in engineering it was almost always the special systems that were of interest. A mathematical foundation for the classification was provided by Gibson and Hunt [76]. The principles underlying the classification, which is $\mathcal{I}$–invariant, are the following [73].

(1) $P\mathfrak{se}(3)$ is partitioned by the pencil of quadric hypersurfaces of constant pitch:

$$Q_h(\mathbf{u}, \mathbf{v}) = \langle \mathbf{u}, \mathbf{v} \rangle - h \langle \mathbf{u}, \mathbf{u} \rangle = 0, \qquad h \in \mathbb{R} \cup \{\infty\}.$$

where we use Q_h to denote both the quadratic form and the associated hypersurface. For $h \neq \infty$, $Q_\infty \subset Q_h$ so properly we should use $\tilde{Q}_h = Q_h - Q_\infty$. For $h \neq \infty$ these quadrics have two rulings: by the α-planes, corresponding to the screws of pitch h whose axes pass through a given point, and by the β-planes, corresponding to those whose axes lie in a given plane. Q_0 corresponds to the classical Klein quadric, representing the set of lines in projective 3–space, in terms of Plücker line coordinates, which the screw coordinates generalize. It plays a special role in that its axis (see Figure 4 (a)(ii)) consists of instantaneously stationary points and these are the only screws having such points.

(2) Classify screw systems of a given dimension by how they meet this pencil of quadrics:

   (a) Type I systems do not lie wholly in a pitch quadric and type II do.

   (b) Subtypes A, B, C, D are classified according to the dimension of their intersection with Q_∞.

   (c) Type I subclasses are further subdivided by the projective type of the pencil of intersections. For example 3–systems intersect the pitch quadrics in a pencil of real conics. Subtype IA systems can be distinguished by whether their three principle pitches, corresponding to singular conics in the pencil, are distinct (IA_1) or whether two coincide (IA_2).

   (d) Further refinement is provided by the signs of the moduli such as principal pitches [77], e.g. type IA_1^{+0-} denotes the subclass with principal pitches $h_\alpha > 0$, $h_\beta = 0$ and $h_\gamma < 0$.

(3) Each Q_h derives from an associated bilinear form which gives rise to a polarity on the set of screws: $Q_h(\$_1, \$_2) = 0$. Polarity with respect to Q_0, called *reciprocity*, has a particular physical significance in that screws can be used to represent both infinitesimal motion and generalized force (force + torque). Reciprocity indicates that a generalized

force on a screw $\$_1$ produces zero rate of work on a body free to move on screw $\$_2$. The set of screws $S^\perp$ reciprocal to a k–system S is itself a $(6 - k)$–system, so one can deduce a classification of $(6 - k)$-systems from that for k-systems, $k = 1, 2$.

In Ref. 77, it is shown that all the classes described above are submanifolds in the relevant Grassmannian and their adjacency diagrams are established. In particular:

Theorem 6.2. *The Hunt–Gibson classification of screw systems and its refinement form Whitney regular stratifications of the relevant Grassmannians.*

The stratifications can be translated across the jet bundle $J^1(M, SE(3))$, within motion germs of each rank, to give a regular stratification. It follows from the transversality theorem for stratified sets [78] that for a residual set in $C^\infty(M, SE(3))$ the 1–jet extension is transverse to the stratification. In particular, generically, we only encounter screw systems up to codimension $\dim M$ in the Grassmannian. For example, for $\dim M = 3$, type IA_1^{+0-} has codimension 1 so one would expect to find a surface in m along which the screw system is of this type; but IIA^0, which is an α–plane in Q_0, has codimension 6 so one does not expect to encounter this 3–system generically.

It should be noted that there exist other classifications, though essentially equivalent, such as that of Rico Martínez and Duffy [79,80] and there does not, at this stage, appear to be an accepted standard.

The relevance of screw systems to the study of kinematic singularities is obvious. For example, a 6–link serial manipulator is in a singular configuration precisely when the screws defined by its joints span a screw system of dimension ≤ 5. Karger has written several papers [81–83] exploring the singularities of serial manipulators. Starting from the product of exponentials formula (3), he highlights the connection with closed loops and the significance of not only the screw system itself but also the Lie algebra it generates.

7. Instantaneous Singular Sets and Applications

The Transversality Theorem 5.1 indicates a connection between a kinematic mapping λ and singularities of the associated family of trajectories τ_λ. However, the equivalence relations do not preserve much of the rigid geometry that is an explicit feature of kinematic mappings. On the other hand, $\mathcal{I}$–equivalence does preserve this geometry so it is natural to study trajectory singularities in the context of screw theory.

Definition 7.1. Given the germ of a motion $\lambda : M, x \to SE(n), 1$, its **instantaneous singular set (ISS) at** x is

$$I_{\lambda,x} = \{\mathbf{w} \in \mathbb{R}^n : \tau_{\lambda,w} \text{ singular at } x\}$$

where $\tau_{\lambda,w} = e_w \circ \lambda$.

For example, for a 3R regional manipulator, the ISS in a given configuration is precisely the union of singular surfaces identified by Stanišić and Engelberth [43]. Indeed, they characterize these surfaces in terms of the principal pitches of the associated 3–system. Their classification can be derived from the following [84] which applies to screw systems of any dimension.

Theorem 7.1. *Let* $\lambda : M, x \to SE(3), 1$ *be a kinematic mapping germ with rank* d *and let* S *be the associated screw system. For a point* $\mathbf{w} \in \mathbb{R}^3$ *let* A_w *be the* α–*plane in the Klein quadric* Q_0 *representing the bundle of lines through* $\mathbf{w}$. *Then* $\mathbf{w} \in I_{\lambda,x}$ *if and only if* $S \cap A_w$ *has projective dimension* $\geq \max(0, d - 3)$.

This follows by applying the Chain Rule to $\tau_{\lambda,w}$:

$$\mathbf{w} \in I_{\lambda,x} \iff \operatorname{rank} T_x \tau_{\lambda,w} < \min\{d, 3\}$$
$$\iff \dim(\operatorname{im} T_x \lambda \cap \ker T_1 e_{\mathbf{w}}) > \max\{0, d - 3\} \tag{5}$$

and noting that the relevant subspaces correspond to S and A_w on projectivization. An important corollary is that the ISS is the union of all the lines corresponding to points of $S \cap A_w$ satisfying the condition of the theorem. Also, the affine dimension of the intersection in (5) is precisely the corank of the singularity of the trajectory of $\mathbf{w}$. Moreover, it was also shown that the ISS of a screw system is identical to the ISS of its reciprocal system.

It is a straightforward exercise to describe the ISS associated to a given class of screw systems. In the simplest cases of a 3–system, IA_1^{+--}, for example, the ISS is an elliptic single-sheeted hyperboloid, and each point on it has a singularity of corank 1. For type IA_1^{+0-}, where the intersection with Q_0 is a line pair (singular conic) rather than a non-singular conic, the ISS is an intersecting pair of planes and on the line of intersection there are two distinguished points which have corank 2 singularities. For type IIA^0, corresponding to an α–plane of lines through a given point in $\mathbb{R}^3$, the whole space is singular but the given point has a corank 3 singularity. From the point of view of singularity theory, such a singularity is highly non-generic.

As a final application, we consider a family of parallel manipulators: those for which three points on the platform are constrained to lie on three given surfaces (not necessarily distinct), as in Figure 5. There are connections with the research of Pottmann and Ravani [85] on the singularities of motions where the constraint is that one surface (e.g. a milling head) is required to be in contact with another. However they use only the line geometry relevant to screws of zero pitch rather than full screw systems.

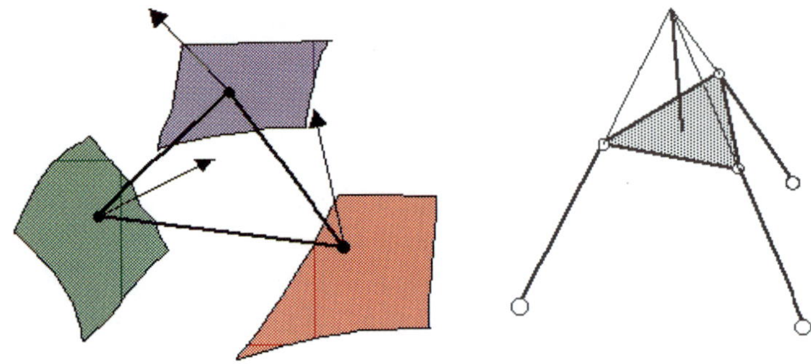

Fig. 5.   3–point constrained parallel manipulator and RCC

Examples are:

- the Darboux motion [10], where three points in a body are constrained to lie in three planes in general position
- the telescope-focussing mechanism of Carretero *et al* [34,35], where three points on the telescope mirror lie on 3 mutually intersecting planes
- a simplified model of the remote centre compliance device [39], illustrated, in which the contact points are constrained to lie on symmetrically placed, congruent spheres.

It was shown in Refs. 66,86 that the screw system in any given configuration is determined by the surface normal lines at the contact points; thought of as screws of pitch 0, they span the reciprocal system. It follows that the normal lines lie in the ISS of the reciprocal system and hence the screw system of the motion itself. This enables us to deduce the screw type from the geometric configuration of the lines. For example, if the normal lines are mutually skew and their directions span $\mathbb{R}^3$ then the screw system type is IA_1^{++-} or IA_1^{+--}. If, however, the normals have independent directions but intersect in a finite point then the type is IIA^0.

Among all 3–dof motions one does not encounter type IIA^0 transversely since it forms a stratum of codimension 6. Moreover it is associated to the existence of a trajectory of corank 3 and hence $\mathcal{A}$–codimension at least 9 (sitting within a 3–parameter family of trajectories). However, among 3–point constrained parallel motions we have the following [86].

Theorem 7.2. *Suppose a 3–point motion has a type IIA^0 screw system. So long as a simple transversality condition is satisfied, then there is an open neighbourhood of contact triangles (and indeed of contact surfaces), containing the given motion, for which the corresponding motion also has a type IIA^0 screw system.*

For a Darboux motion with acute contact triangle and for the simplified RCC, there exist configurations at which the screw system is type IIA^0 and the conditions of Theorem 7.2 are satisfied. Hence these can be regarded as stable occurrences, meaning that the class of surface–constrained motions is not generic in the sense of Gibson and Hobbs.

A similar theorem holds also for type IIB^0 3–systems, which also form a class of codimension 6. This includes the telescope-focussing device. However in that case the transversality condition fails, meaning that the singular phenomenon being utilized is sensitive to the design parameters of the mechanism.

8. Conclusion

Singularities are of great interest and importance in robot manipulator design and control. A number of attempts have been made to apply the methods and perspective of singularity theory. This has resulted in the discovery of some powerful general kinematic theorems and, in some cases, a better understanding of the singularities of specific classes of manipulators. However there is a tension between the engineering and mathematical approaches. Many practical manipulators have special geometries that render statements about generic situations inapplicable. In practice, one often focusses on the non-generic cases, though here singularity theory can provide the right conceptual framework and language and may provide standard local models of the bifurcation set within the manipulator class.

There are several directions for future research in robot manipulator singularities that are likely to be fruitful. These include:

- Further exploration of specific finite–dimensional classes of manipulators with a view to finding transversality theorems in the spirit of, for example, Ref. 87.

- A more detailed understanding of the singularities of serial robot manipulators, where the product of exponentials representation of the kinematic mapping highlights the important interaction with the Lie algebra structure of the screw space.

- Application of algebraic and semi-algebraic singularity theory to the study of parallel manipulators where, frequently, the relevant constraints and mappings can be written in polynomial form. This is linked to generalizations of Kempe's Theorem [88] that (arbitrary large sections of) every plane algebraic curve can be generated as the output curve of a planar mechanism.

- The identification of higher order invariants of the adjoint action. For time dependent motion these were developed by Veldkamp [10,89]. These are also important for understanding the differential-geometric properties of manipulator motions and their trajectories.

- Exploration of the role of symmetry on manipulator singularities. Frequently, the presence of symmetries can give rise to unexpected singularity types. For example, most of the known over-constrained closed loop mechanisms possess symmetries and $\mathbb{Z}_2$–symmetries arise where mechanisms have up–down poses arising from consecutive revolute joints.

- Extension of existing results on singularity avoidance through determination of the topology of the sets of singularities and how these bifurcate under change of design parameters.

- Links to control theory and Lagrangian singularities for the dynamics of manipulators. It is already known that there are subtle links between the geometry of mechanisms and the theory of caustics [87] so it is not unreasonable to expect that there are links between symplectic geometry and the singularities of manipulators.

References

1. IFToMM Commission A, Terminology for Theory of Machines and Mechanisms, *Mechanism and Machine Theory*, **26** (1991) 435–539
2. *Manipulating Industrial Robots—Vocabulary*, ISO 8373, 1994
3. K. H. Hunt, *Kinematic Geometry of Mechanisms* (Clarendon Press, Oxford, 1978)
4. R. Connolly and E. D. Demaine, Geometry and Topology of Polygonal Link-

ages, *Handbook of Discrete and Computational Geometry, 2nd ed.* (CRC Press, Boca Raton, 2004) 197–218

5. ParalleMIC, Terminology Related to Parallel Mechanisms, http://www.parallemic.org/Terminology/General.html

6. S. D. Guest and P. W. Fowler, A Symmetry-Extended Mobility Rule, *Mechanism and Machine Theory*, **40** (2005) 1002–1014

7. J. M. Hervé, Analyse Structurelle des Mécanismes par Groupe des Déplacements, *Mechanism and Machine Theory*, **13** (1978) 437–450

8. J. Beckers, J. Patera, M. Perroud and P. Winternitz, Subgroups of the Euclidean Group and Symmetry Breaking in Nonrelativistic Quantum Mechanics, *J. Math. Phys.*, **18** (1977) 72–83

9. D. Lazard, On the Representation of Rigid–Body Motions and its Application to Generalized Platform Manipulators, *Computational Kinematics*, eds: J. Angeles *et al.* (Kluwer Academic, Dordrecht, 1993) 175–181

10. O. Bottema and B. Roth, *Theoretical Kinematics* (Dover Publications, New York, 1990)

11. J. M. McCarthy, *Introduction to Theoretical Kinematics* (MIT Press, Cambridge MA, 1990)

12. J. Selig, *Geometrical Fundamentals of Robotics* (Springer Verlag, New York, 2005)

13. R. Brockett, Robotic Manipulators and the Product of Exponentials Formula, in *Proc. Mathematical Theory of Networks and Systems, Beer-Sheva, Israel,* ed. P Fuhrman, (1984) 120–129

14. J. Denavit and R. S. Hartenberg, A Kinematic Notation for Lower Pair Mechanisms based on Matrices, *ASME J. Applied Mechanics*, **22** (1955) 215–221

15. R. M. Murray, Z. Li and S. S. Shastry, *A Mathematical Introduction to Robotic Manipulation* (CRC Press, Boca Raton, 1994)

16. T. Koetsier, A Contribution to the History of Kinematics I, *Mechanism and Machine Theory*, **18** (1983) 37–42

17. S. Roberts, On Three–Bar Motion in Plane Space, *Proc. London Math. Soc..* **7** (1875) 14–23

18. C. G. Gibson and P. E. Newstead, On the Geometry of the Planar 4–Bar Mechanism, *Acta Applicandae Mathematicae*, **7** (1986) 113–135

19. M. Golubitsky and V. Guillemin, *Stable Mappings and Their Singularities* (Springer Verlag, New York, 1973)

20. V. E. Gough and S. G. Whitehall, Universal Tyre test Machine, *Proc. 9th Int. Technical Congress FISITA*, (1962) 117–137

21. D. Stewart, A Platform with 6 Degrees of Freedom, *Proc. Inst. Mechanical Engineers, London*, **180** (1965) 371–386

22. H. MacCallion and D. T. Pham, The Analysis of a Six Degree of Freedom Work Station for Mechanized Assembly, *Proc. 5th World Congress on the Theory of Machines and Mechanisms, Montreal*, (1979) 611–616

23. I. Bonev, The True Origins of Parallel Robots, *ParalleMIC Reviews*, **7** (2003) http://www.parallemic.org/Reviews/Review007.html

24. B. Dasgupta and T. S. Mruthyunjaya, The Stewart Platform: a Review, *Mechanism and Machine Theory*, **35** (2000) 15–40

25. C. Gosselin and J. Angeles, Singularity Analysis of Closed-Loop Kinematic Chains, *IEEE Trans. Robotics and Automation*, **6** (1990) 281–290

26. D. Zlatanov, R. G. Fenton and B. Benhabib, Singularity Analysis of Mechanisms and Robots via a Motion–Space Model of the Instantaneous Kinematics, *Proc. IEEE Int. Conf. on Robotics and Automation, San Diego, CA*, (1994) 980–985

27. D. Zlatanov, R. G. Fenton and B. Benhabib, Singularity Analysis of Mechanisms and Robots via a Velocity–Equation Model of the Instantaneous Kinematics, *Proc. IEEE Int. Conf. on Robotics and Automation, San Diego, CA*, (1994) 986–991

28. N. Simaan and M. Shoham, Singularity Analysis of Composite Serial In-Parallel Robots, *IEEE Trans. Robotics and Automation*, **17** (2001) 301–311

29. K. Sugimoto, J. Duffy and K. H. Hunt, Special Configurations of Spatial Mechanisms and Robot Arms, *Mechanism and Machine Theory*, **17** (1982) 119–132

30. F. L. Litvin, Z. Yi, V. Parenti-Castelli and C. Innocenti, Singularities, Configurations and Displacement Functions for Manipulators, *Int. J. Robotics Research*, **5** (1986) 66–74

31. S. L. Wang and K. J. Waldron, A Study of the Singular Configurations of Serial Manipulators, *Trans. ASME J. Mechanisms, Transmissions and Automation in Design*, **109** (1987) 14–20

32. J. Kieffer, Differential Analysis of Bifurcations and Isolated Singularities for Robots and Mechanisms, *IEEE Trans. Robotics and Automation*, **10** (1994) 1–10

33. J. Kieffer and J. Lenarčič, On the Exploitation of Mechanical Advantage Near Robot Singularities, *Proc. 3rd Intl. Workshop on Advances in Robot Kinematics, Ferrara, Italy*, (1992) 65–72

34. J. A. Carretero, M. Nahon, B. Buckham and C. M. Gosselin, Kinematic Analysis of a Three-DoF Parallel Mechanism for Telescope Applications, *Proc. ASME Design Engineering Technical Conf., Sacramento*, ASME, 1997

35. J. A. Carretero, R. P. Podhorodeski and M. Nahon, Architecture Optimization of a Three-DoF Parallel Mechanism , *Proc. ASME Design Engineering Technical Conf., Atlanta* ASME, 1998

36. C. A. Hobbs C, Singularities of Mechanisms with One Degree of Freedom, unpublished

37. J. L. Nevins and D. E. Whitney, Assembly Research, *Automation*, **16** (1980) 595–613

38. P. C. Watson, A Multidimensional System Analysis of the Assembly Process as Performed by a Manipulator, presented at *1st North American Robot Conf., Chicago*, (1976)

39. D. E. Whitney and J. L. Nevins, What is the RCC and what can it do?, *Robot Sensors, Tactile and Non-Vision*, ed. A.Pugh, IFS Publications, (1986) 3–15

40. D. H. Gottlieb, Robots and Fibre Bundles, *Bull. Soc. Math. Belg.*, **38** (1986) 219–223

41. D. R. Baker, Some Topological Problems in Robotics, *The Mathematical Intelligencer*, **12** (1990) 66–76

42. F. L. Litvin and V. Parenti-Castelli, Configurations of Robot Manipulators and Their Identification and the Execution of Prescribed Trajectories, *Trans. ASME J. Mechanisms, Transmissions and automation in Design*, **107** (1985) 170–188

43. M. M. Stanišić and J. W. Engelberth, A Geometric Description of Manipulator Singularities in Terms of Singular Surfaces, *Proc. 1st Int. Meeting of Advances in Robot Kinematics, Ljubljana, Slovenia*, (1988) 132–141

44. K. Tchoń and R. Muszynski, Singularities of Nonredundant Robot Kinematics, *Int. J. Robotics Research*, **16** (1997) 71–89

45. K. Tchoń, Singularities of the Euler Wrist, *Mechanism and Machine Theory*, **35** (2000) 505–515

46. J. P. Merlet, Singular Configurations of Parallel Manipulators and Grassmann Geometry, *Int. J. Robotics Research*, **8** (1992) 45–56

47. R. Di Gregorio and V. Parenti-Castelli, Mobility Analysis of the 3-UPU Parallel Mechanism Assembled for a Pure Translational Motion, *Proc. IEEE/ASME Int. Conf. on Advanced Intelligent Mechatronics, Atlanta, Georgia*, (1999) 520–525

48. D. Zlatanov, I. A. Bonev and C. M. Gosselin, Constraint Singularities of Parallel Mechanisms , *Proc. IEEE Int. Conf. on Robotics and Automation, Washington, DC*, (2002) 496–502

49. D. K. Pai and M. C. Leu, Genericity and Singularities of Robot Manipulators, *IEEE Trans. Robotics and Automation*, **8** (1992) 545–559

50. J. W. Burdick, A Classification of 3R Regional Manipulator Singularities and Geometries, *Mechanism and Machine Theory*, **30** (1995) 71–89

51. M. Baili, P. Wenger and D. Chablat, Classification of One Family of 3R Positioning Manipulators, *Proc. 11th Int. Conf. on Advanced Robotics, Coimbra, Portugal*, (2003)

52. P. Wenger, Classification of 3R Positioning Manipulators, *ASME J. Mechanical Design*, **120** (1998) 327–332

53. P. Wenger and J. El Omri, Changing Posture for Cuspidal Robot Manipulators, *IEEE Int. Conf. on Robotics and Automation, Minneapolis*, (1996) 3173–3178

54. C. G. Gibson and C. A. Hobbs, Simple Singularities of Space Curves, *Math. Proc. Camb. Phil. Soc.*, **113** (1992) 297–310

55. C. T. C. Wall, Geometric Properties of Differentiable Manifolds, *Geometry and Topology, Rio de Janeiro*, Lecture Notes in Mathematics **597**, Springer, Berlin, (1976) 707–774

56. J. Montaldi, On Generic Composites of Maps, *Bull. London Math. Soc.*, **23** (1991) 81–85

57. C. G. Gibson, Kinematic Singularities—A New Mathematical Tool, *Proc. 3rd Int. Workshop on Advances in Robot Kinematics, Ferrara, Italy*, (1992) 209–215.

58. P. S. Donelan, C. G. Gibson and W. Hawes, Trajectory Singularities of General Planar Motions, *Proc. Royal Soc. Edinburgh*, **129A**, (1999) 37–55

59. C. G. Gibson, W. Hawes and C. A. Hobbs, Local Pictures for General Two–Parameter Motions of the Plane, *Advances in Robot Kinematics and Com-*

putational Geometry, Kluwer Academic Publishers, (1994) 49–58.

60. C. G. Gibson and C. A. Hobbs, Local Models for General One–Parameter Motions of the Plane and Space, *Proc. Royal Soc. Edinburgh*, **125A** (1995) 639–656

61. C. G. Gibson and C. A. Hobbs, Singularity and Bifurcation for General Two–Dimensional Planar Motions, *New Zealand J. Math.*, **25** (1996) 141–163

62. C. G. Gibson, C. A.. Hobbs and W. L. Marar, On Versal Unfoldings of Singularities for General Two–Dimensional Spatial Motions, *Acta Applicandae Mathematicae*, **47** (1996) 221–242

63. C. G. Gibson, D. Marsh and Y. Xiang, Singular Aspects of Generic Planar Motions with Two Degrees of Freedom, *Int. J. Robotics Research*, **17** (1998) 1068–1080

64. W. Hawes, Multi–Dimensional Motions of the Plane and Space, Ph. D. Thesis, University of Liverpool (1995)

65. C. A. Hobbs, Kinematic Singularities of Low Dimension, Ph. D. Thesis, University of Liverpool (1993)

66. M. W. Cocke, Natural Constraints on Euclidean Motions, PhD Thesis, University of Liverpool (1998)

67. G. T. Bennett, A New Mechanism, *Engineering*, **76** (1903) 777–778.

68. G. T. Bennett, The Skew Isogram Mechanism, *Proc. London Math. Soc. (2nd series)*, **13** (1913) 151–173.

69. J. Lerbet and K. Hao, Kinematics of Mechanisms to the Second Order— Application to the Closed Mechanisms, *Acta Applicandae Mathematicae*, **59** (1999) 1–19

70. A. Sagle and R. Walde, *Introduction to Lie Groups and Lie Algebras* (Academic Press, New York, 1973)

71. P. S. Donelan, Generic Properties of Euclidean Kinematics, *Acta Applicandae Mathematicae*, **12** (1988) 265–286

72. P. S. Donelan, On the Geometry of Planar Motions, *Quarterly J. Math. Oxford*, **44** (1993) 165–184

73. P. S. Donelan and C. G. Gibson, First–Order Invariants of Euclidean Motions, *Acta Applicandae Mathematicae* **24** (1991) 233–251

74. D. I. Panyushev, Semi-Direct Products of Lie Algebras, Their Invariants and Representations, arXiv:math.AG/0506579

75. R. S. Ball, *The Theory of Screws* (Cambridge University Press, Cambridge, 1900)

76. C. G. Gibson and K. H. Hunt, Geometry of Screw Systems I & II, *Mechanism and Machine Theory*, **25** (1990) 1–27

77. P. S. Donelan and C. G. Gibson, On the Hierarchy of Screw Systems, *Acta Applicandae Mathematicae*, **32** (1993) 267–296

78. C. G. Gibson, K. Wirthmüller, A. A. Du Plessis and E. Looijenga, Topological Stability of Smooth Mappings, Lecture Notes in Mathematics 552 (Springer Verlag, Berlin, 1976)

79. J. M. Rico Martíez and J. Duffy, Orthogonal Spaces and Screw Systems, *Mechanism and Machine Theory*, **27** (1992) 451–458

80. J. M. Rico Martíez and J. Duffy, Classification of Screw Systems I and II,

Mechanism and Machine Theory, **27** (1992) 459–490

81. A. Karger, Classification of Robot–Manipulators with only Singular Configurations, *Mechanism and Machine Theory*, **30** (1995) 727–736

82. A. Karger, Classification of Serial Robot–Manipulators with Non–Removable Singularities, *Trans. ASME J. Mechanical Design*, **118** (1996) 202–208

83. A. Karger, Singularity Analysis of Serial Robot–Manipulators, *Trans. ASME J. Mechanical Design*, **118** (1996) 520–525

84. M. W. Cocke, P. S. Donelan and C. G. Gibson, Instantaneous Singular Sets Associated to Spatial Motions, in *Real and Complex Singularities, São Carlos, 1998*, eds. F. Tari and J. W. Bruce, Res. Notes Math., **412** (Chapman and Hall/CRC Press, Boca Raton, 2000) 147–163

85. H. Pottmann and B. Ravani, Singularities of Motions Constrained by Contacting Surfaces, *Mechanism and Machine Theory*, **35** (2000) 963–984

86. M. W. Cocke, P. S. Donelan and C. G. Gibson, Trajectory Singularities for a Class of Parallel Mechanisms, to appear in *Proc. 8th Intl. Workshop on Real and Complex Singularities, Luminy, France, 2004*

87. J. W. Bruce, P. J. Giblin and C. G. Gibson, On Caustics by Reflexion, *Topology*, **21** (1982) 179–199

88. A. B. Kempe, A Method of Describing Curves of the nth Degree by Linkwork, *Proc. London Math. Soc.*, **7** (1876) 213–216

89. G. R. Veldkamp, Canonical Systems and Instantaneous Invariants in Spatial Kinematics, *J. Mechanisms*, **2** (1967) 329–388

Singularity and stratification theory applied to dynamical systems

Michael Field

Department of Mathematics
Imperial College London, UK, and
University of Houston
Houston, TX 77204-3008, USA
** E-mail: mf@uh.edu*

We outline the theory of equivariant transversality for maps equivariant with respect to a compact Lie group. We indicate some applications to generic equivariant bifurcation theory.

Keywords: equivariant transversality, compact Lie group, dynamical systems

1. Introduction

Transversality theory is a basic technical and theoretical tool in the study of smooth mappings, dynamical systems and bifurcation theory. In this paper we describe a version of transversality theory applicable to the study of maps and vector fields which are equivariant under the smooth action of a compact Lie group G. From a local point of view, we will be outlining a theory for the analysis of solutions of symmetric equations. From the global point of view, we will be describing an intersection theory for G-manifolds.

The foundational theory of transversality for G-manifolds was developed in the mid 1970's by Bierstone [1] and the author [10]. The focus of Bierstone's work was on extending Mather's theory of stable mappings to smooth equivariant maps. As part of that program, Bierstone extended Thom's jet transversality theorem to equivariant maps [2]. On the other hand, Field's motivation was to extend parts of the Smale program to equivariant dynamical systems [11]. Much later it turned out that techniques of equivariant transversality had powerful applications to equivariant bifurcation theory [16,12–14]. Very recently there have also been applications to equivariant reversible systems [8] and equivariant Hamiltonian systems [6]. In this paper, we emphasize applications of equivariant transversality to the

220

bifurcation theory of equivariant vector fields. For a more comprehensive and careful introduction to the theory we refer the reader to the forthcoming monograph [15] which includes the general theory of equivariant transversality and jet transversality as well as applications to equivariant and reversible equivariant dynamical systems and relative equilibria. Part of the motivation for writing this paper was to provide an introduction to some of the main results described in [15] as they apply to bifurcation theory.

1.1. *Equivariant transversality*

Let G be a compact Lie group of transformations acting smoothly (that is, C^∞) on connected differential manifolds M and N. A map $f : M \to N$ is *G-equivariant* if $f(gx) = gf(x)$, for all $g \in G$, $x \in M$. Suppose that Y is a closed G-invariant submanifold of N. We want to describe the 'generic' intersection $f^{-1}(Y)$. If there is no symmetry, then the intersection is generic if f is *transverse* to Y — in symbols $f \pitchfork Y$. That is, $\forall x \in M$, either $f(x) \notin Y$ or $T_x f(T_x M) + T_{f(x)} Y = T_{f(x)} N$. It is easy to describe the local structure of $f^{-1}(Y)$. Every $x \in f^{-1}(Y)$ has an open neighbourhood in $f^{-1}(Y)$ diffeomorphic to an open disk in $\mathbb{R}^n$, where $n = \dim(M) - \dim(N) + \dim(Y)$. In particular,

(a) If $\dim(M) < \dim(N) - \dim(Y)$, then $f^{-1}(Y) = \emptyset$.
(b) $f^{-1}(Y)$ is nonsingular and the local topological type of $f^{-1}(Y)$ is constant.

Neither of these statements need hold in the equivariant context.

Example 1.1. Let $G = \mathbb{Z}_2$ act linearly on $\mathbb{R}^2$ by $(x, t) \mapsto (\pm x, t)$ and on $\mathbb{R}$ by $y \mapsto \pm y$. Let $Y = \{0\} \subset \mathbb{R}$. Suppose that $f : \mathbb{R}^2 \to \mathbb{R}$ is a smooth $\mathbb{Z}_2$-equivariant map

$$f(-x, t) = -f(x, t), \quad (x, t) \in \mathbb{R}^2.$$

Since $f(0, t) \equiv 0$, we may write $f(x, t) = xg(x, t)$, where g is smooth and even in x. Clearly $f^{-1}(Y) \supset \{0\} \times \mathbb{R}$. The map f will be transverse to Y at $(0, t)$ if and only if $g(0, t) \neq 0$. Suppose that $g(0, t_0) = 0$. It follows from the implicit function theorem that if $\frac{\partial g}{\partial t}(0, t_0) \neq 0$, then there will be a curve of solutions to $g(x, t) = 0$ passing through $(0, t_0)$ and distinct from $x = 0$ (in fact perpendicular to $x = 0$ at $(0, t_0)$ — see Figure 1). Consequently, $f^{-1}(Y)$ will be singular at $(0, t_0)$. The singularity cannot be removed by (small) perturbations of f. Indeed, it is easy to construct examples on compact $\mathbb{Z}_2$-manifolds where the singularities in the intersection cannot be removed by any $\mathbb{Z}_2$-equivariant deformation of f.

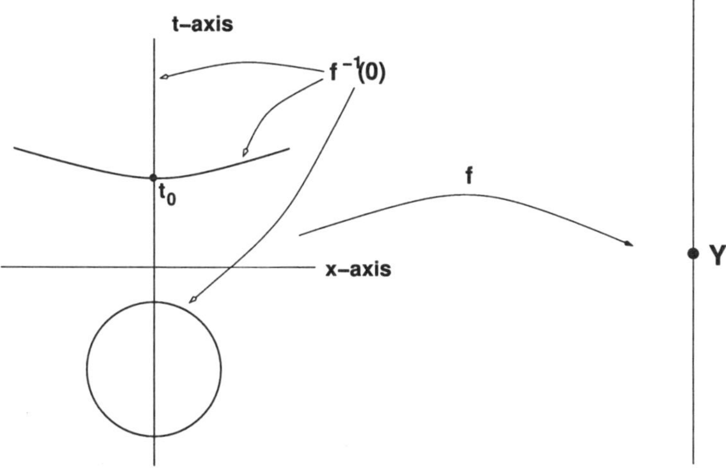

Fig. 1. An example of a generic $\mathbb{Z}_2$-equivariant intersection

Unlike what happens when there is no symmetry, it is not easy to give a simple geometric description of what it means for a map $f : M \to N$ to be G-transverse to a G-invariant submanifold Y of N. However, the problem is certainly local and using slice theory we may reduce to the case where M and N are G-representations and Y is the origin of N. We give the local definition of equivariant transversality in section 3 of this paper. It follows from the definition that if f is G-transversal to Y (we write this $f \pitchfork_G Y$), then the local topological type of $Z = f^{-1}(Y)$ is that of a real algebraic variety. Although the local topological type of Z will be locally constant on an open and dense subset Z_0 of Z, it is not (yet) known whether the local topological type is constant on Z_0 (it is for families of equivariant vector fields). The usual openness and density and isotopy theorems hold for G-transversality. Specifically, if $Y \subset N$ is a closed G-invariant submanifold then the set of maps $f : M \to N$ such that $f \pitchfork_G Y$ is an open and dense subset of the space of all smooth G-equivariant maps from M to N (Whitney C^∞-topology). The definition of G-transversal is open in the sense that if f is G-transversal to Y at $x \in M$ then f will be G-transversal to Y at x' for all x' in some neighbourhood of x in M. Finally, there is an equivariant version of Thom's isotopy theorem. However, isotopies will in general *not* be smooth. (We refer to [1,10,15] for precise statements and proofs.)

222

1.2. *Applications to vector fields and bifurcation theory*

We are interested in studying properties of diffeomorphisms and, in particular, vector fields which are symmetric or *equivariant* with respect to G. Noting that the G action on M extends in the obvious way to a smooth G-action on the tangent bundle TM of M, we say that a vector field $X : M \to TM$ is G-equivariant if

$$X(gx) = gX(x), \quad (x \in M, g \in G).$$

For simplicity, assume for the remainder of the introduction that M is compact and G is finite. Give the space $\mathcal{X} = C_G^\infty(TM)$ of smooth G-equivariant vector fields on M the C^r topology, where $1 \le r \le \infty$. It is not hard to show (see [11,15] for details) that

(1) There is a C^1 open and dense subset $\mathcal{X}_1$ of $C_G^\infty(TM)$ consisting of vector fields all of whose equilibria are hyperbolic.
(2) Given $T > 0$, there is a C^1 open and dense subset $\mathcal{X}_2(T)$ consisting of vector fields with all periodic orbits, period $\le T$, hyperbolic.
(3) There is a residual subset $\mathcal{X}_3$ of $\cap_{T>0}\mathcal{X}_2(T)$ (C^∞-topology) consisting of vector fields such that all invariant manifolds of equilibria and limit cycles meet G-transversally (equivariant version of the Kupka-Smale theorem).

There is an analogous result when G is compact but not finite. In this case, 'equilibria' (resp. 'periodic orbits') is replaced by 'relative equilibria' (resp. 'relative periodic orbits') and 'hyperbolic' by 'normally hyperbolic'.

Suppose that $X : M \times \mathbb{R} \to M$ is a smooth 1-parameter family of equivariant vector fields on M. For $\lambda \in \mathbb{R}$, set $X_\lambda(x) = X(x, \lambda)$, so that $X_\lambda \in \mathcal{X}$, all $\lambda \in \mathbb{R}$. It follows from (1) that for generic families $X_\lambda \in \mathcal{X}_1$ except for λ lying in a discrete subset $\mathcal{B}(X) \subset \mathbb{R}$. If $\lambda_0 \in \mathcal{B}(X)$ there exists at least one equilibrium x_{λ_0} for X_{λ_0} which is not hyperbolic. We want to describe the typical bifurcation behavior of the family X_λ near the bifurcation point $(x_{\lambda_0}, \lambda_0)$. That is, the typical local structure of the germ of $X^{-1}(0)$ at $(x_{\lambda_0}, \lambda_0)$. In the case where there is no group action, it is well-known that generically $X^{-1}(0)$ is a non-singular curve. The only generic bifurcation of equilibria that we see in 1-parameter families is the *saddle-node* bifurcation and this corresponds to a change of stability (index) along the curve of equilibria. Bifurcations of saddle-node type can occur in families of equivariant vector fields. However, these bifurcations are well-understood and involve little in the way of new ideas — essentially everything is reduced via slices to the non-equivariant case. Our focus will be on investigating bifurcations where the symmetry plays an essential role. Typically in these bifurcations we see the appearance of new branches at the bifurcation point that have different (less) symmetry: a symmetry breaking bifurcation. Furthermore, the germ of $X^{-1}(0)$ will generally be singular (this is the case in almost all known examples). We describe some of the basic ideas and indicate the proof of a characteristic genericity and determinacy theorem in section 4. All of what we describe works also for general compact groups G, relative equilibria and (relative) periodic orbits. There is also a theory for equivariant maps (see [14,15]).

We start with a section covering basic definitions and results on smooth G-actions, equivariant maps, stratifications and semialgebraic sets. Much of this section is directed towards experts in dynamical systems who are not familiar with singularity theory and the geometry of stratified sets.

2. Preliminaries and notation

2.1. *Smooth G-actions*

We start by reviewing some facts about smooth actions by compact Lie groups. Proofs and more details may be found in chapter VI of the text by Bredon [7].

Let G be a compact Lie group acting smoothly on the connected differential manifold M. If $x \in M$, let $Gx = \{gx \mid g \in G\}$ denote the G-orbit through x and $G_x = \{g \in G \mid gx = x\}$ denote the isotropy subgroup of (the action of) G at x. Each isotropy group G_x is a closed (therefore Lie) subgroup of G and Gx is (G-equivariantly) diffeomorphic to the compact homogeneous space G/G_x.

Points $x, y \in M$ have the same *isotropy type* if G_x, G_y are conjugate subgroups of G. If $y = gx$, then $G_y = gG_xg^{-1}$ and so all points on the same G-orbit have the same isotropy type. Denote the set of isotropy types for the action of G on M by $\mathcal{O} = \mathcal{O}(M, G)$. If M is compact or G is a linear action on a finite dimensional vector space, then $\mathcal{O}$ is finite.

Given $x \in M$, let $\iota(x) \in \mathcal{O}$ denote the isotropy type of x. If $\tau \in \mathcal{O}$, define $M_\tau = \{x \in M \mid \iota(x) = \tau\}$. In this way we define a partition $\mathcal{M} = \{M_\tau \mid \tau \in \mathcal{O}\}$ into points of the same isotropy type. We refer to $\mathcal{M}$ as the stratification of M by isotropy type or the *orbit stratification* of M. Using slices (see Bredon [7]), it is easy to show that each stratum M_τ is a smooth G-invariant submanifold of M and that $\mathcal{M}$ is a Whitney stratification of M (see subsection 2.5).

We define a partial order $<$ on $\mathcal{O}$ by $\tau < \mu$ if $\exists H \in \tau$, $\exists J \in \mu$ such that $H \subsetneq J$. We remark that this condition holds if $\partial M_\tau \cap M_\mu \neq \emptyset$. The converse is true for linear actions.

There exists a unique minimal isotropy type ν and M_ν is an open and dense subset of M. In the sequel we refer to ν as the *principal* isotropy type and M_ν as the principal stratum. If τ is a *maximal* isotropy type, M_τ is always a closed submanifold of M. Linear actions have a unique maximal isotropy type (G). Nonlinear actions may have many maximal isotropy types.

If H is a subset of G, let $M^H = \{x \in M \mid Hx = x\}$ (in the bifurcation literature, this subspace is often denoted by $\mathrm{Fix}(H)$). The fixed point space M^H is a closed submanifold of M and $M^H = M^{\langle H \rangle}$ ($\langle H \rangle$ is the closure of the subgroup of G generated by H). If $H \in \tau \in \mathcal{O}$, then $M_\tau^H \subset M^H$. The inclusion will be strict unless τ is a maximal isotropy type. We have

$$M_\tau = \cup_{H \in \tau} M_\tau^H.$$

Suppose N is a G-manifold and $f : M \to N$ is G-equivariant. For all $x \in M$, $G_{f(x)} \supset G_x$. It follows that given $H \subset G$ we have

$$f(M^H) \subset N^H.$$

If f is 1:1 then $\mathcal{O}(M) \subset \mathcal{O}(N)$ and f preserves isotropy type.

$$f(M_\tau) \subset N_\tau, \text{ for all } \tau \in \mathcal{O}(M). \tag{1}$$

If f is a diffeomorphism we have equality in (1).

2.2. *Equivariant vector fields*

Suppose that X is an equivariant vector field on M. We list some simple consequences of equivariance and (1).

(1) If $X(x) = 0$, then $X(gx) = 0$, all $g \in G$. (Equilibria occur in group orbits.)
(2) The flow $\phi_t^X = \phi_t$ of X is G-equivariant: $\phi(gx, t) = \phi_t(gx) = g\phi_t(x)$, all $x \in M$, $g \in G$ (we assume flows are complete — defined for all time. This is so if M is compact and can be achieved by time rescaling if M is not compact).
(3) The flow respects the orbit stratification $\mathcal{M}$ and X is tangent to each orbit stratum.
(4) The G-orbit Gx is a *relative equilibrium* of X if X is tangent to Gx (by equivariance, tangent at one point will suffice). Equilibria are always relative equilibria. The converse is only true if G is finite. What we

discuss applies to relative equilibria and non-finite groups — however, we only describe results for equilibria.

2.3. *Representations*

Let V be a real finite dimensional inner product space. An orthogonal representation (V, G) of the compact Lie group G on V is a homomorphism $\rho : G \to O(V)$. We have a corresponding action of G on V by orthogonal transformations. The action is *trivial* if $G\mathbf{v} = \mathbf{v}$ for all $\mathbf{v} \in V$ (that is $V_{(G)} = V$) and *irreducible* if there are no proper G-invariant linear subspaces of V.

In this paper we assume representations are defined over $\mathbb{R}$. Let $L_G(V, V)$ denote the space of linear G-equivariant maps. If (V, G) is an irreducible representation then it follows from Frobenius' theorem that $L_G(V, V)$ is isomorphic (as a division algebra) to either $\mathbb{R}$, $\mathbb{C}$ or $\mathbb{Q}$ (the quaternions).

We shall only consider *absolutely irreducible* representations where $L_G(V, V) \approx \mathbb{R}$ (we refer to [13,15] for the general theory).

Example 2.1. Let $\mathbf{D}_n \subset O(2)$ denote the group of isometries of the regular n-gon. The induced action of $\mathbf{D}_n$ on $\mathbb{R}^2$ is absolutely irreducible for all $n \geq 3$. Similarly, the symmetry groups of the platonic solids and $SO(3)$, $O(3)$ define absolutely irreducible representations on $\mathbb{R}^3$. In Figure 2 we show the orbit stratification of $\mathbb{R}^2$ for the standard action of $\mathbf{D}_4$ on $\mathbb{R}^2$.

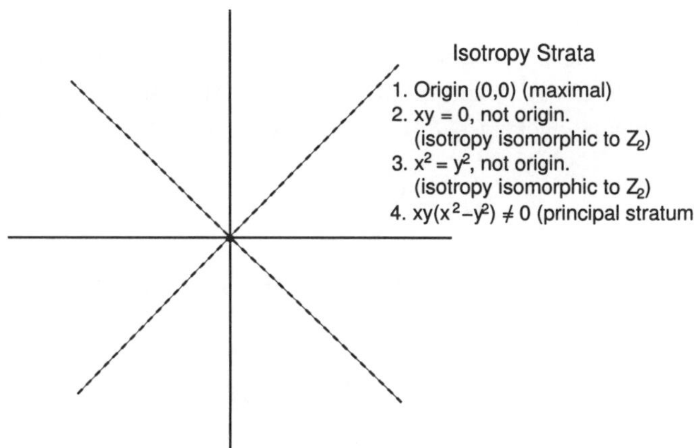

Fig. 2.   Orbit stratification of $\mathbb{R}^2$ when $G = \mathbf{D}_4$

Note that non-zero points on the diagonals $x^2 = y^2$ and axes $xy = 0$ have isotropy isomorphic to $\mathbb{Z}_2$. However, these isotropy groups are not conjugate within $\mathbf{D}_4$ and so define *different* isotropy types.

2.4. *Smooth invariant theory*

Let (V, G), (W, G) be G-representations. Let $P(V)^G$ denote the $\mathbb{R}$-algebra of G-invariant polynomials on V and $P_G(V, W)$ denote the $P(V)^G$-module of G-equivariant polynomial maps from V to W. It follows from the Hilbert basis theorem (using Haar integration) that $P(V)^G$ is finitely generated as an $\mathbb{R}$-algebra and $P_G(V, W)$ is finitely generated as a $P(V)^G$-module (see [24,19]).

Let $\mathcal{F} = \{F_1, \ldots, F_k\}$ be a minimal set of homogeneous generators for the $P(V)^G$-module $P_G(V, W)$. Let $\deg(F_j) = d_j$ and label the generators so that $0 \le d_1 \le \ldots \le d_k$. It follows easily from minimality and the homogeneity of the F_j that the number of generators k and the degrees $d_1, \ldots, d_k$ depend only on the isomorphism class of the representations (V, G), (W, G) (see also Remarks 3.1).

Let $C^\infty(V)^G$ denote the $\mathbb{R}$-algebra of G-invariant smooth functions on V and $C_G^\infty(V, W)$ denote the $C^\infty(V)^G$-module of G-equivariant smooth maps from V to W. It follows from the equivariant version of Stone-Weierstrass approximation theorem that the $C^\infty(V)^G$-submodule of $C_G^\infty(V, W)$ generated by $\mathcal{F}$ is a dense subset of $C_G^\infty(V, W)$ (here, as elsewhere in this section, we always take the C^∞-topology). Since $\mathcal{F}$ consists of a finite set of polynomials, it follows by results of Malgrange [21,27] on closed ideals of differentiable functions that the $C^\infty(V)$-submodule of $C^\infty(V, W)$ generated by $\mathcal{F}$ is closed in the C^∞-topology. Averaging over G using Haar measure, it follows that the $C^\infty(V)^G$-submodule of $C_G^\infty(V, W)$ generated by $\mathcal{F}$ equals $C_G^\infty(V, W)$. That is, every $f \in C_G^\infty(V, W)$ may be written

$$f(x) = \sum_{j=1}^{k} f_j(x) F_j(x),$$

where $f_j \in C^\infty(V)^G$. The coefficient functions f_j will not generally be unique.

Remark 2.1. (1) Although we will not need it here, we recall the basic result on smooth invariants proved by Schwarz [26]. This states if $p_1, \ldots, p_\ell$ is a set of polynomial generators for the $\mathbb{R}$-algebra $P(V)^G$, then every smooth invariant may be written as a smooth function of $p_1, \ldots, p_\ell$. If we write $P = (p_1, \ldots, p_\ell) : V \to \mathbb{R}^\ell$ and let $P^\star : C^\infty(\mathbb{R}^\ell) \to C^\infty(V)^G$ denote the mapping defined by composition with P, then Schwarz's result amounts to showing that $P^\star(C^\infty(\mathbb{R}^\ell)) \subset C^\infty(V)^G$ is a *closed* subspace of $C^\infty(V)$ in the C^∞-topology. $(P^\star(C^\infty(\mathbb{R}^\ell)))$ is dense in $C^\infty(V)^G$ by the Weierstrass approximation theorem). Schwarz's original proof used properties of the G-action. Subsequently, it has been shown that $P^\star(C^\infty(\mathbb{R}^\ell))$ is a closed linear subspace of $C^\infty(V)$ with closed complement whenever P is a proper polynomial map (see for example [5,4,3] and note that rather simple proofs of Mather's extension [23] showing that $P^\star$ has a continuous linear section can be given based on results of Vogt and Wagner [28,29]).

(2) Schwarz's result on smooth invariants can be used to give an alternative proof of the previous result on smooth equivariants. The method depends on an observation of Malgrange and may be found in [24,1].

(3) We have stated our results with the domain of functions and maps equal to V. Similar results hold if we replace V by any nonempty G-invariant open subset of V. It is also not necessary (or always desirable) to assume that polynomial generators are homogeneous.

2.5. *Semialgebraic sets and stratifications*

We start with generalities about semialgebraic sets and their stratifications and conclude by describing the *canonical* "minimum" stratification of a semialgebraic set. We refer the reader to Coste [9], Mather [22], Gibson *et al.* [18] or Risler [25] for proofs and further details about semialgebraic sets.

Definition 2.1. A semialgebraic subset X of $\mathbb{R}^n$ is a finite union of sets of the form

$$\{x \in \mathbb{R}^n \mid p_i(x) = 0, \ q_j(x) > 0\},$$

where $p_i, q_j : \mathbb{R}^n \to \mathbb{R}$ are a finite set of polynomials.

(P0) *The collection of semialgebraic subsets of $\mathbb{R}^n$ is closed under finite union, intersection and complementation.*

(P1) *The closure, interior and frontier of a semialgebraic set $X \subset \mathbb{R}^n$ are semialgebraic. The frontier ∂X of X is of dimension strictly less than that of X.*

(P2) *A semialgebraic subset has finitely many connected components.*

(P3) [Tarski-Seidenberg theorem] *If $P : \mathbb{R}^n \to \mathbb{R}^m$ is a polynomial and $X \subset \mathbb{R}^n$ is semialgebraic, then $P(X)$ is a semialgebraic subset of $\mathbb{R}^m$.*

Stratifications

Recall that a *stratification* S of a subset X of $\mathbb{R}^n$ is a locally finite partition of X into smooth and connected submanifolds of $\mathbb{R}^n$ called *strata*. We denote the union of the i-dimensional strata by S_i, $i \geq 0$. By abuse of notation, we also regard S_i as the set of i-dimensional strata. If X is semialgebraic, we say that S is a *semialgebraic stratification* if each stratum is semialgebraic.

In order to obtain a satisfactory definition of transversality to a stratified set we need to recall the recall some facts about the Whitney regularity conditions.

Definition 2.2. A stratification S of a set $X \subset \mathbb{R}^n$ satisfies Whitney's condition (b) if given any pair $U, V \in S$ then for all $u \in U \cap \overline{V}$ and sequences $(u_i) \subset U$, $(v_i) \subset V$ such that

(1) $u_i \to u$ and $v_i \to u$,
(2) the line joining u_i to v_i converges (in $P^{n-1}(\mathbb{R})$) to a line L, and
(3) the family of tangent planes $T_{v_i} Q$ converges in the Grassmannian of $\dim(V)$ planes to a plane P,

we have $P \supset L$.
If the stratification satisfies Whitney's condition (b), we refer to S as a Whitney stratification.

Remark 2.2. If S satisfies Whitney's condition (b), it follows easily that S satisfies Whitney's condition (a). That is, given a pair $U, V \in S$, $u \in U \cap \overline{V}$ and sequence $(v_i) \subset V$ such that $v_i \to u$ and $T_{v_i} V \to P$, we have $P \supset T_u U$.

(P4) [Frontier condition] *If S is a Whitney stratification of a (semialgebraic) subset of $\mathbb{R}^n$ then the frontier of every stratum is a union of lower dimensional strata.*

If S is a Whitney stratification of a semialgebraic subset of $\mathbb{R}^n$ then a stratum S is 'top-dimensional' if S is not contained in the union of frontiers of other strata.

Filtrations

Let S be a Whitney stratification of a semialgebraic set $X \subset \mathbb{R}^n$. We define the associated filtration of X by dimension to be the filtration (X_i) of X obtained by taking X_i to be the union of all strata of dimension $\leq i$. If T is another Whitney semialgebraic stratification of X, we write $S < T$ if there exists an index j such that $X_j \supsetneq T_j$ and $X_i = T_i$, for $i > j$. We say S is minimal if $\mathcal{S} < \mathcal{T}$ for all Whitney semialgebraic stratifications $\mathcal{T} \neq \mathcal{S}$ of X.

(P5) *Every semialgebraic subset X of $\mathbb{R}^n$ has a canonical minimal stratification.* [22]

(Mather shows that the canonical minimal semialgebraic stratification of a semialgebraic set is minimal amongst all stratifications by smooth manifolds – not just semialgebraic (or semianalytic) stratifications.)

Henceforth, we refer to the stratification of X given by (P5) as the *canonical stratification* of X.

Transversality to stratified sets

Let $\mathcal{S}$ be a Whitney stratification of the closed subset $X \subset \mathbb{R}^n$, M be a differential manifold and $f : M \to \mathbb{R}^n$ be a smooth map. Given $x \in M$, f is transverse to $\mathcal{S}$ at x if either $f(x) \notin X$ or $f(x) \in U \in \mathcal{S}$ and f is transverse to $U \subset \mathbb{R}^n$ at x. It follows from Whitney regularity (in fact (a)-regularity) that if f is transverse to $\mathcal{S}$ at x then f will be transverse to $\mathcal{S}$ at all points in some neighbourhood of x in M.

We say f is transverse to $\mathcal{S}$ if f is transverse to $\mathcal{S}$ at all points of M. In case $\mathcal{S}$ is the canonical stratification of the semialgebraic set X, we often just say f is transverse to X and write $f \pitchfork X$.

Using (b)-regularity it may be shown that if f is transverse to a Whitney stratification $\mathcal{S}$ then an isotopy theorem holds (though isotopies will in general only be continuous). For further details on all of this we refer to [22]. We caution that although the theory works well for the canonical stratification of a semialgebraic set, it is well known that the canonical stratification may sometimes not be the most natural Whitney stratification.

3. Local theory of equivariant transversality

Let $f : M \to N$ be a G-equivariant diffeomorphism and P be a G-invariant closed submanifold of N. Just as in standard transversality theory, it is easy to give a local description of the intersection $f^{-1}(P)$ in terms of solutions to equivariant equations defined on a representation (see [1,10,15] for details). In what follows we assume this reduction and focus on the issue of finding generic conditions for solutions of equivariant equations.

Suppose then that (V, G), (W, G) are finite dimensional real G-representations. Following 2.4, let $\mathcal{F} = \{F_1, \ldots, F_k\}$ be a minimal set of homogenous generators for the $P(V)^G$-module $P_G(V, W)$ and set degree$(F_j) = d_j$, labelling generators so that $0 \leq d_1 \leq d_2 \leq \ldots \leq d_k$.

Lemma 3.1. *Let $\mathcal{F} = \{F_1, \ldots, F_k\}$ be a minimal set of homogenous generators for $P_G(V, V)$. Then any relation of the form*

$$\sum_{j=1}^{k} p_j F_j = 0,$$

where $p_j \in P(V)^G$, implies that $p_j(0) = 0$, $1 \leq j \leq k$. The same result holds if we allow $p_j \in C^\infty(V)^G$.

Proof. If $p \in P(V)^G$, let p^ℓ denote the homogeneous part of p of degree ℓ. For $1 \leq i \leq k$ we have

$$-p_i F_i = \sum_{j \neq i} p_j F_j.$$

Taking the homogeneous parts of degree d_i we see that

$$-p_i(0) F_i = \sum_{j \neq i} p_j^{d_i - d_j} F_j.$$

Hence $p_i(0) = 0$ by the minimality of $\mathcal{F}$. If we allow the coefficients p_i to be smooth invariants, the result follows immediately by taking the d_k-jet of $\sum_{j=1}^{k} p_j F_j$ at the origin and applying the result for polynomials. $\quad\square$

Let $\mathfrak{M} = \{p \in P(V)^G \mid p(0) = 0\}$ and $\mathfrak{M}_\infty = \{f \in C^\infty(V)^G \mid f(0) = 0\}$.

Lemma 3.2.

(1) Any minimal set of homogeneous generators for $P_G(V, W)$ maps to a vector space basis of $P_G(V, W)/\mathfrak{M}P_G(V, W)$.

(2) $C_G^\infty(V, W)/\mathfrak{M}_\infty C_G^\infty(V, W) \approx P_G(V, W)/\mathfrak{M}P_G(V, W)$ (as vector spaces).

Proof. (1) follows from Lemma 3.1, and (2) from 2.4 (smooth invariant theory). $\quad\square$

Remark 3.1. (1) It follows from Lemma 3.2 that the number of polynomials in a minimal set of homogeneous generators for $P_G(V, W)$ depends only on the isomorphism class of the representations V and W.

(2) If $\mathcal{F}$ is a minimal set of homogeneous generators for $P_G(V, W)$, then the set of degrees (counting multiplicities) $\{d_1, \ldots, d_k\}$ depends only on isomorphism class of the representations V and W.

(3) For our purposes it suffices to restrict attention to homogeneous generators. However, when it comes to proving openness of G-transversality it is necessary to allow for sets of inhomogeneous generators (see [1,15]).

Set $\mathbb{U} = P_G(V, W)/\mathfrak{M}P_G(V, W)$ and let $\Pi : C_G^\infty(V, W) \to \mathbb{U}$ be the projection given by Lemma 3.2.

It follows from Lemma 3.2 that $\mathcal{F}$ determines a vector space isomorphism $I_{\mathcal{F}}$ between $\mathbb{U}$ and $\mathbb{R}^k$. Set $\gamma = I_{\mathcal{F}}\Pi : C_G^\infty(V, W) \to \mathbb{R}^k$. If $f = \sum f_j F_j$, then $\gamma(f) = (f_1(0), \ldots, f_k(0))$.

Set $d = d_k$. For $f \in C_G^\infty(V, W)$, let $J^d(f)$ denote the d-jet (Taylor polynomial of degree d) of f at the origin. If $J^d(f) = 0$ then $\gamma(f) = 0$. Hence γ factorizes as

$$C_G^\infty(V, W) \xrightarrow{J^d} P_G^{(d)}(V, W) \xrightarrow{\bar{\gamma}} \mathbb{R}^k,$$

where $\bar{\gamma} = \gamma | P_G^{(d)}(V, W)$. It follows that γ is continuous if we give $C_G^\infty(V, W)$ the C^r-topology, $r \geq d$ (Whitney or uniform convergence on compact sets).

Lemma 3.3. *Suppose V, W are G-representations and $\mathbb{R}^s$ is a trivial G-representation. Every minimal set of homogeneous generators $\mathcal{F}$ for $P_G(V, W)$ defines a minimal set of homogeneous generators for $P_G(V \times \mathbb{R}^s, W)$. (Each $F \in \mathcal{F}$ defines a map $F : V \times \mathbb{R}^s \to W$ by $F(x, t) = F(x)$.)*

Proof. We leave this as an easy exercise for the reader. $\square$

Suppose that V, W are G-representations and $\mathbb{R}^s$ is a trivial G-representation. It follows from Lemmas 3.2, 3.3 that we have a linear map $\Pi^s : C_G^\infty(V \times \mathbb{R}^s, W) \to C^\infty(\mathbb{R}^s, \mathbb{U})$ defined by $\Pi^s(f)(t) = \Pi(f_t) \in \mathbb{U}$. Given $f \in C_G^\infty(V \times \mathbb{R}^s, W)$ we may write

$$f(x, t) = \sum_{j=1}^k f_j(x, t) F_j(x), \quad (f_j \in C^\infty(V \times \mathbb{R}^s)^G).$$

We define $\gamma = \gamma^s = I_{\mathcal{F}}\Pi^s : C_G^\infty(V \times \mathbb{R}^s, W) \to C^\infty(\mathbb{R}^s, \mathbb{R}^k)$ by

$$\gamma(f)(t) = (f_1(0, t), \ldots, f_k(0, t)), \quad (t \in \mathbb{R}^s, f \in C_G^\infty(V \times \mathbb{R}^s, W))$$

When f is fixed we usually write γ_f rather than $\gamma^s(f)$.

Lemma 3.4. *The map $\gamma^s : C_G^\infty(V \times \mathbb{R}^s, W) \to C^\infty(\mathbb{R}^s, \mathbb{R}^k)$ is continuous with respect to the C^∞-topologies on $C^\infty(\mathbb{R}^s, \mathbb{R}^k)$ and $C_G^\infty(V \times \mathbb{R}^s, W)$.*

Proof. (following [1]) Let $\alpha : (C^\infty(V \times \mathbb{R}^s)^G)^k \to C_G^\infty(V \times \mathbb{R}^s, W)$ and $\beta : (C^\infty(V \times \mathbb{R}^s)^G)^k \to C^\infty(\mathbb{R}^s, \mathbb{R}^k)$ be defined by

$$\alpha(f_1, \ldots, f_k) = \sum_{j=1}^{k} f_j F_j,$$

$$\beta(f_1, \ldots, f_k)(t) = (f_1(0, t), \ldots, f_k(0, t)), \ t \in \mathbb{R}^s.$$

Both α and β are continuous (with respect to the C^∞-topologies on function spaces). Since α is a continuous linear surjective map between Fréchet spaces, it follows by the Open Mapping Theorem that α is an open map. Since $\gamma\alpha = \beta$, for all open subsets V in $C^\infty(\mathbb{R}^s, \mathbb{R}^k)$, $\alpha(\beta^{-1}(V)) = \gamma^{-1}(V)$ is open and so γ is continuous. $\qquad\square$

3.1. *The universal variety*

Define the polynomial map $\vartheta \in P_G(V \times \mathbb{R}^k, W)$ by

$$\vartheta(x, t) = \sum_{j=1}^{k} t_j F_j(x), \ ((x, t) \in V \times \mathbb{R}^k).$$

Define $\Sigma = \vartheta^{-1}(0) \subset V \times \mathbb{R}^k$ and note that Σ is a G-invariant algebraic subset of $V \times \mathbb{R}^k$. We sometimes refer to Σ as the *universal variety*, and ϑ as the *universal polynomial* (for the pair (V, W)). We have

$$\Sigma \supset V = V \times \{0\} \subset V \times \mathbb{R}^k, \tag{2}$$

$$\Sigma \supset \mathbb{R}^k = \{0\} \times \mathbb{R}^k \subset V \times \mathbb{R}^k, \text{ if } W^G = \{0\}. \tag{3}$$

Every $f \in C_G^\infty(V \times \mathbb{R}^s, W)$ factorizes through ϑ. Specifically, if $f \in C_G^\infty(V \times \mathbb{R}^s, W)$, then we may write $f(x, s) = \sum_{j=1}^{k} f_j(x, s) F_j(x)$, where $f_j \in C^\infty(V \times \mathbb{R}^s)^G$. Define $\Gamma_f : V \times \mathbb{R}^s \to V \times \mathbb{R}^k$ by $\Gamma_f(x, s) = (x, f_1(x, s), \ldots, f_k(x, s))$. Then

$$f = \vartheta \circ \Gamma_f,$$

$$f^{-1}(0) = \Gamma_f^{-1}(\Sigma).$$

3.2. *The local definition of G-transversality*

Let $\mathcal{S}$ denote the canonical minimal stratification of Σ. Since Σ is algebraic, each stratum of $\mathcal{S}$ is a semialgebraic subset of $V \times \mathbb{R}^k$. Since the stratification is canonical and Σ is G-invariant, it follows that G permutes strata. In particular, group orbits of connected strata are G-manifolds. Our convention will be that if $S \in \mathcal{S}$ is a stratum then S is a G-manifold and S/G

(rather than S) is connected. A smooth map is transverse to Σ if the map is transverse to each stratum of $\mathcal{S}$.

Definition 3.1. Let $f \in C_G^\infty(V, W)$. The map f is G-transversal to $0 \in W$ at $0 \in V$ if $\Gamma_f : V \to V \times \mathbb{R}^k$ is transverse to Σ at $0 \in V$.

Remark 3.2. It follows from the openness property of transversality to a Whitney stratification that if f is G-transversal to $0 \in W$ at $0 \in V$, then $\Gamma_f : V \to V \times \mathbb{R}^k$ is transverse to Σ on some G-invariant neighbourhood U of $0 \in V$. In fact, the G-transversality of f to $0 \in W$ at $0 \in V$ implies the G-transversality of f to $0 \in W$ on a neighbourhood of $0 \in V$. However, we will not discuss this point further here.

We omit the verification that the definition is is independent of choice of minimal set of homogeneous generators for $P_G(V, W)$ (see [15,10]). Granted this independence it still remains to show that the definition is independent of the coefficient functions f_j. We do this by proving that the transversality of Γ_f to Σ at $0 \in V$ is determined by the values of $f_1(0), \dots, f_k(0)$ – which are uniquely determined by the choice of $\mathcal{F}$. The approach we outline has two advantages: it gives a more geometric and natural definition of G-transversality, and it gives immediate applications to equivariant bifurcation theory.

Henceforth we shall assume that $V^G = \{0\}$ and let $s \in \mathbb{N}$. We regard $\mathbb{R}^s$ as embedded in $V \times \mathbb{R}^s$ as $\{0\} \times \mathbb{R}^s$.

Theorem 3.1. *There exists a natural Whitney semialgebraic stratification $\mathcal{A}$ of U with the property that $f \in C_G^\infty(V \times \mathbb{R}^s, W)$ is G-transverse to $0 \in W$ on $K \subset \mathbb{R}^s \subset V \times \mathbb{R}^s$, if and only if $\Pi^s(f) : \mathbb{R}^s \to U$ is transverse to $\mathcal{A}$ along K.*

Remark 3.3. (1) We are restricting attention to G-transversality along sets of points in the domain $V \times \mathbb{R}^s$ with trivial isotropy. This allows us to avoid discussion of openness of G-transversality. Note, however, that we obtain transversality to Σ on an open neighbourhood of K in $V \times \mathbb{R}^s$.
(2) If we choose a minimal set of homogeneous generators $\mathcal{F}$ for $P_G(V, W)$, then $I_\mathcal{F}(\mathcal{C})$ is a Whitney stratification $\mathcal{C}_\mathcal{F}$ of $\mathbb{R}^k$ and $\Pi^s(f) : \mathbb{R}^s \to U$ is transverse to $\mathcal{C}$ along K if and only if $\gamma_f : \mathbb{R}^s \to \mathbb{R}^k$ is transverse to $\mathcal{C}_\mathcal{F}$ along K. Consequently, if $s = 0$, the theorem implies that f is G-transverse to $0 \in W$ at 0 if and only if $\gamma_f(0)$ belongs to the top $(k\text{-})$ dimensional stratum of $\mathcal{C}_\mathcal{F}$. That is, G-transversality is determined by $(f_1(0), \dots, f_k(0))$. Similar remarks hold for $s > 0$. In particular, if $s < k$, then a necessary condition

234

for G-transversality along K is that $\gamma_f|K$ does not take values in the strata of $C_{\mathcal{F}}$ which are of dimension less than $k - s$.

Proof. We sketch the construction of the stratification $\mathcal{A}_{\mathcal{F}}$. The proof that the stratification $I_{\mathcal{F}}^{-1}(\mathcal{A}_{\mathcal{F}})$ of $\mathbb{U}$ is independent of $\mathcal{F}$ is in [15].

Denote the canonical stratification of Σ by $\mathcal{S}$. Extend $\mathcal{S}$ to a Whitney semialgebraic stratification $\mathcal{S}^{\star}$ of $V \times \mathbb{R}^k$ by adding the stratum $(V \times \mathbb{R}^k) \backslash \Sigma$. Define $\mathcal{A}_{\mathcal{F}} = \{S^G \mid S \in \mathcal{S}^{\star},\ S^G \neq \emptyset\}$. It is straightforward to verify that $\mathcal{A}_{\mathcal{F}}$ is a Whitney semialgebraic stratification of $\mathbb{R}^k$. Moreover, it is obvious that $\Gamma_f \pitchfork \Sigma$ along K if and only if $\gamma_f \pitchfork C_{\mathcal{F}}$ along K. $\qquad\square$

Remark 3.4. If $W^G = \{0\}$, then $\Sigma \supset \Sigma^G = \mathbb{R}^k$ and $C_{\mathcal{F}} = \{S^G \mid S \in \mathcal{S},\ S^G \neq \emptyset\}$. If $(W, G) = (V, G)$, then $C_{\mathcal{F}}$ is a *union* of $\mathcal{S}$-strata.

Example 3.1. (1) Let $V = W = \mathbb{R}$ and take the nontrivial representation of $\mathbb{Z}_2$ on V. In this case $\Sigma = \{(x, t) \mid tx = 0\}$ and $C_{\mathcal{F}} = \{\mathbb{R} \setminus \{0\}, \{0\}\}$. If $f \in C_{\mathbb{Z}_2}^{\infty}(V \times \mathbb{R}, W)$ then $f(x, t) = g(x, t)x$. and f is $\mathbb{Z}_2$-transversal to $0 \in W$ along $K \subset \mathbb{R}$ if and only if $g(0, \cdot) : \mathbb{R} \to \mathbb{R}$ is transverse to $0 \in \mathbb{R}$ along K.
(2) Let $O(2)$ act on $V = \mathbb{C}^2$ as $e^{i\theta}(z_1, z_2) = (e^{2i\theta}z_2, e^{i\theta}z_2)$ and on $W = \mathbb{C}$ in the standard way. A minimal set of homogeneous generators $\mathcal{F}$ of $P_{O(2)}(\mathbb{C}^2, \mathbb{C})$ is given by $F_1(z_1, z_2) = z_2$ and $F_2(z_1, z_2) = z_1\bar{z}_2$. The natural stratification $C_{\mathcal{F}}$ of $\mathbb{R}^2$ is given by $\{\{(0, 0)\}, \{t_1 = 0, t_2 \neq 0\}, \{t_1 \neq 0\}\}$. The top-dimensional stratum is *not* a stratum of the canonical stratification $\mathcal{S}$ of Σ. Consequently, even if $V^G = W^G = \{0\}$ it does not follow that the induced stratification of $\mathbb{R}^k$ is a union of strata of $\mathcal{S}$.

4. Applications to equivariant bifurcation theory

We give a simple application of equivariant transversality to equivariant bifurcation theory and conclude with some examples that illustrate some of the phenomena that can be expected for various classes of vector fields.

Suppose that (V, G) is a nontrivial absolutely irreducible representation of the compact Lie group G (see 2.3). Let $C_G^{\infty}(V \times \mathbb{R}, V)$ denote the space of smooth 1-parameter families of G-equivariant vector fields on V. Suppose $X \in C_G^{\infty}(V \times \mathbb{R}, V)$. Since (V, G) is a irreducible representation, $V^G = \{0\}$ and by G-equivariance we have

$$X_\lambda(0) = 0, \quad (\lambda \in \mathbb{R}).$$

We refer to $x = 0$ as the *trivial* solution of X. Since $DX_\lambda(0) \in L_G(V, V)$, we have by absolute irreducibility that

$$DX_\lambda(0) = \sigma(\lambda)I_V,$$

where $\sigma : \mathbb{R} \to \mathbb{R}$ is smooth. Since bifurcations of the trivial solution occur at points where $DX_\lambda(0)$ is singular, it follows that the bifurcation set for the trivial solution is precisely the zero set of σ.

It is natural to impose the generic condition that at bifurcation points λ_0, $\sigma'(\lambda_0) \neq 0$. Under this assumption, bifurcation points are isolated. By a local smooth change of λ-coordinates, we may require that $\sigma(\lambda) = \lambda$. Since we shall only be interested in the local behaviour of the zero set of X near a generic bifurcation point, it is no loss of generality to restrict to the space $\mathcal{V}(V, G)$ of smooth equivariant families on V which are of the form

$$X_\lambda(x) = \lambda x + Q(x, \lambda),$$

where $Q(x, \lambda) = O(\|x\|^2)$ on compact subsets of $V \times \mathbb{R}$. For any $X \in \mathcal{V}(V, G)$ there is a non-degenerate change of stability of the trivial solution at $\lambda = 0$.

Let $\mathcal{G}(V, G) \subset \mathcal{V}(V, G)$ consist of those families which are G-transversal to $0 \in V$ at $(0, 0) \in V \times \mathbb{R}$. If $X \in \mathcal{G}(V, G)$, the germ of $X^{-1}(0)$ is stable (topologically) under perturbation of X.

Suppose that $\mathcal{F} = \{F_1, \ldots, F_k\}$ is a minimal set of homogeneous generators for the $P(V)^G$-module $P_G(V, V)$. Since (V, G) is absolutely irreducible we may take $F_1 = I_V$ and then $d_j \geq 2$, $j \geq 2$. Let $\Sigma \subset V \times \mathbb{R}^k$ denote the zero set of

$$\vartheta(x, t) = \sum_{j=1}^k t_j F_j(x) = t_1 x + \sum_{j=2}^k t_j F_j(x).$$

Let $\Sigma = \cup_{\tau \in \mathcal{O}(V,G)} \Sigma_\tau$ denote the partition of Σ into points of the same isotropy type. It may be shown [12,13] that each Σ_τ is a semialgebraic submanifold of $V \times \mathbb{R}^k$. To simplify our exposition, assume from now on that G is finite. We then have $\dim(\Sigma_\tau) = k$, all $\tau \in \mathcal{O}(V, G)$. In particular, $\Sigma_{(G)} = \Sigma^G = \{0\} \times \mathbb{R}^k$. If $t_1 \neq 0$, then $(0, t_1) \notin \overline{\Sigma_\tau}$, all $\tau \neq (G)$, and so $\{t_1 \neq 0\} \subset \mathbb{R}^k$ is contained in a top (k) dimensional stratum of the minimal stratification of Σ.

Let $\mathcal{A} = \{A_0, \ldots, A_k\}$ denote the natural stratification of $\mathbb{R}^k$ induced from the minimal stratification of Σ. We always have $A_k \supset \{t_1 \neq 0\}$.

If $X \in \mathcal{V}(V, G)$,

$$X(x, \lambda) = f_1(x, \lambda)x + \sum_{j=2}^k f_j(x, t)F_j(x),$$

where $f_j \in C^\infty(V \times \mathbb{R})^G$ and $f_1(0, \lambda) = \lambda$. Hence

$$\gamma_f(\lambda) = (\lambda, f_2(0, \lambda), \ldots, f_k(0, \lambda)).$$

By definition, $\gamma_f \pitchfork \mathcal{A}$ at $0 \in \mathbb{R}$ if and only if $\gamma_f \pitchfork A_j$ at $0 \in \mathbb{R}$, $0 \leq j \leq k$. There are only two ways we can satisfy the condition $\gamma_f \pitchfork A_j$ at $0 \in \mathbb{R}$, $0 \leq j \leq k$.

(1) $\gamma_f(0) \in A_k$ (in particular, $A_k \supsetneq \mathbb{R}^{k-1}$).
(2) $\gamma_f(0) \in A_{k-1}$ (transversality to A_{k-1} is automatic since A_{k-1} is an open subset of $\{t_1 = 0\}$ and $\gamma_f \pitchfork \{t_1 = 0\}$).

If the first condition holds then no new branches of equilibria occur for X as λ passes through zero. I know of no examples where (1) holds.

In either case, the branching pattern for X – the germ of $X^{-1}(0)$ at the origin (see [16,15]) – is entirely determined by $f_2(0, 0), \ldots, f_k(0, 0)$. Consequently, if $X \in \mathcal{V}(V, G)$ and we write $X(x, \lambda) = \lambda x + Q(x, \lambda)$, then the dependence of Q on λ is irrelevant as far as the local homeomorphism type of $X^{-1}(0)$ is concerned. This remark still holds if we take account of stabilities along branches [12,13,15].

Example 4.1. (see [16]) For $n \geq 2$, let $H_n \subset O(n)$ denote the group of signed $n \times n$ permutation matrices. We have $H_n = \Delta_n \rtimes S_n$, where Δ_n is the group of diagonal matrices, entries ± 1 and S_n is the symmetric group on n-symbols. The group H_n is the symmetry group of the n-dimensional hypercube and is a finite reflection group. A *basis* for the $P(\mathbb{R}^n)^{H_n}$-module of equivariants is given by

$$F_j(x_1, \ldots, x_n) = \sum_{i=1}^{n} x_i^{2j+1}, \quad 1 \leq j \leq n.$$

It is shown in [16] that the natural stratification of $\mathbb{R}^n$ has filtration

$$\mathbb{R}^n \supset \mathbb{R}^{n-1} \supset \ldots \supset \mathbb{R} \supset \{0\}.$$

If we write $X(x, \lambda) = (\lambda + g(x, \lambda))x + bF_2(x) + O(\|x\|^5)$, then X is generic if and only if $b \neq 0$. Results for other finite reflection groups may be found in [16,17].

Example 4.2. The minimal number $k = k(V, G)$ of generators for the $P(V)^G$-module $P_G(V, V)$ will usually be (much) larger than the dimension of V. As a result computation of the natural stratification $\mathcal{A}$ of $\mathbb{R}^k$ may be very difficult. Nevertheless, it is often possible to gain a lot of information about the codimension one strata of $\mathcal{A}$. For each $\tau \in \mathcal{O}(V, G)$, $\tau \neq (G)$, let $A_\tau^\star = \mathbb{R}^k \cap \overline{\Sigma_\tau}$. We have

(1) $A^\star_\tau \subset \{t_1 = 0\}$.

(2) If $\tau = (H)$ and V^H is odd-dimensional, then $A^\star_\tau = \{t_1 = 0\}$. (In case $\dim(V^H) = 1$, this amounts to the equivariant branching lemma of Cicogna and Vanderbauwhede. See also Example 4.3.10 [13]).

In [17] results are given that identify a large class of subgroups G of the hyperoctahedral group H_n for which one can compute all $\tau \in \mathcal{O}(\mathbb{R}^n, G)$ such that $\dim(A^\star_\tau) = k - 1$ (the 'symmetry breaking isotropy types'). For example, if $G = \Delta_3 \rtimes S_3 \subset H_3$ (an example studied by Guckenheimer and Holmes [20]), then

$$\mathcal{O}(\mathbb{R}^3, G) = \{(G), (S_3), (\mathbb{Z}_2^2 \subset \Delta_3), (\mathbb{Z}_2 \subset \Delta_3), (e)\}.$$

The isotropy types $(S_3), (\mathbb{Z}_2^2)$ are both maximal (that is, maximal isotropy subgroups of G) and have one-dimensional fixed point spaces. It follows from (2) that $A^\star_{(S_3)} = A^\star_{(\mathbb{Z}_2)} = \mathbb{R}^{k-1}$. Thus far, we have not needed any quantitative information on the equivariants. To proceed further, it suffices to note that there are no quadratic equivariants (since $-I \in G$, all equivariants are odd) and that there are two cubic equivariants in a minimal set of homogeneous generators. These may be taken to be $F_2(x, y, z) = (y^2, z^2, x^2)$ and $F_3(x, y, z) = (z^2, x^2, y^2)$. With these choices, one can easily show that

$$A^\star_{(\mathbb{Z}_2)} = \{t_1 = 0, \; t_2 t_3 \geq 0\}.$$

Clearly $A^\star_{(\mathbb{Z}_2)}$ is of codimension one in $\mathbb{R}^k$ and is not equal to $\{t_1 = 0\}$. This provides the simplest example for which $A^\star_\tau$ is a proper semialgebraic, non algebraic, subset of $\mathbb{R}^k$. We refer to [17,15] for more details and generalizations.

Much of what we have described above for equivariant vector fields can be extended to other classes of vector fields. We conclude with an example of a reversible equivariant vector field with a *forced kernel*.

Example 4.3. (see [15] and also [8]) Let W denote the index 2 finite reflection subgroup $\Delta'_4 \rtimes S_4$ of H_4. As basis for the $\mathbb{R}$-algebra $P(\mathbb{R}^4)^W$ we take (see [16])

$$p_1(x) = \frac{1}{2}\|x\|^2, \; p_2(x) = \frac{1}{4}\sum_{i=1}^4 x_i^4, \; p_3(x) = \frac{1}{6}\sum_{i=1}^6 x_i^6, \; p_4(x) = x_1 x_2 x_3 x_4.$$

Corresponding generators for the equivariants are given by $\phi_i = \mathrm{grad}(p_i)$. It follows from smooth invariant theory that every smooth W-equivariant vector field $X : \mathbb{R}^4 \to \mathbb{R}^4$ may be written (uniquely) in the form

$$X = \sum_{j=1}^4 f_i(p_1, \ldots, p_4) F_i, \tag{4}$$

where $f_i \in C^\infty(\mathbb{R}^4)$. Define two, non-isomorphic, representations of $G = H_4$ on $\mathbb{R}^4$. The first representation of G will be the standard representation $\rho : G = H_4 \to O(4)$ defined previously. For the second representation, let $\sigma : G \to O(1) = \mathbb{Z}_2$ be the representation defined by mapping W to $+1$ and $G \setminus W$ to -1. We then define $\rho_\sigma : G \to O(4)$ by $\rho_\sigma(g) = \sigma(g)\rho(g)$. It is easy to verify that these two representations of G are absolutely irreducible and non-isomorphic. We write the first representation as $(\mathbb{R}^4, G)$, the second as $(\mathbb{R}^4_\sigma, G)$. Obviously every $P \in P_G(\mathbb{R}^4, \mathbb{R}^4_\sigma)$ may be written in the form (4). While $F_4 \in P_G(\mathbb{R}^4, \mathbb{R}^4_\sigma)$, the polynomials $F_1, F_2, F_3 \in P_W(\mathbb{R}^4, \mathbb{R}^4)$ do not lie in $P_G(\mathbb{R}^4, \mathbb{R}^4_\sigma)$. In order that $f_i(p_1, \ldots, p_4)F_i \in P_{H_4}(\mathbb{R}^4, \mathbb{R}^4_\sigma)$, $i \neq 4$, it is necessary and sufficient that $f_i(p_1, \ldots, p_4)(gx) = -f_i(p_1, \ldots, p_4)(x)$, for all g such that $\sigma(g) = -1$. Similarly, $f_4(p_1, \ldots, p_4)F_4 \in P_G(\mathbb{R}^4, \mathbb{R}^4_\sigma)$ only if $f_4(p_1, \ldots, p_4)(gx) = f_4(p_1, \ldots, p_4)(x)$, for all g such that $\sigma(g) = -1$. It follows straightforwardly that if we define

$$\bar{F}_i = p_4 F_i, \; i = 1, 2, 3, \; \bar{F}_4 = F_4,$$

then $\bar{F}_1, \ldots, \bar{F}_4$ generate the $P(\mathbb{R}^4)^G$-module $P_G(\mathbb{R}^4, \mathbb{R}^4_\sigma)$. In particular, by smooth invariant theory, every $X \in C_G^\infty(\mathbb{R}^4, \mathbb{R}^4_\sigma)$ may be written (uniquely) in the form

$$X = \sum_{j=1}^{4} f_i(p_1, p_2, p_3, p_4^2)\bar{F}_i, \tag{5}$$

where $f_i \in C^\infty(\mathbb{R}^4)$. Here we have used the fact that p_1, p_2, p_3, p_4^2 generate $P(\mathbb{R}^4)^{H_4}$. Elements of $C_G^\infty(\mathbb{R}^4, \mathbb{R}^4_\sigma)$ are *reversible* equivariant vector fields (see [8]).

Let $F(x, t) = \sum_{j=1}^{4} t_j \bar{F}_j(x)$. Clearly, $F(x_1, x_2, x_3, x_4, t) = 0$ if any two of x_1, x_2, x_3, x_4 are zero. Hence

$$A_\tau^\star = \mathbb{R}^4, \; \text{if } \tau = (G_{1,0,0,0})), (G_{1,1,0,0})), (G_{1,2,0,0}).$$

It is not hard to compute the remaining $A_\tau^\star$. For example, we have

$$A_{(G_{(1,1,1,0)})}^\star = A_{(G_{(1,1,1,1)})}^\star = \{t_1 = 0\},$$

$$A_{(G_{(1,1,2,2)})}^\star = \{t_1, t_2 = 0\}, \; A_{(G_{(1,2,3,4)})}^\star = \{(0, 0, 0, 0)\}.$$

Acknowledgements

Research supported in part by NSF Grant DMS-0244529 and the Leverhulme Foundation. Thanks also to Imperial College London and the University of Colorado at Boulder for their hospitality while this paper was being written.

References

1. E Bierstone. 'General position of equivariant maps', *Trans. Amer. Math. Soc.* **234** (1977), 447–466.
2. E Bierstone. 'Generic equivariant maps', *Real and Complex Singularities, Oslo 1976*, Proc. Nordic Summer School/NAVF Sympos. Math. (Sijthoff and Noordhoff International Publ.) Leyden (1977), 127–161.
3. E Bierstone and P Milman. 'Composite differentiable functions', *Ann. Math.* **116** (1982), 541–558.
4. E Bierstone and P Milman. 'Local analytic invariants and splitting theorems in differential analysis', *Israel J. Math.* **60** (1987), 257–280.
5. E Bierstone and G Schwarz. 'Continuous linear division and extension of C^∞ functions', *Duke Math. J.* **50** (1), 1983, 233–271.
6. P Birtea, M Puta, T S Ratiu and R M Tudoran. 'On the symmetry breaking phenomenon', preprint 2005.
7. G E Bredon. *Introduction to compact transformation groups* (Pure and Applied Mathematics, **46**, Academic Press, New York and London, 1972).
8. P-L Buono, J S W Lamb and M Roberts. 'Bifurcation and branching of equilibria in reversible equivariant vector fields', *Nonlinearity*, to appear.
9. M Coste. 'Ensembles semi-algébriques', in *Géométrie Algebébrique Réelle et Formes Quadratiques*, Springer Lecture Notes in Math., **959**, 1982, 109–138.
10. M J Field. 'Transversality in G-manifolds', *Trans. Amer. Math. Soc.* **231** (1977), 429–450.
11. M J Field. 'Equivariant dynamical systems', *Trans. Amer. Math. Soc.* **259**(1) (1980), 185–205.
12. M J Field. 'Equivariant Bifurcation Theory and Symmetry Breaking', *J. Dynamics and Diff. Eqns.* **1**(4) (1989), 369–421.
13. M J Field. 'Symmetry breaking for compact Lie groups', *Mem. Amer. Math. Soc.* **574** (1996).
14. M J Field. 'Symmetry breaking for equivariant maps'. In: *Algebraic groups and Lie groups*, Volume in Honour of R. W. Richardson, G I Lehrer et al.(ed.), Cambridge University Press, (1997), 219–253.
15. M J Field. *Dynamics and Symmetry* (to appear: Imperial College Press, series in Pure Mathematics).
16. M J Field and R W Richardson. 'Symmetry Breaking and the Maximal Isotropy Subgroup Conjecture for Reflection Groups', *Arch. for Rational Mech. and Anal.* **105**(1) (1989), 61–94.
17. M J Field and R W Richardson. 'Symmetry breaking and branching patterns in equivariant bifurcation theory II', *Arch. Rational Mech. and Anal.* **120** (1992), 147–190.
18. C Gibson, K Wirthmüller, A A du Plessis and E Looijenga. Springfer Lecture Notes in Math. **553** (1976).
19. M Golubitsky, D G Schaeffer and I N Stewart. *Singularities and Groups in Bifurcation Theory, Vol. II*, (Appl. Math. Sci. **69**, Springer-Verlag, New York, 1988).
20. J Guckenheimer and P Holmes. Structurally stable heteroclinic cycles, *Math. Proc. Camb. Phil. Soc.* **103** (1988), 189-192.

21. B Malgrange. *Ideals of Differentiable Functions*, Oxford Univ. Press, London (1966).

22. J N Mather. 'Stratifications and mappings', *Proceedings of the Dynamical Systems Conference, Salvador, Brazil*, ed. M. Peixoto, (Academic Press, New-York, San Franscisco, London, 1973.)

23. J N Mather. 'Differentiable invariants', *Topology* **16** (1977), 145–155.

24. V Poenaru. *Singularités C^∞ en Prśence de Symétrie*, Springer Lect. Notes Math. **510**, Springer-Verlag, New York and Berlin 1976.

25. J J- Risler. *Real Algebraic and Semi-algebraic Sets*, Hermann, 1990.

26. G W Schwarz. 'Smooth functions invariant under the action of a compact Lie group', *Topology* **14** (1975), 63–68.

27. J C Tougeron. *Idéaux de fonctions différentiable*, Springer-Verlag, New York and Berlin 1972.

28. D Vogt. 'Subspaces and quotient spaces of (s)', *Functional Analysis: Surveys and recent results* (Proc. Conf. Paderborn, 1976), North-Holland Math. Studies **27**, 167–187, Amsterdam: North-Holland 1977.

29. D Vogt and M J Wagner. 'Charakterisierung der Quotientenräume von s und eine Vermutung von Martineau', *Studia Math.* **67** (1980), 225–240.

DIFFERENTIAL GEOMETRY FROM THE VIEW POINT OF LAGRANGIAN OR LEGENDRIAN SINGULARITY THEORY

Shyuichi IZUMIYA

Department of Mathematics, Hokkiado University,
Sapporo, 060-0810, Japan
** E-mail:izumiya@math.sci.hokudai.ac.jp*

This is a half survey on the classical results of extrinsic differential geometry of hypersurfaces in Euclidean space from the view point of Lagrangian or Legendrian singularity theory. Many results in this paper have been already obtained in some articles. However, we can discover some new information of geometric properties of hypersurfaces from this point of view.

Keywords: Lagrangian singularities, Legendrian singularities, hypersurfaces, extrinsic differential geometry.

1. Introduction

In this paper we revise the classical differential geometry from the view point of the theory of Lagrangian or Legendrian singularities. Recently we apply the theory of Lagrangian or Legendrian singularities to the extrinsic differential geometry on submanifolds of pseudo-spheres in Minkowski space [9–19]. As consequences, we have obtained several interesting geometric properties of such submanifolds mainly from the view point of contact with model hypersurfaces (i.e., totally umbilic hypersurfaces). The theory of contact between submanifolds has been systematically developed by Montaldi [24,25] for the study of curves and surfaces in Euclidean space as an application of the theory of singularities of smooth mappings due to Mather [21,22]. However, we have discovered that if we apply the theory of Lagrangian or Legendrian singularities, we might be able to have much more detailed geometric properties through the previous researches [9,14,18]. Although such researches were focused on submanifolds of pseudo-spheres in Minkowski space, this method also supplies new information on submanifolds of Euclidean space.

In §2 we give a quick review on the classical Gaussian differential geom-

"

242

etry of hypersurfaces in Euclidean space. The fundamental concept is the Gauss map of a hypersurface whose Jacobian determinant is the Gauss-Kronecker curvature. Therefore the singularities of the Gauss map is the set of the points where the Gauss-Kronecker curvature vanishes (i.e., the parabolic points). We also have the notion of evolutes (focal sets) and pedal hypersurfaces whose singularities correspond to some important geometric properties (umbilical points, ridge points and parabolic points etc). The height functions family and the distance squared functions family are the fundamental tools for the study of classical differential geometry as applications of singularity theory. The importance of such families were originally pointed out by Thom and the idea of Thom has been first realized by Porteous [26]. In his pioneering work on "generic differential geometry", Terry Wall [30] pointed out that the theory of Lagrangian singularities might be useful for the study of Gauss maps and normal exponential maps (the critical value sets are the evolutes) of hypersurfaces. See also [3–5,20,27,28]. We review the basic properties of the height functions family and the distance squared functions family in §3. We can show that these families are Morse families in the theory of Lagrangian or Legendrian singularities which control the singularities of evolutes, Gauss maps and pedals of hypersurfaces (cf., §4). It was shown by Wall [30] that Gauss map has generically the same types of singularities of as any Lagrangian maps (see also [1], §18.6). We also review the theory of contact between submanifolds due to Montaldi [24,25] in §5. In [14] we have considered the contact of submanifolds with families of hypersurfaces for the study of contact of hypersurfaces with families of hyperspheres in hyperbolic space as an application of Goryunov's result([6], Appendix). This technique is also useful for the study of the contact of hypersurfaces with families of hyperspheres in Euclidean space. We apply Lagrangian or Legendrian singularity theory to these theories of contact and show some new results in §6. §7 is devoted to a more detailed study of the case $n = 3$. We remark that this method also work for a higher codimensional submanifold if we consider the canal hypersurface of the submanifold.

We shall assume throughout the whole paper that all the maps and manifolds are C^∞ unless the contrary is explicitly stated.

2. Hypersurfaces in Euclidean space

In this section we review the classical theory of differential geometry on hypersurfaces in Euclidean space and introduce some singular mappings associated to geometric properties of hypersurfaces.

Let $\boldsymbol{X} : U \to \mathbb{R}^n$ be an embedding, where $U \subset \mathbb{R}^{n-1}$ is an open subset. We denote that $M = \boldsymbol{X}(U)$ and identify M and U through the embedding $\boldsymbol{X}$. The tangent space of M at $p = \boldsymbol{X}(u)$ is

$$T_pM = \langle \boldsymbol{X}_{u_1}(u), \boldsymbol{X}_{u_2}(u), \ldots, \boldsymbol{X}_{u_{n-1}}(u) \rangle_{\mathbb{R}}.$$

For any $\boldsymbol{a}_1, \boldsymbol{a}_2, \ldots, \boldsymbol{a}_{n-1} \in \mathbb{R}^n$, we define

$$\boldsymbol{a}_1 \times \boldsymbol{a}_2 \times \cdots \times \boldsymbol{a}_{n-1} = \begin{vmatrix} \boldsymbol{e}_1 & \boldsymbol{e}_2 & \cdots & \boldsymbol{e}_n \\ a_1^1 & a_2^1 & \cdots & a_n^1 \\ a_1^2 & a_2^2 & \cdots & a_n^2 \\ \vdots & \vdots & \cdots & \vdots \\ a_1^{n-1} & a_2^{n-1} & \cdots & a_n^{n-1} \end{vmatrix},$$

where $\{\boldsymbol{e}_1, \ldots, \boldsymbol{e}_n\}$ is the canonical basis of $\mathbb{R}^n$ and $\boldsymbol{a}_i = (a_1^i, a_2^i, \ldots, a_n^i)$. It follows that we can define the unit normal vector field

$$\boldsymbol{n}(u) = \frac{\boldsymbol{X}_{u_1}(u) \times \cdots \times \boldsymbol{X}_{u_{n-1}}(u)}{\|\boldsymbol{X}_{u_1}(u) \times \cdots \times \boldsymbol{X}_{u_{n-1}}(u)\|}$$

along $\boldsymbol{X} : U \longrightarrow \mathbb{R}^n$. A map $G : U \longrightarrow S_1^n$ defined by $G(u) = \boldsymbol{n}(u)$ is called the *Gauss map* of $M = \boldsymbol{X}(U)$. We can easily show that $D_v\boldsymbol{n} \in T_pM$ for any $p = \boldsymbol{X}(u) \in M$ and $v \in T_pM$. Here D_v denotes the covariant derivative with respect to the tangent vector v. Therefore the derivative of the Gauss map $dG(u)$ can be interpreted as a liner transformation on the tangent space T_pM at $p = \boldsymbol{X}(u)$. We call the linear transformation $S_p = -dG(u) : T_pM \longrightarrow T_pM$ the *shape operator* (or *Weingarten map*) of $M = \boldsymbol{X}(U)$ at $p = \boldsymbol{X}(u)$. We denote the eigenvalue of S_p by κ_p which we call a *principal curvature*. We call the eigenvector of S_p the *principal direction*. By definition, κ_p is a principal curvature if and only if $\det(S_p - \kappa_pI) = 0$. *The Gauss-Kronecker curvature* of $M = \boldsymbol{X}(U)$ at $p = \boldsymbol{X}(u)$ is defined to be $K(u) = \det S_p$.

We say that a point $p = \boldsymbol{X}(u) \in M$ is an *umbilical* point if $S_p = k_p id_{T_pM}$. We also say that M is *totally umbilic* if all points of M are umbilic. Then the following proposition is a well-known result:

Proposition 2.1. *Suppose that $M = \boldsymbol{X}(U)$ is totally umbilic, then κ_p is constant κ. Under this condition, we have the following classification:*
1) *If $\kappa \neq 0$, then M is a part of a hypersphere.*
2) *If $\kappa = 0$, then M is a part of a hyperplane.*

In the extrinsic differential geometry, totally umbilic hypersurfaces are considered to be the model hypersurfaces in Euclidean space. Since the set

244

$\{\boldsymbol{X}_{u_i} \mid (i = 1, \ldots, n-1)\}$ is linearly independent, we induce the Riemannian metric (*first fundamental form*) $ds^2 = \sum_{i=1}^{n-1} g_{ij} du_i du_j$ on $M = \boldsymbol{X}(U)$, where $g_{ij}(u) = \langle \boldsymbol{X}_{u_i}(u), \boldsymbol{X}_{u_j}(u) \rangle$ for any $u \in U$. We define the *second fundamental invariant* by $h_{ij}(u) = \langle -\boldsymbol{n}_{u_i}(u), \boldsymbol{X}_{u_j}(u) \rangle$ for any $u \in U$. We have the following *Weingarten formula*:

$$\boldsymbol{n}_{u_i}(u) = -\sum_{j=1}^{n-1} h_i^j(u) \boldsymbol{X}_{u_j}(u),$$

where $(h_i^j(u)) = (h_{ik}(u))(g^{kj}(u))$ and $(g^{kj}(u)) = (g_{kj}(u))^{-1}$. By the Weingarten formula, the Gauss-Kronecker curvature is given by

$$K(u) = \frac{\det(h_{ij}(u))}{\det(g_{\alpha\beta}(u))}.$$

For a hypersurface $\boldsymbol{X} : U \longrightarrow \mathbb{R}^n$, we say that a point $u \in U$ or $p = \boldsymbol{X}(u)$ is a *flat point* if $h_{ij}(u) = 0$ for all i, j. Therefore, $p = \boldsymbol{X}(u)$ is a flat point if and only if p is an umbilic point with the vanishing principal curvature. We say that a point $p = \boldsymbol{X}(u) \in M$ is a *parabolic point* if $K(u) = 0$. For a hypersurface $\boldsymbol{X} : U \longrightarrow \mathbb{R}^n$, we define the *evolute* of $\boldsymbol{X}(U) = M$ by

$$\mathrm{Ev}_M = \left\{ \boldsymbol{X}(u) + \frac{1}{\kappa(u)}\boldsymbol{n}(u) \middle| \kappa(u) \text{ is a principal curvature at } u \in U \right\}.$$

The evolute is also called the *focal set* of M. We define a smooth mapping $\mathrm{Ev}_\kappa : U \longrightarrow \mathbb{R}^n$ by

$$\mathrm{Ev}_\kappa(u) = \boldsymbol{X}(u) + \frac{1}{\kappa(u)}\boldsymbol{e}(u),$$

where we fix a principal curvature $\kappa(u)$ on U at u with $\kappa(u) \neq 0$. This map gives a parametrization of a component of Ev_M. We also define the *pedal hypersurface* of $M = \boldsymbol{X}(U)$ by

$$\mathrm{Pe}_M : U \longrightarrow \mathbb{R}^n \ ; \ \mathrm{Pe}_M(u) = \langle \boldsymbol{X}(u), \boldsymbol{n}(u) \rangle \boldsymbol{n}(u).$$

Concerning on the pedal hypersurface in $\mathbb{R}^n$, we define the *cylindrical pedal* of $M = \boldsymbol{X}(U)$ by

$$\mathrm{CPe}_M : U \longrightarrow S^{n-1} \times \mathbb{R} \ ; \ \mathrm{CPe}_M(u) = (\boldsymbol{n}(u), \langle \boldsymbol{X}(u), \boldsymbol{n}(u) \rangle).$$

The cylindrical pedal of M is called the *dual* of M in [4,28]. We have the following well-known result:

Proposition 2.2. *Let $M = \boldsymbol{X}(U)$ be a hypersurface in $\mathbb{R}^n$.*
(a) Suppose that there are no parabolic points or flat points, then the following are equivalent:
   (1) M is totally umbilic with $\kappa \neq 0$.
   (2) Ev_M is a point in $\mathbb{R}^n$.
   (3) M is a part of a hypersphere.
(b) The following are equivalent:
   (1) M is totally umbilic with $\kappa = 0$.
   (2) The Gauss map is a constant map.
   (3) M is a part of a hyperplane.

We define a mapping $\Psi : S^{n-1} \times (\mathbb{R} \setminus \{0\}) \longrightarrow \mathbb{R}^n \setminus \{\boldsymbol{0}\}$ by $\Psi(\boldsymbol{v}, r) = r\boldsymbol{v}$. We can easily show that Ψ is a double covering and $\Psi(\mathrm{CPe}_M(u)) = \mathrm{Pe}_M(u)$ under the assumption that $\langle \boldsymbol{X}(u), \boldsymbol{n}(u) \rangle \neq 0$. If necessary, by applying a Euclidean motion in $\mathbb{R}^n$, we have the condition $\langle \boldsymbol{X}(u), \boldsymbol{n}(u) \rangle \neq 0$. Since we consider the geometric properties which are invariant under Euclidean motion, we might assume the above condition. Therefore the singularities of the pedal and the cylindrical pedal of a hypersurface are diffeomorphic. Although the notion of pedals are classically given, we consider the cylindrical pedal instead of the pedal of $M = \boldsymbol{X}(U)$ by the above reason.

3. Height functions and distance squared functions

We now define two kinds of functions families in order to describe the Gauss map, the evolute and the pedal hypersurface of a hypersurface in $\mathbb{R}^n$.

For the purpose, we need some concepts and results in the theory of unfoldings of function germs. We shall give a brief review of the theory in the appendices.

We now define two families of functions

$$H : U \times S^{n-1} \longrightarrow \mathbb{R}$$

by $H(u, \boldsymbol{v}) = \langle \boldsymbol{X}(u), \boldsymbol{v} \rangle$ and

$$D : U \times \mathbb{R}^n \longrightarrow \mathbb{R}$$

by $D(u, \boldsymbol{x}) = \|\boldsymbol{X}(u) - \boldsymbol{x}\|^2$. We call H a *height function* and D *distance squared function*) on $M = \boldsymbol{X}(U)$. We denote that $h_v(u) = H(u, \boldsymbol{v})$ and $d_x(u) = D(u, \boldsymbol{x})$. These two families of functions are introduced by Thom for the study of parabolic points and umbilical points. Actually, Porteous and Montaldi realized Thom's program [23,26,27]. The following proposition follows from direct calculations:

Proposition 3.1. *Let* $\boldsymbol{X} : U \longrightarrow \mathbb{R}^n$ *be a hypersurface. Then*
(1) $(\partial h_v / \partial u_i)(u) = 0$ $(i = 1, \ldots, n-1)$ *if and only if* $\boldsymbol{v} = \pm \boldsymbol{n}(u)$.
(2) $(\partial d_x / \partial u_i)(u) = 0$ $(i = 1, \ldots, n-1)$ *if and only if there exist real numbers* λ *such that* $\boldsymbol{v} = \boldsymbol{x}(u) + \lambda \boldsymbol{n}(u)$.

By Proposition 3.1, we can detect both the catastrophe sets (cf., Appendix A) of H and D as follows:

$$C(H) = \left\{ (u, \boldsymbol{v}) \in U \times S^{n-1} \middle| \boldsymbol{v} = \pm \boldsymbol{n}(u) \right\},$$

$$C(D) = \left\{ (u, \boldsymbol{x}) \in U \times \mathbb{R}^n \middle| \boldsymbol{x} = \boldsymbol{x}(u) + \mu \boldsymbol{n}(u) \right\}.$$

For $\boldsymbol{v} = \boldsymbol{n}(u)$, We also calculate that

$$\frac{\partial^2 H}{\partial u_i \partial u_j}(u, \boldsymbol{v}) = \langle \boldsymbol{X}_{u_i u_j}(u), \boldsymbol{v} \rangle = \mp h_{ij}(u) \text{ on } C(H)$$

and

$$\frac{\partial^2 D}{\partial u_i \partial u_j}(u, \boldsymbol{x}) = 2(\langle \boldsymbol{X}_{u_i u_j}(u), \boldsymbol{X}(u) - \boldsymbol{x} \rangle + \langle \boldsymbol{X}_{u_j}(u), \boldsymbol{X}_{u_j}(u) \rangle)$$

$$= 2(-\lambda h_{ij}(u) + g_{ij}(u)) \text{ on } C(D).$$

Therefore, for any $\boldsymbol{v} = \boldsymbol{n}(u)$, $\det(\mathcal{H}(h_v)(u)) = \det(\partial^2 H / \partial u_i \partial u_j)(u, \boldsymbol{v})) = 0$ if and only if $K(p) = 0$ (i.e., $p = \boldsymbol{X}(u)$ is a parabolic point). Moreover, for any $\boldsymbol{x} = \boldsymbol{X}(u) + \lambda \boldsymbol{n}(u)$, $\det(\mathcal{H}(d_x)(u)) = \det(\partial^2 D / \partial u_i \partial u_j)(u, \boldsymbol{x})) = 0$ if and only if $\kappa(u) = \frac{1}{\lambda}$ is a principal curvature. By the above calculation, we have the following well-known results:

Proposition 3.2. *For any* $p = \boldsymbol{X}(u)$, *we have the following assertions:*
Suppose that $\boldsymbol{v} = \boldsymbol{n}(u)$, *then*
    (a) p *is a parabolic point if and only if* $\det(\mathcal{H}(h_v)(u)) = 0$.
    (b) p *is a flat point if and only if* $\operatorname{rank} \mathcal{H}(h_v)(u) = 0$.
Suppose that p *is not a flat point and* $\boldsymbol{x} = \boldsymbol{X}(u) + (1/\kappa(u))\boldsymbol{n}(u)$ *for a non-zero principal curvature* $\kappa(u)$. *Then*
    (c) p *is an umbilical point if and only if* $\operatorname{rank} \mathcal{H}(d_x)(u) = 0$.

We say that u is a *ridge point* if h_v has the $A_{k \geq 3}$-type singular point at u, where $\boldsymbol{v} \in \mathrm{Ev}_M(U)$. For a function germ $f : (\mathbb{R}^{n-1}, \boldsymbol{x}_0) \longrightarrow \mathbb{R}$, f has A_k-type singular point at $\boldsymbol{x}_0$ if f is $\mathcal{R}^+$-equivalent to the germ $x_1^{k+1} \pm x_2^2 \pm \cdots \pm x_{n-1}^2$. We say that two function germs $f_i : (\mathbb{R}^{n-1}, \boldsymbol{x}_i) \longrightarrow \mathbb{R}$ $(i = 1, 2)$ are $\mathcal{R}^+$-*equivalent* if there exists a diffeomorphism germ $\Phi : (\mathbb{R}^{n-1}, \boldsymbol{x}_1) \longrightarrow (\mathbb{R}^{n-1}, \boldsymbol{x}_2)$ and a real number c such that $f_2 \circ \Phi(\boldsymbol{x}) = f_2(\boldsymbol{x}) + c$. The notion of ridge points was introduced by Porteous [26] as an application of the singularity theory of unfoldings to the evolute and the geometric meaning of ridge points is given as follows: Let $F : \mathbb{R}^n \longrightarrow \mathbb{R}$ be a function and $X : U \longrightarrow \mathbb{R}^n$ a hypersurface. We say that X and $F^{-1}(0)$ have a *corank r contact* at $p = X(u)$ if the Hessian of the function $g(u) = F \circ X(u)$ has corank r at u. We also say that X and $F^{-1}(0)$ have an *A_k-type contact* at $p = X(u)$ if the function $g(u) = F \circ X(u)$ has the A_k-type singularity at u. By definition, if X and $F^{-1}(0)$ have an *A_k-type contact* at $p = X(u)$, then these have a corank 1 contact. For any $r \in \mathbb{R}$ and $a_0 \in \mathbb{R}^n$, we consider a function $F : \mathbb{R}^n \longrightarrow \mathbb{R}$ defined by $F(x) = \|x - a_0\|^2 - r^2$. We denote that

$$S^{n-1}(a, r) = F^{-1}(0) = \{u \in \mathbb{R}^n|\ \|x - a\|^2 = r^2\}.$$

It follows that $S^{n-1}(a, r)$ is a hypersphere with the center a and the radius $|r|$. We put $a = \mathrm{Ev}_\kappa(u)$ and $r = 1/\kappa(u)$, where we fix a principal curvature $\kappa(u)$ on U at u, then we have the following simple proposition:

Proposition 3.3. *Under the above notations, there exists an integer ℓ with $1 \le \ell \le n-1$ such that $M = X(U)$ and $S^{n-1}(a, r)$ have corank ℓ contact at u.*

In the above proposition, $S^{n-1}(a, r)$ is called an *osculating hypersphere* of $M = X(U)$. We also call a the *center of the principal curvature $\kappa(u)$*. By Proposition 3.2, $M = X(U)$ and the osculating hypersphere has corank $n-1$ contact at an umbilic point. Therefore the ridge point is not an umbilic point.

By the general theory of unfoldings of function germs, the bifurcation set B_F is non-singular at the origin if and only if the function $f = F|\mathbb{R}^n \times \{0\}$ has the A_2-type singularity (i.e., the fold type singularity). Therefore we have the following proposition:

Proposition 3.4. *Under the same notations as in the previous proposition, the evolute Ev_M is non-singular at $a = \mathrm{Ev}_\kappa(u)$ if and only if $M = X(U)$ and $S^{n-1}(a, r)$ have A_2-type contact at u.*

All results mentioned in the above paragraphs on the evolute have been shown by Porteous and Montaldi [23,26].

We also define a family of functions $\widetilde{H} : U \times (S^{n-1} \times \mathbb{R}) \longrightarrow \mathbb{R}$ by

$$\widetilde{H}(u, v, r) = \langle X(u), v \rangle - r.$$

We call it the *extended height function* of $M = X(U)$. By the previous calculations, we have

$$D_{\widetilde{H}} = \{\pm \mathrm{CPe}_M(u) \mid u \in U\} \quad \text{and} \quad B_D = \mathrm{Ev}_M.$$

Moreover, the catastrophe map of H is $\pi_{C(H)}(u, \pm n(u)) = \pm n(u) = \pm G(u)$. Therefore, we can identify the Gauss map of $M = \boldsymbol{X}(U)$ with the positive component of the catastrophe map $\pi_{C(H)}$.

4. Evolutes and Cylindrical pedals as Caustics and Wavefronts

In this section we naturally interpret the evolute (respectively, the cylindrical pedal) of a hypersurface as a caustics (respectively, a wave front) in the framework of symplectic (respectively, contact) geometry and consider the geometric meaning of those singularities. In Appendix A (respectively, Appendix B) we give a brief survey of the theory of Lagrangian (respectively, Legendrian) singularities. For notions and basic results on the theory of Lagrangian or Legendrian singularities, please refer to these appendices.

For a hypersurface $\boldsymbol{X} : U \longrightarrow \mathbb{R}^n$, we consider the distance squared function D and the height function H. We have the following propositions:

Proposition 4.1. *Both of the distance squared function $D : U \times \mathbb{R}^n \longrightarrow \mathbb{R}$ and the height function $H : U \times S^{n-1} \longrightarrow \mathbb{R}$ of $M = \boldsymbol{X}(U)$ are Morse families of functions.*

Proof. First we consider the distance squared function.

For any $\boldsymbol{x} = (x_1 \ldots, x_n) \in \mathbb{R}^n$, we have $D(u, \boldsymbol{x}) = \sum_{i=1}^{n}(x_i(u) - x_i)^2$, where $\boldsymbol{X}(u) = (x_1(u), \ldots, x_n(u))$. We will prove that the mapping

$$\Delta D = \left(\frac{\partial D}{\partial u_1}, \ldots, \frac{\partial D}{\partial u_{n-1}}\right)$$

is non-singular at any point. The Jacobian matrix of ΔD is given as follows:

$$\begin{pmatrix} A_{11} & \cdots & A_{1(n-1)} & -2x_{1u_1}(u) & \cdots & -2x_{nu_1}(u) \\ \vdots & \vdots & \vdots & \vdots & \vdots & \vdots \\ A_{(n-1)1} & \cdots & A_{(n-1)(n-1)} & -2x_{1u_{n-1}}(u) & \cdots & -2x_{nu_{n-1}}(u) \end{pmatrix},$$

where $A_{ij} = 2(\langle \boldsymbol{X}_{u_i u_j}(u), \boldsymbol{X}(u) - \boldsymbol{x}\rangle + \langle \boldsymbol{X}_{u_i}(u), \boldsymbol{X}_{u_j}(u)\rangle)$. Since $\boldsymbol{X} : U \longrightarrow \mathbb{R}^n$ is an embedding, the rank of the matrix

$$X = \begin{pmatrix} 2x_{1u_1}(u) & \cdots & -2x_{nu_1}(u) \\ \vdots & \vdots & \vdots \\ 2x_{1u_{n-1}}(u) & \cdots & -2x_{nu_{n-1}}(u) \end{pmatrix}$$

is $n - 1$ at any $u \in U$.

Therefore the rank of the Jacobian matrix of ΔD is $n-1$.

Next we consider the height function. The proof is also given by direct calculations but a bit more carefully than in the previous case. For any $v \in S^{n-1}$, we have $v_1^2 + \cdots + v_n^2 = 1$. Without loss of the generality, we might assume that $v_n > 0$. We have $v_n = \sqrt{1 - v_1^2 - \cdots - v_{n-1}^2}$, so that

$$H(u, v) = x_1(u)v_1 + \cdots + x_{n-1}(u)v_{n-1} + x_n(u)\sqrt{1 - v_1^2 - \cdots - v_{n-1}^2}.$$

We also prove that the mapping

$$\Delta H = \left(\frac{\partial H}{\partial u_1}, \ldots, \frac{\partial H}{\partial u_{n-1}}\right)$$

is non-singular at any point. The Jacobian matrix of ΔH is given as follows:

$$\left(\left. \begin{matrix} \langle \boldsymbol{X}_{u_1 u_1}, v \rangle & \cdots & \langle \boldsymbol{X}_{u_1 u_{n-1}}, v \rangle \\ \vdots & \vdots & \vdots \\ \langle \boldsymbol{X}_{u_{n-1} u_1}, v \rangle & \cdots & \langle \boldsymbol{X}_{u_{n-1} u_{n-1}}, v \rangle \end{matrix} \right| \widetilde{X} \right),$$

where

$$\widetilde{X} = \begin{pmatrix} x_{1u_1} - x_{nu_1}\dfrac{v_1}{v_n} & \cdots & x_{n-1u_1} - x_{nu_1}\dfrac{v_{n-1}}{v_n} \\ \vdots & \vdots & \vdots \\ x_{1u_{n-1}} - x_{nu_{n-1}}\dfrac{v_1}{v_n} & \cdots & x_{n-1u_{n-1}} - x_{nu_{n-1}}\dfrac{v_{n-1}}{v_n} \end{pmatrix}.$$

We will show that the rank of the matrix $\widetilde{X}$ is $n-1$ at $(u, v) \in C(H)$. We denote that $\boldsymbol{a}_i = \begin{pmatrix} x_{iu_1} \\ \vdots \\ x_{iu_{n-1}} \end{pmatrix}$ for $i = 0, \ldots, n$. It should be proven that the rank of the matrix

$$\widetilde{A} = \left(\boldsymbol{a}_1 - \boldsymbol{a}_n\frac{v_1}{v_n}, \ldots, \boldsymbol{a}_{n-1} - \boldsymbol{a}_n\frac{v_{n-1}}{v_n}\right)$$

is $n-1$ at $(u, v) \in C(H)$.

Therefore we have

$$\det\widetilde{A} = (-1)^{n+1}\frac{v_1}{v_n}\det(\boldsymbol{a}_2,\ldots,\boldsymbol{a}_n)$$

$$+\cdots+(-1)^{2n}\frac{v_n}{v_n}\det(\boldsymbol{a}_1,\ldots,\boldsymbol{a}_{n-1})$$

$$= (-1)^{n-1}\left\langle\left(\frac{v_1}{v_n},\ldots,\frac{v_n}{v_n}\right),\boldsymbol{X}_{u_1}\times\cdots\times\boldsymbol{X}_{u_{n-1}}\right\rangle$$

$$= \frac{(-1)^{n-1}}{v_n}\langle\pm\boldsymbol{n},\boldsymbol{X}_{u_1}\times\cdots\times\boldsymbol{X}_{u_{n-1}}\rangle$$

$$= \pm\frac{(-1)^{n-1}}{v_n}\|\boldsymbol{X}_{u_1}\times\cdots\times\boldsymbol{X}_{u_{n-1}}\|\neq 0$$

for $(u,\boldsymbol{v}) = (u,\pm\boldsymbol{n}(u)) \in C(H)$. This completes the proof of the proposition. $\square$

By the method for constructing the Lagrangian immersion germ from Morse family of functions (cf., Appendix A), we can define a Lagrangian immersion germ whose generating family is the distance squared function or the height function of $M = \boldsymbol{X}(U)$ as follows: For a hypersurface $\boldsymbol{X} : U \longrightarrow \mathbb{R}^n$ with $\boldsymbol{X}(u) = (x_1(u),\ldots,x_n(u))$, we define a smooth mapping

$$L(D) : C(D) \longrightarrow T^*\mathbb{R}^n$$

by

$$L(D)(u,\boldsymbol{x}) = \big(\boldsymbol{x},-2(x_1(u)-x_1),\ldots,-2(x_n(u)-x_n)\big),$$

where $\boldsymbol{x} = (x_1,\ldots,x_n) \in \mathbb{R}^n$. Here we have used the triviality of the cotangent bundle $T^*\mathbb{R}^n$. For the $(n-1)$-sphere S^{n-1}, we consider the local coordinate $U_i = \{\boldsymbol{v} = (v_1,\ldots,v_n) \in S^{n-1} \mid v_i \neq 0 \}$. Since $T^*S^{n-1}|U_i$ is a trivial bundle, we define a map

$$L_i(H) : C(H) \longrightarrow T^*S^{n-1}|U_i \ (i = 0,1,\ldots,n)$$

by

$$L_i(H)(u,\boldsymbol{v})=\Big(\boldsymbol{v},x_1(u)-x_i(u)\frac{v_1}{v_i},\ldots,\widehat{x_i(u)-x_i(u)}\frac{v_i}{v_i},\ldots,x_n(u)-x_i(u)\frac{v_n}{v_i}\Big),$$

where $\boldsymbol{v} = (v_1,\ldots,v_n) \in S^{n-1}$ and we denote $(x_1,\ldots,\hat{x}_i,\ldots,x_n)$ as a point in the $(n-1)$-dimensional space such that the i-th component x_i is removed. We can show that if $U_i \cap U_j \neq \emptyset$ for $i \neq j$, then $L_i(H)$ and $L_j(H)$ are Lagrangian equivalent which are given by the local coordinate transformation of S^{n-1} and Lagrangian lift of it. Indeed, we denote that the local coordinate change of S^{n-1} for $i < j$; $\varphi_{ij} : U_i \longrightarrow U_j$, defined by

$$\varphi_{ij}(v_1, \ldots, \hat{v}_i, \ldots, v_n)$$
$$= (v_1, \ldots, v_i = \sqrt{1 - v_1^2 - \cdots - \hat{v}_i^2 - \cdots - v_n^2}, \ldots, \hat{v}_j, \ldots, v_n),$$

and $\tilde{\varphi}_{ij} : T^*S^{n-1} \longrightarrow T^*S^{n-1}$ are Lagrangian lift of φ_{ij} which defined by $\tilde{\varphi}_{ij}(\xi) = (\varphi_{ij*}^{-1})^*\xi$. Then $\tilde{\varphi}_{ij}$ are symplectic diffeomorphism germs (c.f [1]). Also we define diffeomorphism germs $\hat{\sigma}_{ij} : U \times U_i \to U \times U_j$ by $\hat{\sigma}_{ij}(u, v) = (u, \varphi_{ij}(v))$ and $\sigma_{ij} = \hat{\sigma}_{ij}|_{C(H)}$, then $\tilde{\varphi}_{ij} \circ L_i(H) = L_j(H) \circ \sigma_{ij}$ and $\varphi_{ij} \circ \pi = \pi \circ \tilde{\varphi}_{ij}$. Therefore we can define a global Lagrangian immersion, $L(H) : C(H) \longrightarrow T^*S^{n-1}$.

By definition, we have the following corollary of the above proposition:

Corollary 4.2. *Under the above notations, $L(D)$ (respectively, $L(H)$) is a Lagrangian immersion such that the distance squared function $D : U \times \mathbb{R}^n \longrightarrow \mathbb{R}$ (respectively, height function $H : U \times S^{n-1} \longrightarrow \mathbb{R}$) of $M = \boldsymbol{X}(U)$ is a generating family of $L(D)$ (respectively, $L(H)$).*

Therefore, we have the Lagrangian immersion $L(D)$ whose caustics is the evolute of $M = \boldsymbol{X}(U)$. We call $L(D)$ the *Lagrangian lift* of the evolute Ev_M of $M = \boldsymbol{X}(U)$. Moreover, the positive component of the Lagrangian map $\pi \circ L(H)$ can be identified with the Gauss map of $M = \boldsymbol{X}(U)$. We also call $L(H)$ the *Lagrangian lift* of the Gauss map $G : U \longrightarrow S^{n-1}$ of $M = \boldsymbol{X}(U)$.

On the other hand, we consider the extended height function $\widetilde{H} : U \times (S^{n-1} \times \mathbb{R}) \longrightarrow \mathbb{R}$ of $M = \boldsymbol{X}(U)$. We have the following proposition.

Proposition 4.3. *The extended height function $\widetilde{H} : U \times (S^{n-1} \times \mathbb{R}) \longrightarrow \mathbb{R}$ on $M = \boldsymbol{X}(U)$ is a Morse family of hypersurfaces.*

Proof. The proof is given by almost the similar calculation as the case for the height function. For any $v \in S^{n-1}$, we have $v_1^2 + \cdots + v_n^2 = 1$. Without loss of the generality, we also assume that $v_n > 0$. We have $v_n = \sqrt{1 - v_1^2 - \cdots - v_{n-1}^2}$, so that

$$\widetilde{H}(u, v, r) = x_1(u)v_1 + \cdots + x_{n-1}(u)v_{n-1} + x_n(u)\sqrt{1 - v_1^2 - \cdots - v_{n-1}^2} - r.$$

We also prove that the mapping

$$\Delta^*\widetilde{H} = \left(\widetilde{H}, \frac{\partial \widetilde{H}}{\partial u_1}, \ldots, \frac{\partial \widetilde{H}}{\partial u_{n-1}}\right)$$

252

is non-singular at any point in $\Sigma_*(\widetilde{H}) = \Delta^*\widetilde{H}^{-1}(0)$. The Jacobian matrix of $\Delta^*\widetilde{H}$ is given as follows:

$$\left(\begin{array}{ccc} \langle \boldsymbol{X}_{u_1}, \boldsymbol{v} \rangle & \cdots & \langle \boldsymbol{X}_{u_n}, \boldsymbol{v} \rangle \\ \langle \boldsymbol{X}_{u_1 u_1}, \boldsymbol{v} \rangle & \cdots & \langle \boldsymbol{X}_{u_1 u_{n-1}}, \boldsymbol{v} \rangle \\ \vdots & \vdots & \vdots \\ \langle \boldsymbol{X}_{u_{n-1} u_1}, \boldsymbol{v} \rangle & \cdots & \langle \boldsymbol{X}_{u_{n-1} u_{n-1}}, \boldsymbol{v} \rangle \end{array} \right. \left| \ \widetilde{X}\ \right| \left. \begin{array}{c} 1 \\ 0 \\ \vdots \\ 0 \end{array} \right),$$

where

$$\widetilde{X} = \left(\begin{array}{ccc} x_{1u_1} - x_{nu_1}\dfrac{v_1}{v_n} & \cdots & x_{n-1u_1} - x_{nu_1}\dfrac{v_{n-1}}{v_n} \\ \vdots & \vdots & \vdots \\ x_{1u_{n-1}} - x_{nu_{n-1}}\dfrac{v_1}{v_n} & \cdots & x_{n-1u_{n-1}} - x_{nu_{n-1}}\dfrac{v_{n-1}}{v_n} \end{array} \right)$$

It is enough to show that the rank of the matrix $\widetilde{X}$ is $n-1$ at $(u, \boldsymbol{v}, r) \in \Sigma_*(\widetilde{H})$. It has been done in the proof of Proposition 4.1. This completes the proof of the proposition. $\qquad\square$

We can also define a Legendrian immersion germ whose generating family is the extended height function of $M = \boldsymbol{X}(U)$ as follows (cf., Appendix B): For the $(n-1)$-sphere S^{n-1}, we consider the local coordinate $U_i = \{\boldsymbol{v} = (v_1, \ldots, v_n) \in S^{n-1} \mid v_i \neq 0\}$. Since $PT^*(S^{n-1} \times \mathbb{R})|(U_i \times \mathbb{R})$ is a trivial bundle, we define a map

$$\mathcal{L}_i(\widetilde{H}) : \Sigma_*(\widetilde{H})|U \times (U_i \times \mathbb{R}) \longrightarrow PT^*(S^{n-1} \times \mathbb{R})|(U_i \times \mathbb{R}) \ (i = 0, 1, \ldots, n)$$

by

$$\mathcal{L}_i(\widetilde{H})(u, \boldsymbol{v}, r) =$$
$$\left(\boldsymbol{v}, r, [x_1(u) - x_i(u)\frac{v_1}{v_i} : \cdots : \widehat{x_i(u) - x_i(u)\frac{v_i}{v_i}} : \cdots : x_n(u) - x_i(u)\frac{v_n}{v_i} : -1] \right),$$

where $v = (v_1, \ldots, v_n) \in S^{n-1}$ and we denote $(x_1, \ldots, \hat{x}_i, \ldots, x_n)$ as a point in the $(n-1)$-dimensional space such that the i-th component x_i is removed. We can also show that if $U_i \cap U_j \neq \emptyset$ for $i \neq j$, then $\mathcal{L}_i(\widetilde{H})$ and $\mathcal{L}_j(\widetilde{H})$ are Legendrian equivalent which are given by the local coordinate transformation of $S^{n-1} \times \mathbb{R}$ and Legendrian lift of it by exactly the same method as the case for Lagrangian equivalence. Therefore we can define a global Legendrian immersion, $\mathcal{L}(\widetilde{H}) : \Sigma_*(\widetilde{H}) \longrightarrow PT^*(S^{n-1} \times \mathbb{R})$.

By definition, we have the following corollary of the above proposition:

Corollary 4.4. *Under the above notations, $\mathcal{L}(\widetilde{H})$ is a Legendrian immersion such that the extended height function $\widetilde{H} : U \times (S^{n-1} \times \mathbb{R}) \longrightarrow \mathbb{R}$ of $M = \boldsymbol{X}(U)$ is a generating family of $\mathcal{L}(\widetilde{H})$.*

Therefore, we have the Legendrian immersion $\mathcal{L}(\widetilde{H})$ whose wave front is the cylindrical pedal of $M = \boldsymbol{X}(U)$. We call $\mathcal{L}(\widetilde{H})$ the *Legendrian lift* of the cylindrical pedal CPe_M of $M = \boldsymbol{X}(U)$.

5. Contact with model hypersurfaces and families of model hypersurfaces

In [24,25] Montaldi studied the contact of surfaces with hyperplanes or hyperspheres in $\mathbb{R}^n$ ($n = 3, 4$). For the purpose, he has developed a general theory of contact between submanifolds. Let X_i, Y_i ($i = 1, 2$) be submanifolds of $\mathbb{R}^n$ with $\dim X_1 = \dim X_2$ and $\dim Y_1 = \dim Y_2$. We say that the *contact of X_1 and Y_1 at y_1 is of the same type* as the *contact of X_2 and Y_2 at y_2* if there is a diffeomorphism germ $\Phi : (\mathbb{R}^n, y_1) \longrightarrow (\mathbb{R}^n, y_2)$ such that $\Phi(X_1) = X_2$ and $\Phi(Y_1) = Y_2$. In this case we write $K(X_1, Y_1; y_1) = K(X_2, Y_2; y_2)$. It is clear that in the definition $\mathbb{R}^n$ could be replaced by any manifold. In his paper [24], Montaldi gives a characterization of the notion of contact by using the terminology of Singularity theory.

Theorem 5.1. *Let X_i, Y_i ($i = 1, 2$) be submanifolds of $\mathbb{R}^n$ with $\dim X_1 = \dim X_2$ and $\dim Y_1 = \dim Y_2$. Let $g_i : (X_i, x_i) \longrightarrow (\mathbb{R}^n, y_i)$ be immersion germs and $f_i : (\mathbb{R}^n, y_i) \longrightarrow (\mathbb{R}^p, 0)$ be submersion germs with $(Y_i, y_i) = (f_i^{-1}(0), y_i)$. Then $K(X_1, Y_1; y_1) = K(X_2, Y_2; y_2)$ if and only if $f_1 \circ g_1$ and $f_2 \circ g_2$ are $\mathcal{K}$-equivalent. For the definition of the $\mathcal{K}$-equivalence and the basic properties, see Appendix B or [21].*

On the other hand, we now briefly describe the theory of contact with foliations. Here we consider the relationship between the contact of submanifolds with foliations and the $\mathcal{R}^+$-class of functions. Let X_i ($i = 1, 2$) be submanifolds of $\mathbb{R}^n$ with $\dim X_1 = \dim X_2$, $g_i : (X_i, \bar{x}_i) \longrightarrow (\mathbb{R}^n, \bar{y}_i)$ be immersion germs and $f_i : (\mathbb{R}^n, \bar{y}_i) \longrightarrow (\mathbb{R}, 0)$ be submersion germs. For a submersion germ $f : (\mathbb{R}^n, 0) \longrightarrow (\mathbb{R}, 0)$, we denote that $\mathcal{F}_f$ be the regular foliation defined by f; i.e., $\mathcal{F}_f = \{f^{-1}(c)|c \in (\mathbb{R}, 0)\}$. We say that *the contact of X_1 with the regular foliation $\mathcal{F}_{f_1}$ at $\bar{y}_1$ is of the same type as the contact of X_2 with the regular foliation $\mathcal{F}_{f_2}$ at $\bar{y}_2$* if there is a diffeomorphism germ $\Phi : (\mathbb{R}^n, \bar{y}_1) \longrightarrow (\mathbb{R}^n, \bar{y}_2)$ such that $\Phi(X_1) = X_2$ and $\Phi(Y_1(c)) = Y_2(c)$, where $Y_i(c) = f_i^{-1}(c)$ for each $c \in (\mathbb{R}, 0)$. In this case we write $K(X_1, \mathcal{F}_{f_1}; \bar{y}_1) = K(X_2, \mathcal{F}_{f_2}; \bar{y}_2)$. It is also clear that in the definition $\mathbb{R}^n$ could be replaced by any manifold. We apply the method of Goryunov [6] to the case for $\mathcal{R}^+$-equivalences among function germs, so that we have the following:

Proposition 5.2 ([6], Appendix). *Let X_i $(i = 1, 2)$ be submanifolds of $\mathbb{R}^n$ with $\dim X_1 = \dim X_2 = n - 1$ (i.e. hypersurface), $g_i : (X_i, \bar{x}_i) \longrightarrow (\mathbb{R}^n, \bar{y}_i)$ be immersion germs and $f_i : (\mathbb{R}^n, \bar{y}_i) \longrightarrow (\mathbb{R}, 0)$ be submersion germs. Then $K(X_1, \mathcal{F}_{f_1}; \bar{y}_1) = K(X_2, \mathcal{F}_{f_2}; \bar{y}_2)$ if and only if $f_1 \circ g_1$ and $f_2 \circ g_2$ are $\mathcal{R}^+$-equivalent.*

Golubitsky and Guillemin [7] have given an algebraic characterization for the $\mathcal{R}^+$-equivalence among function germs. We denote $C_0^\infty(X)$ is the set of function germs $(X, 0) \longrightarrow \mathbb{R}$. Let J_f be the Jacobian ideal in $C_0^\infty(X)$ (i.e., $J_f = \langle \partial f / \partial x_1, \ldots, \partial f / \partial x_n \rangle_{C_0^\infty(X)}$). Let $\mathcal{R}_k(f) = C_0^\infty(X)/J_f^k$ and $\bar{f}$ be the image of f in this local ring. We say that f satisfies the *Milnor Condition* if $\dim_{\mathbb{R}} \mathcal{R}_1(f) < \infty$.

Proposition 5.3 ([7], Proposition 4.1). *Let f and g be germs of functions at 0 in X satisfying the Milnor condition with $df(0) = dg(0) = 0$. Then f and g are $\mathcal{R}^+$-equivalent if*
(1) The rank and signature of the Hessians $\mathcal{H}(f)(0)$ and $\mathcal{H}(g)(0)$ are equal, and
(2) There is an isomorphism $\gamma : \mathcal{R}_2(f) \longrightarrow \mathcal{R}_2(g)$ such that $\gamma(\bar{f}) = \bar{g}$.

On the other hand, we define the following functions:

$$\mathcal{H} : \mathbb{R}^n \times S^{n-1} \longrightarrow \mathbb{R} \; ; \; \mathcal{H}(\boldsymbol{x}, \boldsymbol{v}) = \langle \boldsymbol{x}, \boldsymbol{v} \rangle,$$
$$\widetilde{\mathcal{H}} : \mathbb{R}^n \times (S^{n-1} \times \mathbb{R}) \longrightarrow \mathbb{R} \; ; \; \mathcal{H}(\boldsymbol{x}, \boldsymbol{v}, r) = \langle \boldsymbol{x}, \boldsymbol{v} \rangle - r,$$
$$\mathcal{D} : \mathbb{R}^n \times \mathbb{R}^n \longrightarrow \mathbb{R} \; ; \; \mathcal{D}(\boldsymbol{y}, \boldsymbol{x}) = \|\boldsymbol{y} - \boldsymbol{x}\|^2.$$

We now consider the contact of hypersurfaces with hyperplane. For any $\boldsymbol{v} \in S^{n-1}$ we denote that $\mathfrak{h}_v(\boldsymbol{x}) = \mathcal{H}(\boldsymbol{x}, \boldsymbol{v})$ and we have a hyperplane $\mathfrak{h}_v^{-1}(r)$. We denote it as $H(\boldsymbol{v}, r)$. For any $u \in U$, we consider the unit normal vector $\boldsymbol{v} = \boldsymbol{n}(u)$ and $r = \langle \boldsymbol{X}(u), \boldsymbol{n}(u) \rangle$, then we have

$$\mathfrak{h}_v \circ \boldsymbol{X}(u) = \mathcal{H} \circ (\boldsymbol{X} \times id_{S^{n-1}})(u, \boldsymbol{v}) = H(u, \boldsymbol{n}(u)) = r.$$

We also have relations that

$$\frac{\partial \mathfrak{h}_v \circ \boldsymbol{X}}{\partial u_i}(u) = \frac{\partial H}{\partial u_i}(u, \boldsymbol{n}(u)) = 0$$

for $i = 1, \ldots, n-1$. This means that the hyperplane $\mathfrak{h}_v^{-1}(r) = H(v, r)$ is tangent to $M = X(U)$ at $p = X(u)$. Therefore, $H(v, r)$ is the *tangent hyperplane* of $M = X(U)$ at $p = X(u)$ (or, u), which we write $H(X(U), u)$. Let v_1, v_2 be unit vectors. If v_1, v_2 are linearly dependent, then corresponding hyperplanes $H(v_1, r_1), H(v_2, r_2)$ are parallel. Then we have the following simple lemma.

Lemma 5.4. *Let $X : U \longrightarrow \mathbb{R}^n$ be a hypersurface. Consider two points $u_1, u_2 \in U$. Then*

(1) $\mathrm{CPe}_M(u_1) = \mathrm{CPe}_M(u_2)$ *if and only if* $H(X(U), u_1) = H(X(U), u_2)$.
(2) $G(u_1) = G(u_2)$ *if and only if* $H(X, u_1)$, $H(X, u_2)$ *are parallel.*

We also consider the family of parallel hyperplanes which contains a tangent hyperplane of $M = X(U)$. Since $\mathfrak{h}_v$ is a submersion, we have a regular foliation $\mathcal{F}_{\mathfrak{h}_v} = \{ H(v, c) \mid c \in (\mathbb{R}, r) \}$ whose leaves are hyperplanes such that the case $c = r$ corresponds to the tangent hyperplane $H(X(U), u)$. It follows that we have a singular foliation germ $(X^{-1}(\mathcal{F}_{\mathfrak{h}_v}), u)$ which we call the *Dupin foliation germ* of $M = X(U)$ at u. We denote it by $\mathcal{DF}(X(U), u)$. We remark that the Dupin foliation germ is diffeomorphic to the germ of the Dupin indicatrices family in the classical sense at a non-parabolic point ([29], page 136).

We consider the function $\mathcal{D} : \mathbb{R}^n \times \mathbb{R}^n \longrightarrow \mathbb{R}$. For any $x \in \mathbb{R}^n \setminus M$, we denote that $\mathfrak{d}_x(y) = \mathcal{D}(y, x)$ and we have a hypersphere $\mathfrak{d}_x^{-1}(r^2) = S^{n-1}(x, r)$. It is easy to show that $\mathfrak{d}_x$ is a submersion. For any $u \in U$, we consider a point $x = X(u) + rn(u) \in \mathbb{R}^n \setminus M$, then we have

$$\mathfrak{d}_x \circ X(u) = \mathcal{D} \circ (X \times id_{\mathbb{R}^n})(u, x) = r,$$

and

$$\frac{\partial \mathfrak{d}_x \circ X}{\partial u_i}(u) = \frac{\partial D}{\partial u_i}(u, x) = 0.$$

for $i = 1, \ldots, n-1$. This means that the hypersphere $\mathfrak{d}_x^{-1}(r) = S^{n-1}(x, r)$ is tangent to $M = X(U)$ at $p = X(u)$. In this case, we call $S^{n-1}(x, r)$ a *tangent hypersphere* at $p = X(u)$ with the center x. However, there are infinitely many tangent hyperspheres at a general point $p = X(u)$ depending on the real number r. If x is a point of the hyperbolic evolute, the tangent hypersphere with the center x is called the *osculating hypersphere* (*focal hypersphere*) at $p = X(u)$ which is uniquely determined. For $x = X(u) + rn(u)$, we also have a regular foliation

$$\mathcal{F}_{\mathfrak{d}_x} = \left\{ S^{n-1}(x, c) \mid c \in (\mathbb{R}, r) \right\}$$

whose leaves are hyperspheres with the center $\boldsymbol{x}$ such that the case $c = r$ corresponding to the tangent hypersphere with radius $|r|$. Moreover, if $r = 1/\kappa(u)$, then $S^{n-1}(\boldsymbol{x}, 1/\kappa(u))$ is the osculating hypersphere. In this case $(\boldsymbol{X}^{-1}(\mathcal{F}_{\eth_x}), u)$ is a singular foliation germ at u which is called a *osculating hyperspherical foliation* of $M = \boldsymbol{X}(U)$ at $p = \boldsymbol{X}(u)$ (or, u). We denote it by $\mathcal{OF}(\boldsymbol{X}(U), u)$.

6. The theory of contact from the view point of Lagrangian or Legendrian singularity theory

In this section we apply Lagrangian or Legendrian singularity theory to the study of contact of hypersurfaces with hyperplanes or hyperspheres.

First we consider the contact of hypersurfaces with hyperplanes. Let $\mathrm{CPe}_{M_i} : (U, u_i) \longrightarrow (S^{n-1} \times \mathbb{R}, (\boldsymbol{v}_i, r_i))$ $(i = 1, 2)$ be two cylindrical pedal germs of hypersurface germs $\boldsymbol{X}_i : (U, u_i) \longrightarrow (\mathbb{R}^n, \boldsymbol{X}_i(u_i))$ and $M_i = \boldsymbol{X}_i(U)$. We say that two map germs $f_i : (\mathbb{R}^n, \boldsymbol{x}_i) \longrightarrow (\mathbb{R}^p, \boldsymbol{y}_i)$ $(i = 1, 2)$ are $\mathcal{A}$-*equivalent* if there exist diffeomorphism germs $\phi : (\mathbb{R}^n, \boldsymbol{x}_1) \longrightarrow (\mathbb{R}^n, \boldsymbol{x}_2)$ and $\psi : (\mathbb{R}^p, \boldsymbol{y}_1) \longrightarrow (\mathbb{R}^p, \boldsymbol{y}_2)$ such that $\psi \circ f_1 = f_2 \circ \phi$. If for both $i = 1, 2$ the regular set of CPe_{M_i} is dense in (U, u_i), it follows from Proposition B.2 that CPe_{M_1} and CPe_{M_2} are $\mathcal{A}$-equivalent if and only if the corresponding Legendrian immersion germs $\mathcal{L}(\widetilde{H}_1) : (U, u_1) \longrightarrow PT^*(S^{n-1} \times \mathbb{R})$ and $\mathcal{L}(\widetilde{H}_2) : (U, u_2) \longrightarrow PT^*(S^{n-1} \times \mathbb{R})$ are Legendrian equivalent, where $\widetilde{H}_i$ is the extended height function germ of $M_i = \boldsymbol{X}_i(U)$. This condition is also equivalent to the condition that two generating families $\widetilde{H}_1$ and $\widetilde{H}_2$ are P-$\mathcal{K}$-equivalent by Theorem B.3.

On the other hand, we consider the case that $\boldsymbol{v}_i = \boldsymbol{n}_i(u)$, $r_i = \langle \boldsymbol{X}_i(u), \boldsymbol{n}_i(u) \rangle$. We denote that $\widetilde{h}_{i,(v_i,r_i)}(u) = \widetilde{H}_i(u, \boldsymbol{v}_i, r_i)$, then we have $\widetilde{h}_{i,(v_i,r_i)}(u) = \mathfrak{h}_{v_i} \circ \boldsymbol{X}_i(u) - r_i$. By Theorem 5.1, $K(\boldsymbol{X}_1(U), H(\boldsymbol{X}_1, u_1), p_1) = K(\boldsymbol{X}_2(U), H(\boldsymbol{X}_2(U), u_2), p_2)$ if and only if $\widetilde{h}_{1,(v_1,r_1)}$ and $\widetilde{h}_{1,(v_2,r_2)}$ are $\mathcal{K}$-equivalent, where $p_i = \boldsymbol{X}(u_i)$. Therefore, we can apply the arguments in Appendix B to our situation. We denote $Q(\boldsymbol{X}, u)$ the local ring of the function germ $\widetilde{h}_{v_0,r_0} : (U, u_0) \longrightarrow \mathbb{R}$, where $(\boldsymbol{v}_0, r_0) = \mathrm{CPe}_M(u_0)$. We remark that we can explicitly write the local ring as follows:

$$Q(\boldsymbol{X}(U), u_0) = \frac{C_{u_0}^\infty(U)}{\langle \langle \boldsymbol{X}(u), \boldsymbol{n}(u_0) \rangle - r_0 \rangle_{C_{u_0}^\infty(U)}},$$

where where $r_0 = \langle \boldsymbol{X}(u_0), \boldsymbol{n}(u_0) \rangle$ and $C_{u_0}^\infty(U)$ is the local ring of function germs at u_0 with the unique maximal ideal $\mathfrak{M}_{u_0}(U)$.

Theorem 6.1. *Let* $\boldsymbol{X}_i : (U, u_i) \longrightarrow (\mathbb{R}^n, p_i)$ $(i = 1, 2)$ *be hypersurfaces germs such that the corresponding Legendrian immersion germs* $\mathcal{L}(\widetilde{H}_i)$: $(U, u_i) \longrightarrow PT^*(S^{n-1} \times \mathbb{R})$ *are Legendrian stable. Then the following conditions are equivalent:*

(1) *Cylindrical pedal germs* CPe_{M_1} *and* CPe_{M_2} *are* $\mathcal{A}$-*equivalent.*

(2) $\widetilde{H}_1$ *and* $\widetilde{H}_2$ *are* P-$\mathcal{K}$-*equivalent.*

(3) $\widetilde{h}_{1,(v_1,r_1)}$ *and* $\widetilde{h}_{1,(v_2,r_2)}$ *are* $\mathcal{K}$-*equivalent, where* $(\boldsymbol{v}_i, r_i) = \mathrm{CPe}_{M_i}(u_i)$.

(4) $K(\boldsymbol{X}_1(U), H(\boldsymbol{X}_1(U), u_1), p_1) = K(\boldsymbol{X}_2(U), H(\boldsymbol{X}_2(U), u_2), p_2)$.

(5) $Q(\boldsymbol{X}_1, u_1)$ *and* $Q(\boldsymbol{X}_2, u_2)$ *are isomorphic as* $\mathbb{R}$-*algebras.*

Proof. By the previous arguments (mainly from Theorem 5.1), it has been already shown that conditions (3) and (4) are equivalent. Other assertions follow from Proposition B.4. $\qquad\square$

As an application of a kind of the transversality theorems, we cam show that the assumption of the theorem is generic in the case when $n \leq 6$. In general we have the following proposition.

Proposition 6.2. *Let* $\boldsymbol{X}_i : (U, u_i) \longrightarrow (\mathbb{R}^n, p_i)$ $(i = 1, 2)$ *be hypersurface germs such that their sets of parabolic points have no interior points as subspaces of* U. *If cylindrical pedal germs* CPe_{M_1}, CPe_{M_2} *are* $\mathcal{A}$-*equivalent, then*

$$K(\boldsymbol{X}_1(U), H(\boldsymbol{X}_1(U), u_1), p_1) = K(\boldsymbol{X}_2(U), H(\boldsymbol{X}_2(U), u_2), p_2).$$

In this case, $(\boldsymbol{X}_1^{-1}(H(\boldsymbol{X}_1(U), u_1)), u_1)$ *and* $(\boldsymbol{X}_2^{-1}(H(\boldsymbol{X}_2(U), u_2), u_2)$ *are diffeomorphic as set germs.*

Proof. The set of parabolic points is the set of singular points of the cylindrical pedal. So the corresponding Legendrian lifts $\mathcal{L}(\widetilde{H}_i)$ satisfy the hypothesis of Proposition B.2. If cylindrical pedal germs CPe_{M_1}, CPe_{M_2} are $\mathcal{A}$-equivalent, then $\mathcal{L}(\widetilde{H}_1)$, $\mathcal{L}(\widetilde{H}_2)$ are Legendrian equivalent, so that $\widetilde{H}_1$, $\widetilde{H}_2$ are P-$\mathcal{K}$-equivalent. Therefore, $\widetilde{h}_{1,(v_1,r_1)}$, $\widetilde{h}_{1,(v_2,r_2)}$ are $\mathcal{K}$-equivalent, where $r_i = \langle \boldsymbol{X}_i(u), \boldsymbol{n}_i(u) \rangle$. By Theorem 5.1, this condition is equivalent to the condition that $K(\boldsymbol{X}_1(U), H(\boldsymbol{X}_1(U), u_1), p_1) = K(\boldsymbol{X}_2(U), H(\boldsymbol{X}_2(U), u_2), p_2)$.

On the other hand, we have $(\boldsymbol{X}_i^{-1}(H(\boldsymbol{X}_i(U), u_i)), u_i) = (\widetilde{h}_{i,(v_i,r_i)}^{-1}(0), u_i)$. It follows from this fact that $(\boldsymbol{X}_1^{-1}(H(\boldsymbol{X}_1(U), u_1)), u_1)$ and $(\boldsymbol{X}_2^{-1}(H(\boldsymbol{X}_2(U), u_2), u_2)$ are diffeomorphic as set germs because the $\mathcal{K}$-equivalence preserve the zero level sets. $\qquad\square$

For a hypersurface germ $\boldsymbol{X} : (U,u) \longrightarrow (\mathbb{R}^n, p)$, we call $(\boldsymbol{X}^{-1}(H(\boldsymbol{X}(U),u)),u)$ the *tangent indicatrix germ* of $M = \boldsymbol{X}(U)$ at u (or p). By Proposition 6.2, the diffeomorphism type of the tangent indicatrix germ is an invariant of the $\mathcal{A}$-classification of the cylindrical pedal germ of $\boldsymbol{X}$. Moreover, by the above results, we can borrow some basic invariants from the singularity theory on function germs. We need $\mathcal{K}$-invariants for function germ. The local ring of a function germ is a complete $\mathcal{K}$-invariant for generic function germs. It is, however, not a numerical invariant. The $\mathcal{K}$-codimension (or, Tyurina number) of a function germ is a numerical $\mathcal{K}$-invariant of function germs [21]. We denote that

$$\text{T-ord}(\boldsymbol{X}(U), u_0) = \dim \frac{C_{u_0}^\infty(U)}{\langle \langle \boldsymbol{X}(u), \boldsymbol{n}(u_0) \rangle - r_0, \langle \boldsymbol{X}_{u_i}(u), \boldsymbol{n}(u_0) \rangle \rangle_{C_{u_0}^\infty}},$$

where $r_0 = \langle \boldsymbol{X}(u_0), \boldsymbol{n}(u_0) \rangle$. Usually $\text{T-ord}(\boldsymbol{x}(U), u_0)$ is called the $\mathcal{K}$-*codimension of* $\widetilde{h}_{(v_0, r_0)}$. However, we call it the *order of contact with the tangent hyperplane* at $\boldsymbol{X}(u_0)$. We also have the notion of corank of function germs.

$$\text{T-corank}(\boldsymbol{X}(U), u_0) = (n-1) - \text{rank}\,\text{Hess}(h_{v_0}(u_0)),$$

where $v_0 = \boldsymbol{n}(u_0)$.

By Proposition 3.2, $\boldsymbol{X}(u_0)$ is a parabolic point if and only if $\text{T-corank}(\boldsymbol{X}(U), u_0) \geq 1$. Moreover $\boldsymbol{X}(u_0)$ is a flat point if and only if $\text{T-corank}(\boldsymbol{X}(U), u_0) = n - 1$.

On the other hand, a function germ $f : (\mathbb{R}^{n-1}, \boldsymbol{a}) \longrightarrow \mathbb{R}$ has the A_k-type singularity if and only if f is $\mathcal{K}$-equivalent to the germ $x_1^{k+1} \pm x_1^2 \cdots \pm x_{n-1}^2$. If $\text{T-corank}(\boldsymbol{X}(U), u_0) = n - 2$, the height function h_{v_0} has the A_k-type singularity at u_0 in generic. In this case we have $\text{T-ord}(\boldsymbol{X}(U), u_0) = k$. This number is equal to the order of contact in the classical sense (cf., [5]). This is the reason why we call $\text{T-ord}(\boldsymbol{X}(U), u_0)$ the order of contact with the tangent hyperplane at $\boldsymbol{X}(u_0)$.

We now consider the contact of hypersurfaces with families of hyperplane. Let $\boldsymbol{X}_i : (U, \bar{u}_i) \longrightarrow (\mathbb{R}^n, p_i)$ $(i = 1, 2)$ be hypersurface germs. We consider height functions $H_i : (U \times S^{n-1}, (\bar{u}_i, \boldsymbol{v}_i)) \longrightarrow \mathbb{R}$ of $\boldsymbol{X}_i(U)$, where $\boldsymbol{v}_i = \boldsymbol{n}(\bar{u}_i)$ respectively. We denote that $h_{i, v_i}(u) = H_i(u, \boldsymbol{v}_i)$, then we have $h_{i, v_i}(u) = \mathfrak{h}_{v_i} \circ \boldsymbol{X}_i(u)$. Then we have the following theorem:

Theorem 6.3. *Let* $\boldsymbol{X}_i : (U, \bar{u}_i) \longrightarrow (\mathbb{R}^n . p_i)$ *be hypersurface germs such that the corresponding Lagrangian immersion germs* $L(H_i) : (C(H_i), (\bar{u}_i, \boldsymbol{v}_i)) \longrightarrow T^* S^{n-1}$ *are Lagrangian stable, where* $\boldsymbol{v}_i = \boldsymbol{n}(\bar{u}_i)$ *re-*

spectively. Then the following conditions are equivalent:

(1) $K(\boldsymbol{X}_1(U), \mathcal{F}_{\mathfrak{h}_{v_1}}; p_1) = K(\boldsymbol{X}_2(U), \mathcal{F}_{\mathfrak{h}_{v_2}}; p_2)$.

(2) h_{1,v_1} *and* h_{2,v_2} *are* $\mathcal{R}^+$*-equivalent.*

(3) H_1 *and* H_2 *are* P-$\mathcal{R}^+$*-equivalent.*

(4) $L(H_1)$ *and* $L(H_2)$ *are Lagrangian equivalent.*

(5) (a) *The rank and signature of the* $\mathcal{H}(h_{1,v_1})(\bar{u}_1)$ *and* $\mathcal{H}(h_{2,v_2})(\bar{u}_2)$ *are equal,*

(b) *There is an isomorphism* $\gamma : \mathcal{R}_2(h_{1,v_1}) \longrightarrow \mathcal{R}_2(h_{2,v_2})$ *such that* $\gamma(\overline{h_{1,v_1}}) = \overline{h_{2,v_2}}$.

Proof. By Proposition 5.2, the condition (1) is equivalent to the condition (2). Since both of $L(H_i)$ are Lagrangian stable, both of H_i are $\mathcal{R}^+$-versal unfoldings of h_{i,v_i} respectively. By the uniqueness theorem on the $\mathcal{R}^+$-versal unfolding of a function germ, the condition (2) is equivalent to the condition (3). By Theorem A.2, the condition (3) is equivalent to the condition (4). It also follows from Theorem A.2 that both of h_i satisfy the Milnor condition. Therefore we can apply Proposition 5.3 to our situation, so that the condition (2) is equivalent to the condition (5). This completes the proof☐

We remark that if $L(H_1)$ and $L(H_2)$ are Lagrangian equivalent, then the corresponding Lagrangian map germs $\pi \circ L(H_1)$ and $\pi \circ L(H_1)$ are $\mathcal{A}$-equivalent. The Gauss map of a hypersurface $\boldsymbol{x}(U) = M$ is considered to be the Lagrangian map germ of $L(H)$ (or, the catastrophe map germ of H_1). Moreover, if $h_{1.v_1}$ and h_{2,v_2} are $\mathcal{R}^+$-equivalent then the level set germs of function germs $h_{1.v_1}$ and h_{2,v_2} are diffeomorphic. Therefore, we have the following corollary.

Corollary 6.4. *Under the same assumptions as those of the above theorem for hypersurface germs* $\boldsymbol{X}_i : (U, \bar{u}_i) \longrightarrow (\mathbb{R}^n, p_i)$ $(i = 1, 2)$, *we have the following: If one of the conditions of the above theorem is satisfied, then*

(1) *The Gauss map germs* G_1, G_2 *are* $\mathcal{A}$*-equivalent.*

(2) *The Dupin foliation germs* $\mathcal{DF}(\boldsymbol{X}_1(U), \bar{u}_1)$, $\mathcal{DF}(\boldsymbol{X}_2(U), \bar{u}_2)$ *are diffeomorphic.*

We also consider the contact of hypersurfaces with families of hyperspheres. Let $\boldsymbol{X}_i : (U, \bar{u}_i) \longrightarrow \mathbb{R}^n, p_i)$ $(i = 1, 2)$ be hypersurface germs. We consider distance squared functions $D_i : (U \times \mathbb{R}^n, (\bar{u}_i, \boldsymbol{x}_i)) \longrightarrow \mathbb{R}$ of $\boldsymbol{X}_i(U)$,

260

where $x_i = \mathrm{Ev}_{\kappa_i}(\bar{u}_i)$. We denote that $d_{i,v_i}(u) = D_i(u, x_i)$, then we have $d_{i,x_i}(u) = \mathfrak{d}_{x_i} \circ X_i(u)$. Then we have the following theorem:

Theorem 6.5. *Let* $X_i : (U, \bar{u}_i) \longrightarrow \mathbb{R}^n, p_i)$ $(i = 1, 2)$ *be hypersurface germs such that the corresponding Lagrangian immersion germs* $L(D_i) : (C(D_i), (\bar{u}_i, x_i)) \longrightarrow T^*\mathbb{R}^n$ *are Lagrangian stable, where* $x_i = \mathrm{Ev}_{\kappa_i}(\bar{u}_i)$ *are centers of the osculating hyperspheres of* $X_i(U)$ *respectively. Then the following conditions are equivalent:*

(1) $K(X_1(U), \mathcal{F}_{\mathfrak{d}_{x_1}}; p_1) = K(X_2(U), \mathcal{F}_{\mathfrak{d}_{x_2}}; p_2)$.

(2) d_{1,x_1} *and* d_{2,x_2} *are* $\mathcal{R}^+$-*equivalent.*

(3) D_1 *and* D_2 *are* P-$\mathcal{R}^+$-*equivalent.*

(4) $L(D_1)$ *and* $L(D_2)$ *are Lagrangian equivalent.*

(5) (a) *The rank and signature of the* $\mathcal{H}(d_{1,x_1})(\bar{u}_1)$ *and* $\mathcal{H}(d_{2,x_2})(\bar{u}_2)$ *are equal,*

  (b) *There is an isomorphism* $\gamma : \mathcal{R}_2(d_{1,x_1}) \longrightarrow \mathcal{R}_2(d_{2,x_2})$ *such that* $\gamma(\overline{d_{1,x_1}}) = \overline{d_{2,x_2}}$.

The proof of the theorem is parallel to those of Theorem 6.3, so that we omit it.

We remark that if $L(D_1)$ and $L(D_2)$ are Lagrangian equivalent, then the corresponding evolutes are diffeomorphic. Since the evolute of a hypersurface $M = X(U)$ is considered to be the caustic of $L(D)$, the above theorem gives a symplectic interpretation for the contact of hypersurfaces with family of hyperspheres (cf., Appendix A). We have the following corollary.

Corollary 6.6. *Under the same assumptions as those of the above theorem for hypersurface germs* $X_i : (U, \bar{u}_i) \longrightarrow (\mathbb{R}^n, p_i)$ $(i = 1, 2)$, *we have the following: If one of the conditions of the above theorem is satisfied, then*

  (1) *The evolutes* Ev_{M_1} *and* Ev_{M_2} *are diffeomorphic as set germs.*

  (2) *The osculating hyperspherical foliation germs* $\mathcal{OF}(X_1(U), \bar{u}_1)$, $\mathcal{OF}(X_2(U), \bar{u}_2)$ *are diffeomorphic.*

7. Surfaces in 3-space

In this section we consider the case $n = 3$. Before we start to consider the case $n = 3$, we study generic properties of hypersurfaces in $\mathbb{R}^n$ for general n. The main tool is a kind of transversality theorems. We consider the space of embeddings $\mathrm{Emb}\,(U,\mathbb{R}^n)$ with Whitney C^∞-topology. We also consider the functions:

$$\mathcal{H} : \mathbb{R}^n \times S^{n-1} \longrightarrow \mathbb{R},$$
$$\widetilde{\mathcal{H}} : \mathbb{R}^n \times (S^{n-1} \times \mathbb{R}) \longrightarrow \mathbb{R},$$
$$\mathcal{D} : \mathbb{R}^n \times \mathbb{R}^n \longrightarrow \mathbb{R}.$$

which are given in §5. We claim that $\mathfrak{h}_v$, $\widetilde{\mathfrak{h}}_{(v,r)}$ and $\mathfrak{d}_x$ are respectively submersions for any $v \in S^{n-1}$, $(v,r) \in S^{n-1} \times \mathbb{R}$ and $x \in \mathbb{R}^n \setminus M$ respectively. where $\mathfrak{h}_v(x) = \mathcal{H}(x,v)$, $\widetilde{\mathfrak{h}}_{(v,r)}(x) = \widetilde{\mathcal{H}}(x,v,r)$ and $\mathfrak{d}_x(y) = \mathcal{D}(y,x)$. For any $X \in \mathrm{Emb}\,(U,\mathbb{R}^n)$, we have

$$H = \mathcal{H} \circ (X \times id_{S^{n-1}}), \ \widetilde{H} = \widetilde{\mathcal{H}} \circ (X \times id_{S^{n-1} \times \mathbb{R}}) \text{ and } D = \mathcal{D} \circ (X \times id_{\mathbb{R}^n}).$$

We also have the ℓ-jet extensions:

$$j_1^\ell H : U \times S^{n-1} \longrightarrow J^\ell(U,\mathbb{R}) \ ; \ j_1^\ell H(u,v) = j^\ell h_v(u),$$
$$j_1^\ell \widetilde{H} : U \times (S^{n-1} \times \mathbb{R}) \longrightarrow J^\ell(U,\mathbb{R}) \ ; \ j_1^\ell \widetilde{H}(u,(v,r)) = j^\ell \widetilde{h}_{(v,r)}(u),$$
$$j_1^\ell D : U \times \mathbb{R}^n \longrightarrow J^\ell(U,\mathbb{R}) \ ; \ j_1^\ell D(u,x) = j^\ell d_x(u).$$

We consider the trivialization $J^\ell(U,\mathbb{R}) \equiv U \times \mathbb{R} \times J^\ell(n-1,1)$. For any submanifold $Q \subset J^\ell(n-1,1)$, we denote that $\widetilde{Q} = U \times \{0\} \times Q$. Then we have the following proposition as a corollary of Lemma 6 in Wassermann [31]. (See also Montaldi [25]).

Proposition 7.1. *Let Q be a submanifold of $J^\ell(n-1,1)$. Then the set*

$$T_Q(F) = \{X \in \mathrm{Emb}\,(U,\mathbb{R}^n) \mid j_1^\ell F \text{ is transversal to } \widetilde{Q}\,\}$$

is a residual subset of $\mathrm{Emb}\,(U,\mathbb{R}^n)$. If Q is a closed subset, then T_Q is open. Here, F is H, $\widetilde{H}$ or D.

As a corollary of the above proposition and classification results of function germs [1], we have the following theorem.

Theorem 7.2. *Suppose that $n \leq 6$. There exists an open dense subset $\mathcal{O} \subset \mathrm{Emb}\,(U,\mathbb{R}^n)$ such that for any $X \in \mathcal{O}$, the germ of the corresponding the germs of the Lagrangian lifts $L(D)$ and $L(H)$ of the evolute Ev_M and the Gauss map G at each point are Lagrangian stable. Moreover the germ of the Legendrian lift $\mathcal{L}(\widetilde{H})$ of the cylindrical pedal CPe_M at each point is Legendrian stable.*

We now stick to the case when $n = 3$. In this case we call $\boldsymbol{X} : U \longrightarrow \mathbb{R}^3$ a *surface*, S^2 a *sphere* and $H(\boldsymbol{X}(U), u)$ the *tangent plane* and etc. By Theorem 7.2 and the classification of function germs [1], we have the following theorem.

Theorem 7.3. *There exists an open dense subset $\mathcal{O} \subset \mathrm{Emb}\,(U, \mathbb{R}^3)$ such that for any $\boldsymbol{X} \in \mathcal{O}$, the following conditions hold:*

(1) The parabolic set $K^{-1}(0)$ is a regular curve. We call such a curve the parabolic curve.

(2) The Gauss map G along the parabolic curve are the folds except at isolated points. At this point G is the cusp.

Here, a map germ $f : (\mathbb{R}^2, \boldsymbol{a}) \longrightarrow (\mathbb{R}^2, \boldsymbol{b})$ is called a fold if it is $\mathcal{A}$-equivalent to the germ (x_1, x_2^2) (cf., Fig. 1) and a cusp if it is $\mathcal{A}$-equivalent to the germ $(x_1, x_2^3 + x_1 x_2)$ (cf., Fig. 1).

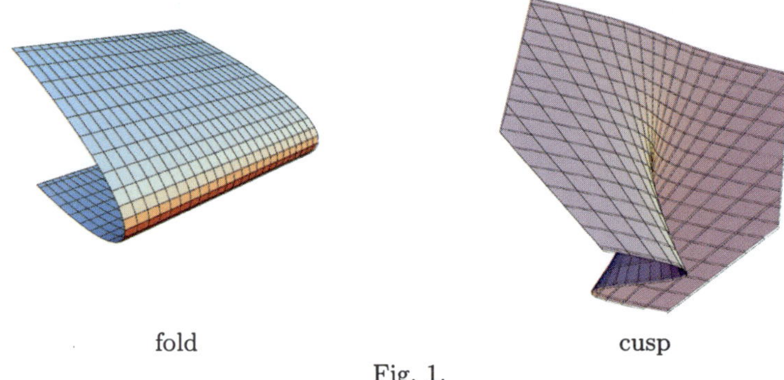

Fig. 1.

(3) A parabolic point $u \in U$ is a fold of the Gauss map G if and only if it is the cuspidaledge of the cylindrical pedal CPe_M.

(4) A parabolic point $u \in U$ is a cusp of the Gauss map G if and only if it is the swallowtail of the cylindrical pedal CPe_M.

Here, a map germ $f : (\mathbb{R}^2, \boldsymbol{a}) \longrightarrow (\mathbb{R}^3, \boldsymbol{b})$ is called a cuspidaledge if it is $\mathcal{A}$-equivalent to the germ (x_1, x_2^2, x_2^3) (cf., Fig. 2) and a swallowtail if it is $\mathcal{A}$-equivalent to the germ $(3x_1^4 + x_1^2 x_2, 4x_1^3 + 2x_1 x_2, x_2)$ (cf., Fig.2).

The assertion (1) and (2) can be interpreted that the Lagrangian lift $L(H)$ of the Gauss map G of $\boldsymbol{X} \in \mathcal{O}$ is Lagrangian stable at each point. Since $\mathcal{L}(\widetilde{H})$ is the Legendrian covering of the Lagrangian map $L(H)$ whose Lagrangian map is the Gauss map G, it has been known that the corre-

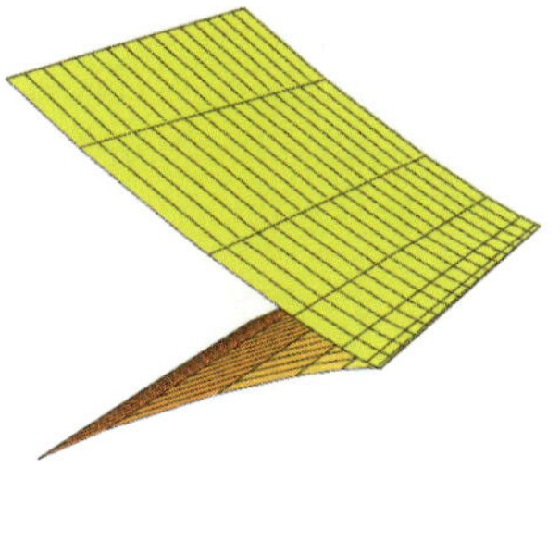

cuspidaledge

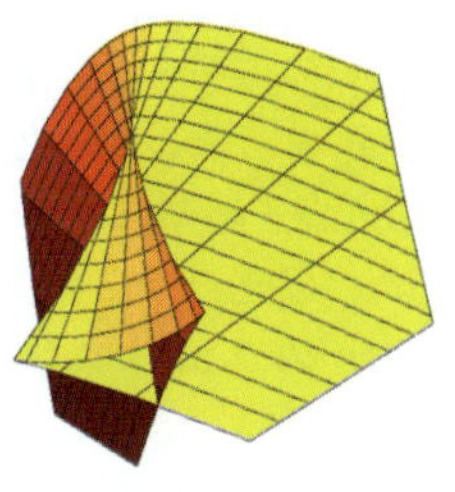
swallowtail

Fig. 2.

sponding singularities of the wavefront of $\mathcal{L}(\widetilde{H})$ are the cuspidaledge or the swallowtail [1]. Therefore we have the assertion (3) and (4).

Following the terminology of Whitney [32], we say that a surface $\boldsymbol{X}$: $U \longrightarrow \mathbb{R}^3$ has the *excellent Gauss map* G if $L(H)$ is a stable Lagrangian immersion germ at each point. In this case, the Gauss map G has only folds and cusps as singularities. Theorem 7.3 asserts that a surface with the excellent Gauss map is generic in the space of all surfaces in $\mathbb{R}^3$. We now consider the geometric meanings of folds and cusps of the Gauss map. We have the following results the main part of which is given by Banchoff et al [2]. However, we add few new information from the view point of Legendrian singularity theory.

Theorem 7.4. *Let* $G : (U, u_0) \longrightarrow (\mathbb{R}^3, v_0)$ *be the excellent Gauss map of a surface* $\boldsymbol{X}$ *and* $h_{v_0} : (U, u_0) \longrightarrow \mathbb{R}$ *be the height function germ at* $v_0 = G(u_0) = \boldsymbol{n}(u_0)$. *Then we have the following:*
(1) u *is a parabolic point of* $\boldsymbol{X}$ *if and only if* $T\text{-}corank(\boldsymbol{X}(U), u_0) = 1$ *(i.e.,* u_0 *is not a flat point of* $\boldsymbol{X}$).
(2) If u_0 *is a parabolic point of* $\boldsymbol{X}$, *then* $\widetilde{h}_{(v_0, r_0)}$ *has the* A_k-*type singularity for* $k = 2, 3$, *where* $\widetilde{h}_{(v_0, r_0)}(u) = h_{v_0}(u) - r_0$.
(3) Suppose that u_0 *is a parabolic point of* $\boldsymbol{X}$. *Then the following conditions are equivalent:*

    (a) The cylindrical pedal CPe_M *is the cuspidaledge at* u_0
    (b) $\widetilde{h}_{(v_0, r_0)}$ *has the* A_2-*type singularity.*
    (c) $T\text{-}ord(\boldsymbol{X}(U), u_0) = 2$.
    (d) Tangent indicatrix $(\boldsymbol{X}^{-1}(H(\boldsymbol{X}(U), u_0), u_0)$ *is a ordinary cusp, where a curve* $C \subset \mathbb{R}^2$ *is called an ordinary cusp if it is diffeomorphic*

to the curve given by $\{(x_1, x_2) \mid x_1^2 - x_2^3 = 0\}$.

(e) *For each* $\varepsilon > 0$, *there exist two distinct points* $u_1, u_2 \in U$ *such that* $|u_0 - u_i| < \varepsilon$ *for* $i = 1, 2$, *both of* u_1, u_2 *are not parabolic points and the tangent planes to* $M = \boldsymbol{x}(U)$ *at* u_1, u_2 *are parallel.*

(f) *The Gauss map* G *is the fold at* u_0.

(4) *Suppose that* u_0 *is a parabolic point of* $\boldsymbol{X}$. *Then the following conditions are equivalent:*

(a) *The cylindrical pedal* CPe_M *is the swallowtail at* u_0

(b) $\widetilde{h}_{(v_0, r_0)}$ *has the* A_3-*type singularity.*

(c) *T-ord*$(\boldsymbol{X}(U), u_0) = 3$.

(d) *Tangent indicatrix* $(\boldsymbol{X}^{-1}(H(\boldsymbol{X}(U), u_0), u_0)$ *is a point or a tachnodal, where a curve* $C \subset \mathbb{R}^2$ *is called a tachnodal if it is diffeomorphic to the curve given by* $\{(x_1, x_2) \mid x_1^2 - x_2^4 = 0\}$.

(e) *For each* $\varepsilon > 0$, *there exist three distinct points* $u_1, u_2, u_3 \in U$ *such that* $|u_0 - u_i| < \varepsilon$ *for* $i = 1, 2, 3$, *both of* u_1, u_2, u_3 *are not parabolic points and the tangent planes to* $M = \boldsymbol{x}(U)$ *at* u_1, u_2, u_3 *are parallel.*

(f) *For each* $\varepsilon > 0$, *there exist two distinct points* $u_1, u_2 \in U$ *such that* $|u_0 - u_i| < \varepsilon$ *for* $i = 1, 2$, *both of* u_1, u_2 *are not parabolic points and the tangent planes to* $M = \boldsymbol{x}(U)$ *at* u_1, u_2 *are equal.*

(g) *The Gauss map* G *is the cusp at* u_0.

Proof. We have shown in §6 that u_0 is a parabolic point if and only if T-corank$(\boldsymbol{X}(U), u_0) \geq 1$. Since $n = 3$, we have T-corank$(\boldsymbol{X}(U), u_0) \leq 2$. Since the extended height function germ $\widetilde{H}$: $(U \times (S^{n-1} \times \mathbb{R}), (u_0, (v_0, r_0))) \longrightarrow \mathbb{R}$ can be considered as a generating family of the Legendrian immersion germ $\mathcal{L}(\widetilde{H})$, $\widetilde{h}_{(v_0, r_0)}$ has only the A_k-type singularities ($k = 1, 2, 3$). This means that the corank of the Hessian matrix of $\widetilde{h}_{(v_0, r_0)}$ at a parabolic point is 1. The assertion (2) also follows. By the same reason, the conditions (3);(a),(b),(c) (respectively, (4); (a),(b),(c)) are equivalent. If the height function germ $\widetilde{h}_{(v_0, r_0)}$ has the A_2-type singularity, it is $\mathcal{K}$-equivalent to the germ $\pm x_1^2 + x_2^3$. Since the $\mathcal{K}$-equivalence preserves the zero level sets, the tangent indicatrix is diffeomorphic to the curve given by $\pm x_1^2 + x_2^3 = 0$. This is the ordinary cusp. The normal form for the A_3-type singularity is given by $\pm x_1^2 + x_2^4$, so the tangent indicatrix is diffeomorphic to the curve $\pm x_1^2 + x_2^4 = 0$. This means that the condition (3),(d) (respectively, (4),(d)) is also equivalent to the other conditions.

Suppose that u_0 is a parabolic point, then the Gauss map has only folds or cusps. If the point u_0 is the fold point, there is a neighborhood of u_0 on which the Gauss map is 2 to 1 except the parabolic curve (i.e, fold curve).

By Lemma 5.4, the condition (3), (e) is satisfied. If the point u_0 is the cusp, the critical value set is the ordinary cusp. By the normal form, we can understand that the Gauss map is 3 to 1 inside region of the critical values. Moreover, the point u_0 is in the closure of the region. This means that the condition (4),(e) holds. We can also observe that near by the cusp point, there are 2 to 1 points which near to the cusp u_0. However, one of those points is always a parabolic point. Since no other singularities appear for in this case, we have the condition (3),(e) (respectively, (4),(e)) characterizes the fold (respectively, the cusp).

If we consider the cylindrical pedal instead of the Gauss map, the only singularities are cuspidaledges or swallowtails. For a swallowtail point u_0, there is a self intersection curve (cf., Fig. 1) approaching to u_0. On this curve, there are two distinct point u_1, u_2 such that $\mathrm{CPe}_M(u_1) = \mathrm{CPe}_M(u_2)$. By Lemma 5.4, this means that the tangent planes to $M = \boldsymbol{x}(U)$ at points u_1, u_2 are equal. Since there are no other singularities in this case, the condition (4),(f) characterizes a swallowtail point of CPe_M. This completes the proof. $\qquad\square$

We now apply Theorem 6.3 to the above theorem and obtain new information from the view point of Lagrangian singularity theory.

Proposition 7.5. *Let $G : (U, u_0) \longrightarrow (\mathbb{R}^3, v_0)$ be the excellent Gauss map of a surface $\boldsymbol{X}$ and $h_{v_0} : (U, u_0) \longrightarrow \mathbb{R}$ be the height function germ at $v_0 = G(u_0) = \boldsymbol{n}(u_0)$. Then the Dupin foliation germ $\mathcal{DF}(\boldsymbol{X}(U), u_0)$ is diffeomorphic to a foliation germ $(\mathcal{F}_f, \boldsymbol{0})$ where f is one of the germs in the following list:*
(1) $x_1^3 + x_2^2$ (fold)
(2) $\pm x_1^4 + x_2^2$ ($\pm$cusp)

By Theorems 7.2, A.2 and the classification of function germs under $\mathcal{R}^+$-codimension ≤ 3, we have the following classification theorem:

Theorem 7.6. *There exists an open dense subset $\mathcal{O} \subset \mathrm{Emb}\,(U, \mathbb{R}^3))$ such that for any $\boldsymbol{X} \in \mathcal{O}$, the corresponding Lagrangian immersion germ $L(D)$ at any point $(u_0, \boldsymbol{x}_0) \in U \times (\mathbb{R}^3 \setminus M)$ is Lagrangian equivalent to a Lagrangian immersion germ $L(F) : (C(F), 0) \longrightarrow T^*\mathbb{R}^3$ whose generating family $F(x_1, x_2, \boldsymbol{q})$ $(\boldsymbol{q} = (q_1, q_2, q_3) \in \mathbb{R}^3)$ is one of the germs in the following list:*
(1) $x_1^3 + x_2^2 + q_1 x_1$ (fold)
(2) $\pm x_1^4 + x_2^2 + q_1 x_1 + q_2 x_1^2$ ($\pm$cusp)
(3) $x_1^5 + x_2^2 + q_1 x_1 + q_2 x_1^2 + q_3 x_1^3$ (swallowtail)

(4) $x_1^3 - x_1 x_2^2 + q_1 x_1 + q_2 x_2 + q_3(x_1^2 + x_2^2)$ *(pyramid)*
(5) $x_1^3 + x_2^3 + q_1 x_1 + q_2 x_1 + q_3 x_1 x_2$ *(purse)*.

We can draw the pictures of the foliation germs $\mathcal{F}_f$ for the germs f in Theorem 7.6:

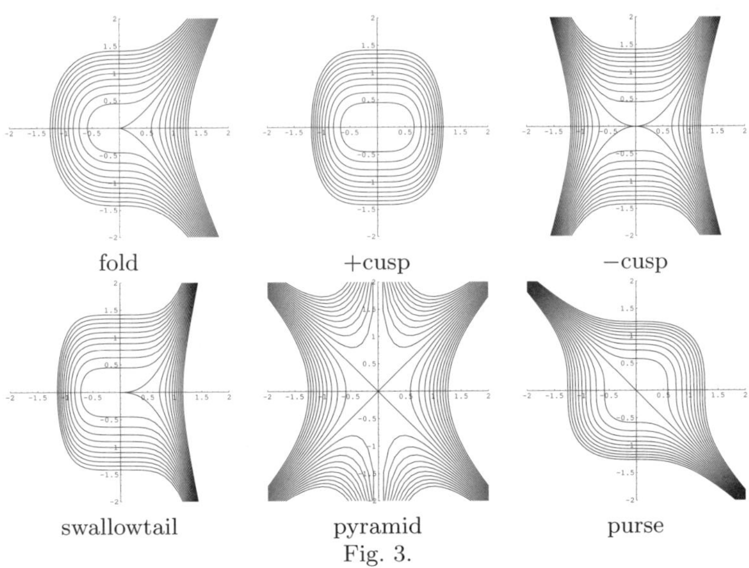

fold	+cusp	−cusp
swallowtail	pyramid	purse

Fig. 3.

We now apply Corollary 6.6 to the above classification theorem. Let $F(x_1, x_2, q)$ be one of the germs in the above list. We write $f(x_1, x_2) = F(x_1, x_2, 0)$. As a corollary of the above classification theorem and Corollary 6.6, we have the following:

Corollary 7.7. *There exists an open dense subset $\mathcal{O} \subset \mathrm{Emb}\,(U, \mathbb{R}^3)$ such that for any $X \in \mathcal{O}$ and any point $(u_0, x_0) \in U \times (\mathbb{R}^3 \setminus M)$, we have the following assertions :*

(1) The evolute germ (Ev_M, x_0) is diffeomorphic to the cuspidaledge, the swallowtail, the pyramid or the purse.

(2) The osculating spherical foliation germ $\mathcal{OF}(X(U), u_0)$ is diffeomorphic to a foliation germ $(\mathcal{F}_f, 0)$ where $F(x_1, x_2, q)$ is one of the germs in the list of Theorem 7.6.

Here, the purse and the pyramid are depicted in Figure 4.

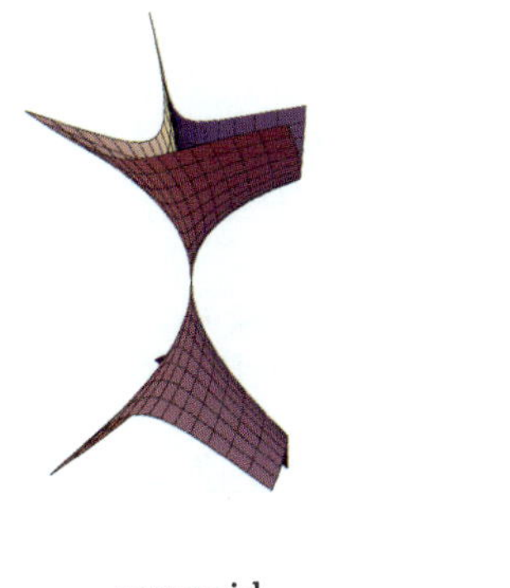

pyramid

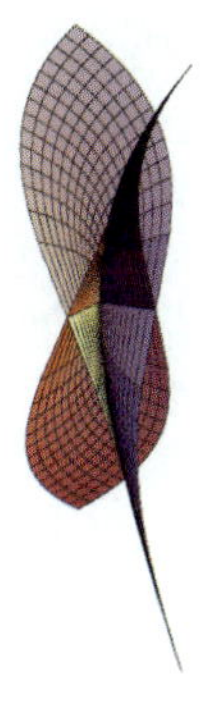

purse

Fig. 4.

We consider the geometric meanings of swallowtails, pyramids and purses of the evolute. By Theorem 7.6, we have the following theorem.

Theorem 7.8. *let $X : U \longrightarrow \mathbb{R}^3$ be an embedding. Suppose that the Lagrangian immersion germ of $L(D)$ at (u_0, x_0) is Lagrangian stable. Then we have the following:*

(1) Suppose that u_0 is not a umbilic point of $M = X(U)$, then the following conditions are equivalent:

(a) The germ of the evolute (Ev_M, x_0) is diffeomorphic to the swallowtail.

(b) The osculating spherical foliation germ $\mathcal{OF}(X(U), u_0))$ is diffeomorphic to $(\mathcal{F}_f, 0)$, where $f(x_1, x_2) = x_1^5 + x_2^2$.

(c) For each $\varepsilon > 0$, there exist two distinct points $u_1, u_2 \in U$ such that $|u_0 - u_i| < \varepsilon$ for $i = 1, 2$ and both of u_1, u_2 are the ridge points of $M = X(U)$.

(d) For each $\varepsilon > 0$, there exist two distinct points $u_1, u_2 \in U$ such that $|u_0 - u_i| < \varepsilon$ for $i = 1, 2$ and the osculating spheres of $M = X(U)$ at u_1, u_2 are equal.

(2) Suppose that u_0 is a umbilic point of $M = X(U)$, then the following conditions are equivalent:

(a) The germ of the evolute (Ev_M, x_0) is diffeomorphic to the pyramid.

(b) The osculating spherical foliation germ $\mathcal{OF}(X(U), u_0))$ is diffeomorphic to $(\mathcal{F}_f, 0)$, where $f(x_1, x_2) = x_1^3 - x_1 x_2^2$.

(c) For each $\varepsilon > 0$, there exist six distinct points $u_i \in U$ such that $|u_0 - u_i| < \varepsilon$ for $i = 1, 2, 3, 4, 5, 6$ and all of u_i are the ridge points of $M = X(U)$.

(3) *Suppose that u_0 is a umbilic point of $M = \boldsymbol{X}(U)$, then the following conditions are equivalent:*

(a) *The germ of the evolute $(Ev_M, \boldsymbol{x}_0)$ is diffeomorphic to the purse.*

(b) *The osculating spherical foliation germ $\mathcal{OF}(\boldsymbol{X}(U), u_0))$ is diffeomorphic to $(\mathcal{F}_f, 0)$, where $f(x_1, x_2) = x_1^3 + x_2^3$.*

(c) *For each $\varepsilon > 0$, there exist two distinct points $u_1, u_2 \in U$ such that $|u_0 - u_i| < \varepsilon$ for $i = 1, 2$ and both of u_1, u_2 are the ridge points of $M = \boldsymbol{X}(U)$.*

(d) *For each $\varepsilon > 0$, there exist four distinct points $u_j^i \in U$ such that $|u_0 - u_j^i| < \varepsilon$ for $i = 1, 2; j = 1, 2$ and each two osculating spheres of $M = \boldsymbol{X}(U)$ at u_1^i, u_2^i $(i = 1, 2)$ are equal.*

Acknowledgments

The work is partially supported by Grant-in-Aid for formation of COE "Mathematics of Nonlinear Structure via Singularities" (Hokkaido University) and Grant-in-Aid for Scientific Research (B) 18340013 JSPS.

Appendix A. The theory of Lagrangian singularities

In this section we give a brief review on the theory of Lagrangian singularities due to [1,33]. We consider the cotangent bundle $\pi : T^*\mathbb{R}^r \longrightarrow \mathbb{R}^r$ over $\mathbb{R}^r$. Let $(u, p) = (u_1, \ldots, u_r, p_1, \ldots, p_r)$ be the canonical coordinate on $T^*\mathbb{R}^r$. Then the canonical symplectic structure on $T^*\mathbb{R}^r$ is given by the *canonical two form* $\omega = \sum_{i=1}^{r} dp_i \wedge du_i$. Let $i : L \longrightarrow T^*\mathbb{R}^r$ be an immersion. We say that i is a *Lagrangian immersion* if $\dim L = r$ and $i^*\omega = 0$. In this case the critical value of $\pi \circ i$ is called the *caustic* of $i : L \longrightarrow T^*\mathbb{R}^r$ and it is denoted by C_L. The main result in the theory of Lagrangian singularities is to describe Lagrangian immersion germs by using families of function germs. Let $F : (\mathbb{R}^n \times \mathbb{R}^r, (\boldsymbol{0}, \boldsymbol{0})) \longrightarrow (\mathbb{R}, 0)$ be an r-parameter unfolding of function germs. We call

$$C(F) = \left\{ (x, u) \in (\mathbb{R}^n \times \mathbb{R}^r, (\boldsymbol{0}, \boldsymbol{0})) \Big| \frac{\partial F}{\partial x_1}(x, u) = \cdots = \frac{\partial F}{\partial x_n}(x, u) = 0 \right\},$$

the *catastrophe set* of F and

$$B_F = \left\{ u \in (\mathbb{R}^r, 0) \Big| \exists (x, u) \in C(F) \text{ s.t. rank} \left(\frac{\partial^2 F}{\partial x_i \partial x_j}(x, u) \right) < n \right\}$$

the *bifurcation set* of F. Let $\pi_r : (\mathbb{R}^n \times \mathbb{R}^r, 0) \longrightarrow (\mathbb{R}^r, 0)$ be the canonical projection, then we can easily show that the bifurcation set of F is the critical value set of $\pi_r|C(F)$. We call $\pi_{C(F)} = \pi|C(F) : (C(F), 0) \longrightarrow \mathbb{R}$ a *catastrophe map* of F. We say that F is a *Morse family of functions* if the map germ

$$\Delta F = \left(\frac{\partial F}{\partial u_1}, \ldots, \frac{\partial F}{\partial u_r} \right) : (\mathbb{R}^n \times \mathbb{R}^r, 0) \longrightarrow (\mathbb{R}^r, 0)$$

is non-singular, where $(x, u) = (x_1, \ldots, x_n, u_1, \ldots, u_r) \in (\mathbb{R}^n \times \mathbb{R}^r, 0)$. In this case we have a smooth submanifold germ $C(F) \subset (\mathbb{R}^n \times \mathbb{R}^r, 0)$ and a map germ $L(F) : (C(F), 0) \longrightarrow T^*\mathbb{R}^r$ defined by

$$L(F)(x, u) = \left(u, \frac{\partial F}{\partial u_1}, \ldots, \frac{\partial F}{\partial u_r} \right).$$

We can show that $L(F)$ is a Lagrangian immersion. Then we have the following fundamental theorem ([1], page 300).

Proposition A.1 *All Lagrangian submanifold germs in $T^*\mathbb{R}^r$ are constructed by the above method.*

Under the above notation, we call F a *generating family* of $L(F)$.

We define an equivalence relation among Lagrangian immersion germs. Let $i : (L, x) \longrightarrow (T^*\mathbb{R}^r, p)$ and $i' : (L', x') \longrightarrow (T^*\mathbb{R}^r, p')$ be Lagrangian immersion germs. Then we say that i and i' are *Lagrangian equivalent* if there exist a diffeomorphism germ $\sigma : (L, x) \longrightarrow (L', x')$, a symplectic diffeomorphism germ $\tau : (T^*\mathbb{R}^r, p) \longrightarrow (T^*\mathbb{R}^r, p')$ and a diffeomorphism germ $\bar{\tau} : (\mathbb{R}^r, \pi(p)) \longrightarrow (\mathbb{R}^r, \pi(p'))$ such that $\tau \circ i = i' \circ \sigma$ and $\pi \circ \tau = \bar{\tau} \circ \pi$, where $\pi : (T^*\mathbb{R}^r, p) \longrightarrow (\mathbb{R}^r, \pi(p))$ is the canonical projection and a symplectic diffeomorphism germ is a diffeomorphism germ which preserves symplectic structure on $T^*\mathbb{R}^r$. In this case the caustic C_L is diffeomorphic to the caustic $C_{L'}$ by the diffeomorphism germ $\bar{\tau}$.

A Lagrangian immersion germ into $T^*\mathbb{R}^r$ at a point is said to be *Lagrangian stable* if for every map with the given germ there is a neighborhood in the space of Lagrangian immersions (in the Whitney C^∞-topology) and a neighborhood of the original point such that each Lagrangian immersion belonging to the first neighborhood has in the second neighborhood a point at which its germ is Lagrangian equivalent to the original germ.

We can interpret the Lagrangian equivalence by using the notion of generating families. We denote $\mathcal{E}_m$ the local ring of function germs $(\mathbb{R}^m, 0) \longrightarrow \mathbb{R}$ with the unique maximal ideal $\mathfrak{M}_m = \{h \in \mathcal{E}_m | h(0) = 0\}$. Let $F, G : (\mathbb{R}^n \times \mathbb{R}^r, 0) \longrightarrow (\mathbb{R}, 0)$ be function germs. We say that F and G are *P-$\mathcal{R}^+$-equivalent* if there exists a diffeomorphism germ $\Phi : (\mathbb{R}^n \times \mathbb{R}^r, 0) \longrightarrow (\mathbb{R}^n \times \mathbb{R}^r, 0)$ of the form $\Phi(x, u) = (\Phi_1(x, u), \phi(u))$ and a function germ $h : (\mathbb{R}^r, 0) \longrightarrow \mathbb{R}$ such that $G(x, u) = F(\Phi(x, u)) + h(u)$.

For any $F_1 \in \mathfrak{M}_{n+r}$ and $F_2 \in \mathfrak{M}_{n'+r}$, F_1, F_2 are said to be *stably P-$\mathcal{R}^+$-equivalent* if they become P-$\mathcal{R}^+$-equivalent after the addition to the arguments to x_i of new arguments y_i and to the functions F_i of nondegenerate quadratic forms Q_i in the new arguments (i.e., $F_1 + Q_1$ and $F_2 + Q_2$ are P-$\mathcal{R}^+$-equivalent).

Let $F : (\mathbb{R}^n \times \mathbb{R}^r, 0) \longrightarrow (\mathbb{R}, 0)$ be a function germ. We say that F is an $\mathcal{R}^+$-*versal deformation* of $f = F|_{\mathbb{R}^n \times \{0\}}$ if

$$\mathcal{E}_n = J_f + \left\langle \frac{\partial F}{\partial u_1}|\mathbb{R}^n \times \{0\}, \ldots, \frac{\partial F}{\partial u_r}|\mathbb{R}^n \times \{0\} \right\rangle_{\mathbb{R}} + \langle 1 \rangle_{\mathbb{R}},$$

where

$$J_f = \left\langle \frac{\partial f}{\partial x_1}, \ldots, \frac{\partial f}{\partial x_n} \right\rangle_{\mathcal{E}_n}.$$

Theorem A.2 *Let $F_1 \in \mathfrak{M}_{n+r}$ and $F_2 \in \mathfrak{M}_{n'+r}$ be Morse families. Then we have the following:*
(1) $L(F_1)$ and $L(F_2)$ are Lagrangian equivalent if and only if F_1, F_2 are stably P-$\mathcal{R}^+$-equivalent.
(2) $L(F)$ is Lagrangian stable if and only if F is a $\mathcal{R}^+$- versal deformation of $F|\mathbb{R}^n \times \{0\}$.

For the proof of the above theorem, see ([1], page 304 and 325). The following proposition describes the well-known relationship between bifurcation sets and equivalence among unfoldings of function germs:

Proposition A.3 *Let $F, G : (\mathbb{R}^n \times \mathbb{R}^r, 0) \longrightarrow (\mathbb{R}, 0)$ be function germs. If F and G are P-$\mathcal{R}^+$-equivalent then there exist a diffeomorphism germ $\phi : (\mathbb{R}^r, 0) \longrightarrow (\mathbb{R}^r, 0)$ such that $\phi(B_F) = B_G$*

Appendix B. The theory of Legendrian singularities

In which we give a quick survey on the Legendrian singularity theory mainly due to Arnol'd-Zakalyukin [1,33]. Almost all results have been known at least implicitly. Let $\pi : PT^*(M) \longrightarrow M$ be the projective cotangent bundle over an n-dimensional manifold M. This fibration can be considered as a Legendrian fibration with the canonical contact structure K on $PT^*(M)$. We now review geometric properties of this space. Consider the tangent bundle $\tau : TPT^*(M) \to PT^*(M)$ and the differential map $d\pi : TPT^*(M) \to N$ of π. For any $X \in TPT^*(M)$, there exists an element $\alpha \in T^*(M)$ such that $\tau(X) = [\alpha]$. For an element $V \in T_x(M)$, the property

$\alpha(V) = \mathbf{0}$ does not depend on the choice of representative of the class $[\alpha]$. Thus we can define the canonical contact structure on $PT^*(M)$ by

$$K = \{X \in TPT^*(M) | \tau(X)(d\pi(X)) = 0\}.$$

For a local coordinate neighborhood $(U, (x_1, \ldots, x_n))$ on M, we have a trivialization $PT^*(U) \cong U \times P(\mathbb{R}^{n-1})^*$ and we call

$$((x_1, \ldots, x_n), [\xi_1 : \cdots : \xi_n])$$

homogeneous coordinates, where $[\xi_1 : \cdots : \xi_n]$ are homogeneous coordinates of the dual projective space $P(\mathbb{R}^{n-1})^*$.

It is easy to show that $X \in K_{(x,[\xi])}$ if and only if $\sum_{i=1}^{n} \mu_i \xi_i = 0$, where $d\tilde{\pi}(X) = \sum_{i=1}^{n} \mu_i \frac{\partial}{\partial x_i}$. An immersion $i : L \to PT^*(M)$ is said to be a *Legendrian immersion* if $\dim L = n$ and $di_q(T_q L) \subset K_{i(q)}$ for any $q \in L$. We also call the map $\pi \circ i$ the *Legendrian map* and the set $W(i) = \mathrm{image}\, \pi \circ i$ the *wave front* of i. Moreover, i (or, the image of i) is called the *Legendrian lift* of $W(i)$.

The main tool of the theory of Legendrian singularities is the notion of generating families. Here we only consider local properties, we may assume that $M = \mathbb{R}^n$. Let $F : (\mathbb{R}^k \times \mathbb{R}^n, \mathbf{0}) \longrightarrow (\mathbb{R}, 0)$ be a function germ. We say that F is a *Morse family* if the mapping

$$\Delta^* F = \left(F, \frac{\partial F}{\partial q_1}, \ldots, \frac{\partial F}{\partial q_k}\right) : (\mathbb{R}^k \times \mathbb{R}^n, \mathbf{0}) \longrightarrow (\mathbb{R} \times \mathbb{R}^k, \mathbf{0})$$

is non-singular, where $(q, x) = (q_1, \ldots, q_k, x_1, \ldots, x_n) \in (\mathbb{R}^k \times \mathbb{R}^n, \mathbf{0})$. In this case we have a smooth $(n-1)$-dimensional submanifold

$$\Sigma_*(F) = \left\{(q, x) \in (\mathbb{R}^k \times \mathbb{R}^n, \mathbf{0}) \ \bigg| \ F(q, x) = \frac{\partial F}{\partial q_1}(q, x) = \cdots = \frac{\partial F}{\partial q_k}(q, x) = 0\right\}$$

and the map germ $\mathcal{L}(F) : (\Sigma_*(F), \mathbf{0}) \longrightarrow PT^*\mathbb{R}^n$ defined by

$$\mathcal{L}(F)(q, x) = \left(x, [\frac{\partial F}{\partial x_1}(q, x) : \cdots : \frac{\partial F}{\partial x_n}(q, x)]\right)$$

is a Legendrian immersion germ. Then we have the following fundamental theorem of Arnol'd-Zakalyukin [1,33].

Proposition B.1 *All Legendrian submanifold germs in $PT^*\mathbb{R}^n$ are constructed by the above method.*

We call F a *generating family* of $\mathcal{L}(F)(\Sigma_*(F))$. Therefore the wave front is

$$W(\mathcal{L}(F)) = \left\{x \in \mathbb{R}^n \ \bigg| \ \exists q \in \mathbb{R}^k \text{s.t.} F(q, x) = \frac{\partial F}{\partial q_1}(q, x) = \cdots = \frac{\partial F}{\partial q_k}(q, x) = 0\right\}.$$

We sometime denote $\mathcal{D}_F = W(\mathcal{L}(F))$ and call it the *discriminant set* of F.

On the other hand, for any map $f : N \longrightarrow P$, we denote by $\Sigma(f)$ the set of singular points of f and $D(f) = f(\Sigma(f))$. In this case we call $f|\Sigma(f) : \Sigma(f) \longrightarrow D(f)$ the *critical part of* the mapping f. For any Morse family $F : (\mathbb{R}^k \times \mathbb{R}^n, \mathbf{0}) \longrightarrow (\mathbb{R}, \mathbf{0})$, $(F^{-1}(0), \mathbf{0})$ is a smooth hypersurface, so we define a smooth map germ $\pi_F : (F^{-1}(0), \mathbf{0}) \longrightarrow (\mathbb{R}, 0)$ by $\pi_F(q, x) = x$. We can easily show that $\Sigma_*(F) = \Sigma(\pi_F)$. Therefore, the corresponding Legendrian map $\pi \circ \mathcal{L}(F)$ is the critical part of π_F.

We now introduce an equivalence relation among Legendrian immersion germs. Let $i : (L, p) \subset (PT^*\mathbb{R}^n, p)$ and $i' : (L', p') \subset (PT^*\mathbb{R}^n, p')$ be Legendrian immersion germs. Then we say that i and i' are *Legendrian equivalent* if there exists a contact diffeomorphism germ $H : (PT^*\mathbb{R}^n, p) \longrightarrow (PT^*\mathbb{R}^n, p')$ such that H preserves fibers of π and that $H(L) = L'$. A Legendrian immersion germ $i : (L.p) \subset PT^*\mathbb{R}^n$ (or, a *Legendrian map* $\pi \circ i$) at a point is said to be *Legendrian stable* if for every map with the given germ there is a neighborhood in the space of Legendrian immersions (in the Whitney C^∞ topology) and a neighborhood of the original point such that each Legendrian immersion belonging to the first neighborhood has in the second neighborhood a point at which its germ is Legendrian equivalent to the original germ.

Since the Legendrian lift $i : (L, p) \subset (PT^*\mathbb{R}^n, p)$ is uniquely determined on the regular part of the wave front $W(i)$, we have the following simple but significant property of Legendrian immersion germs:

Proposition B.2 *Let* $i : (L, p) \subset (PT^*\mathbb{R}^n, p)$ *and* $i' : (L', p') \subset (PT^*\mathbb{R}^n, p')$ *be Legendrian immersion germs such that regular sets of* $\pi \circ i, \pi \circ i'$ *are dense respectively. Then* i, i' *are Legendrian equivalent if and only if wave front sets* $W(i), W(i')$ *are diffeomorphic as set germs.*

This result has been firstly pointed out by Zakalyukin [34]. The assumption in the above proposition is a generic condition for i, i'. Specially, if i, i' are Legendrian stable, then these satisfy the assumption.

We can interpret the Legendrian equivalence by using the notion of generating families. We denote $\mathcal{E}_n$ the local ring of function germs $(\mathbb{R}^n, 0) \longrightarrow \mathbb{R}$ with the unique maximal ideal $\mathfrak{M}_n = \{h \in \mathcal{E}_n \mid h(0) = 0 \}$. Let $F, G : (\mathbb{R}^k \times \mathbb{R}^n, \mathbf{0}) \longrightarrow (\mathbb{R}, \mathbf{0})$ be function germs. We say that F and G are *P-$\mathcal{K}$-equivalent* if there exists a diffeomorphism germ $\Psi : (\mathbb{R}^k \times \mathbb{R}^n, \mathbf{0}) \longrightarrow (\mathbb{R}^k \times \mathbb{R}^n, \mathbf{0})$ of the form $\Psi(x, u) = (\psi_1(q, x), \psi_2(x))$ for $(q, x) \in (\mathbb{R}^k \times \mathbb{R}^n, \mathbf{0})$ such that $\Psi^*(\langle F \rangle_{\mathcal{E}_{k+n}}) = \langle G \rangle_{\mathcal{E}_{k+n}}$. Here $\Psi^* : \mathcal{E}_{k+n} \longrightarrow \mathcal{E}_{k+n}$ is the pull back $\mathbb{R}$-algebra isomorphism defined by $\Psi^*(h) = h \circ \Psi$. If $n = 0$, we simply say these germs are $\mathcal{K}$-*equivalent.*

Let $F : (\mathbb{R}^k \times \mathbb{R}^3, \mathbf{0}) \longrightarrow (\mathbb{R}, \mathbf{0})$ be a function germ. We say that F is a $\mathcal{K}$-*versal deformation of* $f = F|\mathbb{R}^k \times \{\mathbf{0}\}$ if

$$\mathcal{E}_k = T_e(\mathcal{K})(f) + \left\langle \frac{\partial F}{\partial x_1}|\mathbb{R}^k \times \{\mathbf{0}\}, \dots, \frac{\partial F}{\partial x_n}|\mathbb{R}^k \times \{\mathbf{0}\} \right\rangle_{\mathbb{R}},$$

where

$$T_e(\mathcal{K})(f) = \left\langle \frac{\partial f}{\partial q_1}, \dots, \frac{\partial f}{\partial q_k}, f \right\rangle_{\mathcal{E}_k}.$$

(See [21].)

The main result in Arnol'd-Zakalyukin's theory [1,33] is the following:

Theorem B.3 *Let* $F, G : (\mathbb{R}^k \times \mathbb{R}^n, \mathbf{0}) \longrightarrow (\mathbb{R}, 0)$ *be Morse families. Then*
(1) $\mathcal{L}(F)$ *and* $\mathcal{L}(G)$ *are Legendrian equivalent if and only if* F, G *are* P-$\mathcal{K}$-*equivalent.*
(2) $\mathcal{L}(F)$ *is Legendrian stable if and only if* F *is a* $\mathcal{K}$-*versal deformation of* $F \mid \mathbb{R}^k \times \{\mathbf{0}\}$.

Since F, G are function germs on the common space germ $(\mathbb{R}^k \times \mathbb{R}^n, \mathbf{0})$, we do no need the notion of stably P-$\mathcal{K}$-equivalences under this situation (cf., [1]). By the uniqueness result of the $\mathcal{K}$-versal deformation of a function germ, Proposition B.2 and Theorem B.3, we have the following classification result of Legendrian stable germs. For any map germ $f : (\mathbb{R}^n, \mathbf{0}) \longrightarrow (\mathbb{R}^p, \mathbf{0})$, we define the *local ring of* f by $Q(f) = \mathcal{E}_n / f^*(\mathfrak{M}_p)\mathcal{E}_n$.

Proposition B.4 *Let* $F, G : (\mathbb{R}^k \times \mathbb{R}^n, \mathbf{0}) \longrightarrow (\mathbb{R}, 0)$ *be Morse families. Suppose that* $\mathcal{L}(F), \mathcal{L}(G)$ *are Legendrian stable. The the following conditions are equivalent.*
(1) $(W(\mathcal{L}(F)), \mathbf{0})$ *and* $(W(\mathcal{L}(G)), \mathbf{0})$ *are diffeomorphic as germs.*
(2) $\mathcal{L}(F)$ *and* $\mathcal{L}(G)$ *are Legendrian equivalent.*
(3) $Q(f)$ *and* $Q(g)$ *are isomorphic as* $\mathbb{R}$-*algebras,*
where $f = F|\mathbb{R}^k \times \{\mathbf{0}\}$, $g = G|\mathbb{R}^k \times \{\mathbf{0}\}$.

Proof. Since $\mathcal{L}(F)$, $\mathcal{L}(G)$ are Legendrian stable, these satisfy the generic condition of Proposition B.2, so that the conditions (1) and (2) are equivalent. The condition (3) implies that f, g are $\mathcal{K}$-equivalent [21,22]. By the uniqueness of the $\mathcal{K}$-versal deformation of a function germ, F, G are P-$\mathcal{K}$-equivalent. This means that the condition (2) holds. By Theorem B.3, the condition (2) implies the condition (3). $\qquad\square$

References

1. V. I. Arnol'd, S. M. Gusein-Zade and A. N. Varchenko, *Singularities of Differentiable Maps vol. I*. Birkhäuser (1986)
2. T. Banchoff, T. Gaffney and C. McCrory, *Cusps of Gauss mappings*. Research notes in Mathematics, Pitman, **55** (1982)
3. D. Bleeker and L. Wilson, *Stability of Gauss maps*. Illinois J. Math. **22**, (1978), 279–289
4. J. W. Bruce, *The dual of generic hypersurfaces*. Math. Scand., **49** (1981), 36–60
5. J. W. Bruce and P. J. Giblin, *Curves and singularities (second edition)*, Cambridge University press, (1992)
6. V. V. Goryunov, *Projections of Generic Surfaces with Boundaries*, Adv. Soviet Math., **1** (1990), 157–200
7. M. Golubitsky and V. Guillemin, *Contact equivalence for Lagrangian manifold*, Adv. Math., **15** (1975), 375–387
8. M. Golubitsky and V. Guillemin, *Stable Mappings and their Singularities*. Springer GTM.
9. S. Izumiya, D-H. Pei and T. Sano, *Singularities of hyperbolic Gauss maps*. Proceedings of the London Mathematical Society **86** (2003), 485–512
10. S. Izumiya, D-H. Pei, T. Sano and E. Torii, *Evolutes of hyperbolic plane curves*, Acta Mathmatica Sinica **20**, (2004), 543–550
11. S. Izumiya, D-H. Pei and T. Sano, *Horospherical surfaces of curves in Hyperbolic space*, Publ. Math. (Debrecen) **64** (2004),1–13
12. S. Izumiya, D-H. Pei and M. Takahasi, *Curves and surfaecs in Hyperbolic space*, Banach center publications **65**, Geometric singularity theory (2004), 197–123
13. S. Izumiya, D-H. Pei, M. C. Romero-Fuster and M. Takahashi, *On the ridges of submanifolds of codimension 2 in Hyperbolic n-space*, Bull. Braz. Math. Soc. **35** (2) (2004), 177–198
14. S. Izumiya, D-H. Pei and M. Takahashi, *Singularities of evolutes of hypersurfaces in hyperbolic space*, Proceedings of the Edinburgh Mathematical Society **47** (2004), 131–153
15. S. Izumiya, D-H. Pei and M. C. Romero-Fuster, *The geometry of surfaces in Hyperbolic 4-space*, to appear in Israel Journal of Mathematics
16. S. Izumiya, D-H. Pei, M. C. Romero-Fuster and M. Takahashi, *Geometry of submanifolds in hyperbolic n-space*, Journal of London Mathematical Society **71**, (2005) 779–800
17. S. Izumiya and M. C. Romero-Fuster, *The horospherical Gauss-Bonnet type theorem in hyperbolic space*. to appear in J. Math. Soc. Japan **58**, (2006)
18. S. Izumiya, *Legendrian dualities and spacelike hypersurfaces in the light-cone*, preprint
19. S. Izumiya and M. Takahashi. *Spacelike Prallels and Evolutes in Minkowski pseudo-spheres*, preprint
20. E. E. Landis, *Tangential singularities*, Funct. Anal. Appli., **15** (1981), 103–114

21. J. Martinet, *Singularities of Smooth Functions and Maps*, London Math. Soc. Lecture Note Series, Cambridge Univ. Press,**58** (1982)

22. J. N. Mather, *Stability of C^∞-mappings IV:Classification of stable germs by $\mathbb{R}$ algebras*, Publi. Math. I.H.E.S., **37** (1970), 223–248

23. J. A. Montaldi, *Surfaces in 3-space and their contact with circles*, J. Diff. Geom., **23** (1986),109–126

24. J. A. Montaldi, *On contact between submanifolds*, Michigan Math. J., **33** (1986), 81–85

25. J. A. Montaldi, *On generic composites of maps*, Bull. London Math. Soc., **23** (1991), 81–85

26. I. Porteous, *The normal singularities of submanifold*, J. Diff. Geom., vol 5, (1971), 543–564

27. I. Porteous, *Geometric Differentiation* second edition, Cambridge Univ. Press (2001)

28. M. C. Romero Fuster, *Sphere stratifications and the Gauss map.* Proceedings of the Royal Soc. Edinburgh, **95A** (1983), 115–136

29. I. Vaisman , *A first course in Differential Geometry*, Marcel Dekker (1984)

30. C. T. C. Wall, *Geometric properties of gneric differential manifolds*, Geometry and Topology, Rio de Janeiro, 1976, Lect. Notes in Math. **597**, Springer-Verlag, Berlin (1977), 707–774

31. G. Wassermann, *Stability of Caustics*, Math. Ann., **216** (1975), 43–50

32. H. Whitney, *On singularities of mappings of Euclidean spaces I.* Ann. of Math. **62** (1955), 374–410

33. V. M. Zakalyukin, *Lagrangian and Legendrian singularities*, Funct. Anal. Appl., **10** (1976), 23–31

34. V. M. Zakalyukin, *Reconstructions of fronts and caustics depending one parameter and versality of mappings.* J. Sov. Math., **27** (1984), 2713-2735

CAUSTICS AND VISUALIZATION TECHNIQUES

A. JOETS

Laboratoire de Physique des Solides, Bât. 510
Université Paris-Sud, 91405 Orsay cedex, France
E-mail: joets@lps.u-psud.fr

Optical caustics are formed by the focalisation of light rays. They are observed
in nature as well as in experimental physics. They constitute a concrete and
visual realization of the abstract notion of singularity, or better of Lagrangian
singularity, as defined in the modern theory of singularity. Caustics appear naturally in applications to physics, and more particularly in visualization techniques. The main problem is then to extract physical informations from the
observation of the caustic, that is to say to clarify the relation between the
caustic and the refractive index field producing it. In this difficult program,
the first step is to be able to calculate, from a given index field, the structure
of a caustic, i.e. its decomposition into different types of singularities: folds,
cusps, swallowtails, umbilics. We show, using the example of the visualization
of convective structures in nematic liquid crystals, how this first step may be
efficiently realized with the help of Thom-Boardman classes.

Keywords: Caustics; Flow visualization; Nematic liquid crystals.

1. Caustics in Visualization Techniques

Optical caustics are the luminous forms created by the focalization of light
rays. They are almost always observed by interposing a screen transversally
to the mean direction of the rays. Usual examples include the cusp formed
in the tee cup (the screen is the surface of the liquid) and the bright moving
lines one sees under the wavy surface of a swimming pool (the screen is the
bottom of the swimming pool). These cusps and lines are in fact the planar
section of a 2D caustic surface of our 3D-physical space.

A very particular case of caustic is provided by the focus of a thin lens:
the caustic is reduced to a single point. The difficulties encountered by the
opticians to produce a strict convergence at that point, or at least in a
small region, shows that this situation is highly unstable. However, due to
the technical importance of perfect optical systems, many efforts are made
to realize such "degenerate" caustics.

In contrast with the previous exemple, natural systems, i.e. systems not constructed with the aim to reduce caustic surfaces to a focus, produce generic caustic surfaces. A generic caustic has a regular part, the fold-surface A_2, whose trace in the screen is composed of bright lines. These bright lines possibly meet at cusp-points A_3, corresponding to the trace of the cusp-lines of the caustic in the screen. In addition, for special positions of the screen, one may observe three other types of caustic points: swallowtails A_4, elliptic umbilics D_4^- and hyperbolic umbilics D_4^+ [1].

An example of systems producing generic caustics is provided by the shadowgraph method, an important visualization technique used in aerodynamics or in fluid mechanics [2]. The principle is simple. A beam of initially parallel rays is sent through a medium interacting with light. The interaction is described by a refractive index field $N(x, y, z)$. If the medium is homogeneous, the index field is uniform and the rays are not deflected. The transmitted rays are then parallel and a screen cutting the emergent rays shows a uniform lighting. On the other hand, if some physical mechanism (convection, instabilities, etc.) makes the index non uniform, the transmitted rays are no more parallel and the lighting in the screen becomes non uniform. One observes bright zones corresponding to convergent emergent rays and dark zones corresponding to divergent emergent rays. These dark zones are at the origin of the name "shadowgraph method". The emergent beam, as does any set of rays, admits a caustic. In the usual case of small ray deflections, the associated caustic is formed very far from the physical system and it is hardly observable. However, there are now many cases, e.g. experiments with liquid crystals, in which the deflection is strong [3,4]. The caustic is then easily observable.

One then understands that the caustic is an integral part of the image. Its analysis must be included in the visualization techniques of media interacting strongly with light. The aim of this article is to show, according to our practice, how some notions on singularities may be efficiently used in visualization techniques.

2. Caustics as Singularities

Modeling caustics traces back to the discovery of Calculus. Caustics appeared in the literature as "evolutes", "envelopes", "centers of curvature", "focals", etc. Caustics are now understood as Langangian singularities, or more precisely as critical values of a Lagrangian map π [1]. The Lagrangian map applies the Lagrangian submanifold Λ representing the rays in the phase space into the ordinary physical space $R^3 = \{x, y, z\}$: $\pi : \Lambda \to R^3$.

In Λ the rays trajectories do not intersect. The intersection of the rays is recovered by projecting Λ into the ordinary space R^3. Since at a caustic point two infinitely close rays intersect, at the corresponding point in Λ the projection π must have a non trivial kernel. This leads to define a singular point as a point where the rank of π has not the maximum possible value 3. The singular points form the singular set $\Sigma \subset \Lambda$. The caustic K is the image of the singular set: $K = \pi(\Sigma)$.

Three variables are necessary to parametrize the Lagrangian submanifold. For the two first variables, we can take the variables λ, μ describing the initial wave front, or any surface globally transverse to the rays. Each ray is associated with a particular value of λ, μ. The third variable s is, for instance, the distance from the initial wave front. The Lagrangian projection is then a map between spaces of the same dimension 3. It is given by relations of the type $x = x(\lambda, \mu, s)$, etc. The equation for Σ is obtained by saying that π^*, the derivative of π, has a non trivial kernel at the singular point.

It is known that general maps between two 3D-spaces, i.e. maps which are not necessarily Lagrangian, have only three types of singularities: folds A_2 (which form surfaces), cusp-lines A_3 (which form lines) and swallowtails A_4 (point singularities). These singularities are effectively observed in caustics. Ones observes also umbilics D_4 (elliptic umbilics D_4^- and hyperbolic umbilics D_4^+). The appearance of the umbilics is related to the Lagrangian character of the projection, that is to say to the existence of a wave front.

3. Calculating Caustics

In practice, the first problem to solve is finding the projection π, that is to say the functions $x = x(\lambda, \mu, s)$, etc. In the problem of the visualization, this problem is equivalent to calculating the deviation of the rays by the non homogeneous medium. In the medium deflecting the rays, the Fermat principle applies: the optical path $\int N dl$ is extremal. The ray equations are the Euler-Lagrange equations associated with this extremum principle. They constitute a set of ordinary differential equations of second order. Except in very special cases, one cannot find an explicit analytical solution. However the equations may always be numerically integrated, for instance by the Runge-Kutta method. For each ray of parameter λ, μ, one finds (numerically) its exit point $P(\lambda, \mu)$, where it leaves the deflecting medium, and its direction $\vec{r}(\lambda, \mu)$, which is now constant. The projection π is given by $\pi(\lambda, \mu, s)) = P(\lambda, \mu) + s\vec{r}(\lambda, \mu)$. Integrating a great number of trajectories (typically 10^4 or 10^5) one obtains the numerical functions defining the

Lagragian projection π.

The second problem is to calculate the caustic equation. The singular set is given by vanishing the determinant of the Jacobian matrix associated to π:

$$\det|\partial(x, y, z)/\partial(\lambda, \mu, s)| = 0. \tag{1}$$

To obtain the equation (1), the functions x, y, z must be numerically differentiated. Then equation (1) is numerically solved. In fact, an important simplification occurs here. The particular form of π shows that equation (1) is an algebraic equation of second degree in the variable s. Its solution $s_\pm(\lambda, \mu)$ is found explicitly as a function of x, y, z and of their derivatives. Introducing $s_\pm$ in the expressions for x, y, z gives a parametrization of the caustic K, the parameters being λ and μ. The caustic is composed of two sheets K_+ and K_-, which connect at the umbilic points D_4.

At this point appears a third problem: the practical problem of the representation of the caustic in the plane of the paper sheet. Caustics are not regular surfaces. They possess line-singularities (cusps) and point-singularities (swallowtails and umbilics). They have also self-intersection lines which may end to singular points (swallowtails and hyperbolic umbilics). Another complication arises from the fact that the representation of the caustic realizes a mapping from a 2D-surface (the caustic) into a plane (the paper sheet). It is known that these mappings generically possess singularities, the Whitney singularities, that constitute the apparent contour of the caustic [5]. These additional singularities may lead to ambiguities or to mistakes. For example the fold of the apparent contour of the caustic may be taken for a cusp-line of the caustic, and a cusp of the apparent contour may be taken for a swallowtail of the caustic. For theses reasons, the calculation of the generic caustic points given by equation (1) is clearly not sufficient. One has to calculate also the singularities of the caustic itself, that to say the Thom-Boardman classes [1,6,7]. Let us recall that a class Σ^k is the set of points where the projection π has a kernel of dimension k. By definition $\Sigma^{k,i,\dots,j}$ is the class Σ^k of the restriciton of π to $\Sigma^{i,\dots,j}$. The important result is that Σ^1 represents the fold-surface in Λ, $\Sigma^{1,1}$ represents the cusp-lines, $\Sigma^{1,1,1}$ represents the swallowtails and Σ^2 represents the umbilics.

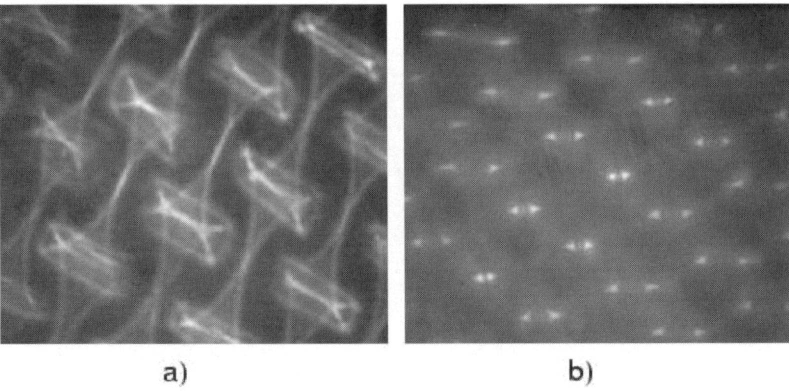

Fig. 1.   Varicose structure observed by the shadowgraph method: a) section of the virtual part of the caustic showing a periodic network of hyperbolic umbilics b) section of the real part of the caustic showing a periodic network of elliptic umbilics.

4. Application to the Visualization of Convective Structures in Nematics

We have realized the program exposed in section 3 in the case of the visualization of convective structures produced in nematic liquid crystals.

Nematic liquid crystals, the only type of liquid crystal considered here, are anisotropic liquids [8]. They are characterized by a new dynamical variable, the director $\vec{n}$, which indicates the local orientation of their rodlike molecules ($\vec{n} \equiv -\vec{n}$, $|\vec{n}| = 1$). The director field is coupled to the other fields: velocity field $\vec{v}$, external fields (electric field, magnetic field,), etc. As a consequence, convective structures may be easily produced by applying, for instance, an alternative electric field across a nematic liquid crystal layer [9]. These electro-convective structures have different symmetries and different time behaviors. In the most simple case the structure is composed of stationary straight rolls. There exists also a biperiodic stationary structure, the varicose structure [10]. From the viewpoint of the singularities, this structure is very interesting, since it possesses the 5 generic types of caustic points.

The convective structures are easily observed by the shadowgraph method. Inside the nematic layer, light is decomposed into an ordinary wave and an extraordinary wave. The ordinary light does not interact with the director field and it is eliminated by using a polarizer. On the contrary, the

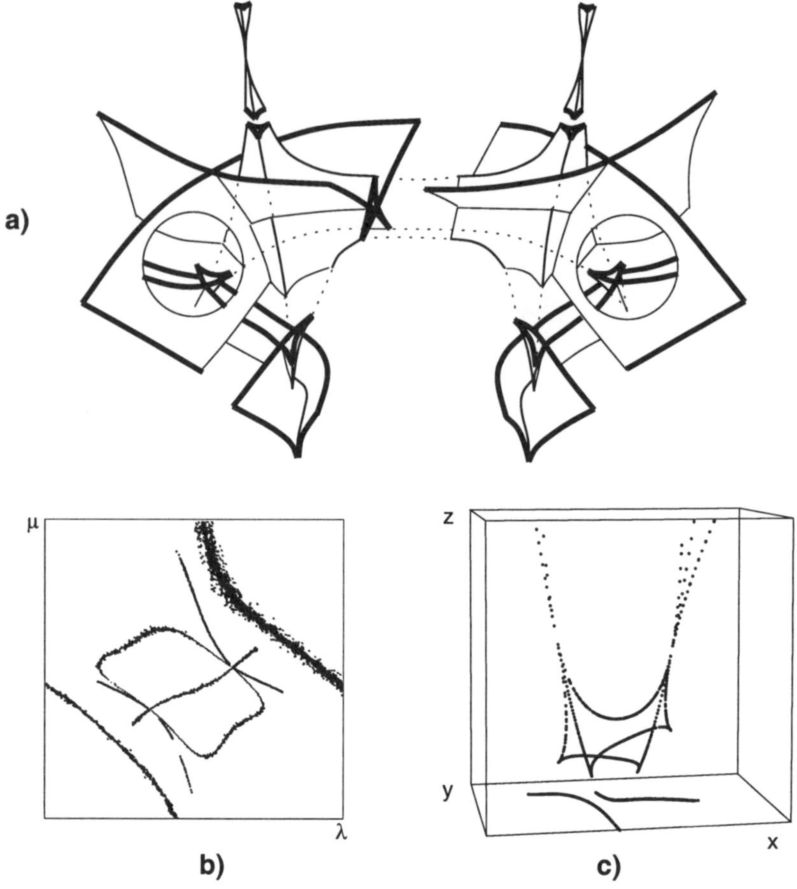

Fig. 2. Detail of the caustic associated with the varicose structure: a) sketch of the caustic surface near the elliptic umbilics, b) the class $\Sigma^{1,1}$, represented in the plane λ, μ, c) the corresponding cusp-lines $\pi(\Sigma^{1,1})$.

extraordinary light interacts with the director field. Its (energy) index N depends on the angle β between the ray direction and the director $\vec{n}$: $N = (n_o^2 \cos^2 \beta + n_e^2 \sin^2 \beta)^{1/2}$, where n_o and n_e are the ordinary refractive index and the extraordinary refractive index [11]. For our nematic compound, the birefringence $n_e - n_o = 1.94 - 1.65 = 0.29$ is high, meaning a strong interaction. Any convective motion induces, due to the coupling between $\vec{v}$ and $\vec{n}$, a distortion of the director field. Consequently the (extraordinary) rays see a varying N and are deviated by the liquid crystal. The envelope of the transmitted rays constitutes the caustic associated to the convective structure. It is composed of two parts: a real part above the layer and a virtual part located below the layer and formed by the prolongations of the transmitted rays in the inverse direction of the light. In fact, the two parts are connected "at infinity". The thickness d of the layer is typically 10 or 100 microns and the caustic is observed by using a polarizing microscope. The focal plane of the microscope plays the role of the screen. It is important to note that our screen is immaterial. It may be placed above, inside, or below the layer and it allows us to observe both real and virtual parts (see Fig. (1)).

Actually the director field $\vec{n}$ cannot be deduced directly from the basic equations of the nemato-hydrodynamics. The calculations are done by starting from a given form for the director distortion compatible with the symmetry of the optical images. For example, for the varicose structure, we write $\vec{n} = (\cos\varphi, 0, \sin\varphi)$, with $\varphi = \varphi_0 \cos(\pi z/d)[\cos(kx + qy) + \varepsilon \cos(kx - qy)]$ ($x - y$ is the plane of the layer, z is the direction of the incoming rays, and ε is a parameter typically equal to 0.5.). The time consuming part of the calculations is the calculation of the Lagrangian projection by the numerical integration of the ray trajectories. As explained above, the important result concerns the set of the caustic singularities. Figure (2) shows an example of the calculation of the class $\Sigma^{1,1}$ (Fig.2-b) and its image $\pi(\Sigma^{1,1})$ (Fig.2-c), i.e. the cusp-lines, associated to the varicose structure. The elliptic umbilic appears in the λ, μ plane as the meeting point of 6 half cusp-lines. The swallowtails, invisible in the λ, μ plane, appear as "cusp-points of cusp-lines". The caustic surface itself forms a very complex surface, having many intersection lines (Fig.2-a). However the skeleton formed by its singular lines allows one to understand immediately its structure.

The complete structure of the caustic associated with any convective structure in nematics can be understood in the same manner, with the help of Thom-Boardman classes. This technique may be applied to other types of visualization: usual fluids, shock waves, combustion, defects, etc. It may also be applied to the fundamental research on caustics. In particular, we applied it in the first experimental determination of a topological invariant associated with a set of rays [12,13].

References

1. V.I. Arnold, S.M. Gusein-Zade and A.N. Varchenko, *Singularities of Differentiable Maps, Vol. I* (Birkhäuser, Boston, 1985).
2. W. Merzkirch, *Flow Visualization* (Academic Press, Orlando, 1987).
3. A. Joets and R. Ribotta, *J. Phys. I France* **4**, 1013–1026 (1994).

4. A. Joets, in *11e Colloque de Visualisation et de Traitement d'images en Mécanique des Fluides*, Ecole Centrale de Lyon, France, 2005.
5. H. Whitney, *Ann. Math.* **62**, 374–410 (1955).
6. R. Thom, *Ann. Inst. Fourier (Grenoble)* **6**, 43–87 (1956).
7. A. Joets and R. Ribotta, *Europhys. Lett.* **29**, 593–598 (1995).
8. P.-G de Gennes, *The Physics of Liquid Crystals* (Clarendon Press, Oxford, 1974).
9. A. Joets and R. Ribotta, *J. Physique (Paris)*, **47**, 595–606 (1986).
10. R. Ribotta and A. Joets, *J. Physique (Paris)* **47**, 739–743 (1986).
11. A. Joets and R. Ribotta, *Opt. Comm.* **107**, 200-204 (1994).
12. Yu. V. Chekanov, *Funct. Anal. Appl.* **20**, 223–226 (1986).
13. A. Joets and R. Ribotta, *Phys. Rev. Lett.* **77**, 1755–1758 (1996).

Singularities and Genericity in Medical Imaging: Old and New

Yannick L. Kergosien

Université de Cergy-Pontoise, Département d'Informatique,
196 rue des Rabats, F–92160 Antony, France
yannick.kergosien@libertysurf.fr

We describe some applications of singularity theory to medical imaging in the spirit of R. Thom's Catastrophe Theory. In the first part we address the interpretation of standard projection radiographs. Some of the signs used by radiologists rely on generic properties of smooth mappings from 2-manifolds to the plane first described by Whitney. The model is extended to situations where the projection is controlled by a small number of parameters. A catastrophe set for that setting can be computed and its generic properties are briefly described. Further applications to visualization and shape classification are mentioned.

The second part addresses issues in interventional imaging and vision. It describes stochastic algorithms which build trees in high dimensional Euclidean spaces with some adaptation to the geometry of a chosen target subset. Such growing trees provide an example to Thom's concept of a generalized catastrophe which helps analyzing their behavior. Some of them produces search trees and is used to approximately identify in real time the pose of a polyhedron from its external contour. A search tree is first grown in a space of shapes of plane curves which are a set of precomputed polygonal outlines of the polyhedron. The tree is then used to find in real time a best match to the outline of the polyhedron in the current pose.

Keywords: singularities, medical imaging, radiology, tree, vision

PART 1 : GENERICITY AND DIAGNOSTIC MEDICAL IMAGING

1. Projection diagnostic radiology and signs for interpretation

Let us consider first the well known case of projection radiology. A part of the patient's body is placed between an X-ray tube and a plane detector (either digital or analog, including a screen or not, these differences are irrelevant here), and the detector is briefly exposed to the X-rays photons that crossed that solid, all of them coming approximately from the same focal point in the tube. After appropriate processing, one is able to see on the plane (let us call it the film) different shades of gray according to the density of incident photons. The radiologist relates the image his eyes see on the film to the 3D geometry and anatomy of the imaged part, then

285

to pathologic conditions, possibly using other clinical or biological information.

Medicine has a long history of trying to analyze and classify diseases and their manifestations in very different contexts. It thus evolved a general framework for analysis in terms of elementary signs (e.g., fever, tremor, or tenderness) subsets of which are associated to diseases or other entities like syndromes. Radiology too has been analyzed in that way. Here we shall only be interested in the geometric part of the interpretation. Let us first describe what is known as the sign (or law) of the tangential incidence. The tissues radiographed have different X-photon absorption characteristics, but it is a crucial fact that (1) they are organized into macroscopic compartments, (2) each of which has a close to constant photon absorbing power, (3) these compartments are bounded by anatomic surfaces, known as the *contrast surfaces*, which (4) are smooth to a good approximation. The eye of the radiologist sees lines on the film where the shades of gray have sharp variations, and *this (lines on film) occurs at, and only at, the points where the ray hitting the detector has been tangent to a contrast surface* (Fig. 1).

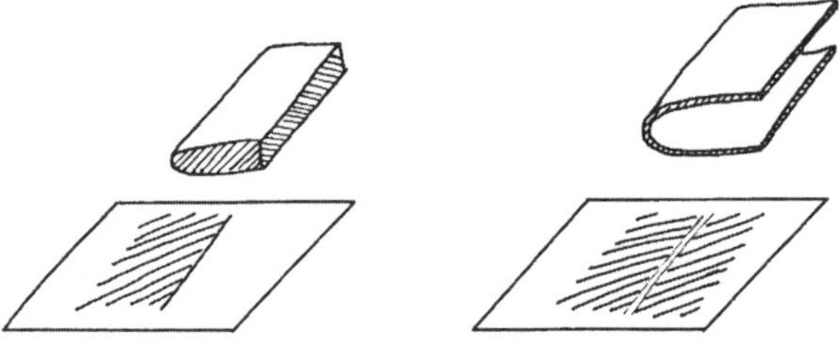

Fig. 1.   Formation of the radiologic image from singular projection of contrast manifolds: simple contrast (left) and double contrast (right) techniques.

This law really has two parts. The "at" part (tangency implies a line) finds its limits if the curvature of the contrast surface is too large, leading to an absence of line, or if the difference between the absorbing powers of adjacent compartments is too small. To artificially enhance contrast and make more anatomical interfaces visible, one can fill a hollow cavity (such as the colon) with a very opaque liquid to obtain a contrast between the cavity and its wall. Injecting a gas is another way of creating contrast with the liquid-like opacity of the wall. Computed imaging has made many of such opacifications unnecessary because much subtler contrast can now be detected, but a variant of these *single contrast* techniques is still very common in vascular imaging where a soluble contrast agent is injected in the blood vessels, thus creating strong contrast between the blood in the lumen and the wall of the vessels. A second technique, called *double contrast* first fills the cavity with a liquid more opaque than the wall, then empties it and injects gas. A thin opaque film is thus created between *two* more transparent volumes (wall on one side, gas on the other side). The lines we refer to on projections are thus either single boundaries with single contrast, or double boundaries (e.g., white curves on a dark background) with double contrast (Fig. 1), but their relations to the contrast surface are very similar. The "only at" part of the law (line implies tangency) can be considered very safe.

1.1. *The silhouette sign for chest roentgenology*

In the early 1970's, before C.T. scanners were available, Benjamin Felson [4] showed to the radiologic community how to use another sign to build complex deductions that were very important to chest imaging. His *silhouette sign* is the fact that "an intra-thoracic radiopacity, if in anatomic contact with a border of the heart or aorta, will obscure that border" (thus leading to a loss of the *silhouette*).

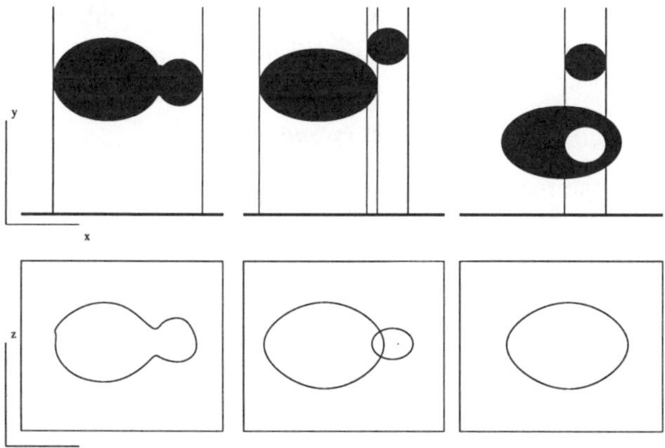

Fig. 2.   The silhouette sign: contact of the solids (top left) erases some of their silhouettes (bottom left). Top : horizontal sections, bottom: projections on a vertical plane. The thin lines figure the direction of projection and indicate the relation of silhouettes to tangencies. Center: in the absence of contact, the silhouettes are preserved. Right: an exception to the silhouette sign, both unlikely and unstable.

A simplified situation is shown in (Fig. 2) for solids which are "smoothed" unions of two solid ellipsoids of comparable opacities (e.g. figuring the heart and a neighboring mass): in a (O,x,y,z) frame, the top parts are horizontal sections by the (O,x,y) plane and the bottom parts are their projections on the vertical (O,x,z)

plane, the thick lines figuring the film seen from above. The two separate bodies give rise to two complete contours which cross, whereas in the connected case no boundary is seen to separate the two projections.

Combined with anatomical knowledge, this sign often permits to precisely locate a lesion, e.g., to decide whether an opacity seen on a postero-anterior chest view is anterior or posterior and in which pulmonary lobe and segment it stands. For instance on the same simplified example, if one knows that the chest geometry does not allow the small mass to fit, with the same projection, anterior to the heart, i.e. with a horizontal section closer to the thick line, then seeing the crossing pattern on the film indicates that the mass is posterior. More generally "an intra-thoracic lesion touching the border of the heart, aorta or diaphragm will obliterate that border on the roentgenogram. An intra-thoracic lesion not anatomically contiguous with a border of one of these structures will not obliterate that border".

These facts could be reproduced experimentally, but the disappearance of the silhouette might also have been produced (Fig. 2 right) by a perfect alignment of silhouettes arising from non-contiguous structures [17] (the experimental evidence remaining debated, though). No consensus emerged about the foundation of the silhouette sign except that the exceptions to it must be clinically very rare.

1.2. *Justifying signs: singularity theory and genericity*

The problematic nature and validity of such signs calling for some formalization, we were led to a more general study of generic sign systems [7] [9]. Assuming the law of tangential incidence to hold, the lines on the film occur at singular values of the projection (along the X-rays) of the contrast manifold to the film. It is natural, following R. Thom's discussions of experimental methodology [15], to require the patterns described in a theory or morphology to be structurally stable, and inquire about the genericity of such stability. Here we require the projection to be topologically stable (in the usual Whitney topology) and we use Whitney's results [18] : such stability is generic, and the the stable local types of projections of a 2-manifold to the plane are either non singular (no curve on the film at that point) or of one of three types (see Fig. 3): (1) the fold (simple line), (2) the transverse crossing (two lines crossing, not tangent), and (3) the cusp (a cusp on the line).

To use the sign system for interpretation, we first assume that the experimental conditions that led to the radiograph we see belong to a generic case. That such assumption is not restrictive would rigorously result from a Bayesian inference if we had built a realistic probability measure on our set of experimental settings and if we had proved that our generic properties are almost sure for that law, a task

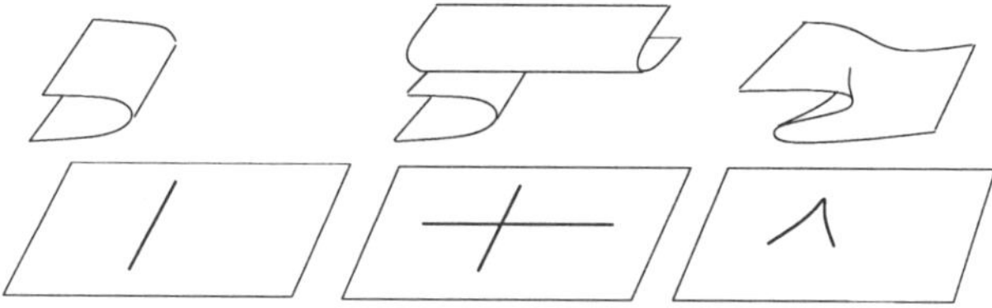

Fig. 3.   The three stable signs: line, crossing, cusp.

that we shall not pursue here. Now knowing that only one of these types can occur locally, we have a rule that associates to each local type of curve seen on the film a local type of 3D embedding for the contrast surface (up to some few symmetries, however). Up to the few ambiguities, such rule constitutes a complete sign system which holds generically. The ambiguities mentioned are common for many sign systems such as natural languages; they are usually resolved by global constraints, prior anatomical knowledge, or what is sometimes called the "context".

One should be aware of the numerous circumstances that made projection radiology possible, that is, interpretable by humans using only qualitative reasoning rather than numerical measurements and computations: Among them, the organization of tissue opacities in macroscopic compartments, the smooth anatomic boundaries of these compartments, the detection of lines by the human eye, the fact that stability of line patterns (singularities) is generic and leads to a small number of recognizable curve types associated to a small number of embedding types with low ambiguity. The interplay between stability, genericity and *paucisemy* (a term by which we mean low ambiguity of signs) is central to the tractability of such qualitative interpretation. It is a reason to try using singularities to build new sign systems in visualization problems.

1.3. *The three stable signs: applications*

The model just described for local signs provides a setting in which the validity of the signs can be discussed. For instance, one can address the issues of the intensity of the contrast, the curvature of the surfaces, the width of contrast lines (involving also the width of the X-ray source), or noise. When applicable, the silhouette sign is related the line sign, but in practice it is often used with the crossing sign: two crossing lines arise from contrast surfaces embedded at different depths. We found some clinical applications to the cusp sign, for instance in bone imaging (Fig. 4) and gastrointestinal imaging (Fig. 5).

After thirty years, projection radiology has lost its unique status: all sorts of computer reconstructed imaging modalities like C.T., M.R.I., U.S., or P.E.T., are

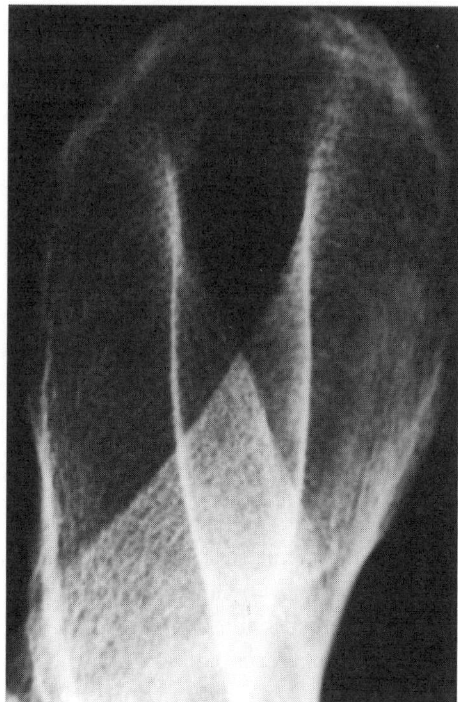

Fig. 4.   Stable projection of the iliac bone with cusps and crossings.

available to provide volume exploration, and they are mostly considered to be sectional imaging. It is thus much less important now to deduce the location of lesions from a single projection, and one might wonder whether these signs are still useful at all. It recently turned out that the huge amounts of data output by these devices have become a problem of their own and that visualization benefits from designs that decouple it from acquisition formats. For instance, the data acquired as slices from abdominal C.T. scanners are now presented for interpretation as virtual colonoscopies or synthetic double contrast barium enema-like simulations: all the sign systems of projection radiology are ready to be used again and to inspire new visualization paradigms.

2. Extending control

If patient positioning is controlled precisely, like during interventional radiology, it is possible to make unstable patterns appear, and the former sign system is not valid anymore. However, if the control is constrained to spaces of low dimension

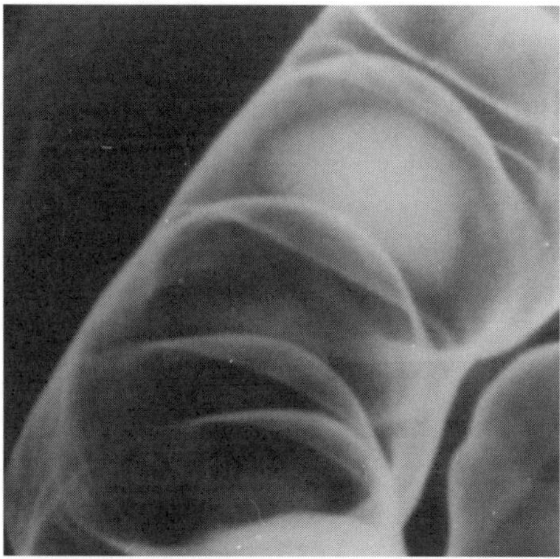

Fig. 5. Some cusps seen on a double contrast image of the colon. The cusps associate into regional lip patterns which can be understood using the knowledge of 1-codimensional events under control (see below).

in general position, it is still possible to extend the former setting and still keep simple sign systems.

2.1. *The formalism of controlled mappings*

Let A, B, and C be smooth manifolds, where C is the control space to parameterize some smooth mappings from A to B. A mapping from A to B controlled by C is a smooth mapping $F : A \times C \to B \dot{\times} C$, where $B \dot{\times} C$ is a fibre bundle with base C and fiber B, also noted, for any $c \in C$, by $f_c : A \to B$ where $f_c = F(\ ,c)$. We can define in the usual way a smooth equivalence between such controls F and F', only requiring that the source and target diffeomorphisms commute with the projections onto C. Smooth stability can then be defined for controls. There is a strong analogy with Elementary Catastrophe Theory even if no gradient dynamical system is used, and we apply a very similar set of tools and methods. We now detect changes of the type of f_c as c moves within C. The set of c's where such changes happen is thus analogous to a catastrophe set. Referring to the radiologic setting, we call it *obturation set* or *collimation set*: it is the set of $c \in C$ such that f_c is not stable as a mapping from A to B. That set is stratified by the codimension of the germ (or multi-germ) at which f_c is unstable.

We call *singular set* of the control the set of points x of A at which f_c is unstable for at least one c in C. In our applications that set is important for effective computations since one can compute it first from simple local differential geometric properties of an embedding of A which determines the mapping control, then deducing the obturation set from it. Another set that we shall use later for applications to visualization and shape coding is the *diagram of contours*, which we define to be the subset of $B \dot{\times} C$ made of the union of all the singular values of the f_c's for $c \in C$.

2.2. *Controlled projections : generic results*

For a simple model of projection radiology under controlled rotations, we consider the set of orthogonal projections from the contrast 2-manifold A embedded in R^3 to the planes B_c of the film, parameterized by the set C of directions, where for each direction of projection the plane is orthogonal to the direction of projection. The control space C is thus the real projective 2-space. Several authors have studied equivalent or related settings, among which [1] [13] [2] [7]. The singular set is then generically the union of parabolic curves, swallow tail curves and some multi-local strata, on which one can distinguish 1-codimensional points (among which the well known lip, beak to beak and swallowtail points) and isolated 2-dimensional points (including godrons, gouttieres, butterflies). That set can be computed numerically for a given surface, i.e., once the embedding of A in R^3 is known. For each 1-codimensional type of singularity of the projection, the corresponding strata of the singular set are computed. The unstable direction at each point of a stratum is then easy to compute, which permits to compute the obturation set from the singular set. For multi-local strata (points of contacts of doubly tangent planes) we computed the obturation strata first as self-intersections of a dual surface.

Global surface configurations, like the fossette [7] [9], can be studied with these methods (Fig. 6). The computations can also address and visualize, for each $c \in C$, the critical set of f_c, to show, as predicted by the theory, that topological changes of the critical set of f_c occur at points of the singular set (and of course for angles in the obturation set).

Another way to use the former results is to detect from the patterns in an image the proximity of the control parameter to a stratum of a certain type such as a swallowtail stratum if a small "swallowtail pattern", i.e. the unfolding of a swallowtail singularity, is seen. This is what is intuitively used by radiologists who know from an image how to rotate a patient to make some patterns appear or disappear.

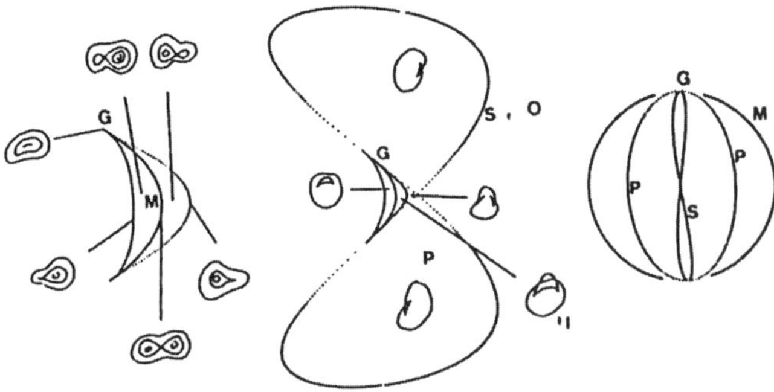

Fig. 6. The fossette: Singular set (right), obturation set for projections (center, with typical projections in stable strata), and obturation set for section stacks (left, with typical sections associated to strata). S: swallowtails, P: parabolic points, M: multi-local strata, G: godron.

3. Sectional imaging

3.1. *Controlled sections and stacks of parallel sections*

Sectional imaging has become the main stream of medical imaging. The formalism of mapping controls and singularity theory can be extended to it. Contrast surfaces are now surfaces limiting the anatomical compartments which produce different intensities or textures on a sectional image, i.e. leading to the recognition of a curve in the sectional image. To get one sectional image, one has to choose a sectional plane. In the sectional plane, curves will be seen at the intersection of the contrast surface with that plane. A generic plane is transversal to the contrast surface. It is convenient to consider stacks of sections by parallel planes and use Morse theory on the corresponding height functions, a generic stack producing a Morse function. We now control the stack by its common normal direction, the controlled mapping being the height function, thus taking $B = R$. The control set is the same as for projections. It turns out that for a given contrast surface, the singular set for sectional stacks is a subset of the singular set for projections, containing the parabolic points and some multi-local strata. The same holds for the obturation set.

294

4. Other applications

4.1. *Data Visualization*

In the former sections we have built a model for projection radiology, an imaging modality evolved before the digital era which used images directly built during the physical process of data acquisition. Nowadays visualization can be decoupled from acquisition, of course temporally, but also conceptually. We already mentioned that image data acquired as sections can be displayed as projections or even virtual endoscopies. Many kinds of data sets, medical or not, wait for efficient visualization procedures. Visualization has to take into account the viewer's aptitudes and habits, and many successful visualizations are adaptations of an existing mode of representation, just like graphic user interfaces have used such metaphors as "cut and paste". Conventional radiology skills and habits are a good source for visualization design. Data reduction through dimension reduction, now a popular concept, occurs in projections and greatly benefits from the tools we evolved for mapping controls [7] [9]. For instance a temporal sequence of maps (in the context of electrocardiography, for a technique called Body Surface Potential Mapping) can be condensed using the diagram of contours already mentioned: Each map is a mapping of a 2-sphere (the surface of the skin) to R (the set of values of an electric potential), these maps being controlled by time (R). The diagram of contours is thus a set of curves in R^2: plot against time the singular values of the potentials, i.e., the values of maxima, minima, and saddles points. One can recognize swallowtail and lip patterns on such diagram and induce events in the sequence of maps, thus providing a first step in the exploration of usually huge sets [7] [9]. Notice that the same process can in principle be applied to sequences of functions on higher dimensional spaces.

4.2. *Shape coding*

The idea of using some special points on curves or surfaces to characterize their shapes is not new. Already Hilbert and Cohn-Vossen showed the parabolic set on the face of a statue [5]. Mapping controls provide a wealth of special subsets for which genericity methods can be used and often provide classifications into a small number of types. Let us describe the simple case of plane curves obtained as smooth embeddings of the unit circle in R^2: the orthogonal projections of a fixed plane curve to straight lines in the same plane can be controlled by the direction of projection (identifying mappings associated to parallel directions). We can build the diagram of contours on a Moebius strip and generically code it with a slight extension of Gauss words for coding immersions [7] to get a classification and some coding for curve shapes. Controlling other mappings from the curve to

R leads to other classifications. In the case of the distance from the curve to a controlled point of the plane, the obturation set sits in R^2 and is the evolute for strictly convex curves, which d'Arcy Thomson already used to precisely describe the shapes of eggs [16].

PART 2 : GENERALIZED CATASTROPHE, ADAPTIVE TREES, AND INTERVENTIONAL IMAGING

5. New imaging problems

Imaging has now become mostly digital and computed. Imaging devices often readily output data as reconstructed volumes, apparently solving many of the problems of geometric interpretation. Assistance is now needed for visualization, fusion of the data from several modalities, morphometry, diagnosis, or interventional imaging. Often some non-linear optimization is eventually used and problems of multiple minima must be addressed. We here discuss an alternative to some of the classical minimization algorithms. It borrows at the same time from classical search tree algorithms and from biologically inspired models of tree growth and branching. We then apply it to problems inspired by interventional imaging and vision.

5.1. *Adaptive branching in nature, generalized catastrophes, simple models*

René Thom introduced the concept of generalized catastrophe [15] as an attempt to describe many of the branching phenomena which can be observed in Nature, like vegetal or vascular trees, river deltas, glass fractures, or sparks. It is important to notice that adaptivity to some boundary condition is an essential feature of these phenomena.

Branching can be reproduced by the physical models of the phenomena which produce it, like, for crystal growth, diffusion limited aggregation models or the Stephan problem. In Biology, reaction-diffusion has been shown by Gierer and Meinhardt to be able to generate branching using two diffusible species. These models, however, lead to technical difficulties when trying to prove very general properties on their branching and adaptivity properties. We shall look for a more tractable model of a phenomenon we call abstract angiogenesis, taking as our central metaphor the growth of a vascular network toward an organ or a tumor.

296

5.2. *Wish list for a model of abstract angiogenesis*

Let us list a set of desirable properties for a model of adaptive braniching. The algorithm should :

- exhibit branching
- adapt to the geometry of targets
- display cooperation and competition behaviors (between different branches, networks, or seeds)
- be able to take place in general spaces
- allow easier formal study of adaptivity and branching
- its numerical complexity should permit simulations.

6. Adaptive trees

We shall describe two different algorithms with the same broad features: They take place in some space like a Euclidean space (more general spaces are possible), where one or more seed points are given together with a target which is simply a subset of the space with a probability measure on it. The target can be thought to grossly model a tumor or an organ to be irrigated by a growing vascular network starting from the seed points. They build (or "grow") an increasing sequence of finite subsets (starting from the seed set) which progressively approach the target and adapt to it. At each step a single point is added (we say accreted) to the current network. The algorithm is stochastic and at each step a point is randomly drawn from the target, which, together with the shape of the current network, will determine the point to be accreted to the network. The sequence of subsets can naturally be given the structure of a growing tree, and the second algorithm will be able to use that tree to perform searches in the target.

6.1. *Algorithm*

Let the algorithm take place in $E = R^n$. The network at time $i \in N$ is called N_i, with the initial network N_0 a finite set (often a single point) called the seed. Define a probability measure on E and call T (the target) its support. Often, and more simply, T can be defined first as a compact submanifold of E and the probability distribution on T is taken to be uniform. Choose $0 < \varepsilon < 1$ to be a fixed small positive constant.

Now iterate the following operations, starting from $i = 1$:

- randomly draw a point a_i from T, using the probability defined on T,
- find b_i, the point of N_{i-1} closest to a_i (in most settings, ties almost surely do not occur, but you can provide a rule for breaking ties)

- compute $b_i' = \varepsilon.a_i + (1 - \varepsilon).b_i$ which is the point to be accreted,
- set $N_i = N_{i-1} \cup \{b_i'\}$. We say that b_i' was accreted to N_{i-1} at b_i, that b_i is the parent of b_i', and that b_i' and b_i are neighbors in any N_{i+k}, $k \in N$.

The number of steps is at user's will or a stopping rule can be defined.

6.2. *Experimental results*

Executing the former algorithm in R^2 with a seed reduced to a point and for different targets shows how the network branches and progressively adapts to very general shapes of targets (Fig. 7). Competition or target sharing between two seed points also occurs according to their positions and the geometry of the target. Using indirect visualization, such as projections on a plane, one can check that the same features hold in higher dimensional spaces (Fig. 8). Varying ε effects the regularity of the branches but the qualitative features of branching and adaptations are relatively insensitive to it.

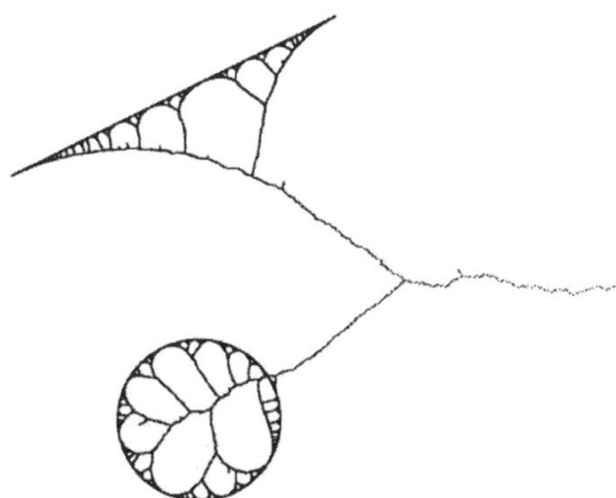

Fig. 7.   Growing a tree to a line segment and a circle in $E = R^2$ with the first algorithm. Abortive branchings can be observed.

6.3. *Abortive bifurcations*

With decreasing ε, and magnifying the network in the neighborhood of a branching point, one can check that the apparent regularity of the branches hides some repeated microscopic branches which did not grow much after they where started. We say that these branches are abortive [8] [10]. It is illuminating to study why

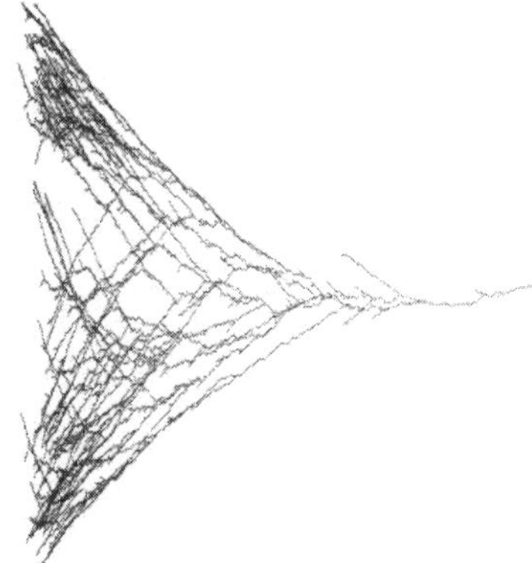

Fig. 8. Growing a tree in $E = R^8$ toward a target 7-cube with the first algorithm. The network has been projected to a 2-plane normal to the hyperplane containing the target

many of these have to occur before a branching can lead to two viable branches. For a point of the network to be able to grow the network (let us assume it is a tip of the network, i.e., it has only one neighbor), it has to be the closest network point for some part of the target which we call its sub-target. Now assume ε small. If that sub-target, as seen from the tip point, spans an angle more than $\pi/2$, it is possible for the points drawn from that very sub-target at two further steps to be accreted at that tip (because a first point can be accreted and still leave some sub-target to its parent), thus producing branching. The two new tips now have their own sub-targets, but the size of these sub-targets, and thus the probability of a growth, evolves with the growths of the two tips. Notice that these two sub-targets are separated by the hyperplane of points equidistant from the tips. If one of the tips grows too much relative to the other, it moves the hyperplane and can take over the other's sub-target and thus kills the other tip. It is possible to write the equations of the expected growth forces on the two tips knowing the current network. The structure of the dynamical system on the couple of tips depends on the distance to the target. Away from the target, the system is unstable and the probability is very high that one of the two tips will kill the other. Closer to the target, the simultaneous survival of two newborn tips becomes stable and the branching does not abort (competition being replaced by sharing). This analysis supports our use of the term generalized catastrophe in reference to Thom since adaptive branching is seen to arise from repeated bifurcations of a dynamical system.

6.4. *Formal results and discussion*

Other simple formal results can be given as to the adaptive properties of the algorithm. Any neighborhood of the target will in probability be eventually entered by the network, uniformly if the target is compact [10].

To fulfill the wish list, we need to address some computational complexity and implementation issues. The main computational burden is to find at each step the element b_i of the network closest to the a_i just drawn from the target. When possible one should remove from the set to be searched the points of the network that cannot grow anymore. This is easy and very effective for some simple target geometries such as a segment in R^2. General targets however lack such simple algorithms. The next algorithm will solve that problem by using search trees, however at the price of giving up some items of our wish list.

7. Adaptive search trees

To prevent abortive branching and speed up the search for closest network points, we now slightly modify the former algorithm by explicitly and irreversibly assigning sub-targets to the points of the network, and even to regions of the space where the network can grow. This disables any competition between tips after branching. Also, one can chose the assignment rule to facilitate the successive searches for a closest point.

7.1. *Algorithm*

At each step the whole space will be partitioned in a finite set of regions, each step finishing with the partition being refined. We start with the whole ambient E as a single region. In each region, a part of the network will grow from a seed in that region toward that part of the target which intersects that region (we call it the sub-target). Branching in that region will trigger splitting of that region, with two new seeds replacing the children of the branching point. Thus we repeat the following steps:

(1) Repeat:

- draw a random point from the target
- determine to which region and thus to which sub-target this point belongs
- perform accretion to the corresponding subnetwork

until some branching occurs, say in the k-th region at a point c which is the parent of two points.

(2) In the k-th region, keep the points between the seed and the two sons s_1, s_2 of the branching point c, remove the other points. Partition that region by the hyperplane equidistant from s_1 and s_2. Take s_1 and s_2 as seeds of these new regions.

Using classical algorithms and data structures inspired from search trees, the computation of the region assigned to each new point drawn from the target can be made fast. See [11] for details.

7.2. *Experimental results*

The performance of this algorithm is satisfactory even for complex targets, and the trees produced for searches are found to be reasonably balanced. For many targets, branching and adaptive behaviors appear mostly similar to those of the first algorithm, with the difference that no abortive bifurcations appear (Fig. 9).

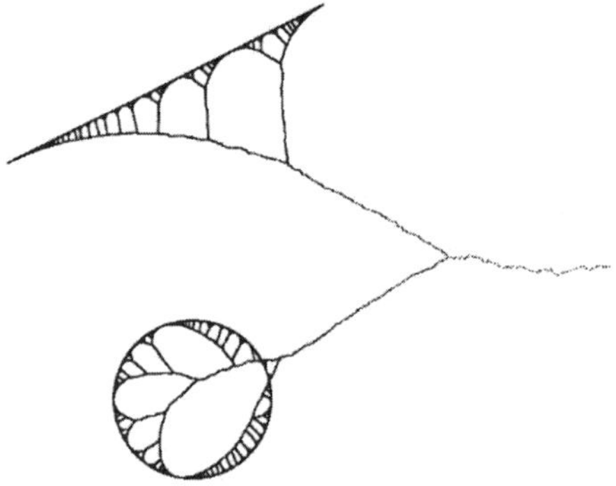

Fig. 9.   Growing a tree with the second algorithm, same target, E, and ε as before. Abortive branchings cannot be observed.

The pattern of successive target splits can be seen better when looking at a projection of the network on the plane where a target is (Fig. 10). However some seed and target configurations definitely challenge the adaptive properties of this second algorithm, which is weaker there, especially when some parts of the target have to be crossed by the network to reach other parts [11]. Difficulties are not limited to target self-hiding but also involve lateral phenomena.

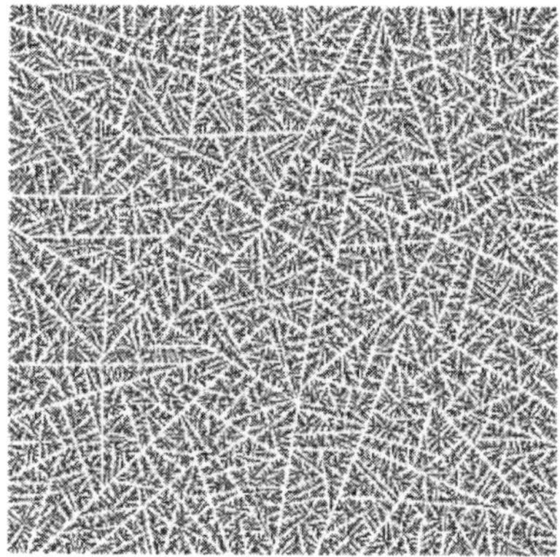

Fig. 10. From a seed not in the target's plane, evolution toward a square target with the second algorithm. What is shown is the projection of the network onto the target's plane. Notice the nesting of regions and the linear separatrices.

8. Growing search trees in shape spaces

The good performance of the second algorithm and its capability to build a tree to search the target can be used in spaces of geometric objects such as polygons, with enough dimensionality to approximate differentiable curves. For most Vision and Interventional Imaging problems, we need to address curve shapes, i.e. equivalence classes of curves under geometric transforms, rather than embeddings [6]. We shall grow search trees in spaces of curve shapes and thus need to modify the former setting slightly.

Let S^1 denote the unit circle. Given two embeddings $e_1, e_2 : S^1 \to R^2$, where the parameter is proportional to curvilinear abscissa, we can build a bijection such as $e_2 \circ e_1^{-1}$ and a distance $d(e_1, e_2)$ integrating along the circle euclidean distances between corresponding points. To get a distance $d_s(e_1, e_2)$ between the shapes of e_1 and e_2, we compute the minimum over ϕ and g of $d(e_1 \circ T_\phi, g \circ e_2)$ where $T_\phi : \alpha \mapsto \alpha + \phi$ is the angle translation by a phase ϕ and g is an element of the group G of plane displacements. In other words d_s is computed like d after matching the curves using plane displacements and phase shifts. We use the values $\hat{\phi}$ and $\hat{g}$ for which the minimum is reached to build a barycenter in matched position $(1 - \varepsilon).\hat{e_1} + \varepsilon.\hat{e_2}$ where $\hat{e_1} = e_1 \circ T_{\hat{\phi}}$ and $\hat{e_2} = \hat{g} \circ e_2$. That formalism applies as well to shapes of plane polygons.

8.1. *Application : curve to surface matching and Interventional Imaging*

Medical imaging has been extended to computer assisted surgery and interventional imaging. One issue of these fields is to monitor in real time the precise location of organs and instruments acting on them during surgical interventions, or biopsies, or radio-frequency ablations. If ultrasound is used for interventional imaging, the precise position of the probe can be of interest, especially if some fusion with data from another imaging modality is sought, for instance from C.T. or M.R.I which could show a tumor better. Solutions are now available using optical or magnetic tracking of the probe. Another avenue for research is to deduce that position from the image, using prior information from the other modality, and also possibly deducing possible deformations of the probed organ.

We shall restrict the clues for such a positioning to visible boundaries of anatomical compartments and address the more abstract problem of matching planes curves to surfaces: given a plane curve which was obtained as the intersection of a known surface by an unknown plane, find a plane which intersects the surface along the same curve, up to plane displacements. More precisely, given a surface in R^3, a plane in R^3, and a curve is that plane, we shall look for a 3D displacement of the surface which makes it intersect the given plane along the given curve. The problem can be addressed using descent searches (and their improvements). in the space of 3D displacements looking for a best fit, but these are plagued by multiple minima. We are going to build a searchable atlas of intersection curves where we record for each curve the matrix of the transformation which makes the surface intersect along that curve. We can then use it to find an approximation of the matrix producing an intersection along a new curve and possibly start other methods from there.

As a target we take the set of plane sections of the given surface, with the probability measure induced from the uniform distribution on the set of displacement matrices which lead to a non void intersection. As a seed, we take an ellipse, and we build a search tree according to the second algorithm, with the slight modifications mentioned earlier: each time we draw a random matrix, we compute the corresponding intersection curve (point of the target), we locate the region in the space of curves to which that curve belongs using the search tree built so far and d_s, then perform accretion in that region after interpolation in matched positions, detect branching and split the region accordingly.

It is interesting to observe the evolution of the curves associated to the leaves of the search tree as it grows: the curves progressively differentiate toward the different shapes obtainable as intersections of the surface. Once computed, the search tree can be used to retrieve approximate matrices from intersection curves in real time.

Fig. 11.   The binary search tree during its growth toward outline curves of a polyhedron, represented with some overlap. The curves correspond to branching points of the tree grown with the second algorithm.

8.2. *Application : Vision*

As an application to computer vision, let us address the classical problem of identifying the pose of a known polyhedron given an outline of it. In the same way as before, we build a search tree toward the target taken as the set of outlines with the distribution induced from the uniform distribution on poses. Here again, as the tree grows, the curves associated to its nodes progressively differentiate toward the different shapes of possible outlines in the target (Fig. 11) In that case, it is possible to test the retrieval of poses from an interactively controlled polyhedron. Real time approximate retrieval can be achieved from a tree built off-line (Fig. 12)

9. Conclusions and Prospects

The applications of singularity theory we described for medical imaging in the first part have been strongly inspired by Catastrophe Theory, even if they do not involve underlying dynamical systems. They use most its idea of structuring (stratifying) a control space by the topological types of the controlled mappings, here related to image patterns, and the paradigm that what is not controlled should be generic. Methods and results are available for projection imaging and sectional

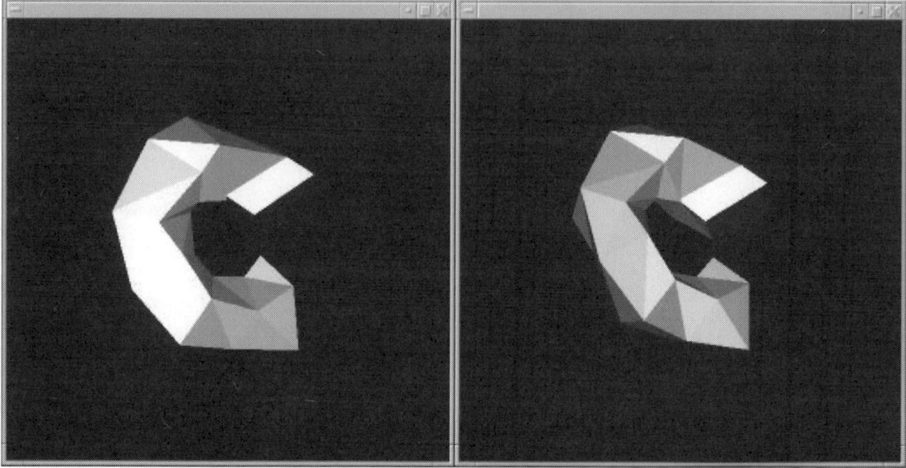

Fig. 12.   Screen copy of a real time test of the search tree. The pose of the polyhedron in the right hand window is controlled with the mouse and the left hand window displays the same polyhedron with a pose recognized from the outline of the controlled polyhedron.

imaging. After computerized imaging enabled sectional imaging to become the main stream of imaging, projections come of interest again for visualizing the data and they are now mostly computed but require the same signs for interpretation. The methodology of generic sign systems extends outside radiology. In the second part we abstracted models built for natural branching phenomena to more general settings. We related a first algorithm to R. Thom's notion of generalized catastrophe. The second algorithm was computationally more efficient and built search trees applicable to subsets of polygonal curves, with foreseen applications in computer vision, interventional imaging, or where other tree methods [3] [14] are already thriving, like for processing of genomic and biometric data.

References

1. Arnold, V. I. : Indices of singular points of 1-forms on a manifold with boundary, convolutions of invariants of reflection groups, and singular projections of smooth surfaces. Russian math. surveys Vol 34 (2) 1979, pp. 1–42.
2. Banchoff T;, Gaffney T., McCrory C.: Cusps of Gauss mappings. Pitman 1982.
3. Breiman L., Friedman J. H., Ohlsen R. A., Stone C. J. : Classification and regression trees. Wadsworth, Belmont, 1984.
4. Felson, B. : Chest roentgenology. W. B. Saunders, New York, 1973.
5. Hilbert, D., Cohn-Vossen, S.: Geometry and the imagination. Chelsea, 1952.
6. Kendall, D.G. : Shape manifolds, Procrustean metrics, and complex projective spaces. Bull. London Math. Soc. **16** (1984), 81–121.
7. Kergosien, Y. L. : Medical exploration of rythmic phenomena: a topological semiology.

Rythms in Biology and Other Fields of Application. M. Cosnard et al. (Eds), Lecture Notes in Biomathematics, Springer 1983.

8. Kergosien, Y. L. : Adaptive ramification and abortive concepts. Neural networks from models to applications (NEURO'88), I.D.S.E.T., Paris, 1988, pp. 439–449.

9. Kergosien, Y. L. : Generic sign systems in Medical Imaging. IEEE Comput. Graph. Appl., **11** (5) (sept. 1991), 46–65.

10. Kergosien, Y.: Adaptive branching in Epigenesis and Evolution. C. R. Biologies **326** (May 2003) 477–485.

11. Kergosien, Y.L.: Adaptive trees and pose identification from external contours of polyedra. DSSCV 2005, O.F. Olsen ed, Springer Lecture Notes in Computer Science 1353, pp. 157–168.

12. Kergosien, Y. L.: The family of the orthogonal projections of a surface and its singularities. C.R. Académie des Sciences, Paris, t. 292, série I–929–932 (1981).

13. Kergosien, Y. L., Thom R. : Sur les points paraboliques des surfaces. C.R. Académie des Sciences, Paris, t. 290, série A–705–710 (1980).

14. Kohonen, T. : Self-organizing maps, Springer, Berlin, 1997.

15. Thom, R. : Stabilité structurelle et morphogénèse: essai d'une théorie générale des modèles, Benjamin, Reading, 1972.

16. Thompson, D'A.W.: On growth and form. Cambridge, 1942.

17. Tuddenham, W.J.: Problems of perception in chest roentgenology: facts and fallacies. Radiol. Clin. N. America. 1: 277, 1963.

18. Whitney, H : On singularities of mappings of Euclidean spaces I: mappings of the plane into the plane. Annals of Math. Vol 62, 1955, PP. 374–410.

Geometric contacts and 2-regularity of surfaces in euclidean space

MARIA DEL CARMEN ROMERO-FUSTER

Departament de Geometria i Topologia, Universitat de València,
46100 Burjassot (València), Spain,
e-mail: carmen.romero@uv.es

We study the problem of second order regularity in Feldman's sense for surfaces immersed in $I\!R^n$ and its connection with the contacts of these surfaces with hyperspheres and hyperplanes of the ambient space. We obtain some obstructions to the second order regularity.

Keywords: Distance squared functions, height functions, 2-regular immersions, asymptotic directions, $\nu-$ principal curvature foliations, umbilicity, convexity.

1. Introduction

E. A. Feldman[8] and W. Pohl[26] introduced the concept of kth-regular immersion of a submanifold M in Euclidean space in terms of maps between osculating bundles. For a curve immersed in 3-space, $2nd$ and $3rd$-order regularity are respectively equivalent to having non vanishing curvature and torsion. It can be shown by standard transversality techniques that the subspace of 2-regular closed curves is open and dense in the set of immersions $Imm(S^1, I\!R^3)$ with the Whitney C^∞-topology. It is also known that the convexity (i.e. when the curve lies on the boundary of its convex hull) is an obstruction for the $3rd$-order regularity of such curves ([28,33]). Families of non convex closed 3-regular curves lying on the torus have been constructed by S. I. R. Costa.[6] In the case of an immersion $f : M \to I\!R^n$ of a surface M in n-space, the $2nd$ order regularity at a point p is equivalent to the second fundamental form having maximum rank at p. An immersion $f : M \to I\!R^n$ is **2-regular** if all the points of M are 2-regular. Feldman[8] proved that the set of 2-regular immersions of any closed surface M in $I\!R^n$ is open and dense, for $n = 3$ and $n \geq 7$. It thus remains to analyze the cases $n = 4, 5, 6$. For $n = 4$, the 2-singular points coincide with the inflection points defined by J. Little[14] and it follows from the results obtained in Ref. 11 that the local convexity is an obstruction for the 2-regularity of closed surface with non vanishing Euler number immersed in $I\!R^4$. The problem of 2-regularity of surfaces immersed in $I\!R^5$ appears to be much more complicated. A *2nd-order* regular immersion of the 2-sphere into $I\!R^5$ is given by the Veronese surface (see Ref. 5). This is a double covering of a projective plane embedded in S^4. But no 2-regular immersions of surfaces with non zero genus into $I\!R^5$ are known so far. This problem has been recently connected in Ref. 19 to the behavior of the family of height functions on the surfaces, and in Ref. 20 to the existence of globally defined special sections of their normal bundle. Our aim in this paper is to provide a survey of results on geometrical properties of surfaces immersed in $I\!R^5$ related its contacts with hyperplanes and hyperspheres, making special emphasis in their connections with the 2-regularity problem. In particular, we define the concepts of essential normal field, degenerate normal field, essential convexity and umbilic curvature that lead to the following obstructions for the 2-regularity on closed (compact without boundary) surfaces immersed in $I\!R^5$:

a) Existence of an essential normal field globally defined (Corollary 4.7).
b) Existence of a degenerate normal field globally defined (Theorem 6.2).
c) Essential convexity (Corollary 7.4).
d) Zeroes of the umbilic curvature (Theorem 8.3).

We also analyze the possibility of having 2-regular immersions of surfaces in S^4, arriving to the following conclusion:

Orientable surfaces cannot be 2-regularly embedded in S^4 (Theorem 8.1).

Finally, we show that

Any 2-regular surface immersed in $I\!R^n, n \geq 7$ can be isometrically immersed into $I\!R^5$ (Corollary 10.2).

2. Basic concepts

2.1. *Second fundamental form of surfaces in $I\!R^n$*

Let $\bar{\nabla}$ be the riemannian conexion of $I\!R^n$ and $M \subseteq I\!R^n$ with $n > 3$, a smooth surface.

If X, Y are vector fields locally defined along M and $\bar{X}, \bar{Y}$ are local extensions on a neighborhood of M in $I\!R^n$, we have the following riemannian conexion on M,

$$\nabla_X Y = (\bar{\nabla}_{\bar{X}} \bar{Y})^\top.$$

The second fundamental form of M is given by:

$$\alpha_M : \mathcal{X}(M) \times \mathcal{X}(M) \longrightarrow \mathcal{N}(M)$$

$$(X, Y) \longmapsto \bar{\nabla}_{\bar{X}} \bar{Y} - \nabla_X Y,$$

$\mathcal{X}(M) =$ Space of tangent vector fields on M,
$\mathcal{N}(M) =$ Space of normal vector fields on M.

For each $p \in M$ and $\nu \in N_p M$, $\nu \neq 0$ we have a bilinear form on $T_p M$ given by

$$H_\nu : T_p M \times T_p M \longrightarrow \mathbb{R}$$

$$(v, w) \longmapsto \langle \alpha(v, w), \nu \rangle,$$

and a quadratic form given by

$$II_\nu : T_p M \longrightarrow \mathbb{R}$$

$$v \longmapsto II_\nu(v) = H_\nu(v, v) = \langle \alpha(v, v), \nu \rangle.$$

known as the **second fundamental form in the direction** ν.

We can take M locally as the image of an embedding $f : \mathbb{R}^2 \to \mathbb{R}^n$. If (x, y) are isothermic coordinates and $\{e_1, e_2, ..., e_n\}$ is an orthonormal frame in a neighbourhood of a point $p = f(0, 0) \in M$, in such a way that $\{e_1, e_2\}$ is the tangent frame determined by these coordinates and $\{e_3, ..., e_n\}$ is a normal frame, then the second fundamental form of M at p is represented by the matrix

$$\alpha_f(p) = \begin{bmatrix} a_1 & b_1 & c_1 \\ & \vdots & \\ a_{n-2} & b_{n-2} & c_{n-2} \end{bmatrix},$$

where
$a_i = \alpha_f(e_1, e_1) \cdot e_{i+2} = \frac{1}{E} \frac{\partial^2 f}{\partial x^2}(p) \cdot e_{i+2}$,
$b_i = \alpha_f(e_1, e_2) \cdot e_{i+2} = \frac{1}{E} \frac{\partial^2 f}{\partial x^2}(p) \cdot e_{i+2}$ and
$c_i = \alpha_f(e_2, e_2) \cdot e_{i+2} = \frac{1}{E} \frac{\partial^2 f}{\partial y^2}(p) \cdot e_{i+2}$,
for $i = 1, \cdots, n - 2$, and $ds^2 = E(dx^2 + dy^2)$ is the first fundamental form (in the isothermic coordinates (x, y)).

2.2. *The curvature ellipse*

Given $p \in M$, consider the unit circle in $T_p M$ parametrized by $\theta \in [0, 2\pi]$. Denote by γ_θ the **normal section** of M in the direction θ, i.e. $\gamma_\theta = M \cap H_\theta$, where $H_\theta = \{\lambda\theta\} \oplus N_p M$. The curvature vector $\eta(\theta)$ of γ_θ at p lies in $N_p M$. Varying θ from 0 to 2π, this vector describes an ellipse in $N_p M$, called **curvature ellipse** of M at p.

The curvature ellipse can be seen as the image of the affine map ([14])

$$\eta : S^1 \subset T_p M \to N_p M,$$

given by

$$\eta(\theta) = H + B \cos 2\theta + C \sin 2\theta,$$

where $H = \frac{1}{2} \sum_{i=1}^{n-2} (a_i + c_i) \cdot e_{i+2}$, $B = \frac{1}{2} \sum_{i=1}^{n-2} (a_i - c_i) \cdot e_{i+2}$ and $C = \sum_{i=1}^{n-2} b_i \cdot e_{i+2}$.

The curvature ellipse may degenerate into a segment at certain points $p \in M$. Such points are called **semiumbilics**. An **inflexion point** is a semiumbilic at which the curvature ellipse is a radial segment (i.e. the three vectors H, B and C are parallel). The inflection points are said to be of **real** or **imaginary type** according the origin belongs or not to the curvature segment. If the curvature ellipse at p degenerates into a point, we say that p is an **umbilic point**. Finally, we say that p is a **flat umbilic** if the curvature ellipse at p degenerates into a point that coincides with the origin p of $N_p M$.

In the case of a surface in $\mathbb{R}^4$, the relative position of the ellipse with respect to the origin p of the normal plane $N_p M$ allows to classify the non-semiumbilic points into hyperbolic, elliptic or parabolic according to p lies outside, on, or inside the ellipse.

We denote by Aff_p the affine hull of the curvature ellipse in $N_p M$ and by E_p the linear subspace of $N_p M$ parallel to Aff_p. Clearly, if p is semiumbilic or an inflexion point, then Aff_p is an affine line in $N_p M$, and reduces to a point at umbilics.

2.3. *The rank of the second fundamental form of surfaces in $\mathbb{R}^5$*

Given an immersion $f : M \to \mathbb{R}^5$, we define the following subsets of M:

$$M_i = \{p \in M : rank(\alpha_f(p)) = i\}, \ i = 0, 1, 2, 3.$$

The **first normal bundle** of M at p is defined as

$$N_p^1 M = (Ker(\alpha_f(p)))^\perp \subseteq N_p M.$$

The second fundamental form at any point $p \in M$ induces a linear map

$$A_p : N_p M \longrightarrow Q$$

$$v \longmapsto II_v.$$

where Q is the linear 3-space of quadratic forms in two variables. Denote by C the cone of degenerate quadratic forms in Q. This induces a cone in $N_p M$,

$$C_p = II_p^{-1}(C) = \{v \in N_p M \mid II_v \text{ is degenerate }\}.$$

Clearly $Ker\alpha_f(p) \subseteq C_p$.

The following lemmas characterize the points p in M_i in terms of the cones C_p and of the relative positions of the curvature ellipse with respect to the origin of the normal space. Their proofs can be found in Ref. 7.

Lemma 2.1. *Given a surface M immersed in $\mathbb{R}^5$ we have the following*

a) If $p \in M_3$ then C_p is a cone in $N_p M$.

b) $p \in M_2$ if and only if either

    i) C_p consists of 2 planes intersecting in $Ker\alpha_f(p)$, or
    ii) C_p is a plane containing the line $Ker\alpha_f(p)$, or
    iii) $C_p = Ker\alpha_f(p)$.

c) $p \in M_1$ if and only if either

    i) C_p coincides with the plane $Ker\alpha_f(p)$, or
    ii) $C_p = N_p M$.

d) $p \in M_0$ if and only if $C_p = N_p M$.

Lemma 2.2.
Given any point $p \in M$, we have the following.

a) If $p \in M_3$ then E_p is a plane in $N_p^1 M$.

b) $p \in M_2$ if and only if either

    i) $Aff_p = N_p^1 M$ is a plane (through the origin p), or
    ii) p is a non radial semiumbilic.

c) $p \in M_1$ if and only if either

    i) $Aff_p = N_p^1 M$ is a line (through the origin p), or
    ii) p is a non flat umbilic.

d) $p \in M_0$ if and only if p is a flat umbilic.

The generic distribution of the subsets M_i on surfaces in $I\!R^5$ is investigated in Ref. 19.

Proposition 2.1 (Mochida, Romero-Fuster and Ruas[19]). *For a generic immersion $f : M \to I\!R^5$, we have*

$$M = M_3 \cup M_2.$$

Here M_2 is a regular simple curve in M containing the semiumbilics as isolated points.

Similar methods applied to surfaces in $I\!R^6$ lead to the conclusion that we also have, generically, that $M = M_3 \cup M_2$. But in this case M_2 is the union of isolated points which are non semiumbilic. In fact, the curvature ellipse determines a plane that passes through the origin of the normal space at such points. Surfaces generically immersed in $I\!R^n, n > 6$ only have points of type M_3.

3. 2-singular points

Let M be a surface and $f : M \to I\!R^n, n \geq 3$ an immersion. Following Feldman $(^{8,10})$ and Pohl $(^{26})$, we say that a point $p \in M$ is **2-regular** provided there is some coordinate system, $\{x, y\}$, for M at p such that the subspace S_p generated by the vectors $\{\frac{\partial}{\partial x}|_p, \frac{\partial \phi}{\partial y}|_p, \frac{\partial^2 \phi}{\partial x^2}|_p, \frac{\partial^2 \phi}{\partial x \partial y}|_p, \frac{\partial^2 \phi}{\partial y^2}|_p\}$ has maximal rank in $I\!R^n$. Otherwise, p is said to be **2-singular**. It can be seen that S_p is the projection of the second osculating space of M at p onto $T_p I\!R^n$. It is easy to show that this concept does not depend on the choice of the coordinate system (x, y) at p. In the case $n \geq 5$, being 2-regular is equivalent to the above vectors being linearly independent. Moreover, dim $S_p = 2+$ rank $\alpha_f(p) \leq 5$. Clearly S_p has maximal dimension if and only if $\alpha_f(p)$ has maximal rank. This implies the following result

Proposition 3.1.

Given a surface M immersed in $I\!R^n$ with $n \geq 5$, a point $p \in M$ is 2-singular if and only if $p \in M_2 \cup M_1 \cup M_0$.

In the case $n = 4$, it is not difficult to see that:

Proposition 3.2.

The 2-singular points of surfaces generically immersed in $I\!R^4$ are inflection points.

In non generic situations the inflection points may degenerate into umbilics, which also are 2-singular points.

In the next sections we characterize the 2-singular points of surfaces immersed in $I\!R^n$ with $n = 4, 5$ in terms of Singularity Theory and Dynamics. From the Singularity Theory viewpoint the 2-singular points are seen to be the corank 2 singularities of height functions. This setting allows to introduce special (degenerated and binormal) normal fields on the surfaces whose associated principal configurations have the 2-singular points as critical points. Then we can use the Poincaré-Hopf formula in order to obtain obstructions to the 2-regularity condition on closed surfaces with non vanishing Euler number. In order to do this we need to ensure that such normal fields are globally defined. We see that this condition can be expressed in terms of convexity properties of the surface.

The analysis of the singularities of distance squared functions leads to the definition of the umbilical curvature function on surfaces in $I\!R^5$. This function, given in terms of the distance of the affine subspace determined by the curvature ellipses to the surface, coincides with the curvature of the hypersphere having corank 2 contact with the surface at the considered point. The zeros of this function are seen to be 2-singular points.

4. Extrinsic dynamics: ν-principal configurations

The **shape operator** associated to a normal field ν locally defined at a point p of a surface M immersed in $I\!R^n$ is given by

$$
\begin{aligned}
S_\nu : T_pM &\longrightarrow T_pM \\
X &\longmapsto S_\nu(X) = -\left(\bar{\nabla}_{\bar{X}}\bar{\nu}\right)^\top,
\end{aligned}
$$

where $\bar{\nu}$ is a local extension of ν at p in $I\!R^n$. The map S_ν is self-adjoint and satisfies,

$$
\langle S_\nu(X), Y \rangle = H_\nu(X, Y), \quad \forall X, Y \in T_pM.
$$

Therefore,

$$
II_\nu(X) = \langle S_\nu(X), X \rangle.
$$

The eigenvectors of S_ν at $p \in M$ are called ν-**principal directions** and provide an orthonormal basis for T_pM. The corresponding eigenvalues, k_1 and k_2 are the ν-**principal curvatures**. A point p at which the two ν-principal curvatures coincide is called ν-**umbilic**. Let $\mathcal{U}_\nu = \{\nu$-umbilic points of $M\}$. The ν-principal directions define two tangent fields on M, which are orthogonal on $M - \mathcal{U}_\nu$. Their corresponding integral lines are the ν-**curvature lines**. These two foliations, together with the subset $\mathcal{U}_\nu$ of critical points form the ν-**principal configuration**. The differential equation of ν-lines of curvature is given by

$$S_\nu(X(p)) = \lambda(p)X(p). \tag{1}$$

The generic behavior of the ν-principal lines in a neighborhood of a ν-umbilic for surfaces immersed in $I\!\!R^4$ has been studied by Ramirez-Galarza and Sánchez-Bringas.[27]

The following result characterizes the critical points of the principal configurations on the surface in terms of the curvature ellipses.

Proposition 4.1 (Moraes and Romero-Fuster[20]). *Given a surface $M \subset I\!\!R^n$ with $n \geq 4$ and a normal field ν locally defined at a point p of M, we have that the point point p is ν-umbilic if and only if $\nu(p) \in E_p^\perp$.*

This allows us to relate the concepts of semiumbilics and critical points of principal configurations (ν-umbilics) as described below for surfaces in $I\!\!R^4$ and $I\!\!R^5$.

Corollary 4.1 (Moraes and Romero-Fuster[20]). *A point p of a surface M immersed in $I\!\!R^4$ is semiumbilic (or umbilic) if and only if there exists a normal field ν, locally defined at p, such that p is a ν-umbilic point.*

It then follows from the Poincaré-Hopf formula that

Corollary 4.2.
Any closed orientable surface with non vanishing Euler number immersed in $I\!\!R^4$ has semiumbilic points.

Since J. A. Little[14] proved that any torus immersed in $I\!\!R^4$ has semiumbilics, we can state the following result.

Corollary 4.3.
Any orientable surface immersed in $I\!\!R^4$ has semiumbilic points.

In the case of surfaces in $I\!\!R^5$ Proposition 4.1 leads to the following result.

Corollary 4.4.
Let M be a surface immersed in $I\!\!R^5$. Suppose that there exists a normal field ν which is orthogonal to E_p at every point. Then M is ν-umbilic.

The set of curvature planes $\{E_p\}_{p\in M}$ determine a rank 2 sub-bundle, EM', of $N^1 M$ over the complement M' of the set of semiumbilic points of M. A normal field on M is said to be **essential** if its restriction to M' lies on the sub-bundle EM'. Principal configurations associated to essential fields are called **essential configurations on** M. The concept of essential normal field on a surface immersed in $\mathbb{R}^5$ is introduced in Ref. 22. Such fields are called essential because they provide all the principal configurations on M in the following sense: It follows from Proposition 4.1 that any normal field η on M can be written as a sum of an essential normal field η_1 (essential component) and a totally umbilic normal field η_2 (umbilic component). Clearly, the principal configuration of η coincides with that of its essential part η_1. Moreover, it follows that the critical points of the essential configurations on M are semiumbilics or umbilics.

The Poincaré-Hopf formula leads to the following.

Corollary 4.5.

Let M be a closed connected orientable surface with non vanishing Euler number immersed in $\mathbb{R}^5$. If M admits some globally defined essential normal field then it has semiumbilics (or umbilics).

Since these are points of type $M_i, i < 3$, which in turn are 2-singular, we obtain the following result.

Corollary 4.6.

Let M be a closed connected orientable surface with non vanishing Euler number immersed in $\mathbb{R}^5$. If M admits some globally defined essential normal field then M cannot be 2-regular.

5. Contacts with hyperplanes

Suppose that M is locally given by an embedding $f : \mathbb{R}^2 \to \mathbb{R}^n$. The **family of height functions** associated to f is defined as

$$\lambda(f) : M \times S^{n-1} \longrightarrow \mathbb{R}$$

$$(p, v) \longmapsto \langle f(p), v \rangle = f_v(p).$$

A point $p = f(x) \in M$ is a singular point of f_v if and only if $v \in N_p M$.

The singular subset

$$\Sigma_{\lambda(f)} = \{(p, v) \in M \times S^{n-1} | \frac{\partial f_v}{\partial x} = 0\}$$

can be viewed as the **canal hypersurface** CM of M in $\mathbb{R}^n$. Let $\Gamma : CM \to S^{n-1}$ be its associated Gauss map. A point $p = f(x) \in M$ is a degenerate singularity of f_v if and only if $v \in \Sigma\Gamma$. In such case we say that v is a **degenerate normal direction** for M at p. If f_v has a non degenerate singularity at x, we say that the hyperplane $H_v = v^\perp$ has a non degenerate contact with M at $p = f(x)$. We say that H_v is a **local support hyperplane** for M at p, provided the surface lies locally at p in one of the half-spaces determined by H_v in $\mathbb{R}^n$. That is, $f_v(y) - f_v(x) \geq 0$, for all y lying in some neighborhood of x. Obviously, this is the case when f_v is a non degenerate Morse function of elliptic type at x.

Proposition 5.1 (See Ref. 25). *Given a surface M immersed in $\mathbb{R}^n$ with $n \geq 4$, a point $p \in M$ and a non null vector $\nu \in N_p M$, the quadratic forms $II_v(p)$ and $Hess(f_v)(p)$ are equivalent (up to local coordinate changes in M).*

Consequently, $v \in N_p M$ is a degenerate normal direction if and only if $v \in C_p$.

The corank 1 degenerate singularities of f_v on a surface M generically immersed in $\mathbb{R}^5$ are of type A_k with $k \leq 5$. As for the corank 2 singularities of f_v on a surface generically immersed in $\mathbb{R}^5$, they are of type $D_4^\pm$ along curves on M and of type D_5 at isolated points; see Ref. 19.

The corank 2 singularities of the height function f_v can also be characterized as follows. A point p is a corank 2 singularity of f_v if and only if (p, v) is a corank 2 singularity of Γ if and only if $v \in Ker A_p = Ker\alpha_f(p)$ if and only if $p \notin M_3$.

Theorem 5.1 (Mochida, Romero-Fuster and Ruas[19]). *For an embedding $f : M \to \mathbb{R}^5$, the following conditions are equivalent.*
a) A point $p \in M$ is 2-singular.
b) The point $p \in M$ is a singularity of corank 2 for some height function f_v on M.

6. Contact directions

The kernel of the Hessian quadratic form $Hess(f_v)(p)$ of the height function in a degenerate normal direction $v \in N_p M$ contains non zero vectors. The corresponding tangent directions are called **contact directions** associated to v.

A normal field b on M such that $b(p)$ is a degenerate normal direction at each point $p \in M$ is said to be a **degenerate normal field** on M. If b

determines height functions of corank one at each point of an open subset V of M, then it has an associated contact directions field whose integral lines define a contact foliation on V.

Given a surface M immersed in $\mathbb{R}^n$, a unit vector $v \in N_p M$ is said to be a **binormal direction** for M if and only if f_v has a singularity of type A_{n-2} or worse (i.e. the $\mathcal{A}$-codimension of f_v is at least $n - 3$) at p. For instance, in the case of a surface immersed in $\mathbb{R}^4$, the binormal directions coincide with the degenerate normal directions. They are introduced in,[18] where it is shown that there are at most two normal directions over each point of a surface. Moreover, generic surfaces can be decomposed into an open region of elliptic points over which there are no binormal directions and an open region of hyperbolic points with two binormal fields. The two regions are separated by a regular curve of parabolic points at which there is a unique binormal direction.

For surfaces immersed in $\mathbb{R}^5$, it can be shown that $p \in M_3$ is an $A_{k \geq 3}$ point of f_v if and only if (p, v) is a Morin singularity of type S_{1_k} (^{12}) of the Gauss map Γ.

Binormal fields are a particular case of degenerate normal fields. The corresponding contact direction fields are called **asymptotic** fields and their integrals **asymptotic lines** of M. The critical points of the asymptotic configurations of surfaces immersed in $\mathbb{R}^4$ are inflection points. Their generic behavior in a neighborhood of inflection points of imaginary type was analyzed by Garcia, Mochida, Romero-Fuster and Ruas.[11] A complete analysis of the generic behavior of asymptotic configurations at their critical points, including generic 1-parameter families of immersions of surfaces in $\mathbb{R}^4$ can be found in Ref. 3.

For surfaces immersed in $\mathbb{R}^5$, we have the following result.

Proposition 6.1 (Mochida, Romero-Fuster and Ruas[19]). *A surface generically immersed in $\mathbb{R}^5$ has at least 1 and at most 5 asymptotic directions at each point.*

These directions determine locally defined asymptotic direction fields. Their differential equations as well as their generic properties are described in detail in a forthcoming paper (Ref. 31).

Given a binormal field b on M, it follows from Proposition 4.1 that the matrix of the shape operator S_b at any point $p \in M$ is equivalent to the hessian matrix $\mathrm{Hess}(f_v)(p)$. This implies that one of the b-principal curvatures is zero at every point. Therefore, the corresponding b-principal directions foliation coincides with the asymptotic lines associated to b.

Given any surface M immersed in $I\!R^5$, let ν be a degenerate normal field on M. Then if a point $p \in M$ is ν-umbilic, $S_\nu(X) = \lambda X, \forall X \in T_p M$. Now, the matrix of S_ν coincides with that of II_ν which is equivalent to $Hess(f_{\nu(p)}(p))$. Since ν is degenerate, we must have $\lambda = 0$, and thus $p \in M_i, i < 3$. Therefore, as a consequence of the Poincaré-Hopf formula, we can assert the following.

Theorem 6.1. *A closed oriented connected surface with non vanishing Euler number immersed in $I\!R^5$ that admits some globally defined degenerate field cannot be 2-regular.*

7. Essential convexity

A hypersurface $M \subset I\!R^n$ is convex at some point p provided its Gaussian curvature is non negative at p, or equivalently, it admits some locally support hyperplane at p. This second assertion generalizes easily to the submanifolds immersed in $I\!R^n$ with codimension higher than one. So we say that a surface immersed in $I\!R^n$ with $n > 3$ is (**locally**) **convex** at a point p if it admits some (locally) support hyperplane at p. This can be expressed in terms of height functions as follows. A surface immersed in $I\!R^n$ with $n > 3$ is convex at a point p if there exists some direction $v \in N_p M$ such that $f_v(x) \geq f_v(p)$, for all x in a neighborhood of p. We then say that M is **strictly convex** at p provided there exists $v \in N_p M$, such that f_v is an elliptic Morse function in a neighborhood of p. A surface immersed in $I\!R^4$ is strictly convex at a point p if and only if p is a hyperbolic point, or equivalently, if and only if M admits exactly two asymptotic directions at p ([18]). It follows that convex surfaces in $I\!R^4$ have globally defined asymptotic direction fields. Since the inflection points are their critical points, and these are also the 2-singular points for such surfaces, we can state the following.

Theorem 7.1. *Closed orientable surfaces with non vanishing Euler number that are convexly immersed in $I\!R^4$ cannot be 2-regular.*

One can show that the stereographic projection takes bijectively inflection points of surfaces in S^3 into umbilic points of their images in $I\!R^3$. Therefore, any torus immersed in $I\!R^3$ without umbilics is mapped by the inverse of the stereographic projection onto a 2-regular surface in $I\!R^4$. Such a surface lies in S^3 and is locally convex.

It can be shown that any surface immersed in $I\!R^n$ with $n \geq 5$ is locally strictly convex at any of its points of type M_3. In view of this fact, it was introduced in Ref. 21 the concept of essential convexity for surfaces in $I\!R^5$. A surface M is said to be **essentially convex** at a point p if there is some normal vector $v \in E_p$, such that f_v defines a non degenerate (Morse) elliptic function at p. That is, M admits locally some support hyperplane H_v at p, perpendicular to the hyperplane $T_pM \oplus E_p$, which has a non degenerate contact with M at p. We observe that the restriction of the natural projection $\pi_p : I\!R^5 \to T_pM \oplus E_p$, to M is a local diffeomorphism in a neighborhood of p, and hence $\pi_p(M)$ is a regular surface in 4-space in some small enough neighborhood of $\pi_p(p)$. Then we have the following characterization of essential convexity.

Proposition 7.1 (Moraes and Romero Fuster[21]). *A surface* $M \subset I\!R^5$ *is essentially convex at* p *if and only if* $\pi_p(M)$ *is locally convex at* p *in the 4-space* $T_pM \oplus E_p$.

A surface that is essentially convex at every point is said to be **essentially convex**. The following result provides a connection between the essential convexity and the extrinsic dynamics on surfaces in 5-space.

Proposition 7.2 (Moraes and Romero Fuster[21]). *If* $M \subset I\!R^5$ *is essentially convex, then it admits two (essential) degenerated normal directions at every point.*

Each one of these degenerate directions determines a corank one singularity of the corresponding height functions at each point. Their kernels determine tangent direction fields globally defined on the surface. Then as a consequence of the Poincaré-Hopf formula we arrive to the following result.

Corollary 7.1. *A closed orientable essentially convex surface with non vanishing Euler number cannot be 2-regular in* $I\!R^5$.

8. Contacts with hyperspheres

Given an immersion $f : I\!R^2 \longrightarrow I\!R^n$ of a surface M in $I\!R^n$ with $n \geq 4$ the **family of distance squared functions** on M is given by:

$$\begin{aligned}
\Phi : I\!R^2 \times I\!R^n &\to I\!R \\
(x, a) &\longmapsto d_a(x) = \|a - f(x)\|^2.
\end{aligned}$$

A point $p = f(x) \in M$ is a singular point of the distance squared function d_a if and only if the vector $a - f(x)$ is normal to M at p. Those points $a \in I\!R^n$ for which d_a has a degenerate singularity at $p \in M$ form the subset F_p called the **focal centers** at p. Such points are the centers of all the

focal hyperspheres of M. The degenerate singularities of corank 1 of d_a on a generically immersed surface M are of type A_k with $k \leq 6$. Those of corank 2, are of type $D_k^{\pm}, k = 4, 5, 6$. The **focal set** of M, made of all the focal centers of M, is stratified according to the above classification. The focal centers for which d_a has corank 2 are called **umbilical foci** of M.

J. Montaldi[16] proved that the singularities of corank 2 of distance squared functions on surfaces immersed in $I\!R^4$ are the semiumbilic points of these surfaces. They form, generically, closed regular curves. The case of surfaces immersed in $I\!R^n$ with $n \geq 5$ is treated in Ref. 7, where the following result concerning the distribution of umbilical foci on such surfaces is proven.

Theorem 8.1.

Given a surface M immersed in $I\!R^5$, if a is an umbilical focus for M at p, then $a \in E_p^{\perp}$. Moreover,

(a) If $p \in M_3$ then there is a unique umbilical focus

$$q(p) = p + \frac{1}{\lambda_p(v)}v,$$

where $v \in E_p^{\perp}$ is a unit vector pointing towards the plane $Aff_p \subset N_pM$ and $\lambda_p(v) = d(p, Aff_p)$.

(b) If $p \in M_2$ is non semiumbilic, then the umbilical focus lies at infinity and the corresponding focal hypersphere becomes a hyperplane.

(c) If $p \in M_2$ is a semiumbilic, then there is a straight line of umbilical foci for M at p lying in the plane $E_p^{\perp} \subset N_pM$.

Given $M \subset I\!R^5$ and $p \in M$, we define the **umbilical curvature** $\kappa_u(p)$ of M at p as the distance $d(p, Aff_p)$ of the affine plane determined by the curvature ellipse to the origin p of the normal space N_pM.

It follows from Theorem 7.1 that $\kappa_u(p)$ coincides with the curvature of the unique hypersphere whose contact with M at p is of corank 2.

Chen and Yano[4] proved that a surface M contained in $I\!R^n$ with $n \geq 4$ lies in a hypersphere if and only if M admits a parallel umbilic normal field. In such case, the direction $\nu(p)$ coincides with that of the radius of the hypersphere at the point p and the radius of the hypersphere is $\frac{1}{\kappa_\nu}$. As a consequence of this one can show the following.

Theorem 8.2 (Costa, Moraes and Romero-Fuster[7]). *Let M be a surface immersed in $I\!R^5$ and suppose that it admits a unit normal field*

ν such that

$$\nu(p) \in E_p^{\perp} \cap N_p^1 M, \ \forall p \in M.$$

Then, provided ν is parallel, the umbilic curvature of M is a constant function and M lies in a 4-sphere of radius $\frac{1}{\kappa_u}$.

It follows from Lemma 2.2 that if $\kappa_u(p) = 0$ then $p \notin M_3$. We obtain the following result.

Theorem 8.3. *The umbilical curvature of a 2-regular surface immersed in $I\!R^5$ never vanishes.*

9. 2-regularity for surfaces in S^4

Corollary 4.3 shows that closed orientable surfaces immersed in $I\!R^4$ have semiumbilics (or umbilics). Semiumbilic and umbilic points of a surface M immersed in $I\!R^4$ are the corank 2 singularities of the distance squared functions ([16]). The stereographic projection,

$$\xi : S^4 - \{P\} \to I\!R^4,$$

transforms bijectively corank 2 singularities of height functions on surfaces in $S^4 \subset I\!R^5$ into corank 2 singularities of distance squared functions on their images into $I\!R^4$ (see Ref. 29). However, the 2-singular points of surfaces immersed in S^4 are either semiumbilic or umbilic (see Ref. 30). Therefore we have the following result.

Theorem 9.1. *No orientable closed 2-regular surface of $I\!R^5$ may be contained in S^4.*

Surfaces contained in S^4 do not need to be essentially convex. In fact, if $M \subset S^4 \subset I\!R^5$ is essentially convex, there exists $v \in E_p$ such that the hyperplane $H_v = v^{\perp}$ locally supports M at p in $I\!R^5$. Now, it is shown in Ref. 30 that $T_p S^4 = T_p M \oplus E_p$. So H_v contains the radial direction and thus passes through the center of S^4. Therefore, $S_p^3 = H_v \cap S^4$ is a 3-sphere of maximal radius in S^4 that locally supports M at p. This means that M is locally convex at p as a submanifold of the 4-sphere.

Consider the restriction to the 2-sphere of the map

$$V : \quad I\!R^3 \quad \longrightarrow \quad I\!R^6$$

$$(x, y, z) \longmapsto (x^2, y^2, z^2, xy, xz, yz).$$

It is not difficult to see that the surface $V(S^2)$, being contained in both a 5-sphere and a hyperplane of $I\!R^6$ is contained in a 4-sphere. Due to the antipodal symmetry of V, its restriction to S^2 defines a 2-regular immersion of the real projective plane into the 4-sphere. Its image is known as the **Veronese surface** and is locally given by

$$\tilde{V}(x,y) = \left(\frac{y\sqrt{4 - x^2 - y^2}}{2}, \frac{x\sqrt{4 - x^2 - y^2}}{2}, \frac{xy}{2}, \frac{x^2 - y^2}{4}, \frac{3x^2 + 3y^2 - 8}{4\sqrt{3}} \right).$$

The curvature ellipse is a circle at every point of $V(S^2)$. One can show that all the height functions on $V(S^2)$ have infinite codimension, and it is therefore a very degenerate immersion from the viewpoint of contacts with hyperplanes. Since the subspace of 2-regular immersions of any surface in $\mathbb{R}^5$ is open (in the Whitney C^∞-topology), there exists a 2-regular immersion (close enough to a $V(S^2)$) in $\mathbb{R}^5$ whose family of height functions is structurally stable. Clearly, such image cannot lie in S^4. Moreover, such 2-regular embeddings of S^2 into $\mathbb{R}^5$ do not admit globally defined essential normal fields, nor degenerate normal fields.

10. Isometric reduction of the codimension and 2-regularity of surfaces in $\mathbb{R}^n$

J. Nash[23,24] proved that any surface with a given riemannian metric can be isometrically immersed as submanifold of $\mathbb{R}^n$, for some n. Now, any isometric immersion f of a surface M into $\mathbb{R}^n$ induces a second fundamental form and a family of principal configurations on M, and the subspace

$$N_p^1 M = (Ker(\alpha_f(p)))^\perp \subseteq N_p M$$

contains all the relevant informations on the second fundamental form $\alpha_f(p)$, at $p \in M$. Since the dimension of $N_p^1 M$ is at most 3, it is natural to formulate the following question.

Question 10.1. When is it possible to isometrically immerse M in $\mathbb{R}^5$ in such a way that the relevant part of the second fundamental form α_f, and therefore, the whole family of principal configurations induced by f is preserved?

In such case, we say that *the codimension of $f(M)$ can be isometrically reduced to 3.*

An answer to the above question is the following.

Theorem 10.1 (Romero-Fuster and Sánchez-Bringas[32]). *Let M be a simply connected surface immersed in $\mathbb{R}^n, n > 5$ and suppose that $N^1 M$*

has constant rank $r \leq 3$. Then the codimension of M can be isometrically reduced to r.

It follows from Proposition 2.4 that a surface M is 2-regular in $\mathbb{R}^n$ in $n > 5$ if and only if $M = M_3$, which implies that $N^1 M$ has constant rank $r = 3$.

Corollary 10.1. *The codimension of any simply connected 2-regular surface immersed in $\mathbb{R}^n$ with $n > 5$ can be isometrically reduced to 3.*

We observe in Lemma 2.2 that if the umbilical curvature of M vanishes identically, every point of M is either a non semiumbilic M_2-point (dim $N_p^1 M = 2$), an inflection point (dim $N_p^1 M = 1$), or a flat umbilic point (dim $N_p^1 M = 0$). So, simply connected surfaces with vanishing umbilical curvature whose first normal space has constant rank admit isometric immersions that preserve their second fundamental form into $\mathbb{R}^4$ (provided $M = M_2$), or $\mathbb{R}^3$ (provided $M = M_1$).

The above results allow us to expect that there is a reasonably large subset of 2-regular immersions of the 2-sphere in $\mathbb{R}^5$. Nevertheless, nothing is known so far about the existence of some 2-regular embedding. The existence of 2-regular immersions of orientable surfaces with non vanishing genus (i.e. non simply connected) into $\mathbb{R}^5$ is an open problem too.

Finally, an interesting question to be considered is whether the h-principle for 2-regular immersions holds: *Is any immersion of S^2 into $\mathbb{R}^5$ regularly homotopic to some 2-regular immersion?* The corresponding problem for 2-regular immersions into $\mathbb{R}^6$ is considered in Ref. 13.

Acknowledgements

This work has been partially supported by DGCYT grant no. BFM2003-0203.

References

1. V.I. Arnold, S.M. Gusein-Zade, A.N. Varchenko, *Singularities of differentiable maps.* Birkhäuser, Boston-Basel-Stuttgart (1985).
2. J.W. Bruce and P.J. Giblin, *Curves and Singularities.* Cambridge University Press, 2nd Edition (1991).
   \bibitem{B-N} J.W. Bruce and A.C. Nogueira, Surfaces in $\mathbb{R}^4$ and duality. *Quart. J. Oxford.*(2) 49 (1998), 433-443.
3. J.W. Bruce and F. Tari, Families of surfaces in $\mathbb{R}^4$. *Proc. Edinb. Math. Soc.* (2) 45 (2002), 181-203.

4. B. Y. Chen and K. Yano, Integral formulas for submanifolds and their applications. *J. Differential Geometry* 5 (1971), 467-477.

5. S.I.R. Costa, *Aplicações não singulares de ordem p*. Doctoral Thesis, University of Campinas, 1982.

6. S.I.R. Costa, On closed twisted curves. *Proc. Amer. Math. Soc.* 1098(1) (1990), 205-214.

7. S.I.R. Costa, S. Moraes and M. C. Romero Fuster, Curvature ellipses and geometric contacts of surfaces immersed in $\mathbb{R}^n$, $n \geq 5$. Preprint (2005).

8. E.A. Feldman, Geometry of immersions I. *Trans. AMS* 120 (1965), 185-224.

9. E.A. Feldman, Geometry of immersions II. *Trans. AMS* 125 (1966), 181-315.

10. E.A. Feldman, On parabolic and umbilic points of immersed surfaces, *Trans. AMS* 127 (1967), 1-28.

11. R A. Garcia, D.K.H. Mochida, M.C. Romero-Fuster and M.A.S. Ruas, Inflection Points and Topology of Surfaces in 4-space, *Trans. AMS*. 352 (2000), 3029-3043.

12. M. Golubitsky and V. Guillemin, *Stable Mappings and Their Singularities*. Grad. Texts in Maths 14, Springer-Verlag (1973).

13. M. Gromov and Y. Eliashberg, Removal of singularities of smooth maps. *Izv. Akad. Nauk SSS Ser. Mat.* 35 (1971), 600-627.

14. J.A. Little, On singularities of submanifolds of higher dimensional euclidean space. *Annali Mat. Pura et Appl.*, (ser. 4A) 83 (1969), 261-336.

15. E.J.N. Looijenga, *Structural stability of smooth families of C^∞-functions*, Ph.D. Thesis, University of Amsterdam, 1974.

16. J.A. Montaldi, *Contact with applications to submanifolds of $\mathbb{R}^n$*. Ph.D. Thesis, University of Liverpool, 1983.

17. J.A. Montaldi, On contact between submanifolds. *Michigan Math. J.* 33 (1986), 195-199.

18. D.K.H. Mochida, M.C. Romero-Fuster and M.A. Ruas, The Geometry of surfaces in 4-space from a contact viewpoint. *Geom. Dedicata* 54, (1995), 323-332.

19. D.K.H. Mochida, M.C. Romero-Fuster and M.A.S. Ruas, Inflection points and nonsingular embeddings of surfaces in $\mathbb{R}^5$. *Rocky Mountain J. Maths.* 33 (2003), 995-1009.

20. S.M. Moraes and M.C. Romero-Fuster, Semiumbilics and normal fields on surfaces immersed in $\mathbb{R}^n, n > 3$. *Rocky Mountain J. Maths* (2005).

21. S.M. Moraes and M.C. Romero-Fuster, Convexity and semiumbilicity for surfaces in $\mathbb{R}^5$. *Differential geometry, Valencia, 2001*, World Sci. Publishing, River Edge, NJ (2002), 222-234.

22. S.M. Moraes, M.C. Romero-Fuster and F. Sánchez-Bringas, Principal configurations and umbilicity of submanifolds in $\mathbb{R}^N$. *Bull. Belg. Math Soc-Simon Stevin* 11, vol 2(2004), 227-245.

23. J. Nash, C^1 isometric imbeddings. *Ann. of Math.* (2) 60 (1954), 38-396.

24. J. Nash, The imbedding problem for Riemannian manifolds. *Ann. of Math.* (2) 63 (1956), 20-63.

25. R. S. Palais and C-L. Terng, *Critical Point Theory and Submanifolds Geometry*, Lecture Notes in Math 1353, Springer-Verlag (1988).

26. W. Pohl, Differential geometry of higher order. *Topology* 1 (1962), 169-211.

27. A. Ramírez-Galarza and F. Sánchez-Bringas, Lines of Curvature near Umbilical Points on Surfaces Immersed in $I\!R^4$, *Annals of Global Analysis and Geometry* 13 (1995), 129-140.

28. M.C. Romero-Fuster, Convexly generic curves in $I\!R^3$. *Geom. Dedicata* 28 (1988), 7-29.

29. M.C. Romero-Fuster, Stereographic Projections and Geometric Singularities. *Matemática Contemporânea* 12 (1997), 167-182.

30. M.C. Romero-Fuster, Semiumbilics and geometrical dynamics on surfaces in 4-spaces. Real and complex singularities, *Contemp. Math.*, 354, Amer. Math. Soc., Providence, RI (2004), 259-276.

31. M.C. Romero-Fuster, M.A.S. Ruas and F. Tari, Asymptotic curves on surfaces in $I\!R^5$. Preprint (2006).

32. M.C. Romero-Fuster and F. Sánchez-Bringas, Isometric reduction of the codimension and 2nd order non degeneracy of submanifolds. Preprint (2005).

33. V.D. Sedykh, Four vertices of a convex space curve. *Bull. London Math. Soc.* 26 (1994), 177-180.

Geometry of resonance tongues

Henk W. Broer

Institute of Mathematics and Computing Science, University of Groningen. P.O. Box 800, 9700 AV Groningen, The Netherlands

Martin Golubitsky

Department of Mathematics, University of Houston. Houston, TX 77204-3476, USA. The work of MG was supported in part by NSF Grant DMS-0244529

Gert Vegter

Institute of Mathematics and Computing Science, University of Groningen. P.O. Box 800, 9700 AV Groningen, The Netherlands

1. Introduction

Resonance tongues arise in bifurcations of discrete or continuous dynamical systems undergoing bifurcations of a fixed point or an equilibrium satisfying certain resonance conditions. They occur in several different contexts, depending, for example, on whether the dynamics is dissipative, conservative, or reversible. Generally, resonance tongues are domains in parameter space, with periodic dynamics of a specified type (regarding period of rotation number, stability, etc.). In each case, the tongue boundaries are part of the bifurcation set. We mention here several standard ways that resonance tongues appear.

1.1. *Various contexts*

Hopf bifurcation from a fixed point. Resonance tongues can be obtained by Hopf bifurcation from a fixed point of a map. This is the context of Section 2, which is based on Broer, Golubitsky, Vegter.[7] More precisely, Hopf bifurcations of maps occur when eigenvalues of the Jacobian of the map at a fixed point cross the complex unit circle away from the strong resonance points $e^{2\pi pi/q}$ with $q \leq 4$. Instead, we concentrate on the *weak* *resonance points* corresponding to roots of unity $e^{2\pi pi/q}$, where p and q are coprime integers with $q \geq 5$ and $|p| < q$. *Resonance tongues* themselves are regions in parameter space near the point of Hopf bifurcation where periodic points of period q exist and *tongue boundaries* consist of points in parameter space where the q-periodic points disappear, typically in a saddle-node bifurcation. We assume, as is usually done, that the critical eigenvalues are simple with no other eigenvalues on the unit circle. Moreover, usually just two parameters are varied; The effect of changing these parameters is to move the eigenvalues about an open region of the complex plane. In the non-degenerate case a *pair of q-periodic orbits* arises or disappears as a single complex parameter governing the system crosses the boundary of a resonance tongue. Outside the tongue there are no q-periodic orbits. In the degenerate case there are two complex parameters controlling the evolution of the system. Certain domains of complex parameter space correspond to the existence of zero, two or even *four q-periodic orbits.*

Lyapunov-Schmidt reduction is the first main tool used in Section 2 to reduce the study of q-periodic orbits in families of planar diffeomorphisms to the analysis of zero sets of families of $\mathbb{Z}_q$-*equivariant* functions on the plane. Equivariant Singularity Theory, in particular the theory of equivariant contact equivalence, is used to bring such families into low-degree polynomial normal form, depending on one or two complex parameters. The discriminant set of such polynomial families corresponds to the resonance tongues associated with the existence of q-periodic orbits in the original family of planar diffeomorphisms.

Hopf bifurcation and birth of subharmonics in forced oscillators
Let

$$\frac{dX}{dt} = F(X)$$

be an autonomous system of differential equations with a periodic solution $Y(t)$ having its Poincaré map P centered at $Y(0) = Y_0$. For simplicity we take $Y_0 = 0$, so $P(0) = 0$. A Hopf bifurcation occurs when eigenvalues of the Jacobian matrix $(dP)_0$ are on the unit circle and resonance occurs when these eigenvalues are roots of unity $e^{2\pi pi/q}$. *Strong resonances* occur when $q < 5$. This is one of the contexts we present in Section 3. Except at strong resonances, Hopf bifurcation leads to the existence of an invariant circle for the Poincaré map and an invariant torus for the autonomous system. This is usually called a Naimark-Sacker bifurcation. At weak resonance points the flow on the torus has very thin regions in parameter space (between the tongue boundaries) where this flow consists of a *phase-locked* periodic solution that winds around the torus q times in one direction (the direction approximated by the original periodic solution) and p times in the other. Section 3.1 presents a Normal Form Algorithm for continuous vector fields, based on the method of Lie series. This algorithm is applied in Section 3.2 to obtain the results summarized in this paragraph. In particular, the analysis of the Hopf normal form reveals the birth or death of an invariant circle in a non-degenerate Hopf bifurcation.

Related phenomena can be observed in periodically forced oscillators. Let

$$\frac{dX}{dt} = F(X) + G(t)$$

be a periodically forced system of differential equations with 2π-periodic forcing $G(t)$. Suppose that the autonomous system has a hyperbolic equilibrium at $Y_0 = 0$; That is, $F(0) = 0$. Then the forced system has a 2π-periodic solution $Y(t)$ with initial condition $Y(0) = Y_0$ near 0. The dynamics of the forced system near the point Y_0 is studied using the *stroboscopic map* P that maps the point X_0 to the point $X(2\pi)$, where $X(t)$ is the solution to the forced system with initial condition $X(0) = X_0$. Note that $P(0) = 0$ in coordinates centered at Y_0. Again resonance can occur as a parameter is varied when the stroboscopic map undergoes Hopf bifurcation with critical eigenvalues equal to roots of unity. Resonance tongues correspond to regions in parameter space near the resonance point where the stroboscopic map has q-periodic trajectories near 0. These q-periodic trajectories are often called *subharmonics of order q*. Section 3.2 presents a Normal Form Algorithm for such periodic systems. The Van der Pol transformation is a tool for reducing the analysis of subharmonics of order q to the study of zero sets of $\mathbb{Z}_q$-equivariant polynomials. In this way we obtain the $\mathbb{Z}_q$-equivariant Takens Normal Form[37] of the Poincaré time-2π-map of the system. After this transformation, the final analysis of the resonance tongues corresponding to the birth or death of these subharmonics bears strong resemblance to the approach of Section 2.

Coupled cell systems. Finally, in Section 4 we report on work in progress by presenting a case study, focusing on a feed-forward network of three coupled cells. Each cell satisfies the same dynamic law, only different choices of initial conditions may lead to different kinds of dynamics for each cell. To tackle such systems, we show how a certain class of dynamic laws may give rise to time-evolutions that are equilibria in cell 1, periodic in cells 2,

330

and exhibting the Hopf-Neĭmark-Sacker phenomenon in cell 3. This kind
of dynamics of the third cell, which is revealed by applying the theory
Normal Form theory presented in Section 3, occurs despite the fact that
the dynamic law of each individual cell is simple, and identical for each cell.

1.2. *Methodology: generic versus concrete systems.*

Analyzing bifurcations in generic families of systems requires different tools
than analyzing a concrete family of systems and its bifurcations. Furthermore, if we are only interested in restricted aspects of the dynamics, like
the emergence of periodic orbits near fixed points of maps, simpler methods might do than in situations where we are looking for complete dynamic
information, like normal linear behavior (stability), or the coexistence of
periodic, quasi-periodic and chaotic dynamics near a Hopf-Neĭmark-Sacker
bifurcation. In general, the more demanding context requires more powerful
tools.

This paper illustrates this 'paradigm' in the context of local bifurcations
of vector fields and maps, corresponding to the occurrence of degenerate
equilibria or fixed points for certain values of the parameter. These degenerate features are encoded by a semi-algebraic stratification of the space of
jets (of some fixed order) of local vector fields or maps. Ideally, each stratum is represented by a 'simple model', or *normal form*, to which all other
systems in the stratum can be reduced by a coordinate transformation (or,
normal form transformation). These 'simple models' are usually low-degree
polynomial systems, equivalent to either the full system or some jet of sufficiently high order. Moreover, generic unfoldings of such degenerate systems
also have simple polynomial normal forms. The guiding idea is that the
interesting features of the system are much more easily extracted from the
normal form than from the original system.

Singularity Theory provides us with algebraic algorithms that compute
such simple polynomial models for *generic (families of) functions*. In Section 2 we apply Equivariant Singularity Theory in this way to determine
resonance tongues corresponding to bifurcations of periodic orbits from
fixed points of maps. However, before Singularity Theory can be applied
we have to use the Lyapunov-Schmidt method to reduce the study of bi-
furcating periodic orbits to the analysis of zero sets of equivariant families
of functions on the plane. In this reduction we loose all other information
on the dynamics of the system.

To overcome this restriction, we apply Normal Form algorithms in the
context of flows[4,36] yielding simple models of generic families of vector

fields (possibly up to terms of high order), without first reducing the system according to the Lyapunov-Schmidt approach. Therefore, all dynamic information is present in the normal form. We follow this approach to study the Hopf-Neĭmark-Sacker phenomenon in *concrete systems*, like the feed-forward network of coupled cells.

1.3. *Related work*

The geometric complexity of resonance domains has been the subject of many studies of various scopes. Some of these, like the present paper, deal with quite universal problems while others restrict to interesting examples. As opposed to this paper, often normal form theory is used to obtain information about the nonlinear dynamics. In the present context the normal forms automatically are $\mathbb{Z}_q$-equivariant.

Chenciner's degenerate Hopf bifurcation. Chenciner[17–19] considers a 2-parameter unfolding of a degenerate Hopf bifurcation. Strong resonances to some finite order are excluded in the 'rotation number' ω_0 at the central fixed point. Chenciner[19] studies corresponding periodic points for sequences of 'good' rationals p_n/q_n tending to ω_0, with the help of $\mathbb{Z}_{q_n}$-equivariant normal form theory. For a further discussion of the codimension k Hopf bifurcation compare Broer and Roussarie.[12]

The geometric program of Peckam *et al.* The research program reflected in Peckam *et al.*[30,31,33,35] views resonance 'tongues' as projections on a 'traditional' parameter plane of (saddle-node) bifurcation sets in the product of parameter and phase space. This approach has the same spirit as ours and many interesting geometric properties of 'resonance tongues' are discovered and explained in this way. We note that the earlier result Peckam and Kevrekidis[34] on higher order degeneracies in a period-doubling uses $\mathbb{Z}_2$ equivariant singularity theory.

Particularly we like to mention the results of Peckam and Kevrekidis[35] concerning a class of oscillators with doubly periodic forcing. It turns out that these systems can have coexistence of periodic attractors (of the same period), giving rise to 'secondary' saddle-node lines, sometimes enclosing a flame-like shape. In the present, more universal, approach we find similar complications of traditional resonance tongues, compare Figure 1 and its explanation in Section 2.4.

Related work by Broer *et al.* Broer *et al.*[15] an even smaller universe of annulus maps is considered, with Arnold's family of circle maps as a limit. Here 'secondary' phenomena are found that are similar to the ones discussed presently. Indeed, apart from extra saddle-node curves inside tongues also many other bifurcation curves are detected.

We like to mention related results in the reversible and symplectic settings regarding parametric resonance with periodic and quasi-periodic forcing terms by Afsharnejad[1] and Broer *et al.*[5,6,8–11,13,14,16] Here the methods use Floquet theory, obtained by averaging, as a function of the parameters.

Singularity theory (with left-right equivalences) is used in various ways. First of all it helps to understand the complexity of resonance tongues in the stability diagram. It turns out that crossing tongue boundaries, which may give rise to instability pockets, are related to Whitney folds as these occur in 2D maps. These problems already occur in the linearized case of Hill's equation. A question is whether these phenomena can be recovered by methods as developed in the present paper. Finally, in the nonlinear cases, application of $\mathbb{Z}_2$- and $\mathbb{D}_2$-equivariant singularity theory helps to get dynamical information on normal forms.

2. Bifurcation of periodic points of planar diffeomorphisms

2.1. *Background and sketch of results*

The types of resonances mentioned here have been much studied; we refer to Takens,[37] Newhouse, Palis, and Takens,[32] Arnold[2] and references therein. For more recent work on strong resonance, see Krauskopf.[29] In general, these works study the complete dynamics near resonance, not just the shape of resonance tongues and their boundaries. Similar remarks can be made on studies in Hamiltonian or reversible contexts, such as Broer and Vegter[16] or Vanderbauwhede.[39] Like in our paper, in many of these references some form of singularity theory is used as a tool.

The problem we address is how to find resonance tongues in the general setting, without being concerned by stability, further bifurcation and similar dynamical issues. It turns out that contact equivalence in the presence of $\mathbb{Z}_q$ symmetry is an appropriate tool for this, when first a Liapunov-Schmidt reduction is utilized, see Golubitsky, Schaeffer, and Stewart.[23,25] The main question asks for the number of q-periodic solutions as a function of parameters, and each tongue boundary marks a change in this number. In the next subsection we briefly describe how this reduction process works. Using equivariant singularity theory we arrive at equivariant normal forms for the reduced system in Section 2.3. It turns out that the standard, nondegen-

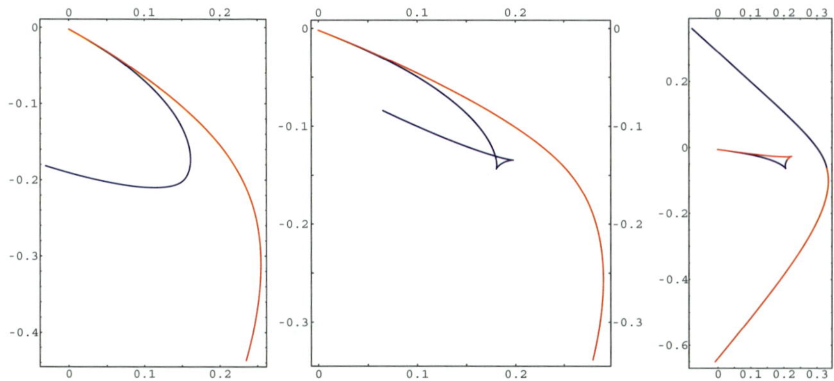

Fig. 1.   Resonance tongues with pocket- or flame-like phenonmena near a degenerate Hopf bifurcation through $e^{2\pi i p/q}$ in a family depending on two complex parameters. Fixing one of these parameters at various (three) values yields a family depending on one complex parameter, with resonance tongues contained in the plane of this second parameter. As the first parameter changes, these tongue boundaries exhibit cusps (middle picture), and even become disconnected (rightmost picture). The small triangle in the rightmost picture encloses the region of parameter values for which the system has four q-periodic orbits.

erate cases of Hopf bifurcation[2,37] can be easily recovered by this method. When $q \geq 7$ we are able to treat a degenerate case, where the third order terms in the reduced equations, the 'Hopf coefficients', vanish. We find pocket- or flame-like regions of four q-periodic orbits in addition to the regions with only zero or two, compare Figure 1. In addition, the tongue boundaries contain new cusp points and in certain cases the tongue region is blunter than in the nondegenerate case. These results are described in detail in Section 2.4.

2.2. *Reduction to an equivariant bifurcation problem*

Our method for finding resonance tongues — and tongue boundaries — proceeds as follows. Find the region in parameter space corresponding to points where the map P has a q-periodic orbit; that is, solve the equation $P^q(x) = x$. Using a method due to Vanderbauwhede (see[39,40]), we can solve for such orbits by Liapunov-Schmidt reduction. More precisely, a q-periodic orbit consists of q points $x_1, \ldots, x_q$ where

$$P(x_1) = x_2, \ldots, P(x_{q-1}) = x_q, P(x_q) = x_1.$$

334

Such periodic trajectories are just zeroes of the map

$$\widehat{P}(x_1, \ldots, x_q) = (P(x_1) - x_2, \ldots, P(x_q) - x_1).$$

Note that $\widehat{P}(0) = 0$, and that we can find all zeroes of $\widehat{P}$ near the resonance point by solving the equation $\widehat{P}(x) = 0$ by Liapunov-Schmidt reduction. Note also that the map $\widehat{P}$ has $\mathbb{Z}_q$ symmetry. More precisely, define

$$\sigma(x_1, \ldots, x_q) = (x_2, \ldots, x_q, x_1).$$

Then observe that

$$\widehat{P}\sigma = \sigma\widehat{P}.$$

At 0, the Jacobian matrix of $\widehat{P}$ has the block form

$$J = \begin{pmatrix} A & -I & 0 & 0 & \cdots & 0 & 0 \\ 0 & A & -I & 0 & \cdots & 0 & 0 \\ & & & \vdots & & & \\ 0 & 0 & 0 & 0 & \cdots & A & -I \\ -I & 0 & 0 & 0 & \cdots & 0 & A \end{pmatrix}$$

where $A = (dP)_0$. The matrix J automatically commutes with the symmetry σ and hence J can be block diagonalized using the isotypic components of irreducible representations of $\mathbb{Z}_q$. (An *isotypic component* is the sum of the $\mathbb{Z}_q$ isomorphic representations. See[25] for details. In this instance all calculations can be done explicitly and in a straightforward manner.) Over the complex numbers it is possible to write these irreducible representations explicitly. Let ω be a qth root of unity. Define V_ω to be the subspace consisting of vectors

$$[x]_\omega = \begin{pmatrix} x \\ \omega x \\ \vdots \\ \omega^{q-1} x \end{pmatrix}.$$

A short calculation shows that

$$J[x]_\omega = [(A - \omega I)x]_\omega.$$

Thus J has zero eigenvalues precisely when A has q^{th} roots of unity as eigenvalues. By assumption, A has just one such pair of complex conjugate q^{th} roots of unity as eigenvalues.

Since the kernel of J is two-dimensional — by the simple eigenvalue assumption in the Hopf bifurcation — it follows using Liapunov-Schmidt reduction that solving the equation $\widehat{P}(x) = 0$ near a resonance point is equivalent to finding the zeros of a reduced map from $\mathbb{R}^2 \to \mathbb{R}^2$. We can, however, naturally identify $\mathbb{R}^2$ with $\mathbb{C}$, which we do. Thus we need to find the zeros of a smooth implicitly defined function

$$g : \mathbb{C} \to \mathbb{C},$$

where $g(0) = 0$ and $(dg)_0 = 0$. Moreover, assuming that the Liapunov-Schmidt reduction is done to respect symmetry, the reduced map g commutes with the action of σ on the critical eigenspace. More precisely, let ω be the critical resonant eigenvalue of $(dP)_0$; then

$$g(\omega z) = \omega g(z). \tag{1}$$

Since p and q are coprime, ω generates the group $\mathbb{Z}_q$ consisting of all q^{th} roots of unity. So g is $\mathbb{Z}_q$-equivariant.

We propose to use $\mathbb{Z}_q$-equivariant singularity theory to classify resonance tongues and tongue boundaries.

2.3. $\mathbb{Z}_q$ singularity theory

In this section we develop normal forms for the simplest singularities of $\mathbb{Z}_q$-equivariant maps g of the form (1). To do this, we need to describe the form of $\mathbb{Z}_q$-equivariant maps, contact equivalence, and finally the normal forms.

The structure of $\mathbb{Z}_q$-equivariant maps. We begin by determining a unique form for the general $\mathbb{Z}_q$-equivariant polynomial mapping. By Schwarz's theorem[25] this representation is also valid for C^∞ germs.

Lemma 2.1. *Every $\mathbb{Z}_q$-equivariant polynomial map $g : \mathbb{C} \to \mathbb{C}$ has the form*

$$g(z) = K(u, v)z + L(u, v)\overline{z}^{q-1},$$

where $u = z\overline{z}$, $v = z^q + \overline{z}^q$, and K, L are uniquely defined complex-valued function germs.

$\mathbb{Z}_q$ contact equivalences. Singularity theory approaches the study of zeros of a mapping near a singularity by implementing coordinate changes that transform the mapping to a 'simple' normal form and then solving the normal form equation. The kinds of transformations that preserve the zeros of a mapping are called contact equivalences. More precisely, two $\mathbb{Z}_q$-equivariant germs g and h are $\mathbb{Z}_q$-*contact equivalent* if

$$h(z) = S(z)g(Z(z)),$$

where $Z(z)$ is a $\mathbb{Z}_q$-equivariant change of coordinates and $S(z) : \mathbb{C} \to \mathbb{C}$ is a real linear map for each z that satisfies

$$S(\gamma z)\gamma = \gamma S(z)$$

for all $\gamma \in \mathbb{Z}_q$.

Normal form theorems. In this section we consider two classes of normal forms — the codimension two standard for resonant Hopf bifurcation and one more degenerate singularity that has a degeneracy at cubic order. These singularities all satisfy the nondegeneracy condition $L(0,0) \neq 0$; we explore this case first.

Theorem 2.1. *Suppose that*

$$h(z) = K(u,v)z + L(u,v)\overline{z}^{q-1}$$

where $K(0,0) = 0$.

(1) Let $q \geq 5$. If $K_u L(0,0) \neq 0$, then h is $\mathbb{Z}_q$ contact equivalent to

$$g(z) = |z|^2 z + \overline{z}^{q-1}$$

with universal unfolding

$$G(z,\sigma) = (\sigma + |z|^2)z + \overline{z}^{q-1}. \tag{2}$$

(2) Let $q \geq 7$. If $K_u(0,0) = 0$ and $K_{uu}(0,0)L(0,0) \neq 0$, then h is $\mathbb{Z}_q$ contact equivalent to

$$g(z) = |z|^4 z + \overline{z}^{q-1}$$

with universal unfolding

$$G(z,\sigma,\tau) = (\sigma + \tau|z|^2 + |z|^4)z + \overline{z}^{q-1}, \tag{3}$$

where $\sigma, \tau \in \mathbb{C}$.

Remark. Normal forms for the cases $q = 3$ and $q = 4$ are slightly different. See[7] for details.

2.4. *Resonance domains*

We now compute boundaries of resonance domains corresponding to universal unfoldings of the form

$$G(z) = b(u)z + \overline{z}^{q-1}. \tag{4}$$

By definition, the *tongue boundary* is the set of parameter values where local bifurcations in the number of period q points take place; and, typically, such bifurcations will be saddle-node bifurcations. For universal unfoldings of the simplest singularities the boundaries of these parameter domains have been called tongues, since the domains have the shape of a tongue, with its tip at the resonance point. Below we show that our method easily recovers resonance tongues in the standard least degenerate cases. Then, we study a more degenerate singularity and show that the usual description of tongues needs to be broadened.

Tongue boundaries of a $p : q$ resonance are determined by the following system

$$\begin{aligned}
\overline{z}G &= 0 \\
\det(dG) &= 0.
\end{aligned} \tag{5}$$

This follows from the fact that local bifurcations of the period q orbits occur at parameter values where the system $G = 0$ has a singularity, that is, where the rank of dG is less than two. Recalling that

$$u = z\overline{z} \quad v = z^q + \overline{z}^q \quad w = i(z^q - \overline{z}^q),$$

we prove the following theorem, which is independent of the form of $b(u)$.

Theorem 2.2. *For universal unfoldings (4), equations (5) have the form*

$$\begin{aligned}
|b|^2 &= u^{q-2} \\
b\overline{b}' + \overline{b}b' &= (q - 2)u^{q-3}
\end{aligned}$$

To begin, we discuss weak resonances $q \geq 5$ in the nondegenerate case corresponding to the situation of Theorem 2.1, part 1. where a $\frac{q}{2} - 1$ cusp forms the tongue-tip and where the concept of resonance tongue remains unchallenged. Using Theorem 2.2, we recover several classical results on the geometry of resonance tongues in the present context of Hopf bifurcation. Note that similar tongues are found in the Arnold family of circle maps,[2] also compare Broer, Simó and Tatjer.[15]

We find some new phenomena in the case of weak resonances $q \geq 7$ in the mildly degenerate case corresponding to the situation of Theorem 2.1, part 2. Here we find 'pockets' in parameter space corresponding to the occurrence of *four* period-q orbits.

The nondegenerate singularity when $q \geq 5$. We first investigate the nondegenerate case $q \geq 5$ given in 2. Here

$$b(u) = \sigma + u$$

where $\sigma = \mu + i\nu$. We shall compute the tongue boundaries in the (μ, ν)-plane in the parametric form $\mu = \mu(u), \nu = \nu(u)$, where $u \geq 0$ is a local real parameter. Short computations show that

$$|b|^2 = (\mu + u)^2 + \nu^2$$
$$b\bar{b}' + \bar{b}b' = 2(\mu + u).$$

Then Theorem 2.2 gives us the following parametric representation of the tongue boundaries:

$$\mu = -u + \frac{q-2}{2} u^{q-3}$$
$$\nu^2 = u^{q-2} - \frac{(q-2)^2}{4} u^{2(q-3)}.$$

In this case the tongue boundaries at $(\mu, \nu) = (0,0)$ meet in the familiar $\frac{q-2}{2}$ cusp

$$\nu^2 \approx (-\mu)^{q-2}.$$

See also Figure 2. It is to this and similar situations that the usual notion

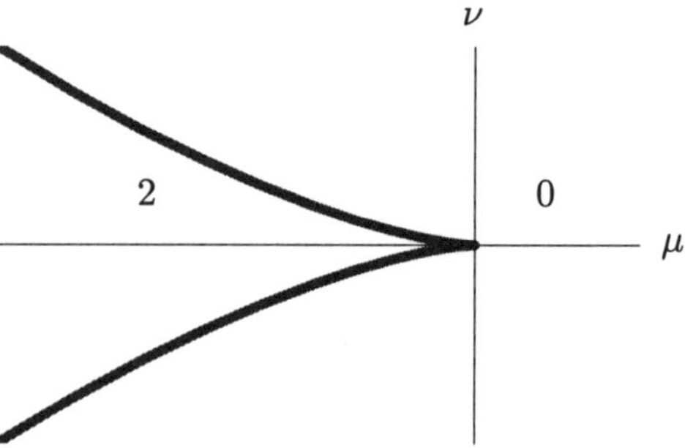

Fig. 2.   Resonance tongue in the parameter plane. A pair of q-periodic orbits occurs for parameter values inside the tongue.

of resonance tongue applies: inside the sharp tongue a pair of period q orbits exists and these orbits disappear in a saddle-node bifurcation at the boundary.

Tongue boundaries in the degenerate case. The next step is to analyze a more degenerate case, namely, the singularity

$$g(z) = u^2 z + \overline{z}^{q-1},$$

w hen $q \geq 7$. We recall from 3 that a universal unfolding of g is given by $G(z) = b(u)z + \overline{z}^{q-1}$, where

$$b(u) = \sigma + \tau u + u^2.$$

Here σ and τ are complex parameters, which leads to a real 4-dimensional parameter space. As before, we set $\sigma = \mu + i\nu$ and consider how the tongue boundaries in the (μ, ν)-plane depend on the complex parameter τ. Broer, Golubitsky and Vegter[7] find an explicit parametric representation of the tongue boundaries in (σ, τ)-space for $q = 7$. Cross-sections of these resonance tongues of the form $\tau = \tau_0$ are depicted in Figure 1 for several constant values of τ_0. A new complication occurs in the tongue boundaries for certain τ, namely, cusp bifurcations occur at isolated points of the fold (saddle-node) lines. The interplay of these cusps is quite interesting and challenges some of the traditional descriptions of resonance tongues when $q = 7$ and presumably for $q \geq 7$.

3. Subharmonics in forced oscillators

As indicated in the introduction, subharmonics of order q ($2q\pi$-periodic orbits) correspond to q-periodic orbits of the Poincaré time-2π-map. However, since the Poincaré map is not known explicitly, applying the method of Section 2 directly is at best rather cumbersome, if not completely infeasible in most cases, especially since we are after a method for computing resonance tongues in *concrete* systems. Therefore we follow an other, more direct approach by introducing a Normal Form Algorithm for time dependent periodic vector fields. This method is explicit, and in principle computes a Normal Form up to any order.

3.1. *A Normal Form Algorithm*

First we present the Normal Form Algorithm in the context of autonomous vector fields. Our approach is an extension of the well-known methods introduced in,[36] and aimed at the derivation of an implementable algorithm. This procedure transforms the terms of the vector field in 'as simple a form as possible', up to a user-defined order. It does so via iteration with respect to the total degree of these terms. In concrete systems we determine this normal form exactly, i.e., making the dependence on the coefficients of the input system explicit. To this end we have to compute the transformed system explicitly up to the desired order. The method of Lie series turns out to be a powerful tool in this context. We first present the key property of the Lie series approach, that allows us to computate the transformed system in a rather straightforward way, up to any desired order.

Lie series expansion. For nonnegative integers m we denote the space of vector fields of total degree m by $\mathcal{H}_m$, and the space of vector fields with vanishing derivatives up to and including order m at $0 \in \mathbb{C}$ by $\mathcal{F}_m$. Note that $\mathcal{F}_m = \prod_{k \geq m} \mathcal{H}_k$.

Proposition 3.1. *Let X and Y be vector fields on $\mathbb{C}$, where X is of the form*

$$X = X^{(1)} + X^{(2)} + \cdots + X^{(N)} \mod \mathcal{F}_{N+1}, \tag{6}$$

with $X^{(n)} \in \mathcal{H}_n$, and $Y \in \mathcal{H}_m$, with $m \geq 2$. Let Y_t, $t \in \mathbb{R}$, be the one-parameter group generated by Y, and let $X^t = (Y_t)_*(X)$. Then

$$X^t = X + \sum_{k=1}^{\lfloor \frac{N-1}{m-1} \rfloor} \frac{(-1)^k}{k!} t^k \operatorname{ad}(Y)^k(X) \mod \mathcal{F}_{N+1} \tag{7}$$

$$= X + \sum_{n=1}^{N} \sum_{k=1}^{\lfloor \frac{N-n}{m-1} \rfloor} \frac{(-1)^k}{k!} t^k \operatorname{ad}(Y)^k(X^{(n)}) \mod \mathcal{F}_{N+1}. \tag{8}$$

Proof. We follow the approach of Takens[36] and Broer et.al.[3,4] to obtain the Taylor series of X^t with respect to t in $t = 0$ using the basic identity

$$\frac{\partial}{\partial t} X^t = [X^t, Y] = -\operatorname{ad}(Y)(X^t).$$

Using this relation, we inductively prove that:

$$\frac{\partial^k}{\partial t^k} X^t = (-1)^k \operatorname{ad}(Y)^k(X^t).$$

Using the latter identity for $t = 0$, we obtain the formal Taylor series

$$X^t = \sum_{k \geq 0} \frac{1}{k!} t^k \left. \frac{\partial^k}{\partial t^k} \right|_{t=0} X^t$$

$$= \sum_{k \geq 0} \frac{(-1)^k}{k!} t^k \operatorname{ad}(Y)^k(X).$$

Since $Y \in \mathcal{H}_m$, the operator $\mathrm{ad}(Y)^k$ increases the degree of each term in its argument by $k(m-1)$. Since the terms of lowest order in X are linear, we see that

$$\mathrm{ad}(Y)^k(X) = 0 \mod \mathcal{F}_{N+1},$$

if $1 + k(m-1) > N$. Therefore,

$$X^t = \sum_{k=0}^{\lfloor \frac{N-1}{m-1} \rfloor} \frac{(-1)^k}{k!} t^k \, \mathrm{ad}(Y)^k(X) \mod \mathcal{F}_{N+1}$$

$$= X + \sum_{k=1}^{\lfloor \frac{N-1}{m-1} \rfloor} \frac{(-1)^k}{k!} t^k \, \mathrm{ad}(Y)^k(X) \mod \mathcal{F}_{N+1},$$

which proves (7). In view of (6) the latter identity expands to

$$X^t = \sum_{n=1}^{N} \sum_{k=0}^{\lfloor \frac{N-1}{m-1} \rfloor} \frac{(-1)^k}{k!} t^k \, \mathrm{ad}(Y)^k(X^{(n)}) \mod \mathcal{F}_{N+1}. \tag{9}$$

Since $\mathrm{ad}(Y)^k(X^{(n)}) \in \mathcal{H}_{n+k(m-1)}$, we see that

$$\mathrm{ad}(Y)^k(X^{(n)}) = 0 \mod \mathcal{F}_{N+1},$$

for $k > \frac{N-n}{m-1}$. Therefore, for fixed index n, the inner sum in (9) can be truncated at $k = \lfloor \frac{N-n}{m-1} \rfloor$, which concludes the proof of (8). $\qquad\square$

The Normal Form Algorithm. Consider a vector field X having a singular point with semisimple linear part S. Our goal is to design an iterative algorithm bringing X into normal form, to some prescribed order N.

Lemma 3.1. *(Normal Form Lemma[36])*
The vector field X can be brought into the normal form

$$X = S + G^{(2)} + \cdots + G^{(m)} \mod \mathcal{F}_{m+1},$$

for any $m \geq 2$, where $G^{(i)} \in \mathcal{H}_i$ belongs to $\mathrm{Ker}\,\mathrm{ad}(S)$.

Proof. Assume that X is of the form

$$X = S + G^{(2)} + \cdots + G^{(m-1)} + X^{(m)} \mod \mathcal{F}_{m+1}, \tag{10}$$

where $X^{(m)} \in \mathcal{H}_m$, and $G^{(i)} \in \mathcal{H}_i$ belongs to $\mathrm{Ker}\,\mathrm{ad}(S)$. If $Y \in \mathcal{H}_m$ and $X^t = (Y_t)_*(X)$, then

$$X^t = X - t\,\mathrm{ad}(Y)(X^{(1)}) \mod \mathcal{F}_{m+1}. \tag{11}$$

This is a direct consequence of Proposition 3.1. See also Takens[36] and Broer et.al.[3,4] Since S is semisimple, we know that

$$\mathcal{H}_m = \operatorname{Ker} \operatorname{ad}(S) + \operatorname{Im} \operatorname{ad}(S),$$

so we write $X^{(m)} = G^{(m)} + B^{(m)}$, where $G^{(m)} \in \operatorname{Ker} \operatorname{ad}(S)$ and $B^{(m)} \in \operatorname{Im} \operatorname{ad}(S)$. If the vector field Y satisfies the *homological equation*

$$\operatorname{ad}(S)(Y) = -B^{(m)}, \tag{12}$$

it follows from (10) and (11) that X^1 is in normal form to order m, since

$$X^1 = S + G^{(2)} + \cdots + G^{(m-1)} + G^{(m)} \quad \operatorname{mod} \mathcal{F}_{m+1}. \qquad \square$$

Our final goal, namely bringing X into normal form to order N, is achieved by repeating this algorithmic step with X replaced by the *transformed vector field* X^1, bringing the latter vector field into normal form to order $m + 1$. Since the homological equation involves the homogeneous terms of X^1 of order $m+1$, we use identity (8) to compute these terms. Furthermore, we enforce uniqueness of the solution Y of (12) by imposing the condition $Y \in \operatorname{Im} \operatorname{ad}(S)$. However, computing just the homogeneous terms of X^1 of order $m + 1$ is not sufficient if $m + 1 < N$, since subsequent steps of the algorithm access the terms of even higher order in the transformed vector field. Therefore, we use (8) to compute these higher order terms.

These steps are then repeated until the final transformed vector field is in normal form to order $N + 1$. This procedure is expressed more precisely in the normal form algorithm in Figure 3.

3.2. *Applications of the Normal Form Algorithm*

The Hopf bifurcation occurs in one-parameter families of planar vector fields having a nonhyperbolic equilibrium with a pair of pure imaginary eigenvalues with nonzero imaginary part. In this bifurcation a limit cycle emerges from the equilibrium as the parameters of the system push the eigenvalues off the imaginary axis. See also Figure 4.

In this context it is easier to express the system in coordinates $z, \bar{z}$ on the complex plane. The linear part of the vector field at the point of bifurcation is then $\dot{z} = i\omega z$. To apply the Normal Form Algorithm, we first derive an expression for the Lie brackets of real vector fields with in these coordinates.

Algorithm (Normal Form Algorithm)
Input: N, S, $X[2..N]$, satisfying
    1. S is a semisimple linear vector field
    2. $X = S + X[2] + \cdots + X[N] \mod \mathcal{F}_{N+1}$,
        with $X[n] \in \mathcal{H}_n$
    ($*$ X is in normal form to order 1 $*$)

for $m = 2$ **to** N **do**
    ($*$ bring X into normal form to order m $*$)
    determine $G \in \operatorname{Ker} \operatorname{ad}(S) \cap \mathcal{H}_m$ and $B \in \operatorname{Im} \operatorname{ad}(S) \cap \mathcal{H}_m$ such that
        $X[m] = G + B$
    determine Y, with $Y \in \operatorname{Im} \operatorname{ad}(S) \cap \mathcal{H}_m$, such that
        $\operatorname{ad}(S)(Y) = -B$
    ($*$ compute terms of order $m+1, \ldots, N$ of transformed vector field $*$)
    for $n = 1$ **to** N **do**
        for $k = 1$ **to** $\lfloor \frac{N-n}{m-1} \rfloor$ **do**

$$X[n + k(m-1)] := X[n + k(m-1)] + \frac{(-1)^k}{k!} \operatorname{ad}(Y)^k(X[n])$$

Fig. 3.   The Normal Form Algorithm.

The Lie-subalgebra of real vector fields. We identify $\mathbb{R}^2$ with $\mathbb{C}$, by associating the point (x_1, x_2) in $\mathbb{R}^2$ with $x_1 + ix_2$ in $\mathbb{C}$. The real vector field X, defined on $\mathbb{R}^2$ by

$$X = Y_1 \frac{\partial}{\partial x_1} + Y_2 \frac{\partial}{\partial x_2},$$

corresponds to the vector field

$$X = Y \frac{\partial}{\partial z} + \overline{Y} \frac{\partial}{\partial \overline{z}}, \tag{13}$$

on $\mathbb{C}$, where $Y = Y_1 + iY_2$.

Example. Taking $Y(z, \overline{z}) = cz^{k+1}\overline{z}^k$, with c a complex constant, the vector field X given by (13) is SO(2)-equivariant. Writing $c = a + ib$, with $a, b \in \mathbb{R}$, and $z = x_1 + ix_2$, we obtain its real form via a straightforward computation:

$$X = (x_1^2 + x_2^2)^k \left(a(x_1 \frac{\partial}{\partial x_1} + x_2 \frac{\partial}{\partial x_2}) + b(-x_2 \frac{\partial}{\partial x_1} + x_1 \frac{\partial}{\partial x_2}) \right).$$

In particular, the real vector field $\omega_N(-x_2 \frac{\partial}{\partial x_1} + x_1 \frac{\partial}{\partial x_2})$ corresponds to the complex vector field $S = i\omega_N(z \frac{\partial}{\partial z} - \overline{z} \frac{\partial}{\partial \overline{z}})$.

We denote the $\dfrac{\partial}{\partial z}$-component of a real vector field X by $X_{\mathbb{R}}$, so:

$$X = X_{\mathbb{R}}\,\frac{\partial}{\partial z} + \overline{X_{\mathbb{R}}}\,\frac{\partial}{\partial \bar{z}}.$$

The real vector fields form a Lie-subalgebra of the algebra of all vector fields on $\mathbb{C}$. The following result justifies this claim.

Lemma 3.2. *Let X and Y be real vector fields on $\mathbb{C}$, and let $f : \mathbb{C} \to \mathbb{C}$ be a smooth function. Then*

$$\overline{X(f)} = X(\bar{f}), \tag{14}$$

and

$$[X,Y] = \langle X,Y\rangle\,\frac{\partial}{\partial z} + \overline{\langle X,Y\rangle}\,\frac{\partial}{\partial \bar{z}},$$

where the bilinear antisymmetric form $\langle \cdot,\cdot \rangle$ is defined by

$$\langle X,Y\rangle = X(Y_{\mathbb{R}}) - Y(X_{\mathbb{R}}).$$

Derivation of the Hopf Normal Form. To derive the Hopf Normal Form, we apply the Normal Form Algorithm to a vector field with linear part

$$S = i\omega_N\Big(z\,\frac{\partial}{\partial z} - \bar{z}\,\frac{\partial}{\partial \bar{z}}\Big).$$

The adjoint action of S on the Lie-subalgebra of real vector fields is given by:

$$\mathrm{ad}(S)(X) = \langle S,X\rangle\,\frac{\partial}{\partial z} + \overline{\langle S,X\rangle}\,\frac{\partial}{\partial \bar{z}},$$

with

$$\langle S,X\rangle = i\omega_N\Big(z\,\frac{\partial X_{\mathbb{R}}}{\partial z} - \bar{z}\,\frac{\partial X_{\mathbb{R}}}{\partial \bar{z}} - X_{\mathbb{R}}\Big).$$

In particular, if $Y = Y_{\mathbb{R}}\,\dfrac{\partial}{\partial z} + \overline{Y_{\mathbb{R}}}\,\dfrac{\partial}{\partial \bar{z}}$ with $Y_{\mathbb{R}} = z^k\bar{z}^l$, then

$$\langle S,Y\rangle = i\omega_N\,(k - l - 1)\,z^k\bar{z}^l.$$

Therefore, the adjoint operator $\mathrm{ad}(S) : \mathcal{H}_m \to \mathcal{H}_m$ has non-trivial kernel for m odd. If $m = 2k + 1$, this kernel consists of the monomial vector field Y with $Y_{\mathbb{R}} = z|z|^{2k}$. These observations lead to the following Normal Form.

Corollary 3.1. *If a vector field on $\mathbb{C}$ has linear part $S = i\omega_N\left(z\dfrac{\partial}{\partial z} - \bar{z}\dfrac{\partial}{\partial \bar{z}}\right)$, then the Normal Form Algorithm brings this vector field into the form*

$$\dot{z} = i\omega z + \sum_{k=1}^{m} c_k z |z|^{2k} + O(|z|^{2m+3}). \tag{15}$$

The nondegenerate Hopf bifurcation. An other application of the Normal Form Algorithm is the computation of the first Hopf coefficient c_1 in (15). Let X be given by

$$\dot{z} = i\omega z + a_0 z^2 + a_1 z\bar{z} + a_2 \bar{z}^2 + b_0 z^3 + b_1 z^2\bar{z} + b_2 z\bar{z}^2 + b_3 \bar{z}^3 + O(|z|^4).$$

The Normal Form Algorithm computes the following normal form for this system:

$$\dot{z} = i\omega z + \left(b_1 - \frac{i}{3\omega}(3a_0 a_1 - 3|a_1|^2 - 2|a_2|^2)\right) z^2\bar{z} + O(|z|^4).$$

This result can also be obtained by a tedious calculation. See, for example, [26, page 155]*

To analyze the emergence of limit cycles we rewrite the Hopf Normal Form

$$\dot{w} = i\omega_N w + wb(|w|^2, \mu) + O(n+1)$$

in polar coordinates as:

$$\dot{r} = r\,\mathrm{Re}\,b(r^2, \mu) + O(n+1)$$
$$\dot{\varphi} = \omega_N + \mathrm{Im}\,b(r^2, \mu) + O(n+1)$$

Limit cycles are obtained by solving $r = r(\mu)$ from the equation $\mathrm{Re}\,b(r^2, \mu) = 0$. The frequency of the limit cycle is then of the form $\omega(\mu) = \omega_N + \mathrm{Im}\,b(r(\mu)^2, \mu)$.

A non-degenerate Hopf bifurcation occurs if the first Hopf coefficient c_1 in (15) is nonzero. Consider, e.g., the simple case $b(u, \mu) = \mu + u$. Putting $\mu = a + i\delta$ we see that the limit cycle corresponds to the trajectory w

$$w_{a,\delta}(t) = \sqrt{-a}\,e^{i(\omega_N + \delta)t} \qquad (a \le 0)$$

This limit cycle exists for $a < 0$, and is repulsive in this case.

---

*The term $|h_{ww}|^2$ in identity (3.4.26) of [26, page 155] should be replaced by $|h_{\bar{w}\bar{w}}|^2$.

346

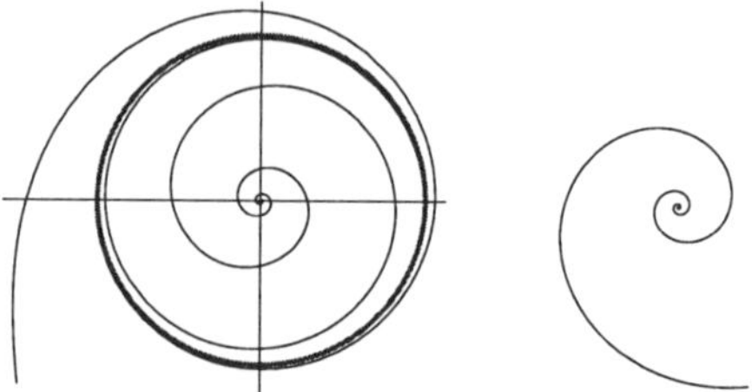

Fig. 4.   Birth or death of a limit cycle via a Hopf bifurcation.

Hopf-Neĭmark-Sacker bifurcations in forced oscillators. We now study the birth or death of subharmonics in forced oscillators depending on parameters. In particular, we consider 2π-periodic systems on $\mathbb{C}$ of the form

$$\dot{z} = F(z, \overline{z}, \mu) + \varepsilon G(z, \overline{z}, t, \mu), \tag{16}$$

obtained from an autonomous system by a small 2π-periodic perturbation. Here ε is a real perturbation parameter, and $\mu \in \mathbb{R}^k$ is an additional k-dimensional parameter. Subharmonics of order q may appear or disappear upon variation of the parameters if the linear part of F at $z = 0$ satisfies a $p : q$-resonance condition which is appropriately detuned upon variation of the parameter μ.

The Normal Form Algorithm of Section 3.1 can be adapted to the derivation of the Hopf-Neĭmark-Sacker Normal form of periodic systems. Consider a 2π-periodic forced oscillator on $\mathbb{C}$ of the form

$$\dot{z} = X_{\mathbb{R}}(z, \overline{z}, t, \mu),$$

where

$$X_{\mathbb{R}}(z, \overline{z}, t, \mu) = i\omega_N z + (\alpha + i\delta)z + zP(z, \overline{z}, \mu) + \varepsilon Q(z, \overline{z}, t, \mu). \tag{17}$$

Here $\mu \in \mathbb{R}^k$, and ε is a small real parameter. Furthermore we assume that P and Q contain no terms that are independent of z and $\overline{z}$ (i.e., $P(0, 0, \mu) = 0$ and $Q(0, 0, t, \mu) = 0$), and that Q does not even contain terms that are linear in z and $\overline{z}$. Any system of the form (16) with linear part $\dot{z} = i\omega_N z$ can be brought into this form after a straightforward initial transformation. See[3] for details. Subharmonics of order q are to be expected if the linear part satisfies a $p : q$-resonance condition, in other words, if the normal frequency ω_N is equal to $\dfrac{p}{q}$ (with p and q relatively prime).

Theorem 3.1. *(Normal Form to order q)*
The system (17) has normal form

$$\dot{z} = i\omega_N z + (\alpha + i\delta)z + zF(|z|^2, \mu) + d\varepsilon\,\overline{z}^{q-1}\,e^{ipt} + O(q+1), \qquad (18)$$

where $F(|z|^2, \mu)$ is a complex polynomial of degree $q - 1$ with $F(0, \mu) = 0$, and d is a complex constant.

Proof. To derive a normal form for the system (17) we consider 2π-periodic vector fields on $\mathbb{C} \times \mathbb{R}/(2\pi\mathbb{Z})$ of the form

$$X = X_{\mathbb{R}}\,\frac{\partial}{\partial z} + \overline{X_{\mathbb{R}}}\,\frac{\partial}{\partial\overline{z}} + \frac{\partial}{\partial t},$$

with 'linear' part

$$S = i\omega_N\Big(z\frac{\partial}{\partial z} - \overline{z}\frac{\partial}{\partial\overline{z}}\Big) + \frac{\partial}{\partial t}.$$

For nonnegative integers m we denote the space of 2π-periodic vector fields of total degree m in $(z, \overline{z}, \mu)$ with vanishing $\dfrac{\partial}{\partial t}$-component by $\mathcal{H}_m$. As before $\mathcal{F}_m = \prod_{k \geq m} \mathcal{H}_k$.

The adjoint action of S on the Lie-subalgebra of real 2π-periodic vector fields with zero $\dfrac{\partial}{\partial t}$-component is given by:

$$\mathrm{ad}(S)(X) = \langle S, X\rangle_{\mathbb{R}}\,\frac{\partial}{\partial z} + \overline{\langle S, X\rangle_{\mathbb{R}}}\,\frac{\partial}{\partial\overline{z}},$$

with

$$\langle S, X\rangle_{\mathbb{R}} = i\omega_N\Big(z\frac{\partial X_{\mathbb{R}}}{\partial z} - \overline{z}\frac{\partial X_{\mathbb{R}}}{\partial\overline{z}} - X_{\mathbb{R}}\Big) + \frac{\partial X_{\mathbb{R}}}{\partial t}.$$

If $X_{\mathbb{R}} = \mu^\sigma z^k\overline{z}^l e^{imt}$, with $|\sigma| + k + l = n$, then

$$\langle S, X\rangle_{\mathbb{R}} = (i\omega_N(k - l - 1) + im)\,X_{\mathbb{R}}.$$

Therefore, the normal form contains time-independent rotationally symmetric terms corresponding to $k = l + 1$ and $m = 0$. Since for $\varepsilon = 0$ the system is in Hopf Normal Form, all non-rotationally symmetric terms contain a factor ε, so $|\sigma| > 0$ for these terms. It is not hard to see that for $n \leq q$ and $|\sigma| > 0$ the only non-rotationally symmetric term corresponds to $k = 0$, $l = q - 1$, $m = p$, and $|\sigma| = 1$. Therefore, this non-symmetric term is of the form

$$d\varepsilon\,\overline{z}^{q-1}\,e^{ipt},$$

for some complex constant d. $\qquad\square$

3.3. *Via covering spaces to the Takens Normal Form*

Existence of $2\pi q$-periodic orbits. The Van der Pol transformation.
Subharmonics of order q of the 2π-periodic forced oscillator (17) correspond
to q-periodic orbits of the Poincaré time 2π-map $P : \mathbb{C} \to \mathbb{C}$. These periodic
orbits of the Poincaré map are brought into one-one correspondence with
the zeros of a vector field on a q-sheeted cover of the phase space $\mathbb{C} \times
\mathbb{R}/(2\pi\mathbb{Z})$ via the *Van der Pol transformation*, cf.[16] This transformation
corresponds to a q-sheeted covering

$$\Pi : \mathbb{C} \times \mathbb{R}/(2\pi q\mathbb{Z}) \to \mathbb{C} \times \mathbb{R}/(2\pi\mathbb{Z}),$$
$$(z,t) \mapsto (ze^{itp/q}, t \ (\mathrm{mod}\ 2\pi\mathbb{Z})) \tag{19}$$

with cyclic Deck group of order q generated by

$$(z,t) \mapsto (ze^{2\pi ip/q}, t - 2\pi).$$

The Van der Pol transformation $\zeta = ze^{-i\omega_N t}$ lifts the forced oscillator (16)
to the system

$$\dot{\zeta} = (\alpha + i\delta)\zeta + \zeta P(\zeta e^{i\omega_N t}, \overline{\zeta}e^{-i\omega_N t}, \mu) + \varepsilon Q(\zeta e^{i\omega_N t}, \overline{\zeta}e^{-i\omega_N t}, t, \mu) \tag{20}$$

on the covering space $\mathbb{C} \times \mathbb{R}/(2\pi q\mathbb{Z})$. The latter system is $\mathbb{Z}_q$-equivariant.
A straightforward application of (20) to the normal form (18) yields the
following normal form for the lifted forced oscillator.

Theorem 3.2. *(Equivariant Normal Form of order q)*
*On the covering space, the lifted forced oscillator has the $\mathbb{Z}_q$-equivariant
normal form:*

$$\dot{\zeta} = (\alpha + i\delta)\zeta + \zeta F(|\zeta|^2, \mu) + d\varepsilon \overline{\zeta}^{q-1} + O(q+1), \tag{21}$$

where the $O(q+1)$ terms are $2\pi q$-periodic.

Resonance tongues for families of forced oscillators. Bifurcations of
q-periodic orbits of the Poincaré map P on the base space correspond to
bifurcations of fixed points of the Poincaré map $\tilde{P}$ on the q-sheeted covering
space introduced in connection with the Van der Pol transformation (19).
Denoting the normal form system (18) on the base space by $\mathcal{N}$, and the
normal form system (21) of the lifted forced oscillator by $\tilde{\mathcal{N}}$, we see that

$$\Pi_* \tilde{\mathcal{N}} = \mathcal{N}.$$

The Poincaré mapping $\tilde{P}$ of the normal form on the covering space now is
the $2\pi q$-period mapping

$$\tilde{P} = \tilde{\mathcal{N}}^{2\pi q} + O(q+1),$$

where $\tilde{\mathcal{N}}^{2\pi q}$ denotes the $2\pi q$-map of the (planar) vector field $\tilde{\mathcal{N}}$. Following the Corollary to the Normal Form Theorem of [16, page 12], we conclude for the original Poincaré map P of the vector field X on the base space that

$$P = R_{2\pi\omega_N} \circ \tilde{\mathcal{N}}^{2\pi} + O(q+1),$$

where $R_{2\pi\omega_N}$ is the rotation over $2\pi\omega_N = 2\pi p/q$, which precisely is the Takens Normal Form[37] of P at $(z,\mu) = (0,0)$.

Our interest is with the q-periodic points of P_μ, which correspond to the fixed points of $\tilde{P}_\mu$. This fixed point set and the boundary thereof in the parameter space $\mathbb{R}^3 = \{a,\delta,\varepsilon\}$ is approximately described by the discriminant set of

$$(a + i\delta)\zeta + \zeta\tilde{F}(|\zeta|^2,\mu) + \varepsilon d\bar{\zeta}^{q-1},$$

which is the truncated right hand side of (21). This gives rise to the bifurcation equation that determine the boundaries of the resonance tongues. The following theorem implies that, under the conditions that $d \neq 0 \neq F_u(0,0)$, the order of tangency at the tongue tips is generic. Here $F_u(0,0)$ is the partial derivative of $F(u,\mu)$ with respect to u.

Theorem 3.3. *(Bifurcation equations modulo contact equivalence) Assume that $d \neq 0$ and $F_u(0,0) \neq 0$. Then the polynomial (21) is $\mathbb{Z}_q$-equivariantly contact equivalent with the polynomial*

$$G(\zeta,\mu) = (a + i\delta + |\zeta|^2)\zeta + \varepsilon\bar{\zeta}^{q-1}. \tag{22}$$

The discriminant set of the polynomial $G(\zeta,\mu)$ is of the form

$$\delta = \pm\varepsilon(-a)^{(q-2)/2} + O(\varepsilon^2). \tag{23}$$

Proof. The polynomial (22) is a universal unfolding of the germ $|\zeta|^2\zeta + \varepsilon\bar{\zeta}^{q-1}$ under $\mathbb{Z}_q$ contact equivalence. See[7] for a detailed computation. The tongue boundaries of a $p : q$ resonance are given by the bifurcation equations

$$G(\zeta,\mu) = 0,$$
$$\det(dG)(\zeta,\mu) = 0.$$

As in [7, Theorem 3.1] we put $u = |z|^2$, and $b(u,\mu) = a + i\delta + u$. Then $G(\zeta,\mu) = b(u,\mu)\zeta + \varepsilon\bar{\zeta}^{q-1}$. According to (the proof of) [7, Theorem 3.1], the system of bifurcation equations is equivalent to

$$|b|^2 = \varepsilon^2 u^{q-2},$$
$$b\bar{b}' + \bar{b}b' = (q-2)\varepsilon^2 u^{q-3},$$

where $b' = \dfrac{\partial b}{\partial u}(u, \mu)$. A short computation reduces the latter system to the equivalent

$$(a + u)^2 + \delta^2 = \varepsilon^2 u^{q-2},$$

$$a + u = \frac{1}{2}(q - 2)\varepsilon^2 u^{q-3}.$$

Eliminating u from this system of equations yields expression (23) for the tongue boundaries. $\qquad\square$

The discriminant set of the equivariant polynomial (22) forms the boundary of the resonance tongues. See Figure 5. At this surface we expect

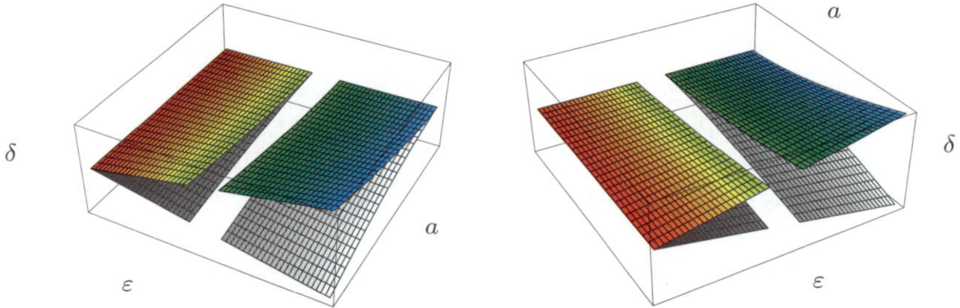

Fig. 5.   Resonance zones for forced oscillator families: the Hopf-Neĭmark-Sacker phenomenon.

the Hopf-Neĭmark-Sacker bifurcation to occur; here the Floquet exponents of the linear part of the forced oscillator cross the complex unit circle. This bifurcation gives rise to an invariant 2-torus in the 3D phase space $\mathbb{C} \times \mathbb{R}/(2\pi\mathbb{Z})$. Resonances occur when the eigenvalues cross the unit circle at roots of unity $e^{2\pi i p/q}$. 'Inside' the tongue the 2-torus is phase-locked to subharmonic periodic solutions of order q.

4. Generic Hopf-Neĭmark-Sacker bifurcations in feed forward systems?

Coupled Cell Systems. A coupled cell system is a network of dynamical systems, or cells, coupled together. This network is a finite directed graph with nodes representing cells and edges representing couplings between these cells. See, e.g., Golubitsky, Nicol and Stewart.[22]

We consider the three-cell feed-forward network in Figure 6, where the first cell is coupled externally to itself. The network has the form of a *coupled*

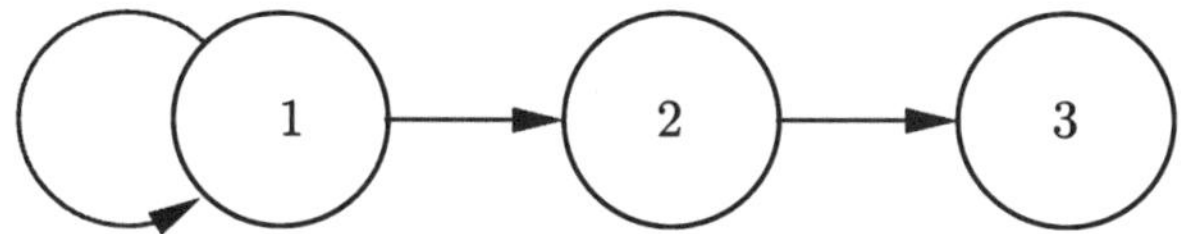

Fig. 6.   Three-cell linear feed-forward network

cell system

$$\dot{x}_1 = f(x_1, x_1)$$
$$\dot{x}_2 = f(x_2, x_1)$$
$$\dot{x}_3 = f(x_3, x_2)$$

with $x_j \in \mathbb{R}^2$.

Under certain conditions these networks have time-evolutions that are equilibria in cell 1 and periodic in cells 2 and 3. Elmhirst and Golubitsky[20] describe a curious phenomenon: the amplitude growth of the periodic signal in cell 3 is to the power $\frac{1}{6}$ rather than to the power $\frac{1}{2}$ with respect to the bifurcation parameter in the Hopf bifurcation. See also Section 3.2.

For technical reasons we assume that the function f, describing the dynamics of each cell, is $\mathbb{S}^1$-symmetric in the sense that

$$f(e^{i\theta} z_2, e^{i\theta} z_1) = e^{i\theta} f(z_2, z_1), \tag{24}$$

for all real θ. Here we identify the two-dimensional phase space of each cell with $\mathbb{C}$ by writing $z_j = x_{j1} + i x_{j2}$. Identity (24) is a special assumption, that we will try to relax in future research. However, Elmhirst and Golubitsky[20] verify that this symmetry condition holds to third order after a change of coordinates. We also assume that the dynamics of each cell depends on external parameters λ, μ, to be specified later on.

Dynamics of the first and second cell. The $\mathbb{S}^1$-symmetry (24) implies that $f_{\lambda,\mu}(0,0) = 0$. Note that from now on we make the dependence of f on the parameters explicit in our notation. Assume that the linear part of $f_{\lambda,\mu}(z_1, z_1)$ at $z_1 = 0$ has eigenvalues with negative real part. Then *the first cell has a stable equilibrium at $z_1 = 0$.*

The second cell has dynamics

$$\dot{z}_2 = f_{\lambda,\mu}(z_2, z_1) = f_{\lambda,\mu}(z_2, 0),$$

where we use that the first cell is in its stable equilibrium. Golubitsky and Stewart[24] introduce a large class of functions $f_{\lambda,\mu}$ for which the *second cell undergoes a Hopf bifurcation.* For this class of cell dynamics, and for linear feed-forward networks of increasing length, there will be 'repeated Hopf' bifurcation, reminiscent of the scenarios named after Landau-Lifschitz and Ruelle-Takens.

To obtain more precise information on the Hopf bifurcation in the dynamics of the second cell we consider a special class of functions $f_{\lambda,\mu}$ satisfying (24). In particular, we require that

$$f_{\lambda,\mu}(z_2, 0) = (\lambda + i - |z_2|^2)\, z_2, \tag{25}$$

giving a supercritical Hopf bifurcation at $\lambda = 0$. The stable periodic solution, occurring for $\lambda > 0$, has the form

$$z_2(t) = \sqrt{\lambda}\, e^{it}.$$

Dynamics of the third cell. The main topic of our research is the generic dynamics of the *third cell*, given simple time-evolutions of the first two cells. Here we like to know what are the correspondences and differences with the general ODE setting. In particular this question regards the coexistence of periodic, quasi-periodic and chaotic dynamics.

In co-rotating coordinates the dynamics of the third cell becomes time-independent. To see this, set $z_3 = e^{it}y$, and use the $\mathbb{S}^1$-symmetry to derive

$$\begin{aligned} ie^{it}y + e^{it}\dot{y} = &= f_{\lambda,\mu}(e^{it}y, \sqrt{\lambda}\, e^{it}) \\ &= e^{it}\, f_{\lambda,\mu}(y, \sqrt{\lambda}). \end{aligned}$$

Therefore, the dynamics of the third cell is given by

$$\dot{y} = -iy + f_{\lambda,\mu}(y, \sqrt{\lambda}). \tag{26}$$

Equation (26) is autonomous, so the present setting might exhibit Hopf bifurcations, but it is still too simple to produce resonance tongues. Indeed, all (relative) periodic motion in (26) will lead to parallel (quasi-periodic) dynamics and the Hopf-Neĭmark-Sacker phenomenon. Therefore, we now perturb the basic function $f = f_{\lambda,\mu}(z_2, z_1)$, to

$$F_{\lambda,\mu,\varepsilon}(z_2, z_1) := f_{\lambda,\mu}(z_2, z_1) + \varepsilon P(z_2, z_1).$$

In cells 1 and 2 any choice of the perturbation term $P(z_2, z_1)$ gives the dynamics

$$\begin{aligned} \dot{z}_1 &= F_{\lambda,\mu,\varepsilon}(z_1, z_1) = f_{\lambda,\mu}(z_1, z_1) + \varepsilon P(z_1, z_1), \\ \dot{z}_2 &= F_{\lambda,\mu,\varepsilon}(z_2, 0), \end{aligned}$$

with the same conclusions as before, namely a steady state $z_1 = 0$ in cell 1 and a periodic state $z_2 = \sqrt{\lambda}\, e^{it}$ in cell 2 (when $\lambda > 0$). For these two conclusions it is sufficient that

$$P(z_2, 0) \equiv 0.$$

Turning to the third cell we again put $y = e^{-it}z_3$, and so get a perturbed reduced equation

$$\dot{y} = -iy + f_{\lambda,\mu}(y, \sqrt{\lambda}) + \varepsilon e^{-it} P(y\, e^{it}, \sqrt{\lambda}\, e^{it}). \tag{27}$$

The third cell therefore has forced oscillator dynamics with driving frequency 1. The question about generic dynamics regards the possible coexistence of periodic and quasi-periodic dynamics. We aim to investigate (27) for Hopf-Neĭmark-Sacker bifurcations, which are expected along curves $\mathcal{H}_\varepsilon$ in the (λ, μ)-plane of parameters. We expect to find periodic tongues (See also Figure 5) and quasiperiodic hairs, like in Broer et al.[15] This is the subject of ongoing research. The machinery of Section 3.2 should provide us with sufficiently powerful tools to investigate this phenomenon for a large class of coupled cell systems.

5. Conclusion and future work

We have presented several contexts in which bifurcations from fixed points of maps or equilibria of vector fields lead to the emergence of periodic orbits. For each context we present appropriate normal form techniques, illustrating the general paradigm of 'simplifying the system before analyzing it'. In the context of generic families we apply generic techniques, based on Lyapunov-Schmidt reduction and $\mathbb{Z}_q$-equivariant contact equivalence. In this way we recover standard results on resonance tongues for nondegenerate maps, but also discover new phenomena in unfoldings of mildly degenerate systems. Furthermore, we present an algorithm for bringing concrete families of dynamical systems into normal form, without losing information in a preliminary reduction step, like the Lyapunov-Schmidt method. An example of such a concrete system is a class of feedforward networks of coupled cell systems, in which we expect the Hopf-Neĭmark-Sacker-phenomenon to occur.

With regard to further research, our methods can be extended to other contexts, in particular, to cases where extra symmetries, including time reversibility, are present. This holds both for Lyapunov-Schmidt reduction and $\mathbb{Z}_q$ equivariant singularity theory. In this respect Golubitsky, Marsden, Stewart, and Dellnitz,[21] Knobloch and Vanderbauwhede,[27,28] and Vanderbauwhede[38] are helpful.

Furthermore, there is the issue of how to apply our results to a concrete family of dynamical systems. Golubitsky and Schaeffer[23] describe methods for obtaining the Taylor expansion of the reduced function $g(z)$ in terms of the Poincaré map P and its derivatives. These methods may be easier to apply if the system is a periodically forced second order differential equation, in which case the computations again may utilize parameter dependent Floquet theory. We also plan to turn the Singularity Theory methods of Section 2 into effective algorithms, along the lines of our earlier work.[9]

Finally, in this paper we have studied only degeneracies in tongue boundaries. It would also be interesting to study low codimension degeneracies in the dynamics associated to the resonance tongues. Such a study will require tools that are more sophisticated than the singularity theory ones that we have considered here.

References

1. Z. Afsharnejad. Bifurcation geometry of mathieu's equation. *Indian J. Pure Appl. Math.*, 17:1284–1308, 1986.
2. V.I. Arnold. *Geometrical Methods in the Theory of Ordinary Differential Equations.* Springer-Verlag, 1982.
3. B.L.J. Braaksma, H.W. Broer, and G.B. Huitema. Toward a quasi-periodic bifurcation theory. In *Mem. AMS*, volume 83, pages 83–175. 1990.
4. H.W. Broer. Formal normal form theorems for vector fields and some consequences for bifurcations in the volume preserving case. In *Dynamical Systems and Turbulence*, volume 898 of *LNM*, pages 54–74. Springer-Verlag, 1980.
5. H.W. Broer, S.-N. Chow, Y. Kim, and G. Vegter. normally elliptic hamiltonian bifurcation. *ZAMP*, 44:389–432, 1993.
6. H.W. Broer, S.-N. Chow, Y. Kim, and G. Vegter. The hamiltonian double-zero eigenvalue. In *Normal Forms and Homoclinic Chaos, Waterloo 1992*, volume 4 of *Fields Institute Communications*, pages 1–19, 1995.
7. H.W. Broer, M. Golubitsky, and G. Vegter. The geometry of resonance tongues: A singularity theory approach. *Nonlinearity*, 16:1511–1538, 2003.
8. H.W. Broer, I. Hoveijn, G.A. Lunter, and G. Vegter. Resonances in a spring-pendulum: algorithms for equivariant singularity theory. *Nonlinearity*, 11:1–37, 1998.
9. H.W. Broer, I. Hoveijn, G.A. Lunter, and G. Vegter. *Bifurcations in Hamiltonian systems: Computing singularities by Gröbner bases*, volume 1806 of *Springer Lecture Notes in Mathematics.* Springer-Verlag, 2003.
10. H.W. Broer and M. Levi. Geometrical aspects of stability theory for hill's equations. *Arch. Rational Mech. Anal.*, 13:225–240, 1995.
11. H.W. Broer, G.A. Lunter, and G. Vegter. Equivariant singularity theory with distinguished parameters, two case studies of resonant hamiltonian systems. *Physica D*, 112:64–80, 1998.
12. H.W. Broer and R. Roussarie. Exponential confinement of chaos in the bi-

furcation set of real analytic diffeomorphisms. In B. Krauskopf H.W. Broer and G. Vegter, editors, *Global Analysis of Dynamical Systems, Festschrift dedicated to Floris Takens for his 60th birthday*, pages 167–210. IOP, Bristol and Philadelphia, 2001.

13. H.W. Broer and C. Simó. Hill's equation with quasi-periodic forcing: resonance tongues, instability pockets and global phenomena. *Bol. Soc. Bras. Mat.*, 29:253–293, 1998.

14. H.W. Broer and C. Simó. Resonance tongues in hill's equations: a geometric approach. *J. Diff. Eqns*, 166:290–327, 2000.

15. H.W. Broer, C. Simó, and J.-C. Tatjer. Towards global models near homoclinic tangencies of dissipative diffeomorphisms. *Nonlinearity*, 11:667–770, 1998.

16. H.W. Broer and G. Vegter. Bifurcational aspects of parametric resonance. In *Dynamics Reported, New Series*, volume 1, pages 1–51. Springer-Verlag, 1992.

17. A. Chenciner. Bifurcations de points fixes elliptiques, i. courbes invariantes. *Publ. Math. IHES*, 61:67–127, 1985.

18. A. Chenciner. Bifurcations de points fixes elliptiques, ii. orbites périodiques et ensembles de Cantor invariants. *Invent. Math.*, 80:81–106, 1985.

19. A. Chenciner. Bifurcations de points fixes elliptiques, iii. orbites périodiques de "petites" périodes et élimination résonnantes des couples de courbes invariantes. *Publ. Math. IHES*, 66:5–91, 1988.

20. T. Elmhirst and M. Golubitsky. Nilpotent hopf bifurcations in coupled cell networks. *SIAM J. Appl. Dynam. Sys.*, (To appear).

21. M. Golubitsky, J.E. Marsden, I. Stewart, and M. Dellnitz. The constrained liapunov-schmidt procedure and periodic orbits. In W. Langford J. Chadam, M. Golubitsky and B. Wetton, editors, *Pattern Formation: Symmetry Methods and Applications*, volume 4 of *Fields Institute Communications*, pages 81–127. American Mathematical Society, 1996.

22. M. Golubitsky, M. Nicol, and I. Stewart. Some curious phenomena in coupled cell networks. *J. Nonlinear Sci.*, 14(2):119–236, 2004.

23. M. Golubitsky and D.G. Schaeffer. *Singularities and Groups in Bifurcation Theory: Vol. I*, volume 51 of *Applied Mathematical Sciences*. Springer-Verlag, New York, 1985.

24. M. Golubitsky and I. Stewart. Synchrony versus symmetry in coupled cells. In *Equadiff 2003: Proceedings of the International Conference on Differential Equations*, pages 13–24. World Scientific Publ. Co., 2005.

25. M. Golubitsky, I.N. Stewart, and D.G. Schaeffer. *Singularities and Groups in Bifurcation Theory: Vol. II*, volume 69 of *Applied Mathematical Sciences*. Springer-Verlag, New York, 1988.

26. J. Guckenheimer and Ph. Holmes. *Nonlinear oscillations, dynamical systems, and bifurcations of vector fields*, volume 42 of *Applied Mathematical Sciences*. Springer-Verlag, New York, Heidelberg, Berlin, 1983.

27. J. Knobloch and A. Vanderbauwhede. Hopf bifurcation at k-fold resonances in equivariant reversible systems. In P. Chossat, editor, *Dynamics. Bifurcation and Symmetry. New Trends and New Tools.*, volume 437 of *NATO ASI*

Series C, pages 167–179. Kluwer Acad. Publ., 1994.

28. J. Knobloch and A. Vanderbauwhede. A general method for periodic solutions in conservative and reversible systems. *J. Dynamics Diff. Eqns.*, 8:71–102, 1996.

29. B. Krauskopf. Bifurcation sequences at 1:4 resonance: an inventory. *Nonlinearity*, 7:1073–1091, 1994.

30. R.P. McGehee and B.B. Peckham. Determining the global topology of resonance surfaces for periodically forced oscillator families. In *Normal Forms and Homoclinic Chaos*, volume 4 of *Fields Institute Communications*, pages 233–254. AMS, 1995.

31. R.P. McGehee and B.B. Peckham. Arnold flames and resonance surface folds. *Int. J. Bifurcations and Chaos*, 6:315–336, 1996.

32. S.E. Newhouse, J. Palis, and F. Takens. Bifurcation and stability of families of diffeomorphisms. *Publ Math. I.H.E.S*, 57:1–71, 1983.

33. B.B. Peckham, C.E. Frouzakis, and I.G. Kevrekidis. Bananas and banana splits: a parametric degeneracy in the hopf bifurcation for maps. *SIAM. J. Math. Anal.*, 26:190–217, 1995.

34. B.B. Peckham and I.G. Kevrekidis. Period doubling with higher-order degeneracies. *SIAM J. Math. Anal.*, 22:1552–1574, 1991.

35. B.B. Peckham and I.G. Kevrekidis. Lighting arnold flames: Resonance in doubly forced periodic oscillators. *Nonlinearity*, 15:405–428, 2002.

36. F. Takens. Singularities of vector fields. *Publ. Math. IHES*, 43:48–100, 1974.

37. F. Takens. Forced oscillations and bifurcations. In B. Krauskopf H.W. Broer and G. Vegter, editors, *Global Analysis of Dynamical Systems, Festschrift dedicated to Floris Takens on his 60th birthday*, pages 1–61. IOP, Bristol and Philadelphia, 2001.

38. A. Vanderbauwhede. Hopf bifurcation for equivariant conservative and time-reversible systems. *Proc. Royal Soc. Edinburgh*, 116A:103–128, 1990.

39. A. Vanderbauwhede. Branching of periodic solutions in time-reversible systems. In H.W. Broer and F. Takens, editors, *Geometry and Analysis in Non-Linear Dynamics*, volume 222 of *Pitman Research Notes in Mathematics*, pages 97–113. Pitman, London, 1992.

40. A. Vanderbauwhede. Subharmonic bifurcation at multiple resonances. In *Proceedings of the Mathematics Conference*, pages 254–276, Singapore, 2000. World Scientific.

GENERIC SINGULARITIES OF SURFACES

Y. YOMDIN

Department of Mathematics
The Weizmann Institute of Science
Rehovot 76100, Israel
E-mail:yosef.yomdin@weizmann.ac.il

We suggest an approach to a description of the hierarchy of singularities of surfaces in $\mathbb{R}^3$, which uses as the starting point the level surfaces $Y(c) = \{F(x_1, x_2, x_3) = c\}$ of smooth functions $F(x_1, x_2, x_3)$ of three variables. In our setting we explicitly allow *singular* level surfaces $Y(c)$, corresponding to the critical values c of F.

In order to obtain as a special case of our definition sharp edges and corners of the surfaces, as well as their "smoothed" versions, we consider functions $F(x_1, x_2, x_3)$ of a special product form $F = F_1 F_2 \ldots F_m$. We modify also the notion of a "genericity" or of a "general position" for such products: we say that $F = F_1 F_2 \ldots F_m$ is in general position if F_1, F_2, $\ldots$, F_m are generic smooth functions, and in particular, they are in a general position with respect to one another. Under these assumptions the product $F = F_1 F_2 \ldots F_m$ usually *is not a generic smooth function in the sense of the classical Singularity Theory*. Indeed, F has non-isolated singularities along the crossing curves C_{ij} of the surfaces $C_i = \{F_i = 0\}$ and $C_j = \{F_j = 0\}$, $i, j = 1, \ldots, m$. From the point of view of the *classification of singularities of smooth functions in the standard setting* this is a very degenerate situation, appearing only in "codimension infinity".

1. Summary

1.1. *Motivations for the setting of the problem*

Probably, there is no "canonical" mathematical setting for the treatment of singularities of surfaces in $\mathbb{R}^3$. In most of applications we intend to call "singular" the points where our surface is not smooth. Typically, these are sharp ridges and vertices (corners) on the surface. Respectively, "near-singular" are those points of the surface where at least one of its main curvatures is "large".

However, to use a differential-geometric definition (through the curvatures) for a classification of surface singularities is technically rather difficult. This approach would also require a separate treatment of non-smooth singularities, i.e. of sharp ridges and vertices on the surface, and of smooth "almost singular" points.

Another mathematical possibility, which looks better fitted to the requirement of including both the smooth parts, as well as sharp ridges and vertices, is to consider graphs of *maxima functions of smooth families*, or, better, graphs of differences of such maxima functions. Here we can easily get a sharp edge: the function $g(x_1, x_2) = \max\{f_1(x_1, x_2), f_2(x_1, x_2)\}$ has an edge along the curve $\{f_1(x_1, x_2) = f_2(x_1, x_2)\}$. A sharp corner we get for $g(x_1, x_2) = \max\{f_1(x_1, x_2),\ f_2(x_1, x_2),\ f_3(x_1, x_2)\}$ at the points where all the three functions $f_1(x_1, x_2),\ f_2(x_1, x_2),\ f_3(x_1, x_2)$ take an equal value.

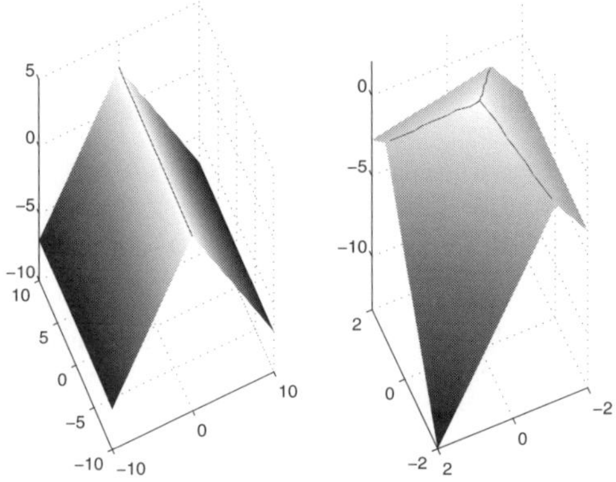

The advantage of this approach is that there is a bunch of results on the classification of singularities of maxima and minimax functions and their differences (see [3,6,7,11,14]). However, some of the typical (stable) singularities in this classification do not look natural from the point of view of the study of the surface geometry. This concerns, for example, the typical singularity of the maxima of two-parametric families $g(x, y) = \max_t(-t^4 + xt^2 + yt)$. This specific singularity does not look relevant for surfaces (at least for those which did not appear explicitly as the envelops of certain smooth families).

Another problem of the minimax approach is the representation of smooth ridges and corners (obtained by a "low-pass" smoothening of the sharp ones). Although such patterns can be, in principle, represented in the minimax form, this representation is neither easy nor natural.

There are other possible approaches to the problem of a description of surface singularities. Algebraic-geometric treatment of surfaces provides, in particular, a natural hierarchy of singularities, which differs from the one suggested below. Another set of relevant problems appears in Image Processing. In particular, it would be interesting to compare our description with the results of [4–6]. Approximation of surfaces with a triangulated mesh (see [13] and references there) provides another natural approach to the treatment of of surface singularities and near-singularities.

1.2. *The suggested setting*

We suggest below an alternative approach, which uses as the starting point the level surfaces $Y(c) = \{F(x_1, x_2, x_3) = c\}$ of smooth functions $F(x)$ of three variables $(x_1, x_2, x_3) = x$, of a special "product" form. Our approach is motivated by the following consideration: surfaces usually appear as the boundaries of three-dimensional bodies in $B \subset \mathbb{R}^3$. Let us assume that a connected body $B \subset \mathbb{R}^3$ is defined by the inequalities $F_1(x) \geq 0, \ldots, F_m(x) \geq 0$. For example, this is always the case for the surfaces produced by the Computer Assisted Design - Computer Assisted Manufacturing (CAD-CAM) systems, widely used in engineering. The interior $\hat{B}$ is exactly one of the connected components G_0^i of the set $G_0 = \{F(x) > 0\}$, where $F = F_1 F_2 \ldots F_m$. So our surface is the boundary of $\hat{B} = G_0^i$, and it is a part of the level surface $Y(0) = \{F(x) = 0\}$

If we want to smooth out sharp edges and corners of our surface, one of possibilities is to shift it slightly inside the body B by taking the appropriate component of the surface $Y(\epsilon) = \{F(x, y, z) = \epsilon\}$, where ϵ is a small positive number.

Following this example, we propose as a mathematical model of a "general" surface a level surface $Y(c) = \{F(x) = c\}$ of a smooth function $F(x)$ of the product form as above.

1.3. *The notion of a "general position"*

In order to apply the techniques of Singularity Theory to the description of typical singularities of the surfaces as above, we first of all need an appropriate notion of a "genericity" or of a "general position" for the smooth functions of a product form $F = F_1 F_2 \ldots F_m$. We refer the reader to [1,2,9,10,15] for accurate definitions and for a discussion of this very important notion. In this paper we give only its informal explanation. Assume that our functions F are allowed to vary inside a certain functional space $\mathcal{F}$. The space $\mathcal{F}$ may

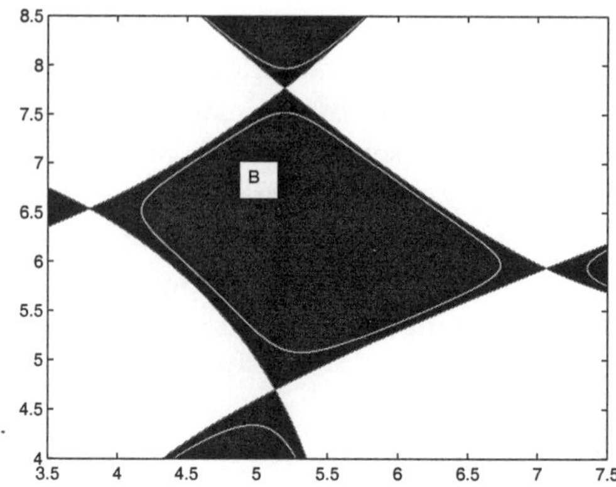

strongly change from one example to another. In particular, it may be the space of C^k functions, of C^∞ or analytic functions, or of polynomials of a given degree with the uniform bound on the coefficients. Consider a certain property P of functions F (like the property to have only non-degenerate critical points).

The property P is generic, with respect to the functional space $\mathcal{F}$, if it is satisfied for "almost all" functions in $\mathcal{F}$.

Another intuitive explanation is that *the property P is satisfied with a probability 1 for a randomly picked $F \in \mathcal{F}$.*

In this case we say also that P is satisfied for F in "general position". The usual in Singularity Theory way to formulate a result about genericity of a certain property P is: "For a generic F (or for F in "general position") the following property P is satisfied ...". We mostly use below this form.

If the functional space $\mathcal{F}$ is finite-dimensional and compact, like in the case of polynomials of a given degree with the uniform bound on the coefficients, the above intuitive explanations are, in fact, accurate mathematical definitions. In the case of C^k, C^∞ or analytic functions, more detailed definitions are necessary, that can be found in the references above.

Consider now, for a given functional space $\mathcal{F}$, the space W of the functions $F(x, y, z)$ of the product form: $F = F_1 F_2 \ldots F_m$, with $F_i \in \mathcal{F}$, $i = 1, \ldots, m$.

We say that the property P of $F = F_1 F_2 \ldots F_m$ is generic, with respect to the functional space W, if it is satisfied for $F = F_1 F_2 \ldots F_m$ with "almost all" functions $F_1, \ldots, F_m \in \mathcal{F}$ (or, if it is satisfied with a probability 1 for a randomly and independently picked $F_1, \ldots, F_m \in \mathcal{F}$.) In this case we say *also that P is satisfied for $F = F_1 F_2 \ldots F_m$ in a general position.*

Under this definition we see that all the usual generic properties of the functions F_i by themselves can be assumed to be satisfied. We can equally assume that these functions are in a general position with respect to one another, in the usual sense of Singularity Theory. However, it is important to stress that many *generic in the usual setting of the classical Singularity Theory properties of F are not generic anymore in the "product" setting.* In particular, the product $F = F_1 F_2 \ldots F_m$ generically has non-isolated singularities along the crossing curves C_{ij} of the surfaces $C_i = \{F_i = 0\}$ and $C_j = \{F_j = 0\}$, $i, j = 1, \ldots, m$. From the point of view of the *classification of singularities of smooth functions in the standard setting* this is a very degenerate situation, appearing only in "codimension infinity". In other words, it cannot appear *in generic families of functions, with any finite number of parameters.*

1.4. *Normal Forms*

Let us remind that a "normal form" is the simplest form to which a given object can be brought by the allowed "normalizing transformations". Of course, in each specific case this informal definition is replaced by an appropriate formal one. The "lists of normal forms" are among the main "outputs" of Singularity Theory, and the quantitative version of normal forms plays an important role in our approach. In the present paper the allowed "normalizing transformations" are the smooth coordinate changes in the source space $\mathbb{R}^3$.

We do not stress below the notion of a "quantitative normal form" (which involves, in particular, the explicit bounds on the size of the coordinate neighborhoods and on the derivatives of the normalizing transformations), although it appears implicitly in several results below.

1.5. *How to use our approach in applications?*

There are several problems which have to be settled before the initial mathematical treatment of surface singularities presented in this paper can be implemented in applied algorithms.

First of all, *how to identify a singularity (or, more importantly, a near-singularity) of the type suggested in this paper, in empiric surfaces?*

Even a purely mathematical version of (one possible setting of) this problem looks difficult: let the surface $Y = Y(c)$ be a level surface $Y(c) = \{F(x_1, x_2, x_3) = c\}$ of a smooth function $F(x_1, x_2, x_3)$. However, we do not assume F to be of the "product" form, as above.

Is it enough to use the normal forms of this paper, and how, for a given F, to identify such singularities and near-singularities?

Another important problem is *to find an efficient representation of the surface singularities and near-singularities, based on their normal forms.* This representation should combine computational efficiency, flexibility and high accuracy of the provided approximation with robustness and noise resistance. The local normal forms described in this paper, are only the initial mathematical building blocks for such a representation.

Generally, it would be very important to find a comprehensive scheme for an efficient representation of smooth and singular surfaces, incorporating the normal forms based representation of the surface singularities and near-singularities, and combining it with a compact and flexible representation of smooth parts.

We plan to present some results in these directions in separate publications.

Let us mention also the important problem of making the "general position" arguments used in this paper (and throughout Singularity Theory) numerically and computationally meaningful. This is a deep problem, which we discuss here only very shortly, in the concluding remarks of Section 4. For some initial discussions and results in this direction see [16,19].

The author would like to thank Dvir Haviv for a careful reading of this paper and for his remarks and suggestions.

2. Singularities of a generic product function

To simplify the presentation, we shall consider only compact surfaces without boundary. Accordingly, the definitions below are arranged in such a way that they imply, in particular, that all the considered level surfaces $Y(c) = \{F(x_1, x_2, x_3) = c\}$ are contained strictly inside the unit ball in $\mathbb{R}^3$.

Consider the class W of smooth (C^k, $k \geq 3$, or C^∞) functions $F(x)$, $x = (x_1, x_2, x_3)$, defined on the unit ball $\mathbb{B}_1 \subset \mathbb{R}^3$, having the form $F = F_1 F_2 \ldots F_m$, where $F_1(x), \ldots, F_m(x)$ are smooth functions on $\mathbb{B}_1$. We always assume in addition that each $F \in W$ is bounded from below by 1 on the boundary sphere $S^1 \subset \mathbb{B}_1$.

Definition 2.1. A surface $Y \subset \mathbb{B}_1$ is a level set $Y = Y(c) = \{F(x) = c\}$ with $F \in W$ and $c \in \mathbb{R}$, $c < 1$.

A shell $S \subset \mathbb{B}_1$ is the boundary of one of the connected components B of the set $\{x = (x_1, x_2, x_3) \in \mathbb{B}_1, \ F(x) < c\}$.

Notice that this definition is not "uniform" with respect to c: the value $c = 0$ plays a special role, in particular, since the function $G = F_1 \cdots F_m - c$ for $c \neq 0$ does not have the product structure anymore. Consequently, below we mostly consider the level set $Y(c)$ for $c = 0$ or for $c = \epsilon$ with ϵ small.

Another important remark is that Definition 2.1 can be naturally generalized in the following way: for each $i = 1, \ldots, m$ we take the connected components of $C_i \setminus \cup_{j \neq i} C_j$ (where, as above, $C_i = \{F_i = 0\}$), and then form unions of some of these components or of their closures. The description of the local structure of such unions (in a generic situation) can be obtained from the results of this paper by simple combinatorial arguments.

Let us remind that we denote by $C_i = \{F_i = 0\}$ the zero surfaces of the functions F_i, $i = 1, \ldots, m$, by $C_{ij} = \{F_i = 0\} \cap \{F_j = 0\}$, $i, j = 1, \ldots, m$, $i \neq j$, the zero curves of the couples of these functions, and by $w_{ijl} = \{F_i = 0\} \cap \{F_j = 0\} \cap \{F_l = 0\}$ - their "triple zeroes".

To clarify the description of the generic singularities below we have to remind the notion of transversality.

Definition 2.2. The intersection of smooth submanifolds $Z_1, \ldots, Z_n$ of $\mathbb{R}^n$ at the point x is called transversal, if the tangent spaces TZ_i to Z_i at x, $i = 1, \ldots, n$, span the entire space $\mathbb{R}^n$.

The intersection of $Z_1, \ldots, Z_n$ at x is transversal if and only if there exists a new local coordinate system $y_1, \ldots, y_n$, centered at $x \in \mathbb{R}^n$, in which Z_i become the coordinate subspaces (defined by the vanishing of some of the coordinates $y_1, \ldots, y_n$) and spanning together the entire space $\mathbb{R}^n$. This can be proved using the Implicit Function Theorem, and we give this proof (in our special situation) in the proof of Theorem 3.1 below.

Implicit Function Theorem provides, in particular, *a normal form of a differential mapping at its regular point*. One of many possible formulations is the following:

Theorem 2.1. *Let $f : \mathbb{R}^n \to \mathbb{R}^m, n \geq m$, be a C^k-mapping, $k \geq 1$, given in a coordinate form by $y_1 = f_1(x_1, \ldots, x_n), \ldots, y_m = f_m(x_1, \ldots, x_n)$ and let the differential $df(0)$ of f at the origin $0 \in \mathbb{R}^n$ be non-degenerate (i.e. it has the maximal possible rank m). Then the functions $y_1, \ldots, y_m$ can be completed to a new C^k-coordinate system $y_1, \ldots, y_m, y_{m+1}, \ldots, y_n$ in a neighborhood of the origin.*

Another important result we have to remind here is the Morse theorem and the notion of the Morse singular point. One part of this theorem is the following:

Theorem 2.2. *Let $f(x_1,...,x_n)$ be a smooth function in a neighborhood of the origin in $\mathbb{R}^n$. Assume that the origin is a critical point of f, i.e. $\mathrm{grad} f(0) = 0$, and the Hessian $H(f)$, i.e. the matrix of the second partial derivatives of f, is non-degenerate at the origin. Then there is a new coordinate system $y_1,...,y_n$, centered at the origin, such that*

$$f(y_1,...,y_n) = y_1^2 + y_2^2 + ... + y_l^2 - y_{l+1}^2 - y_{l+2}^2 - ... - y_n^2 + \ const.$$

Morse singular points of f are those with the Hessian $H(f)$ non-degenerate. Equivalently, at the Morse points the function f can be written in the above form in an appropriate coordinate system. In $\mathbb{R}^3$ we get $\pm f = y_1^2 + y_2^2 + y_3^2 + \ const$ or $\pm f = y_1^2 - y_2^2 - y_3^2 + \ const$.

The following lemma describes the singular structure of a generic function $F \in W$:

Lemma 2.1. *For a generic function $F \in W$ the critical set $\Sigma(F)$ consists of isolated non-degenerate (Morse) points w_i with $F(w_i) \neq 0$, of smooth curves C_{ij}, and of isolated triple points w_{ijl}, being the intersections of the zero surfaces C_i, C_j and C_l (and of the curves C_{ij}, C_{il} and C_{jl}). At the curves C_{ij} the zero surfaces C_i and C_j intersect transversally, and at the triple points w_{ijl} the corresponding triples of the zero surfaces C_i, C_j and C_l intersect transversally.*

Proof. Consider first singular points of F with the singular value zero. Such points may belong either to the parts of the zero surfaces C_i outside of the intersection curves C_{ij}, or to the curves C_{ij}, or they coincide with the triple points w_{ijl}. A priori, non-empty intersections of more than three zero surfaces C_i are also possible.

However, for generic smooth functions F_i we can assume, by the standard results of Singularity Theory, that all the zero surfaces C_i, $i = 1, \ldots, m$ are regular, i.e. that $\mathrm{grad}\, F_i(x) \neq 0$ for any $x = (x_1, x_2, x_3) \in C_i$. We can assume also that all their mutual intersections are transversal. In particular, this implies that generically at the curves C_{ij} the zero surfaces C_i and C_j intersect transversally, and at the triple points w_{ijl} the corresponding triples of the zero surfaces C_i, C_j and C_l intersect transversally. It follows also that generically there are no non-empty intersections of more than three zero surfaces C_i. Then the following easy calculation shows that the points belonging to the parts of the zero surfaces $F_i = 0$ outside of the intersection curves C_{ij}, are in fact regular points of F: for $F = F_1 \ldots F_m$, $\mathrm{grad}\, F = (\mathrm{grad}\, F_1) \cdot F_2 \ldots F_m + (\mathrm{grad}\, F_2) \cdot F_1 F_3 \ldots F_m + \cdots + (\mathrm{grad}\, F_m) \cdot F_1 \ldots F_{m-1}$. If at a certain point $x \in C_i$ we have $F_i(x) = 0$, $F_j(x) \neq 0$ for $j \neq i$, and

$grad\ F_i(x) \neq 0$, then the above formula shows that $grad\ F(x) \neq 0$. The same formula shows also that the points of the curves C_{ij} and the triple points w_{ijl} are indeed singular points of F.

In particular, there are no critical points of F on its zero level surface, besides the points of C_{ij} and the triple points w_{ijl}. Notice that in three dimensional space these singular points of F are definitely more degenerate than the Morse points. Indeed, in each case there are adjacent one dimensional strata of singular points, while the singularity at the Morse point is always isolated. The formula for the $grad\ F$ given above shows also that at the triple point all the second order derivatives vanish, so the Hessian of F cannot be nonzero.

It remains to show only that for a generic $F \in W$ all the critical points w with the critical value $F(w) \neq 0$ are non-degenerate (Morse) points. But at such points each of the factors F_j, $j = 1, \ldots, m$ does not vanish. Then we can use essentially the same proof as in [12]. Namely, consider all the linear functions l on $\mathbb{R}^3$. Denoting by $\tilde{l}$ the function $\frac{l}{F_2 F_3 \ldots F_m}$ which is smooth near w we have $F_l = (F_1 + \tilde{l})F_2 F_3 \ldots F_m = F + l$. We observe that F_l has degenerated critical points if and only if l is the critical value of the mapping $grad\ F : \mathbb{B}_1 \to \mathbb{R}^3$. Now applying Sard theorem we show that for almost all l locally near w the function F_l has only Morse singularities. This completes the proof of Lemma 2.1. $\qquad\square$

Remark 2.1. We formulate all the results in the present paper only for surfaces in three-dimensional space. Most of these results remain true for hypersurfaces in any dimension $n \geq 2$, just the statements become much less transparent for $n > 3$. In particular, in the plane, a generic product function has a zero set consisting of smooth curves, transversally crossing one another at double points. Notice that these double points correspond in dimension $n = 2$, in contrast to the case $n \geq 3$, to the Morse points of the index one of the product function F.

Remark 2.2. Our definition of a surface (Definition 2.1 above) and our notion of the "general position" for the product functions exclude the possibility for a generic surface to have "ridges" and "corners" *together* with the Morse type singularities. Indeed, by Lemma 2.1, the level surfaces $Y(c) = \{F = c\}$ for $c \neq 0$ may contain Morse points, but cannot have corners (all the singular points of F with the critical value $c \neq 0$ are isolated). On the other hand, the zero level surface $Y(0)$ may have "ridges" and "corners", but not Morse points. By changing slightly the definitions we can have both the types of singularities on generic surfaces. Consider the products F of the form $F = (F_1 - c_1) \cdots (F_i - c_i) \cdots (F_m - c_m)$. Defining our surfaces as the level sets of such products, we get generically ridges, corners, and Morse points on the same surface. However, varying freely the constants c_i, we may get Morse points on the ridges, as well as other "higher codimension" configurations. We can exclude this by some additional assumptions, but the statement of the results becomes much less transparent.

3. Stable singularities of surfaces

In this section we prove the stability of all the possible types of local singularities of generic surfaces, and produce their normal forms. Consider the surface $Y(0) = \{F(x_1, x_2, x_3) = 0\}$, $F(0) = 0$, in a neighborhood of the origin. We distinguish the following four special cases:

(1) F at the origin has a regular point.
(2) F at the origin has a a non-degenerate singularity (Morse point).
(3) F at the origin has the form $F = F_1 F_2$, with F_1 and F_2 vanishing at the origin and the gradients $grad\ F_1(0)$ and $grad\ F_2(0)$ are linearly independent.
(4) F at the origin has the form $F = F_1 F_2 F_3$, with F_1, F_2 and F_3 vanishing at the origin and the gradients $grad\ F_1(0)$, $grad\ F_2(0)$ and $grad\ F_3(0)$ are linearly independent.

Consider now the following four "model surfaces" (normal forms):
1. $Y_1 = \{y_1 = 0\}$, $S_1 = Y_1$.
2. $Y_2^0 = \{0\}$, $Y_2 = \{y_1^2 - y_2^2 - y_3^2 = 0\}$, $S_2 = Y_2 \cap \{y_1 \geq 0\}$.
3. $Y_3 = \{y_1 y_2 = 0\}$, $S_3 = Y_3 \cap \{y_1 \geq 0,\ y_2 \geq 0\}$.
4. $Y_4 = \{y_1 y_2 y_3 = 0\}$, $S_4 = Y_4 \cap \{y_1 \geq 0,\ y_2 \geq 0,\ y_3 \geq 0\}$.

Before we can state the main theorem of this section, we need also a definition of "structural stability":

Definition 3.1. A singularity of a surface $Y(c) = \{F = c\}$ or of a shell $S(c) \subset Y(c)$ at a certain its point x is called (structurally) stable, if there is a neighborhood U of x in $\mathbb{R}^3$ with the following property: on each surface $\tilde{Y}(c) = \{\tilde{F} = c\}$ (or on a shell $\tilde{S}(c) \subset \tilde{Y}(c)$), with $\tilde{F}$ sufficiently close to F, there is a point $\tilde{x}$ and a neighborhood $\tilde{U}$ of $\tilde{x}$ such that the couples $(U, Y(c))$ and $(\tilde{U}, \tilde{Y}(c))$ are diffeomeorphic. (Respectively, for the shells, the couples $(U, S(c))$ and $(\tilde{U}, \tilde{S}(c))$ are diffeomeorphic.)

In our "product" setting we have also to explain what does it mean that $\tilde{F}$ is sufficiently close to F. We understand this in the following way: $F =

$F_1 \cdot \ldots \cdot F_m$, $\tilde{F} = \tilde{F}_1 \cdot \ldots \cdot \tilde{F}_m$, and each $\tilde{F}_i$ is sufficiently close to F_i in the C^k-norm.

Now we are ready to prove the stability and classification theorem:

Theorem 3.1. *For a generic function $F \in W$ and for each $c \in \mathbb{R}$ the surface $Y(c) = \{F(x_1, x_2, x_3) = c\}$ (respectively, the shell $S(c) \subset Y(c)$ at each of its points has singularities only of the form (1)-(4) above. These singularities are stable with respect to small perturbations of $F \in W$. In a neighborhood of each of the points of the type (i), $i = 1, 2, 3, 4$, there is a smooth coordinate system y_1, y_2, y_3 such that in the new coordinates the surface $Y(c)$ has the normal form Y_i (for the Morse points (type (2)) also the normal form Y_2^0 is possible.) The normal forms of the shell singularities are S_i, respectively.*

Proof. The first part of the required result follows from Lemma 1.1. Indeed, if $c \in \mathbb{R}$ is a regular value of F then at each point of Y_c this surface has the form (1). If $c \in \mathbb{R}, c \neq 0$ is a singular value of F, then all the singularities of Y_c are Morse points by Lemma 2.1. This corresponds to the case (2). Finally, for $c = 0$ the singular points of Y_0 are either the points of C_{ij} or the triple points w_{ijl}. In the first case we take $\tilde{F}_1 = F_i$, $\tilde{F}_2 = F_1 \ldots F_{i-1} F_{i+1} \ldots F_m$. Then $F = \tilde{F}_1 \tilde{F}_2$, and it is easy to check that the gradients of $\tilde{F}_1$, $\tilde{F}_2$ are linearly independent. This corresponds to the case (3). At the triple point w_{ijl} we take $\tilde{F}_1 = F_i$, $\tilde{F}_2 = F_j$, $\tilde{F}_3 = F_1 \ldots F_{i-1} F_{i+1} \ldots F_{j-1} F_{j+1} \ldots F_m$. Then $F = \tilde{F}_1 \tilde{F}_2 \tilde{F}_3$, and it is easy to check that the gradients of $\tilde{F}_1$, $\tilde{F}_2$ and $\tilde{F}_3$ are linearly independent. This corresponds to the case (4).

The stability of the properties (1)-(4) defining these four types of singularities follows from the fact that transversality condition, as well as the condition of the Hessian to be non-degenerate, are *open* and so they persist small perturbations. We complete the proof of the structural stability of these singularities (and, in particular, the existence of the diffeomorphism between the original and the perturbed singularities) after the proof of the reduction to the normal forms.

To prove the existence of the "normalizing" coordinate system y_1, y_2, y_3 in cases (1), (3), (4), we use the Implicit Function Theorem (Theorem 2.1 above). In the case (1) we have just one function F, which is non-degenerate at the origin. Consequently, we can take it as the first coordinate y_1.

In the cases (3) and (4) we take as the new coordinates y_1, y_2 (respectively, y_1, y_2, y_3) the functions F_1, F_2 (respectively, F_1, F_2, F_3).

Finally, in the case (2) we use the Morse Theorem (Theorem 2.2 above). As we apply this theorem to our function F we get (after multiplying the equation, if necessary, by -1) either $F = y_1^2 + y_2^2 + y_3^2$, in which case the surface degenerates to the point Y_2^0, or $F = y_1^2 - y_2^2 - y_3^2$, which gives Y_2. This completes the proof of the reduction to the normal form for the case of surfaces.

As the shells are concerned, we just notice that the condition for the part of the level surface $Y(c)$ to be the boundary of one of the the connected components of $\{F < c\}$ cuts out from the normal forms Y_i exactly the normal forms S_i, as defined above.

Now to complete the proof of the structural stability of our singularities, we just notice that also after a perturbation they satisfy the same conditions (1)-(4), and consequently, they can be brought *to the same normal forms* by the appropriate change of coordinates. The composition of the transformation of the original singularity to its normal form, and then from the normal form to the deformed singularity (and back) provides the required diffeomorphisms.

This completes the proof of Theorem 3.1. $\qquad\qquad\square$

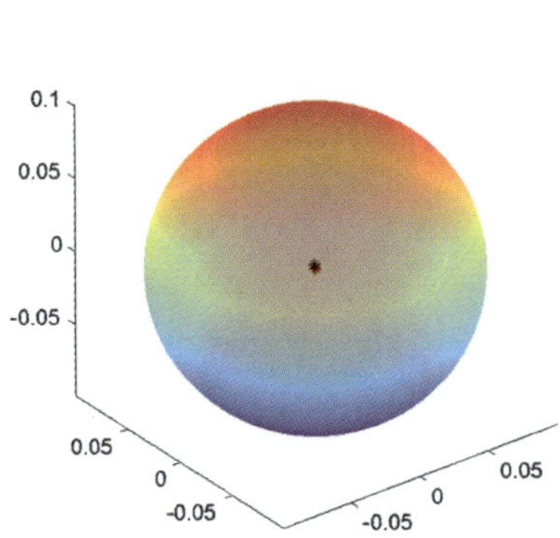

The key point in the applications of Theorem 3.1 is the existence of the inverse coordinate transformation from the new coordinates y_1, y_2, y_3 to the old ones x_1, x_2, x_3:

$$x_1 = \Psi_1(y_1, y_2, y_3), \quad x_2 = \Psi_2(y_1, y_2, y_3), \quad x_3 = \Psi_3(y_1, y_2, y_3). \qquad (1)$$

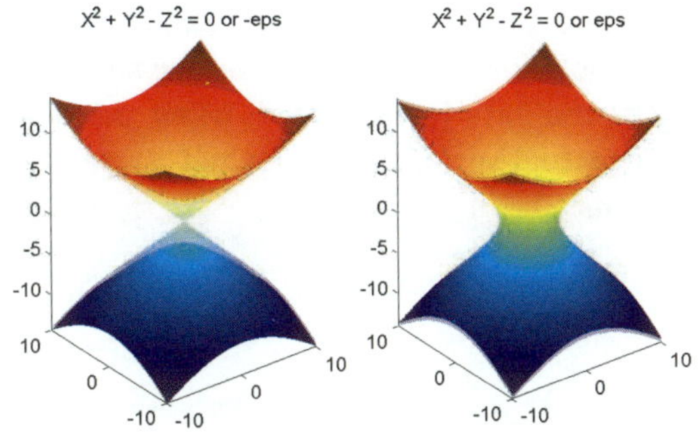

X^2 + Y^2 - Z^2 = 0 or -eps
X^2 + Y^2 - Z^2 = 0 or eps

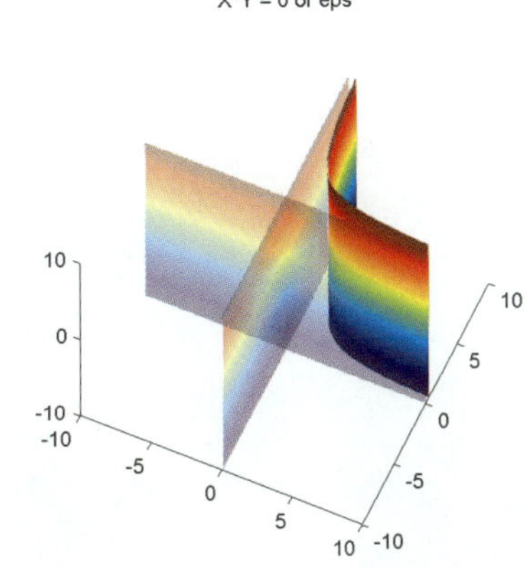

X*Y = 0 or eps

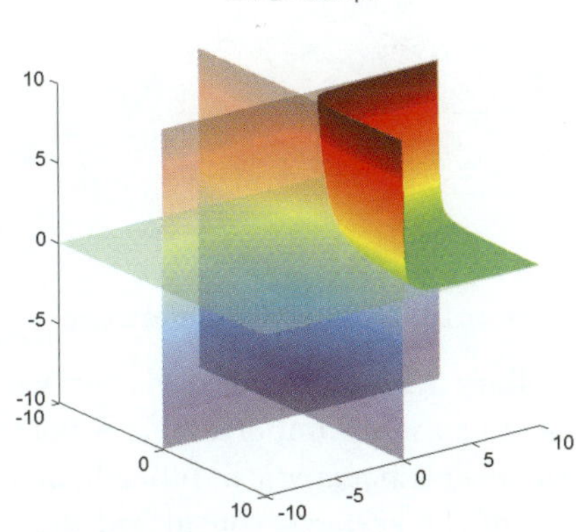

X*Y*Z = 0 or eps

370

and a possibility to find this transformation explicitly (see [17] where the second order Taylor polynomials (jets) of the normalizing transformations are given in terms of the data jets). Indeed, using the expressions (3.1) we can parametrize our actual surface Y (or a shell S) near its singular points by the "Normal Forms" Y_1, Y_2, Y_3, Y_4 ($S_1 - S_4$, respectively):

Corollary 3.1. *Under the assumptions of Theorem 3.1 in a neighborhood of each of its points the surface $Y(c)$ can be parametrized as follows:*

$$x_1 = \Psi_1(y_1,\ y_2,\ y_3),\ x_2 = \Psi_2(y_1,\ y_2,\ y_3),\ x_3 = \Psi_3(y_1,\ y_2,\ y_3),$$

with $(y_1,\ y_2,\ y_3) \in \tilde{Y}$, where $\tilde{Y} = Y_1$, Y_2, Y_2^0, Y_3 or Y_4 for the singular points of the types (1), (2), (3) or (4), respectively. In a neighborhood of each of its points the shell $S(c)$ can be parametrized via the same expressions as above, with $(y_1,\ y_2,\ y_3) \in \tilde{S}$, where $\tilde{S} = S_1$, S_2, S_3 or S_4 for the singular points of the types (1), (2), (3) or (4), respectively.

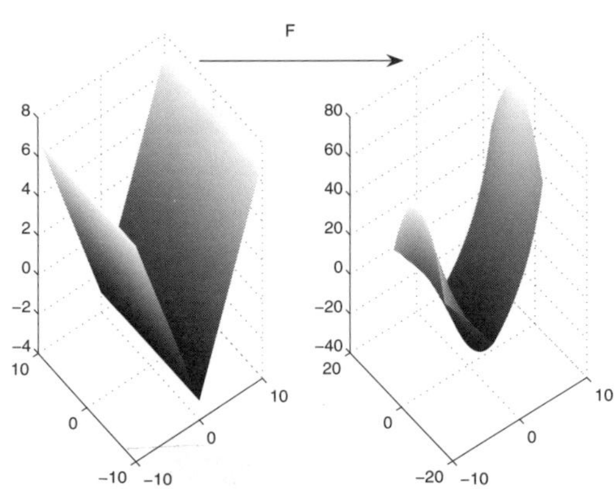

4. Near-singularities and Organizing Center

Theorem 3.1 and Corollary 3.1. describe generic surface singularities as they appear in our setting. Even more important is a possibility to completely describe their generic *near-singularities*. Indeed, in most of applications the edges and vertices of the surfaces considered *are not absolutely sharp*. They appear rather as a result of a certain "smoothenning" of the sharp prototypes.

It would be highly desirable to associate each "near-singularity" with a certain virtual "absolute singularity" and to use the Normal form of the last for the description of the first. This is a special case of a well known and very important *problem of finding the "Organizing Center"* in a terminology of R. Thom ([15]). The results of this section show that indeed in our setting each near-singularity can be associated with a "true singularity" of one of the types described above; hence we can rigorously define and apply in our context the notion of the Organizing Center.

First of all, let us define a notion of a "controlled neighborhood":

Definition 4.1. The controlled neighborhood $U(x)$ of a generic singular point x of a surface (or of a shell) is its neighborhood, covered by the coordinate system $(y_1,\ y_2,\ y_3)$, as defined in Theorem 3.1.

Now we can formulate our first "quantitative" result:

Theorem 4.1. *Let $F \in W$ be a generic function. There exists a constant $K = K(F)$ such that for any $c \in \mathbb{R}$ and $Y(c) = \{F = c\}$ the following is true: each point $x \in Y(c)$, where the sum of the absolute values of the main curvatures of $Y(c)$ at x exceeds K, belongs to a controlled neighborhood of one of the singular points of F.*

Proof. We give only a sketch of the proof. First of all, we need a notion of a "near-critical" point.

Definition 4.2. The point x is called γ-critical for F, if

$$\|grad\ F(x)\| \leq \gamma.$$

The value $F(x)$ of f at its γ-critical point x is called a γ-critical value of F.

We do not formalize the notions of a near-critical point and near-critical value, applying this name to the γ-critical points and γ-critical values of F, for γ small. The following proposition relates the curvature of the level surface at a point x with the "degree of regularity" of this point:

Proposition 4.1. *Assume that all the derivatives up to order 3 of F are uniformly bounded by K. There is an explicit function $H(K, \gamma)$, tending to infinity, as γ tends to zero, such that for $\|grad\ F(x)\| = \gamma > 0$ the absolute values of the main curvatures of $Y(c) = \{F = c\}$ at x are bounded from above by $H(K, \gamma)$.*

Thus, the curvature of the level surface $Y(c) = \{F = c\}$ at x may be high only if x is a near-critical point and $c = F(x)$ is a near-critical value of F.

The geometry of near critical points and values of smooth functions has been studied in many recent publications (see [16,17,19] and references there). In particular, the "Quantitative Sard Theorem" proved in [16,19] shows that if the function F has enough continuous derivatives, then the size of its γ-critical values tend to zero as γ tends to zero.

Moreover, the following result is proved, via the Quantitative Sard Theorem, in [17]:

Theorem 4.2. *Let a C^k function $f_0 : B^n \to \mathbb{R}$ be given, $k > n$, with all the derivatives up to order k uniformly bounded by K. There is an explicit function $\eta(K, \epsilon) > 0$ such that for any given $\epsilon > 0$, we can find a linear function h with $\|h\| \leq \epsilon$, such that $f = f_0 + h$ has the following two properties:*

(1) All the critical points are non-degenerated Morse points.
(2) Each point $x \in B^n$ with the norm of the grad $f(x)$ smaller than $\eta(K, \epsilon)$ belongs to one of the controlled neighborhoods of the singular points x_i of f.

Theorem 4.2 implies the desired result of Theorem 4.1 for the Morse singularities of F. Indeed, since we consider only the generic functions F, we may assumed that the property of Theorem 4.2 is satisfied for F (otherwise, we perturb it by adding an appropriate linear h). Now, if the curvature of the level surface $Y(c) = \{F = c\}$ at x is high, then by Proposition 4.1, the norm of the *grad $F(x)$* is small. As this norm becomes smaller than η from Theorem 4.2, the point x must enter one of the controlled neighborhoods of the Morse points of F.

As for the "product" singularities of the zero level surface of F, applying the Quantitative Sard Theorem, we can prove a result similar to Theorem 4.2 also for this type of singularities. Then the same considerations as above settle also the near-critical points of F approaching its zero level surface. This completes the proof of Theorem 4.1. $\qquad\square$

Corollary 4.1. *At each point $x \in Y(c)$ where the sum of the absolute values of the main curvatures of $Y(c)$ at x exceeds K the surface $Y(c)$ has the form either*

(1) $F(x_1, x_2, x_3) = \epsilon$, where F at the origin has a a non-degenerate singularity (Morse point), or

(2) $F(x_1, x_2, x_3) = \epsilon$, where F at the origin has the form $F = F_1 F_2$. Here F_1 and F_2 vanish at the origin and the gradients $\mathrm{grad}\, F_1(0)$ and $\mathrm{grad}\, F_2(0)$ are linearly independent, or

(3) $F(x_1, x_2, x_3) = \epsilon$, where F at the origin has the form $F = F_1 F_2 F_3$. Here F_1, F_2 and F_3 vanish at the origin the gradients $\mathrm{grad}\, F_1(0)$, $\mathrm{grad}\, F_2(0)$ and $\mathrm{grad}\, F_3(0)$ are linearly independent.

In each of this cases ϵ is assumed to be a sufficiently small positive constant.

Proof. This follows directly from Theorem 4.1 and Theorem 3.1, describing the generic singularities of $Y(c)$. The case of the regular point of $Y(c)$ is naturally excluded here, since the curvatures of the level surfaces in a neighborhood of a regular point are uniformly bounded (for instance, via Proposition 4.1). $\qquad\square$

Let us use the following notations:
1. $Y_2^+(\epsilon) = \{y_1^2 + y_2^2 - y_3^2 = \epsilon\}$, $S_2^+(\epsilon) = Y_2^+(\epsilon)$.
2. $Y_2^-(\epsilon) = \{y_1^2 + y_2^2 - y_3^2 = -\epsilon\}$, $S_2^-(\epsilon) = Y_2^-(\epsilon) \cap \{y_3 > 0\}$.
3. $Y_3(\epsilon) = \{y_1 y_2 = \epsilon\}$, $S_3(\epsilon) = Y_3(\epsilon) \cap \{y_1 > 0,\ y_2 > 0\}$.
4. $Y_4(\epsilon) = \{y_1 y_2 y_3 = \epsilon\}$, $S_4(\epsilon) = Y_4(\epsilon) \cap \{y_1 > 0,\ y_2 > 0,\ y_3 > 0\}$.

Corollary 4.2. *At each regular point $x \in Y(c)$ of the surface $Y(c)$, where the sum of the absolute values of the main curvatures of $Y(c)$ at x exceeds K, this surface can be parametrized as follows:*

$$x_1 = \Psi_1(y_1,\ y_2,\ y_3),\ x_2 = \Psi_2(y_1,\ y_2,\ y_3),\ x_3 = \Psi_3(y_1,\ y_2,\ y_3),$$

with $(y_1,\ y_2,\ y_3) \in \tilde{Y}(\epsilon)$, where $\tilde{Y}(\epsilon) = Y_2^{\pm}(\epsilon)$, $Y_3(\epsilon)$, $Y_4(\epsilon)$ for the singular points of the types (2), (3) or (4), respectively.

At each regular point $x \in S(c)$ of the shell $S(c)$, where the sum of the absolute values of the main curvatures of $S(c)$ at x exceeds K, this shell can be parametrized as follows:

$$x_1 = \Psi_1(y_1,\ y_2,\ y_3),\ x_2 = \Psi_2(y_1,\ y_2,\ y_3),\ x_3 = \Psi_3(y_1,\ y_2,\ y_3),$$

with $(y_1,\ y_2,\ y_3) \in \tilde{S}(\epsilon)$, where $\tilde{S}(\epsilon) = S_2^{\pm}(\epsilon)$, $S_3(\epsilon)$, $S_4(\epsilon)$ for the singular points of the types (2), (3) or (4), respectively.

Remark 4.1. More accurate quantitative results can be obtained here. They are motivated by the following question:

Is it possible to make the notion of a "general position" quantitative, and, in particular, applicable in numerical computations?

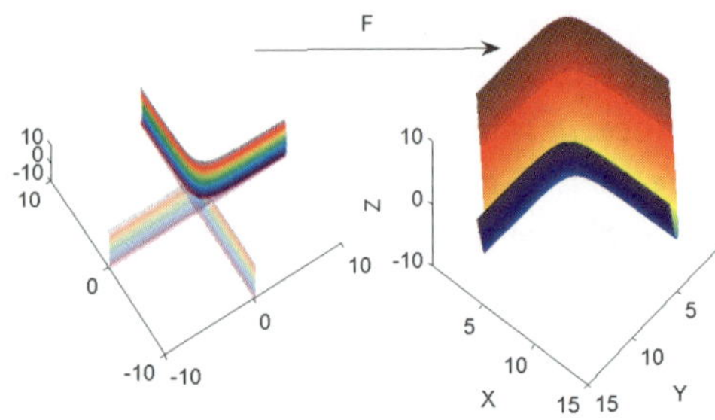

There are several possible ways to achieve this goal. One way is to take the following interpretation of a general position: the property P of F is satisfied for F in a general position, if it can be achieved by an arbitrarily small perturbation of the initial function F_0. A typical form of a quantitative result in this direction is:

For each positive ϵ there exists a perturbation $F_h = F + h$ of F with $\|h\| \leq \epsilon$, such that for the perturbed function F_h the property P is satisfied with the explicit estimates of the relevant parameters in terms of ϵ.

Another interpretation of the genericity of the property P - that it is satisfied for a randomly picked F with probability 1 - leads to the following form of a corresponding quantitative result:

For each probability p, $0 \leq p < 1$, and for a randomly picked F, with the probability at least p the property P is satisfied with the explicit estimates of the relevant parameters in terms of p.

Some examples of quantitative results in Singularity Theory, having this form, one can find in [17]. For the surface singularities the most important parameters of the "general position" are the size of the controlled neighborhood of the singularity, the size of the derivatives of the normalizing change of variables, and the bound on the curvature of the point, implying this point to be covered by one of the controlled neighborhoods. All these parameters can be included into a quantitative framework, as above. We plan to present some results in this direction separately.

Another important question is to describe higher-codimension singularities of $F \in W$, and not only the stable ones, as in this paper.

References

1. J. M. Boardman, Singularities of differential maps. *Publ. Math. I.H.E.S.* **33**, (1967), 21-57.
2. Th. Bröcker; L. Lander, Differentiable Germs and Catastrophes, London Math. Society Lecture Notes Series **17**, Cambridge University Press, 1975.
3. L. N. Bryzgalova, Singularities of a maximum of a function depending on parameters. *Funct. Anal. Appl.* **11**, (1977), 49-50.
4. J. Damon, Generic structure of two-dimensional images under Gaussian blurring, *SIAM J. Appl. Math.* **59** (1999), no. 1, 97-138.
5. J. Damon, Properties of Ridges and Cores for Two-Dimensional Images, *J. Math. Imaging Vision* **10** (1999), no. 2, 163-174.
6. J. Damon, Smoothness and geometry of boundaries associated to skeletal structures I: sufficient conditions for smoothness. *Ann. de l'Institut Fourier* **53**, No. 6 (2003), 1941-1985.
7. A. A. Davydov; V. M. Zakalyukin, Point singularities of a conditional minimum on a three-dimensional manifold. *Tr. Mat. Inst. Steklova* **220** (1998), *Optim. Upr., Differ. Uravn. i Gladk. Optim.*, 113-129.
8. J. H. G. Fu, Curvatures measures of subanalytic sets, *Amer. J. Math.* **116** (1994), 819-880.
9. V. Golubitski; V. Guillemin, Stable mappings and their singularities. *Graduate Texts in Math.* **14**, (1973).
10. J. Guckenheimer, Catastrophes and Partial Differential Equations, *Ann. Inst. Fourier* **23** (1973), 31-59.
11. V. I. Matov, Topological classification of germs of maximum and minimax functions generic families. *Uspekhi Mat. Nauk* **37**, 4, (1982), 167-168.
12. J. Milnor, Morse Theory, *Princeton Univ. Press*, Princeton, N.J. 1963.
13. J. M. Morvan, B. Thibert, On the approximation of a smooth surface with a triangulated mesh, *Comp. Geometry* **23** (2002) 337-352.
14. V. D. Sedyh, On the topology of singularities of Maxwell sets, *Moscow Math. Journal* **3**, No. 3, (2003), 1097-1112.
15. R. Thom, Stabilité structurelle et morphogénese, *W. A. Benjamin, Inc.*, 1972. English edition: R. Thom, Structural Stability and Morphogenesis, *Benjamin-Cumming, Inc.*, London-Amsterdam-Sydney-Tokyo, 1975.
16. Y. Yomdin, The geometry of critical and near-critical values of differentiable mappings, *Math. Ann.* **264**, (1983), n. 4. 495-515.
17. Y. Yomdin, Some quantitative results in Singularity Theory, to appear.
18. Y. Yomdin, C^k and analytic reparametrization of semialgebraic sets, preprint.
19. Y. Yomdin, G. Comte, Tame Geometry with Applications in Smooth Analysis, *Lecture Notes in Mathematics*, **1834**, Springer, Berlin, Heidelberg, New York, 2004.

PART III

Geometry and Topology of Singularities

HODGE-RIEMANN RELATIONS FOR POLYTOPES
A GEOMETRIC APPROACH

GOTTFRIED BARTHEL AND LUDGER KAUP

Fachbereich Mathematik und Statistik
Universität Konstanz, Fach D 203
D-78457 Konstanz, Deutschland
E-mail: Gottfried.Barthel@uni-konstanz.de
E-mail: Ludger.Kaup@uni-konstanz.de

J.-P. BRASSELET

IML/CNRS, Luminy Case 907
F-13288 Marseille Cedex 9, France
E-mail: jpb@iml.univ-mrs.fr

KARL-HEINZ FIESELER

Matematiska Institutionen
Uppsala Universitet
SE-751 06 Uppsala, Sverige
E-mail: Karl-Heinz.Fieseler@math.uu.se

The key to the Hard Lefschetz Theorem for combinatorial intersection cohomology of polytopes is to prove the Hodge-Riemann bilinear relations. In these notes, we strive to present an easily accessible proof. The strategy essentially follows the original approach of [Ka], applying induction à la [BreLu2], but our guiding principle here is to emphasize the geometry behind the algebraic arguments by consequently stressing polytopes rather than fans endowed with a strictly convex conewise linear function. It is our belief that this approach makes the exposition more transparent since polytopes are more appealing to our geometric intuition than convex functions on a fan.

Keywords: Combinatorial intersection cohomology; polytopes; Hard Lefschetz.

1. Introduction

The proof of the Hard Lefschetz Theorem for the "Combinatorial Intersection Cohomology" of polytopes given in [Ka] was the keystone in a long endeavour of several research groups to verify that Stanley's generalized ("toric") h-vector for polytopes has the conjectured properties: The theorem (usually referred to as "**HLT**" in the sequel) implies that the generalized h-vector agrees with the vector of even degree Intersection Cohomology Betti numbers and that this vector enjoys the unimodality property (in addition to symmetry and non-negativity).

The HLT is an easy consequence of the so-called bilinear "Hodge-Riemann relations" ("HR relations" or "**HRR**" for short); and since the latter, being a "positivity result", reflect convexity in a more appropriate way than the HLT, the focus has shifted towards proving these relations. The first proof of the HRR given in [Ka] has been rather involved. The task of making it more easily accessible has been taken up in different articles, cf. [BreLu$_2$] and [BBFK$_3$]. With the present notes, we further pursue this direction: Being convinced that polytopes are closer to our geometric intuition, we present an approach that stresses geometric operations on polytopes rather than algebraic operations on strictly convex conewise linear functions.

Let us briefly recall the setup, referring to section 4 for further details: To an n-dimensional polytope P in an n-dimensional real vector space V, one associates its outer normal fan $\Delta = \Delta(P)$ in the dual vector space V^*, and a conewise linear strictly convex function ψ. The "combinatorial intersection cohomology" $IH(\Delta)$ is a finite-dimensional real vector space with even grading $\bigoplus_{k=0}^{n} IH^{2k}(\Delta)$. There is a perfect pairing

$$\cap : IH^q(\Delta) \times IH^{2n-q}(\Delta) \longrightarrow \mathbb{R} ,$$

the "intersection product", so Poincaré duality holds on $IH(\Delta)$.

On $IH(\Delta)$, multiplication with ψ induces an endomorphism

$$L : IH^q(\Delta) \longrightarrow IH^{q+2}(\Delta)$$

called the **Lefschetz operator**. The key result of [Ka] (see also BreLu$_2$) reads as follows:

Combinatorial Hard Lefschetz Theorem (HLT)

Theorem 1.1. *For each $k \geq 0$, the iterated Lefschetz operator*

$$L^k : IH^{n-k}(\Delta) \longrightarrow IH^{n+k}(\Delta)$$

is an isomorphism.

By Poincaré duality, it suffices to prove that each map L^k be injective or surjective.

Using the intersection product, the Hard Lefschetz Theorem can be restated in a different framework: For each $k \geq 0$, the mapping L^k yields a bilinear form

$$s_k \colon IH^{n-k}(\Delta) \times IH^{n-k}(\Delta) \longrightarrow \mathbb{R}, \quad (\xi, \eta) \longmapsto \xi \cap L^k(\eta),$$

called the k-th **Hodge-Riemann bilinear form**, or **"HR-form"** for short. It is symmetric since L is self-adjoint with respect to the intersection product. In this set-up, the HLT is equivalent to the non-degeneracy of all forms s_k.

Beyond non-degeneracy, the HR relations provide explicit formulæ for the signatures of these pairings (see Proposition 1.2). For a proof, one considers the **primitive intersection cohomology**

$$IP^{n-k}(\Delta) \ := \ \ker\bigl(L^{k+1} : IH^{n-k}(\Delta) \longrightarrow IH^{n+k+2}(\Delta)\bigr)$$

(with $k \equiv n \mod 2$). In fact, assuming the HLT, there is an s_k-orthogonal decomposition

$$IH^{n-k}(\Delta) \ = \ L\bigl(IH^{n-k-2}(\Delta)\bigr) \oplus IP^{n-k}(\Delta).$$

More generally, we see:

Proposition 1.1. *If the HLT holds for the Lefschetz operator L on the intersection cohomology of the fan $\Delta = \Delta(P)$, then, for each k, the intersection cohomology splits as an orthogonal direct sum*

$$IH^{n-k}(\Delta) \ = \ \bigoplus_{j \geq 0} L^j\bigl(IP^{n-k-2j}(\Delta)\bigr).$$

Now for each $q \leq n-2$, the restricted operator L provides an isometric embedding $IH^q(\Delta) \hookrightarrow IH^{q+2}(\Delta)$ with respect to the pertinent HR-forms. Hence, in order to determine the signature of s_k, it suffices to consider the restrictions of the Hodge Riemann forms s_{k+2j} to the corresponding primitive subspaces $IP^{n-k-2j}(\Delta)$. Our aim is to give a proof of the following result:

Hodge-Riemann bilinear relations (HRR)

Theorem 1.2. *For each $k \geq 0$ (with $k \equiv n \mod 2$), the Hodge-Riemann bilinear form s_k is $(-1)^{(n-k)/2}$-definite on $IP^{n-k}(\Delta)$.*

The HR relations imply the HLT, since the HR forms are readily seen to be non-degenerate by descending induction on k: That follows from $IP^0(\Delta) = IH^0(\Delta)$ for $k = n$, whereas for $k < n$, we assume that s_{k+2} is non-degenerate. Then so is the restriction of s_k to $L\big(IH^{n-k-2}(\Delta)\big)$. This implies

$$IH^{n-k}(\Delta) \;=\; L\big(IH^{n-k-2}(\Delta)\big) \oplus L\big(IH^{n-k-2}(\Delta)\big)^{\perp},$$

and it now suffices to prove $L\big(IH^{n-k-2}(\Delta)\big)^{\perp} = IP^{n-k}(\Delta)$. The inclusion "$\supset$" follows from the fact that L is $\cap$-self-adjoint, while "$\subset$" is a consequence of Poincaré duality for the complementary dimensions $n-k-2$ and $n+k+2$.

From Proposition 1.1, we immediately obtain a reformulation of the HRR in which the primitive cohomology does not enter explicitly:

Proposition 1.2. *The HRR are equivalent to the HLT together with the additional condition that the Hodge-Riemann bilinear forms s_k on $IH^{n-k}(\Delta)$ satisfy the "**HR-equation**"*

$$\mathrm{sign}(s_k) \;=\; \mathrm{sign}(s_{k+2}) + (-1)^{(n-k)/2}\big(b_{n-k} - b_{n-k-2}\big),$$

where $b_q := \dim_{\mathbb{R}} IH^q(\Delta)$ denotes the qth intersection cohomology Betti number of the fan Δ.

2. Outline of the proof of the Hodge-Riemann relations

The HR relations are known to hold if the polytope P is simple. The first proof has been given in [Mc]; a simplified version followed in [Ti]. This result is the basis for the proof of the general case by a twofold induction: The "outer loop" is on the dimension $n := \dim(P)$. For the more involved "inner loop", following [BreLu$_2$], we associate to P an integer $\mu := \mu(P) \geqq 0$ that measures how far P is from being simple: It counts those faces, here called "normally stout" (see Definition 3.5), that witness non-simplicity, with $\mu = 0$ characterizing simple polytopes. The inner induction on μ requires three main steps:

Cutting off (see Section 3):

> Given a face $F \prec P$, we consider an affine hyperplane H intersecting the relative interior of P that is sufficiently near and parallel to a supporting hyperplane for the face F. Let $P = G \cup R$ be the corresponding decomposition of P into the "germ $G = G_P(F)$ of P along the face F" and the "residual polytope" R.
>
> If the face F is normally stout and of minimal dimension, then $\mu(R) < \mu(P)$, so the HRR hold for R by induction hypothesis. For the investigation of the germ G, it is important that the face F

itself is a simple polytope and that it is "normally trivial" in P, cf. Lemma 3.4.

HRR for special n-polytopes (see Section 5):

From the assumption that the HRR hold for lower-dimensional polytopes, we prove the validity of the HRR for the following special n-polytopes:

5.1 A pyramid $P = \Pi(Q)$ with an $(n-1)$-dimensional base Q.

5.2 A product $P = S \times P_0$, where S is simple and where $\dim P_0 < n$.

Furthermore, we establish the following **Gluing property**:

5.3 The HRR hold for an n-polytope P that can be cut "transversally" into two polytopes P_1 and P_2 such that the HRR hold for both pieces.

Deformation of the germ G into a product (see Section 6):

There is a continuous family $(Q_t)_{t \in [0,1]}$ of pairwise combinatorially equivalent polytopes with $Q_1 = G$ and $Q_0 = F \times \Pi(L)$ with the pyramid $\Pi(L)$ over a "link" $L = L_P(F)$ of F in P. Then the HRR are valid for $Q_1 = G$ if and only that holds for Q_0.

Using these results, the HRR for the polytope P are proved as follows: By induction hypothesis, they hold for the lower dimensional polytopes F and L. Hence, by the results of Subsections 5.1 and 5.2 stated above, they also hold for $Q_0 = F \times \Pi(L)$, and thus, for the germ G. Eventually, the gluing result of Subsection 5.3 applied with $P_1 = R$ and $P_2 = G$ as introduced in the Step "Cutting off" yields the HRR for the initial polytope P.

We recall the definition and basic properties of combinatorial intersection cohomology as needed later on, see Section 4.

3. Cutting off

In this section, we explain how a polytope can be made simple by successively cutting off faces containing non-simple points. In that process, we have to make sure at each step that we get closer to the class of simple polytopes. A measure for the "distance" of a polytope P to that class is the number $\mu(P)$ of its "normally stout" faces, see Definition 3.5.

We first introduce this series of basic notions:

Definition 3.1. In the vector space V, let H be an affine hyperplane, U_1, U_2 the two open connected components of its complement, P, a polytope

384

of dimension d, and F, a proper face of P. We say:

(1) A subset A of V *lies strictly on one side* of H if it is included in exactly one U_i.

(2) H is a *cutting hyperplane* for P if it intersects the relative interior of P, i.e., $H \cap \overset{\circ}{P} \neq \emptyset$. Then $H \cap P$ is the *cut facet* of the two d-polytopes $P_i := P \cap \overline{U}_i$. It is obvious that

$$P = P_1 \cup P_2 \quad \text{and} \quad P_1 \cap P_2 = H \cap P$$

hold.

(3) H cuts P *transversally* if no vertex of P lies on H.

(4) A hyperplane H as in (3) is *sufficiently near* to F (or a "nearby hyperplane") if F lies strictly on one side of H, whereas all the vertices of P not contained in F lie on the other side.

(5) A hyperplane H as in (4) "*cuts off*" the face F if in addition, it is parallel to a supporting hyperplane H_0 for that face F, which means $P \cap H_0 = F$.

(6) In the "cut decomposition" $P = P_1 \cup P_2$ of P according to (2), given by a hyperplane H as in (5), the part P_i that includes F is called the *germ of P along F* and denoted by $G := G_P(F)$. The other part P_j is the *corresponding residual polytope*, denoted by $R := R_P(F)$. We thus have

$$P = G_P(F) \cup R_P(F). \tag{1}$$

In the sequel, such a cutting off decomposition $P = G \cup R$ for a proper face F of P by a hyperplane plays an important role: It allows a "*divide et impera*" approach to the HRR problem.

Definition 3.2. Let F be a proper face of a polytope P in V. A *link* $L = L_P(F)$ of F in P is any polytope obtained in the following way:

(1) If F is just a vertex $\{\mathbf{a}\}$ of P, then $L_P(F) := L_P(\mathbf{a})$ is the cut facet $P \cap H$ for H as in Definition 3.1 (2).

(2) If $\dim F > 0$, we choose an affine subspace N in V which is *transversal to F*, i.e., complementary to the affine span $\mathrm{aff}(F)$ and intersecting the relative interior of F. Then

$$L_P(F) := L_{P \cap N}(F \cap N) \ (= N \cap P \cap H). \tag{2}$$

Given such $F \precneqq P$, the combinatorial types of a germ and of a link are independent of all choices made in the construction. — In the literature, a link of a vertex often is called a *"vertex figure"*, and a link $L_P(F)$ of a face, a *"face figure"* or a *"quotient polytope"*, sometimes denoted by P/F.

We recall the notion of the *join* of two "relatively skew" polytopes.

Definition 3.3. Let Q_1, Q_2 be disjoint polytopes in V such that $\dim \mathrm{aff}(Q_1 \cup Q_2) = \dim Q_1 + \dim Q_2 + 1$. Then their *join* $Q_1 * Q_2$ is the convex hull of $Q_1 \cup Q_2$ in V.

We note that $Q * \emptyset = Q$, and $Q_1 * Q_2 = Q_2 * Q_1$. We remark that the join $Q_1 * Q_2$ is the disjoint union of Q_1, Q_2, and all open segments (x, y) joining points $x \in Q_1$ and $y \in Q_2$. We further mention that all faces of the join are of the form $F_1 * F_2$, where each F_i is a (possibly empty) face of Q_i, and that a link $L_{Q_1 * Q_2}(Q_1)$ is combinatorially equivalent to Q_2. — We denote by $\Pi(P) := P * \{\mathbf{a}\}$ for $\mathbf{a} \notin \mathrm{aff}(P)$ the *pyramid with apex* $\mathbf{a}$ *and base* P. An iterated pyramid $\Pi^i(P)$ for $i > 0$ is thus a join $P * S_{i-1}$ with an $(i-1)$-simplex S_{i-1}, whereas $\Pi^0(P) = P$.

We now study the local geometry near a face F of a polytope P in V. To that end, we fix a vertex $\mathbf{a} \in F$ and a nearby cutting hyperplane H for it. In the following definition, we use the link $L_P(\mathbf{a}) := P \cap H$ of $\mathbf{a}$ in P, and the fact that its face $L_F(\mathbf{a}) := F \cap H$ is a link of $\mathbf{a}$ relative to F.

Definition 3.4. A proper face F of the polytope P is called
  (1) *normally trivial (in P) at the vertex* $\mathbf{a}$ if the link $L_P(\mathbf{a})$ is the join $L_F(\mathbf{a}) * S_{\mathbf{a}}$ with a suitable "complementary" face $S_{\mathbf{a}}$ of $L_P(\mathbf{a})$, and
  (2) *normally trivial (in P)* if it is normally trivial at each of its vertices.

We remark that each vertex of a polytope P is normally trivial as a face. If $\mathbf{a}$ is a simple vertex of P, then each face F containing $\mathbf{a}$ is normally trivial at $\mathbf{a}$: A link $L = L_P(\mathbf{a})$ of $\mathbf{a}$ in P is a simplex, so for the face $F' = L_F(\mathbf{a})$ of L, there is a unique complementary face; the latter being again a simplex, any link of F in P is a simplex. If $F \lneqq P$ is an edge or a facet of a three-dimensional polytope, then the converse holds: Normal triviality at a vertex $\mathbf{a}$ is equivalent to $\mathbf{a}$ being simple.

More generally, for a face $F \lneqq P$ that is normally trivial at the vertex $\mathbf{a}$, there is a unique face $F'_{\mathbf{a}} \lneqq P$ "cutting out" the complementary face $S_{\mathbf{a}}$ in the link L, i.e., satisfying $S_{\mathbf{a}} = L_P(\mathbf{a}) \cap F'_{\mathbf{a}}$. That face is complementary to F at $\mathbf{a}$, i.e., we have $F \cap F'_{\mathbf{a}} = \{\mathbf{a}\}$, $\dim F + \dim F'_{\mathbf{a}} = \dim P$, and each edge emanating from $\mathbf{a}$ either lies in F or in $F'_{\mathbf{a}}$. Shifting the affine span of $F'_{\mathbf{a}}$ to the relative interior of F yields a transversal subspace N to F as in Definition 3.2. As a consequence, the polytope $S_{\mathbf{a}}$ has the same combinatorial type as $L_P(F)$, so that type does not depend on the vertex $\mathbf{a} \in F$.

Normal triviality of a face yields a combinatorial local product structure:

Remark 3.1. Let F be a normally trivial proper face of P with link $L = L_P(F) = N \cap P \cap H$ as in Formula (2). We denote by $\pi\colon V \to N$ the (affine) projection onto N along $\mathrm{aff}(F)$, i.e., collapsing $\mathrm{aff}(F)$ to a single point $\mathbf{v}_0$. For the germ $G = G_P(F)$, it induces a surjective map

$$\pi|_G\colon G \longrightarrow \Pi(L)$$

onto the pyramid $\Pi(L) := G \cap N$ over L with apex $\mathbf{v}_0$, mapping vertices onto vertices. Moreover, we obtain a bijection between the vertices of G and the vertices of $F \times \Pi(L)$ as follows: A vertex $\mathbf{u}$ lying on the "ridge" F of the "hip roof" G is mapped to $(\mathbf{u}, \mathbf{v}_0)$, and a vertex $\mathbf{v}$ lying on the "bottom facet" $G \cap H$ (i.e., on the cut facet), being the end point of an edge emanating from a unique vertex $\mathbf{u} \in F$, is mapped to $(\mathbf{u}, \pi(\mathbf{v}))$. That map yields a *combinatorial equivalence* between the polytopes G and $F \times \Pi(L)$.

If the link of a face is not a pyramid, then no vertex lying on that face is a simple point of the ambient polytope. This observation motivates the interest in the following concept, essential for the inner loop, cf. [BreLu$_2$, 2.7]:

Definition 3.5. Let F be a non-empty face of a polytope P.

    (1) The polytope P is called *stout* if it is not the pyramid over one of its facets.

    (2) The face F is called *normally stout in P* if one (and thus any) link $L = L_P(F)$ is stout.

Equivalently, a polytope P is stout if for each facet F, there are at least two vertices of P not lying on F. Hence, "stoutness" only depends on the combinatorial type and $\dim P \geq 2$ for a stout polytope P. In particular, a normally stout face $F \prec P$ always has codimension at least 3.

A general polytope can be built up from a unique stout "core":

Lemma 3.1. *If a polytope P is not a simplex, then it has exactly one non-empty maximal stout face $B \preceq P$. In particular, P is the iterated pyramid*

$$P = \Pi^c(B) = B * S_{c-1}$$

(with $c := \mathrm{codim}_P B \geq 0$) over that "base face". Moreover, if P is not stout (i.e., $c > 0$), then the complementary simplex S_{c-1} is the unique minimal normally stout face of P.

Proof. If P is stout, then $B = P$, $c = 0$, and there is nothing to show. The general case is seen by induction on $n := \dim P \geq 2$, with the case $n = 2$ already being settled. For $n \geq 3$, we may thus assume that P is a pyramid $\Pi(F) = F * \{\mathbf{a}\}$ over one of its facets $F \prec P$. By induction hypothesis, the statement holds for that facet F. Since all faces of P containing the apex $\mathbf{a}$ are pyramids, every stout face already lies in F. Hence, the unique maximal stout face B of F also is the unique maximal stout face of P. $\square$

The fundamental role played by normally stout faces in the present approach to the HRR is that they witness non-simplicity, cf. [BreLu$_2$, 2.9]:

Lemma 3.2. *A polytope is simple if and only if it has no normally stout faces.*

Proof. If a polytope is simple, then the links of all its faces are simplices, so no face is normally stout. On the other hand, a non-simple n-polytope P has a vertex $\mathbf{a} \in P$ that is incident to at least $n+1$ edges. A link $L = L_P(\mathbf{a})$ of that vertex is thus an $(n-1)$-polytope with more than n vertices, so it is not a simplex. Hence, as seen above, it can be (uniquely) written as an iterated pyramid $L = \Pi^c(B)$ (for some $c \geq 0$) over a stout base face $B \preceq L$. If $c = 0$, i.e., $L = B$, then the vertex $\mathbf{a}$ already is a normally stout face of P. Otherwise we have $L = B * S_{c-1}$ with a (non-empty) simplex $S_{c-1} \npreceq L$ that is normally stout in L. Then the unique face $F \prec P$ cutting out that simplex $S_{c-1} \prec L$, i.e., such that $S_{c-1} = F \cap L$, is normally stout in P. $\square$

We may thus measure how "far" a polytope is from being simple:

Definition 3.6. The *defect* $\mu(P) \in \mathbb{N}$ of a polytope P is defined as the number of normally stout faces of P.

We can restate Lemma 3.2 in these terms: *A polytope P is simple if and only if its defect vanishes, i.e., $\mu(P) = 0$.* — Pursuing the idea sketched at the beginning of this section, we now show that cutting off a minimal normally stout face brings us closer to "simplicity":

Lemma 3.3. *Let $F \npreceq P$ be a normally stout face of minimal dimension, and let R denote the residual polytope obtained by cutting off the face F from P. Then the defect satisfies*

$$\mu(R) = \mu(P) - 1.$$

Proof. No proper face $F_0 \not\gtrsim G \cap R$ is normally stout in R, since $G \cap R \prec R$, as a cut facet, is normally trivial in R and thus $L_R(F_0) = \Pi(L_{G \cap R}(F_0))$. On the other hand, there is a bijection between the faces of P not contained in F and the faces of R not contained in $G \cap R$. Since corresponding faces have the same links and no proper face of F is normally stout in P by the minimality of F, we obtain $\mu(R) = \mu(P) - 1$. $\qquad\square$

Corollary 3.1. *By finitely many successive cut-offs, every polytope is transformed into a simple one.*

Proof. This follows from the above result together with the fact that a polytope P with $\mu(P) = 0$ is simple, cf. Lemma 3.2. $\qquad\square$

So, finally, we are left with the problem to show that the HRR for the residual polytope R obtained by cutting off a minimal normally stout face imply the HRR for the polytope P itself. To that end, we have to study the "cut-off" part, namely, a germ of that face. With Remark 3.1 at our disposal, the following result turns out to be of crucial importance, cf. also [BreLu$_2$, 2.12]:

Lemma 3.4. *A normally stout face $F \prec P$ of minimal dimension is normally trivial and is itself a simple polytope.*

Proof. We let $d := \dim F$, the minimal dimension of any normally stout face. The case $d = 0$ being trivial, we may assume $d > 0$. Since an arbitrary vertex $\mathbf{a} \in F$ is neither simple nor normally stout in P, its link may be written in the form $L_P(\mathbf{a}) = B * S_{c-1}$, where B is stout and $c \geq 1$. The normally stout faces $F' \not\gtrsim P$ containing $\mathbf{a}$ correspond bijectively to the normally stout faces of $L_P(\mathbf{a})$ via $F' \mapsto F' \cap L_P(\mathbf{a})$. Since $F' = F$ has minimal dimension, and S_{c-1} is the unique normally stout face of $L_P(\mathbf{a})$ having minimal dimension, it follows that $L_F(\mathbf{a}) = F \cap L_P(\mathbf{a}) = S_{c-1}$, i.e., the point $\mathbf{a}$ is a simple vertex of F, and with $S_{\mathbf{a}} := B \not\gtrsim L_P(\mathbf{a})$ in Definition 3.4, the face F is seen to be normally trivial in P at $\mathbf{a}$. $\qquad\square$

4. Intersection Cohomology of Fans

In this section, we briefly recall the construction of the intersection cohomology of a (quasi-convex) fan Δ, referring to [BBFK$_2$] or [BreLu$_1$] for details. All complete fans considered in the sequel occur as outer normal fans $\Delta(P)$ for a polytope $P \subset V$. Hence, we systematically consider fans in the dual V^* of a given vector space V. We are not going to deal with non-polytopal complete fans.

4.A The fan space: Motivated by the coarse "toric topology" on a toric variety given by torus-invariant open sets, we consider a fan Δ in V^* as a finite topological space with the subfans as open subsets. The "affine" fans

$$\langle\sigma\rangle := \{\sigma\} \cup \partial\sigma \preceq \Delta \quad \text{with boundary fan} \quad \partial\sigma := \{\tau \in \Delta \; ; \; \tau \npreceq \sigma\}$$

form a basis of the fan topology by open sets that cannot be covered by smaller ones. Here $\preceq$ means that a cone is a face of another cone or that a set of cones is a subfan of some other fan. In fact, by abuse of notation, we often write σ instead of $\langle\sigma\rangle$, if there is no danger of confusion.

4.B Sheaves: Sheaf theory on a fan (space) Δ is particularly simple since a presheaf given on the basis uniquely extends to a sheaf. In order to simplify notation, given a sheaf $\mathcal{F}$ on Δ, we write

$$F_\Lambda := \mathcal{F}(\Lambda)$$

for the set of sections on the open subset (i.e., subfan) $\Lambda \preceq \Delta$. Such a sheaf $\mathcal{F}$ is flabby if and only if each restriction homomorphism

$$\varrho^\sigma_{\partial\sigma} : F_\sigma \to F_{\partial\sigma}$$

is surjective.

Here are the two most important examples:

(1) **The structure sheaf** $\mathcal{A}$ of Δ is defined by

$$A_\sigma := S(\mathrm{span}(\sigma)^*) \, ,$$

the graded algebra of real-valued polynomial functions on the subspace $\mathrm{span}(\sigma) \subset V^*$ or rather on σ itself, the homomorphisms $\varrho^\sigma_\tau : A_\sigma \to A_\tau$ for $\tau \preceq \sigma$ being the restriction of functions. Hence, for $\Lambda \preceq \Delta$, the global sections in A_Λ are the Λ-conewise polynomial functions $|\Lambda| \to \mathbb{R}$. The grading is chosen to be twice the standard grading, e.g. conewise linear functions get the degree 2.

The structure sheaf $\mathcal{A}$ is flabby if and only if Δ is a simplicial fan.

(2) **The "equivariant" intersection cohomology sheaf** $\mathcal{E}$ – called "minimal extension sheaf" in [BBFK$_2$] – is the "smallest" flabby sheaf of graded $\mathcal{A}$-modules on Δ such that E_σ is a finitely generated free A_σ-module for every cone $\sigma \in \Delta$, and $E_o = A_o = \mathbb{R}$ for the zero cone $o := \{0\}$.

Let us explain the minimality condition in "smallest": Let

$$A := S(V) = S\big((V^*)^*\big)$$

denote the (even-graded) algebra of polynomial functions on the vector space V^* (so in particular, $A_\sigma = A$ for an n-cone σ, and for any fan Λ both, A_Λ and E_Λ are graded A-modules in a natural way). Furthermore, let

$$\mathfrak{m} := A^{>0}$$

denote the unique homogeneous maximal ideal of the graded algebra A. Then, given a graded A-module, we define its *reduction* $\overline{M}$, a graded real vector space, by

$$\overline{M} := (A/\mathfrak{m}) \otimes_A M .$$

Since $\mathcal{E}$ is flabby, the induced restriction homomorphism

$$\overline{\varrho}^{\,\sigma}_{\partial\sigma} : \overline{E}_\sigma \to \overline{E}_{\partial\sigma}$$

is also surjective. Requiring it to be even an isomorphism means minimizing the rank of the free A_σ-module E_σ. Note that, on the other hand, the surjectivity of $\overline{\varrho}^{\,\sigma}_{\partial\sigma}$ already implies that of $\varrho^{\sigma}_{\partial\sigma}$.

The above conditions determine $\mathcal{E}$ up to isomorphy of graded $\mathcal{A}$-modules, and in particular we see that $\mathcal{E} \cong \mathcal{A}$ if and only if Δ is simplicial.

4.C The intersection cohomology $IH(\Delta)$ of a complete (or, more generally, a "quasi-convex") fan is defined as the graded vector space

$$IH(\Delta) := \overline{E}_\Delta .$$

4.D Quasi-convex fans: We call a fan *quasi-convex* if it is purely n-dimensional, i.e., all maximal cones are n-dimensional, and the support $|\partial\Delta|$ of its boundary subfan is a real homology manifold or empty. Here $\partial\Delta \preceq \Delta$ is the subfan generated by those $(n-1)$-cones which are a facet of exactly one n-cone in Δ. In fact, quasi-convex fans Δ are characterized by the fact that E_Δ is a (finitely generated) free A-module, cf. [BBFK$_2$] 4.1 and 4.4.

So in particular, fans with convex or "co-convex" support (i.e., $V^* \setminus |\Delta|$ is convex) as well as stars of cones in a complete fan provide examples of such fans. Furthermore, if Λ is a quasi-convex subfan of the complete fan Δ, we denote by Λ^c its (quasi-convex) "complementary" subfan in Δ, generated by the n-cones in $\Delta \setminus \Lambda$.

4.E Outer normal fan and Lefschetz Operator: Any n-polytope P in V induces a fan $\Delta = \Delta(P)$ in V^* together with a strictly convex Δ-conewise linear function $\psi \colon V^* \to \mathbb{R}$ as follows: For any facet $F \preceq_1 P$, we choose an "outer normal vector" $\mathbf{n}_F \in V^* \setminus \{0\}$, i.e., $\mathbf{n}_F|_F \equiv \mathrm{const} \geq \mathbf{n}_F|_P$, and denote by $\nu(F) := \mathbb{R}_{\geq 0} \cdot \mathbf{n}_F$ the associated "outer normal ray" of the facet F. To any face $G \preceq P$, we associate a cone $\sigma(G) \subset V^*$ as follows:

$$\sigma(G) := \sum_{G \preceq F \preceq_1 P} \nu(F) \ .$$

Note in particular that $\sigma(P) = o := \{0\} \subset V^*$, the zero cone. Then the *outer normal fan* $\Delta(P)$ is defined as

$$\Delta(P) := \{\sigma(G); G \preceq P\} \ .$$

We remark that $\Delta(P)$ is simplicial if and only if P is simple.

For $i = 1, \ldots, r$, let $\mathbf{v}_i$ denote the vertices of P. The corresponding $\sigma_i := \sigma(\{\mathbf{v}_i\})$ are the maximal (n-dimensional) cones in Δ. Moreover, we denote by $\psi_i \in (V^*)^*$ the image of $\mathbf{v}_i$ with respect to the biduality isomorphism $V \to (V^*)^*$. Then

$$\psi|_{\sigma_i} := \psi_i|_{\sigma_i}$$

defines a strictly convex conewise linear function $\psi \in A^2_{\Delta(P)}$. For ease of notation, we let $\Delta := \Delta(P)$. The multiplication map

$$\mu_\psi \colon E_\Delta \longrightarrow E_\Delta \, , \ f \longmapsto \psi f$$

induces a degree 2 map

$$L := \overline{\mu}_\psi \colon \overline{E}_\Delta = IH\big(\Delta(P)\big) \longrightarrow \overline{E}_\Delta = IH\big(\Delta(P)\big) \ ,$$

the "**Lefschetz operator**". We remark that $\Delta(P+\mathbf{a}) = \Delta(P)$ for $\mathbf{a} \in V$ with the same Lefschetz operator, since the correponding strictly convex functions only differ by the "globally linear" function $\mathbf{a} \in V \cong (V^*)^* = A^2$.

If $\mathrm{aff}(P) \neq V$, the above constructions apply *mutatis mutandis* in order to give a fan $\Delta(P)$ in $V^*/\mathrm{aff}_0(P)^\perp$, with the subspace $\mathrm{aff}_0(P) := \mathrm{aff}(P) - \mathbf{a}$, $\mathbf{a} \in \mathrm{aff}(P)$, as well as a Lefschetz operator on $IH\big(\Delta(P)\big)$.

4.F The intersection product (cf. [BBFK$_3$] for details): For a sheaf $\mathcal{F}$ on a quasi-convex fan Δ, we apply this notation: The module $F_{(\Delta,\partial\Delta)} \subset F_\Delta$ of "sections with compact support on Δ" is defined as

$$F_{(\Delta,\partial\Delta)} := \ker(\varrho^\Delta_{\partial\Delta}) = \{f \in F_\Delta; f|_{\partial\Delta} = 0\} \ .$$

392

For every fan Λ including Δ, trivial extension of sections thus provides a natural inclusion $F_{(\Delta,\partial\Delta)} \subset F_\Lambda$. In order to discuss the intersection product, we have to fix a volume form $\omega \in \det V := \bigwedge^n V$ on V^*. If the fan Δ is simplicial, we can, after [Bri], define a graded A-linear "evaluation map"

$$\varepsilon : A_{(\Delta,\partial\Delta)} \longrightarrow A[-2n]$$

as follows: For each n-cone σ, we denote by $g_\sigma \in A^{2n}_{(\sigma,\partial\sigma)} \subset A_\sigma = A$ the unique non-trivial function $g_\sigma \geq 0$ given as the product of those linear forms in $A^2 \cong V$ the wedge product of which agrees, up to sign, with ω. Then the map ε is the composite

$$E_{(\Delta,\partial\Delta)} \cong A_{(\Delta,\partial\Delta)} \subset \bigoplus_{\sigma\in\Delta^n} A_\sigma \to Q(A) \,, \ f = (f_\sigma)_{\sigma\in\Delta^n} \mapsto \sum_{\sigma\in\Delta^n} \frac{f_\sigma}{g_\sigma} \,, \quad (3)$$

mapping $A_{(\Delta,\partial\Delta)}$ onto $A \subset Q(A)$. We remark that any (graded) A-linear map $A_{(\Delta,\partial\Delta)} \to A[-2n]$ is a scalar multiple of ε, and that a multiplication of ω with a scalar $\lambda \in \mathbb{R}$ results in a multiplication of ε with $|\lambda|$.

The intersection product then is the composite

$$\cap : A_\Delta \times A_{(\Delta,\partial\Delta)} \xrightarrow{\text{mult}} A_{(\Delta,\partial\Delta)} \xrightarrow{\varepsilon} A[-2n]$$

of the multiplication and the evaluation map $\varepsilon \colon A_{(\Delta,\partial\Delta)} \to A[-2n]$. In the general case, the definition uses the dual sheaf $\mathcal{DE}$ of $\mathcal{E}$, cf. [BBFK$_3$]. The module of sections of $\mathcal{DE}$ over a cone $\sigma \in \Delta$ is

$$(\mathcal{DE})_\sigma := \mathrm{Hom}(E_{(\sigma,\partial\sigma)}, A_\sigma) \otimes \det V_\sigma,$$

with $V_\sigma := V/\mathrm{lin}(\sigma)^\perp \cong \mathrm{lin}(\sigma)^*$. The determinant factor produces a degree shift ($V_\sigma = A^2_\sigma$ being of weight 2) and plays an important role in the definition of the restriction homomorphisms $(\mathcal{DE})_\sigma \to (\mathcal{DE})_\tau$ for $\tau \preceq \sigma$. Here it is necessary to fix an orientation of $\mathrm{lin}(\sigma)$ for every cone $\sigma \in \Delta$, with the n-cones getting the orientation defined by the volume form $\omega \in \det V$. Then the defining formula holds even globally:

$$(\mathcal{DE})_\Delta \cong \mathrm{Hom}\big(E_{(\Delta,\partial\Delta)}, A\big) \otimes \det V \cong \mathrm{Hom}\big(E_{(\Delta,\partial\Delta)}, A[-2n]\big) \,,$$

where the second isomorphy uses the isomorphism $\det V \cong \mathbb{R}$ given by $\omega \mapsto 1$. Furthermore there are natural isomorphisms $\mathcal{E} \cong \mathcal{DE}$ – in fact, the naturality is obtained only with the HLT for fans in lower dimensions – and $E_\Delta \cong (\mathcal{DE})_\Delta$. Hence, we finally obtain the intersection product

$$\cap : E_\Delta \times E_{(\Delta,\partial\Delta)} \longrightarrow A[-2n] \,,$$

which uniquely extends to a map

$$\cap : E_\Delta \times E_\Delta \longrightarrow Af^{-1}[-2n],$$

where $f \in A$ is a minimal square free product of linear forms in $A^2 = V$ with $f|_{\partial\Delta} = 0$. If we apply that to the subfans $\langle\sigma\rangle$ generated by n-cones $\sigma \in \Delta$, we obtain a formula representing the intersection product of two sections $f \in E_\Delta$ and $g \in E_{(\Delta,\partial\Delta)}$ as a sum of local contributions:

$$f \cap g = \sum_{\sigma \in \Delta^n} f_\sigma \cap g_\sigma \in A$$

with $f_\sigma := f|_\sigma$ and $g_\sigma := g|_\sigma$. The reader should keep in mind that in general $f_\sigma \cap g_\sigma \in Q(A)$ does not belong to A.

There is another way to obtain the intersection product, cf. [BBFK$_3$, 4]: One takes a simplicial refinement $\iota : \Sigma \to \Delta$ and realizes $\mathcal{E}$ as a direct summand of $\iota_*(\mathcal{A})$, where $\mathcal{A}$ denotes the structure sheaf of the fan Σ, cf. [BBFK$_2$, 2.5] – the corresponding inclusion then is also called a **direct embedding**. Then the composition

$$E_\Delta \times E_{(\Delta,\partial\Delta)} \hookrightarrow A_\Sigma \times A_{(\Sigma,\partial\Sigma)} \xrightarrow{\text{mult}} A_{(\Sigma,\partial\Sigma)} \xrightarrow{\varepsilon} A[-2n]$$

of the induced embeddings and the intersection product on Σ provides the intersection product on Δ.

A third possibility is to mimic the multiplication of functions, cf. [BBFK$_3$, 4]: One chooses an "internal intersection product", i.e., any symmetric $\mathcal{A}$-bilinear sheaf homomorphism $\beta : \mathcal{E} \times \mathcal{E} \to \mathcal{E}$ extending the multiplication of functions on the 2-skeleton (its construction involves choices and is not natural). On the other hand, there is a distinguished section $1 \in E_\Delta$. Its image with respect to the isomorphism

$$E_\Delta \xrightarrow{\cong} (\mathcal{D}\mathcal{E})_\Delta \cong \mathrm{Hom}_A(E_{(\Delta,\partial\Delta)}, A[-2n])$$

provides an evaluation map $\varepsilon : E_{(\Delta,\partial\Delta)} \to A[-2n]$. Then the composite

$$E_\Delta \times E_{(\Delta,\partial\Delta)} \xrightarrow{\beta} E_{(\Delta,\partial\Delta)} \xrightarrow{\varepsilon} A[-2n]$$

once again yields the intersection product!

5. HRR for special n-polytopes

5.1. *HRR for pyramids*

Proposition 5.1. *If the HRR hold for polytopes in dimension $d < n$, they also hold for any n-dimensional pyramid $P = \Pi(Q)$ over some $(n-1)$-polytope Q.*

Proof. We may assume that the pyramid is of the form $\Pi(Q) = Q * \{0\}$ with apex at the origin $0 \in V$. Let $\Delta := \Delta(\Pi(Q))$ and denote by $\sigma := \sigma(\{0\}) \in \Delta$ the cone corresponding to the apex 0 of the pyramid. Then the complementary fan $\Delta_0 := \Delta \setminus \{\sigma\} = \langle\sigma\rangle^c$ satisfies

$$\Delta_0 = \mathrm{st}(\nu(Q)) = \partial\sigma + \nu(Q) := \partial\sigma + \langle\nu(Q)\rangle$$

with the outer normal ray $\nu(Q)$ of $Q \preceq_1 P = \Pi(Q)$; moreover, we have $\psi|_\sigma = 0$ and thus $\psi \in A^2_{(\Delta_0,\partial\Delta_0)} \subset A^2_\Delta$. We look at the exact sequence

$$0 \longrightarrow E_{(\sigma,\partial\sigma)} \longrightarrow E_\Delta \longrightarrow E_{\Delta_0} \longrightarrow 0 \ .$$

It even splits, since E_{Δ_0} is free, the fan Δ_0 being quasi-convex. By reduction, we thus obtain the corresponding exact sequence

$$0 \longrightarrow IH(\sigma,\partial\sigma) \longrightarrow IH(\Delta) \longrightarrow IH(\Delta_0) \longrightarrow 0 \ ;$$

moreover $IH^q(\sigma,\partial\sigma) = 0$ holds for $q \leq n$, since HLT holds for fans in any dimension $d < n$, cf. [BBFK$_2$, 1.8]. Hence, for $k \geq 0$, the restriction from Δ to Δ_0 induces the first isomorphism in

$$IH^{n-k}(\Delta) \xrightarrow{\cong} IH^{n-k}(\Delta_0) \cong IH^{(n-1)-(k-1)}(\Delta(Q)) \ .$$

Let us comment here on the second isomorphy: The outer normal fan $\Delta(Q)$ lies in $W := V^*/(\mathbb{R}\cdot n_Q)$, and the quotient projection $\pi\colon V^* \to W$ induces a fan map $\Delta_0 \to \Delta(Q)$. Then, with $B := S(W^*) \subset A = S((V^*)^*)$, we have

$$E_{\Delta_0} \cong A \otimes_B E_{\Delta(Q)} \ ,$$

whence the last isomorphism. The dual picture looks as follows:

$$IH^{n+k}(\Delta) \cong IH^{n+k}(\Delta_0,\partial\Delta_0) \cong IH^{n+k-2}(\Delta_0) \cong IH^{(n-1)+(k-1)}(\Delta(Q)) \ .$$

Here the second isomorphism is the **"Thom isomorphism"**, the isomorphism induced by:

$$E_{\Delta_0} \xrightarrow{\cong} E_{(\Delta_0,\partial\Delta_0)} \ , \quad f \mapsto \psi f \ .$$

Replacing k with $k+2$, we obtain the isomorphism

$$IH^{n+k+2}(\Delta) \cong IH^{(n-1)+(k-1)+2}(\Delta(Q)) \ .$$

For $k > 0$, these isomorphisms transform

$$L^k : IH^{n-k}(\Delta) \longrightarrow IH^{n+k}(\Delta)$$

into

$$L^{k-1} : IH^{(n-1)-(k-1)}(\Delta(Q)) \longrightarrow IH^{(n-1)+(k-1)}(\Delta(Q)) \ .$$

This gives the HLT for Δ.

Now let us look at the HRR: The homomorphism

$$L^{k+1} : IH^{n-k}(\Delta) \longrightarrow IH^{n+k+2}(\Delta)$$

corresponds to

$$L^{(k-1)+1} : IH^{(n-1)-(k-1)}\big(\Delta(Q)\big) \longrightarrow IH^{(n-1)+(k-1)+2}\big(\Delta(Q)\big) .$$

So

$$IP^{n-k}(\Delta) \cong IP^{(n-1)-(k-1)}\big(\Delta(Q)\big) \quad \text{for } k > 0 ,$$

while for $k = 0$ there is no contribution: $IP^n(\Delta) = 0$ since L^0 is the identity. Now the above isomorphism respects the Hodge-Riemann forms, if we endow $V^*/(\mathbb{R}{\cdot}\mathbf{n}_Q)$ with the volume form η given as follows: It satisfies $q^*(\eta) \wedge \psi_\tau = \omega$ with the fixed volume form ω of V^*, the quotient map $q \colon V^* \to V^*/\mathbb{R}\mathbf{n}_Q$, and $\psi_\tau = \psi|_\tau \in A^2 = (V^*)^*$ with an n-cone $\tau \in \Delta_0$. So the HRR hold for $\Pi(Q)$, since they do for Q. $\qquad\square$

5.2. *The Künneth formula*

We want to show that the product $S \times P_0$ of a "HRR polytope" P_0 with a simple factor S again has the "HRR property". We start with discussing the intersection cohomology, endowed with the intersection product.

Proposition 5.2. *Let $P = S \times P_0$ be a polytope in $V \times W$ with a simple factor S, and let $\Delta = \Sigma \oplus \Delta_0$ be the corresponding decomposition of the respective outer normal fans. Then there is a natural isomorphism*

$$IH(\Delta) \xrightarrow{\;\cong\;} IH(\Sigma) \otimes_{\mathbb{R}} IH(\Delta_0)$$

of graded vector spaces endowed with the intersection forms.

Proof. We let $A = S(V)$ and $B = S(W)$ denote the algebra of polynomials on V^* and on W^*, respectively. Disregarding the intersection products, the isomorphism is seen as follows: Since S is simple, the fan Σ is simplicial. Hence, assigning to a cone $\delta = \sigma \times \delta_0$ in $\Delta = \Sigma \oplus \Delta_0$ the A_δ-module

$$E_\delta := A_\sigma \otimes_{\mathbb{R}} E_{\delta_0}$$

defines a minimal extension sheaf on Δ, as follows from an iterated application of Lemma 1.5 in [BBFK$_2$]. Since the tensor product with A_σ yields an exact functor, we obtain

$$E_{\sigma \times \Delta_0} \cong A_\sigma \otimes E_{\Delta_0},$$

for each $\sigma \in \Sigma$. Using analogously the tensor product with E_{Δ_0}, it follows

$$E_\Delta \cong A_\Sigma \otimes E_{\Delta_0}.$$

Since both, A_Σ and E_{Δ_0}, are free modules over their base rings A and B, respectively, the latter isomorphism descends to the level of intersection cohomology.

It remains to check the compatibility with the intersection products. We first assume that the fan Δ_0 is simplicial, too. In that case, up to suitable shifts, the tensor product of the evaluation maps $A_\Sigma \to A$ and $A_{\Delta_0} \to B$ defines the evaluation map

$$A_\Delta \cong A_\Sigma \otimes A_{\Delta_0} \longrightarrow A \otimes B,$$

associated to the product of the respective volume forms on V and on W. This implies the compatibility.

If Δ_0 is non-simplicial, we choose a simplicial subdivision $\iota \colon \widehat{\Delta}_0 \to \Delta_0$ and a direct embedding $\mathcal{E} \hookrightarrow \iota_*(\widehat{\mathcal{A}})$ of the intersection cohomology sheaf $\mathcal{E}$ on Δ_0 into the direct image of the structure sheaf $\widehat{\mathcal{A}}$ on $\widehat{\Delta}_0$. It induces a direct embedding on $\Delta = \Sigma \times \Delta_0$. Since these embeddings provide the respective intersection products on E_{Δ_0} and on E_Δ, the requested compatibility holds. $\qquad\square$

To show the HRR property, we need some purely algebraic considerations. In that framework, it is convenient to make degrees symmetric to zero by a shift: Instead of $IH(\Delta)$, with even grading from 0 to $2n$ and endowed with both, the intersection pairing and the Lefschetz operator, we consider the following

Abstract HR setup: Let

$$W := \bigoplus_{k=-m}^{m} W^k$$

be a finite dimensional graded vector space endowed with the following:

- A non-degenerate symmetric bilinear form, also called the "intersection form",

$$\langle \_, \_ \rangle : W \times W \longrightarrow \mathbb{C}$$

  of total degree 0 satisfying

$$\langle W^{-k}, W^k \rangle \subset i^k \mathbb{R},$$

- the structure of a graded module over the polynomial ring $\mathbb{R}[L]$ with $\deg L = 2$ such that the "Lefschetz operator" μ_L given by multiplication with L is self-adjoint with respect to the above form.

For convenience of notation we simply write L instead of μ_L. These data give rise to "HR-forms" $s_k(x, y) := (x, L^k y)$ on W^{-k}, and furthermore, to "primitive subspaces"

$$P(W^{-k}) := \ker(L^{k+1} \colon W^{-k} \to W^{k+2}).$$

Definition 5.1. A graded $\mathbb{R}[L]$-module W endowed with such a structure is called an **HR-module** if the restriction of $i^k s_k$ to the primitive subspace $P(W^{-k})$ is positive definite.

We note that, obviously, an HR-module satisfies the "numerical Poincaré duality" $\dim W^k = \dim W^{-k}$.

The *simple* HR-modules A_m (for $m \in \mathbb{N}$) are defined as

$$A_m^k := \begin{cases} \mathbb{R} & \text{for } -m \leq k \leq m \text{ with } k \equiv m \mod 2, \\ 0 & \text{otherwise,} \end{cases}$$

where the intersection form maps $(1, 1) \in A_m^{-k} \times A_m^k$ to $\langle 1, 1 \rangle = (-i)^m$, and the "Lefschetz operator" $L \colon A_m^k \to A_m^{k+2}$ maps 1 to 1, whenever that makes sense.

Remark 5.1.

i) Every HR-module is isomorphic to a direct sum of modules A_m.
ii) A graded $\mathbb{R}[L]$-submodule $U \subset W$ of an HR-module W is again an HR-module if and only if it satisfies $\dim U^{-k} = \dim U^k$.

The link with intersection cohomology of polytopes is provided as follows:

Remark 5.2. Let P be an n-polytope and put $\Delta := \Delta(P)$. Endow the graded $\mathbb{R}$-vector space $W(P) := IH(\Delta)[n]$ (i.e., having weight spaces $W^k(P) := IH^{n+k}(\Delta)$ for $-n \leq k \leq n$) with the intersection form, multiplied by $(-i)^n$, and put L to be the Lefschetz operator. Then the polytope P satisfies the HRR if and only if $W(P)$ is an HR-module. Furthermore $W(S_n) \cong A_n$ holds for the n-simplex S_n.

We now state and prove the "Künneth theorem" for HR-modules.

Proposition 5.3. *Let W, W' be HR-modules. Then both, $W \oplus W'$ and $W \otimes W'$ are HR-modules, where the action of L on $W \otimes W'$ is given by $L(x \otimes y) := Lx \otimes y + x \otimes Ly$.*

Proof. The first part of the statement being obvious, we only have to consider the tensor product. Since both, W and W' are direct sums of modules of type A_n and the tensor product commutes with direct sums, it suffices to look at $A_n \otimes A_m$. As a first step, a direct computation shows that, for $n \geq 1$,

$$A_n \otimes A_1 \cong A_{n+1} \oplus A_{n-1}$$

is an HR-module. By induction on n, it follows that

$$(A_1)^{\otimes n} \cong A_n \oplus R$$

also is an HR-module, where the "remainder" R is a direct sum of terms A_m with $m < n$ of the same parity as n. Hence the graded vector space

$$B := A_n \otimes A_m$$

is an $\mathbb{R}[L]$-submodule of the HR-module $C := (A_1)^{\otimes(n+m)}$. Satisfying numerical Poincaré duality $\dim B^{-k} = \dim B^k$, it also is an HR-module. $\square$

Corollary 5.1. *If, in the situation of Proposition 5.2, the polytope P_0 satisfies the HRR, then so does $P = S \times P_0$.*

Proof. According to Proposition 5.2 and in the notation of Remark 5.2, the graded $\mathbb{R}[L]$-module $W(P)$ can be written as

$$W(P) \cong W(S) \otimes W(P_0).$$

The claim now follows from Proposition 5.3 and the HRR for the simple polytope S. $\square$

5.3. *Transversal Cuttings*

Proposition 5.4. *If the affine hyperplane H cuts P transversally into the polytopes P_1 and P_2, i.e., H has nonempty intersection with the relative interior of P and does not contain vertices of P, then the validity of the HRR for P_1 and P_2 and for lower dimensional polytopes implies the HRR for P.*

Proof. Let us write

$$F_i := P \cap H \preceq P_i,$$

using the index $i = 1, 2$ in order to indicate when $P \cap H$ should be considered as a facet of P_i.

(A) Fans involved. For the fans

$$\Delta := \Delta(P) \ \text{ and } \ \Delta_i := \Delta(P_i) \ \text{ with } \ i = 1, 2,$$

we consider the "intermediate" polytope Q cut out from P by H and a nearby parallel hyperplane. Its outer normal fan $\Delta(Q)$ is obtained by gluing together the stars

$$\Lambda_i := \mathrm{st}\big(\nu(F_i)\big) \preceq \Delta_i$$

of the outer normal ray of the "cut" facet with respect to the fans Δ_i in order to produce a new complete fan

$$\Lambda := \Lambda_1 \cup \Lambda_2 = \Delta(Q).$$

We let $\Phi := \Delta(F_1) = \Delta(F_2)$ denote the outer normal fan of $F := F_1 = F_2$ in $W := V^*/(\mathbb{R} \cdot \mathbf{n}_{F_i})$, noting that $\mathbf{n}_{F_2} = -\mathbf{n}_{F_1}$.

(B) Gluing of IH. Let $\mathcal{G}$ and $\mathcal{E}$ denote the respective intersection cohomology sheaves on Φ and Λ. The projection $\pi \colon V^* \to W$ induces a map of fans $\Lambda \to \Phi$, and we have $\mathcal{E} \cong \pi^*(\mathcal{G})$. In particular, writing $B := S(W^*) \subset A := S((V^*)^*)$, there is a natural injection

$$A \otimes_B G_\Phi \hookrightarrow E_\Lambda,$$

which after restriction to the subfans $\Lambda_i \preceq \Lambda$ gives isomorphisms

$$E_{\Lambda_1} \xleftarrow{\cong} A \otimes_B G_\Phi \xrightarrow{\cong} E_{\Lambda_2}.$$

We use the resulting A-module isomorphism

$$S : E_{\Lambda_1} \xrightarrow{\cong} E_{\Lambda_2}$$

to define an exact sequence

$$0 \longrightarrow K \longrightarrow E_{\Delta_1} \oplus E_{\Delta_2} \longrightarrow E_{\Lambda_2} \longrightarrow 0 , \tag{4}$$

the second nontrivial map sending (f_1, f_2) to $S(f_1|_{\Lambda_1}) - f_2|_{\Lambda_2}$. Since E_{Λ_2} is a free A-module and $IH(\Lambda_i)$ is isomorphic to $IH(\Phi)$, the sequence (4) descends to an exact sequence

$$0 \longrightarrow \overline{K} \longrightarrow IH(\Delta_1) \oplus IH(\Delta_2) \longrightarrow IH(\Phi) \longrightarrow 0 . \tag{5}$$

Furthermore we use the exact sequence

$$0 \longrightarrow D \longrightarrow K \longrightarrow E_\Delta \longrightarrow 0 \,, \tag{6}$$

where the second nontrivial map is given by a gluing of sections: The fan Δ is the union

$$\Delta = \Lambda_1^c \cup \Lambda_2^c$$

of the complementary subfans $\Lambda_i^c \preceq \Delta_i$ of $\Lambda_i \preceq \Delta_i$, i.e.,

$$\Delta_i = \Lambda_i \cup \Lambda_i^c$$

with the quasi-convex fans Λ_i and Λ_i^c intersecting only in their common boundary fan and, in particular, $|\Lambda_i| = |\Lambda_j^c|$ for $j \neq i$. Now for a pair $(f_1.f_2) \in K \subset E_{\Delta_1} \oplus E_{\Delta_2}$ we define its image as the section $f \in E_\Delta$ satisfying $f|_{\Lambda_i^c} := f_i|_{\Lambda_i^c}$. The kernel D then satisfies

$$D = \{(h, S(h)); \ h \in E_{(\Lambda_1, \partial\Lambda_1)}\} \cong E_{(\Lambda_1, \partial\Lambda_1)} \,.$$

(C) Gluing of the intersection product. The exact sequence (6) yields an isomorphism

$$E_\Delta \cong K/D,$$

in fact that quotient representation holds even with respect to the intersection pairings on E_Δ and $K \subset E_{\Delta_1} \oplus E_{\Delta_2}$: Consider two pairs $(f_1, f_2), (g_1, g_2) \in K$ and denote by $f, g \in E_\Delta$ their respective images in E_Δ. Then in $Q(A)$ we obtain

$$\begin{aligned}
(f_1, f_2) \cap (g_1, g_2) &= f_1 \cap g_1 + f_2 \cap g_2 \\
&= f_1 \cap_{\Lambda_1^c} g_1 + f_1 \cap_{\Lambda_1} g_1 + f_2 \cap_{\Lambda_2} g_2 + f_2 \cap_{\Lambda_2^c} g_2 \\
&= f_1 \cap_{\Lambda_1^c} g_1 + f_2 \cap_{\Lambda_2^c} g_2 \\
&= f \cap g \,,
\end{aligned} \tag{7}$$

since the middle terms in (7) add up to 0. This can be seen as follows: The restrictions $E_{\Delta_i} \to E_{\Lambda_i}$, $i = 1, 2$, combine to a map

$$E_{\Delta_1} \oplus E_{\Delta_2} \supset K \longrightarrow E_\Lambda \subset E_{\Lambda_1} \oplus E_{\Lambda_2}$$

with image $A \otimes_B G_\Phi$. Denote by $\hat{f}, \hat{g}$ the respective images in E_Λ of the pairs $(f_1, f_2), (g_1, g_2) \in K$. Then

$$\hat{f} \cap \hat{g} = f_1 \cap_{\Lambda_1} g_1 + f_2 \cap_{\Lambda_2} g_2.$$

So what we finally have to prove is that

$$A \otimes_B G_\Phi \subset E_\Lambda$$

is an isotropic subspace. To that end, we may even assume that Φ and thus also Λ is simplicial (keeping $\Phi \cong \partial \Lambda_i$ in mind): Otherwise, we may take a simplicial refinement $\iota \colon \widehat{\Phi} \to \Phi$ and a direct embedding $\mathcal{G} \to \iota_*(\mathcal{A})$, where $\mathcal{A}$ is the structure sheaf of $\widehat{\Phi}$. Then there is an induced simplicial refinement $\widehat{\Lambda} \to \Lambda$ and embedding $E_\Lambda \hookrightarrow A_{\widehat{\Lambda}}$ respecting the intersection product. The resulting inclusion $G_\Phi \subset A_{\widehat{\Phi}}$ yields $A \otimes_B G_\Phi \subset A \otimes_B A_{\widehat{\Phi}}$, so our claim holds for Λ if it does for $\widehat{\Lambda}$.

So let us now consider the case where Φ and hence also Λ is simplicial. In that situation, it suffices to check that the evaluation map $\varepsilon \colon E_\Lambda = A_\Lambda \to A[-2n]$ vanishes on $A \otimes_B G_\Phi$. We use the formula (3) in section 4.F for ε:

$$E_\Lambda \cong A_\Lambda \subset \bigoplus_{\sigma \in \Lambda^n} A_\sigma \longrightarrow Q(A), \quad f = (f_\sigma)_{\sigma \in \Lambda^n} \longmapsto \sum_{\sigma \in \Lambda^n} \frac{f_\sigma}{g_\sigma}.$$

But the n-cones in Λ may be grouped in pairs $\sigma_i \in \Lambda_i, i = 1, 2$ with images $\pi(\sigma_1) = \pi(\sigma_2)$. Then for $f \in A \otimes_B G_\Phi = A \otimes_B A_\Phi$, we have

$$f_{\sigma_1} = f_{\sigma_2} \,, \quad \text{while} \quad g_{\sigma_1} = -g_{\sigma_2} \,,$$

and thus $\varepsilon(f) = 0$.

(D) The HR relations for P. Since

$$\overline{D} \cong IH^*(\Lambda_i, \partial \Lambda_i) \cong IH^{*-2}(\Lambda_i) \cong IH^{*-2}(\Phi)$$

and the third module E_Δ in the exact sequence (6) is free, there is an associated exact sequence

$$0 \longrightarrow IH^{*-2}(\Phi) \longrightarrow \overline{K} \longrightarrow IH(\Delta) \longrightarrow 0,$$

realizing $IH(\Delta) = \overline{E}_\Delta$ in the same way as before E_Δ. Furthermore it is compatible with the natural Lefschetz operators on all three terms; we shall denote them simply by L in all cases.

Now let $\zeta \in IH^{n-k}(\Delta)$, $\zeta \neq 0$, be a primitive class, i.e., $L^{k+1}(\zeta) = 0$. We can lift it to a pair $\xi = (\xi_1, \xi_2) \in \overline{K}$. We may actually assume that the classes $\xi_i \in IH^{n-k}(\Delta_i)$ are again primitive, i.e., $\xi_i \in IP^{n-k}(\Delta_i)$: Because of $L^{k+1}(\zeta) = 0$, we have

$$L^{k+1}(\xi) = \eta \in IH^{n+k}(\Phi) \subset \overline{K}^{n+k+2}.$$

The Lefschetz operator $L^{k+1} \colon IH^{n-k-2}(\Phi) \to IH^{n+k}(\Phi)$ is composed of

$$IH^{n-k-2}(\Phi) = IH^{(n-1)-(k+1)}(\Phi) \xrightarrow{\cong} IH^{(n-1)+(k+1)}(\Phi) = IH^{n+k}(\Phi)\,,$$

so it is an isomorphism – by the assumption, HRR and thus HLT holds for the lower dimensional polytope $P_1 \cap P_2$. We may thus replace ξ with $\xi - L^{-(k+1)}(\eta)$.

So now let both ξ_1 and ξ_2 be primitive. Since $\zeta \neq 0$, we have $\xi_i \neq 0$ for at least one index i. Thus

$$(-1)^{(n-k)/2}\zeta \cap L^k(\zeta) = (-1)^{(n-k)/2}\xi_1 \cap L^k(\xi_1) + (-1)^{(n-k)/2}\xi_2 \cap L^k(\xi_2) > 0,$$

since the HRR hold for P_1 and P_2. $\qquad\qquad\square$

6. Deformation

Proposition 6.1. *The germ $G = G_P(F)$ of a normally trivial face $F \prec P$ can be deformed into the product $F \times \Pi(L)$, where $L := L_P(F)$ denotes a link of F in P.*

Proof. In the terminology of Remark 3.1, we assume that $\overset{\circ}{F} \cap N = \{0\}$; so the affine span $U := \mathrm{aff}(F)$ as well as N are linear subspaces, and $V = U \oplus N$. Furthermore we write $N = W \oplus \mathbb{R}$, such that $H = U \times W \times \{1\}$, and $L_P(F) = L \times \{1\}$ with a polytope $L \subset W$. Denote by $\mathbf{u}_1, \ldots, \mathbf{u}_r \in U$ the vertices of F and by $\mathbf{w}_1, \ldots, \mathbf{w}_s \in W$ those of L. Then G has vertices $(\mathbf{u}_i, 0, 0)$ and $(\mathbf{u}_i + \mathbf{u}_{ij}, \mathbf{w}_j, 1)$ with suitable vectors $\mathbf{u}_{ij} \in U$.

We now let $G_t := G \cap (U \times W \times [0, t])$ be the truncated germ, and consider on

$$V = U \oplus N = U \oplus (W \oplus \mathbb{R})$$

the linear isomorphism

$$F_t := \mathrm{id}_U \oplus t^{-1}\mathrm{id}_N.$$

Then the family $(Q_t)_{0 < t \leq 1}$ of polytopes $Q_t := F_t(G_t)$ extends to $t = 0$ with $Q_0 = F \times \Pi(L)$. In fact, the polytope Q_t has vertices $(\mathbf{u}_i, 0, 0)$ and $(\mathbf{u}_i + t\mathbf{u}_{ij}, \mathbf{w}_j, 1)$. $\qquad\qquad\square$

Theorem 6.1. *Let $F \prec P$ be a normally trivial face of an n-dimensional polytope P, and assume that F itself is a simple polytope. Then the HRR hold for $G = G_P(F)$ if they hold for lower dimensional polytopes.*

Proof. For $\dim F = 0$, the germ $G = \Pi\big(L_P(F)\big)$ is a pyramid, so we may apply Proposition 5.1. Now let $\dim F > 0$. Let us first give a

Survey of proof: We consider the deformation Q_t, $0 \leq t \leq 1$, of Proposition 6.1, which deforms $Q_1 = G$ into $Q_0 = F \times \Pi(L)$. The HLT and the HRR hold on Q_0 according to the Künneth formula. Then we show that the HLT holds on Q_t for all $t \in (0,1]$, cf. Proposition 6.3, hence the HR-forms on $IH\big(\Delta(Q_t)\big)$ are non-degenerate for any $t \in [0,1]$. Since the combinatorial type of the polytopes Q_t is constant along the deformation, the Betti numbers are so, too. In fact, both $IH^{n-k}\big(\Delta(Q_t)\big)$ and the k-th HR-form s_k^t on it depend continuously on $t \in [0,1]$, cf. Proposition 6.2. Since the HR-forms are non-degenerate, their signature is independent of the parameter $t \in [0,1]$. Hence the HR-equations of Proposition 1.2 hold for all t, since they do for $t = 0$.

(A) The deformation on the fan level:

The case $t > 0$: For $t > 0$ the fan $\Delta_t := \Delta(Q_t)$ is "linearly equivalent" to $\Delta_1 = \Delta(G)$, i.e., Δ_t is the image of Δ_1 with respect to a linear isomorphism of the vector space V^*, namely the inverse $(F_t^*)^{-1}$ of the dual F_t^* of the map $F_t: V \to V$ transforming G_t into Q_t in the proof of Proposition 6.1. It provides an isomorphism

$$F_t^* : \Delta_t = \Delta(Q_t) \longrightarrow \Delta(G_t) = \Delta(G) = \Delta(Q_1) = \Delta_1 \ .$$

Behaviour near 0: We replace the linear isomorphism $(F_t^*)^{-1}$ mapping Δ_1 onto Δ_t with a Δ_0-conewise linear isomorphism $S_t: V^* \to V^*$, such that

$$S_t(\Delta_0) = \Delta_t \ .$$

The construction of S_t is as follows: We consider the subfan $\Gamma \preceq \Delta(G)$ generated by the cones $\sigma(F_0)$, where $F_0 \preceq G$ is a minimal face projecting onto the entire pyramid $\Pi(L)$, i.e., $\pi(F_0) = \Pi(L)$ with the projection $\pi: V = U \oplus N \to N$. The support $|\Gamma|$ is the graph of a map $H: U^* \to N^*$. In fact that map is Φ-conewise linear for the outer normal fan $\Phi := \Delta(F)$ of the polytope $F \subset U$, and

$$S_t : U^* \oplus N^* \longrightarrow U^* \oplus N^*, \ (\mathbf{x},\mathbf{y}) \mapsto \big(\mathbf{x},\mathbf{y} + tH(\mathbf{x})\big)$$

then defines a Δ_0-conewise linear isomorphism with the desired properties. Note here that Δ_0 is the product

$$\Delta_0 = \Phi \times \Lambda$$

of the (simplicial) fan $\Phi := \Delta(F)$ in U^* and the fan $\Lambda := \Delta\big(\Pi(L)\big)$ in N^*.

(B) Pull back isomorphisms: Both $(F_t^*)^{-1}$ and S_t act on the global sections of the structure sheaf and the intersection cohomology sheaf, respectively, by pull back. Let us write

$$A_t := A_{\Delta_t} \ , \ E_t := E_{\Delta_t}.$$

Then $(F_t^*)^{-1}$ induces isomorphisms

$$A_t \longrightarrow A_1 \ , \ E_t \longrightarrow E_1$$

of A-modules; in particular $IH(\Delta_t) \cong IH(\Delta_1)$ in a natural way. For S_t, the corresponding maps

$$A_t \longrightarrow A_0 \ , \ E_t \longrightarrow E_0 \ ,$$

both denoted by S_t^*, are only isomorphisms of graded vector spaces: This is due to the fact that for the subalgebra $A \subset A_t$ of "global polynomials" we in general have $S_t^*(A) \not\subset A \subset A_0$. So we can not any longer identify $IH(\Delta_t)$ in a reasonable way with $IH(\Delta_0)$.

(C) Continuity properties: That the relevant data depend continuously on $t \in (0,1]$ follows now immediately from the fact that the strictly convex function on $\Delta_1 = \Delta(G)$ given by the vertices of G_t is continuous in t. Near 0, however, there is no natural trivialization of the family $IH(\Delta_t)$; instead we represent $IH(\Delta_t)$ as a factor space E/M_t of a constant bigger vector space E with respect to a varying subspace M_t:

Proposition 6.2. *There is a finite dimensional graded vector space E and continuous families of*

> *(1) subspaces $M_t \subset E$ of constant dimension, such that in a natural way*
>
> $$IH(\Delta_t) \cong E/M_t \ ,$$
>
> *(2) endomorphisms $\hat{L}_t \colon E \to E$ with $\hat{L}_t(M_t) \subset M_t$ inducing the Lefschetz operator of $\Delta_t = \Delta(Q_t)$,*
> *(3) symmetric bilinear forms $\beta_t \colon E \times E \to \mathbb{R}$ with $\beta_t(M_t, E) = 0$, inducing the intersection product.*

Proof. Let us start with

The vector spaces E and M_t: We write $\Delta := \Delta_0$ and take:

$$E := \bigoplus_{q=0}^{2n} E_0^q$$

and

$$M_t := S_t^*(\mathfrak{m}E_t) \cap E \subset E.$$

The subspaces M_t can be represented in the form $\Phi_t(\mathfrak{m}^{<2n} \otimes E)$ with the continuous family of linear maps

$$\Phi_t : \mathfrak{m}^{<2n} \otimes E \longrightarrow E, \quad g \otimes f \mapsto S_t^*(g)f;$$

furthermore, since Δ_t and Δ are combinatorially equivalent, we get that $\dim M_t = \dim E - \dim IH(\Delta_t)$ is independent of t. The map $\hat{L}_t$ is multiplication with $S_t^*(\psi_t) := \psi_t \circ S_t$, except on the highest weight subspace E_Δ^{2n}, where it vanishes. Here ψ_t denotes the strictly convex function belonging to Q_t.

Continuity of the intersection product: We now analyse a general fan Δ in V^* and its images $\Delta_t := S_t(\Delta)$ under a continuous family of Δ-conewise linear isomorphisms $S_t \colon V^* \to V^*$. The bilinear form we consider is

$$\beta_t : E \times E \subset E_\Delta \times E_\Delta \xrightarrow{(S_t^{-1})^*} E_t \times E_t \xrightarrow{\cap} A[-2n] \longrightarrow A^0 \cong \mathbb{R}$$

where the last arrow is (up to the shift) the projection $A = A^0 \oplus \mathfrak{m} \to A^0$.

Let us first look at the simplest case:

The fan Δ is simplicial. In this situation, we have $\mathcal{E} = \mathcal{A}$. For r, s with $r + s = 2n$, we have to verify that the map

$$A_0^r \times A_0^s \xrightarrow{(S_t^{-1})^*} A_t^r \times A_t^s \xrightarrow{\cap} A^0 = \mathbb{R}$$

depends continuously on $t \in [0, 1]$. But that map may be rewritten as

$$A_0^r \times A_0^s \xrightarrow{\mathrm{mult}} A_0^{2n} \xrightarrow{(S_t^{-1})^*} A_t^{2n} \xrightarrow{\varepsilon_t} A^0 = \mathbb{R} \;,$$

using the fact that $(S_t^{-1})^*$ commutes with the multiplication of functions. Here ε_t denotes the restriction of the evaluation map

$$A_t \longrightarrow A[-2n]$$

to the $2n$-th weight space A_t^{2n}. It is well defined after having fixed a volume form on V. So, eventually we have to check that the map

$$A_0^{2n} \xrightarrow{(S_t^{-1})^*} A_t^{2n} \xrightarrow{\varepsilon_t} \mathbb{R}$$

depends continuously on $t \in [0, 1]$. We take any n-cone $\sigma_0 \in \Delta = \Delta_0$, set $\sigma_t := S_t(\sigma_0) \in \Delta_t$ and choose a non-negative function $f_t \in A^{2n}_{(\sigma_t, \partial \sigma_t)} \subset A^{2n}_t$, the product of linear forms $\in (V^*)^*$, whose $\wedge$-product is up to sign the volume form on V. Denote by $T_t \colon V^* \to V^*$ the linear map, which coincides with S_t on σ_0. Then we have $S_t^*(f_t) = \det(T_t)\, f_0$, and the map $\varepsilon_t \circ (S_t^{-1})^* \colon A^{2n}_0 \to \mathbb{R}$ can be thought of as $\det(T_t)^{-1}$ times the projection operator $A^{2n}_0 \longrightarrow \mathbb{R}f_0 \subset A^{2n}_0$ with kernel $M^{2n}_t \subset E^{2n} = A^{2n}_0$, since $\varepsilon_t(f_t) = 1$. That yields the desired continuity with respect to $t \in [0, 1]$.

The general case: We take a simplicial refinement $\Sigma \overset{\iota}{\longrightarrow} \Delta$ and consider an enbedding

$$\mathcal{E} \hookrightarrow \iota_*(\mathcal{A}) \quad \text{and} \quad E_\Delta \hookrightarrow A_\Sigma$$

as in section 4.F. For $\Sigma_t := S_t(\Sigma)$ it induces embeddings

$$E_t := E_{\Delta_t} \hookrightarrow A_t := A_{\Sigma_t} \,,$$

which according to [BBFK$_3$] respect the intersection pairings. Then the intersection product takes the form

$$E^r_0 \times E^s_0 \longrightarrow A^r_0 \times A^s_0 \overset{(S_t^{-1})^*}{\longrightarrow} A^r_t \times A^s_t \overset{\cap}{\longrightarrow} A^0 = \mathbb{R} \,,$$

so the simplicial case applies. $\qquad\qquad\qquad\qquad\qquad\qquad\qquad\qquad\square$

(D) The proof of HLT for the polytopes Q_t with $0 < t \le 1$: Finally we show

Proposition 6.3. *The HLT holds for the polytopes Q_t with $t \in (0, 1]$, in particular the HR-forms are non-degenerate for all $t \in [0, 1]$.*

Proof. Since the linear isomorphism $F_t^* \colon V^* \to V^*$ induces an isomorphism

$$IH(\Delta_t) \overset{\cong}{\longrightarrow} IH(\Delta_1)$$

we may replace Δ_t with $\Delta := \Delta_1 = \Delta(G) = \Delta(G_t)$. Denote $\psi := \psi_t$ the function given by the vertices of G_t and $L = L_t$ the corresponding Lefschetz operator. Because of Poincaré duality it suffices to prove that $L^k \colon IH^{n-k}(\Delta) \to IH^{n+k}(\Delta)$ is injective for $k > 0$. We have

$$\Delta = \Theta \cup \Theta_0 \,,$$

with the subfan $\Theta = \mathrm{st}(\sigma(F))$ corresponding to the ridge $F \preceq G$ of the "hip roof" G, and the subfan $\Theta_0 := \mathrm{st}(\nu(G \cap H))$ associated to the bottom or cut facet $G \cap H$. The rays of Δ not contained in $\sigma(F)$ are the outer normal rays $\varrho_i := \nu(F_i)$ of the facets $F_0, \ldots, F_r$ of G not containing F – say, $F_0 := G \cap H \prec G$ is the bottom of G. From the fact that F is simple, we deduce that all these facets are normally trivial in G: This uses the fact that G is combinatorially equivalent to $F \times \Pi(L)$, cf. Remark 3.1, and that under that equivalence, the facets $F_1, \ldots, F_r$ correspond to facets of F times the pyramid $\Pi(L)$, whereas for the cut facet F_0 the claim is obvious. As a particular consequence, each n-cone $\sigma \succeq \varrho_i$ is the sum of $\varrho_i = \nu(F_i)$ and the unique opposite facet of σ corresponding to the unique edge starting in the vertex the cone σ is associated with, and not contained in the facet F_i.

As a consequence, there are (unique) functions $\psi_i \in A_\Delta^2$ vanishing outside $\Theta_i := \mathrm{st}\big(\nu(F_i)\big)$ with $\psi_i = \psi$ on the ray $\nu(F_i)$ for $i = 0, \ldots, r$. On the other hand, we may assume $0 \in F$ resp. $\psi|_{\sigma(F)} = 0$. So altogether we have

$$\psi = \sum_{i=0}^{r} \psi_i \ .$$

Now assume $k \equiv n \bmod(2), 0 < k \le n$ and $\xi \in IH^{n-k}(\Delta)$ with $L^k(\xi) = 0$. For $\Theta_i := \mathrm{st}\big(\nu(F_i)\big)$ we show $0 = \xi|_{\Theta_i} \in IH(\Theta_i)$. First of all, the equality $L^k(\xi) = \psi^k \xi = 0$ implies

$$0 = \xi \cap \psi^k \xi = \sum_{i=0}^{r} \xi \cap \psi_i \psi^{k-1} \xi = \sum_{i=0}^{r} \xi_i \cap \varphi_i^{k-1} \xi_i \, ,$$

where

- $\xi_i := \xi|_{\Theta_i} \in IH^{n-k}(\Theta_i) \cong IH^{(n-1)-(k-1)}\big(\Delta(F_i)\big)$,
- the strictly convex conewise linear functions $\varphi_i \in A_{\Delta(F_i)}^2$ are the pullback of $\psi|_{\partial\Theta_i}$ with respect to the inverse mapping of $\pi|_{\partial\Theta_i} \colon \partial\Theta_i \xrightarrow{\cong} \Delta(F_i)$, where π is the projection $\pi \colon V^* \to V^*/\mathrm{lin}\big(\nu(F_i)\big)$,
- the intersection product $\xi_i \cap \varphi_i^{k-1} \xi_i$ refers to $\Delta(F_i)$.

The equality $\xi \cap \psi_i \psi^{k-1} \xi = \xi_i \cap \varphi_i^{k-1} \xi_i$ is obtained as follows: If we use the internal product approach to the intersection product, we obtain a commutative diagram (using the obvious fact that $n-k = (n-1) - (k-1)$)

$$
\begin{array}{ccc}
(\xi_i, \xi_i) & & (\xi, \psi_i\xi)|_{\Theta_i} \\
\in & & \in \\
IH^{n-k}\big(\Delta(F_i)\big) \times IH^{n-k}\big(\Delta(F_i)\big) & \longrightarrow & IH^{n-k}(\Theta_i) \times IH^{n-k+2}(\Theta_i, \partial\Theta_i) \\
\downarrow & & \downarrow \\
IH^{2n-2}\big(\Delta(F_i)\big) & \xrightarrow{\cong} & IH^{2n}(\Theta_i, \partial\Theta_i)
\end{array}
$$

408

where the left vertical arrow denotes the HR-form in degree $k-1$ for $\Delta(F_i)$, and the right one is the intersection product composed in one argument with multiplication by ψ^{k-1}. The second component of the upper horizontal homomorphism and the lower one are pull back to the fan Θ_i followed by the Thom isomorphism (cf. the proof of Proposition 5.1), i.e., multiplication with ψ_i.

As a consequence of $L^k(\xi) = 0$, we obtain $\xi_i \in IP^{(n-1)-(k-1)}(\Delta(F_i))$. Hence, the HRR for $\Delta(F_i)$ yield that all summands are either non-negative or non-positive. So necessarily $\xi_i \cap \varphi_i^{k-1}\xi_i = 0$ and thus $\xi_i = 0$ for $i = 0, \ldots, r$. In particular $\xi_0 = 0$, and the exact sequence

$$0 \longrightarrow IH^{n-k}(\Theta, \partial\Theta) \longrightarrow IH^{n-k}(\Delta) \longrightarrow IH^{n-k}(\Theta_0) \longrightarrow 0$$

tells us that $\xi \in IH^{n-k}(\Theta, \partial\Theta) \subset IH^{n-k}(\Delta)$. Consequently, in order to conclude $\xi = 0$, it suffices to prove that $L^k|_{IH^{n-k}(\Theta, \partial\Theta)}$ is injective or, equivalently, that dually $L^k \colon IH^{n-k}(\Theta) \to IH^{n+k}(\Theta)$ is surjective for $k > 0$. To that end, we show that the graded vector space $IH(\Theta)/L^k(IH(\Theta))$ has weights at most $n + k - 1$.

The fan Θ is of the form

$$\Theta = S(\Phi \times \sigma(0)) \subset U^* \oplus N^*$$

with the cone $\sigma(0) \subset N^*$ associated to the apex $0 \in \Pi(L)$ of the pyramid $\Pi(L) \subset N$ and $S := S_1$. Using the induced vector space isomorphism

$$S^* \colon E_\Theta \longrightarrow E_{\Phi \times \sigma(0)} \cong A_\Phi \otimes E_{\sigma(0)}$$

we may write

$$IH(\Theta) \cong A_\Phi \otimes E_{\sigma(0)}/S^*(\mathfrak{m})(A_\Phi \otimes E_{\sigma(0)})$$

with $S^*(\mathfrak{m}) \subset A_{\Phi \times \sigma(0)}$. Since $\psi|_{\sigma(F)} = 0$ (because of $0 \in F$) and $\Theta = \mathrm{st}(\sigma(F))$ (remember that $\sigma(F) = o \times \sigma(0) \subset U^* \oplus N^*$), we can write $\psi|_\Theta = \chi \circ p$ with the projection $p \colon U^* \oplus N^* \to U^*$ and a function $\chi \in A_\Phi^2$. Now in order to compute $IH(\Theta)/L^k(IH(\Theta))$ we have to regard on $A_\Phi \otimes E_{\sigma(0)}$ the "twisted" A-module structure obtained from that of E_Θ by pull back via S, with other words, a function $f \in A$ acts on $A_\Phi \otimes E_{\sigma(0)}$ by "standard" multiplication with $f \circ S \in A_{\Phi \times \sigma(0)}$.

We now write $A = C \otimes D$ with the respective polynomial algebras $C := S((U^*)^*)$ on U^* and $D := S((N^*)^*)$ on N^*. Then $g = g(\mathbf{x}) \in C$ acts on the first factor A_Φ only, while $h \in D$ acts by standard multiplication with $h(\mathbf{y} + H(\mathbf{x}))$. We now have to divide by the submodule obtained by multiplication with $\mathfrak{m}_C$, χ^k, and $\mathfrak{m}_D$. Looking first at $\mathfrak{m}_C$ and χ^k gives

$$IH(\Phi)/L^k(IH(\Phi)) \otimes E_{\sigma(0)} \ ,$$

a D-module. The graded vector space $F := IH(\Phi)/L^k\big(IH(\Phi)\big)$ has weights at most $s + k - 1$ for $s := \dim U^* < n$, according to the HLT for F. It admits a descending filtration by the D-submodules

$$F^{\geq i} \otimes E_{\sigma(0)}, \; 0 \leq i \leq s + k \, ,$$

with free successive quotients

$$F^{\geq i} \otimes E_{\sigma(0)}/F^{\geq i+1} \otimes E_{\sigma(0)} \;\cong\; (F^{\geq i}/F^{\geq i+1}) \otimes E_{\sigma(0)} \, ,$$

since $h \in D$ acts only on the second factor of the right hand side, the twist being factored out. The short exact sequences

$$0 \longrightarrow F^{\geq i+1} \otimes E_{\sigma(0)} \longrightarrow F^{\geq i} \otimes E_{\sigma(0)} \longrightarrow F^{\geq i} \otimes E_{\sigma(0)}/F^{\geq i+1} \otimes E_{\sigma(0)} \longrightarrow 0$$

remain exact after reduction mod $\mathfrak{m}_D$: The third terms being free D-modules, they are split. So, since the third non-trivial term has weights less than $i + t$ with $t := \dim N^*$ according to [BBFK$_2$, 1.7], we see by descending induction on i that the reduction of $F^{\geq i} \otimes E_{\sigma(0)}$ has weights at most $(s + k - 1) + t = n + k - 1$. The case $i = 0$ yields the claim. $\qquad\square$

This ends the proof of both, Proposition 6.3 and Theorem 6.1.

References

BBFK$_1$. G. BARTHEL, J.-P. BRASSELET, K.-H. FIESELER AND L. KAUP, *Equivariant Intersection Cohomology of Toric Varieties*, Algebraic Geometry, Hirzebruch 70, 45–68, Contemp. Math. **241**, Amer. Math. Soc., Providence, R.I., 1999.

BBFK$_2$. G. BARTHEL, J.-P. BRASSELET, K.-H. FIESELER AND L. KAUP, *Combinatorial Intersection Cohomology for Fans*, Tôhoku Math. J. **54** (2002), 1–41.

BBFK$_3$. G. BARTHEL, J.-P. BRASSELET, K.-H. FIESELER AND L. KAUP, *Combinatorial Duality and Intersection Product: A Direct Approach*, Tôhoku Math. J. **57** (2005), 273 – 292.

BreLu$_1$. P. BRESSLER AND V. LUNTS, *Intersection cohomology on nonrational polytopes*, Compos. Math. **135** (2003), 245–278.

BreLu$_2$. P. BRESSLER AND V. LUNTS, *Hard Lefschetz theorem and Hodge-Riemann relations for intersection cohomology of nonrational polytopes*, (pr)e-print `math.AG/0302236 v2` (46 pages), 2003.

Bri. M. BRION, *The structure of the polytope algebra*, Tôhoku Math. J. **49** (1997), 1–32.

Ka. K. KARU, *Hard Lefschetz Theorem for Nonrational Polytopes*, Invent. Math. **157** (2004), 419–447.

Mc. P. McMULLEN, *On simple Polytopes*, Invent. Math. **113** (1993), 419–444.

St. R. STANLEY, *Generalized h-vectors, intersection cohomology of toric varieties and related results*, M. NAGATA, H. MATSUMURA, eds., *Commutative Algebra and Combinatorics*, 187–213, Adv. Stud. Pure Math. **11**, Kinokunia, Tokyo, and North Holland, Amsterdam/New York, 1987.

Ti. V. A. TIMORIN, *An analogue of the Hodge-Riemann relations for simple convex polytopes*, Russian Mathematical Surveys **54.2** (1999), 381–426

ON RATIONAL CUSPIDAL PLANE CURVES, OPEN SURFACES AND LOCAL SINGULARITIES

J. FERNÁNDEZ DE BOBADILLA

Departamento de Matemáticas Fundamentales
Facultad de Ciencias U.N.E.D.
c/ Senda del Rey 9, 28040 Madrid, Spain
E-mail:javier@mat.uned.es

I. LUENGO* and A. MELLE-HERNÁNDEZ**

Facultad de Matemáticas
Universidad Complutense
Plaza de Ciencias 3
E-28040, Madrid, Spain
** E-mail: iluengo@mat.ucm.es*
*** Email: amelle@mat.ucm.es*

A. NÉMETHI

Department of Mathematics
Ohio State University
Columbus, OH 43210,USA;
Rényi Institute of Mathematics,
Budapest, Hungary
E-mail: nemethi@renyi.hu,
nemethi@math.ohio-state.edu

Dedicated to Jean-Paul Brasselet on the Occasion if His 60th Birthday

Let C be an irreducible projective plane curve in the complex projective space $\mathbb{P}^2$. The classification of such curves, up to the action of the automorphism group $PGL(3,\mathbb{C})$ on $\mathbb{P}^2$, is a very difficult open problem with many interesting connections. The main goal is to determine, for a given d, whether there exists a projective plane curve of degree d having a fixed number of singularities of given topological type. In this note we are mainly interested in the case when C is a rational curve. The aim of this article is to present some of the old conjectures and related problems, and to complete them with some results and new conjectures from the recent work of the authors.

Keywords: Rational curves, logarithmic Kodaira dimension, Nagata-Coolidge problem, Flenner-Zaidenberg rigidity conjecture, surface singularities, $\mathbb{Q}$-homology spheres, Seiberg-Witten invariant, graded roots, Heegaard Floer homology, Ozsváth-Szabó invariants

1. Introduction

Let C be an irreducible projective plane curve in the complex projective space $\mathbb{P}^2$. The classification of such curves, up to the action of the automorphism group $PGL(3, \mathbb{C})$ on $\mathbb{P}^2$, is a very difficult open problem with many interesting connections. The main goal is to determine, for a given d, whether there exists a projective plane curve of degree d having a fixed number of singularities of given topological type. In this note we are mainly interested in the case when C is a rational curve.

The problem remains very difficult even if we aim much less, e.g. the determination of the maximal number of cusps among all the rational cuspidal plane curves (a problem proposed by F. Sakai in [14]), — this number is expected to be small. In [53] K. Tono recently proved that it is less than 9; the maximal number of cusps known by the authors is 4; and, in fact, it is expected to be 4. The referee pointed out to us that in mid-90-s S. Orevkov found a bound bigger than 8 but he never published his result.

This remarkable problem of classification is not only important for its own sake, but it is also connected with crucial properties, problems and conjectures in the theory of open surfaces, and in the classical algebraic geometry.

For instance, the open surface $\mathbb{P}^2 \setminus C$ is $\mathbb{Q}$-acyclic if and only if C is a rational cuspidal curve. On the other hand, regarding these surfaces, Flenner and Zaidenberg in [8] formulated the *rigidity conjecture*. This says that every $\mathbb{Q}$-acyclic affine surfaces Y with logarithmic Kodaira dimension $\bar{\kappa}(Y) = 2$ must be rigid. This conjecture for $Y = \mathbb{P}^2 \setminus C$ would imply the projective rigidity of the curve C in the sense that every equisingular deformation of C in $\mathbb{P}^2$ would be projectively equivalent to C. (Notice that if C has at least three cusps then $\bar{\kappa}(\mathbb{P}^2 \setminus C) = 2$ by [59]; and, in fact, all curves with $\bar{\kappa}(\mathbb{P}^2 \setminus C) < 2$ are classified, see below.) Many known examples support the rigidity conjecture, see [7–10]. Zaidenberg in [14] also conjectured that the set of shapes of the Eisenbud-Neumann splice diagrams is finite for all $\mathbb{Q}$-acyclic affine surfaces Y with $\bar{\kappa}(Y) = 2$ (this is stronger than the existence of a uniform bound for the number of cusps for rational cuspidal curves).

Another related, very old, famous open problem has its roots in early algebraic geometry, and wears the name of Coolidge and Nagata, see [3, 30]. It predicts that every rational cuspidal curve can be transformed by a Cremona transformation into a line.

The aim of this article is to present some of these conjectures and related problems, and to complete them with some results and new conjectures from the recent work of the authors.

In section 3 we present the Nagata-Coolidge problem. The main theme of section 4 is Orevkov's conjecture [41], which formulates an inequality involving the degree d and numerical invariants of local singularities. In a different formulation, this is equivalent with the positivity of the virtual dimension of the space of curves with fixed degree and certain local type of singularities which can be geometrically realized. (This was used, as a 'first test', by the second author to check that some singularities might be realized or not.) The equivalence of the two inequalities is proved via some properties of the numerical, local and global deformation invariants; this material is presented in sections 2 and 3.

Section 5 starts with some classification results: the classification of projective plane curves with $\bar{\kappa}(\mathbb{P}^2 \setminus C) < 2$ by the work of Kashiwara, Lin, Miyanishi, Sugie, Tsunoda, Tono, Wakabayashi, Yoshihara, Zaidenberg (among others). Also, we recall the classification (of the authors) of rational unicuspidal curves whose cusp has one Puiseux pair. The end of this section deals with the *rigidity conjecture* of Flenner and Zaidenberg.

In section 6 we present the author's 'compatibility property' (a sequence of inequalities), conjecturally satisfied by the degree and local invariants of the singularities of a rational cuspidal curve. It turns out that in the unicuspidal case, the inequalities are true if and only if they are (in fact) equalities. One of the reformulations (valid in the unicuspidal case) of this conjectural series of identities is the *semigroup distribution property*, which is a very precise compatibility property connecting the semigroup of the local singularity and the degree of the curve. We also present one of its equivalent statements suggested to us by Campillo. Section 7 explains the relation of the semigroup distribution property (for the unicuspidal case) with the 'Seiberg-Witten invariant conjecture' (of the forth author and Nicolaescu [36]), which basically leads us to this compatibility property. The last section present another connection with the Seiberg-Witten theory (based on the articles [32–34]), but now exploiting the relation with the Heegaard-Floer homology (introduced and studied by Ozsváth and Szabó [43]). Here, a crucial intermediate object is the 'graded root' (introduced by the forth author [32]), which provides a completely different interpretation of the semigroup distribution property.

The authors thank A. Campillo, K. Tono and M.G. Zaidenberg for helpful information, comments and interesting discussions. The authors thank to the referee for her/his very useful remarks and comments.

2. Invariants of plane curve singularities and deformations

One of our main goals is to characterise the local embedded topological types of local singular germs $(C, p_i) \subset (\mathbb{P}^2, p_i)$ which can be realized by a projective plane curve C of degree d. (Here, the points $\{p_i\}$ are not fixed.) In this section we collect some results about the (deformation of) local invariants (C, p_i).

2.1. *Germs of plane curve singularities*

Let (C, p) be the germ at p of a reduced curve $C \subset \mathbb{P}^2$ and let $f \in \mathcal{O}_{\mathbb{P}^2, p}$ be a function defining the singularity (C, p) in some local coordinates x and y.

2.1.1. *Semi-universal deformation*

Let $\phi : \mathcal{C}_{(C,p)} \to \mathcal{S}_{(C,p)}$ be a *semi-universal deformation* of (C, p). The base space $\mathcal{S}_{(C,p)}$ is smooth and its tangent space is isomorphic to the vector space $\mathcal{O}_{C,p}/(f_x, f_y)$. In particular its dimension is equal to the *Tjurina number* $\tau(C, p) := \dim_{\mathbb{C}}(\mathcal{O}_{\mathbb{P}^2, p}/I^{ea}(C, p))$, where $I^{ea}(C, p)$ is the ideal (f, f_x, f_y) (see [24].)

2.1.2. *Equisingular deformations*

Let $\mathcal{S}^{es}_{(C,p)} \subset \mathcal{S}_{(C,p)}$ be the smooth subgerm of a *semi-universal equisingular deformation* of (C, p), see [62], Theorem 7.4. Let $T_\varepsilon := Spec(\mathbb{C}[\varepsilon]/(\varepsilon^2))$ be the base space of *first order infinitesimal* deformations. The *equisingularity ideal* of (C, p) is the ideal

$$I^{es}(C, p) = \{g \in \mathcal{O}_{\mathbb{P}^2, p} \mid f + \varepsilon g \text{ is equisingular over } T_\varepsilon\}.$$

It contains $I^{ea}(C, p)$. Moreover the tangent space of $\mathcal{S}^{es}_{(C,p)} \subset \mathcal{S}_{(C,p)}$ is isomorphic to the vector space $I^{es}(C, p)/I^{ea}(C, p)$. In particular the codimension of $\mathcal{S}^{es}_{(C,p)}$ in $\mathcal{S}_{(C,p)}$ is the topological invariant

$$\tau^{es}(C, p) := \dim_{\mathbb{C}}(\mathcal{O}_{\mathbb{P}^2, p}/I^{es}(C, p)). \tag{1}$$

The equisingular stratum $\mathcal{S}^{es}_{(C,p)}$ coincides with the μ-*constant stratum* of $\mathcal{S}_{(C,p)}$, where $\mu = \mu(C, p)$ is the Milnor number of (C, p).

2.1.3. *δ-constant deformations*

Let $n : \mathcal{O}_{C,p} \hookrightarrow \tilde{\mathcal{O}}_{C,p} := \prod_{i=1}^{r} \mathbb{C}\{t\}$ be the normalisation map, where r is the number of local irreducible components of (C,p). Let $\delta(C,p)$ be the dimension of the cokernel $\tilde{\mathcal{O}}_{C,p}/n(\mathcal{O}_{C,p})$. Let $\mathrm{cond}(\mathcal{O}_{C,p})$ be the conductor ideal of $\mathcal{O}_{C,p}$, that is the annihilator of the $\mathcal{O}_{C,p}$-module $\tilde{\mathcal{O}}_{C,p}/n(\mathcal{O}_{C,p})$.

The *δ-constant stratum* $\mathcal{S}^{\delta}_{(C,p)} \subset \mathcal{S}_{(C,p)}$ is the locus of points where δ is constant, see [4,48]. In general, $\mathcal{S}^{\delta}_{(C,p)}$ is not longer smooth at C, but it is locally irreducible. Even more, the tangent cone of $\mathcal{S}^{\delta}_{(C,p)}$ at C is always a linear space which is identified with the vector space $\mathrm{cond}(\mathcal{O}_{C,p})/I^{ea}(C,p)$, [4], Theorem 4.15. In particular, the codimension of $\mathcal{S}^{\delta}_{(C,p)}$ in $\mathcal{S}_{(C,p)}$ is equal to

$$\mathrm{codim}\,(\mathcal{S}^{\delta}_{(C,p)} \subset \mathcal{S}_{(C,p)}) = \delta(C,p). \tag{2}$$

2.1.4. *Relations among strata*

S. Diaz and J.Harris [4] showed the inclusion of ideals

$$I^{ea}(C,p) \subset I^{es}(C,p) \subset \mathrm{cond}(\mathcal{O}_{C,p}) \subset \mathcal{O}_{C,p}.$$

Since the equisingular stratum $\mathcal{S}^{es}_{(C,p)}$ is contained in the δ-constant stratum $\mathcal{S}^{\delta}_{(C,p)}$ then

$$\bar{M}(C,p) := \mathrm{codim}\,(\mathcal{S}^{es}_{(C,p)} \subset \mathcal{S}^{\delta}_{(C,p)}) = \tau^{es}(C,p) - \delta(C,p) \geq 0. \tag{3}$$

In fact, $\bar{M}(C,p)$ is a topological invariant of the singularity (see next paragraph). S. Orevkov and M. Zaidenberg used $\bar{M}(C,p)$ in a different situation, cf. [41,42].

2.2. *Local invariants and embedded resolution*

The minimal embedded resolution of (C,p) is obtained via a sequence of blowing-ups at infinitely near points to p. The topological numerical invariants $\delta(C,p), \mu(C,p), \tau^{es}(C,p)$ and $\bar{M}(C,p)$ can be written in terms of the multiplicity sequence $[m_1^p, \ldots, m_{k_p}^p]$ of (C,p) associated with this minimal resolution. This is the set of multiplicities of the strict transform of C at all the infinitely near points along all the branches. Let r_p denote the number of branches of C at p. (We do not omit the 1's; for this notation we follow [27] and [8].)

Using Milnor's formula $\mu(C,p) = 2\delta(C,p) - r_p + 1$ (see e.g. Theorem

416

6.5.9 in [61]) one gets

$$\mu(C,p) + r_p - 1 = 2\delta(C,p) = \sum_{i=1}^{k_p} m_i^p(m_i^p - 1). \tag{4}$$

A blow-up in the minimal good resolution is called *inner* (or *subdivisional*) if its center is at the intersection point of two exceptional curves of the resolution process (such a center is called a *satellite point* in [60]). If the center is situated on exactly one exceptional divisor, then the blow-up is called *outer* (or *sprouting*). Notice that the first blow-up is neither inner nor outer. A center is called a *free infinitely near point* if it is either outer or it is p, the center of the very first blow up. It is convenient to count this first infinitely near point by two. Let ω_p, resp. ρ_p, denote the number of inner, respectively of outer, blow-ups. Then $k_p - 1 = \omega_p + \rho_p$ and the number of free infinitely near points is $L_p := 2 + \rho_p = k_p - \omega_p + 1$.

C.T.C. Wall in Theorem 8.1 of [60] proved the following formula for $\tau^{es}(C,p)$ (see also Proposition 11.5.8 in [61])

$$\tau^{es}(C,p) = \sum_{i=1}^{k_p} \frac{(m_i^p - 1)(m_i^p + 2)}{2} + \omega_p - 1 = \sum_{i=1}^{k_p} \frac{m_i^p(m_i^p + 1)}{2} - L_p. \tag{5}$$

Different proofs of this formula have been also given by J.F. Mattei [28], E. Shustin [46] or Theo de Jong [20].

The parametric codimension is equal to

$$\bar{M}(C,p) = \sum_{i=1}^{k_p}(m_i^p - 1) + \omega_p - 1 = -L_p + \sum_{i=1}^{k_p} m_i^p. \tag{7}$$

For other equivalent formulae of $\bar{M}(C,p)$ see [42].

3. Global invariants and existence problem

3.1. *Expected dimension*

Let $\mathbb{P}^N$, where $N = d(d+3)/2$, be the Hilbert scheme parametrising algebraic projective plane curves of degree d. The locus $V_{d,g}$ in $\mathbb{P}^N$ of reduced and irreducible curves of degree d and genus g is irreducible (see e.g. [16]). The locus of reduced and irreducible curves of degree d and genus g having only nodes as singularities is smooth of dimension $3d - 1 + g$ and is dense in $V_{d,g}$.

Recall that our goals is to characterise the local embedded topological types of local singular germs $(C,p_i) \subset (\mathbb{P}^2, p_i)$ which can be realized by a projective curve C of degree d. Assume that a projective reduced plane curve C of degree d exists with fixed topological types $S_1, \ldots, S_\nu$. Let $V(S_1, \ldots, S_\nu)$ denote the locally closed subscheme of reduced curves on $\mathbb{P}^2$ having exactly ν-singularities with topological types $S_1, \ldots, S_\nu$.

Greuel and Lossen in Theorem 3.6 of [13] proved that the dimension of $V(S_1, \ldots, S_\nu)$ at C satisfies

$$\dim(V(S_1, \ldots, S_\nu), C) \geq C^2 + 1 - p_a(C) - \tau^{es}(C) = \frac{d(d+3)}{2} - \tau^{es}(C), \quad (8)$$

where $\tau^{es}(C) := \sum_{i=1}^{\nu} \tau^{es}(C, p_i)$. The right hand side

$$\operatorname{expdim}(V(S_1, \ldots, S_\nu), C) := \frac{d(d+3)}{2} - \tau^{es}(C)$$

is called *expected dimension* of $V(S_1, \ldots, S_\nu)$ at C. The study of the locally closed subscheme $V(S_1, \ldots, S_\nu)$ of $\mathbb{P}^N$ and its properties have been studied intensively in the literature (see for instance works by Artal Bartolo, Diaz, Greuel, Harris, Karras, Lossen, Shustin, Tannenbaum, Wahl, Zariski among many others).

3.2. *The action of $PGL(3, \mathbb{C})$*

Consider a reduced curve C of degree d in $\mathbb{P}^2$ and the action of the group $PGL(3) := PGL(3, \mathbb{C})$ on the space $\mathbb{P}^N$, which parametrises plane curves of degree d. The orbit of a curve C is a quasi-projective variety of dimension $\dim PGL(3) - \dim \operatorname{Stab}_{PGL(3)}(C)$. For a general curve, the dimension of the orbit is 8, that is, its stabiliser is 0-dimensional.

Since the topological types $S_1, \ldots, S_\nu$ of singularities of C and the degree d of C remain constant under the $PGL(3)$-action, we consider the *virtual dimension* of $V(S_1, \ldots, S_\nu)$ at C defined by

$$\operatorname{expdim}(V(S_1, \ldots, S_\nu), C) - (8 - \dim \operatorname{Stab}_{PGL(3)}(C)). \quad (9)$$

Curves with small orbits have been studied and classified by P. Allufi and C. Faber in [1]. According to this classification, C is always a configuration of rational curves. Moreover, C consists of irreducible components of the form below, with arbitrary multiplicities. We reproduce here their list together with the dimension of the stabiliser $\operatorname{Stab}_{PGL(3)}(C)$.

(1) C consists of a single line; $\dim \operatorname{Stab}_{PGL(3)}(C) = 6$.
(2) C consists of 2 (distinct) lines; $\dim \operatorname{Stab}_{PGL(3)}(C) = 4$.
(3) C consists of 3 or more concurrent lines; $\dim \operatorname{Stab}_{PGL(3)}(C) = 3$.
(4) C is a triangle (consisting of 3 lines in general position); $\dim \operatorname{Stab}_{PGL(3)}(C) = 2$.

418

(5) C consists of 3 or more concurrent lines, together with 1 other (non-concurrent) line; $\dim \mathrm{Stab}_{PGL(3)}(C) = 1$.

(6) C consists of a single conic; $\dim \mathrm{Stab}_{PGL(3)}(C) = 3$.

(7) C consists of a conic and a tangent line; $\dim \mathrm{Stab}_{PGL(3)}(C) = 2$.

(8) C consists of a conic and 2 (distinct) tangent lines; $\dim \mathrm{Stab}_{PGL(3)}(C) = 1$.

(9) C consists of a conic and a transversal line and may contain either one of the tangent lines at the 2 points of intersection or both of them; $\dim \mathrm{Stab}_{PGL(3)}(C) = 1$.

(10) C consists of 2 or more bitangent conics (conics in the pencil $y^2 + \lambda xz$) and may contain the line y through the two points of intersection as well as the lines x and/or z, tangent lines to the conics at the points of intersection; again, $\dim \mathrm{Stab}_{PGL(3)}(C) = 1$.

(11) C consists of 1 or more (irreducible) curves from the pencil $y^b + \lambda z^a x^{b-a}$, with $b \geq 3$, and may contain the lines x and/or y and/or z; $\dim \mathrm{Stab}_{PGL(3)}(C) = 1$.

(12) C contains 2 or more conics from a pencil through a conic and a double tangent line; it may also contain that tangent line. In this case, $\dim \mathrm{Stab}_{PGL(3)}(C) = 1$.

3.3. *The Coolidge-Nagata conjecture*

Let C_1 and C_2 be two curves in $\mathbb{P}^2$. The pairs $(\mathbb{P}^2, C_1)$ and $(\mathbb{P}^2, C_2)$ are *birationally equivalent* if there exist a birational map $\sigma : \mathbb{P}^2 \dashrightarrow \mathbb{P}^2$ such that the proper image of C_1 by f coincides with C_2. Traditionally, the map σ is called a *Cremona transformation*.

Let $\pi : V \to \mathbb{P}^2$ be an embedded resolution of singularities of C, let $\bar{C}$ be the strict transform of C by π and let K_V be the canonical divisor of V. One can show that $h^0(m(K_V + \bar{C}))$ and the Kodaira dimension $\kappa(K_V + \bar{C}, V)$ are birational invariants of C as plane curves (see [19], [22]). Thus, one defines $\kappa[C] := \kappa(K_V + \bar{C}, V)$.

Let D be the reduced total preimage of C by the embedded resolution π. One can also show that $h^0(m(K_V + D))$ and the Kodaira dimension $\kappa(K_V + D, V)$ are birational invariants of the open surface $Y := \mathbb{P}^2 \setminus C$ (see [19]). Thus one defines the logarithmic Kodaira dimension $\bar{\kappa}(Y) := \kappa(K_V + D, V)$ of Y.

One of the open problems regarding projective plane curves is the following famous *Coolidge-Nagata problem/conjecture*, cf. [3] and [30]:

Conjecture 3.1 (Coolidge-Nagata conjecture). *Every rational cuspidal curve can be transformed by a Cremona transformation into a straight line.*

In [3], J.I. Coolidge proved:

Theorem 3.1 (Coolidge). *A rational curve can be transformed into a straight line by a Cremona transformation if and only if all the conditions for special adjoints of any index are incompatible.*

A *special adjoint of index m* is an effective divisor in the complete linear system $|mK_V + \bar{C}|$. In fact one can check that $\kappa[C] = -\infty$ if and only if C has no special adjoints. One has the following criterium due to N.M. Kumar and M.P. Murthy (cf. Corollary 2.4 in [22]), cf. also with Iitaka (Proposition 12 in [19]):

Theorem 3.2. *Let C be an irreducible rational plane curve. The following conditions are equivalent:*

a) *the curve C can be transformed into a straight line by a Cremona transformation,*

b) $\kappa[C] = -\infty$,

c) $|2K_V + \bar{C}| = \emptyset$,

d) $|2(K_V + \bar{C})| = \emptyset$.

In [22] N.M. Kumar and M.P. Murthy also showed that a sufficient condition is $\bar{C}^2 \geq -3$.

The Nagata-Coolidge problem has been solved for cuspidal rational plane curves with logarithmic Kodaira dimension of the complement $\bar{\kappa}(\mathbb{P}^2 \setminus C) < 2$ (using the classification listed in subsection 5.1) and for all known curves with $\bar{\kappa} = 2$ (see below, cf. also with Corollary 5.1).

4. On rational plane curves

From now on we are interested in rational projective plane curves. Let C be a reduced rational projective plane curve of degree d in the complex projective plane with singular points $\{p_j\}_{j=1}^{\nu}$. Let S_j be the topological type of the singularity (C, p_j).

The main theme of this section is Orevkov's conjecture [41]. Since this conjecture is not true for a small number of curves (with positive dimensional stabiliser $Stab_{PGL(3)}(C)$, all of them with $\bar{\kappa}(Y) < 2$), we correct the original version by adding a contribution provided by this stabiliser; and we also present some equivalent reformulations.

Conjecture 4.1 (Virtual dimension conjecture). *The virtual dimension* $\mathrm{virtdim}(V(S_1,\ldots,S_\nu),C)$ *of reduced rational projective plane curve is non-negative:*

$$expdim(V(S_1,\ldots,S_\nu),C) - 8 + \dim Stab_{PGL(3)}(C)) \geq 0. \tag{10}$$

This version was also conjectured independently by I. Luengo.

Since C is rational, if the multiplicity sequence of the singularity (C,p) is $[m_1^p,\ldots,m_k^p]$, then (the genus-formula reads as):

$$(d-1)(d-2) = \sum_{p\in\mathrm{Sing}(C)}\sum_{i=1}^{k} m_i^p(m_i^p-1) = \sum_{p\in\mathrm{Sing}(C)}\left(\sum_{i=1}^{k}(m_i^p)^2 - \sum_{i=1}^{k}m_i^p\right). \tag{11}$$

Eliminating d^2 from (9) and (5), one gets

$$\text{virtual dim} = 3d-9 - \sum_{p\in\mathrm{Sing}(C)}\sum_{i=1}^{k}m_i^p + \sum_{P\in\mathrm{Sing}(C)} L_p + \dim\mathrm{Stab}_{PGL(3)}(C).$$

Substituting (7) in this equality, one gets

$$\text{virtual dim} = 3d-9 - \sum_{p\in\mathrm{Sing}(C)} \bar{M}(C,p) + \dim\mathrm{Stab}_{PGL(3)}(C). \tag{12}$$

In particular, (10) is equivalent to

$$3d-9 - \sum_{p\in\mathrm{Sing}(C)} \bar{M}(C,p) + \dim\mathrm{Stab}_{PGL(3)}(C) \geq 0. \tag{13}$$

S. Orevkov in [41] conjectured the following inequality.

Conjecture 4.2 (Orevkov's conjecture). *For a rational cuspidal curve*

$$\sum_{p\in Sing(C)} \bar{M}(C,p) \leq 3d-9. \tag{14}$$

Orevkov's conjecture is stated for any rational curve (without restrictions). In such a case, the sum is over all irreducible branches at each singular point. Note that (14) is not true for the curve C defined by $x^2y + z^3 = 0$ since $\bar{M}(C,p) = 1$ at its singular point. Nevertheless, $\dim\mathrm{Stab}_{PGL(3)}(C) = 1$ and

(10) and (13) hold. We believe that the correct statement of the conjecture is (10), or equivalently (13). Nevertheless, one can prove (cf. Lemma 5.1) that for irreducible, cuspidal, rational plane curve with with $\bar{\kappa}(Y) = 2$ the statements of (10) and (14) are equivalent.

4.1. *The $\bar{C}^2$–Conjecture*

Let $\pi : V \to \mathbb{P}^2$ be the minimal good embedded resolution of $C \subset \mathbb{P}^2$, and let $\bar{C}$ be the strict transform of C and $D = \pi^{-1}(C)$ be the reduced preimage of C as above. One of the integers which plays a special role in the classification problem is the self-intersection of $\bar{C}$ in V. It equals

$$\bar{C}^2 = d^2 - \sum_{p \in \mathrm{Sing}(C)} \sum_{i=1}^{k} (m_i^p)^2. \tag{15}$$

From (10) and (15) one gets

$$3d = \bar{C}^2 + 2 + \sum_{p \in \mathrm{Sing}(C)} \sum_{i=1}^{k} m_i^p.$$

Thus, via (12):

$$\text{virtual dim} = \bar{C}^2 - 7 + \sum_{p \in \mathrm{Sing}(C)} L_p + \dim \mathrm{Stab}_{PGL(3)}(C). \tag{16}$$

Therefore, the conjecturelly inequality (10) is equivalent to:

$$\bar{C}^2 - 7 + \sum_{p \in \mathrm{Sing}(C)} L_p + \dim \mathrm{Stab}_{PGL(3)}(C) \geq 0. \tag{17}$$

5. Cuspidal rational plane curves and the Rigidity Conjecture

Let C be an irreducible curve of degree d in the complex projective plane. One of the main invariants of such curves is the logarithmic Kodaira dimension $\bar{\kappa} = \bar{\kappa}(Y)$, where $Y := \mathbb{P}^2 \setminus C$. The following result of Wakabayashi [59] is crucial in the classification procedure.

Theorem 5.1 (Wakabayashi). *Let C be an irreducible curve of degree d in $\mathbb{P}^2$.*

(1) If $g(C) \geq 1$ and $d \geq 4$ then $\bar{\kappa}(\mathbb{P}^2 \setminus C) = 2$.
(2) If $g(C) = 0$ and C has at least 3 cuspidal points then $\bar{\kappa}(\mathbb{P}^2 \setminus C) = 2$.

(3) If $g(C) = 0$ and C has at least 2 singular points and one of them is locally reducible then $\bar{\kappa}(\mathbb{P}^2 \setminus C) = 2$.
(4) If $g(C) = 0$ and C has 2 cuspidal points then $\bar{\kappa}(\mathbb{P}^2 \setminus C) \geq 0$.

5.1. *Logarithmic Kodaira dimension and classification*

Recall that the open surface $Y := \mathbb{P}^2 \setminus C$ is $\mathbb{Q}$-acyclic if and only if C is a rational cuspidal curve. From now on we assume that C is a rational cuspidal plane curve of degree d, in particular C is irreducible. Let $(C, p_i)_{i=1}^{\nu}$ be the collection of local plane curve singularities, all of them locally irreducible.

(a) If $\bar{\kappa} = -\infty$ then $\nu = 1$ by [59]. Moreover, all these curves are classified by M. Miyanishi and T. Sugie [29] (see also H. Kashiwara [21]). The family contains as an important subfamily the Abhyankar-Moh-Suzuki curves (see [11]).

(b) The case $\bar{\kappa} = 0$ cannot occur by a result of Sh. Tsunoda [54], see also Orevkov's article [41].

(c) If $\bar{\kappa} = 1$ then by the above result of Wakabayashi [59] one has $\nu \leq 2$. In the case $\nu = 1$, K. Tono provides the possible equations of the curves [52]. (Notice that Tsunoda's classification in [55] is incomplete.). On the other hand, by another result of Tono [51], the case $\nu = 2$ corresponds exactly to the Lin-Zaidenberg bicuspidal rational plane curves.

Lemma 5.1. *If C is an irreducible, cuspidal, rational plane curve with with $\bar{\kappa}(Y) = 2$ then $\dim Stab_{PGL(3)}(C) = 0$.*

Proof. According with the above classification of Aluffi and Faber, C is a rational cuspidal plane curve with $\dim Stab_{PGL(3)}(C) > 0$ if and only if C is a generic member of the pencil $y^d + \lambda z^a x^{d-a}$ with $d \geq 3$ and $(d, a) = 1$. If $a = d - 1$ then C is of Abhyankar-Moh-Suzuki type (with $\bar{\kappa} = -\infty$). Otherwise $a \neq d - 1$ and C is of Lin-Zaidenberg type (with $\bar{\kappa} = 1$). $\square$

5.2. *Rational unicuspidal plane curves with one Puiseux pair*

From a different point of view, one can classify triples (d, a, b) such that there exists a unicuspidal rational plane curve C of degree d whose singularity has only one Puiseux pair of type (a, b), where $1 < a < b$. Let $\{\varphi_j\}_{j \geq 0}$ denote the Fibonacci numbers $\varphi_0 = 0$, $\varphi_1 = 1$, $\varphi_{j+2} = \varphi_{j+1} + \varphi_j$.

Theorem 5.2 ([12]). *The Puiseux pair (a, b) can be realized by a unicuspidal rational curve of degree d if and only if (d, a, b) appears in the* *following list.*

> *(a)* $(a,b) = (d-1, d)$;
> *(b)* $(a,b) = (d/2, 2d-1)$, *where d is even;*
> *(c)* $(a,b) = (\varphi_{j-2}^2, \varphi_j^2)$ *and* $d = \varphi_{j-1}^2 + 1 = \varphi_{j-2}\varphi_j$, *where j is odd and* ≥ 5;
> *(d)* $(a,b) = (\varphi_{j-2}, \varphi_{j+2})$ *and* $d = \varphi_j$, *where j is odd and* ≥ 5;
> *(e)* $(a,b) = (\varphi_4, \varphi_8 + 1) = (3, 22)$ *and* $d = \varphi_6 = 8$;
> *(f)* $(a,b) = (2\varphi_4, 2\varphi_8 + 1) = (6, 43)$ *and* $d = 2\varphi_6 = 16$.

In the first four cases $\bar{\kappa} = -\infty$ and they can be realized by some particular curves which appear in Kashiwara's classification [21]. The last two sporadic cases have $\bar{\kappa} = 2$ and were found by Orevkov and Artal-Bartolo, cf. [41].

5.3. *More classification results*

Another aproach is to classify rational cuspidal curves C such that the highest multiplicity of the singular points m is close to the degree d. Flenner and Zaidenberg classified the curves with $m = d - 2$ in [9] and $m = d - 3$ in [10]. The case $m = d - 4$ is partially solved by Fenske [7]. Note that m can not be too small because in [27] it is proved that $d < 3m$ solving a conjecture of Yoshihara [63]. Let $\alpha = \frac{(3+\sqrt{5})}{2}$, Orevkov [41] gives two families of curves with $\alpha m < d$ and conjectured that those families gives the only curves verifyng $\alpha m < d$.

5.4. *The rigidity conjecture of Flenner and Zaidenberg*

Let Y be an $\mathbb{Q}$-acyclic affine surface, and fix one of its 'minimal logarithmic compactifications' (V, D). This means that V is a smooth projective surface with a normal crossing divisor D, such that $Y = V \setminus D$, and (V, D) is minimal with these properties.

The sheaf of the logarithmic tangent vectors $\Theta_V\langle D\rangle$ controls the deformation theory of the pair (V, D), cf. [8]. E.g., $H^0(V, \Theta_V\langle D\rangle)$ is the set of infinitesimal automorphisms, $H^1(V, \Theta_V\langle D\rangle)$ is the space of infinitesimal deformations, and $H^2(V, \Theta_V\langle D\rangle)$ is the space of obstructions. Iitaka showed in [17] that if $\bar{\kappa}(Y) = 2$ then the automorphism group of the surface Y is finite (this also provides a different proof of Lemma 5.1). Therefore $h^0(\Theta_V\langle D\rangle) = 0$. In [8,9,66] Flenner and Zaidenberg conjectured the following

Conjecture 5.1 (Rigidity conjecture). *Every $\mathbb{Q}$-acyclic affine surfaces Y with logarithmic Kodaira dimension $\bar{\kappa}(Y) = 2$ is rigid and has unobstructed deformations. That is,*

$$h^1(\Theta_V\langle D\rangle) = 0 \qquad and \qquad h^2(\Theta_V\langle D\rangle) = 0. \tag{18}$$

In particular, the Euler characteristic $\chi(\Theta_V\langle D\rangle) = h^2(\Theta_V\langle D\rangle) - h^1(\Theta_V\langle D\rangle)$ must vanish.

In [8], [9] and [10] the conjecture was verified for most of the known examples. In [8] unobstructedness was proved for all $\mathbb{Q}$-acyclic surfaces of non log-general type. In [65] it is proved that a rigid rational cuspidal curve has at most 9 cusps.

This can be applied in our situation as follows. Consider a projective curve C, and write $Y := \mathbb{P}^2 \setminus C$. The $\mathbb{Q}$-acyclicity of Y is equivalent to the fact that C is rational and cuspidal. For V one can take the minimal embedded resolution of the pair $(\mathbb{P}^2, C)$.

The conjecture for $Y = \mathbb{P}^2 \setminus C$ implies the projective rigidity of the curve C. This means that every equisingular deformation of C in $\mathbb{P}^2$ would be projectively equivalent to C. Thus $V(S_1, \ldots, S_\nu)$ has expected dimension 8 (see Section 4).

In Corollary 2.5 of [9], Flenner and Zaidenberg show that for any cuspidal rational plane curve

$$\chi(\Theta_V\langle D\rangle) = K_V(K_V + D) = -3(d-3) + \sum_{p\in\mathrm{Sing}(C)} \bar{M}(C,p). \tag{19}$$

By (12) and Lemma 5.1 then

$$\text{virtual dim} = -\chi(\Theta_V\langle D\rangle). \tag{20}$$

The vanishing of $\chi(\Theta_V\langle D\rangle)$ implies any of the equivalent equalities (10), (13) or (17). On the other hand, if (10), (13) or (17) hold, then

$$\chi(\Theta_V\langle D\rangle) \leq 0. \tag{21}$$

Proposition 5.1 (Tono). *For cuspidal rational plane curves with $\bar{\kappa} = 2$ the following inequality holds*

$$\chi(\Theta_V\langle D\rangle) \geq 0. \tag{22}$$

Proof. (22) follows from the article [53] of K. Tono in the following way. F. Sakai in [45] introduce the invariant $\gamma_2 := h^0(2K_V + D)$. Lemma 4.1 in [53] states that if the pair (V, D) satisfies the following three conditions (for details see [loc. cit.])

(A1) $\bar{\kappa}(V \setminus D) = 2$,

(A2) (V, D) is almost minimal, and

(A3) D contains neither a rod consisting of (-2)-curves nor a fork consisting of (-2)-curves,

then

$$\gamma_2 = K_V(K_V + D) + \frac{D(D + K_V)}{2} + \chi(\mathcal{O}_V).$$

(The main point here is that by a vanishing theorem $h^1(2K_V + D) = 0$, by an easy argument $h^2(2K_V + D) = 0$ too, hence $\gamma_2 = \chi(2K_V + D)$ can be computed by Riemann-Roch.)

One can check that in our case the minimal embedded resolution satisfies these conditions. Moreover, $\chi(\mathcal{O}_V) = 1$ and (since D is a rational tree, the adjunction formula implies) $K_V D + D^2 = -2$. Thus $\gamma_2 = K_V(K_V + D)$. Therefore, via (19), one has:

$$\chi(\Theta_V\langle D\rangle) = h^0(2K_V + D) \geq 0. \qquad \square$$

Corollary 5.1. *Let C be an irreducible, cuspidal, rational plane curve with $\bar{\kappa}(\mathbb{P}^2 - C) = 2$. The following conditions are equivalent:*

(i) $\chi(\Theta_V\langle D\rangle) = 0$,

(ii) $virtdim(V(S_1, \ldots, S_\nu), C) \geq 0$, i.e. (10) holds, where S_j is the topological type of the corresponding uni-branch singularity (C, p_j).

(iii) $\chi(\Theta_V\langle D\rangle) \leq 0$.

In such a case, the curve C can be transformed by a Cremona transformation of $\mathbb{P}^2$ into a straight line (i.e., the Coolidge-Nagata problem has a positive answer).

Proof. $(i) \Rightarrow (ii) \Rightarrow (iii)$ follows from (20) and (21). $(iii) \Rightarrow (i)$ follows from Proposition 5.1 and (21). Finally, the characterisation (c) of Theorem 3.2 shows that C can be transform into a straight line by a Cremona transformation. Indeed, $h^0(2K_V + D) = \chi(\Theta_V\langle D\rangle) = 0$, but $\mathcal{O}_V(2K_V + \bar{C})$ is a subsheaf of $\mathcal{O}_V(2K_V + D)$, hence $h^0(2K_V + \bar{C}) = 0$ as well. $\qquad \square$

6. The semigroup distribution property

6.1. *Compatibility property*

The characterisation problem of the realization of prescribed topological types of singularities has a long and rich history providing many interesting compatibility properties connecting local invariants of the germs $\{(C, p_i)\}_i$ with some global invariants of C — like its degree, or the log-Kodaira dimension of $\mathbb{P}^2 \setminus C$, etc. (For a — non-complete — list of some of these restrictions, see e.g. [11,12].)

In [11] we proposed a new compatibility property — valid for rational cuspidal curves C. Its formulation is surprisingly very elementary. Consider a collection $(C, p_i)_{i=1}^{\nu}$ of locally irreducible plane curve singularities (i.e. cusps), let $\Delta_i(t)$ be the characteristic polynomial of the monodromy action associated with (C, p_i), and $\Delta(t) := \prod_i \Delta_i(t)$. Its degree is 2δ, where δ is the sum of the delta-invariants $\delta(C, p_i)$ of the singular points. Then $\Delta(t)$ can be written as $1 + (t-1)\delta + (t-1)^2 Q(t)$ for some polynomial $Q(t)$. Let c_l be the coefficient of $t^{(d-3-l)d}$ in $Q(t)$ for any $l = 0, \ldots, d-3$.

Conjecture 6.1 (Conjecture A [11]). *Let $(C, p_i)_{i=1}^{\nu}$ be a collection of local plane curve singularities, all of them locally irreducible, such that $2\delta = (d-1)(d-2)$ for some integer d. If $(C, p_i)_{i=1}^{\nu}$ can be realized as the local singularities of a degree d (automatically rational and cuspidal) projective plane curve then*

$$c_l \leq (l+1)(l+2)/2 \quad \text{for all} \quad l = 0, \ldots, d-3.$$

In fact, the integers $n_l := c_l - (l+1)(l+2)/2$ are symmetric: $n_l = n_{d-3-l}$; and $n_0 = n_{d-3} = 0$. We also mention that examples with strict inequality occur, cf. [11] (in all these examples known by the authors $\nu > 1$).

The main result of [11] is :

Theorem 6.1 ([11]). *If $\bar{\kappa}(\mathbb{P}^2 \setminus C)$ is ≤ 1, then the above conjecture A is true (in fact with $n_l = 0$).*

There is an additional surprising phenomenon in the above conjecture. Namely, in the *unicuspidal* case one can show the following.

Proposition 6.1 ([11]). *If $\nu = 1$ then $c_l \geq (l+1)(l+2)/2$ for $0 \leq l \leq d-3$.*

Therefore, Conjecture A in this case can be reformulated as follows:

Conjecture 6.2 (Conjecture B1). *With the notations of 6.1, if $\nu = 1$, then $n_l = 0$ for all $l = 0, \ldots, d-3$, that is*

$$c_l = (l+1)(l+2)/2 \quad \text{for all} \quad l = 0, \ldots, d-3.$$

In fact, if $\nu = 1$, we can do more. Recall that the characteristic polynomial Δ of $(C,p) \subset (\mathbb{P}^2, p)$ is a complete (embedded) topological invariant of this germ, similarly as the semigroup $\Gamma_{(C,p)} \subset \mathbb{N}$. In the next discussion we will replace Δ by $\Gamma_{(C,p)}$. Recall that the semigroup $\Gamma_{(C,p)} \subset \mathbb{N}$ consists of all possible intersection multiplicities $I_p(C, h)$ at the point p for all $h \in \mathcal{O}_{(\mathbb{C}^2, p)}$.

Hence, one can reformulate conjecture B1 in terms of the semigroup of the germ (C,p) and the degree d. It turns out that the of vanishing of the coefficients n_l (as in conjecture B1.) is replaced by a very precise and mysterious distribution of the elements of the semigroup with respect to the intervals $I_l := ((l-1)d, ld]$:

Conjecture 6.3 (Conjecture B2). *Assume that $\nu = 1$. Then for any $l > 0$, the interval I_l contains exactly $\min\{l+1, d\}$ elements from the semigroup $\Gamma_{(C,p)}$.*

In other words, for every rational unicuspidal plane curve C of degree d, the above conjecture is equivalent to the identity

$$D(t) \equiv 0, \tag{DP}$$

where:

$$D(t) := \sum_{k \in \Gamma_{(C,p)}} t^{\lceil k/d \rceil} - \left(1 + 2t + \cdots + (d-1)t^{d-2} + d(t^{d-1} + t^d + t^{d+1} + \cdots) \right).$$

For the equivalences of conjectures B1 and B2, see Theorem 7.1. Here we only mention a key relation between the coefficients c_l and the semigroup $\Gamma_{(C,p)}$.

First, consider the identity (cf. [15]) $\Delta(t) = (1-t) \cdot L(t)$, where $L(t) = \sum_{k \in \Gamma_{(C,p)}} t^k$ is the Poincaré series of $\Gamma_{(C,p)}$. Write $\Delta(t) = 1 - P(t)(1-t)$ for some polynomial $P(t)$, then $L(t) + P(t) = 1/(1-t) = \sum_{k \geq 0} t^k$. In particular, $P(t) = \sum_{k \in \mathbb{N} \setminus \Gamma_{(C,p)}} t^k$. Then

$$Q(t) = \frac{P(t) - \delta}{t - 1} = \sum_{k \notin \Gamma_{(C,p)}} \frac{t^k - 1}{t - 1} = \sum_{k \notin \Gamma_{(C,p)}} (1 + t + \cdots + t^{k-1}).$$

Hence $c_l = \#\{k \notin \Gamma_{(C,p)} : k > (d-3-l)d\}$. Since $k \in \Gamma_{(C,0)}$ if and only if $\mu - 1 - k \notin \Gamma_{(C,p)}$ for any $0 \leq k \leq \mu - 1$, one gets $c_l = \#\{k \in \Gamma_{(C,p)} ; k \leq ld\}$.

6.2. *An equivalent formulation*

The following equivalent formulation was suggested by A. Campillo.

Theorem 6.2. *Let C be a unicuspidal rational plane curve of degree d. The curve C satisfies the semigroup compatibility property (DP) (i.e. conjectures B1 and/or B2) if and only if the elements of the semigroup $\Gamma_{(C,p)}$ in $[0, ld]$ are realized by projective (possibly non-reduced) curves of degree l for $l \leq d - 3$.*

Proof. The proof of the 'if' part is easy. For the 'only if' part fix a projective coordinate system $[X : Y : Z]$ such that the affine chart $Z \neq 0$ contains the singular point p. Let V be the vector space of polynomials of degree l in variables $(X/Z, Y/Z)$. Its dimension is $N := (l + 1)(l + 2)/2$, which, in fact, equals the number of elements of the semigroup in the interval $[0, ld]$. Denote these elements by $0 = s_1, ..., s_N$, ordered in an increasing way.

Consider the decreasing filtration of vector spaces $V_1 \supset V_2 \supset \cdots \supset V_N$, defined by

$$V_i := \{f \in V : I_p(C, f) \geq s_i\}.$$

First, we verify that $\dim(V_i/V_{i+1})$ is at most 1. Indeed, assume that $I_p(C, f_i) = I_p(C, f_2) = I$. Let $n : (\mathbb{C}, 0) \to (C, p)$ be the normalisation of (C, p), and write $f_i \circ n(t) = a_i t^I + \cdots$ with $a_i \neq 0$, for $i = 1$ and 2. Then $I_p(C, a_2 f_1 - a_1 f_2) > I$. Since there is no semigroup element between s_i and s_{i+1}, the inequality $\dim(V_i/V_{i+1}) \leq 1$ follows.

Next, notice that to prove the theorem it is enough to show that each dimension $\dim(V_i/V_{i+1})$ is exactly 1.

But, if $\dim(V_i/V_{i+1}) = 0$ for some i then $\dim(V_N)$ is at least 2. Since for any $f \in V_N$ one has $I_p(C, f) \geq s_N$, $\dim(V_N) \geq 2$ would imply (by similar argument as above) the existence of an $f \in V_N$ with $I_p(C, f) > s_N$. Since $I_p(C, f)$ is an element of the semigroup and the last element of the semigroup in the interval $[0, ld]$ is s_N, we get that $I_p(C, f) > ld$, which contradicts the irreducibility of C by Bézout Theorem. $\qquad\square$

6.3. *A counterexample to an 'extended' version*

In [11] we formulated the following conjecture, as an extension of the conjecture B2. to an 'if and only if' statement.

Conjecture 6.4 ('Conjecture' C). *The local topological type $(C, p) \subset (\mathbb{P}^2, p)$ can be realized by a degree d unicuspidal rational curve if and only if the property (DP) is valid.*

In the sequel we present a counterexample to the 'if' part (i.e. to the 'extension').

If the germ $(C,0)$ has g Newton pairs $\{(p_k, q_k)\}_{k=1}^g$ with $\gcd(p_k, q_k) = 1$, $p_k \geq 2$ and $q_k \geq 1$ (and by convention, $q_1 > p_1$), define the integers $\{a_k\}_{k=1}^g$ by $a_1 := q_1$ and $a_{k+1} := q_{k+1} + p_{k+1}p_k a_k$ for $k \geq 1$. Then its Eisenbud-Neumann splice diagram decorated by the numerical data $\{(p_k, a_k)\}_{k=1}^g$ has the following shape [5]:

Consider now the local singularity whose Eisenbud-Neumann splice diagram is decorated by two pairs $(p_1, a_1) = (2, 7)$ and $(p_2, a_2) = (4, 73)$. A local equation for such singularity can be $(x^2 - y^7)^4 + x^{33}y = 0$. Its multiplicity sequence is $[8_3, 4_6, 1_4]$. A minimal set of generators of its semigroup $\Gamma_{(C,p)}$ is given by $\langle 8, 28, 73 \rangle$. Its Milnor number is $16 \cdot 15$, hence a possible unicuspidal plane curve C of degree 17 might exist with such local singularity. Moreover the distribution property (DP) of the semigroup is also satisfied. Nevertheless, such a curve C does not exist. To prove this, one can either use Cremona transformations to transform C into another curve for which one sees that it does not exist, or one uses Varchenko's semi-continuity criterium for the spectrum of the singularity [57,58]. Here we will follow the second argument.

The spectrum of the irreducible singularity $(C,0)$ can be computed from the Newton pairs of the singularity. The forth author provided such a formula in [31]. It is convenient to consider the spectrum $Sp(C,0) = \sum_r n_r(r)$ as an element of $\mathbb{Z}[\mathbb{Q} \cap (0,2)]$. We write $Sp_{(0,1)}(C,0)$ for the collection of spectral elements situated in the interval $(0,1)$.

Theorem 6.3. *If the irreducible germ $(C,0)$ has g Newton pairs $\{(p_k, q_k)\}_{k=1}^g$ then*

$$Sp_{(0,1)}(C,0) = \sum_{k=1}^g S_k \quad \text{where} \quad S_k = \sum \left(\frac{i/a_k + j/p_k + t}{p_{k+1}p_{k+2}\cdots p_g} \right),$$

where the second sum is over $0 < i < a_k$, $0 < j < p_k$, $i/a_k + j/p_k < 1$ and $0 \leq t \leq p_{k+1}p_{k+2}\cdots p_g - 1$ (if $k = g$ then $S_g = \sum(l/a_g + k/p_g)$ where the sum is over $0 < l < a_g$, $0 < k < p_g$, $l/a_g + k/p_g < 1$).

If the local singular type $\{(C,p)\}$ can be realized by a degree d plane curve C, then (C,p) is in the deformation of the 'universal' plane germ $(U,0) := (x^d + y^d, 0)$. In particular, the collection of all spectral numbers $Sp(C,p)$ of the local plane curve singularity (C,p) satisfies the semi-continuity property compared with the spectral numbers of $(U,0)$ for any interval $(\alpha, \alpha+1)$. Since the spectral numbers of $(U,0)$ are of type l/d, the semi-continuity property for intervals $(-1 + l/d, l/d)$ $(l = 2, 3, \ldots, d-1)$ reads as follows:

$$\#\{\alpha \in Sp(C,p) \ : \ \alpha < l/d\} \leq (l-2)(l-1)/2. \tag{23}$$

In our case, for $d = 17$ and $l = 12$, using Theorem 6.3 we get

$$\#\{\alpha \in Sp(C,p) \ : \ \alpha < 12/17\} - (12-2)(12-1)/2 = 1,$$

which contradicts (23). Thus the rational unicuspidal plane curve C of degree 17 with such singularity cannot exist.

Thus, in the realization problem, the above case $(p_1, a_1; p_2, a_2; d)$ cannot be eliminated by the semigroup distribution property (DP), but it can be eliminated by the semi-continuity of the spectrum. However it is not true that the semi-continuity implies (DP). For a more precise discussion see [11].

7. The semigroup compatibility property and normal surface singularities

7.1. *Superisolated singularities*

The theory of normal surface singularities (in fact, of isolated hypersurface surface singularities) 'contains' in a canonical way the theory of complex projective plane curves via the family of *superisolated* singularities. These singularities were introduced by the second author in [25], see also [2] for a survey on them. A hypersurface singularity $f : (\mathbb{C}^3, 0) \to (\mathbb{C}, 0)$, $f = f_d + l^{d+1}$ (where f_d is homogeneous of degree d and l is linear) is superisolated if the projective plane curve $C := \{f_d = 0\} \subset \mathbb{P}^2$ is reduced, and none of its singularities $\{p_i\}_{i=1}^{\nu}$ is situated on $\{l = 0\}$. The equisingular type of f depends only on f_d, i.e. only on the projective curve $C \subset \mathbb{P}^2$. In particular, all the invariants (of the equisingular type) of f can be determined from the invariants of the pair $(\mathbb{P}^2, C)$.

In the next discussion we follow [11,26]. There is a standard procedure which provides the plumbing graph of the link M of f from the embedded resolution graphs of (C, p_i)'s and the integer d. The point is that the link M

is a rational homology sphere if and only if C is rational and cuspidal. In this section, we will assume that these conditions are satisfied. Let $\mu_i = \mu(C, p_i)$ and Δ_i be the Milnor number and the characteristic polynomial of the local plane curve singularities (C, p_i). Set $2\delta := \sum_i \mu_i$, $\Delta := \prod_i \Delta_i$, and $\bar{\Delta}(t) := t^{-\delta}\Delta(t)$.

Let (V, D) be the minimal embedded resolution of the pair $(\mathbb{P}^2, C)$ as above. The minimal plumbing graph of M (or, equivalently, the minimal good resolution graph of the surface singularity $\{f = 0\}$) can be obtained from the dual graph of D by decreasing the decoration (self-intersection) of $\bar{C}$ by $d(d+1)$. In the language of topologists, if C is unicuspidal ($\nu = 1$), then $M = S^3_{-d}(K)$ (i.e. M is obtained via surgery of the 3-sphere S^3 along K with surgery coefficient $-d$), where $K \subset S^3$ is the local knot of (C, p). One can also verify that $H_1(M, \mathbb{Z}) = \mathbb{Z}_d$.

Another topological invariant of f is the following one. Let $Z \to (\{f = 0\}, 0)$ be the minimal good resolution, K_Z be the canonical divisor of Z and $\#$ the number of irreducible components of the exceptional divisor (which equals the number of irreducible components of D). Then $K_Z^2 + \#$ is a well-defined invariant of f, which, in fact, can be computed from the link M (or, from its graph) as well. In our case, surprisingly, in this invariant of the link M all the information about the local types (C, p_i) are lost: $K_Z^2 + \# = 1 - d(d-2)^2$, it depends only on d.

The same is true for the Euler characteristic $\chi(F)$, or for the signature $\sigma(F)$ of the Milnor fiber F of f, or about the geometric genus p_g of f. In fact, it is well-known that for any hypersurface singularity, any of p_g, $\sigma(F)$ and $\chi(F)$ determines the remaining two modulo $K_Z^2 + \#$. E.g., one has the relation:

$$8p_g + \sigma(F) + K_Z^2 + \# = 0. \tag{24}$$

In our case, for the superisolated singularity f, one has $p_g = d(d-1)(d-2)/6$, hence the smoothing invariants $\chi(F)$ and $\sigma(F)$ depend only on the degree d.

7.2. *Normal surfaces whose link is a rational homology sphere*

For a normal surface singularity with rational homology sphere link (and with some additional analytic restriction, e.g. complete intersection or Gorenstein property) there is a subtle connection between the Seiberg-Witten invariants of its link M and some analytic/smoothing invariants. The hope is that the geometric genus (or, equivalently, $\chi(F)$ or $\sigma(F)$, see

(24) and the discussion nearby), can be recovered from the link. The starting point is an earlier conjecture of Neumann and Wahl [39]:

Conjecture 7.1 (Casson invariant conjecture). *For any isolated complete intersection whose link M is an integral homology sphere we have the equality $\sigma(F) = 8\lambda(M)$, where $\lambda(M)$ is the Casson invariant of the link.*

Notice that the link of a hypersurface superisolated singularity is never an integral homology sphere. The generalised conjecture, applied to rational homology spheres (Conjecture SWC below) was proposed by the forth author in a joint work with L. Nicolaescu in [36] involving the Seiberg-Witten invariant of the link. It was verified for rather large number of non-trivial special families (rational and elliptic singularities, suspension hypersurface singularities $f(x, y) + z^n$ with f irreducible, singularities with good $\mathbb{C}^*$ action) [32,35–38]. But the last three authors of the present article have shown in [26] that the conjecture fails in general. The counterexamples were provided exactly by superisolated singularities and/or their universal abelian covers, see also Stevens paper [47] where he computes explicit equations for the universal abelian covers. Nevertheless, in the next paragraph we will recall this conjecture (in its original form), since this have guided us to the semigroup compatibility property, and we believe that it hides a deep mathematical substance (even if at this moment it is not clear for what family we should expect its validity).

Let $\mathbf{sw}_M(can)$ be the Seiberg-Witten invariant of the link M associated with the canonical $spin^c$ structure (this is induced by the complex structure of $\{f = 0\} \setminus \{0\}$, and it can be identified combinatorially from the graph of M; in this article we will not discuss the invariants associated with the other $spin^c$ structures).

Conjecture 7.2 ('Conjecture' SWC [36]). *For a $\mathbb{Q}$-Gorenstein surface singularity whose link M is a rational homology sphere one has*

$$\mathbf{sw}_M(can) - (K_Z^2 + \#)/8 = p_g.$$

In particular, if the singularity is Gorenstein and admits a smoothing, then $-\mathbf{sw}_M(can) = \sigma(F)/8$ (cf. (24)).

If M is an integral homology sphere then $\mathbf{sw}_M(can) = -\lambda(M)$. If M is a rational homology sphere then by a result of Nicolaescu [40], $\mathbf{sw}_M(can) = \mathcal{T}_M - \lambda(M)/|H_1(M, \mathbb{Z})|$, where $\lambda(M)$ is the Casson-Walker invariant of

M (normalised as in [23]), and $\mathfrak{T}_M$ denotes the sign refined Reidemeister-Turaev torsion (associated with the canonical $spin^c$ structure) [56].

7.3. *Seiberg-Witten invariant of a superisolated singularity*

In our present situation, when M is the link of a superisolated singularity f, one shows, cf. [26] (using the notations of 7.1), that

$$\mathfrak{T}_M = \frac{1}{d} \sum_{\xi^d=1 \neq \xi} \frac{\Delta(\xi)}{(\xi-1)^2} \quad \text{and} \quad \lambda(M) = -\frac{\bar{\Delta}(t)''(1)}{2} + \frac{(d-1)(d-2)}{24}.$$

$$(25)$$

Therefore, since p_g and $K_Z^2 + \#$ depend only on d, the SWC imposes serious restriction on the local invariant Δ. This condition, for some cases when the number of singular points of C is ≥ 2, is not satisfied (hence SWC fails, cf. [26]); nevertheless, as we will see, the SWC identity in the unicuspidal case is equivalent with Conjecture B2 of section 6 about the distribution property of the semigroup. In order to explain this, let us *assume that C is unicuspidal*, and consider (motivated by (25))

$$R(t) := \frac{1}{d} \sum_{\xi^d=1} \frac{\Delta(\xi t)}{(1-\xi t)^2} - \frac{1-t^{d^2}}{(1-t^d)^3}.$$

Similarly,

$$N(t) := \sum_{l=0}^{d-3} \left(c_l - \frac{(l+1)(l+2)}{2}\right) t^{d-3-l}; \quad \text{and} \quad D(t) := \sum_{k\in\Gamma_{(C,p)}} t^{\lceil k/d \rceil} - \frac{1-t^d}{(1-t)^2}$$

Notice that this $D(t)$ agrees with the one defined in (DP), section 6. In [11] the following facts are verified:

$$R(t) = D(t^d)/(1-t^d) = N(t^d). \tag{26}$$

$$N(t) \text{ (hence } R(t) \text{ too) has non-negative coefficients.} \tag{27}$$

$$R(1) = \mathbf{sw}_M(can) - \frac{K^2 + \#}{8} - p_g. \tag{28}$$

Therefore, in this case, we have the equivalence of the 'Seiberg-Witten invariant conjecture' with the 'semigroup distribution property':

Theorem 7.1. *Assume that C is unicuspidal and rational (that is, $\nu = 1$). Then the following facts are equivalent:*

434

(a) $R(1) = 0$, i.e. Conjecture SWC is true (for the above germ f);
(b) $R(t) \equiv 0$;
(c) $N(t) \equiv 0$, i.e. Conjecture B1 is true;
(d) $D(t) \equiv 0$, i.e. Conjecture B2 is true.

8. The semigroup distribution property and Heegaard Floer Homology

8.1. *Graded roots*

The presentation of this section is based on some recent results of the forth author in [32–34]. In the sequel we assume that C is *unicupidal*, and we keep the notations of the previous section.

There is another way to compute the Seiberg-Witten invariant of the link M via its Heegaard-Floer homology. For any oriented rational homology 3-sphere M the Heegaard Floer homology $HF^+(M)$ was introduced by Ozsváth and Szabó in [43] (cf. also with their long list of articles). $HF^+(M)$ is a $\mathbb{Z}[U]$-module with compatible $\mathbb{Q}$-grading. Moreover, $HF^+(M)$ has a natural direct sum decomposition (compatible with the $\mathbb{Q}$-grading) corresponding to the *spin^c*-structures of M: In this article we write $HF^+(M, can)$ for the Heegaard-Floer homology associated with the canonical *spin^c* structure.

For some (negative definite) plumbed rational homology 3-spheres M, one can compute the Heegaard Floer homology of $HF^+(M, can)$ of M (equivalently, of $-M$) in a purely combinatorial way from the plumbing graph G. This is true for all the 3-manifolds discussed in this section. This is done via some intermediate objects, the *graded root* associated with G (in fact, one has a graded root corresponding to each *spin^c*-structure of M, but here we will discuss only the 'canonical' one). The theory of graded roots, from the point of view of singularity theory, is rather interesting by itself, and we plan to exploit further this connection in the future.

Next, we provide a short presentation of abstract graded roots (cf. [32]).

Definition 8.1 (Definition of the 'abstract graded root' (R, χ)).
Let R be an infinite tree with vertices $\mathcal{V}$ and edges $\mathcal{E}$. We denote by $[u, v]$ the edge with end-points u and v. We say that R is a graded root with grading $\chi : \mathcal{V} \to \mathbb{Z}$ if
(a) $\chi(u) - \chi(v) = \pm 1$ for any $[u, v] \in \mathcal{E}$;
(b) $\chi(u) > \min\{\chi(v), \chi(w)\}$ for any $[u, v]$, $[u, w] \in \mathcal{E}$;
(c) χ is bounded below, $\chi^{-1}(n)$ is finite for any $n \in \mathbb{Z}$, and $\#\chi^{-1}(n) = 1$ if $n \gg 0$.

8.2. *Examples*

(1) For any integer $n \in \mathbb{Z}$, let R_n be the tree with $\mathcal{V} = \{v^k\}_{k \geq n}$ and $\mathcal{E} = \{[v^k, v^{k+1}]\}_{k \geq n}$. The grading is $\chi(v^k) = k$.

(2) Let I be a finite index set. For each $i \in I$ fix an integer $n_i \in \mathbb{Z}$; and for each pair $i, j \in I$ fix $n_{ij} = n_{ji} \in \mathbb{Z}$ with the next properties: (i) $n_{ii} = n_i$; (ii) $n_{ij} \geq \max\{n_i, n_j\}$; and (iii) $n_{jk} \leq \max\{n_{ij}, n_{ik}\}$ for any $i, j, k \in I$. For any $i \in I$ consider R_{n_i} with vertices $\{v_i^k\}$ and edges $\{[v_i^k, v_i^{k+1}]\}$, $(k \geq n_i)$. In the disjoint union $\coprod_i R_{n_i}$, for any pair (i, j), identify v_i^k and v_j^k, resp. $[v_i^k, v_i^{k+1}]$ and $[v_j^k, v_j^{k+1}]$, whenever $k \geq n_{ij}$, and take the induced χ.

(3) Any map $\tau : \{0, 1, \ldots, r\} \to \mathbb{Z}$ produces a starting data for construction (2). Indeed, set $I = \{0, \ldots, r\}$, $n_i := \tau(i)$ $(i \in I)$, and $n_{ij} := \max\{n_k : i \leq k \leq j\}$ for $i \leq j$. Then the root constructed in (2) using this data will be denoted by (R_τ, χ_τ).

8.2.1. *Examples of graded roots*

Here are two (typical) graded roots (cf. with Example 8.7):

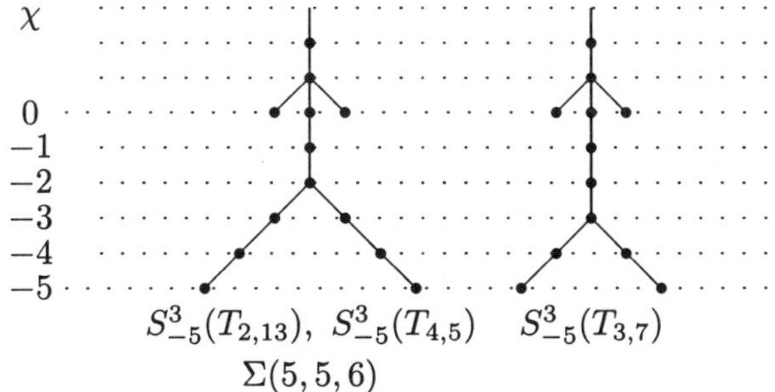

8.3. *The canonical graded root* (R, χ) *of* M. *[32]*

Next, we define for any (negative definite, plumbed) rational homology sphere M a graded root.

We fix a plumbing graph G and denote by L the corresponding lattice: the free $\mathbb{Z}$-module of rank # with fixed basis $\{A_j\}_j$, and bilinear form $(A_i, A_j)_{i,j}$. (In our case, a possible choice is the dual resolution graph and the corresponding intersection form associated with the minimal good resolution $Z \to (\{f = 0\}, 0)$.) Set $L' = \mathrm{Hom}_{\mathbb{Z}}(L, \mathbb{Z}) \subset L \otimes \mathbb{Q}$. Let $K_Z \in L'$ be the canonical cycle defined by $K_Z(A_j) + A_j^2 + 2 = 0$ for any j. Then define $\chi : L \to \mathbb{Z}$ by (the Riemann-Roch formula) $\chi(x) := -(K_Z(x) + x^2)/2$.

436

The definition of the graded root captures the position of the lattice points in the different ellipsoids $\chi^{-1}(n)$. For any $n \in \mathbb{Z}$, one constructs a finite 1-dimensional simplicial complex $\bar{L}_{\leq n}$ as follows. Its 0-skeleton is $L_{\leq n} := \{x \in L : \chi(x) \leq n\}$. For each x and j, with both x and $x + A_j \in L_{\leq n}$, we consider a unique 1-simplex with endpoints at x and $x + A_j$ (e.g., the segment $[x, x+A_j]$ in $L \otimes \mathbb{R}$). We denote the set of connected components of $\bar{L}_{\leq n}$ by $\pi_0(\bar{L}_{\leq n})$. For any $v \in \pi_0(\bar{L}_{\leq n})$, let C_v be the corresponding connected component of $\bar{L}_{\leq n}$.

Next, we define (R, χ) as follows. The vertices $\mathcal{V}(R)$ are $\cup_{n \in \mathbb{Z}} \pi_0(\bar{L}_{\leq n})$. The grading $\mathcal{V}(R) \to \mathbb{Z}$, still denoted by χ, is $\chi | \pi_0(\bar{L}_{\leq n}) = n$. If $v_n \in \pi_0(\bar{L}_{\leq n})$, and $v_{n+1} \in \pi_0(\bar{L}_{\leq n+1})$, and $C_{v_n} \subset C_{v_{n+1}}$, then $[v_n, v_{n+1}]$ is an edge of R. All the edges are obtained in this way.

8.4. *Example*

Recall that the link of the superisolated singularity f (where C is rational and unicuspidal of degree d) is the surgery manifold $S^3_{-d}(K)$, where $K \subset S^3$ is the local knot of (C, p). The graded root of M can be represented by a function τ as in Example 8.2(3) associated with the Alexander polynomial Δ of $K \subset S^3$, see [33]. Similarly as in section 6, set $\mu = 2\delta$ for the degree of Δ (which equals $(d-1)(d-2)$), and write $\Delta(t)$ as $1 + \delta(t-1) + (t-1)^2 Q(t)$ for some polynomial $Q(t) = \sum_{i=0}^{\mu-2} \alpha_i t^i$. Set $c_l := \alpha_{(d-3-l)d}$ (cf. with 6.1). Then define $\tau : \{0, 1, \ldots, 2d - 4\} \to \mathbb{Z}$ by

$$\tau(2l) = \frac{l(l-1)}{2}d - l(\delta - 1), \qquad \tau(2l + 1) = \tau(2l + 2) + c_{d-3-l}.$$

Then $(R, \chi) = (R_\tau, \chi_\tau)$.

8.5. *Example*

Let $\Sigma(d, d, d + 1)$ be the Seifert 3-manifold $(d, d, d + 1)$; equivalently, the link of the Brieskorn singularity $x^d + y^d + z^{d+1} = 0$. Its graded root also can be represented by the 'τ-construction' (for the more general situation of Seifert manifolds, see [32]).

For any $0 \leq l \leq d-3$ define $c_l^u := (l+1)(l+2)/2$, and $2\delta := (d-1)(d-2)$. Then define $\tau^u : \{0, 1, \ldots, 2d - 4\} \to \mathbb{Z}$ by

$$\tau^u(2l) = \frac{l(l-1)}{2}d - l(\delta - 1), \qquad \tau^u(2l + 1) = \tau^u(2l + 2) + c_{d-3-l}^u.$$

Then $(R, \chi) = (R_{\tau^u}, \chi_{\tau^u})$.

Notice the shocking similarities of Example 8.4 and Exmaple 8.5: the graded roots associated with $S^3_{-d}(K)$ and $\Sigma(d, d, d+1)$ coincide exactly when $c_l =$

c_l^u for all l.

To any graded root, one can associate a natural graded $\mathbb{Z}[U]$-module.

8.6. *The $\mathbb{Z}[U]$-module associated with a graded root*

Consider the $\mathbb{Z}[U]$-module $\mathbb{Z}[U, U^{-1}]$, and (following [44]) denote by $\mathcal{T}_0^+$ its quotient by the submodule $U \cdot \mathbb{Z}[U]$. It is a $\mathbb{Z}[U]$-module with grading $\deg(U^{-h}) = 2h$.

Now, fix a graded root (R, χ). Let $\mathbb{H}(R, \chi)$ be the set of functions $\phi :$ $\mathcal{V} \to \mathcal{T}_0^+$ with the property that whenever $[v, w] \in \mathcal{E}$ with $\chi(v) < \chi(w)$ one has $U \cdot \phi(v) = \phi(w)$. Then $\mathbb{H}(R, \chi)$ is a $\mathbb{Z}[U]$-module via $(U\phi)(v) = U \cdot \phi(v)$. Moreover, $\mathbb{H}(R, \chi)$ has a grading: $\phi \in \mathbb{H}(R, \chi)$ is homogeneous of degree $h \in \mathbb{Z}$ if for each $v \in \mathcal{V}$ with $\phi(v) \neq 0$, $\phi(v) \in \mathcal{T}_0^+$ is homogeneous of degree $h - 2\chi(v)$.

In the sequel, the following notation is useful: If P is a $\mathbb{Q}$-graded $\mathbb{Z}[U]$-module with h-homogeneous elements P_h, then for any $r \in \mathbb{Q}$ we denote by $P[r]$ the same module graded in such a way that $P[r]_{h+r} = P_h$.

Theorem 8.1 ([32,44]). *Assume that M is either $S^3_{-d}(K)$ or $\Sigma(d, d, d+1)$. Then*

$$HF^+(-M, can) = \mathbb{H}(R, \chi)[-(K_Z^2 + \#)/4].$$

In other words, for these 3-manifolds, the Heegaard-Floer homology can be recovered from the graded root via $\mathbb{H}(R, \chi)$ modulo a shift in grading by $-(K_Z^2 + \#)/4$. (The shift in the above two examples are different; in the case of $S^3_{-d}(K)$ one has $K_Z^2 + \# = 1 - d(d - 2)^2$, while for $\Sigma(d, d, d + 1)$ one has $K_Z^2 + \# = -d(d - 1)(d - 3)$.)

Now, Conjecture B1 and the above discussion/examples read as follows:

Theorem 8.2. *Assume that $\nu = 1$. Then the following facts are equivalent:*
   (a) Conjecture B1 is true,
   (b) The canonical graded roots of $S^3_{-d}(K)$ and $\Sigma(d, d, d + 1)$ are the same.
   (c) The canonical Heegaard-Floer homologies of $-S^3_{-d}(K)$ and $-\Sigma(d, d, d + 1)$ are the same modulo a shift in the grading, namely:

$$HF^+(-S^3_{-d}(K), can)[1 - d(d-2)^2] = HF^+(-\Sigma(d, d, d+1), can)[-d(d-1)(d-3)].$$

Proof. The equivalence $(a) \Leftrightarrow (b)$ is clear from the above discussion, while $(b) \Leftrightarrow (c)$ can be deduced by a direct computation, or from an easy for-

438

mula which provides $\mathbb{H}(R_\tau, \chi_\tau)$ from τ, cf. [32]. (Nevertheless, see another argument below.) $\qquad\square$

Remark 8.1. Regarding the Seiberg-Witten invariant of $M = S^3_{-d}(K)$, one has

$$\mathbf{sw}_M(can) - \frac{K_Z^2 + \#}{8} = \sum_{l \geq 0} \tau(2l+1) - \tau(2l+2) = \sum_{l \geq 0} c_l, \qquad (29)$$

and there is a similar formula for $M = \Sigma(d, d, d+1)$ with the obvious replacements.

Therefore, to the equivalences of Theorem 8.2 one can add:

(d) $\mathbf{sw}_M(can) - \dfrac{K_Z^2 + \#}{8}\big|_{M=S^3_{-d}(K)} = \mathbf{sw}_M(can) - \dfrac{K_Z^2 + \#}{8}\big|_{M=\Sigma(d,d,d+1)}.$

Since the Conjecture SWC is true for the Brieskorn singularity $f_{BR} := x^d + y^d + z^{d+1}$ (cf. [37]), and the geometric genus of the superisolated singularity f equals the geometric genus of f_{BR} (both equal $d(d-1)(d-2)/6$), this last identity (d) is also equivalent with the validity of the SWC for f — a fact already proved in Theorem 7.1.

(Notice also that the expression $\mathbf{sw}_M(can) - (K_Z^2 + \#)/8$ can be deduced from $\mathbb{H}$, a fact which implies $(c) \Rightarrow (d)$, while $(d) \Rightarrow (a)$ follows from (29).)

8.7. *Example*

Assume that $d = 5$ and C is unicuspidal whose singular point has only one Puiseux pair (a, b) with $a < b$. Then by the genus formula the possible values of (a, b) are $(4, 5)$, $(3, 7)$ and $(2, 13)$. It turns out that the first and the third cases can be realized, while the second not. The corresponding graded roots (together with the root of $\Sigma(5, 5, 6)$) are drawn in the above figure (8.2.1).

Acknowledgments

The first three authors are grateful for the warm hospitality of the Alfréd Rényi Institute of Mathematics (Budapest, Hungary) where they spent a fruitful period in an ideal mathematical environment.

The first author is supported by Ramón y Cajal contract. The first three authors thank Marie Curie Fellowship for the Transfer of Knowledge supporting their visit at the Rényi Institute, Budapest, Hungary; the first three authors are partially supported by the Spanish grant MTM2004-08080-C02-01; the last author is also supported by the Marie Curie Fellowship and OTKA Grant.

References

1. P. Aluffi and C. Faber, Plane curves with small linear orbits II, *International Journal of Mathematics*, **11** (2000), 591–608.

2. E. Artal Bartolo, I. Luengo and A. Melle-Hernández, Superisolated Surface Singularities, Proceedings of the Conference "Singularities and Computer Algebra" on Occasion of Gert-Martin Greuel's 60th Birthday, LMS Lecture Notes 324, 13–39.

3. J.L. Coolidge, *A treatise of algebraic plane curves*, Oxford Univ. Press. Oxford, (1928).

4. S. Diaz and J. Harris, Ideals associated to deformations of singular plane curves. *Trans. Amer. Math. Soc.* **309** (1988), no. 2, 433–468.

5. D. Eisenbud and W. Neumann, *Three-Dimensional Link Theory and Invariants of Plane Curve Singularities*, Ann. of Math. Studies 110, Princeton University Press, (1985).

6. T. Fenske, Rational 1- and 2-cuspidal plane curves, *Beiträge Algebra Geom.* **40** (1999), no. 2, 309–329.

7. T. Fenske, Rational cuspidal plane curves of type $(d, d-4)$ with $\chi(\Theta_V \langle D \rangle) \leq 0$. *Manuscripta Math.* **98** (1999), no. 4, 511–527.

8. H. Flenner and M. Zaidenberg, $\mathbb{Q}$-acyclic surfaces and their deformations, *Contemporary Math.* **162** (1994), 143–208.

9. H. Flenner and M. Zaidenberg, On a class of rational cuspidal plane curves. *Manuscripta Math.* **89** (1996), no. 4, 439–459.

10. H. Flenner and M. Zaidenberg, Rational cuspidal plane curves of type $(d, d-3)$. *Math. Nachr.* **210** (2000), 93–110.

11. J. Fernández de Bobadilla, I. Luengo, A. Melle-Hernández and A. Némethi, On rational cuspidal projective plane curves, *Proc. London Math. Soc.* (3) **92** (2006), no. 1, 99–138.

12. J. Fernández de Bobadilla, I. Luengo, A. Melle-Hernández and A. Némethi, Classification of rational unicuspidal projective curves whose singularities have one Puiseux pair, to appear in Real and Complex Singularities (Luminy, 2004).

13. G-M. Greuel and C. Lossen, Equianalytic and equisingular families of curves on surfaces, *Manuscripta Math.* **91** (1996) 323–342.

14. R.V. Gurjar, S. Kaliman, N.M. Kumar, M. Miyanishi, R. Russell, F. Sakai, D. Wright and M. Zaidenberg, Open problems on open algebraic varieties. arXiv:math.AG/9506006.

15. S.M. Gusein-Zade, F. Delgado and A. Campillo, On the monodromy of a plane curve singularity and the Poincaré series of the ring of functions on the curve, *Functional Analysis and its Applications*, **33**(1) (1999), 56-67.

16. J. Harris and I.Morrison, *Moduli of curves*, Graduate Texts in Mathematics, 187. Springer-Verlag, New York, 1998.

17. S. Iitaka, On logarithmic Kodaira dimension of algebraic varities, *Complex Analysis and Algebraic Geometry*, Cambridge Univ. Press, Cambridge, (1977), 175–190.

18. S. Iitaka, On irreducible plane curves, *Saitama Math. J.* , **1** (1983) 47–63.

19. S. Iitaka, Birational geometry of plane curves, *Tokyo J. Math.* **22** (1999) 289–321.

20. Th. de Jong, Equisingular Deformations of Plane Curve and of Sandwiched Singularities, arXiv:math.AG/0011097.

21. H. Kashiwara, Fonctions rationelles de type (0,1) sur le plan projectif complexe, *Osaka J. Math.* **24** (1987) 521–577.

22. N.M. Kumar and M.P. Murthy, Curves with negative self intersection on rational surfaces, *J. Math. Kyoto Univ.* **22** (1983) 767–777.

23. C. Lescop, Global Surgery Formula for the Casson-Walker Invariant, *Annals of Math. Studies*, vol. **140**, Princeton University Press, 1996.

24. E. Looijenga, *Isolated singular points of complete intersections*. London Math. Soc. Lecture Notes Series **77**, (1983).

25. I. Luengo, The μ-constant stratum is not smooth, *Invent. Math.*, **90** (1) (1987), 139–152.

26. I. Luengo, A. Melle-Hernández, and A. Némethi, Links and analytic invariants of superisolated singularities, *Journal of Algebraic Geometry*, **14** (2005), 543-565.

27. T. Matsuoka and F. Sakai, The degree of rational cuspidal curves, *Math. Ann.*, **285** (1989), 233–247.

28. J.F. Mattei, Modules de feuilletages holomorphes singuliers. I. quisingularit, *Invent. Math.* **103** (1991), no. 2, 297–325.

29. M. Miyanishi and T. Sugie, On a projective plane curve whose complement has logarithmic Kodaira dimension $-\infty$, *Osaka J. Math.* , **18** (1981), 1–11.

30. M. Nagata, On rational surfaces I. Irreducible curves of arithmetic genus 0 and 1, *Memoirs of College of Science,* Univ. of Kyoto, Series A, Vol. **XXXII** (3) (1960), 351-370.

31. A. Némethi, On the spectrum of curve singularities, *Proceedings of the Singularity Conference*, Oberwolfach, July 1996; Progress in Mathematics, **Vol. 162**, 93-102, Birkhäuser (1998).

32. A. Némethi, On the Ozsváth-Szabó invariant of negative definite plumbed 3-manifolds, *Geometry and Topology* **9** (2005), 991-1042.

33. A. Némethi, On the Heegaard Floer homology of $S^3_{-d}(K)$ and unicuspidal rational plane curves, *Geometry and topology of manifolds*, Editors: H.U. Boden, I.Hambleton, A.J. Nicas and B.D. Park, Fields Inst. Commun., **47**, Amer. Math. Soc., Providence, RI, 2005, 219–234.

34. A. Némethi, On the Heegaard Floer homology of $S^3_{-p/q}(K)$, submitted (math.GT/0410570).

35. A. Némethi, Line bundles associated with normal surface singularities, submitted (math.AG/0310084).

36. A. Némethi and L.I. Nicolaescu, Seiberg-Witten invariants and surface singularities, *Geometry and Topology*, Volume **6** (2002), 269-328.

37. A. Némethi and L.I. Nicolaescu, Seiberg-Witten invariants and surface singularities II (singularities with good $\mathbf{C}^*$-action), *J. London Math. Soc.* **(2) 69** (2004), no. 3, 593–607.

38. A. Némethi and L.I. Nicolaescu, Seiberg-Witten invariants and surface singularities splicings and cyclic covers, *Selecta Math.* New Series **11** Nr. 3-4, (2005), 399–451.

39. W. Neumann and J. Wahl, Casson invariant of links of singularities, *Com-*

ment. Math. Helv. **65**, 58-78, (1991).

40. L.I. Nicolaescu, Seiberg-Witten invariants of rational homology 3-spheres, *Comm. in Contemp. Math.*, **6** (2004), 833-866.

41. S.Yu. Orevkov, On rational cuspidal curves, I. Sharp estimate for degree via multiplicity, *Math. Ann.* **324** (2002), 657-673.

42. S.Yu. Orevkov and M.G. Zaidenberg, On the number of singular points of plane curves, In: *Algebraic Geometry. Proc. Conf., Saintama Univ.*, March 15–17, 1995, alg-geom/9507005.

43. P.S. Ozsváth and Z. Szabó, Holomorphic disks and topological invariants for closed three-manifolds. *Ann. of Math.* **(2) 159** (2004), no. 3, 1027–1158.

44. P.S. Ozsváth and Z. Szabó, On the Floer homology of plumbed three-manifolds, *Geom. Topol.* **7** (2003), 185-224.

45. F. Sakai, Kodaira dimensions of complements of divisors, *Complex Analysis and Algebraic Geometry*, Iwanami Shoten, Tokyo, 1977, 239–257.

46. E.I. Shustin, On manifolds of singular algebraic curves, *Selecta Math. Soviet.* **10** (1991), no. 1, 27–37.

47. J. Stevens, Universal abelian covers of superisolated singularities, math.AG/0601669.

48. B. Teissier, Résolution simultanée—I, II, *Séminaire sur les Singularités des Surfaces*. Edited by Michel Demazure, Henry Charles Pinkham and Bernard Teissier. Lecture Notes in Mathematics, 777. Springer, Berlin, (1980). 71–146

49. B. Teissier and O. Zariski, *Le problème des modules pour les branches planes*, (Hermann, Paris, 1986, Appendice).

50. K. Tono, Defining equations of certain rational cuspidal plane curves, *Manuscripta Math.* **103** (2000) 47–62.

51. K. Tono, Defining equations of certain rational cuspidal plane curves, doctoral thesis, Saitama University, 2000.

52. K. Tono, On rational unicuspidal plane curves with $\bar{\kappa} = 1$, *RIMS-Kôkyûroku* **1233** (2001) 82–89.

53. K. Tono, On the number of cusps of cuspidal plane curves, *Math. Nachr.* **278** (2005) 216–221.

54. Sh. Tsunoda, The complements of projective plane curves, *RIMS-Kôkyûroku* **446** (1981) 48–56.

55. Sh. Tsunoda, The Structure of Open Algebraic Surfaces and Its Application to Plane Curves, *Proc. Japan Acad. Ser. A* **57** (1981) 230–232.

56. V.G. Turaev, Torsion invariants of $Spin^c$-structures on 3-manifolds, *Math. Res. Letters*, **4** (1997), 679-695.

57. A.N. Varchenko, On the change of discrete characteristics of critical points of functions under deformations, *Uspekhi Mat. Nauk*, **38:5** (1983), 126-127.

58. A.N. Varchenko, Asymtotics of integrals and Hodge structures. Science rewievs: current problems in mathematics 1983, **22**, 130-166; *J. Sov. Math.* (1984), Vol. 27.

59. I. Wakabayashi, On the Logarithmic Kodaira Dimension of the Complement of a Curve in $\mathbb{P}^2$, *Proc. Japan Acad.*, **54**, Ser. A, (1978), 167162.

60. C.T.C. Wall, Notes on the classification of singularities, *Proc. London Math. Soc.* **48**(3) (1984), no. 3, 461–513.

61. C.T.C. Wall, *Singular points of plane curves*, London Mathematical Society Student Texts, 63. Cambridge University Press, Cambridge, 2004.

62. J. Wahl, Equisingular deformations of plane algebroid curves, *Trans. Amer. Math. Soc.* **193** (1974), 143–170.

63. H. Yoshihara, A problem concerning rational plane curves. (Japanese) *Sûgaku* **31** (1979), no. 3, 256–261.

64. M.G. Zaidenberg and V.Yu Lin, An irreducible simply connected algebraic curve in $\mathbb{C}^2$ is equivalent to a quasihomogeneous curve, *Soviet Math. Dokl.* **28** (1983) 200–204.

65. M.G. Zaidenberg and S.Yu Orevkov, On rigid rational cuspidal plane curves, *Russian Math. Surveys* **51** (1996), no. 1, 179–180

66. M.G. Zaidenberg, Selected problems, arkiv:AG/0501457.

A HOMOLOGICAL APPROACH TO SINGULAR REDUCTION IN DEFORMATION QUANTIZATION

Martin Bordemann

Laberatoire de Mathématique, Informatique et Applications,
Université de Haute Alsace,
4, rue des Frères Lumière,
68093 Mulhouse, France
E-mail: Martin.Bordemann@uha.fr

Hans-Christian Herbig and Markus J. Pflaum

Fachbereich Mathematik, Goethe-Universität,
Robert-Mayer-Straße 10,
D-60054 Frankfurt am Main, Germany
E-mail: herbig@math.uni-frankfurt.de
E-mail: pflaum@math.uni-frankfurt.de

Dedicated to Jean-Paul Brasselet on the occasion of his 60th birthday

We use the method of homological quantum reduction to construct a deformation quantization on singular symplectic quotients in the situation, where the coefficients of the moment map define a complete intersection. Several examples are discussed, among others one where the singularity type is worse than an orbifold singularity.

1. Introduction

In hamiltonian mechanics, reducing the number of degrees of freedom of a hamiltonian system by exploiting its symmetry is a standard method to determine the dynamics of the system. Within the language of symplectic geometry, regular reduction has been introduced independently by Meyer and Marsden–Weinstein and is usually called Marsden–Weinstein reduction. In [11] and, subsequently, in [3] it was shown that Marsden-Weinstein reduction has an analog in deformation quantization (see [7] for an overview on deformation quantization) in case the hamiltonian group action satisfies certain regularity conditions. This quantum reduction was used to obtain differentiable star products on regular symplectic quotient spaces. The general approach followed in [3] is known as the BRST-method and goes back to works of Batalin, Fradkin and Vilkoviski (for an overview and references on classical homological reduction see [24]).

In the following, we will see that the above method, suitably modified, works also for cases of *singular* reduction, where the singular behavior of the moment map is "not too bad". This will yield continuous star products on the corresponding singular quotient spaces. Let us be more specific about the premises to be made. We will consider a hamiltonian action of a connected and compact Lie group G acting on a symplectic manifold M with equivariant moment map $J : M \to \mathfrak{g}^*$, where $\mathfrak{g}^*$ is the dual space of the Lie algebra $\mathfrak{g}$ of G. Let $Z := J^{-1}(0)$ be the zero set of J, it will be also called *constraint surface*. Due to the equivariance of J, the constraint surface is an invariant subset. Let us denote by $I(Z) \subset \mathcal{C}^\infty(M)$ the vanishing ideal of Z. We will assume that the moment map satisfies the following conditions:

(a) the components of J generate $I(Z)$ (generating hypothesis),
(b) the Koszul complex on J in the ring $\mathcal{C}^\infty(M)$ is acyclic (cf. Section 3).

Substantial work has been done in [1] in order to understand the generating hypothesis. Using local normal coordinates for the moment map this issue is reduced to a problem in algebraic geometry (cf. also Section 2). Note that the generating hypothesis puts severe restrictions on the geometry of Z: it implies that $I(Z)$ is a Poisson subalgebra. Using Dirac's terminology we say: Z is *first class*. If the Koszul complex is acyclic, one also says J is a *complete intersection* (see e.g. [4]). Misleadingly, the physicist's denotation is here: "J is irreducible". The question whether a variety is a complete intersection is fundamental in commutative algebra, but there the most interesting techniques to determine that rely on the assumption that the base ring is *noetherian*, as opposed to the ring of smooth functions on a manifold which is the base ring in our considerations. So we have to find alternatives and attack this problem directly by providing a simple crititerion for J to be a complete intersection (cf. Theorem 3.1). The proof may be interesting in its own right.

If zero is a singular value of the moment map, the constraint surface Z is not a smooth manifold, but, according to [23], a stratified space. A continuous function f on Z is said to be *smooth* if there is a smooth function $F \in \mathcal{C}^\infty(M)$ such that $f = F_{|Z}$. The algebra of smooth functions $\mathcal{C}^\infty(Z)$ is isomorphic to $\mathcal{C}^\infty(M)/I(Z)$. It is naturally a Fréchet algebra, since it is the quotient of a Fréchet algebra by a closed ideal. In [23] Sjamaar and Lerman could show that the orbit space of Z under the action of G is a stratified symplectic space. The Poisson algebra of smooth functions on it is naturally isomorphic to the Poisson algebra $C^\infty(Z)^G/I(Z)^G$. Since Z is first class, $C^\infty(Z)^{\mathfrak{g}}$ carries a canonical Poisson structure, which is referred to as the *Dirac reduced algebra*. Since G is compact and connected, these Poisson algebras are isomorphic. If the conditions a) and b above are true, this Poisson algebra is identified with the zeroth cohomology of the classical BRST-algebra (cf. Section 4).

According to [3], it is relatively easy to find a formal deformation of the classical BRST-algebra into a differential graded associative algebra such that the cohomology is essentially unchanged (see Section 5 and 6), and thus yielding a deformation of the classical reduced Poisson structure. In [3] some efforts have been made to provide explicit formulas for contracting homotopies of the Koszul resolution, which have certain technical properties. Using these formulas and techniques from homological perturbation theory, it was shown that, in the regular case, a differentiable reduced star product can be found. Here we use the extension theorem and the division theorem of [2] to provide *continuous* contracting homotopies that satisfy similar technical assumptions.

In this way, we obtain the main result of this article. Given a hamiltonian action of a compact connected Lie g roup on a symplectic manifold such that the moment map satisfies conditions (a) and (b) above, then there exists a *continuous* formal deformation of the Dirac reduced algebra, i.e. a continuous star product on the singular reduced space (see Corollary 6.1). Since it is clear, that a situation, where both conditions (a) and (b) are true, is rather special, we start the discussion by giving some examples (cf. Section 2). Needless to say, this will show that the theory does not reduce to the regular situation. But, more importantly, there are examples where the reduced spaces are not orbifolds, but genuine stratified symplectic spaces. To the authors knowledge, this is the first known instance of such a space admitting a deformation quantization. Homological reduction therefore provides a construction method for formal deformation quantizations which works for more general singular symplectic spaces than the Fedosov type construction introduced in [19] for orbifolds.

We have included an appendix providing basic notions of homological perturbation theory and two variants of the well known basic perturbation lemma (see e.g. [?,16]), which are less universal but fit our purposes. The perturbation lemma A.1 is also implicit in Fedosov's construction [10].

Througout this paper we shall use the following conventions. Unless otherwise stated, all complexes are *cochain* complexes in the category of

446

$\mathbb{K}$-vector spaces, $\mathbb{K}$ being $\mathbb{R}$ or $\mathbb{C}$. The shift $V[j]$ of a graded vector space $V = \oplus_i V^i$ is defined by $V[j]^i := V^{i+j}$. If not said otherwise, maps of graded vector spaces are of degree zero. Concerning symplectic structure, moment maps, star products etc. we adopt the conventions of [3]. The formal parameter $\nu = i\lambda$ stands for $i\hbar$.

Acknowledgements. The authors would like to thank Markus Hunziker for stimulationg discussions and drawing our attention to important references concerning commuting varieties. We thank Paweł Domański for explaining the notion of a split, and Richard Cushman, Marc Gotay, Nolan Wallach and Patrick Erdelt for helpful advice. M.P. and H.-C.H. gratefully acknowledge support by Deutsche Forschungsgemeinschaft. H.-C.H. also acknowledges a travel stipend by Hermann Willkomm-Stiftung.

2. Examples

Before we start to explain the general machinery let us provide some examples of hamiltonian G-spaces, which satisfy the generating and the complete intersection hypothesis. In general, it is not at all a trivial matter to check, whether the generating hypothesis is true. The following is based on results of the seminal article [1]. We begin the discussion with the most simple case, where G is a torus.

2.1. *Hamiltonian torus actions*

In [1] it was proven that for a moment map $J : M \to \mathfrak{g}^*$ of a torus action to generate the vanishing ideal $I(Z)$, $Z = J^{-1}(0)$, it is necessary and sufficient, that the following *nonpositivity condition* applies: for all $\xi \in \mathfrak{g}$ and $z \in Z$ one has either

(i) $J(\xi) = 0$ in a neighborhood $U \subset M$ of z, or
(ii) in every neighborhood $U \subset M$ of z the function $J(\xi)$ takes strictly positive as well as strictly negative values.

This nonpositivity condition and Theorem 3.1 make it easy to provide first nontrivial examples.

2.1.1. *Zero Angular Momentum for m particles in $\mathbb{R}^2$.*

We consider the system of m particles in $\mathbb{R}^2$ with zero total angular momentum (see e.g. [17, Section 5] and [14, Section 6]). More precisely, the phase space is $M := (T^*\mathbb{R}^2)^m$ and we let $SO(2,\mathbb{R}) \cong S^1$ act on it by lifting the diagonal action, i.e.,

$$SO(2) \times M \to M$$

$$(g, (\boldsymbol{q}_1, \boldsymbol{p}^1, \ldots, \boldsymbol{q}_m, \boldsymbol{p}^m)) \mapsto (g\boldsymbol{q}_1, g\boldsymbol{p}^1, \ldots, g\boldsymbol{q}_m, g\boldsymbol{p}^m),$$

where $\boldsymbol{q}_i = (q_i^1, q_i^2)^t$ and $\boldsymbol{p}^i = (p_1^i, p_2^i)^t$ for $i = 1, \ldots, m$. The moment map $J : M \to \mathfrak{so}(2) = \mathbb{R}$ is given by $J(\boldsymbol{q}, \boldsymbol{p}) = \sum_{i=1}^m q_i^1 p_2^i - q_i^2 p_1^i$. In [17] the reduced space is described as a branched double cover of the closure of a certain coadjoint orbit of $\mathfrak{sp}(m, \mathbb{R})$. The moment map J obviously satisfies the nonpositivity condition above. Since $Z = J^{-1}(0)$ is of codimension 1, this implies that the Koszul complex (cf. Section 3) is a resolution of $C^\infty(Z)$.

2.1.2. An S^1-action with a worse-than-orbifold quotient.

The following example is taken from [6, p.125]. Consider the S^1-action on $\mathbb{C}^4$, endowed with symplectic form $\omega = \frac{i}{2} \sum_k dz_k \wedge d\bar{z}_k$, given by $e^{i\vartheta} \cdot (z_1, z_2, z_3, z_4) := (e^{i\vartheta} z_1, e^{i\vartheta} z_2, e^{-i\vartheta} z_3, e^{-i\vartheta} z_4)$. The moment map for the action is

$$J(z_1, z_2, z_3, z_4) = \frac{1}{2}(|z_3|^2 + |z_4|^2 - |z_1|^2 - |z_2|^2).$$

The constraint surface Z is the real cone $C(S^3 \times S^3)$, and by a topological argument (see [6]), the reduced space $C(S^3 \times_{S^1} S^3)$ can not be an orbifold. Since J clearly satisfies the nonpositivity condition above, it generates the vanishing ideal $I(Z)$. Again, we conclude that the Koszul complex is a resolution of $C^\infty(Z)$.

2.1.3. A T^2-action on $\mathbb{C}^4$.

We consider example 7.7 from [1]. The action is given by $T^2 \times \mathbb{C}^4 \to \mathbb{C}^4$, $((\vartheta_1, \vartheta_2), (z_1, z_2, z_3, z_4)) \mapsto (e^{i(\alpha\vartheta_1 + \beta\vartheta_2)} z_1, e^{-i\vartheta_2} z_2, e^{i\vartheta_1} z_3, e^{-i\vartheta_2} z_4)$ for $\alpha, \beta \in \mathbb{Z}$. A moment map for the action is $J : \mathbb{C}^4 \to \mathbb{R}^2$, $J(z_1, z_2, z_3, z_4) := \frac{1}{2}(-\alpha|z_1|^2 - |z_3|^2, -\beta|z_1|^2 + |z_2|^2 - |z_4|^2)$. J satisfies the nonpositivity condition for $\alpha < 0$. An elementary calculation gives that also condition b) of Theorem 3.1 is true. Consequently, the corresponding Koszul complex is a resolution of the space of smooth functions on $Z := J^{-1}(0)$.

2.2. Hamiltonian actions of nonabelian Lie groups

As the nonpositivity condition, in the case of nonabelian group actions, is only *necessary* for the ideal $I(Z) \subset C^\infty(M)$ to be generated by $J_1, \ldots, J_\ell$, the reasoning here is usually more intricate. In [1] it was proven that the latter is the case iff in every normal coordinate system the ideal I generated by the moment map in the real polynomial ring $\mathbb{R}[x^1, \ldots, x^{2n}]$ is *real* in the sense of real algebraic geometry (cf. [1, Theorem 6.3]). Recall that an ideal I in $\mathbb{R}[x^1, \ldots, x^m]$ is real, if it coincides with its *real radical*

$$\sqrt[\mathbb{R}]{I} := \Big\{ f \in \mathbb{R}[x^1, \ldots, x^m] \,\Big|\, f^{2i} + \sum_{j=0}^{k} g_j^2 \in I$$

$$\text{for some } i \text{ and } g_1, \ldots, g_k \in \mathbb{R}[x^1, \ldots, x^m] \Big\}.$$

In [1] we find the following criterion for such an ideal to be real.

Theorem 2.1. *Let I be an ideal in $\mathbb{R}[x^1, \ldots, x^m]$. Then I is real, if and only if the following two conditions hold:*

(i) $I_{\mathbb{C}} := I \otimes_{\mathbb{R}} \mathbb{C}$ is radical in $\mathbb{C}[x^1, \ldots, x^m]$, and
(ii) for every irreducible component $W \subset \mathbb{C}^m$ of the (complex) locus of $I_{\mathbb{C}}$

$$\dim_{\mathbb{R}}(W \cap \mathbb{R}^m) = dim_{\mathbb{C}}(W).$$

In other words, in order to know whether the ideal I is real, it is enough to gain detailed insight into the complex algebraic geometry behind the scene (e.g. knowing the primary decomposition of $I_{\mathbb{C}}$). Regardless the fact that the varieties in question are cones, there is no straightforward way to provide this information. A basic example, which one is tempted to consider is zero angular momentum of one particle in dimension n. Since the components of the moment map can be written as the 2×2-minors of a $2 \times n$-matrix, the ideal $I_{\mathbb{C}}$ is prime, and the complex locus is of dimension $n+1$ by a theorem of Hochster [13]. It follows easily, that the ideal I is real. Unfortunately, this example is not a complete intersection for $n \geq 3$. The only class of nonabelian examples, which the authors are aware of, where generating and complete intersection hypothesis are true at the same time, is the following.

2.2.1. *Commuting Varieties*

Let S the space of symmetric $n \times n$-matrices with real entries. We let $SO(n)$ act on S by conjugation and we lift this action to an action of $SO(n)$ on the cotangent bundle $T^*S = S \times S$. This action is hamiltonian with the moment map

$$J : S \times S \to \wedge^2 \mathbb{R}^n = \mathfrak{so}(n)^*$$

$$(Q, P) \mapsto [Q, P],$$

where we have identified $\mathfrak{so}(n)^*$ with the space $\wedge^2\mathbb{R}^n$ of antisymmetric $n \times n$-matrices. The complex locus $Z_{\mathbb{C}}$ defined by the these $\frac{1}{2}n(n-1)$ quadratic equations is an instance of what is called a *commuting variety*. In [5] it was shown that $Z_{\mathbb{C}}$ is irreducible of codimension $\frac{1}{2}n(n-1)$, and the ideal generated by the coefficients of J in the complex polynomial ring is prime. Let $S_{\mathrm{reg}} \subset S$ be the open subset of symmetric matrices with pairwise distinct eigenvalues. Since the action of $SO(n)$ on T^*S_{reg} is locally free, it follows that $Z \cap T^*S_{\mathrm{reg}}$ is of codimension $\frac{1}{2}n(n-1)$ likewise. As a consequence of Theorem 2.1, the components of J generate the vanishing ideal $I(Z)$ in $C^\infty(T^*S)$. It is easy to see, that T_zJ is surjective for $z \in Z \cap T^*S_{\mathrm{reg}}$. By Theorem 3.1 below, the Koszul complex is a resolution of the space of smooth function on Z. Using invariant theory, the reduced space was identified in [17] as the quotient $(\mathbb{R}^n \times \mathbb{R}^n)/S_n$, the symmetric group S_n acting diagonally. Note that the results of [5] have been generalized to moment maps of the isotropy representations of symmetric spaces of maximal rank [18].

3. Koszul resolution

Given a smooth map $J : M \to \mathbb{R}^\ell =: V^*$ we consider the Koszul holomogical complex of the sequence of ring elements $J_1, \ldots, J_\ell \in C^\infty(M)$, but we will view it later artificially as a cochain complex. In other words, we define the space of (co)chains to be $K^i := K_{-i}(M, J) := S^i_{C^\infty(M)}(V[1])$, i.e. the free (super)symmetric $C^\infty(M)$-algebra generated by the graded vector space $V[1]$, where we consider V to be concentrated in degree zero. $K_\bullet$ may also be viewed as the space of sections of the trivial vector bundle over M with fibre $\wedge^\bullet V$. Denoting by $e^1, \ldots, e^\ell$ the canonical bases of the dual space V of $V^* = \mathbb{R}^\ell$, we define the Koszul differential $\partial := \sum_a J_a i(e^a)$, where the $i(e^a)$ are the derivations extending the dual pairing. We will say, in accordance with [4], that $J_1, \ldots, J_\ell \in C^\infty(M)$ is a *complete intersection*, if the homology of the Koszul complex vanishes in degree $\neq 0$.

Now we would like to have a simple geometric criterion for J to be a complete intersection. We achieve this goal only after knowing that J generates the vanishing ideal (which is sometimes difficult to decide).

Theorem 3.1. *Let M be an analytic manifold and $J : M \to \mathbb{R}^\ell$ an analytic map, such that the following conditions are true*

(i) $(J_1, \ldots, J_\ell)$ generate the vanishing ideal of $Z := J^{-1}(0)$ in $C^\infty(M)$,

(ii) the regular stratum $Z_r := \{z \in Z \mid T_zJ$ is surjective$\}$ is dense in $Z := J^{-1}(0)$.

Then the Koszul complex $K := K(M, J)$ is acyclic and $H_0 = C^\infty(Z)$.

Proof. We will show that the Koszul complex $K(C_x^\omega(M), J)$ is acyclic for the ring $C_x^\omega(M)$ of germs in x of real analytic functions. Then it will follow that the Koszul complex $K(C^\infty(M), J)$ is acyclic, since the ring of germs of smooth functions $C_x^\infty(M)$ is flat over $C_x^\omega(M)$ (see [25, p.118]), and the sheaf of smooth functions on M is fine. Since $C_x^\omega(M)$ is noetherian, Krull's intersection theorem says that $\cap_{r \geq 0} I_x^r = 0$, where I_x is the ideal of germs of analytic functions vanishing on Z. According to [4, A X.160], it is therefore sufficient to show that $H_1(C_x^\omega(M), J) = 0$. Note that since J generates the vanishing ideal of Z in $C^\infty(M)$, it also generates the vanishing ideal of Z in $C_x^\omega(M)$. This can easily be seen using M. Artin's approximation theorem (see e.g. [21]). Suppose $f = \sum_a f^a e_a \in K_1$ is a cycle, i.e. $\partial f = \sum_a J_a f^a = 0$. Since the restriction to Z of the Jacobi matrix $D(\sum_a J_a f^a)$ vanishes, we conclude (using condition (ii)) that $f_{|Z}^a = 0$ for all $a = 1, \ldots, \ell$. Since J generates the vanishing ideal, we find an $\ell \times \ell$-matrix $F = (F^{ab})$ with smooth (resp. analytic) entries such that $f^a = \sum_b F^{ab} J_b$. It remains to be shown, that this matrix can be choosen to be *antisymmetric*. We have to distinguish two cases. If $x \notin Z$, the claim is obvious, since then one can take for example $F^{ab} := (\sum_a J_a^2)^{-1}(J_b f^a - J_a f^b)$. So let us consider the other case $x \in Z$. We then introduce some formalism to avoid tedious symmetrization arguments. Let E denote the free $k := C_x^\omega(M)$-module on ℓ generators, and consider the Koszul-type complex $SE \otimes \wedge E$. Generators of the symmetric part will be denoted by $\mu_1, \ldots, \mu_\ell$, generators of the Grassmann part by $e_1, \ldots, e_\ell$, respectively. We have two derivations $\delta := \sum_a e_a \wedge \frac{\partial}{\partial \mu_a} : S^n E \otimes \wedge^m E \to S^{n-1} E \otimes \wedge^{m+1} E$, and $\delta^* := \sum_a \mu_a i(e^a) : S^n E \otimes \wedge^m E \to S^{n+1} E \otimes \wedge^{m-1} E$. They satisfy the well known identities: $\delta^2 = 0$, $(\delta^*)^2 = 0$ and $\delta \delta^* + \delta^* \delta = (m+n)\,\mathrm{id}$. Furthermore, we introduce the two commuting derivations $i_J := \sum_a J_a i(e^a)$ and $d_J = \sum_a J_a \frac{\partial}{\partial \mu_a}$. They obey the identities $i_J^2 = 0$, $[i_J, \delta] = d_J$, $[d_J, \delta^*] = i_J$ and $[i_J, \delta^*] = 0 = [d_J, \delta]$. We interpret the cycle f above as being in $E \otimes k$ and the matrix F as a member of $E \otimes E$. We already know that $d_J f = 0$ implies $f = i_J F$. This argument may be generalized as follows: if $a \in S^n E \otimes k$ obeys $d_J^n a = 0$, then there is an $A \in S^n E \otimes E$ such that $a = i_J A$. The proof is easily provided by taking all n-fold partial derivatives of $d_J^n a = 0$, evaluating the result on Z and using conditions (i) and (ii). We now claim that there is a sequence of $F_{(n)} \in S^{n+1} E \otimes E$, $n \geq 0$, such that $F = F_{(0)}$, $\delta^* F_{(n)} = (n+2)i_J F_{(n+1)}$ and

$$f = d_J^n i_J F_{(n)} + i_J \delta^* \underbrace{\left(\sum_{i=0}^{n-1} \frac{1}{i+2} d_J^i \delta F_{(i)} \right)}_{=:B_{n-1}} \quad \text{for all } n \geq 1. \tag{1}$$

We prove this by induction. Setting $B_{-1} := 0$, we may start the induction with $n = 0$, where nothing has to be done. Suppose now, that the claim is true for $F_{(0)}, \ldots, F_{(n)}$. We obtain $f = \frac{1}{n+2} d_J^n i_J \big(\delta \delta^* F_{(n)} + \delta^* \delta F_{(n)} \big) + i_J \delta^* B_{n-1} = \frac{1}{n+2} d_J^{n+1} \delta^* F_{(n)} + i_J \delta^* B_n$, where we made use of the relations $[d_J^n i_J, \delta^*] = 0$ and $[d_J^n i_J, \delta] = d_J^{n+1}$. Since $0 = d_J f = d_J^{n+2} \delta^* F_{(n)}$, we find an $F_{(n+1)}$ such that $\frac{1}{n+2} \delta^* F_{(n)} = i_J F_{(n+1)}$, and the claim is proven. Finally, we want to take the limit of equation (1) as n goes to ∞. For this limit to make sense, we have to change the ring to the ring of formal power series. Let us denote this change of rings by $\hat{\ } : C_x^\omega(M) \to \mathbb{K}[[x^1, \ldots, x^n]]$. Since by Krull's intersection theorem $\cap_{r \geq 0} \hat{I}^r = 0$ ($\hat{I}$ the ideal generated by $\hat{J}_1, \ldots, \hat{J}_\ell$), we obtain a formal solution of the problem: $\hat{f} = i_{\hat{J}} \delta^* B_\infty$, where $B_\infty := \sum_{i=0}^\infty \frac{1}{i+2} d_{\hat{J}}^i \delta \hat{F}_{(i)}$ is well defined since $\hat{I}$ contains the maximal ideal. Applying M. Artin's approximation theorem yields an analytic solution, and we are done. $\qquad\square$

The above reasoning can be considered to be folklore, as the subtlety of finding an *antisymmetric* source term is often swept under the rug in semirigorous arguments. The next theorem though is a consequence of rather deep analytic results. The problem of splitting the Koszul resolution in the context of Fréchet spaces was also addressed in [8] from a different perspective.

Theorem 3.2. *Let M be a smooth manifold, $J : M \to \mathbb{R}^\ell$ be smooth map such that around every $m \in M$ there is a local chart in which J is real analytic. Moreover, assume that the Koszul complex $K = K(M, J)$ is a resolution of $C^\infty(Z)$, $Z = J^{-1}(0)$. Then there are a prolongation map* prol $: C^\infty(Z) \to C^\infty(M)$ *and contracting homotopies* $h_i : K_i \to K_{i+1}$, $i \geq 0$, *which are continuous in the respective Fréchet topologies, such that*

$$(C^\infty(Z), 0) \underset{\text{prol}}{\overset{\text{res}}{\leftrightarrows}} (K, \partial), h \tag{2}$$

is a contraction, i.e. res *and* prol *are chain maps and* res prol $=$ id *and* id $-$ prol res $= \partial h + h \partial$. *If necessary, these can be adjusted in such a way,* *that the side conditions (see Appendix A) $h_0 \, \mathrm{prol} = 0$ and $h_{i+1} \, h_i = 0$ are fulfilled. If, moreover, a compact Lie group G acts smoothly on M, G is represented on $\mathbb{R}^\ell$ and $J : M \to \mathbb{R}^\ell$ is equivariant, then prol and h can additionally be chosen to be equivariant.*

Proof. A closed subset $X \subset \mathbb{R}^n$ is defined to have the *extension property*, if there is a continuous linear map $\lambda : \mathcal{C}^\infty(X) \to \mathcal{C}^\infty(\mathbb{R}^n)$, such that $\mathrm{res} \, \lambda = \mathrm{id}$. The extension theorem of E. Bierstone and G. W. Schwarz, [2, Theorem 0.2.1] says that Nash subanalytic sets (and hence closed analytic sets) have the extension property. Using a partition of unity, we get a continuous linear map $\lambda : \mathcal{C}^\infty(Z) \to \mathcal{C}^\infty(M)$, such that $\mathrm{res} \, \lambda = \mathrm{id}$. In the same reference, one finds a "division theorem" (Theorem 0.1.3.), which says that for a matrix $\varphi \in \mathcal{C}^\omega(\mathbb{R}^n)^{r,s}$ of analytic functions the image of $\varphi : \mathcal{C}^\infty(\mathbb{R}^n)^s \to \mathcal{C}^\infty(\mathbb{R}^n)^r$ is closed, and there is a continuous split $\sigma : \mathrm{im}\varphi \to \mathcal{C}^\infty(\mathbb{R}^n)^s$ such that $\varphi\sigma = \mathrm{id}$. Using a partition of unity, we conclude that there are linear continuous splits $\sigma_i : \mathrm{im} \, \partial_{i+1} \to K_{i+1}$ for the Koszul differentials $\partial_{i+1} : K_{i+1} \to K_i$ for $i \geq 0$, i.e. $\partial_{i+1} \, \sigma_i = \mathrm{id}$. We observe that $\mathrm{im} \, \lambda \oplus \mathrm{im} \, \partial_1 = K_0$, since for every $x \in K_0$ the difference $x - \lambda \, \mathrm{res} \, x$ is a boundary due to exactness and the sum is apparantly direct. Similarly, we show that $\mathrm{im} \, \sigma_i \oplus \mathrm{im} \, \partial_{i+2} = K_{i+1}$ for $i \geq 0$. The next step is to show that $\mathrm{im} \, \sigma_i$ is a *closed* subspace of K_0. Therefor we assume that $(x_n)_{n \in \mathbb{N}}$ is a sequence in $\mathrm{im} \, \partial_{i+1}$ such that $\sigma_i(x_n)$ converges to $y \in K_{i+1}$. Then $x_n = \partial_{i+1}\sigma_i(x_n)$ converges to $\partial_{i+1}y$, since ∂_{i+1} is continuous. Since $\partial_{i+1}y$ is in the domain of σ_i, we obtain that $\sigma_i(x_n)$ converges to $\sigma_i\partial_{i+1}y = y \in \mathrm{im} \, \sigma_i$. Similarly, we have that $\mathrm{im} \, \lambda$ is a closed subspace of K_0. Altogether, it is feasible to extend σ_i to a linear continuous map $K_i \to K_{i+1}$ (cf. [20, p.133]). If necessary, λ and σ_i can be made equivariant by averaging over G, since res and ∂ are equivariant. We observe that we have $\lambda \, \mathrm{res}_{|\mathrm{im}\lambda} = \mathrm{id}$ and $\lambda \, \mathrm{res}_{|\mathrm{im}\partial_1} = 0$ and analogous equations in higher degrees. We now replace λ by $\mathrm{prol} := \lambda - \partial_1\sigma_0\lambda$ and σ_i by $h_i := \sigma_i - \partial_{i+2}\sigma_{i+1}\sigma_i$ for $i \geq 0$. These maps share all of the above mentioned properties with λ and σ_i. Additionally, we have $\partial_1 h_{0|\mathrm{im(prol)}} = 0$ and $\partial_{i+2}h_{i+1|\mathrm{im}(h_i)} = 0$ for $i \geq 0$. This concludes the construction of (2). The side conditions can be achieved by algebraic manipulations (see Appendix A). Note that these modifications do not ruin the equivariance $\square$

A crucial property of the Koszul resolution is that it is a differential graded commutative algebra. In the present context, where the constraint functions are the components of a moment map, it has the following extra feature. The Lie algebra $\mathfrak{g}$ acts on it by even derivations, extending the actions on $\mathfrak{g}$ and on $\mathcal{C}^\infty(M)$.

4. Classical homological reduction

The BRST-algebra is defined to be $\mathscr{A} := S_{\mathcal{C}^\infty(M)}\big(\mathfrak{g}[1] \oplus \mathfrak{g}^*[-1]\big)$, i.e. the free graded commutative $\mathcal{C}^\infty(M)$-algebra generated by $\mathfrak{g}$ (of degree -1) and $\mathfrak{g}^*$ (of degree 1). We adopt the usual convention to call the elements of $\mathfrak{g}^*$ and $\mathfrak{g}$ *ghosts* and *antighosts*, respectively. We will frequently refer to a basis $e_1, \ldots, e_\ell$ and $e^1, \ldots, e^\ell$ of $\mathfrak{g}$ and $\mathfrak{g}^*$, respectively (we will use latin indices: $a, b, \ldots$). There is an even graded Poisson bracket on $\mathscr{A}$ extending that on M, which is uniquely defined by the requirements $\{\alpha, x\} = 2\langle \alpha, x \rangle$ and $\{f, x\} = 0 = \{f, \alpha\}$ for all $x \in \mathfrak{g}$, $\alpha \in \mathfrak{g}^*$ and $f \in \mathcal{C}^\infty(M)$. With the Lie bracket and the moment map we build an element $\theta := -\frac{1}{4}\sum_{a,b,c} f_{ab}^c \, e^a e^b e_c + \sum_a J_a e^a \in \mathscr{A}^1$, where the f_{ab}^c are the structure constants of $\mathfrak{g}$. An easy calculation yields $\{\theta, \theta\} = 0$, hence $\mathscr{D} := \{\theta, ?\}$ is a differential. Summing up, we obtain a differential graded Poisson algebra $(\mathscr{A}, \{, \}, \mathscr{D} = \{\theta, ?\})$, we call θ the *BRST-charge* and $\mathscr{D}$ the *classical BRST-differential*.

Closer examination shows that $\mathscr{D} = \delta + 2\partial$ is a linear combination of two supercommuting differentials. Here, δ is the codifferential of the Lie algebra cohomology corresponding to the $\mathfrak{g}$-module $S_{\mathcal{C}^\infty(M)}(\mathfrak{g}[1])$, this representation will be denoted by L, and $\partial = \sum_a J_a i^a$ is the extension of the Koszul differential. We view $\mathscr{D}$ as a perturbation (see Appendix A) of the acyclic differential ∂.

We extend the restriction map res to a map res : $\mathscr{A} \to S_{\mathcal{C}^\infty(Z)}(\mathfrak{g}^*[-1])$ by setting it zero for all terms containing antighosts and restricting the coefficients. In the same fashion, we extend prol to a map $S_{\mathcal{C}^\infty(Z)}(\mathfrak{g}^*[-1]) \to \mathscr{A}$ extending the coefficients.

Since the moment map J is G-equivariant, G acts on $Z = J^{-1}(0)$. Hence $\mathcal{C}^\infty(Z)$ is a $\mathfrak{g}$-module, this representation will be denoted by L^z. Note that $L_X^z = \text{res } L_X \text{ prol}$ for all $X \in \mathfrak{g}$. We identify $S_{\mathcal{C}^\infty(Z)}(\mathfrak{g}^*[-1])$ with the space of cochains of Lie algebra cohomology $C^\bullet\big(\mathfrak{g}, \mathcal{C}^\infty(Z)\big)$. Let us denote $d : C^\bullet\big(\mathfrak{g}, \mathcal{C}^\infty(Z)\big) \to C^{\bullet+1}\big(\mathfrak{g}, \mathcal{C}^\infty(Z)\big)$ the codifferential of Lie algebra cohomology coresponding to L^z. Since res is a morphism of $\mathfrak{g}$-modules we obtain $d \text{ res} = \text{res } \delta$.

Theorem 4.1. *There are $\mathbb{K}$-linear maps $\Phi : C^\bullet\big(\mathfrak{g}, \mathcal{C}^\infty(Z)\big) \to \mathscr{A}^\bullet$ and $H : \mathscr{A}^\bullet \to \mathscr{A}^{\bullet-1}$ which are continuous in the respective Fréchet topologies such that*

$$\big(C^\bullet(\mathfrak{g}, \mathcal{C}^\infty(Z)), d\big) \underset{\Phi}{\overset{\text{res}}{\leftrightarrows}} (\mathscr{A}^\bullet, \mathscr{D}), H \tag{1}$$

is a contraction.

Proof. Apply lemma A.1 to the perturbation $\mathscr{D}_\nu$ of 2∂. Explicitly, we get $H := \frac{1}{2}h\sum_{j=0}^{\ell}(-\frac{1}{2})^j(h\delta + \delta h)^j$ and $\Phi = \mathrm{prol} - H(\delta\ \mathrm{prol} - \mathrm{prol}\ d)$, which are obviously Fréchet continuous. Note that from $h\,\mathrm{prol} = 0$ and $h^2 = 0$ it follows that $H\Phi = 0$ and $H^2 = 0$. If prol is chosen to be equivariant, then the expression for Φ simplifies to $\Phi = \mathrm{prol}$. In the same way one gets $H = \frac{1}{2}h$, if h is equivariant. $\qquad\square$

Corollary 4.1. *There is a graded Poisson structure on* $\mathrm{H}^\bullet(\mathfrak{g}, C^\infty(Z))$. *If* $[a], [b]$ *are the cohomology classes of* $a, b \in C^\bullet(\mathfrak{g}, C^\infty(Z))$, *then the bracket is given given by* $\{[a], [b]\} := [\mathrm{res}\{\Phi(a), \Phi(b)\}]$. *The restriction of this bracket to* $\mathrm{H}^0(\mathfrak{g}, C^\infty(Z)) = C^\infty(Z)^{\mathfrak{g}}$ *coincides with the Dirac reduced Poisson structure.*

5. The quantum BRST-algebra

In this section we will introduce the quantum BRST algebra, which is $\mathbb{K}[[\nu]]$-dg algebra $(\mathscr{A}^\bullet[[\nu]], *, \mathscr{D}_\nu)$ deforming the classical dg Poisson algebra $(\mathscr{A}^\bullet, \{,\}, \mathscr{D})$. The exposition parallels that of [3]. In order to define a graded product $*$ on $\mathscr{A}[[\nu]]$, we use on one hand a Clifford multiplication $x \cdot y := \mu\big(\mathrm{e}^{-2\nu\sum_a i^a \otimes i_a}(x \otimes y)\big)$ for $x, y \in S_{\mathbb{K}}(\mathfrak{g}[1] \oplus \mathfrak{g}[-1])$. Here μ denotes the supercommutative multiplication, i^a and i_a are the left derivations extending the dual pairing with e^a and e_a, respectively and $\otimes$ denotes the graded tensor product. On the other hand, we will need a star product $\star$ on M, which is compatible with the $\mathfrak{g}$-action in the following sense

$$\mathbb{J}(X) \star \mathbb{J}(Y) - \mathbb{J}(Y) \star \mathbb{J}(X) = \nu\mathbb{J}([X,Y]) \text{ for all } X, Y \in \mathfrak{g}, \tag{1}$$

where $\mathbb{J} = J + \sum_{i\geq 1}\nu^i J_{(i)} \in \mathscr{A}^1[[\nu]]$ is a deformation of the moment map J. In other words, $\star$ is *quantum covariant* for the *quantum moment map* $\mathbb{J}$. For $f, g \in C^\infty(M)$ and $x, y \in S(\mathfrak{g}[1] \oplus \mathfrak{g}^*[-1])$ we define $(fx) * (gy) := (f \star g)(x \cdot y)$. Note that $*$ is graded. The next step is to quantize the BRST-charge. Luckily, we are done with (see e.g. [15])

$$\theta_\nu := -\frac{1}{4}\sum_{a,b,c} f_{ab}^c\, e^a e^b e_c + \sum_a \mathbb{J}_a\, e^a + \frac{1}{2}\nu\sum_a f_{ab}^b e^a \in \mathscr{A}^1[[\nu]],$$

since a straightforward calculation yields $\theta_\nu * \theta_\nu = 0$. We define the *quantum BRST differential* to be $\mathscr{D}_\nu := \frac{1}{\nu}\mathrm{ad}_*(\theta_\nu)$.

Before we take a closer look, at $\mathscr{D}_\nu$ let us introduce some terminology. We define the (superdifferential) operators $\delta_\nu, \mathscr{R}, q, u : \mathscr{A}^\bullet \to \mathscr{A}^{\bullet+1}$,

$$\delta_\nu(f) := -\frac{1}{2}\sum_{a,b,c} f^c_{ab}\, e^a e^b\, i_c(f) + \sum_{a,b,c} f^c_{ab}\, e^a e_c\, i_b(f) + \sum_a e^a\,\frac{1}{\nu}[\mathbb{J}_a, f]_*,$$

$$\mathscr{R}(f) := \sum_a i^a\, f * \mathbb{J}_a, \qquad\qquad \text{``right multiplication''}$$

$$q(f) := -\frac{1}{2}\sum_{a,b,c} f^c_{ab}\, e_c\, i^a i^b(f), \qquad\qquad \text{``quadratic ...''}$$

$$u(f) := \sum_{a,b} f^b_{ab}\, i^a(f), \qquad\qquad \text{``unimodular term''}$$

for $f \in \mathscr{A}$. Note that δ_ν is the coboundary operator of Lie algebra cohomology corresponding to the representation

$$\mathbb{L}_X : S_{\mathcal{C}^\infty(M)}(\mathfrak{g}[1])[[\nu]] \to S_{\mathcal{C}^\infty(M)}(\mathfrak{g}[1])[[\nu]],$$

$$af \mapsto (\mathrm{ad}_X(a))f + a\,\nu^{-1}(\mathbb{J}(X) \star f - f \star \mathbb{J}(X)), \quad (2)$$

where $X \in \mathfrak{g}$, $a \in S_{\mathbb{K}}(\mathfrak{g}[1])$ and $f \in \mathcal{C}^\infty(M)[[\nu]]$. Finally, we set

$$\partial_\nu := \mathscr{R} + \nu\Big(\frac{1}{2}u - q\Big).$$

This operator will be called the *deformed* or *quantum Koszul differential*. Note that ∂_ν is a homomorphism of $\mathcal{C}^\infty(M)[[\nu]]$-left-modules. As a side remark, ∂_ν may also be interpreted as a differential of Lie algebra homology of a certain representaion of $\mathfrak{g}$. This point of view was adopted in [22].

Theorem 5.1. *The quantum BRST differential $\mathscr{D}_\nu = \delta_\nu + 2\partial_\nu$ is a linear combination of two supercommuting differentials δ_ν and ∂_ν.*

Proof. Straightforward calculation. $\qquad\qquad\qquad\qquad\qquad\qquad\square$

6. Quantum reduction

The main idea, which we follow in order to compute the quantum BRST cohomology (i.e. the cohomology of $(\mathscr{A}[[\nu]], \mathscr{D}_\nu)$), is to provide a deformed version of the contraction (1). This will be done by applying Lemma A.2 to the contraction (2) for the perturbation ∂_ν of ∂ and then applying Lemma A.1 for the perturbation $\mathscr{D}_\nu$ of $2\partial_\nu$. We will also need to examine a deformed version of the representation L^z of $\mathfrak{g}$ on $\mathcal{C}^\infty(Z)$.

Proposition 6.1. *If we choose h_0 such that $h_0\,\mathrm{prol} = 0$, then there are deformations of the restriction map $\mathrm{res}_\nu = \mathrm{res} + \sum_{i\geq 1} \nu^i\,\mathrm{res}_i : \mathcal{C}^\infty(M) \to$*

$\mathcal{C}^\infty(Z)[[\nu]]$ *and of the contracting homotopies* $h_{\nu i} = h_i + \sum_{j\geq 1} \nu^j\, h_i^j :$
$K_i[[\nu]] \to K_{i+1}[[\nu]]$, *which are a formal power series of Fréchet continuous maps and such that*

$$\left(\mathcal{C}^\infty(Z)[[\nu]], 0\right) \underset{\text{prol}}{\overset{\text{res}_\nu}{\leftrightarrows}} \left(K[[\nu]], \partial_\nu\right), h_\nu \tag{1}$$

is a contraction with $h_{\nu 0}\, \text{prol} = 0$. *Explicitly, we have*

$$\text{res}_\nu := \text{res}\,(\text{id} + (\partial_{\nu 1} - \partial_1)h_0)^{-1}.$$

If we choose h *to be* $\mathfrak{g}$-*equivariant, the same is true for* h_ν.

Proof. Apply lemma A.2 to the perturbation ∂_ν of ∂. $\square$

We now define the quantized representation $\mathbb{L}^z$ of $\mathfrak{g}$ on $\mathcal{C}^\infty(Z)[[\nu]]$ by setting

$$\mathbb{L}_X^z := \text{res}_\nu\, \mathbb{L}_X\, \text{prol} \quad \text{for } X \in \mathfrak{g}.$$

That this is in fact a representation, follows easily from the observation $\mathbb{L}_X \partial_\nu - \partial_\nu \mathbb{L}_X = 0$ for all $X \in \mathfrak{g}$ (this is a consequence of Theorem 5.1), and from h_ν being a contracting homotopy. In the same fashion as in Section 4, we define $d_\nu : C^\bullet(\mathfrak{g}, \mathcal{C}^\infty(Z)[[\nu]]) \to C^{\bullet+1}(\mathfrak{g}, \mathcal{C}^\infty(Z)[[\nu]])$ to be the differential of Lie algebra cohomology of the representation $\mathbb{L}^z$, i.e. $d_\nu\, \text{res}_\nu = \text{res}_\nu\, \delta_\nu$. In the same manner, we extend res_ν and h_ν as in Section 4 to maps $\text{res}_\nu : \mathscr{A} \to C(\mathfrak{g}, \mathcal{C}^\infty(Z)[[\nu]])$ and $h_\nu : \mathscr{A}^\bullet[[\nu]] \to \mathscr{A}^{\bullet-1}[[\nu]]$.

Theorem 6.1. *There are* $\mathbb{K}[[\nu]]$-*linear maps* $\Phi_\nu : C^\bullet\left(\mathfrak{g}, \mathcal{C}^\infty(Z)\right) \to \mathscr{A}^\bullet[[\nu]]$ *and* $H_\nu : \mathscr{A}^\bullet \to \mathscr{A}^{\bullet-1}[[\nu]]$, *which are series of Fréchet continuous maps such that*

$$\left(C^\bullet\left(\mathfrak{g}, \mathcal{C}^\infty(Z)[[\nu]]\right), d_\nu\right) \underset{\Phi_\nu}{\overset{\text{res}_\nu}{\leftrightarrows}} (\mathscr{A}^\bullet[[\nu]], \mathscr{D}_\nu), H_\nu$$

is a contraction.

Proof. Since the requisite condition $\text{res}_\nu\, h_\nu = 0$ is obviously fulfilled, we apply Lemma A.1 to the perturbation $\mathscr{D}_\nu$ of $2\partial_\nu$. Explicitly, this means that $H_\nu := \frac{1}{2}h_\nu \sum_{j=0}^\ell (-\frac{1}{2})^j (h_\nu \delta_\nu + \delta_\nu h_\nu)^j$ and $\Phi_\nu = \text{prol} - H_\nu(\delta_\nu\, \text{prol} - \text{prol}\, d_\nu)$, which are obviously series of Fréchet continuous maps. Note that from $h_0\, \text{prol} = 0$ and $h^2 = 0$, we get $H_\nu \Phi_\nu = 0$ and $H_\nu^2 = 0$. If prol is chosen to be equivariant, then the expression for Φ

simplifies to $\Phi_\nu = \text{prol}$. If h and (hence h_ν) is equivariant, then it follows that $H_\nu = \frac{1}{2}h_\nu$. $\qquad\qquad\square$

We use this contraction to transfer the associative algebra structure from $\mathscr{A}[[\nu]]$ to the Lie algebra cohomology $\mathrm{H}^\bullet\big(\mathfrak{g}, C^\infty(Z)[[\nu]]\big)$ of the representation $\mathbb{L}^z$ by setting

$$[a] * [b] := \big[\mathrm{res}_\nu\big(\Phi_\nu(a) * \Phi_\nu(b)\big)\big] \tag{2}$$

where $[a], [b]$ denote the cohomology classes of $a, b \in C^\bullet(\mathfrak{g}, C^\infty(Z)[[\nu]])$. But in fact that is *not exactly*, what we want to accomplish. The primary obstacle on the way to the main result, Corollary 6.1, is that, in general, we have $\mathrm{H}^\bullet\big(\mathfrak{g}, C^\infty(Z)[[\nu]]\big) \neq \mathrm{H}^\bullet\big(\mathfrak{g}, C^\infty(Z)\big)[[\nu]]$. An example where this phenomenon occurs was given in [3, section 7]. One way out is to sharpen the compatibility condition (1). We require, that $\mathbb{J} = J$ and

$$J(X) \star f - f \star J(X) = \nu\{J(X), f\} \quad \text{for all } X \in \mathfrak{g}, f \in C^\infty(M).$$

This property is also referred to as *strong invariance* of the star product $\star$ with respect to the Lie algebra action. For proper group actions a strongly invariant star product can always be found (see [9]). Of course, now the representations $\mathbb{L}$ and L coincide and we get $\delta = \delta_\nu$. But with some mild restrictions on the contracting homotopy h of the Koszul resolution we also have the following.

Lemma 6.1. *If h_0 is $\mathfrak{g}$-equivariant and $h_0\,\mathrm{prol} = 0$, then $\mathbb{L}^z = \mathrm{L}^z$.*

Proof. For $X \in \mathfrak{g}$ we have $\mathbb{L}^z_X = \mathrm{res}_\nu\, L_X\, \mathrm{prol} = \mathrm{res}\,(\mathrm{id} + (\partial_{\nu 1} - \partial_1)h_0)^{-1}L_X\, \mathrm{prol}$. Since L_X commutes with $\partial_{\nu 1}$, ∂_1 and h_0 the last expression can be written as $\mathrm{res}\, L_X(\mathrm{id} + (\partial_{\nu 1} - \partial_1)h_0)^{-1}\, \mathrm{prol} = \mathrm{res}\, L_X\, \mathrm{prol}$. $\square$

Corollary 6.1. *With the assumptions made above the product defined by equation (2) makes $\mathrm{H}^\bullet\big(\mathfrak{g}, C^\infty(Z)\big)[[\nu]]$ into a graded associative algebra. For the subalgebra $\mathrm{H}^0\big(\mathfrak{g}, C^\infty(Z)\big)[[\nu]] = \big(C^\infty(Z)\big)^{\mathfrak{g}}[[\nu]]$ this formula simplifies to*

$$f * g := \mathrm{res}_\nu\big(\mathrm{prol}(f) * \mathrm{prol}(g)\big) \quad \text{for } f, g \in \big(C^\infty(Z)\big)^{\mathfrak{g}}. \tag{3}$$

Since $\big(C^\infty(Z)\big)^{\mathfrak{g}}[[\nu]]$ is $\mathbb{K}[[\nu]]$-linearly isomorphic to the algebra of smooth functions on the symplectic stratified space M_{red}, we obtain an associative product on $C^\infty(M_{\mathsf{red}})[[\nu]]$ which gives rise to a continuous Hochschild cochain.

Proof. It remains to show (3). Let us denote by $\mathscr{A}^+$ the kernel of the augmentation map $\mathscr{A} \to C^\infty(M)$. Equation (3) follows from the fact that $(\mathscr{A}^+ \cap \mathscr{A}^0)[[\nu]]$ is a two-sided ideal in $\mathscr{A}^0[[\nu]]$. $\qquad\square$

Finally, if $H^1(\mathfrak{g}, C^\infty(Z))$ vanish, it is possible to find a topologically linear isomorphism between the spaces of invariants for the classical and the deformed representation.

Corollary 6.2. *Let G be a compact, connected semisimple Lie group acting on the Poisson manifold M in a Hamiltonian fashion. Assume that the equivariant moment map J satisfies the generating and the complete intersection hypothesis. Then for a star product $*$ on M with quantum moment map $\mathbb{J}$ there is an invertible sequence of continuous maps*

$$S = \sum_{i \geq 0} \nu^i \, S_i : H^0(\mathfrak{g}, C^\infty(Z))[[\nu]] = C^\infty(Z)^{\mathfrak{g}}[[\nu]] \to H^0(\mathfrak{g}, C^\infty(Z)[[\nu]])$$

such that the formula

$$f \star g := S^{-1}\big(S(f) * S(g)\big) = S^{-1}\Big(\mathrm{res}_\nu \big(\Phi_\nu(S(f)) * \Phi_\nu(S(g))\big)\Big)$$

defines a continuous formal deformation of the Poisson algebra $C^\infty(Z)^{\mathfrak{g}}$ into an associative algebra.

Proof. According to Viktor L. Ginzburg (see [12, Theorem 2.13]) we have for any compact, connected Lie group G with a smooth representation on a Fréchet space W an isomorphism $H^\bullet(\mathfrak{g}, W) \cong H^\bullet(\mathfrak{g}, \mathbb{K}) \otimes W^{\mathfrak{g}}$. In particular, this implies that if $\mathfrak{g}$ is semisimple, the first and the second cohomology groups of the $\mathfrak{g}$-module $C^\infty(Z)$ vanish. Note, that, since G is compact, the space of invariants $C^\infty(Z)^{\mathfrak{g}} \subset C^\infty(Z)$ has a closed complement. This can be taken to be the kernel of the averaging projection. Using these observations it is straight forward to construct S by a standard inductive argument (see e.g. [3, p.140]). $\qquad\square$

Appendix A. Two perturbation lemmata

We consider (cochain) complexes in an additive $\mathbb{K}$-linear category $\mathscr{C}$ (e.g. the category of Fréchet spaces). A *contraction* in $\mathscr{C}$ consists of the following data

$$(X, d_X) \underset{i}{\overset{p}{\leftrightarrows}} (Y, d_Y), h_Y, \tag{A.1}$$

where i and p are chain maps between the chain complexes (X, d_X) and (Y, d_Y), $h_Y : Y \to Y[-1]$ is a morphism, and we have $pi = \mathrm{id}_X$, $d_Y h_Y + h_Y d_Y = \mathrm{id}_Y - ip$. The contraction is said to satisfy the *side conditions* (sc1–3), if moreover, $h_Y^2 = 0$, $h_Y i = 0$ and $ph_Y = 0$ are true. It was observed in [16], that in order to fulfill (sc2) and (sc3), one can replace h_Y by $h_Y' := (d_Y h_Y + h_Y d_Y)\, h_Y\, (d_Y h_Y + h_Y d_Y)$. If one wants to have in addition (sc1) to be satisfied, one may relapce h_Y' by $h_Y'' := h_Y' d_Y h_Y'$. Let $C := \mathrm{Cone}(p)$ be the mapping cone of p, i.e. $C = X[1] \oplus Y$ is the complex with differential $d_C(x, y) := (d_X x + (-1)^{|y|} py, d_Y y)$. The homology of C is trivial, because $h_C(x, y) := (0, h_Y y + (-1)^{|x|} ix)$ is a contracting homotopy, i.e. $d_C h_C + h_C d_C = \mathrm{id}_C$, if (sc3) is true.

Let us now assume that the objects X and Y carry complete descending filtrations and the structure maps are filtration preserving. Moreover, pretend that we have found a *perturbation* $D_Y = d_Y + t_Y$ of d_Y, i.e. $D_Y^2 = 0$ and $t_Y : Y \to Y[1]$, called the *initiator*, has the property that $t_Y h_Y + h_Y t_Y$ raises the filtration. Since, in general, $t_X := pt_Y i$ needs not to be a perturbation of d_X, we impose that as an extra condition: we *assume* that $D_X = d_X + t_X$ is a differential. Setting $t_C := (t_X, t_Y)$, we will get a perturbation $D_C := d_C + t_C$ of d_C, if we have in addition $t_X p = pt_Y$ (this will imply that $(d_X + t_X)^2 = 0$). Then an easy calculation yields that $H_C := h_C(D_C h_C + h_C D_C)^{-1} = h_C(\mathrm{id}_C + t_C h_C + h_C t_C)^{-1}$ is well defined and satisfies $D_C H_C + H_C D_C = \mathrm{id}_C$. Defining the morphism $I : X \to Y$, $H_C(x, 0) =: (0, (-1)^{|x|} Ix)$ and the homotopy $H_Y : Y \to Y[-1]$, $H_C(0, y) =: (0, H_Y y)$ we get the following

Lemma A.1 (*Perturbation Lemma – Version 1*). *If the contraction (A.1) satisfies (sc3) and $D_Y = d_Y + t_Y$ is a perturbation of d_Y such that $t_X p = pt_Y$, then*

$$(X, D_X) \overset{p}{\underset{I}{\leftrightarrows}} (Y, D_Y), H_Y, \tag{A.2}$$

is a contraction fulfilling (sc3). Moreover, we have $H_Y = h_Y(\mathrm{id}_Y + t_Y h_Y + h_Y t_Y)^{-1}$ and $Ix = ix - H_Y(t_Y ix - it_X x)$. If all side conditions are true for (A.1), then they are for (A.2), too.

Starting with the mapping cone $K = \mathrm{Cone}(i)$, i.e. the complex $K = Y[1] \oplus X$ with the differential $d_K(y, x) = (d_Y y + (-1)^{|x|} ix, d_X x)$, we may give a version of the above argument arriving at a contraction with all data perturbed except i. More precisely, we have a homotopy $h_K(y, x) :=$

$(h_Y y, (-1)^{|y|} p y)$, for which $d_K h_K + h_K d_K = \mathrm{id}_K$ follows from (sc2). Mimicking the above argument, we get a differential $D_K := d_K + t_K$ with $t_K := (t_Y, t_X)$, if $t_Y i = i t_X$ (this will imply $D_X^2 = 0$). Assuming (A.1) to satisfy (sc2), $H_K := h_K (D_K h_K + h_K D_K)^{-1}$ will become a contracting homotopy $D_K H_K + H_K D_K = \mathrm{id}_K$. Defining $P : Y \to X$ and $H_Y' : Y \to Y[-1]$ by $H_K(y, 0) = H_K(y, x) =: (H_Y' y, (-1)^{|y|} P y)$ we get the following

Lemma A.2 (*Perturbation Lemma – Version 2*). *If the contraction (A.1) satisfies (sc2) and $D_Y = d_Y + t_Y$ is a perturbation of d_Y such that $t_Y i = i t_X$, then*

$$(X, D_X) \overset{P}{\underset{i}{\leftrightarrows}} (Y, D_Y), H_Y', \tag{A.3}$$

is a contraction fulfilling (sc2). Moreover, we have $H_Y' = h_Y (\mathrm{id}_Y + t_Y h_Y + h_Y t_Y)^{-1}$ and $P = p (\mathrm{id} + t_Y h_Y + h_Y t_Y)^{-1}$. If all side conditions are true for (A.1), then they are for (A.3), too.

Bibliography

1. Judith M. Arms, Mark J. Gotay, and George Jennings, *Geometric and algebraic reduction for singular momentum maps*, Adv. Math. **79** (1990), no. 1, 43–103.
2. Edward Bierstone and Gerald W. Schwarz, *Continuous linear division and extension of C^∞ functions*, Duke Math. J. **50** (1983), no. 1, 233–271.
3. Martin Bordemann, Hans-Christian Herbig, and Stefan Waldmann, *BRST cohomology and phase space reduction in deformation quantization*, Comm. Math. Phys. **210** (2000), no. 1, 107–144.
4. Nicolas Bourbaki, *Éléments de mathématique*, Masson, Paris, 1980, Algèbre. Chapitre 10. Algèbre homologique.
5. J. P. Brennan, M. V. Pinto, and W. V. Vasconcelos, *The Jacobian module of a Lie algebra*, Trans. Amer. Math. Soc. **321** (1990), no. 1, 183–196.
6. Richard Cushman and Reyer Sjamaar, *On singular reduction of Hamiltonian spaces*, Symplectic geometry and mathematical physics (Aix-en-Provence, 1990) (P. Donato, C. Duval, J. Elhadad, and G. M. Tuynman, eds.), Progr. Math., vol. 99, Birkhäuser Boston, Boston, MA, 1991, pp. 114–128.
7. Giuseppe Dito and Daniel Sternheimer, *Deformation quantization: genesis, developments and metamorphoses*, Deformation quantization (Strasbourg, 2001), IRMA Lect. Math. Theor. Phys., vol. 1, de Gruyter, Berlin, 2002, pp. 9–54.
8. P. Domański and B. Jakubczyk, *Linear continuous division for exterior and interior products*, Proc. Amer. Math. Soc. **131** (2003), no. 10, 3163–3175 (electronic).

9. Boris Fedosov, *Deformation quantization and index theory*, Mathematical Topics, vol. 9, Akademie Verlag, Berlin, 1996.

10. ______, *A simple geometrical construction of deformation quantization*, J. Differential Geom. **40** (1994), no. 2, 213–238.

11. ______, *Non-abelian reduction in deformation quantization*, Lett. Math. Phys. **43** (1998), no. 2, 137–154.

12. Viktor L. Ginzburg, *Equivariant Poisson cohomology and a spectral sequence associated with a moment map*, Internat. J. Math. **10** (1999), no. 8, 977–1010.

13. Melvin Hochster, *Topics in the homological theory of modules over commutative rings*, Published for the Conference Board of the Mathematical Sciences by the American Mathematical Society, Providence, R.I., 1975.

14. Johannes Huebschmann, *Lie-Rinehart algebras, descent, and quantization*, Galois theory, Hopf algebras, and semiabelian categories, Fields Inst. Commun., vol. 43, Amer. Math. Soc., Providence, RI, 2004, pp. 295–316.

15. Bertram Kostant and Shlomo Sternberg, *Symplectic reduction, BRS cohomology, and infinite-dimensional Clifford algebras*, Ann. Physics **176** (1987), no. 1, 49–113.

16. Larry Lambe and Jim Stasheff, *Applications of perturbation theory to iterated fibrations*, Manuscripta Math. **58** (1987), no. 3, 363–376.

17. Eugene Lerman, Richard Montgomery, and Reyer Sjamaar, *Examples of singular reduction*, Symplectic geometry (Coventry, 1990) (Dietmar Salamon, ed.), London Math. Soc. Lecture Note Ser., vol. 192, Cambridge Univ. Press, Cambridge, 1993, pp. 127–155.

18. Dmitrii I. Panyushev, *The Jacobian modules of a representation of a Lie algebra and geometry of commuting varieties*, Compositio Math. **94** (1994), no. 2, 181–199.

19. Markus J. Pflaum, *On the deformation quantization of symplectic orbispaces*, Diff. Geometry and its Applications **19** (2003), 343–368.

20. Walter Rudin, *Functional analysis*, second ed., International Series in Pure and Applied Mathematics, McGraw-Hill Inc., New York, 1991.

21. Jesús M. Ruiz, *The basic theory of power series*, Advanced Lectures in Mathematics, Friedr. Vieweg & Sohn, Braunschweig, 1993.

22. Alexey Sevostyanov, *Reduction of quantum systems with arbitrary first class constraints and Hecke algebras*, Comm. Math. Phys. **204** (1999), no. 1, 137–146.

23. Reyer Sjamaar and Eugene Lerman, *Stratified symplectic spaces and reduction*, Ann. of Math. (2) **134** (1991), no. 2, 375–422.

24. Jim Stasheff, *Homological reduction of constrained Poisson algebras*, J. Differential Geom. **45** (1997), no. 1, 221–240.

25. Jean-Claude Tougeron, *Idéaux de fonctions différentiables*, Springer-Verlag, Berlin, 1972, Ergebnisse der Mathematik und ihrer Grenzgebiete, Band 71.

Differentiability and Composite Functions

S. BROMBERG

Departamento de Matemáticas, Universidad Autónoma Metropolitana-Iztapalapa,
México, D.F., MEXICO
E-mail: stbs@xanum.uam.mx

S. LOPEZ DE MEDRANO

Instituto de Matemáticas, Universidad Nacional Autónoma de México
México, D.F., MEXICO
E-mail: santiago@matem.unam.mx

Pour J-P. Brasselet à l'occasion de son 60ème anniversaire.

The aim of this article is to study the increase of the class of differentiability
of a function when composed with a smooth mapping. For some important
special cases, we give the exact bound for that gain.

Keywords: Differentiability class; composite functions; blow-up; quasihomogeneous maps; Newton map.

1. Introduction

Consider a C^∞ map $\Theta : M^n \to \mathbb{R}^p$ where M is a smooth manifold. Let
$X := \Theta(M)$.

In this article we address the following question:

Given a continuous function $f : \mathbb{R}^p \to \mathbb{R}$ such that $\hat{f} := f \circ \Theta$ is of class
C^K, what can be said about the differentiability class of f?

Some simple precisions are pertinent: First, no conclusion about the differentiability of f outside X is possible from information about $\hat{f}$. Therefore, we shall consider only a continuous function $f : X \to \mathbb{R}$ and ask whether it belongs or not to $C^s(X)$, which is the space of functions on X that can be extended to C^s functions on $\mathbb{R}^p$.

Second, for f to belong to $C^s(X)$ there is an obvious necessary condition on the s−jet of the function $\hat{f}$:

(F_s) : *For every $a \in X$, the s−jet of $\hat{f}$ in $\Theta^{-1}(a)$ comes from the s−jet* *of some C^s function in the neighborhood of a.*

We will see that this necessary *formal* condition is not sufficient and that a higher differentiability class of $\hat{f}$ is required (to begin with).

To reformulate our question[a] we define a function $K(s)$ (depending only on Θ) to be the smallest value of K satisfying:

(*) For every continuous $f : X \to \mathrm{IR}$ such that $\hat{f}$ is of class C^K and satisfies condition (F_s), f is in $C^s(X)$.

Our problem is to find the value of $K(s)$, which we solve in some cases:

(a) the case where Θ is a blow-down transformation over the real or complex numbers, where our results show that $K(s) = 2s$ (Theorems 2.1 and 2.2);

(b) For quasihomogeneous maps $\Theta : \mathrm{IR}^n \to \mathrm{IR}^p$ satisfying certain conditions we can give a lower bound for $K(s)$ (Theorem 3.1). In some important cases we can combine this bound with known results to determine $K(s)$, and thus show that those results are best possible. We illustrate this with the Newton map given by the elementary symmetric functions on n variables, for which we show $K(s) = ns$, using a theorem of Barbançon [1];

(c) the case where Θ is the mapping of the plane given in a complex coordinate by $z \to z^m$, where we show that the lower bound in b) is not the actual value. Actually, no value of K, not even ∞, satisfies condition (*) above. We give nevertheless a positive result involving not the $s-$jet but the $ms-$jet of $\hat{f}$ (Theorem 3.2).

This discussion started with the analysis of the map sending polar coordinates in the plane into Cartesian ones, a particular case of a) above which we will recall.

Related questions were considered by Bierstone, Milman and Pawłucki (Theorem 1.2 in [2]), where certain constants ℓ_s are introduced that have some resemblance with our $K(s)$, but with several differences. The main ones are:

(1) the interest in that article is to show the existence of such finite constants for a certain type of analytic mappings and not to find their optimal value in any specific example, and

---

[a]In many of the cases considered in this article the mappings are surjective ($X = \mathrm{IR}^p$) and regular outside $\Theta^{-1}(0)$, so the statements of the theorems will be simpler.

(2) their definition involves the stronger formal condition (F_K), that is, on the whole K−jet of the C^K function $\hat{f}$ and not only on its s−jet.

Our results give bounds for the optimal values of ℓ_s under some conditions. In cases a), b), c) above we can give the exact optimal values, due to the fact that our examples happen to satisfy (F_K).

It can be shown that, for a general semiproper and generically submersive analytic mapping Θ, the optimal value of ℓ_s is less or equal to $K(s)$. In cases a) and b) above they are equal, but in case c) they differ widely. We are just beginning to look for some general conditions under which those values agree or the lower bound in b) is optimal.

We thank the referee for pointing this article to us and for his careful and constructive reading of our first vague manuscript.

2. Differentiability and blow-up

Let $\Bbbk$ be either the real or complex number field. We will consider the projective space $P_{\Bbbk}^{n-1}$ of lines through the origin in $\Bbbk^n$ and $\xi_{\Bbbk}$ the tautological bundle over $P_{\Bbbk}^{n-1}$ whose total space $E(\xi_{\Bbbk})$ consists of the pairs (v, ℓ) where ℓ is in $P_{\Bbbk}^{n-1}$ and v is a vector in ℓ. Consider the blow-down mapping $\Theta : E(\xi_{\Bbbk}) \to \Bbbk^n$ given by $(v, \ell) \mapsto v$ which is a diffeomorphism outside $\Theta^{-1}(0) = P_{\Bbbk}^{n-1}$.

Theorem 2.1. *Let $f : \Bbbk^n \to I\!R$ be such that the function $\hat{f} := f \circ \Theta$ satisfies*

(i) $\hat{f}$ is of class C^{2s}
(ii) The s−jet of $\hat{f}$ is zero at the points of $\Theta^{-1}(0)$.

Then f is of class C^s and the s−jet of f is zero at the origin.

This result is optimal: There are functions f such that $\hat{f}$ is of class C^{2s-1} and the $(2s-1)$−jet of $\hat{f}$ is zero at the points of $\Theta^{-1}(0)$ but such that f is only of class C^{s-1}.

Proof. We start by proving the case $\Bbbk = I\!R$: Since the double cover of $E(\xi_{\Bbbk})$ is $I\!R \times S^{n-1}$ we can consider the composition $\tilde{\Theta}$ of Θ with the covering map, which is given by $\tilde{\Theta}(r, u) = ru$. The differentiability properties of functions in $I\!R^n$ after composition with Θ or with $\tilde{\Theta}$ are the same, so we can work with the latter. To avoid cumbersome notation we will drop the $\sim$ in $\tilde{\Theta}$ and still write $\hat{f}$ for the new composite function. Notice that now, since $\hat{f}(0, u) = 0$, the condition on the s−jet of $\hat{f}$ is really a condition on the partial derivatives of $\hat{f}$ with respect to r only.

We begin with a Lemma:

Lemma 2.1. *Let f be a function on $I\!R^n$ of class C^m which is m–flat at the origin and let H be a homogeneous function of degree $-q$ which is defined and of class C^{m-q} outside the origin, where q and $m - q \geq 0$. Then fH is of class C^{m-q} and $(m - q)$–flat at the origin.*

Proof of the Lemma. We will prove the Lemma by induction on $k :=$ $m - q$. For $k = 0$ it follows from the continuity of H on the unit sphere.

Assuming the statement of the Lemma for some k, suppose f of class C^m and m–flat at the origin and H homogeneous of degree $-q$ and of class C^{m-q} outside the origin, where $m - q = k + 1$. Then, writing Leibniz's rule for the partial derivatives of the product fH one verifies that each summand satisfies the induction hypothesis and is therefore of class C^k and k–flat at the origin. Then fH is of class C^{k+1} and $(k+1)$–flat at the origin.

$\square$

To prove the theorem for $\Theta(r, u) = ru$ observe first that $f(x) = \hat{f}(\|x\|, x/\|x\|)$ for $x \neq 0$, and $f(0) = 0$.

We proceed by induction on s. For $s = 0$, the assertion follows because Θ is proper.

Suppose the theorem valid for some s and suppose $\hat{f}$ is of class C^{2s+2} and $(s + 1)$–flat for $r = 0$. Let $\varphi_1(x) = \|x\|$, $\varphi_2(x) = x/\|x\|$ which are homogeneous of degrees 1 and 0, respectively. Now,

$$f(x) = \hat{f}(\varphi_1(x), \varphi_2(x)),$$

$$\frac{\partial f}{\partial x_j} = \frac{\partial \hat{f}}{\partial r}(\varphi_1, \varphi_2)\frac{\partial \varphi_1}{\partial x_j} + \frac{\partial \hat{f}}{\partial u}(\varphi_1, \varphi_2)\frac{\partial \varphi_2}{\partial x_j}.$$

Since $\partial \hat{f}/\partial r$ is of class C^{2s+1} and s–flat in $r = 0$, $\partial \hat{f}/\partial r(\varphi_1, \varphi_2)$ is of class C^s and s–flat at 0 by the induction hypothesis. By Lemma 2.1, it remains so when multiplied by a homogeneous function of degree 0 and the first summand of $\partial f/\partial x_j$ above is of class C^s and s–flat at 0.

On the other hand, $\partial \hat{f}/\partial u$ is also of class C^{2s+1} and s–flat in $r = 0$, so again by the induction hypothesis $\partial \hat{f}/\partial u(\varphi_1, \varphi_2)$ is of class C^s and s–flat at 0. But in the second summand above this term is multiplied by a homogeneous function of degree -1 and Lemma 2.1 applied directly would not give the expected answer.

Looking more closely we can see that $\partial \hat{f}/\partial u$ is actually $(s + 1)$–flat in $r = 0$ because, from $\hat{f}(0, u) = 0$ and the fact that $\hat{f}$ is $(s+1)$–flat at $r = 0$, any partial derivative of order $s+2$ of $\hat{f}$ involving u is 0 for $r=0$. Then,

$$\frac{\partial \hat{f}}{\partial u}(r,u) = r\hat{g}(r,u)$$

where $\hat{g}$ is of class C^{2s} and s-flat in $r=0$. Hence, by the induction hypothesis, $\hat{g}(\varphi_1,\varphi_2)$ is of class C^s and s-flat at 0. Now it is possible to apply Lemma 2.1 to

$$\frac{\partial \hat{f}}{\partial u}(\varphi_1,\varphi_2)\frac{\partial \varphi_2}{\partial x_j} = \hat{g}(\varphi_1,\varphi_2)\varphi_1\frac{\partial \varphi_2}{\partial x_j},$$

to conclude that the second summand of $\partial f/\partial x_j$ is of class C^s and s-flat at 0, since $\varphi_1 \partial \varphi_2/\partial x_j$ is homogeneous of degree 0.

Therefore both summands are of class C^s and s-flat at 0, so is $\partial f/\partial x_j$ and f itself is of class C^{s+1} and $(s+1)$-flat at 0, thus completing the induction step.

A function satisfying the condition in the second part of the theorem is the following:

$$f(x,y) = \frac{x^{2s+4}}{x^4 + \|y\|^2}$$

where we denote points in $\mathbb{R}^n$ as pairs (x,y) with $x \in \mathbb{R}$ and $y \in \mathbb{R}^{n-1}$.

The function f is quasihomogeneous in (x,y) with weights $(1,2)$ and degree $2s$. It is therefore of class C^{s-1} and $(s-1)$-flat at 0. Since its partial derivatives of order s with respect to y are quasihomogeneous of degree 0 and not constant, they are not continuous at 0 and therefore f is not of class C^s.

Then

$$\hat{f}(r,(x,y)) = \frac{r^{2s+2}x^{2s+4}}{r^2x^4 + \|y\|^2}$$

(for $(x,y) \in S^{n-1}$) which is of class C^∞ and $(2s-1)$-flat at $r=0$, except possibly where the denominator vanishes, i.e. when $r=0$, $y=0$ and $x=\pm 1$. Since it is even in (x,y) it is enough to consider what happens around the point $(0,(1,0))$.

To this end, consider the local parametrization $\Phi : \mathbb{R} \times \mathbb{R}^{n-1} \to \mathbb{R} \times S^{n-1}$ defined by:

$$\Phi(t,y) = (t\sqrt{1+\|y\|^2}, (\frac{1}{\sqrt{1+\|y\|^2}}, \frac{y}{\sqrt{1+\|y\|^2}}))$$

whose image is $\mathbb{R} \times S^{n-1} \cap \{x>0\}$ and whose inverse is $(r,(x,y)) \mapsto (rx, y/x)$. Then the local form of Θ in the chart Φ is $(t,y) \mapsto (t,ty)$

The local form of $\hat{f}$ in the chart Φ is

$$(t,y) \mapsto \hat{f}(\Phi(t,y)) = \frac{t^{2s+2}}{t^2 + \|y\|^2}$$

which is a homogeneous function of degree $2s$ and therefore of class C^{2s-1} and $(2s-1)$–flat at $t = 0$. Therefore, $\hat{f}$ is everywhere of class C^{2s-1} and $(2s-1)$–flat at $r = 0$.

Being even in (x,y), $\hat{f}$ factors through $E(\xi_\mathbb{k})$ and also gives an example for the blow-down map.

To prove the result when $\mathbb{k}$ is the complex field, notice that the question about the differentiability of compositions with Θ is, as previously, unchanged if we compose further with the natural S^1-bundle map $\mathbb{C} \times S^{2n-1} \to E(\xi_\mathbb{C})$ so we can substitute Θ by the composition. Then, if the composite map $\hat{f} : \mathbb{C} \times S^{2n-1} \to \mathbb{R}$ is of class C^{2s} so is the restriction to $\mathbb{R} \times S^{2n-1}$ and we can apply to the latter the proof for the case $\mathbb{k} = \mathbb{R}$ and conclude that f is of class C^s.

The example for this case can be given in a similar way as in the real case. $\qquad\qquad\square$

Other blow-up transformations, such as the one given in [4], can be treated in the same way.

The idea of using quasihomogeneous functions to give examples that increase their degree of differentiability when composed with a transformation can be used in other situations like those described in the following section.

In Theorem 2.1 the condition on the s–jet of f at the origin is not necessary; to state the exact condition we reformulate the theorem, in the real non-projective situation only, as follows:

Theorem 2.2. *Let $f : \mathbb{R}^n \to \mathbb{R}$ be such that the function $\hat{f}(r,u) := f(ru)$ is of class C^{2s}. Then f is of class C^s if, and only if, $\dfrac{\partial^j \hat{f}}{\partial r^j}(0,u)$ is a homogeneous polynomial of degree j on u, for $0 \le j \le s$.*

Proof. Suppose first that f is of class C^s. Then

$$f(x) = \sum_{|\alpha| \le s} c_\alpha x^\alpha + o(\|x\|^s).$$

Hence

$$\hat{f}(r, u) = \sum_{|\alpha| \le s} c_\alpha u^\alpha r^{|\alpha|} + o(r^s) = \sum_{j=0}^{s} P_j(u) r^j + o(r^s).$$

And $P_j(u)$, which is the claimed partial derivative, is a homogeneous polynomial of degree j.

Conversely, if

$$\hat{f}(r, u) = \sum_{j=0}^{s} P_j(u) r^j + o(r^s),$$

Then,

$$\hat{f}(r, u) - \sum_{j=0}^{s} P_j(u) r^s$$

is $2s$–flat at $r = 0$ and by Theorem 2.1 $f(x) - \sum_{j=0}^{s} P_j(x)$ is of class C^s and so is f. $\qquad \square$

Remark 2.1. The condition on the polynomial dependence of the jet of $\hat{f}$ on $\Theta^{-1}(0)$ (as well as the Fourier polynomial condition in Theorem 2.3 below) are clearly equivalent to condition (F_s) of the introduction in this case, since Θ is regular outside $\Theta^{-1}(0)$.

Remark 2.2. Even though we ask that $\hat{f}$ is of class C^{2s}, only restrictions on the s–jet are necessary. The residue of order $> s$ of $\hat{f}$ can be arbitrary, as can been seen by applying Theorem 2.1 to the terms of higher order. In the notations of the introduction and of [2] we have shown that for the blow-down maps

$$optimal\ \ell_s = K(s) = 2s.$$

Remark 2.3. The case $\Bbbk = \mathbb{R}$ and $n = 2$ of the above theorems can be reformulated in terms of polar coordinates (See [3]):

Theorem 2.3. *Let* $f : \mathbb{R}^2 \to \mathbb{R}$ *be such that the function* $\hat{f}(r, \theta) := f(r\cos\theta, r\sin\theta)$ *is of class* C^{2s}. *Then* f *is of class* C^s *if, and only if, the j-th derivative of* $\hat{f}$ *with respect to* r *at* $r = 0$ *is a homogeneous Fourier polynomial of degree* j *in* θ *for all* $j \le s$.

This result is optimal: There are functions f *such that* $\hat{f}$ *is of class* C^{2s-1} *and the* $(2s-1)$–*jet of* $\hat{f}$ *is zero when* $r = 0$ *but such that* f *is only of class* C^{s-1}.

The question of the relation between the degree of differentiability of a function with respect to its degree of differentiability when expressed in polar coordinates was raised in a first version of the Ph.D. thesis by Samaniego [8]. This was used to give a first answer to a question of Marc Chaperon regarding the degree of differentiability of the manifold of periodic points of a certain family of transformations of a Banach space. The fact that the degree of differentiability can drop drastically when passing from polar to Cartesian coordinates forced the search for a different proof. In [8] and [5] a better result was proved by avoiding the passage through polar coordinates when considering differentiability questions.

The questions about differentiability and polar coordinates, as well as the extensions considered in the present article, seem nevertheless interesting in themselves, and do not seem to have been explored in the literature.

The C^∞ version of this theorem was obtained by G. Glaeser [6] and by Kazdan and Warner [7]. We thank Alain Chenciner and Daniel Meyer for pointing to us those references.

Remark 2.4. The corresponding results are not true for the infinite dimensional situation, due essentially to the fact that the corresponding maps are not proper: In ℓ_2, for any K there are analytic functions $\hat{f}$, K−flat at $r = 0$ and such that f is only continuous.

The same happens in the finite dimensional case for the improper map $(t, x) \to (t, tx)$ which is the local form of the map Θ.

3. Differentiability and quasihomogeneous maps

3.1. *General bounds for a class of quasihomogeneous maps*

Consider a quasihomogeneous map $\Theta : \mathbb{R}^n \to \mathbb{R}^p$ with weights $m_1, ..., m_n$ and components of degrees $d_1, ..., d_p$, where m_i and d_i are positive integers. Let $m = max\{m_1, ..., m_n\}$, $d = max\{d_1, ..., d_p\}$ and $g = d/m$. For a real x denote by $[[x]]$ the largest integer strictly smaller than x. (Thus, if x is an integer, $[[x]] = x - 1$.)

Assume that $\Theta^{-1}(0) = 0$ and that X has non-empty interior.

Theorem 3.1. *There are functions $f : \mathbb{R}^n \to \mathbb{R}$ of class C^∞ outside the origin such that $\hat{f}$ is of class $C^{[[gs]]}$ and the $[[gs]]$−jet of $\hat{f}$ is zero at 0 but such that f is onle of class C^{s-1}.*

Proof. It is based on the following Lemma:

Lemma 3.1. *There is a positively quasihomogeneous function $f : \mathbb{R}^p \to \mathbb{R}$*

with weights $d_1, ..., d_p$ and any given degree which is of class C^∞ *outside the origin, but not a polynomial.*

(To show this take any non-polynomial C^∞ function on the unit sphere of $\mathbb{R}^p$ and extend it to the whole space in a quasihomogeneous way.)

For degree ds the differentiability class of this function is $s-1$, since the d-th partial derivative with respect to one of the variables is quasihomogeneous of degree 0 and not constant, therefore not continuous at 0. Its limit can be taken along a path in the interior of X so it also follows that f is not in $C^s(X)$. But the composite function $\hat{f}$ is quasihomogeneous of degree ds and all its partial derivatives of order strictly less than $ds/m = gs$ are quasihomogeneous of positive degree and therefore they are continuous and vanish at the origin. Outside the origin $\hat{f}$ is of class C^∞, since $\Theta^{-1}(0) = 0$. The theorem follows. $\square$

Remark 3.1. For a homogeneous mapping, $m = 1$, the function $\hat{f}$ is of class C^{ds-1}.

Remark 3.2. The maps $(x, y) \to (x^{2m+1} - xy, y)$ provide interesting examples where gs is not an integer.

Remark 3.3. In the notations of the introduction and of [2] we have shown that, for Θ satisfying the above hypotheses, $[[gs]] + 1$ is a lower bound for both the optimal ℓ_s and $K(s)$.

3.2. *Some special homogeneous maps. The Newton map*

In some special interesting cases there are known results about the differentiability of f. Combined with our lower bound above they can give precise results. We illustrate this with the Newton mapping of order n which is homogeneous with degrees $1, \dots, n$.

A theorem of Barbançon in [1] shows that if $\hat{f}$ is of class ns then f is in $C^s(X)$. In this case no condition on the jet of $\hat{f}$ has to be given explicitly, since it is implicit in the symmetry of $\hat{f}$.

The above construction gives examples showing that this is best possible and therefore for this map

$$optimal\ \ell_s = K(s) = 2s.$$

In other words, we have shown that for all n there is a symmetric function of differentiability class $ns - 1$ that cannot be expressed as a C^s function of the elementary symmetric functions, thus showing that Barbançon's theorem is best possible. (In [1] only examples for $n = 2$ are given.)

3.3. *Holomorphic mappings of the plane*

Consider for $m > 1$ an integer the holomorphic map $\Theta_m : \mathrm{IR}^2 \to \mathrm{IR}^2$ given in complex notation by

$$\Theta_m(z) = z^m.$$

We will see that in this case the analog of Theorem 2.1 is not valid because we need to consider a higher order jet:

Theorem 3.2. *Let* $f : \mathrm{IR}^2 \to \mathrm{IR}$ *be such that the function* $\hat{f} := f \circ \Theta_m$ *satisfies*

(i) $\hat{f}$ *is of class* C^{ms},
(ii) The $ms-$*jet of* $\hat{f}$ *is zero at the origin.*

Then f *is of class* C^s *and* $s-$*flat at the origin.*
*   There are functions* f *not of class* C^s *such that* $\hat{f}$ *is of class* C^∞ *and the* $s-$*jet of* $\hat{f}$ *is zero at the origin.*

Proof. We will prove the assertion by induction on s. For $s = 0$ the assertion states that $f(w)$ is continuous when $\hat{f}(z) = f(z^m)$ is so, which is trivial. Assuming the result true for s, let $\hat{f}$ be of class $\mathrm{C}^{m(s+1)}$ and $m(s+1)-$flat at the origin.

Using the complex notation $\partial/\partial z = (1/2)(\partial/\partial x - i\partial/\partial y)$ we have:

$$\frac{\partial \hat{f}}{\partial z}(z) = \frac{\partial f}{\partial w}(z^m) m z^{m-1}.$$

So,

$$\widehat{\frac{\partial f}{\partial w}}(z) = \frac{\partial \hat{f}}{\partial z}(z) \frac{1}{m z^{m-1}}$$

Since $\partial \hat{f}/\partial z$ is of class $\mathrm{C}^{m(s+1)-1}$ and $1/(m z^{m-1})$ is a homogeneous function of degree $-(m-1)$, Lemma 2.1, applied componentwise, implies that the left-hand side is of class C^{ms} and $ms-$flat at the origin. By the induction hypothesis, applied also componentwise, $\partial f/\partial w$ is of class C^s and $s-$flat at the origin. Therefore f itself is of class C^{s+1} and $(s + 1)-$flat at the origin, and the induction step is done.

The example is as follows:

For m even take

$$f(w) = |w| Re(w^{s-1}).$$

This function is homogeneous of degree s and is not a polynomial. It is therefore not of class C^s. But

$$\hat{f}(z) = |z|^m Re(z^{m(s-1)})$$

is a polynomial and therefore actually analytic. Being homogeneous of degree ms it is $(ms - 1)$–flat at the origin.

For m odd take

$$f(w) = |w|^{1-1/m} Re(w^{s-1}).$$

This function is homogeneous of degree $s - 1/m$. It is therefore not of class C^s. But

$$\hat{f}(z) = |z|^{m-1} Re(z^{ms-m})$$

is a polynomial and therefore actually analytic. Being homogeneous of degree $ms - 1$ it is $(ms - 2)$–flat at the origin and $(ms - 2) \geq s$. $\quad\square$

In this case the formal necessary condition for f being C^s involves more than the s–jet of $\hat{f}$. Actually, the whole ms–jet of $\hat{f}$ is determined by the s–jet of f and has to be very special.

In the notations of the introduction and of [2] we have shown that in these cases the optimal ℓ_s is (at least when m is even) ms but that $K(s)$ cannot even be taken to be ∞. $\quad\square$

References

1. G. Barbançon, *Théorème de Newton pour les fonctions de classe C^r*, Ann. Scient. Éc. Norm. Sup. **5**, 435 (1972).
2. E. Bierstone, P. D. Milman and W. Pawłucki, *Composite differentiable functions*, Duke Math. J. **83** (3), 607 (1996).
3. S. Bromberg, S. López de Medrano and J.L. Samaniego, *A remark on differentiability in polar coordinates*, (to appear).
4. M. Chaperon and F. Coudray, *Invariant manifolds, conjugacies and blow-up*, Ergodic Theory and Dynamical Systems **17** 783 (1997).
5. M. Chaperon, S. López de Medrano and J.L. Samaniego, *On sub-harmonic bifurcations*, C.R.Acad. Sci. Paris, Ser.I **340** 827 (2005).
6. G. Glaeser, *Fonctions composées différentiables*, Ann. Math. **77** 193 (1963).
7. J.L. Kazdan and F.W. Warner, *Curvature functions for open 2-manifolds*, Ann. Math. **99** 203 (1974).
8. J.L. Samaniego, *Sobre el grado de diferenciabilidad de la variedad de puntos p–periódicos de una familia de transformaciones de un espacios de Banach*, (Ph.D. thesis, UNAM, August 2005).

CONTACT STRUCTURES AND NON-ISOLATED SINGULARITIES

CLÉMENT CAUBEL

Univ. Paris 7 Denis Diderot, Inst. de Maths.-UMR CNRS 7586,
équipe "Géométrie et dynamique"
case 7012, 2, place Jussieu, 75251 Paris cedex 05, France.
E-mail: caubel@math.jussieu.fr

A contact manifold is *smoothly Milnor fillable* if it is contactomorphic to the contact boundary of the Milnor fibre of a germ with no blowing up. In this note, this notion is compared with the previously defined *singular Milnor fillability*, both on topological and contact levels. In particular, an example where the two notions give rise to non isomorphic contact structures on the same 3-manifold is given.

Keywords: Singularities, contact structures

1. Introduction

Let $(X, x) \subset (\mathbb{C}^N, \mathbf{0})$ be a germ of complex analytic variety having an isolated singularity. Since [V], one knows how to associate to (X, x) a well defined contact structure $\xi(X)$ on its boundary $M(X) := X \cap S_\varepsilon^{2N-1}$, $\varepsilon \ll 1$. In [CNP], we defined a contact manifold to be *Milnor fillable* if it is contactomorphic to such a contact boundary. We then got the following result: any oriented 3-manifold admits at most one Milnor fillable contact structure up to contactomorphism.

For a singularist, this result may seem somewhat frustrating: among all the different contact structures on a given 3-manifold, which are the subject of intense research for twenty years, there is only one (at most!) which is naturally related to singularity theory!

The goal of this paper is then to introduce another family of contact manifolds linked to singularity theory: the boundaries of the *Milnor fibres* of some (possibly) *non isolated* singularities. These will be the *smoothly Milnor fillable* contact manifolds (see definition 3.1), to be compared with the *singularly* Milnor fillable ones defined earlier.

476

In section 2, I introduce and compare different notions of fillability for
manifolds only, regardless of any contact structure. Examples prove that
the consideration of both smoothly and singularly Milnor fillable manifolds
enriches the family of manifolds arising from singularity theory, but the
question whether the smooth Milnor fillability is weaker than the singular
one is still open.

Section 3 introduces the smooth Milnor fillability for contact manifolds.
I then show on an example that there can be at least two non isomorphic
contact structures on the same 3-manifold which both arise from singularity
theory (see proposition 3.1).

The paper ends with the study of the following problem, raised by
P. Popescu-Pampu: are there naturally defined Milnor open books in the
boundary of the Milnor fibre of a non-isolated singularity which support the
natural contact structure? I provide here a positive answer to this question
under natural conditions on the considered singularity. The Milnor open
books were an essential ingredient in the proof of the main result of [CNP],
but in this new situation they seem much less tractable: perhaps one should
make (or wait) some progress along the lines of [MP] in the comprehension
of the topology of the boundaries of Milnor fibres.

2. Smooth Milnor fillability

2.1. *Definitions*

In [CNP], a manifold is said to be *(singularly) Milnor fillable* if it is diffeomorphic to the boundary $M(X)$ of some germ (X, x) of normal complex analytic variety with an isolated singularity. We give here a definition
which takes into account the case of *non-isolated* singularities. The point
is to associate to such a non-isolated singularity the boundary of one of its
smoothings. We consider the following wide family of singularities, where
this procedure is natural (it was first introduced in [HMS]).

Definition 2.1. Let $f : (\mathbb{C}^{n+p}, \mathbf{0}) \to (\mathbb{C}^p, \mathbf{0})$ be a holomorphic germ defining a complete intersection singularity (non necessarily isolated). We say
that *f has no blowing-up (in codimension 0)* if its zero set admits a stratification satisfying Thom's (a_f)-condition: for any sequence (p_n) of regular
points of f tending to a point p in a stratum $\Sigma \subset f^{-1}(\mathbf{0})$, the limiting
tangent space $\lim T_{p_n} f^{-1}(f(p_n))$ to the fibres, when it exists, contains the
tangent space $T_p \Sigma$ to the stratum.

Example 2.1. a) The function germs $f : (\mathbb{C}^{n+1}, \mathbf{0}) \to (\mathbb{C}, 0)$, the germs defining isolated complete intersection singularities, or the germs (f, ℓ) : $(\mathbb{C}^{n+p}, \mathbf{0}) \to (\mathbb{C}^p \times \mathbb{C}^q, \mathbf{0})$ with f with no blowing-up and ℓ a generic linear projection w.r.t. f, are all examples of germs with no blowing-up.
b) The map $(x, y, z) \mapsto (xy - z^2, x)$ has blowing-up.

If f has no blowing-up, it defines a generalized Milnor fibration in the following sense (see [Ca]): for all $0 < \eta \ll \varepsilon \ll 1$, the restriction

$$\phi_{\varepsilon, \eta}(f) : B_\varepsilon^{2n+2p} \cap f^{-1}(B_\eta^{2p} \setminus \mathrm{Disc}\, f) \xrightarrow{f} B_\eta^{2p} \setminus \mathrm{Disc}\, f$$

is a smooth fibration. Here $\mathrm{Disc}\, f$ denotes the *discriminant* of f, that is, the image by f of its critical locus $\mathrm{Crit}\, f$. The fibre $F_{\varepsilon, s}(f) := \phi_{\varepsilon, \eta}(f)^{-1}(s)$, for any $s \in B_\eta^{2p} \setminus \mathrm{Disc}\, f$, is the *Milnor fibre* of f: this is a n-dimensional Stein manifold with boundary, whose diffeomorphism type does not depend on $0 < |s| \ll \varepsilon \ll 1$, nor on the choice of local analytic coordinates. Its boundary $M(f) = M_{\varepsilon, s}(f) := \partial F_{\varepsilon, s}(f)$ is the *Milnor boundary* of f. It is a smooth closed oriented $(2n - 1)$-dimensional real manifold, which only depends on the analytic type of f.

Definition 2.2. The $(2n - 1)$-dimensional manifold M is *smoothly Milnor fillable* if it is diffeomorphic to the Milnor boundary $M(f)$ of some germ $f : (\mathbb{C}^{n+p}, \mathbf{0}) \to (\mathbb{C}^p, \mathbf{0})$ with no blowing up.

2.2. *Singular vs. smooth Milnor fillability*

The two notions of Milnor fillability we have just given are in fact really different, as the following example shows.

Example 2.2. a) The 3-torus $\mathbb{T}^3$ is smoothly Milnor fillable but not singularly. It is indeed the Milnor boundary of the polynomial $f(x, y, z) = xyz$, but it cannot be the boundary of a surface singularity, as D. Sullivan first showed (see [Du] and [S]).
b) For any $k \geq 0$, the connected sum $\#_k S^2 \times S^1$ is the Milnor boundary of the polynomial $h_k(x, y, z) := x^{k+1} + y^2$. Since the boundaries of surface singularities are known to be irreducible ([Neu]), these connected sums are not singularly Milnor fillable. We will come back to this example later.
c) For any $k \geq 1$, the lens space $L(2k, 1)$ is Milnor fillable in both senses: it is indeed the Milnor boundary of the polynomial $f_k(x, y, z) := xy^k + z^2$ (see [MPW]) and the boundary of the Hirzebruch-Jung singularity $X_{2k,1}$, whose embedding dimension is $2k + 1$. We will also come back to this example later.

Conversally, if a normal surface singularity is the zero set of an isolated complete intersection singularity, it is then clear that its boundary is Milnor fillable in both senses. But there are restrictions on the topology of such singularities, or more generally on *smoothable* normal surface singularities. For instance, any circle bundle over the torus $\mathbb{T}^2$ with Euler number $e < -9$ is the boundary of a non-smoothable normal surface singularity (see [W]). But nothing prevents *a priori* this manifold from being the Milnor boundary of a *non-isolated* surface singularity. Thus one may ask the following question:

Question 2.1. Does the singular Milnor fillability imply the smooth Milnor fillability?

2.3. *Smooth Milnor fillability vs. Stein fillability*

More generally, it is clear from the definition that any smoothly Milnor fillable manifold is *Stein fillable*, that is, is diffeomorphic to the boundary of a Stein domain in an affine space. It is also true for singularly Milnor fillable 3-manifolds (see [CNP]).

Now Gompf in [Go] gives a characterisation of Stein fillable 3-manifolds in terms of surgery diagrams. In particular, any manifold obtained by surgery along a given link in the 3-sphere and with sufficiently negative coefficients is Stein fillable. This, combined with the work of Thurston [T], gives a wealth of Stein fillable 3-manifolds admitting a hyperbolic metric (e.g. any manifold obtained by surgery along the figure-eight knot in the 3-sphere and with sufficiently negative coefficient).

But it is shown in [MP] that the Milnor boundaries of functions $f :$ $(\mathbb{C}^3, 0) \to (\mathbb{C}, 0)$ always admit a Waldhausen decomposition, and the proof given there may be adapted to larger ambient dimensions. This would prove that there are a lot of Stein fillable 3-manifolds which are not Milnor fillable in both senses.

Remark 2.1. Considering the 3-dimensional case only, we have mentioned a topological characterisation of Stein fillability. The singular Milnor fillability also admits such a characterisation in terms of plumbing (see [Neu]). A natural problem is then to find an analogous statement for smooth Milnor fillability.

3. Smoothly Milnor fillable contact manifolds

3.1. *Definition*

Let $f : (\mathbb{C}^{n+p}, \mathbf{0}) \to (\mathbb{C}^p, \mathbf{0})$ be a germ of map with no blowing up. We have just showed how to associate to f a well-defined Milnor boundary $M(f)$. But, like in the case of isolated singularities developed in [CNP], we can endow this boundary with a well defined contact structure.

For this, notice that $M(f)$ is defined as a smooth level set of the function $\rho : \mathbf{z} \mapsto |\mathbf{z}|^2$ in the complex manifold $f^{-1}(s)$, with s a sufficiently small regular value of f. Since ρ is a plurisubharmonic function, the complex hyperplane distibution $\xi(f) := TM(f) \cap i.TM(f) \subset TM(f)$ indeed defines a contact structure on $M(f)$, and this contact structure does not depend on the choices, nor on the analytic coordinates (see [CNP] and [CT] for details in related cases). This contact manifold $(M(f), \xi(f))$ is the *contact boundary* of the germ $f : (\mathbb{C}^{n+p}, \mathbf{0}) \to (\mathbb{C}^p, \mathbf{0})$.

Definition 3.1. The $(2n - 1)$-dimensional contact manifold (M, ξ) is *(smoothly) Milnor fillable* if it is contactomorphic to the contact boundary $(M(f), \xi(f))$ of some germ $f : (\mathbb{C}^{n+p}, \mathbf{0}) \to (\mathbb{C}^p, \mathbf{0})$ with no blowing up.

3.2. *Non unicity*

The preceding notion is related to the (singular) Milnor fillability for contact manifolds, which just means for a given contact manifold to be isomorphic to the contact boundary of some isolated singularity. In [CNP], the following main result is proved: *any singularly Milnor fillable 3-manifold admits a unique singularly Milnor fillable contact structure up to contactomorphism.* One may then ask if this unicity result can be extended to the non-isolated case. In other words, for a given Milnor fillable 3-manifold in both senses, is there a smoothly Milnor fillable contact structure on it which is different from the singularly Milnor fillable one? The following proposition gives an answer in a particular case.

Proposition 3.1. *Fix any $k \geq 2$. Denote by $X_{2k,1}$ the cyclic quotient singularity corresponding to the scalar multiplication by $\exp(i\pi/k)$ in $\mathbb{C}^2$ (this is the cone over the rational normal curve of degree $2k$ in $\mathbb{C}P^{2k}$). Put also $f_k(x, y, z) := xy^k + z^2$. Then the contact boundaries $(M(X_{2k,1}), \xi(X_{2k,1}))$ and $(M(f_k), \xi(f_k))$, which are both diffeomorphic to the lens space $L(2k, 1)$, are not contactomorphic.*

480

Proof. We first sketch why $M(f_k) \simeq L(2k, 1)$ (see [MPW] for a more general case). Replacing round spheres by polydiscs we can make the following identification:

$$M(f_k) \simeq F(f_k) \cap \partial(B^4_{\varepsilon_1} \times B^2_{\varepsilon_2}) = M_1 \cup M_2$$

where

$$M_1 := \{(x, y, z) \mid xy^k = \eta - z^2, \; |x|^2 + |y|^2 = \varepsilon_1, \; |z|^2 \leq \varepsilon_2\}$$

and

$$M_2 := \{(x, y, z) \mid xy^k = \eta - z^2, \; |x|^2 + |y|^2 \leq \varepsilon_1, \; |z|^2 = \varepsilon_2\},$$

with $0 < \eta \ll \varepsilon_2 \ll \varepsilon_1 \ll 1$. Now M_2 can be parameterized by y in an annulus and $\arg z \in S^1$: it is then a thickened torus, fibered by circles corresponding to the different values of $(|y|, \arg z)$. M_1 is a branched covering of a neighborhood of the link of xy^k in $S^3_{\varepsilon_1}$. It has two components:
– the first, about $\{x = 0\}$, corresponding to large values of $|y|$, is a solid torus parameterized by z in a disc and $\arg y$;
– the second, about $\{y = 0\}$, is a Seifert bundle over the disc parameterized by z with two singular fibres corresponding to $z = \pm\sqrt{\eta}$, each having a neighborhood saturated like that of $\{y = 0\} \cap S^3_{\varepsilon_1}$ by the links $\{xy^k = \sigma\} \cap S^3_{\varepsilon_1}$, $0 < |\sigma| \ll \varepsilon_1$.

Glueing all theses foliated parts, we see that $M(f_k)$ is a Seifert manifold with base S^2, Euler number 0 and two singular fibres of type $(k, 1)$, which proves the result (see Theorem 4.4 in [JN]).

Now, we will distinguish the two contact boundaries $(M(X_{2k,1}), \xi(X_{2k,1}))$ and $(M(f_k), \xi(f_k))$. Notice first that the homotopy class of $\xi(X_{2k,1})$ as an oriented plane field is well defined *up to isotopy*, and not only up to isomorphism. This is a consequence of [CNP] and of [P], where it is proved that the natural plumbing decomposition of any singularly Milnor fillable 3-manifold is well defined up to isotopy. This said, it now suffices to show that the Chern classes $c_1(\xi(X_{2k,1}))$ and $c_1(\xi(f_k))$ are different elements of $H^2(L(2k, 1))$, since they will then lie in two different orbits of the action of $\mathrm{Diff}_+ L(2k, 1)$ on this cohomology group (the orbit of $c_1(\xi(X_{2k,1}))$ being reduced to one point).

First, the complex distribution $\xi(f_k)$ on $M(f_k)$ is induced by the complex structure on the Milnor fibre $F(f_k)$ of f_k, which is stably trivial. Thus $c_1(\xi(f_k)) = 0$ in $H^2(L(2k, 1)) \simeq \mathbb{Z}/2k\mathbb{Z}$.

Now, let $\widetilde{X} \to X_{2k,1}$ be a minimal good resolution. On $M(X_{2k,1}) \simeq \partial\widetilde{X}$, we have $T\widetilde{X}|_{M(X_{2k,1})} = \xi(X_{2k,1}) \oplus \mathbb{C}.\nu$, where ν denotes an outward normal field to $M(X_{2k,1})$ in $\widetilde{X}$. This shows that $c_1(\xi(X_{2k,1})) = c_1(T\widetilde{X}|_{M(X_{2k,1})})$. But $\widetilde{X}$ is a disc bundle over the exceptional divisor $E \simeq S^2$, with Chern class $-2k$ in $H^2(E) \simeq \mathbb{Z}$. Hence, following the homotopy equivalence $\widetilde{X} \simeq E$,

$$c_1(\widetilde{X}) = c_1(T\widetilde{X}|_E) \text{ in } H^2(\widetilde{X}) \simeq H^2(E) \simeq \mathbb{Z},$$

and furthermore $T\widetilde{X}|_E = TE \oplus N_{E/\widetilde{X}}$, which shows that

$$c_1(\widetilde{X}) = c_1(TE) + c_1(N_{E/\widetilde{X}}) = 2 + (-2k) = 2(1 - k).$$

If $i : M(X_{2k,1}) \hookrightarrow \widetilde{X}$ denotes the inclusion, then $i^* : H^2(\widetilde{X}) \to H^2(M(X_{2k,1}))$ corresponds to the natural projection $\mathbb{Z} \to \mathbb{Z}/2k\mathbb{Z}$, and thus

$$c_1(\xi(X_{2k,1})) = c_1(T\widetilde{X}|_{M(X_{2k,1})}) = i^*(c_1(\widetilde{X})) = 2 \neq 0 \text{ in } \mathbb{Z}/2k\mathbb{Z},$$

which ends the proof. $\qquad\qquad\qquad\qquad\qquad\qquad\qquad\qquad\qquad\qquad\qquad\square$

Remark 3.1. An analogous reasoning shows that whenever an isolated surface singularity X satisfies $K_{\widetilde{X}}^2 \notin \mathbb{Z}$, where $K_{\widetilde{X}}$ denotes the purely exceptional divisor having the same numerical properties as the canonical divisor of the good resolution $\widetilde{X} \to X$, then its contact boundary cannot be isomorphic to any smoothly Milnor fillable contact manifold.

3.3. *Milnor open books on smoothly Milnor fillable contact 3-manifolds*

The main result of [CNP] we have recalled earlier was proved using *Milnor open books*, that is, the natural open book decompositions of the boundary $M(X)$ associated to the functions $g : (X, x) \to (\mathbb{C}, 0)$ with an isolated singularity. These open books indeed carry (in the sense of Giroux [Gi], see [Et]) the natural contact structure $\xi(X)$, and thus give a natural way to study it.

We now give an analogous statement for some contact Milnor boundaries.

Proposition 3.2. *Let $f : (\mathbb{C}^{n+p}, 0) \to (\mathbb{C}^p, 0)$ be a germ of map with no blowing up. If the germ $g : (\mathbb{C}^{n+p}, 0) \to (\mathbb{C}, 0)$ defines with f an isolated complete intersection singularity in $\mathbb{C}^{n+p}$, then for $1 \gg \varepsilon \gg |s| > 0$ the restriction*

$$\arg g_{\varepsilon,s} : M_{\varepsilon,s}(f) \setminus (M_{\varepsilon,s}(f) \cap V(g)) \to S^1$$

of the argument of g to the Milnor boundary $M_{\varepsilon,s}(f) = f^{-1}(s) \cap S_\varepsilon^{2n+2p-1}$ of f defines an open book decomposition on it which carries its natural contact structure. This is the Milnor open book of g in the Milnor boundary of f.

Remark 3.2. The existence of a convenient g in the last statement imposes in fact quite strong restrictions on f: for instance, the critical locus of f must be at most one-dimensional. But for the surface case – which is the main context of interest – this statement seems reasonably general: it encompasses in particular the function case $f : (\mathbb{C}^3, \mathbf{0}) \to (\mathbb{C}, 0)$.

Moreover, like in the isolated singularity case, the fact that the open book carries the contact structure is a direct consequence of the proof that it is indeed an open book. Thus the hypotheses we have made on f are here merely to guarantee that any sufficiently general g w.r.t. f defines a Milnor open book in the Milnor boundary of f, regardless of any contact property.

Proof. We adapt the main steps of the proof of the corresponding result (Theorem 3.9) in [CNP] to our case. See that paper for further details and notations.

First, since $\Phi := (f, g)$ defines an isolated complete intersection singularity, there is a $\varepsilon_0 > 0$ such that, for all $0 < \varepsilon < \varepsilon_0$, there is a $\eta > 0$ such that all the fibers $\Phi^{-1}(s, t)$ with $|s| < \eta$ and $|t| < \eta$ cut the sphere $S_\varepsilon^{2n+2p-1}$ transversally. This implies that, for all $s \in B_\eta^{2p} \setminus \operatorname{Disc} f$, there is a η such that all the fibres $g^{-1}(t)$ with $|t| < \eta$ cut the Milnor boundary $M_{\varepsilon,s}(f)$ transversally. This settles our problem "near the binding".

Second, we can choose ε_0 small enough, so that for all $z \in V(f) \cap B_{\varepsilon_0}^{2n+2p} \setminus V(g)$ the following implication holds:

$$\underbrace{\exists (\lambda, c) \in \mathbb{C} \times \mathbb{C}^p \mid \nabla g(z) = i.\lambda.g(z)\nabla\rho(z) + \sum c_j.\nabla f_j(z)}_{(*)} \Rightarrow |\arg \lambda| < \frac{\pi}{4},$$

where ∇ denotes the gradient in $\mathbb{C}^{n+p}$ (this is the Proposition 3.8 of [CNP], adapted to our context). Indeed, suppose that there in an analytic path $\gamma : [0, \sigma[\to V(f)$ so that $\gamma(0) = \mathbf{0}$, $g(\gamma(\tau)) \neq 0$ and $\gamma(\tau)$ satisfies $(*)$ for all $\tau > 0$ but with $|\arg \lambda| \geq \pi/4$. There are two cases:
1) If the image of γ is in the critical locus $\operatorname{Crit} f$, then ∇f vanishes and $(*)$ becomes $\exists \lambda \mid \nabla g = i\lambda.g.\nabla\rho$. But the Milnor fibration theorem for g in $\mathbb{C}^{n+p}$ proves that in this case $|\arg \lambda| < \pi/4$, which is impossible.
2) If the image of γ is in the smooth locus $V(f)_{\text{smooth}}$, then one can project all the gradients on this smooth complex manifold, get the intrinsic version of $(*)$ and use Proposition 3.8 of [CNP] to show that it is also impossible.

Finally, suppose that there is a $\varepsilon < \varepsilon_0$ such that there is a sequence $(p_k) \to p_\infty$ in $S_\varepsilon^{2n+2p-1}$ with $|f(z_k)| \to 0$ and $(*)$ true for all the z_k, but with $|\arg \lambda| \geq \pi/4$. Then z_∞ must satisfy $(*)$ and $|\arg \lambda| \geq \pi/4$ but with $z_\infty \in V(f) \cap S_\varepsilon^{2n+2p-1}$. Now $g(z_\infty) \neq 0$, for otherwise we would have

$\mathrm{rk}(\nabla f(z_\infty), \nabla g(z_\infty), \nabla \rho(z_\infty)) < 3$ which is impossible in $S_\varepsilon \cap \Phi^{-1}(0,0)$ after the first paragraph of this proof. But the second paragraph then proves that $|\arg \lambda| < \pi/4$, which is also impossible. $\qquad\square$

Example 3.1. Take any reduced function $h : (\mathbb{C}^2, \mathbf{0}) \to (\mathbb{C}, 0)$, and consider it as a function in three variables: $f(x, y, z) := h(x, y)$. To identify the Milnor boundary $M(f)$ of f, we do exactly like in the proof of 3.1: we can split $M(f)$ as a union $M(f) \simeq T \cup C$, where

$$T := \{(x, y, z) \mid h(x, y) = \eta, \ (x, y) \in B^4_{\varepsilon_1}, \ z \in S^1_{\varepsilon_2}\}$$

and

$$C := \{(x, y, z) \mid h(x, y) = \eta, \ (x, y) \in S^3_{\varepsilon_1}, \ z \in B^2_{\varepsilon_2}\}$$

with $0 < \eta \ll \varepsilon_2 \ll \varepsilon_1 \ll 1$. Now T is exactly the mapping torus of the identity map of the Milnor fibre $F(h) = \{(x, y) \in B^4_{\varepsilon_1} \mid h(x, y) = \eta\}$ of h in $\mathbb{C}^2$, and C is just the product of its boundary with the disc $B^2_{\varepsilon_2}$. These two parts matching naturally, this shows that $M(f)$ admits an open book decomposition whose page is $F(h)$ and whose monodromy is the identity. This implies that

$$M(f) \simeq \#_{\mu(h)} S^2 \times S^1$$

where $\mu(h)$ is the Milnor number of h, equal to $2g(F(h)) + \#\pi_0(\partial F(h)) - 1$ (see [Et, 2.10]).

Moreover, the third coordinate $z : (\mathbb{C}^3, \mathbf{0}) \to (\mathbb{C}, 0)$ defines with f an isolated complete intersection singularity. Therefore Proposition 3.2 applies, and in fact the Milnor open book of z in $M(f)$ is exactly the one we have just described. But in this case, since the smoothly Milnor fillable contact structure on $M(f)$ is tight and since there is only one tight contact structure on the connected sums $\#_k S^2 \times S^1$ (a consequence of [El] and [Co]), it is not necessary to know this open book to determine the contact structure it carries.

Remark 3.3. The same work can be made in higher dimension, giving that *the Milnor fibre of any isolated hypersurface singularity, along with its identity map, is a Milnor open book in the Milnor boundary of a non-isolated hypersurface singularity.*

Acknowledgements

I would like to thank Patrick Popescu-Pampu for asking me the question which motivated this paper, and Anne Pichon for valuable discussions and for providing the function f_k from her joint work [MPW] with F. Michel and C. Weber. I'm also grateful to Christine Lescop for having pointed out a simplification in the original proof of Proposition 3.1.

References

Ca.   Caubel, Clément *Variation of the Milnor fibration in pencils of hypersurface singularities.* Proc. London Math. Soc. (3) 83 (2001), no. 2, 330–350.

CNP.   Caubel, Clément; Némethi, András; Popescu-Pampu, Patrick. *Milnor open books and Milnor fillable contact 3-manifolds.* math.SG/0409160. To appear in Topology.

CT.   Caubel, Clément; Tibăr, Mihai *The contact boundary of a complex polynomial.* Manuscripta Math. 111 (2003), no. 2, 211–219.

Co.   Colin, Vincent. *Chirurgies d'indice un et isotopies de sphères dans les variétés de contact tendues.* C. R. Acad. Sci. Paris Sér. I Math. 324 (1997), 659–663.

Du.   Durfee, Alan H. *Knot invariants of singularities.* Algebraic geometry (Proc. Sympos. Pure Math., Vol. 29, Humboldt State Univ., Arcata, Calif., 1974), pp. 441–448. Amer. Math. Soc., Providence, R.I., 1975.

El.   Eliashberg, Yakov. *Contact 3-manifolds twenty years since J. Martinet's work.* Ann. Inst. Fourier 42 (1992), 165–192.

Et.   Etnyre, John. *Lectures on open book decompositions and contact structures.* math.SG/0409402

Gi.   Giroux, Emmanuel *Géométrie de contact: de la dimension trois vers les dimensions supérieures.* Proceedings of the International Congress of Mathematicians, Vol. II (Beijing, 2002), 405–414, Higher Ed. Press, Beijing, 2002.

Go.   Gompf, Robert E. *Handlebody construction of Stein surfaces.* Ann. of Math. (2) 148 (1998), no. 2, 619–693.

HMS.   Henry, J. P.; Merle, M.; Sabbah, C. *Sur la condition de Thom stricte pour un morphisme analytique complexe.* Ann. Sci. École Norm. Sup. (4) 17 (1984), no. 2, 227–268.

JN.   Jankins, Mark; Neumann, Walter D. *Lectures on Seifert manifolds.* Brandeis Lecture Notes, 2. Brandeis University, Waltham, MA, 1983. i+84+27 pp.

MP.   Michel, Françoise; Pichon, Anne *On the boundary of the Milnor fibre of non-isolated singularities.* Int. Math. Res. Not. 2003, no. 43, 2305–2311.

MPW.   Michel, Françoise; Pichon, Anne; Weber, Claude *The boundary of the Milnor fiber of Hirzebruch surface singularities.* Preprint (2005).

NN.   Némethi, András; Nicolaescu, Liviu I. *Seiberg-Witten invariants and surface singularities.* Geom. Topol. 6 (2002), 269–328

Neu.   Neumann, Walter D. *A calculus for plumbing applied to the topology of complex surface singularities and degenerating complex curves.* Trans. Amer. Math. Soc. 268 (1981), no. 2, 299–344.

P.   Popescu-Pampu, Patrick. *The geometry of continued fractions and the topology of surface singularities.* math.GT/0506432, Proceedings of the third Franco-Japanese Conference on Singularities, Sapporo, 2004 (to appear).

S.   Sullivan, Denis.*On the intersection ring of compact three manifolds.* Topology 14 (1975), 275-277.

T.   Thurston, William P. *Three-dimensional manifolds, Kleinian groups and hyperbolic geometry.* Bull. Amer. Math. Soc. (N.S.) 6 (1982), no. 3, 357–381.

V.   Varchenko, A.N. *Contact structures and isolated singularities.* Mosc. Univ. Math. Bull. **35** no.2 (1980), 18–22.

W.   C.T.C. Wall, *Quadratic forms and normal surface singularities*. Quadratic forms and their applications (Dublin, 1999), 293–311, Contemp. Math., 272, Amer. Math. Soc., Providence, RI, 2000.

ON LOCAL REDUCTION THEOREMS FOR SINGULAR SYMPLECTIC FORMS ON A 4-DIMENSIONAL MANIFOLD

W. DOMITRZ

Faculty of Mathematics and Information Science, Warsaw University of Technology,
Plac Politechniki 1, 00-661 Warszawa, Poland
E-mail: domitrz@mini.pw.edu.pl

We study local invariants of singular symplectic forms with structurally smooth Martinet hypersurfaces on a 4-dimensional manifold M. We prove that the equivalence class of a germ at $p \in M$ of a singular symplectic form ω is determined by the Martinet hypersurface, the canonical orientation of it, the pullback of the singular symplectic form to it and the 2-dimensional kernel of ω at p. We also show which germs of closed 2-forms on a 3-dimensional submanifold can be realizable as pullbacks of singular symplectic forms to structurally smooth Martinet hypersurfaces.

Keywords: Symplectic forms; Singularities; Normal forms

1. Introduction

Let ω be a closed 2-form on a $2n$-dimensional manifold M. ω is a symplectic form on M if for any $p \in M$

$$\omega^n|_p = \omega \wedge \cdots \wedge \omega|_p \neq 0. \tag{1}$$

By the Darboux Theorem there exists a system of local coordinates $(p_1, \cdots, p_n, q_1, \cdots, q_n)$ around any point $p \in M$ such that

$$\omega = \sum_{i=1}^{n} dp_i \wedge dq_i.$$

If the set of points $p \in M$, where ω does not satisfy (1), is nowhere dense we call ω a *singular symplectic form*.

In this paper we study local invariants of singular symplectic forms on a 4-dimensional manifold.

Because our consideration is local, we may assume that ω is a germ of a $\mathbb{K}$-analytic or smooth closed 2-form on $\mathbb{K}^4$ for $\mathbb{K} = \mathbb{R}$ or $\mathbb{K} = \mathbb{C}$. Then

$\omega^2 = f\Omega$, where f is a function-germ at 0 and Ω is a germ at 0 of a volume form on $\mathbb{K}^4$.

The *Martinet hypersurface* $\Sigma_2 = \Sigma_2(\omega)$ is the following set

$$\{p \in \mathbb{K}^4 : \omega^2|_p = 0\} = \{f = 0\}.$$

We assume that $f(0) = 0$ and $df_0 \neq 0$. Then Σ_2 is called *structurally smooth at 0*. In dimension 4 such situation is generic (see [12]).

Let ω be a germ of a singular symplectic form with a structurally smooth Martinet hypersurface at 0. It is obvious that Σ_2 is an invariant of ω. It is also obvious that the pullback of ω to Σ_2 is an invariant of ω. In this paper we consider the following problem.

Do the Martinet hypersurface Σ_2 and the pullback of ω to Σ_2 form a complete set of invariants?

The starting point of this paper is the articles [8,9] where an affirmative answer to the above question is given for all local *singular contact structures* excluding degenerations of infinite codimension. B. Jakubczyk and M. Zhitomirskii show that local $\mathbb{C}$-analytic singular contact structures on $\mathbb{C}^3$ with structurally smooth Martinet hypersurfaces are diffeomorphic if their Martinet hypersurfaces and restrictions of singular structures to them are diffeomorphic. In the $\mathbb{R}$-analytic category a complete set of invariants contains, in general, one more independent invariant. It is a canonical orientation on the Martinet hypersurface. The same is true for smooth local singular contact structures $P = (\alpha)$ on $\mathbb{R}^3$ provided $\alpha|_S$ is either not flat at 0 or $\alpha|_S = 0$. The authors also study local singular contact structures in higher dimensions. They find more subtle invariants of a singular contact structure $P = (\alpha)$ on $\mathbb{K}^{2n+1}$: a line bundle L over the Martinet hypersurface S, a canonical partial connection Δ_0 on the line bundle L at $0 \in \mathbb{K}^{2n+1}$ and a 2-dimensional kernel $ker(\alpha \wedge (d\alpha)^{n-1})|_0$. They also consider the more general case when S has singularities.

For the first occurring singularities of singular symplectic forms on a 4-dimensional manifold the answer for the above question follows from Martinet's normal forms of types Σ_{20} and Σ_{220} (see [11,12,15]). In fact it is proved that the Martinet hypersurface Σ_2 and a characteristic line field on Σ_2 (i.e. $\{X$ *is a smooth vector field* $: X \rfloor (\omega|_{T\Sigma_2}) = 0\}$) form a complete set of invariants. Since $(\omega|_{T\Sigma_2})|_0 \neq 0$ for Σ_{20}-singularity, then its characteristic line field is generated by a non-vanishing vector field. But for Σ_{220}-singularity both $\omega|_{T\Sigma_2}$ and the characteristic line vanish at 0 (see [11,15]).

In this paper we assume that $\omega|_{T\Sigma_2}$ vanishes at 0 (if $\omega|_{T\Sigma_2}$ does not vanish at 0 then ω is a symplectic singular form of type Σ_{20} and these problems for this singularity are solved in [12]). We show that a complete set of invariants for local $\mathbb{C}$-analytic singular symplectic forms on $\mathbb{C}^4$ with structurally smooth Martinet hypersurfaces consists of the Martinet hypersurface, the pullback of the singular symplectic form to it and the 2-dimensional kernel of the singular symplectic form at 0 (Theorem 3.1). The same is true for local $\mathbb{R}$-analytic and smooth singular symplectic forms on $\mathbb{R}^4$ with structurally smooth Martinet hypersurfaces if we add to the invariants the canonical orientation of the Martinet hypersurface (Theorem 3.2). These results are obtained as corollaries of Theorem 2.1 on 'normal' forms of singular symplectic forms with a given pullback to the Martinet hypersurface. Another corollary of Theorem 2.1 is a realization theorem (Theorem 2.2), where we show which closed 2-forms on $\mathbb{K}^3$ vanishing at 0 can be obtained as a pullback of a singular symplectic form to its Martinet hypersuface.

In section 4 (see Theorems 4.1, 4.2) we also prove that an equivalence class of a $\mathbb{K}$-analytic singular symplectic form ω on $\mathbb{K}^4$ with a structurally smooth Martinet hypersurface is determined only by the Martinet hypersurface, its canonical orientation (only if $\mathbb{K} = \mathbb{R}$) and the pullback of the singular form to it if ω satisfies the following condition :

$$\forall X \ (X \ is \ a \ \mathbb{K} - analytic \ vector \ field \ and \ X \rfloor (\omega|_{T\Sigma_2}) = 0) \implies X|_0 = 0.$$

The same statement holds for local smooth singular symplectic forms ω on $\mathbb{R}^4$ with structurally smooth Martinet hypersurfaces if the two generators of the ideal generated by coefficients of $\omega|_{T\Sigma_2}$ form a regular sequence of length 2 (Theorem 4.3).

The local invariants of singular symplectic forms in higher dimensions and with singular Martinet hypersurfaces will be studied in [4].

2. The normal form and realization theorems

The main result of this section is Theorem 2.1. In this theorem a 'normal' form of ω with the given pullback to the Martinet hypersurface is presented and sufficient conditions for the equivalence of germs of singular symplectic forms with the same pullback to the common Martinet hypersurface are found. We also show which germs of closed 2-forms on $\mathbb{K}^3$ vanishing at 0 can be obtained as a pullback of a germ of a singular symplectic form on $\mathbb{K}^4$ to its structurally smooth Martinet hypersurface. All results of this section hold in $\mathbb{C}$-analytic, $\mathbb{R}$-analytic and (C^∞) smooth categories.

Let Ω be a germ of a volume form on $\mathbb{K}^4$. Let ω_0 and ω_1 be two germs of singular symplectic forms on $\mathbb{K}^4$ with structurally smooth Martinet hypersurfaces at 0. It is obvious that if there exists a diffeomorphism-germ of $\mathbb{K}^4$ at 0 such that $\Phi^*\omega_1 = \omega_0$ then $\Phi(\Sigma_2(\omega_0)) = \Sigma_2(\omega_1)$. Therefore we assume that these singular symplectic forms have the same Martinet hypersurface.

If the singular symplectic forms are equal on their common Martinet hypersurface then we obtain the following result (see see [7]).

Proposition 2.1. *Let ω_0 and ω_1 be two germs at 0 of singular symplectic forms on $\mathbb{K}^4$ with the common structurally smooth Martinet hypersurface Σ_2.*

If $\frac{\omega_1^2}{\omega_0^2}|_0 > 0$ for $\mathbb{K} = \mathbb{R}$ ($\Re e\left(\frac{\omega_1^2}{\omega_0^2}|_0\right) > 0$ or $\Im m\left(\frac{\omega_1^2}{\omega_0^2}|_0\right) \neq 0$ for $\mathbb{K} = \mathbb{C}$) and $\omega_0|_{T_{\Sigma_2}\mathbb{K}^4} = \omega_1|_{T_{\Sigma_2}\mathbb{K}^4}$ then there exists a diffeomorphism-germ $\Phi : (\mathbb{K}^4, 0) \to (\mathbb{K}^4, 0)$ such that

$$\Phi^*\omega_1 = \omega_0$$

and $\Phi|_{\Sigma_2} = Id_{\Sigma_2}$.

Proof. We present the proof in $\mathbb{R}$-analytic and smooth categories. The proof in the $\mathbb{C}$-analytic category is similar. Firstly we simplify the forms ω_0 and ω_1. We find a local coordinate system (p_1, p_2, p_3, p_4) such that $\omega_0^2 = p_1\Omega$, $\omega_1^2 = p_1(A + g)\Omega$, where $\Omega = dp_1 \wedge dp_2 \wedge dp_3 \wedge dp_4$, g is a function-germ, $g(0) = 0$ and $A > 0$ (see [12]). In this coordinate system $\omega_i = \sum_{1 \leq j < k \leq 4} f_{i,j,k} dp_j \wedge dp_k$, where $f_{i,j,k}$ is a function-germ on $\mathbb{K}^4$ for $i = 0, 1$ and $1 \leq j < k \leq 4$. We can decompose $f_{i,j,k}$ in the following way $f_{i,j,k}(p_1, p_2, p_3, p_4) = p_1 g_{i,j,k}(p_1, p_2, p_3, p_4) + h_{i,j,k}(p_2, p_3, p_4)$, where $g_{i,j,k}$ is a function-germ and $h_{i,j,k}$ is a function-germ that does not depend on p_1 for $i = 0, 1$ and $1 \leq j < k \leq 4$. Let $\alpha_i = \sum_{1 \leq j < k \leq 4} g_{i,j,k} dp_j \wedge dp_k$ and $\tilde{\omega}_i = \sum_{1 \leq j < k \leq 4} h_{i,j,k} dp_j \wedge dp_k$. Then we have $\omega_i = p_1\alpha_i + \tilde{\omega}_i$ for $i = 0, 1$. By assumptions we have $\tilde{\omega}_0|_{T_{\Sigma_2}\mathbb{K}^4} = \tilde{\omega}_1|_{T_{\Sigma_2}\mathbb{K}^4}$. It implies that $\tilde{\omega}_0 = \tilde{\omega}_1$, because $h_{i,j,k}$ does not depend on p_1. We denote $\tilde{\omega}_1 = \tilde{\omega}_0$ by $\tilde{\omega}$. Then $\omega_i = p_1\alpha_i + \tilde{\omega}$ for $i = 0, 1$.

Further on we use the Moser homotopy method (see [14]). Let $\omega_t = t\omega_1 + (1 - t)\omega_0$, for $t \in [0; 1]$.

We want to find a family of diffeomorphisms Φ_t, $t \in [0; 1]$ such that $\Phi_t^*\omega_t = \omega_0$, for $t \in [0; 1]$, $\Phi_0 = Id$. Differentiating the above homotopy equation by t, we obtain

$$d(V_t \lrcorner \omega_t) = \omega_0 - \omega_1 = p_1(\alpha_0 - \alpha_1),$$

where $V_t = \frac{d}{dt}\Phi_t$. We need to solve the above equation for V_t. Now we prove the following lemmas.

Lemma 2.1 ([2]). *Let γ be a germ of a 2-form on $\mathbb{R}^4$ and θ be a germ of a 1-form on $\mathbb{R}^4$. If $p_1\gamma + dp_1 \wedge \theta$ is a germ of a closed 2-form on $\mathbb{R}^4$ then there exists a germ of a 1-form δ such that $p_1\gamma + dp_1 \wedge \theta = d(p_1\delta)$.*

Proof. $p_1\gamma + dp_1 \wedge \theta$ is closed, therefore there exists a 1-form ξ such that $d\xi = p_1\gamma + dp_1\wedge\theta$. There exist a germ of a 1-form ξ_1 on $\mathbb{R}^4$, a function-germ g on $\mathbb{R}^4$ and a germ of 1-form ξ_2 on $\{p_1 = 0\}$ such that $\xi = p_1\xi_1 + gdp_1 + \pi^*\xi_2$, where $\pi : \mathbb{R}^4 \ni (p_1, p_2, p_3, p_4) \mapsto (p_2, p_3, p_4) \in \{p_1 = 0\}$. The pullback of $d\xi$ to $\{p_1 = 0\}$ vanishes. It implies that $d\xi_2 = 0$. Thus $d(p_1\xi_1 + gdp_1) = d(\xi - \pi^*\xi_2) = p_1\gamma + dp_1 \wedge \theta$. It implies that $d(p_1(\xi_1 - dg)) = p_1\gamma + dp_1 \wedge \theta$, which finishes the proof of Lemma 2.1. $\qquad\square$

Lemma 2.2. *Let α be a germ of a 2-form on $\mathbb{R}^4$. If $p_1\alpha$ is a germ of a closed 2-form on $\mathbb{R}^4$ then there exists a germ of a 1-form β such that $p_1\alpha = d(p_1^2\beta)$.*

Proof. By Lemma 2.1 there exists a germ of a 1-form γ such that $p_1\alpha = d(p_1\gamma) = dp_1 \wedge \gamma + p_1 d\gamma$. It implies that $dp_1 \wedge \gamma|_{T_{\{p_1=0\}}\mathbb{R}^4} = 0$. Hence there exist a germ of a 1-form δ and a smooth function-germ f such that $\gamma = p_1\delta + f dp_1$. If we take $\beta = \delta - \frac{df}{2}$ then

$$p_1\alpha = d(p_1\gamma - d(\frac{p_1^2 f}{2})) = d(p_1^2\beta),$$

which finishes the proof of Lemma 2.2. $\qquad\square$

Let us notice that $p_1(\alpha_0 - \alpha_1) = \omega_1 - \omega_0$ is closed. By the above lemma it is enough to solve for V_t the equation

$$V_t \rfloor \omega_t = p_1^2\beta. \tag{2}$$

Now we calculate $\Sigma_2(\omega_t)$. It is easy to see that

$$\omega_i^2 = (p_1\alpha_i + \tilde{\omega})^2 = \tilde{\omega}^2 + p_1(2\alpha_i \wedge \tilde{\omega} + p_1\alpha_i^2).$$

But $\omega_i^2|_{T_{\{p_1=0\}}\mathbb{R}^4} = 0$. This clearly forces $\tilde{\omega}^2|_{T_{\{p_1=0\}}\mathbb{R}^4} = 0$. It implies that $\tilde{\omega}^2 = 0$, because coefficients of $\tilde{\omega}$ do not depend on p_1. By the above formula we get

$$2\alpha_0 \wedge \tilde{\omega} = \Omega - p_1\alpha_0^2$$

and

$$2\alpha_1 \wedge \tilde{\omega} = (A + g)\Omega - p_1\alpha_1^2$$

492

The above formulas imply the following formula

$$\omega_t^2 = (p_1(t\alpha_1 + (1-t)\alpha_0) + \tilde{\omega})^2 =$$
$$= p_1(1 + t(A + g - 1))\Omega + \tag{3}$$
$$+ p_1^2\left((t\alpha_1 + (1-t)\alpha_0)^2 - t\alpha_1^2 - (1-t)\alpha_0^2\right).$$

From (3) we obtain

$$\omega_t^2 = p_1(1 + t(A + g - 1) + p_1 h_t)\Omega, \tag{4}$$

where h_t is a function-germ. Let us notice that $(1 + t(A + g(0) - 1)) \neq 0$ for $A > 0$ and for $t \in [0,1]$. Since $V_t \rfloor \omega_t^2 = 2(V_t \rfloor \omega_t) \wedge \omega_t$ and $\Sigma_2(\omega_t) = \{p_1 = 0\}$ is nowhere dense, equation (2) is equivalent to the following equation

$$V_t \rfloor \omega_t^2 = 2p_1^2 \beta \wedge \omega_t. \tag{5}$$

Combining (5) with (4) we obtain

$$V_t \rfloor (1 + t(A + g - 1) + p_1 h_t)\Omega = 2p_1 \beta \wedge \omega_t \tag{6}$$

But if $A > 0$ then $(1 + t(A - 1)) \neq 0$ for $t \in [0; 1]$. Therefore we can find a germ of smooth (or $\mathbb{R}$-analytic) vector field V_t that satisfies (6). $V_t|_{\Sigma_2} = 0$, because the right hand side of (6) vanishes on Σ_2. Hence there exists a diffeomorphism Φ_t such that $\Phi_t^* \omega_t = \omega_0$ for $t \in [0,1]$ and $\Phi_t|_{\Sigma_2} = Id_{\Sigma_2}$. This completes the proof of Theorem 2.1.

Now we define

$$\iota : \Sigma_2 = \{p_1 = 0\} \ni (p_2, p_3, p_4) \mapsto (0, p_2, p_3, p_4) \in \mathbb{K}^4$$

and

$$\pi : \mathbb{K}^4 \ni (p_1, p_2, p_3, p_4) \mapsto (p_2, p_3, p_4) \in \Sigma_2 = \{p_1 = 0\}.$$

If $rank\iota^*\omega|_0$ is 2 then ω is equivalent to Σ_{20} Martinet's singular form (see [12]). Therefore we study singular symplectic forms such that $rank\iota^*\omega|_0 = 0$.

In the next theorem we describe all germs of singular symplectic forms ω on $\mathbb{K}^4$ with structurally smooth Martinet hypersurfaces at 0 and $rank\iota^*\omega|_0 = 0$. We also find the sufficient conditions for equivalence of singular symplectic forms of this type.

Theorem 2.1. *Let ω be a germ of a singular symplectic form on $\mathbb{K}^4$ with a structurally smooth Martinet hypersurface at 0.*

(a) If $rank\iota^*\omega|_0 = 0$ then there exists a germ of a diffeomorphism $\Phi : (\mathbb{K}^4, 0) \to (\mathbb{K}^4, 0)$ such that

$$\Phi^*\omega = d\left(p_1\pi^*\alpha\right) + \pi^*\sigma,$$

where $\sigma = \iota^* \Phi^* \omega$ is a germ of a closed 2-form on $\{p_1 = 0\}$ and α is a germ of a contact form on $\{p_1 = 0\}$ such that $\alpha \wedge \sigma = 0$.

(b)Moreover if $\omega_0 = d\left(p_1 \pi^* \alpha_0\right) + \pi^* \sigma$ and $\omega_1 = d\left(p_1 \pi^* \alpha_1\right) + \pi^* \sigma$ are two germs of singular symplectic forms satisfying the above conditions and

(1) $\frac{\alpha_1 \wedge d\alpha_1}{\alpha_0 \wedge d\alpha_0}\big|_0 > 0$ if $\mathbb{K} = \mathbb{R}$,
(2) $\alpha_1|_0 \wedge \alpha_0|_0 = 0$,

then there exists a germ of a diffeomorphism $\Psi : (\mathbb{K}^4, 0) \to (\mathbb{K}^4, 0)$ such that

$$\Psi^* \omega_1 = \omega_0.$$

Remark 2.1. Assumption (1) is only needed in $\mathbb{R}$-analytic and smooth categories. In the $\mathbb{C}$-analytic category we have

$$\Phi^*\left(d\left(p_1 \pi^* \alpha\right) + \pi^* \sigma\right) = d\left(p_1 \pi^* i\alpha\right) + \pi^* \sigma,$$

where Φ is the following diffeomorphism $\Phi(p_1, p_2, p_3, p_4) = (ip_1, p_2, p_3, p_4)$ and $i^2 = -1$. It is obvious that $\Phi|_{\Sigma_2} = Id_{\Sigma_2}$, where $\Sigma_2 = \{p_1 = 0\}$ and $i\alpha \wedge d(i\alpha) = -\alpha \wedge d\alpha$.

Proof. By Lemma 2.1 there exists a 1-form γ such that $\omega = d(p_1\gamma) + \pi^* \sigma$. It is clear that we can write γ in the following form $\gamma = \pi^* \alpha + p_1 \delta + g dp_1$, where α is a germ of a 1-form on $\{p_1 = 0\}$, g is a function-germ and δ is a germ of a 1-form. Then

$$d(p_1(p_1\delta + g dp_1)) = p_1(2dp_1 \wedge \delta + p_1 d\delta + dg \wedge dp_1).$$

By Lemma 2.2 we have $\omega = d(p_1\pi^* \alpha) + \pi^* \sigma + d(p_1^2 \theta)$.

It is easy to see that

$$\omega^2 = 2dp_1 \wedge \pi^* \alpha \wedge \pi^* \sigma + 4p_1 dp_1 \wedge \theta \wedge \pi^* \sigma$$
$$+ 2p_1 dp_1 \wedge \pi^* \alpha \wedge d\pi^* \alpha + p_1^2 v\Omega,$$

where v is a function-germ at 0. We have $\alpha \wedge \sigma = 0$, because $\omega^2|_{T_{\{p_1=0\}}\mathbb{K}^4} = 0$. From $\sigma|_0 = 0$, we have

$$\omega^2 = 2p_1 dp_1 \wedge \pi^* \alpha \wedge d\pi^* \alpha + p_1 g\Omega,$$

where g is a function-germ vanishing at 0. From the above we obtain that

$$\alpha \wedge d\alpha|_0 \neq 0.$$

Let

$$\omega_0 = d\left(p_1 \pi^* \alpha\right) + \pi^* \sigma.$$

494

Then

$$\omega_0^2 = 2p_1 dp_1 \wedge \pi^*\alpha \wedge d\pi^*\alpha + p_1 h\Omega,$$

where h is a smooth function-germ at 0 such that $h(0) = 0$. One can check that

$$\omega_0|_{T_{\{p_1=0\}}\mathbb{K}^4} = dp_1 \wedge \pi^*\alpha + \pi^*\sigma = \omega|_{T_{\{p_1=0\}}\mathbb{K}^4}.$$

Therefore by Proposition 2.1 there exists a germ of a diffeomorphism $\Theta :$ $(\mathbb{K}^4, 0) \to (\mathbb{K}^4, 0)$ such that $\Theta^*\omega = \omega_0$ and $\Theta|_{\{p_1=0\}} = Id_{\{p_1=0\}}$.

This finishes the proof of part (a).

Now we prove part (b).

Assumption (2) implies that there exists $B \neq 0$ such that $\alpha_1|_0 = B\alpha_0|_0$. If $B \neq 1$ then $\Phi^*\omega_0 = d(p_1\pi^*(B\alpha_0)) + \pi^*\sigma$ where Φ is a diffeomorphism-germ of the form $\Phi(p) = (Bp_1, p_2, p_3, p_4))$. Thus we may assume that $B = 1$.

We use the Moser homotopy method. Let $\alpha_t = t\alpha_1 + (1-t)\alpha_0$ and $\omega_t = d(p_1\pi^*\alpha_t) + \pi^*\sigma$ for $t \in [0,1]$. It is easy to check that $\alpha_t \wedge \sigma = 0$.

Now we look for germs of diffeomorphims Φ_t such that

$$\Phi_t^*\omega_t = \omega_0, \ for \ t \in [0;1], \ \Phi_0 = Id. \tag{7}$$

Differentiating the above homotopy equation by t, we obtain

$$d(V_t \rfloor \omega_t) = d(p_1\pi^*(\alpha_0 - \alpha_1)),$$

where $V_t = \frac{d}{dt}\Phi_t$. Therefore we have to solve for V_t the following equation

$$V_t \rfloor \omega_t = p_1\pi^*(\alpha_0 - \alpha_1). \tag{8}$$

We calculate the Martinet hypersurface of ω_t. $\omega_t^2 = 2p_1 dp_1 \wedge \pi^*(\alpha_t \wedge d\alpha_t)$, because $\sigma^2 = 0$, $d\alpha_t^2 = 0$ and $\alpha_t \wedge \sigma = 0$.

$\alpha_0|_0 = \alpha_1|_0$ and there exists $A > 0$ such that $(\alpha_1 \wedge d\alpha_1)|_0 = A(\alpha_0 \wedge d\alpha_0)|_0$. It implies that

$$\alpha_t \wedge d\alpha_t|_0 = (tA + (1-t))(\alpha_0 \wedge d\alpha_0)|_0.$$

Therefore

$$dp_1 \wedge \pi^*(\alpha_t \wedge d\alpha_t)|_0 \neq 0 \tag{9}$$

for $t \in [0;1]$. Thus $\Sigma_2(\omega_t) = \{p_1 = 0\}$.

Since $V_t \rfloor \omega_t^2 = 2(V_t \rfloor \omega_t) \wedge \omega_t$ and $\Sigma_2(\omega_t) = \{p_1 = 0\}$ is nowhere dense, equation (8) is equivalent to

$$V_t \rfloor \omega_t^2 = 2p_1\pi^*(\alpha_0 - \alpha_1) \wedge \omega_t.$$

Therefore we have to solve the following equation

$$V_t \rfloor (2dp_1 \wedge \pi^*(\alpha_t \wedge d\alpha_t)) = 2\pi^*(\alpha_0 - \alpha_1) \wedge \omega_t. \tag{10}$$

Hence by (9) we can find a smooth solution V_t of (10) and $V_t|_0 = 0$, because $\alpha_1|_0 = \alpha_0|_0$ Therefore there exist germs of diffeomorphisms Φ_t, which satisfy (7). For $t = 1$ we have $\Phi_1^* \omega_1 = \omega_0$. $\qquad\square$

We call a germ of a closed 2-form σ on $\mathbb{K}^3$ *realizable with a structurally smooth Martinet hypersurface* if there exists a germ of a singular symplectic form ω on $\mathbb{K}^4$ such that $\Sigma_2(\omega) = \{0\} \times \mathbb{K}^3$ is structurally smooth and $\omega|_{T\Sigma_2(\omega)} = \sigma$.

From Martinet's normal form of type Σ_{20} we know that all germs of closed 2-forms on $\mathbb{K}^3$ of the rank 2 are realizable with a structurally smooth Martinet hypersurface (see [12]). From part (a) of the Theorem 2.1 we obtain the following realization theorem of closed 2-forms on $\mathbb{K}^3$ of rank 0 at $0 \in \mathbb{K}^3$.

Theorem 2.2. *Let σ be a germ of a closed 2-form on $\mathbb{K}^3$ and $\mathrm{rank}\,\sigma|_0 = 0$. σ is realizable with a structurally smooth Martinet hypersurface if and only if there exists a germ of a contact form α on $\mathbb{K}^3$ such that $\alpha \wedge \sigma = 0$.*

3. The canonical orientation and the 2-dimensional kernel of ω at 0

In $\mathbb{R}$-analytic and smooth categories assumption (1) of Theorem 2.1 means that ω_0 and ω_1 determine the same orientation. The orientation may be defined invariantly. Let ω be a germ of a singular symplectic structure on $\mathbb{R}^4$ with a structurally smooth Martinet hypersurface Σ_2 at 0. Then $\Sigma_2 = \{f = 0\}$ and $df|_0 \neq 0$. We define the volume form Ω_{Σ_2} on Σ_2 which determines the orientation of Σ_2 in the following way

$$\Omega_{\Sigma_2} \wedge df = \frac{\omega^2}{f}.$$

This definition is analogous to the definition in [8] proposed by V. I. Arnol'd. It is easy to see that this definition of the orientation does not depend on the choice of f such that $\Sigma_2 = \{f = 0\}$ and $df|_0 \neq 0$. We call this orientation of Σ_2 the *canonical orientation of Σ_2*.

Assumption (2) of Theorem 2.1 can be also expressed invariantly. We call a subspace $\ker \omega|_0 = \{v \in T_0\mathbb{K}^4 : v \rfloor \omega|_0 = 0\}$ *the kernel of ω at 0*. It is easy to see that $\ker \omega|_0$ is 2-dimensional subspace of $T_0\Sigma_2$ if $\omega|_{T_0\Sigma_2} = 0$. $\ker \omega|_0$ can be also described as a kernel of a non-vanishing 1-form on Σ_2. Let Y be a germ of a vector field on $\mathbb{K}^4$ that is transversal to Σ_2 at 0. Let $\iota : \Sigma_2 \hookrightarrow \mathbb{K}^4$ be an inclusion. Then the kernel of $\iota^*(Y\rfloor\omega)|_0$ is a 2-dimensional linear subspace of $T_0\Sigma_2$. By Theorem 2.1 it is easy to check that this definition does not depend on the choice of Y and that the subspace $\ker\iota^*(Y\rfloor\omega)|_0$ is $\ker\omega|_0$. Assumption (2) of Theorem 2.1 means that $\ker\omega_0|_0 = \ker\omega_1|_0$, which is equivalent to $\ker\iota^*(Y\rfloor\omega_1)|_0 = \ker\iota^*(Y\rfloor\omega_0)|_0$. Now we formulate part (b) of Theorem 2.1 invariantly.

In the $\mathbb{C}$-analytic category ω is determined by the restriction to $T\Sigma_2$ and the 2-dimensional kernel of ω at 0.

Theorem 3.1. *Let ω_0 and ω_1 be germs of $\mathbb{C}$-analytic singular symplectic forms on $\mathbb{C}^4$ with a common structurally smooth Martinet hypersurface Σ_2 at 0 and $rank\,\iota^*\omega_0|_0 = rank\,\iota^*\omega_1|_0 = 0$.*

If $\iota^*\omega_0 = \iota^*\omega_1$ and $\ker\omega_0|_0 = \ker\omega_1|_0$ then there exists a germ of a $\mathbb{C}$-analytic diffeomorphism $\Psi : (\mathbb{C}^4, 0) \to (\mathbb{C}^4, 0)$ such that

$$\Psi^*\omega_1 = \omega_0.$$

In the $\mathbb{R}$-analytic and smooth categories ω is determined by the restriction to $T\Sigma_2$, the kernel of ω at 0 and the canonical orientation of Σ_2.

Theorem 3.2. *Let ω_0 and ω_1 be germs of smooth ($\mathbb{R}$-analytic) singular symplectic forms on $\mathbb{R}^4$ with a common structurally smooth Martinet hypersurface Σ_2 at 0 and $rank\,\iota^*\omega_0|_0 = rank\,\iota^*\omega_1|_0 = 0$.*

If $\iota^*\omega_0 = \iota^*\omega_1$, $\ker\omega_0|_0 = \ker\omega_1|_0$ and ω_0, ω_1 define the same canonical orientation of Σ_2 then there exists a germ of a smooth ($\mathbb{R}$-analytic) diffeomorphism $\Psi : (\mathbb{R}^4, 0) \to (\mathbb{R}^4, 0)$ such that

$$\Psi^*\omega_1 = \omega_0.$$

4. Determination by the restriction of ω to $T\Sigma_2$ and the canonical orientation

In this section we find conditions in the $\mathbb{C}$-analytic category for the determination of the equivalence class of a singular symplectic form by its pullback to the Martinet hypersurface (Theorem 4.1). The same conditions are valid for the determination of the equivalence class of a singular symplectic form by its pullback to the Martinet hypersurface and the canonical orientation in the $\mathbb{R}$-analytic category (Theorem 4.2). In the smooth category we need a stronger condition to obtain an analogous result.

Theorem 4.1. *Let ω_0 and ω_1 be germs of $\mathbb{C}$-analytic singular symplectic* *forms on $\mathbb{C}^4$ with a common structurally smooth Martinet hypersurface Σ_2 at 0 and $\mathrm{rank}\iota^*\omega_0|_0 = \mathrm{rank}\iota^*\omega_1|_0 = 0$.*

If $\iota^*\omega_0 = \iota^*\omega_1 = \sigma$ and there does not exist a germ of a $\mathbb{C}$-analytic vector field X on Σ_2 at 0 such that $X\rfloor\sigma = 0$ and $X|_0 \neq 0$ then there exists a germ of a $\mathbb{C}$-analytic diffeomorphism $\Psi : (\mathbb{C}^4, 0) \to (\mathbb{C}^4, 0)$ such that

$$\Psi^*\omega_1 = \omega_0.$$

Theorem 4.2. *Let ω_0 and ω_1 be germs of $\mathbb{R}$-analytic singular symplectic forms on $\mathbb{R}^4$ with a common structurally smooth Martinet hypersurface Σ_2 at 0 and $\mathrm{rank}\iota^*\omega_0|_0 = \mathrm{rank}\iota^*\omega_1|_0 = 0$.*

If $\iota^*\omega_0 = \iota^*\omega_1 = \sigma$, ω_0 and ω_1 define the same canonical orientation of Σ_2 and there does not exist a germ of an $\mathbb{R}$-analytic vector field X on Σ_2 at 0 such that $X\rfloor\sigma = 0$ and $X|_0 \neq 0$ then there exists a germ of an $\mathbb{R}$-analytic diffeomorphism $\Psi : (\mathbb{R}^4, 0) \to (\mathbb{R}^4, 0)$ such that

$$\Psi^*\omega_1 = \omega_0.$$

Proof. We present the proof of Theorem 4.2. The proof of Theorem 4.1 is similar.

By Theorem 2.1 we obtain $\omega_0 = d(p_1\pi^*\alpha_0) + \sigma$ and $\omega_1 = d(p_1\pi^*\alpha_1) + \sigma$, where α_0, α_1 are germs of analytic contact forms on $\Sigma_2 = \{p_1 = 0\}$ such that $\alpha_0 \wedge \sigma = \alpha_1 \wedge \sigma = 0$ and $\alpha_0 \wedge d\alpha_0$, $\alpha_1 \wedge d\alpha_1$ define the same orientation on Σ_2.

α_0 is a contact form, therefore $\alpha_0|_0 \neq 0$. We can find a coordinate system (x, y, z) on Σ_2 such that $\alpha_0 = f_0 dx + g_0 dy + h_0 dz$, where f_0, g_0 and h_0 are function-germs on Σ_2 and $h_0(0) \neq 0$. Let $\sigma = a\, dy \wedge dz + b\, dz \wedge dx + c\, dx \wedge dy$, where a, b, c are function-germs on Σ_2 vanishing at 0. $\alpha_0 \wedge \sigma = 0$, thus we get $c = -\frac{f_0}{h_0}a - \frac{g_0}{h_0}b$.

Let $\alpha_1 = f_1 dx + g_1 dy + h_1 dz$, where f_1, g_1, h_1 are functions-germs on Σ_2. From $\alpha_1 \wedge \sigma = 0$ we obtain the equation

$$a\left(f_1 - \frac{h_1}{h_0}f_0\right) + b\left(g_1 - \frac{h_1}{h_0}g_0\right) = 0 \tag{11}$$

and $a(0) = b(0) = 0$.

Let l be the greatest common divisor of a and b ($GCD(a, b)$). Then $a = la_1$ and $b = lb_1$, where a_1 and b_1 are germs of analytic functions on Σ_2 and $GCD(a_1, b_1) = 1$. Thus $\sigma = l(a_1 dy \wedge dz + b_1 dz \wedge dx - (\frac{f_0}{h_0}a_1 + \frac{g_0}{h_0}b_1)dx \wedge dy)$. If $a_1 \neq 0$ or $b_1 \neq 0$ then a germ of an analytic vector field $X = a_1\frac{\partial}{\partial x} + b_1\frac{\partial}{\partial y} - (\frac{f_0}{h_0}a_1 + \frac{g_0}{h_0}b_1)\frac{\partial}{\partial z}$ does not vanish at 0. It is easy to see that $X\rfloor\sigma = 0$. Therefore $a_1(0) = b_1(0) = 0$.

498

Thus the equation (11) has the following form

$$la_1(f_1 - \frac{h_1}{h_0}f_0) = -lb_1(g_1 - \frac{h_1}{h_0}g_0)$$

and $GCD(a_1, b_1) = 1$.

Therefore $f_1 - \frac{h_1}{h_0}f_0 = b_1 r$ and $g_1 - \frac{h_1}{h_0}g_0 = -a_1 r$, where r is a function-germ on Σ_2 at 0.

Then $\alpha_1 = \frac{h_1}{h_0}(f_0 dx + g_0 dy + h_0 dz) + r(b_1 dx - a_1 dy)$. $\alpha_1|_0 \neq 0$ and $a_1(0) = b_1(0) = 0$ thus $h_1(0) \neq 0$.

Hence $\alpha_1|_0 = \frac{h_1(0)}{h_0(0)}\alpha_0|_0$.

It is easy to see that $\omega_i^2 = 2p_1 dp_1 \wedge \pi^*(\alpha_i \wedge d\alpha_i)$ for $i = 0, 1$. Therefore by assumptions of the theorem we have $\alpha_1 \wedge d\alpha_1 = A\alpha_0 \wedge d\alpha_0$, where $A > 0$.

Thus ω_0 and ω_1 satisfy the assumptions of Theorem 2.1. Then there exists a germ of an analytic diffeomorphism $\Psi : (\mathbb{R}^4, 0) \to (\mathbb{R}^4, 0)$ such that

$$\Psi^*\omega_1 = \omega_0. \qquad \square$$

Now we find the normal form of a germ of a singular symplectic form on $\mathbb{K}^4$ at 0 which does not satisfy the assumptions of the above theorem. The following result is also true in the smooth category.

Proposition 4.1. *Let ω be a germ of a $\mathbb{K}$-analytic singular symplectic form on $\mathbb{K}^4$ with a structurally smooth Martinet hypersurface at 0 and $rank \iota^*\omega|_0 = 0$.*

If there exists a germ of a $\mathbb{K}$-analytic vector field X on Σ_2 at 0 such that $X \rfloor \sigma = 0$ and $X|_0 \neq 0$ then there exists a germ of a $\mathbb{K}$-analytic diffeomorphism $\Psi : (\mathbb{K}^4, 0) \to (\mathbb{K}^4, 0)$ such that

$$\Psi^*\omega = d(p_1(dx + Cdy + zdy)) + g(x,y)dx \wedge dy$$

or

$$\Psi^*\omega = d(p_1(dy + Cdx + zdx)) + g(x,y)dx \wedge dy,$$

where $C \in \mathbb{K}$ and g is a $\mathbb{K}$-analytic function-germ on $\mathbb{K}^4$ at 0 that does not depend on p_1 and z.

Proof. By Theorem 2.1 we may assume that $\omega = d(p_1\pi^*\alpha) + \pi^*\sigma$, where $\sigma = \iota^*\omega$ and α is a germ of an analytic contact form on $\Sigma_2 = \{p_1 = 0\}$ such that $\alpha \wedge \sigma = 0$. Let X be a germ of an analytic vector field on Σ_2 at 0 such that $X \rfloor \sigma = 0$ and $X|_0 \neq 0$. Then we may choose a coordinate system on Σ_2 such that $X = \frac{\partial}{\partial z}$. In this system the closed 2-form σ has the following form $\sigma = h(x,y)dx \wedge dy$, where h is an analytic function-germ on Σ_2 at 0 that does not depend on z. In this coordinate system $\alpha = a(x,y,z)dx + b(x,y,z)dy$, because $\alpha \wedge \sigma = 0$. Therefore ω has the following form

$$\omega = d(p_1(a(x,y,z)dx + b(x,y,z)dy)) + h(x,y)dx \wedge dy. \qquad (12)$$

$a(0) \neq 0$ or $b(0) \neq 0$, because $\alpha_0 \neq 0$. Assume that $a(0) \neq 0$. Then by a diffeomorphism of the form

$$\Phi : (\mathbb{K}^4, 0) \to (\mathbb{K}^4, 0); (p_1, x, y, z) \mapsto \left(\frac{p_1}{a(x,y,z)}, x, y, z \right)$$

we obtain $\Phi^*\omega = d(p_1(dx + b_1(x,y,z)dy)) + h(x,y)dx \wedge dy$, where $b_1(x,y,z) = \frac{b(x,y,z)}{a(x,y,z)}$.

But $\alpha = dx + b_1(x,y,z)dy$ is a germ of a contact form on Σ_2. Therefore

$$\alpha \wedge d\alpha|_0 = \frac{\partial b_1}{\partial z}(0)dx \wedge dz \wedge dy \neq 0.$$

Thus $\frac{\partial b_1}{\partial z}(0) \neq 0$.

Then by a diffeomorphism of the form

$$\Phi : (\mathbb{K}^4, 0) \to (\mathbb{K}^4, 0); (p_1, x, y, z) \mapsto (p_1, x, y, b_1(x,y,z) - b_1(0))$$

we obtain $\Phi^*\omega = d(p_1(dx + Cdy + zdy)) + h(x,y)dx \wedge dy$, where $C = b_1(0)$.

If $a(0) = 0$ in (12) then $b(0) \neq 0$ and we obtain $\Psi^*\omega = d(p_1(dy + Cdx + zdx)) + g(x,y)dx \wedge dy$, by the analogous coordinate changes. $\qquad \square$

Now we need some notions from commutative algebra (see Appendix 1 of [8], [3]) to formulate the result in the smooth category. We recall that a sequence of elements $a_1, \cdots, a_r$ of a proper ideal I of a ring R is called *regular* if a_1 is a nonzerodivisor of R and a_i is a nonzerodivisor of $R/ < a_1, \cdots, a_{i-1} >$ for $i = 2, \cdots, r$. Here $< a_1, \cdots, a_i >$ denotes the ideal generated by $a_1, \cdots, a_i$. The *length* of a regular sequence $a_1, \cdots, a_r$ is r.

The *depth* of the proper ideal I of the ring R is the supremum of lengths of regular sequences in I. We denote it by $depth(I)$. If $I = R$ then we define $depth(I) = \infty$.

Let σ be a germ of a smooth ($\mathbb{K}$-analytic) closed 2-form on $\Sigma_2 = \mathbb{K}^3$ and $rank\,\sigma|_0 = 0$. In the local coordinate system (x,y,z) on Σ_2 we have $\sigma = ady \wedge dz + bdz \wedge dx + cdx \wedge dy$, where a, b, c are smooth ($\mathbb{K}$-analytic) function-germs on Σ_2. By $I(\sigma)$ we denote the ideal of the ring of smooth ($\mathbb{K}$-analytic) function-germs on Σ_2 generated by a, b, c i.e. $I(\sigma) = < a, b, c >$. It is easy to see that $I(\sigma)$ does not depend on the local coordinate system on Σ_2. σ satisfies the condition $\alpha \wedge \sigma = 0$, where α is a germ of a contact form on $\mathbb{K}^3$. It implies that $I(\sigma)$ is generated by two function-germs.

500

In the $\mathbb{K}$-analytic category if $depthI(\sigma) \geq 2$ then the two generators of $I(\sigma)$ form a regular sequence of length 2 (see [3]). One can easily check that it implies that there does not exist a germ of a $\mathbb{K}$-analytic vector field on Σ_2 such that $X\rfloor\sigma = 0$ and $X|_0 \neq 0$. The inverse implication is not true in general. Now we formulate the following result in the smooth category.

Theorem 4.3. *Let ω_0 and ω_1 be germs of smooth singular symplectic forms on $\mathbb{R}^4$ with a common structurally smooth Martinet hypersurface Σ_2 at 0 and $rank\iota^*\omega_0|_0 = rank\iota^*\omega_1|_0 = 0$.*

If $\iota^*\omega_0 = \iota^*\omega_1 = \sigma$, ω_0 and ω_1 define the same canonical orientation of Σ_2 and the two generators of the ideal $I(\sigma)$ form a regular sequence of length 2 then there exists a germ of a smooth diffeomorphism $\Psi : (\mathbb{R}^4, 0) \to (\mathbb{R}^4, 0)$ such that

$$\Psi^*\omega_1 = \omega_0.$$

Proof. The proof is similar to the proof of Theorem 4.2. By Theorem 2.1 we obtain $\omega_0 = d(p_1\pi^*\alpha_0) + \sigma$ and $\omega_1 = d(p_1\pi^*\alpha_1) + \sigma$, where α_0, α_1 are germs of smooth contact forms on $\Sigma_2 = \{p_1 = 0\}$ such that $\alpha_0 \wedge \sigma = \alpha_1 \wedge \sigma = 0$ and $\alpha_0 \wedge d\alpha_0$, $\alpha_1 \wedge d\alpha_1$ define the same orientation on Σ_2.

α_0 is a contact form therefore $\alpha_0|_0 \neq 0$. We can find a coordinate system (x, y, z) on Σ_2 such that $\alpha_0 = f_0 dx + g_0 dy + h_0 dz$, where f_0, g_0 and h_0 are function-germs on Σ_2 and $h_0(0) \neq 0$. Let $\sigma = ady \wedge dz + bdz \wedge dx + cdx \wedge dy$, where a, b, c are function-germs on Σ_2 vanishing at 0. $\alpha_0 \wedge \sigma = 0$, thus we get $c = -\frac{f_0}{h_0}a - \frac{g_0}{h_0}b$. Thus $I(\sigma) = <a, b, c> = <a, b>$.

Let $\alpha_1 = f_1 dx + g_1 dy + h_1 dz$, where f_1, g_1, h_1 are functions-germs on Σ_2. From $\alpha_1 \wedge \sigma = 0$ we obtain the equation

$$a(f_1 - \frac{h_1}{h_0}f_0) + b(g_1 - \frac{h_1}{h_0}g_0) = 0 \tag{13}$$

and $a(0) = b(0) = 0$.

By assumptions a, b is a regular sequence.

Therefore $f_1 - \frac{h_1}{h_0}f_0 = br$ and $g_1 - \frac{h_1}{h_0}g_0 = -ar$, where r is a smooth function-germ on Σ_2 at 0.

Then proceeding in the same way as in the proof of Theorem 4.2 we get the result. $\square$

Acknowledgements

The author wishes to express his thanks to S. Janeczko, B. Jakubczyk and M. Zhitomirskii for many helpful conversations and remarks during writing this paper. The author is also grateful to the organizers of the Singularity 5 weeks programme at CIRM Luminy for hospitality. The author thanks the referee of this paper for many useful comments, especially for a simpler description of the 2-dimensional kernel of ω at 0. .

References

1. V. I. Arnold, S. M. Gusein-Zade, A. N. Varchenko, *Singularities of Differentiable Maps*, Vol. 1, Birhauser, Boston, 1985.
2. V. I. Arnold, A.B. Givental, *Symplectic Geometry, Itogi Nauki, Contemporary Problems in Mathematics, Fundamental Directions*, **4**,(1985),5-139 (Russian edn)
3. D. Eisenbud, *Commutative algebra with a view toward algebraic geometry*, Springer-Verlag, New York, 1994.
4. W. Domitrz, *Local invariants for singular symplectic structures*, in preparation.
5. W. Domitrz, S. Janeczko, *Equivalence of lagrangian germs in the presence of a surface*, Banach Center Publications, Vol. 39, (1997), 31-37.
6. W. Domitrz, S. Janeczko, *Normal forms of symplectic structures on the stratified spaces*, Colloquium Mathematicum, vol. LXVIII, No.1, (1995), 101-119.
7. W. Domitrz, S. Janeczko, Z. Pasternak-Winiarski, *Geometry and representation of the singular symplectic forms*, Banach Center Publications, Vol. 62, (2004), 57-71.
8. B. Jakubczyk, M. Zhitomirskii, *Local reduction theorems and invariants for singular contact structures*, Ann. Inst. Fourier (Grenoble) 51 (2001), no. 1, 237–295.
9. B. Jakubczyk, M. Zhitomirskii, *Odd-dimensional Pfaffian equations: reduction to the hypersurface of singular points*, C. R. Acad. Sci. Paris Sr. I Math. 325 (1997), no. 4, 423–428.
10. S. Janeczko, A. Kowalczyk, *On singularities in the degenerated symplectic geometry*, Hokkaido Mathematical Journal, Vol. 19(1990), 103-123.
11. M. Golubitsky and D. Tischler, *An example of moduli for singular symplectic forms*, Inventiones Math., 38 (1977), 219-225.
12. J. Martinet, *Sur les singularités des formes différentielles*, Ann. Inst. Fourier (Grenoble), 20 (1970), 95-178.
13. J. Martinet, *Singularities of Smooth Functions and Maps*, Cambridge Univ. Press, Cambridge, 1982.
14. J. Moser, *On volume elements on manifold*, Trans. Amer. Math. Soc., 120 (1965), 280-296.
15. R. Roussarie, *Modèles locaux de champs et de formes*, Astérisque, 30, (1975), 1-181.
16. A. Weinstein, *Lectures on Symplectic Manifolds*, CBMS Regional Conf. Ser. in Math. 29, Amer. Math. Soc., Providence, R.I., 1977.

DEVISSAGE DE LA FORME DE SEIFERT ENTIERE DES GERMES DE COURBE PLANE A DEUX BRANCHES

Philippe DU BOIS* et Emmanuel ROBIN

*LAREMA - UMR 6093 - Faculté des Sciences, Université d'Angers,
Angers, 49045, France*
** E-mail : pdubois@univ-angers.fr*

Nous proposons une méthode de dévissage de la forme de Seifert d'un germe de courbe plane. Sous certaines hypothèses techniques, nous expliquons comment trouver le(s) type(s) topologique(s) des germes associés à la forme de Seifert d'un germe de courbe plane à deux branches. Réciproquement, nous démontrons que deux germes de courbe plane à deux branches, qui sont "isomères", ont des formes de Seifert entières isomorphes. La filtration par le poids sur l'homologie entière de la fibre de Milnor est l'ingrédient clé de la démonstration.

A devissage method for the Seifert form of a plane curve germ is proposed. Assuming certain technical hypotheses, it is explained how one can find the topological type(s) of germs associated with the Seifert form of a given plane curve germ with two branches. Conversely, two plane curve germs with two branches, which are "isomeric", are shown to have isomorphic integral Seifert forms. The weight filtration on the integral homology of the Milnor fiber is the key ingredient of the proof.

Keywords: Plane curve germs. Monodromy. Milnor fiber. Seifert form. Units of cyclotomic fields.

Introduction

Soit $f : (\mathbf{C}^2, 0) \to (\mathbf{C}, 0)$ un germe de fonction analytique à singularité isolée et $f^{-1}(0)$ le germe de courbe plane associé. On note $K(f) := S^3_\varepsilon \cap f^{-1}(0)$ l'entrelacs de la singularité où $\varepsilon \ll 1$. Les composantes de $K(f)$ sont appelées *composantes de bord* de la fibre de Milnor F de f. Soit h une monodromie géométrique de F et h_* le morphisme induit par h sur $H_1(F, \mathbf{Z})$. On note $A(f)$ la *forme de Seifert* de f, on rappelle que $A(f)$ est la forme bilinéaire définie sur le $\mathbf{Z}$-module libre $H_1(F, \mathbf{Z})$ par $A(f)(\alpha, \beta) = \mathcal{L}(i_+(\alpha), \beta)$, où $\mathcal{L}(.,.)$ est le nombre d'enlacement dans la sphère S^3_ε, et $i_+(\alpha)$ est le cycle obtenu en poussant α hors de F dans une direction normale positive induite par l'orientation de F. Les principales propriétés de $A(f)$ sont les suivantes : la forme $A(f)$ est unimodulaire et h_*-équivariante, elle détermine la forme d'intersection S et le morphisme h_* par les égalités $S(x,y) = A(x,y) - A(y,x) = A(x,y) - A(x, h_*(y))$.

L'*arbre de désingularisation avec multiplicités* $T(f)$ du germe f est l'arbre dual de la résolution minimale de f, pondéré par les multiplicités des composantes irréductibles du diviseur exceptionnel de la résolution. Le sommet correspondant au premier diviseur qui apparait lors de la résolution par éclatements successifs sera noté #1 la multiplicité e_1 de ce sommet est égale à la multiplicité à l'origine du germe f, notée $\nu_0(f)$. La *valence $v(i)$* d'un sommet i de $T(f)$ est le nombre de voisins de i et e_i sa multiplicité. On dit que i est un *sommet de rupture* si $v(i) \geq 3$. Le *halo de multiplicité centrale* e_i est le n-uplet $\mathcal{H}_i := (e_i; (\eta_{ij})_{j \in V(i)})$ où $V(i)$ est l'ensemble des voisins de i et η_{ij} est l'entier de l'intervalle $[0, e_i - 1]$ congru à e_j modulo e_i. Une arête d'extrémités un sommet de rupture i et un autre sommet j est une *arête sortante* (resp. une *arête entrante*) pour le sommet i si $e_i < e_j$ (resp. $e_i > e_j$).

L'arbre de désingularisation avec multiplicités est équivalent au type topologique du germe. Si le germe est irréductible, d'après Burau (1933), le polynôme caractéristique de h_* détermine le type topologique de f. Par contre, si f est réductible, la forme de Seifert ne détermine pas le type topologique, voir (DBM94). On étudie ici la question suivante : *partant de la forme de Seifert A d'un germe de courbe plane, peut-on retrouver le(s) type(s) topologique(s) des germes de courbe plane dont la forme de Seifert est isomorphe à A ?*

Pour répondre à cette question, nous donnons une méthode de dévissage de $A(f)$ en deux étapes. Dans la première étape, on se demande si on peut déterminer à partir de $A(f)$ les données numériques suivantes : les halos des sommets de rupture, ainsi que les nombres d'enlacement des composantes de bord. Ensuite dans la seconde étape, on se demande quels sont les germes déterminés par ces données numériques. Nous suivons ainsi une démarche réciproque de celle suivie en (DBM94).

D'après (K96), la restriction de $A(f)$ à $\mathrm{Ker}(t-1)$ détermine les nombres d'enlacement des composantes de bord. Nous décrirons au §1 une méthode pour déterminer les halos de valence 3 à partir de la forme de Seifert. Cette méthode utilise la théorie des corps cyclotomiques.

À partir du §2, nous restreignons notre étude au cas des germes de courbe plane à deux branches. Sous certaines hypothèses techniques portant sur les unités cyclotomiques associées à la forme de Seifert, les propositions 2.3, 2.4 et 2.5 montrent que la forme de Seifert d'un tel germe permet de trouver comment les deux branches se séparent, en termes de la résolution minimale de $f^{-1}(0)$. On déterminera au passage la multiplicité du germe. Enfin, la proposition 2.6 donne une formule très simple qui calcule le nombre d'intersection des deux branches de f en termes des multiplicités dans $T(f)$, ce qui entraîne que ce nombre d'intersection est déterminé par la collection des halos de $T(f)$.

Le §3 est consacré à la reconstruction de l'arbre réduit $TR(f)$ de l'arbre de désingularisation $T(f)$ (*cf.* déf. 3.1) à partir de la forme de Seifert. Nous démontrerons (théorème 3.1) que, sous les mêmes hypothèses techniques, si deux germes de courbe plane à deux branches ont des formes de Seifert entières isomorphes, ils sont isomères (*cf.* déf. 3.2) . Voir (R99) pour la première version de cette étude.

Le §4 est consacré au calcul des paires de Zariski d'un germe de courbe plane à partir de l'arbre réduit $TR(f)$. Si le germe n'a pas d'isomères, et sous les mêmes hypothèses techniques, on en déduira (théorème 4.1) que la donnée de la forme de Seifert, à isomorphisme près, détermine le type topologique du germe. Nous démontrerons (théorème 4.2) que, si la reconstruction de l'arbre réduit de l'arbre de désingularisation $T(f)$ peut être effectuée de plusieurs manières, chaque reconstruction provient effectivement d'un germe de courbe plane, c'est-à-dire que chaque arbre réduit, isomère de l'arbre $TR(f)$, est l'arbre réduit de l'arbre de désingularisation $T(g)$ d'un germe de courbe plane, $T(g)$ est donc un isomère de $T(f)$; de plus, des germes isomères ont la même multiplicité, et le nombre d'intersection des deux branches ne dépend pas de l'isomère choisi.

Réciproquement, nous démontrerons (théorème 5.2) que deux germes de courbe plane à deux branches qui sont isomères ont des formes de Seifert entières isomorphes. Voir (DB03) pour l'annonce de ce résultat.

Enfin, le §6 est consacré au calcul d'un exemple suivant la méthode développée au cours de l'article.

Le point de départ de ce travail est la thèse du deuxième auteur (R99), complétée par (DB03).

1. La filtration par le poids et les halos

La forme de Seifert induit sur l'homologie de la fibre de Milnor, la filtration par le poids M et une structure isométrique (le couple (S, t)). Le gradué central de la filtration, $\mathrm{Gr}^{M}_{-1} H_1(F, \mathbf{Z})$, est la somme directe de facteurs associés aux halos de l'arbre de désingularisation. Nous étudierons d'abord

(Sec. 1.1) le problème de l'unicité de cette décomposition en somme directe. L'ingrédient clé est ici le calcul des résultants des polynômes cyclotomiques, dû à T. Apostol, (A70).

La structure isométrique sur $H_1(F, \mathbf{Z})$ induit une structure isométrique, notée (S_{-1}, t), sur le gradué central de la filtration. La décomposition en somme directe de $\mathrm{Gr}_{-1}^M H_1(F, \mathbf{Z})$ associée aux halos est orthogonale pour S_{-1}. Nous étudierons ensuite (Sec. 1.2) les facteurs de cette somme directe (correspondants aux halos de valence 3), en leur associant des unités cyclotomiques, dans le but de retrouver les halos à partir de ces unités.

1.1. *Décomposition de certains $\mathbf{Z}[t, t^{-1}]$-modules*

La filtration par le poids M sur l'homologie $H_1(F, \mathbf{Z})$ de la fibre de Milnor F du germe f est définie dans (DBM92) et utilisée dans (DBM93) et (DBM94), par les formules suivantes, où la monodromie homologique h_* est notée comme la multiplication par t, et e désigne un exposant de la monodromie.

$$M_0(H_1(F, \mathbf{Z})) = H_1(F, \mathbf{Z}), \ M_{-1}(H_1(F, \mathbf{Z})) = \mathrm{Ker}(t^e - 1),$$

$$M_{-2}(H_1(F, \mathbf{Z})) = \Big(\big(\mathrm{Ker}(t - 1) + \mathrm{Im}(t^e - 1) \big) \otimes \mathbf{Q} \Big) \cap H_1(F, \mathbf{Z}).$$

Dans (DBM94), on associe à chaque halo $\mathcal{H}_i$ une surface D_i qui est un revêtement cyclique d'ordre e_i de $\mathbf{P}^1$, ramifié en $v(i)$ points dont les entiers du revêtement sont les η_{ij}. On a l'isomorphisme de $\mathbf{Z}[t, t^{-1}]$-modules suivant :

$$\big(\mathrm{Gr}_{-1}^M H_1(F, \mathbf{Z}), S_{-1} \big) \cong \overset{\perp}{\underset{i}{\bigoplus}} \big(H_1(D_i, \mathbf{Z}), S_i \big),$$

où la somme directe (orthogonale pour S) est effectuée sur les halos de rupture, S_{-1} et S_i désignant les formes bilinéaires (unimodulaires) induites par la forme d'intersection S.

Considérons un sommet de rupture i de valence 3, notons $\mathcal{H}_i := (e_i; \eta_1, \eta_2, \eta_3)$ le halo associé et $H_1(D_i, \mathbf{Z})$ le facteur correspondant de la somme directe. Posons $m_{ij} = \mathrm{pgcd}(e_i, \eta_j)$, $r_i = \mathrm{pgcd}(m_{ij})$ et

$$\Lambda_i(t) = \frac{(t^{e_i} - 1)(t^{r_i} - 1)^2}{(t^{m_{i1}} - 1)(t^{m_{i2}} - 1)(t^{m_{i3}} - 1)},$$

on a l'isomorphisme de $\mathbf{Z}[t, t^{-1}]$-modules suivant :

$$H_1(D_i, \mathbf{Z}) \cong \frac{\mathbf{Z}[t, t^{-1}]}{(\Lambda_i(t))} := \mathbf{Z}_{\Lambda_i},$$

Soit i un sommet de rupture de valence $v(i) > 3$, la structure de $\mathbf{Z}[t, t^{-1}]$-module de $H_1(D_i, \mathbf{Z})$ est plus complexe que dans le cas de la valence 3, voir (DBM94, 1.17). Notons pour le moment que le polynôme caractéristique (resp. minimal) de l'action de t sur $H_1(D_i, \mathbf{Z})$ est Λ_i (resp. le polynôme réduit associé Λ_i^{red}), où :

$$\Lambda_i(t) = \frac{(t^{e_i} - 1)^{v(i)-2}(t^{r_i} - 1)^2}{(t^{m_{i1}} - 1)(t^{m_{i2}} - 1)\cdots(t^{m_{iv(i)}} - 1)}\,.$$

Les propriétés suivantes des $\Lambda_i(t)$ sont immédiates.

Lemme 1.1. *Soit i un sommet de rupture de $T(f)$, soit $J(i)$ l'ensemble des indices des polynômes cyclotomiques qui divisent Λ_i, i.e.*

$$J(i) = \{j \mid \Phi_j(t) \mid \Lambda_i(t)\} = \{j \in \mathbf{N} \mid j \mid e_i,\, j \nmid m_1,\, j \nmid m_2, \ldots, j \nmid m_{j(i)}\},$$

considérons le graphe $G_{J(i)}$ dont les sommets sont les $j \in J(i)$ et dont les arêtes joignent deux sommets j et j' s'il existe un nombre premier p tel que $j' = pj$. Le graphe $G_{J(i)}$ est connexe.

Lemme 1.2. *Soit i un sommet de rupture de $T(f)$, soit p et q des nombres premiers distincts et c un entier, $c > 0$. Supposons que $\Phi_{npq}(t)$ et $\Phi_n(t)$ divisent $\Lambda_i(t)$, alors $\Phi_{np}(t)$ et $\Phi_{nq}(t)$ divisent $\Lambda_i(t)$. Supposons que $\Phi_n(t)$ et $\Phi_{np^c}(t)$ divisent $\Lambda_i(t)$, alors $\Phi_{np}(t), \ldots, \Phi_{np^{c-1}}(t)$ divisent $\Lambda_i(t)$.*

On dit que que deux éléments A et B de $\mathbf{Z}[t, t^{-1}]$ sont *fortement premiers entre eux* s'il existe U et V dans $\mathbf{Z}[t, t^{-1}]$ tels que $AU + BV = 1$. On notera cette relation $(A, B)_{\mathbf{Z}} = 1$.

Proposition 1.1. *Considérons les polynômes cyclotomiques Φ_a et Φ_b, $a < b$. Les conditions suivantes sont équivalentes :*

(i) b/a n'est pas une puissance d'un nombre premier,
(ii) le résultant $R(\Phi_a, \Phi_b)$ de Φ_a et Φ_b est égal à 1,
(iii) Φ_a et Φ_b sont fortement premiers entre eux.

De plus, si $a > 1$ et s'il existe un nombre premier p et un entier $c > 0$ tels que $b = ap^c$, alors le résultant de Φ_a et Φ_b est :

$$\mathrm{res}(\Phi_a, \Phi_b) = \mathrm{res}(\Phi_a, \Phi_{ap^c}) = p^{\varphi(a)}.$$

Démonstration. Voir (A70) pour (i) $\Leftrightarrow$ (ii), ainsi que pour le calcul du résultant, et (VW31, §30, formule 4) pour (ii) $\Leftrightarrow$ (iii). $\qquad\square$

508

Lemme 1.3. *Soit $A(t), B(t) \in \mathbf{Z}[t]$ deux polynômes unitaires, notons (A, B) l'idéal de $\mathbf{Z}[t, t^{-1}]/A \cdot B$ engendré par A et B, l'indice du $\mathbf{Z}$-module (A, B) dans $\mathbf{Z}[t, t^{-1}]/A \cdot B$ est égal au résultant $\mathrm{res}(A, B)$ de A et B.*

Démonstration. Notons $\alpha = \deg(A)$ et $\beta = \deg(B)$, le $\mathbf{Z}$-module (A, B) est engendré par $A(t), tA(t), \ldots, t^{\beta-1}A(t), B(t), tB(t), \ldots, t^{\alpha-1}B(t)$. Par suite, l'indice cherché est égal au déterminant de Sylvester associé à A et B, et donc au résultant $\mathrm{res}(A, B)$. $\qquad\square$

Proposition 1.2. *Soit i un sommet de rupture de $T(f)$, de valence 3, soit g_i un générateur de $H_1(D_i, \mathbf{Z})$, de sorte que $H_1(D_i, \mathbf{Z}) = \mathbf{Z}[t, t^{-1}] \cdot g_i \cong \mathbf{Z}[t, t^{-1}]/\Lambda_i(t)$. Si $\Phi_a\Phi_b$ divise Λ_i, avec $a \neq b$, l'indice $\mathrm{ind}_i^{a,b}$ de $H_1(D_i, \mathbf{Z}) \cap \bigl(\ker(\Phi_a(t)) \oplus \ker(\Phi_b(t))\bigr)$ dans $H_1(D_i, \mathbf{Z}) \cap \ker(\Phi_a(t) \cdot \Phi_b(t))$ est alors égal à $\mathrm{res}(\Phi_a, \Phi_b)$. Si on suppose de plus que Φ_a et Φ_b ne sont pas fortement premiers entre eux, on a donc $\mathrm{ind}_i^{a,b} > 1$.*

Démonstration. On a, pour $\Psi = \Phi_a$, Φ_b ou $\Phi_a\Phi_b$:

$$H_1(D_i, \mathbf{Z}) \cap \ker(\Psi(t)) = \mathbf{Z}[t, t^{-1}] \cdot (\Lambda_i(t)/\Psi(t))g_i \cong \mathbf{Z}[t, t^{-1}]/\Psi(t),$$

le $\mathbf{Z}[t, t^{-1}]$-module $H_1(D_i, \mathbf{Z}) \cap \ker(\Phi_a(t) \cdot \Phi_b(t))$ est donc engendré par $\Lambda_i(t)/\bigl(\Phi_a(t)\Phi_b(t)\bigr)g_i$, et l'image de $H_1(D_i, \mathbf{Z}) \cap \bigl(\ker(\Phi_a(t)) \oplus \ker(\Phi_b(t))\bigr)$ dans le précédent est engendrée, en tant que $\mathbf{Z}[t, t^{-1}]$-module, par $\Phi_a(t) \cdot \bigl(\Lambda_i(t)/\bigl(\Phi_a(t)\Phi_b(t)\bigr)\bigr)g_i, \Phi_b(t) \cdot \bigl(\Lambda_i(t)/\bigl(\Phi_a(t)\Phi_b(t)\bigr)\bigr)g_i$, le résultat est donc donné par le lemme 1.3. $\qquad\square$

Proposition 1.3. *Soit i un sommet de rupture de $T(f)$, de valence 4, supposons que $\Phi_a\Phi_b$ divise Λ_i, où $a \mid b$ et $a \neq b$, et que $\mathrm{res}(\Phi_a, \Phi_b) > 1$. L'indice $\mathrm{ind}_i^{a,b}$ de $H_1(D_i, \mathbf{Z}) \cap \bigl(\ker(\Phi_a(t)) \oplus \ker(\Phi_b(t))\bigr)$ dans $H_1(D_i, \mathbf{Z}) \cap \ker(\Phi_a(t) \cdot \Phi_b(t))$ est alors > 1.*

Démonstration. Soit $\Lambda_i(t)$ le polynôme caractéristique de la monodromie sur $H_1(D_i, \mathbf{Z})$, le polynôme minimal est $\Lambda_i^{\mathrm{red}}(t)$ (*cf.* lemme 1.1). Définissons $A, B \in \mathbf{Z}[t]$ par $A(t)B^2(t) = \Lambda_i(t)$ et $A(t)B(t) = \Lambda_i^{\mathrm{red}}(t)$, on a l'isomorphisme suivant, associé à une matrice de présentation diagonale :

$$H_1(D_i, \mathbf{Q}) \cong \mathbf{Q}[t]/B(t) \oplus \mathbf{Q}[t]/(A(t)B(t)).$$

D'après (DBM94, 1.17), on peut trouver une famille de 2 générateurs $(g_{i,1}, g_{i,2})$ du $\mathbf{Z}[t, t^{-1}]$-module $H_1(D_i, \mathbf{Z})$, de telle sorte que la matrice de présentation associée M soit triangulaire supérieure. En comparant avec la forme réduite obtenue sur $\mathbf{Q}$, on voit qu'il existe $A_1, A_2, C \in \mathbf{Z}[t]$ tels que $A(t) = A_1(t)A_2(t)$ et

$$M = \begin{pmatrix} B(t)A_1(t) & B(t)C(t) \\ 0 & B(t)A_2(t) \end{pmatrix}.$$

Si Φ_a divise Λ_i, deux cas se présentent :

(i) $\Phi_a(t)$ ne divise aucun des $t^{m_j} - 1$, $j = 1, \ldots, 4$, et alors Φ_a divise B,

(ii) $\Phi_a(t)$ divise l'un des $t^{m_j} - 1$, $j = 1, \ldots, 4$, ceci pour un unique j, et alors Φ_a divise A.

Si Φ_a et Φ_b divisent Λ_i et si a divise b, trois cas se présentent :

(i) Φ_a et Φ_b divisent B,

(ii) $\Phi_b(t)$ ne divise aucun des $t^{m_j} - 1$ et $\Phi_a(t)$ divise l'un d'entre eux, alors Φ_a divise A et Φ_b divise B,

(iii) Φ_a et Φ_b divisent A.

Dans le cas (i), le produit $\Phi_a\Phi_b$ est en facteur dans la matrice M et on vérifie facilement que :

$$H_1(D_i, \mathbf{Z}) \cap \ker(\Phi_a(t)\Phi_b(t)) \cong \mathbf{Z}[t, t^{-1}]/(\Phi_a(t)\Phi_b(t)) \oplus \mathbf{Z}[t, t^{-1}]/(\Phi_a(t)\Phi_b(t)),$$

par suite, l'indice cherché vérifie $\text{ind}_i^{a,b} = \text{res}(\Phi_a, \Phi_b)^2 > 1$.

Dans le cas (ii), numérotons les voisins du sommet i de telle sorte que $\Phi_a(t)$ divise $t^{m_3} - 1$, choisissons les générateurs donnés dans (*loc. cit.*) en accord avec cette numérotation, on trouve alors que $B(t)A_2(t)$ divise $(t^{e_i} - 1)/(t^{m_3} - 1)$, par suite Φ_a ne divise pas BA_2 et donc divise A_1. On vérifie facilement que :

$$H_1(D_i, \mathbf{Z}) \cap \ker(\Phi_a(t)\Phi_b(t)) \cong \mathbf{Z}[t, t^{-1}]/(\Phi_a(t)\Phi_b(t)) \oplus \mathbf{Z}[t, t^{-1}]/(\Phi_b(t)),$$

par suite, l'indice cherché vérifie $\text{ind}_i^{a,b} = \text{res}(\Phi_a, \Phi_b) > 1$.

Dans le cas (iii), on peut de même choisir les générateurs de telle sorte que $\Phi_a\Phi_b$ divise A_1, on trouve ici :

$$H_1(D_i, \mathbf{Z}) \cap \ker(\Phi_a(t)\Phi_b(t)) \cong \mathbf{Z}[t, t^{-1}]/(\Phi_a(t)\Phi_b(t)),$$

par suite, l'indice cherché vérifie $\text{ind}_i^{a,b} = \text{res}(\Phi_a, \Phi_b) > 1$. $\qquad\square$

Théorème 1.1. *Soit $f : (\mathbf{C}^2, 0) \to (\mathbf{C}, 0)$ un germe de fonction analytique à singularité isolée, supposons que les sommets de rupture de l'arbre $T(f)$ sont de valence 3 ou 4, et que les polynômes Λ_i, $i \in \Re := \{i \mid v(i) \geq 3\}$ sont 2 à 2 premiers entre eux, on peut alors retrouver la famille $(\Lambda_i)_{i \in \Re}$ à partir de la forme de Seifert $A(f)$ en procédant comme suit. Soit Φ_a et*

Φ_b deux polynômes cyclotomiques distincts tels que $\Phi_a\Phi_b$ divise $\prod_{i\in\Re}\Lambda_i$ et $\mathrm{res}(\Phi_a,\Phi_b) > 1$, alors Φ_a et Φ_b divisent le même Λ_i si, et seulement si, les sous $\mathbf{Z}[t,t^{-1}]$-modules de $H_1(F,\mathbf{Z})$ suivants, $\ker(\Phi_a(t)) \oplus \ker(\Phi_b(t))$ et $\ker(\Phi_a(t)\cdot\Phi_b(t))$, sont distincts.

Si tous les sommets de rupture sont de valence 3, ceci donne directement les $(\Lambda_i)_{i\in\Re}$. S'il existe des sommets de rupture de valence 4, le procédé donne les $(\Lambda_i^{\mathrm{red}})_{i\in\Re}$, on en déduit facilement les $(\Lambda_i)_{i\in\Re}$. On notera que si f est un germe à deux branches, $T(f)$ admet au plus un sommet de valence 4 et n'admet aucun sommet de valence > 4.

Démonstration. D'après les propositions 1.2 et 1.3, si $\Phi_a\Phi_b$ divise Λ_{i_0}, l'indice de $\ker(\Phi_a(t)) \oplus \ker(\Phi_b(t)) = H_1(D_{i_0},\mathbf{Z}) \cap \big(\ker(\Phi_a(t)) \oplus \ker(\Phi_b(t))\big)$ dans $\ker(\Phi_a(t)\cdot\Phi_b(t)) = H_1(D_{i_0},\mathbf{Z})\cap\ker(\Phi_a(t)\cdot\Phi_b(t))$ est > 1. Si Φ_a divise Λ_{i_1} et Φ_b divise Λ_{i_2}, $i_1 \neq i_2$, on a les égalités suivantes, d'où le théorème :

$$\ker(\Phi_a(t)) \oplus \ker(\Phi_b(t)) =$$

$$\big(H_1(D_{i_1},\mathbf{Z}) \cap \ker(\Phi_a(t))\big) \oplus \big(H_1(D_{i_2},\mathbf{Z}) \cap \ker(\Phi_b(t))\big) =$$

$$\big(H_1(D_{i_1},\mathbf{Z}) \cap \ker(\Phi_a(t) \cdot \Phi_b(t))\big) \oplus \big(H_1(D_{i_2},\mathbf{Z}) \cap \ker(\Phi_a(t) \cdot \Phi_b(t))\big) =$$

$$\ker(\Phi_a(t) \cdot \Phi_b(t)). \qquad\qquad \square$$

Remarque 1.1. Si l'on ne suppose plus que les polynômes Λ_i, $i \in \Re := \{i \mid v(i) \geq 3\}$ sont 2 à 2 premiers entre eux, on ne peut pas espérer prolonger le théorème 1.1 en toute généralité, ainsi que le montre l'exemple suivant.

Supposons que $\Phi_{np}\Phi_{nq}\Phi_{npq}^2$ divise $\prod_{i\in\Re}\Lambda_i$, où p et q sont des nombres premiers distincts, et n est un nombre entier, et que les calculs d'indice comme dans les propositions 1.2 et 1.3 nous assurent qu'il existe i_1 et i_2 tels que $\Phi_{np}\Phi_{npq}$ divise Λ_{i_1} et $\Phi_{nq}\Phi_{npq}$ divise Λ_{i_2} ; la méthode proposée ne permet pas de distinguer les cas $i_1 = i_2$ et $i_1 \neq i_2$. Cependant, l'utilisation du lemme 1.2 permet d'utiliser les calculs d'indice pour prolonger le résultat d'unicité du théorème 1.1, voir l'exemple au §6 ou la proposition suivante.

Proposition 1.4. Supposons que le produit $\Phi_n\Phi_{np}^2\Phi_{npq}$, où p et q sont des nombres premiers distincts, apparaît dans la décomposition de $\prod_{i\in\Re}\Lambda_i$ en produit de puissances de polynômes cyclotomiques distincts, supposons de plus que l'on ait démontré (en utilisant la prop. 1.2) qu'il existe i_0 et i_1 tels que $\Phi_{np}\Phi_{npq} \mid \Lambda_{i_0}$ et $\Phi_n\Phi_{np} \mid \Lambda_{i_1}$. Alors, si Φ_{nq} ne divise pas $\prod_{i\in\Re}\Lambda_i$, on a $i_0 \neq i_1$; si $\prod_{i\in\Re}\Lambda_i$ est divisible par Φ_{nq}, mais pas par Φ_{nq}^2, on a

$i_0 = i_1$ *si, et seulement si, l'indice de* $\ker(\Phi_n) \oplus \ker(\Phi_{nq}) \oplus \ker(\Phi_{npq})$ *dans* $\ker(\Phi_n \Phi_{nq} \Phi_{npq})$ *est égal à* $\mathrm{res}(\Phi_{nq}, \Phi_n \Phi_{npq}) = q^{\varphi(n)} p^{\varphi(nq)}$.

1.2. *Invariant complet de la structure isométrique associée à un halo de valence 3 et unités cyclotomiques*

Nous allons donner un invariant complet de la structure isométrique $(H_1(D_i, \mathbf{Z}), S_i, t)$, et en déduire une méthode pour déterminer un halo de valence 3 à partir de cet invariant. Nous omettrons l'indice i pour alléger l'écriture.

Soit h_1 et h_2 des éléments de $H_1(D, \mathbf{Z})$, on notera $\mathbf{S}(h_1, h_2)$ l'élément de $\mathbf{Z}[[t, t^{-1}]]$ suivant :

$$\mathbf{S}(h_1, h_2) := \sum_{n \in \mathbf{Z}} S(h_1, t^n h_2) t^n.$$

On trouve immédiatement les égalités $\mathbf{S}(th_1, h_2) = \mathbf{S}(h_1, t^{-1}h_2) = t\mathbf{S}(h_1, h_2)$. Décomposons la série $\mathbf{S}(h_1, h_2)$ de la façon suivante : $\mathbf{S}(h_1, h_2) = \mathbf{S}_-(h_1, h_2) + \mathbf{S}_+(h_1, h_2)$ où les degrés des termes de $\mathbf{S}_+(h_1, h_2)$ sont minorés, et ceux de $\mathbf{S}_-(h_1, h_2)$ majorés. On a $\Lambda(t)\mathbf{S}(h_1, h_2) = \mathbf{S}(\Lambda(t)h_1, h_2) = 0$, d'où $\Lambda(t)\mathbf{S}_+(h_1, h_2) = -\Lambda(t)\mathbf{S}_-(h_1, h_2)$ et, par suite, $\Lambda(t)\mathbf{S}_+(h_1, h_2) \in \mathbf{Z}[t, t^{-1}]$. La série $\mathbf{S}_+(h_1, h_2)$ étant définie à l'addition d'un polynôme près, $U(h_1, h_2) := \Lambda(t)\mathbf{S}_+(h_1, h_2)$ est ainsi un élément bien défini de $\mathbf{Z}_\Lambda$. Réciproquement, $U(h_1, h_2)$ détermine la série $\mathbf{S}(h_1, h_2)$ par l'égalité suivante, grâce à la relation de périodicité $S(h_1, t^n h_2) = S(h_1, t^{n+e} h_2)$:

$$\frac{t^e - 1}{\Lambda(t)} U(h_1, h_2) = (t^e - 1) \sum_{n=0}^{+\infty} S(h_1, t^n h_2) t^n = - \sum_{n=0}^{e-1} S(h_1, t^n h_2) t^n.$$

Soit maintenant g un générateur du $\mathbf{Z}[t, t^{-1}]$-module $H_1(D, \mathbf{Z})$, soit $v \in \mathbf{Z}_\Lambda^\times$, où $\mathbf{Z}_\Lambda^\times$ désigne le groupe des unités de $\mathbf{Z}_\Lambda$, on a l'égalité $U(vg, vg) = v(t)v(t^{-1})U(g, g)$. Notons $N_+(v) = v(t)v(t^{-1})$ et $U = U(t)$ la classe de $U(g, g)$ dans $\mathbf{Z}_\Lambda^\times / N_+(\mathbf{Z}_\Lambda^\times)$.

Théorème 1.2 (Robin, 1999). *Avec les notations ci-dessus, $U(g, g)$ est une unité de $\mathbf{Z}_\Lambda$, sa classe U dans $\mathbf{Z}_\Lambda^\times / N_+(\mathbf{Z}_\Lambda^\times)$ est un invariant complet de la structure isométrique $(H_1(D, \mathbf{Z}), S, t)$. De plus, on trouve :*

$$U = t^r \frac{t^{\eta_1} - 1}{t^{m_1} - 1} \times \frac{t^{\eta_2} - 1}{t^{m_2} - 1} \times \frac{t^{\eta_3} - 1}{t^{m_3} - 1}.$$

Démonstration. Voir (DBM94, 1.16) pour la méthode et (R99, 4.2.4) pour un calcul détaillé. L'invariant U est une unité, car la structure isométrique est unimodulaire. $\qquad\square$

La forme d'intersection S étant antisymétrique, il peut être préférable de caractériser la structure isométrique ci-dessus par une unité symétrique en t et t^{-1}. Pour cela, on remarque que le degré du polynôme Λ est pair, disons $\deg(\Lambda) = 2d$, plus précisement, $\Lambda(t)$ est un produit de polynômes cyclotomiques 2 à 2 distincts, et ce produit n'est divisible ni par $\Phi_1(t) = t - 1$, ni par $\Phi_2(t) = t + 1$. On a de plus, pour tout halo de valence 3, $\eta_{i1} + \eta_{i2} + \eta_{i3} = e_i$. On pose alors $U'(t) = t^{-d}U(t)$ et on vérifie immédiatement que :

$$U(t^{-1}) = t^{-2d}U(t) \quad \text{et} \quad U'(t^{-1}) = U'(t).$$

Considérons le morphisme d'anneaux $\mathbf{Z}_\Lambda \to \mathbf{C}$ défini par $t \mapsto \exp(2\pi i/e)$. L'image $U'(\exp(2\pi i/e))$ de $U'(t)$ par ce morphisme est donc un nombre réel, dont le signe ne change pas si on remplace $U'(t)$ par $U'(t)v(t)v(t^{-1})$, où $v(t)$ est une unité de $\mathbf{Z}_\Lambda$. Le calcul de l'argument de $U'(\exp(2\pi i/e))$ montre que ce nombre réel est positif. Par suite, si $U(t)$ est une unité de $\mathbf{Z}_\Lambda$ associée à une structure isométrique comme ci-dessus, alors $-U(t)$ ne peut pas être associé à une structure isométrique dont le polynôme annulateur est le même polynôme $\Lambda(t)$.

1.3. *Détermination d'un halo de valence 3*

La question que nous nous posons maintenant est de déterminer un halo de valence 3 à partir de la donnée de l'unité associée $U \in \mathbf{Z}_\Lambda^\times/N_+(\mathbf{Z}_\Lambda^\times)$. Nous allons pour cela utiliser les applications naturelles de l'anneau $\mathbf{Z}_\Lambda$ dans les corps cyclotomiques $\mathbf{K}_n := \mathbf{Q}[t]/\Phi_n(t)$ tels que Φ_n divise Λ. Soient E_n le groupe des unités de $\mathbf{K}_n$, $C_n := \{\pm t^b \prod_a (t^a - 1)\} \cap E_n$ le groupe des unités circulaires et W_n le groupe des racines de l'unité. On notera $\widetilde{E}_n = E_n/W_n$ et $\widetilde{C}_n = C_n/W_n$, on rappelle que $\widetilde{E}_n$ et $\widetilde{C}_n$ sont des groupes abéliens libres de rang $\frac{1}{2}\varphi(n) - 1$. De plus, pour tout $u \in E_n$, il existe $\zeta \in W_n$ tel que $u\bar{u} = \zeta u^2$, en effet, si on pose $\zeta = \bar{u}/u$, on trouve que ζ est un entier algébrique dont tous les conjugués sont de module 1, par suite ζ est une racine de l'unité. L'image de U dans $\mathbf{K}_n$, qu'on note toujours U, donne donc un élément bien défini de $M_n := \widetilde{C}_n/(\widetilde{C}_n \cap \widetilde{E}_n^2)$, qui est un espace vectoriel de dimension $\frac{1}{2}\varphi(n) - 1$ sur $\mathbf{Z}/2\mathbf{Z}$.

On trouvera dans l'article de R. Gold et J. Kim (GK89) la construction d'un ensemble minimal de générateurs de C_n, ce qui donne une base de l'espace vectoriel M_n. Notons γ_a l'image de $t^a - 1$ dans $\widetilde{C}_n$, les relations de Bass, rappelées ci-dessous (voir (B66), (E72) ou (W83, 8.9)) permettent alors de travailler dans $\widetilde{C}_n$ ou dans M_n :

$$\gamma_a = \gamma_{n-a} \quad \text{et} \quad \gamma_{km} = \prod_{j=0}^{m-1} \gamma_{k+jn/m}, \text{ pour } m \mid n \text{ et } km \not\equiv 0 \;(\text{mod } n).$$

D'après (S78), $[\widetilde{E}_n : \widetilde{C}_n] = 2^b h_n^+$ où h_n^+ est le nombre de classes d'idéaux du sous-corps réel maximal de $\mathbf{K}_n$ et où $b = 0$ si n a 1, 2 ou 3 facteurs premiers et $b > 0$ sinon. Nous obtenons alors des résultats qui dépendent du nombre de facteurs premiers de la multiplicité e.

Soit e le produit de s puissances de nombres premiers distincts, soient η_1, η_2, η_3 trois entiers de l'intervalle $[1, e-1]$ tels que $\eta_1 + \eta_2 + \eta_3 \equiv 0 \;(\text{mod } e)$, notons, pour $j = 1, 2$ ou 3, $m_j = \text{pgcd}(e, \eta_j)$, $r = \text{pgcd}(m_1, m_2, m_3)$ et

$$\Lambda(t) = \frac{(t^e - 1)(t^r - 1)^2}{(t^{m_1} - 1)(t^{m_2} - 1)(t^{m_3} - 1)}.$$

Par définition, le *problème $\mathcal{U}_s$ associé à e et Λ est* : "*déterminer les entiers η_1, η_2, η_3 à partir de la classe de*

$$U = t^r \frac{t^{\eta_1} - 1}{t^{m_1} - 1} \times \frac{t^{\eta_2} - 1}{t^{m_2} - 1} \times \frac{t^{\eta_3} - 1}{t^{m_3} - 1}$$

dans le groupe $\mathbf{Z}_\Lambda^\times / N_+(\mathbf{Z}_\Lambda^\times)$".

Théorème 1.3 (Robin, 1999). *(R99, th. 4.23, 4.43, 4.46, 4.54) Soit e un entier positif. On considère les halos de valence 3 dont la multiplicité centrale est égale à e.*

(i) *Dans le cas où $e = p^m$ avec p premier ou $e = 2p^m$ avec p premier impair, si $h_{p^m}^+$ est impair, alors la structure isométrique $(H_1(D, \mathbf{Z}), S, t)$ détermine le halo associé.*

(ii) *Supposons que $e = p_1^a p_2^b \not\equiv 2 \;(\text{mod } 4)$ avec p_1, p_2 premiers distincts. Soit $I_e = \{1 \leq a < e/2 \mid (a, e) = 1\}$. Supposons vérifiées les conditions suivantes : h_e^+ est impair, la relation $\prod_{a \in I_e}(t^a - 1) \in W_e$ est la seule relation de la famille $(t^a - 1)_{a \in I_e}$ et $m_1 + m_2 + m_3 < \frac{1}{4}\varphi(e)$ (noter que cette condition porte sur le polynôme Λ). Alors la structure isométrique $(H_1(D, \mathbf{Z}), S, t)$ détermine le halo associé.*

(iii) *Supposons que $e = \prod_{i=1}^{s} p_i^{a_i}$ où $s \geq 3$ et les p_i sont des nombres premiers distincts. Supposons qu'on sache résoudre les problèmes $\mathcal{U}_1$ associés à $p_i^{a_i}$ $(1 \leq i \leq s)$ et les problèmes $\mathcal{U}_2$ associés à $p_i^{a_i} p_j^{a_j}$ $(1 \leq i < j \leq s)$, ceci pour n'importe quel polynôme Λ. Alors la structure isométrique $(H_1(D, \mathbf{Z}), S, t)$ détermine le halo associé en dehors de la situation* *particulière suivante : quitte à échanger les m_j, on a $m_1 = m_2$, et à une permutation près des p_i, si l'on pose $m_3 = \prod_{i=1}^{s} p_i^{\gamma_i}$, il existe un entier q, $1 \leq q \leq s - 2$, tel que, pour $1 \leq i \leq q$, on a $\gamma_i < a_i$ et $\eta_1 \equiv \eta_2$ (mod $p_i^{a_i}$) et, pour $q+1 \leq i \leq s$, on a $\gamma_i = a_i$. Cependant, on détermine l'entier η_3 dans cette situation particulière.*

On obtient le point (i) car la famille $\left((t^a - 1)/(t - 1)\right)_{1 < a < p^m/2\,,\,(a,p)=1}$ est une base du $\mathbf{Z}/2\mathbf{Z}$-espace vectoriel $\widetilde{C}_e / \widetilde{C}_e^2$. Ce résultat n'est plus vrai lorsque e est composée. Mais lorsque e a deux facteurs premiers, on a une méthode similaire si la famille $(t^a - 1)_{a \in I_e}$ a une seule relation. La troisième hypothèse provient de l'existence de cette relation car celle-ci implique deux écritures de l'invariant U dans cette famille génératrice. On obtient le point (iii) en appliquant le théorème des restes chinois. Ce théorème n'englobe pas toutes les multiplicités e. Cependant, h_e^+ étant toujours supposé impair, la même méthode s'applique dans tous les cas pour déterminer le halo en partant de l'unité U, mais nous n'avons pas de résultat général quant à l'unicité du halo correspondant à une unité. Si h_e^+ est de la forme $2^a b$ où b est impair et $a \leq 8$, on peut encore appliquer notre méthode, voir (R99, 4.60). D'après les tables numériques, voir (W83), cette hypothèse est vérifiée pour tout e inférieur à 300. Le cas de la structure isométrique associée à un halo i_0 de valence 4 ne semble pas pouvoir être traité de façon praticable : nous avons une grande latitude dans le choix des générateurs $(g_{i_0,1}, g_{i_0,2})$, et ceux-ci ont une forme peu propice au calcul des séries $\mathbf{S}(g_{i_0,1}, g_{i_0,1})$, $\mathbf{S}(g_{i_0,1}, g_{i_0,2})$ et $\mathbf{S}(g_{i_0,2}, g_{i_0,2})$. Nous utiliserons uniquement le polynôme Λ_{i_0}, ce qui donne e_{i_0} et les $m_{i_0,j} = \mathrm{pgcd}(e_{i_0}, \eta_{i_0,j})$, où $j = 1, \ldots, 4$, sans chercher à retrouver le halo $\mathcal{H}_{i_0} := (e_{i_0}; \eta_{i_0,1}, \ldots, \eta_{i_0,4})$.

Nous avons donc une décomposition en somme directe orthogonale (pour S) de $\mathrm{Gr}_{-1}^{M}(H_1(F, \mathbf{Z}))$, associée aux sommets de rupture de $T(f)$, et les structures isométriques sur les facteurs de cette décomposition ont, dans le cas des sommets de valence 3, une forme très particulière, donnée par une unité cyclotomique qui se déduit simplement du halo associé. Ceci excuse ou justifie l'hypothèse $\mathcal{HT}$ que nous allons faire pour continuer notre travail.

2. L'arbre de désingularisation au voisinage du sommet #1

Nous allons maintenant restreindre notre étude au cas des germes de courbe plane à deux branches. Dans toute la suite, $f : (\mathbf{C}^2, 0) \to (\mathbf{C}, 0)$ désignera un germe de fonction analytique à singularité isolée, définissant un germe de courbe plane à deux branches. On se donne la forme de Seifert $A(f)$ du germe f. Sous l'hypothèse $\mathcal{HT}$ ci-dessous, nous allons trouver, en partant de $A(f)$, comment les deux branches de f se séparent, *cf.* prop. 2.3, 2.4 et 2.5. Nous déterminerons au passage la multiplicité du germe. Enfin, la proposition 2.6 donne une formule très simple qui calcule le nombre d'intersection des deux branches de f en termes des multiplicités dans $T(f)$, ce qui entraîne que ce nombre d'intersection est déterminé par la collection des halos de $T(f)$.

L'hypothèse de travail $\mathcal{HT}$

Nous supposerons dorénavant que la décomposition en somme directe orthogonale (pour S) du $\mathbf{Z}[t, t^{-1}]$-module $\mathrm{Gr}^M_{-1} H_1(F, \mathbf{Z})$ est unique, et que, dans cette décomposition, la structure isométrique de chaque facteur associé à un sommet de rupture de valence 3 définit un unique triplet (η_1, η_2, η_3) ; on rappelle qu'un tel facteur est isomorphe à $\mathbf{Z}[t, t^{-1}]/(\Lambda_i(t))$ pour un certain $\Lambda_i(t)$. La filtration M étant définie en termes de la forme de Seifert (via la monodromie homologique), ainsi que les structures isométriques sur les facteurs directs de $\mathrm{Gr}^M_{-1} H_1(F, \mathbf{Z})$, notre hypothèse porte donc sur la forme de Seifert et sur les unités cyclotomiques que nous lui avons associé.

Définition 2.1. *On notera φ et φ' des développements de Puiseux des branches du germe f. Les paires de Zariski de φ (resp. φ') seront notées $p_1/q_1, \ldots, p_g/q_g$ (resp. $p'_1/q'_1, \ldots, p'_{g'}/q'_{g'}$). L'exposant de coïncidence de φ et φ' est le nombre rationnel suivant :*

$$\mathfrak{C}(\varphi, \varphi') = \max(\mathrm{val}(\sigma(\varphi) - \tau(\varphi'))),$$

où $\sigma(\varphi)$ (resp. $\tau(\varphi')$) parcourt l'ensemble des développements de Puiseux de la branche définie par φ (resp. φ'), cf. (MW85, 3.2). Un nombre rationnel r est un exposant permis pour φ si les développements de Puiseux $\varphi(x)$ et $x^r + \varphi(x)$ ont les mêmes paires caractéristiques. Il s'ensuit que r est dans $(q_1 \cdots q_g)^{-1}\mathbf{N}$. L'exposant de coïncidence est permis pour l'une au moins des deux branches, on conviendra qu'il est permis pour la branche φ. On définit l'entier c par la double inégalité suivante :

$$\frac{p_1}{q_1} + \frac{p_2}{q_1 q_2} + \cdots + \frac{p_c}{q_1 \cdots q_c} \leq \mathfrak{C}(\varphi, \varphi') < \frac{p_1}{q_1} + \frac{p_2}{q_1 q_2} + \cdots + \frac{p_{c+1}}{q_1 \cdots q_{c+1}}.$$

On a donc $\mathfrak{C}(\varphi, \varphi') \in \dfrac{1}{q_1 \cdots q_c}\mathbf{N}$ et $e_1 = \nu_0(f) = q_1 \cdots q_g + q'_1 \cdots q'_{g'}$. On reprend les notations données aux §0 et §1. On notera $\Gamma(1, \varphi)$ (resp. $\Gamma(1, \varphi')$) la géodésique de $T(f)$ joignant #1 à la flèche symbolisant φ (resp. φ'). On notera som$(T(f))$ l'ensemble des sommets de $T(f)$, m_{ij} le pgcd des

516

multiplicités des sommets i et j, extrémités de l'arête (ij) de $T(f)$; ce pgcd est constant le long d'un segment géodésique, on rappelle qu'un segment géodésique *est la réunion des arêtes situées entre deux sommets de rupture consécutifs sur une géodésique de $T(f)$, ou entre un sommet de rupture et un sommet de valence 2 portant une flèche symbolisant φ ou φ', consécutifs sur une géodésique de $T(f)$. Une* branche morte *est la réunion des arêtes situées entre un sommet de rupture et un sommet de valence 1 distinct du sommet #1. Le pgcd m_{ij} est constant le long d'une branche morte ; de plus, il est égal à la multiplicité du sommet de valence 1 de la branche. De même, si le sommet #1 est de valence 1, le m_{ij} est constant, et égal à e_1, le long de la réunion des arêtes situées entre #1 et le sommet de rupture de plus petite multiplicité ; dans ce cas, on appellera* branche géodésique d'extrémité #1 *cette réunion d'arêtes. Enfin, on appellera* sommet de séparation *le sommet de rupture en lequel se séparent les géodésiques de $T(f)$ qui joignent le sommet #1 aux sommets qui portent les flèches symbolisant les composantes de f.*

Le lemme suivant donne la multiplicité sortante après un sommet de rupture. Il se déduit de (MW85, 5.4.1 et 6.6.4), voir aussi (BK86). Il va nous permettre de distinguer entre branche sortante et branche entrante. Nous donnons l'énoncé pour un germe ayant un nombre quelconque de branches.

Lemme 2.1. *Soit $g : (\mathbf{C}^2, 0) \to (\mathbf{C}, 0)$ un germe de fonction analytique à singularité isolée, ayant un nombre de branches quelconque, soit $T(g)$ l'arbre de désingularisation avec multiplicités de g. Soit i un sommet de rupture de $T(g)$, ou le sommet #1 s'il est de valence 2. Soit i^Λ le sommet voisin de i sur une arête sortante de i. Notons $(\chi^\ell)_{\ell \in \Lambda}$ des développements de Puiseux des branches de g telles que la géodésique $\Gamma(1, \chi^\ell)$ de $T(g)$ joignant #1 à la flèche symbolisant χ^ℓ passe par l'arête qui porte le sommet i^Λ ; notons $q_{k_\ell}, \ldots, q_{g_\ell}$ les dénominateurs des paires de Zariski de χ^ℓ qui interviennent, dans la suite d'éclatements donnant la désingularisation minimale de g, après l'éclatement qui crée le diviseur représenté par i. On a alors :*

$$e_i^\Lambda \equiv \sum_{\ell \in \Lambda} q_{k_\ell} \cdots q_{g_\ell} \quad (\mathrm{mod}\, e_i).$$

Proposition 2.1. *Soit $f : (\mathbf{C}^2, 0) \to (\mathbf{C}, 0)$ un germe de fonction analytique à singularité isolée, définissant un germe de courbe plane à deux branches. Considérons l'arbre de désingularisation $T(f)$ du germe f.*

(i) *Le long d'une branche morte, le pgcd m_{ij} est constant, et sa valeur est strictement supérieure à $e_1 = \nu_0(f)$.*

(ii) *Si $T(f)$ présente une branche géodésique d'extrémité #1, le pgcd m_{ij} est constant le long de cette branche, et sa valeur est égale à $e_1 = \nu_0(f)$.*

(iii) *Le long d'un segment géodésique, le pgcd m_{ij} est constant, et sa valeur est strictement inférieure à $e_1 = \nu_0(f)$.*

Démonstration. La multiplicité d'un sommet de valence 1 extrémité de branche morte est strictement supérieure à e_1, le point (i) se déduit donc de la déf. 2.1. Le point (ii) est donné plus haut. Si un segment géodésique commence par une arête sortante pour le sommet de rupture i, le lemme 2.1 donne la majoration suivante : $\mathrm{pgcd}(e_i, e_i^\Lambda) < e_1 = q_1 \cdots q_g + q_1' \cdots q_{g'}'$. Ceci donne le point (iii) à l'exception du cas suivant. Si l'exposant de coïncidence $\mathfrak{C}(\varphi, \varphi')$ est permis pour la branche φ, mais non pour la branche φ', considérons le segment géodésique de l'arbre $T(f)$ dont les extrémités, $i(c)$ et $i'(c)$, sont associées à la c-ième paire de Zariski de la branche φ, pour $i(c)$, ou φ', pour $i'(c)$. Autrement dit, $i(c)$ est le sommet où les géodésiques $\Gamma(1, \varphi)$ et $\Gamma(1, \varphi')$ se séparent, et $i'(c)$ est le sommet de rupture suivant sur $\Gamma(1, \varphi')$ (ou, à défaut, le sommet portant la flèche φ'). Les arêtes situées aux extrémités de ce segment géodésique sont toutes deux des arêtes entrantes (pour le sommet $i(c)$ ou pour le sommet $i'(c)$). Notons $i(c)^{\varphi'}$ le sommet voisin de $i(c)$ dans la direction de la flèche associée à φ', notons $e_{i(c)}^{\varphi'}$ sa multiplicité et $m_{i(c)\varphi'} = \mathrm{pgcd}(e_{i(c)}, e_{i(c)}^{\varphi'})$, alors, d'après le lemme 2.17 de (R99), le pgcd $m_{i(c)\varphi'}$ le long du segment géodésique considéré divise $q_c' \cdots q_{g'}'$. On a donc encore $m_{i(c)\varphi'} < e_1$. $\qquad\square$

D'après (AC75, th. 4), le polynôme caractéristique de l'action de la monodromie sur $H_1(F, \mathbf{Z})$ est, dans le cas d'un germe à deux branches :

$$\Delta(t) = (t - 1) \prod_i (t^{e_i} - 1)^{v(i) - 2},$$

où l'on effectue le produit sur l'ensemble des sommets de $T(f)$. La forme de Seifert nous donne donc, via le polynôme caractéristique Δ, l'élément $\sum_i (v(i) - 2)[e_i]$ de $\mathbf{Z}[\mathbf{N}]$. Rappelons que $T(f)$ comporte au plus un sommet de valence 4 dans le cas d'un germe à deux branches. Si c'est le cas, notons i_0 ce sommet, le polynôme caractéristique de l'action induite par la monodromie sur $H_1(D_{i_0}, \mathbf{Z})$ est alors, en notant $\mathcal{H}_{i_0} = (e_{i_0}; \eta_1, \eta_2, \eta_3, \eta_4)$ le halo correspondant, puis $m_{i_0 j} = \mathrm{pgcd}(e_{i_0}, \eta_j)$ et $r_{i_0} = \mathrm{pgcd}(m_{i_0 j} \mid 1 \leq$

$j \le 4$) :

$$\Lambda_{i_0}(t) = \frac{(t^{e_{i_0}} - 1)^2 (t^{r_{i_0}} - 1)^2}{(t^{m_{i_0 1}} - 1)(t^{m_{i_0 2}} - 1)(t^{m_{i_0 3}} - 1)(t^{m_{i_0 4}} - 1)}.$$

Par ailleurs, l'hypothèse $\mathcal{HT}$ nous donne la liste des polynômes Λ_i associés aux sommets de rupture. Vu que, pour tout sommet de rupture i, on a, pour $j = 1, 2, 3$ (ou $j = 1, 2, 3, 4$), $e_i > m_{ij} \ge r_i$, ceci donne l'expression $\sum_{i, v(i) \ge 3} (v(i) - 2)[e_i]$, où l'on somme sur les seuls sommets de rupture. On trouve ainsi la liste $SV(3)$ des sommets de valence au moins égale à 3 :

$$SV(3) = \{(i, e_i, v(i)) \mid i \in \mathrm{som}(T(f)), v(i) \ge 3\}.$$

La comparaison avec $\sum_i (v(i) - 2)[e_i]$ nous donne ensuite la liste $SV(1)$ des sommets de valence 1 :

$$SV(1) = \{(i, e_i) \mid i \in \mathrm{som}(T(f)), v(i) = 1\}.$$

D'après la proposition 2.1, l'ensemble des m_{ij} associés aux segments géodésiques est disjoint de l'ensemble $\{e_i \mid v(i) = 1\}$, qui est égal à l'ensemble des m_{ij} associés aux branches mortes et à l'éventuelle branche géodésique d'extrémité #1. En considérant de nouveau la définition de Λ_i et la liste de ces polynômes, on trouve donc la proposition suivante.

Proposition 2.2. *Si $T(f)$ n'admet pas de sommet de valence 4, on a* $\mathrm{card}SV(1) = \mathrm{card}SV(3)$, *et si $T(f)$ admet un sommet de valence 4, on a* $\mathrm{card}SV(1) = \mathrm{card}SV(3) + 1$.

Si l'hypothèse $\mathcal{HT}$ est vérifiée, le polynôme Δ et la liste des polynômes Λ_i permettent de déterminer l'application

$$\sigma : SV(1) \to SV(3),$$

qui à un sommet de valence 1, extrémité d'une branche morte, associe le sommet de valence 3 ou 4 qui est l'autre extrémité de celle-ci, et au sommet #1, s'il est de valence 1, associe le sommet de valence 3 ou 4 extrémité de la branche géodésique d'extrémité #1.

Si f est le produit de deux germes lisses transverses, on a $H_1(F, \mathbf{Z}) \cong \mathbf{Z}[t, t^{-1}]/(t-1)$, et $SV(1) = SV(3) = \emptyset$. Nous supposerons dorénavant que f n'est pas le produit de deux germes lisses transverses, on a alors $SV(3) \ne \emptyset$. En utilisant la description de $T(f)$ en fonction des développements de Puiseux des branches de f, donnée dans (BK86, p. 698–704) ou dans (MW85, 6.5), on déduit de la proposition 2.2 les informations suivantes.

Proposition 2.3. *Cas A. Si l'application σ n'est pas injective, le sommet #1 est de valence 1, le sommet $\sigma(1)$ porte une branche morte, la multiplicité de f est donnée par :*

$$\nu_0(f) = e_1 = \min(e_i \mid i \in \mathrm{som}(T(f)), v(i) \neq 2),$$

les paires de Zariski des branches de f vérifient :

$$\frac{p_1}{q_1} = \frac{p_1'}{q_1'} \leq \mathfrak{C}(\varphi, \varphi').$$

et nous sommes dans l'un des 3 cas suivants.

A.1. L'application σ est surjective. Dans ce cas, $T(f)$ admet un sommet de valence 4 (avec 2 arêtes sortantes) et σ induit une bijection de $SV(1) \setminus \sigma^{-1}(\{\sigma(1)\})$ sur $SV(3) \setminus \{\sigma(1)\}$.

A.2. L'application σ n'est pas surjective et l'exposant de coïncidence $\mathfrak{C}(\varphi, \varphi')$ est permis pour les deux branches de f. Alors, les branches se séparent en un sommet de valence 3 associé à une paire non-caractéristique, ce sommet est le seul sommet de valence 3 de $T(f)$ sans branche morte, et il admet 2 arêtes sortantes.

A.3. L'application σ n'est pas surjective et l'exposant de coïncidence $\mathfrak{C}(\varphi, \varphi')$ est permis pour la branche φ mais non pour la branche φ'. Alors, les branches se séparent en un sommet de valence 3 associé à une paire caractéristique pour φ, mais non pour φ' ; ce sommet est le seul sommet de valence 3 de $T(f)$ sans branche morte, et il admet une unique arête sortante, portée par la géodésique $\Gamma(1, \varphi)$.

Cas B. Si l'application σ est injective, $T(f)$ n'admet pas de sommet de valence 4, l'application σ est bijective et nous sommes dans l'un des 3 cas suivants.

B.1. Les deux branches de f sont transverses. Dans ce cas, tous les sommets de valence 1 sont extrémités de branche morte et ceux de valence 3 admettent une unique arête sortante.

B.2. Le sommet #1 est de valence 1 et $\mathfrak{C}(\varphi, \varphi')$ est permis pour les deux branches. Celles-ci se séparent sur une paire non-caractéristique. De plus, le sommet de valence 3 extrémité de la branche géodésique d'extrémité #1 admet 2 arêtes sortantes.

B.3. Le sommet #1 est de valence 1 et $\mathfrak{C}(\varphi, \varphi')$ est permis pour la branche φ, mais non pour la branche φ'. De plus, le sommet de valence 3 extrémité de la branche géodésique d'extrémité #1 admet une unique arête sortante.

Voyons comment distinguer arête sortante et arête entrante. Soit $\mathcal{H}_i := (e_i; \eta_{i_1}, \eta_{i_2}, \eta_{i_3})$ le halo associé à un sommet de valence 3, et e_{i_j}, $j = 1, 2$

520

ou 3 la multiplicité du sommet i_j, voisin du sommet i, on a deux cas de figure :

ou bien i_j est sur une arête sortante pour le sommet i, par suite $e_{i_j} > e_i$, et, d'après le lemme 2.1, $\eta_{i_j} < e_1$,

ou bien i_j est sur une arête entrante pour le sommet i, par suite $e_{i_j} < e_i$, et donc $\eta_{i_j} = e_{i_j} \geq e_1$. Dans les cas $A.2$ et $A.3$, on trouve une meilleure majoration quand i est distinct de $\sigma(1)$, en effet, si i est le sommet où se séparent $\Gamma(1,\varphi)$ et $\Gamma(1,\varphi')$), on a, toujours pour une arête entrante : $\eta_{i_j} = e_{i_j} \geq e_{\sigma(1)} > 2e_1$.

Dans le cas A, e_1 est donné par la prop. 2.3, on trouve donc la proposition suivante.

Proposition 2.4. *Soit $\mathcal{H}_i := (e_i; \eta_1, \eta_2, \eta_3)$ le halo associé au sommet de rupture sur lequel se séparent $\Gamma(1,\varphi)$ et $\Gamma(1,\varphi')$. Quitte à renuméroter les voisins de i, le critère suivant permet de distinguer entre les cas $A.2$ et $A.3$:*

dans le cas $A.2$, on a $0 < \eta_2, \eta_3 < e_1 < 2e_1 < \eta_1 < e_i$;

dans le cas $A.3$, on a $0 < \eta_3 < e_1 < 2e_1 < \eta_1, \eta_2 < e_i$.

Il nous reste à distinguer les sous-cas du cas B. Dans les cas $B.1$ et $B.3$, chaque sommet de rupture de $T(f)$ correspond soit à une paire de Zariski de φ, soit à une paire de Zariski de φ' ; on a donc $g + g'$ sommets de rupture, avec $g \geq 1$ et $g' \geq 0$. Dans le cas $B.2$, le sommet de rupture de plus petite multiplicité est le sommet de séparation des branches de f, les autres sommets de rupture correspondent soit à une paire de Zariski de φ, soit à une paire de Zariski de φ' ; on a donc $g + g' + 1$ sommets de rupture, avec $g \geq 0$ et $g' \geq 0$.

On notera $i(j)$ (resp. $i'(j)$) le sommet de rupture associé à la j-ème paire de Zariski de φ (resp. φ'), ceci pour $j = 1, \ldots, g$ (resp. $j = 1, \ldots, g'$). Dans le cas $B.2$, le sommet de séparation sera noté $i(0)$. On notera $e_{i(j)}$ (resp. $e'_{i(j)}$) la multiplicité du sommet $i(j)$ (resp. $i'(j)$), $a_{i(j)}$ (resp. $a'_{i(j)}$) la multiplicité du sommet de valence 1 associé, *i.e.* $a_{i(j)} = e_{\sigma^{-1}(i(j))}$ et $a'_{i(j)} = e_{\sigma^{-1}(i'(j))}$. On désignera par $\eta_{i(j)}$ le plus petit des $\eta_{i(j),k}$, $k = 1, 2$ ou 3 ; si $j > 0$, $\eta_{i(j)}$ est donc associé à l'unique arête sortante de $i(j)$, et si $j = 0$, $\eta_{i(0)}$ est associé à l'une des deux arêtes sortantes de $i(0)$. On notera de même $\eta'_{i(j)}$ l'homologue de $\eta_{i(j)}$ pour la branche φ'.

Lemme 2.2. *Si le sommet $i(j)$ (resp. $i'(j)$) porte une branche morte, on*

a :

$$\frac{e_{i(j)}}{a_{i(j)}} = q_j \quad (resp. \quad \frac{e'_{i(j)}}{a'_{i(j)}} = q'_j).$$

Dans le cas B.2 (resp. B.3), le sommet $i(0)$ (resp. $i(1)$) est la deuxième extrémité de la branche géodésique d'extrémité #1, et on a :

$$\frac{e_{i(0)}}{a_{i(0)}} = \mathfrak{C}(\varphi, \varphi') > 1 \quad (resp. \frac{e_{i(1)}}{a_{i(1)}} = p_1 > q_1).$$

Démonstration. Voir le calcul des multiplicités dans l'arbre de désingularisation dans (BK86), dans (MW85, 5.4.1 et 6.6.4) ou ci-dessous le lemme 2.4. $\qquad\square$

Supposons que nous sommes dans le cas B, posons :

$$\nu_0^{(1)}(f) = 2 + \sum_{j=0/1}^{j=g} \eta_{i(j)}\Big(\frac{e_{i(j)}}{a_{i(j)}} - 1\Big) + \sum_{j=1}^{j=g'} \eta'_{i(j)}\Big(\frac{e'_{i(j)}}{a'_{i(j)}} - 1\Big),$$

$$\nu_0^{(2)}(f) = \min(e_i \mid i \in \mathrm{som}(T(f)), v(i) = 1).$$

De façon précise, $\nu_0^{(1)}(f) - 2$ est la somme des termes indiqués, étendue à l'ensemble des sommets de valence 3 de $T(f)$, et cette somme contient ou non un terme d'indice $j = 0$, d'où la notation $j = 0/1$ pour noter $j = 0$ ou 1. Nous supposons que l'hypothèse $\mathcal{HT}$ est vérifiée, nous pouvons donc calculer $\nu_0^{(1)}(f)$ et $\nu_0^{(2)}(f)$ à partir des données.

Proposition 2.5. *Dans le cas B, $\nu_0^{(1)}(f)$ et $\nu_0^{(2)}(f)$ sont distincts, et la multiplicité $\nu_0(f)$ est donnée par :*

$$\nu_0(f) = \min(\nu_0^{(1)}(f), \nu_0^{(2)}(f)).$$

Dans le cas B.1, on a $\nu_0(f) = \nu_0^{(1)}(f)$. Dans les cas B.2 et B.3, on a $\nu_0(f) = \nu_0^{(2)}(f)$. Quitte à renuméroter les voisins du sommet $\sigma(1)$ de $T(f)$, où $\sigma(1)$ est le sommet de valence 3 image par σ du sommet de valence 1 de plus petite multiplicité, et en notant provisoirement $\sigma(1) = i(0/1)$, le critère suivant permet de distinguer entre les cas B.2 et B.3 :

dans le cas B.2, on a $0 < \eta_2, \eta_3 < \nu_0(f) \leq \eta_1 < e_{i(0/1)}$, i.e. $\sigma(1)$ admet deux arêtes sortantes ;

dans le cas B.3, on a $0 < \eta_3 < \nu_0(f) \leq \eta_1, \eta_2 < e_{i(0/1)}$, i.e. $\sigma(1)$ admet deux arêtes entrantes.

522

Démonstration. Dans le cas *B.1*, on a, d'après les lemmes 2.1 et 2.2 :

$$\nu_0^{(1)}(f) = 2 + \sum_{j=1}^{j=g} \eta_{i(j)}\Big(\frac{e_{i(j)}}{a_{i(j)}} - 1\Big) + \sum_{j=1}^{j=g'} \eta'_{i(j)}\Big(\frac{e'_{i(j)}}{a'_{i(j)}} - 1\Big),$$

$$\nu_0^{(1)}(f) = 2 + \sum_{j=1}^{j=g} q_{j+1}\cdots q_g(q_j - 1) + \sum_{j=1}^{j=g'} q'_{j+1}\cdots q'_{g'}(q'_j - 1),$$

$$\nu_0^{(1)}(f) = q_1 \cdots q_g + q'_1 \cdots q'_{g'} = \nu_0(f) < \nu_0^{(2)}(f).$$

Dans le cas *B.2*, on a, d'après les lemmes 2.1 et 2.2, et puisque $\mathfrak{C}(\varphi, \varphi') > 1$:

$$\nu_0^{(1)}(f) = 2 + \sum_{j=0}^{j=g} \eta_{i(j)}\Big(\frac{e_{i(j)}}{a_{i(j)}} - 1\Big) + \sum_{j=1}^{j=g'} \eta'_{i(j)}\Big(\frac{e'_{i(j)}}{a'_{i(j)}} - 1\Big) =$$

$$= 2 + \eta_{i(0)}\big(\mathfrak{C}(\varphi, \varphi') - 1\big) \sum_{j=1}^{j=g} q_{j+1}\cdots q_g(q_j - 1) + \sum_{j=1}^{j=g'} q'_{j+1}\cdots q'_{g'}(q'_j - 1),$$

$$\nu_0^{(1)}(f) > q_1 \cdots q_g + q'_1 \cdots q'_{g'} = \nu_0(f) = e_1 = \nu_0^{(2)}(f).$$

Dans le cas *B.3*, on a, d'après les lemmes 2.1 et 2.2, et puisque $p_1 > q_1$:

$$\nu_0^{(1)}(f) = 2 + \sum_{j=1}^{j=g} \eta_{i(j)}\Big(\frac{e_{i(j)}}{a_{i(j)}} - 1\Big) + \sum_{j=1}^{j=g'} \eta'_{i(j)}\Big(\frac{e'_{i(j)}}{a'_{i(j)}} - 1\Big),$$

$$\nu_0^{(1)}(f) = 2 + q_2 \cdots q_g(p_1 - 1) + \sum_{j=2}^{j=g} q_{j+1}\cdots q_g(q_j - 1) + \sum_{j=1}^{j=g'} q'_{j+1}\cdots q'_{g'}(q'_j - 1),$$

$$\nu_0^{(1)}(f) = p_1 q_2 \cdots q_g + q'_1 \cdots q'_{g'} > q_1 \cdots q_g + q'_1 \cdots q'_{g'} = \nu_0(f) = e_1 = \nu_0^{(2)}(f).$$

On distingue ensuite entre les cas *B.2* et *B.3* en utilisant le lemme 2.2. $\square$

Donnons une méthode de calcul des multiplicités dans un arbre de désingularisation, en commençant par le calcul de la multiplicité d'intersection de deux germes irréductibles. Soit g et h deux germes irréductibles, notons $m = \nu_0(g)$ et $n = \nu_0(h)$ les multiplicités à l'origine de ces germes, supposons que $\{x = 0\}$ n'est pas dans le cône tangent à $\{gh = 0\}$ et choisissons un développement de Puiseux χ (resp. ψ) pour g (resp. h), de sorte que $\{g = 0\}$ est paramétré par $x = t^m$, $y = \chi(t)$ et

$\{h = 0\}$ est paramétré par $x = u^n$, $y = \psi(u)$. Quitte à multiplier par un inversible de $\mathbf{C}\{x, y\}$, on peut supposer que g et h sont dans $\mathbf{C}\{x\}[y]$ et que :

$$g(x, y) = \prod_{\alpha^m = 1} \big(y - \chi(\alpha t)\big) \quad \text{et} \quad h(x, y) = \prod_{\beta^n = 1} \big(y - \psi(\beta u)\big).$$

Le résultant $R_y(g, h)$ de g et h est l'élément suivant de $\mathbf{C}\{x\}$, en notant que $x = t^m = u^n$:

$$R_y(g, h) = \prod_{\alpha^m = 1} h(t^m, \chi(\alpha t)) = \prod_{\alpha^m = \beta^n = 1} \big(\chi(\alpha t) - \psi(\beta u)\big).$$

Lemme 2.3. *Avec les mêmes notations, supposons que g et h ne sont pas proportionnels, la multiplicité d'intersection en 0 de ces deux germes est donnée par l'égalité suivante :*

$$\nu_0(g, h) = \mathrm{val}_x\big(R_y(g, h)\big) = \mathrm{val}_x\Big(\prod_{\alpha^m = \beta^n = 1} \big(\chi(\alpha t) - \psi(\beta u)\big)\Big).$$

Démonstration. La définition générale $\nu_0(g, h) = \dim\big(\mathbf{C}\{x, y\}/(g, h)\big)$, valable sans condition d'irréductibilité sur g et h, donne, si g est irréductible, et pour tout α tel que $\alpha^m = 1$, $\nu_0(g, h) = \mathrm{val}_t\big(h(t^m, \chi(\alpha t))\big)$. On a donc :

$$m\,\nu_0(g, h) = m\,\mathrm{val}_t\big(h(t^m, \chi(\alpha t))\big) = \mathrm{val}_t\big(R_y(g, h)\big) = m\,\mathrm{val}_x\big(R_y(g, h)\big).$$

Ceci donne la première égalité. La deuxième égalité s'ensuit, si on suppose de plus h irréductible. $\qquad\square$

Lemme 2.4. *Soit $f : (\mathbf{C}^2, 0) \to (\mathbf{C}, 0)$ un germe de fonction analytique à singularité isolée et $T(f)$ l'arbre de désingularisation avec multiplicités de f. Soit i un sommet de $T(f)$ et γ_i une curvette de f associée au sommet i. La multiplicité e_i du sommet i est donnée par l'égalité suivante :*

$$e_i = \nu_0(f, \gamma_i).$$

L'arbre $T(f)$ donne sans calcul un développement de Puiseux d'une curvette associée au sommet i, on peut donc utiliser les deux lemmes précédents pour déterminer explicitement e_i.

Proposition 2.6. *Soit $f = f' \cdot f''$ un germe de fonction analytique à singularité isolée définissant un germe de courbe plane à deux branches, et $T(f)$ l'arbre de désingularisation avec multiplicités de f. Le nombre*

524

d'intersection des deux branches f' et f'' est donné par la formule suivante, où le produit est étendu aux sommets de l'arbre $T(f)$:

$$\nu_0(f', f'') = \prod_i e_i^{v(i)-2}.$$

Démonstration. On utilise les notations de la déf 2.1, on note donc φ (resp. φ') un développement de Puiseux de f' (resp. f''), $p_1/q_1, \ldots, p_g/q_g$ les paires de Zariski de φ, $p'_1/q'_1, \ldots, p'_g/q'_{g'}$ les paires de Zariski de φ' et $\mathcal{L} = \nu_0(f', f'')$, qui est aussi le nombre d'enlacement des deux composantes de bord de la fibre de Milnor. Nous donnons une démonstration pour chacun des cas énumérés prop. 2.3. Dans chacun des cas, les résultats du lemme 2.2 seront utilisés pour regrouper, dans le produit des $e_i^{v(i)-2}$, les multiplicités des sommets, extrémités de chaque branche morte ou de la branche géodésique d'extrémité #1, si ce sommet est de valence 1.

(i) Le cas *B.1*. Les branches sont transverses. On a directement

$$\mathcal{L} = q_1 \cdots q_g \, q'_1 \cdots q'_{g'} = \prod_i e_i^{v(i)-2}.$$

(ii) Le cas *B.2*. Les branches se séparent sur une paire non caractéristique $c = \mathfrak{C}(\varphi, \varphi')$, telle que $c < p_1/q_1$ et $c < p'_1/q'_1$. On peut supposer pour le calcul que :

$$\varphi = x^{p_1/q_1} + \cdots \quad \text{et} \quad \varphi' = x^c + x^{p'_1/q'_1} + \cdots$$

Chacun des termes qui interviennent dans le produit donné au lemme 2.3 est de valuation c, on trouve donc le résultat demandé :

$$\mathcal{L} = q_1 \cdots q_g \, q'_1 \cdots q'_{g'} \, \mathfrak{C}(\varphi, \varphi') = \prod_i e_i^{v(i)-2}.$$

(iii) Le cas *B.3*, en supposant de plus la branche f'' lisse. Les branches se séparent sur la première paire caractéristique de φ, $p_1/q_1 = \mathfrak{C}(\varphi, \varphi')$. On peut supposer que :

$$\varphi = x^{p_1/q_1} + \cdots \quad \text{et} \quad \varphi' = x^d, \quad \text{avec} \quad d > p_1/q_1.$$

Chacun des termes qui interviennent dans le produit donné au lemme 2.3 est de valuation p_1/q_1, on trouve donc le résultat demandé :

$$\mathcal{L} = q_1 \cdots q_g \, p_1/q_1 = p_1 q_2 \cdots q_g = \prod_i e_i^{v(i)-2}.$$

(iv) Le cas $B.3$, en supposant de plus la branche f'' non lisse. Les deux branches se séparent sur la première paire caractéristique de φ, $p_1/q_1 = \mathfrak{C}(\varphi, \varphi')$, et $p_1/q_1 < p_1'/q_1'$. On peut supposer que :

$$\varphi = x^{p_1/q_1} + \cdots \quad \text{et} \quad \varphi' = x^{p_1'/q_1'} + \cdots .$$

Chacun des termes qui interviennent dans le produit donné au lemme 2.3 est de valuation p_1/q_1, on trouve donc le résultat demandé :

$$\mathcal{L} = q_1 \cdots q_g\, q_1' \cdots q_{g'}'\, p_1/q_1 = p_1 q_2 \cdots q_g\, q_1' \cdots q_{g'}' = \prod_i e_i^{v(i)-2}.$$

(v) Le cas $A.1$. Les deux branches ont c paires de Zariski en commun et se séparent sur la paire caractéristique p_c/q_c. Notons e_1 la multiplicité du sommet #1, e_s celle du sommet de séparation (de valence 4), g_s et g_s' deux curvettes associées à ce sommet et en position générale. On peut supposer que les développements de Puiseux $\varphi, \varphi', \chi_s$ et χ_s' de f', f'', g_s et g_s' sont de la forme suivante :

$$\varphi = x^{p_1/q_1}(1 + \cdots (1 + x^{p_c/q_1 \cdots q_c} + \cdots)),$$

$$\varphi' = x^{p_1/q_1}(1 + \cdots (1 + 2x^{p_c/q_1 \cdots q_c} + \cdots)),$$

$$\chi_s = x^{p_1/q_1}(1 + \cdots (1 + 3x^{p_c/q_1 \cdots q_c}))$$

$$\text{et} \quad \chi_s' = x^{p_1/q_1}(1 + \cdots (1 + 4x^{p_c/q_1 \cdots q_c})).$$

On trouve successivement, en utilisant les lemmes 2.3 et 2.4 :

$$e_s = \nu_0(g_s, f') + \nu_0(g_s, f'') = (q_{c+1} \cdots q_g + q_{c+1}' \cdots q_{g'}')\, \nu_0(g_s, g_s'),$$

$$e_1 = q_1 \cdots q_c\,(q_{c+1} \cdots q_g + q_{c+1}' \cdots q_{g'}')$$

$$\text{et} \quad \prod_i e_i^{v(i)-2} = \frac{e_s}{e_1} q_1 \cdots q_c\, q_{c+1} \cdots q_g\, q_{c+1}' \cdots q_{g'}',$$

$$\text{donc,} \quad \prod_i e_i^{v(i)-2} = q_{c+1} \cdots q_g\, q_{c+1}' \cdots q_{g'}'\, \nu_0(g_s, g_s') = \nu_0(f', f'') = \mathcal{L}.$$

(vi) Le cas $A.2$. Les deux branches ont c paires de Zariski en commun et se séparent sur une paire non caractéristique. Les notations e_1, e_s, g_s et g_s' gardent le sens précédent. On peut supposer que les développements de Puiseux $\varphi, \varphi', \chi_s$ et χ_s' de f', f'', g_s et g_s' sont de la forme suivante, où $p_c < a$ et $aq_{c+1} < p_c q_{c+1} + p_{c+1}$:

$$\varphi = x^{p_1/q_1}(1 + \cdots (1 + x^{p_c/q_1 \cdots q_c} + x^{(p_c q_{c+1} + p_{c+1})/q_1 \cdots q_{c+1}} + \cdots)),$$

526

$$\varphi' = x^{p_1/q_1}(1 + \cdots (1 + x^{p_c/q_1\cdots q_c} + x^{a/q_1\cdots q_c} +$$

$$x^{(p_c q'_{c+1} + p'_{c+1})/q_1\cdots q_c q'_{c+1}} + \cdots)),$$

$$\chi_s = x^{p_1/q_1}(1 + \cdots (1 + x^{p_c/q_1\cdots q_c} + 2x^{a/q_1\cdots q_c}))$$

$$\text{et} \quad \chi'_s = x^{p_1/q_1}(1 + \cdots (1 + x^{p_c/q_1\cdots q_c} + 3x^{a/q_1\cdots q_c})).$$

Le calcul donné en $A.1$ s'applique sans changement au cas $A.2$. Noter cependant que la valeur de $\nu_0(g_s, g'_s)$ dépend du cas considéré, puisque $\nu_0(g_s, g'_s) = a q_1 \cdots q_c$, le cas $A.1$ correspondant à $a = p_1$.

(vii) Le cas $A.3$. Premier sous-cas : les deux branches ont $(c - 1)$ paires de Zariski en commun, $(c - 1) > 0$, et se séparent sur la paire p_c/q_c, caractéristique pour φ, mais non pour φ', qui a $(c - 1)$ paires de Zariski. Les notations e_1, e_s, g_s et g'_s gardent le sens précédent. On peut supposer que les développements de Puiseux φ, φ', χ_s et χ'_s de f', f'', g_s et g'_s sont de la forme suivante, où $dq_c > p_{c-1}q_c + p_c$:

$$\varphi = x^{p_1/q_1}(1 + \cdots (1 + x^{p_c/q_1\cdots q_c} + \cdots)),$$

$$\varphi' = x^{p_1/q_1}(1 + \cdots (1 + x^{p_{c-1}/q_1\cdots q_{c-1}} + x^{d/q_1\cdots q_{c-1}})),$$

$$\chi_s = x^{p_1/q_1}(1 + \cdots (1 + 2x^{p_c/q_1\cdots q_c}))$$

$$\text{et} \quad \chi'_s = x^{p_1/q_1}(1 + \cdots (1 + 3x^{p_c/q_1\cdots q_c})).$$

On trouve successivement :

$$e_s = \nu_0(g_s, f') + \nu_0(g_s, f'') = q_{c+1}\cdots q_g\, \nu_0(g_s, g'_s) + \nu_0(g_s, f''),$$

$$e_1 = q_1 \cdots q_{c-1}(1 + q_c \cdots q_g),$$

$$\nu_0(g_s, g'_s) = \frac{p_c}{q_1\cdots q_c}(q_1\cdots q_c)^2 = p_c q_1 \cdots q_c,$$

$$\nu_0(g_s, f'') = \frac{p_c}{q_1\cdots q_c}q_1 \cdots q_c\, q_1 \cdots q_{c-1} = p_c q_1 \cdots q_{c-1},$$

$$\mathcal{L} = \nu_0(f', f'') = q_{c+1}\cdots q_g\, \nu_0(g_s, f'') = \frac{e_s}{e_1}\frac{q_1\cdots q_g}{q_c} = \prod_i e_i^{v(i)-2}.$$

(viii) Le cas $A.3$. Deuxième sous-cas : les deux branches ont $(c-1)$ paires de Zariski en commun, $(c-1) > 0$, et se séparent sur la paire p_c/q_c, caractéristique pour φ, mais non pour φ', qui a au moins c paires de Zariski. Les notations e_1, e_s, g_s et g'_s gardent le sens précédent. On peut supposer que les développements de Puiseux φ, φ', χ_s et χ'_s de f', f'', g_s et g'_s sont de la forme suivante, où $p_c/q_c < p'_c/q'_c$:

$$\varphi = x^{p_1/q_1}(1 + \cdots (1 + x^{p_{c-1}/q_1\cdots q_{c-1}}(1 + x^{p_c/q_1\cdots q_c} + \cdots))),$$

$$\chi_s = x^{p_1/q_1}(1 + \cdots (1 + 2x^{p_c/q_1\cdots q_c})),$$

$$\varphi' = x^{p_1/q_1}(1 + \cdots (1 + x^{p_{c-1}/q_1\cdots q_{c-1}}(1 + x^{p'_c/q_1\cdots q_{c-1}q'_c} + \cdots))),$$

$$\chi'_s = x^{p_1/q_1}(1 + \cdots (1 + 3x^{p_c/q_1\cdots q_c})).$$

On trouve successivement :

$$e_s = \nu_0(g_s, f') + \nu_0(g_s, f'') = q_{c+1}\cdots q_g\,\nu_0(g_s, g'_s) + \nu_0(g_s, f''),$$

$$e_1 = q_1\cdots q_{c-1}(q_c\cdots q_g + q'_c\cdots q'_{g'}),$$

$$\nu_0(g_s, g'_s) = p_c q_1\cdots q_c, \quad \nu_0(g_s, f'') = p_c q_1\cdots q_{c-1}\,q'_c\cdots q'_{g'},$$

$$\mathcal{L} = \nu_0(f', f'') = q_{c+1}\cdots q_g\,\nu_0(g_s, f'')$$

$$\mathcal{L} = \frac{e_s}{e_1}\frac{q_1\cdots q_g\,q'_c\cdots q'_{g'}}{q_c} = \prod_i e_i^{v(i)-2}. \qquad \square$$

3. Reconstruction de l'arbre réduit de l'arbre $T(f)$

Soit $f : (\mathbf{C}^2, 0) \to (\mathbf{C}, 0)$ un germe de fonction analytique à singularité isolée, définissant un germe de courbe plane à deux branches. On se donne la forme de Seifert $A(f)$ et on suppose l'hypothèse $\mathcal{HT}$ vérifiée. On connait donc l'ensemble des halos de valence 3 de $T(f)$ et, s'il existe un halo $\mathcal{H}_i$ de valence 4, on connait e_i et les $m_{i,j} = \text{pgcd}(e_i, \eta_{i,j})$, pour $j = 1, \ldots, 4$ (*cf.* fin du §1). Si $\mathcal{H}_i = (e_i; \eta_{i1}, \eta_{i2}, \eta_{i3})$ est le halo d'un sommet de rupture de valence 3 de $T(f)$, l'étude effectuée en 2 donne les réponses aux questions suivantes :

η_{ij} est-il porté par une arête entrante ou par une arête sortante ?

dans le cas d'une arête entrante, celle-ci est-elle une arête portée par une branche morte, une arête reliant le sommet de rupture au sommet #1 ou l'arête du type particulier rencontré (en un seul exemplaire) dans les seuls cas *A.3* et *B.3* ?

Nous dirons qu'un halo est *orienté* si on sait distinguer l'arête entrante qui relie le sommet de rupture au sommet #1. La proposition 3.1 montre que l'orientation des halos dans $T(f)$ est fixée. La proposition 3.3 donne la disposition relative des halos dans l'arbre $T(f)$. Dans certains cas, la reconstruction n'est pas unique, ce qui conduit à la définition de germes isomères, *cf.* déf. 3.2. Le théorème 3.1 montre que si deux germes de courbe plane à deux branches ont des formes de Seifert isomorphes, et si l'hypothèse $\mathcal{HT}$ est vérifiée, ces germes sont isomères.

Proposition 3.1 (Orientation des halos). *(i) Soit $\mathcal{H}_i = (e_i; \eta_1, \eta_2, \eta_3)$ le halo d'un sommet de rupture de valence 3 de $T(f)$. On peut renuméroter les voisins de i de sorte que les propriétés suivantes soient vérifiées.*

(a) Si le sommet central i de $\mathcal{H}_i$ n'est pas le sommet de séparation entre les branches de f, il admet :

une arête entrante, associée à η_1, portée par la géodésique qui relie les sommets i et #1, et caractérisée par les inégalités $\eta_1 \geq \nu_0(f) \geq \mathrm{pgcd}(e_i, \eta_1)$,

une arête entrante, associée à η_2, portée par une branche morte, et caractérisée par les inégalités $\eta_2 \geq \mathrm{pgcd}(e_i, \eta_2) > \nu_0(f)$,

et une arête sortante, associée à η_3, et caractérisée par l'inégalité $\eta_3 < \nu_0(f)$.

(b) Si le sommet central i de $\mathcal{H}_i$ est le sommet de séparation entre les branches de f, et est associé à une paire non-caractéristique (cas A.2 et B.2), il admet :

une arête entrante, associée à η_1, portée par la géodésique qui relie les sommets i et #1, et caractérisée par l'inégalité $\eta_1 \geq \nu_0(f)$,

et deux arêtes sortantes (indiscernables), associées à η_2 et η_3, caractérisées par les inégalités $\eta_2 < \nu_0(f)$ et $\eta_3 < \nu_0(f)$.

(c) Si le sommet central i de $\mathcal{H}_i$ est le sommet de séparation entre les branches de f, et est associé à une paire caractéristique pour φ, mais non pour φ' (cas A.3 et B.3), il admet :

une arête entrante, associée à η_1, portée par la géodésique qui relie les sommets i et #1, et caractérisée par les inégalités $\eta_1 \geq \nu_0(f)$ et $\mathrm{pgcd}(\eta_1, e_i) > \mathrm{pgcd}(\eta_2, e_i)$,

une arête entrante, associée à η_2, portée par la géodésique qui relie le sommet i au sommet symbolisant la branche φ', et caractérisée par les inégalités $\eta_2 > \nu_0(f)$ et $\mathrm{pgcd}(\eta_2, e_i) < \mathrm{pgcd}(\eta_1, e_i)$,

529

et une arête sortante, associée à η_3, et caractérisée par l'inégalité $\eta_3 < \nu_0(f)$.

(ii) Soit $\mathcal{H}_i = (e_i; \eta_1, \eta_2, \eta_3, \eta_4)$ le halo du sommet de rupture de valence 4 de $T(f)$, s'il en existe un. On rappelle que les $m_j := \mathrm{pgcd}(e_i, \eta_j)$ sont connus, mais pas les η_j (on note m_j au lieu de m_{ij} pour alléger les notations). On peut renuméroter les voisins de i de sorte que i admet :

une arête entrante, associée à η_1, portée par la géodésique qui relie les sommets i et $\#1$,

une arête entrante, associée à η_2, portée par une branche morte,

et deux arêtes sortantes (indiscernables), associées à η_3 et η_4, où la numérotation est définie par la condition : $m_2 > m_1 > \max(m_3, m_4)$.

Démonstration. La proposition récapitule les résultats relatifs à l'orientation des halos, obtenus plus haut. L'inégalité qui permet de distinguer entre les deux arêtes entrantes dans le cas *(i.c)* provient, d'une part, de l'égalité $\mathrm{pgcd}(\eta_1, e_i) = q_c \cdots q_g + q'_c \cdots q'_{g'}$, voir pour ce calcul les références données lemme 2.2, et, d'autre part, de la relation de divisibilité $\mathrm{pgcd}(\eta_2, e_i) \mid q'_c \cdots q'_{g'}$ de (R99, 2.17) déjà utilisée prop. 2.1.

Dans le cas *(ii)*, on vérifie facilement qu'on a, avec les notations du §2, les inégalités :

$$m_1 = q_c(q_{c+1} \cdots q_g + q'_{c+1} \cdots q'_{g'}), \quad m_2 > p_c(q_{c+1} \cdots q_g + q'_{c+1} \cdots q'_{g'}),$$

$$m_3 \leq q_{c+1} \cdots q_g, \quad m_4 \leq q'_{c+1} \cdots q'_{g'}. \qquad \square$$

Définition 3.1. *Soit f un germe de fonction analytique à singularité isolée, définissant un germe de courbe plane et soit $T(f)$ son arbre de désingularisation. On appellera arbre réduit de l'arbre $T(f)$, l'arbre $TR(f)$ obtenu à partir de $T(f)$ en effaçant les sommets de valence 2, et en ajoutant, pour tout sommet de rupture i de $T(f)$ et pour tout voisin j de i dans $T(f)$, un sommet pondéré par η_{ij} sur l'arête (ij). Notons que la multiplicité e_k d'un sommet k de valence 1 vérifie $e_k = \mathrm{pgcd}(e_i, \eta_{ij_k})$, où $i = \sigma(k)$ est le sommet de rupture associé à k comme dans la prop. 2.2, et j_k est le voisin de i sur l'arête dirigée vers k. Les multiplicités des sommets de valence 1 sont ainsi déterminées par les halos. De même, la donnée de l'arbre réduit $TR(f)$ détermine la position des flèches qui symbolisent dans $T(f)$ les composantes de f, de la manière suivante : les halos tels que $\eta_3 = 1$ (resp. $\eta_2 = \eta_3 = 1$ ou $\eta_3 = \eta_4 = 1$) portent une flèche (resp. deux flèches) ; si la méthode indiquée ne positionne qu'une des deux flèches, nous sommes dans le cas A.3 ou B.3 et de plus la branche qui admet φ' pour développement de Puiseux n'a pas de* *branche morte au delà du sommet de séparation, alors, la flèche symbolisant φ' se place au bout de l'arête issue du sommet de séparation et portant la composante η_2 du halo de ce sommet.*

Proposition 3.2. *Les halos qui sont voisins sur $T(f)$ doivent vérifier les conditions suivantes, en numérotant les voisins des sommets de rupture comme indiqué dans la proposition 3.1 :*

(i) *Si le halo avec branche morte $\mathcal{H} = (e, \eta_1, \eta_2, \eta_3)$ a pour voisin immédiat dans $T(f)$, dans la direction du sommet #1, le halo, de valence 3, $\mathcal{H}^* = (e^*, \eta_1^*, \eta_2^*, \eta_3^*)$ ou le halo, de valence 4, $\mathcal{H}^* = (e^*, \eta_1^*, \eta_2^*, \eta_3^*, \eta_4^*)$, et si le segment géodésique, qui relie ces deux halos, commence du côté de $\mathcal{H}^*$ par une arête sortante qui porte la composante η_3^* de ce halo, on a les règles de compatibilité suivantes :*

$$\frac{e\eta_3}{\mathrm{pgcd}(e, \eta_2)} = \eta_3^*, \quad \mathrm{pgcd}(e, \eta_1) = \mathrm{pgcd}(e^*, \eta_3^*) \quad \text{et} \quad e^* < \mathrm{pgcd}(e, \eta_2).$$

(ii) *Si le halo avec branche morte $\mathcal{H}$ est de valence 4, $\mathcal{H} = (e, \eta_1, \eta_2, \eta_3, \eta_4)$, les conditions sur $\mathcal{H}^*$ restant les mêmes, les règles de compatibilité sont les suivantes :*

$$\frac{e(\eta_3 + \eta_4)}{\mathrm{pgcd}(e, \eta_2)} = \eta_3^*, \ \mathrm{pgcd}(e, \eta_1) = \mathrm{pgcd}(e^*, \eta_3^*) \text{ et } e^* < \mathrm{pgcd}(e, \eta_2).$$

(iii) *Si le halo avec branche morte $\mathcal{H} = (e, \eta_1, \eta_2, \eta_3)$ a pour voisin immédiat dans $T(f)$, dans la direction du sommet #1, le halo $\mathcal{H}^* = (e^*, \eta_1^*, \eta_2^*, \eta_3^*)$, et si le segment géodésique, qui relie ces deux halos, commence du côté de $\mathcal{H}^*$ par une arête entrante qui porte la composante η_2^* de ce halo, nous sommes dans le cas A.3 ou B.3, le halo $\mathcal{H}^*$ n'est autre que le halo du sommet de séparation et on a la règle de compatibilité suivante :*

$$\mathrm{pgcd}(e, \eta_1) = \mathrm{pgcd}(e^*, \eta_2^*) \text{ et } e^* > \mathrm{pgcd}(e, \eta_2).$$

Démonstration. Ce résultat traduit la compatibilité des multiplicités sortantes (*cf.* lemme 2.1), rappelle que le pgcd est constant le long des arêtes d'un même segment géodésique et utilise le fait que la multiplicité du sommet de valence 1 situé au bout d'une branche morte est égale au pgcd des multiplicités le long de cette branche. $\qquad\square$

Définition 3.2. *Considérons l'arbre réduit $TR(f)$ de l'arbre de désingularisation $T(f)$ d'un germe de courbe plane à deux branches, numérotons les composantes η_{ij} des halos comme l'indique la prop. 3.1, et considérons la famille des halos de $TR(f)$. On appellera* arbre réduit *un arbre obtenu* *en reliant les halos de cette famille, en suivant les règles indiquées dans la prop. 3.2. On appellera* peuplier *un sous-arbre de l'arbre réduit, obtenu en reliant des halos de valence 3 avec branche morte (choisis dans la famille des halos de $TR(f)$), avec la condition que le halo de multiplicité centrale maximale dans le peuplier admette 1 pour multiplicité sortante, c'est-à-dire, soit de la forme $\mathcal{H}_M = (e_M; \eta_{M1}, \eta_{M2}, 1)$. Le poids $\varpi(\mathcal{P})$ d'un peuplier $\mathcal{P}$ est défini à partir de son halo de multiplicité centrale minimale $\mathcal{H}_m = (e_m; \eta_{m1}, \eta_{m2}, \eta_{m3})$ par l'égalité suivante :*

$$\varpi(\mathcal{P}) = \left(\eta_{m3} \times \frac{e_m}{\mathrm{pgcd}(e_m, \eta_{m2})}, \mathrm{pgcd}(e_m, \eta_{m1})\right).$$

Il est utile de remarquer le point suivant : si $\mathcal{P}$ est un peuplier composé de p halos, on obtient de nouveaux peupliers en enlevant à $\mathcal{P}$ le halo de plus petite multiplicité centrale, ceci 1, 2, ..., ou $p-1$ fois. Deux arbres réduits sont dits isomères *si l'on passe de l'un à l'autre en effectuant une ou plusieurs fois l'échange de peupliers de même poids. On dit que deux germes de courbe plane, définis par des germes de fonction analytique f_1 et f_2 à singularité isolée, sont* isomères *si leurs arbres réduits $TR(f_1)$ et $TR(f_2)$ sont isomères. Dans ce cas, on dira aussi que $T(f_1)$ et $T(f_2)$ sont isomères. Les définitions ci-dessus gardent un sens dans le cas d'un germe de courbe plane, défini par une fonction analytique g à singularité isolée avec un nombre arbitraire de branches, si l'on suppose que les sommets de rupture de $T(g)$ sont tous de valence 3. Voir la Rem. 5.1 pour une utilisation de cette notion.*

Proposition 3.3. *Soit f un germe de fonction analytique à singularité isolée, définissant un germe de courbe plane à deux branches, les arbres réduits construits à partir de la famille des halos de $T(f)$ sont isomères.*

Démonstration. Nous partons de la famille $\mathcal{H}(T(f))$ des halos de $T(f)$. Nous effectuons l'étude vue au §2 pour savoir dans lequel des cas *A.1, A.2, A.3* ou *B.1, B.2, B.3*, nous nous trouvons, et nous orientons les halos.

Si l'arbre $T(f)$ ne possède aucun halo avec branche morte tel que $\eta_3 = 1$, on attache au sommet de séparation une flèche dans les cas *A.3* et *B.3*, ou deux flèches dans les cas *A.1, A.2* et *B.2*, et on passe au raccordement des halos situés en dessous du sommet de séparation, voir la fin de la démonstration.

Si l'arbre $T(f)$ possède un halo $\mathcal{H} = (e; \eta_1, \eta_2, \eta_3)$ avec branche morte tel que $\eta_3 = 1$, nous construisons un peuplier comme suit : on cherche un halo $\mathcal{H}^* = (e^*; \eta_1^*, \eta_2^*, \eta_3^*)$ avec branche morte vérifiant les conditions de compatibilité indiquées en 3.2.i. S'il n'existe pas un tel halo, le halo $\mathcal{H}$

532

est le peuplier cherché. Dans le cas contraire, il existe au plus deux halos vérifiant les conditions 3.2.i, parce que la multiplicité sortante du halo, qui est le voisin immédiat de $\mathcal{H}$ dans $T(f)$ dans la direction du sommet #1, ne peut se trouver que deux fois dans l'arbre de désingularisation d'un germe à deux branches; le cas échéant, on choisit pour $\mathcal{H}^*$ un des deux halos possibles et on raccorde alors $\mathcal{H}$ et $\mathcal{H}^*$ en traçant une arête entre η_1 et η_3^*.

En répétant la même opération, on construit ainsi un premier peuplier, puis s'il existe un deuxième halo $\mathcal{H}'$ avec branche morte tel que $\eta_3' = 1$, un deuxième peuplier, en partant de $\mathcal{H}'$. Si la construction fait apparaître deux peupliers, elle peut donner plusieurs arbres réduits distincts (si chaque peuplier est la réunion de n halos, on peut trouver jusqu'à 2^n arbres réduits), mais on passe d'un arbre réduit à un autre en effectuant une ou plusieurs fois l'échange de peupliers de même poids.

La Figure 1 montre comment les halos $\mathcal{H}$ et $\mathcal{H}'$ peuvent se raccorder à $\mathcal{H}^* = (e^*; \eta_j^*)$ et $\mathcal{H}'^* = (\varepsilon^*; \eta_j'^*)$ de deux manières différentes si les conditions 3.2.i le permettent. On a posé $a = \mathrm{pgcd}(e, \eta_2)$, $a' = \mathrm{pgcd}(e', \eta_2')$ et représenté les branches mortes issues de e et e'.

$$
\begin{array}{ccccc}
\eta_2 \bullet \eta_3 & \eta_3' \bullet \eta_2' & & \eta_2' \bullet \eta_3' & \eta_3 \bullet \eta_2 \\
a \bullet \bullet \bullet e & e' \bullet \bullet \bullet a' & \quad\text{et}\quad & a' \bullet \bullet \bullet e' & e \bullet \bullet \bullet a \\
\bullet\, \eta_1 & \eta_1' \,\bullet & & \bullet\, \eta_1' & \eta_1 \,\bullet \\[4pt]
\eta_3^* \,\bullet & \bullet\, \eta_3'^* & & \eta_3^* \,\bullet & \bullet\, \eta_3'^* \\
e^* \,\bullet & \bullet\, \varepsilon^* & & e^* \,\bullet & \bullet\, \varepsilon^*
\end{array}
$$

Figure 1 - Échange des halos $\mathcal{H} = (e; \eta_1, \eta_2, \eta_3)$ et $\mathcal{H}' = (e'; \eta_1', \eta_2', \eta_3')$

Dans le cas *B.1*, les deux branches de f sont transverses, et $TR(f)$ est obtenu en raccordant les deux peupliers construits ci-dessus (resp. le peuplier et une flèche) par une arête qui relie les (la) composante(s) η_1 des (du) halo(s) de plus petite multiplicité centrale de chacun des deux peupliers (resp. du peuplier avec la flèche). Dans les autres cas, l'étude indique quel est le sommet de séparation.

Dans les cas *A.1, A.2* et *B.2*, on raccorde les deux peupliers, ou le peuplier et une flèche, ou les deux flèches, au sommet de séparation e_s, en respectant les règles de compatibilité 3.2.i ou ii; en particulier, l'arête sortante, associée à la composante η_{sj} de $\mathcal{H}_s$, est attachée à un peuplier dont le poids admet η_{sj} pour première composante, ou à une flèche si $\eta_{sj} = 1$. Noter que la première composante du poids d'un peuplier n'est autre que la multiplicité sortante (du sommet e_s) le long de l'arête de raccordement, si c'est une arête sortante.

Dans les cas *A.3* et *B.3*, on raccorde les deux peupliers, ou le peuplier et une flèche, ou les deux flèches, au sommet de séparation e_s, en respectant les règles de compatibilité 3.2.i pour le raccordement à l'arête sortante (associée à la composante η_{s3} de $\mathcal{H}_s$) et les règles de compatibilité 3.2.iii pour le raccordement à l'arête entrante (associée à la composante η_{s2} de $\mathcal{H}_s$).

Noter d'une part, que dans chaque cas, on peut construire au moins un arbre réduit, parce que nous sommes partis d'un arbre de désingularisation $T(f)$ et, d'autre part, que les choix qui interviennent lors de la construction concernent l'échange de peupliers de même poids au dessus du sommet de séparation.

L'isomérie éventuellement rencontrée dans les cas *A.3* et *B.3* est représentée Figure 2.

Les halos restants se raccordent dans l'arbre réduit dans l'ordre décroissant des multiplicités centrales, du halo associé au sommet de séparation jusqu'au halo de plus petite multiplicité centrale ; de plus, l'orientation des halos décrite dans la prop. 3.1 détermine leurs dispositions relatives de façon unique. □

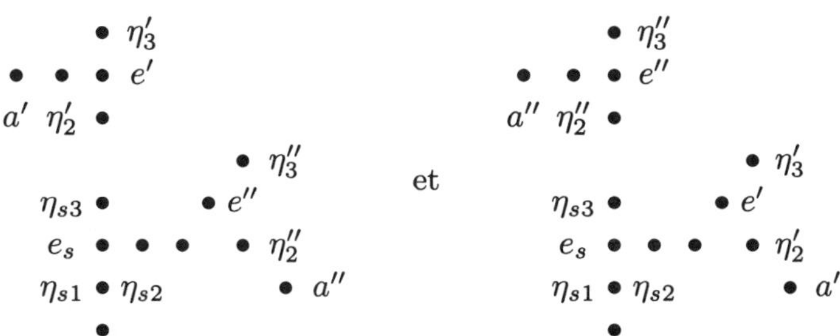

Figure 2 - Échange des halos $\mathcal{H}' = (e'; \eta_1', \eta_2', \eta_3')$ et $\mathcal{H}'' = (e''; \eta_1'', \eta_2'', \eta_3'')$

Théorème 3.1. *Soit f_1 et f_2 des germes de fonction analytique à singularité isolée, définissant des germes de courbe plane à deux branches, supposons que leurs formes de Seifert $A(f_1)$ et $A(f_2)$ sont isomorphes et que l'hypothèse $\mathcal{HT}$ est satisfaite, alors les arbres $T(f_1)$ et $T(f_2)$ sont isomères.*

Démonstration. Les hypothèses entraînent que la liste des halos de $T(f_1)$ et celle de $T(f_2)$ sont identiques, la prop. 3.3 donne donc le résultat. □

4. Calcul des paires de Zariski et fin de la reconstruction de l'arbre de désingularisation

Considérons de nouveau la situation étudiée au §3 : f désigne un germe de fonction analytique à singularité isolée, définissant un germe de courbe plane à deux branches, on se donne la forme de Seifert $A(f)$ et on suppose l'hypothèse $\mathcal{HT}$ satisfaite. Nous allons calculer les paires de Zariski du germe f et des germes isomères, s'il en existe, à partir de la forme de Seifert $A(f)$, au moyen de l'arbre réduit $TR(f)$.

Si f n'a pas d'isomères, ce qui est le cas général, on en déduira que la donnée de la forme de Seifert détermine le type topologique du germe f, *cf.* th. 4.1.

Si, au contraire, l'étude donnée en 3 conduit à plusieurs arbres réduits (qui sont donc isomères), nous démontrerons, *cf.* th. 4.2, que chaque arbre réduit construit à partir de $A(f)$ est l'arbre réduit de l'arbre de désingularisation d'un germe de courbe plane à deux branches défini par un germe de fonction analytique à singularité isolée g : les arbres $T(f)$ et $T(g)$ sont donc isomères. De plus, des germes isomères ont la même multiplicité et le nombre d'intersection des deux branches ne dépend pas de l'isomère choisi.

Considérons un germe de courbe plane à deux branches, défini par un germe de fonction analytique à singularité isolée f, et l'arbre de désingularisation $T(f)$ de f. Notons, comme au §2, φ et φ' des développements de Puiseux des branches de f. On supposera que la droite $\{x = 0\}$ n'est pas dans le cône tangent au germe de courbe $f^{-1}(0)$, ce qui entraîne que les paires de Zariski de φ et φ' sont strictement supérieures à 1. Les lemmes 4.1 à 4.8 donnent le calcul des paires de Zariski associées aux halos $\mathcal{H} = (e; \eta_j)$ et $\mathcal{H}' = (e'; \eta'_j)$ de $T(f)$ représentés Figures 3, 4 et 5, chaque lemme correspondant à un arbre de même numéro. Dans les cas 4.1 et 4.4, le halo $\mathcal{H}$ peut aussi être un halo de valence 4, la figure doit dans ce cas être modifiée en remplaçant le sommet η_3 par deux sommets, de multiplicités η_3 et η_4.

Le lemme 4.9 donne le calcul de l'exposant de coïncidence dans le cas *A.2*. Cet exposant est égal à 1 dans le cas *B.1*, il est donné par le lemme 2.2 dans le cas *B.2*. Dans les cas *A.1*, *A.3* et *B.3*, il est donné en fonction des paires de Zariski par la formule suivante, *cf.* §2 pour les notations :

$$\mathfrak{C}(\varphi, \varphi') = \frac{p_1}{q_1} + \frac{p_2}{q_1 q_2} + \cdots + \frac{p_c}{q_1 \cdots q_c}.$$

Pour la démonstration des lemmes, voir le calcul des multiplicités dans l'arbre de désingularisation dans (BK86) ou (MW85, 5.4.1 et 6.6.4), voir aussi (R99, lemmes 3.12 à 3.19) ou le lemme 2.4 ci-dessous.

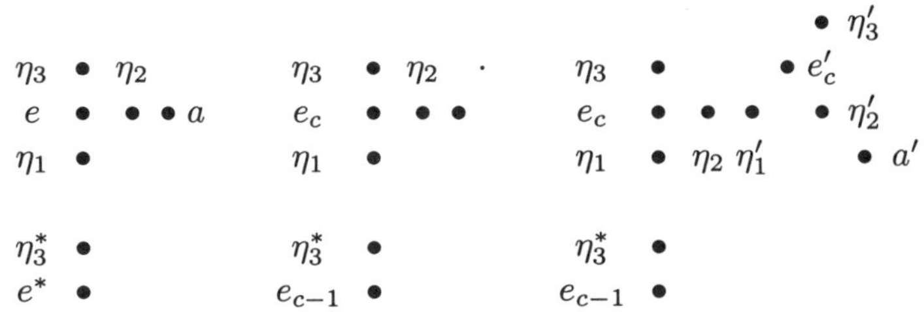

$$
\begin{array}{cccc}
& & & \bullet\ \eta_3' \\
\eta_3\ \bullet\ \eta_2 \qquad \eta_3\ \bullet\ \eta_2\ \cdot \qquad \eta_3\ \bullet & & \bullet\ e_c' \\
e\ \bullet\ \bullet\ \bullet\ a \qquad e_c\ \bullet\ \bullet\ \bullet \qquad e_c\ \bullet\ \bullet\ \bullet & & \bullet\ \eta_2' \\
\eta_1\ \bullet \qquad\qquad \eta_1\ \bullet \qquad\qquad \eta_1\ \bullet\ \eta_2\ \eta_1' & & \bullet\ a' \\
\\
\eta_3^*\ \bullet \qquad\qquad \eta_3^*\ \bullet \qquad\qquad \eta_3^*\ \bullet \\
e^*\ \bullet \qquad\qquad e_{c-1}\ \bullet \qquad\qquad e_{c-1}\ \bullet \\
4.1 \qquad\qquad\qquad 4.2 \qquad\qquad\qquad 4.3
\end{array}
$$

Figure 3

Lemme 4.1. *Considérons un halo avec branche morte $\mathcal{H} = (e; \eta_j)$, de valence 3 ou 4, et le halo $\mathcal{H}^* = (e^*; \eta_j^*)$ situé dans l'arbre réduit immédiatement en-dessous de $\mathcal{H}$. Notons $a = \mathrm{pgcd}(e, \eta_2)$ la multiplicité de l'extrémité de la branche morte. La paire de Zariski p/q associée à $\mathcal{H}$ est donnée par les formules suivantes :*

$$
q = \frac{e}{a}, \qquad p = \frac{e - qe^*}{\eta_3^*}.
$$

Si $\mathcal{H}^*$ est de valence 4, η_3^* n'est pas connu directement, mais il est donné par l'égalité $\eta_3^* = \eta_3 e/a$.

Lemme 4.2. *Dans le cas A.3, considérons le halo $\mathcal{H}_c = (e_c; \eta_j)$ du sommet de séparation, et supposons de plus que la branche associée à φ' a $c-1$ paires de Zariski, c'est-à-dire que la flèche associée à φ' se raccorde en l'arbre réduit à l'extrémité de l'arête portant la composante η_2. Notons $\mathcal{H}_{c-1} = (e_{c-1}; \eta_j^*)$ le halo associé à la $(c-1)$-ième paire de Zariski (commune à φ et φ' par hypothèse). La paire de Zariski p_c/q_c, associée à $\mathcal{H}_c$, est donnée par les formules suivantes :*

$$
q_c = \frac{\eta_3^* - 1}{\eta_3}, \qquad p_c = \frac{e_c - q_c e_{c-1}}{\eta_3^*}.
$$

Lemme 4.3. *Dans le cas A.3, considérons le halo $\mathcal{H}_c = (e_c; \eta_j)$ du sommet de séparation, et supposons de plus que la branche associée à φ' a au moins c paires de Zariski. Notons $\mathcal{H}_c' = (e_c'; \eta_j')$ le halo qui se raccorde à l'arbre réduit en l'extrémité de l'arête portant la composante η_2. Notons $\mathcal{H}_{c-1} = (e_{c-1}; \eta_j^*)$ le halo associé à la $(c-1)$-ième paire de Zariski (commune à φ et φ' par hypothèse), notons enfin $a' = \mathrm{pgcd}(e_c', \eta_2')$. Les*

536

paires de Zariski p_c/q_c, associée à $\mathcal{H}_c$, et p'_c/q'_c, associée à $\mathcal{H}'_c$, sont données par les formules suivantes :

$$q_c = \frac{a'\eta_3^* - e'_c\eta'_3}{a'\eta_3}\,, \quad q'_c = \frac{e'_c}{a'}\,, \quad p_c = \frac{e_c - q_ce_{c-1}}{\eta_3^*}\,, \quad p'_c = \frac{a' - e_{c-1} - p_c\eta_3}{\eta'_3}\,.$$

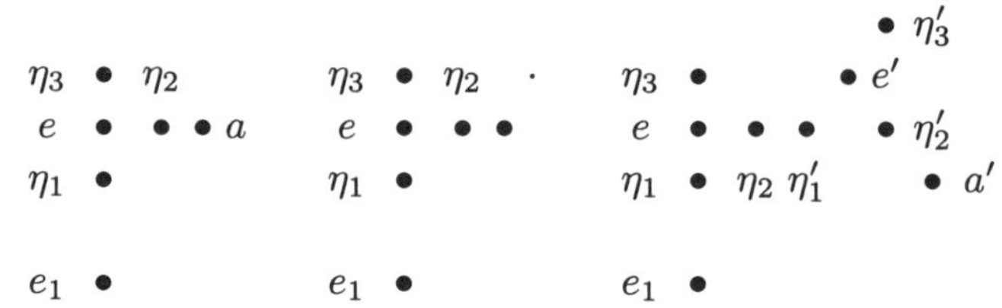

Figure 4

Lemme 4.4. *Dans les cas A.1, A.2 et A.3, considérons le halo $\mathcal{H} = (e; \eta_j)$ de plus petite multiplicité centrale. Notons $a = \mathrm{pgcd}(e, \eta_2)$ et e_1 la multiplicité du sommet #1 (de valence 1). La paire de Zariski p_1/q_1 associée à $\mathcal{H}$ est donnée par les formules suivantes :*

$$q_1 = \frac{e}{a}\,, \qquad p_1 = \frac{e}{e_1}\,.$$

Lemme 4.5. *Dans le cas B.3, considérons le halo $\mathcal{H}_1 = (e_1; \eta_j)$ du sommet de séparation, et supposons de plus que la branche φ' est lisse. Notons e_1 la multiplicité du sommet #1 (de valence 1). La paire de Zariski p_1/q_1 associée à $\mathcal{H}_1$ est donnée par les formules suivantes :*

$$q_1 = \frac{e_1 - 1}{\eta_3}\,, \qquad p_1 = \frac{e}{e_1}\,.$$

Lemme 4.6. *Dans le cas B.3, considérons le halo $\mathcal{H}_1 = (e_1; \eta_j)$ du sommet de séparation, et supposons de plus que la branche φ' n'est pas lisse. Notons $\mathcal{H}'_1 = (e'_1; \eta'_j)$ le halo qui se raccorde à l'arbre réduit à l'extrémité de l'arête portant la composante η_2. Notons $a' = \mathrm{pgcd}(e'_1, \eta'_2)$. Les paires de Zariski p_1/q_1, associée à $\mathcal{H}_1$, et p'_1/q'_1, associée à $\mathcal{H}'_1$, sont données par les formules suivantes :*

$$p_1 = \frac{e}{e_1}\,, \qquad q'_1 = \frac{e'}{a'}\,, \qquad q_1 = \frac{e_1 - q'_1\eta'_3}{\eta_3}\,, \qquad p'_1 = \frac{a' - p_1\eta_3}{\eta'_3}\,.$$

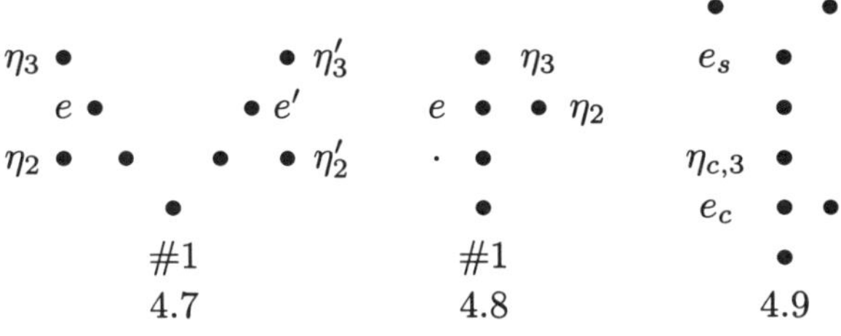

Figure 5

Lemme 4.7. *Dans le cas B.1, supposons que l'arbre $T(f)$ comporte deux peupliers. Notons $\mathcal{H} = (e; \eta_j)$ et $\mathcal{H}' = (e'; \eta_j)$ les deux halos reliés par une arête au sommet #1, notons $a = \mathrm{pgcd}(e, \eta_2)$ et $a' = \mathrm{pgcd}(e', \eta_2')$. Les paires de Zariski p/q, associée à $\mathcal{H}$, et p'/q', associée à $\mathcal{H}'$, sont données par les formules suivantes (on rappelle que l'on a supposé $p/q > 1$ et $p'/q' > 1$) :*

$$q = \frac{e}{a}, \qquad q' = \frac{e'}{a'}, \qquad p = \frac{a - q'\eta_3'}{\eta_3}, \qquad p' = \frac{a' - q\eta_3}{\eta_3'}.$$

Lemme 4.8. *Dans le cas B.1, supposons que l'arbre $T(f)$ comporte un seul peuplier. Notons $\mathcal{H} = (e; \eta_j)$ le halo relié par une arête au sommet #1, notons $a = \mathrm{pgcd}(e, \eta_2)$. La paire de Zariski p/q, associée à $\mathcal{H}$ est donnée par les formules suivantes :*

$$q = \frac{e}{a}, \qquad p = \frac{a - 1}{\eta_3}.$$

Lemme 4.9. *Dans le cas A.2, l'exposant de coïncidence $\mathfrak{C}(\varphi, \varphi')$ entre les deux branches de f se calcule comme suit. On note $\mathcal{H}_s = (e_s; \eta_{s,j})$ le halo du sommet de séparation, et $\mathcal{H}_c = (e_c; \eta_{c,1}, \eta_{c,2}, \eta_{c,3})$ le halo du sommet de rupture (avec branche morte) qui correspond à la c-ième paire de Zariski, cf. §2 et Figure 5 pour les notations, on définit l'entier γ, $\gamma > 0$, par l'égalité suivante :*

$$\mathfrak{C}(\varphi, \varphi') = \frac{p_1}{q_1} + \frac{p_2}{q_1 q_2} + \cdots + \frac{p_c}{q_1 \cdots q_c} + \frac{\gamma}{q_1 \cdots q_c}. \quad \textit{On a alors : } \gamma = \frac{e_s - e_c}{\eta_{c,3}}.$$

Théorème 4.1. *Soit f_1 et f_2 des germes de fonction analytique à singularité isolée, définissant des germes de courbe plane à deux branches. Supposons que les formes de Seifert sur $H_1(F(f_a), \mathbf{Z})$, $a = 1$ ou 2, sont isomorphes, que l'hypothèse $\mathcal{HT}$ est satisfaite, et que f_1 n'a pas d'isomères, alors f_1 et f_2 ont le même type topologique.*

Démonstration. D'après l'étude faite au §3, les hypothèses du théorème entraînent que les arbres réduits $TR(f_1)$ et $TR(f_2)$ sont identiques. D'après les lemmes 4.1 à 4.8, il s'ensuit que les branches des germes f_1 et f_2 ont les mêmes paires de Zariski. De plus, d'après le lemme 4.9, l'exposant de coïncidence entre les deux branches de f_1 est égal à l'exposant de coïncidence entre les deux branches de f_2. Par suite, les germes f_1 et f_2 ont le même type topologique. Sans utiliser l'exposant de coïncidence, on peut aussi remarquer que la forme de Seifert détermine directement le nombre d'intersection des deux branches f_1' et f_1'' (resp. f_2' et f_2'') de $f_1 = f_1' \cdot f_1''$ (resp. $f_2 = f_2' \cdot f_2''$), par la formule $\nu_0(f_1', f_1'') = -A(b, b)$, où b est un générateur de $\mathrm{Ker}(t - 1) \subset H_1(F, \mathbf{Z})$, on a donc :

$$\nu_0(f_1', f_1'') = \nu_0(f_2', f_2''). \qquad \square$$

Théorème 4.2. *Soit f un germe de fonction analytique à singularité isolée, définissant un germe de courbe plane à deux branches. Supposons que l'arbre réduit $TR(f)$ admet des isomères. Alors chaque arbre réduit isomère de l'arbre $TR(f)$ est l'arbre réduit de l'arbre de désingularisation d'un germe de courbe plane à deux branches défini par un germe de fonction analytique à singularité isolée g, $T(g)$ est donc un isomère de $T(f)$. De plus, si les germes à deux branches définis par les fonctions analytiques $f = f' \cdot f''$ et $g = g' \cdot g''$ sont isomères, ils ont la même multiplicité, et le nombre d'intersection des deux branches est le même dans les deux cas, autrement dit :*

$$\nu_0(f) = \nu_0(g) \quad \text{et} \quad \nu_0(f', f'') = \nu_0(g', g'').$$

Démonstration. On part d'un germe de courbe plane à deux branches, défini par un germe de fonction analytique à singularité isolée f, on suppose avoir trouvé un arbre réduit TR', isomère de $TR(f)$. On veut montrer qu'il existe un germe de courbe plane à deux branches, défini par un germe de fonction analytique g tel que $TR' = TR(g)$. On supposera que TR' s'obtient à partir de $TR(f)$ par un unique échange de peupliers de même poids. Le cas général s'en déduit en itérant le cas d'un échange unique. D'après la prop. 3.3, deux cas se présentent, voir les Figures 1 et 2, dont nous reprenons les notations.

Premier cas : le cas représenté Figure 1. Dans l'arbre $T(f)$, les paires de Zariski p/q, associée à $\mathcal{H}$, et p'/q', associée à $\mathcal{H}'$ sont données par les formules suivantes :

$$q = \frac{e}{a}, \quad p = \frac{e - qe^*}{\eta_3^*} = \frac{a - e^*}{\eta_3}, \quad q' = \frac{e'}{a'}, \quad p' = \frac{e' - q'\varepsilon^*}{\eta_3^*} = \frac{a' - \varepsilon^*}{\eta_3'}.$$

Si le germe g existe, nous noterons π/χ la paire de Zariski associée à $\mathcal{H}$ et π'/χ' la paire associée à $\mathcal{H}'$ dans $T(g)$; celles-ci doivent vérifier les formules suivantes :

$$\chi = \frac{e}{a}, \quad \pi = \frac{a - \varepsilon^*}{\eta_3}, \quad \chi' = \frac{e'}{a'}, \quad \pi' = \frac{a' - e^*}{\eta_3'}.$$

On doit donc avoir $\chi = q$, $\chi' = q'$, $\pi = p + \dfrac{e^* - \varepsilon^*}{\eta_3}$ et $\pi' = p' + \dfrac{\varepsilon^* - e^*}{\eta_3'}$. Les multiplicités sortantes des halos $\mathcal{H}^*$ et $\mathcal{H}'^*$ sont égales à η_3^*, il s'ensuit que e^* et ε^* sont divisibles par η_3^*, ainsi que les multiplicités des sommets situés dans $T(f)$ sur la géodésique qui joint les deux sommets de valence 3 au centre des halos $\mathcal{H}^*$ et $\mathcal{H}'^*$. Ceci entraîne que e^* et ε^* sont divisibles par $\eta_3 = \eta_3^*/q$ et $\eta_3' = \eta_3^*/q'$ et que les nombres π et π' sont des nombres entiers. De plus, les conditions $\varepsilon^* < a$ et $e^* < a'$ prouvent que π et π' sont positifs.

On remarquera plus précisément que $\pi - p$ (resp. $\pi' - p'$) est un multiple entier de q (resp. q') :

$$\pi - p = q\,\frac{e^* - \varepsilon^*}{\eta_3^*} \quad \text{et} \quad \pi' - p' = q'\frac{\varepsilon^* - e^*}{\eta_3^*}.$$

Autrement dit, les développements en fraction continue de p/q et de π/q ne se distinguent que par leurs parties entières, et il en est de même pour les développements de p'/q' et de π'/q' :

$$\frac{p}{q} = h_0 + \cfrac{1}{h_1 + \cfrac{1}{h_2 + \cdots}} \quad \text{et} \quad \frac{\pi}{q} = \bar{h}_0 + \cfrac{1}{h_1 + \cfrac{1}{h_2 + \cdots}}.$$

On notera $\mathrm{ez}(h)$ les exposants caractéristiques de Zariski d'un germe irréductible h. Supposons que les exposants des branches f' et f'' du germe $f = f' \cdot f''$ sont :

$$\mathrm{ez}(f') = (\frac{p_1}{q_1}, \frac{p_2}{q_2}, \ldots, \frac{p_a}{q_a}, \frac{p}{q}, \frac{p_{a+2}}{q_{a+2}}, \ldots, \frac{p_g}{q_g}),$$

$$\mathrm{ez}(f'') = (\frac{p_1'}{q_1'}, \frac{p_2'}{q_2'}, \ldots, \frac{p_b'}{q_b'}, \frac{p'}{q'}, \frac{p_{b+2}'}{q_{b+2}'}, \ldots, \frac{p_{g'}'}{q_{g'}'}).$$

Considérons un germe $g = g' \cdot g''$ à deux branches dont le type topologique est défini comme suit. D'une part, les exposants caractéristiques de Zariski des branches g' et g'' sont :

$$\mathrm{ez}(g') = (\frac{p_1}{q_1}, \frac{p_2}{q_2}, \ldots, \frac{p_a}{q_a}, \frac{\pi'}{q'}, \frac{p_{b+2}'}{q_{b+2}'}, \ldots, \frac{p_{g'}'}{q_{g'}'}),$$

540

$$\mathrm{ez}(g'') = \left(\frac{p'_1}{q'_1}, \frac{p'_2}{q'_2}, \ldots, \frac{p'_b}{q'_b}, \frac{\pi}{q}, \frac{p_{a+2}}{q_{a+2}}, \ldots, \frac{p_g}{q_g}\right),$$

et, d'autre part, l'exposant de coïncidence entre les branches de g est égal à l'exposant de coïncidence entre les branches de f. La méthode de calcul donnée dans (MW85, 6.6) permet alors de vérifier que les arbres TR' et $TR(g)$ sont identiques, le point clé étant la remarque sur les développements en fraction continue de p/q et de π/q (resp. de p'/q' et de π'/q'), qui assure que le halo associé à π/q (resp. π'/q') dans $T(g)$ est identique au halo $\mathcal{H}$ (resp. $\mathcal{H}'$) de $T(f)$.

Les arbres $T(f)$ et $T(g)$ étant isomères, on a, d'après la prop. 2.6, $\nu_0(f', f'') = \nu_0(g', g'')$. On a enfin, par construction, $\eta_3^* = q q_{a+2} \cdots q_g = q' q'_{b+2} \cdots q'_{g'}$, ce qui donne :

$$\nu_0(f) = q_1 q_2 \cdots q_g + q'_1 q'_2 \cdots q'_{g'} = \eta_3^* (q_1 q_2 \cdots q_a + q'_1 q'_2 \cdots q'_b) = \nu_0(g).$$

Le germe g construit ci-dessus satisfait donc les propriétés demandées.

Deuxième cas : le cas représenté Figure 2. Dans l'arbre $T(f)$, les paires de Zariski p'/q', associée à $\mathcal{H}'$, et p''/q'', associée à $\mathcal{H}''$ sont données par les formules suivantes, où p_s/q_s est la paire de Zariski associée au sommet de séparation et e_{s-1} est la multiplicité du sommet de rupture qui précède le sommet de séparation, s'il en existe un, et 0 sinon :

$$q' = \frac{e'}{a'}, \quad p' = \frac{e' - q'e_s}{\eta_{s3}} = \frac{a' - e_s}{\eta'_3}, \quad q'' = \frac{e''}{a''}, \quad p'' = \frac{a'' - e_{s-1} - p_s\eta_{s3}}{\eta''_3}.$$

Si le germe g existe, nous noterons π'/χ' la paire de Zariski associée à $\mathcal{H}'$ et π''/χ'' la paire associée à $\mathcal{H}''$ dans $T(g)$; celles-ci doivent vérifier les formules suivantes :

$$\chi'' = \frac{e''}{a''}, \quad \pi'' = \frac{e'' - \chi''e_s}{\eta_{s3}} = \frac{a'' - e_s}{\eta''_3}, \quad \chi' = \frac{e'}{a'}, \quad \pi' = \frac{a' - e_{s-1} - p_s\eta_{s3}}{\eta'_3}.$$

On doit donc avoir $\chi' = q'$, $\chi'' = q''$, $\pi' = p' + \dfrac{e_s - e_{s-1} - p_s\eta_{s3}}{\eta'_3}$ et $\pi'' = p'' - \dfrac{e_s - e_{s-1} - p_s\eta_{s3}}{\eta''_3}$. La condition imposée sur les halos donne en particulier $\eta_{s3} = q'\eta'_3 = q''\eta''_3$, par suite, les deux branches de f ont une multiplicité en 0 multiple de η_{s3}. On trouve ainsi que e_s et e_{s-1} sont multiples de η_{s3}. Les nombres π' et π'' sont donc entiers. Plus précisément, on voit que $\pi' - p'$ est un multiple entier de q', et $\pi'' - p''$ un multiple entier de q'' :

$$\pi' - p' = q'\left(\frac{e_s - e_{s-1}}{\eta_{s3}} - p_s\right) \quad \text{et} \quad \pi'' - p'' = -q''\left(\frac{e_s - e_{s-1}}{\eta_{s3}} - p_s\right).$$

Supposons que les exposants caractéristiques de Zariski des branches f' et f'' du germe $f = f' \cdot f''$ sont :

$$\mathrm{ez}(f') = (\frac{p_1}{q_1}, \frac{p_2}{q_2}, \ldots, \frac{p_s}{q_s}, \frac{p'}{q'}, \frac{p_{s+2}}{q_{s+2}}, \ldots, \frac{p_g}{q_g}),$$

$$\mathrm{ez}(f'') = (\frac{p_1}{q_1}, \frac{p_2}{q_2}, \ldots, \frac{p_{s-1}}{q_{s-1}}, \frac{p''}{q''}, \frac{p'_{s+1}}{q'_{s+1}}, \ldots, \frac{p'_{g'}}{q'_{g'}}).$$

Considérons un germe $g = g' \cdot g''$ à deux branches dont le type topologique est défini comme suit. D'une part, les exposants caractéristiques de Zariski des branches g' et g'' sont :

$$\mathrm{ez}(g') = (\frac{p_1}{q_1}, \frac{p_2}{q_2}, \ldots, \frac{p_s}{q_s}, \frac{\pi''}{q''}, \frac{p'_{s+1}}{q'_{s+1}}, \ldots, \frac{p'_{g'}}{q'_{g'}}),$$

$$\mathrm{ez}(g'') = (\frac{p_1}{q_1}, \frac{p_2}{q_2}, \ldots, \frac{p_{s-1}}{q_{s-1}}, \frac{\pi'}{q'}, \frac{p_{s+2}}{q_{s+2}}, \ldots, \frac{p_g}{q_g}),$$

et, d'autre part, l'exposant de coïncidence entre les branches de g est égal à l'exposant de coïncidence $\mathfrak{C}(\varphi, \varphi') = p_s/q_s$ entre les branches de f. On termine la démonstration comme dans le premier cas. $\qquad\square$

5. Formes de Seifert de germes isomères

Comme plus haut, f désigne un germe de fonction analytique à singularité isolée, définissant un germe de courbe plane à deux branches, et F sa fibre de Milnor. Nous allons démontrer dans cette section que $M_{-2}H_1(F, \mathbf{Z})$ et $\mathrm{Gr}_0^M(H_1(F, \mathbf{Z}))$ sont des $\mathbf{Z}[t, t^{-1}]$-modules cycliques, *cf.* théorème 5.1. On en déduira que, si deux germes de courbe plane à deux branches sont isomères, leurs formes de Seifert sont isomorphes, *cf.* théorème 5.2.

5.1. *Lemmes préliminaires*

On notera $\Gamma(\varphi, \varphi')$ la géodésique de $T(f)$ joignant les flèches symbolisant φ et φ' ; les *sommets* de $\Gamma(\varphi, \varphi')$ seront les sommets de rupture, le sommet #1 s'il est sur $\Gamma(\varphi, \varphi')$ (c'est-à-dire si les deux branches sont transverses) et l'éventuel sommet de valence 2 portant la flèche associée à φ' de $T(f)$ portés par cette géodésique ; les *arêtes* de $\Gamma(\varphi, \varphi')$ seront les segments géodésiques de $T(f)$ portés par cette géodésique.

Les sommets (resp. les arêtes) seront pondérés par les e_i et r_i (resp. les m_{ij}) calculés dans $T(f)$. Les sommets de $\Gamma(\varphi, \varphi')$ seront renumérotés en suivant leur position sur la géodésique, de 1 (correspondant au sommet de rupture sur lequel s'attache la flèche associée à φ) à $N+1$ (associé à φ'). Les e_i et r_i seront numérotés par le nouveau numéro du sommet correspondant et les m_{ij} par le numéro du segment géodésique, comme indiqué Figure 6 ci-dessous.

$$r_1 \; m_1 \; r_2 \; m_2 \; \ldots \; m_{N-1} \; r_N \; m_N \; r_{N+1}$$
$$\bullet \qquad \bullet \qquad \cdots \qquad \bullet \qquad \bullet$$

Figure 6 - $\Gamma(\varphi, \varphi')$

Si $\mathfrak{C}(\varphi, \varphi')$ est un exposant permis pour φ et pour φ' (cas *A.1*, *A.2*, *B.1* et *B.2*), on posera :

$$\psi = x^{p_1/q_1}(1 + x^{p_2/q_1 q_2}(1 + \cdots + x^{p_c/q_1 \cdots q_c}) \ldots) \text{ et } \psi = x \text{ si } c = 0.$$

Si $\mathfrak{C}(\varphi, \varphi')$ est un exposant permis pour φ et non pour φ' (cas *A.3* et *B.3*), on posera :

$$\psi = x^{p_1/q_1}(1 + x^{p_2/q_1 q_2}(1 + \cdots + x^{p_{c-1}/q_1 \cdots q_{c-1}}) \ldots).$$

L'arbre $T(f)$ est alors la réunion du sous-arbre $T^\top(f)$, constitué de la géodésique de $T(f)$ joignant les flèches symbolisant φ et φ' et des branches mortes qui y sont attachées, et d'un sous-arbre $T^\perp(f)$ isomorphe à l'arbre $T(\psi)$ de désingularisation d'un germe ayant ψ pour développement de Puiseux, les deux sous-arbres étant rattachés par un segment géodésique qui joint le sommet de rupture de $T^\perp(f)$ ayant la plus grande multiplicité à celui de $T^\top(f)$ ayant la plus petite multiplicité, ou au sommet #1 si $c = 0$; ces deux sommets sont distincts, à l'exception du cas où $T(f)$ admet un sommet de rupture de valence 4, qui est alors l'unique sommet commun à $T^\perp(f)$ et $T^\top(f)$. Suivant la méthode de calcul donnée dans (BK86, p. 682-708) ou (MW85, 6.6), on voit que les multiplicités des sommets de $T^\perp(f)$ se déduisent de celles des sommets de $T(\psi)$ par multiplication par $(q_{c+1} \cdots q_g + q'_{c+1} \cdots q'_{g'})$, si $\mathfrak{C}(\varphi, \varphi')$ est un exposant permis pour φ et pour φ', et par multiplication par $(q_c \cdots q_g + q'_c \cdots q'_{g'})$ sinon. Soit g un germe de fonction analytique à singularité isolée, définissant un germe de courbe plane ayant un nombre quelconque de branches, et soit $F(g)$ sa fibre de Milnor. On peut calculer les $\mathbf{Z}[t, t^{-1}]$-modules $M_{-2}H_1(F(g), \mathbf{Z})$ et $\mathrm{Gr}_0^M(H_1(F(g), \mathbf{Z}))$ en considérant le graphe $G(g)$, revêtement ramifié de l'arbre $T(g)$, construit comme suit : le sommet (i) de $T(g)$ a pour image réciproque r_i sommets, l'arête (ij) de $T(g)$ a pour image réciproque m_{ij} arêtes, chaque flèche de $T(g)$ a pour image réciproque un segment de $G(g)$. L'ensemble des extrémités extérieures de ces segments sera noté ∂G, les points de ∂G sont donc en bijection avec l'ensemble des flèches de $T(g)$ ou avec l'ensemble des branches de g. De plus, l'action de la monodromie sur

$G(g)$ est un isomorphisme du revêtement. On a alors les isomorphismes de $\mathbf{Z}[t, t^{-1}]$-modules suivants, d'après (DBM92, 5.5 et 6.5) :

$$M_{-2}H_1(F(g), \mathbf{Z}) \cong H^1(G(g), \partial G, \mathbf{Z}) \text{ et } \mathrm{Gr}_0^M(H_1(F(g), \mathbf{Z})) \cong H^1(G(g), \mathbf{Z}).$$

Comme observé dans (*loc. cit.*), les branches mortes de $T(g)$ n'apportent pas de contribution au calcul des $\mathbf{Z}[t, t^{-1}]$-modules étudiés ici ; en effet, une branche morte, dont les extrémités sont les sommets (j), de valence 1, et (k), de valence ≥ 3, a pour image réciproque dans $G(g)$, e_j segments attachés à $G(g)$ par les sommets qui forment l'image réciproque du sommet (k). L'image réciproque d'une branche morte est donc contractile.

Revenons maintenant au cas d'un germe à deux branches et aux notations du §5.

Lemme 5.1. *Soit* $f : (\mathbf{C}^2, 0) \to (\mathbf{C}, 0)$ *un germe de fonction analytique à singularité isolée, définissant un germe de courbe plane à deux branches. L'image réciproque de* $T^{\perp}(f)$ *dans* $G(f)$ *est contractile. Par suite, les arêtes de* $T^{\perp}(f)$ *n'apportent pas de contribution au calcul des* $\mathbf{Z}[t, t^{-1}]$-*modules* $M_{-2}H_1(F, \mathbf{Z})$ *et* $\mathrm{Gr}_0^M(H_1(F, \mathbf{Z}))$.

Démonstration. Il reste à voir que si (i) est un sommet de rupture de $T(f)$ situé sur $T^{\perp}(f)$ et (j) le sommet suivant sur l'unique arête sortante du sommet (i), on a l'égalité $r_i = m_{ij}$. Ceci provient de la description, donnée plus haut, des multiplicités des sommets de $T^{\perp}(f)$ en fonction de celles des sommets de $T(\psi)$, et du fait bien connu que la monodromie d'un germe de courbe plane irréductible (ici le germe dont ψ est un développement de Puiseux) est unipotente.

Notons (l) le sommet de $\Gamma(\varphi, \varphi')$ dont la multiplicité e_l est minimale parmi les sommets de $\Gamma(\varphi, \varphi')$; à la renumérotation près, (l) est associé au sommet de rupture de $T^{\top}(f)$, noté $(\widetilde{l})$, sur lequel se rattache $T^{\perp}(f)$, ou au sommet #1 s'il est de valence 2. Le lemme 2.1 donne les informations suivantes sur les multiplicités de certains voisins de (l).

Si $\mathfrak{C}(\varphi, \varphi')$ est un exposant permis pour φ et pour φ', le sommet $(\widetilde{l})$ admet deux arêtes sortantes. La multiplicité e_l^{φ} (resp. $e_l^{\varphi'}$) du sommet de $T(f)$ voisin de $(\widetilde{l})$ sur l'arête sortante dirigée vers φ (resp. φ') vérifie :

$$e_l^{\varphi} \equiv q_{c+1} \cdots q_g \,(\mathrm{mod}\, e_l), \quad e_l^{\varphi'} \equiv q'_{c+1} \cdots q'_{g'} \,(\mathrm{mod}\, e_l).$$

Si $\mathfrak{C}(\varphi, \varphi')$ est un exposant permis pour φ, mais non pour φ', le sommet $(\widetilde{l})$ admet une unique arête sortante (vers φ) et le sommet $(\widetilde{l+1})$ de $T(f)$, associé au sommet $(l+1)$ sur $\Gamma(\varphi, \varphi')$, admet une unique arête sortante (vers φ'). La multiplicité e_l^φ (resp. $e_{l+1}^{\varphi'}$) du sommet de $T(f)$ voisin de $(\widetilde{l})$ (resp. $(\widetilde{l+1})$) sur l'arête sortante dirigée vers φ (resp. φ') vérifie :

$$e_l^\varphi \equiv q_{c+1} \cdots q_g \;(\mathrm{mod}\, e_l)\,, \quad e_{l+1}^{\varphi'} \equiv q'_{c+1} \cdots q'_{g'} \;(\mathrm{mod}\, e_{l+1})\,. \qquad \square$$

Lemme 5.2. *On a la relation de divisibilité suivante entre les m_i le long de $\Gamma(\varphi, \varphi')$: si, pour un certain couple (i, j), $1 \le i < j \le N$, l'entier a divise m_i et m_j, alors, pour tout k, $i \le k \le j$, a divise m_k.*

Démonstration. Si $i + 1 = j$, il n'y a rien à démontrer. On supposera donc que $i + 1 < j$. Deux cas se présentent, suivant la position de (l) par rapport aux arêtes $(i\ i+1)$ et $(j\ j+1)$. Si $k \le l$ (resp. $k \ge l+1$), on notera e_k^φ (resp. $e_k^{\varphi'}$) la multiplicité du sommet de $T(f)$ voisin du sommet de rupture numéroté k dans $\Gamma(\varphi, \varphi')$ dans la direction de la flèche associée à φ (resp. φ').

Premier cas : $j \le l$ ou $l \le i$. Si $j \le l$, l'hypothèse $a \mid m_i$ et $a \mid m_j$ entraine que a divise e_i, e_{i+1}, e_j et e_{j+1} ainsi que le produit $q_b \cdots q_g$ des dénominateurs des paires de Zariski associées aux sommets de rupture situés entre (1) et (i) inclus, en effet le lemme 2.1 donne $e_{i+1}^\varphi \equiv q_b \cdots q_g \;(\mathrm{mod}\, e_{i+1})$, par suite, a divise e_j et e_j^φ, puisque $e_j^\varphi \equiv q_{b+1+i-j} \cdots q_b \cdots q_g \;(\mathrm{mod}\, e_j)$. Ceci entraine que a divise m_{j-1}, d'où le résultat demandé. On procède de même si $l \le i$. Noter que si $\mathfrak{C}(\varphi, \varphi')$ n'est pas un exposant permis pour φ' et si $l = i$, le lemme 2.1 ne s'applique pas au sommet $(i) = (l)$ dans la direction de φ', mais il nous suffit de savoir que $e_{i+1}^{\varphi'}$ est divisible par a.

Deuxième cas : $i < l < j < N$. L'hypothèse entraine ici que a divise e_i, e_{i+1}, e_j et e_{j+1} ainsi que le produit $q_b \cdots q_g$ (resp. $q'_d \cdots q'_{g'}$) des dénominateurs des paires de Zariski associées aux sommets de rupture situés entre (1) et (i) inclus (resp. entre $(j+1)$ et $(N+1)$ inclus), on a en effet, par le lemme 2.1, $e_{i+1}^\varphi \equiv q_b \cdots q_g \;(\mathrm{mod}\, e_{i+1})$ et $e_j^{\varphi'} \equiv q'_d \cdots q'_{g'} \;(\mathrm{mod}\, e_j)$. Par suite, a divise $e_{\#1} = (q_1 \cdots q_g + q'_1 \cdots q'_{g'})$, et a divise les multiplicités des sommets de $T(f)$ portés par la géodésique qui joint $\#1$ à $(\widetilde{l})$ (le sommet de rupture de $T(f)$ renuméroté (l) dans $\Gamma(\varphi, \varphi')$). Le cas où le sommet $(\widetilde{l})$ est de valence 4 se traite comme le premier cas ci-dessus ; sinon, le sommet $(\widetilde{l})$ a trois voisins dans $T(f)$, dont nous noterons les multiplicités e_l^φ (multiplicité du voisin de $(\widetilde{l})$ dans la direction de φ), $e_l^{\varphi'}$ (dans la direction de φ') et $e_l^\perp$ (dans la direction de $\#1$). On vient de voir que $a \mid e_l^\varphi$ et $a \mid e_l^\perp$, mais $e_l^\perp + e_l^\varphi + e_l^{\varphi'} \equiv 0 \;(\mathrm{mod}\, e_l)$, donc aussi $e_l^\perp + e_l^\varphi + e_l^{\varphi'} \equiv 0 \;(\mathrm{mod}\, a)$. Ceci montre que $a \mid e_l^{\varphi'}$, puis, d'une part $a \mid m_l$, car $m_l = \mathrm{pgcd}(e_l, e_l^{\varphi'})$, ce qui donne, pour $i < k < l$, $a \mid m_k$ (en appliquant le premier cas entre i et l) et d'autre part, $a \mid e_{l+1}$, car $m_l = \mathrm{pgcd}(e_l, e_l^{\varphi'}, e_{l+1})$, ce qui donne, pour $l < k$, $a \mid m_k$ (en appliquant le premier cas entre l et j). $\qquad\square$

5.2. *Matrice de présentation de* $M_{-2}H_1(F, \mathbf{Z})$

Nous allons travailler avec des produits de polynômes cyclotomiques fortement premiers entre eux, on rappelle (*cf.* prop. 1.1) qu'on écrit $(A, B)_{\mathbf{Z}} = 1$ pour indiquer que A et B de $\mathbf{Z}[t, t^{-1}]$ sont fortement premiers entre eux.

On posera, pour $1 \leq i \leq N$, $\alpha_i = (t^{m_i} - 1)/(t^{r_i} - 1)$, $\beta_i = t^{m_i} - 1$ et $\gamma_i = (t^{m_i} - 1)/(t^{r_{i+1}} - 1)$. Vu que les sommets de rupture situés aux extrémités de $\Gamma(\varphi, \varphi')$ portent une flèche dans l'arbre $T(f)$, on a $r_1 = r_{N+1} = 1$, ce qui donne l'égalité $\alpha_1 \alpha_2 \cdots \alpha_N = \gamma_1 \gamma_2 \cdots \gamma_N$. On posera aussi $\beta_0 = \beta_{N+1} = t - 1$, $\gamma_0 = \alpha_{N+1} = 1$, $u_0 = 1$, $v_1 = 0$, $u_N = 0$ et $v_{N+1} = 1$ de sorte que $u_0 \gamma_0 + v_1 \alpha_1 = 1$ et $u_N \gamma_1 \cdots \gamma_N + v_{N+1} \alpha_{N+1} = 1$.

Lemme 5.3. *Les polynômes α_i et γ_i, $1 \leq i \leq N$, vérifient les relations suivantes :*

$$(\gamma_1, \alpha_2)_{\mathbf{Z}} = 1, \ (\gamma_1 \gamma_2, \alpha_3)_{\mathbf{Z}} = 1, \ldots, \ (\gamma_1 \gamma_2 \cdots \gamma_{N-1}, \alpha_N)_{\mathbf{Z}} = 1.$$

Il existe donc des éléments $u_1, \ldots, u_{N-1}$ et $v_2, \ldots, v_N$ de $\mathbf{Z}[t, t^{-1}]$ tels que :

$$u_1 \gamma_1 + v_2 \alpha_2 = 1, \ u_2 \gamma_1 \gamma_2 + v_3 \alpha_3 = 1, \ldots, \ u_{N-1} \gamma_1 \gamma_2 \cdots \gamma_{N-1} + v_N \alpha_N = 1.$$

Démonstration. Procédant comme dans le lemme 5.2, on trouve d'abord que, pour tout i, $1 < i < N$, on a $r_i = \mathrm{pgcd}(m_{i-1}, m_i)$, (noter que le sommet $(\tilde{\imath})$ de $T(f)$, renuméroté en i dans $\Gamma(\varphi, \varphi')$, a 3 voisins dans $T(f)$, sauf si $T(f)$ admet un sommet de valence 4 et si $i = l$). Il suffit de démontrer que, pour tout i et j, $1 \leq i < j \leq N$, on a $(\gamma_i, \alpha_j)_{\mathbf{Z}} = 1$, c'est-à-dire que $\gamma_i = (t^{m_i} - 1)/(t^{r_{i+1}} - 1)$ et $\alpha_j = (t^{m_j} - 1)/(t^{r_j} - 1)$ sont fortement premiers entre eux. Supposons par l'absurde qu'il existe des entiers $a \geq 2$, $n \geq 1$ et un nombre premier p tels que $\Phi_a \mid \alpha_j$ et $\Phi_{ap^n} \mid \gamma_i$, on aurait alors $a \mid m_j$ et $ap^n \mid m_i$, donc $a \mid m_i$, et aussi, pour tout k, $i \leq k \leq j$, $a \mid m_k$, ce qui donne en particulier, par la remarque ci-dessus, a divise $r_{i+1}, \ldots, r_j$. Il s'ensuit que Φ_a ne divise pas α_j : contradiction. L'hypothèse $\Phi_{ap^n} \mid \alpha_j$ et $\Phi_a \mid \gamma_i$ conduit de même à une contradiction. $\qquad\square$

Passons maintenant au calcul du $\mathbf{Z}[t, t^{-1}]$-module $M_{-2}H_1(F, \mathbf{Z})$. En utilisant (DBM92, 6.6) et le lemme 5.1, on voit que la matrice M suivante est une matrice de présentation de ce module :

$$
M = \begin{pmatrix}
\beta_0 & \gamma_0 & 0 & 0 & 0 & \cdots & 0 & 0 & 0 & 0 \\
0 & \alpha_1 & \beta_1 & \gamma_1 & 0 & \cdots & 0 & 0 & 0 & 0 \\
0 & 0 & 0 & \alpha_2 & \beta_2 & \cdots & 0 & 0 & 0 & 0 \\
\cdots\cdots\cdots\cdots\cdots & & & & & \cdots & \cdots & & \cdots \\
0 & 0 & 0 & 0 & 0 & \cdots & \alpha_N & \beta_N & \gamma_N & 0 \\
0 & 0 & 0 & 0 & 0 & \cdots & 0 & 0 & \alpha_{N+1} & \beta_{N+1}
\end{pmatrix}.
$$

En multipliant successivement M à gauche par les matrices $P_i \in \mathrm{GL}(N+2-i, \mathbf{Z}[t,t^{-1}])$ suivantes, $0 \le i \le N$:

$$
P_i = \begin{pmatrix}
u_i & v_{i+1} & 0 \\
-\alpha_{i+1} & \gamma_0\gamma_1\gamma_2\cdots\gamma_i & 0 \\
0 & 0 & I_{N-i}
\end{pmatrix},
$$

où I_{N-i} désigne la matrice unité d'ordre $N - i$, et en effectuant les simplifications, on trouve les matrices de présentation suivantes :

$$
M_1 = \begin{pmatrix}
\beta_0\alpha_1 & \gamma_1 & \cdots & 0 & 0 \\
0 & \alpha_2 & \cdots & 0 & 0 \\
\cdots & \cdots\cdots & \cdots & & \cdots \\
0 & 0 & \cdots & \gamma_N & 0 \\
0 & 0 & \cdots & \alpha_{N+1} & \beta_{N+1}
\end{pmatrix}, \
M_2 = \begin{pmatrix}
\beta_0\alpha_1\alpha_2 & \gamma_1\gamma_2 & \cdots & 0 & 0 \\
0 & \alpha_3 & \cdots & 0 & 0 \\
\cdots & \cdots\cdots & \cdots & \cdots & \cdots \\
0 & 0 & \cdots & \gamma_N & 0 \\
0 & 0 & \cdots & \alpha_{N+1} & \beta_{N+1}
\end{pmatrix},
$$

$$
M_N = \begin{pmatrix}
\beta_0\alpha_1\cdots\alpha_N & \gamma_1\cdots\gamma_N & 0 \\
0 & \alpha_{N+1} & \beta_{N+1}
\end{pmatrix}, \
M_{N+1} = \left((t-1)\alpha_1\cdots\alpha_N \right).
$$

On a ainsi déterminé la structure du $\mathbf{Z}[t,t^{-1}]$-module $M_{-2}H_1(F,\mathbf{Z})$. Le $\mathbf{Z}[t,t^{-1}]$-module $\mathrm{Gr}_0^M H_1(F,\mathbf{Z})$ est donné par un calcul analogue.

Théorème 5.1. *Soit F la fibre de Milnor d'un germe de courbe plane à deux branches, on a les isomorphismes de $\mathbf{Z}[t,t^{-1}]$-modules suivants :*

$$
M_{-2}H_1(F,\mathbf{Z}) \cong \mathbf{Z}[t,t^{-1}]/\left(\frac{(t^{m_1}-1)\cdots(t^{m_N}-1)}{(t^{r_2}-1)\cdots(t^{r_N}-1)}\right),
$$

$$
\mathrm{Gr}_0^M H_1(F,\mathbf{Z}) \cong \mathbf{Z}[t,t^{-1}]/(\alpha_1\cdots\alpha_N) = \mathbf{Z}[t,t^{-1}]/\left(\frac{(t^{m_1}-1)\cdots(t^{m_N}-1)}{(t^{r_1}-1)\cdots(t^{r_N}-1)}\right).
$$

5.3. *Germes isomères*

Rappelons la définition 3.2 : deux germes de courbe plane, définis par des germes de fonctions analytiques à singularité isolée f_1 et f_2, sont *isomères* si l'on peut passer de $T(f_1)$ à $T(f_2)$ par une suite d'échanges de peupliers de même poids. Dans ce cas, les sommets de $T(f_1)$ qui portent une (ou deux) flèche(s) sont des sommets de rupture avec branche morte.

Si f_1 et f_2 sont des germes à deux branches, la relation d'isomérie entre f_1 et f_2 signifie que l'on passe de $T(f_1)$ à $T(f_2)$ en effectuant une ou plusieurs fois l'opération élémentaire d'échange de sous-arbres qui fait passer du graphe $\Gamma(\varphi_1, \varphi_1')$ au graphe $\Gamma(\varphi_2, \varphi_2')$ comme indiqué Figure 7 ci-dessous, en respectant les conditions ci-dessous. Soit $\eta_{j+1}^{(a)}$ le produit des dénominateurs des paires de Zariski des sommets de rupture $(\widetilde{1}), (\widetilde{2}), \ldots, (\widetilde{j})$ de $T(f_1)$, si $a = 1$, ou des sommets de rupture $(\widetilde{N+1}), (\widetilde{N}), \ldots, (\widetilde{k})$ de $T(f_2)$, si $a = 2$. Soit $\eta_{k-1}^{(a)}$ le produit des dénominateurs des paires de Zariski des sommets de rupture $(\widetilde{k}), \ldots, (\widetilde{N}), (\widetilde{N+1})$ de $T(f_1)$, si $a = 1$, ou des sommets de rupture $(\widetilde{j}), \ldots, (\widetilde{2}), (\widetilde{1})$ de $T(f_2)$, si $a = 2$. Le nombre $\eta_{j+1}^{(a)}$ n'est autre que la multiplicité sortante du sommet $(\widetilde{j+1})$ de l'arbre $T(f_a)$ dans la direction de φ_a ; de même, le nombre $\eta_{k-1}^{(a)}$ est la multiplicité sortante du sommet $(\widetilde{k-1})$ de l'arbre $T(f_a)$ dans la direction de φ_a', si l'arête associée est sortante. Ces nombres sont aussi les premières composantes des poids des peupliers correspondants.

Les relations imposées sont les suivantes :
(i) $j+1 \le l \le k-1$ et, si $\mathfrak{C}(\varphi, \varphi')$ est un exposant permis pour φ et pour φ', $j+1 < k-1$ (de sorte que $T(f_1)$ et $T(f_2)$ sont distincts),
(ii) $\eta_{j+1}^{(1)} = \eta_{j+1}^{(2)} = \eta_{k-1}^{(1)} = \eta_{k-1}^{(2)}$.

$$e_1 \; m_1 \; e_2 \; \ldots \; e_j \; m_j \; e_{j+1} \; \ldots \; e_{k-1} \; m_{k-1} \; e_k \; \ldots \; e_N \; m_N \; e_{N+1}$$

$$e_{N+1} \; m_N \; e_N \; \ldots \; e_k \; m_j \; e_{j+1} \; \ldots \; e_{k-1} \; m_{k-1} \; e_j \; \ldots \; e_2 \; m_1 \; e_1$$

Figure 7 - $\Gamma(\varphi_1, \varphi_1')$ et $\Gamma(\varphi_2, \varphi_2')$

Les sommets (resp. les arêtes) de $\Gamma(\varphi_2, \varphi_2')$ seront numérotés par les indices des e_i (resp. m_i) correspondants dans le diagramme. Comme dans la démonstration du lemme 5.2, la condition $\eta_{j+1} = \eta_{k-1}$ (on omet l'exposant (1) ou (2)) entraine que $m_j = \eta_{j+1} = \eta_{k-1} = m_{k-1}$; on en déduit les égalités $r_{j+1}^{(1)} = r_{j+1}^{(2)}$, $r_{k-1}^{(1)} = r_{k-1}^{(2)}$ et donc, pour tout i, $1 \le i \le N+1$, $r_i^{(1)} = r_i^{(2)}$.

Théorème 5.2. *Soit f_1 et f_2 deux germes de fonction analytique à singularité isolée, définissant des germes de courbe plane à deux branches. Si les arbres $T(f_1)$ et $T(f_2)$ sont isomères, les formes de Seifert sur $H_1(F(f_a), \mathbf{Z})$, $a = 1$ et 2 sont isomorphes.*

548

Démonstration. Par définition de la relation d'isomérie, la collection des halos des sommets de rupture de $T(f_1)$ est identique à celle de $T(f_2)$ et, d'après la proposition 2.6, le nombre d'intersection des deux branches de f_1 est égale à celle des deux branches de f_2. Notons φ_1 et φ_1' (resp. φ_2 et φ_2') des développements de Puiseux des branches de f_1 (resp. f_2), la relation entre les arbres pondérés $\Gamma(\varphi_1, \varphi_1')$ et $\Gamma(\varphi_2, \varphi_2')$ montre que les $\mathbf{Z}[t, t^{-1}]$-modules $M_{-2}H_1(F(f_1), \mathbf{Z})$ et $M_{-2}H_1(F(f_2), \mathbf{Z})$ sont des $\mathbf{Z}[t, t^{-1}]$-modules isomorphes, d'après le théorème 5.1.

Notons $x_0^1 = x_{N+1}^1$ la classe de la composante de bord de $F(f_1)$ associée à φ_1, orientée comme le bord de $F(f_1)$. Notons x_i^1 un cycle de recollement entre les parties de $F(f_1)$ associées aux sommets de rupture renumérotés (i) et $(i+1)$ sur $\Gamma(\varphi_1, \varphi_1')$, voir (DBM92) ou (DBM94) pour cette construction. Pour $1 \leq i \leq N$, on peut choisir x_i^1 parmi les $t^a x_i^1$ et choisir son orientation de sorte que la matrice de présentation de $M_{-2}H_1(F(f_1), \mathbf{Z})$ décrive ce $\mathbf{Z}[t, t^{-1}]$-module par générateurs et relations comme suit : les générateurs sont $x_0^1, \ldots, x_{N+1}^1$, les relations sont, pour $0 \leq i \leq N+1$, $\beta_i x_i^1 = 0$, et, pour $0 \leq i \leq N$, $\gamma_i x_i^1 = \alpha_{i+1} x_{i+1}^1$. On notera de même x_i^2, pour $0 \leq i \leq N+1$, les cycles de recollement correspondants dans $H_1(F(f_2), \mathbf{Z})$, en utilisant, pour numéroter x_i^2, l'indice de l'entier m_i associé à l'arête correspondante dans $\Gamma(\varphi_2, \varphi_2')$.

On supposera pour simplifier l'exposition qu'on passe de $T(f_1)$ à $T(f_2)$ par un unique échange de sous-arbres et que $\Gamma(\varphi_1, \varphi_1')$ et $\Gamma(\varphi_2, \varphi_2')$ sont comme indiqué plus haut ; le cas général consiste en plusieurs pas du même calcul. On choisira le générateur x^a de $M_{-2}H_1(F(f_a), \mathbf{Z})$, $a = 1$ ou 2, de telle sorte que les cycles de recollement x_i^a soient donnés par $x_i^a = \gamma_1 \cdots \gamma_{i-1} \alpha_{i+1} \cdots \alpha_N x^a$. On a alors immédiatement $x_j^a = x_{k-1}^a$.

La détermination de la forme de Seifert des germes f_1 et f_2 utilise les données suivantes.

(i) Le nombre d'intersection des deux branches du germe,

(ii) Les relations entre les cycles de recollement ci-dessus, codées dans la donnée de $(m_1, r_1, m_2, \ldots, r_N, m_N)$,

(iii) La description de l'image par la monodromie de $M_{-1}H_1(F(f_a), \mathbf{Z})$, $a = 1$ ou 2, codée dans la donnée des halos de $T(f_a)$,

(iv) Le polynôme de twist $\mathrm{Tw}_e(t)$, défini en (DBM94, 2.21 et 4.5) et associé au choix d'un générateur x.

Les résultats de (DBM94), la définition de l'isomérie et la proposition 2.6 montrent que les germes f_1 et f_2 ne sont pas distingués par les trois premiers points. Comparons maintenant les polynômes de twist associés aux deux germes. On note ν (resp. ν') la multiplicité de la branche associée à φ_1 ou φ_2, (resp. φ_1' ou φ_2'), les branches notées φ_a (resp. φ_a') ayant par hypothèse la même multiplicité pour $a = 1$ et 2. On note η_j (resp. η_k) la valeur du η_i correspondant à l'arête sortante du sommet (j) (resp. du sommet (k)) de $\Gamma(\varphi_1, \varphi_1')$ vers φ_1 (resp. φ_2) ou de $\Gamma(\varphi_2, \varphi_2')$ vers φ_2' (resp. φ_1'). Cette notation est licite par définition de l'isomérie, $i.e.$ la valeur de η_j (resp. η_k) est la même pour $a = 1$ ou 2. Avec ces notations, en désignant par y^a le relevé dans $H_1(F(f_a), \mathbf{Z})$ d'un générateur de $\mathrm{Gr}^0 H_1(F(f_a), \mathbf{Z})$ choisi comme expliqué dans ($loc.\ cit.$, 4.5), et en notant S la forme d'intersection sur $H_1(F(f_a), \mathbf{Z})$, on a les résultats suivants.

Dans le cas du germe f_1, la contribution des arêtes $(j\ j+1)$ et $(k-1\ k)$ à $\mathrm{Tw}_e(t) \cdot x^1$ est :

$$\sum_{h=1}^{m_j} S(t^h x_j^1, y^1) \frac{e m_j}{\nu} \left(\frac{\nu}{\eta_{j+1} e_{j+1}} - \frac{\nu}{\eta_j e_j} \right) t^h x_j^1 +$$

$$\sum_{h=1}^{m_{k-1}} S(t^h x_{k-1}^1, y^1) \frac{e m_{k-1}}{\nu'} \left(\frac{\nu'}{\eta_{k-1} e_{k-1}} - \frac{\nu'}{\eta_k e_k} \right) t^h x_{k-1}^1.$$

Dans le cas du germe f_2, la contribution des arêtes $(k\ j+1)$ et $(k-1\ j)$ à $\mathrm{Tw}_e(t) \cdot x^2$ est :

$$\sum_{h=1}^{m_j} S(t^h x_j^2, y^2) \frac{e m_j}{\nu} \left(\frac{\nu}{\eta_{j+1} e_{j+1}} - \frac{\nu}{\eta_k e_k} \right) t^h x_j^2 +$$

$$\sum_{h=1}^{m_{k-1}} S(t^h x_{k-1}^2, y^2) \frac{e m_{k-1}}{\nu'} \left(\frac{\nu'}{\eta_{k-1} e_{k-1}} - \frac{\nu'}{\eta_j e_j} \right) t^h x_{k-1}^2.$$

Vu que x_j^a et x_{k-1}^a sont égaux, ainsi que m_j et m_{k-1}, ces deux contributions sont donc égales. De plus, la contribution des autres arêtes ne distingue pas les deux germes. Par suite, les polynômes de twist des deux germes sont égaux.

Le théorème 4.6 de $loc.\ cit.$ nous donne alors le résultat indiqué. $\qquad \square$

Remarque 5.1. *L'article (DBM93) donne un exemple de deux germes de courbe plane à trois branches, isomères, dont les formes de Seifert ne sont pas isomorphes, ce qui permet de construire des nœuds algébriques (de grande dimension) cobordants et non isotopes.*

6. Un exemple

Nous allons mettre en œuvre la méthode décrite dans l'article sur un exemple. On se donne la forme de Seifert $A(f)$ d'un germe de courbe plane à deux branches, défini par un germe de fonction analytique à singularité isolée $f = f' \cdot f''$; la question est alors de trouver le (ou les) type(s) topologique(s) des germes dont la forme de Seifert est isomorphe à $A(f)$, à partir des données suivantes, qui se déduisent de $A(f)$: le polynôme caractéristique $\Delta(t)$ de l'action de la monodromie sur $H_1(F, \mathbf{Z})$, la décomposition du $\mathbf{Z}[t, t^{-1}]$-module $\mathrm{Gr}_{-1}^M H_1(F, \mathbf{Z})$ en somme directe orthogonale pour la forme d'intersection S, les unités associées aux structures isométriques sur les facteurs de cette décomposition et le nombre d'enlacement des composantes de bord de F, $i.e.$ $\nu_0(f', f'')$.

6.1. *Les données*

Les données relatives à l'exemple proposé sont les suivantes.

(i) Le polynôme caractéristique de l'action de la monodromie sur $H_1(F, \mathbf{Z})$ est :

$$\Delta(t) = (t - 1)\frac{(t^{80} - 1)(t^{172} - 1)(t^{348} - 1)(t^{350} - 1)}{(t^{16} - 1)(t^{86} - 1)(t^{87} - 1)(t^{175} - 1)},$$

(ii) La décomposition de $\mathrm{Gr}_{-1}^M H_1(F, \mathbf{Z})$ en somme directe est :

$$\mathrm{Gr}_{-1}^M H_1(F, \mathbf{Z}) \cong$$

$$\mathbf{Z}[t, t^{-1}]/\Lambda_2(t) \oplus \mathbf{Z}[t, t^{-1}]/\Lambda_4(t) \oplus \mathbf{Z}[t, t^{-1}]/\Lambda_6(t) \oplus \mathbf{Z}[t, t^{-1}]/\Lambda_8(t),$$

$$\text{où } \Lambda_2(t) = \frac{(t^{80} - 1)}{(t^{16} - 1)}, \ \Lambda_4(t) = \frac{(t^{172} - 1)(t^2 - 1)}{(t^{86} - 1)(t^4 - 1)},$$

$$\Lambda_6(t) = \frac{(t^{348} - 1)(t - 1)}{(t^{87} - 1)(t^4 - 1)}, \ \Lambda_8(t) = \frac{(t^{350} - 1)(t - 1)}{(t^{175} - 1)(t^2 - 1)}.$$

(iii) Les unités $U_2 \in \mathbf{Z}[t, t^{-1}]/\Lambda_2$, $U_4 \in \mathbf{Z}[t, t^{-1}]/\Lambda_4$, $U_6 \in \mathbf{Z}[t, t^{-1}]/\Lambda_6$ et $U_8 \in \mathbf{Z}[t, t^{-1}]/\Lambda_8$, associées comme dans le th. 1.2 aux 4 halos de valence 3 qui correspondent à la décomposition en somme directe, donnent les unités circulaires suivantes, après passage aux corps cyclotomiques indiqués (on rappelle que ces unités sont définies à multiplication par le carré d'une unité près) :

$$U_2' = \frac{t^{32} - 1}{t^{16} - 1} \cdot \frac{t^{28} - 1}{t^4 - 1} \in \mathbf{Q}(t)/\Phi_{80}(t), \ U_4' = \frac{t^{84} - 1}{t^4 - 1} \in \mathbf{Q}(t)/\Phi_{172}(t),$$

$$U_6' = \frac{t^{88} - 1}{t^4 - 1} \in \mathbf{Q}(t)/\Phi_{348}(t), \ U_8' = \frac{t^{174} - 1}{t^2 - 1} \in \mathbf{Q}(t)/\Phi_{350}(t).$$

(iv) Le nombre d'enlacement est : $\nu_0(f', f'') = 80$.

On retrouve les polynômes Λ_2, Λ_4, Λ_6 et Λ_8 en utilisant le théorème 1.1, la remarque 1.1 et le lemme 1.2 (ce dernier indique qu'un des Λ_i est divisible par $\Phi_5\Phi_{10}\Phi_{20}$). Ceci montre que l'arbre $T(f)$ compte 4 sommets de rupture, tous de valence 3. Plus précisement, l'application σ, définie prop. 2.2, nous indique que les arbres cherchés ont 4 sommets de valence 1, que nous numéroterons 1, 3, 5 et 7, de multiplicités $e_1 = 16$, $e_3 = 86$, $e_5 = 87$ et $e_7 = 175$, et 4 sommets de valence 3, que nous numéroterons 2, 4, 6 et 8, de multiplicités $e_2 = 80$, $e_4 = 172$, $e_6 = 348$ et $e_8 = 350$, les sommets de valence 3 étant associés à ceux de valence 1 par : $\sigma(1) = 2$, $\sigma(3) = 4$, $\sigma(5) = 6$ et $\sigma(7) = 8$. Nous sommes donc dans le cas B.

Les sommets de valence 3 sont ordonnés sur la géodésique $\Gamma(\varphi, \varphi')$ dans l'ordre e_6, e_2, e_4, e_8 ou dans l'ordre inverse, en effet, les multiplicités des composantes du diviseur exceptionnel croissent suivant leur ordre d'apparition lors de la désingularisation, et on a ici $e_5 > e_2$ et $e_3 > e_2$. Les polynômes Λ_4, Λ_6 et Λ_8 permettent alors de calculer les nombres m_{ij}, pour $i = 4, 6$ ou 8 et $j = 1, 2$ ou 3 : $(m_{41}, m_{42}, m_{43}) = (4, 86, 2)$, $(m_{61}, m_{62}, m_{63}) = (4, 87, 1)$, $(m_{81}, m_{82}, m_{83}) = (2, 175, 1)$, ce qui donne enfin $(m_{21}, m_{22}, m_{23}) = (16, 4, 4)$.

6.2. *Calcul des unités*

On notera, comme en Sec. 1.3, $M_n := \widetilde{C}_n/(\widetilde{C}_n \cap \widetilde{E}_n^2)$ le groupe des unités circulaires du corps cyclotomique $\mathbf{Q}[t]/\Phi_n(t))$, modulo le sous-groupe des carrés d'unités. Dans les cas considérés ici, on a $M_n = \widetilde{C}_n/(\widetilde{C}_n^2)$, parce que 80, $172 = 2^2 \cdot 43$, $348 = 2^2 \cdot 3 \cdot 29$ et $350 = 2 \cdot 5^2 \cdot 7$ ont 2 ou 3 facteurs premiers et les h_n^+ valent 1.

D'après Gold et Kim (GK89) on peut choisir pour base de l'espace vectoriel M_{80} (espace vectoriel de dimension $15 = \frac{1}{2}\varphi(80) - 1$ sur $\mathbf{Z}/2\mathbf{Z}$) la famille suivante, où l'on note γ_a l'image de $t^a - 1$ et on conserve l'écriture multiplicative en passant à M_{80} :

$$\left(\gamma_1, \gamma_3, \gamma_7, \gamma_9, \gamma_{13}, \gamma_{17}, \gamma_{19}, \gamma_{21}, \gamma_{23}, \gamma_{27}, \gamma_{37}, \frac{\gamma_{32}}{\gamma_{16}}, \frac{\gamma_{15}}{\gamma_5}, \frac{\gamma_{25}}{\gamma_5}, \frac{\gamma_{35}}{\gamma_5}\right).$$

Les relations de Bass (*cf.* Sec. 1.3) donnent successivement dans M_{80} :

$$\frac{\gamma_{32}}{\gamma_{16}} = \frac{\gamma_{48}}{\gamma_{16}} = \frac{\gamma_{36}}{\gamma_4} = \gamma_8 = \gamma_{24}, \ \frac{\gamma_{24}}{\gamma_8} = 1, \ \frac{\gamma_{28}}{\gamma_4} = 1, \ \gamma_{29} = \gamma_3 \cdot \gamma_{13} \cdot \gamma_{19} \cdot \frac{\gamma_{15}}{\gamma_5} \cdot \frac{\gamma_{35}}{\gamma_5},$$

552

$$\gamma_{31} = \gamma_1 \cdot \gamma_3 \cdot \gamma_7 \cdot \gamma_{13} \cdot \gamma_{23} \cdot \gamma_{27} \cdot \gamma_{37} \cdot \frac{\gamma_{32}}{\gamma_{16}} \cdot \frac{\gamma_{15}}{\gamma_5}, \quad \gamma_{11} = \gamma_{21} \cdot \gamma_{27} \cdot \gamma_{37} \cdot \frac{\gamma_{25}}{\gamma_5},$$

$$\gamma_{33} = \gamma_3 \cdot \gamma_7 \cdot \gamma_{13} \cdot \gamma_{17} \cdot \gamma_{23} \cdot \gamma_{27} \cdot \gamma_{37} \cdot \frac{\gamma_{32}}{\gamma_{16}}, \quad \gamma_{39} = \gamma_7 \cdot \gamma_9 \gamma_{23} \cdot \frac{\gamma_{25}}{\gamma_5} \cdot \frac{\gamma_{35}}{\gamma_5}$$

$$\text{et enfin } \frac{\gamma_{12}}{\gamma_4} = \gamma_1 \cdot \gamma_3 \cdot \gamma_7 \cdot \gamma_9 \cdot \gamma_{17} \cdot \gamma_{19} \cdot \gamma_{21} \cdot \gamma_{37} \cdot \gamma_{25} \gamma_5 \cdot \frac{\gamma_{35}}{\gamma_5}.$$

Vu que $(m_{21}, m_{22}, m_{23}) = (16, 4, 4)$, les η_{2j} sont des multiples de 4, et les valeurs de $\gamma_a / \gamma_{(80,a)}$ qui interviennent dans le calcul de U_2 figurent dans la liste suivante :

$$\frac{\gamma_4}{\gamma_4} = \frac{\gamma_{16}}{\gamma_{16}} = \frac{\gamma_{28}}{\gamma_4} = \frac{\gamma_{52}}{\gamma_4} = \frac{\gamma_{64}}{\gamma_{16}} = \frac{\gamma_{76}}{\gamma_4} = 1, \quad \frac{\gamma_{12}}{\gamma_4} = \frac{\gamma_{68}}{\gamma_4}, \quad \frac{\gamma_{32}}{\gamma_{16}} = \frac{\gamma_{36}}{\gamma_4} = \frac{\gamma_{48}}{\gamma_{16}},$$

où l'on a utilisé la relation $\gamma_a = \gamma_{80-a}$. La liste des η_{2j} possibles est donc :

$$(4, 12, 16, 28, 32, 36, 44, 48, 52, 64, 68, 76).$$

On cherche alors les triplets d'éléments de cette liste qui vérifient les conditions :

$$\eta_{21} + \eta_{22} + \eta_{23} = e_2 = 80, \quad \gamma_{\eta_{21}} \cdot \gamma_{\eta_{22}} \cdot \gamma_{\eta_{23}} \cdot \gamma_{16} = U_2', \quad \text{pgcd}(\eta_{21}, \eta_{22}, \eta_{23}) = 4.$$

On trouve deux halos possibles, $\mathcal{H}_2^1 = (80; 48, 28, 4)$ et $\mathcal{H}_2^2 = (80; 16, 28, 36)$. Un calcul dans M_{40} montre que $\mathcal{H}_2^2$ doit être écarté, en effet, les unités associées s'écrivent, en notant U_2'' (resp. γ_a') l'image de U_2 (resp. $t^a - 1$) dans M_{40} :

$$U_2'' = U(\mathcal{H}_2^1) = \gamma_8' \cdot \gamma_{16}' \cdot \gamma_{12}' \cdot \gamma_4' \quad \text{et} \quad U(\mathcal{H}_2^2) = \gamma_{12}' \cdot \gamma_4',$$

et les relations de Bass donnent ici

$$\gamma_{16}' = \gamma_8' \cdot \gamma_{12}' \quad \text{et} \quad \gamma_{12}' = \gamma_3' \cdot \gamma_7' \cdot \gamma_{13}' \cdot \gamma_{17}' \neq 1,$$

ce qui distingue les deux halos. On a donc trouvé $\mathcal{H}_2 = (80; 48, 28, 4)$.

Le cas des halos $\mathcal{H}_4$ et $\mathcal{H}_8$ est aisé, en effet, d'une part, les polynômes Λ_4 et Λ_8 donnent $\eta_{42} = 86$ et $\eta_{82} = 175$ et, d'autre part, les multiplicités sortantes sont connues et donnent $\eta_{43} = 2$ et $\eta_{83} = 1$. Dans le cas du halo $\mathcal{H}_6$, on sait que $\eta_{63} = 1$ et $\eta_{62} = 87$ ou $174 = 2 \cdot 87$ ou $261 = 3 \cdot 87$, donc $\eta_{61} = 260$ ou 173 ou 86, et seul 260 convient car $\text{pgcd}(348, \eta_{61}) = 4$. On trouve donc les halos suivants, ce qui est conforme aux données U_4, U_6 et U_8 :

$$\mathcal{H}_4 = (172; 86, 84, 2), \quad \mathcal{H}_6 = (348; 260, 87, 1), \quad \mathcal{H}_8 = (350; 175, 174, 1).$$

La proposition 2.5 permet maintenant de distinguer entre $B.1$, $B.2$ et $B.3$. En effet, avec les mêmes notations, on trouve : $\nu_0^{(2)}(f) = 16$ et

$$\nu_0^{(1)}(f) = 2 + 4(\frac{80}{16} - 1) + 2(\frac{172}{86} - 1) + (\frac{348}{87} - 1) + (\frac{350}{175} - 1) = 24,$$

ce qui exclut le cas $B.1$ et indique que la multiplicité du germe est 16. Les composantes (η_1, η_2, η_3) du halo de séparation $\mathcal{H}_1$, numérotées comme dans la prop. 3.1, sont :

$$\eta_3 = 4 < \nu_0(f) = 16 < \eta_2 = 28 < \eta_3 = 48,$$

ce qui exclut le cas $B.2$: les germes cherchés sont dans le cas $B.3$; en particulier, chaque sommet de valence 3 admet une unique arête sortante.

6.3. *Les solutions du problème*

En conclusion, la reconstruction de l'arbre réduit peut se faire de deux manières.

Premier cas : $\mathcal{H}_4$ est rattaché à $\mathcal{H}_2$ le long de l'arête sortante (associée à η_3) et $\mathcal{H}_6$ est rattaché à $\mathcal{H}_2$ le long de l'arête entrante (associée à η_2).

Deuxième cas : $\mathcal{H}_6$ est rattaché à $\mathcal{H}_2$ le long de l'arête sortante (associée à η_3) et $\mathcal{H}_4$ est rattaché à $\mathcal{H}_2$ le long de l'arête entrante (associée à η_2).

Dans les deux cas, $\mathcal{H}_8$ est rattaché à $\mathcal{H}_4$ le long de l'arête sortante. On calcule alors les paires de Zariski en utilisant les lemmes 4.1 et 4.6. Dans le premier cas, on trouve les paires suivantes : $(5/3, 3/2, 3/2)$ pour une branche et $(67/4)$ pour l'autre, le type topologique est celui défini par les développements de Puiseux suivants, ou par l'arbre $T(f_1)$ ci-dessous :

$$\varphi_1 = x^{5/3}(1 + x^{3/6}(1 + x^{3/12})) \quad \text{et} \quad \varphi_1' = x^{67/4}.$$

Dans le deuxième cas, on trouve les paires suivantes : $(5/3, 7/4)$ pour une branche et $(33/2, 3/2)$ pour l'autre, le type topologique est celui défini par les développements de Puiseux suivants, ou par l'arbre $T(f_2)$ ci-dessous :

$$\varphi_2 = x^{5/3}(1 + x^{7/12}) \quad \text{et} \quad \varphi_2' = x^{33/2}(1 + x^{3/4}).$$

Dans les deux cas, la proposition 2.6 donne $\nu_0(f', f'') = 80$, en accord avec la donnée (iv).

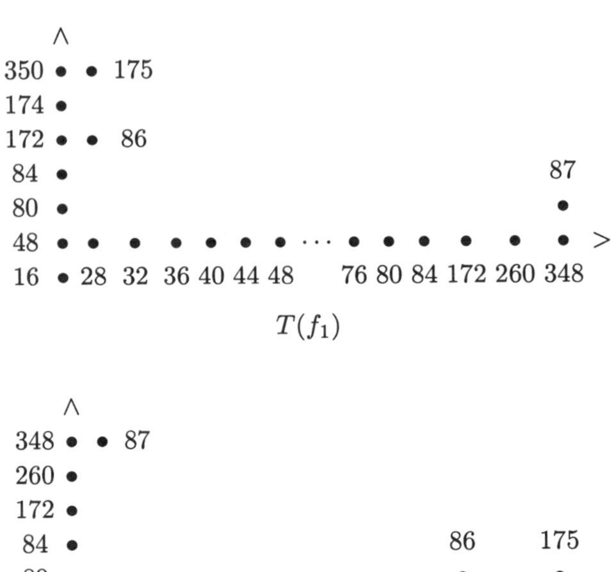

$$T(f_1)$$

$$T(f_2)$$

References

1. Du Bois Ph., Michel F., *The integral Seifert form does not determine the topology of plane curve germs*. Journal of Algebraic Geometry, **3**, 1994, 1-38
2. Kaenders R., *The Seifert Form of a Plane Curve Singularity determines its Intersection Multiplicities*, Indag. Mathem., **7** (1996), no. 2, 185-197
3. Robin E., *Sur la forme de Seifert d'un germe de courbe plane*, Thèse de doctorat, Université d'Angers, 1999 et *Dévissage de la forme de Seifert d'un germe de courbe plane*. C. R. Acad. Sc. Paris, **329**, 1999, Série I, 863–866
4. Du Bois Ph., *Sur la forme de Seifert entière des germes de courbe plane à deux branches*. C. R. Acad. Sc. Paris, **336**, 2003, Série I, 757-762
5. Apostol T., *Resultants of cyclotomic polynomials*. Proc. Amer. Math. Soc., **24**, 1970, 457-462
6. Du Bois Ph., Michel F., *Filtration par le poids et monodromie entière*. Bull. Soc. Math. France, **120**, 1992, 129-167
7. Du Bois Ph., Michel F., *Cobordism of algebraic knots via Seifert forms*. Inventiones math., **111**, 1993, 151-169
8. van der Waerden B., *Moderne Algebra*, Springer Verlag, Berlin, 1931
9. Gold R., Kim J., *Bases for cyclotomic units*. Compositio Mathematica, **71** (1989), 13–28

10. Bass H., *Generators and relations for cyclotomic units*. Nagoya Math. J., **27** (1966), 401–407

11. Ennola V., *On relations between cyclotomic units*. J. Number Theory, **4** (1972), 236–247

12. Washington L., *Introduction to cyclotomic fields*, G.T.M. Springer-Verlag **83**

13. Sinnott W., *On the Stickelberger ideal and the circular units of a cyclotomic field*, Ann. of Math., **108** (1978), 107-134

14. Michel F., Weber C., *Topologie des germes de courbe planes à plusieurs branches*. Prépublication de l'Université de Genève, 1985

15. Brieskorn E., Knörrer H., *Plane algebraic Curves*. Birkhäuser Verlag, 1986

16. A'Campo N., *La fonction zéta d'une monodromie*, Comment. Math. Helvetica, **50** (1975), 233-248

Soumis le 6 Juin 2005 et, sous forme révisée, le 27 Mars 2006

CHERN OBSTRUCTIONS FOR COLLECTIONS OF 1-FORMS ON SINGULAR VARIETIES

W. EBELING*

*Institut für Algebraische Geometrie, Leibniz Universität Hannover,
Postfach 6009, D-30060 Hannover, Germany
* E-mail: ebeling@math.uni-hannover.de*

S. M. GUSEIN-ZADE

*Faculty of Mechanics and Mathematics, Moscow State University,
Moscow, 119992, Russia
E-mail: sabir@mccme.ru*

Dedicated to Jean-Paul Brasselet on the occasion of his 60th birthday

We introduce a certain index of a collection of germs of 1-forms on a germ
of a singular variety which is a generalization of the local Euler obstruction
corresponding to Chern numbers different from the top one.

Keywords: singular variety, 1-form, index.
Mathematics Subject Classification: 14B05, 58A10, 55S99.

Introduction

The aim of this paper is to bring together some ideas of [3] and [6]. A
germ of a vector field or of a 1-form on the complex affine space $\mathbb{C}^n$ at
the origin not vanishing in a punctured neighbourhood of it has a topological invariant — the Poincaré–Hopf index. The sum of the Poincaré–
Hopf indices of the singular points of a vector field on a compact complex
manifold is equal to the Euler characteristic of the manifold. There are
several generalizations of this notion to vector fields and/or to 1-forms on
complex analytic varieties with singularities (isolated or not) started by
M.-H. Schwartz: [1,3–5,8,12,13, . . .]. For the case of an isolated complete
intersection singularity there is defined an index which is sometimes called
the GSV index: [5,8,13]. Another generalization which makes sense not only
for isolated complete intersection singularities and also not only for vari-

eties with isolated singularities is the so called local Euler obstruction: [2,3] (its analogue for 1-forms was considered in [7]). One can say that in some sense all these invariants correspond to the Euler characteristic, which, for a compact complex analytic manifold M^n, coincides with the top Chern number $\langle c_n(M), [M] \rangle$.

A generalization of the GSV-index corresponding to other Chern numbers (different from the top one) was introduced and studied in [6]. It is defined for a collection of germs of 1-forms on an isolated complete intersection singularity. For a collection of 1-forms on a projective complex complete intersection with isolated singularities, the sum of these indices of the singular points is equal to plus-minus the corresponding Chern number of a smoothing of the variety.

Here we define and study an index of a collection of germs of 1-forms on a germ of a singular variety which is an analogue of the local Euler obstruction corresponding to a Chern number different from the top one.

1. Special points of 1-forms

Let $(X^n, 0) \subset (\mathbb{C}^N, 0)$ be the germ of a purely n-dimensional reduced complex analytic variety at the origin (generally speaking with a non-isolated singularity). Let $\mathbf{k} = \{k_i\}$, $i = 1, \ldots, s$, be a fixed partition of n (i.e., k_i are positive integers, $\sum\limits_{i=1}^{s} k_i = n$). Let $\{\omega_j^{(i)}\}$ $(i = 1, \ldots, s, \; j = 1, \ldots, n - k_i + 1)$ be a collection of germs of 1-forms on $(\mathbb{C}^N, 0)$ (not necessarily complex analytic; it suffices that the forms $\omega_j^{(i)}$ are complex linear functions continuously depending on a point of $\mathbb{C}^N$). Let $\varepsilon > 0$ be small enough so that there is a representative X of the germ $(X, 0)$ and representatives $\omega_j^{(i)}$ of the germs of 1-forms inside the ball $B_\varepsilon(0) \subset \mathbb{C}^N$.

Definition 1.1. A point $P \in X$ is called a *special* point of the collection $\{\omega_j^{(i)}\}$ of 1-forms on the variety X if there exists a sequence $\{P_m\}$ of points from the non-singular part X_{reg} of the variety X converging to P such that the sequence $T_{P_m} X_{\mathrm{reg}}$ of the tangent spaces at the points P_m has a limit L as $m \to \infty$ (in the Grassmann manifold of n-dimensional vector subspaces of $\mathbb{C}^N$) and the restrictions of the 1-forms $\omega_1^{(i)}, \ldots, \omega_{n-k_i+1}^{(i)}$ to the subspace $L \subset T_P \mathbb{C}^N$ are linearly dependent for each $i = 1, \ldots, s$.

Definition 1.2. The collection $\{\omega_j^{(i)}\}$ of 1-forms has an *isolated special point* on the germ $(X, 0)$ if it has no special points on X in a punctured neighbourhood of the origin.

Remark 1.1. If the 1-forms $\omega_j^{(i)}$ are complex analytic, the property to have an isolated special point is a condition on the classes of these 1-forms in the module

$$\Omega_{X,0}^1 = \Omega_{\mathbb{C}^N,0}^1 / \{ f \cdot \Omega_{\mathbb{C}^N,0}^1 + df \cdot \mathcal{O}_{\mathbb{C}^N,0} | f \in \mathcal{J}_X \}$$

of germs of 1-forms on the variety X ($\mathcal{J}_X$ is the ideal of germs of holomorphic functions vanishing on X).

Remark 1.2. For the case $s = 1$ (and therefore $k_1 = n$), i.e. for one 1-form ω, there exists a notion of a *singular* point of the 1-form ω on X (see, e.g., [7]). It is defined in terms of a Whitney stratification of the variety X. A point $x \in X$ is a *singular* point of the 1-form ω on the variety X if the restriction of the 1-form ω to the stratum of X containing x is equal to zero at the point x. (One should consider points of all zero-dimensional strata as singular ones.) One can easily see that a special point of the 1-form ω on the variety X is singular, but not vice versa. (E.g. the origin is a singular point of the 1-form dx on the cone $\{x^2 + y^2 + z^2 = 0\}$, but not a special one.) On a smooth variety these two notions coincide.

The notion of a *non-degenerate* special (singular) point of a collection of germs of 1-forms on a smooth variety was introduced in [6]. The index of a non-degenerate point of a collection of germs of holomorphic 1-forms is equal to 1.

Let

$$\mathcal{L}^{\mathbf{k}} = \prod_{i=1}^{s} \prod_{j=1}^{n-k_i+1} \mathbb{C}_{ij}^{N*}$$

be the space of collections of linear functions on $\mathbb{C}^N$ (i.e. of 1-forms with constant coefficients).

Proposition 1.1. *There exists an open and dense subset $U \subset \mathcal{L}^{\mathbf{k}}$ such that each collection $\{\ell_j^{(i)}\} \in U$ has only isolated special points on X and, moreover, all these points belong to the smooth part X_{reg} of the variety X and are non-degenerate.*

Proof. Let $Y \subset X \times \mathcal{L}^{\mathbf{k}}$ be the closure of the set of pairs $(x, \{\ell_j^{(i)}\})$ where $x \in X_{\mathrm{reg}}$ and the restrictions of the linear functions $\ell_1^{(i)}, \ldots, \ell_{n-k_i+1}^{(i)}$ to the tangent space $T_x X_{\mathrm{reg}}$ are linearly dependent for each $i = 1, \ldots, s$. Let $\pi : Y \to \mathcal{L}^{\mathbf{k}}$ be the projection to the second factor. One has $\operatorname{codim} Y = \sum_{i=1}^{s} k_i = n$ and therefore $\dim Y = \dim \mathcal{L}^{\mathbf{k}}$. Moreover, $Y \setminus (X_{\mathrm{reg}} \times \mathcal{L}^{\mathbf{k}})$ is a proper subvariety of Y and therefore its dimension is strictly smaller than $\dim \mathcal{L}^{\mathbf{k}}$. A generic point of the space $\mathcal{L}^{\mathbf{k}}$ is a regular value of the map π which means that it has only finitely many preimages, all of them belong to $X_{\mathrm{reg}} \times \mathcal{L}^{\mathbf{k}}$ and the map π is non-degenerate at them. This implies the statement. $\qquad \square$

Corollary 1.1. *Let $\{\omega_j^{(i)}\}$ be a collection of 1-forms on X with an isolated special point at the origin. Then there exists a deformation $\{\widetilde{\omega}_j^{(i)}\}$ of the collection $\{\omega_j^{(i)}\}$ whose special points lie in X_{reg} and are non-degenerate. Moreover, as such a deformation one can use $\{\omega_j^{(i)} + \lambda \ell_j^{(i)}\}$ with a generic collection $\{\ell_j^{(i)}\} \in \mathcal{L}^{\mathbf{k}}$, $\lambda \neq 0$ small enough.*

Corollary 1.2. *The set of collections of holomorphic 1-forms with a non-isolated special point at the origin has infinite codimension in the space of all holomorphic collections.*

2. Local Chern obstructions

Let $\{\omega_j^{(i)}\}$ be a collection of germs of 1-forms on $(X, 0)$ with an isolated special point at the origin. Let $\nu : \widehat{X} \to X$ be the Nash transformation of the variety $X \subset B_\varepsilon(0)$ defined as follows. Let $G(n, N)$ be the Grassmann manifold of n-dimensional vector subspaces of $\mathbb{C}^N$. There is a natural map $\sigma : X_{\mathrm{reg}} \to B_\varepsilon(0) \times G(n, N)$ which sends a point $x \in X_{\mathrm{reg}}$ to $(x, T_x X_{\mathrm{reg}})$. The Nash transform $\widehat{X}$ of the variety X is the closure of the image $\mathrm{Im}\, \sigma$ of the map σ in $B_\varepsilon(0) \times G(n, N)$, ν is the natural projection. The Nash bundle $\widehat{T}$ over $\widehat{X}$ is a vector bundle of rank n which is the pullback of the tautological bundle on the Grassmann manifold $G(n, N)$. Let $T\mathbb{C}^N|_X$ be the restriction to X of the tangent bundle $T\mathbb{C}^N$ of $\mathbb{C}^N$. There is a natural bundle map from the Nash bundle $\widehat{T}$ to $T\mathbb{C}^N|_X$ which is an embedding on fibres:

$$
\begin{array}{ccc}
\widehat{T} & \longrightarrow & T\mathbb{C}^N|_X \\
\downarrow & & \downarrow \\
\widehat{X} & \xrightarrow{\;\nu\;} & X
\end{array}
$$

This is an isomorphism of $\widehat{T}$ and $T X_{\mathrm{reg}} \subset T\mathbb{C}^N|_X$ over the non-singular part X_{reg} of X.

The collection of 1-forms $\{\omega_j^{(i)}\}$ gives rise to a section $\widehat{\omega}$ of the bundle

$$\widehat{\mathbb{T}} = \bigoplus_{i=1}^{s} \bigoplus_{j=1}^{n-k_i+1} \widehat{T}_{i,j}^{*}$$

where $\widehat{T}_{i,j}^{*}$ are copies of the dual Nash bundle $\widehat{T}^{*}$ over the Nash transform $\widehat{X}$ numbered by indices i and j. Let $\widehat{\mathbb{D}} \subset \widehat{\mathbb{T}}$ be the set of pairs $(x, \{\alpha_j^{(i)}\})$ where $x \in \widehat{X}$ and the collection $\{\alpha_j^{(i)}\}$ of elements of $\widehat{T}_x^{*}$ (i.e. of linear functions on $\widehat{T}_x$) is such that $\alpha_1^{(i)}, \ldots, \alpha_{n-k_i+1}^{(i)}$ are linearly dependent for each $i = 1, \ldots, s$. The image of the section $\widehat{\omega}$ does not intersect $\widehat{\mathbb{D}}$ outside of the preimage $\nu^{-1}(0) \subset \widehat{X}$ of the origin. The map $\widehat{\mathbb{T}} \backslash \widehat{\mathbb{D}} \to \widehat{X}$ is a fibre bundle. The fibre $W_x = \widehat{\mathbb{T}}_x \setminus \widehat{\mathbb{D}}_x$ of it is $(2n-2)$-connected, its homology group $H_{2n-1}(W_x; \mathbb{Z})$ is isomorphic to $\mathbb{Z}$ and has a natural generator: see, e.g., [6]. The latter fact implies that the fibre bundle $\widehat{\mathbb{T}} \setminus \widehat{\mathbb{D}} \to \widehat{X}$ is homotopically simple in dimension $2n-1$, i.e. the fundamental group $\pi_1(\widehat{X})$ of the base acts trivially on the homotopy group $\pi_{2n-1}(W_x)$ of the fibre, the last one being isomorphic to the homology group $H_{2n-1}(W_x; \mathbb{Z})$: see, e.g., [15].

Definition 2.1. The *local Chern obstruction* $\mathrm{Ch}_{X,0} \{\omega_j^{(i)}\}$ of the collections of germs of 1-forms $\{\omega_j^{(i)}\}$ on $(X, 0)$ at the origin is the (primary, and in fact the only) obstruction to extending the section $\widehat{\omega}$ of the fibre bundle $\widehat{\mathbb{T}} \backslash \widehat{\mathbb{D}} \to \widehat{X}$ from the preimage of a neighbourhood of the sphere $S_\varepsilon = \partial B_\varepsilon$ to $\widehat{X}$, more precisely its value (as an element of $H^{2n}(\nu^{-1}(X \cap B_\varepsilon), \nu^{-1}(X \cap S_\varepsilon); \mathbb{Z})$) on the fundamental class of the pair $(\nu^{-1}(X \cap B_\varepsilon), \nu^{-1}(X \cap S_\varepsilon))$.

The definition of the local Chern obstruction $\mathrm{Ch}_{X,0} \{\omega_j^{(i)}\}$ can be re-formulated in the following way. Let $\mathcal{D}_X^{\mathbf{k}} \subset \mathbb{C}^N \times \mathcal{L}^{\mathbf{k}}$ be the closure of the set of pairs $(x, \{\ell_j^{(i)}\})$ such that $x \in X_{\mathrm{reg}}$ and the restrictions of the linear functions $\ell_1^{(i)}, \ldots, \ell_{n-k_i+1}^{(i)}$ to $T_x X_{\mathrm{reg}} \subset \mathbb{C}^N$ are linearly dependent for each $i = 1, \ldots, s$. (For $s = 1$, $\mathbf{k} = \{n\}$, $\mathcal{D}_X^{\mathbf{k}}$ is the (non-projectivized) conormal space of X [16].) The collection $\{\omega_j^{(i)}\}$ of germs of 1-forms on $(\mathbb{C}^N, 0)$ defines a section $\breve{\omega}$ of the (trivial) fibre bundle $\mathbb{C}^N \times \mathcal{L}^{\mathbf{k}} \to \mathbb{C}^N$. Then

$$\mathrm{Ch}_{X,0} \{\omega_j^{(i)}\} = (\breve{\omega}(\mathbb{C}^N) \circ \mathcal{D}_X^{\mathbf{k}})_0$$

where $(\cdot \circ \cdot)_0$ is the intersection number at the origin in $\mathbb{C}^N \times \mathcal{L}^{\mathbf{k}}$. This description can be considered as a generalization of an expression of the local Euler obstruction as a microlocal intersection number defined in [9], see also [11, Sections 5.0.3 and 5.2.1].

Remark 2.1. On a smooth manifold X the local Chern obstruction $\mathrm{Ch}_{X,0}\{\omega_j^{(i)}\}$ coincides with the index $\mathrm{ind}_{X,0}\{\omega_j^{(i)}\}$ of the collection $\{\omega_j^{(i)}\}$ defined in [6].

Remark 2.2. The local Euler obstruction is defined for vector fields as well as for 1-forms. One can see that vector fields are not well adapted to a definition of the local Chern obstruction. A more or less direct version of the definition above for vector fields demands to consider vector fields on a singular variety $X \subset \mathbb{C}^N$ to be sections $v = v(x)$ of $T\mathbb{C}^N|_X$ such that $v(x) \in T_x X \subset T_x \mathbb{C}^N$ ($\dim T_x X$ is not constant). (Traditionally vector fields tangent to smooth strata of the variety X are considered.) There exist only continuous (non-trivial, i.e. with $s > 1$) collections of such vector fields "on X" with isolated special points, but not holomorphic ones.

Remark 2.3. The definition of the local Chern obstruction $\mathrm{Ch}_{X,0}\{\omega_j^{(i)}\}$ may also be formulated in terms of a collection $\{\omega^{(i)}\}$ of germs of 1-forms with values in vector spaces L_i of dimensions $n - k_i + 1$. Therefore (via differentials) it is also defined for a collection $\{f^{(i)}\}$ of germs of maps $f^{(i)} :$ $(\mathbb{C}^N, 0) \to (\mathbb{C}^{n-k_i+1}, 0)$ (just as the Euler obstruction is defined for a germ of a function).

Being a (primary) obstruction, the local Chern obstruction satisfies the law of conservation of number, i.e. if a collection of 1-forms $\{\widetilde{\omega}_j^{(i)}\}$ is a deformation of the collection $\{\omega_j^{(i)}\}$ and has isolated special points on X, then

$$\mathrm{Ch}_{X,0}\{\omega_j^{(i)}\} = \sum \mathrm{Ch}_{X,Q}\{\widetilde{\omega}_j^{(i)}\}$$

where the sum on the right hand side is over all special points Q of the collection $\{\widetilde{\omega}_j^{(i)}\}$ on X in a neighbourhood of the origin. With Corollary 1.1 this implies the following statements.

Proposition 2.1. *The local Chern obstruction* $\mathrm{Ch}_{X,0}\{\omega_j^{(i)}\}$ *of a collection* $\{\omega_j^{(i)}\}$ *of germs of holomorphic 1-forms is equal to the number of special points on X of a generic (holomorphic) deformation of the collection.*

This statement is an analogue of Proposition 2.3 in [14].

Proposition 2.2. *If a collection $\{\omega_j^{(i)}\}$ of 1-forms on a compact (say, projective) variety X has only isolated special points, then the sum of the local Chern obstructions of the collection $\{\omega_j^{(i)}\}$ at these points does not depend on the collection and therefore is an invariant of the variety.*

It is possible to consider this sum multiplied by $(-1)^n$ as a version of the corresponding Chern number of the singular variety X (or, more accurately, taking into account the similarity with Mather classes [10], Mather-Chern number).

Let $(X, 0)$ be an isolated complete intersection singularity. As it was mentioned above, a collection of germs of 1-forms $\{\omega_j^{(i)}\}$ on $(X, 0)$ with an isolated special point at the origin has an index $\mathrm{ind}_{X,0}\{\omega_j^{(i)}\}$ which is an analogue of the GSV–index of a vector field: [5]. The fact that both the Chern obstruction and the index satisfy the law of conservation of number and they coincide on a smooth manifold yields the following statement.

Proposition 2.3. *For a collection $\{\omega_j^{(i)}\}$ of germs of 1-forms on an isolated complete intersection singularity $(X, 0)$ the difference*

$$\mathrm{ind}_{X,0}\{\omega_j^{(i)}\} - \mathrm{Ch}_{X,0}\{\omega_j^{(i)}\}$$

does not depend on the collection and therefore is an invariant of the germ of the variety.

Since, by Proposition 1.1, $\mathrm{Ch}_{X,0}\{\ell_j^{(i)}\} = 0$ for a generic collection $\{\ell_j^{(i)}\}$ of linear functions on $\mathbb{C}^N$, one has the following statement.

Corollary 2.1. *One has*

$$\mathrm{Ch}_{X,0}\{\omega_j^{(i)}\} = \mathrm{ind}_{X,0}\{\omega_j^{(i)}\} - \mathrm{ind}_{X,0}\{\ell_j^{(i)}\}$$

for a generic collection $\{\ell_j^{(i)}\}$ of linear functions on $\mathbb{C}^N$.

Acknowledgments

This research was partially supported by the DFG-programme "Global methods in complex geometry" (Eb 102/4–3) and grants RFBR–04–01–00762 and NWO–RFBR 047.011.2004.026.

We are grateful to the referee for useful remarks.

References

1. Ch. Bonatti, X. Gómez-Mont: The index of holomorphic vector fields on singular varieties I. Astérisque **222**, 9–35 (1994).
2. J.-P. Brasselet, Lê Dũng Tráng, J. Seade: Euler obstruction and indices of vector fields. Topology **39**, 1193–1208 (2000).
3. J.-P. Brasselet, D. Massey, A. J. Parameswaran, J. Seade: Euler obstruction and defects of functions on singular varieties. J. London Math. Soc. (2) **70**, 59–76 (2004).

4. J.-P. Brasselet, M.-H. Schwartz: Sur les classes de Chern d'un ensemble analytique complexe. In: Caractéristique d'Euler-Poincaré, Astérisque **82–83**, 93–147 (1981).

5. W. Ebeling, S. M. Gusein-Zade: Indices of 1-forms on an isolated complete intersection singularity. Moscow Math. J. **3**, 439–455 (2003).

6. W. Ebeling, S. M. Gusein-Zade: Indices of vector fields or 1-forms and characteristic numbers. Bull. London Math. Soc. **37**, 747–754 (2005).

7. W. Ebeling, S. M. Gusein-Zade: Radial index and Euler obstruction of a 1-form on a singular variety. Geom. Dedicata **113**, 231–241 (2005).

8. X. Gómez-Mont, J. Seade, A. Verjovsky: The index of a holomorphic flow with an isolated singularity. Math. Ann. **291**, 737–751 (1991).

9. M. Kashiwara, P. Schapira: Sheaves on Manifolds. Springer-Verlag, 1990.

10. R. MacPherson: Chern classes for singular varieties. Annals of Math. **100**, 423–432 (1974).

11. J. Schürmann: Topology of Singular Spaces and Constructible Sheaves. Birkhäuser, 2003.

12. M.-H. Schwartz: Classes caractéristiques définies par une stratification d'une variété analytique complexe. C. R. Acad. Sci. Paris Sér. I Math. **260**, 3262–3264, 3535–3537 (1965).

13. J. A. Seade, T. Suwa: A residue formula for the index of a holomorphic flow. Math. Ann. **304**, 621–634 (1996).

14. J. Seade, M. Tibăr, A. Verjovsky: Milnor numbers and Euler obstruction. Bull. Braz. Math. Soc. (N.S.) **36**, no. 2, 275–283 (2005).

15. N. Steenrod: The Topology of Fibre Bundles. Princeton Math. Series, Vol. **14**, Princeton University Press, Princeton, N. J., 1951.

16. B. Teissier: Variétés polaires. II. Multiplicités polaires, sections planes, et conditions de Whitney. In: Algebraic geometry (La Rábida, 1981), Lecture Notes in Math., Vol. **961**, Springer, Berlin, 1982, pp. 314–491.

THE FINITE GENERATION OF THE MONOID OF EFFECTIVE DIVISOR CLASSES ON PLATONIC RATIONAL SURFACES

G. FAILLA

*Department of Mathematics, University of Messina,
Messina, 98166/Sicily, Italy
E-mail: gfailla@dipmat.unime.it*

M. LAHYANE*

*Departamento de Álgebra, Geometría y Topología,
Universidad de Valladolid, Calle Prado de la Magdalena s/n,
Valladolid, 47005/Castilla y León, Spain
* E-mail: lahyane@agt.uva.es
www.cie.uva.es/algebra/*

G. MOLICA BISCI

*DIMET, University of Reggio Calabria,
Reggio Calabria, 89100/Reggio Calabria, Italy
E-mail: giovanni.molica@ing.unirc.it*

On the occasion of Jean-Paul BRASSELET's 60th birthday.

We prove the finite generation of the monoid of effective divisor classes on a Platonic rational surface, then derive some consequences. We also show the vanishing of the irregularity of any numerically effective divisor, solving thus the Riemann-Roch Problem for numerically effective divisors. Platonic rational surfaces provide new evidence to a speculation of Felix Klein about the interaction between geometry and discrete mathematics.

Keywords: Smooth rational surfaces; Anticanonical divisor; Anticanonical rational surfaces; Points in general position; Picard group; Blowing-up; Monoid of effective divisor classes.

1. Introduction

In ([1, Theorem 4a, page 283]), Masayoshi Nagata proved that the surface S obtained by blowing up the projective plane at $r > 8$ points in general position has an infinite number of (-1)-curves, hence its monoid of effective divisor classes modulo algebraic equivalence is not finitely generated (see also [2, Fact, page 426] and [3, Exercise 4.15, page 409]). Here a (-1)-curve on S means that it is smooth, rational and of self-intersection -1. It follows that the configuration of the points should be special in order to ensure the finite generation of the monoid of effective divisor classes modulo algebraic equivalence on the surface obtained by blowing up these points. In ([4, Theorem (1.1), page 271]), Eduard Looijenga studied, among other things, smooth projective rational surfaces having a triangle anticanonical divisor. Here an anticanonical divisor $-K_S$ on a smooth projective rational surface S is said to be a triangle if it is effective and has only three irreducible components, all of them are smooth rational curves intersecting each other transversally and the intersection diagram is a triangle. In particular, he proved that such surface has the projective plane as a minimal model and at each step of contracting exceptional curves till reaching the projective plane, the obtained surface remain to have a triangle anticanonical divisor. In particular, the image of its anticanonical divisor $-K_S$ in the projective plane is a cubic curve with three irreducible components.

In this work we deal mainly with the smooth projective rational surfaces having a triangle anticanonical divisor, allowing in particular that some components to be (-1)-curves. Such assumption was not allowed by Looijenga in his further analysis. More precisely, we consider the surface obtained by blowing up the projective plane at $(p + q + r)$ points which are on the 3 edges of a triangle, say p points on one edge, q points on one of the two other edges and r points on the remaining edge such that these points are smooth for the cubic defined by the triangle and such that the nonnegative integers satisfy either the equality $pqr = 0$ or the inequality $pqr - pq - pr - qr < 0$. This smooth projective rational surface will be denoted by $S_{(p,q,r)}$ and we refer to it as a rational surface of type (p, q, r), or simply a *Platonic rational surface*. Classical examples of these kind of surfaces are the ones in which the integers either both p, q or r vanishes. On the other hand, one may observe that for certain values of p, q and r, the anticanonical complete linear system $|-K_{S_{(p,q,r)}}|$ of $S_{(p,q,r)}$ is a singleton whose element is a reduced divisor having three irreducible components, all of which are smooth rational curves of strictly negative self-intersection.

When the nonnegative integer p, q and r are larger than or equal to one, the triplet (p, q, r) is a Platonic one, i.e., the inequality $\frac{1}{p} + \frac{1}{q} + \frac{1}{r} > 1$ holds[a].

---

[a]Note that this is not the classical definition of Platonic numbers.

In [5], Felix Christian Klein gave an interaction between geometry and discrete mathematics. Our surfaces $S_{(p,q,r)}$ may be also considered as another geometric realization of a nonorientable graph studied by Igor Dolgachev in [6].

The aim of this work is to prove the finite generation of the monoid of effective divisor classes $M(S_{(p,q,r)})$ on $S_{(p,q,r)}$. I.e., we have:

Theorem 1.1. *With the same notation as above, the monoid $M(S_{(p,q,r)})$ of effective divisor classes modulo algebraic equivalence on $S_{(p,q,r)}$ is finitely generated.*

From the singularity theory, this result may be interpreted as follows: the number of integral exceptional curves on $S_{(p,q,r)}$ up to the automorphism of the surface is finite.

As another consequence, the following known result is recovered, see [7, Lemma 3.1.1.]:

Corollary 1.1. *The monoid of effective divisor classes on the surface obtained by the blow up the projective plane either at all collinear points or at the smooth points of a degenerate conic is finitely generated.*

Remark 1.1. For certain values of p, q and r, the theorem gives new smooth projective rational surfaces X having a canonical divisor K_X of strictly negative self-intersection and for which the monoid of effective divisor classes $M(X)$ is finitely generated. Surfaces X with $K_X^2 \geq 0$ are very well understood by now, see [2,4,8–13].

The following lemma is useful, its proof is postponed to Section 3. We recall that a divisor on $S_{(p,q,r)}$ is numerically effective if it meets every integral curve on $S_{(p,q,r)}$ nonnegatively.

Lemma 1.1. *With the same notation as above, there is no nonzero numerically effective divisor D on $S_{(p,q,r)}$ satisfying the equality $K_{S_{(p,q,r)}}.D = 0$, $K_{S_{(p,q,r)}}$ being a canonical divisor on $S_{(p,q,r)}$.*

Here we study the vanishing problem of the first cohomology group of an arbitrary numerically effective divisor on a Platonic rational surface.

Theorem 1.2. *With the same notation as above, if D is a numerically effective divisor on $S_{(p,q,r)}$, then*

$$h^1(S_{(p,q,r)}, O_{S_{(p,q,r)}}(D)) = 0,$$

$O_{S_{(p,q,r)}}(D)$ being an invertible sheaf associated to the divisor D.

Proof. Apply [14, Theorem III.1, page 1197] and the above Lemma 1.1.$\square$

A straightforward consequence of the Theorem 1.2 is a solution to the Riemann-Roch Problem for any numerically effective divisor on a Platonic rational surface.

Corollary 1.2. *Let D be a numerically effective divisor on a Platonic rational surface Z. Then*

$$h^0(Z, O_Z(D)) = 1 + \frac{1}{2}(D^2 - D.K_Z),$$

$O_Z(D)$ being an invertible sheaf associated to the divisor D, and K_Z being a canonical divisor on Z.

Proof. Apply Theorem 1.2 and the below Lemma 2.1 and Lemma 2.3. $\square$

The plan of this paper is as follows: in section 2, we give some standard facts about smooth rational surfaces and fix our notation. Section 3 is devoted to proving Theorem 1.1 and Lemma 1.1.

2. Preliminaries

Let X be a smooth projective rational surface defined over an algebraically closed field of arbitrary characteristic. A canonical divisor on X, respectively the Picard group $Pic(X)$ of X will be denoted by K_X and $Pic(X)$ respectively. There is an intersection form on $Pic(X)$ induced by the intersection of divisors on X, it will be denoted by a dot, that is, for x and y in $Pic(X)$, $x.y$ is the intersection number of x and y (see [3,15]).

The following result known as the Riemann-Roch theorem for smooth projective rational surfaces is stated using the Serre duality.

Lemma 2.1. *Let D be a divisor on a smooth projective rational surface X having an algebraically closed field of arbitrary characteristic as a ground field. Then the following equality holds:*

$$h^0(X, O_X(D)) - h^1(X, O_X(D)) + h^0(X, O_X(K_X - D)) = 1 + \frac{1}{2}(D^2 - D.K_X),$$

$O_X(D)$ being an invertible sheaf associated to the divisor D.

Here we recall some standard results, see [14] and [3]. A divisor class x modulo algebraic equivalence on a smooth projective rational surface Z is effective respectively numerically effective, nef in short, if an element of x is an effective, respectively numerically effective, divisor on Z. Here a divisor D on Z is nef if $D.C \geq 0$ for every integral curve C on Z. Now, we start with some properties which follow from successive iterations of blowing up closed points of a smooth projective rational surface.

Lemma 2.2. *Let $\pi^\star : Pic(X) \to Pic(Y)$ be the natural group homomorphism on Picard groups induced by a given birational morphism $\pi : Y \to X$ of smooth projective rational surfaces. Then $\pi^\star$ is an injective intersection-form preserving map of free abelian groups of finite rank. Furthermore, it preserves the dimensions of cohomology groups, the effective divisor classes and the numerically effective divisor classes.*

Proof. See [14, Lemma II.1, page 1193]. $\qquad\qquad\square$

Lemma 2.3. *Let x be an element of the Picard group $Pic(X)$ of a smooth projective rational surface X. The effectiveness or the nefness of x implies the noneffectiveness of $k_X - x$, where k_X denotes the element of $Pic(X)$ which contains a canonical divisor on X. Moreover, the nefness of x implies also that the self-intersection of x is greater than or equal to zero.*

Proof. See [14, Lemma II.2, page 1193]. $\qquad\qquad\square$

We also need the following result, we recall that a (-1)-curve, respectively a (-2)-curve, is a smooth rational curve of self-intersection -1, respectively -2.

Lemma 2.4. *The monoid of effective divisor classes modulo algebraic equivalence on a smooth projective rational surface X having an effective anticanonical divisor is finitely generated if and only if X has only a finite number of (-1)-curves and a finite number of (-2)-curves.*

Proof. See [16, Corollary 4.2, page 109]. $\qquad\qquad\square$

3. Proofs of Theorem 1.1 and Lemma 1.1

To give a proof of the result stated in Theorem 1.1 of section one, we need to give explicitly the Picard lattice of the surface $S_{(p,q,r)}$, i.e., to give a suitable basis of the Picard group $Pic(S_{(p,q,r)})$ and the values of the quadratic form on this basis.

Firstly, the integral basis

$$(\mathcal{E}_0; -\mathcal{E}_1^{L_1}, \ldots, -\mathcal{E}_p^{L_1}; -\mathcal{E}_1^{L_2}, \ldots, -\mathcal{E}_q^{L_2}; -\mathcal{E}_1^{L_3}, \ldots, -\mathcal{E}_r^{L_3}),$$

570

is defined by:

- $\mathcal{E}_0$ is the class of a line on the projective plane which does not pass through any of the assigned points $P_1, \ldots, P_p; Q_1, \ldots, Q_q; R_1, \ldots, R_r$ in consideration,
- $\mathcal{E}_i^{L_1}$ is the class of the exceptional divisor corresponding to the i^{-th} point blown-up P_i for every $i = 1, \ldots, p$,
- $\mathcal{E}_j^{L_2}$ is the class of the exceptional divisor corresponding to the j^{-th} point blown-up Q_j for every $j = 1, \ldots, q$,
- $\mathcal{E}_k^{L_3}$ is the class of the exceptional divisor corresponding to the k^{-th} point blown-up R_k for every $k = 1, \ldots, r$.

Then it follows that the class of a divisor on $S_{(p,q,r)}$ will be represented by the $(1 + p + q + r)$-tuple $(a; b_1^{L_1}, \ldots, b_p^{L_1}; b_1^{L_2}, \ldots, b_q^{L_2}; b_1^{L_3}, \ldots, b_r^{L_3})$,

Secondly, the quadratic form on $Pic(S_{(p,q,r)})$ is given by the fact that the basis elements are pairwise orthogonal and by the following equalities:

- $(\mathcal{E}_i^{L_1})^2 = (\mathcal{E}_j^{L_2})^2 = (\mathcal{E}_k^{L_3})^2 = -\mathcal{E}_0^2 = -1$ for every $i = 1, \ldots, p$, $j = 1, ..., q$ and $k = 1, \ldots, r$.

Remark 3.1. we observe that if the class $(a; b_1^{L_1}, \ldots, b_p^{L_1}; b_1^{L_2}, \ldots, b_q^{L_2}; b_1^{L_3}, ..., b_r^{L_3})$ is effective, then it represents the class of a projective plane curve of degree a and having at least multiplicity $b_1^{L_1}, ..., b_p^{L_1}$ (respectively, $b_1^{L_2}, ..., b_q^{L_2}$ and $d_1^{L_3}, ..., d_r^{L_3}$) at the points $P_1, ..., P_p$ (respectively $Q_1, ..., Q_q$ and $R_1, ..., R_r$). Also we note by assumption that the classes $\mathcal{E}_0, \mathcal{E}_i^{L_1}, \mathcal{E}_j^{L_2}, \mathcal{E}_k^{L_3}$ are all the classes of smooth rational curves on $S_{(p,q,r)}$ for every $i = 1, ..., p, j = 1, \ldots, q$, and $k = 1, \ldots, r$.

To prove Theorem 1.1, it is enough from Lemma 2.4 to prove that the set of (-1)-curves and the set of (-2)-curves are both finite. To do so, we first show that the set of (-2)-curves is finite. So let V be a general (-2)-curve on $S_{(p,q,r)}$. This means, by assumption, that it is not a fixed component of the complete linear system $|-K_{S_{(p,q,r)}}|$. Let π be the natural projection from $S_{(p,q,r)}$ to $\mathbb{P}^2$ and let $(a; b_1^{L_1}, \ldots, b_p^{L_1}; b_1^{L_2}, \ldots, b_q^{L_2}; b_1^{L_3}, \ldots, b_r^{L_3})$ be the $(1 + p + q + r)$-tuple representing the class of V in the Picard group $Pic(S_{(p,q,r)})$ relative to the integral basis $(\mathcal{E}_0; -\mathcal{E}_1^{L_1}, \ldots, -\mathcal{E}_p^{L_1}; -\mathcal{E}_1^{L_2}, \ldots, -\mathcal{E}_q^{L_2}; -\mathcal{E}_1^{L_3}, \ldots, -\mathcal{E}_r^{L_3})$.

It follows that the degree $a = V.\mathcal{E}_0$ is larger than or equal to one. From the two equalities $V^2 = -2$ and $V.K_{S_{(p,q,r)}} = 0$, one may obtain the following equalities:

$$(b_1^{L_1})^2 + \ldots + (b_p^{L_1})^2 + (b_1^{L_2})^2 + \cdots + (b_q^{L_2})^2 + (b_1^{L_3})^2 + \ldots + (b_r^{L_3})^2 = a^2 + 2, \quad (1)$$

and

$$b_1^{L_1} + \cdots + b_p^{L_1} + b_1^{L_2} + \cdots + b_q^{L_2} + b_1^{L_3} + \cdots + b_r^{L_3} = 3a. \quad (2)$$

From the equality 2, one may obtain the following three equalities:

$$b_1^{L_1} + \cdots + b_p^{L_1} = a, \quad (3)$$

and

$$b_1^{L_2} + \cdots + b_q^{L_2} = a, \quad (4)$$

and

$$b_1^{L_3} + \cdots + b_r^{L_3} = a. \quad (5)$$

It follows that if either p, q or r vanishes, then a also vanishes. Hence there is at most two (-2)-curves, generically there is no (-2)-curve at all. Consequently, we assume that p, q and r do not vanish.

We claim that the integer a is bounded. To see this, we argue as follows. Define $x_i^{L_1}$, $y_j^{L_2}$ and $z_k^{L_3}$ for every $i = 1, \ldots, p, j = 1, \ldots, q$ and $k = 1, \ldots, r$ as follows.

$$x_i^{L_1} = \left(b_i^{L_1} - \frac{a}{p} \right), \quad (6)$$

and

$$y_j^{L_2} = \left(b_j^{L_2} - \frac{a}{q} \right), \quad (7)$$

and

$$z_k^{L_3} = \left(b_k^{L_3} - \frac{a}{r} \right), \quad (8)$$

Then the equations 3, 4 and 5 become respectively:

$$x_1^{L_1} + \cdots + x_p^{L_1} = 0, \quad (9)$$

572

and

$$y_1^{L_2} + \cdots + y_q^{L_2} = 0, \tag{10}$$

and

$$z_1^{L_3} + \cdots + z_r^{L_3} = 0, \tag{11}$$

Whereas the equation 1 gives the following equation:

$$\sum_{i=1}^{i=p}(x_i^{L_1})^2 + \sum_{j=1}^{j=q}(y_j^{L_2})^2 + \sum_{k=1}^{k=r}(z_k^{L_3})^2 = 2 + a^2\left(1 - \frac{1}{p} - \frac{1}{q} - \frac{1}{r}\right), \tag{12}$$

which implies by our assumption that the nonnegative integer a is bounded.

Now we proceed to prove that the set of (-1)-curves on $S_{(p,q,r)}$ is also finite. Indeed, let U be a general (-1)-curve on $S_{(p,q,r)}$. This means, by assumption, that it is not a fixed component of the complete linear system $|-K_{S_{(p,q,r)}}|$ and is different from some well known (-1)-curves which are finite in number.

Let $(a; b_1^{L_1}, \ldots, b_p^{L_1}; b_1^{L_2}, \ldots, b_q^{L_2}; b_1^{L_3}, \ldots, b_r^{L_3})$ be the $(1 + p + q + r)$-tuple corresponding to the class of U in the Picard group $Pic(S_{(p,q,r)})$ relative to the integral basis $(\mathcal{E}_0; -\mathcal{E}_1^{L_1}, \ldots, -\mathcal{E}_p^{L_1}; -\mathcal{E}_1^{L_2}, \ldots, -\mathcal{E}_q^{L_2}; -\mathcal{E}_1^{L_3}, \ldots, -\mathcal{E}_r^{L_3})$. Since U is general, it follows that the degree $a = U.\mathcal{E}_0$ is greater than or equal to one. From the two equalities $U^2 = -1$ and $U.K_{S_{(p,q,r)}} = -1$, one may obtain the following two equalities:

$$(b_1^{L_1})^2 + \ldots + (b_p^{L_1})^2 + (b_1^{L_2})^2 + \cdots + (b_q^{L_2})^2 + (b_1^{L_3})^2 + \ldots + (b_r^{L_3})^2 = a^2 + 1, \tag{13}$$

and

$$(a - b_1^{L_1} - \cdots - b_p^{L_1}) + (a - b_1^{L_2} - \cdots - b_q^{L_2}) + (a - b_1^{L_3} - \cdots - b_r^{L_3}) = 1, \tag{14}$$

Hence either the following case which we refer to as the case 1,

$$(b_1^{L_1})^2 + \ldots + (b_p^{L_1})^2 + (b_1^{L_2})^2 + \ldots + (b_q^{L_2})^2 + (b_1^{L_3})^2 + \ldots + (b_r^{L_3})^2 = a^2 + 1, \tag{15}$$

and

$$b_1^{L_1} + \cdots + b_p^{L_1} = a - 1, \tag{16}$$

and

$$b_1^{L_2} + \cdots + b_q^{L_2} = b_1^{L_3} + \cdots + b_r^{L_3} = a. \tag{17}$$

or the following case which we refer to as the case 2

$$(b_1^{L_1})^2 + \ldots + (b_p^{L_1})^2 + (b_1^{L_2})^2 + \ldots + (b_q^{L_2})^2 + (b_1^{L_3})^2 + \ldots + (b_r^{L_3})^2 = a^2 + 1, \tag{18}$$

and

$$b_1^{L_2} + \cdots + b_q^{L_2} = a - 1, \tag{19}$$

and

$$b_1^{L_1} + \cdots + b_p^{L_1} = b_1^{L_3} + \cdots + b_r^{L_3} = a. \tag{20}$$

or the following case which we refer to as the case 3

$$(b_1^{L_1})^2 + \ldots + (b_p^{L_1})^2 + (b_1^{L_2})^2 + \ldots + (b_q^{L_2})^2 + (b_1^{L_3})^2 + \ldots + (b_r^{L_3})^2 = a^2 + 1, \tag{21}$$

and

$$b_1^{L_3} + \cdots + b_r^{L_3} = a - 1, \tag{22}$$

and

$$b_1^{L_1} + \cdots + b_p^{L_1} = b_1^{L_2} + \cdots + b_q^{L_2} = a. \tag{23}$$

holds.

It follows that if either p, q or r vanishes, then a also vanishes. So we may consider the integers p, q and r to be all not equal to zero. Assume that we are in the case 1, and consider the new scalars $(\alpha_i^{L_1})_{i \in \{1,\ldots,p\}}$, $(\beta_j^{L_2})_{j \in \{1,\ldots,q\}}$ and $(\gamma_k^{L_3})_{k \in \{1,\ldots,r\}}$ defined by $\alpha_i^{L_1} = b_i^{L_1} - \left(\frac{a-1}{p}\right)$ for every $i = 1, \ldots, p$, $\beta_j^{L_2} = b_j^{L_2} - \left(\frac{a}{q}\right)$ for every $j = 1, \ldots, q$, and $\gamma_k^{L_3} = b_k^{L_3} - \left(\frac{a}{r}\right)$ for every $k = 1, \ldots, r$.

Then the equations 15, 16 and 17 give

$$\sum_{i=1}^{i=p} (\alpha_i^{L_1})^2 + \sum_{j=1}^{j=q} (\beta_j^{L_2})^2 + \sum_{k=1}^{k=r} (\gamma_k^{L_3})^2 = 1 + a^2 - \frac{(a-1)^2}{p} - \frac{a^2}{q} - \frac{a^2}{r}, \tag{24}$$

and

$$\alpha_1^{L_1} + \cdots + \alpha_p^{L_1} = 0, \qquad (25)$$

and

$$\beta_1^{L_2} + \cdots + \beta_q^{L_2} = 0. \qquad (26)$$

and

$$\gamma_1^{L_3} + \cdots + \gamma_r^{L_3} = 0. \qquad (27)$$

It follows then from the equation 24 that a is bounded. With the same method, we prove the boundness of a in the case 2 and in the case 3.

Proof of Lemma 1.1.

Let D be a numerically effective divisor on $S_{(p,q,r)}$ such that $D.K_{S_{(p,q,r)}} = 0$. We would like to prove that D is the zero divisor. For let $(a; b_1^{L_1}, \ldots, b_p^{L_1}; b_1^{L_2}, \ldots, b_q^{L_2}; b_1^{L_3}, \ldots, b_r^{L_3})$ be the $(1 + p + q + r)$-tuple representing the class of D in the Picard group $Pic(S_{(p,q,r)})$ relative to the integral basis $(\mathcal{E}_0; -\mathcal{E}_1^{L_1}, \ldots, -\mathcal{E}_p^{L_1}; -\mathcal{E}_1^{L_2}, \ldots, -\mathcal{E}_q^{L_2}; -\mathcal{E}_1^{L_3}, \ldots, -\mathcal{E}_r^{L_3})$.

Our primarily task is to prove that a vanishes. Indeed, since $D.K_{S_{(p,q,r)}} = 0$, one may obtain the three following equalities.

$$b_1^{L_1} + \cdots + b_p^{L_1} = b_1^{L_2} + \cdots + b_q^{L_2} = b_1^{L_3} + \cdots + b_r^{L_3} = a. \qquad (28)$$

It follows that if either p, q or r vanishes, then a also vanishes. Now assume that $pqr \neq 0$ and consider the scalars $(x_i^{L_1})_{i \in \{1,\ldots,p\}}$, $(x_j^{L_2})_{j \in \{1,\ldots,q\}}$ and $(x_k^{L_3})_{k \in \{1,\ldots,r\}}$ defined by $x_i^{L_1} = b_i^{L_1} - \left(\frac{a}{p}\right)$ for every $i = 1, \ldots, p$, $x_j^{L_2} = b_j^{L_2} - \left(\frac{a}{q}\right)$ for every $j = 1, \ldots, q$, and $x_k^{L_3} = b_k^{L_3} - \left(\frac{a}{r}\right)$ for every $k = 1, \ldots, r$. It then follows the equalities:

$$x_1^{L_1} + \cdots + x_p^{L_1} = x_1^{L_2} + \cdots + x_q^{L_2} = x_1^{L_3} + \cdots + x_r^{L_3} = 0. \qquad (29)$$

On the other hand, the inequality $D^2 \geq 0$ (see the above Lemma 2.3) gives

$$a^2\left(1 - \frac{1}{p} - \frac{1}{q} - \frac{1}{r}\right) - \sum_{i=1}^{i=p}(x_i^{L_1})^2 - \sum_{j=1}^{j=q}(x_j^{L_2})^2 - \sum_{k=1}^{k=r}(x_k^{L_3})^2 \geq 0. \qquad (30)$$

Hence a vanishes by the Platonic assumption. Consequently D is nothing than the zero divisor (since $D^2 \geq 0$). $\qquad\square$

Remark 3.2. For other kinds of rational surfaces for which the finite generation of the monoid of effective divisor classes holds, one may look at the recent works in [17,18].

Acknowledgements

The authors are highly indebted to the referee for her/his careful reading of the manuscript, suggestions to make the paper more readable and above all to her/his encouragements to study the truth of the result involved in Theorem 1.2 and also to inform them about related research topics. This work has its incarnation during the second and the third weeks of the five weeks in singularity at Luminy 2005. The second author would like to thank warmly the organizers of such meeting, in particular deep thanks to Professors Jean-Paul Brasselet, David Trotman, Anne Pichon, Claudio Murolo, Nicolas Dutertre, Maurice Bourguel from the computer section, and all the library staff. Many thanks also to Professor Gaetana Restuccia for making our stay at the Mathematics Department of the University of Messina a very nice one. This work was supported by G.N.S.A.G.A at the Mathematics Department of Messina University (Messina, Italy), and was partially supported by a grant number MEC 2004 MTM 00958 from the Department of "Álgebra, Geometría y Topología" of the Valladolid University (Valladolid, Spain).

References

1. M. Nagata, *On rational surfaces, II, Memoirs of the College of Science, University of Kyoto, Series A* **33** (1960), no. 2, 271–293.
2. J. Rosoff, *Effective divisor classes and blowings-up of* $\mathbb{P}^2$, *Pacific Journal of Mathematics* **89 (2)** (1980), 419–429.
3. R. Hartshorne, *Algebraic Geometry,* (Graduate Texts in Mathematics, Springer Verlag, 1977).
4. E. Looijenga, *Rational surfaces with an anticanonical cycle, Annals of Mathematics* **114** (1981), no. 2, 267–322.
5. Felix Christian Klein, *Lectures on the Icosahedron and the Solution of Equations of the Fifth Degree,* (1884).
6. I. Dolgachev, *Weyl groups and cremona transformations, Proceedings of Symposia in Pure Mathematics* Volume **40** Part 1 (1983), 283–294.
7. B. Harbourne, *Free resolutions of fat point ideals on* $\mathbb{P}^2$, *Journal of Pure and Applied Algebra* 125 (1998) 213–234.

8. B. Harbourne, *Blowings-up of* $\mathbb{P}^2$ *and their blowings-down, Duke Mathematical Journal* **52**:1 (1985), 129–148.

9. R. Miranda, U. Persson, *On Extremal Rational Elliptic Surfaces, Mathematische Zeitschrift* **193** (1986), 537–558.

10. B. Harbourne, *Rational surfaces with* $K^2 > 0$, *Proceedings of the American Mathematical Society* Volume 124, Number 3, March 1996.

11. M. Lahyane, *Exceptional curves on rational surfaces having* $K^2 \geq 0$, *C. R. Acad. Sci. Paris, Ser. I* **338** (2004) 873–878.

12. M. Lahyane, *Rational surfaces having only a finite number of exceptional curves, Mathematische Zeitschrift* Volume **247**, Number 1, 213–221 (May 2004).

13. M. Lahyane, *Exceptional curves on smooth rational surfaces with* $-K$ *not nef and of self-intersection zero, Proceedings of the American Mathematical Society* **133** (2005) 1593–1599.

14. B. Harbourne, *Anticanonical rational surfaces, Transactions of the American Mathematical Society* Volume **349** (1997), Number 3, 1191–1208.

15. W. Barth, C. Peters, A. Van de Ven. *Compact Complex Surfaces*, (Berlin, Springer 1984).

16. M. Lahyane, B. Harbourne, *Irreducibility of* -1-*classes on anticanonical rational surfaces and finite generation of the effective monoid, Pacific Journal of Mathematics* Volume **218**, Number 1 (2005), pp. 101–114.

17. G. Failla, M. Lahyane, G. Molica Bisci, *On the finite generation of the monoid of effective divisor classes on rational surfaces of type* (n, m), *Atti dell' Accademia Peloritana dei Pericolanti Classe di Scienze Fisiche, Matematiche e Naturali* Vol. **LXXXIV**, C1A0601001 (2006), 1–9.

18. G. Failla, M. Lahyane, G. Molica Bisci, *Rational surfaces of Kodaira type IV.* To appear in *Bollettino dell'Unione Matematica Italiana, Sezione B*.

AN APPLICATION OF RESOLUTION OF SINGULARITIES: COMPUTING THE TOPOLOGICAL ζ-FUNCTION OF ISOLATED SURFACE SINGULARITIES IN $(\mathbb{C}^3,0)$

ANNE FRÜHBIS-KRÜGER*

Fachbereich Mathematik
University of Kaiserslautern
67653 Kaiserslautern, Germany

Introduction

The existence of a resolution of singularities over a field of characteristic zero has been known since the famous work of Hironaka [8] more than 40 years ago. But it was not until the late 1980s that a constructive approach to this problem had been found (see e.g. [2], [10]) which in turn allowed the development and implementation of algorithms for resolution of singularities (see e.g. [4], [7]). On this basis it is now possible to compute several invariants directly, whose definition relies on knowledge about a resolution. The aim of this article is to describe the computation of one of these invariants, the topological ζ-function, based on an implementation of embedded resolution of singularities in the computer algebra system SINGULAR ([9], [7]). In this article, we only discuss the case of surfaces, but calculations in higher dimensions or the computation of different ζ-functions like the Hodge-ζ-function are also possible provided that there is an algorithm to compute the coefficients appearing in the respective sums.

To make the article mostly self-contained, we start with a very brief description of the notions and basic results in the field of resolution of singularities and with the definition of the invariant which we want to compute: the topological ζ-function. This is followed by a detailed description of how to tackle all arising computational tasks in the case of surfaces. As the final result of a resolution algorithm is represented by means of a collection

---

*supported in part by the dfg-schwerpunkt "globale methoden in der komplexen geometrie".

578

of affine charts (which allow gluing), the first problem to be solved is the identification of the exceptional divisors. After this, we consider how multiplicities of the exceptional divisors in the total transform can be computed and how the relevant Euler characteristics can be determined. Throughout the article, all calculations are illustrated by an example which is discussed in detail.

1. Resolution of Singularities and the Topological ζ-Function

In this section, we shall briefly recall well-known results and constructions about resolution of singularities in characteristic zero and, in particular, about algorithmic resolution of singularities; then we continue by recalling the notion of the topological ζ-function. More detailed presentations of the theoretical background of desingularization can, for instance, be found in [2], [3] or [5]; the practical aspects are covered in [7] and in [4]. For the topological ζ-function, we refer to [6] for details.

As the problem of desingaritzation can be rephrased as the problem of resolving singularities of a given variety or scheme X by means of an appropriate finite sequence of blow-ups, the first notions which we would like to recall are clearly the different kinds of transforms under a blow-up:

Definition 1. Let $I \subset \mathcal{O}_W$ be a sheaf of ideals on a smooth algebraic variety W and let $\phi : \tilde{W} \longrightarrow W$ be a blow-up map at a smooth center C with exceptional divisor H. Then

$$\phi^*(I) = I\mathcal{O}_{\tilde{W}}$$

and

$$I_{\text{strict}} = (\phi^*(I) :_{\mathcal{O}_W} \mathcal{I}(H)^\infty)$$

are called the total transform and the strict transform of I respectively. The weak transform of I is defined as the ideal I_{weak} such that the total transform may be factorized as

$$\phi^*(I) = \mathcal{I}(H)^k I_{\text{weak}}$$

(for a suitable k) and H is not a component of $V(I_{\text{weak}})$.

Geometrically speaking, the strict transform is obtained from the total transform by dropping all components which lie inside the exceptional divisor H, whereas the weak transform originates from dropping only the component which coincides with the exceptional divisor.

After fixing the notation for the transforms under a blow-up, we can now turn our considerations to resolution type theorems. There exist several variants of these which are closely related to each other – some requiring weak transforms, some strict transforms. For the application which we would like to consider in this article, we can restrict our considerations to the following theorem of embedded resolution of singularities[a]:

Theorem 1. *(Hironaka) Let X be a subscheme of a smooth algebraic scheme W. Then there exists a sequence*

$$W = W_0 \xleftarrow{\pi_1} W_1 \xleftarrow{\pi_2} \cdots \xleftarrow{\pi_r} W_r$$

of blow-ups $\pi_i : W_i \longrightarrow W_{i-1}$ at smooth centers $C_{i-1} \subset W_{i-1}$ such that

- a *The exceptional divisor of the induced morphism $W_i \longrightarrow W$ has only normal crossings and C_i has normal crossings with it.*
- b *Let $X_i \subset W_i$ be the strict transform of X. All centers C_i are disjoint from $Reg(X) \subset X_i$, the set of points where X is smooth.[b]*
- c *X_r is smooth and has normal crossings with the exceptional divisor of the morphism $W_r \longrightarrow W$.*
- d *The morphism $(W_r, X_r) \longrightarrow (W, X)$ is equivariant under group actions.*

This theorem does not give any hint on how to find the appropriate centers in an algorithmic way, it merely states that they exist – and so does the original proof of Hironaka. The usual approach to determining these centers is to assign to each point of the given X_i an appropriate invariant and use the locus of the maximal value of the invariant as the center for the next blow-up π_{i+1}. There are several ways of defining such an invariant of which two have been implemented up to now: the invariant of Villamayor[5], implemented by Bodnar and Schicho[4], and an variant thereof introduced in [7].

---

[a]The theorem is due to H.Hironaka[8]; it is stated in a version close to [5] resp. [11]

[b]This is not a typographical error, it is really $Reg(X)$, not $Reg(X_i)$. This condition simply ensures that the sequence of blow-ups is an isomorphism on $Reg(X)$.

Common to each of these variants of resolution of singularities is that not the value of the invariant at each point is computed, but the locus of maximal value. Moreover, calculations take place in affine charts by which the scheme X resp. the smooth ambient space W is covered. In particular, the center of a blow-up can consist of components which lie in different charts and there may not even be a chart which contains the whole center.

Computational Remark:(Computational effect of the use of charts)
The fact that there might not be a chart containing the whole center is by far not the only problem arising due to the representation of the relevant objects by means of charts. By far more cumbersome for practical use of an implementation of desingularization is the fact that upon each blow-up a number of new charts arises. As these usually overlap in large parts and as in subsequent steps centers for upcoming blow-ups will be computed in each chart, there is often a large number of blow-ups which are done not just once or twice but many times amounting to a large increase in time and memory consumption.
This shows that steps like passing to open covers need to be avoided, if possible, and that all charts arising from blow-ups which do not contribute relevant information should be dropped. What should be regarded as relevant information in this context, however, can differ significantly depending on what should eventually be computed from the resolution data. In the application described in this article, for instance, it turns out that we can only drop charts which do not contain any points of the total transform outside the open subset already covered by the other charts of the blow-up. This is caused by the fact that we also need information on the intersection loci of the exceptional divisors for computing the corresponding Euler characteristics as we see below.

Throughout the article, we use the following rather small example to illustrate the different computational aspects of determining the topological ζ-function from the output of an algorithmic resolution of singularities:

Example: We consider an embedded resolution of the isolated hypersurface singularity[c]

$$V(x^2 y^3 + y^5 + x^4 + z^2) \subset \mathbb{A}^3_{\mathbb{C}},$$

---

[c]Its singularity at the origin is of type W_{12} in Arnold's list ([1]), its Milnor number is 12 and its δ–invariant is 6.

which is sketched (by means of its tree of charts) in figure 1. The tree of charts of this resolution consists of 26 charts, fifteen of which are final charts, and there are (in total) 7 exceptional divisors

As our goal is the computation of the topological ζ-function from a given tree of charts of a resolution process, we now recall the definition of this ζ-function:

Definition 2. Let $f \in \mathbb{C}[x_1, \ldots, x_n]$ be a non-zero polynomial defining a hypersurface V and let $\pi : X \longrightarrow \mathbb{C}^n$ be an embedded resolution of V. Denote by E_i, $i \in I$, the irreducible components of the divisor $\pi^{-1}(f^{-1}(0))$. To fix notation, we define for each subset $J \subset I$

$$E_J := \cap_{j \in J} E_j \text{ and } E_J^* := E_J \setminus \cup_{j \notin J} E_{J \cup \{j\}}$$

and denote for each $j \in I$ the multiplicity of E_j in the divisor of $f \circ \pi$ by $N(E_j)$. We further set $\nu(E_j) - 1$ to be the multiplicity of E_j in the divisor $\pi^*(dx_1 \wedge \ldots \wedge dx_n)$. Using this notation, we define

(1) The global Denef-Loeser ζ - function of f is

$$Z_{top}^{(d)}(f, s) := \sum_{\substack{J \subset I \text{s.th.} \\ d \mid N(E_j) \forall j \in J}} \chi(E_J^*) \prod_{j \in J} (\nu(E_j) + N(E_j)s)^{-1} \in \mathbb{Q}(s).$$

(2) Intersecting the E_J^* with the preimage of zero in the above formula leads to the local Denef-Loeser zeta function

$$Z_{top,0}^{(d)}(f, s) := \sum_{\substack{J \subset I \text{s.th.} \\ d \mid N(E_j) \forall j \in J}} \chi(E_J^* \cap \pi^{-1}(0)) \prod_{j \in J} (\nu(E_j) + N(E_j)s)^{-1} \in \mathbb{Q}(s).$$

Here, it is important to observe that in the above context the irreducible components are taken over $\mathbb{C}$, while practical calculations usually take place over $\mathbb{Q}$ and further passing to components taken over $\mathbb{C}$ is rather expensive. The following lemma shows that considering $\mathbb{Q}$-irreducible components already allows the computation of the ζ-function:

Lemma 1. *Let D_l , $l \in L$, be the $\mathbb{Q}$-irreducible components of the divisor $\pi^{-1}(f^{-1}(0))$. For each subset $J \subset L$ define D_J and D_J^* as above. Then*

$$Z_{top}^{(d)}(f, s) = \sum_{\substack{J \subset L \text{s.th.} \\ d \mid N(D_j) \forall j \in J}} \chi(D_J^*) \prod_{j \in J} (\nu(D_j) + N(D_j)s)^{-1} \in \mathbb{Q}(s)$$

$$Z_{top,0}^{(d)}(f, s) = \sum_{\substack{J \subset I \text{s.th.} \\ d \mid N(D_j) \forall j \in J}} \chi(D_J^* \cap \pi^{-1}(0)) \prod_{j \in J} (\nu(D_j) + N(D_j)s)^{-1} \in \mathbb{Q}(s).$$

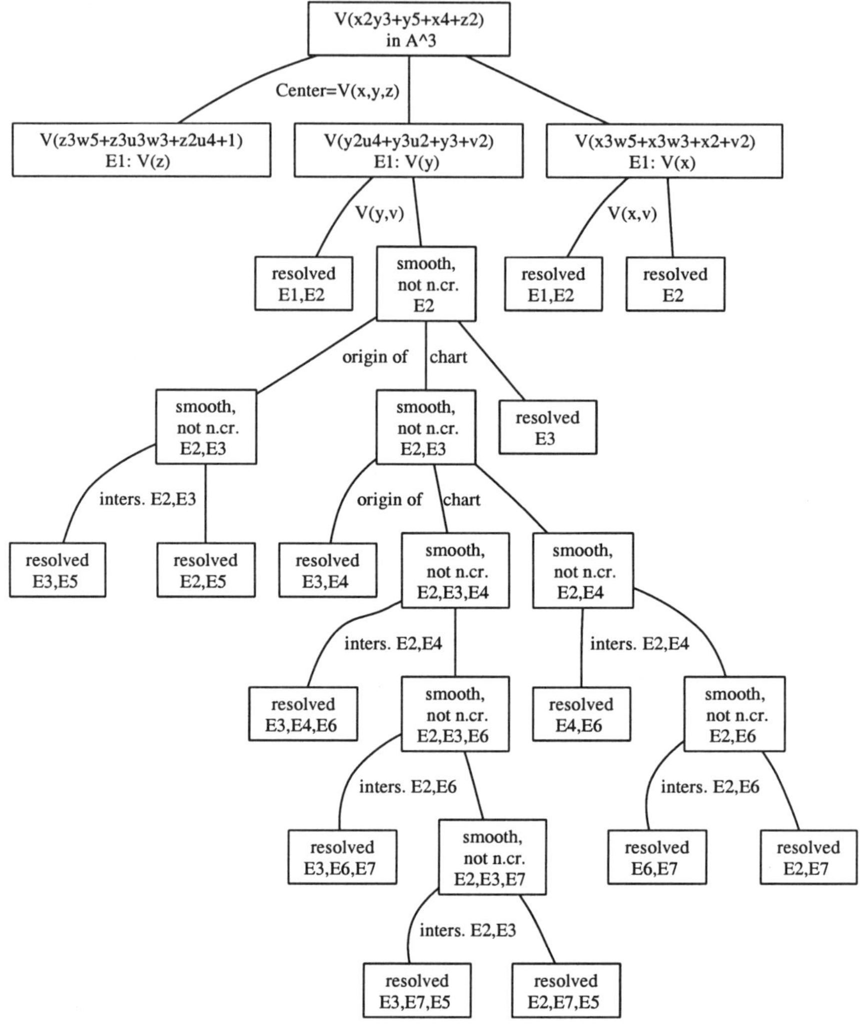

Figure 1. Tree of Charts for a resolution process of the variety $V(x^2y^3 + y^5 + x^4 + z^2) \subset \mathbb{A}^3_{\mathbb{C}}$. The first step is listed explicitly, for the subsequent steps only a very brief description of the situation (including the list of exceptional divisors which are visible in the respective chart) and of the centers is given. Here the exceptional divisors are labeled E1,...,E7, each box corresponds to a chart and the charts in the subsequent line which are connected to one in the previous line arise from a blow-up at a center which is noted between the connecting lines for this blow-up. All centers are contained in the strict transform of the original variety and the notation 'inters. E1,E2' denotes the intersection locus of the exceptional divisors labeled E1 and E2. The labeling of exceptional divisors, which is used here, requires the identification of exceptional divisors appearing in different charts which is one of the computational tasks discussed below.

Proof: The D_l are smooth and therefore disjoint unions of some of the E_j. This implies that for $M = \{l_1, \ldots, l_q\} \subset L$ we obtain $\chi(D_M^*) = \sum \chi(E_J^*)$, the sum is on all subsets $J = \{j_1, \ldots, j_q\} \subset I$ such that E_{j_i} is a component of D_{l_i}. Moreover, it is easy to see that for a component E_j of D_l always $N(E_j) = N(D_l)$ and $\nu(E_j) = \nu(D_l)$.

$\square$

2. The Computational Tasks

The very brief summary of the relevant definitions and of the structure of a tree of charts , which was given in the previous section, already shows the practical tasks that need to be tackled for computing the topological ζ-function of a given (surface) singularity: Given a tree of charts of a resolution of singularities of the original surface, the first task is the identification of exceptional divisors appearing in more than one chart; subsequently, the multiplicities $N(E_i)$ and $\nu(E_i)$ and the relevant Euler characteristics need to be computed. Eventually, all these data have to be combined to obtain the topological ζ-function.[d]

Identification of Exceptional Divisors

Our first task is to find a way to identify points resp. subvarieties which appear in more than one chart; in particular we need to decide whether two given exceptional divisors living in two different charts actually belong to the same exceptional divisor of the global object. To this end, we will move through the tree of charts arising during the resolution process, first blowing-down from the first chart to the one in which the history of the two charts in question branched, and then blowing-up again to the other chart with which we want to compare (cf. figure 1).

As blow-ups are isomorphisms away from the center, this process of successively blowing-down and then blowing-up again does not cause any problems for points which do not lie on an exceptional divisor at all or only lie on exceptional divisors, which already exist in the chart at which the history of the considered charts branched. If, however, the point lies on an exceptional divisor which arises later, then blowing-down beyond the moment of birth of this divisor will inevitably lead to incorrect results, because

---

[d]Treating similar problems like the computation of the topological ζ-function in higher dimensions or of the Hodge-ζ-function is in principle possible along these lines as long as an algorithmic approach for determining the relevant coefficients is available.

this blow-up map is not an isomorphism. To avoid this problem, we need to represent the point on the exceptional divisor as the locus of intersection of the exceptional divisor with an auxiliary variety which is not contained in the exceptional divisor. More formally speaking, we use the following simple fact from commutative algebra:

Lemma 2. *Let $I \subset K[x_1, \ldots, x_n]$ be a prime ideal, $J \subset K[x_1, \ldots, x_n]$ another ideal such that $I + J$ is equidimensional and $ht(I) = ht(I + J) - r$ for some integer $0 < r < n$. Then there exist polynomials $p_1, \ldots, p_r \in I + J$ and a polynomial $f \in K[x_1, \ldots, x_n]$ such that*

$$\sqrt{I + J} = \sqrt{(I + (p_1, \ldots, p_r)) : f}.$$

In our situation, the ideal I is, of course, the ideal of the intersection of the exceptional divisors in which the point or subvariety $V(J)$ is contained. As any sufficiently general set of polynomials $p_1, \ldots, p_r \in J \setminus (I \cap J)$ leading to the correct height of $I + (p_1, \ldots, p_r)$ will do and as the only truly restricting condition on f is that it has to exclude all extra components of $I + (p_1, \ldots, p_r)$, we also have enough freedom of choice of the $p_1, \ldots, p_r, f$ to achieve that none of them is contained in any further exceptional divisor that might be in our way when blowing-down.

Having solved the problem of identifying points which exist in more than one chart, we can now determine which exceptional divisor in one chart coincides with which one in another chart by simply comparing the centers leading to these exceptional divisors. To this end, we start at the root of the tree of charts of the resolution and work our way up to the final charts. The criteria for identifying the centers are quite simple[e]: first of all, the centers cannot be the same, if the corresponding values of the governing function do not agree, secondly, the centers cannot be the same if the exceptional divisors, in which they are contained, are not the same and, in the last step, the remaining candidates are compared explicitly by mapping them through the resolution tree as described above.

Example:(Example 1 revisited)
In the tree of charts of our example, the most obvious identifications are

---

[e]Here we assume that the tree of charts arose from a resolution process which was governed by an invariant as it is the case for the resolution algorithms of Villamayor-Encinas and of Bierstone-Milman.

the divisors E1, E3 and E4 as the earlier ones are in the history of the respective charts leading to the later ones[f] and as there is clearly no further chart giving rise to a zero-dimensional center. For all other identifications there is at least one comparison where we need to move through the tree; the comparison which has the highest number of blow-ups/blow-downs is the one for E5 (appearing on the left side in the 5th line and again in the 8th line with the branching point in the resolution history in 3rd line of the tree).

As a rather simple example of the movement through the tree, we consider the divisor E7. To this end, we need to compare the centers of the two blow-ups leading from the sixth row to the seventh in figure 1. In the first of the two charts (of the sixth row), which give rise to these centers, we have

strict transform	$V(x_1^2 x_3^2 y_0 + x_1 x_3^3 + x_1 x_3 + y_0)$
exceptional divisors	E2$= V(y_0)$, E3$= V(x_3)$, E6$= V(x_1)$
center	$V(x_1, y_0)$

where the variables x_1 and x_3 are the ones which were preserved during the preceding blow-up and y_0 is a new variable which arose in that blow-up. In the other chart we have

strict transform	$V(x_1 x_3^2 + x_1^2 y_0 + x_1 + y_0)$
exceptional divisors	E2$= V(y_0)$, E6$= V(x_1)$
center	$V(x_1, y_0)$

where the x_i and y_j have the same meaning as above. In each of these two charts, E6 is the exceptional divisor which arose from the preceding blow-up. Since we have to move back through the tree beyond the moment of birth of E6, we have to apply lemma 2 which can be done in an obvious way here: We can choose the polynomials as $p_1 = y_0$ and $f = 1$. As p_1 happens to be the equation of the exceptional divisor E2 and as this divisor already exists at the chart where the history of our two charts branched, it is now obvious that the centers in the two charts are actually parts of the same center.

---

[f]From the point of view of the governing invariant, this implies that they cannot correspond to the same value of the the governing invariant which drops upon each blow-up

N and ν

This is a rather easy task compared to the other ones. Performing this computation by hand, we would simply pass through the tree and extract the relevant information directly. But as the trees can become rather large, it is desirable to determine the multiplicities from the final charts. To this end, we only need to pass to some final chart, in which we can see the exceptional divisor in question. There the multiplicity $N(E_i)$ (for the given exceptional divisor E_i) can then be determined by finding the highest exponent j, such that the quotient $I(E_i)^j : J$ is still the whole ring, where J denotes the ideal of the total transform of the original variety. $\nu(E_i)$ can be computed in a similar way, taking into account the appropriate Jacobian determinant.

Example:(Example 1 revisited)
In our example, we obtain the following multiplicities:

	E1	E2	E3	E4	E5	E6	E7
N	2	4	5	10	10	15	20
ν	3	4	6	11	10	15	19

Euler characteristic

For the computation of the Euler characteristics $\chi(E_J^*)$, we have to proceed in several steps: First, we compute the Euler characteristics $\chi(E_J)$ for sets J consisting of 2 or 3 elements in a direct way[g]; then we consider the $\chi(E_i)$ at the birth of the exceptional divisors E_i and their changes under subsequent blow-ups. Finally we determine $\chi(E_J^*)$ from these data.

For the first step, we observe that E_J is a set of points, whenever J consists of three elements, and hence computation of the Euler characteristic boils down to counting points. This, in turn, involves the correct identification of the respective points which appear in more than one chart – a task which has already been discussed.

For $\#J = 2$, we are dealing with curves and can, hence, compute the Euler characteristic by means of the following formula:

$$\chi(E_i \cap E_j) = 2 - 2g(E_i \cap E_j),$$

---

[g]Note that we are only dealing with surfaces in our situation and hence the sets J cannot consist of more than 3 elements.

where g denotes the geometric genus. For computing this latter invariant, it is a well known method to use a (sufficiently general) projection to a plane curve such that the degree of the curve is preserved and the normalizations of the original and the projected curve coincide. As the geometric genus of a curve is the arithmetic genus of its normalization, the geometric genus of this plane curve yields the desired value. For the plane curve, in turn, the δ-invariant can be determined by analysis of the appearing singularities[h] and the arithmetic genus can be found be considering the constant term of the Hilbert polynomial. Hence also the geometric genus is easily accessible, since the difference between arithmetic and geometric genus is exactly the δ-invariant of the curve (sum over the δ-invariants of all appearing singularities).

In the second step we need to determine the Euler characteristics of all exceptional divisors. To this end, we first consider each exceptional divisor at the time of its birth, using the formulae:

$$\chi(E) = k \cdot \chi(\mathbb{P}^2) = 3k \qquad \text{if the center is a set of } k \text{ points}$$
$$\chi(E) = \chi(\mathbb{P}^1 \times C) = 4 - 4g(C) \quad \text{if the center is a curve } C$$

But, of course, subsequent blow-ups have an effect on the Euler characteristic of the exceptional divisor. More precisely, given an exceptional divisor E in Z and a single blow-up $\pi : Y \longrightarrow Z$ at a center C_1 (where E intersects C_1 transversally), the Euler characteristic changes according to the following equation:

$$\chi(\overline{E}) - \chi(\overline{E} \cap D) = \chi(E) - \chi(E \cap C)$$

where D denotes the new exceptional divisor arising from the blow-up π and $\overline{E}$ denotes the strict transform of E under this blow-up. This allows computation of the Euler characteristics of the exceptional divisors by keeping track of the blow-ups affecting the respective divisor and changing the value accordingly.

Finally, we need to pass from the $\chi(E_J)$ to the $\chi(E_J^*)$, which is done in the following way:

---

[h]This analysis of singularities usually involves two different steps: for nodes and cusps, whose δ-invariant is one, it suffices to count the singularities, for the other singularities the use of a Puiseux expansion is necessary.

if $\#J = 3$: $\chi(E_J^*) = \chi(E_J)$

if $\#J = 2$: $\chi(E_J^*) = \chi(E_J) - \sum_{k,k\notin J} \chi(E_{J\cup\{k\}})$

if E_i exceptional divisor:
$$\chi(E_i^*) = \chi(E_i) - \sum_{j,j\neq i} \chi(E_{\{i,j\}}) + \sum_{j,k\neq i} \chi(E_{\{i,j,k\}})$$

if E_i component of strict transform:
$$\chi(E_i^*) = 0$$

Example: (Example 1 continued)

In our example, we will not show all the necessary calculations, since they all follow the steps outlined above. Instead, we focus on one particular set and compute its Euler characteristic explicitly: We consider $\chi(E_3^*)$. To this end, we first need to determine the values of the Euler characteristics for sets J where $3 \in J$ and $\#J > 1$. For simplicity of notation, the strict transform of the original surface is denoted by E_8. Counting the number of points in the intersection locus of the respective exceptional divisors, we obtain the following list for $\#J = 3$ (omitting those sets J for which the intersection locus is empty):

J	$\{2,5,7\}$	$\{2,7,8\}$	$\{3,4,6\}$	$\{3,5,7\}$	$\{3,6,7\}$	$\{5,7,8\}$
$\chi(E_J)$	1	2	1	1	1	1

For $\#J = 2$, we compute the geometric genera of the respective intersection loci, which are curves, and then pass to the corresponding Euler characteristics to obtain the following values[i]. For readers convenience, we have also subtracted the sum of the appropriate entries from the previous table from each of the resulting values and listed the results in the last line.

J	$\{1,2\}$	$\{2,5\}$	$\{2,7\}$	$\{2,8\}$	$\{3,4\}$	$\{3,5\}$	$\{3,6\}$
$g(E_J)$	0	0	0	-1	0	0	0
$\chi(E_J)$	2	2	2	4	2	2	2
$\chi(E_J^*)$	2	1	-1	2	1	1	0

J	$\{3,7\}$	$\{4,6\}$	$\{5,7\}$	$\{5,8\}$	$\{6,7\}$	$\{7,8\}$
$g(E_J)$	0	0	0	0	0	0
$\chi(E_J)$	2	2	2	2	2	2
$\chi(E_J^*)$	0	1	-1	1	1	-1

---

[i]Again sets J which have empty intersection are omitted

As the tables for the sets J, $\#J > 1$, are computed now, we can proceed to compute the Euler characteristic of E_3, the exceptional divisor which we are interested in: It arises from the third blow-up in the tree of charts, leading from the 3rd to the 4th row. The center of this blow-up is clearly a single point, which implies that the Euler characteristic of E_3 at this moment (i.e. in the 4th row of the tree) is

$$\chi(E_3) = 3 \cdot 1 = 3.$$

But, of course, the subsequent blow-ups influence the Euler characteristic of the strict transform of E_3 at the end of the resolution process. More precisely, there are three possible situations all of which occur in this case: First of all, the center can be a curve which is contained in E_3. In this case, the Euler characteristic of the center and the one of the intersection locus of the strict transform[j] of E_3 and the new exceptional divisor coincide; hence $\chi(E_3)$ is not changed. This behavior occurs in the blow-up giving rise to E_5.

In the second case, the intersection locus of E_3 and the center is a set of points - either because the center is itself zero-dimensional or because the 1-dimensional center intersects E_3 in a set of points. Here, the Euler characteristic of the intersection locus of the center with E_3 is exactly the number points and the Euler characteristic of the intersection locus after the blow-up is $2 - 2 \cdot g(E_3 \cap E_{\mathrm{new}})$. In our example, we see this situation in the blow-ups leading to E_4, E_6 and E_7. In the first of these three blow-ups the center is itself one point, in the other two the center meets E_3 in a single point; in all of these cases the geometric genus of the intersection locus after the blow-up is 0. Hence, we know that each of these three blow-ups changes the Euler characteristic of E_3 as follows:

$$\chi(\overline{E_3}) = \chi(E_3) - 1 + (2 - 2 \cdot 0) = \chi(E_3) + 1.$$

The third possible case is that the center does not meet the exceptional divisor and hence the exceptional divisor and its Euler characteristic stay unchanged.

Applying these considerations to the calculation of $\chi(E_3)$, we obtain

$$\chi(E_3) = 3 + 3 \cdot 1 = 6.$$

The following table contains the Euler characteristics for all exceptional divisors:

---

[j]By abuse of notation, we also denote the respective strict transforms of E_3 by the same identifier.

590

i	1	2	3	4	5	6	7	8
$\chi(E_i)$	3	6	6	3	4	4	5	-
$\chi(E_i^*)$	1	-1	1	0	-1	0	1	0

Having computed all these data, it is now easy to combine it to obtain the topological ζ-function of our singularity:

$$Z_{top}^{(1)}(f, s) = \frac{8s + 19}{20s^2 + 39s + 19}.$$

Computational Remark: Combining these data to obtain the desired topological ζ-function involves a rather large number of terms each of which is a quotient of two polynomials. Moreover, we know a common denominator for all these terms a priori anyway. Therefore all simplifications by canceling common factors in enumerator and denominator should be done at the very end of the computations, whereas only the enumerators[k] need to be considered during the intermediate steps.

References

1. Arnold,V., Gusein-Zade,S., Varchenko,A.: *Singularities of Differentiable Maps I*, Birkhäuser (1985)
2. Bierstone,E., Milman,P.:*Canonical Desingularization in Characteristic Zero by Blowing up the Maximum Strata of a Local Invariant*, Invent.Math. **128** (1997), pp. 207–302
3. Bierstone,E., Milman,P.:*Desingularization Algorithms I: The Role of Exceptional Divisors*, Mosc. Math. J. **3** (2003), pp. 751-805
4. Bodnar,G., Schicho,J.:*A Computer Program for the Resolution of Singularities*, in Resolution of Singularities (eds. H.Hauser, J.Lipman, F. Oort, A. Quiros), Progr. in Math. **181** (2000), pp. 231–238
5. Bravo,A., Encinas,S., Villamayor,O.:*A Simplified Proof of Desingularisation and Applications*, Rev. Math. Iberoamericana **21** (2005), pp. 349–458
6. Denef,J., Loeser,F.:*Caractéristiques de Euler-Poincaré, fonctions zeta locales, et modifications analytiques*, J. Amer. Math. Soc. 4 (1992), pp. 705–720
7. Frühbis-Krüger, A., Pfister, G.: *Practical Aspects of Algorithmic Resolution of Singularities*, preprint http://www.mathematik.uni-kl.de/~zca/ Reports_on_ca/33/paper_full.ps.gz
8. Hironaka,H.:*Resolution of Singularities of an Algebraic Variety over a Field of Characteristic Zero*, Annals of Math. **79** (1964), pp. 109–326

---

[k]Of course, these should be the enumerators after passing to the common denominator.

9. Greuel,G.-M., Pfister,G., Schönemann,H.: SINGULAR 3.0, `http://www.singular.uni-kl.de/`

10. Villamayor, O.: *Constructiveness of Hironaka's resolution*, Ann.Scien.Ec. Norm.Sup. 4eme serie **22** (1989), pp. 1-32

Wlodarczyk, J.: *Simple Hironaka Resolution in Characteristic Zero*, J.

11. Amer. Math. Soc. **18** (2005), pp. 779-822

Global properties of integrable implicit Hamiltonian systems

Takuo Fukuda

Department of Mathematics, College of Humanities and Sciences,
Sakurajousui 3-25-40, Setagaya-ku, Tokyo, Japan
E-mail: fukuda@math.chs.nihon-u.ac.jp

Stanislaw Janeczko

Institute of Mathematics, Polish Academy of Sciences, Śniadeckich 8, 00-956
Warszawa, Poland,
and Faculty of Mathematics and Information Science,
Warsaw University of Technology,
Pl. Politechniki 1, 00-661 Warszawa, Poland
E-mail: janeczko@impan.gov.pl

The generalized Hamiltonian dynamics of an implicit Hamiltonian system considered as a Lagrangian variety in the symplectic tangent bundle is studied. Global properties of compact, smoothly integrable Lagrangian immersions with fold singularities are investigated. It is proved that the number of intersection points of an immersion with the zero section of the bundle is estimated by a doubled sum of the self-intersection numbers. Examples of the sphere and the compact orientable surface of genus 2 were explicitly constructed.*

Dedicated to Jean-Paul Brasselet for his 60th birthday

1. Introduction and main results

1.1. *Implicit Hamiltonian systems*

Let $(\mathbb{R}^{2n}, \omega)$ be a Euclidean symplectic manifold endowed with the symplectic structure $\omega = \sum_{i=1}^{n} dy_i \wedge dx_i$, where (x, y) are the standard coordinates on $\mathbb{R}^{2n}$. By the canonical isomorphism between the tangent and cotangent bundles, $T\mathbb{R}^{2n} \ni u \to \omega(u, \bullet) \in T^*\mathbb{R}^{2n}$ the tangent bundle $T\mathbb{R}^{2n}$ is also a

---

*AMS(2000) subject classification. Primary: 57R45, 58F05 Secondary: 58C27, 70H05, 34A26

symplectic manifold with the natural symplectic structure

$$\bar{\omega} = \sum_{i=1}^{n} (d\dot{y}_i \wedge dx_i - d\dot{x}_i \wedge dy_i),$$

where $(x, y, \dot{x}, \dot{y})$ are coordinates on $T\mathbb{R}^{2n} \equiv \mathbb{R}^{2n} \times \mathbb{R}^{2n}$.

Let $\pi : T\mathbb{R}^{2n} \to \mathbb{R}^{2n}, \pi(x, y, \dot{x}, \dot{y}) = (x, y)$, denote the projection to the base space $\mathbb{R}^{2n}$. A smooth Lagrangian submanifold L of $T\mathbb{R}^{2n}$, i.e. $\bar{\omega}|_L = 0$ and $\dim L = 2n$, is called an *implicit Hamiltonian system*. Let $L \subset T\mathbb{R}^{2n}$ be an implicit Hamiltonian system. By $\Sigma(L)$ we denote the set of singular points of $\pi|_L : L \to \mathbb{R}^{2n}$. Then $L - \Sigma(L)$ is a symplectic manifold endowed with the symplectic structure $(\pi|_{L-\Sigma(L)})^* \omega$.

Throughout this paper, we make a generic assumption that the set of fold singular points of $\pi|_L$ is dense in $\Sigma(L)$.

Let $L \subset T\mathbb{R}^{2n}$ be an implicit Hamiltonian system. *Solutions* of L are smooth curves $\alpha : (a, b) \to \mathbb{R}^{2n}$ such that $(\alpha(t), \dot{\alpha}(t)) \in L$ for every $t \in (a, b)$. A point $p \in L$ is called an *integrable point* of L if there exists a solution $\alpha : (a, b) \to \mathbb{R}^{2n}$ of L such that $(\alpha(0), \dot{\alpha}(0)) = p$, $0 \in (a, b)$. An implicit Hamiltonian system L is *integrable* if it consists only of integrable points. A point $p \in L$ is called a *smoothly integrable point* of L if there exists a neighborhood U of p in L and a family $\{\alpha_p \mid (a, b) \to \mathbb{R}^{2n} : p \in U\}$ of solutions of L such that $(\alpha_p(0), \dot{\alpha}_p(0)) = p$ and the family depends smoothly on p and t. An implicit Hamiltonian system L is said to be *smoothly integrable* if all the points of L are smoothly integrable. See Definition 2.1 for a precise description of smooth integrability.

1.2. *Main theorems*

Since P.A.M. Dirac [5], generalized Hamiltonian dynamics became the natural subject. Local criteria for smooth integrability of generalized Hamiltonian systems were proved in [8,10,11]. In this paper we study global properties of such systems. The set of singular points $\Sigma(L)$ encloses integral curves and the configuration of $\Sigma(L)$ together with the ordinary zeroes of dynamical system controls the geometry of solutions. In particular, if $\dim L = 2$, in many cases they determine the geometry of solutions. We will see this fact from the following theorems.

Theorem 1.1. *Let $L \subset T\mathbb{R}^{2n}$ be a smoothly integrable implicit Hamiltonian system. Suppose that the set of fold singular points of $\pi|_L$ is dense* *in $\Sigma(L)$. Then there exists a unique smooth vector field ξ on L with the following properties:*

1)

$$d\pi_{(x,y,\dot{x},\dot{y})}(\xi(x,y,\dot{x},\dot{y})) = \sum_{i=1}^{n} \dot{x}_i \frac{\partial}{\partial x_i} + \dot{y}_i \frac{\partial}{\partial y_i}.$$

2) A curve $\alpha : (a,b) \to \mathbb{R}^{2n}$ is a solution of the implicit Hamiltonian system L if and only if there exists an integral curve $\gamma : (a,b) \to L$ of ξ such that $\pi \circ \gamma = \alpha$ or α is an envelope of such solutions.

3) Integral curves of ξ preserve the singular point set $\Sigma(L)$, i.e. if $\gamma : (a,b) \to L$ is an integral curve of ξ and if $\gamma(c) \in \Sigma(L)$ for some $c \in (a,b)$, then $\gamma(t) \in \Sigma(L)$ for all $t \in (a,b)$.

Let us extend the notion of implicit Hamiltonian systems to include images of Lagrangian immersions (cf. [4,6]). For a Lagrangian immersion $i : \widehat{L} \to T\mathbb{R}^{2n}$, its image $L = i(\widehat{L})$ is called an immersed Lagrangian submanifold of $T\mathbb{R}^{2n}$. We also consider immersed Lagrangian submanifolds of $T\mathbb{R}^{2n}$ as differential equations and call them also implicit Hamiltonian systems (or immersed implicit Hamiltonian systems).

Analogously Theorem 1.1 also holds for Lagrangian immersions:

Theorem 1.2. *Let $i : \widehat{L} \to T\mathbb{R}^{2n}$ be a smoothly integrable Lagrangian immersion such that the set of fold points of $\pi \circ i : \widehat{L} \to \mathbb{R}^{2n}$ is dense in the singular point set $\Sigma(\widehat{L})$ of $\pi \circ i : \widehat{L} \to \mathbb{R}^{2n}$. Then there exists a unique smooth vector field $\widehat{\xi}$ on $\widehat{L}$ with the following properties:*

1) For any point $\widehat{p} \in \widehat{L}$ with $i(\widehat{p}) = (x,y,\dot{x},\dot{y})$, we have

$$d(\pi \circ i)_{\widehat{p}}\widehat{\xi}(\widehat{p}) = \sum_{i=1}^{n} \dot{x}_i \frac{\partial}{\partial x_i} + \dot{y}_i \frac{\partial}{\partial y_i}.$$

2) A curve $\alpha : (a,b) \to \mathbb{R}^{2n}$ is a solution of the implicit Hamiltonian system $L = i(\widehat{L})$ if and only if there exists an integral curve $\widehat{\gamma} : (a,b) \to \widehat{L}$ of $\widehat{\xi}$ such that $\pi \circ i \circ \widehat{\gamma} = \alpha$ or α is an envelope of such solutions.

3) Integral curves of $\widehat{\xi}$ preserve the singular point set $\Sigma(\widehat{L})$, i.e. if $\widehat{\gamma} : (a,b) \to \widehat{L}$ is an integral curve of $\widehat{\xi}$ and if $\widehat{\gamma}(c) \in \Sigma(\widehat{L})$ for some $c \in (a,b)$, then $\widehat{\gamma}(t) \in \Sigma(\widehat{L})$ for all $t \in (a,b)$.

As a corollary of Theorem 1.2 we have

596

Theorem 1.3.

Let $\widehat{L}$ be a $2n-$dimensional compact manifold and let $i : \widehat{L} \to T\mathbb{R}^{2n}$ be a smoothly integrable Lagrangian immersion such that the set of fold singular points of $\pi \circ i : \widehat{L} \to \mathbb{R}^{2n}$ is dense in the singular point set $\Sigma(\widehat{L})$ of $\pi \circ i : \widehat{L} \to \mathbb{R}^{2n}$. Then $L = i(\widehat{L})$ intersects the zero section of the tangent bundle $T\mathbb{R}^{2n}$ at least in $|\chi(\widehat{L})| = 2|\sharp(i(\widehat{L}))|$ points. Here $\sharp(i(\widehat{L}))$ denotes the sum of the local self-intersection numbers of $i : \widehat{L} \to T\mathbb{R}^{2n}$ and $\chi(\widehat{L})$ denotes the Euler characteristic of $\widehat{L}$.

2. Smooth integrability

2.1. *Integrability condition in terms of a generating family*

Let L be a Lagrangian submanifold of $T\mathbb{R}^{2n}$. Then by the Hörmander-Arnold-Weinstein construction, L can be locally described by a generating family (cf. [1,16]).

Lemma 2.1. *([16]) Let $p = (x_0, y_0, \dot{x}_0, \dot{y}_0)$ be a singular point of $\pi|_L$. Then there exist a neighborhood $\mathcal{O}$ of $p = (x_0, y_0, \dot{x}_0, \dot{y}_0)$ in $T\mathbb{R}^{2n}$ and a smooth function $F : \mathbb{R}^{2n} \times \mathbb{R}^k \to \mathbb{R}$ (Morse family) defined in a neighborhood W of $(x_0, y_0, 0)$ such that*

$$L \cap \mathcal{O} = \{(x, y, \frac{\partial F}{\partial y}(x, y, \lambda), -\frac{\partial F}{\partial x}(x, y, \lambda)) \mid \frac{\partial F}{\partial \lambda_\ell}(x, y, \lambda) = 0, (x, y, \lambda) \in W\},$$

$$\mathrm{rank}\left(\frac{\partial^2 F}{\partial x_j \partial \lambda_\ell}, \frac{\partial^2 F}{\partial y_i \partial \lambda_\ell}\right)(x_0, y_0, 0) = k, \quad \left(\frac{\partial^2 F}{\partial \lambda_s \partial \lambda_r}\right)(x_0, y_0, 0) = 0.$$

where $1 \leq \ell, s, r \leq k$, $1 \leq i, j \leq n$.

Set

$$\widetilde{L} = \{(x, y, \lambda) \in W \mid \frac{\partial F}{\partial \lambda_\ell}(x, y, \lambda) = 0, \quad \ell = 1, \ldots, k\}$$

and define a map $\phi : \widetilde{L} \to L \cap \mathcal{O}$ by $\phi(x, y, \lambda) = (x, y, \frac{\partial F}{\partial y}(x, y, \lambda), -\frac{\partial F}{\partial x}(x, y, \lambda))$. Let $\widetilde{\pi} : \mathbb{R}^{2n} \times \mathbb{R}^k \to \mathbb{R}^{2n}$ denote the projection $\widetilde{\pi}(x, y, \lambda) = (x, y)$. Then $\phi : \widetilde{L} \to L \cap \mathcal{O}$ is a diffeomorphism and the following diagram commutes.

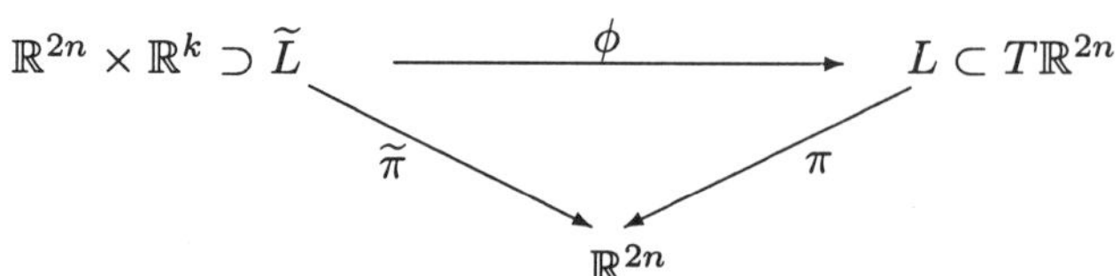

Consider a vector field $\widetilde{\xi}$ on $\mathbb{R}^{2n} \times \mathbb{R}^k$ of the form

$$\widetilde{\xi}(x,y,\lambda) = \sum_{i=1}^{n} \left(\frac{\partial F}{\partial y_i}(x,y,\lambda)\frac{\partial}{\partial x_i} - \frac{\partial F}{\partial x_i}(x,y,\lambda)\frac{\partial}{\partial y_i} \right) + \sum_{\ell=1}^{k} \mu_\ell(x,y,\lambda)\frac{\partial}{\partial \lambda_\ell}.$$

Definition 2.1. A singular point $p = (x_0, y_0, \dot{x}_0, \dot{y}_0) \in L \cap \mathcal{O}$ is a *smoothly integrable point* of L if there exist smooth functions $\mu_\ell(x,y,\lambda)$, $\ell = 1, \ldots, k$, such that $\widetilde{\xi}$ is tangent to $\widetilde{L}$. An implicit Hamiltonian system L is *smoothly integrable* if every singular point of L is smoothly integrable.

This definition is justified as follows. If for some smooth functions $\mu_\ell(x,y,\lambda)$, $\ell = 1, \ldots, k$, $\widetilde{\xi}$ is tangent to $\widetilde{L}$, then for integral curves $\widetilde{\gamma} : (a,b) \rightarrow \widetilde{L}$ of $\widetilde{\xi}$, $\widetilde{\pi} \circ \widetilde{\gamma} : (a,b) \rightarrow \mathbb{R}^{2n}$ are solutions of the implicit Hamiltonian system L.

Consider the family $\mathcal{G}$ of solutions of $L \cap \mathcal{O}$ derived from the integral curves of $\widetilde{\xi}$, $\mathcal{G} = \{\widetilde{\pi} \circ \widetilde{\gamma} \mid \widetilde{\gamma} : (a,b) \rightarrow \widetilde{L} \text{ is an integral curve of } \widetilde{\xi}\}$.

Lemma 2.2. *If for some smooth functions $\mu_\ell(x,y,\lambda)$, $\ell = 1, \ldots, k$, the vector field $\widetilde{\xi}(x,y,\lambda)$ is tangent to $\widetilde{L}$, then the implicit Hamiltonian system $L \cap \mathcal{O}$ is integrable and any solution of this system is either an element of $\mathcal{G}$ or an envelope of elements of $\mathcal{G}$.*

Here a curve $\gamma_0 : (a,b) \rightarrow \mathbb{R}^{2n}$ is an *envelope* of elements of $\mathcal{G}$ if no part of γ_0 is contained in $\mathcal{G}$ and if for any $t_0 \in (a,b)$, there exists a solution $\gamma \in \mathcal{G}$, $\gamma : (t_0 - \epsilon, t_0 + \epsilon) \rightarrow \mathbb{R}^{2n}$, different from γ_0 at any point other than t_0, such that γ is tangent to γ_0 at t_0 (cf. [14,15]).

Proof. Let $p = (x_0, y_0, \dot{x}_0, \dot{y}_0)$ be an arbitrary point of $L \cap \mathcal{O}$ and let $(x_0, y_0, \lambda_0) \in \widetilde{L}$ be a unique point such that $\phi(x_0, y_0, \lambda_0) = p$. Then there exists an integral curve $\widetilde{\gamma} : (-\epsilon, \epsilon) \rightarrow \widetilde{L}$ of $\widetilde{\xi}$ with $\widetilde{\gamma}(0) = (x_0, y_0, \lambda_0)$. The curve $\gamma = \widetilde{\pi} \circ \widetilde{\gamma} : (a,b) \rightarrow \mathbb{R}^{2n}$ is a solution of the implicit Hamiltonian system $L \cap \mathcal{O}$ such that $(\gamma(0), \dot{\gamma}(0)) = p$. Thus p is an integrable point of $L \cap \mathcal{O}$. Hence $L \cap \mathcal{O}$ is integrable.

Now let $\gamma_0 : (a,b) \rightarrow \mathbb{R}^{2n}$ be a solution of $L \cap \mathcal{O}$ no part of which is contained in $\mathcal{G}$ and let $t_0 \in (a,b)$. Since in the above argument, the integrability of $L \cap \mathcal{O}$ was guaranteed by elements of $\mathcal{G}$, there exists a solution $\gamma : (t_0 - \epsilon, t_0 + \epsilon) \rightarrow \mathbb{R}^{2n}$, $\gamma \in \mathcal{G}$, such that $(\gamma(0), \dot{\gamma}(0)) = (\gamma_0(0), \dot{\gamma}_0(0))$. Thus γ is tangent to γ_0 at $\gamma_0(0)$. Since no part of γ_0 is contained in $\mathcal{G}$, γ is different from γ_0 at any point of $(t_0 - \epsilon, t_0 + \epsilon)$ other than t_0. Thus γ_0 is an envelope of $\mathcal{G}$. This completes the proof of Lemma 2.2. $\square$

Lemma 2.3. *([10]) Let $L \subset T\mathbb{R}^{2n}$ be an implicit Hamiltonian system and let $p = (x_0, y_0, \dot{x}_0, \dot{y}_0)$ be a singular point of L generated by a Morse family*

598

$F : \mathbb{R}^{2n} \times \mathbb{R}^k \to \mathbb{R}$. *Then p is a smoothly integrable point of L if and only if the linear equation*

$$\left(\frac{\partial^2 F}{\partial \lambda_s \partial \lambda_r} \right)(x,y,\lambda) \begin{pmatrix} \mu_1 \\ \vdots \\ \mu_k \end{pmatrix} = \begin{pmatrix} \{\frac{\partial F}{\partial \lambda_1}, F\}(x,y,\lambda) \\ \vdots \\ \{\frac{\partial F}{\partial \lambda_k}, F\}(x,y,\lambda) \end{pmatrix} \tag{1}$$

has a smooth solution $\mu(x,y,\lambda) = (\mu_1(x,y,\lambda), \ldots, \mu_k(x,y,\lambda))$ defined in a neighborhood of $(x_0, y_0, 0)$ in $\widetilde{L}$, where $\{\cdot, \cdot\}$ denotes the Poisson bracket on $(\mathbb{R}^{2n}, \omega)$.

Let us denote by $[\frac{\partial^2 F}{\partial \lambda_s \partial \lambda_r}](x,y,\lambda)$ the $k \times k$ cofactor matrix of $(\frac{\partial^2 F}{\partial \lambda_s \partial \lambda_r})(x,y,\lambda)$ at a point (x,y,λ), and let

$$\begin{pmatrix} \overline{\{\frac{\partial F}{\partial \lambda_1}, F\}}(x,y,\lambda) \\ \vdots \\ \overline{\{\frac{\partial F}{\partial \lambda_k}, F\}}(x,y,\lambda) \end{pmatrix} = \left[\frac{\partial^2 F}{\partial \lambda_s \partial \lambda_r} \right](x,y,\lambda) \begin{pmatrix} \{\frac{\partial F}{\partial \lambda_1}, F\}(x,y,\lambda) \\ \vdots \\ \{\frac{\partial F}{\partial \lambda_k}, F\}(x,y,\lambda) \end{pmatrix}.$$

Then we have

Lemma 2.4. *([8]) Let the assumptions be the same as in Lemma 2.3. Then the linear equation (1) has a smooth solution $\mu(x,y,\lambda) = (\mu_1(x,y,\lambda), \ldots, \mu_k(x,y,\lambda))$ defined in a neighborhood of $(x_0, y_0, 0)$ in $\widetilde{L}$ if and only if*

$$\overline{\{\frac{\partial F}{\partial \lambda_1}, F\}}, \ldots, \overline{\{\frac{\partial F}{\partial \lambda_k}, F\}} \in \langle \det \left(\frac{\partial^2 F}{\partial \lambda_s \partial \lambda_r} \right), \frac{\partial F}{\partial \lambda_1}, \ldots, \frac{\partial F}{\partial \lambda_k} \rangle_{\mathcal{E}(x,y,\lambda)},$$

where $\mathcal{E}(x,y,\lambda)$ denotes the space of smooth function-germs at $(x_0, y_0, 0)$.

Proof. Multiplying (1) by the cofactor matrix of the Hessian matrix of F with respect to λ we have (cf. [8,13])

$$\det \left(\frac{\partial^2 F}{\partial \lambda_s \partial \lambda_r} \right)(x,y,\lambda) \begin{pmatrix} \mu_1 \\ \vdots \\ \mu_k \end{pmatrix} = \left[\frac{\partial^2 F}{\partial \lambda_s \partial \lambda_r} \right](x,y,\lambda) \begin{pmatrix} \{\frac{\partial F}{\partial \lambda_1}, F\}(x,y,\lambda) \\ \vdots \\ \{\frac{\partial F}{\partial \lambda_k}, F\}(x,y,\lambda) \end{pmatrix}$$

$$= \begin{pmatrix} \overline{\{\frac{\partial F}{\partial \lambda_1}, F\}}(x,y,\lambda) \\ \vdots \\ \overline{\{\frac{\partial F}{\partial \lambda_k}, F\}}(x,y,\lambda) \end{pmatrix}.$$

Thus (1) has a smooth solution $\mu = (\mu_1, \dots, \mu_k)$ defined on

$$\widetilde{L} = \{\frac{\partial F}{\partial \lambda_1} = \dots = \frac{\partial F}{\partial \lambda_k} = 0\}$$

if and only if

$$\overline{\{\frac{\partial F}{\partial \lambda_1}, F\}}, \dots, \overline{\{\frac{\partial F}{\partial \lambda_k}, F\}} \in \langle \det\left(\frac{\partial^2 F}{\partial \lambda_s \partial \lambda_r}\right), \frac{\partial F}{\partial \lambda_1}, \dots, \frac{\partial F}{\partial \lambda_k}\rangle_{\mathcal{E}(x,y,\lambda)}.$$

$\square$

2.2. *Normal forms of fold singularities*

For the proof of Theorem 1.1 we need the normal forms of fold singularities (cf. [7,12]).

Lemma 2.5. *Let $L \subset T\mathbb{R}^{2n}$ be a Lagrangian submanifold and let $(x_0, y_0, \dot{x}_0, \dot{y}_0) \in L$ be a fold singular point of $\pi|_L : L \to \mathbb{R}^{2n}$. Then the germ $(L, (x_0, y_0, \dot{x}_0, \dot{y}_0))$ is symplectomorphic to the germ of a Lagrangian submanifold generated by a function-germ at $(x, y, \lambda) = (0, 0, 0)$ of the form*

$$F(x, y, \lambda) = \lambda^3 + y_1\lambda + a(x, y).$$

Proof. Since (x_0, y_0) is a fold singular point of $\pi|_L$, Lemma 2.1 shows that $(L, (x_0, y_0, \dot{x}_0, \dot{y}_0))$ is symplectomorphic to the germ of Lagrangian submanifold generated by

$$F(x, y, \lambda) = \lambda^3 + a_1(x, y)\lambda + a_0(x, y).$$

Since $(x_0, y_0, \dot{x}_0, \dot{y}_0)$ is a fold singular point, we see that $da_1(x_0, y_0) \neq 0$. Then, preserving the symplectic structure, we may assume that $a_1(x, y) = y_1$. Thus F has the form

$$F(x, y, \lambda) = \lambda^3 + y_1\lambda + a_0(x, y).$$

$\square$

Lemma 2.6. *The implicit Hamiltonian system germ generated by a function-germ $F(x, y, \lambda) = \lambda^3 + y_1\lambda + a_0(x, y)$ at $(x, y, \lambda) = (0, 0, 0)$ is smoothly integrable if and only if F has the form*

$$F(x, y, \lambda) = \lambda^3 + y_1\lambda + y_1 a(x, y) + b(x_2, \dots, x_n, y)$$

for some smooth function-germs $a(x, y)$ and $b(x_2, \dots, x_n, y)$.

Proof. The implicit Hamiltonian system L is integrable if and only if $\overline{\{\frac{\partial F}{\partial \lambda}, F\}} = \{\frac{\partial F}{\partial \lambda}, F\} \in \langle \frac{\partial F}{\partial \lambda}, \frac{\partial^2 F}{\partial \lambda^2} \rangle_{\mathcal{E}(x,y,\lambda)}$. Since $\frac{\partial F}{\partial \lambda} = 3\lambda^2 + y_1$, and $\frac{\partial^2 F}{\partial \lambda^2} = 6\lambda$, we have $\langle \frac{\partial F}{\partial \lambda}, \frac{\partial^2 F}{\partial \lambda^2} \rangle_{\mathcal{E}(x,y,\lambda)} = \langle y_1, \lambda \rangle_{\mathcal{E}(x,y,\lambda)}$. On the other hand, $\{\frac{\partial F}{\partial \lambda}, F\} = \{y_1, y_1 + a_0(x, y)\} = \frac{\partial a_0}{\partial x_1}(x, y)$. The condition that

$$\{\frac{\partial F}{\partial \lambda}, F\} = \{y_1, y_1 + a_0(x, y)\} = \frac{\partial a_0}{\partial x_1}(x, y) \in \langle y_1, \lambda \rangle_{\mathcal{E}(x,y,\lambda)}$$

is equivalent to $\frac{\partial a_0}{\partial x_1}(x, y) \in \langle y_1 \rangle_{\mathcal{E}(x,y)}$. Thus, L is smoothly integrable if and only if $a_0(x, y)$ has the form

$$a_0(x, y) = y_1 a(x, y) + b(x_2, \ldots, x_n, y)$$

for some smooth functions $a(x, y)$ and $b(x_2, \ldots, x_n, y)$. This completes the proof. $\square$

Corollary 2.1. *Let* $(L, (0, 0, \dot{x}_0, \dot{y}_0)) \subset T\mathbb{R}^{2n}$ *be a fold singularity germ of the Lagrangian submanifold generated by a function-germ at* $(x, y, \lambda) = (0, 0, 0)$ *of the form*

$$F(x, y, \lambda) = \lambda^3 + y_1 \lambda + y_1 a(x, y) + b(x_2, \ldots, x_n, y).$$

Then integral curves of the tangent vector field $\widetilde{\xi}$ *of* $\widetilde{L}$ *preserve the singular point set* $\Sigma(\widetilde{L})$ *of* $\widetilde{\pi} \mid_{\widetilde{L}} \colon \widetilde{L} \to \mathbb{R}^{2n}$.

Proof. The tangent vector field $\widetilde{\xi}$ has the form

$$\widetilde{\xi}(x, y, \lambda) = \sum_{i=1}^{n} \left(\frac{\partial F}{\partial y_i}(x, y, \lambda) \frac{\partial}{\partial x_i} - \frac{\partial F}{\partial x_i}(x, y, \lambda) \frac{\partial}{\partial y_i} \right) + \mu(x, y, \lambda) \frac{\partial}{\partial \lambda}.$$

and μ is a solution of the equation $\frac{\partial^2 F}{\partial \lambda^2} \mu = \{\frac{\partial F}{\partial \lambda}, F\}$ on $\widetilde{L} = \{\frac{\partial F}{\partial \lambda} = 0\}$. As a result we get $\mu = -\frac{1}{2}\lambda \frac{\partial a}{\partial x_1}(x, y)$ on $\widetilde{L}$. On the other hand

$$\Sigma(\widetilde{L}) = \{(x, y, \lambda) \mid \frac{\partial F}{\partial \lambda} = \frac{\partial^2 F}{\partial \lambda^2} = 0\} = \{(x, y, \lambda) \mid \lambda = y_1 = 0\},$$

and we have $\widetilde{\xi}\lambda = \widetilde{\xi}y_1 = 0$ on $\Sigma(\widetilde{L})$. Hence $\widetilde{\xi}$ is tangent to $\Sigma(\widetilde{L})$. This completes the proof. $\square$

3. Proofs of main theorems

3.1. *Proof of Theorem 1.1*

Locally, there always exists a smooth vector field ξ satisfying the conditions of Theorem 1.1 as follows.

In a small neighborhood U of a regular point of $\pi|_L : L \to \mathbb{R}^{2n}$, since $\pi(L \cap U)$ is an open subset of $\mathbb{R}^{2n}$ and $\pi|_{L\cap U} : L \cap U \to \pi(L \cap U) \subset \mathbb{R}^{2n}$ is a diffeomorphism, there is a unique vector field $\xi_{L\cap U}$ satisfying conditions 1) and 2). Note that in this case a solution of L cannot be an envelope of other solutions.

In a small neighborhood U of a singular point of $\pi|_L : L \to \mathbb{R}^{2n}$, there exists a Hörmander-Arnold-Weinstein generating family $F(x, y, \lambda)$ satisfying the conditions of Lemma 2.1. Since L is smoothly integrable, by definition, there exists a smooth tangent vector field $\widetilde{\xi}$ on $\widetilde{L}$ of the form

$$\widetilde{\xi}(x, y, \lambda) = \sum_{i=1}^{n}\left(\frac{\partial F}{\partial y_i}(x, y, \lambda)\frac{\partial}{\partial x_i} - \frac{\partial F}{\partial x_i}(x, y, \lambda)\frac{\partial}{\partial y_i}\right) + \sum_{\ell=1}^{k}\mu_\ell(x, y, \lambda)\frac{\partial}{\partial \lambda_\ell}.$$

Let $\phi : \widetilde{L} \to L \cap U$ be the diffeomorphism defined in §2.1. Then $d\phi(\widetilde{\xi})$ satisfies conditions 1) and 2). Let us note again that, in this case, it may happen that a family of solutions α of the form $\pi \circ \gamma = \alpha$ has an envelope. Then of course this envelope is also a solution of L.

Since such a vector field is unique in the set of regular points and the regular points are dense, there exists a unique vector field ξ satisfying 1) and 2).

From Corollary 2.1, ξ preserves the set of fold singular points and the fold points are dense in the singular point set $\Sigma(L)$. Therefore ξ preserves $\Sigma(L)$. This completes the proof of Theorem 1.1. $\square$

3.2. *Proof of Theorem 1.2*

Lagrangian immersions are locally Lagrangian embeddings, so from Theorem 1.1 we have Theorem 1.2. Indeed, let $p \in \widehat{L}$. Since $i : \widehat{L} \to T\mathbb{R}^{2n}$ is a Lagrangian immersion, there exists an open neighborhood $\widehat{U}$ of p in $\widehat{L}$ such that $i|_{\widehat{U}} : \widehat{U} \to T\mathbb{R}^{2n}$ is a Lagrangian embedding and $L_{\widehat{U}} = i(\widehat{U})$ is a smoothly integrable implicit Hamiltonian system whose set of fold singular points of $\pi|_{L_{\widehat{U}}}$ is dense in $\Sigma(L_{\widehat{U}})$. By Theorem 1.1, there exists a unique smooth vector field $\xi_{\widehat{U}}$ on $L_{\widehat{U}}$ satisfying the conditions in Theorem 1.1. Since $i\,|_{\widehat{U}}: \widehat{U} \to L_{\widehat{U}}$ is a diffeomorphism, there is a unique smooth vector field $\widehat{\xi}_{\widehat{U}}$ on $\widehat{U}$ such that $di(\widehat{\xi}_{\widehat{U}}) = \xi_{\widehat{U}}$. Then, for $\widehat{U}$, $\widehat{\xi}_{\widehat{U}}$ is the unique vector field which satisfies the conditions of Theorem 1.2. Gluing such unique vector fields $\widehat{\xi}_{\widehat{U}}$, we obtain a unique global vector field $\widehat{\xi}$ on $\widehat{L}$ that satisfies the conditions of Theorem 1.2. $\quad\square$

3.3. *Proof of Theorem 1.3*

Let $\widehat{\xi}$ be the unique vector field on $\widehat{L}$ which satisfies the conditions in Theorem 1.2. From the Euler-Poincaré-Hopf formula, $\widehat{\xi}$ has at least $|\chi(\widehat{L})|$ equilibrium (i.e. singular) points.

Let $\widehat{p} \in \widehat{L}$ be a singular point of $\widehat{\xi}$: $\widehat{\xi}(\widehat{p}) = 0$. By Theorem 1.2, for $\widehat{p} \in \widehat{L}$ with $i(\widehat{p}) = (x, y, \dot{x}, \dot{y})$, we have

$$d(\pi \circ i)_{\widehat{p}}\widehat{\xi}(\widehat{p}) = \sum_{i=1}^{n} \dot{x}_i \frac{\partial}{\partial x_i} + \dot{y}_i \frac{\partial}{\partial y_i}.$$

Since $\widehat{\xi}(\widehat{p}) = 0$, the left hand side of the above equality is 0 and we have $(\dot{x}, \dot{y}) = (0, 0)$. Thus $L = i(\widehat{L})$ intersects the zero section of $T\mathbb{R}^{2n}$ at $i(\widehat{p})$. Hence $L = i(\widehat{L})$ intersects the zero section of the tangent bundle $T\mathbb{R}^{2n}$ in at least $|\chi(\widehat{L})|$ points. The equality $|\chi(\widehat{L})| = 2|\sharp(i(\widehat{L}))|$ comes from M. Audin's theorem (see [3], p. 594, cf. also [2,9]). This completes the proof of Theorem 1.3. $\quad\square$

4. Examples of smoothly integrable global Lagrangian immersions

4.1. *A method for constructing smoothly integrable global Lagrangian immersions*

Before giving a concrete example, we give a method for constructing smoothly integrable global Lagrangian immersions.

Let W^{2n} be a $2n-$dimensional compact smooth manifold with boundary embedded in $\mathbb{R}^{2n}$. Let ∂W denote its boundary. Let $(x, y) = (x_1, \ldots, x_n, y_1, \ldots, y_n)$ be the standard coordinates of $\mathbb{R}^{2n}$. Then there exists a smooth function $\alpha : \mathbb{R}^{2n} \to \mathbb{R}$ satisfying the following conditions:

$$\alpha^{-1}(0) = \partial W, \tag{2}$$

$$\mathrm{grad}\,\alpha(x, y) \neq 0, \qquad \forall (x, y) \in \partial W, \tag{3}$$

$$\alpha(x, y) < 0, \qquad \forall (x, y) \in W - \partial W, \tag{4}$$

$$\alpha(x, y) > 0, \qquad \forall (x, y) \notin W. \tag{5}$$

Consider a global Morse family $F : \mathbb{R}^{2n} \times \mathbb{R} \to \mathbb{R}$ of the form

$$F(x, y, \lambda) = \frac{1}{3}\lambda^3 + \alpha(x, y)\lambda + \alpha(x, y)a(x, y),$$

where $a(x, y)$ is an arbitrary smooth function. Set

$$\widetilde{L} = \{(x, y, \lambda) \in \mathbb{R}^{2n} \times \mathbb{R} \mid \frac{\partial F}{\partial \lambda} = \lambda^2 + \alpha(x, y) = 0\}$$

and define a map $i : \widetilde{L} \to T\mathbb{R}^{2n}$ by

$$i(x, y, \lambda) = (x, y, \frac{\partial F}{\partial y}(x, y, \lambda), -\frac{\partial F}{\partial x}(x, y, \lambda)).$$

Let

$$\pi : T\mathbb{R}^{2n} \to \mathbb{R}^{2n} \quad \text{and} \quad \widetilde{\pi} : \mathbb{R}^{2n} \times \mathbb{R} \to \mathbb{R}^{2n}$$

denote the canonical projections as in the previous sections. Then we have

$$\pi \circ i = \widetilde{\pi}|_{\widetilde{L}} : \widetilde{L} \to \mathbb{R}^{2n}.$$

Let $\widetilde{\Sigma}$ denote the singular point set of $\pi \circ i = \widetilde{\pi}|_{\widetilde{L}} : \widetilde{L} \to \mathbb{R}^{2n}$.

Prop 4.1.

1) $\widetilde{L}$ is a compact smooth manifold.
2) $i : \widetilde{L} \to T\mathbb{R}^{2n}$ is a smoothly integrable Lagrangian immersion.
3) The critical value set of $\pi \circ i = \widetilde{\pi}|_{\widetilde{L}}$ is ∂W.
4) Singularities of $\pi \circ i = \widetilde{\pi}|_{\widetilde{L}} : \widetilde{L} \to \mathbb{R}^{2n}$ are fold singularities, hence the Lagrangian singularities of $i(\widetilde{L})$ are fold singularities.
5) Solutions of the implicit Hamiltonian system $i(\widetilde{L}) \subset T\mathbb{R}^{2n}$ preserve ∂W.

The properties 1)-5) can be easily verified, noticing the following fact: At a singular point $(x_0, y_0, \lambda_0) \in \widetilde{\Sigma}$ of $\pi \circ i = \widetilde{\pi}|_{\widetilde{L}} : \widetilde{L} \to \mathbb{R}^{2n}$, by the assumption (3) that grad $\alpha(x, y) \neq 0$ on ∂W, there exists a local symplectic coordinate system $(u_1, \ldots, u_n, v_1, \ldots, v_n)$ around (x_0, y_0) such that $v_1 = \alpha$. Then

$$F(x, y, \lambda) = \frac{1}{3}\lambda^3 + \alpha(x, y)\lambda + \alpha(x, y)a(x, y) = \frac{1}{3}\lambda^3 + v_1\lambda + v_1 a(x, y)$$

is a Morse family of a smoothly integrable fold singularity and this family generates the germ of a Lagrangian submanifold $i((\widetilde{L}, (x_0, y_0, \lambda_0)))$.

604

4.2. *Example: A compact orientable surface with genus two*

As a smooth manifold W with boundary in Proposition 4.1, we take (see Figure 1)

$$W = \{(x,y) \in \mathbb{R}^2 \mid (x^2+y^2-16)((x-2)^2+y^2-1)(((x+2)^2+y^2-1) \leq 0\}.$$

Then

$$\partial W = \{(x,y) \in \mathbb{R}^2 \mid (x^2+y^2-16)((x-2)^2+y^2-1)(((x+2)^2+y^2-1) = 0\}.$$

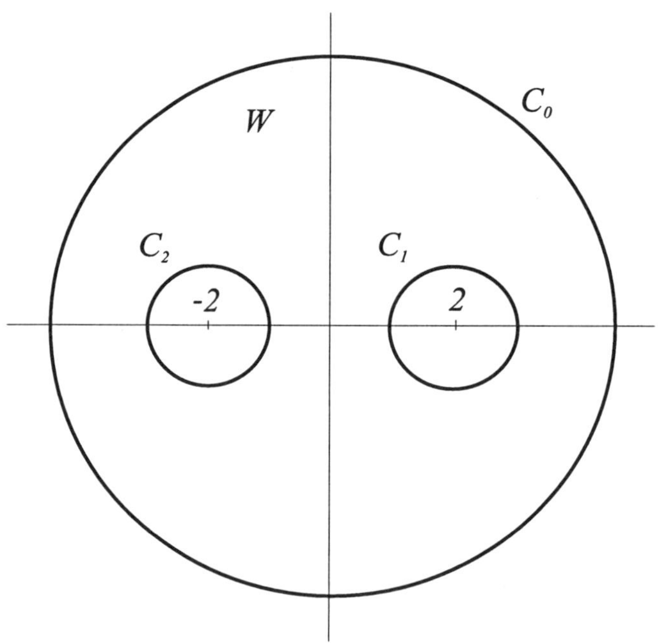

Fig. 1.   Projection of the surface with genus two

As a function $\alpha : \mathbb{R}^2 \to \mathbb{R}$ satisfying the conditions (2), (3), (4) and (5) in §4.1, we choose

$$\alpha(x,y) = (x^2 + y^2 - 16)((x - 2)^2 + y^2 - 1)(((x + 2)^2 + y^2 - 1),$$

and set

$$F(x, y, \lambda) = \frac{1}{3}\lambda^3 + \alpha(x, y)\lambda + c\alpha(x, y), \quad c \gg 1,$$

c being large enough so that $\lambda + c > 0$ on

$$\widetilde{L} = \{(x, y, \lambda) \mid \frac{\partial F}{\partial \lambda}(x, y, \lambda) = \lambda^2 + \alpha(x, y) = 0\}.$$

Then $\widetilde{L}$ is a compact orientable surface with genus 2. By Proposition 4.1, the mapping

$$i : \widetilde{L} \to T\mathbb{R}^2, \qquad i(x, y, \lambda) = (x, y, \frac{\partial F}{\partial y}(x, y, \lambda), -\frac{\partial F}{\partial x}(x, y, \lambda)),$$

is a smoothly integrable Lagrangian immersion.

Equilibrium points of $i(\widetilde{L})$ as an implicit Hamiltonian system are the images, under the Lagrangian immersion $i : \widetilde{L} \to T\mathbb{R}^2$, of the equilibrium points of the vector field $\widetilde{\xi}$ tangent to $\widetilde{L}$, which is defined by

$$\widetilde{\xi}(x, y, \lambda) = \frac{\partial F}{\partial y}(x, y, \lambda)\frac{\partial}{\partial x} - \frac{\partial F}{\partial x}(x, y, \lambda)\frac{\partial}{\partial y} + \mu(x, y, \lambda)\frac{\partial}{\partial \lambda}.$$

We express $\widetilde{L}$ as the union of

$$\widetilde{L}_+ = \{(x, y, \lambda) \in \widetilde{L} \mid \lambda \geq 0\} \quad \text{and} \quad \widetilde{L}_- = \{(x, y, \lambda) \in \widetilde{L} \mid \lambda \leq 0\}.$$

We consider the projections to $\mathbb{R}^2$ of the restricted vector fields $\widetilde{\xi}_{\widetilde{L}_+}$ and $\widetilde{\xi}_{\widetilde{L}_-}$:

$$\bar{\xi}_+(x, y) = \frac{\partial F}{\partial y}(x, y, \sqrt{-\alpha(x, y)})\frac{\partial}{\partial x} - \frac{\partial F}{\partial x}(x, y, \sqrt{-\alpha(x, y)})\frac{\partial}{\partial y},$$

$$\bar{\xi}_-(x, y) = \frac{\partial F}{\partial y}(x, y, -\sqrt{-\alpha(x, y)})\frac{\partial}{\partial x} - \frac{\partial F}{\partial x}(x, y, -\sqrt{-\alpha(x, y)})\frac{\partial}{\partial y}.$$

Then we have : The equilibrium points of $\bar{\xi}_+$ and $\bar{\xi}_-$ coincide and they are

$$(0, 0), \quad (0, \pm\sqrt{\frac{29}{3}}) \quad \text{and} \quad (\pm\sqrt{13}, 0).$$

In total, there are 10 equilibrium points of $\widetilde{\xi}$. The indices of $\bar{\xi}_+$ and $\bar{\xi}_-$ at the same equilibrium points coincide and they are

$$\begin{cases} -1 & \text{at } (0, 0), \\ +1 & \text{at } (0, \pm\sqrt{\frac{29}{3}}), \\ -1 & \text{at } (\pm\sqrt{13}, 0). \end{cases}$$

Thus, the sum of the indices of the equilibrium points of $\bar{\xi}$ is equal to

$$2 \times (-1) + 4 \times (+1) + 4 \times (-1) = -2 = \chi(\widetilde{L}).$$

Now we investigate the self-intersection points of the immersion $i : \widetilde{L} \to T\mathbb{R}^2$. Set

$$\lambda_+(x, y) = +\sqrt{-\alpha(x, y)}, \qquad \lambda_-(x, y) = -\sqrt{-\alpha(x, y)}.$$

The Lagrangian immersion $i : \widetilde{L} \to T\mathbb{R}^2$ intersects itself exactly at the images of the equilibrium points:

$$i(0, 0, \lambda_+(0, 0)) = i(0, 0, \lambda_-(0, 0)),$$

$$i(0, \pm\sqrt{\frac{29}{3}}, \lambda_+(0, \pm\sqrt{\frac{29}{3}})) = i(0, \pm\sqrt{\frac{29}{3}}, \lambda_-(0, \pm\sqrt{\frac{29}{3}}))$$

$$i(\pm\sqrt{13}, 0, \lambda_+(\pm\sqrt{13}, 0)) = i(\pm\sqrt{13}, 0, \lambda_-(\pm\sqrt{13}, 0)).$$

The indices of the self-intersections are

$$\begin{cases} +1 & \text{at} \quad i(0, 0, \lambda_+(0, 0)) = i(0, 0, \lambda_-(0, 0)), \\ -1 & \text{at} \quad i(0, \pm\sqrt{\frac{29}{3}}, \lambda_+(0, \pm\sqrt{\frac{29}{3}})) = i(0, \pm\sqrt{\frac{29}{3}}, \lambda_-(0, \pm\sqrt{\frac{29}{3}})), \\ +1 & \text{at} \quad i(\pm\sqrt{13}, 0, \lambda_+(\pm\sqrt{13}, 0)) = i(\pm\sqrt{13}, 0, \lambda_-(\pm\sqrt{13}, 0)). \end{cases}$$

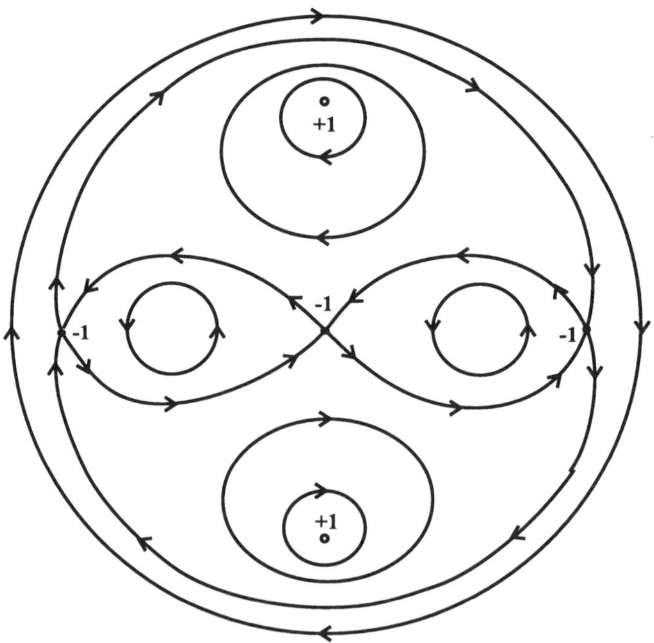

Fig. 2.   Phase curves determined by the configuration of singularities

Thus, twice the sum of the indices of the local self-intersections of $i : \widetilde{L} \to T\mathbb{R}^2$ is equal to

$$2 \times \{1 \times (+1) + 2 \times (-1) + 2 \times (+1)\} = 2 = -\chi(\widetilde{L}),$$

as Audin's theorem asserts (cf. [3] and Figure 2).

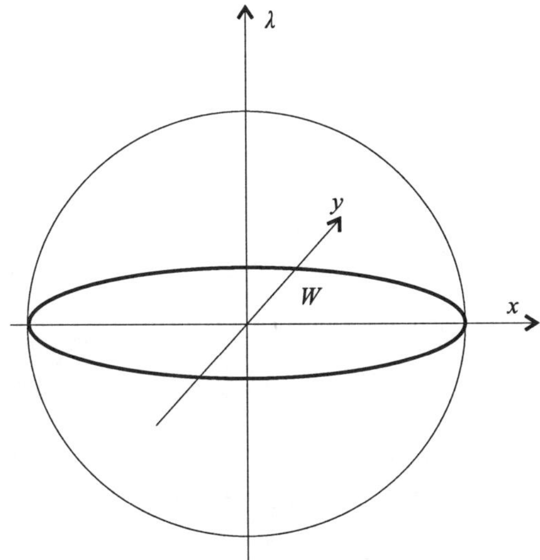

Fig. 3.   Immersion of a sphere

4.3. *Example: An integrable Lagrangian immersion of a sphere*

Now according to Proposition 4.1, we take (see Figure 3)

$$W = \{(x,y) \in \mathbb{R}^2 \mid x^2 + y^2 - 1 \le 0\}, \qquad \partial W = \{(x,y) \in \mathbb{R}^2 \mid x^2 + y^2 - 1 = 0\}.$$

As a function $\alpha : \mathbb{R}^2 \to \mathbb{R}$ satisfying the conditions (2), (3), (4) and (5) in § 4.1, we choose $\alpha(x,y) = x^2 + y^2 - 1$ and set

$$F(x,y,\lambda) = \frac{1}{3}\lambda^3 + \alpha(x,y)\lambda + 2x\alpha(x,y).$$

Then we have

$$\widetilde{L} = \{(x,y,\lambda) \mid \frac{\partial F}{\partial \lambda}(x,y,\lambda) = \lambda^2 + \alpha(x,y) = \lambda^2 + x^2 + y^2 - 1 = 0\}.$$

608

Thus $\widetilde{L}$ is the standard unit sphere S^2. By Proposition 4.1, the mapping

$$i : \widetilde{L} \to T\mathbb{R}^2, \qquad i(x, y, \lambda) = (x, y, \frac{\partial F}{\partial y}(x, y, \lambda), -\frac{\partial F}{\partial x}(x, y, \lambda))$$

is a smoothly integrable Lagrangian immersion.

Equilibrium points of $i(\widetilde{L})$ as an implicit Hamiltonian system are the images, under the Lagrangian immersion $i : \widetilde{L} \to T\mathbb{R}^2$, of the equilibrium points of the vector field $\widetilde{\xi}$ tangent to $\widetilde{L}$, which is defined by

$$\widetilde{\xi}(x, y, \lambda) = \frac{\partial F}{\partial y}(x, y, \lambda)\frac{\partial}{\partial x} - \frac{\partial F}{\partial x}(x, y, \lambda)\frac{\partial}{\partial y} + \mu(x, y, \lambda)\frac{\partial}{\partial \lambda},$$

where μ is uniquely chosen so that $\widetilde{\xi}$ is tangent to $\widetilde{L}$.

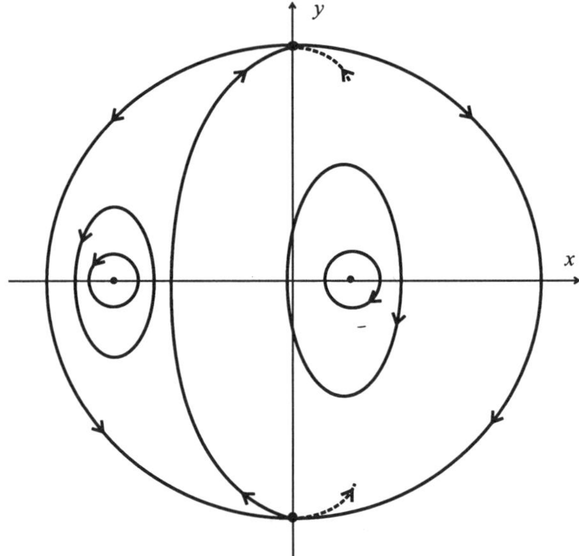

Fig. 4.   Phase curves of $\bar{\xi}_+(x, y)$, $\lambda > 0$

Then we have

Fact   *There are 6 equilibrium points of $\widetilde{\xi}$ and they are*

$$\text{for} \quad \lambda > 0, \qquad (-\sqrt{\tfrac{1}{2}}, 0, \sqrt{\tfrac{1}{2}}), \quad (\sqrt{\tfrac{1}{5}}, 0, \sqrt{\tfrac{4}{5}}),$$

$$\text{for} \quad \lambda < 0, \qquad (\sqrt{\tfrac{1}{2}}, 0, -\sqrt{\tfrac{1}{2}}), \quad (-\sqrt{\tfrac{1}{5}}, 0, -\sqrt{\tfrac{4}{5}}),$$

$$\text{for} \quad \lambda = 0, \qquad (0, +1, 0), \qquad (0, -1, 0).$$

The indices of $\widetilde{\xi}$ at the equilibrium points are

$$
\begin{cases}
+1 & \text{at} \quad (-\sqrt{\tfrac{1}{2}}, 0, \sqrt{\tfrac{1}{2}}) \quad \text{and} \quad (\sqrt{\tfrac{1}{5}}, 0, \sqrt{\tfrac{4}{5}}), \\
+1 & \text{at} \quad (\sqrt{\tfrac{1}{2}}, 0, -\sqrt{\tfrac{1}{2}}) \quad \text{and} \quad (-\sqrt{\tfrac{1}{5}}, 0, -\sqrt{\tfrac{4}{5}}), \\
-1 & \text{at} \quad (0, 1, 0) \quad \text{and} \quad (0, -1, 0).
\end{cases}
$$

Thus, the sum of the indices of the equilibrium points of $\widetilde{\xi}$ is equal to

$$
4 - 2 = 2 = \chi(\widetilde{L} = S^2).
$$

With this information, knowing the values of $\widetilde{\xi}$ at several points and the zero locus $F^{-1}(0)$, we have the corresponding phase portraits of $\bar{\xi}_{+}, \bar{\xi}_{-}$ and $\widetilde{\xi}$ (see Figures 4, 5 and 7).

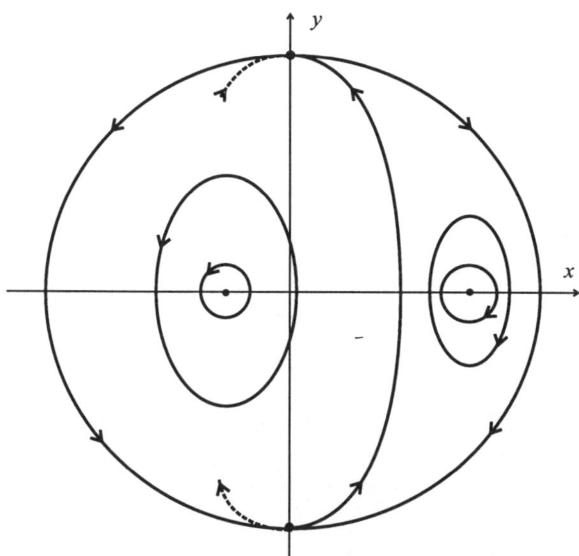

Fig. 5.   Phase curves of $\bar{\xi}_{-}(x, y)$, $\lambda < 0$

Now we investigate the self-intersection points of the immersion $i : \widetilde{L} \to T\mathbb{R}^2$.

Fact   *The Lagrangian immersion $i : \widetilde{L} \to T\mathbb{R}^2$ intersects itself exactly at one point $i(0, 0, 1) = i(0, 0, -1) = (0, 0, 0, 2)$ (Figure 6). The index of self-intersection at this point is equal to -1.*

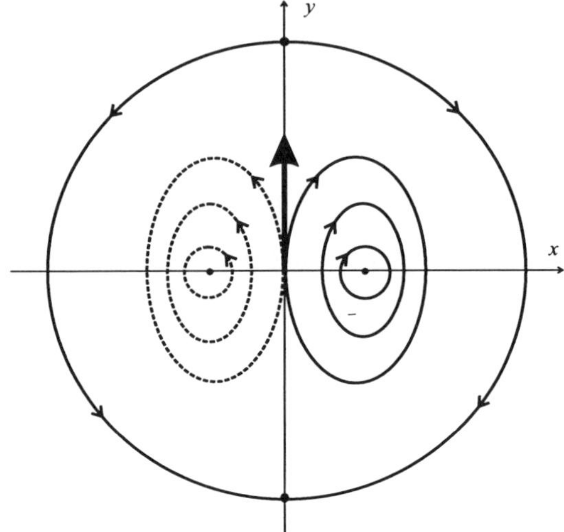

Fig. 6.   Self-intersection point

Thus, twice the sum of the indices of the local self-intersections is equal to $-2 = -\chi(\widetilde{L})$, as Audin's theorem asserts (cf. [3]).

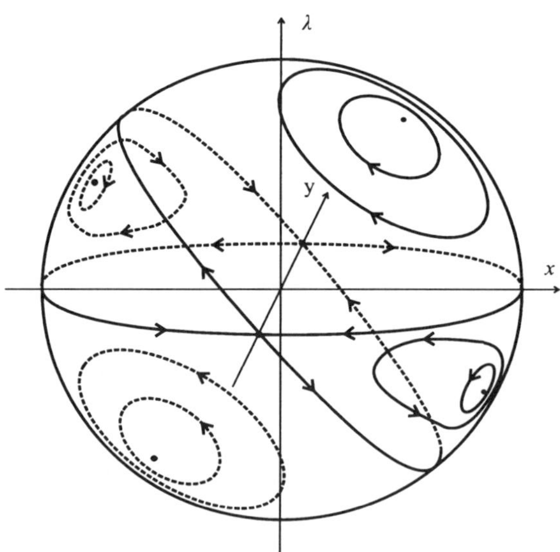

Fig. 7.   Phase curves of $\widetilde{\xi}$ on the sphere $\widetilde{L}$

Acknowledgements. The authors are grateful to Goo Ishikawa for helpful discussions and suggestions.

References

1. V. I. Arnold, S. M. Gusein-Zade, A. N. Varchenko, *Singularities of Differentiable Maps*, Vol. 1, Birkhäuser, Boston, 1985.
2. V.I. Arnol'd, *First steps in symplectic topology*, Russian Math. Surveys **41**, (6), (1986), 1-21.
3. M. Audin, *le fibrés normaux d'immersions en dimension double, points doubles d'immersions lagrangiennes et plongements totalement réels*, Comment. Math. Helv. **63**, No. 4, (1988), 593-623.
4. M. Audin, J. Lafontaine, *Holomorphic Curves in Symplectic Geometry*, Progress in Math. Vol. 117, Birkhäuser, 1994.
5. P.A.M. Dirac, *Generalized Hamiltonian Dynamics* , Canadian J. Math. **2**, (1950), 129-148.
6. I. Ekeland, H. Hofer, *Symplectic topology and Hamiltonian dynamics*, Math. Zeitschrift **200**, (1989),355-378.
7. T. Fukuda, *Local topological properties of differentiable mappings I*, Invent. Math. **65**, (1981), 227-250.
8. T. Fukuda, S. Janeczko, *Singularities of implicit differential systems and their integrability*, Banach Center Publications, **65**, (2004), 23-47.
9. A.B. Givental, *Lagrangian embeddings of surfaces and unfolded Whitney umbrella*, Funktsional Anal. i Prilozhen. **20-3**, (1986), 35-41.
10. S. Janeczko, *On implicit Lagrangian differential systems*, Annales Polonici Mathematici, **74**, (2000),133-141
11. S. Janeczko, F. Pelletier, *Singularities of implicit differential systems and Maximum Principle*, Banach Center Publications, **62,** (2004), 117-132.
12. J. Martinet, *Singularities of Smooth Functions and Maps*, Cambridge Univ. Press, Cambridge, 1982.
13. J.N. Mather, *Solutions of generic linear equations*, Dynamical Systems, (1972),185-193.
14. F. Takens, *Implicit differential equations: some open problems*, Springer Lecture Notes in Mathematics, Vol. **535**, (1976), 237-253.
15. R. Thom, *Sur les équations différentielles multiformes et leurs intégrales singuliéres*, Colloque E. Cartan, Paris, 1971.
16. A. Weinstein, *Lectures on Symplectic Manifolds*, CBMS Regional Conf. Ser. in Math., 29, AMS, Providence, R.I. 1977.

MOTIVIC VANISHING CYCLES AND APPLICATIONS

Gil Guibert

Lycée Chaptal, 45 boulevard des Batignolles, 75008 Paris, France
E-mail: guibert9@wanadoo.fr

We present here a survey on the motivic Milnor fiber introduced by Denef-Loeser and some recent results about it. In particular, we give an application of this notion to a conjecture of Steenbrink concerning the spectrum of certain hypersurface singularities. The result uses iterated motivic vanishing cycles.

Keywords: motivic Milnor fiber, iterated vanishing cycles, motivic convolution, Hodge spectrum.

1. Motivic Milnor fiber

1.1. *Arc spaces*

Let X a connected algebraic variety of pure dimension d over a field k. We denote as usual by $\mathcal{L}_n(X)$ the space of arcs of order n, also known as the n-th jet space on X (see [4] or [10] for a survey). It is a k-scheme whose K-points, for K a field containing k, is the set of morphisms $\varphi :$ $\operatorname{Spec} K[t]/t^{n+1} \to X$. There are canonical morphisms $\mathcal{L}_{n+1}(X) \to \mathcal{L}_n(X)$ which are $\mathbf{A}_k^d$-bundles when X is smooth of pure dimension d (here $\mathbf{A}_k^d$ stands for the d-dimensional affine space $\operatorname{Spec} k[X_1, \ldots, X_d]$).

The arc space $\mathcal{L}(X)$ is defined as the projective limit of this system. We denote by $\pi_n : \mathcal{L}(X) \to \mathcal{L}_n(X)$ the canonical morphism. In particular, we have a canonical morphism $\pi_0 : \mathcal{L}(X) \to \mathcal{L}_0(X) \simeq X$. There is a canonical action of the multiplicative group $\mathbf{G}_m = \mathbf{A}_k^1 \setminus \{0\}$ on $\mathcal{L}_n(X)$ and on $\mathcal{L}(X)$ given by $a \cdot \varphi(t) = \varphi(at)$.

For an element φ in $K[[t]]$ or in $K[t]/t^{n+1}$, we denote by $\operatorname{ord}_t(\varphi)$ the valuation of φ and by $\operatorname{ac}(\varphi)$ its first non zero coefficient, with the convention $\operatorname{ac}(0) = 0$.

614

1.2. *Grothendieck rings*

We fix here a variety S over k. We will consider varieties Y over $S \times \mathbf{G}_m^r$ with good $\mathbf{G}_m^r$-action (in the sense of [4]), such that the morphism to S has invariant fibers under the action of $\mathbf{G}_m^r$ and that the morphism $\pi : Y \to \mathbf{G}_m^r$ is diagonally monomial of weight $\mathbf{n}$ in $\mathbf{N}_{>0}^r$, that is $\pi(\lambda x) = \lambda^{\mathbf{n}} \pi(x)$ for all λ in $\mathbf{G}_m^r$ and x in Y. We will note $\mathrm{Var}_{S \times \mathbf{G}_m^r}$ the category of these varieties with the corresponding morphisms.

If $\mathbf{n} = \mathbf{km}$ (the multiplication in $\mathbf{N}_{>0}^r$ being componentwise), we have a natural functor

$$\theta_{\mathbf{n}}^{\mathbf{m}} : \mathrm{Var}_{S \times \mathbf{G}_m^r}^{\mathbf{G}_m^r, \mathbf{m}} \longrightarrow \mathrm{Var}_{S \times \mathbf{G}_m^r}^{\mathbf{G}_m^r, \mathbf{n}}, \tag{1}$$

sending $X \to S \times \mathbf{G}_m^r$ to the same object, but with the action $\lambda \mapsto \lambda x$ on X replaced by $\lambda \mapsto \lambda^{\mathbf{k}} x$. We define the category $\mathrm{Var}_{S \times \mathbf{G}_m^r}^{\mathbf{G}_m^r}$ as the colimit of the inductive system of categories $\mathrm{Var}_{S \times \mathbf{G}_m^r}^{\mathbf{G}_m^r, \mathbf{n}}$. We define $K_0(\mathrm{Var}_{S \times \mathbf{G}_m^r}^{\mathbf{G}_m^r})$ as the free abelian group on isomorphism classes of objects $Y \to S \times \mathbf{G}_m^r$ in $\mathrm{Var}_{S \times \mathbf{G}_m^r}^{\mathbf{G}_m^r}$, modulo the relations

$$[Y \to S \times \mathbf{G}_m^r] = [Y' \to S \times \mathbf{G}_m^r] + [Y \backslash Y' \to S \times \mathbf{G}_m^r] \quad (\text{"additivity relation"}) \tag{2}$$

for Y' closed $\mathbf{G}_m^r$-invariant in Y and, for $f : Y \to S \times \mathbf{G}_m^r$ in $\mathrm{Var}_{S \times \mathbf{G}_m^r}^{\mathbf{G}_m^r, \mathbf{n}}$,

$$[Y \times \mathbf{A}_k^n \to S \times \mathbf{G}_m^r, \sigma] = [Y \times \mathbf{A}_k^n \to S \times \mathbf{G}_m^r, \sigma'] \tag{3}$$

if σ and σ' are two liftings of the same $\mathbf{G}_m^r$-action on Y to affine actions, the morphism $Y \times \mathbf{A}_k^n \to S \times \mathbf{G}_m^r$ being the composition of f with the projection on the first factor.

Fiber product over $S \times \mathbf{G}_m^r$ with diagonal action induces a product in the category $\mathrm{Var}_{S \times \mathbf{G}_m^r}^{\mathbf{G}_m^r, \mathbf{n}}$, which allows to endow $K_0(\mathrm{Var}_{S \times \mathbf{G}_m^r}^{\mathbf{G}_m, \mathbf{n}})$ with a natural ring structure. By external product, there is a natural structure of $K_0(\mathrm{Var}_k)$-module on $K_0(\mathrm{Var}_{S \times \mathbf{G}_m^r}^{\mathbf{G}_m^r, \mathbf{n}})$.

We denote by $\mathbf{L}$ the class of the affine line in $K_0(\mathrm{Var}_k)$ and by $\mathbf{L}_{S \times \mathbf{G}_m^r}$ the element $\mathbf{L} \cdot 1_{S \times \mathbf{G}_m^r}$ in the module $K_0(\mathrm{Var}_{S \times \mathbf{G}_m^r}^{\mathbf{G}_m^r, \mathbf{n}})$. When the context is clear we will simply write $\mathbf{L}$ instead of $\mathbf{L}_{S \times \mathbf{G}_m^r}$. We define $\mathcal{M}_{S \times \mathbf{G}_m^r}^{\mathbf{G}_m^r, \mathbf{n}}$ as the localisation of $K_0(\mathrm{Var}_{S \times \mathbf{G}_m^r}^{\mathbf{G}_m^r, \mathbf{n}})$ with respect to the multiplicative system $\{\mathbf{L}^n , n \in \mathbf{N}\}$, namely

$$\mathcal{M}_{S \times \mathbf{G}_m^r}^{\mathbf{G}_m^r, \mathbf{n}} = K_0(\mathrm{Var}_{S \times \mathbf{G}_m^r}^{\mathbf{G}_m^r, \mathbf{n}})[\mathbf{L}^{-1}]$$

If $f : S \to S'$ is a morphism of varieties, composition with f leads for each n to a push-forward morphism $f_! : \mathcal{M}_{S \times \mathbf{G}_m^r}^{\mathbf{G}_m^r, \mathbf{n}} \to \mathcal{M}_{S' \times \mathbf{G}_m^r}^{\mathbf{G}_m^r, \mathbf{n}}$, while fiber product leads to a pull-back morphism $f^* : \mathcal{M}_{S' \times \mathbf{G}_m^r}^{\mathbf{G}_m^r,\mathbf{n}} \to \mathcal{M}_{S \times \mathbf{G}_m^r}^{\mathbf{G}_m^r,\mathbf{n}}$ (these morphisms may already be defined at the K_0-level) ; by passing to the colimit we get functors $f_!$ and f^* defined on $\mathcal{M}_{S \times \mathbf{G}_m^r}^{\mathbf{G}_m^r}$ and $\mathcal{M}_{S' \times \mathbf{G}_m^r}^{\mathbf{G}_m^r}$ respectively.

More generally if W is a constructible subset of Y stable by the $\mathbf{G}_m^r$-action, we shall call a morphism $\pi : W \to \mathbf{G}_m^r$ *piecewise monomial* if there is a finite partition of W into locally closed $\mathbf{G}_m^r$-invariant subsets on which the restriction of π is a monomial morphism. To such a W endowed with a morphism $(p, \pi) : W \to S \times \mathbf{G}_m^r$ such that the fibers of $p : W \to S$ are $\mathbf{G}_m^r$-invariant and $\pi : W \to \mathbf{G}_m^r$ is piecewise monomial, we assign by additivity a class $[(p, \pi) : W \to S \times \mathbf{G}_m^r]$ in $\mathcal{M}_{S \times \mathbf{G}_m^r}^{\mathbf{G}_m^r}$.

1.3. *Motivic zeta function and motivic Milnor fiber*

Let us start by recalling some basic constructions introduced by Denef and Loeser.

Let X be a smooth variety over k of pure dimension d and $f : X \to \mathbf{A}_k^1$. We set $X_0(f)$ for the zero locus of f, and consider, for $n \geq 1$, the variety

$$\mathcal{X}_n(f) := \Big\{ \varphi \in \mathcal{L}_n(X) \ \Big| \ \mathrm{ord}_t f(\varphi) = n \Big\}. \tag{4}$$

Note that since $n \geq 1$, this is a $X_0(f)$-variety via the morphism π_0. Note also that $\mathcal{X}_n(f)$ is invariant by the $\mathbf{G}_m$-action on $\mathcal{L}_n(X)$. Furthermore f induces a morphism from $\mathcal{X}_n(f)$ to $\mathbf{G}_m$, assigning to a point φ in $\mathcal{L}_n(X)$ the coefficient $\mathrm{ac}(f(\varphi))$ of t^n in $f(\varphi)$, which we shall also denote by $\mathrm{ac}(f)(\varphi)$. This morphism is diagonally monomial of weight n with respect to the $\mathbf{G}_m$-action on $\mathcal{X}_n(f)$ since $\mathrm{ac}(f(a.\varphi)) = a^n \mathrm{ac}(f(\varphi))$, so we can consider the class $[\mathcal{X}_n(f)]$ of $\mathcal{X}_n(f)$ in $\mathcal{M}_{X_0(f) \times \mathbf{G}_m}^{\mathbf{G}_m}$.

We now consider the motivic zeta function

$$Z_f(T) := \sum_{n \geq 1} [\mathcal{X}_n(f)] \, \mathbf{L}^{-nd} \, T^n \tag{5}$$

in $\mathcal{M}_{X_0(f) \times \mathbf{G}_m}^{\mathbf{G}_m}[[T]]$. Note that $Z_f = 0$ if $f = 0$ on X.

Denef and Loeser showed, using a resolution of singularities of f, that $Z_f(T)$ lies is the sub-$\mathcal{M}_{X_0(f) \times \mathbf{G}_m}^{\mathbf{G}_m}$ module of $\mathcal{M}_{X_0(f) \times \mathbf{G}_m}^{\mathbf{G}_m}[[T]]$ generated by rational series of the form $\dfrac{\mathbf{L}^a T^b}{1 - \mathbf{L}^a T^b}$ (*with* (a, b) *in* $\mathbf{Z} \times \mathbf{N}$). We will note $\mathcal{M}_{X_0(f) \times \mathbf{G}_m}^{\mathbf{G}_m}[[T]]_{sr}$ this module.

Hence we can consider the so called *motivic nearby cycle class*

$$\mathcal{S}_f := -\lim_{T \mapsto \infty} Z_f(T), \text{ in } \mathcal{M}^{\mathbf{G}_m}_{X_0(f) \times \mathbf{G}_m}. \tag{6}$$

We shall also consider in this paper the *motivic vanishing cycles class* defined as

$$\mathcal{S}_f^{\phi} := (-1)^{d-1}(\mathcal{S}_f - [\mathbf{G}_m \times X_0(f)]) \tag{7}$$

in $\mathcal{M}^{\mathbf{G}_m}_{X_0(f) \times \mathbf{G}_m}$. Here $\mathbf{G}_m \times X_0(f)$ is endowed with the standard $\mathbf{G}_m$-action on the first factor and the trivial $\mathbf{G}_m$-action on the second factor.

If x is any closed point of $X_0(f)$ and i_x is the inclusion of x in X, we set $\mathcal{S}_{f,x} = i_x^*(\mathcal{S}_f^{\phi})$: this is the *motivic Milnor fiber of f at x*. We will also consider $\mathcal{S}_{f,x}^{\phi} = i_x^*(\mathcal{S}_f^{\phi})$.

2. Motivation

Denef and Loeser ([2] corollary 4.3.1, see also [4]) proved that usual invariant of the singularity of f at x can be recovered from the motivic Milnor fiber. For example we have the following result :

Theorem 2.1. *Assuming the previous notations and designing by $sp(f,x)$ the Steenbrink's Hodge spectrum of f at x (see [14], [15] or [4] subsection 3.1.3), we have*

$$sp(f,x) = \text{Sp}(\mathcal{S}_{g,x}^{\phi}) \tag{8}$$

where $Sp : \mathcal{M}^{\mathbf{G}_m}_{\mathbf{G}_m} \longrightarrow \mathbf{Z}[\mathbf{Q}]$ is a suitable group morphism.

It is then quite natural to use arc spaces to compute invariants of singularities of hypersurfaces and to look for motivic analogue of Milnor fiber-theoretical results. For example Denef and Loeser have shown ([3] Theorem 5.2.2, see also [10]) the following Thom-Sebastiani property for the motivic Milnor fiber :

Theorem 2.2. *Let X and Y be smooth varieties of pure dimension, and consider functions $f : X \to \mathbf{A}_k^1$ and $g : Y \to \mathbf{A}_k^1$. Denote by $f \oplus g$ the function on $X \times Y$ sending (u,v) to $f(u) + g(v)$. Then for any closed point (x,y) in $X \times Y$, we have*

$$\mathcal{S}_{f \oplus g,(x,y)}^{\phi} = \mathcal{S}_{f,x}^{\phi} * \mathcal{S}_{g,y}^{\phi} \text{ in } \mathcal{M}^{\mathbf{G}_m}_{\mathbf{G}_m}. \tag{9}$$

where $*$ is a binary operator on $\mathcal{M}^{\mathbf{G}_m}_{\mathbf{G}_m}$ called convolution product.

They recovered from this result the usual Thom-Sebastiani theorems.

In a similar way we consider the following situation, studied by Iomdin, Steenbrink, Siersma and Saito (in [8], [15], [13] and [12] respectively) : let f a regular function on a complex algebraic variety X, let x a closed point in $X_0(f)$, and assume that the singular locus of f is a curve Γ, having r local components Γ_ℓ, $1 \leq \ell \leq r$, in a neighborhood of x. We denote by m_ℓ the multiplicity of Γ_ℓ. Let g be a generic linear form vanishing at x (that is, a function g vanishing at x whose differential at x is a generic linear form). For N large enough, the function $f + g^N$ has an isolated singularity at x. In a neighborhood of the complement Γ_ℓ° to $\{x\}$ in Γ_ℓ, we may view f as a family of isolated hypersurface singularities parametrized by Γ_ℓ°. The cohomology of the Milnor fiber of this hypersurface singularity is naturally endowed with the action of two commuting monodromies: the monodromy of the function and the monodromy of a generator of the local fundamental group of Γ_ℓ°. We denote by $\alpha_{\ell,j}$ the *exponents* (see [15] for a definition) of that isolated hypersurface singularity (recall that they form a finite set of rational numbers) and by $\beta_{\ell,j}$ the corresponding rational numbers in $[0,1)$ such that the complex numbers $\exp(2\pi i \beta_{\ell,j})$ are the eigenvalues of the monodromy along Γ_ℓ°. The following result was conjectured by Steenbrink [15] and proved by Saito [12].

Theorem 2.3 (Saito). *For $N \gg 0$,*

$$\mathrm{Sp}(f + g^N, x) - \mathrm{Sp}(f, x) = \sum_{\ell,j} t^{\alpha_{\ell,j} + (\beta_{\ell,j}/m_\ell N)} \frac{1-t}{1 - t^{1/m_\ell N}}. \tag{10}$$

If we want to state a motivic analogue of this formula, we have first to define a motivic analogue of the second term of the previous equality. Let us begin by giving some generalisation of the motivic vanishing cycle construction.

3. Motivic vanishing cycles morphism on the Grothendieck group

We will now briefly recall how to extend $\mathcal{S}_g$ to the whole Grothendieck group $\mathcal{M}_X$ in such a way that $\mathcal{S}_g([X \to X])$ is equal to $\mathcal{S}_g$. Such an extension has also been considered with a different approach by Bittner in [1]. We refer to [6] (section 3.7) for details.

3.1. *A modified zeta function*

Let X be a smooth variety of pure dimension d and let U be a dense open in X. Consider a function $g : X \to \mathbf{A}^1_k$. We start by defining $\mathcal{S}_g([U \to X])$.

We denote by F the closed subset $X \setminus U$ and by $\mathcal{I}_F$ the ideal of functions vanishing on F.

Fix $\gamma \geq 1$ a positive integer. We will consider the modified zeta function $Z^\gamma_{g,U}(T)$ defined as follows. For $n \geq 1$, we consider the constructible set

$$\mathcal{X}^{\gamma n}_n(g,U) := \left\{ \varphi \in \mathcal{L}_{\gamma n}(X) \;\middle|\; \mathrm{ord}_t g(\varphi) = n, \mathrm{ord}_t \varphi^*(\mathcal{I}_F) \leq \gamma n \right\}. \tag{11}$$

As in 1.3, we consider the morphism $\mathcal{X}^{\gamma n}_n(g,U) \to \mathbf{G}_m$ induced by $\varphi \mapsto \mathrm{ac}(g(\varphi))$. It is piecewise monomial, so we can consider the class $[\mathcal{X}^{\gamma n}_n(g,U)]$ in $\mathcal{M}^{\mathbf{G}_m}_{X_0(g) \times \mathbf{G}_m}$. We set

$$Z^\gamma_{g,U}(T) := \sum_{n \geq 1} [\mathcal{X}^{\gamma n}_n(g,U)] \, \mathbf{L}^{-\gamma n d} \, T^n \tag{12}$$

in $\mathcal{M}^{\mathbf{G}_m}_{X_0(g) \times \mathbf{G}_m}[[T]]$. Note that for $U = X$, $Z^\gamma_{g,U}(T)$ is equal to $Z_g(T)$ for every γ, since in this case, $[\mathcal{X}^{\gamma n}_n(g,U)] \mathbf{L}^{-\gamma n d} = [\mathcal{X}_n(g)] \mathbf{L}^{-nd}$.

This equality follows from the fact that when $U = X$ and X is smoooth, $\mathcal{X}_n(g)$ is the image of $\mathcal{X}^{\gamma n}_n(g,U)$ by the canonical projection from $\mathcal{L}_{\gamma n}(X)$ to $\mathcal{L}_n(X)$ which is a $\mathbf{A}^{(\gamma n - n)d}_k$-bundle in this case.

Note also that $Z^\gamma_{g,U}(T) = 0$ if g is identically zero on X.

This modified zeta function allows to construct the motivic nearby cycle class of a function on an open set ; after that, we can extend this construction to the whole Grothendieck group of varieties over X. The following results are respectively proposition 3.8 and theorem 3.9 of [6] :

Prop 3.1. Let U be a dense open in the smooth variety X of pure dimension d with a function $g : X \to \mathbf{A}^1_k$. There exists γ_0 such that for every $\gamma > \gamma_0$ the series $Z^\gamma_{g,U}(T)$ lies in $\mathcal{M}^{\mathbf{G}_m}_{X_0(g) \times \mathbf{G}_m}[[T]]_{\mathrm{sr}}$ and $\lim_{T \mapsto \infty} Z^\gamma_{g,U}(T)$ is independent of $\gamma > \gamma_0$. We set $\mathcal{S}_{g,U} = - \lim_{T \mapsto \infty} Z^\gamma_{g,U}(T)$.

Theorem 3.1 (Extension to the Grothendieck group). *Let X be a variety with a function $g : X \to \mathbf{A}^1_k$. There exists a unique $\mathcal{M}_k$-linear group morphism*

$$\mathcal{S}_g : \mathcal{M}_X \longrightarrow \mathcal{M}^{\mathbf{G}_m}_{X_0(g) \times \mathbf{G}_m} \tag{13}$$

such that, for every proper morphism $p : Z \to X$, with Z smooth, and every dense open subset U in Z,

$$\mathcal{S}_g([U \to X]) = p_!(\mathcal{S}_{g \circ p, U}). \tag{14}$$

This result admits an equivariant version for varieties with $\mathbf{G}_m^r$-action ([6] Theorem 3.12).

4. Iterated vanishing cycles and convolution

4.1. *Iterated vanishing cycles*

Now we consider a smooth variety X of pure dimension d with two functions $f : X \to \mathbf{A}_k^1$ and $g : X \to \mathbf{A}_k^1$. The motivic Milnor fiber $\mathcal{S}_f$ lies in $\mathcal{M}_{X_0(f) \times \mathbf{G}_m}^{\mathbf{G}_m}$. We still denote by g the function $X_0(f) \times \mathbf{G}_m \to \mathbf{A}_k^1$ obtained by composition of g with the projection $X_0(f) \times \mathbf{G}_m \to X$. Hence, thanks to previous construction (in the equivariant context), we may consider the image

$$\mathcal{S}_g(\mathcal{S}_f) = \mathcal{S}_g(\mathcal{S}_f([X \to X])) \tag{15}$$

of $\mathcal{S}_f = \mathcal{S}_f([X \to X])$ by the nearby cycles morphism

$$\mathcal{S}_g : \mathcal{M}_{X_0(f) \times \mathbf{G}_m}^{\mathbf{G}_m} \longrightarrow \mathcal{M}_{(X_0(f) \cap X_0(g)) \times \mathbf{G}_m^2}^{\mathbf{G}_m^2} \tag{16}$$

We will refer to this virtual variety as the *motivic iterated vanishing cycle class* of g and f.

4.2. *Convolution*

Let us denote by a and b the coordinates on each factor of $\mathbf{G}_m^2$. Let X be a variety. We denote by $i : X \times (a + b)^{-1}(0) \to X \times \mathbf{G}_m^2$ the inclusion of the antidiagonal and by j the inclusion of its complement. We consider the morphism

$$a + b : X \times \mathbf{G}_m^2 \setminus (a + b)^{-1}(0) \longrightarrow X \times \mathbf{G}_m \tag{17}$$

which is the identity on the X-factor and is equal to $a+b$ on the $\mathbf{G}_m^2 \setminus (a+b)^{-1}(0)$-factor. We denote by pr_1 and pr_2 the projection of $X \times \mathbf{G}_m \times (a+b)^{-1}(0)$ on $X \times \mathbf{G}_m$ and $X \times (a + b)^{-1}(0)$, respectively.

If A is an object in $\mathcal{M}_{X \times \mathbf{G}_m^2}$, the object

$$\Psi_\Sigma^0(A) := -(a + b)_! j^*(A) + \mathrm{pr}_{1!} \mathrm{pr}_2^* i^*(A) \tag{18}$$

lives in $\mathcal{M}_{X \times \mathbf{G}_m}$.

By [6] (section 5.1), when A is in $\mathcal{M}_{X \times \mathbf{G}_m^2}^{\mathbf{G}_m^2, (n,m)}$, $\Psi_\Sigma^0(A)$ can be naturally viewed as an object in $\mathcal{M}_{X \times \mathbf{G}_m}^{\mathbf{G}_m, nm}$. By passing to the the colimit, Ψ_Σ^0 lifts to a $\mathcal{M}_k$-linear group morphism $\Psi_\Sigma : \mathcal{M}_{X \times \mathbf{G}_m^2}^{\mathbf{G}_m^2} \to \mathcal{M}_{X \times \mathbf{G}_m}^{\mathbf{G}_m}$.

620

5. The main theorem

Let us consider again a smooth variety X of pure dimension d with two functions f and g from X to $\mathbf{A}_k^1$. Let us denote by i_1 and i_2 the inclusion of $(X_0(f) \cap X_0(g)) \times \mathbf{G}_m$ in $X_0(f) \times \mathbf{G}_m$ and $X_0(f + g^N) \times \mathbf{G}_m$, respectively.

We can now state the main result (see [6] Theorem 5.7) :

Theorem 5.1. *Let X be a smooth variety of pure dimension d, and f and g be two functions from X to $\mathbf{A}_k^1$. For every $N >> 0$, the equality*

$$i_1^* \mathcal{S}_f^\phi - i_2^* \mathcal{S}_{f+g^N}^\phi = \Psi_\Sigma(\mathcal{S}_{g^N}(\mathcal{S}_f^\phi)) \tag{19}$$

holds in $\mathcal{M}_{(X_0(f) \cap X_0(g)) \times \mathbf{G}_m}^{\mathbf{G}_m}$.

Theorem 5.1 has the following local corollary:

Corollary 5.1. *Let X be a smooth variety of pure dimension d, and f and g be two functions from X to $\mathbf{A}_k^1$. Let x be a closed point of $X_0(f) \cap X_0(g)$. For every $N >> 0$, the equality*

$$\mathcal{S}_{f,x}^\phi - \mathcal{S}_{f+g^N,x}^\phi = \Psi_\Sigma(\mathcal{S}_{g^N,x}(\mathcal{S}_f^\phi)) \tag{20}$$

holds in $\mathcal{M}_{\mathbf{G}_m}^{\mathbf{G}_m}$.

The motivic Thom-Sebastiani Theorem of [3] , [10] and [4] may be deduced from Theorem 5.1.

6. Spectrum and the Steenbrink conjecture

6.1.

We now assume $k = \mathbf{C}$.

Denoting by $\mathrm{HS}^{2-\mathrm{mon}}$ the abelian category of Hodge structures endowed with two commuting automorphisms of finite order and by $K_0(\mathrm{HS}^{2-\mathrm{mon}})$ the corresponding Grothendieck ring, one has a natural ring morphism, called the Hodge characteristic (see subsection 3.1.2 of [4] for a precise definition)

$$\chi_h : \mathcal{M}_{\mathbf{G}_m^2}^{\mathbf{G}_m^2} \longrightarrow K_0(\mathrm{HS}^{2-\mathrm{mon}}) \tag{21}$$

As explained in the section 6.1 of [6], we can define a generalised Hodge spectrum hsp from $K_0(\mathrm{HS}^{2-\mathrm{mon}})$ to $\mathbf{Z}[(\mathbf{Q}/\mathbf{Z})^2 \times \mathbf{Z}]$. We shall denote by Sp the composite morphism of abelian groups

$$\mathrm{Sp} := (\mathrm{hsp} \circ \chi_h) : \mathcal{M}_{\mathbf{G}_m^2}^{\mathbf{G}_m^2} \longrightarrow \mathbf{Z}[(\mathbf{Q}/\mathbf{Z})^2 \times \mathbf{Z}]. \tag{22}$$

Now if $g : X \to \mathbf{A}^1$ is another function vanishing at x, we shall define the following generalised Hodge-Steenbrink spectrum of f and g at x :

$$\mathrm{Sp}(f, g, x) := \mathrm{Sp}(\mathcal{S}_{g,x}(\mathcal{S}_f^\phi)). \tag{23}$$

Let us denote by δ_N the morphism of abelian groups $\mathbf{Z}[(\mathbf{Q}/\mathbf{Z})^2 \times \mathbf{Z}] \to \mathbf{Z}[\mathbf{Q}]$ sending $t^a u^b v^c$ to $t^{s(a)+s(b)/N+c}$, where s is a section of the restriction to $[0,1)$ of the projection $\pi : \mathbf{Q} \to \mathbf{Q}/\mathbf{Z}$. We have the following result (proposition 6.7 of [6]) :

Proposition 6.1. *For every positive integer N, the spectrum of $\Psi_\Sigma(\mathcal{S}_{g^N,x}(\mathcal{S}_f^\phi))$ is equal to*

$$\mathrm{Sp}(\Psi_\Sigma(\mathcal{S}_{g^N,x}(\mathcal{S}_f^\phi))) = \frac{1-t}{1-t^{\frac{1}{N}}} \delta_N(\mathrm{Sp}(f, g, x)). \tag{24}$$

Hence, we deduce immediately the following statement from Corollary 5.1.

Theorem 6.1. *Let X be a smooth variety of pure dimension d, and f and g be two functions from X to $\mathbf{A}^1$. Let x be a closed point of $X_0(f) \cap X_0(g)$. Then, for $N >> 0$,*

$$\mathrm{Sp}(f, x) - \mathrm{Sp}(f + g^N, x) = \frac{1-t}{1-t^{\frac{1}{N}}} \, \delta_N(\mathrm{Sp}(f, g, x)). \tag{25}$$

7. A computation : Motivic Milnor fiber of a non-degenerate composite

Consider p algebraic varieties over k denoted $X_1, \ldots, X_p$, each one endowed with a morphism $f_j : X_j \to \mathbf{A}^1_k$. We shall denote by f the function induced on the product $X = \prod_j X_j$ by the f_j's. We denote by $X_0(\mathbf{f})$ the set of zeroes of f in X.

Let $P \in k[y_1, \ldots, y_p]$ be a polynomial, which we assume to be *non-degenerate* with respect to its Newton polyhedron in the sense of [9]. We present here a computation of the motivic nearby cycles class on the open set $U = X \setminus X_0(f)$ of the composed function $P(\mathbf{f})$ (see section 3) as a sum over the set of *compact faces* δ of the Newton polyhedron of P. For every such δ, let us denote by P_δ the corresponding quasi-homogeneous polynomial (see [9] for precise definitions). We associate to such a quasi-homogeneous polynomial a convolution operator Ψ_{P_δ}, which in the special case where P_δ is the polynomial $\Sigma = y_1 + y_2$ is nothing but the operator Ψ_Σ considered in section 4.2. For such a compact face δ, one may also define generalized nearby cycles $\mathcal{S}_{\mathbf{f}}^{\sigma(\delta)}$, constructed as the limit, as $T \mapsto \infty$, of certain truncated motivic zeta functions :

$$Z_{\mathbf{f}}^{\sigma(\delta),\ell}(\mathbf{T}) := \sum_{\mathbf{n}\in\sigma(\delta)} [\mathcal{X}_{\mathbf{n}}(\mathbf{f})]\,\mathbf{L}^{-s(\mathbf{n})d}\,T^{\ell(\mathbf{n})} \tag{26}$$

in $\mathcal{M}_{X_0(\mathbf{f})\times\mathbf{G}_m{}^p}^{\mathbf{G}_m}[[T]]$.

Here ℓ represent the piecewise linear form associated with the Newton polyhedron of P.

Our main result states :

$$i^*\mathcal{S}_{P(\mathbf{f}),U} = \sum_{\delta\in\Gamma^\emptyset} \Psi_{P_\delta}(\mathcal{S}_{\mathbf{f}}^{\sigma(\delta)}). \tag{27}$$

Here $\Gamma^\emptyset$ denotes the set of compact faces of the Newton polyhedron of P not contained in any coordinate hyperplane, $\mathcal{S}_{P(\mathbf{f}),U}$ refers to the extension of $\mathcal{S}_{P(\mathbf{f})}$ constructed in section 3 (and in [1]), and i^* denotes restriction to $X_0(\mathbf{f})$.

References

1. F. Bittner, *On motivic zeta functions and the motivic Milnor fiber*, [arXiv:math.AG/0307033].
2. J. Denef, F. Loeser, *Motivic Igusa zeta functions*, J. Algebraic Geom. **7**, (1998), 505–537.
3. J. Denef, F. Loeser, *Motivic exponential integrals and a motivic Thom-Sebastiani Theorem*, Duke Math. J. **99** (1999), 285–309.
4. J. Denef, F. Loeser, *Geometry on arc spaces of algebraic varieties*, Proceedings of 3rd European Congress of Mathematics, Barcelona 2000, Progress in Mathematics **201** (2001), 327–348, Birkhaüser.
5. G. Guibert, *Espaces d'arcs et invariants d'Alexander*, Comment. Math. Helv. **77** (2002), 783–820.
6. G. Guibert, F. Loeser, M. Merle, *Iterated vanishing cycles, convolution, and a motivic analogue of a conjecture of Steenbrink*, Duke Mathematical Journal **132** (2006), 409–457.
7. G. Guibert, F. Loeser, M. Merle, *Nearby cycles and composition with a non-degenerate polynomial*, IMRN **31** (2005), 1874-1888.
8. I.N. Iomdin, *Complex surfaces with a one-dimensional set of singularities* (Russian), Sibirsk. Mat. Ž. **15** (1974), 1061–1082, 1181, English translation: Siberian Math. J. **15** (1974), no. 5, 748–762 (1975).
9. A.G. Kouchnirenko, *Polydres de Newton et nombres de Milnor*, Invent.math. **32** (1976),1–31.

10. E. Looijenga, *Motivic Measures, Astérisque* **276** (2002), 267–297, Séminaire Bourbaki, exposé 874.

11. A. Némethi, J. Steenbrink, *Spectral pairs, mixed Hodge modules, and series of plane curve singularities, New York J. Math.* **1** (1994/95), 149–177.

12. M. Saito, *On Steenbrink's conjecture*, Math. Ann. **289** (1991), 703–716.

13. D. Siersma, *The monodromy of a series of hypersurface singularities*, Comment. Math. Helv. **65** (1990), 181–197.

14. J. Steenbrink, *Mixed Hodge structures on the vanishing cohomology*, in Real and Complex Singularities, Sijthoff and Noordhoff, Alphen aan den Rijn, 1977, 525–563.

15. J. Steenbrink, *The spectrum of hypersurface singularities*, Actes du Colloque de Théorie de Hodge (Luminy, 1987). Astérisque No. 179-180, (1989) **11**, 163–184.

COMPLEMENTS OF HYPERSURFACES AND EQUISINGULARITY

H. A. HAMM

Mathematisches Institut, Westf. Wilhelms-Universität,
Einsteinstr. 62, 48149 Münster, Germany
** E-mail: hamm@math.uni-muenster.de*

In this article we study the question of equisingularity for germs of complex spaces each of which is the complement of some hypersurface. This is motivated by the question of global equisingularity of complex affine varieties but will be related to classical local equisingularity questions, too.

Keywords: Equisingularity, Whitney stratification, Bekka condition, Thom condition.

1. Introduction

In the course of the topological investigation of singularities which has been intensified in the second half of the twentieth century it was natural to study mappings up to topological triviality, too: this is the question of equisingularity. An important step was the proof of Thom's isotopy theorem. In the seventies equisingularity has been investigated intensively in the framework of local singularity theory, asking how far it can be controlled by numerical invariants and how it is connected with notions from stratification theory. Later on one encountered a similar local question of a somewhat different nature when passing to a corresponding global equisingularity problem - topological triviality of polynomial mappings. Usually such a mapping is compactified first in order to reduce the global problem to a local one. In this way we are led to the question of local triviality outside a hypersurface (here: the hypersurface at infinity), cf. [15], [14].

In this paper we do not restrict to the special situation which arises in the study of polynomial mappings. On the other side we will for simplicity only consider the case of isolated singularities (in the stratified sense), by imposing a suitable transversality condition.

After that we will compare with the case which has been treated earlier, here the initial situation creates a somewhat different point of view. On this occasion we will take care of new developments and results: the fact that in the complex case Whitney regularity implies Thom's a_f-condition [3], [21], the possibity of finding continuous controlled vector fields ([16], [6]), and Bekka's regularity condition for stratifications [1].

Furthermore we will give a simple example which shows that the questions of triviality outside resp. including the hypersurface do not lead to the same answer (Example 3.2).

Now let us fix the situation which will be studied throughout this paper and indicate the most important results.

The general assumption will be the following:

Assumption 1.1. *Let U be an open Stein neighbourhood of 0 in $\mathbb{C}^{m+1}$, $g : \mathbb{C}^{m+1} \longrightarrow \mathbb{C} : (z_1, \ldots, z_{m+1}) \mapsto z_{m+1}$ the projection onto the last coordinate. Let X be a closed analytic subset of U of pure dimension n and Y a hypersurface in X such that $0 \in Y$. We assume that $X \setminus Y$ is smooth.*

In connection with this assumption we will keep the following notations: For $\epsilon > 0$ put $B_\epsilon := \{z \in \mathbb{C}^{m+1} \, | \, ||z|| \le \epsilon\}$, $X_\epsilon := X \cap B_\epsilon$, $Y_\epsilon := Y \cap B_\epsilon$, $D_\epsilon := \{t \in \mathbb{C} \, | \, |t| \le \epsilon\}$. Note that $X_\epsilon \setminus Y$ is a C^∞ manifold with boundary if $0 < \epsilon \ll 1$. For $\epsilon > 0, t \in \mathbb{C}$ put $X_{\epsilon,t} := X \cap B_\epsilon \cap \{g = t\}$, $Y_{\epsilon,t} := Y \cap B_\epsilon \cap \{g = t\}$.

The aim is to compare the spaces $X_{\epsilon,t} \setminus Y_{\epsilon,t} = X \cap B_\epsilon \cap \{g = t\} \setminus Y$ for different values of t: Are they diffeomorphic to each other? If yes and if Y is the zero level set of some holomorphic function $f : U \longrightarrow \mathbb{C}$: is there a diffeomorphism which is fibre-preserving with respect to the mapping $\phi := \frac{f}{|f|}$? By the way, diffeomorphy will be denoted by $\approx$, whereas homeomorphy will be denoted by $\sim$.

Let us fix a Whitney-regular stratification of (X, Y); we may assume that the connected components of $X \setminus Y$ are strata. We make the following assumption:

Assumption 1.2. *$\{g = 0\}$ intersects the strata of X in some punctured neighbourhood of 0 transversally.*

Theorem 1.1. *We start from the assumptions 1.1 and 1.2. Furthermore, let $f : U \longrightarrow \mathbb{C}$ be holomorphic, and assume that $Y = \{z \in X \, | \, f(z) = 0\}$.*

Then the following conditions are equivalent:

a_1) $\chi(X_{\epsilon,t}) = \chi(Y_{\epsilon,t})$, $0 < |t| \ll \epsilon \ll 1$,

a_2) $\chi(X_{\epsilon,t} \setminus Y_{\epsilon.t}) = 0$, $0 < |t| \ll \epsilon \ll 1$,

a_3) $\chi(X_\epsilon \cap \{f = \tau\}) = \chi(X_{\epsilon,t} \cap \{f = \tau\})$, $0 < |t| \le \beta, 0 < |\tau| \ll \beta \ll \epsilon \ll 1$,

b) $(f,g) : X_\epsilon \setminus Y \longrightarrow \mathbb{C}^2$ *has no critical values* (c,d) *with* $|c| + |d| \ll \epsilon$,

c_1) *there is a commutative diagram of the form*

$$
\begin{array}{ccc}
X_\epsilon \cap \{|g| \le \beta\} \setminus Y & & \\
\quad \approx \downarrow & \searrow (\phi,g) & \\
(X_\epsilon \cap \{g = 0\} \setminus Y) \times D_\beta & \xrightarrow{(\phi,pr_2)} & S_1 \times D_\beta
\end{array}
$$

with $0 < \beta \ll \epsilon \ll 1$,

c_2) *there is a commutative diagram of the form*

$$
\begin{array}{ccc}
X_\epsilon \cap \{0 < |f| \le \alpha, |g| \le \beta\} & & \\
\quad \approx \downarrow & \searrow (f,g) & \\
(X_{\epsilon,0} \cap \{0 < |f| \le \alpha\}) \times D_\beta & \xrightarrow{(f,pr_2)} & (D_\alpha \setminus \{0\}) \times D_\beta
\end{array}
$$

where $\alpha > 0$, $\beta > 0$, $\max\{\alpha,\beta\} \ll \epsilon \ll 1$.

The condition c_2) is an assertion about the local triviality of $g|(X \setminus Y)$, where the trivialization is to be compatible with f. The condition c_1) is a variant where f is replaced by ϕ.

The theorem above will be proved in the next paragraph (see Theorem 2.1) where we will discuss similar results, too. In the subsequent paragraphs we will also discuss the question of local triviality of $g|X$. Particularly accessible is the case $\dim X = 2$. Among other statements we will show (cf. Theorem 4.2):

Theorem 1.2. *Let us start from the assumptions 1.1 and 1.2 again. Furthermore assume that* $\dim X = 2$, *that* Y *is smooth, and that* 0 *is not a critical point of* $g|Y$. *Then the following conditions are equivalent:*

a) $\chi(cl_0 X) = 1$, *where* $cl_0 X =$ *"complex link" (cf. [7]),*

b) there is a commutative diagram of the form

$$
\begin{array}{ccc}
(X_\epsilon \cap \{|g| \le \beta\}, Y \cap \{|g| \le \beta\}) & & \\
\quad \sim \downarrow & \searrow g & \\
(X_{\epsilon,0}, 0) \times D_\beta & \xrightarrow{pr_2} & D_\beta
\end{array}
$$

where $0 < \beta \ll \epsilon \ll 1$,

c) (X,Y) *satisfies Bekka's c-condition, cf. [1].*

In this case there is at least a weakly holomorphic continuous function $f : X \to \mathbb{C}$ with $Y = X \cap \{f = 0\}$. For the notions involved cf. §4.

2. On local triviality of $g|(X \setminus Y)$

In this paragraph we keep the following additional assumption:

Assumption 2.1. *There is a holomorphic function $f : U \longrightarrow \mathbb{C}$ such that $Y = \{z \in X \mid f(z) = 0\}$.*

For the proof of Theorem 1.1 transversality statements are useful. Let us recall the following: If M is a complex manifold, $h : M \to \mathbb{C}$, h C^∞-differentiable, there is a decomposition $dh = \partial h + \bar{\partial} h$ into forms of type $(1, 0)$ resp. $(0, 1)$. If h is holomorphic, $dh = \partial h$; if h is real-valued we have $\bar{\partial} h = \overline{\partial h}$. Let $\psi : \mathbb{C}^{m+1} \longrightarrow \mathbb{R}\colon z \mapsto ||z||^2$, $0 < \epsilon \ll 1$, $\alpha, \beta > 0$.

Lemma 2.1. *Under the assumptions 1.1, 1.2, and 2.1, we have:*

a) $(f, g)|X \cap \partial B_\epsilon \setminus Y$ has no critical points in $\{|f| \le \alpha, |g| \le \beta\}$, $\max(\alpha, \beta) \ll \epsilon$,

b) $(f, g)|X$ has no critical points in $B_\epsilon \cap \{|f| \ge \alpha, |g| \le \beta\}$, $\alpha > 0$, $0 < \beta \ll \min(\alpha, \epsilon)$,

c) $(\phi, g)|X \setminus Y$ has no critical points in $B_\epsilon \cap \{|f| \ge \alpha, |g| \le \beta\}$, $0 < \beta \ll \min(\alpha, \epsilon)$,

d) $(\phi, g)|X \cap \partial B_\epsilon \setminus Y$ has no critical points in $\{|g| \le \beta\}$, $0 < \beta \ll \epsilon$,

e) $g|X \setminus Y$ has no critical points in B_ϵ,

f) $(f, g)|X$ has no critical points in $B_\epsilon \cap \{0 < |f| \le \alpha, |g| \ge \beta\}$, $0 < \alpha \ll \min(\beta, \epsilon)$,

g) $(f, g)|X \setminus Y$ has in some neighbourhood of $B_\epsilon \cap Y \setminus \{0\}$ no critical points.

Proof. a) First, by the Curve Selection Lemma [13] there is an $\epsilon_0 > 0$ such that ϵ^2 is not a critical value of $\psi|S$ whenever $0 < \epsilon \le \epsilon_0$ and S is a stratum of X.

Shrinking U if necessary we may assume that $S \cap \{g = 0\} \setminus \{0\}$ is smooth, for each stratum S of Y, by Assumption 1.2. By the Curve Selection Lemma again there is an $\epsilon_1 > 0$, $\epsilon_1 \le \epsilon_0$, such that ϵ^2 is not a critical value of $\psi|S \cap \{g = 0\} \setminus \{0\}$ whenever $0 < \epsilon \le \epsilon_1$. Fix $\epsilon, 0 < \epsilon \le \epsilon_1$. We just showed that 0 is not a critical value of $g|S \cap \partial B_\epsilon$. By continuity, there is a $\beta > 0$ such that t is not a critical value of $g|S \cap \partial B_\epsilon$ as soon as $|t| \le \beta$. So $\{g = t\}$ intersects the space $S \cap \partial B_\epsilon$ transversally, $|t| \le \beta$.

Now S was supposed to be contained in $Y = \{g = 0\}$. We know that our stratification satisfies Thom's a_f-condition because Whitney regularity implies Thom's a_f-condition in the complex case [3]. Therefore, there is an $\alpha > 0$ such that $\{g = t\}$ intersects the space $\{f = \tau\} \cap X \cap \partial B_\epsilon$ transversally for $|t| \leq \beta, |\tau| \leq \alpha$, too: Suppose that this is not true. Then there is a sequence (z_ν) of points in $X \cap \partial B_\epsilon$ such that $z_\nu \to z^* \in Y$, $|g(z^*)| \leq \beta$, $T_{z_\nu}(\{f = f(z_\nu)\} \cap X) \to T$ with $T \supset T_{z^*}S$, where S is the stratum of Y which contains z^* and where the spaces $\{f = f(z_\nu)\} \cap X$ and $\partial B_\epsilon \cap \{g = g(z_\nu)\}$ are not transversal. This is a contradiction to the transversality of $\partial B_\epsilon \cap \{g = g(z^*)\}$ to S which was shown just before.

b) $(X \setminus Y) \cap \{g = 0\}$ is smooth by hypothesis. If $\epsilon > 0$ is sufficiently small, the restriction of f onto this space has no critical point in B_ϵ because of the curve selection lemma. Choose $\alpha > 0$. Then $\{g = 0\}$ intersects the spaces $X \cap \{f = \tau\}$ with $|\tau| \geq \alpha$ within B_ϵ transversally, too. The same holds for $\{g = t\}$, $|t| \ll \alpha$. This implies our assertion.

c) follows from b).

d) First we have because of the curve selection lemma: If z is a critical point of $f|(X \setminus Y) \cap \partial B_\epsilon \cap \{g = 0\}$, i.e. there is $\lambda \in \mathbb{C}$ with $\frac{d_z f}{f(z)} = \lambda \partial_z \psi$ on $T_z(X \cap \{g = 0\})$, then $\lambda \neq 0$. Furthermore, $|\arg \lambda| < \frac{\pi}{4}$, where arg is normalized by $-\pi < \arg \leq \pi$: Otherwise the curve selection lemma would imply the existence of a real-analytical curve p in $X \cap \{g = 0\}$ with $p(0) = 0$ and $\frac{d_{p(t)} f}{f(p(t))} = \lambda(t) \partial_{p(t)} \psi$ on $T_{p(t)}(X \cap \{g = 0\})$, $\lambda(t) \neq 0$, $|\arg \lambda(t)| \geq \frac{\pi}{4}$. Let $f(p(t)) = at^m + \ldots$, $\lambda(t) = bt^n + \ldots$, $p(t) = ct^k + \ldots$; here $m, k > 0$. Application to $\dot{p}(t)$ gives: $mt^{-1} + \ldots = b||c||^2 kt^{n+2k-1} + \ldots$. This leads to a contradiction, because $|\arg b| \geq \frac{\pi}{4}$.

Assume that $\phi|(X \setminus Y) \cap \partial B_\epsilon \cap \{g = 0\}$ has a critical point z: then there is a real μ with $d_z(\operatorname{Im} \log f) = \mu d_z \psi$, so $\partial_z(\operatorname{Im} \log f) = \mu \partial_z \psi$ on $T_z(X \cap \{g = 0\})$. Now $\partial_z(\operatorname{Im} \log f) = -\frac{i}{2} \frac{d_z f}{f(z)}$, so $\frac{d_z f}{f(z)} = 2i\mu \partial_z \psi$ on $T_z(X \cap \{g = 0\})$, contradiction.

Let $\alpha > 0$. Then $\phi|(X \setminus Y) \cap \partial B_\epsilon \cap \{g = t\}$ has no critical point in $\{|f| \geq \alpha\}$ for $|t| \ll \beta$, too. Because of a) we get the assertion.

e) follows from the curve selection lemma and the transversality of $\{g = 0\}$ and $X \setminus Y$ near 0.

f) First we have for each stratum S of Y : $g|S$ has no critical points in $B_\epsilon \setminus \{0\}$, so in $B_\epsilon \cap \{|g| \geq \beta\}$. This implies our assertion because of Thom's a_f-condition.

g) This follows from the transversality of the sets $\{g = t\}$ to the strata of Y outside 0 and Thom's a_f-condition. $\qquad\square$

Furthermore we need in certain cases of non-transversality an assertion about the sign:

Lemma 2.2. *Under the same assumptions as in Lemma 2.1, we have:*

a) *If $z \in X_\epsilon \cap \{g = t, |f| \geq \alpha\}$ there is no $\lambda \leq 0$ with $d_z|f| = \lambda d_z\psi$ on $T_z(X \cap \{g = t\})$, $|t| \ll \min(\alpha, \epsilon)$.*

b) *If $z \in X_\epsilon \setminus Y$, $g(z) \neq 0$ there is no $\lambda \leq 0$ with $d_z|g| = \lambda d_z\psi$ on $T_z X$.*

c) *If $z \in X_\epsilon \cap \{f = \tau, |g| \geq \beta\}$ there is no $\lambda \leq 0$ with $d_z|g| = \lambda d_z\psi$ on $T_z X \cap \{f = \tau\}$, $0 < |\tau| \ll \min(\beta, \epsilon)$.*

Proof. b) follows from the curve selection lemma.

a) We get by the curve selection lemma: For $t = 0$ there is no $\lambda \leq 0$ with $d_z|f| = \lambda d_z\psi$ on $T_z(X \cap \{g = t\})$. This implies the assertion.

c) By the curve selection lemma we get as in b): If S is a stratum of Y there is for $z \in S \cap B_\epsilon \cap \{g \neq 0\}$ no $\lambda \leq 0$ with $d_z|g| = \lambda d_z\psi$ on $T_z S$. Because of Thom's a_f-condition we get our assertion. $\qquad\square$

We apply Lemma 2.1 and 2.2 in order to modify the conditions a_1) and a_3) from Theorem 1.1.

Lemma 2.3. *Again we start from the assumptions of Lemma 2.1.*

a) $\chi(X_{\epsilon,t}) = \chi(X_\epsilon \cap \{|f| \leq \alpha, g = t\})$, $0 < |t| \ll \alpha \ll \epsilon \ll 1$,

b) $\chi(X_\epsilon \cap \{f = \tau\}) = \chi(X_\epsilon \cap \{f = \tau, |g| \leq \beta\})$, $0 < |t| \leq \beta, 0 < |\tau| \ll \beta \ll \epsilon \ll 1$.

Proof. a) $X_\epsilon \cap \{|f| \leq \alpha, g = t\}$ is a deformation retract of $X_{\epsilon,t}$: here we consider the function $|f|$ and use Lemma 2.1b) as well as 2.2a).

b) $X_\epsilon \cap \{f = \tau, |g| \leq \beta\}$ is a deformation retract of $X_\epsilon \cap \{f = \tau\}$: this can be seen using $|g|$, Lemma 2.1f) as well as Lemma 2.2c). $\qquad\square$

Now we prove Theorem 1.1, modifying a_1 and a_3 according to Lemma 2.3 and adding equivalent conditions c_1' and c_2' :

Theorem 2.1. *Under the assumptions 1.1,1.2, and 2.1, the following conditions are equivalent:*

a_1) $\chi(X_{\epsilon,t} \cap \{|f| \leq \alpha\}, Y_{\epsilon,t}) = 0,\ 0 < |t| \ll \alpha \ll \epsilon \ll 1$,

a_2) $\chi(X_{\epsilon,t} \setminus Y) = 0,\ 0 < |t| \ll \epsilon \ll 1$,

a_3) $\chi(X_\epsilon \cap \{f = \tau, |g| \leq \beta\}, X_{\epsilon,t} \cap \{f = \tau\}) = 0,\ 0 < |t| \leq \beta, 0 < |\tau| \ll \beta \ll \epsilon \ll 1$,

b) $(f,g) : X_\epsilon \setminus Y \longrightarrow \mathbb{C}^2$ *has no critical values* (c,d) *with* $|c| + |d| \ll \epsilon$,

c_1') *there is a commutative diagram of the form*

$$
\begin{array}{ccc}
X_\epsilon \cap \{|g| \leq \beta\} \setminus Y & & \\
\approx \downarrow & \searrow g & \\
(X_{\epsilon,0} \setminus Y) \times D_\beta & \xrightarrow{pr_2} & D_\beta
\end{array}
$$

where $0 < \beta \ll \epsilon \ll 1$,

c_1) *there is a commutative diagram of the form*

$$
\begin{array}{ccc}
X_\epsilon \cap \{|g| \leq \beta\} \setminus Y & & \\
\approx \downarrow & \searrow (\phi, g) & \\
(X_{\epsilon,0} \setminus Y) \times D_\beta & \xrightarrow{(\phi, pr_2)} & S_1 \times D_\beta
\end{array}
$$

where $0 < \beta \ll \epsilon \ll 1$,

c_2') *there is a commutative diagram of the form*

$$
\begin{array}{ccc}
X_\epsilon \cap \{0 < |f| \leq \alpha, |g| \leq \beta\} & & \\
\approx \downarrow & \searrow g & \\
(X_{\epsilon,0} \cap \{0 < |f| \leq \alpha\}) \times D_\beta & \xrightarrow{pr_2} & D_\beta
\end{array}
$$

where $\max\{\alpha, \beta\} \ll \epsilon \ll 1$,

c_2) *there is a commutative diagram of the form*

$$
\begin{array}{ccc}
X_\epsilon \cap \{0 < |f| \leq \alpha, |g| \leq \beta\} & & \\
\approx \downarrow & \searrow (f, g) & \\
(X_{\epsilon,0} \cap \{0 < |f| \leq \alpha\}) \times D_\beta & \xrightarrow{(f, pr_2)} & (D_\alpha \setminus \{0\}) \times D_\beta
\end{array}
$$

where $\max\{\alpha, \beta\} \ll \epsilon \ll 1$.

Proof. Let Γ be the closure of the set of all critical points $z \in X_\epsilon \setminus Y$ of $(f,g)|X \setminus Y$ in X. We consider the mapping $(f,g) : B_\epsilon \cap \Gamma \cap \{|f| \leq \alpha, |g| \leq \beta\} \longrightarrow D_\alpha \times D_\beta,\ \max(\alpha, \beta) \ll \epsilon$.

On the boundary of B_ϵ there is because of Lemma 2.1g) no point of $\Gamma \cap \{|f| \leq \alpha, |g| \leq \beta\}$, so $B_\epsilon \cap \Gamma \cap \{|f| < \alpha, |g| < \beta\}$ is an analytic set which is mapped properly onto $\{(s,t)\,|\,|s| < \alpha, |t| < \beta\}$. By Sard's theorem we conclude that the image is a curve if $\Gamma \neq \emptyset$. The fibres are compact and analytic, therefore zero-dimensional; so $\Gamma = \emptyset$ or $\dim \Gamma = 1$ ("polar curve"). Furthermore the fibres of $(f,g)|X_\epsilon$ above points of $\Gamma \setminus \{0\}$ have at most isolated singularities.

We want to show the following implications:

(i) $a_1) \iff a_2) \iff a_3) \iff b)$

(ii) $b) \implies c_1) \implies c_1') \implies a_2)$

(iii) $b) \implies c_2) \implies c_2') \implies a_2)$.

(i): $a_3) \iff b)$: We have $\chi := \chi(X_\epsilon \cap \{f = \tau, |g| \leq \beta\}, X_{\epsilon,t} \cap \{f = \tau\}) = \pm \sum_{x \in \Gamma : f(x) = \tau} \mu_x$, where μ_x is the reduced Milnor number of (f,g) in x. So $\chi = 0 \iff \{x \in \Gamma \mid f(x) = \tau\} = \emptyset \iff \Gamma = \emptyset$.

$a_2) \iff b)$: Let $0 < |t| \ll \alpha \ll \epsilon \ll 1$ and $\partial X_{\epsilon,t} := X_{\epsilon,t} \cap \partial B_\epsilon$. Then $\partial X_{\epsilon,t} \setminus Y$ has a Whitney stratification with real strata of odd dimension, so $\chi(\partial X_{\epsilon,t} \setminus Y) = 0$ by Sullivan [17]. This implies $\chi(X_{\epsilon,t} \setminus Y) = \chi(X_{\epsilon,t} \setminus Y, \partial X_{\epsilon,t} \setminus Y)$. Therefore $a_2)$ is equivalent to the condition that $\chi(X_{\epsilon,t} \setminus Y, \partial X_{\epsilon,t} \setminus Y) = 0$.
Here we use $-|f|$ as a kind of Morse function on $X_{\epsilon,t} \setminus Y$ modulo $\partial X_{\epsilon,t} \setminus Y$ in the sense of [8]. Now $|f| \, | \, \partial X_{\epsilon,t}$ has no critical points with value in $]0, \alpha]$, cf. Lemma 2.1a). The critical points of $|f| \, | \, \partial X_{\epsilon,t}$ with value $> \alpha$ do not contribute because of Lemma 2.2a) . Furthermore $|f| \, | \, X \cap \{g = t\} \setminus Y$ has because of Lemma 2.1b) in $X_{\epsilon,t} \setminus (\partial X_{\epsilon,t} \cup Y)$ only critical points with value $< \alpha$. These may be degenerate. For each critical point x we have to attach μ_x cells of dimension $n-1$, in total $\sum_{x \in \Gamma : g(x) = t} \mu_x$ such cells. This number is zero if and only if Γ is void. Note that $\Gamma \neq \emptyset \Rightarrow \Gamma \not\subset \{g = 0\}$.

$a_2) \iff a_1)$: The number in $a_1)$ can be computed with the help of $|f| \, | \, X_{\epsilon,t} \cap \{|f| \leq \alpha\}$. It is $\pm \sum_{x \in \Gamma : g(x) = t} \mu_x$. Because of the considerations in the proof $a_2) \iff b)$ we get the assertion.

So (i) has been shown. As for (ii):

$b) \implies c_1)$: We consider the mapping $(f,g)|X_\epsilon$. Near Y we lift the constant vector fields $(0,1)$ resp. $(0,i)$ locally, cf. $b)$ and Lemma 2.1a). Outside some neighbourhood of Y we lift the vector fields $(0,1)$ resp. $(0,i)$ locally with respect to (ϕ, g), because by Lemma 2.1c),d) $(\phi, g)|X \setminus Y$ and $(\phi, g)|X \cap \partial B_\epsilon \setminus Y$ have no critical points z with $|g(z)| \leq \beta$ there. After this we use a partition of unity.

$c_1) \implies c_1')$: trivial.

$c_1' \implies a_2)$: First, $\chi(X_{\epsilon,t} \setminus Y) = \chi(X_{\epsilon,0} \setminus Y)$. But $\chi(X_{\epsilon,0} \setminus Y) = 0$, because $\partial X_{\epsilon,0} \setminus Y$ is a deformation retract of $X_{\epsilon,0} \setminus Y$ and admits a stratification with odd-dimensional strata, so it has Euler characteristic 0 by Sullivan [17].

Ad (iii): b) $\Longrightarrow c_2$): Lift the vector fields $(0,1)$ resp. $(0,i)$ locally with respect to $(f,g)|X_\epsilon$ (cf. b) and Lemma 2.1a)) and use a partition of unity.

c_2) $\Longrightarrow c_2'$) $\Longrightarrow a_2$): trivial, cf. proof of $c_1' \Longrightarrow a_2$). $\qquad\qquad\square$

Remarks on Theorem 2.1:

1. c_1) implies, because $\phi : X_{\epsilon,0} \setminus Y \longrightarrow S_1$ defines a C^∞-fibre bundle (cf. Lemma 2.1c),d)):

$(\phi,g) : X_\epsilon \cap \{|g| \leq \beta\} \setminus Y \longrightarrow S_1 \times D_\beta$ defines a C^∞-fibre bundle.

Equally c_2) implies, since $f : X_{\epsilon,0} \cap \{0 < |f| \leq \alpha\} \longrightarrow D_\alpha \setminus \{0\}$ defines a C^∞-fibre bundle (cf. b) and Lemma 2.1a)): $(f,g) : X_\epsilon \cap \{|g| \leq \beta, 0 < |f| \leq \alpha\} \longrightarrow (D_\alpha \setminus \{0\}) \cap D_\beta$ defines a C^∞-fibre bundle.

2. It has been assumed that $\{g = 0\}$ intersects the strata of X transversally outside 0. The situation is of course much simpler if we have transversality at 0, too: because of Thom's a_f-condition the condition b) from Theorem 2.1 is then fulfilled automatically, cf. Theorem 3.1.

3. By Milnor [13] it seems natural to look at ∂B_ϵ instead of B_ϵ in the case of c_1), c_1'), or at $|f| = \alpha$ instead of $0 < |f| \leq \alpha$ in the case of c_2) and c_2'). The corresponding assertions are then always true, however, where in the latter case we have to assume that $\beta \ll \alpha$:

Lemma 2.4. *Let us start from the same assumptions as in Theorem 2.1.*

a) There is a commutative diagram of the form

$$X \cap \partial B_\epsilon \cap \{|g| \leq \beta\} \setminus Y$$
$$\approx\downarrow \qquad\qquad\qquad \searrow (\phi,g)$$
$$(X \cap \partial B_\epsilon \cap \{g = 0\} \setminus Y) \times D_\beta \xrightarrow{(\phi,pr_2)} S_1 \times D_\beta$$

where $0 < \beta \ll \epsilon \ll 1$,

b) there is a commutative diagram of the form

$$X_\epsilon \cap \{|f| = \alpha, |g| \leq \beta\}$$
$$\approx\downarrow \qquad\qquad \searrow (f,g)$$
$$(X_{\epsilon,0} \cap \{|f| = \alpha\}) \times D_\beta \xrightarrow{(f,pr_2)} \alpha \cdot S_1 \times D_\beta$$

where $0 < \beta \ll \alpha \ll \epsilon \ll 1$.

Proof. a) Because of Lemma 2.1a), $(f,g)|X \cap \partial B_\epsilon \setminus Y$ has no critical points in $\{|f| \leq \alpha, |g| \leq \beta\}$, $0 < \beta \ll \alpha \ll 1$. So the constant vector fields $(0,1)$ and $(0,i)$ can be lifted with respect to (f,g) in a neighbourhood of this set locally. At the same time we get liftings with respect to (ϕ,g), of course.

634

Furthermore, $(\phi, g)|X \cap \partial B_\epsilon \setminus Y$ has no critical point in $\{|g| \leq \beta\}$ by Lemma 2.1d), so we can lift the vector fields above in neighbourhoods which are contained in $\{|f| > \frac{\alpha}{2}\}$. Using a partition of unity we obtain in total vector fields v_1 and v_2 with $d\phi(v_j) = 0$, $df(v_j) = 0$ for $|f| \leq \frac{\alpha}{2}$, $dg(v_1) = 1$, $dg(v_2) = i$. These vector fields are (each) integrable.

b) Because of Lemma 2.1a),b), $(f, g)|X \setminus Y$ and $(f, g)|X \cap \partial B_\epsilon \setminus Y$ have no critical points in $\{|g| \leq \beta, |f| = \alpha\}$ with $0 < \beta \ll \alpha \ll \epsilon$. As above we argue with liftings of the vector fields $(0, 1)$ resp. $(0, i)$. $\qquad\square$

By the way, the Euler characteristics which are used in Theorem $2.1a_2$) resp. a_3) are different in general; Theorem 2.1 only says that they can only vanish simultaneously.

Example 2.1. Let $X := \mathbb{C}^2$, $f(z_1, z_2) := z_1^2 + z_2^2$, $g(z_1, z_2) := z_2$. Then the hypotheses of Theorem 2.1 are fulfilled but the conditions a_2) and a_3) are not satisfied. Indeed, we have $\sum_{x \in \Gamma \,:\, g(x) = t_0} \mu_x = 1$, $\sum_{x \in \Gamma \,:\, f(x) = w_0} \mu_x = 2$, $t_0 \neq 0$, $w_0 \neq 0$, which also shows that the Euler characteristics are different.

The numbers used in Theorem 2.1 a_1) and a_2) coincide, however (up to sign), as the proof above shows. This can also been seen using formal arguments:

As above, let $\partial X_{\epsilon,t} := X_{\epsilon,t} \cap \partial B_\epsilon$.

In a_1) we consider the Euler characteristic of $(X_{\epsilon,t}, Y \cap X_{\epsilon,t})$, in a_2) (up to sign) the one of $(X_{\epsilon,t} \setminus Y, \partial X_{\epsilon,t} \setminus Y)$, cf. the proof above. The corresponding cohomology groups are dual to each other:

Let $j_t : X_{\epsilon,t} \setminus Y \to X_{\epsilon,t}$ be the inclusion. Then: $H^\nu(X_{\epsilon,t}, Y \cap X_{\epsilon,t}; \mathbb{C})$
$= \mathbb{H}^\nu(X_{\epsilon,t}, (j_t)_! \mathbb{C}_{X_{\epsilon,t} \setminus Y}) = \mathbb{H}^\nu(X_{\epsilon,t} \setminus \partial X_{\epsilon,t}, (j_t)_! \mathbb{C}_{X_{\epsilon,t} \setminus Y})$,
$H^\mu(X_{\epsilon,t} \setminus Y, \partial X_{\epsilon,t} \setminus Y; \mathbb{C}) = \mathbb{H}^\mu_c(X_{\epsilon,t} \setminus \partial X_{\epsilon,t}, (j_t)_* \mathbb{C}_{X_{\epsilon,t} \setminus Y})$.

Here $(j_t)_!$ and $(j_t)_*$ are considered as functors in the derived category. The two cohomology groups are therefore for $\mu + \nu = 2n - 2$ dual to each other. We can reformulate this using the functor of nearby cycles: Let $j : X \setminus Y \to X$ and $k : \{0\} \to X \cap \{g = 0\}$ be the inclusions. Then:

$\mathbb{H}^\nu(X_{\epsilon,t}, (j_t)_! \mathbb{C}_{X_{\epsilon,t} \setminus Y}) = H^\nu(k^* \Psi_g j_! \mathbb{C}_{X \setminus Y})$,
$\mathbb{H}^\mu_c(X_{\epsilon,t}, (j_t)_* \mathbb{C}_{X_{\epsilon,t} \setminus Y}) = H^\mu(k^! \Psi_g j_* \mathbb{C}_{X \setminus Y})$.

Finally the complexes $k^* j_! \mathbb{C}_{X \setminus Y}$ and $k^! j_* \mathbb{C}_{X \setminus Y}$ are acyclic, which implies the duality of

$H^\nu(k^* \Phi_g j_! \mathbb{C}_{X \setminus Y})$ and $H^\mu(k^! \Phi_g j_* \mathbb{C}_{X \setminus Y})$ for $\mu + \nu = 2n$.

Because of our transversality hypothesis $\Phi_g j_* \mathbb{C}_{X \setminus Y}$ is concentrated upon 0, so we can replace $k^!$ here by k^* and arrive finally at $H^\mu(k^* \Phi_g j_* \mathbb{C}_{X \setminus Y}) = H^\mu(X_\epsilon \setminus Y, X_{\epsilon,t} \setminus Y; \mathbb{C})$.

So the condition a_2) can be reformulated as follows:

$\chi(X \cap B_\epsilon \setminus Y, X \cap B_\epsilon \cap \{g = t\} \setminus Y) = 0$.

Since the scope is only the Euler characteristic this can be shown in a much simpler way directly using the result of Sullivan quoted above according to which the Euler characteristic of $X \cap \partial B_\epsilon \setminus Y$ vanishes, so the one of $X_\epsilon \setminus Y$, too.

3. Transition from $g|X \setminus Y$ to $g|X$

a) Arbitrary dimension

In connection with the condition c_2) from Theorem 2.1 we may ask whether there is a trivialization that can be extended to the special fibre of f, i.e. whether there a commutative diagram of the form

$$X_\epsilon \cap \{|f| \leq \alpha, |g| \leq \beta\}$$
$$\sim\downarrow \qquad\qquad \searrow (f,g)$$
$$(X_{\epsilon,0} \cap \{|f| \leq \alpha\}) \times D_\beta \xrightarrow{(f,pr_2)} D_\alpha \times D_\beta$$

where $\max\{\alpha, \beta\} \ll \epsilon \ll 1$.

Remember that $\sim$ denotes homeomorphy.

It is easy to give a sufficient condition. Recall that we started in connection with Assumption 1.2 from a stratification which is Whitney-regular:

Theorem 3.1. *Assume that the assumptions 1.1 and 1.2 are satisfied, as well as assumption 2.1, i.e.* $Y = \{z \in X \mid f(z) = 0\}$, *where* $f : U \longrightarrow \mathbb{C}$ *is holomorphic. Furthermore assume that* 0 *is not a critical point for the restriction of* g *on the stratum of* Y *through* 0 *(which implies that* $\{0\}$ *is not a stratum). Then there is a commutative diagram of the form*

$$X_\epsilon \cap \{|f| \leq \alpha, |g| \leq \beta\}$$
$$\sim\downarrow \qquad\qquad \searrow (f,g)$$
$$(X_{\epsilon,0} \cap \{|f| \leq \alpha\}) \times D_\beta \xrightarrow{(f,pr_2)} D_\alpha \times D_\beta$$

where $\max\{\alpha, \beta\} \ll \epsilon \ll 1$.

Proof. The function $g|X$ has no critical points in B_ϵ. So $(f,g)|X_\epsilon \setminus Y$ has no critical values in $D_\alpha \times D_\beta$, because of Thom's a_f-condition. As the proof of Lemma 2.1a) shows, $g|Y \cap \partial B_\epsilon$ has no critical values in D_β, too, and according to Lemma 2.1a) $(f,g)|X \cap \partial B_\epsilon \setminus Y$ has no critical values in $D_\alpha \times D_\beta$. Now we may find continuous controlled vector fields on X which constitute a lifting of the constant vector fields 1 resp. i with respect

636

to g along $X \cap B_\epsilon$, cf. [16], [6]. We project these vector fields outside of Y onto the tangent space to the fibre of $f|X$ resp. - near ∂B_ϵ - onto the tangent space to the fibres of $(g, \psi)|X$. Because of Thom's condition we get continuous vector fields. By some partition of unity we get continuous vector fields that are parallel to the fibres of f . By integration we obtain the assertion. $\qquad\square$

If we renounce to the compatibility with f we may drop the hypothesis that Y is defined by some holomorphic function and still have the assertion of Theorem 3.1; the following is known:

Theorem 3.2. *Let us start from assumptions 1.1 and 1.2. Furthermore assume that 0 is not a critical point for the restriction of g to the stratum of Y through 0. Then there is a commutative diagram of the form*

$$(X_\epsilon \cap \{|g| \leq \beta\}, Y_\epsilon \cap \{|g| \leq \beta\})$$
$$\sim\downarrow \qquad\qquad\qquad \searrow g$$
$$(X_{\epsilon,0}, Y_{\epsilon,0}) \times D_\beta \qquad \xrightarrow{pr_2} D_\beta$$

where $0 < \beta \ll \epsilon \ll 1$.

Proof. Let $\epsilon > 0$ be sufficiently small. Then the restriction of g to the strata has no critical points in B_ϵ, and ∂B_ϵ intersects the sets $g^{-1}(\{0\}) \cap S$, S stratum of X, transversally. The same holds if we replace $g^{-1}(\{0\})$ by $g^{-1}(\{t\})$, $|t| \leq \beta$, $0 < \beta \ll \epsilon$. So $g|X_\epsilon$ is a proper stratified submersion above D_β, the assertion follows therefore from Thom's isotopy theorem. $\square$

Now let us turn back to the situation studied before, so for the rest of this paragraph we return to Assumption 2.1:

$Y = \{z \in X \mid f(z) = 0\}$, *where* $f : U \longrightarrow \mathbb{C}$ *is some holomorphic function.*

It would be exaggerated to connect the possibility of a local trivialization automatically with the hypothesis of Theorem 3.1, as shown by the famous example of Briançon-Speder (recall that we started in Theorem 3.1 from a stratification which is Whitney-regular):

Example 3.1. [4] Let $X := \mathbb{C}^4$, $f(z_1, z_2, z_3, t) = f_t(z_1, z_2, z_3) := z_3^5 + tz_2^6 z_3 + z_2^7 z_1 + z_1^{15}$. For each t, f_t is weighted homogeneous of degree 15 with respect to the weights $1, 2, 3$ with isolated singularity. Let $\phi(z) := |z_1|^{12} + |z_2|^6 + |z_3|^4$, $\Phi(z, t) := \phi(z)$. Then (f, Φ) is outside $\{0\} \times \mathbb{C}$ a submersion. The mapping $g|\Phi^{-1}(\{1\})$ can be trivialized in a way which is compatible with f, i.e. there is some homeomorphism ("trivialization") $h = (h_1, h_2) : \Phi^{-1}(\{1\}) \to (\Phi^{-1}(\{1\}) \cap \{g = 0\}) \times \mathbb{C}$ such that $h_2 = g|\Phi^{-1}(\{1\})$ and $f \circ h_1 = f|\Phi^{-1}(\{1\})$. Therefore we can extend h to a trivialization of g which is compatible with $f : (z, t) \mapsto ((\phi(z)^{\frac{1}{12}} \circ h_1(\phi(z)^{-\frac{1}{12}} \circ z, 0), t)$ for $\phi(z) > 0$, $(0, t) \mapsto ((0, 0), t)$. Here $c \circ z := (cz_1, c^2 z_2, c^3 z_3)$.

On the other side, $(Y, \{0\} \times \mathbb{C})$ is not Whitney-regular according to [4].

In general the answer on the initial question is "no", as the following example shows:

Example 3.2. Let $X := \{(z_1, w_1, z_2, w_2) \in \mathbb{C}^4 \mid z_1 = z_2, w_1 = w_2 \text{ or } z_1 = -z_2, w_1 = -w_2\}$, $f(z_1, w_1, z_2, w_2) := z_1$, $g(z_1, w_1, z_2, w_2) := w_2$.
The fibre of $(f, g) : X \to \mathbb{C}^2$ above $(0, 0)$ consists of one point, the other fibres consist of two points. Outside $f = 0$ we have the following uniquely determined trivialization $h : X \setminus Y \to ((X \setminus Y) \cap g^{-1}(\{0\})) \times \mathbb{C}$:
$(z_1, w_1, z_2, w_2) \mapsto ((z_1, 0, z_2, 0), w_2)$.
Inverse mapping: $((z_1, 0, z_2, 0), t) \mapsto (z_1, t, z_2, t)$ if $z_1 = z_2$,
$((z_1, 0, z_2, 0), t) \mapsto (z_1, -t, z_2, t)$ if $z_1 = -z_2$.
But an extension is impossible, of course.

What is the reason? We may look for different answers.

First, the space X is the standard example for a space which cannot be realized as a complete intersection. In [9] we introduced the notion $rHd_{\mathbb{Q}}$ of rational homological depth; the index $\mathbb{Q}$ makes clear that we mean homology with rational coeffizients.
Recall that $rHd_{\mathbb{Q}}(X) \geq r$ means that for each k the following holds: the set of all $x \in X$ with $H^k(X, X \setminus \{x\}; \mathbb{Q}) \neq 0$ is contained in some analytic subset of dimension $\leq k - r$.
Obviously, $rHd_{\mathbb{Q}}(X) \leq \dim X$. For a complete intersection we have that $rHd_{\mathbb{Q}}(X) = \dim X$. By the way, from $rHd_{\mathbb{Q}}(X) = \dim X$ we get $rHd_{\mathbb{Q}}(Y) = \dim Y$.
In the example above we have however: $rHd_{\mathbb{Q}}(X) < \dim X$, cf. Lemma 3.2.

Other possibility: If the extension is possible, the spaces $X_{\epsilon,t}$ and $Y_{\epsilon,t}$, $|t| \ll \epsilon$, are contractible, just as $X_{\epsilon,0}$ and $Y_{\epsilon,0}$. In the example above, however, $Y_{\epsilon,t}$ consists of exactly 2 points for $0 < |t| \ll \epsilon$.

The situation is clear in the case $n = 2$, as we will see. For $n > 2$ it is unclear whether there is an example with $rHd_{\mathbb{Q}}(X) = \dim X$. That the situation is not so simple can be seen from

Theorem 3.3. *Let us start from the assumptions 1.1, 1.2, and 2.1. Fur-*

thermore assume that $rHd_{\mathbb{Q}}X = n = \dim X$. Then the conditions of Theorem 2.1 are equivalent to

d) $Y_{\epsilon,t}$ and $X_{\epsilon,t}$ are contractible, $0 < |t| \ll \epsilon$.

Proof. $b) \Longrightarrow d)$: The spaces in $d)$ are transformed by attaching cells of dimension $n - 1$ resp. n into contractible ones and have because of $b)$ the same homotopy type. Therefore there are added no cells in fact, i.e. the spaces are already contractible.

$d) \Longrightarrow a_1)$: trivial. $\qquad\qquad\square$

Of course, $d)$ may be reformulated as follows: the homotopy type of $Y_{\epsilon,t}$ does not depend on t, $|t| \ll \epsilon$, similarly for $X_{\epsilon,t}$. Further equivalent conditions are obtained by

Lemma 3.1. *Let us start from Assumption 1.1,1.2, and 2.1.*

a) If $rHd_{\mathbb{Q}}X = n$ we have for $0 < |t| \ll \epsilon$:
  $X_{\epsilon,t}$ *contractible* $\Longleftrightarrow \chi(X_{\epsilon,t}) = 1 \Longleftrightarrow \chi(X_\epsilon, X_{\epsilon,t}) = 0.$

b) In the case $n = 1$ we have:
  $X_{\epsilon,t}$ *contractible* $\Longleftrightarrow \chi(X_{\epsilon,t}) = 1 \Longleftrightarrow X_{\epsilon,t}$ *consists of exactly one point*
  $\Longleftrightarrow g : X \cap \{|g| < \beta\} \longrightarrow \{|t| < \beta\}$ *is biholomorphic.*

Proof. a) $X_{\epsilon,t}$ is transformed by attaching cells of dimension n into a contractible space, cf. proof of Theorem 3.3. The rest ist clear.

b) Suppose that $X_{\epsilon,t}$ consists of exactly one point: Let $\pi : \tilde{X} \longrightarrow X$ be the normalization of X. Then the fibre of $g \circ \pi$ above $t \neq 0$ consists of exactly one point, so $g \circ \pi$ and therefore g is biholomorphic. The other implications are clear. $\qquad\qquad\square$

b) The case $n = 2$

Now consider the case $n = 2$, i.e. $\dim X = 2$: First we want to reformulate the condition $rHd_{\mathbb{Q}}X = 2$. A complex space X is called connected in $x \in X$ if x admits a fundamental system of neighbourhoods U such that $U \setminus \{x\}$ is connected.

Lemma 3.2. *Let us make assumptions 1.1, 1.2, and 2.1. Assume that $\dim X = 2$ and that $\{0\}$ is the only 0-dimensional stratum of X. Then the following assertions are equivalent:*

a) $rHd_{\mathbb{Q}}X = 2$,

b) X is connected in 0 .

Proof. a) $\Rightarrow$ b): $\{x \mid H_1(X, X \setminus \{x\}; \mathbb{Q}) \neq 0\}$ is contained in some analytic subset of dimension ≤ -1 . Therefore $\tilde{H}_0(X \setminus \{0\}; \mathbb{Q}) = H_1(X, X \setminus \{0\}; \mathbb{Q}) = 0$.

b) $\Rightarrow$ a): Let $z \in Y \setminus \{0\}$ and let H be a hyperplane through z which is sufficiently general. Then the neighbourhood boundary of $Y \cap H$ with respect to z is non-void, because X has been supposed to be purely 2-dimensional. Furthermore the neighbourhood boundary $\partial B_\epsilon \cap X$ of X with respect to 0 is connected and non-void: the former holds by b), the latter because X has been supposed to be purely 2-dimensional. $\square$

Theorem 3.4. *If the assumptions 1.1,1.2, and 2.1 hold and if $n = 2$ the following conditions are equivalent:*

a) *$Y \cap \{g = t\}$ consists for $|t| \ll \epsilon$ of exactly one point, and the conditions from Theorem 2.1 hold,*

b) *$\chi(Y \cap \{g = t\}) = \chi(X_{\epsilon,t}) = 1$, $0 < |t| \ll \epsilon$,*

b') *$Y \cap \{g = t\}$ and $X_{\epsilon,t}$ are contractible, $0 < |t| \ll \epsilon$,*

c) *$rHd_{\mathbb{Q}} X = 2$ and the equivalent conditions from Theorem 2.1 (or from Theorem 3.2) hold,*

d) *there is a commutative diagram of the form*

$$X_\epsilon \cap \{|f| \leq \alpha, |g| \leq \beta\}$$
$$\sim \downarrow \qquad\qquad \searrow (f,g)$$
$$(X_{\epsilon,0} \cap \{|f| \leq \alpha\}) \times D_\beta \overset{(f,pr_2)}{\longrightarrow} D_\alpha \times D_\beta$$

where $\max(\alpha, \beta) \ll \epsilon \ll 1$.

Proof. We show: $a) \Rightarrow d) \Rightarrow b') \Rightarrow b) \Rightarrow a)$ as well as $b' \Leftrightarrow c)$.

a) $\Rightarrow$ d): By Lemma 3.1b), $g|Y \cap \{|g| \leq \beta\} \longrightarrow D_\beta$ is biholomorphic: In fact, replace (X, Y) by $(Y, \{0\})$ in Lemma 3.1b); of course Assumption 2.1 still holds in our case. By Theorem 2.1 there is a diffeomorphism h' : $X_\epsilon \cap \{0 < |f| \leq \alpha, |g| \leq \beta\} \longrightarrow (X_{\epsilon,0} \cap \{0 < |f| \leq \alpha\}) \times D_\beta$ which is compatible with (f, g). We extend h' to a bijective mapping $h : X_\epsilon \cap \{|f| \leq \alpha, |g| \leq \beta\} \longrightarrow (X_{\epsilon,0} \cap \{|f| \leq \alpha\}) \times D_\beta$ by $y \mapsto (0, g(y))$ for $y \in Y$. Then h is continuous: It is sufficient to check the continuity of the first component h_1 of h in $y \in Y$. Let $0 < \epsilon' \ll 1$, then the inverse image of the compact set $(X \cap \{|f| \leq \alpha, g = 0, ||z|| \geq \epsilon'\}) \times D_\beta$ under h' is compact; the complement is an open neighbourhood of y which is mapped under h_1 into the open neighbourhood $X \cap \{|f| \le \alpha, g = 0, ||z|| < \epsilon'\}$ of 0 in $X_{\epsilon,0} \cap \{|f| \le \alpha\}$. As for the continuity of h^{-1}, it is sufficient to check it in $(0, t)$. Let $y \in Y$ with $g(y) = t$. Let V be an open neighbourhood of y, without loss of generality of the form $V' \cap \{|g - t| < \beta'\}$, V' being a neighbourhood of $Y \cap \{|g| \le \beta\}$ in $X_\epsilon \cap \{|f| \le \alpha, |g| \le \beta\}$. The complement of V' is mapped under h_1 onto a compact subset of $X \cap \{0 < |f| \le \alpha, g = 0\} \cap B_\epsilon$, let U_1 be its complement. Then $U_1 \times \{t' \,|\, |t' - t| < \beta'\}$ is mapped under h^{-1} into V .

$d) \Rightarrow b')$: $Y \cap \{g = t\}$ and $X_{\epsilon,t}$ are homeomorphic to $Y \cap \{g = 0\}$ resp. $X_{\epsilon,0}$ and therefore contractible.

$b') \Rightarrow b)$: trivial.

$b) \Rightarrow a)$: Because of Lemma 3.1 $Y \cap B_\epsilon \cap \{g = t\}$ consists of one point. Furthermore, $\chi(X_{\epsilon,t}, Y \cap B_\epsilon \cap \{g = t\}) = 0$, $0 < |t| \ll \epsilon$, so the condition $a_1)$ from Theorem 2.1 is fulfilled.

$b') \Rightarrow c)$: Obviously condition $a_1)$ from Theorem 2.1 is fulfilled. Now $rHd_\mathbb{Q}X = 2$: Because of Lemma 3.2 we must show that X is connected in 0. Let $0 < \beta \ll \epsilon$. We can replace the neighbourhood boundaries of X with respect to 0 by $X \cap \{|g| = \beta, ||z|| \le \epsilon$ or $|g| \le \beta, ||z|| = \epsilon\}$. Since $X_{\epsilon,t}$ is connected and non-void for $|t| = \beta$ the same holds for $X_\epsilon \cap \{|g| = \beta\}$. Since every point of $X \cap \partial B_\epsilon \cap \{|g| \le \beta\}$ can be combined inside this space with a point of $X \cap \partial B_\epsilon \cap \{|g| = \beta\}$ we get the assertion.

$c) \Rightarrow b')$: Condition $d)$ from Theorem 3.3 is fulfilled. $\qquad\square$

Now we make

Assumption 3.1. $Y_0 := U \cap (\{0\} \times \mathbb{C})$ *is contained in* Y *for* $0 < \beta \ll 1$.

In fact, this assumption is natural if we look at equisingularity of a family of space germs.

For $|t| \le \beta$ let $\chi_t := \chi(\{(\xi, t) \in X \,|\, ||\xi|| \le \epsilon', f(\xi, t) = \tau\})$ where for given (!) t the following holds: $0 < |\tau| \ll \epsilon' \ll 1$. Then we can use in the case $X = \mathbb{C}^2$ the following lemma in order to add further equivalent conditions to Theorem 3.4, see the Remark below:

Lemma 3.3. *Let us start from the assumptions 1.1, 1.2, 2.1, and 3.1. For* $\dim X = 2$ *the following assertions are equivalent:*

a) $Y \cap \{g = t\} = \{(0, t)\}$, $|t| \le \beta$, *i.e. the germs of* Y_0 *and* Y *in* 0 *coincide,*
b) $Y \cap \{g = t\}$ *consists for* $|t| \le \beta$ *of exactly one point,*

c) χ_t does not depend on t, $|t| \leq \beta$.

Proof. a) $\Leftrightarrow$ b): clear.

a) $\Rightarrow$ c): Obviously χ_t is because of a) the number of inverse images of (τ, t) with respect to (f, g), where $\tau \neq 0$ is sufficiently small for fixed t . So it is the mapping degree of (f, g).

c) $\Rightarrow$ a): More irreducible components of Y would imply that the mapping degree of (f, g) coincides with $\chi(t)$ for $t = 0$ and is larger than $\chi(t)$ for $t \neq 0$. $\qquad\square$

Remark: For the case $X = \mathbb{C}^2$ we get under the additional assumption 3.1 the result that the conditions from Theorem 3.4 are equivalent to

e) χ_t does not depend on t.

Namely for this X we have $\chi(X_{\epsilon,t}) = 1$, $0 < |t| \ll \epsilon$, so that we can work with condition b) from Theorem 3.4.

For arbitrary X this is not true:

Example 3.3. Let $X := \{z \in \mathbb{C}^3 \mid z_2^2 = z_1(z_1 - z_3)\}$, $f(z) := z_1$. Obviously $\chi_t \equiv 2$, indeed $Y = \{0\} \times \mathbb{C}$, in accordance with Lemma 3.3. The equivalent conditions from Theorem 3.3 are not fulfilled, however: for instance $\Gamma \neq \emptyset$, and we have no topological triviality. This can be read off e.g. from the multiplicity: Let m_t be the multiplicity of X in $(0, t)$. Then $m_0 = 2$, $m_t = 1$ für $t \neq 0$. Indeed X has an isolated singularity in 0.

The invariant χ_t depends not only on g but also on f :

Example 3.4. Let $X := \{z \in \mathbb{C}^3 \mid z_2 = z_1^2\}$.
If $f(z) = z_1$ (and equally in the general case) $\chi_t = 1$;
if $f(z) = z_2$, $\chi_t = 2$.

For the validity of Theorem 3.4 we need in the case $Y_0 \subset Y$ (i.e. under assumption 3.1) that $Y_0 = Y$ because of Theorem 3.4b') . So we assume in the following section that Y is smooth.

4. Local triviality of $g|X$ in the case $\dim X = 2$, Y smooth

Let us assume that $\dim X = 2$. In contrast to §2 and §3 we give up the assumption 2.1 that Y is the zero level set of some holomorphic function f. Of course we keep assumptions 1.1 and 1.2.

In order to have local triviality in the sense of Theorem 3.2 it is because of Lemma 3.1b) (with Y instead of X) necessary that the mapping $g|Y$ is locally biholomorphic near 0. Therefore we assume in this paragraph without loss of generality:

Assumption 4.1. $Y = U \cap (\{0\} \times \mathbb{C})$

Let $h : U \to \mathbb{C}$ be holomorphic. For the moment we suppose that h is a linear form such that $h|Y \not\equiv 0$ and $h|T \not\equiv 0$ for all T of the form $T = \lim_{k \to \infty} T_{p_k} X$, $p_k \to 0$, $p_k \in X \setminus Y$. Let $\epsilon > 0$ be sufficiently small, $0 < |s| \ll \epsilon$, $cl_0 X := X \cap B_\epsilon \cap \{h = s\}$. It is well-known that the homeomorphy type of $cl_0 X$ ("complex link") does not depend on the choice of h, ϵ, s, cf. [7].

a) The case $X = $ hypersurface

The situation is particularly transparent if X is a hypersurface, cf. the list in [19] p.623f., too:

Theorem 4.1. *(cf. [11]): Let us make assumptions 1.1,1.2, and 4.1, and assume that X is a hypersurface: $X = \{h = 0\} \subset U \subset \mathbb{C}^3$, $h : U \to \mathbb{C}$ holomorphic and without critical points outside Y. Then the following assertions are equivalent:*

a) *The Milnor number of $h|\{g = t\}$ in $(0,t)$ does not depend on t, $|t| \ll \epsilon$,*

b) *(X, Y) is Whitney-regular,*

c) *there is a commutative diagram of the form*

$$(B_\epsilon \cap \{|h| \le \alpha, |g| \le \beta\}, Y \cap \{|g| \le \beta\})$$
$$\sim \downarrow \qquad\qquad\qquad\qquad \searrow (h,g)$$
$$(B_\epsilon \cap \{|h| \le \alpha, g = 0\}, Y \cap \{g = 0\})) \times D_\beta \xrightarrow{(h, pr_2)} D_\alpha \times D_\beta$$

where $\max\{\alpha, \beta\} \ll \epsilon \ll 1$,

d) *there is a commutative diagram of the form*

$$X_\epsilon \cap \{|g| \le \beta\}$$
$$\sim \downarrow \qquad \searrow g$$
$$X_{\epsilon,0} \times D_\beta \xrightarrow{pr_2} D_\beta$$

where $0 < \beta \ll \epsilon \ll 1$,

e) *$\chi(X_{\epsilon,t})$ does not depend on t, $|t| \ll \epsilon \ll 1$.*

Proof. a) $\Rightarrow$ b): cf. [11].

b) $\Rightarrow$ c): Altogether we have a Whitney-regular stratification of U. The assertion now follows as in the proof of Theorem 3.1.

c) $\Rightarrow$ d), d) $\Rightarrow$ e): trivial.

c) $\Rightarrow$ e): clear, since $\chi(X_{\epsilon,t}) = \chi(X_{\epsilon,t} \cap \{|h| \leq \alpha\}) = \chi(X_{\epsilon,0} \cap \{|h| \leq \alpha\}) = 1$ for $0 < |t| \ll \alpha \ll \epsilon \ll 1$.

e) $\Rightarrow$ a): Let $0 < \epsilon' \ll |t| \ll \epsilon$. Then $\chi(X \cap \{(\xi,t) \,|\, \|\xi\| \leq \epsilon'\}) = 1$. By hypothesis $\chi(X_{\epsilon,t}) = 1$, so $\chi(X \cap B_\epsilon \cap \{(\xi,t) \,|\, \|\xi\| \geq \epsilon'\}) = 0$. Therefore $\chi(B_\epsilon \cap \{(\xi.t) \,|\, \|\xi\| \geq \epsilon', h(\xi,t) = \tau\}) = 0$, too, $0 < \tau \ll \epsilon'$. So we get: Milnor number of $h|\{g = t\}$ in $(0,t) = \pm\chi(B_\epsilon \cap \{g = t, h = \tau\}) = \pm\chi(B_\epsilon \cap \{g = 0, h = \tau\}) = $ Milnor number of $h|\{g = 0\}$ in $(0,0)$. $\qquad\square$

Further equivalent conditions can be obtained using Theorem 4.2 and 4.3, starting from c) and d), resp. using Theorem 4.4, starting from b).

Note that under the conditions of Theorem 4.1 Y can be defined by some holomorphic function, see Theorem 4.4a) below.

b) Equisingularity and Bekka's condition

The conditions which will be studied now will automatically imply that Y is defined by some function which is continuous and weakly holomorphic. For this notion see [10].

Note that Theorem 2.1 still holds if we suppose about f only that f is continuous and weakly holomorphic (pass to the normalization of X).

Let us recall Bekka's c-condition [1]:

(X,Y) fulfills this condition if there is a C^1-function $\phi : X \to \mathbb{R}$ with $\phi^{-1}(\{0\}) = Y$ such that Thom's a_ϕ-condition is fulfilled and $\phi|X \setminus Y$ is a submersion.

Theorem 4.2. *Under the assumptions 1.1, 1.2, and 4.1, the following assertions are equivalent:*

a) *There is a commutative diagram of the form*

$$(X_\epsilon \cap \{|g| \leq \beta\}, Y \cap \{|g| \leq \beta\})$$
$$\sim\downarrow \qquad\qquad \searrow g$$
$$(X_{\epsilon,0}, 0) \times D_\beta \xrightarrow{pr_2} D_\beta$$

where $0 < \beta \ll \epsilon \ll 1$,

b) *for all $g_1 : U \to \mathbb{C}$ holomorphic, $g_1(0) = 0$, $g_1|Y$ submersive, a) holds with g_1 instead of g,*

c) (X, Y) fulfills Bekka's c-condition,

d) $\chi(X_{\epsilon,t}) = 1$, $0 < |t| \ll \epsilon \ll 1$,

e) $cl_0 X$ is contractible,

f) $\chi(cl_0 X) = 1$.

In this case there is a weakly holomorphic continuous function $f : U \to \mathbb{C}$ with $Y = X \cap \{f = 0\}$.

Proof. a) $\Rightarrow$ b): Our assumptions enable us to apply [5] Th. 5.2.2. Indeed, the mapping $g|X$ is flat after shrinking U if necessary: By hypothesis the space $\{g = 0\}$ intersects $X \backslash Y$ transversally, so the fibres of $g|X$ are reduced, and $g|X$ is open. Therefore, by [2] V Theorem 2.13 $g|X$ is flat. So we have the hypothesis of the theorem quoted abeve.

According to the implication (3) $\Rightarrow$ (4) of this theorem there is a weak simultaneous resolution $n : \tilde{X} \to X$. Here $\tilde{X}$ is smooth; let $\tilde{Y} := n^{-1}(Y)$ and $\tilde{f} : \tilde{X} \to \mathbb{C}$ be holomorphic, $\tilde{f}^{-1}(\{0\}) = \tilde{Y}$. Of course, $n|\tilde{X} \backslash \tilde{Y} \to X \backslash Y$ is biholomorphic. so there is exactly one function $f : X \to \mathbb{C}$ with $f \circ n = \tilde{f}$. Obviously f is continuous and weakly holomorphic.

Let g_1 be chosen according to b) and let $r > 0$ be sufficiently small. Now (X, Y) fulfills generically, in particular along $|g_1| = r$, Thom's a_f-condition. Since $g_1|Y$ is submersive along $|g_1| = r$, $|g_1| = r$ is transversal to all T of the form $T = \lim_k T_{p_k}(X \cap f^{-1}(f(p_k)))$, where (p_k) is a a sequence in $X \backslash Y$ with $\lim_k p_k \in Y \cap \{|g| = r\}$. Therefore, if $\alpha > 0$ is sufficiently small and if $\tilde{g}$ is sufficient near g_1 with respect to the topology of compact convergence the space $|\tilde{g}| = r$ is transversal to the spaces $X \cap \{f = \tau\}$ with $0 < |\tau| \le \alpha$. Let C be the closure of the critical set of $(f, \tilde{g})|X \cap \{|f| < \alpha, |\tilde{g}| < r\}$, then we get: $C \cap \{|f| < \alpha\}$ is analytic and is mapped by $(f, \tilde{g})$ onto some analytic subset D of $\{(\tau, t)| |\tau| < \alpha, |t| < r\}$. Because of Sard's theorem D is at most one-dimensional, the same holds for C.

Now choose $0 < \tau_0 < \alpha$ and $|t_0| < r$ such that $g_1 = t_0$ intersects the space $X \cap \{f = \tau_0\}$ transversally. If $\tilde{g}$ is sufficiently near g_1 the Euler characteristic of $X \cap f^{-1}(\tau_0) \cap \{|\tilde{g}| \le r\}$ does not depend on $\tilde{g}$, the same holds for the one of $X \cap \{f = \tau_0, \tilde{g} = t_0\}$. If we consider $\tilde{g}|X \cap \{f = \tau_0\}$ we get: whether there are critical points of $\tilde{g}|X \cap \{f = \tau_0\}$ in $\{|\tilde{g}| \le r\}$ is independent of $\tilde{g}$. Furthermore the result does not change if we replace τ_0 by τ with $0 < |\tau| \le \tau_0$. Therefore, if $g|(X, Y)$ is (not) locally trivial, the same holds for $\tilde{g}|(X, Y)$, $\tilde{g}$ near g. This implies our assertion.

b) $\Rightarrow$ c): Choose f as in a) $\Rightarrow$ b). Then (X, Y) fulfills Thom's a_f-condition: Otherwise there is a sequence (p_k) in $X \backslash Y$ which converges to

0 such that $T_{p_k}(X \cap f^{-1}(f(p_k)))$ converges to some $T \neq T_0 Y$. Then there is a linear form g with $g|T = 0$ so that $g|Y$ is submersive. Choose τ_0 as in the proof a) $\Rightarrow$ b). Arbitrarily near g there is - because of the choice of g - some $\tilde{g}$ and $0 < |\tau| < \tau_0$ so that $\tilde{g}|X \cap \{f = \tau\}$ has critical points . On the other hand there cannot be such points because of the considerations above, contradiction.

Now we can give a C^∞-function $\phi : X \to \mathbb{R}$ which shows that (X,Y) Bekka's c-condition is fulfilled: Extend $|f| \, |X \setminus Y \to]0, \infty[$ to a C^∞-function $\lambda : U \setminus Y \to]0, \infty[$; we can achieve that λ can be extended continuously by 0 to U . More precisely: let $A(\epsilon') := \max\{|f(z)| \, ; \, |z| = \epsilon', z \in X\}$; then $A(\epsilon') \to 0$ for $\epsilon' \to 0$. Extend $|f|$ to a neighbourhood W of $X \setminus Y$ in $U \setminus Y$; then $W' := \{z \in W \, | \, |f(z)| < 2A(|z|)\}$ is such a neighbourhood, too. Using some partition of unity we can find a C^∞-function $\lambda : U \setminus Y \longrightarrow \mathbb{R}$ with $0 < \lambda(z) \leq A(|z|)$. Obviously λ can be extended continuously by 0 to U . Put $\phi|U \setminus Y := e^{-1/\lambda}$, $\phi|Y := 0$.

c) $\Rightarrow$ a): follows by integrating some suitable vector field, cf. Bekka [1].

a) $\Rightarrow$ d): clear.

d) $\Rightarrow$ a): By [5] Th. 5.2.2 (3) $\Rightarrow$ (6) (the hypothesis of this theorem has already been verified above, see a) $\Rightarrow$ b)) there is a commutative diagram of the form

$$
\begin{array}{ccc}
X_\epsilon \cap \{|g| \leq \beta\} & & \\
\sim\downarrow \qquad & \searrow g & \\
X_{\epsilon,0} \times D_\beta & \xrightarrow{pr_2} & D_\beta
\end{array}
$$

where $0 < \beta \ll \epsilon \ll 1$. Inspection of the proof shows that even a) holds.

b) $\Rightarrow$ e): clear, since g_1 can be chosen generically.

e) $\Rightarrow$ f): trivial.

f) $\Rightarrow$ a): First, because of d) $\Rightarrow$ a) we obtain the assertion for a generically chosen g_1 instead of of g. Because of a) $\Rightarrow$ b) we get the assertion.

The existence of the function f in question has been shown in the proof a) $\Rightarrow$ b). $\qquad\qquad\square$

In particular, the equivalence of the conditions a), c) and f) corresponds to Theorem 1.2.

Further equivalent conditions can be obtained with the help of

Theorem 4.3. *Again let us start from the assumptions 1.1, 1.2, and 4.1. Let $f : U \to \mathbb{C}$ be a weakly holomorphic continuous function with $Y = X \cap \{f = 0\}$. Then the following assertions are equivalent:*

a) The equivalent assertions of Theorem 2.1 hold,

b) the equivalent assertions of Theorem 4.2 hold,

c) there is a commutative diagram of the form

$$X_\epsilon \cap \{|f| \le \alpha, |g| \le \beta\}$$
$$\sim\downarrow \qquad\qquad \searrow (f,g)$$
$$(X_{\epsilon,0} \cap \{|f| \le \alpha\}) \times D_\beta \xrightarrow{(f,pr_2)} D_\alpha \times D_\beta$$

where $\max(\alpha, \beta) \ll \epsilon \ll 1$,

d) (X, Y) *fulfills Thom's* a_f*-condition.*

Proof. a) $\Rightarrow$ c): with Theorem 3.4, since $\chi(Y \cap \{g = t\}) = 1$, $|t| \ll \epsilon$; Theorem 3.4 is still valid if f is only weakly holomorphic and continuous.

c) $\Rightarrow$ b): clear.

b) $\Rightarrow$ a): Obviously Theorem 4.2d) is fulfilled.

b) $\Leftrightarrow$ d): cf. proof of Theorem 4.2 b) $\Rightarrow$ c). $\qquad\qquad\qquad$ $\square$

Using normalization we can add further equivalent conditions, see. [5].

c) Equisingularity and Whitney condition

As Example 4.1 will show we cannot conclude Whitney-regularity in the case of Theorem 4.2; here a stronger hypothesis is necessary. It will imply automatically that Y is defined by some holomorphic function, see Theorem 4.4a) below.

Theorem 4.4. *Under the assumptions 1.1, 1.2, and 4.1, the following assertions are equivalent:*

a) There is a commutative diagram the form

$$(X_\epsilon \cap \{|g| \le \beta\}, Y \cap \{|g| \le \beta\})$$
$$\sim\downarrow \qquad\qquad \searrow g$$
$$(X_{\epsilon,0}, 0) \times D_\beta \xrightarrow{pr_2} D_\beta$$

where $0 < \beta \ll \epsilon \ll 1$,

and if f *is a linear form with* $f|Y = 0$ *and* f *is sufficiently general the germs of* $X \cap \{f = 0\}$ *and* Y *in* 0 *coincide,*

b) $cl_0 X$ *ist contractible, and the multiplicity* $m_t(X)$ *of* X *in* $(0, t)$ *is constant along* Y ,

c) (X, Y) *is Whitney-regular.*

Proof. a) $\Rightarrow$ b): Because of Theorem 4.2 we may use instead of g an arbitrary linear form g such that $g|Y$ is submersive. If we choose f and g general enough we have $\chi_t = m_t =$ multiplicity of X in $(0, t)$, so the latter is constant. The rest follows from Theorem 4.2.

b) $\Rightarrow$ c): The first condition says that $\chi(X_\epsilon \cap \{g = t\})$ is constant for general linear g , so the assertion follows by [12].

c) $\Rightarrow$ a): Let f be a linear form chosen general enough. If $T \neq T_0 Y$ is a generalized tangent space to X in 0 (cf. [7]) we have $f|T \not\equiv 0$, cf. [20]. Suppose that the germs of Y and $X \cap \{f = 0\}$ in 0 do not coincide: then there are sequences (ξ_k), (t_k) which converge to 0 such that $(\xi_k, t_k) \in X \setminus Y$, $f(\xi_k, t_k) = 0$, $T_{(\xi_k, t_k)} \to T$, $\frac{\xi_k}{||\xi_k||} \to v$. Because of Whitney's b)-condition we have $(v, 0) \in T$ but: $d_0 f(v, 0) = d_0 f(0, 1) = 0$, therefore $f|T \equiv 0$, contradiction. $\qquad\square$

Example 4.1. Let $\Phi : \mathbb{C}^2 \longrightarrow \mathbb{C}^4$ be defined by $(\tau, t) \mapsto (\tau^2, \tau^3, t\tau, t)$. Let $X := \Phi(\mathbb{C}^2)$, $Y := \{0\} \times \mathbb{C} \subset X$, $f : \mathbb{C}^4 \longrightarrow \mathbb{C}: z \mapsto z_1$. Obviously $X = \{(z_1, z_2, z_3, t) \in \mathbb{C}^4 \,|\, z_1^3 = z_2^2, t^2 z_1 = z_3^2, z_1 z_3 = t z_2\}$; we have $\Phi^{-1}(z, t) = (\frac{z_2}{z_1}, t)$ for $(z, t) \in X$, $z_1 \neq 0$.

Of course, the conditions of Theorem 4.3 are fulfilled: we have $Y = X \cap \{f = 0\}$. A topological trivialization is induced by $(\tau, t) \mapsto ((\tau, 0), t)$, i.e.: $(z_1, z_2, z_3, t) \mapsto ((z_1, z_2, 0, 0), t)$ for $(z, 0) \in X$. Therefore Theorem 4.2a) is fulfilled.

On one hand, $\frac{\partial \Phi}{\partial t}(\tau, t) = (0, 0, \tau, 1)$, for $(\tau, t) \to (0, t_0)$ the direction converges to $(0, 0, 0, 1) \in T_0 Y$. Therefore Thom's a_f-condition is fulfilled, i.e. condition d) from Theorem 4.3.

On the other hand, the conditions from Theorem 4.4 are violated: Now let f be a general linear form with $f|Y \equiv 0$: $f = az_1 + bz_2 + cz_3$ with $c \neq 0$. Then $f(\Phi(\tau, t)) = a\tau^2 + b\tau^3 + ct\tau$. Therefore, for $\tau \neq 0$ we have: $f(\Phi(\tau, t)) = 0 \Leftrightarrow (\tau, t)$ is a solution of the equation $b\tau^2 + a\tau + ct = 0$. Obviously there are arbitrarily near $(0, 0)$ solutions with $\tau \neq 0$. So a) is violated.

We have $\frac{\partial \Phi}{\partial \tau}(t, t) = (2t, 3t^2, t, 0)$, the direction converges for $t \to 0$ to that of $(2, 0, 1, 0)$. Furthermore, $\frac{\partial \Phi}{\partial t}(t, t) = (0, 0, t, 1)$, the direction converges to that of $(0, 0, 0, 1)$. So the corresponding generalized tangent space T to X in 0 is spanned by $(2, 0, 1, 0)$ and $(0, 0, 0, 1)$. On the other hand, the direction of $\Phi(t, t) - \Phi(0, t) = (t^2, t^3, t^2, 0)$ converges to that of $v = (1, 0, 1, 0)$. Because of $v \notin T$ Whitney's condition b) is violated, therefore condition c) of Theorem 4.4.

Obviously X is smooth outside 0 , so $m_t = 1$ for $t \neq 0$. Since Whitney's

648

b)-condition is violated, X cannot be smooth (with the reduced structure), so $m_0 \neq 1$. Therefore condition b) from Theorem 4.4 is violated, too.

References

1. K.Bekka: c-régularité et trivialité topologique. In: Singularity Theory and its Applications, Warwick 1989, part I. Springer Lecture notes in Math. **1462**, 42-62 (1991).

2. C.Bănică, O.Stănăşilă: Algebraic methods in the global theory of complex spaces. Ed. Academiei: Bucureşti / John Wiley & Sons: London 1976.

3. J.Briançon, P.Maisonobe, M.Merle: Localisation des systèmes différentiels, stratifications de Whitney et condition de Thom. Invent. Math. **117**, 531-550 (1994).

4. J.Briançon, J.-P.Speder: La trivialité topologique n'implique pas les conditions de Whitney. C.R.Acad.Sci.Paris Sér. A **280**, 365-367 (1975).

5. R.-O.Buchweitz, G.-M.Greuel: The Milnor number and deformations of complex curve singularities. Invent. Math. **58**, 241-281 (1980).

6. A.A.du Plessis: Continuous controlled vector fields. In: Singularity theory, Liverpool 1996. London Math. Soc. Lecture note Ser. **263**, 189-197 (1999).

7. M.Goresky, R.MacPherson: Stratified Morse theory. Springer-Verlag: Berlin 1988.

8. H.A.Hamm: Zum Homotopietyp q-vollständiger Räume. J. reine angew. Math. **364**, 1-9 (1986).

9. H.A.Hamm, Lê D.T.: Rectified homotopical depth and Grothendieck conjectures. The Grothendieck Festschrift, Vol. II, pp. 311-351. Birkhäuser: Boston 1990.

10. L.Kaup, B.Kaup: Holomorphic Functions of Several Variables. De Gruyter: Berlin 1983.

11. Lê D.T., C.P.Ramanujam: The invariance of Milnor's number implies the invariance of the topological type. Amer.J.Math. **98**, 67-78 (1976).

12. Lê D.T., B. Teissier: Cycles évanescents, sections planes et conditions de Whitney. II. In: Proc. Summer Inst. on Singularities (Arcata 1981), vol. II, pp. 65-103. Proc. Symp. Pure Math. **40**. AMS: Providence, R.I. 1983.

13. J.Milnor: Singular points of complex hypersurfaces. Annals of Math. Studies

61. Princeton: Princeton Univ. Press 1968.

14. A.Parusiński: A note on singularities at infinity of complex polynomials. Symplectic singularities and gauge fields, Banach Center Publ. **39**, 131-141 (1997).

15. D.Siersma, M.Tibăr: Singularities at infinity and their vanishing cycles. Duke Math. J. **80**, 771-783 (1995).

16. M.Shiota: Geometry of Subanalytic and Semialgebraic Sets. Birkhäuser: Boston 1997.

17. D.Sullivan: Combinatorial Invariants of Analytic Spaces. In: Proc. Liverpool Singularities Symposium I (1969/70). Springer Lecture notes in Math. **192**, 165-168 (1971).

18. B.Teissier: Résolution simultanée. In: Sém. sur les singularités des surfaces, Palaiseau 1976/77. Springer Lecture notes in Math. **777**, 229-245 (1980).

19. B.Teissier: The Hunting of Invariants in the Geometry of Discriminants. In: Real and Complex Geometry, Proc. Nordic Summer School/NAVF, Oslo 1976, pp. 565-677. Sijthoff&Noordhoff: Alphen aan den Rijn 1977.

20. B.Teissier: Variétés polaires II. Multiplicités polaires, sections planes et conditions de Whitney. In: Algebraic Geometry, La Rábida 1981. Springer Lecture notes in Math. **961**, 314-491 (1982).

21. M.Tibăr: Topology at infinity of polynomial mappings and Thom regularity condition. Compositio Math. **111**, 89-109 (1998).

A General Image Computing Spectral Sequence

Kevin Houston
School of Mathematics, University of Leeds,
Leeds, LS2 9JT, U.K.

e-mail address: k.houston@leeds.ac.uk

A general existence theorem for a spectral sequence that calculates the homology of the image of a certain type of finite and proper map is given. The E^1 terms of this sequence are the alternating homology groups of the multiple point spaces of the map.
AMS Mathematics Subject Classification 2000: 55T05.

1. Introduction

Calculating the homology of the image of a map seems to be a problem more difficult than calculating the homology of the fibre of a map. However, significant advances have been made recently in the topology of singularities through the use of a spectral sequence calculating the homology of an image, e.g., [6], [7] and [8].

In [7] a spectral sequence for calculating the rational cohomology of the image of a finite and proper continuous map was introduced. This was further refined to integer homology for algebraic maps between compact semi-algebraic spaces in [6] using a semi-simplicial resolution similar to that introduced by Vassiliev in his new knot invariant theory, [16], and used in a number of other areas, see [17] for a survey. The E^1 terms of this spectral sequence were given by the alternating homology, (homology arising from anti-symmetric chains under the action S_k, the group of permutations on k objects) of the multiple point spaces of the map. The kth multiple point space of a map f, denoted $D^k(f)$, is the closure in the product of k copies of the source, of the k-tuples of pairwise distinct points having the same image under f.

The key to application, such as in [6], [7], [8], [9] and [10], lies in the description of multiple points spaces as fibres of well understood maps. In general, the multiple point spaces are better behaved than the image. For example, the image may be a highly singular object but the multiple point spaces are non-singular. This is the case for stable maps from surfaces to three-space, (see [9]). And thus, because it is easier to calculate the homology of a fibre, we stand a chance of calculating the homology of the image. Furthermore, in certain situations the sequence has been used, in conjunction with fundamental group information, to provide a homotopy description of spaces, see [8].

We further generalise the existence of the sequence in [6] by the following. (i) Any coefficient group is allowed, (it is not obvious from the statement in [6] that we can change coefficient in the usual way); (ii) the spaces X and Y need no longer be compact nor embeddable in $\mathbb{R}^N$ for some large N; (iii) the maps are no longer restricted to algebraic ones; (iv) one can calculate relative homology groups; (iv) different types of support and different types of homology are easily incorporated.

Furthermore, a description of the differential of the spectral sequence is given; its existence is merely implicit in [6]. The description is also an alternative to that for the corresponding one in the rational cohomology version of [7].

As an example of the type of spectral sequence one can get from the main theorem, Theorem 5.1, we give the following. Let $f : X \to Y$ be a finite analytic map between compact subanalytic spaces and $\widetilde{X} \subset X$ be a subanalytic subspace. Then, there exists a spectral sequence

$$E_{p,q}^1 = H^{alt}(D^{p+1}(f), D^{p+1}(f|\widetilde{X}); G) \implies H_*(f(X), f(\widetilde{X}); G),$$

for any coefficient group G.

The theorem is proved by constructing a semi-simplicial resolution of the image of the map. A simplex in this resolution corresponds to a point of the image and its dimension depends on the number of preimages of the point. The same type of resolution in [6] was produced by first embedding the source into some $\mathbb{R}^N$. That such an embedding exists is not required in the construction here. (Anyone intimately familiar with Vassiliev's original construction will apppreciate this could be done).

The point of the resolution is that it has the same homology as the image and it can be filtered by the union of simplices of dimension less than k. Thus, the natural spectral sequence resulting from the filtration calculates the homology of the image. The relative terms of the filtration can be described in terms of the alternating homology of the multiple point spaces.

In section 2 the notions of alternating and alternated homology are explained and a condition ensuring the two notions coincide is given. Section

3 introduces multiple point spaces and the alternating homology of these are investigated. In section 4 the semi-simplicial resolution of the image of a map is constructed, its homology is investigated and then related to the alternating homology of the multiple point spaces of the map. Section 5 contains the statements of a number of different versions of the spectral sequence, some of which have been used in papers such as [10] and [11].

There are many applications of the sequence outside mainstream singularity theory. For example, some of Vassiliev's theory of knots can be rewritten using multiple point spaces and the sequence (in fact Vassiliev's resolution is where the resolution used by Goryunov originated). Another example of where the sequence can be applied is to a subspace arrangement. This space can be thought of as the image of a map, just take the inclusion of the disjoint subspaces into the ambient space. Again, the resulting semi-simplicial resolution is the same as the one used by Vassiliev in [16] for his proof of Goresky and Macpherson's Theorem on arrangements. Many other examples are of course possible, such as quotient spaces arising from finite group actions.

The author is grateful to Victor Goryunov, David Mond and Terry Wall for comments made upon some of the material in this paper. This paper was begun when the author was supported by a grant from the EPSRC (Grant number GR/K/29227).

2. Alternating Homology

2.1. *Algebra*

Suppose the group H acts on the abelian group M, (we call M an H-module). Let sign $: H \to \{\pm 1\}$ be a homomorphism from H to the group $\{\pm 1\} \cong \mathbb{Z}_2$. Define the operator $\mathrm{Tw}^H : M \to M$ by

$$\mathrm{Tw}^H = \sum_{h \in H} \mathrm{sign}(h)h.$$

We drop the reference to sign as it will be obvious from context which particular homomorphism is being used.

Definition 2.1. Suppose $c \in M$. Then,

(i) c is called H-twisting if $h.c = \mathrm{sign}(h)c$ for all $h \in H$;
(ii) c is called H-twisted if there exists $e \in M$ such that $c = \sum_{h \in H} \mathrm{sign}(h)h.e$.

It is easy to show that an H-twisted chain is H-twisting but the converse is not true.

The H-twisting and H-twisted elements form submodules of M. Denote by $M^{\langle H \rangle}$ (resp. $\mathrm{Tw}^H M$) the submodule of H-twisting (resp. H-twisted) elements of M.

Suppose (C_*, ∂) is a complex of abelian groups such that H acts upon C_n for all n and the action commutes with ∂ , i.e. $\partial(h.e) = h.(\partial e)$ for all $e \in C_*$ and $h \in H$.

Proposition 2.1. *The set of modules $C_n^{\langle H \rangle}$ (resp. $\mathrm{Tw}^H C_n$) and the homomorphisms*

$$\partial_n^{\langle H \rangle} := \partial_n | C_n^{\langle H \rangle} \quad (resp. \, \mathrm{Tw}^H \partial_n := \partial_n | \, \mathrm{Tw}^H C_n)$$

form a subcomplex of (C_*, ∂), denoted $(C_*^{\langle H \rangle}, \partial^{\langle H \rangle})$ (resp. $(\mathrm{Tw} \, C_*, \mathrm{Tw}^H \partial)$).

Both results are easy to prove, the key lies in the fact that the boundary homomorphism commutes with elements of H.

Definition 2.2. The homology of $(C_*^{\langle H \rangle}, \partial^{\langle H \rangle})$ is called the H-twisting homology of C_*. The homology of $(\mathrm{Tw}^H(C_*), \mathrm{Tw}^H \partial)$ is called the H-twisted homology of C_*.

Suppose D_* is a subcomplex of C_* inheriting the H-action. We can define *relative H-twisting* (resp. *H-twisted*) *homology* to be the homology of the complex

$$(C, D)_*^{\langle H \rangle} := C_*^{\langle H \rangle} / D_*^{\langle H \rangle},$$

resp.

$$\mathrm{Tw}^H(C, D)_* := \mathrm{Tw}^H C_* / \mathrm{Tw}^H D_*.$$

Such a definition allows long exact sequences of twisting (resp. twisted) homology to arise from short exact sequences of complexes.

Note that for a general H-submodule $N \subset M$ we do not have $\mathrm{Tw}^H(M/N) = \mathrm{Tw}^H M / \mathrm{Tw}^H N$. However, if a short exact sequence is split exact, then the twisting and tswisted submodules do fit into a short exact sequence:

Lemma 2.1. *A split short exact sequence of S_k-modules*

$$0 \to M'' \to M \to M' \to 0,$$

leads to short exact sequences of alternating and alternated groups:

$$0 \to M''^{\langle H \rangle} \to M^{\langle H \rangle} \to M'^{\langle H \rangle} \to 0,$$

respectively,

$$0 \to \mathrm{Tw}^H M'' \to \mathrm{Tw}^H M \to \mathrm{Tw}^H M' \to 0.$$

The proof is quite standard.

If sign $: H \to \{\pm 1\}$ is the trivial map, then we define the *symmetrised elements of M*, $\mathrm{Sym}\, M$, to be the set

$$\mathrm{Sym}\, M := \mathrm{Tw}^H M = \left\{ \sum_{\sigma \in H} \sigma(m) \text{ for all } m \in M \right\}.$$

This definition will be in a crucial step of the main theorem, but is also useful if one is interested in the H-invariant homology of some complex.

2.2. *Subgroups of S_k*

Denote the group of permutations of degree k by S_k and equip it with the usual sign representation: the sign of an element is negative if and only if the element is a product of an odd number of simple transpositions.

This case will be so important to us in the setting of twisted and twisting homology that we define $\mathrm{Alt}^H = \mathrm{Tw}^H$. If $H = S_k$, then we use the notation Alt.

Definition 2.3. Suppose $c \in M$. Then,

(i) c is called H-alternating if $h.c = \mathrm{sign}(h)c$ for all $h \in H$;
(ii) c is called H-alternated if there exists $e \in M$ such that $c = \mathrm{Alt}^H(e)$.

If $H = S_k$, then such elements are simply called alternating and alternated respectively. The H-alternating and H-alternated elements form submodules of M. Denote by $M^{alt\langle H \rangle}$ (resp. $\mathrm{Alt}^H M$) the submodule of H-alternating (resp. H-alternated) elements of M. If $H = S_k$, then we use the notation M^{alt}.

Even in this case of twisting $\mathrm{Alt}^H M$ may not be a proper submodule of $M^{alt\langle H \rangle}$ as simple well-known examples show. However, in the case of multiple point spaces, the two notions do coincide, see Theorem 2.1.

We can define H-alternating and H-alternated homology to be just special cases of twisting and twisted homology respectively. If $H = S_k$, then we call the homologies alternating and alternated respectively.

Introducing twisting and then specialising to alternating may seem like an unnecessary complication, however we wish to be able to deal with situations such as calculating the twisting homology of an alternating complex.

2.3. *Topology*

In this section the emphasis is moved from algebra to topology by concentrating on the specific case of the alternating homology of chain complexes arising from topological spaces with a group action.

Some basic definitions and results on the action of groups on topological spaces, in particular for CW-complexes, are needed.

Suppose H is a group which acts on the topological spaces X, Y and Z. In this case we call X, etc., an H-space.

Definition 2.4. Suppose $f : X \to Y$ is a continuous map, then we call f an H-equivariant map (or an H-map) if $h.f(x) = f(h.x)$ for all $h \in H$.

Definition 2.5. Let $f_0, f_1 : X \to Y$ be two continuous H-equivariant maps. We call f_0 and f_1 H-homotopic if there exists an H-equivariant map, called an H-homotopy from f_0 to f_1,

$$F : X \times [0,1] \to Y,$$

such that $F(x,0) = f_0(x)$ and $F(x,1) = f_1(x)$. The interval $[0,1]$ is given the trivial H-action and $X \times [0,1]$ the diagonal action. Each map $f_t := x \mapsto F(x,t)$ is then an H-map.

It is trivial that being H-homotopic is an equivalence relation and that we have a category of H-spaces and H-homotopy classes.

Suppose that J and L are CW-complexes upon which H acts.

Definition 2.6. The action on L is called cellular if the following conditions are satisfied.

(i) For any e, an open cell of L, $g.e$ is an open cell of L for all $g \in H$, i.e. cells are mapped to cells.
(ii) If $g.e = e$ for an open cell e, then $g(x) = x$ for all $x \in e$.

An H-equivariant map f is called *cellular* if $f(X_n) \subset Y_n$ for all n, where the subscript denotes the union of all cells of dimension less than or equal to n.

We have a useful (though non-trivial) theorem that, if $f : X \to Y$ is an H-equivariant map, then it is H-homotopic to a cellular map. The proof, see for example [3], is an adaptation of the corresponding theorem in the study of CW-complexes.

Consider the integral cellular complex of L, denoted $C_*(L)$. The action of H on L induces an action on $C_*(L)$. The naturally induced map $h_\# : C_*(L) \to C_*(L)$ on chains, induced from $h \in H$, commutes with the differential of $C_*(L)$ and so we can apply the definitions and theorems of the previous section. For an H-equivariant cellular map $f : J \to L$ we have $f_\# h_\# = h_\# f_\#$.

In this way cellular maps induce maps on homology such that H-homotopic maps induce the same map, etc. So, for example, if two spaces are S_k-homotopically equivalent, then their alternating homologies are isomorphic.

Suppose K is group with a representation $\chi : K \to \{\pm 1\}$ that acts on the cellular chain complex $C_*(L)$

Definition 2.7. Define the twisting homology of L with respect to K to be the homology of the K-twisting chains (denoted $C_*(L)^{\langle K \rangle}$). Denote this homology by $H_*^K(L)$.

Definition 2.8. Suppose that X is an S_k-space and has the S_k-homotopy type of a cellular S_k-complex, i.e. X is S_k-homotopically equivalent to an S_k-complex L. Suppose that H is a subgroup of S_k and that K is a group also acting cellularly on X. Then, we define the H-alternating homology of X twisted with respect to K with coefficient group G to be

$$H_*^{alt\langle H \rangle \times K}(X; G) := H_* \left(\left[(C_*(L))^{alt\langle H \rangle} \right]^{\langle K \rangle} \otimes G \right).$$

Similarly, we can define the H-alternated homology of X twisted with respect to K:

$$H_*^{\mathrm{Alt}^H \times K}(X; G) := H_* \left(\left[\mathrm{Alt}^H (C_*(L)) \right]^{\langle K \rangle} \otimes G \right).$$

If $H = S_k$ and K is the trivial group, then the notation $H^{alt}(X; G)$ is used. This is the notation used in [10] and [11].

The H-alternating and H-alternated homologies of a topological space may not coincide as simple examples will show. However, in the case were the action arises from the permutation of copies of X in $X^k := X \times \cdots \times X$ the notions will coincide. First, we need a couple of definitions.

Definition 2.9. Suppose that X is a topological space and let S_k act on X^k in the obvious way: permutation of copies of X. Let H be a subgroup of S_k.

The union of fixed point sets of all simple transpositions is called the diagonal of X^k with respect to H and is denoted $\mathrm{Diag}^H(X^k)$. Any cell fixed by a simple transposition is said to lie in $\mathrm{Diag}^H(X^k)$.

658

If $H = S_k$, then the space

$$\mathrm{Diag}(X^k) = \{(x_1, \ldots, x_k) \in X^k \,|\, x_i = x_j \ \text{ for some } i \neq j\}$$

is called the diagonal of X^k.

The next theorem is important for later theorems; its conclusion is most definitely false for more general finite group actions.

Theorem 2.1. *Let H be a subgroup of S_k and K be any group. Suppose X is a topological space and Y and Z are H and K-invariant subspaces of X^k with $Z \subset Y$ and H acts by restriction of the action of S_k on X^k. Let Y have a cellular decomposition such that Z is a subcomplex, and the groups H and K act cellularly. Suppose further that they act cellularly on* $\mathrm{Diag}^H(X^k) \cap Y$.

Then, the following are true.

(i) No cell lying in $\mathrm{Diag}^H(X^k)$ is part of an H-alternating chain. Thus,

$$\left(C_*(\mathrm{Diag}^H(X^k) \cap Y; \mathbb{Z})\right)^{alt\langle H \rangle} = \mathrm{Alt}^H\left(C_*(\mathrm{Diag}^H(X^k) \cap Y; \mathbb{Z})\right) = 0.$$

(ii) Any H-alternating chain of the cellular chain complex $C_*(Y; \mathbb{Z})$ is H-alternated, so $(C_i(Y))^{alt\langle H \rangle} \cong \mathrm{Alt}^H(C_i(Y))$.

(iii) $H_i^{alt\langle H \rangle \times K}(Y, Z; G) \cong H_i^{\mathrm{Alt}^H \times K}(Y, Z; G)$.

Proof. (i) Since $\mathrm{Alt}^H M$ is a submodule of $M^{alt\langle H \rangle}$ for any module M, the statement needs only to be proved for alternating chains. Let $c \in C_i(\mathrm{Diag}^H(X^k) \cap Y; \mathbb{Z})^{alt\langle H \rangle}$ be an H-alternating chain. Then, we can assume $c = \sum_{\lambda \in \Lambda} m_\lambda e_\lambda$, such that the e_λ are distinct, and lie in the diagonal with respect to H.

For a cell $e_1 \in \{e_\lambda\}$ we can consider its H-orbit, $\{e_1, \ldots, e_n\}$, for some n, and the corresponding coefficients in c, $\{m_1, \ldots, m_n\}$. Then $c = \sum_{i=1}^n m_i e_i + c'$ where c' contains no cells in the orbit of e_1.

Since $e_1 \subset \mathrm{Diag}^H(X^k)$, we have $e_1 \subset \{x_i = x_j\} \subset X^k$ for some $i \neq j$. Take $h \in H$ such that h only exchanges the ith and the jth coordinates. Then, because h is a simple transposition, $\mathrm{sign}(h) = -1$, and obviously $h(e_1) = e_1$ and $h(e_i) \neq e_1$ for $i \neq 1$.

So,

$$h(c) = m_1 h(e_1) + m_2 h(e_2) + \cdots + m_n h(e_n) + h(c')$$
$$= m_1 e_1 + m_2 h(e_2) + \cdots + m_n h(e_n) + h(c').$$

But if c is alternating, then $h(c) = \text{sign}(h)c = -c$ and this would imply that
$m_1 h(e_1) + m_2 h(e_2) + \cdots + m_n h(e_n) + h(c') = -(m_1 e_1 + \cdots + m_n e_n) - c'$.
Equating coefficients then gives that $m_1 = -m_1$, which can only happen if
$m_1 = 0$. Similarly all the other coefficients are zero and hence $c = 0$. We
can apply this reasoning to c' and through repetition deduce that $c = 0$.

(ii) By (i) we need only consider chains made up of cells that lie in
regular orbits of H (i.e. are not in the diagonal) and can again concentrate
on the orbit of one cell, e_1 say. If e_1 is in such an alternating chain, c, then
all the other cells in the orbit of e_1 are in c, hence we may assume that
$c = m_1 e_1 + \cdots + m_n e_n$ for some coefficients $m_i \in \mathbb{Z}, n = |H|$ and $e_i \neq e_j$
for $i \neq j$. Furthermore, $h_i(e_1) = e_i$ for some unique $h_i \in H, i = 1, \ldots, |H|$.
Thus,

$$\begin{aligned}
h_i(c) &= h_i(m_1 e_1 + \cdots + m_n e_n) \\
&= m_1 h_i(e_1) + \cdots + m_n h_i(e_n) \\
&= m_1 e_i + m_2 h_i(e_2) + \cdots + m_n h_i(e_n).
\end{aligned}$$

But $h_i(c) = \text{sign}(h_i)c$ as c is an alternating chain, and so

$$h_i(c) = \text{sign}(h_i)(m_1 e_1 + \cdots + m_n e_n).$$

Equating coefficients gives $m_1 = \text{sign}(h_i)m_i$, so $m_i = \text{sign}(h_i)m_1$.
Thus,

$$\begin{aligned}
c &= m_1 e_1 + \cdots + m_n e_n \\
&= m_1 \text{sign}(h_1)h_1(e_1) + \cdots + m_1 \text{sign}(h_n)h_n(e_1) \\
&= m_1 \text{Alt}^H(e_1).
\end{aligned}$$

(iii) Since the action is cellular and the alternated and alternating chains
are equal for Y and for Z by (ii), the corresponding quotient chain complexes are equal, hence the result. $\qquad\square$

We end with lemma which allows us to describe alternation of products
of groups. This will be important when applying Corollary 5.4.

Lemma 2.2. *Let H_1 and H_2 be two subgroups of S_k such that $h_1 h_2 = h_2 h_1$
for all $h_1 \in H_1$ and $h_2 \in H_2$. Suppose S_k acts on the module M. Define
the action of $H_1 \times H_2$ on M_2 by $(h_1, h_2).m = h_1.(h_2.m)$. (Note that order
is unimportant here.) Define $\text{sign}(h_1, h_2) = \text{sign}(h_1)\text{sign}(h_2)$.*

Then,

$$\mathrm{Alt}^{H_1}\mathrm{Alt}^{H_2}(m) = \mathrm{Alt}^{H_1 \times H_2}(m) \ \textit{for all } m \in M.$$

Proof.

$$\mathrm{Alt}^{H_1}\mathrm{Alt}^{H_2}(m) = \mathrm{Alt}^{H_1}\left(\sum_{h_2 \in H_2} \mathrm{sign}(h_2)h_2(m)\right)$$

$$= \sum_{h_1 \in H_1} \mathrm{sign}(h_1)h_1\left(\sum_{h_2 \in H_2} \mathrm{sign}(h_2)h_2(m)\right)$$

$$= \sum_{h_1 \in H_1}\sum_{h_2 \in H_2} \mathrm{sign}(h_1)\,\mathrm{sign}(h_2)h_1(h_2(m))$$

$$= \sum_{h_1 \in H_1}\sum_{h_2 \in H_2} \mathrm{sign}(h_1 h_2)(h_1, h_2)(m)$$

$$= \sum_{(h_1,h_2) \in H_1 \times H_2} \mathrm{sign}(h_1, h_2)(h_1, h_2)(m)$$

$$= \mathrm{Alt}^{H_1 \times H_2}(m). \qquad \square$$

3. Multiple Point Spaces of Finite Maps

The most important spaces to which we apply alternating homology will be multiple point spaces, the definition of which we now recall. Suppose $f : X \to Y$ is a continuous map of topological spaces.

Definition 3.1. The kth multiple point space of the map f is defined to be

$$D^k(f) = \mathrm{closure}\{(x_1, \ldots x_k) \in X^k | f(x_1) = \cdots = f(x_k) \text{ for } x_i \neq x_j \text{ with } i \neq j\}.$$

The reference to f in $D^k(f)$ can be dropped if no confusion will result. This definition has a number of advantages over divisions of X and Y by counting preimages of f. One is that $D^k(f)$ has more symmetry: the group of permutations on k objects acts on the space by permuting the set of coordinates x_1, x_2, etc. Another is that singularities of $D^k(f)$ are generally simpler than those in the image, for example, they may even be non-singular, see [13].

For each i and k we have a map $\varepsilon_{i,k} : D^k(f) \to D^{k-1}(f)$ which 'forgets' the ith coordinate of $D^k(f)$,

$$\varepsilon_{i,k}(x_1, \ldots x_{i-1}, x_i, x_{i+1}, \ldots, x_k) = (x_1, \ldots x_{i-1}, x_{i+1}, \ldots, x_k).$$

There is also a well defined map $\varepsilon_k : D^k(f) \to Y$ given by $\varepsilon_k(x_1, \ldots, x_k) = f(x_i)$ for any i.

Definition 3.2. The kth image multiple point space of a map f is the set $\varepsilon_k(D^k(f))$ and is denoted $M_k(f)$.

The following theorem is very useful, it was used implicitly in many theorems in [8].

Theorem 3.1. *Suppose that $f : X \to Y$ is a finite and proper continuous map. Let $\widetilde{X}$ be a subset of X and suppose that $D^k(f) \cap \widetilde{X}^k$ and $D^k(f|\widetilde{X})$ are S_k-cellular. Then, the natural map*

$$H_i^{alt}(D^k(f|\widetilde{X}); G) \to H_i^{alt}(D^k(f) \cap \widetilde{X}^k; G)$$

is an isomorphism for all i.

Proof. The set $D^k(f) \cap \widetilde{X}^k$ is a union of $D^k(f|\widetilde{X})$ and (possibly) spaces lying in the diagonal. So by Theorem 2.1 the homologies are the same. $\square$

Now let us consider equivariant maps and the induced action on multiple point spaces. Suppose $f : X \to Y$ is a finite and proper H-equivariant map for some group H. Then, H acts naturally on $D^k(f)$, for all k, by $h.(x_1, \ldots, x_k) = (h.x_1, \ldots, h.x_k)$ for $h \in H$.

Proposition 3.1. *Suppose that the group H is finite and $f : X \to Y$ is a finite H-equivariant analytic map between compact subanalytic space, such that the action of H is analytic on X and Y. Then, H and the natural S_k-action act cellularly on $D^k(f)$.*

Proof. First, $D^k(f)$ is subanalytic. This is because $D^k(f)$ is the closure of the subanalytic set defined by $\{(x_1, \ldots, x_k) \in X^k | f(x_i) = f(x_j)$ for all $x_i \neq x_j, i \neq j\}$ and the closure of a subanalytic set is subanalytic.

Since locally X is embeddable in $\mathbb{R}^N$ for some N, the action of S_k on X^k is analytic and hence the fixed point sets in $D^k(f)$ are subanaytic. The analytic action of H on X and Y imply that the fixed points sets of H are also subanalytic.

For all k, $D^k(f)$ can be Whitney stratified with the fixed points sets of H and S_k forming substratifications, see p. 43 of [5]. As $D^k(f)$ is compact, it is triangulable, with the substratifications forming subtriangulations, see [4]. This provides the cellular actions on $D^k(f)$. $\square$

Proposition 3.2. *Suppose that $f : X \to Y$ is a simplicial map with a simplicial action of H. Then, H acts cellularly on $D^k(f)$.*

Proof. This follows from simple but lengthy use of the definitions. $\square$

The next proposition will be used in the description of the differential of the spectral sequence. We assume that we are in the situation of Theorem 2.1, where, for a map $f : X \to B$, the set $D^k(f)$, for all k, is such that it satisfies the conditions for Y in that theorem.

Proposition 3.3. *Suppose $f : X \to Y$ is a finite continuous map. Then,*

(i) $\varepsilon_{i,k_*} : C_*(D^k(f)) \to C_*(D^{k-1}(f))$ *maps alternated chains to alternated chains;*
(ii) $(-1)^i \varepsilon_{i,k_*} = (-1)^j \varepsilon_{j,k_*} : \mathrm{Alt}\, C_*(D^k(f)) \to \mathrm{Alt}\, C_*(D^{k-1}(f))$.

Proof. (i) Let $\sigma \in S_{k-1}$, then σ can be represented in the following way,

$$\sigma = \begin{pmatrix} 1 & 2 & \dots & k-1 \\ \sigma(1) & \sigma(2) & \dots & \sigma(k-1) \end{pmatrix}.$$

We can define an element $\tilde{\sigma}$ of S_k via the following

$$\tilde{\sigma} = \begin{pmatrix} 1 & 2 & \dots & i-1 & i & i+1 & \dots & k \\ \sigma(1) & \sigma(2) & \dots & \sigma(i-1) & i & \sigma(i) & \dots & \sigma(k-1) \end{pmatrix}.$$

Then obviously $\mathrm{sign}(\sigma) = \mathrm{sign}(\tilde{\sigma})$ and $\sigma \varepsilon_{i,k} = \varepsilon_{i,k} \tilde{\sigma}$. Thus, if c is an alternating chain in $C_*(D^k(f))^{\langle K \rangle}$ (and so is K-alternated by Theorem 2.1), then we have

$$\begin{aligned} \sigma_* \varepsilon_{i,k*}(c) &= \varepsilon_{i,k*}(\tilde{\sigma}_*(c)) \\ &= \varepsilon_{i,k*}(\mathrm{sign}(\tilde{\sigma})c) \\ &= \mathrm{sign}(\tilde{\sigma}) \varepsilon_{i,k*}(c) \\ &= \mathrm{sign}(\sigma) \varepsilon_{i,k*}(c). \end{aligned}$$

Thus $\varepsilon_{i,k*}(c)$ is a K-alternating chain.

(ii) We can assume that $i < j$. Let σ be the element of S_k that moves the $i+1$ to j numbers inclusive to the left and puts the ith number into the jth position. Then it is easy to see that $\mathrm{sign}(\sigma) = (-1)^{j-i}$ and $\varepsilon_{i,k} = \varepsilon_{j,k}\sigma$. The effect of σ_* on K-alternating chains is to multiply by $\mathrm{sign}(\sigma)$. Thus

$$\varepsilon_{i,k*} = \varepsilon_{j,k*}\, \mathrm{sign}(\sigma),$$

and so

$$\varepsilon_{i,k*} = (-1)^{j-i}\varepsilon_{j,k*}. \qquad \square$$

4. Semi-simplicial Resolution of the Image

Throughout this section we suppose that X and Y are topological spaces with X Hausdorff and both satisfy the first axiom of countability. (That is, the neighbourhood system of every point has a countable base, for example a metric space. This is used to ensure that the closure of a set can be constructed from the limits of sequences. See [12] p. 72).

Suppose that $f : X \to Y$ is a surjective, finite and proper continuous map. Let $\widetilde{X}$ be a subspace of X and $\widetilde{Y} = f(\widetilde{X})$.

For a continuous finite map the concept of proper is equivalent to the map being closed, i.e. the image of a closed set is closed, ([1] Theorem I.10.2.1). If X is Hausdorff and Y is locally compact, then the condition of being proper is equivalent to the preimage of compact sets being compact, ([1] Proposition I.10.3.7). If B is a subset of Y, then the map $f|_{f^{-1}(B)} f^{-1}(B) \to B$ is proper, ([1] Exercise I.10.2).

4.1. *Semi-simplicial resolution*

We shall construct a space W which has the same homology as Y, but which can be filtered naturally, thus giving a spectral sequence, the E^1 terms of which we describe in the next section. The space W is constructed through the substitution of a point of Y by a simplex, the dimension of which depends on the number of preimages of the point. The previous method for construction of this space involved embedding the source into $\mathbb{R}^N$ for large N. Part of the aim of the following theorems is to avoid this embedding requirement.

We begin with a lemma.

Lemma 4.1. *Suppose f is as above and that the number of preimages in X of any point in Y is k, a constant. Consider the kth multiple point space, $D^k(f)$ and the continuous surjective map $g : D^k(f) \to Y$ defined by $g(x_1, \ldots, x_k) = f(x_j)$ for any j. Let $\{E_\alpha\}$ be the set of orbits of connected components of $D^k(f) \backslash \mathrm{Diag}$ and let $Y_\alpha = g(E_\alpha)$.*

Then, the following statements hold.

(i) The union of all the Y_α is Y.
(ii) For $\alpha \neq \beta$, $Y_\alpha \cap Y_\beta = \emptyset$.
(iii) The map $f|f^{-1}(Y_\alpha) \to Y_\alpha$ is a covering of Y_α for all α.

Proof. Condition (i) is proved by the following. Consider $b \in Y$ then $f^{-1}(b) = \{x_1, \ldots, x_k\}$ and $(x_1, \ldots, x_k) \in D^k(f) \backslash \mathrm{Diag}$ so $g(x_1, \ldots, x_k) = b$.

For part (ii) of the theorem suppose that $g(E_\alpha) \cap g(E_\beta) \neq \emptyset$ for some $\alpha \neq \beta$. Then, there exists a y such that $y = g(x) = g(x')$ where $x \neq x'$ such that $x = (x_1, \ldots, x_k)$ and $x' = (x'_1, \ldots, x'_k)$ are in different connected components of $D^k(f) \backslash \text{Diag}$. The set $\{x_1, \ldots, x_k, x'_1, \ldots, x'_k\}$ has at least $k + 1$ distinct members so $(x''_1, \ldots, x''_k, x''_{k+1}) \in D^{k+1}(f)$; this contradicts the number of elements in $f^{-1}(y)$.

We prove (iii) by the following. Let Y_α be a set defined above, with $b \in Y_\alpha$ and $a_j \in f^{-1}(b)$, for $j = 1, \ldots, k$. Then, for every $a_j \in f^{-1}(b)$ there exists a neighbourhood $V_j \subset f^{-1}(Y_\alpha)$ of a_j upon which $f|V \to f(V)$ is injective. Assume not, then there exist two sequences of points z_i and z'_i with $f(z_i) = f(z'_i)$ for $z_i \neq z'_i$ such that the limit of the two sequences is a. For each z_i we can find another $k - 2$ points in X that are contained in $f^{-1}(f(z_i))$. Thus $(z_i, z'_i, z''_i, \ldots, z_i^{(k-1)})$ is a sequence of points in $D^k(f)$ which lies in some E_β and converges to a point $c = (a, a, \lim_{i \to \infty} z''_i, \ldots, \lim_{i \to \infty} z_i^{(k-1)})$. But this limit is in the diagonal of $D^k(f)$, i.e. not in E_α, thus its image under g is not in Y_α, contradicting $g(c) \in Y_\alpha$.

As X is Hausdorff we can take neighbourhoods, W_j, of $a_j \in f^{-1}(b)$ that are mutually disjoint and so that $W_j \cap V_j$ is neighbourhood upon which the map f is injective. The set $U = \text{Interior}(\cap_j(\text{Closure} f(W_j \cap V_j)))$ is thus an open set such that $f^{-1}(U)$ is a union of mutually disjoint sets $\widetilde{U}_j$ with $f(U_j) = U$. The map $f|\widetilde{U}_j \to U$ is injective for each j and since f is proper this implies that the restriction to $\widetilde{U}_j$ must be a homeomorphism. Thus f restricts to a covering. $\qquad\square$

Definition 4.1. Suppose we have a simplex Δ_k. Then the *cellular collapse map of Δ_k into Δ_{k-1} along the edge p_1 to p_2* is the map that identifies all points on lines in Δ_k parallel to the line p_1 to p_2.

Note that we can then collapse Δ_{k-1} into Δ_{k-2} along some edge. Thus, we can collapse Δ_k into Δ_j for $j < k$ by collapsing Δ_k along various edges, a process we shall use in the proof of the next theorem. Note also that collapses obviously commute.

Theorem 4.1. *For f as above there exists a topological space W and a proper map $\pi : W \to Y$ such that the following hold.*

(i) *For every $y \in Y$ with the cardinality of $f^{-1}(y)$ equal to k, we have $\pi^{-1}(y)$ is a $(k - 1)$-simplex.*

(ii) *The set of vertices formed by the simplices in W is homeomorphic to X. The composition of π and this homeomorphism is equal to f.*

Proof.

Figure 1 gives a schematic view of how the space W is constructed. The map shown is the folding in half of a closed interval with an extra, isolated, point identified with the fold point.

Let $Y_k = \{y \in Y | \# f^{-1}(y) = k\}$ and $X_k = f^{-1}(Y_k)$. The restriction map $f_k : X_k \to Y_k$ fulfills the conditions of Lemma 4.1 (it is proper because it is the restriction of f to a preimage set) and hence there exists a decomposition of Y_k by $Y_{k,i}$ so that $f_{k,i} : X_{k,i} \to Y_{k,i}$ is a covering.

We can obviously define a bundle of $(k-1)$-simplices over $Y_{k,i}$ with the set of vertices homeomorphic to the set $X_{k,i}$. In general this bundle is not trivial. Denote this bundle by $\pi_{k,i} : Z_{k,i} \to Y_{k,i}$. Obviously $Z_{k,i}$ is homotopically equivalent to $Y_{k,i}$.

Let $Z = \amalg_{i,j} Z_{k,i}$ be the disjoint union of the bundles. We shall 'close' the bundles over $Y_{k,i}$ so that we can glue together the closures to get W. Let $\widetilde{Y}_{k,i}$ denote the closure of $Y_{k,i}$ in Y. Suppose $y \in \widetilde{Y}_{k,i}$. Then there exists a sequence of points $y_r \in Y_{k,i}$ with limit y. In $D^k(f_{k,i})$ there exists a sequence of points $(x'_r, x''_r, \ldots, x_r^{(k)})$ with $f(x_r^{(i)}) = y_r$ for all r and $1 \leq i \leq k$. For every y_r we get a simplex in $Z_{k,i}$. We define the closure, $\widetilde{Z}_{k,i}$, of $Z_{k,i}$ to be the union of $Z_{k,i}$ with the simplices formed by the limits of a sequence of simplices. If sequences in $D^k(f_{k,i})$ have the same limit, then the simplices are identified. Thus, we get a map $\widetilde{\pi} : \widetilde{Z}_{k,i} \to Y_{k,i}$ for all k and i.

Let $\widetilde{Z} = \amalg \widetilde{Z}_{k,i}$. We will now identify the simplices in $\widetilde{Z}$ to produce a new space W. Suppose $y \in \widetilde{Y}_{k,i}$ for some k and i. Then there exists a sequence $(x'_r, x''_r, \ldots, x_r^{(k)}) \in D^k(f_{k,i})$ with $\lim_{r \to \infty} f(x_r^{(t)}) = y$. Let $c_j = \lim_{r \to \infty} x_r^{(j)}$. We identify the simplex in $\widetilde{Z}_{k,i}$ over y with any other simplex in $(\widetilde{\pi}_{j,s})^{-1}(y)$ for some j and s by the following. If the c_j are distinct, then we simply identify the simplices by the natural inclusion of vertices. If the c_j are not distinct then we collapse the vertices that are the same and fit this into the simplex $(\widetilde{\pi}_{j,s})^{-1}(y)$ by using the remaining vertices.

The resulting space will be denoted W and the map $\pi : W \to Y$ is defined in the obvious way. Since π is a closed map and the fibres are simplices, it is a proper map.

Part (ii) is obvious from the construction of W. $\qquad\square$

Definition 4.2. The space W is called the *semi-simplicial resolution of Y*.

Goryunov states in [6] that, if X is embeddable in $\mathbb{R}^N$, then we can construct a semi-simplicial resolution. This condition is actually too strict, for if $\varepsilon_{2,1}(D^2(f))$ can be embedded in $\mathbb{R}^N$ for some N, then we can construct a resolution in the same manner as [6]. The semi-simplicial resolution pro-

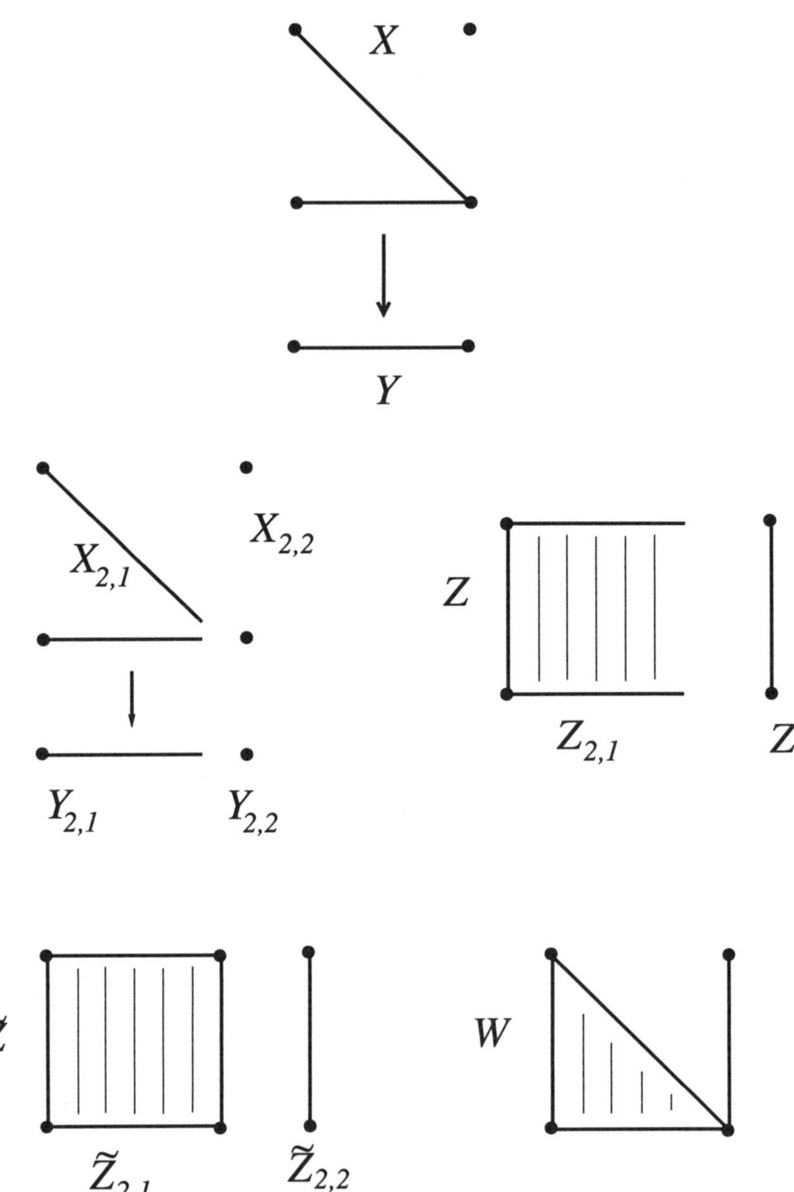

Fig. 1.   Spaces involved in the resolution.

cess detailed there can be carried out for the set $\varepsilon_{2,1}(D^2(f))$ and then X is attached along this double point space by a straight-forward identification.

4.2. *Homology of the resolution*

We can filter W by taking W_k to be the union of all $(k-1)$-simplices in W. Thus, $W_1 = X$ and, for some m, $W_m = W$. This latter space has the same homology as Y, as we shall see. It is incorrectly concluded in [6] that W is homotopically equivalent to Y, this is because in Vassiliev's original paper Whitehead's theorem is incorrectly applied, see [16]. However, that W and Y are homotopically equivalent if W and Y are finite complexes can be proved using Theorem 11.1 of [2]. Presumably the homotopy conclusion is true for the non-finite case, but all we really need to do is relate the homologies of W and Y as in the following.

Lemma 4.2. *Suppose that Y is triangulable so that above each open cell π is a trivialisable bundle and that K acts cellularly on Y for this triangulation. Then, the map $\pi_* : H_i^K(W; G) \to H_i^K(Y; G)$ is an isomorphism for all i.*

Proof. The proof follows that of the proof of Lemma 1 on p. 86 of [16]. Filter Y by the skeletons of the triangulation and W by the preimages of the skeletons under the map π. The filtration of W generates a spectral sequence $E_{p,q}^r(W)$ converging to the homology of W (twisted with respect to K). By construction the $E_{p,q}^1(W)$ term vanishes for $q > 0$ and the complex $\{E_{*,0}^1(W), d_1\}$ is isomorphic to the cellular differential complex of the cellular decomposition of Y (twisted with respect to K), the isomorphism being given by π. $\square$

4.3. *Homology of the filtration*

In this section we prove that the E^1 term of the spectral sequence arising from the filtration of W can be described in terms of the alternating homology of the multiple point spaces of the map.

Suppose $f : X \to Y$ is a surjective finite and proper H-equivariant continuous map and that W is the semi-simplicial resolution of the image. Then H acts on W by taking the simplex with vertices x_1 to x_k to the one with vertices $h.x_1$ to $h.x_n$, where $h \in H$. Note that the map π is H-equivariant.

Definition 4.3. The action of H on W is called compatible with W if, for every $y \in Y$ and $h \in H$, such that $h.y = y$, h fixes the simplex $\pi^{-1}(y)$ pointwise.

Obviously, a trivial action is compatible with any semi-simplicial resolution.

A non-trivial and crucial example of the definition is given by the following. Suppose that $g : X \to Y$ is a finite and proper continuous map and that H is a group that acts on $D^k(g)$. Then, H acts on $D^{k+1}(g)$ by $h.(x_1, \ldots, x_k, x_{k+1}) = (h(x_1, \ldots, x_k), x_{k+1})$. Define $\varepsilon = \varepsilon_{k+1,k+1} : D^{k+1}(g) \to \varepsilon_{k+1,k+1}(D^k(g))$. It is easy to show that ε is a surjective, finite and proper H-equivariant continuous map.

Lemma 4.3. *The action of H is compatible with the semi-simplicial resolution of the image of ε.*

Proof. Let $(z_1, \ldots, z_k)$ be a point in the image of ε and let $\varepsilon^{-1}(z_1, \ldots, z_k) = \{w_1, \ldots, w_l\}$. Then, $w_i = (z_1, \ldots, z_k, \widetilde{z}_i)$ for some $\widetilde{z}_i \in X$, $i = 1, \ldots, l$.

If $h.(z_1, \ldots, z_k) = (z_1, \ldots, z_k)$, then, for all i, $h.w_i = w_i$ and so the simplex in the semi-simplicial resolution is fixed by h. $\qquad\square$

An non-example is $g : [-1, 1] \to [0, 1]$ defined by $g(x) = x^2$, with an action of $S_2 = \{id, \sigma\}$ on $[-1, 1]$ given by $\sigma x = -x$ and trivial on $[0, 1]$. Then, g is S_2-equivariant, and for any $y \in (0, 1]$ the action induced on the resolution space reverses the orientation of the 1-simplex.

We are now in a position to state the relationship between the relative terms of the induced chain filtration and the alternating chains on multiple point spaces.

Proposition 4.1. *Suppose that, for some k, S_k and H act cellulary on $D^k(f)$ with $D^k(f) \cap \mathrm{Diag}$ a subcomplex and that H is compatible with W. Then, there is a natural H-cellular structure on the pair (W_k, W_{k-1}) and there exists a natural isomorphism*

$$\mathrm{Alt}^{S_k} C_{i-k+1}(D^k(f))^{\langle H \rangle} \to C_i(W_k, W_{k-1})^{\langle H \rangle}$$

for all i.

Proof. The proof follows the structure of the proof of Proposition 1.2.1 in [6]. We will drop the reference to f. Let Δ_k denote the k-simplex oriented by a choice of order of the vertices. Define a map $h_k : D^k \times \Delta_{k-1} \to W_k$ by sending $(x_1, \ldots, x_k) \times (t_1, \ldots, t_k)$ to the simplex with vertices $x_1, \ldots, x_k$.

That is, the point of D^k gives a simplex in W and the point of Δ_{k-1} gives a point within that simplex.

Since $h_k(\text{Diag} \times \Delta_{k-1}) \subset W_{k-1}$ and $h_k(D^k \times \partial\Delta_{k-1}) \subset W_{k-1}$ we have a well defined map of pairs,

$$h_k : (D^k, \text{Diag}) \times (\Delta_{k-1}, \partial\Delta_{k-1}) \to (W_k, W_{k-1}).$$

This map is obviously proper and on the complement of the second terms is actually a covering of $W_k - W_{k-1}$, its fibre is the orbit of a point in $D^k \times \Delta_{k-1}$.

We can look at the elements of $C_i(D^k, \text{Diag}) \times (\Delta_{k-1}, \partial\Delta_{k-1}))$ symmetrized with respect to the trivial representation of S_k. The map induced on chains by h_k,

$$h_{k\#} : \left(\text{Sym}\, C_i((D^k, \text{Diag}) \times (\Delta_{k-1}, \partial\Delta_{k-1}))\right)^{\langle H \rangle} \to C_i(W_k, W_{k-1})^{\langle H \rangle}$$

is an isomorphism for all i as the following shows.

The cell structure on D^k and the standard one on Δ_{k-1} give a cellular structure on $D^k \times \Delta_{k-1}$. The group S_k acts on the product; however, it is not a cellular action in our sense as the fixed point set of a cell, under the action of an element of the group, may be a proper subset of the cell. However, the cellular structure on $D^k \times \Delta_{k-1}$ gives a structure on the pair

$$(D^k, \text{Diag}) \times (\Delta_{k-1}, \partial\Delta_{k-1}).$$

Since h_k is a map on the pairs which is essentially a factorisation of the complement by S_k, we get a structure on (W_k, W_{k-1}). The induced map on relative chains is a $k!$-fold homomorphism that provides the isomorphism.

We shall prove now that there is an H-equivariant isomorphism

$$\text{Alt}^{S_k} C_{i-k+1}(D^k, \text{Diag}) \to \text{Sym}\, C_i((D^k, \text{Diag}) \times (\Delta_{k-1}, \partial\Delta_{k-1}))$$

for all i. We have a chain isomorphism

$$\sum_{p+q=i} C_p(D^k, \text{Diag}) \otimes C_q(\Delta_{k-1}, \partial\Delta_{k-1}) \to C_i((D^k, \text{Diag}) \times (\Delta_{k-1}, \partial\Delta_{k-1}))$$

which respects the actions on the spaces, (if H was not compatible, then there would not necessarily be an isomorphism). Thus we get an isomorphism on the S_k-symmetric parts. The group $C_q(\Delta_{k-1}, \partial\Delta_{k-1})$ is trivial except for $q = k - 1$ when it is isomorphic to $\mathbb{Z}$ with the the natural sign action of S_k. These two facts give an isomorphism:

$$\text{Sym}\left(C_{i-k+1}(D^k, \text{Diag}) \otimes \mathbb{Z}\right) \xrightarrow{\cong} \text{Sym}\, C_i\left((D^k, \text{Diag}) \times (\Delta_{k-1}, \partial\Delta_{k-1})\right).$$

The former group is equal to the group

$$\left\{ \sum_{\sigma \in S_k} \sigma(c \otimes a) \text{ for all } c \in C_{i-k+1}(D^k, \mathrm{Diag}), a \in \mathbb{Z} \right\}$$

$$= \left\{ \sum_{\sigma \in S_k} \sigma(c) \otimes \sigma(a) \text{ for all } c \in C_{i-k+1}, a \in \mathbb{Z} \right\}$$

$$= \left\{ \sum_{\sigma \in S_k} \sigma(c) \otimes (\mathrm{sign}(\sigma)a) \text{ for all } c \in C_{i-k+1}, a \in \mathbb{Z} \right\}$$

$$= \left\{ \left(\sum_{\sigma \in S_k} \mathrm{sign}(\sigma)\sigma(c) \right) \otimes a \text{ for all } c \in C_{i-k+1}, a \in \mathbb{Z} \right\}$$

$$= \left\{ \sum_{\sigma \in S_k} \mathrm{sign}(\sigma)\sigma(c) \text{ for all } c \in C_{i-k+1} \right\} \otimes \mathbb{Z}$$

$$\cong \mathrm{Alt}^{S_k} C_{i-k+1}(D^k(f), \mathrm{Diag}).$$

Since the action of H on Δ_{k-1} is trivial, the map is H-equivariant.

The diagonal is a subcomplex of $D^k(f)$ and H acts cellularly, so the short exact sequence

$$0 \to C_i(\mathrm{Diag}) \to C_i(D^k) \to C_i(D^k, \mathrm{Diag}) \to 0$$

is split exact, and this implies by Lemma 2.1 that

$$0 \to \mathrm{Alt}^{S_k} C_i(\mathrm{Diag}) \to \mathrm{Alt}^{S_k} C_i(D^k) \to \mathrm{Alt}^{S_k} C_i(D^k, \mathrm{Diag}) \to 0$$

is also exact. The third map is in fact an isomorphism because $\mathrm{Alt}^H C_i(\mathrm{Diag}) = 0$ by Theorem 2.1.

Thus, we have a chain of isomorphisms giving the isomorphism of the theorem. $\qquad \square$

5. Spectral Sequences

In this section we make the statements on the existence of spectral sequences which calculate the different types of homology of images of maps. First we need a definition of the maps for which the sequences exist.

Definition 5.1. A continuous map $f : X \to Y$ is a good map, if

 (i) it is finite and proper with X Hausdorff,
(ii) X and $f(X)$ satisfy the first axiom of countability;
(iii) S_k acts cellularly on $D^k(f)$ and $D^k(f) \cap \mathrm{Diag}$ is a subcomplex;

(iv) the semi-simplicial resolution of $f(X)$ can be triangulated so that over each open cell the projection $\pi : W \to f(X)$ is a trivialisable bundle.

There exists a large supply of good maps as the following two propositions show.

Proposition 5.1. *A surjective finite analytic map* $f : X \to Y$ *between compact subanalytic spaces is a good map.*

Proof. For the properties of subanalytic sets see, for example, [14].

The first two conditions follow from the fact that as X is compact and Y is locally compact the map is proper, see [1] Proposition I.10.3.7.

For the third condition one uses Proposition 3.1. The reasoning in the proof provides the S_k-cellular decomposition of the pair $(D^k(f), D^k(f) \cap \text{Diag})$.

Stratify each $D^k(f)$ with a Whitney stratification such that each fixed point set under an element of S_k is a union of strata. Now, suppose that m is the largest integer such that $D^m(f) \neq \emptyset$. Then the map $\varepsilon_{m,m} : D^m(f) \to D^{m-1}(f)$ can be stratified so that the image multiple point spaces are union of strata. Proceeding in this way for all $\varepsilon_{k,k} : D^k(f) \to D^{k-1}(f)$ we can stratify $f : X \to Y$ so that above each stratum A of Y the preimage $f^{-1}(A)$ is a union of strata and the number of preimages is constant. Using this stratification one can triangulate Y so that over each open cell the map $\pi : W \to Y$ is a trivialisable bundle. $\qquad\square$

Proposition 5.2. *Suppose that* $f : X \to Y$ *is a finite and proper simplicial map between locally finite simplicial complexes. Then,* f *is a good map.*

Proof. Conditions (i) and (ii) are satisfied because simplicial complexes are Hausdorff ([15] p111) and being locally finite is equivalent to satisfying the first axiom of countability ([15] p119).

The space $D^k(f)$ is simplicial and the projection $\varepsilon_{k,k} : D^k(f) \to D^{k-1}(f)$ is a simplicial map. The diagonal will be a subcomplex of $D^k(f)$ by construction so condition (iii) is satified.

The fourth condition is true by construction of W in this case. $\qquad\square$

We now come to the main theorem, the existence of a spectral that computes the homology of the image of a map using the multiple point spaces.

Theorem 5.1. *Let* $f : X \to Y$ *be a continuous map,* $\widetilde{X}$ *a topological subspace of* X, *and* H *be a group. Suppose*

(i) $f : X \to Y$ and $f|\widetilde{X}$ are H-equivariant and good maps;

(ii) that $D^k(f|\widetilde{X})$ is a subcomplex of $D^k(f)$ for all $k \geq 1$;

(iii) that H is compatible with the semi-simplicial resolutions of $f(X)$ and $f(\widetilde{X})$;

(iv) H acts cellularly on $D^k(f)$ for all k, with $D^k(\widetilde{X})$ as a subcomplex.

Then, there exists a spectral sequence

$$E^{1,H}_{p,q} = H^{alt\langle S_{p+1}\rangle \times H}_q(D^{p+1}(f), D^{p+1}(f|\widetilde{X}); G) \implies H^H_*(f(X), f(\widetilde{X}); G),$$

where G is a coefficient group. If ε_{i,k_*} is H-equivariant, then the differential $d_{k-1,q}$ is induced from the map $\varepsilon_{r,k} : D^k(f) \to D^{k-1}(f)$ for any $r = 1, \ldots, k$.

Proof. Let m be the maximal number of points in $f^{-1}(y)$ for $y \in Y$. Let W be the semi-simplicial resolution of $f(X)$, this can be filtered by

$$X = W_1 \subset W_2 \subset \cdots \subset W_{m-1} \subset W_m,$$

where W_k is the union of all simplices of dimension less than k. Through restricting to simplices that have vertices in $\widetilde{X}$ one also gets a semi-simplicial resolution of $\widetilde{Y}$, denote this by $\widetilde{W}$. This leads to a filtration on the the chains $C_i(W, \widetilde{W})$.

From Proposition 4.1 we obtain two isomorphisms

$$\left(\mathrm{Alt}^{S_k} C_{i-k+1}(D^k(f))\right)^{\langle H\rangle} \to C_i(W_k, W_{k-1})^{\langle H\rangle}$$

and

$$\left(\mathrm{Alt}^{S_k} C_{i-k+1}(D^k(f|\widetilde{X}))\right)^{\langle H\rangle} \to C_i(\widetilde{W}_k, \widetilde{W}_{k-1})^{\langle H\rangle},$$

so there exists an isomorphism

$$\mathrm{Alt}^{S_k} C_{i-k+1}(D^k(f))^{\langle H\rangle} / \mathrm{Alt}^{S_k} C_{i-k+1}(D^k(f|\widetilde{X}))^{\langle H\rangle}$$
$$\to C_i(W_k, W_{k-1})^{\langle H\rangle} / C_i(\widetilde{W}_k, \widetilde{W}_{k-1})^{\langle H\rangle}.$$

As $D^k(f|\widetilde{X})$ is an H-invariant subcomplex of $D^k(f)$ we get that the first group is isomorphic to $\mathrm{Alt}^{S_k} C_{i-k+1}(D^k(f), D^k(f|\widetilde{X}))^{\langle H\rangle}$.

By the three-by-three lemma and the fact that the short exact sequences are split we get (by Lemma 2.1) a natural isomorphism

$$C_i(W_k, W_{k-1})^{\langle H\rangle} / C_i(\widetilde{W}_k, \widetilde{W}_{k-1})^{\langle H\rangle} \cong C_i(W_k, \widetilde{W}_k)^{\langle H\rangle} / C_i(W_{k-1}, \widetilde{W}_{k-1})^{\langle H\rangle}.$$

The right-hand side terms form the groups in the E^1 page of the spectral sequence associated to the filtration of $C_i(W, \widetilde{W})^{\langle H \rangle}$. Thus, the group $E^1_{p,q}$ is isomorphic to

$$H_q \left(\mathrm{Alt}^{S_{p+1}} C_*(D^{p+1}(f), D^{p+1}(f|\widetilde{X}))^{\langle H \rangle} \otimes G \right),$$

where G is a general coefficient group. By Theorem 2.1 this group is isomorphic to

$$H_q^{alt\langle S_{p+1}\rangle \times H}(D^{p+1}(f), D^{p+1}(f|\widetilde{X}); G).$$

Since $D^{m+1} = \emptyset$, the spectral sequence is bounded, and so by Lemma 4.2 it converges to $H_*^H(f(X), f(\widetilde{X}); G)$.

The statement on the differential follows from Proposition 3.3 and a simple (if lengthy) diagram chase. $\square$

As a corollary one gets the following useful spectral sequence.

Corollary 5.1. *For a good map $f : X \to Y$ and a subspace $\widetilde{X} \subset X$, with $D^k(f|\widetilde{X})$ a subcomplex of the S_k-cellular $D^k(f)$, there exists a spectral sequence*

$$E^1_{p,q} = H_q^{alt}(D^{p+1}(f), D^{p+1}(f|\widetilde{X}); G) \implies H_*(f(X), f(\widetilde{X}); G),$$

for any coefficient group G.

Proof. Take H trivial in the theorem. $\square$

In light of Theorem 5.1 another obvious corollary is the one mentioned in the introduction. One can deduce from this the sequence in Corollary 1.2.2 of [6].

Corollary 5.2. *Let $f : X \to Y$ be a finite surjective algebraic mapping between compact semialgebraic spaces. Then, there exists a spectral sequence*

$$E^1_{p,q} = H_q^{alt}(D^{p+1}(f); \mathbb{Z}) \implies H_*(Y; \mathbb{Z}).$$

The spectral sequence originally arose as a rational cohomological sequence in [7]. Here, we can immediately deduce a rational homological version.

Corollary 5.3. *Suppose $f : X \to Y$ is a good map. Then, there exists a spectral sequence*

$$E^1_{p,q} = H_q(D^{p+1}(f); \mathbb{Q})^{alt} \implies \widetilde{H}_*(f(X); \mathbb{Q}).$$

Proof. By the theorem there exists a spectral sequence with $E^1_{p,q}$ term equal to $H^{alt}_q(D^{p+1}(f); \mathbb{Q})$. This group is isomorphic to $H_q\left(\frac{1}{k!} \mathrm{Alt}\left(C_*\left(D^{p+1}(f); \mathbb{Q}\right)\right)\right)$. Since it is idempotent, the functor $\frac{1}{k!} \mathrm{Alt}$ passes through the homology functor, and the resulting group is isomorphic to $H_q(D^{p+1}(f); \mathbb{Q})^{alt}$. $\qquad\square$

The next corollary is useful in proving results on the image multiple point spaces of a map, see [10] and [11].

Corollary 5.4. *Suppose that f is a good map and that H acts on $D^k(f)$ by permutation of (some of the) copies of X in X^k. Assume that $\varepsilon_{k+1,k+1}$ and $\widetilde{\varepsilon}_{k+1,k+1} := \varepsilon_{k+1,k+1}|D^{k+1}(f|\widetilde{X})$ are good maps for some k and that $D^k(f)$ has $D^k(f|\widetilde{X})$ as a subcomplex. Then, there exists a spectral sequence,*

$$E^{1,H}_{p,q} = H^{alt\langle H \times S_{p+1}\rangle}_q(D^{p+k+1}(f), D^{p+k+1}(f|\widetilde{X}); G)$$

$$\implies H^{alt\langle H\rangle}_*(\varepsilon_{k+1,k+1}(D^{k+1}(f)), \widetilde{\varepsilon}_{k+1,k+1}(D^{k+1}(f|\widetilde{X})); G).$$

Proof. First, note that, by Lemma 4.3, H is compatible with W and $\widetilde{W}$, the semi-simplicial resolutions of the images of $\varepsilon_{k+1,k+1}$ and $\widetilde{\varepsilon}_{k+1,k+1}$ respectively. Since $\varepsilon_{k+1,k+1}$ and $\widetilde{\varepsilon}_{k+1,k+1}$ are good there exists a spectral sequence

$$E^1_{p,q} = H^{alt\langle S_{p+1}\rangle \times H}(D^{p+1}(\varepsilon_{k+1,k+1}), D^{p+1}(\widetilde{\varepsilon}_{k+1,k+1}); G)$$

converging to

$$H^{alt\langle H\rangle}_*(\varepsilon_{k+1,k+1}(D^{k+1}(f)), \varepsilon_{k+1,k+1}(D^{k+1}(f|\widetilde{X})); G)$$

By [7] page 55 $D^{p+k+1}(f)$ is S_k-homeomorphic to $D^{p+1}(\varepsilon_{k+1,k+1})$, and similarly for $\widetilde{\varepsilon}_{k+1,k+1}$. By Theorem 2.1 and Lemma 2.2 the group

$$H^{alt\langle S_{p+1}\rangle \times H}_q(D^{p+k+1}(f), D^{p+k+1}(f|\widetilde{X}); G)$$

is isomorphic to $H^{alt\langle S_{p+1} \times H\rangle}_q(D^{p+k+1}(f), D^{p+k+1}(f|\widetilde{X}); G)$. $\qquad\square$

One can also prove theorems on different types of support, for example compact support. In general, calculation of homology with supports depends on the family of supports being preserved in the isomorphism of Theorem 4.1; the precise details are left to the interested reader.

References

1. N. Bourbaki, Elements of Mathematics, General Topology, Chapters 1-4, Springer Verlag, (1989).

2. M. Cohen, Simplicial structures and transverse cellularity, Ann. of Math., 85, (1967), 218-245.
3. T. tom Dieck, Transformation Groups, de Gruyter, Studies in Mathematics 8, 1987.
4. M. Goresky, Triangulation of stratified objects, Proc. Amer. Math. Soc. 261, (1978), 193-200.
5. M. Goresky and R. MacPherson, Stratified Morse Theory, Springer-Verlag Berlin (1988).
6. V.V. Goryunov, Semi-simplicial resolutions and homology of images and discriminants of mappings, Proc. London Math. Soc., 70 (1995), 363-385.
7. V.V. Goryunov and D. Mond, Vanishing cohomology of singularities of mappings, Compositio Math., 89 (1993), 45-80.
8. K. Houston, Local topology of images of finite complex analytic maps, Topology, 36, (1997), 1077-1121.
9. K. Houston, Images of maps with one dimensional double point set, Topology and its Applications, 91, (1999), 197-219.
10. K. Houston, Bouquet and join theorems for disentanglements, Inventiones Mathematicae 147 (2002), 471-485.
11. K. Houston, A note on good real perturbations of singularities, Mathematical Proceedings of the Cambridge Philosophical Society, 132 (2002), 301-310.
12. J. Kelley, General Topology, D. Van Nostrand Company, (1955).
13. W.L. Marar and D. Mond, Multiple point schemes for corank 1 maps, J. London Math. Soc. (2) 39 (1989), 553-567.
14. M. Shiota, Geometry of Subanalytic and Semialgebraic Sets, Birkhäuser, (1997).
15. E. Spanier, Algebraic Topology, McGraw-Hill, (1966).
16. V.A. Vassiliev, Complements of Discriminants of Smooth Maps: Topology and Applications, AMS Translations of Mathematical Monographs, Vol. 98, AMS, Providence, Rhode Island, (1992).
17. V.A. Vassiliev, Resolutions of discriminants and topology of their complements, in New Developments in Singularity Theory, ed. D. Siersma, C.T.C. Wall, and V. Zakalyukin, Kluwer Academic Publishers, (2001).

Generic sections of Singularities

Lê Dũng Tráng

For an equidimensional germ of complex analytic space (X, x) embedded in $\mathbb{C}^N$, we give an explicit inductive construction of generic hyperplanes section of the singularity.

Introduction.

In our paper [5] with B. Teissier we study the limit of tangent hyperplanes of equidimensional germs (X, x) of complex analytic spaces. Among interesting results of that paper, we show that the hyperplane intersections of a germ (X, x) by hyperplanes which are not limit of tangent hyperplanes of (X, x), are Whitney equisingular, i.e. they belong to an analytic family $\varphi \mathcal{X} \to S$ for which the deformation space $\mathcal{X}$ has a stratification satisfying Whitney conditions (see e.g. [4] for the definition) whose strata are by φ locally trivial fibrations over the space of parameters S.

In particular for many algebraic properties or topological properties, hyperplane sections by a hyperplane which is not a limit of tangent hyperplanes are general enough to give the generic property. This is why we shall call generic hyperplane of (X, x) a hyperplane which is not a limit of tangent hyperplanes of (X, x) in the smooth space in which (X, x) is embedded.

Answering a question of F. Michel we give here the way to construct these generic hyperplanes by induction on the dimension of the space.

1. Polar varieties and Limits of Tangents

Let (X, x) be an equidimensional germ of reduced complex analytic space. Let X be a representative of (X, x) closed in an open neighbourhood U of x in $\mathbb{C}^N$ and defined by the analytic equations $F_i : U \to \mathbb{C}$, $i = 1, \ldots, r$ which generate the ideal of X in U. For $2 \leq X \leq \dim X$ consider the germ $\pi_k : (X, x) \to (\mathbb{C}^k, 0)$ of the restriction to X of an affine projection of $p_k : \mathbb{C}^N \to \mathbb{C}^k$.

In [3] we have noticed that there is an open dense Zariski subset Ω_k of the space of affine projections of $p_k : \mathbb{C}^N \to \mathbb{C}^k$ for which $p_k(x) = 0$ such that, for any $p_k \in \Omega_k$:

- the critical locus $C(p_k)$ of the restriction of p_k to the non-singular part X^0 of X is either empty or reduced of dimension $k - 1$;
- the closure $\Gamma(p_k)$ of $C(p_k)$ is either empty or reduced of dimension $k - 1$;
- the analytic family of $(\Gamma(p_k), x)$, for $p_k \in \Omega_k$, is Whitney equisingular;
- the image $\Delta(p_k)$ of $\Gamma(p_k)$ by p_k is reduced and the induced map from $\Gamma(p_k)$ onto $\Delta(p_k)$ is generically one to one over an open neighbourhood of 0 in $\Delta(p_k)$.

We have called the germ $(\Gamma(p_k), x)$ a polar variety of dimension $k - 1$ of (X, x). By defining $\Delta(p_k)$ using a sufficiently small representative X of (X, x), the germ $(\Delta(p_k), x)$ is a polar discriminant of dimension $k - 1$ of (X, x).

In [5] we have defined a limit of tangent hyperplanes of (X, x) as an affine hyperplane of $\mathbb{C}^N$ which contains a limit of tangent space to X at x, i.e. an affine supspace of $\mathbb{C}^N$ which is the limit of a sequence of tangent spaces to the non-singular part X^0 at a sequence of points tending to x. Then, we gave the following characterization of limits of tangent hyperplanes of (X, x):

Theorem 1.1. *A hyperplane H of $\mathbb{C}^N$ is a limit of tangent hyperplanes only if it is a limit of tangent hyperplanes of the tangent cones of the polar varieties $\Gamma(p_k)$ of (X, x) at x for $2 \le k \le \dim X + 1$.*

In fact in [5] we prove a more precise theorem. We define the aureole of (X, x) as the set of components of the tangent cones of the polar varieties $\Gamma(p_k)$ of (X, x) at x which do not depend on the projection $p_k \in \Omega_k$, for $2 \le k \le \dim X + 1$. Then, a hyperplane H of $\mathbb{C}^N$ is a limit of tangent hyperplanes if and only if it is a limit of tangent hyperplanes of the cones in the aureole of (X, x).

Remark. Among the cones in the aureole, there are the components of the tangent cone of (X, x). In particular the hyperplanes of $\mathbb{C}^N$ which are limits of tangent hyperplanes of the reduced tangent cone $|C_{X,x}|$ of (X, x) are limits of tangent hyperplanes of (X, x).

An important result from [5] is:

Theorem 1.2. *Let H, H' be affine hyperplanes of $\mathbb{C}^N$ which contain x. If*

H and H′ are not limit of tangent hyperplanes of (X, x), they are transverse to the strata of a Whitney stratification of X in an open neighbourhood V of x in $\mathbb{C}^N$ and there is a stratified germ of homeomorphism of (H, x) onto $(H′, x)$ which sends $(H \cap X, x)$ onto $(H′ \cap X, x)$.

We express the property of the preceding theorem by saying that If H and $H′$ are not limit of tangent hyperplanes of (X, x), the sections $(H \cap X, x)$ and $(H′ \cap X, x)$ are Whitney equisingular.

As consequence of the preceding theorem, in particular we have the following generalization of a theorem of B. Teissier (see [7]):

Corollary 1.1. *Let $H, H′$ be affine hyperplanes of $\mathbb{C}^N$ which contain x and are not limit of tangent hyperplanes of (X, x), they contain respectively planes P and $P′$ such that the reduced germ of curves $(|P \cap X|, x)$ and $(|P′ \cap X|, x)$ are Whitney equisingular and their Milnor numbers in the sense of Buchweitz-Greuel are equal.*

We define

Definition 1.1. An affine hyperplane H of $\mathbb{C}^N$ which contains x and is not a limit of tangent hyperplanes of (X, x) is called a generic hyperplane for (X, x) in $\mathbb{C}^N$.

If H is a generic hyperplane for (X, x), the intersection $(H \cap X, x)$ is often called a generic hyperplane section of (X, x).

2. Constructing generic hyperplanes

From the above definition one can construct generic hyperplanes for an equidimensional reduced germ (X, x) of analytic space using affine projections.

This will use the following lemma:

Lemma 2.1. *Let L be an affine space of $\mathbb{C}^N$ which contains x and which only intersects the tangent cone $C_{X,x}$ of (X, x) at $\{x\}$. Then, there is a hyperplane of $\mathbb{C}^N$ which is not a limit of tangent hyperplanes of $C_{X,x}$ at x and which contains L.*

Proof. Consider $\tilde{L}$ and $\tilde{C}$ the projective subvarieties of $\mathbb{P}^{N-1}$ respectively associated to L and $C_{X,x}$.

Since L only intersects the tangent cone $C_{X,x}$ at $\{x\}$, the intersection $\tilde{L} \cap \tilde{C}$ is empty. Of course, the codimension of $\tilde{L}$ is $\geq \dim C_{X,x}$. Consider

$\tilde{L}_1$ containing $\tilde{L}$ and of codimension equal to $\dim C_{X,x}$ in $\mathbb{P}^{N-1}$. If we prove the preceding lemma for $\tilde{L}_1$, *a fortiori* it is proved for $\tilde{L}$.

Now assume that $\mathrm{codim}\tilde{L} = \dim C_{X,x}$. Then, the conical projection centered at $\tilde{L}$:

$$\mathbb{P}^{N-1} \setminus \tilde{L} \to \mathbb{P}^{\dim \tilde{C}}$$

onto $\mathbb{P}^{\dim \tilde{C}}$ restricted to $\tilde{C}$ is a finite map $\pi : \tilde{C} \to \mathbb{P}^{\dim \tilde{C}}$. In particular the map π is proper and stratified, i.e. one can find algebraic Whitney stratifications $\mathcal{S}$ and $\mathcal{S}'$ of $\tilde{C}$ and $\mathbb{P}^{\dim \tilde{C}}$ such that π is stratified, i.e. for any stratum S_i of the Whitney stratification $\mathcal{S}$ of $\tilde{C}$, there is a Whitney stratum $S'_{\alpha(i)}$ of the Whitney stratification $\mathcal{S}'$ such that π induces a surjective submersive map of S_i onto $S'_{\alpha(i)}$ (see [1] §3, Theorem).

Let m be a general point of $\mathbb{P}^{\dim \tilde{C}}$, i.e. a point of the biggest stratum of $\mathbb{P}^{\dim \tilde{C}}$. Through m there is a hyperplane $\tilde{H}_0$ which is transverse to all the strata of $\mathcal{S}'$. This is consequence to the fact that a projective hyperplane of $\mathbb{P}^{\dim \tilde{C}}$ is transverse to a stratum of $\mathcal{S}'$ if it does not contain a tangent projective subspace of that stratum. The space of hyperplanes of $\mathbb{P}^{\dim \tilde{C}}$ which contain a projective subspace tangent to a stratum S' has dimension $\dim \tilde{C} - 1$. Since the dimension of the space of projective hyperplanes of $\mathbb{P}^{\dim \tilde{C}}$ is $\dim \tilde{C}$ and the number of strata of $\mathcal{S}'$ is finite, there is an open Zariski dense subset Ω of the space of projective hyperplanes of $\mathbb{P}^{\dim \tilde{C}}$, such that $\tilde{H}_0 \in \Omega$ is transverse to all the strata of $\mathcal{S}'$ in $\mathbb{P}^{\dim \tilde{C}}$.

Now, a projective hyperplane $\tilde{H}$ of $\mathbb{P}^{N-1}$, containing $\tilde{L}$ and the inverse image $\pi^{-1}(\tilde{H}_0)$ of a hyperplane $\tilde{H}_0$ of $\mathbb{P}^{\dim \tilde{C}}$ which contains m and which is transverse to all the strata of $\mathcal{S}'$, is transverse to all limit of tangent spaces of $\tilde{C}$, because, a point $x \in \tilde{H} \cap \tilde{C}$ belongs to some stratum $S(x)$ of $\mathcal{S}$ and, $\tilde{H}_0$ being transverse to its image $\pi(S(x)) := S' \in \mathcal{S}'$, $\tilde{H}$ is transverse to $S(x)$ in $\mathbb{P}^{N-1}$, since π induces a submersive map of $S(x)$ onto S'. The assertion that the projective hyperplane $\tilde{H}$ of $\mathbb{P}^{N-1}$, containing $\tilde{L}$ and the inverse image $\pi^{-1}(\tilde{H}_0)$ is transverse to all limit of tangent spaces of $\tilde{C}$ comes from the fact (see [5]) that all limits of tangent spaces of $\tilde{C}$ at x contain the tangent space $T_{S(x),x}$ of $S(x)$ at x.

Now, consider the cone H at x over $\tilde{H}$, it is an affine hyperplane of $\mathbb{C}^N$ through x which is not a limit of tangent hyperplane of the cone $C_{X,x}$ at x.

The above proof can be summarized as following:

Corollary 2.1. *Let L be an affine subspace of $\mathbb{C}^N$ of codimension $\dim C_{X,x}$ in $\mathbb{C}^N$, which contains x and intersects $C_{X,x}$ only at x, then there is an* *open dense Zariski subset in the space of affine hyperplane through x which contains L in which all hyperplanes are generic hyperplane for (X, x) in $\mathbb{C}^N$.*

Proof. It suffices to use the result proved in [5] that limits of tangent hyperplanes of (X, x) are precisely the limits of tangent hyperplanes to the cones of the Aureole of X at x. Just notice that the cones of the aureole are in finite number and contained in the tangent cone of X at x. So any affine space L satisfying the hypothesis of the corollary satisfies the hypothesis of the preceding lemma for all the components of the Aureole. Then, through L there is an Zariski open dense set of hyperplanes which are generic for (X, x) (and equivalently to the cones of the Aureole at the vertex x) in $\mathbb{C}^N$.

The preceding results give a natural way to find generic hyperplane for (X, x) in $\mathbb{C}^N$.

If (X, x) is a curve, the aureole is given by the lines of the tangent cone. Any hyperplane transverse to these lines is generic for (X, x) in $\mathbb{C}^N$.

If (X, x) is a surface, the aureole is given by the tangent cone and the exceptional tangents, which are lines of the tangent cone. Consider an affine space through x of codimension 2 in $\mathbb{C}^N$ intersecting $C_{X,x}$ only at $\{x\}$:

$$L \cap C_{X,x} = \{x\}.$$

A hyperplane of $\mathbb{C}^N$ containing L which is not a limit of tangent $C_{X,x}$ at x and is transverse to the exceptional tangents is a generic hyperplane for (X, x) in $\mathbb{C}^N$. The geometric way given above to choose that generic hyperplane was to consider the projection p of $\mathbb{C}^N$ onto $\mathbb{C}^2$ parallel to L and its restriction π to (X, x). Then a hyperplane through x equal to the inverse image by p of a line transverse to the discriminant of π is generic for (X, x) in $\mathbb{C}^N$.

Observe that the limits of tangent hyperplanes to the germ of surface (X, x) is a hypersurface $\mathcal{L}(X, x)$ of $\mathbb{P}^{N-1}$ and its intersection with the smooth subspace $\mathcal{V}_L$ of hyperplanes containing L is a hypersurface of $\mathcal{V}_L$. So, let H be the hyperplane $p^{-1}(\ell)$ through x, inverse image by p of a line ℓ transverse to the discriminant of the projection $\pi : (X, x) \to \mathbb{C}^2$. Suppose that it is a limit of tangent hyperplanes. Then, H is in $\mathcal{L}(X, x) \cap \mathcal{V}_L$. Since $\mathcal{L}(X, x) \cap \mathcal{V}_L$ is connected, H is a limit of hyperplanes H_n in $\mathcal{L}^0(X, x) \cap \mathcal{V}_L$, where $\mathcal{L}^0(X, x)$ is the non-singular part of $\mathcal{L}(X, x)$. Such hyperplanes H_n are tangent to X at non-singular points x_n of X and, since H_n contains L, the point x_n lies in the critical locus of the projection of (X, x) onto $\mathbb{C}^2$ parallel to L, i.e. the polar curve of (X, x) relatively to L. The image of H_n

by this projection is a line tangent at the image of x_n to the discriminant to the projection of (X, x) onto $\mathbb{C}^2$ parallel to L. The limit of the lines image of H_n is a line of the tangent cone of this discriminant. This would contradict the fact that the image of H is a line transverse to this discriminant. So the considered hyperplane H is a generic hyperplane of (X, x) in $\mathbb{C}^N$.

In higher dimensions, Corollary 2.1 yields a similar result:

Proposition 2.1. *Let L an affine subspace of $\mathbb{C}^N$ of codimension $\dim C_{X,x}$ in $\mathbb{C}^N$, which contains x and intersects $C_{X,x}$ only at x. Consider the projection p of $\mathbb{C}^N$ onto $\mathbb{C}^{\dim(X,x)}$ parallel to L. We denote by Δ_L the discriminant of π, the projection p restricted to (X, x). Let H_0 be a hyperplane of $\mathbb{C}^{\dim(X,x)}$ which is not a limit of tangent hyperplanes of the discriminant Δ_L. Then, its inverse image $p^{-1}(H_0)$ by p is a generic hyperplane of (X, x) in $\mathbb{C}^N$.*

Since the dimension of Δ_L at $p(x)$ is $\dim(X, x) - 1$, this gives an inductive way to get generic hyperplanes of (X, x) in $\mathbb{C}^N$.

Theorem 2.2.4 of [5] gives a way to understand this induction by comparing the polar varieties of dimension $\leq \dim(X, x) - 1$ of (X, x) and $(\Delta_L, p(x))$.

References

1. H. Hironaka, Stratification and flatness, in Real and complex singularities, Nordic Summer School, Oslo 1976, Sijthoff and Noordhoff, Alphen an den Rijn, Netherlands, 1977.
2. Lê Dũng Tráng, Topological use of polar curves, *Algebraic Geometry*, Proc. Sympos. Pure Math., Vol. 29, Humboldt State Univ., Arcata, Calif., 1974, 507–512.
3. Lê Dũng Tráng, B. Teissier, Variétés polaires locales et classes de Chern singulière, Annals of Maths **114** (1981), 457-491.
4. Lê Dũng Tráng, B. Teissier,Cycles évanescents, sections planes et conditions de Whitney II, in Singularities, Proc. Symp. Pure Math. **40**, part 2 (1983), 65-103.
5. Lê Dũng Tráng, B. Teissier, Limites d'espaces tangents en géométriee analytique, Comm. Math. Helv. **63** (1988), 540-578.
6. J. Milnor, Singular points of complex hypersurfaces, Ann. Math. Studies, Princeton Univ. Press, Princeton, N.J., 1968.
7. B. Teissier, Cycles évanescents, sections planes et conditions de Whitney, Astérisque **7-8**, (1973), 285-362.

MORSE-SMALE-WITTEN COMPLEX FOR GRADIENT-LIKE VECTOR FIELDS ON STRATIFIED SPACES

URSULA LUDWIG

Mathematisches Institut der Universität Freiburg
Eckerstrasse 1
D-79104 Freiburg
E-mail: ursula.ludwig@math.uni-freiburg.de

The aim of this article is to generalise the notion of gradient-like vector fields to stratified spaces and to establish the associated Morse-Smale-Witten complex. The Morse-inequalities which relate the number of singular points of the vector field to the homology of the stratified space are a consequence of the existence of this complex.

1. Introduction

The aim of this article is to generalise the notion of gradient-like vector fields to stratified spaces and to establish the associated Morse-Smale-Witten complex. The Morse-inequalities which relate the number of singular points of the vector field to the homology of the stratified space are a consequence of the existence of this complex.

Let us first recall some main features of the theory of gradient-like vector fields on a smooth compact manifold M of dimension n. Recall first that by a theorem of Smale [24] each gradient-like vector field ξ is indeed the (negative) gradient of a Morse function $f : M \to \mathbb{R}$ with respect to a (generic) metric g, i.e. $\xi = -\nabla_g f$. Critical points of the Morse function are exactly the singular points of the vector field. Denote by $c_k(f) = c_k(\xi)$ the number of critical points of f of index k. Let b_k be the kth Betti number for the homology with $\mathbb{Z}_2$ coefficients. The (strong) Morse inequalities relate the number of critical points of the function f, or equivalently of its gradient vector field, to the singular homology of the manifold:

$$\sum_{i=0}^{k}(-1)^i c_k \geq \sum_{i=0}^{k}(-1)^i b_k, \text{ for all } k < n,$$

and

$$\sum_{i=0}^{n}(-1)^i c_k = \sum_{i=0}^{n}(-1)^i b_k.$$

The strong Morse inequalities are known to be equivalent to the existence of a complex (C_k, ∂_k), with free chain group $C_k = \mathbb{Z}_2^{c_k}$ ($k \in \{0, \ldots, n\}$), such that the homology of this complex is isomorphic to the singular homology of the manifold ([6,13,23,26]). By an idea of Witten [30] one gets a geometric realization of the boundary ∂_k by counting trajectories between critical points of the negative gradient flow, i.e. the flow induced by $-\nabla_g f$ where g is a generic metric on M.

Note that the last Morse-inequality is the Poincaré-Hopf index theorem for the gradient vector field $\nabla_g f$.

Let us mention here that in most presentations the Morse-Smale-Witten complex is associated to a pair (f, g) where f is a Morse function and g is a generic metric. Indeed the complexes associated to two pairs (f, g) and (f', g') are isomorphic if $\xi = -\nabla_g f = -\nabla_{g'} f'$ and thus the Morse-Smale-Witten complex actually depends only on the gradient vector field. This point of view is treated in [6] and will be useful for our purposes. There the departure point is a gradient-like vectorfield ξ. As an intermediate step one constructs a Lyapunov function for ξ, i.e. a function which is decreasing along the flow lines of ξ. Both f and f' are Lyapunov functions for ξ but for the approach in [6] it is not important to consider the Riemannian metric. In this article we want to generalise the above constructions to abstract stratified spaces as introduced by [11] and [27]. First we have to generalise the notion of a gradient-like vector field to the stratified case. On stratified spaces the study of so-called radial vector fields is of some interest. Radial vector fields on Whitney-stratified spaces have first been introduced by Marie-Hélène Schwartz in [18] in order to define characteristic classes on singular spaces. In [19] she showed that the Poincaré-Hopf index theorem can be generalised to radial vector fields and that the theorem may fail to hold for arbitrary vector fields on singular spaces (see also [1] for a short overview). Further generalisations of the Poincaré-Hopf index theorem to more general spaces resp. to more general vector fields have been shown in the sequels [10,22].

Since as remarked before the Poincaré-Hopf index theorem and the Morse-inequalities are related to each other, it is useful to incorporate a radiality condition into a definition of gradient-like vector fields on stratified spaces. More precisely we will define gradient-like vector fields to be radial vector fields with particularly "nice" singularities (def. 3.3) and such that the restriction to each stratum is gradient-like in the smooth sense. Moreover they satisfy a stratified version of the Morse-Smale transversality condition (def. 6.2).

The paper is organised as follows. In section 2 we recall the notion of an abstract stratified space as well as the notion of stratified vector fields. In section 3 we say what we mean by a (stratified) radial vector field (the notion is slightly different from the ones in [10,19,22]) and by a singular point (of the stratified vector field) of standard form.

In section 4 we develop a local model for the flow near a singularity of standard form. This implies (as analogue of thm. 3.3.1 in [19] on the preservation of the index for radial vector fields) that the so-called Conley index of a singular point of standard form is well defined.

In section 6 we define the notion of a gradient-like vector field on a stratified space and show the existence of such vector fields. Using the local model we can prove good properties of the stable and unstable set for a singular point of a gradient-like vector field (section 5 and 6): The unstable set is a submanifold of the stratum of X in which the point lies, the stable set is a substratified space, actually it is a π-fibre space as defined in [8].

In section 7 we study the trajectory spaces between singular points of a gradient-like vector field. We are particularly interested in their asymptotic structure. This is the essential ingredient for the definition of the Morse-Smale-Witten complex in section 8. The main result of the paper is thm. 8.2:

Theorem. *Let X be a compact stratified space. Let ξ be a stratified gradient-like vector field on X. Then there is an isomorphism between the singular homology of X (with coefficients in $\mathbb{Z}_2$) and the homology of the Morse-Smale-Witten complex associated to ξ.*

The Morse-inequalities for a stratified gradient-like vector field are a corollary of this theorem.

Note that this result is closely related to Goresky's results on π-fibre Morse functions ([8], section 9). π-fibre Morse-functions are functions which are increasing normally to each stratum. In 9.3 of [8] the existence of a π-fibre Morse function f on a Whitney stratified space is stated (without proof). Cor. 9.4 of [8] shows that X has the homotopy type of a CW-complex, where k-cells are in one-to-one correspondence with critical points of f of index k. In order to obtain the deformation retracts needed for this result controlled vector fields are used (locally). We will refer to this result in lemma 8.1.

The approach here differs from that in [8] since we start from a "nice" vector field ξ. As an intermediate step we need to construct a so-called Lyapunov function for ξ, which has the property that it is decreasing along the flow lines of ξ. In particular the Lyapunov function of a stratified gradient-like vector field ξ is normally increasing to each stratum in the neighbourhood of singular points.

Note that on a compact space the flow induced by ξ is defined for all time. The advantage of studying the vector field instead of the Morse function is that we can make use of methods from dynamical systems. In particular we can define stable/unstable sets for singular points as well as the trajectory spaces between singular points. The ingredients needed for the definition of the so-called Morse-Smale-Witten complex are the singular points of the vector field as well as the trajectory spaces between singular points of index difference 1 and 2. The attaching maps of the cells in the CW-complex associated to the vector field ξ are related to the trajectory spaces in a precise way (see lemma 8.2) and this fact is used to prove the main theorem.

That it should be possible to generalise the Morse-Smale theory to the stratified context has also been conjectured in [8].

Let us also mention the famous book [7], which treats more general Morse functions on Whitney stratified spaces. In particular on complex spaces Morse inequalities for intersection homology are established.

2. Preliminaries

Abstract stratified spaces. In this article we will deal with stratified spaces as introduced by [11] and Thom [27]. We recall here the basic notions of an abstract stratified space as well as the notions of stratified vector field, controlled stratified vector field, isomorphism of stratified spaces.

Let X be a topological space, Hausdorff, locally compact, paracompact and with countable basis of the topology. A stratification $\mathcal{S}$ of the topological space X is a locally finite family of disjoint locally closed subsets $S \subset X$ called strata such that $X = \bigcup_{S \in \mathcal{S}} S$. Moreover the strata S are smooth manifolds without boundary in the induced topology.

A tubular system for a stratified space is a family of triples

$$\{(T_S, \pi_S, \rho_S)\}_{S \in \mathcal{S}},$$

where T_S is an open neighbourhood of S, $\pi_S : T_S \to S$ is a continuous retraction and $\rho : T_S \to \mathbb{R}_{\geq 0}$ is continuous and such that $\rho_S^{-1}(0) = S$.

Definition 2.1. (see [28]) An abstract stratified space is a topological space X with a stratification $\mathcal{S}$ and a tubular system $\{(T_S, \pi_S, \rho_S)\}_{S \in \mathcal{S}}$ satisfying the following conditions:

(1) For each pair of strata (S, R), $T_S \cap R \neq \emptyset$ implies that $S \leq R$, i.e. $S \subset \overline{R}$.

(2) For each pair of strata (S, R) with $S < R$ (i.e. $S \leq R$, $S \neq R$) the map

$$(\pi_S, \rho_S) : R \cap T_S \to S \times \mathbb{R}_{>0}$$

is smooth and submersive.

(3) The tubular system is controlled, i.e. for each pair of strata (S, R) with $S < R$ and all $x \in T_S \cap T_R$ the following conditions are satisfied:

$$(C1)\ \pi_S \pi_R(x) = \pi_S(x),$$
$$(C2)\ \rho_S \pi_R(x) = \rho_S(x).$$

Because the stratification is locally finite and because of condition (2), two strata of the same dimension are not comparable and thus (by taking the union of all strata of the same dimension) we can assume without loss of generality, that the stratification $\mathcal{S} = \{X_i\}_{i=1,\ldots,n}$ on X is such that $\dim X_1 < \ldots < \dim X_n$. Moreover we will usually make the assumption that $\overline{X_n} = X$.

A consequence of Thom's First Isotopy Lemma (see e.g. [28] thm. 2.6) is that these spaces have locally a cone-like structure, i.e. let $p \in S$ be a point in a stratum S then there exists a neighbourhood of p in X isomorphic (by an isomorphism of stratified spaces as in def. 2.3) to $\mathbb{R}^{\dim S} \times \mathrm{cone}(L)$. Here $L \simeq \pi_S^{-1}(p) \cap \rho_S^{-1}(\epsilon)$ is the normal link at p. The normal link is independent of the choice of $p \in S$ and $\epsilon > 0$ small enough. By taking the obvious restrictions one can easily see that L is itself an abstract stratified space of smaller depth, where the depth of X is defined as follows:

$$\mathrm{depth}(X) = \max\{m \mid \text{ there exist a chain of strata } S_0 < S_1 < \ldots < S_m\}.$$

Example 2.1. The most prominent stratifications are the so-called Whitney stratifications of an analytic subset of $\mathbb{R}^N$ (see e.g. [7] for the definition). Whitney stratified spaces admit a (non-unique) structure of an abstract stratified space ([27] or [17] thm. 3.6.9).

Controlled stratified vector fields. A stratified vector field ξ on an abstract stratified space X is a family $\{\xi_S : S \to TS\}_{S \in \mathcal{S}}$ of smooth vector fields on each stratum.

Definition 2.2. (see [28]) A stratified vector field on X is called controlled if for all pairs of strata (S, R) with $S < R$ and all $x \in T_S \cap R$ the following conditions are satisfied:

$$(C3)\ d\pi_S \xi_R(x) = \xi_S(\pi_S(x)),$$
$$(C4)\ d\rho_S \xi_R(x) = 0.$$

A vector field satisfying only the control condition $(C3)$ is called weakly controlled.

There is no canonical notion of continuity for stratified vector fields but using the control conditions $(C3)$ and $(C4)$ one can still obtain a (local) continuous, stratum-wise smooth flow Φ (see [28], (2.3)). Moreover the homeomorphism Φ_t is a isomorphism of stratified spaces:

Definition 2.3. A homeomorphism $f : X \to Y$ between stratified spaces is an isomorphism of stratified spaces if:

   (i) f sends each stratum of X diffeomorphically to a stratum of Y.
   (ii) For each stratum S of X one has $f(T_S) = T_{f(S)}$ and for all $x \in T_S$ the following conditions are satisfied:

$$(C5)\ \pi_{f(S)} \circ f(x) = f \circ \pi_S(x),$$
$$(C6)\ \rho_S(x) = \rho_{f(S)} \circ f(x).$$

We will call $f : X \to Y$ a stratified homeomorphism if in the above definition the restriction of f to each stratum is only a homeomorphism.

A function $f : X \to M$ into a smooth manifold M is called controlled if for each stratum S and all $x \in T_S$:

$$f(\pi_S(x)) = f(x).$$

The function f is called a stratified controlled submersion if it is controlled and the restriction to each stratum of X is submersive.

One has the following important lifting property for stratified vector fields [28](2.4):

Proposition 2.1. *Let X be an abstract stratified space, M a manifold and $f : X \to M$ a controlled submersion. Then there exists for every smooth* *vector field $\eta : M \to TM$ a controlled vectorfield ξ on X such that for each stratum S of X*

$$Df_{|S} \circ \xi_S = \eta \circ f_{|S}.$$

In particular, given a stratification by strata $\{X_i\}$ with $\dim X_1 < \ldots < \dim X_n$ and a smooth vector field ξ_1 on the smallest stratum X_1 there exists a controlled stratified vector field $\xi = \{\xi_i\}_{i \in I}$ on X, called controlled lift of ξ_1.

3. Radial vector fields

Radial vector fields on Whitney stratified spaces have first been introduced by Marie-Hélène Schwartz [18] and were used to define Chern classes on singular spaces. In [19] the Poincaré-Hopf index theorem has been generalised to radial vector fields. Further generalisations of the Poincaré-Hopf index theorem to more general spaces or more general vector fields can be found in [10] and [22]. See the later reference also for a comparison of the different notions of radiality appearing in the literature. The essential property of radial vectorfields in the above literature is that they allways point *out* of a neighbourhood of the stratum. Since we will work with the flow induced by the vectorfield it is more convenient for us to call "radial" a vector field which is pointing *towards* the stratum.

Definition 3.1. A stratified vector field will be called totally radial if for each stratum S of X there exists a non negative real number A_S such that for all pairs of strata $S < R$ and all $x \in T_S$

$$d\rho_S \xi_R(x) \leq -A_S \rho_S(x).$$

Weakly controlled totally radial vector fields are locally integrable (see [5], prop. 2.5.1.).

Definition 3.2. Let X be an abstract stratified space with a tubular system $\{(T_S, \pi_S, \rho_S)\}_{S \in \mathcal{S}}$. A weakly controlled vector field ξ on T_S is called radial (with respect to the stratum S) if it satisfies

  (i) $\xi_{|S} = 0$ and
  (ii) $\xi_{|T_S - S}$ is a controlled lift of the vector field $-t\frac{\partial}{\partial t}$ on $\mathbb{R}$ along the controlled submersion $\rho_S : T_S - S \to \mathbb{R}$.

Remark 3.1. From the definition of an abstract stratified space it follows that $\rho_S : T_S - S \to \mathbb{R}$ is a stratified controlled submersion. Thus by prop. 2.1

there exists a (non unique) controlled vector field ξ_{rad} on $T_S - S$ satisfying

$$d\rho\xi_{rad} = -t\frac{\partial}{\partial t}.$$

A point $p \in X$ is a singularity of the stratified vector field ξ if $\xi(p) = 0$. We will study stratified vector fields having only nice singularities. Let us first recall that a singular point p of a smooth vector field ξ_S on a smooth manifold S is called hyperbolic if the set of eigenvalues of the linearisation $A := D\xi(p)$ is disjoint from the imaginary axis. The number of eigenvalues of A with positive real part is called the index of the hyperbolic singular point. A hyperbolic singular point is called of standard form if all eigenvalues of A are real.

Definition 3.3.

(i) Let ξ be a stratified vector field on an abstract stratified space X. A singularity $p \in S$ of ξ will be called hyperbolic (resp. of standard form) if the following conditions are satisfied:

- (a) p is a hyperbolic singularity (resp. a singularity of standard form) of the smooth vector field ξ_S.
- (b) There is an open neighbourhood of p such that in this neighbourhood ξ is a radial lift of ξ_S, i.e. ξ can be written as

$$\xi = \xi_{||} + \xi_{rad}$$

where $\xi_{||}$ is a controlled lift of ξ_S and ξ_{rad} is radial with respect to the stratum S.

(ii) A stratified vector field will be called a stratified vector field with hyperbolic singularities (resp. with singularities of standard form) if it is weakly controlled, totally radial and all singular points are hyperbolic (resp. of standard form). We will shortly call such vector fields stratified vector fields w.h.s. (resp. w.s.s.).

Stratified hyperbolic singularities of a vector field are isolated, and if the space is compact one can deduce therefore the finiteness of singular points for stratified vector fields w.h.s.

One can prove the following existence result, the proof of which is postponed to section 6:

Proposition 3.1. *There exists a stratified vector field w.s.s. on each compact abstract stratified space X.*

Note that by integrating a stratified vector field w.h.s. on a compact stratified space one gets a continuous, stratum-wise smooth flow $\Phi : X \times \mathbb{R} \to X$, such that Φ_t satisfies the control condition $(C5)$.

The α- resp. ω-set of a point $x \in X$ is defined as follows:

$$\alpha(x) = \{q \in X \mid \Phi(t_n, x) \to q \text{ for some sequence } t_n \to -\infty\},$$
$$\omega(x) = \{q \in X \mid \Phi(t_n, x) \to q \text{ for some sequence } t_n \to \infty\}.$$

One denotes by L_α resp. L_ω the set of all α- resp. ω-points of the flow Φ. Note that all singular points of a vector field w.s.s. are in $L_\alpha \cup L_\omega$, but the converse need not hold (see [16], example 2 in section 1).

Using the control conditions and the total radiality of a stratified vector field w.h.s. one can deduce that in infinite time flow lines can only go from a larger into a smaller stratum, i.e. for a point $x \in R$ one has:

$$\alpha(x) \subset R \text{ and } \omega(x) \subset \bigcup_{S \leq R} S.$$

4. Local model

In this section we develop a local model for the flow of a stratified vector field in the neighbourhood of a hyperbolic singular point. In the second part we define the index of a hyperbolic singular point.

Let $p \in S$ be a hyperbolic singular point of a stratified vector field w.h.s. ξ. We assume in this section that X is compact and thus the flow is defined for all time. Denote by $d := \dim S$ and by $A := D\xi_S(p)$ the linearisation of ξ_S at p (on the smooth stratum S). After a choice of a basis of T_pS one can identify A with a linear map $A : \mathbb{R}^d \to \mathbb{R}^d$.

Proposition 4.1. *Let ξ be a stratified vector field on an abstract stratified space X. Let $p \in S$ be a hyperbolic singular point of ξ. Denote by Φ the flow associated to ξ. Then one can find an open neighbourhood $V(p) \subset X$ of p and an open neighbourhood $U(0) \subset \mathrm{cone}(L) \times \mathbb{R}^d$ of 0 and a stratified homeomorphism*

$$h : V(p) \to U(0)$$

such that the relation

$$h(\Phi(x, t)) = \Psi(h(x), t)$$

is satisfied for all $(x, t) \in X \times \mathbb{R}$ such that $\Phi(x, t) \in V(p)$ and $\Psi(h(x), t) \in U(0)$.

The flow

$$\Psi : \big(\mathrm{cone}(L) \times \mathbb{R}^d\big) \times \mathbb{R} \to \mathrm{cone}(L) \times \mathbb{R}^d$$

is given by $\Psi\left((r,l,z),t\right) = (re^{-t}, l, e^{tA}z)$, *where* (r,l) *are the coordinates in* $\mathrm{cone}(L)$ *(r the radial coordinate) and z is the coordinate in $\mathbb{R}^d$.*

Proof. By Thom's First Isotopy Lemma there exists an open neighbourhood $V(p)$ and a stratified isomorphism

$$h' : V(p) \simeq \mathrm{cone}(L) \times \mathbb{R}^d,$$

such that $h'(p) = 0$. The tubular retraction $\pi : \mathrm{cone}(L) \times \mathbb{R}^d \to \mathbb{R}^d$ is given (in these coordinates) by $\pi(r,l,z) = z$ and the distance from the stratum $\{0\} \times \mathbb{R}^d$ is given by $\rho(r,l,z) = r$.

The stratified vector field $\xi' = h'^{-1*}\xi$ is a weakly controlled vector field on $\mathrm{cone}(L) \times \mathbb{R}^d$ with a hyperbolic singularity $h'(p) = 0$. The conditions of definition 3.3 (i) imply the following form of the vector field ξ' near the singular point:

$$\xi'(r,l,z) = -r\frac{\partial}{\partial r} + \xi'_l(r,l,z) + \xi_z(z),$$

where $\xi_l(r,-,z)$ is a family of controlled vector fields on L depending smoothly on r and z for $r \neq 0$ and such that $\xi_l(0,-,z) \equiv 0$. Moreover ξ_z is a smooth vector field on $\mathbb{R}^d$ with hyperbolic singularity at 0 and linearisation A at 0.

The induced flow is a continuous, stratum-wise smooth flow and is of the form $\Psi'\left((r,l,z),t\right) = (re^{-t}, \Psi'_l\left((r,l,z),t\right), \Psi'_z(z,t))$, where $\Psi'_l\left((r,l,z),t\right)$ respects the stratification of the normal link and satisfies the control conditions $(C5)$ and $(C6)$.

From the smooth theory of hyperbolic vector fields (see e.g. [16], proof of prop. 4.10) we know that the flow for the vector field ξ_z on $\mathbb{R}^d$ is locally topologically conjugate to the flow of its linearisation A. This implies that the flow Ψ' is locally topologically conjugate to the flow $\Psi''(r,l,z) = \big(re^{-t}, \Psi''_l\left((r,l,z),t\right), e^{tA}z\big)$. This means that there exist open neighbourhoods $W'(0) \subset \mathrm{cone}(L) \times \mathbb{R}^d$ and $W''(0) \subset \mathrm{cone}(L) \times \mathbb{R}^d$ and a homeomorphism $g : W'(0) \to W''(0)$ such that

$$g(\Psi'(x,t)) = \Psi''(g(x),t).$$

Moreover Ψ'' and Ψ are topologically conjugate via the stratified isomorphism

$$
\begin{aligned}
h'' : \mathrm{cone}(L) \times \mathbb{R}^d &\longrightarrow \mathrm{cone}(L) \times \mathbb{R}^d, \\
h''(r,l,z) &= \Psi''\left(\big(1,l,\tfrac{z}{e^{-A\log(r)}}\big), -\log r\right).
\end{aligned}
$$

The conjugation map from the assertion of the proposition is now given by $h := h''^{-1} \circ g \circ h'$ with the obvious choices of the open neighbourhoods. $\square$

Remark 4.1. By imposing a non-resonance condition on the eigenvalues of the linearisation A one can actually show that in the above proposition the conjugation map h is an isomorphism of stratified spaces.

Morse-Conley index for the singular point. An essential result towards the study of the Poincaré-Hopf index theorem for radial vector fields is that the index of the vector field at a singular point p equals the index of the restriction to the stratum in which p lies ([19], [10], [22]).
Prop. 4.2 is in the spirit of these results, while by index here we mean the so-called Conley index. Conley theory is an important generalisation of Morse theory to flows on metric spaces. Let us recall the definition of the Conley index for a flow invariant set I (cf. [4]):

Definition 4.1. Let Φ be a continuous flow on a metric space X. Let I be a flow invariant set. A compact pair (N_1, N_0) in X, $N_0 \subset N_1$, is called an index pair for I if the following holds:

(i) I is the maximal flow invariant set in $\overline{N_1 - N_0}$.
(ii) N_0 is positively invariant relative to N_1, i.e. if $x \in N_0$ and $\Phi(x, [0, t]) \subset N_1$ for $t > 0$ then $\Phi(x, [0, t]) \subset N_0$.
(iii) N_0 is an exit set for N_1: If $x \in N_1$ and $\Phi(x, t_1) \notin N_1$ for some positive t_1, then there exists $t_0 \geq 0$ such that $\Phi(x, [0, t_0]) \subset N_1$ and $\Phi(x, t_0) \in N_0$.

Two index pairs for the invariant set I are homotopically equivalent. The homotopy type of the pointed space $(N_1/N_0, *)$ is called (homotopical) Conley index of I.

One can show that the Conley index of a (smooth) hyperbolic singular point of index k is the homotopy type of the pair $(S^k, *)$.
Let $p \in S$ be a singular point of a stratified vector field w.h.s. ξ. Then by definition p is a (smooth) hyperbolic point for the restriction ξ_S. Let k be the index of ξ_S at p.

Proposition 4.2. *Let ξ be a stratified vector field w.h.s.. Let $p \in S$ be a singular point of ξ. Let (N_1, N_0) be an index pair for p in S (for the restricted flow). Then, for $\rho_0 > 0$ small enough $(N_1^X, N_0^X) := (\pi_S)^{-1}(N_1, N_0) \cap \{\rho_S(x) \leq \rho_0\}$ is an index pair for p in X.*

The two pairs (N_1, N_0) and (N_1^X, N_0^X) are homotopic and the homotopy type of $(N_1^X / N_0^X, *)$ is the homotopy type of $(S^k, *)$.
In particular the index $\mu(p) := k$ of ξ at p is well defined.

Proof. Since the tubular projection $\pi_S : T_S \to S$ is continuous and N_1^X and N_0^X are closed and contained in a compact set, they are themselves compact. In the sequel we will use the fact that π_S commutes with the flow Φ (see section 2). For a set A we will denote by $Inv(A)$ the maximal Φ-invariant set contained in A:

$$Inv(A) := \{x \in A \mid \Phi(x, \mathbb{R}) \subset A\}.$$

We show $Inv(\overline{N_1^X - N_0^X}) = \{p\}$: The inclusion $\supset$ is evident. Let now $x \neq p$, $x \notin S$ be a point in $Inv(\overline{N_1^X - N_0^X})$. Since p is a stratified hyperbolic point, there exists a $t(x)$ such that $\rho(\Phi(x, t)) > \rho_0$ for all $t < t(x)$. This is a contradiction to $x \in Inv(\overline{N_1^X - N_0^X})$.
We show that N_0^X is an exit set for N_1^X : Let $x \in N_1^X$ with $\Phi(x, t) \notin N_1^X$ for a time $t > 0$. Two cases can arise. 1st case: $\pi_S(\Phi(x, t)) \notin N_1$. Since $\pi_S(x) \in N_1$ and N_0 is an exit set for N_1, there exists a time t_0 such that $\Phi(\pi_S(x), [0, t_0]) = \pi_S(\Phi(x, [0, t_0])) \in N_1$ and $\Phi(\pi_S(x), t_0) \in N_0$. Then (using in addition $d\rho_S \xi < 0$) one gets that $\Phi(x, [0, t_0]) \in N_1^X$ and $\Phi(x, t_0) \in N_0^X$ and the assumption follows. 2nd case: $\pi_S(\Phi(x, t)) \in N_1$, i.e. the inequality $\rho_S(\Phi(x, t)) > \rho_0$ must hold. This is a contradiction to $d\rho_S \xi < 0$ and $\rho_S(x) < \rho_0$.
We show that N_0^X is positively invariant relative to N_1^X: Let $x \in N_0^X$ and $\Phi(x, [0, t]) \in N_1^X$. Then $\pi_S(\Phi(x, [0, t])) = \Phi(\pi_S(x), [0, t]) \in N_1$ and since N_0 is positively invariant relative to N_1, $\pi_S(\Phi(x, [0, t])) \in N_0$ follows. Thus we have $\Phi(x, [0, t]) \in N_0^X$.
From the smooth theory we know already that the pair (N_1, N_0) is homotopic to the pair $(S^k, *)$. Thus the last part of the statement follows from the existence of an isomorphism of stratified spaces $(N_1^X, N_0^X) \simeq (N_1, N_0) \times \mathrm{cone}(L)$ and the contractibility of $\mathrm{cone}(L)$. $\qquad\square$

5. Stable/unstable manifold

Let X be a compact abstract stratified space. Let ξ be a stratified vector field w.h.s. with induced flow Φ. The stable resp. unstable set of a singular point $p \in S$ is defined as:

$$W^s(p) := \{y \in X \mid \lim_{t \to \infty} \Phi(y, t) = p\}$$

resp.

$$W^u(p) := \{y \in X \mid \lim_{t \to -\infty} \Phi(y, t) = p\}.$$

Since the flow of a vector field w.h.s. can not leave a stratum in positive time (see end of section 3), the unstable set $W^u(p)$ is completely contained in the stratum S, and the following proposition is a consequence of the unstable manifold theorem of the smooth theory (compare e.g. [16], section 2.6):

Proposition 5.1. *Let p be a singular point of index k of a stratified vector field w.h.s. ξ. Then the unstable set $W^u(p)$ is a k-dimensional immersed smooth submanifold of S.*

Proposition 5.2. *Let $p \in S$ be a singular point of a stratified vector field w.h.s. ξ. Then $W^s(p) \cap S$ is an immersed smooth submanifold of S, whereas for all strata R with $S < R$ the restriction $W^s(p) \cap R$ is an embedded smooth submanifold of R.*

Proof. Since the flow satisfies the control condition $(C5)$ in a tubular neighbourhood, $x \in \pi_S^{-1}(W^s(p) \cap S)$ is a necessary condition for $\lim_{t \to \infty} \Phi(x, t) = p$. Now let $x \in \pi_S^{-1}(W^s(p) \cap S)$. By $(C5)$ and $d\rho\xi = -t\frac{\partial}{\partial t}$ (in a neighbourhood of the singular point) it follows that $x \in W^s(p)$. We have shown that

$$W^s(p) \cap V(p) = \pi_S^{-1}(W^s(p) \cap S) \simeq (W^s(p) \cap S) \times \text{cone}(L)$$

where the open neighbourhood $V(p)$ has been chosen according to Thom's Isotopy Lemma. Thus the assumption holds locally in a neighbourhood $V(p)$.

Let $R \geq S$ be a stratum and let $x \in W^s(p) \cap R$, then there is $t \in \mathbb{R}$ such that $x(t) \in W^s(p) \cap R \cap V(p)$. If (U, h) is a manifold chart near $x(t)$, then $(\Phi(-t, U), \Phi(-t, -) \circ h)$ is a chart near x. This shows that $W^s(p) \cap R$ is an immersed submanifold of R for each stratum $R \geq S$. Note that from the total radiality of ξ we can deduce that the intersection $W^s(p) \cap R$ for $R > S$ is indeed an embedded submanifold of R. The important additional property needed is that flow lines leaving a singular point don't come back in positive time (compare e.g. [9]). In the present situation this property is satisfied since flow lines only go from larger into smaller strata. $\qquad\square$

Remark 5.1. At this point we have shown that the stable set of a hyperbolic singularity is almost a stratified space with the obvious tubular system induced by restriction. The only property in def. 2.1 which eventually fails to hold is that the smallest stratum $W^s(p) \cap S$ is not always a manifold. For gradient-like vector fields however $W^s(p)$ is a substratified space of X, one can even show that it is a so-called π-fibre space (see [8] for the definition).

6. Stratified gradient-like vector fields

In this section we will define the notion of a stratified gradient-like vector field and show the existence of such vector fields. From now on we will always assume that the stratified space X is compact.

Before giving the definition of a stratified gradient-like vector field we need to explain the notion of non-wandering points: A point $x \in X$ is a wandering point of Φ if there is an open neighbourhood V of x and a number $t_0 \in \mathbb{R}$ such that $\Phi(t, V) \cap V = \emptyset$ for $|t| > t_0$. The set of non-wandering points for the flow Φ is denoted by Ω_Φ.

Note that the set Ω_Φ is compact and invariant under the flow. Moreover $L_\alpha \cup L_\omega \subset \Omega_\Phi$ and topological equivalence preserves the set of non-wandering points.

Definition 6.1. The trajectory space between two hyperbolic singular points $p \in R$ and $q \in S$ of a stratified vector field w.h.s. ξ is defined as

$$\mathcal{M}_{pq}(\xi) := \{x(t) \mid t \in \mathbb{R}, \ \dot{x}(t) = \xi(x(t)), \lim_{t \to -\infty} x(t) = p, \lim_{t \to \infty} x(t) = q\}.$$

The trajectory space $\mathcal{M}_{pq}$ is a flow invariant set and equals the intersection of the stable set of q and the unstable set of p

$$\mathcal{M}_{pq}(\xi) = W^u(p) \cap W^s(q).$$

We have already seen that the unstable manifold of a point $p \in R$ is itself always completely contained in the stratum R. Thus also the trajectory space $\mathcal{M}_{pq}$ must be contained in R, and it can be non empty only if the point q lies in a smaller stratum S, $S \leq R$. Thus for a stratified vector field w.h.s. the Morse-Smale condition (see below) makes also sense:

Definition 6.2. A stratified vector field w.s.s. ξ on an abstract stratified space X will be called gradient-like if:

(1) The Morse-Smale condition is satisfied: For all singular points $p \in R$ and $q \in S$ ($S \leq R$) the manifolds $W^u(p)$ and $W^s(q) \cap R$ intersect transversally, i.e. for all points $x \in W^u(p) \cap W^s(q) \subset R$ we have

$$T_x W^u(p) + T_x \left(W^s(q) \cap R \right) = T_x R.$$

(2) The set of non-wandering points Ω_Φ is equal to the set of singular points.

Note that if X is a smooth compact manifold the notion of a gradient-like vector field given above is slightly weaker than the conditions (1)-(4) in [24]. From the implicit function theorem it follows that for a gradient-like vector field the trajectory space $\mathcal{M}_{pq}$ is a submanifold of R of dimension $\mu(p,q) := \mu(p) - \mu(q)$.

The additive group $\mathbb{R}$ acts smoothly freely and properly on the manifold $\mathcal{M}_{pq}$ by

$$\begin{aligned}
\mathbb{R} \times \mathcal{M}_{pq} &\longrightarrow \mathcal{M}_{pq}, \\
(\tau, \gamma(t)) &\mapsto \gamma_\tau(t)
\end{aligned}$$

where γ_τ denotes the shifted curve $\gamma_\tau(t) = \gamma(t + \tau)$ (compare [20], 2.4.1). (Note that $\gamma(\mathbb{R}) \subset \mathcal{M}_{pq}$ implies that $\gamma_\tau(\mathbb{R}) \subset \mathcal{M}_{pq}$.)

The quotient of $\mathcal{M}_{pq}$ by this action, the so called unparametrised trajectory space, is a manifold of dimension $\mu(p) - \mu(q) - 1$ which we denote by

$$\widetilde{\mathcal{M}_{pq}} := \mathcal{M}_{pq}/\mathbb{R}.$$

We will show the existence of stratified gradient-like vector fields:

Existence of stratified gradient-like vector fields. Let us assume that X has a stratification $\{X_i\}_{i \in I}$ with $\dim X_1 < \ldots < \dim X_n$ and $\overline{X_n} = X$. For a real number $\epsilon > 0$ and a stratum S of X we denote by T_S^ϵ the tubular neighbourhood

$$T_S^\epsilon := \{x \in T_S \mid \rho_S(x) < \epsilon\}.$$

Lemma 6.1. *Let X be a compact abstract stratified space. Then there is an $\epsilon > 0$ such that for all $\delta < \epsilon$ and all $i \in I$ the sets*

$$\widetilde{X_i} = \left(\widetilde{X_i}^\delta\right) := X_i - \left(T_{X_1}^\delta \cup \ldots \cup T_{X_{i-1}}^\delta\right)$$

are compact smooth manifolds possibly with boundary and corners.

Proof. The existence of removal data for compact Whitney stratified subsets of $\mathbb{R}^N$ has been shown in [21], pg. 5. One can apply this result to an abstract stratified space X by choosing a realisation of X as a Whitney stratified space according to the result in [14] (see also [15], [25]). $\qquad\square$

Such an ϵ is called removal datum for X. Note that on non compact stratified spaces removal data need not exist.

698

Proposition 6.1. *There exist stratified gradient-like vector fields on a compact abstract stratified space.*

Proof. *Step 1: Construction of a gradient-like vector field on X_1:* By smooth theory (compare e.g. [16]) there exists a smooth Morse function on the compact manifold X_1 such that its gradient vector field (with respect to a generic metric) ξ_1 satisfies the Morse-Smale condition. In particular ξ_1 is gradient-like in the sense of def. 6.2. Let ϵ_1 be a removal data for X. Let us denote also by ξ_1 a radial lift of the vector field to the tubular neighbourhood $T_{X_1}^{\epsilon_1}$.

For $1 \leq i \leq n$ denote by $\epsilon_i = \left(\frac{3}{4}\right)^{i-1} \epsilon_1$. The proof of the theorem is now completed by inductive application of step 2 followed by step 3.

Step 2_k: Assume that the gradient-like vector field ξ has been constructed on $\cup_{j<k} T_{X_j}^{\epsilon_{k-1}}$ such that the following estimates hold for a positive constant $C > 0$:

$$d\rho_1 \xi < -C\rho_1 \quad on \quad T_{X_1}^{\epsilon_{k-1}} \cup \bigcup_{2 \leq j < k} \left(T_{X_j}^{\epsilon_{k-1}} \cap T_{X_1}^{\epsilon_2} \right),$$

$$d\rho_2 \xi < -C\rho_2 \quad on \quad \bigcup_{2 \leq j < k} \left(T_{X_j}^{\epsilon_{k-1}} \cap T_{X_2}^{\epsilon_3} \right) - T_{X_1}^{\epsilon_3},$$

$$\ldots,$$

$$d\rho_i \xi < -C\rho_i \quad on \quad \bigcup_{i \leq j < k} \left(T_{X_j}^{\epsilon_{k-1}} \cap T_{X_i}^{\epsilon_{i+1}} \right) - \bigcup_{j<i} T_{X_j}^{\epsilon_{i+1}}, \ for \ i \leq k-1$$

and

$$d\rho_i \xi = 0 \quad on \quad \bigcup_{j<k} T_{X_j}^{\epsilon_{k-1}}, \ for \ i \geq k.$$

Then one can extend ξ to a stratified gradient-like vector field on $X_k \cup \bigcup_{j<k} T_{X_j}^{\epsilon_k}$ such that all singular points in X_k lie outside of $\bigcup_{j<k} T_{X_j}^{\epsilon_k}$ and the above estimates still hold on $\bigcup_{j<k} T_{X_j}^{\epsilon_k}$.

Note that even if $X_k \cup \bigcup_{j<k} T_{X_j}^{\epsilon_{k-1}}$ is not necessarily a stratified space the meaning of "stratified gradient-like vector field" is obvious here.

For simplicity we will chose the notation $\epsilon = \epsilon_{k-1}$ for the rest of the proof of this step. By Φ we denote the flow induced by the vector field ξ. Denote by $\widetilde{X_k} = \widetilde{X_k}^{\tilde{\epsilon}}$ the manifold with corners constructed as in lemma 6.1 where $\frac{4}{5}\epsilon < \tilde{\epsilon} < \epsilon$. Note that $\tilde{\epsilon}$ has been chosen such that $\partial\widetilde{X_k} \subset \bigcup_{j<k} T_{X_j}^{\epsilon} - \bigcup_{j<k} T_{X_j}^{4/5\epsilon}$. One can construct a smoothing $\widehat{X_k}$ of $\widetilde{X_k}$ such that $\widehat{X_k}$ is a smooth manifold with boundary $W_1 := \partial\widehat{X_k} \subset \bigcup_{j<k} T_{X_j}^{\epsilon} - \bigcup_{j<k} T_{X_j}^{4/5\epsilon}$ and ξ is transversal to W_1.

Because of the compactness of W_1 and the radial estimates of ξ on $\bigcup_{j<k} T^\epsilon_{X_j}$ there exists a finite time $T \in \mathbb{R}$ such that $W_2 := \Phi(W_1, T) \subset \bigcup_{j<k} T^{\epsilon/5}_{X_j}$.

Denote by $W := \Phi(W_1, (0, T))$ and by $A := \bigcup_{j<k} \overline{T^{3\epsilon/4}_{X_j}} - \bigcup_{j<k} T^{\epsilon/4}_{X_j}$. Note that $A \subset W$.

Define a smooth function $f : \overline{W} \to \mathbb{R}$ by $f_{|W_1} \equiv -1$ and $f_{|W_2} \equiv -2$ and $f(\Phi(x,t)) = -1 - \frac{t}{T}$ for $x \in W_1$. Moreover, by construction the vector field ξ is always transversal to the levelsets of f. Therefore one can find a Riemannian metric on the manifold $\widehat{X_k} \cup W$ such that the vector field ξ is the negative gradient vector field of f with respect to this metric. Using lemma 4.15 of [20] we can construct a Morse function $\widetilde{f}$ on the manifold $\widehat{X_k} \cup W$ such that $\widetilde{f}_{|A} = f_{|A}$ on the closed set A. One can even assume (by density of the Morse-Smale condition for gradient vector fields) that $-\nabla \widetilde{f}$ is a gradient-like vector field on $\widehat{X_k} \cup W$ and as in the proof of thm. A in [24] one can assume that $\widetilde{f}$ takes different values for different critical points. We define:

$$\xi_k = \begin{cases} \xi & \text{on } \left(\bigcup_{j<k} T^{3\epsilon/4}_{X_j}\right) \cap X_k, \\ -\nabla \widetilde{f} & \text{on } X_k - \bigcup_{j<k} T^{3\epsilon/4}_{X_j}. \end{cases}$$

Note that since $\xi = -\nabla \widetilde{f}$ on A the vector field ξ_k is well defined and smooth. Also by construction ξ_k is gradient-like and all singular points as well as all intersections of stable/unstable manifolds lie outside of $\bigcup_{j<k} T^{3\epsilon/4}_{X_j}$ and are transversal.

In the sequel of the proof of step 2_k we will denote by ξ the stratified vector field on $X_k \cup \bigcup_{j<k} T^{\epsilon_k}_{X_j}$ defined as before on the tubes of the lower strata and defined by $\xi = \xi_k$ on X_k. By construction ξ is weakly controlled, totally radial and has only singularities of standard form.

We have to show that after possibly perturbating ξ_k in $X_k - \bigcup_{j<k} T^{\epsilon_k}_{X_j}$ all intersections $W^s(p) \cap W^u(q)$ with $p \in \bigcup_{j<k} X_j$ and $q \in X_k$ are transversal. This is done by a successive application of the following lemma which is similar to Lemma 1.2 in [24]. Let $\{q_1, \ldots, q_l\}$ be the set of critical points of $\widetilde{f}$ (= singular points of ξ on X_k). This set is ordered by $\overline{q_1} := \widetilde{f}(q_1) < \ldots < \overline{q_l} := \widetilde{f}(q_l)$.

Lemma 6.2. *Given a sufficiently small $\delta > 0$ and $j \in \{1, \ldots, l\}$, there is a C^1-approximation ξ'_k of ξ_k such that $\xi'_k = \xi_k$ outside of $\widetilde{f}^{-1}(\overline{q_j} - 3\delta, \overline{q_j} - \delta)$ and such that in the ξ'-system all intersections $W^s(p) \cap W^u(q_j)$ with $p \in \bigcup_{j<k} X_j$ as well as all intersections $W^s(q_i) \cap W^u(q_j)$, $i < j$, are transversal.*

To complete step 2, the only thing remaining to show is that all nonsingular points of ξ in $X_k \cup \bigcup_{j<k} T_{X_j}^{\epsilon_k}$ are wandering: By construction ξ is a gradient vector field on $\widehat{X}_k \cup W$ and thus in particular all nonsingular points in $\widehat{X}_k \cup W$ are wandering. The vector field ξ has not been changed on $\bigcup_{j<k} T_{X_j}^{\epsilon_k}$ and thus all nonsingular points in $\bigcup_{j<k} T_{X_j}^{\epsilon_k}$ are wandering by induction hypothesis.

Step 3_k: Assume that the gradient-like vector field ξ has been constructed on $X_k \cup \bigcup_{j<k} T_{X_j}^{\epsilon_k}$ according to step 2_k. We will now construct a gradient-like vector field ξ' on $\bigcup_{j\leq k} T_{X_j}^{\epsilon_k}$ such that the first $k-1$ radial estimates of step 2_k still hold on $\bigcup_{j<k} T_{X_j}^{\epsilon_k}$ and on $T_{X_k}^{\epsilon_k} - \bigcup_{j<k} T_{X_j}^{3\epsilon_k/4}$ one has

$$d\rho_k \xi' < -C\rho_k.$$

Moreover $d\rho_i \xi' = 0$ on $\bigcup_{j\leq k} T_{X_j}^{\epsilon_k}$ for $i > k$.

We set $\epsilon = \epsilon_k$ in the proof of step 3_k.

Denote by ξ_k also a radial lift (on $T_{X_k}^\epsilon$) of the vector field constructed in step 2_k. Set $\xi' = \varphi\xi_k + (1-\varphi)\xi$ where φ is a continuous, stratum-wise smooth, controlled function on $T_{X_k}^\epsilon \cup \bigcup_{j<k} T_{X_j}^\epsilon$ with $\varphi \equiv 1$ on $\bigcup_{j<k} T_{X_j}^{\epsilon/2}$, $\varphi \equiv 0$ on $T_{X_k}^\epsilon - \bigcup_{j<k} T_{X_j}^\epsilon$ and $\varphi < 1$ on $T_{X_k}^\epsilon - \bigcup_{j<k} T_{X_j}^{3\epsilon/4}$. (For the existence of a function φ satisfying the above properties see [28], lemma 1.3.)

Note that the vector field ξ' is weakly controlled (see [28], pg. 10). By construction all singular points of ξ' lie in $\bigcup_{j\leq k} X_j$ and are of standard form. All intersections of stable and unstable manifolds are transversal (by step 2_k).

Using the control condition $d\rho_k \xi = 0$ (on $X_k \cup \bigcup_{j<k} T_{X_j}^{\epsilon_k}$) we have the following radial estimate on $T_{X_k}^\epsilon - \bigcup_{j<k} T_{X_j}^{3\epsilon/4}$:

$$d\rho_k \xi' = (1-\varphi)d\rho_k \xi_k + \varphi d\rho_k \xi = (1-\varphi)d\rho_k \xi_k < -C\rho_k.$$

Let $i < k$. For all $x \in \bigcup_{i\leq j\leq k} \left(T_{X_j}^\epsilon \cap T_{X_i}^{\epsilon_{i+1}}\right) - \bigcup_{j<i} T_{X_j}^{\epsilon_{i+1}}$ we have the estimate:

$$\begin{aligned}
d\rho_i \xi'(x) &= \varphi(x)d\rho_i \xi_k(x) + (1-\varphi(x))d\rho_i \xi(x) \\
&= \varphi(x)d\rho_i d\pi_k \xi_k(x) + (1-\varphi(x))d\rho_i \xi(x) \text{ by condition } (C2), \\
&= \varphi(x)d\rho_i \xi(\pi_k(x)) + (1-\varphi(x))d\rho_i \xi(x) \text{ by weak control of } \xi_k \\
&\quad \text{and construction of } \xi_k, \\
&< -C\varphi(x)\rho_i(\pi_k(x)) - C(1-\varphi(x))\rho_i(x) = -C\rho_i(x),
\end{aligned}$$

where the last inequality follows from the induction hypothesis (radial estimates) for ξ.

Obviously for $i > k$ we have for all $x \in \bigcup_{j \leq k} T^\epsilon_{X_j}$:

$$d\rho_i \xi' = (1 - \varphi) d\rho_i \xi_k + \varphi d\rho_i \xi = 0.$$

The above radial estimates show in particular that ξ' is totally radial.

It remains to show that all points in $T^\epsilon_{X_k} - \left(X_k \cup \bigcup_{j<k} T^\epsilon_{X_j} \right)$ are wandering: Note first that by using the radial estimates one can show easily that for $x \in \bigcup_{j \leq k} T^\epsilon_{X_j}$ and for all positive times t one has $\Phi(x,t) \in \bigcup_{j \leq k} T^\epsilon_{X_j}$. Moreover there is an open neighbourhood V_x of x and a time $t(x) < 0$ such that $\Phi(V_x, t(x))$ lies outside of $\bigcup_{j \leq k} T^\epsilon_{X_j}$ and therefore $\Phi(V_x, t) \cap V_x = \emptyset$ for all $t < t(x)$.

Thus we need only to check that V_x can be removed from itself in positive time. The vector field ξ is weakly controlled and therefore for $x \in T^\epsilon_{X_k}$ we have:

$$\lim_{t \to \infty} \pi_k \left(\Phi(x, t) \right) = \lim_{t \to \infty} \Phi(\pi_k(x), t).$$

Thus the ω-limit of x has to be one of the singular points in $\bigcup_{j \leq k} X_j$. Assume that there is a $t \in \mathbb{R}$ such that $\Phi(x, t) \in \bigcup_{j<k} T^\epsilon_{X_j}$. Then by induction hypothesis $\Phi(x, t)$ is wandering and therefore x is wandering. Assume now that $\Phi(x, t) \notin \bigcup_{j<k} T^\epsilon_{X_j}$ for all $t \in \mathbb{R}$. After possibly shrinking V_x one can assume that $V_x \subset T^\epsilon_{X_k} - \left(X_k \cup \bigcup_{j<k} T^{3\epsilon/4}_{X_j} \right)$ and there exists a $\delta > 0$ such that $\rho_k(V_x) \subset (\rho_k(x) - \delta, \rho_k(x) + \delta)$. By the radial estimate $d\rho_k \xi < -C\rho_k$ there is a $T \in \mathbb{R}$ such that for all t with $t > T$ and all $y \in V_x$:

$$| \rho_k(\Phi(y, t)) - \rho_k(x) | > \delta.$$

Therefore $\Phi(V_x, t) \cap V_x = \emptyset$ for all t with $| t | > \max\{T, | t(x) |\}$. $\qquad \square$

In the following section the existence of Lyapunov functions for gradient-like vector fields will be useful:

Definition 6.3. A continuous, stratum-wise smooth function $f : X \to \mathbb{R}$ will be called Lyapunov function for the stratified vector field ξ if the following conditions are satisfied:

(i) $df_x(\xi) < 0$ for all non singular points x,

(ii) $df_p = 0$ if and only if p is a singular point of ξ.

A Lyapunov function f is called self-indexing if $f(p) = k$ whenever p is a singular point of index k of ξ.

The condition (i) of the above definition ensures that the Lyapunov function is strictly decreasing along flow lines.

Let ξ be a stratified gradient-like vector field. Since for each stratum S the smooth vector field ξ_S is a gradient-like vector field, there exists a Lyapunov function for ξ_S [12]. Actually this function can be chosen to be a Morse function on S. In particular for each singular point $p \in S$ of ξ there is an open neighbourhood $U(p) \subset S$ and a smooth Morse function $f_p : U(p) \to \mathbb{R}$ which satisfies (i) and (ii) of def. 6.3.

Proposition 6.2. *Let ξ be a stratified gradient-like vector field on a compact abstract stratified space X. Then there exists a self-indexing Lyapunov function f for ξ such that for each singular point $p \in S$ of ξ there is an open neighbourhood $U(p) \subset S$ such that $f(x) = f_p(\pi_S(x)) + \rho_S^2(x)$ for all $x \in \pi_S^{-1}(U(p))$.*

Idea of proof. The proof of this proposition is a generalisation of the proof of thm. B in [24] and will be omitted here.

From the existence of a Lyapunov function one can deduce as indicated in rem. 5.1 that the unstable manifold of a singular point of a gradient-like vector field is indeed an embedded submanifold of S whereas the stable set is an abstract stratified space. (Actually it is a π-fibre space, see [8] for a definition). $\qquad\qquad\square$

7. The space of trajectories between singular points

Compactness and Gluing. The trajectory spaces $\mathcal{M}_{pq}$ between two singular points p and q of a stratified gradient-like vector field ξ are manifolds without boundary of dimension $\mu(p) - \mu(q)$.

In this section we will analyse the asymptotic structure of the space of trajectories of a stratified gradient-like vector field ξ. We show two compactness results. The space of trajectories is compact up to broken trajectories. The converse result is the gluing operation. Note that in the functional analytic proof of these compactness results in the smooth case [20], at this point the interplay of $H^{1,2}$- and C_{loc}^{∞}-convergence comes into play. Here however we just study the topological boundary of the set of the space of trajectories. In our proofs we only exploit the local model at singular points given in proposition 4.1. This approach was worked out in [29] for the case of a manifold. In the smooth case the local model is a consequence of the Hartman-Grobman theorem.

The main result of this section is cor. 7.1 which describes the compactification of trajectory spaces between singular points of index difference 1 and 2. This result will be needed in section 8 in order to define the Morse-Smale complex.

In this chapter we will assume that X is a compact stratified space.

For a point $y_i \in X$ we will denote by $y_i(\overline{\mathbb{R}})$ the set of points lying on the trajectory of ξ (including the endpoints). By d we will denote the metric distance on X induced by a stratified (adapted) Riemannian metric as defined in [2]. Note that the topology induced by d is equivalent to the original topology on X. For a subset $A \subset X$ we denote by $U_\epsilon(A)$ the open ϵ-neighbourhood of A. According to prop. 6.2 there exists a self-indexing Lyapunov function for ξ, which will be denoted by $f : X \to \mathbb{R}$.

Definition 7.1. Let $\{x_n\}$ be a sequence of trajectories in the trajectory space $\widetilde{\mathcal{M}_{pq}}$. We say that the sequence converges to a broken trajectory of order k if there exist singular points $p_0 := p, p_1, ..., p_k := q$ and trajectories $y_i \in \widetilde{\mathcal{M}_{p_i,p_{i+1}}}, i = 0, ..., k-1$ such that for all $\epsilon > 0$ there exists $N_0 \in \mathbb{N}$ and for all $n > N_0$ the trajectories x_n lie in an ϵ neighbourhood of the broken trajectory, i.e.

$$x_n(t) \in U_\epsilon(\cup_i y_i(\overline{\mathbb{R}})) \text{ for all } t \in \mathbb{R}.$$

Our first compactness result for the trajectory space states:

Proposition 7.1. *The trajectory space $\widetilde{\mathcal{M}_{pq}}$ is compact up to k-order broken trajectories, where $k \leq \mu(p) - \mu(q) - 1$.*

For the proof of prop. 7.1 the following three lemmas are needed. Lemma 7.1 explains how to continue a broken trajectory.

Lemma 7.1. *Let $p' \in S$ be a singular point of a stratified gradient-like vector field. Let $x_i \in R, (S < R)$ be a sequence of points in a neighbourhood $U(p')$ with $\lim x_i = x \in W^s(p')$. Then there exists a point $y \in W^u(p')$ with $y \in \cup_i x_i(\overline{\mathbb{R}})$.*

Proof. Note that the statement of the lemma holds also in the case $S = R$. If the singular point as well as the sequence $\{x_i\}_{i \in I}$ lie in the stratum S, we can apply the smooth theory due to the fact that the flow does not leave a stratum in finite time (compare section 3). Let $x_i \in R$ where $S < R$. Since the statement is a pure topological one, it is enough to show it for a topologically conjugate situation. Using the local model developed in prop. 4.1 we consider a neighbourhood $V(p') \simeq \mathbb{R}^{\dim S} \times \text{cone}(L)$ and the flow on $\mathbb{R}^{\dim S} \times \text{cone}(L)$ given by $((r,l,z),t) \to (re^{-t}, l, z^u e^{tA^u}, z^s e^{tA^s})$. The z-coordinates have been chosen here according to the strictly expanding and contracting directions of the linear map A, i.e. A^u is a diagonal matrix with positive entries and A^s is a diagonal matrix with negative entries. The distances occuring in the sequel of the proof are taken with respect to the stratified Riemannian metric $dr^2 + r^2 g_L + g$, where g_L is a Riemannian metric on the stratified space L as in [2] and g is a metric on $\mathbb{R}^{\dim S}$ compatible with the splitting of A into expanding and contracting directions as in [16], prop. 10. We denote by $\pi^s : \mathrm{cone}(L) \times \mathbb{R}^{\dim S} \to \mathrm{cone}(L) \times \mathbb{R}^{\dim S - \mu(p')}$ the projection $\pi^s(r, l, z) = (r, l, z^s)$ and by $\pi^u : \mathrm{cone}(L) \times B_\epsilon^{\dim S} \to B_\epsilon^{\mu(p')}$ the projection $\pi^u(r, l, z) = z^u$. Since the vector field ξ is weakly controlled, both projections $\pi^{u/s}$ commute with the flow Φ (compare section 2). Let us assume that the lemma does not hold. Then for each point $y = y^u \in W^u(p')$ there is a sufficiently small ball $S_\delta^u \subset W^u(p')$ around y such that no flow lines pass through $D := S_\delta^u \times S_\delta^s \times \mathrm{cone}_\delta(L)$. Here one denotes by $S_\delta^s := \{z^s || z^s| < \delta\}$, $S_\delta^u := \{z^u \mid |z^u - y| < \delta\}$ and by $\mathrm{cone}_\delta(L) := \{(r, l) \mid r < \delta\}$ and we may assume that $d < d(y, 0)/2$.

Note first that using the fact that e^{tA^s} is a strict contraction for $t > 0$ we can find a time $\tau > 0$ such that:

$$d(\pi^s(\Phi(x_i, \tau)), 0) = d((r_i e^{-\tau}, l, z_i^s e^{\tau A^s}), 0) < \delta \text{ for all } i \in I.$$

Since $\lim x_i \in W^s(p')$ there is an $i_0 \in \mathbb{N}$ large enough such that moreover the following estimate is satisfied

$$d(\pi^u(\Phi(x_{i_0}, \tau)), 0) = d(z_{i_0}^u e^{\tau A^u}, 0) < d(y, 0) - \delta.$$

Since e^{tA^s} is a strict contraction for $t > 0$ we have the following estimate:

$$d(\pi^s(\Phi(x_{i_0}, t)), 0) = d((r_{i_0} e^{-t}, l, z_{i_0}^s e^{tA^s}), 0) < \delta \text{ for all } t > \tau.$$

Moreover, since e^{tA^u} is a strict expansion for all $t > 0$, there exists a time t_0 such that we have

$$d(\pi^u(\Phi(x_{i_0}, t)), 0) = d(z_{i_0}^u e^{tA^u}, 0) > d(y, 0) + \delta \text{ for all } t \geq t_0.$$

By continuity of the flow Φ and the projection π^u we can apply the mean value theorem and find a time $t_1 \in [\tau, t_0]$ such that

$$d(\pi^u(\Phi(x_{i_0}, t_1)), 0) = d(z_{i_0}^u e^{A^u t_1}, 0) = d(y, 0).$$

Thus the flow line $x_{i_0}(\overline{\mathbb{R}})$ passes through D in contradiction to what we assumed. This proves the lemma. $\qquad\square$

The following lemma states that the continued trajectory stays uniformly close to the broken trajectory through y:

Lemma 7.2. *Let ξ be a stratified vector field w.h.s.. Moreover let ξ_S be a Lipschitz continuous vector field on S. Then there exist constants $C > 0$*

and $K > 0$ such that for all $x \in S$ and $y \in R \cap T_S$ with $S \leq R$ the following estimate holds:

$$d(\Phi(x,t), \Phi(y,t)) \leq Ce^{K|t|}d(x,y),$$

where d is the metric induced by a stratified Riemannian metric as in [2].

Proof. For $x, y \in S$ the estimate of the lemma holds as is shown in [16], lemma 4.8. Let $x \in S$ and $y \in R \cap T_S$, $R > S$. The triangle inequality and the weak control of ξ yield:

$$d(\Phi(x,t), \Phi(y,t)) \leq d(\Phi(x,t), \Phi(\pi(y),t)) + d(\Phi(\pi(y),t), \Phi(y,t)).$$

By total radiality of ξ there is a nonnegative constant A_S such that $d\rho_S\xi \leq -A_S\rho$ on T_S and we therefore get:

$$d(\Phi(x,t), \Phi(y,t)) \leq Ce^{k|t|}d(x, \pi(y)) + Ce^{A_S|t|}d(\pi(y), y),$$

where k is the Lipschitz constant of ξ_S. $\qquad\square$

Lemma 7.3 helps us to know where the broken trajectory stops:

Lemma 7.3. *Let $x \in \overline{M_{pq}}$ and $f(x) = f(q)$ (resp. $f(x) = f(p)$), then $x = q$ (resp. $x = p$).*

Proof. Since $x \in \overline{M_{pq}}$, we have especially that $x \in \overline{W^u(q)}$. The result follows since the Lyapunov function is strictly decreasing along flow lines.

$\qquad\square$

Proof of prop. 7.1. Let x_n be a sequence in M_{pq} having no convergent subsequence (in M_{pq}). With the compactness of X we can find a subsequence which converges in X. The limit point x_0 lies in the adherence $\overline{M_{pq}}$. Then all points of the trajectory through x_0 are also in $\overline{M_{pq}}$. The trajectory through x_0 connects two singular points p' and q' and moreover, since f is decreasing along flow lines, $f(p'), f(q') \in [f(p), f(q)]$. If $f(p') = f(p)$ resp. $f(q') = f(q)$ we are done because lemma 7.3 yields $p' = p$ resp. $q' = q$. Otherwise we have to continue the broken trajectory with lemma 7.1. The procedure stops after finitely many steps because X is compact and therefore the vector field w.h.s. ξ has only finitely many singular points.

Thus we obtain singular points $p_0, ..., p_k$ as well as broken trajectories $\{y_i\}_{i=0,...k}$ connecting them. There must be $\mu(p_0) > \mu(p_1) > ... > \mu(p_k)$ and thus the broken trajectory is of order $k < \mu(p, q)$. Note that these points can possibly lie in different strata but the total radiality of the vector field implies that $S(p_0) \geq ... \geq S(p_k)$.

It remains to show the uniform convergence of the trajectories as required in the def. 7.1: Uniform convergence is clear in the neighbourhood of singular points. Outside fixed neighbourhoods of the singular points it is a consequence of the estimate in lemma 7.2, which shows that on compact time intervals the orbits through x_n converge to the broken orbit through x_0. $\qquad\square$

The converse result of prop. 7.1 is the so-called gluing.

Proposition 7.2. *Let X_1, X_2, X_3 be strata with $X_1 \subset \overline{X_2} \subset \overline{X_3}$. Let $q \in X_1$, $p' \in X_2$, $p \in X_3$ be singular points of the stratified gradient-like vector field ξ. Let the relative indices be $\mu(p, p') = 1$ and $\mu(p', q) = 1$. Then there exists a $\rho_0 > 0$ and an embedding*

$$\mathcal{G} : \widetilde{\mathcal{M}_{pp'}} \times \widetilde{\mathcal{M}_{p'q}} \times (\rho_0, \infty) \to \widetilde{\mathcal{M}_{pq}}.$$

Moreover if x_n is a sequence of flow lines in $\widetilde{\mathcal{M}_{pq}}$, which converges to a broken trajectory in $\widetilde{\mathcal{M}_{pp'}} \times \widetilde{\mathcal{M}_{p'q}}$ in the sense defined in 7.1, then there is a N_0 such that $x_n \in \mathrm{im}(\mathcal{G})$ for all $n > N_0$. The compactification of $\widetilde{\mathcal{M}_{pq}}$ in the sense of definition 7.1 corresponds to the limit $\lim \rho = \infty$.

Proof. Denote by $d := \dim X_2$. *Step 1: Let us first explain the definition of the gluing map $\mathcal{G}$:* Let $(u, v) \in \widetilde{\mathcal{M}_{pp'}} \times \widetilde{\mathcal{M}_{p'q}}$ be a broken trajectory. We will construct the gluing map by generalising the geometric construction (see [29]) to the stratified case. Let

$$D(p) \subset W^u(p) \subset X_3$$

be a $\mu(p')$ disc transversal to the trajectory u. Lemma 7.2 assures that for large enough times t we have

$$D_t^u := \Phi(D(p), t) \subset U(p'),$$

where $U(p')$ is a small neighbourhood of the singular point p'. Because of the total radiality of the vector field the disc $D_t^u \subset W^u(p)$ is completely contained in the stratum X_3 and because of the Morse-Smale condition D_t^u is transversal to $W^s(p')$ (in X_3).

Note first that the trajectory v lies completely in X_2. Let

$$D_{X_2}(q) \subset W^s(q) \cap X_2$$

be a disc of dimension $d - \mu(p')$ transversal to the trajectory v. Set $D(q) := \pi_{X_2}^{-1}(D_{X_2}(q))$. The disc $D_{X_2, t}^s := \Phi(-t, D_{X_2}(q))$ is transversal to $W^u(p')$ in X_2. Since the flow is weakly controlled, the space

$$D_t^s := \Phi(-t, D^s(q)) = \pi_{X_2}^{-1}(D_{X_2, t}^s)$$

is an abstract stratified space (with the obvious tubular system given by restriction). For t large enough we have again by lemma 7.2

$$D_t^s \subset U(p').$$

The gluing map $\mathcal{G}$ is defined as follows:

$$\mathcal{G}(u, v, t) = \Pi(\text{trajectory through } D_t^u \cap D_t^s),$$

where $\Pi : \mathcal{M}_{pq} \to \widetilde{\mathcal{M}_{pq}}$ is the canonical projection.

Step 2: The thus defined gluing map $\mathcal{G}$ is well defined: It remains to show that there exists $\rho_0 > 0$ such that for $t > \rho_0$ the intersection $D_t^u \cap D_t^s$ is transversal and consists of exactly one point p_t which depends C^1 on t.
We can choose the neighbourhood $U(p')$ such that there exists a stratified isomorphism

$$U(p') \simeq \text{cone}(L(p')) \times \mathbb{R}^d$$

and the flow is given in these coordinates by $\Phi((r, l, z), t) = (re^{-t}, \Phi_l((r, l, z), t), \Phi_z((0, z), t).)$
The $\mu(p')$-disc D_t^u is transversal to $W^s(p')$ and can thus be written as the graph of a map

$$\varphi_t^u : \mathbb{R}^{\mu(p')} \to \text{cone}(L(p')) \times \mathbb{R}^{d-\mu(p')}.$$

Equally we can write D_t^s as graph of a map

$$\varphi_t^s : \text{cone}(L(p')) \times \mathbb{R}^{d-\mu(p')} \to \mathbb{R}^{\mu(p')},$$

with $\varphi_t^s(r, l, z) = \varphi_t^s(r, l', z)$ for all $l, l' \in L(p')$.
Denote by $\pi : \text{cone}(L(p')) \times \mathbb{R}^d \to \mathbb{R}^{\geq 0} \times \mathbb{R}^d, (r, l, z) \mapsto (r, z)$.
Let us now study the smooth flow

$$\Phi_\pi : (R^{\geq 0} \times \mathbb{R}^d) \times \mathbb{R} \longrightarrow \mathbb{R}^{\geq 0} \times \mathbb{R}^d,$$
$$((r, z), t) \mapsto (re^{-t}, \Phi((0, z), t)).$$

The flow Φ_π has a hyperbolic fixed point of index $\mu(p')$ at 0. Thus we are able to apply the smooth theory developed in [29].
$D_{\pi,t}^u := \text{gr}(\pi \circ \varphi_t^u)$ is a $\mu(p')$-disc transversal to the stable manifold $W^s(0, \Phi_\pi)$ and $D_{\pi,t}^s := \text{gr}\,\varphi_t^s(\ ,l,\ )$ $(l \in L$ fixed but arbitrary) is a $(d+1-\mu(p'))$-disc transversal to $W^u(0, \Phi_\pi)$. According to the λ-Lemma ([16], lemma 7.1) for t large enough $D_{\pi,t}^{u/s}$ gets arbitrarily C^1-close to $W^{s/u}(0, \Phi_\pi)$. Thus the two discs have a unique (transverse) intersection which depends C^1 on t.
The unique intersection point $D_{\pi,t}^u \cap D_{\pi,t}^s$ can be written as $(\pi \circ \varphi_t^u(x), x) = (\pi(y), \varphi_t^s(y))$ where $x \in \mathbb{R}^{\mu(p')}$ and $y \in \text{cone}(L) \times \mathbb{R}^{d-\mu(p')}$. Then also D_t^s

and D_t^u have a unique intersection point $p_t := (\varphi_t^u(x), x)$. The intersection again depends C^1 on the time parameter since all maps involved as well as the flow Φ are stratum-wise smooth and the flow does not leave the stratum in finite time.

Since the flow Φ is strata preserving and its restriction to each stratum is smooth one can directly apply the argument of [29] to show the fact that the gluing map is an embedding. $\qquad\square$

Note that, as they are completely contained in one stratum, the trajectory spaces are very much like in the smooth case. The only difference is that in the case of an abstract stratification parts of the broken trajectories can lie in a smaller stratum.

The following result for the trajectory spaces in index difference 1 and 2 are crucial for the study of the Morse-Smale-Witten complex in paragraph 8:

Corollary 7.1. *Let p and q be singular points of a stratified gradient-like vector field.*

(1) If $\mu(p) - \mu(q) = 1$ the trajectory space $\widetilde{\mathcal{M}_{pq}}$ is compact.

(2) If $\mu(p) - \mu(q) = 2$ each connected component of the trajectory space $\mathcal{M}_{pq}$ is either compact or can be compactified by (exactly) two broken trajectories.

Proof. (1) There are no broken trajectories of order 0 and thus by prop. 7.1 the trajectory space is compact. (2) Assume that the connected component of $\widetilde{\mathcal{M}_{pq}}$ is not compact, i.e. it is diffeomorphic to the open interval $(0, 1)$. According to prop. 7.1 this connected component can be compactified by 1 resp. 2 broken trajectories, the compactification being homeomorphic to $[0, 1]$ resp. S^1. The second case is excluded by prop. 7.2, more precisely by the uniqueness of the intersection point in the definition of the gluing map. $\qquad\square$

8. Morse-Smale-Witten complex for a stratified gradient-like vector field

In this section we will build the Morse-Smale-Witten complex associated to a stratified gradient-like vector field ξ on a compact stratified space X. The main result of this section shows that the homology of this complex is isomorphic to the singular homology of the stratified space.

As explained in section 3 the flow induced by a stratified vector field w.s.s. goes only from larger into smaller strata. This property is essential in order that each singular point contributes to the complex.

The Morse-Smale complex (C_*, ∂_*). We denote by C_k, the k-th chain group, the $\mathbb{Z}_2$-vector field

$$C_k = \bigoplus_{x \in \mathrm{Crit}_k(\xi)} \mathbb{Z}_2 x,$$

where $\mathrm{Crit}_k(\xi)$ is the set of singular points of index k of the stratified gradient-like vector field ξ. Since X is compact and singular points are isolated, the sum appearing in the definition of the chain group is finite. The boundary operator is defined as follows:

$$\partial_{k+1} : C_{k+1} \longrightarrow C_k,$$
$$p \mapsto \sum_{\mu(q)=k} n(p,q)q,$$

where $n(p,q)$ is the number modulo 2 of trajectories between the singular points p and q, i.e. the cardinality of $\widetilde{\mathcal{M}_{pq}}$.

Theorem 8.1.

> *(1) The boundary operator $\partial_{k+1} : C_{k+1} \to C_k$ is well-defined, i.e. $n(p,q) < \infty$ for all singular points p and q.*
> *(2) (C_*, ∂_*) is a complex, i.e. $\partial \circ \partial = 0$.*

Proof. The proof of the well definedness of the Morse-Smale complex follows from cor. 7.1. $\square$

The homology of the Morse-Smale complex is denoted by $H_*(X, \xi, \mathbb{Z}_2)$ and is isomorphic to the singular homology of the stratified space by the following theorem:

Theorem 8.2. *Let X be a compact stratified space. Let ξ be a stratified gradient-like vector field on X. Then there is an isomorphism:*

$$H_*(X, \xi, \mathbb{Z}_2) \simeq H_{sing}(X, \mathbb{Z}_2).$$

We need two lemmas to prove thm. 8.2. The first is the stratified analogue of [6] (theorem 2.3):

Lemma 8.1. *Let ξ be a stratified gradient-like vector field. There exists a CW-complex Y unique up to cell equivalence and a homotopy equivalence*

$g : X \to Y$ such that for each singular point p of index k $g(W^u(p))$ is contained in the base of a single k-cell. In this way g establishes a one-to-one correspondence between singular points of ξ of index k and k-cells of Y. Moreover, the partial order $<$ on singular points defined by $q \leq p$ if and only if $W^s(q) \cap W^u(p) \neq \emptyset$ corresponds to the partial order on the cells of Y.

Proof. Let f be a self-indexing Lyapunov function for ξ as in proposition 6.2. Denote by X_k the sublevelset $f^{-1}([0, k + 1/2])$.

Let $p \in S$ be a singular point of ξ of index k. Then as shown in proposition 5.1 the unstable manifold $W^u(p)$ lies completely in S. Moreover since f is self-indexing the intersection $W^u(p) \cap f^{-1}(c)$ with $c \in (k - 1, k)$ is an $(k - 1)$-sphere (see [13]) and $D(p) := W^u(p) \cap \overline{(X - X_{k-1})}$ is a k-disc.

The construction of Y will be done by induction on the skeleton. Suppose that the homotopy equivalence $g_{k-1} : X_{k-1} \to Y_{k-1}$ has already been defined, where Y_{k-1} is the $(k - 1)$-skeleton of Y. There is a deformation retract

$$\mathrm{retr}_k : X_k \to X_{k-1} \cup \bigcup_{p \in \mathrm{Crit}_k(\xi)} D(p).$$

Outside of neighbourhoods of the singular points the deformation is done by means of the strata preserving flow induced by ξ. For a singular point $p \in S$ denote by $U(p) \subset S$ an open neighbourhood as in prop. 6.2 such that $f(x) = f_p(\pi_S(x)) + \rho_S^2(x)$ for all $x \in \pi^{-1}(U(p))$, where f_p is a Morse function on $U(p)$ with critical point p. In a neighbourhood of a critical point the construction of the deformation retract is explained in [8] thm. 9.3 and cor. 9.4.

Y_k is formed by adding k-cells to the complex Y_{k-1}, one for each singular point of index k. The attaching map is $\mathrm{retr}_{k|\partial D(p)}$. The homotopy equivalence $g_k : X_k \to Y_k$ is the composition of retr_k and g_{k-1}.

The above construction does depend only on the flow and the choice of coordinates in the neighbourhood $\pi^{-1}(U(p))$.

The assumption on the preservation of the partial orders follows as in the proof of 2.3. in [6]. $\qquad\qquad\square$

The next lemma relates the attaching maps of the complex Y constructed above to the trajectory spaces of ξ. Let again $f : X \to \mathbb{R}$ be a self-indexing Lyapunov function for ξ. Since the flow lines of ξ intersect each level set of f only once there is a natural identification

$$\widetilde{\mathcal{M}_{pq}} \simeq W^u(p) \cap W^s(q) \cap f^{-1}(c)$$

where $c \in (f(q), f(p))$.

Lemma 8.2. *Let p, q be singular points of a stratified gradient-like vector field ξ with $\mu(p) = k+1$ and $\mu(q) = k$. Let Y be the CW-complex associated to ξ. Denote by α the attaching map of the cell corresponding to p on the cell corresponding to q. Then the number of points of $\widetilde{\mathcal{M}_{pq}}$ equals (mod 2) the degree of α.*

Proof. The proof of lemma 8.2 is as in thm. 3.3 of [6]. We recall the main ideas: By construction of the CW-complex Y the attaching map α is up to homotopy the composition of the two deformation retracts:

$$\mathrm{retr} : S^k \simeq W^u(p) \cap f^{-1}(k + \epsilon) \to X_{k-1} \cup W^u(q)$$

and

$$h : X_{k-1} \cup W^u(q) \to S^k.$$

Here $(S^k, *)$ denotes the one point compactification of $W^u(q) \simeq D^k$ and X_{k-1} is mapped to $*$ under h.

Because of the Morse-Smale condition, the point q is a regular point for $h \circ \mathrm{retr}$ and $(h \circ \mathrm{retr})^{-1}(q) = \widetilde{\mathcal{M}_{pq}}$ is the Pontyagin manifold associated to $(h \circ \mathrm{retr})$. Therefore the number of points in $\widetilde{\mathcal{M}_{pq}}$ equals mod 2 the degree of $h \circ \mathrm{retr}$. $\qquad\square$

Proof of thm. 8.2. The singular homology of the stratified space X is isomorphic to the homology of the CW-complex Y constructed in lemma 8.1. As the degrees of the attaching maps are the only ingredients needed to calculate the homology of a CW-complex [3], the result follows from the previous lemma. $\qquad\square$

Denote by $c_k(\xi)$ the number of singular points of index k of ξ and by $b_k = \dim H_k(X, \mathbb{Z}_2)$ the kth Betti number of X. We obtain the Morse inequalities as a consequence of thm. 8.2:

Corollary 8.1. *Let X be a compact stratified space and ξ a gradient-like vector field. Then the Morse inequalities hold:*

$$\sum_{i=0}^{k} (-1)^i c_k(\xi) \geq \sum_{i=0}^{k} (-1)^i b_k, \text{ for all } k < n.$$

and

$$\sum_{i=0}^{n} (-1)^i c_k(\xi) = \sum_{i=0}^{n} (-1)^i b_k.$$

712

References

1. Jean-Paul Brasselet. Radial vector fields and the Poincaré-Hopf theorem. In *Real algebraic geometry and ordered structures (Baton Rouge, LA, 1996)*, volume 253 of *Contemp. Math.*, pages 25–30. Amer.Math.Soc., Providence, RI, 2000.

2. Jean-Paul Brasselet, Gilbert Hector, and Martin Saralegi. $\mathcal{L}^2$-cohomologie des espaces stratifiés. *Manuscripta Math.*, 76(1):21–32, 1992.

3. Glen E. Bredon. *Topology and geometry*, volume 139 of *Graduate Texts in Mathematics*. Springer-Verlag, New York, 1997. Corrected third printing of the 1993 original.

4. Charles Conley. *Isolated invariant sets and the Morse index*, volume 38 of *CBMS Regional Conference Series in Mathematics*. American Mathematical Society, Providence, R.I., 1978.

5. Andrew du Plessis and Terry Wall. *The geometry of topological stability*, volume 9 of *London Mathematical Society Monographs. New Series*. The Clarendon Press Oxford University Press, New York, 1995. Oxford Science Publications.

6. John M. Franks. Morse-Smale flows and homotopy theory. *Topology*, 18(3):199–215, 1979.

7. Mark Goresky and Robert MacPherson. *Stratified Morse theory*, volume 14 of *Ergebnisse der Mathematik und ihrer Grenzgebiete (3)*. Springer-Verlag, Berlin 1988.

8. R. Mark Goresky. Whitney stratified chains and cochains. *Trans. Amer. Math. Soc.*, 267(1):175–196, 1981.

9. Jürgen Jost. *Riemannian geometry and geometric analysis*. Universitext. Springer-Verlag, Berlin, third edition, 2002.

10. H. C. King and D. Trotman. Poincaré-Hopf Theorems on Singular Spaces. *Prépublication LATP/URA225, 94-01*, 1994.

11. J. N. Mather. *Notes on Topological Stability*. Mimeographed Notes. Harvard, 1970.

12. K. R. Meyer. Energy functions for Morse Smale systems. *Amer. J. Math.*, 90:1031-1040, 1968.

13. John Milnor. *Lectures on the h-cobordism theorem*. Notes by L. Siebenmann and J. Sondow. Princeton University Press, Princeton, N.J., 1965.

14. Hiroko Natsume. The realization of abstract stratified sets. *Kodai Math. J.*, 3(1):1–7, 1980.

15. Laurent Noirel and David Trotman. Subanalytic and semialgebraic realisations of abstract stratified sets. In *Real algebraic geometry and ordered structures (Baton Rouge, LA, 1996)*, volume 253 of *Contemp. Math.*, pages 203–207. Amer. Math. Soc., Providence, RI, 2000.

16. Jacob Palis, Jr. and Wellington de Melo. *Geometric theory of dynamical systems*. Springer-Verlag, New York, 1982. An introduction, Translated from the Portuguese by A. K. Manning.

17. Markus J. Pflaum. *Analytic and geometric study of stratified spaces*, volume 1768 of *Lecture Notes in Mathematics*. Springer-Verlag, Berlin, 2001.

18. Marie-Héléne Schwartz. Classes caractéristiques définies par une stratification

d'une variété analytique complexe. *C. R. Acad. Sci. Paris*, 260:3262–3264, 3535–3537, 1965.

19. Marie-Héléne Schwartz. *Champs radiaux sur une stratification analytique*, volume 39 of *Travaux en Cours*. Hermann, Paris, 1991.

20. Matthias Schwarz. *Morse homology*, volume 111 of *Progress in Mathematics*. Birkhäuser Verlag, Basel, 1993.

21. Masahiro Shiota. *Geometry of subanalytic and semialgebraic sets*, volume 150 of *Progress in Mathematics*. Birkhäuser Boston Inc., Boston, MA, 1997.

22. Stéphane Simon. Champs totalement radiaux sur une structure de Thom-Mather. *Ann. Inst. Fourier (Grenoble)*, 45(5):1423-1447, 1995.

23. Stephen Smale. Morse inequalities for a dynamical system. *Bull. Amer. Math. Soc.*, 66:43–49, 1960.

24. Stephen Smale. On gradient dynamical systems. *Ann. of Math. (2)*, 74:199–206, 1961.

25. Michael Teufel. Abstract prestratified sets are (*b*)-regular. *J. Differential Geom.*, 16(3):529–536 (1982), 1981.

26. René Thom. Sur une partition en cellules associée à une fonction sur une variété. *C. R. Acad. Sci. Paris*, 228:973–975, 1949.

27. René Thom. Ensembles et morphismes stratifiés, *Bull. Amer. Math. Soc.*, 75:240–284,1969.

28. Andrei Verona. *Stratified mappings—structure and triangulability*, volume 1102 of *Lecture Notes in Mathematics*. Springer-Verlag, Berlin, 1984.

29. Joa Weber. Der Morse-Witten Komplex. *Diplomarbeit, Berlin*, 1993.

30. Edward Witten. Supersymmetry and Morse theory. *J. Differential Geom.*, 17(4):661–692 (1983), 1982.

SOME OBSTRUCTED EQUISINGULAR FAMILIS OF CURVES ON SURFACES IN $\mathbb{P}^3$

Thomas Markwig

Fachbereich Mathematik
Universität Kaiserslautern
Erwin–Schrödinger–Straße
D — 67663 Kaiserslautern
Tel. +496312052732
Fax +496312054795
E-mail: keilen@mathematik.uni-kl.de
www.mathematik.uni-kl.de/~keilen

Very few examples of obstructed equsingular families of curves on surfaces other than $\mathbb{P}^2$ are known. Combining results from [1] and [2] with an idea from [3] we give in the present paper series of examples of families of irreducible curves with simple singularities on surfaces in $\mathbb{P}^3$ which are not T–smooth, i.e. do not have the expected dimension, (Sec. 2) and we compare this with conditions (showing the same asymptotics) which ensure the existence of a T–smooth component (Sec. 3).

Keywords: Equisingular families of curves, simple singularities

1. Introduction

Below we are going to construct two series of equisingular families of curves on surfaces in $\mathbb{P}^3$ over $\mathbb{C}$. In both examples the families are obstructed in the sense that they do not have the expected dimension. However, while in the first example at least the existence of such curves was expected, the families in the second example were expected to be empty. It would be interesting to see if the equisingular families contain further components which are well–behaved. However, the families which we construct fail to satisfy the numerical conditions for the existence of such a component given in Sec. 3 by a factor of two. We do not know whether the families are reducible or not, or if they are smooth.

2. Examples of obstructed families

Throughout this section Σ will denote a smooth projective surface in $\mathbb{P}^3$ of degree $n \geq 2$, and H will be a hyperplane section of Σ. $S = \{S_1, \ldots, S_s\}$ will be a finite set of *simple* singularity types, that is the S_i are of type A_k (given by $x^2 - y^{k+1} = 0$, $k \geq 1$), D_k (given by $x^2 y - y^{k-1} = 0$, $k \geq 4$), or E_k (given by $x^3 - y^4 = 0$, $x^3 - xy^3 = 0$, or $x^3 - y^5 = 0$ for $k = 6, 7, 8$ respectively). In general, for positive integers $r_1, \ldots, r_s$ and d we denote by $V_{|dH|}^{irr}(r_1 S_1, \ldots, r_s S_s)$ the family of irreducible curves in the linear system $|dH|$ with precisely $r = r_1 + \ldots + r_s$ singular points, r_i of which are of the type S_i, $i = 1, \ldots, s$, where S_i may be any analytic type of an isolated singularity. $V_{|dH|}^{irr}(r_1 S_1, \ldots, r_s S_s)$ is called *T–smooth* or *not obstructed* if it is smooth of the expected dimension

$$\operatorname{expdim}\left(V_{|dH|}^{irr}(r_1 S_1, \ldots, r_s S_s)\right) = \dim |dH| - \sum_{i=1}^{s} r_i \cdot \tau(S_i)$$

$$= \frac{nd^2 + (4n - n^2)d}{2} + \frac{n^3 - 6n^2 + 11n - 6}{6} - \sum_{i=1}^{s} r_i \cdot \tau(S_i),$$

where $\tau(S) = \dim_{\mathbb{C}} \mathbb{C}\{x, y\}/\langle \frac{\partial f}{\partial x}, \frac{\partial f}{\partial y}, f \rangle$ is the Tjurina number of the singularity type S given by the local equation $f = 0$. Note that $\tau(A_k) = \tau(D_k) = \tau(E_k) = k$.

In this note we give examples of such equisingular families of curves which are obstructed in the sense that they have dimension larger than the expected one. We use the idea by which Chiantini and Ciliberto in [3] showed the existence of obstructed families of nodal curves.

Let us fix a plane P in $\mathbb{P}^3$, a point p outside P, and a curve C of degree $d > 1$ in P. If we intersect the cone $K_{C,p}$ over C with vertex p with Σ, this gives a curve $C' = K_{C,p} \cap \Sigma$ in $|dH|$ which is determined by the choice of C and p (see Lemma 4.1). In particular, if C varies in an N-dimensional family in P, then C' varies in an N-dimensional family on Σ, and if C is irreducible, then for a general choice of p the curve C' will be irreducible as well (see Lemma 4.2). Moreover, if C has a singular point q of (simple) singularity type S and Σ meets the line joining p and q transversally in n points, then C' will have a singularity of the same type in each of these points.

Example 2.1. Fix the set $S = \{S_1, \ldots, S_s\}$ and let $m = \max\{\tau(S) \mid S \in S\}$. Suppose that $n > 2m + 4$ and $d >> n$, and let $r_1, \ldots, r_s \geq 0$ be such that

$$\frac{d^2 + (4-n)d + 2}{2} \;\leq\; \sum_{i=1}^{s} r_i \cdot \tau(\mathcal{S}_i) \;\leq\; \frac{d^2 + (4-n)d + 2}{2} + m - 1.$$

Then

$$\sum_{i=1}^{s} r_i \cdot \tau(\mathcal{S}_i) \leq \frac{d^2 + (4-n)d + 2}{2} + m - 1 \leq \frac{d^2}{2} - m \cdot d - 3.$$

Hence, by [1] Remark 3.3.5 the family $V = V_d^{irr}(r_1\mathcal{S}_1, \ldots, r_s\mathcal{S}_s)$ of irreducible plane curves C of degree d with precisely $r = r_1 + \ldots + r_s$ singular points, r_i of which are of type $\mathcal{S}_i$, is non-empty, and we may estimate its dimension:

$$\dim(V) \geq \operatorname{expdim}(V) = \frac{d(d+3)}{2} - \sum_{i=1}^{s} r_i \cdot \tau(\mathcal{S}_i)$$

$$\geq \frac{d(d+3)}{2} - \frac{d^2 + (4-n)d + 2}{2} - m + 1 = \frac{n-1}{2} \cdot d - m.$$

By the above construction we see that hence the family of curves C' satisfies

$$\dim\left(V_{|dH|}^{irr}(nr_1\mathcal{S}_1, \ldots, nr_s\mathcal{S}_s)\right) \geq \frac{n-1}{2} \cdot d - m.$$

However, the expected dimension of this family is

$$\operatorname{expdim}\left(V_{|dH|}^{irr}(nr_1\mathcal{S}_1, \ldots, nr_s\mathcal{S}_s)\right)$$

$$= \frac{nd^2 + (4n-n^2)d}{2} + \frac{n^3 - 6n^2 + 11n - 6}{6} - \sum_{i=1}^{s} n \cdot r_i \cdot \tau(\mathcal{S}_i)$$

$$\leq \frac{nd^2 + (4n-n^2)d}{2} + \frac{n^3 - 6n^2 + 11n - 6}{6} - n \cdot \left(\frac{d^2 + (4-n)d + 2}{2}\right)$$

$$= \frac{n^3 - 6n^2 + 5n - 6}{6}.$$

For $d >> n$, more precisely for

$$d > \frac{n^3 - 6n^2 + 5n - 6 + 6m}{3n - 3},$$

the expected dimension will be smaller than the actual dimension, which proves that the family is obstructed.

In particular, if $S = \{\mathcal{S}\}$, $\mathcal{S} \in \{A_k, D_k, E_k\}$, and

$$r = \left\lceil \frac{d^2 + (4-n)d + 2}{2k} \right\rceil,$$

then $V_{|dH|}(nr\mathcal{S})$ is obstructed, once $d >> n > 3k + 4$.

718

Note that in the previous example

$$\mathrm{expdim}\left(V_{|dH|}(nr_1\mathcal{S}_1,\ldots,nr_s\mathcal{S}_s)\right) \geq \frac{n^3 - 6n^2 + 5n - 6}{6} - n\cdot(m-1) > 0,$$

that is, the existence of curves in $|dH|$ with the given singularities was expected. This not so in the following example.

Example 2.2. Let k be an *even*, positive integer, $m \geq 1$, $d = 2(k+1)^m$, and

$$r = \frac{3\cdot(k+1)\cdot\left((k+1)^{2m}-1\right)}{(k+1)^2 - 1}.$$

Hirano proved in [2] the existence of an irreducible plane curve of degree d with precisely r singular points all of type A_k. Thus the above construction shows that

$$V_{|dH|}^{irr}(nrA_k)$$

is non-empty. However, the expected dimension is

$$\mathrm{expdim}\left(V_{|dH|}^{irr}(nrA_k)\right) = \frac{nd^2 + (4n - n^2)d}{2} + \frac{n^3 - 6n^2 + 11n - 6}{6} - knr$$

$$= \left(2 - \frac{3\cdot(k^2 + k)}{k^2 + 2k}\right)\cdot(k+1)^{2m} + o\left((k+1)^m\right),$$

which is negative for m sufficiently large, since

$$\frac{3\cdot(k^2 + k)}{k^2 + 2k} > 2.$$

This shows that $V_{|dH|}^{irr}(nrA_k)$ is obstructed for sufficiently large k.

3. Some remarks on conditions for T–smoothness

Unless otherwise specified in this section Σ will be an arbitrary smooth projective surface, H a very ample divisor on Σ, and $\mathcal{S}_1,\ldots,\mathcal{S}_s$ arbitrary (not necessarily different) topological or analytical singularity types. As in Sec. 2 we denote for $d \geq 0$ by $V_{|dH|}^{irr}(\mathcal{S}_1,\ldots,\mathcal{S}_s)$ the equisingular family of irreducible curves in $|dH|$ with precisely s singular points of types $\mathcal{S}_1,\ldots,\mathcal{S}_s$, and again the expected dimension is

$$\mathrm{expdim}\left(V_{|dH|}^{irr}(\mathcal{S}_1,\ldots,\mathcal{S}_s)\right) = \dim|dH| - \sum_{i=1}^{s}\tau(\mathcal{S}_i).$$

$V_{|dH|}^{irr}(\mathcal{S}_1,\ldots,\mathcal{S}_s)$ is called T–smooth if it is smooth of the expected dimension.

By [4] Theorem 1.2 and 2.3 (which is a slight improvement of [5] Theorem 3.3 and Theorem 4.3) there is a curve $C \in V_{|dH|}^{irr}(\mathcal{S}_1, \ldots, \mathcal{S}_s)$ if

- $d \cdot H^2 - g(H) \geq m_i + m_j$, and
- $h^1\big(\Sigma, \mathcal{J}_{X(\underline{m};\underline{z})/\Sigma}((d-1)H)\big) = 0$ for $\underline{z} \in \Sigma^r$ very general,

where $\underline{m} = (m_1, \ldots, m_s)$ with $m_i = e^*(\mathcal{S}_i)$, a certain invariant which only depends on $\mathcal{S}_i$. Moreover, $V_{|dH|}^{irr}(\mathcal{S}_1, \ldots, \mathcal{S}_s)$ is T–smooth at this curve C (see e.g. [6] Theorem 1). Finally, by [7] Theorem 1.1 there is a number $d(m)$ depending only on $m = \max\{m_1, \ldots, m_s\}$, such that for all $d \geq d(m)$ and for $\underline{z} \in \Sigma^r$ very general the map

$$H^0\big(\Sigma, \mathcal{O}_\Sigma((d-1)H)\big) \longrightarrow H^0\big(\Sigma, \mathcal{O}_{X(\underline{m};\underline{z})/\Sigma}((d-1)H)\big)$$

has maximal rank. In particular, if

$$\dim |(d-1)H| \geq \deg\big(X(\underline{m};\underline{z})\big) = \sum_{i=1}^{s} \frac{m_i \cdot (m_i + 1)}{2},$$

then $h^1\big(\Sigma, \mathcal{J}_{X(\underline{m};\underline{z})/\Sigma}((d-1)H)\big) = 0$. This proves the following Proposition.

Proposition 3.1. *Let $S = \{\mathcal{S}_1, \ldots, \mathcal{S}_s\}$ be a finite set of pairwise different topological or analytical singularity types. Then there exists a number $d(S)$ such that for all $d \geq d(S)$ and $r_1, \ldots, r_s \geq 0$ satisfying*

$$\sum_{i=1}^{s} r_i \cdot \frac{e^*(\mathcal{S}_i) \cdot \big(e^*(\mathcal{S}_i) + 1\big)}{2} < \dim |(d-1)H| \tag{1}$$

the equisingular family $V_{|dH|}^{irr}(r_1\mathcal{S}_1, \ldots, r_s\mathcal{S}_s)$ has a non-empty T–smooth component.

In [8] upper bounds for $e^*(\mathcal{S})$ are given. For a non-simple analytical singularity type we have

$$e^*(\mathcal{S}) = e^a(\mathcal{S}) \leq 3\sqrt{\mu(\mathcal{S})} - 2$$

where $\mu(\mathcal{S})$ is the Milnor number of $\mathcal{S}$, and for any topological singularity type

$$e^*(\mathcal{S}) = e^s(\mathcal{S}) \leq \frac{9}{\sqrt{6}} \cdot \sqrt{\delta(\mathcal{S})} - 1,$$

where $\delta(\mathcal{S})$ is the delta invariant of $\mathcal{S}$.

For *simple* singularity types there are the better bounds

$\mathcal{S}$	$e^*(\mathcal{S})$	$\mathcal{S}$	$e^*(\mathcal{S})$
A_1	2	D_4	3
A_2	3	D_5	4
$A_k, k = 3, \ldots, 7$	4	$D_k, k \leq 6, \ldots, 10$	5
$A_k, k = 8, \ldots, 10$	5	$D_k, k \leq 11, \ldots, 13$	6
$A_k, k \geq 1$	$\leq 2 \cdot \lfloor \sqrt{k+5} \rfloor$	$D_k, k \geq 1$	$\leq 2 \cdot \lfloor \sqrt{k+7} \rfloor + 1$
E_6	4	E_7	4
E_8	5		

In particular, if $S = \{\mathcal{S}_1, \ldots, \mathcal{S}_s\}$ is a finite set of *simple* singularities, then there is a $d(S)$ such that for all $d \geq d(S)$ and all $r_1, \ldots, r_s \geq 0$ satisfying

$$2 \cdot \sum_{i=1}^{s} r_i \cdot \left(\tau(\mathcal{S}_i) + o\left(\sqrt{\tau(\mathcal{S}_i)}\right) \right) \leq \dim |dH| \tag{2}$$

the family $V_{|dH|}^{irr}(nr_1\mathcal{S}_1, \ldots, nr_s\mathcal{S}_s)$ has a non-empty T–smooth component.

The families in Example 2.1 fail to satisfy this condition roughly by the factor 2. We thus cannot conclude that these families are reducible as we could in a similar situation in [9].

However, if we compare Condition 1 respectively 2 to the conditions in [10] or [11] which ensure that the equisingular family is T–smooth at *every* point, the latter basically invole the square of the Tjurina number and are therefore much more restrictive. This, of course, was to be expected.

4. Some remarks on cones

In this section we collect some basic properties on cones used for the construcion in Sec. 2, in particular the dimension counts.

For points $p_1, \ldots, p_r \in \mathbb{P}^3$ we will denote by $\overline{p_1 \ldots p_r}$ the linear span in $\mathbb{P}^3$ of $p_1, \ldots, p_r$, i.e. the smallest linear subspace containing $p_1, \ldots, p_r$.

Let $P \subset \mathbb{P}^3$ be a plane, $C \subset P$ a curve, and $p \in \mathbb{P}^3 \setminus P$ a point. Then we denote by

$$K_{C,p} = \bigcup_{q \in C} \overline{qp}$$

the cone over C with vertex p. Note that

$$K_{C,p} = \bigcup_{q \in K_{C,p}} \overline{qp}$$

and that

$$K_{C,p} \cap P = C.$$

We first show that C and p fix the cone uniquely except when C is a line.

Lemma 4.1. *Let $P \subset \mathbb{P}^3$ be a plane, and $C \subseteq P$ be an irreducible curve which is not a line.*

Then for $p, p' \in \mathbb{P}^3$ with $p \neq p'$ we have that $K_{C,p} \neq K_{C,p'}$.

Proof. Suppose there are points $p \neq p'$ such that $K_{C,p} = K_{C,p'}$. Choose a point $x \in C \setminus \overline{pp'}$ and let $E = \overline{xpp'}$. Then for any point $y \in \overline{xp} \subset K_{C,p} = K_{C,p'}$ we have

$$\overline{yp'} \subset K_{C,p'},$$

and thus $E = \bigcup_{y \in \overline{xp}} \overline{yp'} \subset K_{C,p'}$. This, however, implies that the line

$$l = E \cap P \subseteq K_{C,p'} \cap P = C$$

is contained in C, and since C is irreducible we would have $C = l$ in contradiction to our assumption that C is not a line. Hence, $K_{C,p} \neq K_{C,p'}$ for $p \neq p'$. $\qquad\square$

Finally we show that for a general p the cone $K_{C,p}$ intersects Σ in an irreducible curve.

Lemma 4.2. *Let $\Sigma \subset \mathbb{P}^3$ be a smooth projective surface, $P \subset \mathbb{P}^3$ be a plane such that $P \neq \Sigma$, and $C \subseteq P$ an irreducible curve which is not a line and not contained in Σ. Then for $p \in \mathbb{P}^3 \setminus P$ general $K_{C,p} \cap \Sigma$ is irreducible.*

Proof. Consider the linear system $\mathcal{L}$ in $\mathbb{P}^3$ which is given as the closure of

$$\{K_{C,p} \mid p \in \mathbb{P}^3 \setminus P\},$$

and set for $q \in \mathbb{P}^3 \setminus P$

$$\mathcal{L}_q = \{D \in \mathcal{L} \mid q \in D\}.$$

First we show that for $q' \in C$ and $q \notin P$

$$\bigcap_{p \in \overline{qq'}} K_{C,p} = C \cup \overline{qq'}. \tag{3}$$

Choose pairwise different point $p_1, \ldots, p_n \in \overline{qq'} \setminus \{q, q'\}$. Suppose that there is a $z \in \bigcap_{i=1}^{n} K_{C,p_i} \setminus (C \cup \overline{qq'})$. Since $z \in K_{C,p_i}$ there is a unique intersection point

$$x_i = \overline{zp_i} \cap C,$$

and these points $x_1, \ldots, x_n$ are pairwise different, since $z \notin \overline{qq'} = \overline{p_ip_j}$ for $i \neq j$. However,

$$x_i \in \overline{zp_i} \subset \overline{zp_ip_j} = \overline{zqq'}$$

and $x_i \in C \subset P$, so that

$$q', x_1, \ldots, x_n \in P \cap \overline{zqq'}$$

and $q', x_1, \ldots, x_n$ are pairwise different collinear points on C. Since C is irreducible but not a line, this implies $\deg(C) \geq n + 1$. In particular, if $n \geq \deg(C)$, then

$$\bigcap_{i=1}^{n} K_{C,p_i} = C \cup \overline{qq'},$$

which implies (3).

Note that by (3) for $q \in \mathbb{P}^3 \setminus P$

$$\bigcap_{D \in \mathcal{L}_q} D \subseteq \bigcap_{K_{C,p} \in \mathcal{L}_q} K_{C,p} = \bigcap_{q' \in C} \bigcap_{p \in \overline{qq'}} K_{C,p} = \bigcap_{q' \in C} (C \cup \overline{qq'}) = C \cup \{q\},$$

and thus

$$\bigcap_{D \in \mathcal{L}} D \subseteq \bigcap_{q \in \mathbb{P}^3 \setminus P} \bigcap_{D \in \mathcal{L}_q} D = C. \tag{4}$$

Consider now the linear systems

$$\mathcal{L}_\Sigma = \{D \cap \Sigma \mid D \in \mathcal{L}\} \quad \text{and} \quad \mathcal{L}_{q,\Sigma} = \{D \cap \Sigma \mid D \in \mathcal{L}_q\} = \{D \in \mathcal{L}_\Sigma \mid q \in D\}.$$

Suppose that $\mathcal{L}_\Sigma$ does not contain any irreducible curve. By (4) and since $C \not\subset \Sigma$ the linear system $\mathcal{L}_\Sigma$ has no fixed component. Thus by Bertini's Theorem $\mathcal{L}_\Sigma$ must be composed with a pencil $\mathcal{B}$, and since for a general point $q \in \Sigma$ the pencil $\mathcal{B}$ contains only one element, say $\widetilde{C}$, through q, the linear system $\mathcal{L}_{q,\Sigma}$ has a fixed component $\widetilde{C}$. But then

$$\widetilde{C} \subseteq \bigcap_{D \in \mathcal{L}_q} D \cap \Sigma = C \cap \Sigma.$$

However, $C \cap \Sigma$ is zero-dimensional, while $\widetilde{C}$ has dimension one.

This proves that $\mathcal{L}_\Sigma$ contains an irreducible element, and thus its general element is irreducible. In particular, for $p \in \mathbb{P}^3 \setminus P$ general $K_{C,p} \cap \Sigma$ is irreducible. $\qquad \square$

References

1. Eric Westenberger, *Families of hypersurfaces with many prescribed singularities*, Ph.D. thesis, TU Kaiserslautern, 2004.
2. Atsuko Hirano, *Constructions of plane curves with cusps*, Saitama Math. J. **10** (1992), 21–24.
3. Luca Chiantini and Ciro Ciliberto, *On the severi varieties of surfaces in* $\mathbb{P}^3$, J. Algebraic Geom. **8** (1999), no. 1, 67–83.
4. Thomas Keilen, *Families of curves with prescribed singularities*, Ph.D. thesis, Universität Kaiserslautern, 2001.
5. Thomas Keilen and Ilya Tyomkin, *Existence of curves with prescribed singularities*, Trans. Amer. Math. Soc. **354** (2002), no. 5, 1837–1860.
6. Eugenii Shustin, *Lower deformations of isolated hypersurface singularities*, Algebra i Analiz **10** (1999), no. 5, 221–249.
7. James Alexander and André Hirschowitz, *An asymptotic vanishing theorem for generic unions of multiple points*, Inventiones Math. **140** (2000), no. 2, 303–325.
8. Eugenii Shustin, *Analytic order of singular and critical points*, Trans. Amer. Math. Soc. **356** (2003), no. 3, 953–985.
9. Thomas Keilen, *Reducible families of curves with ordinary multiple points on surfaces in* $\mathbb{P}^3$, Comm. in Alg. **34** (2006), no. 5, 1921–1926.
10. Gert-Martin Greuel, Christoph Lossen, and Eugenii Shustin, *Castelnuovo function, zero-dimensional schemes, and singular plane curves*, J. Algebraic Geom. **9** (2000), no. 4, 663–710.
11. Thomas Keilen, *Smoothness of equisingular families of curves*, Trans. Amer. Math. Soc. **357** (2005), no. 6, 2467–2481.

On the Alexander invariants of hypersurface complements

Laurentiu Maxim

Department of Mathematics, University of Illinois at Chicago,
851 S. Morgan St., Chicago, Illinois, 60607, USA
E-mail: lmaxim@math.uic.edu
&
Institute of Mathematics of the Romanian Academy,
P.O.Box 1-764, Bucharest, Romania, RO-70700

We survey few of the recent developments in the study of Alexander-type invariants associated to complex hypersurface complements, and point out the dependence of such invariants on the local type and position of singularities of the hypersurface.

1. Introduction

In this mostly expository note, we survey few of the recent developments in the study of the topology of hypersurface complements. Most of the results outlined here are contained in [33] and [14].

The study of plane singular curves is a subject going back to the work of Zariski, who observed that the position of singularities has an influence on the topology of the curve, and this phenomena can be detected by the fundamental group of the complement. However, the fundamental group of a plane curve complement is in general highly non-commutative, thus difficult to handle. On the other hand, Alexander invariants of the complement are more manageable, and turn out to be also sensitive to the position of singularities.

Alexander invariants appeared first in the classical knot theory, where it was noted that in order to study a knot, it is useful to consider the topology of its complement. By analogy with knot theory, Libgober [20–23] introduced and studied Alexander-type invariants for the total linking number infinite cyclic cover of complements to affine complex hypersurfaces. For hypersurfaces with only isolated singularities, he showed that there is essentially only one interesting global Alexander invariant, which depends on the local type and the position of singularities. In [5], twisted Alexander invariants of plane algebraic curves are shown to have a similar property, and examples of curves with trivial Alexander polynomial, but non-trivial twisted Alexander polynomials are given.

It is a natural question to ask how such invariants behave if the hypersurface is allowed to have more general singularities. If the hypersurface is reducible, one has to distinguish between Alexander-type invariants associated to an infinite cyclic cover of the complement on one hand, and to the universal abelian cover on the other hand. The relation between invariants of the total linking number infinite cyclic cover and the topology of the polynomial function defining the (affine) hypersurface, mainly reflected by the monodromy and the vanishing cycles, was also considered in [11].

In [33], we extend Libgober's results on the infinite cyclic Alexander invariants to the case of hypersurfaces with non-isolated singularities and in general position at infinity. It turns out that the infinite cyclic Alexander modules of the complement can be realized as intersection homology modules of the ambient projective space (obtained by adding the hyperplane at infinity), with a certain local coefficient system defined on the complement. This new approach allows the use of techniques from homological algebra (e.g., derived categories and perverse sheaves) in showing that most of these Alexander modules are torsion over the ring of complex Laurent polynomials. Moreover, the associated global Alexander polynomials are entirely determined by the local topological information encoded by the link pairs of singular strata of the hypersurface. Similar methods can be used to obtain obstructions on the eigenvalues of the monodromy operators associated to the Milnor fiber of a projective hypersurface arrangement. The Alexander invariants of the total linking number infinite cyclic cover are further studied in [12], where it is shown that there is a natural mixed Hodge structure on the Alexander modules of the complement.

The analogy with link complements in S^3 is reflected in [28], where invariants of the universal abelian cover of a plane curve complement are considered. In was observed in [25] that such invariants depend on the local type of singularities, and they are calculated in terms of position of singularities of the curve in the plane. More general situations are studied in [13,26,27]. In [14], we show that the universal abelian invariants of complements to hypersurfaces with any kind of singularities, are also determined by the corresponding local invariants associated with singular strata of the hypersurface.

2. Infinite cyclic Alexander invariants of the complement

In this section we study Alexander invariants associated to the total linking number infinite cyclic cover of a hypersurface complement. We first recall Libgober's results for the case of hypersurfaces with only isolated singularities, and then show how to extend his results to hypersurfaces with arbitrary singularities and in general position at infinity (for complete details, see [33]).

2.1. *Preliminaries*

To fix notations for the rest of the paper, let V be a reduced hypersurface in $\mathbb{CP}^{n+1}$, defined by a degree d homogeneous equation: $f = 0$. Let f_i, $i = 1, \cdots, s$ be the irreducible factors of f and $V_i = \{f_i = 0\}$ the corresponding irreducible components of V. Throughout this paper, we will assume that V is *in general position at infinity*, that is, we choose a generic hyperplane H_∞ (transversal to all singular strata in a stratification of V) which we call 'the hyperplane at infinity'. Let $\mathcal{U}$ be the (affine) hypersurface complement: $\mathcal{U} = \mathbb{CP}^{n+1} - (V \cup H_\infty)$.

Then $H_1(\mathcal{U}) \cong \mathbb{Z}^s$ (cf. [9], (4.1.3), (4.1.4)), generated by meridian loops γ_i about the non-singular part of each irreducible component V_i, $i = 1, \cdots, s$. Moreover, if γ_∞ denotes a meridian about the hyperplane at infinity, then there is a relation in $H_1(\mathcal{U})$: $\gamma_\infty + \sum d_i \gamma_i = 0$, where $d_i = deg(V_i)$.

We consider the infinite cyclic cover $\mathcal{U}^c$ of $\mathcal{U}$ defined by the kernel of the total linking number homomorphism $Lk : \pi_1(\mathcal{U}) \to \mathbb{Z}$, which maps all meridian generators to 1, and thus any loop α to the linking number $\mathrm{lk}(\alpha, V \cup -dH_\infty)$ of α with the divisor $V \cup -dH_\infty$ in $\mathbb{CP}^{n+1}$. By definition, the *infinite cyclic Alexander modules of the hypersurface complement* are the homology groups $H_i(\mathcal{U}^c; \mathbb{C})$, regarded as $\Gamma := \mathbb{C}[t, t^{-1}]$-modules, where t acts as the canonical covering transformation.

Note that Γ is a principal ideal domain, hence any torsion Γ-module M of finite type has a well-defined associated order (see [35]). This is called the *Alexander polynomial* of M, and is denoted by $\Delta_M(t)$. We regard the trivial module as a torsion module whose associated polynomial is 1. It is easy to see that if $f : M \to N$ is an epimorphism of Γ-modules and M is torsion of finite type, then N is also torsion of finite type and $\Delta_N(t)$ divides $\Delta_M(t)$.

In studying the Alexander modules of the complement we first note that, since $\mathcal{U}$ has the homotopy type of a finite CW complex of dimension

$\leq n+1$ (e.g., see [9] (1.6.7), (1.6.8)), all the associated Alexander modules are of finite type over Γ, but in general not over $\mathbb{C}$. It also follows that $H_i(\mathcal{U}^c; \mathbb{C}) = 0$ for $i > n+1$, and $H_{n+1}(\mathcal{U}^c; \mathbb{C})$ is free over Γ. Thus, of particular interest are the Alexander modules $H_i(\mathcal{U}^c; \mathbb{C})$ for $i < n+1$.

In [21], Libgober showed that if V is a hypersurface with only isolated singularities, then $\tilde{H}_i(\mathcal{U}^c; \mathbb{Z}) = 0$ for $i < n$, and $H_n(\mathcal{U}^c; \mathbb{C})$ is a torsion Γ-module. Moreover, if $\Delta_n(t)$ is the polynomial associated to the torsion module $H_n(\mathcal{U}^c; \mathbb{C})$, then $\Delta_n(t)$ divides (up to a power of $(t-1)$) the product $\prod_{x \in \text{Sing}(V)} \Delta_x(t)$ of the Alexander polynomials of link pairs around the isolated singular points $x \in V$. This shows the dependence of the Alexander polynomial on the local type of singularities of V. If V is a rational homology manifold, then $\Delta_n(1) \neq 0$. As in the case of a homogeneous isolated hypersurface singularity germ, the zeros of $\Delta_n(t)$ are roots of unity of order $d = deg(V)$ and $H_n(\mathcal{U}^c; \mathbb{C})$ is a semi-simple Γ-module (cf. [21]).

As shown by Zariski, the fundamental group of curve complements in $\mathbb{CP}^2$ is sensitive also to the position of singularities. (Note that by a Zariski-Lefchetz type theorem, cf. [9] p. 25, the class of fundamental groups to curve complements coincides with the class of fundamental groups of the complements to hypersurfaces in a projective space.) In [21–24], Libgober observed that the Alexander invariant of an irreducible curve in $\mathbb{CP}^2$ (or more generally, the 'first non-trivial' Alexander invariant of a hypersurface complement) exhibits a similar property.
As an example, let $C \subset \mathbb{CP}^2$ be a sextic with only cusps singularities. Then by the above divisibility result, the global Alexander polynomial $\Delta_1(t)$ of the curve C is either 1 or a power of $t^2 - t + 1$. The influence of the position of singularities can be seen as follows: if C has only 6 cusps then ([21,22]):

(1) if C is in 'special position', i.e., the 6 cusps are on a conic, then $\Delta_1(t) = t^2 - t + 1$.
(2) if C is in 'general position', i.e., the cusps are not on a conic, then $\Delta_1(t) = 1$.

Note. Libgober's divisibility theorem ([21], Theorem 4.3) holds for hypersurfaces with isolated singularities, *including at infinity*. However, for non-generic H_∞ and for hypersurfaces with more general singularities, the Alexander modules $H_i(\mathcal{U}^c; \mathbb{C})$ ($i \leq n$) are not torsion in general. Their Γ-rank is calculated in [11]. One of the main results in [33] asserts that if V is a reduced hypersurface in general position at infinity, then the modules $H_i(\mathcal{U}^c; \mathbb{C})$ are torsion Γ-modules for all $i \leq n$. We will discuss this aspect and related results in the next section.

2.2. *Intersection homology approach*

Our approach to the study of the infinite cyclic Alexander invariants of the complement makes use of intersection homology theory ([3,15,16]) and the foundational work of Cappell-Shaneson [4] on the study of pseudomanifolds, pl-embedded in codimension two into a manifold. We will use freely the background material from these references (but see also [32], §2, for a quick overview).

Following [4], it is possible to think of a n-dimensional projective hypersurface V as the singular locus of $\mathbb{CP}^{n+1}$, which is now regarded as a filtered space stratified by V and the strata of its singularities. This yields a regular stratification of the pair $(\mathbb{CP}^{n+1}, V)$. Due to the transversality assumption we may also consider the induced stratification for the pair $(\mathbb{CP}^{n+1}, V \cup H_\infty)$. Let $\mathcal{L}$ be a locally constant sheaf on $\mathcal{U}$, with stalk $\Gamma := \mathbb{C}[t, t^{-1}]$ and action by an element $\alpha \in \pi_1(\mathcal{U})$ determined by multiplication by $t^{\mathrm{lk}(\alpha, V \cup -dH_\infty)}$. Then, for any perversity $\bar{p}$, the intersection homology complex $\mathcal{IC}_{\bar{p}}^\bullet := \mathcal{IC}_{\bar{p}}^\bullet(\mathbb{CP}^{n+1}, \mathcal{L})$ is defined by Deligne's axiomatic construction as in [3,16]. (Through this section, we make use of the indexing convention of [16].) The *intersection Alexander modules of the hypersurface* V are then defined as hypercohomology groups of the middle-perversity intersection homology complex:

$$IH_i^{\bar{m}}(\mathbb{CP}^{n+1}; \mathcal{L}) := \mathbb{H}^{-i}(\mathbb{CP}^{n+1}; \mathcal{IC}_{\bar{m}}^\bullet), \quad i \in \mathbb{Z}.$$

Note that, in our setting, the following *superduality isomorphism* holds (cf. [4], Theorem 3.3):

$$\mathcal{IC}_{\bar{m}}^\bullet \cong \mathcal{D}\mathcal{IC}_{\bar{l}}^{\bullet\, op}[2n + 2] \tag{1}$$

(here $\mathcal{D}(\mathcal{A}^\bullet)$ is the Verdier-dual to the complex $\mathcal{A}^\bullet$, and A^{op} is the Γ-module obtained from the Γ-module A by composing all module structures with the involution $t \to t^{-1}$.) Recall that the middle and logarithmic perversities are defined by: $\bar{m}(s) = [(s - 1)/2]$ and $\bar{l}(s) = [(s + 1)/2]$.

The assumption on the position with respect to the hyperplane at infinity is crucial in proving the following technical but important fact:

Lemma 2.1. *([33]) If $i : V \cup H_\infty \hookrightarrow \mathbb{CP}^{n+1}$ is the inclusion, then $i^* \mathcal{IC}_{\bar{m}}^\bullet$ is quasi-isomorphic to the zero complex, i.e. the cohomology stalks of the complex $\mathcal{IC}_{\bar{m}}^\bullet$ vanish at points in $V \cup H_\infty$.*

As a corollary, we obtain the intersection homology realization of the infinite cyclic Alexander modules of hypersurface complements:

Theorem 2.1. *([33]) There is an isomorphism of Γ-modules:*

$$IH_*^{\bar{m}}(\mathbb{CP}^{n+1}; \mathcal{L}) \cong H_*(\mathcal{U}; \mathcal{L}) \cong H_*(\mathcal{U}^c; \mathbb{C}).$$

So the intersection Alexander modules of the hypersurface are isomorphic to the infinite cyclic Alexander modules of the hypersurface complement.

At this point, we can use freely the language of derived categories, derived functors etc., in order to describe the infinite cyclic Alexander invariants. Our first main application is the following (recall that V is assumed to be transversal to the hyperplane at infinity):

Theorem 2.2. *([33]) For any $i \leq n$, the group $H_i(\mathcal{U}^c; \mathbb{C})$ is a finitely generated torsion Γ-module.*

Proof. First, by the superduality isomorphism (1) and Lemma 2.1 we obtain the quasi-isomorphism: $i^! \mathcal{IC}_{\bar{l}}^\bullet \overset{q.i.}{\cong} 0$. Therefore:

$$IH_i^{\bar{l}}(\mathbb{CP}^{n+1}; \mathcal{L}) \cong H_i^{BM}(\mathcal{U}; \mathcal{L}) \cong 0 \quad \text{if} \quad i < n+1, \tag{2}$$

where H_*^{BM} stands for the Borel-Moore homology. The vanishing in (2) follows by Artin's theorem ([36], Example 6.0.6) applied to the $(n+1)$-dimensional affine variety $\mathcal{U}$.

Now, recall that the *peripheral complex*, $\mathcal{R}^\bullet$, associated to the finite local type embedding $V \cup H_\infty \subset \mathbb{CP}^{n+1}$, is a torsion complex (i.e. the cohomology stalks $\mathcal{H}^q(\mathcal{R}^\bullet)_x$ are finite dimensional $\mathbb{C}$-vector spaces, for all $x \in \mathbb{CP}^{n+1}$) and its hypercohomology fits into a long exact sequence (for more details, see [4], p. 339-340):

$$\cdots \to \mathbb{H}^q(\mathbb{CP}^{n+1}; \mathcal{IC}_{\bar{m}}^\bullet) \to \mathbb{H}^q(\mathbb{CP}^{n+1}; \mathcal{IC}_{\bar{l}}^\bullet) \to \mathbb{H}^q(\mathbb{CP}^{n+1}; \mathcal{R}^\bullet) \to$$
$$\to \mathbb{H}^{q+1}(\mathbb{CP}^{n+1}; \mathcal{IC}_{\bar{m}}^\bullet) \to \cdots$$

By the hypercohomology spectral sequence, the groups $\mathbb{H}^*(\mathbb{CP}^{n+1}; \mathcal{R}^\bullet)$ are finite dimensional complex vector spaces, hence torsion Γ-modules. Thus, our claim follows from the above long exact sequence and the vanishing for the logarithmic complex $\mathcal{IC}_{\bar{l}}^\bullet$ in (2). $\qquad\square$

Note that if $i \leq n$, the Γ-module $H_i(\mathcal{U}^c; \mathbb{C})$ is actually a finite dimensional complex vector space, thus its order coincides with the characteristic polynomial of the $\mathbb{C}$-linear map induced by a generator of the group of covering transformations (see [35]). It is shown in [12,33] that this map is $\mathbb{C}$-diagonalizable, thus the Γ-module $H_i(\mathcal{U}^c; \mathbb{C})$ is semi-simple.

Definition 2.1. For $i \leq n$, we denote by $\Delta_i(t)$ the polynomial associated to the torsion Γ-module $H_i(\mathcal{U}^c; \mathbb{C})$, and call it the i-th global Alexander polynomial of the hypersurface V.

These polynomials are well-defined up to multiplication by ct^k ($c \in \mathbb{C}$, $k \in \mathbb{Z}$).

As a consequence of Theorem 2.2, for hypersurfaces in general position at infinity we may now calculate the rank of the free Γ-module $H_{n+1}(\mathcal{U}^c; \mathbb{C})$ in terms of the Euler characteristic of the complement:

Corollary 2.1.

$$rank_\Gamma H_{n+1}(\mathcal{U}^c; \mathbb{C}) = (-1)^{n+1}\chi(\mathcal{U}).$$

Another interesting property of the infinite cyclic Alexander invariants is that they depend on the degree d of the hypersurface. More precisely:

Theorem 2.3. (*[33], Theorem 4.1)*
For $i \leq n$, all zeros of the global Alexander polynomial $\Delta_i(t)$ are roots of unity of order d.

This is a generalization of a similar result obtained by Libgober in the case of hypersurfaces with only isolated singularities (cf. [21], Corollary 4.8).

The last two theorems, 2.2 and 2.3, show a striking similarity between the case of hypersurfaces in general position at infinity and that of a homogeneous singularity germ (in which case the total linking number infinite cyclic cover may be replaced by the Milnor fiber). We will discuss this relation in some detail in the next section.

But perhaps the most important consequence of Theorem 2.1 is the dependence of the infinite cyclic cover Alexander invariants on the local type of singularities of the hypersurface. This is in the spirit of early results of Zariski for the fundamental group of the complement, and those of Libgober for the Alexander invariants of hypersurfaces with only isolated singularities (see [20–24]).
We first need some notation. Let $\mathcal{S}$ be a Whitney stratification of V, and consider the induced Whitney stratification of the pair $(\mathbb{CP}^{n+1}, V)$, with $\mathcal{S}$ the set of singular strata. If $S \in \mathcal{S}$ is an s-dimensional stratum of $(\mathbb{CP}^{n+1}, V)$, then a point $p \in S$ has a distinguished neighborhood W in $(\mathbb{CP}^{n+1}, V)$, which is homeomorphic in a stratum-preserving way to

$\mathbb{C}^s \times c^\circ(S^{2n-2s+1}(p), L(p))$, for $S^{2n-2s+1}(p)$ a small sphere at p in a normal slice for S and $L(p) = S^{2n-2s+1}(p) \cap V$. The link pair $(S^{2n-2s+1}(p), L(p))$ has constant topological type along the stratum S, which we denote by $(S^{2n-2s+1}, L)$.

Now fix an arbitrary irreducible component of V, say V_1. For $S \in \mathcal{S}$, an s-dimensional stratum contained in V_1, let $(S^{2n-2s+1}, L)$ be its link pair in $(\mathbb{CP}^{n+1}, V)$. This is a (possibly singular) algebraic link, and has an associated local Milnor fibration ([34]):

$$F^s \hookrightarrow S^{2n-2s+1} - L \to S^1$$

with fibre F^s and monodromy homeomorphism $h^s : F^s \to F^s$. Let $\Delta_r^s(t) = \det(tI - (h^s)_* : H_r(F^s) \to H_r(F^s))$ be the r-th (local) Alexander polynomial associated to S. Then we have the following divisibility result:

Theorem 2.4. *([33], Theorem 4.2)*
Fix $i \leq n$, and let V_1 be a fixed irreducible component of V. Then the prime divisors of $\Delta_i(t)$ are among the divisors of the local polynomials $\Delta_r^s(t)$ associated to strata $S \subset V_1$, such that $n - i \leq s = dimS \leq n$, and with r satisfying $2n - 2s - i \leq r \leq n - s$. Moreover, if V is a rational homology manifold and has no codimension one singularities (e.g., V is normal), then $\Delta_i(1) \neq 0$.

Note. It follows that the zero-dimensional strata of V may only contribute to $\Delta_n(t)$, the one-dimensional singular strata may only contribute to $\Delta_n(t)$ and $\Delta_{n-1}(t)$, and so on. This observation is crucial in obtaining obstructions on the eigenvalues of the monodromy operators of a hypersurface arrangement (see Theorem 2.6 of the next section). It is not clear at this point what is the role played by the position of singularities in the study of Alexander invariants for hypersurfaces with non-isolated singularities. The position itself is yet to be understood.

The proof of Theorem 2.4 uses the intersection homology realization of the infinite cyclic Alexander modules, together with the superduality isomorphism of Cappell-Shaneson and the properties of the associated peripheral complex. In the case of hypersurfaces with only isolated singularities and in general position at infinity, the theorem can be refined as follows (compare [21], Theorem 3.1):

Theorem 2.5. *If V has only isolated singularities, then $\Delta_n(t)$ divides (up* *to a power of $(t-1)$) the product*

$$\prod_{p \in V_1 \cap Sing(V)} \Delta_p(t)$$

of the local Alexander polynomials of links of the singular points p of V which are contained in V_1.

We will sketch a proof of the isolated singularities case, the general case being treated in a similar manner.

Proof. (sketch)
Assume V has only isolated singularities. If $j : \mathbb{CP}^{n+1} - V_1 \hookrightarrow \mathbb{CP}^{n+1}$ and $i : V_1 \hookrightarrow \mathbb{CP}^{n+1}$ denote the inclusion maps, then the hypercohomology long exact sequence associated to the distinguished triangle:

$$i_* i^! \mathcal{IC}_{\bar{m}}^\bullet \to \mathcal{IC}_{\bar{m}}^\bullet \to j_* j^* \mathcal{IC}_{\bar{m}}^\bullet \overset{[1]}{\to}$$

yields:

$$\to \mathbb{H}_{V_1}^{-n}(\mathbb{CP}^{n+1}; \mathcal{IC}_{\bar{m}}^\bullet) \to IH_n^{\bar{m}}(\mathbb{CP}^{n+1}; \mathcal{L}) \to \mathbb{H}^{-n}(\mathbb{CP}^{n+1} - V_1; j^* \mathcal{IC}_{\bar{m}}^\bullet) \to$$

But $\mathbb{CP}^{n+1} - V_1$ is a $(n+1)$-dimensional affine variety, thus by Artin's vanishing theorem we obtain

$$\mathbb{H}^{-n}(\mathbb{CP}^{n+1} - V_1; j^* \mathcal{IC}_{\bar{m}}^\bullet) \cong IH_n^{\bar{m}}(\mathbb{CP}^{n+1} - V_1; \mathcal{L}) \cong 0.$$

So $IH_n^{\bar{m}}(\mathbb{CP}^{n+1}; \mathcal{L})$ is a quotient of

$$\mathbb{H}_{V_1}^{-n}(\mathbb{CP}^{n+1}; \mathcal{IC}_{\bar{m}}^\bullet) \overset{Sd}{\cong} \mathbb{H}^{-n-1}(V_1; i^* \mathcal{IC}_{\bar{l}}^{\bullet \, op}), \tag{3}$$

the isomorphism (3) being a consequence of the Cappell-Shaneson superduality isomorphism (1).

Now let $\Sigma_0 := V_1 \cap \text{Sing}(V)$ and consider the long exact sequence:

$$\to \mathbb{H}_c^{-n-1}(V_1 - \Sigma_0; \mathcal{IC}_{\bar{l}}^{\bullet \, op}) \to \mathbb{H}^{-n-1}(V_1; \mathcal{IC}_{\bar{l}}^{\bullet \, op}) \to \mathbb{H}^{-n-1}(\Sigma_0; \mathcal{IC}_{\bar{l}}^{\bullet \, op}) \to \tag{4}$$

By local calculation and superduality for link pairs ([4], Corollary 3.4), we have that:

$$\mathbb{H}^{-n-1}(\Sigma_0; \mathcal{IC}_{\bar{l}}^{\bullet \, op}) \cong \oplus_{p \in \Sigma_0} H_n(S_p^{2n+1} - S_p^{2n+1} \cap V; \Gamma) \tag{5}$$

where $(S_p^{2n+1}, S_p^{2n+1} \cap V)$ is the link pair of the singular point $p \in \Sigma_0$, and Γ denotes the induced local coefficient system on the link complement. By the hypercohomology spectral sequence, the modules $\mathbb{H}_c^*(V_1 - \Sigma_0; \mathcal{IC}_{\bar{l}}^{\bullet \, op})$

are annihilated by powers of $t - 1$. The theorem follows by observing that for $p \in \Sigma_0$ we have an isomorphism of Γ-modules:

$$H_n(S_p^{2n+1} - S_p^{2n+1} \cap V; \Gamma) \cong H_n(\widetilde{S_p^{2n+1} - S_p^{2n+1}} \cap V; \mathbb{C}) \cong H_n(F_p, \mathbb{C}),$$

where $\widetilde{S_p^{2n+1} - S_p^{2n+1}} \cap V$ is the total linking number infinite cyclic cover of the algebraic link complement, and F_p is the local Milnor fiber at p. The module structure on the group $H_n(F_p, \mathbb{C})$ is induced by the action of the local monodromy homeomorphism at p. $\square$

Example 2.1. Let V be a degree d reduced projective hypersurface, in general position at infinity, such that V is a rational homology manifold with no codimension one singularities. Assume that the local monodromies of link pairs of strata contained in some irreducible component V_1 of V have orders which are relatively prime to d (e.g., the transversal singularities along strata of V_1 are Brieskorn-type singularities, having all exponents relatively prime to d). Then, by Theorem 2.3 and Theorem 2.4, it follows that $\Delta_i(t) = 1$ for all $1 \leq i \leq n$.

2.3. *The Milnor fiber of a projective hypersurface arrangement*

As an application of the previous results, by a conning construction we obtain obstructions on the eigenvalues of the monodromy operators acting on the homology of the Milnor fiber of a projective hypersurface arrangement.

Let $Y = \{f = 0\}$ be a reduced degree d hypersurface in $\mathbb{CP}^n$, defining a projective hypersurface arrangement $\mathcal{A} = (Y_i)_{i=1,s}$, where Y_i are the irreducible components of Y. Associated to the homogeneous polynomial f there is the global Milnor fibration $f : \mathcal{U} = \mathbb{C}^{n+1} - f^{-1}(0) \to \mathbb{C}^*$, whose fiber $F = f^{-1}(1)$ is called the Milnor fiber of the arrangement $\mathcal{A}$. The monodromy homeomorphism $h : F \to F$ of the Milnor fibre is explicitly described by the mapping $h(x) = \tau \cdot x$, where $\tau = \exp(2\pi i/d)$. Denote by $P_q(t)$ the characteristic polynomial of the monodromy operator $h_q : H_q(F) \to H_q(F)$. Since $h^d = id$, the zeros of $P_q(t)$ are roots of unity of order d.

Note that $\mathcal{U}$ is the complement of a central arrangement $A = \{f^{-1}(0)\}$ in $\mathbb{C}^{n+1}$, namely the cone on $\mathcal{A}$. Moreover, it's easy to see that the projective completion of A in $\mathbb{CP}^{n+1}$ is in general position at infinity. The key observation for what follows is that the Milnor fiber F is homotopy equivalent to the infinite cyclic cover $\mathcal{U}^c$ of $\mathcal{U}$, corresponding to the kernel of the total linking number homomorphism and, with this identification, the monodromy homeomorphism h corresponds to a generator of the group of covering transformations (see [9], p. 106-107).

Theorem 2.4, when applied to the projective cone on Y (i.e., the hypersurface $V = \{f = 0\} \subset \mathbb{CP}^{n+1}$), translates into divisibility results for the characteristic polynomials of the monodromy operators of F, thus showing the dependence of the monodromy of the arrangement $\mathcal{A}$ on the local monodromy operators associated with singular strata in a stratification of Y. With the notations from Theorem 2.4, we can now state the following:

Theorem 2.6. *([33])*
Fix an arbitrary component of the arrangement, say Y_1, and let $\mathcal{Y}$ be the set of (open) singular strata of a stratification of the pair $(\mathbb{CP}^n, Y)$. Then for fixed $q \leq n - 1$, a dth root of unity λ is a zero of $P_q(t)$ only if λ is a zero of one of the local polynomials $\Delta_r^s(t)$ associated with strata $V \in \mathcal{Y}$ of complex dimension s, for $n - q - 1 \leq s \leq n - 1$, such that $V \subset Y_1$ and $2(n - 1) - 2s - q \leq r \leq n - s - 1$.

This theorem provides obstructions on the eigenvalues of the monodromy operators, similar to those obtained by Libgober in the case of hyperplane arrangements [30], or Dimca in the case of curve arrangements [10]. For the special case of an arrangement with only normal crossing singularities along one of its components, we deduce from Theorem 2.6 the following result (compare [8], Corollary 16):

Corollary 2.2. *Let $\mathcal{A} = (Y_i)_{i=1,s}$ be a hypersurface arrangement in $\mathbb{CP}^n$, and fix one irreducible component, say Y_1. Assume that $\bigcup_{i=1,s} Y_i$ is a normal crossing divisor at any point $x \in Y_1$. Then the monodromy action on $H_q(F; \mathbb{C})$ is trivial for $q \leq n - 1$.*

3. Universal abelian Alexander invariants of the complement

In [28] (and later in [25]), Libgober introduced new topological invariants of the complement to plane algebraic curves: the sequence of characteristic varieties. These invariants were also considered in E. Hironaka's doctoral thesis, but see also [18]. Characteristic varieties were originally used to obtain information about all abelian covers of the complex projective plane, branched along a curve (see [25], §1.3). In the context of complex hyperplane arrangements, characteristic varieties of the first homology group of the universal abelian cover of the complement are considered in [6,7,29], and studied in relation with the cohomology support loci of rank one local systems defined on the complement (see also [1] for the study of the latter).

Here we consider (co)homological universal abelian invariants of complements to arbitrary hypersurfaces, as they are described in [14].

3.1. *Definition of Characteristic varieties*

In this section, characteristic varieties are defined, first for general noetherian modules, then in the context of complex hypersurface complements.

Let R be a commutative ring with unit, which is Noetherian and a unique factorization domain. Let A be a finitely generated R-module, and M a $(m \times n)$ presentation matrix of A associated to an exact sequence: $R^m \to R^n \to A \to 0$.

Definition 3.1. The i-th elementary ideal $\mathcal{E}_i(A)$ of A is the ideal in R generated by the $(n - i) \times (n - i)$ minor determinants of M, with the convention that $\mathcal{E}_i(A) = R$ if $i \geq n$, and $\mathcal{E}_i(A) = 0$ if $n - i > m$.

Definition 3.2. The support $\mathrm{Supp}(A)$ of A is the reduced sub-scheme of $\mathrm{Spec}(R)$ defined by the order ideal $\mathcal{E}_0(A)$. Equivalently, if P is a prime ideal of R then $P \in \mathrm{Supp}(A)$ if and only if the localized module A_P is non-zero.

The support $\mathrm{Supp}(A)$ is also called the *first characteristic variety* of A, and we define the i-th *characteristic variety* $V_i(A)$ of A to be the reduced sub-scheme of $\mathrm{Spec}(R)$ defined by the (i-th *Fitting ideal*) ideal $\mathcal{E}_{i-1}(A)$.

All definitions above are independent (up to multiplication by a unit of R) of the choices involved, thus the characteristic varieties are invariants of the R-isomorphism type of A.

Now let V be a reduced hypersurface in $\mathbb{CP}^{n+1}$, and H_∞ be the hyperplane at infinity. As in § 2, we let $\mathcal{U}$ be the complement $\mathbb{CP}^{n+1} - (V \cup H_\infty)$. We denote by $\mathcal{U}^{ab}$ the universal abelian cover of $\mathcal{U}$, or equivalently, the covering associated to the kernel of the homomorphism:

$$Lk^{ab} : \pi_1(\mathcal{U}) \to \mathbb{Z}^s, \quad \alpha \mapsto (\mathrm{lk}(\alpha, V_1 \cup -d_1 H_\infty), \cdots, \mathrm{lk}(\alpha, V_s \cup -d_s H_\infty)).$$

The group of covering transformations of $\mathcal{U}^{ab}$ is isomorphic to $\mathbb{Z}^s$ and acts on the covering space. Let Γ_s be the group ring $\mathbb{C}[\mathbb{Z}^s]$, which is identified with the ring of complex Laurent polynomials in s variables, $\mathbb{C}[t_1, t_1^{-1}, \cdots, t_s, t_s^{-1}]$. Note that Γ_s is a regular Noetherian domain, and in particular it is factorial. As a group ring, Γ_s has a natural involution, denoted by an overbar, sending each t_i to $\bar{t}_i := t_i^{-1}$.

Define a local coefficient system $\mathcal{L}^{ab}$ on $\mathcal{U}$, with stalk Γ_s and action of a loop $\alpha \in \pi_1(\mathcal{U})$ determined by multiplication by $\prod_{j=1}^{s} (t_j)^{\mathrm{lk}(\alpha, V_j \cup -d_j H_\infty)}$. In particular, the action of the meridian γ_i is given by multiplication by t_i. We let $\bar{\mathcal{L}}^{ab}$ be the local system obtained from $\mathcal{L}^{ab}$ by composing all module structures with the natural involution of Γ_s.

Definition 3.3. The universal homology k-th Alexander module of $\mathcal{U}$ is by definition $A_k(\mathcal{U}) := H_k(\mathcal{U}, \mathcal{L}^{ab})$, that is, the group $H_k(\mathcal{U}^{ab}; \mathbb{C})$ considered as a Γ_s-module via the action of covering transformations. Similarly, the universal cohomology k-th Alexander module of $\mathcal{U}$ is defined as $A^k(\mathcal{U}) := H^k(\mathcal{U}; \bar{\mathcal{L}}^{ab})$.

Remark 3.1. If C_* is the cellular complex of $\mathcal{U}^{ab}$, as $\mathbb{Z}[\mathbb{Z}^s]$-modules, and if $C_*^0 := C_* \otimes \mathbb{C}$ denotes the complexified complex, then: $A_k(\mathcal{U}) = H_k(C_*^0)$ and $A^k(\mathcal{U}) = H_k(\mathrm{Hom}_{\Gamma_s}(C_*^0, \Gamma_s))$.

As in §2, the modules $A^k(\mathcal{U})$ and resp. $A_k(\mathcal{U})$ are trivial for $k > n + 1$. Moreover, $A_{n+1}(\mathcal{U})$ is a torsion-free Γ_s-module.

It is easy to see that the universal abelian Alexander modules are of finite type over Γ_s. Hence their characteristic varieties are well-defined. The associated characteristic varieties, in particular the supports, become subvarieties of the s-dimensional torus $\mathbb{T}^s = (\mathbb{C}^*)^s$, which is regarded as the set of closed points in $\mathrm{Spec}(\Gamma_s)$. More precisely, for $\lambda = (\lambda_1, \cdots, \lambda_s) \in \mathbb{T}^s$, we denote by m_λ the corresponding maximal ideal in Γ_s, and by $\mathbb{C}_\lambda$ the quotient $\Gamma_s/m_\lambda \Gamma_s$. This quotient is isomorphic to $\mathbb{C}$ and the canonical projection $\rho_\lambda : \Gamma_s \to \Gamma_s/m_\lambda \Gamma_s = \mathbb{C}_\lambda$ corresponds to replacing t_j by λ_j for $j = 1, \cdots, s$. If A is a Γ_s-module, we denote be A_λ the localization of A at the maximal ideal m_λ. For $A = \Gamma_s$, we use the simpler notation Γ_λ when there is no danger of confusion. Note that if A is of finite type, then $A = 0$ if and only if $A_\lambda = 0$ for all $\lambda \in \mathbb{T}^s$. It follows that

$$\mathrm{Supp}(A) = \{\lambda \in \mathbb{T}^s; A_\lambda \neq 0\}$$

In particular $A_0(\mathcal{U}) = \mathbb{C}_\mathbf{1}$, where $\mathbf{1} = (1, \cdots, 1)$. Hence $\mathrm{Supp}(A_0(\mathcal{U})) = \{\mathbf{1}\}$.

We denote by $V_{i,k}(\mathcal{U})$ the i-th characteristic variety associated to the homological Alexander module $A_k(\mathcal{U})$, and by $V^{i,k}(\mathcal{U})$ that associated to the cohomological Alexander module $A^k(\mathcal{U})$. Note that for each universal Alexander module, its characteristic varieties form a decreasing filtration of the character torus $\mathbb{T}^s$. This follows from the fact that for a noetherian R-

module A of finite type, the elementary ideals form an increasing filtration of R.

All definitions in this section work also in the local setting, i.e., when $\mathcal{U}$ is a complement of a hypersurface germ in a small ball.

Remark 3.2. The invariants defined above originate in the classical knot theory (see [17]), where it follows directly from definition that the support of the universal homological Alexander module of a link complement in S^3 is the set of zeros of the multivariable Alexander polynomial. In the case of irreducible hypersurfaces, where the infinite cyclic and universal abelian cover coincide, the support of an Alexander module is simply the zero set of the associated one-variable polynomial.

3.2. *Further study of Supports*

The results mentioned here are taken from [14].

First note that the cohomology modules may be related to the homology modules by the Universal Coefficient spectral sequence (see [17], p. 20, or [19], Theorem 2.3).

$$\mathrm{Ext}^q_{\Gamma_s}(A_p(\mathcal{U}),\Gamma_s) \Rightarrow A^{p+q}(\mathcal{U}). \tag{6}$$

Relations between the corresponding characteristic varieties are consequences of the spectral sequence obtained by localizing at any $\lambda \in \mathbb{T}^s$:

$$\mathrm{Ext}^q_{\Gamma_\lambda}(A_p(\mathcal{U})_\lambda,\Gamma_\lambda) \Rightarrow A^{p+q}(\mathcal{U})_\lambda. \tag{7}$$

If for a fixed $\lambda \in \mathbb{T}^s$, we define

$$k(\lambda) = \min\{m \in \mathbb{N}; A_m(\mathcal{U})_\lambda \neq 0\}, \tag{8}$$

then the spectral sequence (7) yields the following:

Proposition 3.1.
For any $\lambda \in \mathbb{T}^s$, $A^k(\mathcal{U})_\lambda = 0$ *for* $k < k(\lambda)$ *and*

$$A^{k(\lambda)}(\mathcal{U})_\lambda = Hom(A_{k(\lambda)}(\mathcal{U})_\lambda,\Gamma_\lambda). \tag{9}$$

As a simple application of these facts, we obtain the following:

Example 3.1.
(i) Let $\mathcal{U}$ be the complement of a normal crossing divisor germ in a small ball. Then the universal abelian covering $\mathcal{U}^{ab}$ is contractible, so $A_0(\mathcal{U}) = \mathbb{C}_1$ and $A_k(\mathcal{U}) = 0$ for $k > 0$. Moreover, for any $\lambda \neq 1$, the cohomology Alexander modules satisfy $A^k(\mathcal{U})_\lambda = 0$, for any k.

(ii) Let $(Y, 0)$ be an isolated non-normal crossing singularity at the origin of $\mathbb{C}^{n+1}$ (shortly, INNC), that is, each component of Y is nonsingular outside the origin and, moreover, the union of components in a neighborhood of a point outside the origin is a normal crossing divisor. Let $\mathcal{U}$ be its complement in a small open ball centered at the origin in $\mathbb{C}^{n+1}$ and assume that $n \geq 2$. Then the universal abelian cover $\mathcal{U}^{ab}$ is $(n-1)$-connected (see [26]). More precisely, it is a bouquet of n-spheres (see [13]), hence $A_0(\mathcal{U}) = \mathbb{C}_1$ and $A_k(\mathcal{U}) = 0$ for $k \neq n$. Moreover, for $\lambda \neq 1$, we obtain $A^k(\mathcal{U})_\lambda = 0$ for $k < n$.

In relation with the infinite cyclic Alexander invariants, we note that $\mathcal{U}^{ab} \to \mathcal{U}^c$ is a covering map, and there is a spectral sequence:

$$E^2_{p,q} = Tor^{\Gamma_s}_p(A_q(\mathcal{U}), \Gamma_1) \Rightarrow H_{p+q}(\mathcal{U}^c; \mathbb{C}), \tag{10}$$

where the Γ_s-module structure on Γ_1 is defined by sending each t_i to t. For $a \in \mathbb{T}^1 = \{(t, t, ..., t) \in \mathbb{T}^s\}$, we get by localization a new Künneth spectral sequence, namely

$$E^2_{p,q} = Tor^{\Gamma_a}_p(A_q(\mathcal{U})_a, \Gamma_{1,a}) \Rightarrow H_{p+q}(\mathcal{U}^c; \mathbb{C})_a. \tag{11}$$

In particular we obtain:

Proposition 3.2.
  For any $a \in \mathbb{T}^1$, $H_k(\mathcal{U}^c; \mathbb{C})_a = 0$ for $k < k(a)$ and

$$A_{k(a)}(\mathcal{U})_a \otimes_{\Gamma_a} \Gamma_{1,a} = H_{k(a)}(\mathcal{U}^c; \mathbb{C})_a. \tag{12}$$

In connection with the (co)homology support loci of rank one local systems on $\mathcal{U}$, we mention the following: Let $\lambda = (\lambda_1, \cdots, \lambda_s) \in \mathbb{T}^s$ and denote by $\mathcal{L}_\lambda$ the local coefficient system on $\mathcal{U}$ with stalk $\mathbb{C} = \mathbb{C}_\lambda$ and action of a loop $\alpha \in \pi_1(\mathcal{U})$ determined by multiplication by $\prod_{j=1}^s (\lambda_j)^{\mathrm{lk}(\alpha, V_j \cup -d_j H_\infty)}$. One can define new *topological characteristic varieties* by setting

$$V^t_{i,k}(\mathcal{U}) = \{\lambda \in \mathbb{T}^s; \dim_{\mathbb{C}} H_k(\mathcal{U}, \mathcal{L}_\lambda) > i\}$$

and similarly for cohomology. It is known that

$$H_k(\mathcal{U}, \mathcal{L}_\lambda) = H_k(C^0_* \otimes_{\Gamma_s} \mathbb{C}_\lambda).$$

Therefore, by the Künneth spectral sequence, we get

$$E^2_{p,q} = Tor^{\Gamma_s}_p(A_q(\mathcal{U}), \mathbb{C}_\lambda) \Rightarrow H_{p+q}(\mathcal{U}, \mathcal{L}_\lambda). \tag{13}$$

740

Since the localization is exact, the base change for Tor under $\Gamma_s \to \Gamma_\lambda$ yields a new spectral sequence

$$E^2_{p,q} = Tor^{\Gamma_\lambda}_p(A_q(\mathcal{U})_\lambda, \mathbb{C}_\lambda) \Rightarrow H_{p+q}(\mathcal{U}, \mathcal{L}_\lambda). \qquad (14)$$

This is used in proving the next result:

Proposition 3.3. ([14]) *For any point $\lambda \in \mathbb{T}^s$, one has the following:*

(i) $min\{m \in \mathbb{N}, \ H_m(\mathcal{U}, \mathcal{L}_\lambda) \neq 0\} = min\{m \in \mathbb{N}, \ \lambda \in Supp(A_m(\mathcal{U}))\}$.

(ii) $dimH_{k(\lambda)}(\mathcal{U}, \mathcal{L}_\lambda) = max\{m \in \mathbb{N}, \ \lambda \in V_{m,k(\lambda)}(\mathcal{U})\}$.

3.3. *Dependence on the local data*

In [25], characteristic varieties of plane curve complements are described in terms of local type of singularities and dimensions of linear systems which are attached to the configuration of singularities of the curve. In [14] we obtain a different type of dependence on the local data, leading to vanishing results.

Again, we assume that the hyperplane at infinity H_∞ is transversal in the stratified sense to the hypersurface V. Under this assumption, the universal cohomological Alexander invariants of the complement are entirely determined by the degrees of the irreducible components on one hand, and by the local topological information encoded by the singularities of V on the other hand. In particular, these invariants depend on the local type of singularities of the hypersurface.

First, we need some notations. For $x \in V$, we let $\mathcal{U}_x = \mathcal{U} \cap B_x$, for B_x a small open ball at x in $\mathbb{CP}^{n+1}$. Denote by $\mathcal{L}^{ab}_x$ the restriction of the local coefficient system $\mathcal{L}^{ab}$ to $\mathcal{U}_x$. Then:

Theorem 3.1. ([14]) *Let $\lambda = (\lambda_1, \cdots, \lambda_s) \in \mathbb{T}^s$ and $\epsilon \in \mathbb{Z}_{\geq 0}$. Fix an irreducible component V_1 of V, and assume that $\lambda \notin Supp(H^q(\mathcal{U}_x, \bar{\mathcal{L}}^{ab}_x))$ for all $q < n + 1 - \epsilon$ and all points $x \in V_1$. Then $\lambda \notin Supp(A^q(\mathcal{U}))$ for all $q < n + 1 - \epsilon$.*

The assumption on the hyperplane at infinity, together with the universal coefficient spectral sequence imply that the modules $H^*(\mathcal{U}_x, \bar{\mathcal{L}}^{ab}_x)$ in Theorem 3.1 can be expressed in terms of the local universal homological Alexander modules $A_*(\mathcal{U}'_x)$, where $\mathcal{U}' := \mathbb{CP}^{n+1} - V$ and $\mathcal{U}'_x := \mathcal{U}' \cap B_x$. The latter depend only on the hypersurface singularity germ (V, x), and are defined as in § 3.1. For complete details, see [14].

The following consequence of Theorem 3.1 and of Example 3.1 is similar to some results in [13,26,27].

Corollary 3.1. *(i) (Case $\epsilon = 0$) With the notation in the above theorem, assume in addition that V is a normal crossing divisor at any point of the component V_1. Then $Supp(A^k(\mathcal{U})) \subset \{\mathbf{1}\}$ for any $k < n + 1$.*

(ii) (Case $\epsilon = 1$) With the notation in the above theorem, assume in addition that V is an INNC divisor at any point of the component V_1. Then $Supp(A^k(\mathcal{U})) \subset \{\mathbf{1}\}$ for any $k < n$.

The dependence on the degrees of components of V is reflected by the following result, a generalization of a similar result from [28]:

Theorem 3.2. *([14]) For $k \leq n$, $Supp(A^k(\mathcal{U}))$ is contained in the zero set of the polynomial $t_1^{d_1} \cdots t_s^{d_s} - 1$, thus has positive codimension in $\mathbb{T}^s$.*

The positive codimension property of supports in the universal abelian case should be regarded as the analogue of the torsion property in the infinite cyclic case (cf. [12,33]).

Remark 3.3. In proving the results of this section, the general theory of perverse sheaves is used (cf. [2,10,31,36]). The use of intersection homology as in §2 is constrained by lacking the superduality isomorphism (1), which only holds over a Dedekind domain.

Acknowledgements

We would like to thank the organizers of the *Singularities* conference, C.I.R.M.-Luminy 2005, where part of this work was presented.

References

1. Arapura, D.: *Geometry of cohomology support loci for local systems. I*, J. Algebraic Geom. **6** (1997), 563–597.
2. Beilinson, A. A., Bernstein, J., Deligne, P., *Faisceaux pervers*, Astérisque **100** (1982).
3. Borel, A. et. al., *Intersection cohomology*, Progress in Mathematics, vol. **50**, Birkhauser, Boston, 1984.
4. Cappell, S., Shaneson, J., *Singular spaces, characteristic classes, and intersection homology*, Annals of Mathematics, **134** (1991), 325–374.
5. Cogolludo, J. I., Florens, V., *Twisted Alexander polynomials of plane algebraic curves*, arXiv: math:GT/0504356.
6. Cohen, D. C., Suciu, A. I., *Alexander invariants of complex hyperplane arrangements*, Trans. Amer. Math. Soc, Vol. **351** (10), 4043–4067, (1999).

7. Cohen, D. C., Suciu, A. I., *Characteristic varieties of arrangements*, Math. Proc. Cambridge Philos. Soc. **127** (1999), 33–54.

8. Cohen, D.C., Dimca, A., Orlik, P., *Nonresonance conditions for arrangements*, Ann. Inst. Fourier, Grenoble, **53** (6), 2003, 1883–1896.

9. Dimca, A., *Singularities and Topology of Hypersurfaces*, Universitext, Springer-Verlag, 1992.

10. Dimca, A., *Sheaves in Topology*, Universitext, Springer-Verlag, 2004.

11. Dimca, A., Nemethi, A., *Hypersurface complements, Alexander modules and Monodromy*, in *Real and Complex Singularities*, 19–43, Contemp. Math. **354**, Amer. Math. Soc., Providence, RI (2004).

12. Dimca, A., Libgober, A. *Regular functions transversal at infinity*, arXiv: math.AG/0504128, to appear in Tohoku Math. J.

13. Dimca, A., Libgober, A., *Local topology of reducible divisors*, arXiv: math.AG/0303215

14. Dimca, A., Maxim, L., *Multivariable Alexander invariants of hypersurface complements*, arXiv: math.AT/0506324, to appear in Trans. Amer. Math. Soc.

15. Goresky, M., MacPherson, R., *Intersection homology theory*, Topology **19** (1980), 135–162.

16. Goresky, M., MacPherson, R., *Intersection homology II*, Invent. Math. **72** (1983), 77–129.

17. Hillman, J. A., *Alexander ideals of links*, LNM **895**, Springer 1981.

18. Hironaka, E., *Alexander stratifications of character varieties*, Annales de l'institut Fourier, **47** (2), 1997, 555–583.

19. Levine, J., *Knot Modules, I*, Trans. Amer. Math. Soc., Vol **229**, 1–50.

20. Libgober, A., *Homotopy groups of the complements to singular hypersurfaces* Bulletin of the AMS, **13** (1), 1985.

21. Libgober, A., *Homotopy groups of the complements to singular hypersurfaces, II*, Annals of Mathematics, **139** (1994), 117–144.

22. Libgober, A., *Alexander polynomials of plane algebraic curves and cyclic multiple planes*, Duke Math. J., **49** (1982), 833–851.

23. Libgober, A., *Alexander invariants of plane algebraic curves*, Singularities, Proc. Symp. Pure Math., Vol. **40** (2), 1983, 135–143.

24. Libgober, A., *Position of singularities of hypersurfaces and the topology of their complements*, Algebraic Geometry, 5. J. Math. Sci. 82 (1996), no. 1, 3194–3210.

25. Libgober, A., *Characteristic varieties of algebraic curves*, in: C. Ciliberto et al.(eds), Applications of Algebraic Geometry to Coding Theory, Physics and Computation, 215–254, Kluwer, 2001.

26. Libgober, A., *Isolated non-normal crossing*, in Real and Complex Singularities, 145–160, Contemporary Mathematics, **354**, 2004.

27. Libgober, A., *Homotopy groups of complements to ample divisors*, arXiv: math.AG/0404341

28. Libgober, A., *On the homology of finite abelian covers*, Topology and its applications, **43** (1992) 157-166.

29. Libgober, A., Yuzvinsky, S., *Cohomology of the Orlik-Solomon algebras and local systems*, Compositio Math. **121** (2000), no. 3, 337–361.

30. Libgober, A., *Eigenvalues for the monodromy of the Milnor fibers of arrangements*, Trends in Singularities, 141–150, 2002, Birkhauser Verlag.
31. Massey, D. B., *Introduction to perverse sheaves and vanishing cycles,* in *Singularity Theory*, ICTP 1991, Ed. D.T. Le, K. Saito, B. Teissier, 487–509.
32. Maxim, L., *Alexander invariants of hypersurface complements*, Thesis, University of Pennsylvania, 2005.
33. Maxim, L., *Intersection homology and Alexander modules of hypersurface complements*, Comm. Math. Helvetici **81** (1), 2006, 123-155.
34. Milnor, J., *Singular points of complex hypersurfaces*, Annals of Mathematical Studies 61, vol. 50, Princeton University Press, Princeton, 1968.
35. Milnor, J., *Infinite cyclic coverings*, Topology of Manifolds, Boston 1967.
36. Schurmann, J., *Topology of Singular Spaces and Constructible Sheaves*, Birkhauser, Monografie Matematyczne. Vol. **63**, 2003.

THE BOUNDARY OF THE MILNOR FIBER
OF HIRZEBRUCH SURFACE SINGULARITIES

Dedicated to Jean-Paul Brasselet for his first 60 years

FRANÇOISE MICHEL

*Laboratoire de Mathématiques Emile Picard, Université Paul Sabatier,
118 route de Narbonne, 31062 Toulouse Cedex 04, France
E-mail: fmichel@picard.ups-tlse.fr*

ANNE PICHON

*Institut de Mathématiques de Luminy UMR 6206, Université de la Méditerranée,
Case 907, 163 avenue de Luminy,13288 Marseille Cedex 9, France
E-mail: pichon@iml.univ-mrs.fr*

CLAUDE WEBER

*Section de Mathématiques, Université de Genève
CP 64, 1211 Genève 4, Suisse
E-mail: Claude.Weber@math.unige.ch*

We give the first (as far as we know) complete description of the boundary of
the Milnor fiber for some non-isolated singular germs of surfaces in $\mathbf{C}^3$. We
study irreducible (i.e. $gcd\,(m, k, l) = 1$) non-isolated (i.e. $1 \leq k \leq l$ and $2 \leq l$)
Hirzebruch hypersurface singularities in $\mathbf{C}^3$ given by the equation $z^m - x^k y^l = 0$. We show that the boundary L of the Milnor fiber is always a Seifert manifold
and we give an explicit description of the Seifert structure. From it, we deduce
that:

    1) L is never diffeomorphic to the boundary of the normalization.

    2) L is a lens space iff $m = 2$ and $k = 1$.

    3) When L is not a lens space, it is never orientation preserving diffeomorphic to the boundary of a normal surface singularity.

Mathematics subject classification: 14J17 32S25 57M25

1. Introduction

In [MP] the authors prove, among other facts, that the boundary L of the
Milnor fiber of a non-isolated hypersurface singularity in $\mathbf{C}^3$ is a Waldhausen manifold (non-necessarily "reduziert").

745

In this paper, we apply the general method of [MP] to the study of Hirzebruch singularities, defined by the equation

$$z^m - x^k y^l = 0$$

We assume that the germ is irreducible, which amounts to ask that $gcd(m, k, l) = 1$. We also assume that $1 \leq k \leq l$ to avoid redundancies and that $m \geq 2$ in order to have a genuine singularity.

Hirzebruch proved in [H] that the boundary $\tilde{L}$ of the normalization is a lens space and he gave an explicit description of the minimal resolution as a bamboo-shaped graph of rational curves. See also [HNK]. We call "bamboo" a connected graph whose vertices have at most two neighbours. We briefly recall this result in section 2.

We prove in theorem 3.1 that L is always a Seifert manifold. Its canonical star-shaped plumbing graph is described in theorem 4.2.

When $m \geq 3$ or when $m = 2$ and $k \geq 2$ the plumbing graph for L is never a bamboo of rational curves. A little computation shows then (see corollary 4.3) that L is never orientation-preserving diffeomorphic to the boundary of a normal surface singularity.

When $m = 2$ and $k = 1$, the plumbing graph is a bamboo of rational curves. But it is different from the Hirzebruch one. Indeed, the corresponding lens spaces do not have the same fundamental group.

In [MP] it is stated that the boundary L_t of the Milnor fiber of a non-isolated hypersurface singularity in $\mathbf{C}^3$ is never diffeomorphic to the boundary $\tilde{L}_0$ of the normalization. The sketch of proof given in [MP] does not treat the case when L_t is homeomorphic to a lens space. Here, we prove this result for any Hirzebruch surface singularity (even when L_t is a lens space), in a very explicit way as we are able to compare the two corresponding plumbing graphs.

The more general case of germs having equation $z^m - g(x,y) = 0$ is treated in [MPW]. The proofs we present here are self-contained, i.e. independent from [MP] and from [MPW].

The first named author had the idea to study Hirzebruch singularities while reading Egbert Brieskorn's beautiful article [B].

We thank Walter Neumann for very pleasant discussions during the meeting and for bringing our attention to the computation of the invariant

e_0 for Seifert manifolds.

This paper was written with the help of the Fonds National Suisse de la Recherche Scientifique.

2. Plumbing graphs

The 3-dimensional manifolds we consider are compact and oriented. In many cases, they are oriented as the boundary of a complex surface. To describe these manifolds, we use plumbing graphs and we follow [N] as closely as possible. Recall that a vertex of a plumbing graph carries two weights: the genus g of the base space and the Euler number $e \in \mathbf{Z}$. In this paper we always have $g \geq 0$ i.e. the base surfaces are orientable. Particuliarly useful are the bamboos for lens spaces and the star-shaped graphs for "general" Seifert manifolds.

The lens space $L(n,q)$ is defined as the quotient of the sphere $S^3 \subset \mathbf{C^2}$ (oriented as the boundary of the unit 4-ball, equiped with the complex orientation) by the action $C_{n,q}$ of the n-th roots of unity given by $\zeta(z_1, z_2) = (\zeta z_1, \zeta^q z_2)$ with $0 < q < n$ and $gcd(n,q) = 1$. The canonical plumbing graph for $L(n,q)$ is the bamboo of rational curves with Euler numbers, from left to right, $(e_1, e_2, ..., e_u)$ defined as $e_i = -b_i$. The integers b_i are defined by $b_i \geq 2$ together with

$$\frac{n}{q} = b_1 - \cfrac{1}{b_2 - \cfrac{1}{b_3 - \cfrac{1}{\ddots - \cfrac{1}{b_u}}}}$$

As in [N], we summarize the continued fraction expansion as $[b_1, b_2, ...b_u]$.

The Seifert manifolds (with unique Seifert foliation) are described by a star-shaped graph. See [N] corollary 5.7. All vertices, except possibly the central one, have genus zero and Euler number $e \leq -2$.

We now consider Hirzebruch singularity $z^m - x^k y^l = 0$. The boundary $\tilde{L}$ of its normalization is the lens space $L(n,q)$ where n and q are computed as follows. Let $d_k = gcd(m,k)$ and $d_l = gcd(m,l)$. Then

$$n = \frac{m}{d_k d_l}$$

To get q let λ_0 be the smallest integral positive solution of the equation

$$\lambda l \equiv -k d_l \ (\mathrm{mod}\, m)$$

in the unknown λ. This solution λ_0 is divisible by d_k and we have

$$q = \frac{\lambda_0}{d_k}$$

The special case $d_k = 1 = d_l$ is more pleasant. Then

$$n = m \text{ and } q = \lambda_0$$

where λ_0 is the smallest positive solution of the equation $\lambda l \equiv -k$ (mod m). See [BPV].

The description we give below in theorem 4.1 for the boundary L of the Milnor fiber is in sharp contrast with the classical result (essentially Hirzebruch's thesis) about the boundary $\tilde{L}$ of the normalisation. For instance, if m is fixed, $\tilde{L}$ depends only on the residue classes $(\mathrm{mod}\, m)$ of k and l. This is not the case for L. See section 5 below for an example.

3. Vertical monodromies

Let $f(x, y, z) = z^m - x^k y^l$ be an irreducible germ (i.e. $gcd(m, k, l) = 1$) of the hypersurface in $\mathbf{C}^3$ with a singular point (i.e. $2 \leq m$) at the origin. Recall that we assume that $1 \leq k \leq l$ to avoid redundancies.

In this paper, we use for technical reasons a polydisc $B(\alpha) = B_\alpha^2 \times B_\alpha^2 \times B_\epsilon^2$ with $0 < \alpha \leq \epsilon$ and $\alpha^{k+l} < \epsilon^m$ in place of the standard Milnor ball $B_\epsilon^6 = \{P \in \mathbf{C}^3 \text{ with } |P| \leq \epsilon\}$. The equation of f being quasi-homogeneous, for any $B(\alpha)$ there exists η with $0 < \eta \ll \alpha$ such that the restriction of f on $B(\alpha) \cap f^{-1}(B_\eta^2 \setminus \{0\})$ is a locally trivial fibration on $(B_\eta^2 \setminus \{0\})$ and such that this fibration does not depend on α up to isomorphism. Let S be the boundary of $B(\alpha)$. The condition $\alpha^{k+l} < \epsilon^m$ implies that we may choose η with $0 < \eta \ll \alpha$ such that $L_t = f^{-1}(t) \cap S$ is contained in $\{ (x, y, z) \in \mathbf{C}^3$ such that $|x| = \alpha$ or $|y| = \alpha \}$ for all t with $0 \leq |t| \leq \eta$. For such a η, if $t \in B_\eta^2 \setminus 0$ we say that $F_t = B(\alpha) \cap f^{-1}(t)$ is "the" **Milnor fiber** of f and that $L_t = F_t \cap S$ is "the" **boundary of the Milnor fiber** of f. From now on, we write $L = L_t$ for a chosen t such that $0 < |t| \leq \eta$.

We will now describe L as the union of $M' = L \cap \{|x| = \alpha\}$ and $M'' = L \cap \{|y| = \alpha\}$.

Theorem 3.1. The boundary L of the Milnor fiber of $z^m - x^k y^l$ is a Seifert manifold. Moreover, the projection on the z-axis is constant on each Seifert leaf.

Note. We prefer to use the words "Seifert leaf" instead of "Seifert fiber" in order to avoid confusion with the other types of fibers used here, such as the Milnor fibers.

Proof of theorem 3.1. Let $\varphi : M' \to \mathbf{C}^3$ be defined by $\varphi(x, y, z) = (x, z, f(x, y, z))$. Hence we have $\varphi(M') \subset S_\alpha^1 \times B_\epsilon^2 \times \{t\}$. The singular locus $\Sigma(\varphi)$ of φ satisfies the equation $\frac{\partial f}{\partial y} = 0$ i.e. $l x^k y^{l-1} = 0$. But we have $M' \subset \{|x| = \alpha\}$. Hence we have $\Sigma(\varphi) = \cup_{i=1}^m (S_\alpha^1 \times \{0\} \times \{z_i\})$ where $z_i^m = t$.

The set of singular values $\Delta(\varphi) = \varphi(\Sigma(\varphi))$ of the map φ is the union of the m circles $S_\alpha^1 \times \{z_i\} \times \{t\}$ where $z_i^m = t$.

We fill $\varphi(M')$ with the circles $S_\alpha^1 \times \{c\} \times \{t\}$ where $c \in B_\epsilon^2$ and $|c^m - t| \leq \alpha^{k+l}$. As $\Delta(\varphi)$ is the union of m of these circles , we pull-back this (trivial) fibration of $\varphi(M')$ in circles to obtain a Seifert foliation on M'. The Seifert leaves are defined as the components of the intersection $M' \cap \{z = c\}$.

Replacing φ by the restriction to M'' of the morphism $(x, y, z) \mapsto (y, z, f(x, y, z))$ we see that, in a symmetric way, the intersections $M'' \cap \{z = c\}$ fill M'' with a Seifert foliation in circles. The Seifert leaves of M' and of M'' are defined by the same equation $L \cap \{z = c\}$, so they coincide on $T = M' \cap M''$. **End of proof of theorem 3.1.**

Remark. The referee gave us an alternative proof of theorem 3.1 : there is a fixed point free action of the circle on L given by $s.(x, y, z) = (s^{-l}x, s^k y, z)$ for $s \in S^1$. This action equips L with a Seifert fibration. We give the above proof of 3.1 because we want to illustrate the general method given in [MP], which enables one to construct a Waldhausen decomposition of the boundary of the Milnor fiber for any germ f with a 1-dimensional singular locus.

Let $\pi_x : M' \to S_\alpha^1$ (resp $\pi_y : M'' \to S_\alpha^1$) be the restriction to M' (resp M'') of the projection on the x-axis (resp the y-axis). Let $a \in S_\alpha^1$. Now let $G' = \pi_x^{-1}(a)$ and $G'' = \pi_y^{-1}(a)$.

Theorem 3.2. π_x and π_y are locally trivial differentiable fibrations over S_α^1. Moreover:
1) The fibers of π_x (resp π_y) are diffeomorphic to the Milnor fiber of the plane curve germ $z^m - y^l$ (resp $z^m - x^k$).

2) The fibers of π_x (resp π_y) meet transversally the Seifert leaves of M' (resp M'') constructed in the proof of theorem 3.1.

Proof of theorem 3.2. The singular locus of π_x is defined by $lx^k y^{l-1} = 0$ and $mz^{m-1} = 0$. But, if $(x, y, z) \in M'$ we have $|x| = \alpha$ and $z^m - x^k y^l = t$ with $0 < |t|$. So π_x has no singular point. It is easy to see that the restriction of π_x to $\partial M'$ is a submersion onto S^1_α. As M' is a compact differentiable manifold, π_x is a differentiable fibration. The situation is symmetric for π_y.

Now, we have chosen $a \in S^1_\alpha$ and t such that $0 < |t| \le \eta$ where η is very small. By definition we have $G' = \{(a, y, z) \ with \ z^m - a^k y^l = t \ and \ (y, z) \in S^1_\alpha \times B^2_\epsilon\}$ and also $G'' = \{(x, a, z) \ with \ z^m - x^k a^l = t \ and \ (x, z) \in S^1_\alpha \times B^2_\epsilon\}$. Hence, the assertion 1) is obvious.

To prove 2) let b be any l^{th} root of $(a^{-k}(c^m - t))$ and let $P = (a, b, c) \in G'$. The Seifert leaf containing P is parametized by $(e^{i\theta} a, e^{-i\theta \frac{k}{l}} b, c)$ with, say, $\theta \in \mathbf{R}$. Hence, the Seifert leaves are oriented and transverse to the hyperplane $H_a = \{x = a\}$ for all $a \in S^1_\alpha$. The situation is symmetric for M''. **End of proof of theorem 3.2.**

Remarks. 1. If $k = l = 1$ the germ f has an isolated singular point at the origin. In this case, theorem 3.2 shows that G' and G'' are discs and that M' and M'' are solid torii. Hence L is a lens space, diffeomorphic to $L_0 = \tilde{L}$.

2. If we assume that $\dim \Sigma(f) = 1$ then we have $l \ge 2$ and the x-axis $D' = \{(x, 0, 0) \ with \ x \in \mathbf{C}\}$ is a component of $\Sigma(f)$. Then, theorem 3.2 implies that G' is never diffeomorphic to a disc and that M' is not a solid torus. When $D' \subset \Sigma(f)$ we say in [MP] that M' is the **vanishing zone** around D'. When $k \ge 2$ then $D'' = \{(0, y, 0) \ with \ y \in \mathbf{C}\}$ is the second component of $\Sigma(f)$ and M'' is the vanishing zone around D''.

We now proceed to the definition of the vertical monodromy. Let $h' : G' \to G'$ be the diffeomorphism defined by the first return along the (oriented) leaves of M'. Theorem 3.2 implies that h' is a monodromy for the fibration π_x.

Definition. We call h' the **vertical monodromy** for D'.

Likewise, the first return along the (oriented) Seifert leaves of M'' is a diffeomorphism $h'' : G'' \to G''$. We call it the vertical monodromy for D''.

In conclusion, we know that M' is the mapping torus of h' acting on G' and that M'' is the mapping torus of h'' acting on G''. We wish now to describe in details the vertical monodromies.

Notations. Let $d = gcd(k,l)$; $\bar{l} = \frac{l}{d}$; $\bar{k} = \frac{k}{d}$; $d_l = gcd(m,l)$; $d_k = gcd(m,k)$.

Remark. As f is assumed to be irreducible, we have $gcd(m,k,l) = 1$ and $\bar{k}$ is prime to d_l (resp $\bar{l}$ is prime to d_k). Moreover, G' has d_l boundary components and G'' has d_k boundary components.

Theorem 3.3. The vertical monodromy h' (resp h'') has finite order $\bar{l}$ (resp $\bar{k}$). Moreover:
1. If $\bar{l} \geq 2$ (resp $\bar{k} \geq 2$) then h' (resp h'') has exactly m fixed points and any non-fixed point has order $\bar{l}$ (resp $\bar{k}$).
2. At each fixed point h' (resp h'') acts locally as a rotation of angle $-(\bar{k}/\bar{l})2\pi$ (resp $-(\bar{l}/\bar{k})2\pi$).

Proof of theorem 3.3. As in the proof of theorem 3.2, we consider $P = (a,b,c) \in G'$. We have seen that $(e^{i\theta}a, e^{-i\theta\frac{k}{l}}b, c)$ for, say, $\theta \in \mathbf{R}$ is a parametrization of the Seifert leaf which contains P. Hence

$$(\star) \qquad h'(P) = (a, e^{-2i\pi\frac{k}{l}}b, c)$$

As $k/l = \bar{k}/\bar{l}$ with $\bar{k}$ prime to $\bar{l}$, we see that h' has order $\bar{l}$ on each $P = (a,b,c)$ with $b \neq 0$.

Then, if $\bar{l} \geq 2$, it is clear that $h'(P) = P$ iff $b = 0$. Then $c^m = t$ and h' has exactly m fixed points, i.e. the points $\{(a,0,z_i)\}$ where $z_i^m = t$.

The formula $(\star)$ implies directly the last statement of theorem 3.3. **End of proof of theorem 3.3**.

Corollary 3.4. The intersection $T = M' \cap M''$ is a torus.

Proof of corollary 3.4. Indeed, T is the mapping torus of h' acting on the d_l boundary components of G'. As $\bar{k}$ is prime to $\bar{l}$ the formula $(\star)$ in the proof of theorem 3.3 implies that h' permutes transitively the boundary components of G'. **End of proof of corollary 3.4.**

Remark. G', G'' and $F_t = f^{-1}(t) \cap B$ are oriented by the complex structure. L is oriented as the boundary of F_t and this orientation induces one on M' and M''.

Theorem 3.5. Orient $T = M' \cap M''$ as the boundary of M''. Orient $\partial G'$ (resp $\partial G''$) as the boundary of G' (resp G''). Then the intersection number on T of $\partial G'$ with $\partial G''$ is equal to $-m$.

Proof of theorem 3.5. Let $\pi : L \to B_\alpha^2 \times B_\alpha^2$ be the restriction on L of the projection $(x, y, z) \mapsto (x, y)$. The restriction of π to $T = M' \cap M''$ is a regular covering of order m. Moreover, we have $\pi(G') = \{a\} \times S_\alpha^1$ and $\pi(G'') = S_\alpha^1 \times \{a\}$. The complex structure of $\mathbf{C}^2$ induces an orientation on $B_\alpha^2 \times B_\alpha^2$. Let $S_\alpha^1 \times S_\alpha^1 = \pi(T)$ be oriented as the boundary of $B_\alpha^2 \times S_\alpha^1$. The intersection number of $\{a\} \times S_\alpha^1$ with $S_\alpha^1 \times \{a\}$ in $S_\alpha^1 \times S_\alpha^1$ is equal to (-1). The covering projection π being compatible with orientations, this proves that the intersection number we are looking for is equal to $(-m)$. **End of proof of theorem 3.5.**

4. The Seifert structure on the boundary of the Milnor fiber

Theorem 4.1. The Seifert invariants (associated to the Seifert structure described in section 3) for the boundary L of the Milnor fiber of a Hirzebruch singularity are as follows:

1. The genus g of the base space is equal to $(m - 1)(d - 1)$ where $d = gcd(k, l)$.

2. The integral Euler number e is equal to m.

3. Let $\bar{l} = \frac{l}{d}$ and $\bar{k} = \frac{k}{d}$. Then L has $2m$ (possibly) exceptional leaves.

There are m of them with Seifert invariants (α', β') defined by $\alpha' = \bar{l}$ and β' given by $(-\bar{k})\beta' \equiv 1 \bmod \bar{l}$ and $0 < \beta' < \bar{l}$ in normalized form.

There are m of them with Seifert invariants (α'', β'') defined by $\alpha'' = \bar{k}$ and β'' given by $(-\bar{l})\beta'' \equiv 1 \bmod \bar{k}$ and $0 < \beta'' < \bar{k}$.

Comments. 1. The singularity is isolated iff $k = l = 1$. Of course in this case we have $\tilde{L} = L$. The theorem above says that L has no exceptional leaf, that $g = 0$ and that $e = m$. Hence L is the lens space $L(m, m - 1)$. We are happy to see that this agrees with Hirzebruch's result.
Assume from now on that $1 \leq k$ and that $2 \leq l$.
2. Under this hypothesis L is a lens space iff $m = 2$ and $k = 1$. (Quick proof: To get a lens space we need $g = 0$ and the theorem says that this is equivalent to $d = 1$. Then we can admit at most two exceptional leaves. Hence $k = 1$ and $m = 2$). The lens space is $L(2l, 1)$. On the other hand $\tilde{L} = L(1, 1) = S^3$ when l is even and $\tilde{L} = L(2, 1) = P^3(\mathbf{R})$ when l is odd.
3. If $3 \leq m$ or if $m = 2$ and $2 \leq k$ then at least one of the two following statements is true:

i) g is strictly positive

ii) L has strictly more than two exceptional leaves.

We describe the canonical plumbing graph in the next theorem. Its proof follows immediately from theorem 4.1 and from the recipes in [N].

Theorem 4.2. We assume that $\gcd(m, k, l) = 1$ and that $1 \leq k \leq l$. The boundary of the Milnor fiber of $f(x, y, z) = z^m - x^k y^l$ has the following plumbing graph :

1. If $k = l = 1$ the canonical plumbing graph is a bamboo of rational curves, having $(m - 1)$ vertices with Euler number equal to (-2). This is the singularity A_{m-1}.

Assume from now on that $2 \leq l$.

2. If $k = 1$ and $m = 2$ the plumbing graph has just one vertex with $g = 0$ and $e = -2l$.

3. Assume either that $3 \leq m$ or that $m = 2$ and $2 \leq k$. Then the canonical plumbing graph is never a bamboo of rational curves. More precisely:

3a. If $k = l$ the graph has just one vertex with $g = (m - 1)(d - 1)$ and $e = m$. Notice that g is strictly positive because $d = k = l > 1$.

3b. If k divides l but $k \neq l$ the graph is star-shaped with m branches. The central vertex has $g = (m - 1)(d - 1)$ and $e = 0$. Each branch has just one vertex (tied to the central vertex by an edge). Its weights are $g = 0$ and $e = -\frac{l}{k}$.

3c. If k does not divide l then the graph is star-shaped with $2m$ branches. The central vertex has $g = (m - 1)(d - 1)$ and $e = -m$.

There are m branches which are a bamboo of rational curves with $e_i' = -b_i'$ and b_i' defined by $b_i' \geq 2$ and

$$\frac{\alpha'}{\alpha' - \beta'} = [b_1', ..., b_u']$$

The vertex carrying the number 1 is joined to the central vertex by an edge.

There are also m branches which are a bamboo of rational curves with $e_i'' = -b_i''$ and b_i'' defined by $b_i'' \geq 2$ and

$$\frac{\alpha''}{\alpha'' - \beta''} = [b_1'', ..., b_v'']$$

Again, the vertex carrying the number 1 is joined to the central vertex by an edge.

Corollary 4.3. If L is not a lens space, it is never orientation preserving diffeomorphic to the boundary of a normal surface singularity.

Proof of corollary 4.3. L is not a lens space iff we are in case 3. We claim that the intersection form associated to the canonical plumbing graph is never negative definite. In cases 3a and 3b this is obvious since the self-intersection of the central vertex is ≥ 0.

Let us suppose that we are in case 3c. We compute the rational Euler number e_0 of the Seifert structure on L. By definition

$$e_0 = e - \sum \frac{\beta_i}{\alpha_i}$$

From theorem 4.1 we deduce that

$$e_0 = m - m\frac{\beta'}{\bar{l}} - m\frac{\beta''}{\bar{k}}$$

Hence:

$$\bar{k}\bar{l}e_0 = m(\bar{k}\bar{l} - \beta'\bar{k} - \beta''\bar{l})$$

We shall prove later in this section that $(\bar{k}\bar{l} - \beta'\bar{k} - \beta''\bar{l}) = 1$. See lemma 4.6.

Hence

$$e_0 = \frac{m}{\bar{k}\bar{l}} > 0$$

The conclusion follows from [N] Corollary 6 p.300. **End of proof of corollary 4.3.**

Proof of theorem 4.1. We shall compute the Seifert invariants from the data provided by the theorems proved in section 3.

We first determine the genus g. The Euler characteristic $\chi(G')$ is equal to $(-ml + m + l)$. The classical formula for ramified coverings implies that the Euler characteristic χ' of the quotient of G' by the action generated by h' is equal to $(-md + d + m)$. An analogous computation shows that $\chi'' = \chi'$. Hence the Euler characteristic χ of the base space of the Seifert foliation is equal to $2(-md + d + m)$ and we get $g = (m-1)(d-1)$.

The computation of the Seifert invariants (α, β) is routine if we use the dictionary which translates Nielsen invariants into Seifert's.

It is sufficient for us to consider the following special case. Suppose that the angle of rotation at a fixed point of a monodromy h of finite order acting on an oriented surface is equal to $\frac{\omega}{\lambda}2\pi$ with $gcd(\omega, \lambda) = 1$. Define σ as the integer which satisfies $0 < \sigma < \lambda$ and $\omega\sigma \equiv 1 \pmod{\lambda}$. In the mapping torus of h, the Seifert invariant (α, β) for the exceptional leaf which corresponds to the fixed point is given by $\alpha = \lambda$ and $\beta = \sigma$ in normalized form. See [M]. The result follows now immediately from theorem 3.3.

The delicate part of the proof is to determine the Euler number e. As we feel that this invariant is rather elusive, we prefer to deal with closed objects.

Let $\hat{G}'$ be the closed surface obtained from G' by attaching a disc on each of its $d_l = gcd(m, l)$ boundary components. We have seen (in the proof of Corollary 3.4) that the monodromy h' permutes them transitively. Let $\hat{h}'$ be "the" finite order extension of h' on $\hat{G}'$. There is exactly one orbit of $\hat{h}'$ which corresponds to the center of these discs. Its Nielsen invariant $\sigma/\bar{l}$ is given by

$$\frac{\sigma}{\bar{l}} \equiv -m\frac{\beta'}{\bar{l}} \quad in \quad \mathbf{Q} \bmod \mathbf{Z}$$

because the sum of all Nielsen quotients is equal to zero in $\mathbf{Q} \bmod \mathbf{Z}$ for a closed surface.

Let $\hat{M}'$ be the mapping torus of $\hat{h}'$ acting on $\hat{G}'$. It is a closed Seifert manifold. It has m exceptional leaves with Seifert invariant (α', β') and one with Seifert invariant $(\hat{\alpha}', \hat{\beta}')$ which we choose to be defined as

$$\frac{\hat{\beta}'}{\hat{\alpha}'} = -m\frac{\beta'}{\alpha'}$$

where $\hat{\beta}'$ and $\hat{\alpha}'$ are by necessity chosen to be relatively prime. This choice has the advantage that the Euler number $\hat{e}'$ for $\hat{M}'$ is equal to zero, because the rational Euler number for $\hat{M}'$ is equal to zero, as $\hat{M}'$ is the mapping torus of a finite order monodromy acting on a closed surface. See [P].

We proceed along the same path with G'' and h'' to get a closed Seifert manifold $\hat{M}''$ with analogously defined Seifert invariants.

We now state a lemma about glueings of Seifert manifolds. The statement is painful (sorry!).

Lemma 4.4. Let V' and V'' be two closed oriented Seifert manifolds. Let H_0' be a leaf in V' and let H_0'' be one in V''. Let N' be a foliated closed tubular neighborhood of H_0' in V' and let N'' be one for H_0'' in V''.

Let s' be a section in V' (as usual possibly outside some discs in the base space) giving rise to an Euler number e' for V' and a Seifert invariant

756

(a', b') for H'_0. In a similar manner, let s'' be a section in V'' giving rise to the Euler number e'' for V'' and to the Seifert invariant (a'', b'') for H''_0.

Let $\check{V}' = V' \setminus Int(N')$ and $\check{V}'' = V'' \setminus Int(N'')$. Let V be such that $V = \check{V}' \cup \check{V}''$ and $\check{V}' \cap \check{V}'' = \partial\check{V}' \cap \partial\check{V}''$. This intersection is a torus and we write T for it. Suppose that the leaves H' from V' and H'' from V'' coincide on T (hence V is Seifert foliated).

Let m' be a meridian for N' on T and let m'' be one for N''. Let $IN(m', m'')$ be the intersection number of m' and m'' on T, where T is oriented as the boundary of $\check{V}''$.

Then the Euler number e for V (corresponding to a section s essentially built from s' and s'') is given by the equality $e = e' + e'' + \bar{e}$ where $\bar{e}$ is computed from the equation

$$IN(m', m'') = a'b'' + a''b' + a'a''\bar{e}$$

Proof of lemma 4.4. As the section s is built from s' and s'' it follows from the definition of the Euler number as an obstruction (evaluated on a fundamental cycle) that e is the sum of e' and e'' plus a contribution coming from the fact that s' and s'' do not necessarily match along the torus T. The formula of theorem 3.5 will determine that contribution.

Following Seifert conventions we have

$$m' = a's' + b'H' \quad with \quad a' > 0 \quad and \quad m'' = a''s'' + b''H'' \quad with \quad a'' > 0$$

By hypothesis, we have $H' = H'' = H$. Let us choose an orientation (arbitrarily) for H. From Seifert conventions, this choice orients s' and s'' via $IN(s', H) = +1$ on T oriented as $\partial N'$ and $IN(s'', H) = +1$ on T oriented as $\partial N''$. This orients m' on $T = \partial N'$ via $a' > 0$ and m'' on $T = \partial N''$ via $a'' > 0$.

Notice that a change of orientation of H induces a change of orientation on both m' and m'' and hence the intersection number $IN(m', m'')$ does not change. Let us compute that intersection number.

$$IN(m', m'') = IN((a's' + b'H), (a''s'' + b''H))$$
$$= a'a''IN(s', s'') + a'b''IN(s', H) + a''b'IN(H, s'') + b'b''IN(H, H)$$

We have:

1) $IN(H, H) = 0$ because the intersection form is alternating.

2) $IN(s', H) = +1$ from Seifert conventions, because T is oriented as the boundary of $\check{V}''$ which is the same as being oriented as the boundary of N'.

3) $IN(H, s'') = +1$ because $IN(s'', H) = +1$ if T is oriented as the boundary of N'' and two sign changes occur from the last equality to get the first one.

4) $IN(s', s'') = \bar{e}$. To see that the sign is correct, one way to argue is to go back to the definition of Euler numbers. Another way is to remark that this is the good sign in order to be sure that the sum $e' + e'' + \bar{e}$ remains constant under changes of s' (or s'') near the fiber H'_0 (or H''_0).

End of proof of lemma 4.4.

We now use lemma 4.4 to complete the determination of e. To make the argument simpler let us assume that

$$(d_k = gcd(m, k) = 1 \ ; \ d_l = gcd(m, l) = 1 \ ; \ d = gcd(k, l) = 1)$$

Recall that in this case $\hat{M}'$ has m exceptional leaves with Seifert invariant $\alpha' = l$ and β' defined by $0 < \beta' < l$ and $(-k)\beta' \equiv 1 \pmod{l}$. $\hat{M}'$ has one more exceptional leaf with Seifert invariant $(\hat{\alpha}', \hat{\beta}')$ defined by

$$\frac{\hat{\beta}'}{\hat{\alpha}'} = -m\frac{\beta'}{l}$$

As $gcd(m, l) = 1$ we have that $\hat{\alpha}' = l$. We have already seen that $e' = 0$.

Similarly, $\hat{M}''$ has m exceptional leaves with invariant $\alpha'' = k$ and β'' defined by $0 < \beta'' < k$ and $(-l)\beta'' \equiv 1 \pmod{k}$. $\hat{M}''$ has one more exceptional leaf with invariant $(\hat{\alpha}'', \hat{\beta}'')$ defined by

$$\frac{\hat{\beta}''}{\hat{\alpha}''} = -m\frac{\beta''}{k}$$

We have $\hat{\alpha}'' = k$ because $gcd(m, k) = 1$ and $e'' = 0$.

As $gcd(m, l) = 1$ the boundary $\partial G'$ is connected and $\partial G''$ is connected because $gcd(m, k) = 1$. As a consequence, the intersection number $IN(\partial G', \partial G'')$ is equal to $IN(m', m'')$ UP TO SIGN.

Lemma 4.5. We have the equality $IN(m', m'') = -IN(\partial G', \partial G'')$.

Proof of lemma 4.5. The result comes from a comparison between the orientation of meridians coming from Seifert conventions and the orientation coming from $\partial G'$ (or $\partial G''$). What happens is that for one meridian both orientations agree and that for the other one they disagree. Which one it is depends on the orientation selected for H. **End of proof of lemma 4.5.**

We go on with the determination of the Euler number. The formula

$$IN(m', m'') = a'b'' + a''b' + a'a''\bar{e}$$

of lemma 4.4 translates into

$$m = l(-m\beta'') + k(-m\beta') + kl\bar{e}$$

Hence we have

$$(\dagger) \quad m(1 + l\beta'' + k\beta') = kl\bar{e}$$

Lemma 4.6. We have the equality: $(\star\star)$ $1 + l\beta'' + k\beta' = kl$.

From lemma 4.6 and formula $(\dagger)$ we deduce that $\bar{e} = m$ and hence that $e = m$ because $e' = 0 = e''$. This completes the computation of e.

Proof of lemma 4.6. By definition we have

$$l\beta'' \equiv -1 \pmod{k} \quad and \quad k\beta' \equiv -1 \pmod{l}$$

Because $gcd(k, l) = 1$ we deduce that

$$l\beta'' + k\beta' \equiv -1 \pmod{kl}$$

In other words there exists an integer q such that

$$1 + l\beta'' + k\beta' = qkl$$

As $0 < \beta' < l$ and $0 < \beta'' < k$ the only possibility is $q = 1$. **End of proof of lemma 4.6.**

By carefully dividing by adequate gcd's an analogous argument works without assuming that $(d_k = gcd(m, k) = 1$; $d_l = gcd(m, l) = 1$; $d = gcd(k, l) = 1)$. **End of proof of theorem 4.1.**

Comment. The referee has brought our attention to another way to compute the Seifert invariants. The boundary L is a m-fold cyclic cover of S^3 branched over the intersection of S^3 with $x^k y^l = t$. This proves directly that L is a Seifert manifold. Standard methods can be used to get the invariants, particularly the very useful "fonctoriality of the rational Euler number" of [NR] (thm 1.2.).

We have known this other way from the beginning in fact, we used it to check our computations. But we have given the preference to the point of view adopted here because we wanted to illustrate the general method of [MP] and because the ramified cover approach seems impossible to generalize beyond the singularities $z^m - g(x, y)$.

5. Examples

Example 1. Let us consider the Hirzebruch singularity $z^{12} - x^5 y^{11} = 0$

The boundary $\tilde{L}$ of the normalization is the lens space $L(12, 5)$. Its plumbing graph is a bamboo of three rational curves with Euler numbers successively $\{-3, -2, -3\}$.

The Seifert structure of the boundary L of the Milnor fiber is as follows: ($g = 0$ and $e = 12$). L has 24 exceptional leaves. There are 12 of them with Seifert invariant ($\alpha = 11$, $\beta = 2$) and 12 of them with Seifert invariant ($\alpha = 5$, $\beta = 4$).

The plumbing graph of L is star-shaped. The central vertex has weights $g = 0$ and $e = -12$. There are 24 bamboos of rational curves attached to the central vertex. Among them, 12 have Euler numbers equal successively to $\{-2, -2, -2, -2, -3\}$ and 12 of them have just one vertex with Euler number equal to $\{-5\}$.

Example 2. Let us consider the Hirzebruch singularity $z^{12} - x^{17} y^{11} = 0$. In order to make the comparison between examples 1 and 2 easier, we drop the restriction $k \leq l$.

The boundary $\tilde{L}$ of the normalization is the same as in example 1, because 5 is congruent to 17 (mod 12).

But the boundaries L of the Milnor fibers are different. In fact, the Seifert invariants for the exceptional leaves differ. L has 12 leaves with Seifert invariant ($\alpha = 11$, $\beta = 9$) and 12 leaves with Seifert invariant ($\alpha = 17$, $\beta = 3$).

The plumbing graph of L is again star-shaped, as it should be. The

760

central vertex has again weights $g = 0$ and $e = -12$. There are 24 bamboos of rational curves attached to the central vertex. Among them, 12 have Euler numbers equal successively to $\{-6, -2\}$ and 12 of them have Euler numbers successively equal to $\{-2, -2, -2, -2, -3, -2\}$.

References

B.   E. Brieskorn, *Singularities in the work of Friedrich Hirzebruch*. Surv. Diff. Geom. VII, Int. Press, Sommerville, MA (2000), 17-60.

BPV.   W. Barth, C. Peters, A. Van de Ven, *Compact Complex Surfaces*. Ergebnisse der Mathematik und ihrer Grenzgebiete **3**, Band 4, Springer Verlag (1984).

H.   F. Hirzebruch, *Über vierdimensionale Riemannsche Flächen mehrdeutiger analytischeer Funktionen von zwei Veränderlichen*. Math. Ann. **126** (1953), 1-22.

HNK.   F. Hirzebruch, W. D. Neumann, S. S. Koh, *Differentiable manifolds and quadratic forms*. Math. Lecture Notes, vol 4, Dekker, New-York (1972).

J.   H. Jung, *Darstellung der Funktionen eines algebraischen Körpers zweier unabhängigen Veränderlichen (x, y) in der Umgebung einer Stelle $(x - a, y - b)$*. Jour. reine u. angew. Mathematik **133** (1908), 289-314.

M.   J. Montesinos, *Classical tessellations and three-manifolds*. Universitext, Springer Verlag, Berlin (1987).

MP.   F. Michel, A. Pichon, *On the boundary of the Milnor fiber of non-isolated singularities*. IMRN **43** (2003), 2305-2311.

MPW.   F. Michel, A. Pichon, C. Weber, *The boundary of the Milnor fiber for some non-isolated germs of complex surfaces*. math.AG/0605123.

N.   W. D. Neumann, *A calculus for plumbing applied to the topology of complex surface singularities and degenerating complex curves*. Trans. AMS **268** (1981), 299-344.

NR.   W. D. Neumann, F. Raymond, *Seifert manifolds, plumbing, -invariant and orientation reversing maps*. Algebraic and geometric topology (Proc. Sympos., Univ. California, Santa Barbara, Calif., 1977), Lecture Notes in Math., 664, Springer, Berlin, 1978. pp. 163–196

P.   A. Pichon, *Fibrations sur le cercle et surfaces complexes*. Ann. Inst. Fourier (Grenoble) **51** (2001), 337-374.

A SURVEY ON STRATIFIED TRANSVERSALITY*

CLAUDIO MUROLO

*Centre de Mathématiques et Informatique, Université de Provence,
Laboratoire d'Analyse, Topologie et Probabilités UMR 6632
39, rue Joliot-Curie — 13453 — Marseille — France
murolo@cmi.univ-mrs.fr — http://www.cmi.univ-mrs.fr/˜murolo/*

In this paper we give a survey on transversality theorems for stratified spaces
which have appeared in the literature in the last 30 years having interest for
their geometric applications to geometric homology theories.

Keywords: Stratified sets and maps, Stratified transversality, Homology

1. Introduction

We recall that a *stratification* of a topological space A is a locally finite partition Σ of A into C^1 connected manifolds (called the *strata* of Σ) satisfying the *frontier condition* : if X and Y are disjoint strata such that X intersects the closure of Y, then X is contained in the closure of Y. We write then $X < Y$ and $\partial Y = \overline{Y} - Y$ so that $\overline{Y} = Y \sqcup \left(\sqcup_{X<Y} X \right)$ and $\partial Y = \sqcup_{X<Y} X$ ($\sqcup$ = disjoint union).

The pair $\mathcal{X} = (A, \Sigma)$ is called a *stratified space* (or *stratified object*) with *support* A and *stratification* Σ. The union of the strata of dimension $\leq k$ is called the *k-skeleton*, denoted by A_k, inducing a stratified space $\mathcal{X}_k = (A_k, \Sigma_{|A_k})$. A substratified space (or *substratified object* also denoted S.S.O.) of $\mathcal{X}$ is a stratified space $\mathcal{W} = (W, \Sigma_{\mathcal{W}})$, where W is a subset of A, such that each stratum in $\Sigma_{\mathcal{W}}$ is contained in a single stratum of $\mathcal{X}$.

A *stratified map* $f : \mathcal{X} \to \mathcal{X}'$ between stratified spaces $\mathcal{X} = (A, \Sigma)$ and $\mathcal{X}' = (B, \Sigma')$ is a continuous map $f : A \to B$ which sends each stratum X of $\mathcal{X}$ into a unique stratum X' of $\mathcal{X}'$, such that the restriction $f_X : X \to X'$ is smooth. We call such a map f a *stratified homeomorphism* if f is a global homeomorphism and each f_X is a diffeomorphism.

---

* A Zia Carmela

762

A stratified vector field on $\mathcal{X}$ is a family $\zeta = \{\zeta_X\}_{X \in \Sigma}$ of vector fields, such that ζ_X is a smooth vector field on the stratum X.

Extra conditions may be imposed on the stratification Σ, such as to be an *abstract stratified set* in the sense of Thom–Mather [12, 21, 22] or, when A is a subset of a C^1 manifold, to satisfy conditions (a) or (b) of Whitney [21, 22, 42], or (c) of K. Bekka [3] or, when A is a subset of a C^2 manifold, to satisfy conditions (w) of Kuo–Verdier [41], or (L) of Mostowski [38].

We send the reader to the original papers and to the above references for their definitions and main properties.

Let $\mathcal{X} = (A, \Sigma)$ be a stratified space. For stratified transversality we mean the problem of deforming a substratified object $\mathcal{W}$ of $\mathcal{X}$ via a stratified isotopy $\Phi : \mathcal{X} \times I \to \mathcal{X}$ to a substratified object $\mathcal{W}' = \Phi_1(\mathcal{W})$ of $\mathcal{X}$ which is transverse to $\mathcal{V}$ in $\mathcal{X}$, or more generally with respect to a fixed stratified map $g : \mathcal{Y} \to \mathcal{X}$, for some stratified space $\mathcal{Y}$.

This problem was solved by Clint McCrory for stratified polyhedra [23, 24] ; his result is essential to the foundations of intersection homology [16].

For Whitney (b)-regular stratifications Mark Goresky gave a transversality theorem ([15], 5.3) valid only for π-*fibre* substratified objects $\mathcal{W}$, and *controlled maps* $g : \mathcal{Y} \to \mathcal{X}$ [21, 22]. These two properties mean respectively that the support W of $\mathcal{W}$ is locally, near each point x of A, a union of fibres of a projection $\pi_S : T_S \to S$ where S is the stratum of $\mathcal{X}$ containing x and a similar property at the level of the fibre of g. This result is essential in proving the main theorems of Goresky about representing the cohomology by stratified objects ([15] 4.7 and 6.2).

More recently in [31] and [32], A. du Plessis, D. Trotman and myself gave two different proofs that *"after stratified isotopy of $\mathcal{X}$, a stratified subspace $\mathcal{W}$ of $\mathcal{X}$, or a stratified map $h : \mathcal{Z} \to \mathcal{X}$, can be made transverse to a fixed stratified map $g : \mathcal{Y} \to \mathcal{X}$"* : the second (historically the first, [29] 1997) using time-dependent vector field techniques and the family of geodesics introduced by Mather [20] and the first (historically the second) by adapting some ideas indicated at the end of Goresky's thesis ([14], 1976) and in the 1987 book [17]. The authors of [31, 32] obtained a generalisation of Goresky's theorem, with less restrictive hypotheses and which applies to all abstract stratified sets and (w)-regular nice stratifications, hence for any (b)-, (c)- or (L)-regular nice stratification, and which allows one to develop further Goresky's geometric homology theory [29]. This stratified transversality theorem holds for the most important types of regular stratifications, and for every stratified map without assuming control conditions.

In particular, we obtain an analogue of Goresky's theorem for stratified maps $g : \mathcal{Y} \to \mathcal{X}$ which are not necessarily controlled and for substratified objects $\mathcal{W}$ which are not necessarily π-fibre.

The analogous theorem for two stratified maps was also obtained.

We present moreover various applications of Goresky's transversality theorem and of its generalisations in [31, 32], pointing out some related problems which are still open.

I thank David Trotman who suggested I write this survey.

2. The PL-Transversality Theorem

In the context of PL-stratified spaces, the problem of putting in transverse position two substratified polyedra of a stratified polyedron $\mathcal{X}$ was solved in 1977 by Clint McCrory [24]. This theorem states :

Theorem 2.1. *Let $\mathcal{X}$ be a stratified polyedron and A, B, C closed subpolye-dra with $B \supseteq C$. There exists a PL isotopy $H : \mathcal{X} \times I \to I$ such that :*
  i) $|H_t(x) - x| < \epsilon$ for all x and $t \in I$;
  ii) $H_t(x) = x$ for all $x \in C$ and all $t \in I$;
  iii) A and $H_1(B - C)$ are in general position in $\mathcal{X}$.

Such a theorem, which uses a simplicial technique of Zeeman, was first proved without the property ii) by McCrory in his Ph.D. thesis ([23] p. 98, 1972) and previously again without the property i) in Akin's Ph.D. thesis ([2] p. 471, 1969).

Recall historically that in 1895 and 1899, in his famous papers *Analysis Situs* and *Complément à l'Analysis Situs* which founded modern algebraic topology, H. Poincaré studied the intersection of an i-cycle and a j-cycle in a compact oriented n-manifold $\mathcal{X}$ in the case of complementary dimension ($i + j = n$) then, in 1926, the theory was extended by S. Lefschetz to arbitray i and j. Fifty years later in 1980, in their celebrated paper [16] Mark Goresky and Robert MacPherson introduced *Intersection Homology Theory* in which they generalize to a class of singular spaces, the PL-pseudomanifolds, the Poincaré–Lefschetz cup product pairing $\cap : H_i(X) \times H_j(X) \to H_{i+j-n}(X)$.

In this context, McCrory's PL-transversality theorem (in a relative version), played a fundamental role in defining the Goresky–MacPherson inter-section pairing which extends the Poincaré–Leschetz map and was in this way essential to the foundation of intersection homology theory.

3. Transversality for Whitney stratifications

After the years 1965–70, during which H. Whitney laid the foundations of Whitney stratification theory [42] and R. Thom and J. Mather ([39] and

[21, 22]) those of *abstract stratified sets* as the larger class of desirable singular spaces, M. Goresky was preparing his Ph.D. thesis directed by R. MacPherson. The goal of his thesis, *Geometric Cohomology and Homology of Stratified Objects* [14], was to introduce new geometric homology and cohomology theories, for a stratified space $\mathcal{X}$, in which cycles and cocycles of $\mathcal{X}$ could be *represented* by substratified spaces with the same type of singularity as $\mathcal{X}$. In this theory a well adapted stratified transversality theorem would allow to find the geometric cap product as intersection of a cycle and a cocycle in transverse position and many other very nice geometric interpretations of the algebraic operations (see also §4).

We will talk later on about the stratified transversality results, statements and techniques which one can find in the thesis of Goresky, an exciting source of interesting results and nice and useful ideas.

We will present first the geometric homology theories published in 1981 in *Whitney Stratified Chains and Cochains* [15], in a revised version with respect to the thesis of 1976, and the stratified transversality theorem underlying this revised theory of 1981.

In this 1981 paper, Goresky re-defines his geometric homology and cohomology theories only for Whitney (b)-regular stratifications (and not for abstract stratified sets) and gives a new completely revised version of the previous stratified transversality statements and proofs.

For $\mathcal{X} = (A, \Sigma)$ a (b)-regular stratified space of support $A \subseteq \mathbb{R}^n$, Goresky introduces the *homology and cohomology sets* $WH_k(\mathcal{X})$ and $WH^k(\mathcal{X})$ (called *Whitney homology and cohomology theory* from [27]): the elements of $WH_k(\mathcal{X})$ and $WH^k(\mathcal{X})$ are equivalence classes of Whitney substratified k-cycles and k-cocycles of $\mathcal{X}$ with respect to a *Whitney stratified cobordism*.

A *Whitney stratified k-cycle* $\xi = (V, z)$ of $\mathcal{X}$ is a compact (b)-regular k-substratified object V of $\mathcal{X}$ together with an *orientation of V*, that is an element $z = \sum_j n_j V_j^k$ of the free abelian group $C_k(V)$ on $\mathbb{Z}$ on the oriented k-strata $\{V_j^k\}_j$ of V whose *boundary* is 0.

A cobordism between two stratified k-cycles ξ and ξ' is defined as a (b)-regular $(k+1)$-S.S.O. ζ of $\mathcal{X} \times I$ (I stratified by $\{\{0\},]0, 1[, \{1\}\}$) with boundary $\partial\zeta = \xi \times \{0\} - \xi' \times \{1\}$ (modulo *reduction*[a]).

A *substratified object W of $\mathcal{X}$* satisfies the π-*fibre condition* (with respect

---

[a]The reduction of a chain or of a cochain $\xi = (V, z)$ with $z = \sum_j n_j V_j^k$ is the chain or the cochain $\xi_/ = (V_{/z}, z)$ where $V_{/z} = \overline{\cup_{n_j \neq 0} V_j^k} = \cup_{n_j \neq 0} \overline{V_j^k}$. Every chain and cochain is identified with its reduction.

to a fixed system of control data of $\mathcal{X}$, $\mathcal{F} = \{(\pi_S, \rho_S, T_S) | S \text{ stratum of } \mathcal{X}\})$ if there exists an $\epsilon > 0$ such that for every stratum S of $\mathcal{X}$ one has : $\pi_S^{-1}(W) \cap T_S(\epsilon) = W \cap T_S(\epsilon)$. This condition allows Goresky to define for each k-costratum D (i.e. a connected component of the union of all the strata embedded in $\mathcal{X}$ with codimension k) a tubular neighbourood T_D of D in A. Thinking of T_D as a normal fiber bundle of D in $\mathcal{X}$, an orientation of its unit sphere bundle defines a *coorientation* of D.

A *Whitney stratified k-cocycle* $\theta = (\mathcal{W}, c)$ of $\mathcal{X}$ is then defined as a π-fibre Whitney substratified object $\mathcal{W}$ of $\mathcal{X}$, embedded in $\mathcal{X}$ with codimension k together with a *k-coorientation of $\mathcal{W}$*, that is an element $c = \sum_s n_s D_s^k$ of the free abelian group $C^k(\mathcal{W})$ on the oriented k-costrata $\{D_s^k\}_s$ of $\mathcal{W}$, and whose *coboundary* is 0. A *cobordism* between two stratified k-cocycles θ and θ' is defined as a (b)-regular $(k+1)$-π-fibre S.S.O. θ of $\mathcal{X} \times [0,1]$ with boundary $\delta\theta = \theta \times \{0\} - \theta' \times \{1\}$ (modulo *reduction*(*)).

The fundamental reason for which such homology and cohomology sets exist is that every (b)-regular S.S.O. $\mathcal{V}$ admits a system of control data [21]. This again works by considering for $\mathcal{X}$ and $\mathcal{V}$ abstract S.O. (see [14] and [29, Chapter IV, p. 134] for details).

To simplify the notations, we will omit the orientation z of a cycle ξ and the coorientation c of a cocycle θ and will write $\mathcal{V}$ or V (the support of $\mathcal{V}$) for ξ and $\mathcal{W}$ or W (support of $\mathcal{W}$) for θ.

Goresky introduced the two *homology and cohomology representation maps*

$$R_k : WH_k(\mathcal{X}) \to H_k(A) \quad \text{and} \quad R^k : WH^k(\mathcal{X}) \to H^k(A)$$

analogues of the Thom–Steenrod map between the differential bordism and the singular homology of $\mathcal{X} = (A, \Sigma)$.

The stratified transversality theorem underlying this theory is Goresky's *"Transversality Lemma"* (5.3 [15]) that we include with its original proof and notations:

Theorem 3.1. *Suppose $\mathcal{X}_1$ and $\mathcal{X}_2$ are Whitney stratified subsets of two manifolds M_1 and M_2 (respectively). Fix a system of control data on $\mathcal{X}_1$ and $\mathcal{X}_2$ and let $f : \mathcal{X}_1 \to \mathcal{X}_2$ be a stratified map. Suppose $\mathcal{Y} \subseteq \mathcal{X}_2$ is a geometric cocycle. Suppose either (a) f is controlled or (b) f is the restriction of a smoth map $\tilde{f} : M_1 \to M_2$. Then $\mathcal{Y}$ is cobordant to a cocycle $\mathcal{Y}' \subseteq \mathcal{X}_2$ such that f is transverse to $\mathcal{Y}'$.*

Proof. Assume by induction on k that $\mathcal{Y}$ is cobordant to a cocycle $\mathcal{Y}_k$ such that f is transverse to $\mathcal{Y}_k \cap (\mathcal{X}_2)_k$ where $(\mathcal{X}_2)_k$ denotes the k-skeleton of

$\mathcal{X}_2$. We will find a controlled vector field η on $\mathcal{X}_2$ with (controlled) flow $F_t : \mathcal{X}_2 \to \mathcal{X}_2$ such that $\mathcal{Y}_{k+1} = F_1(\mathcal{Y}_k)$ satisfies the induction hypothesis.

Under either assumption (a) or (b) above there is a neigbourhood U of $(\mathcal{X}_2)_k$ such that f is transverse to $\mathcal{Y}_k \cap U$.

Let $S = (\mathcal{X}_2)_k - (\mathcal{X}_2)_{k-1}$, by Sard's Theorem [13] there is a controlled vector field η_S on S with time 1 flow $F_S : S \to S$ such that η_S vanishes near $(\mathcal{X}_2)_k$ and such that f is transverse to $F_S(\mathcal{Y}_k)$. Take η to be any controlled lift [12], [21] of η_S. $\qquad\qquad\qquad\qquad\qquad\qquad\qquad\square$

This proof was for a long time and until 2000, not understandable to me (and to my knowledge to various other mathematicians). Thus I talk in the introduction of my Ph.D. thesis [29] 1997, of a *"mistake in the Goresky proof . . ."*. But after my joint work with D. Trotman and A. du Plessis [31] I no longer think there is a mistake.

The main reason for which this proof was obscure to me was the fact that if we would *first* obtain a transversalizing map $F_S : S \to S$ such that $\mathcal{Y}_{k+1} = F_S(\mathcal{Y}_k)$ is transverse to f, then *after* there is no way to replace the choice of F_S in some open dense set of diffeomorphisms to obtain that *(*): "F_S is also the time 1 flow F_1 of a vector field (not depending on time)"*.

This impossibility comes from a theorem of C. Freifeld [10] who remarked first a phenomenon better explained later by J. Milnor [25] 1980 that *"in the infinite dimensional space $Diff(S, S)$ the maps having the property (*) above, i.e. lying in a one parameter group of diffeomorphisms of S, do not fill a neighbourhood of the identity 1_S"* (more about this difficulty in the introduction of [29]).

On the other hand one could say : in the aim of Goresky the map F_S has to be obtained *at the same time as* the vector field η_S of which F_S is the time 1 flow.

But then how to do it ? In this sense the reference given by Goresky [13] on Sard's theorem is really not clear and very far from orienting the reader to an understandable continuation of the proof. Also remark that for a vector field η to have a flow defined $\forall t \in I$ is equivalent to asking that its flow is defined for every $t \in \mathbb{R}$ (i.e. that η is complete).

Such difficulties, motivated me jointly with A. du Plessis and D. Trotman in 2001 and 2005 [31, 32] to find two new and different generalisations of this stratified transversality theorem : the first by considering (for the η_S) time-dependent vector fields (whose flows are not necessarily one parameter groups !) and the second by putting together some methods sketched in the appendix of Goresky's thesis [14] and in his 1987 book [17] which uses a very fine idea of Abraham [1] to apply Sard's theorem on the space of vector fields on a manifold S and about which we will come back later.

After having established the transversality theorem underlying his whole theory Goresky deduces first of all the important theorem which gives the bijectivity of the cohomological representation.

Theorem 3.2. *For every Whitney stratification $\mathcal{X}$, the cohomology representation map $R^k : WH^k(\mathcal{X}) \to H^k(\mathcal{X})$ is a set bijection.*

The corresponding homology theorem below does not follow from the transversality theorem, however the proof (of the relative to the boundary version) was important in proving Theorem 3.2 above so we like to recall it underlining that it was proved by Goresky only for $\mathcal{X}$ the trivial stratification of a manifold.

Theorem 3.3. *If $\mathcal{X} = \{M\}$ is a trivial stratification of a manifold possibly with boundary, the homology representation map $R_k : WH_k(\mathcal{X}) \to H_k(M)$ is a set bijection.*

The same statement for $\mathcal{X}$ an arbitrary (b)-regular stratification :

Conjecture 3.1. *If $\mathcal{X}$ is a Whitney stratification the homology representation map $R_k : WH_k(\mathcal{X}) \to H_k(M)$ is a set bijection.*

remains a famous problem of Goresky (thesis [14] 1976 and [15] 1981) still unsolved. On the other hand, one easily sees that, if one proves the celebrated conjecture :

Conjecture 3.2. *Every Whitney stratification $\mathcal{X}$ admits a Whitney triangulation.*

this will also give a solution of the above conjecture 3.1 of Goresky.

Then, using also the following proposition :

Proposition 3.1. *Suppose $f : \mathcal{X}_1 \to \mathcal{X}_2$ is a controlled stratified map and $\mathcal{Y}$ is a codimension k geometric cocycle in $\mathcal{X}_2$ such that f is transverse to $\mathcal{Y}$. Then $f^{-1}(\mathcal{Y})$ is a π-fibre subset of $\mathcal{X}_1$ which admits a canonical Whitney stratification. The coorientation of $\mathcal{Y}$ pulls back to a coorientation on $f^{-1}(\mathcal{Y})$ which then becomes a geometric cocycle $f^{-1}(\mathcal{Y})$ and it represents the cohomology class $f^*([\mathcal{Y}])$ in $H^k(\mathcal{X}_1)$.*

Goresky's transversality theorem, again makes richer this geometric theory. It also allows Goresky to prove first the following theorem on the cohomology cup product :

Theorem 3.4. *Suppose $\mathcal{Y}_1$ and $\mathcal{Y}_2$ are geometric cocycles in a Whitney object $\mathcal{X}$. Then $\mathcal{Y}_2$ is cobordant to a cocycle $\mathcal{Y}'_2$ which is transverse to $\mathcal{Y}_1$. In this case $\mathcal{Y}_1 \cap \mathcal{Y}'_2$ is a geometric cocycle with the product coorientation and $[\mathcal{Y}_1 \cap \mathcal{Y}'_2] = [\mathcal{Y}_1] \cup [\mathcal{Y}'_2]$.*

and then the proposition stating that also the cap product has, thanks to the transversality theorem, a nice geometric meaning :

Proposition 3.2. *Suppose $\mathcal{Y}$ is a geometric k-cocycle in $\mathcal{X}$ and $\mathcal{Z}$ is a geometric p-cycle. Then $\mathcal{Y}$ is cobordant to a cocycle $\mathcal{Y}'$ which is transverse to $\mathcal{Z}$. Using the product orientation on $(p-k)$-strata of $\mathcal{Y}' \cap Z$ (which all have the form (p-costratum of $\mathcal{Y}$) $\cap$ (k-stratum of $\mathcal{Z}$)), $\mathcal{Y}' \cap \mathcal{Z}$ becomes a geometric cycle and it represents the cap product $[\mathcal{Y}] \cap [\mathcal{Z}] \in H_{p-k}(\mathcal{X})$.*

4. Further geometric applications of the Goresky Theorem

Using again the transversality theorem of Goresky, in 1994 [27], I improved the Goresky theories by introducing in the homology and cohomology *sets* $WH_*(\mathcal{X})$ and $WH^*(\mathcal{X})$ a geometric sum operation which geometrically means *transverse union* of cycle and/or of cocycles below denoted by $\cup_t$ (see §5.2 for the rigorous definitions of $\cup_t$ and $\cap_t$). This was done in the same spirit as for the *Moving Lemma* in the Chow Group theory for the algebraic cycles of an algebraic manifold [11] and in such a way that the sets $WH_k(\mathcal{X})$ and $WH^k(\mathcal{X})$ become abelian groups and the representation maps R_k and R^k group isomorphisms.

Again in the homology case the full theorem was obtained only when $\mathcal{X} = \{M\}$ reduces to a smooth manifold.

In [27] I complete the Goresky theories with a slight algebraization and by introducing the coefficients in an abelian group G and in [27, 28] showing that the most important cohomology operations, *the Steenrod squares* and *the Steenrod p-powers* (p an odd prime) can be realized through a geometric construction based on transversality methods. Then starting from [27, 28] I refer to WH_* and WH^* as *Whitney Homology and Cohomology theories*.

All geometric applications of the Goresky transversality theorem (rewriting more nicely his theorems on the cup and cap product), together with the improvements in [27, 28] can be summarized as follows :

Theorem 4.1. *If $\mathcal{X}$ is a smooth manifold, the sum operation defined in $WH_k(\mathcal{X})$ by :*

$$+ \quad : \quad WH_k(\mathcal{X}) \times WH_k(\mathcal{X}) \to WH_k(\mathcal{X}) \quad , \quad [\mathcal{V}_1] + [\mathcal{V}_2] = [\mathcal{V}_1 \cup_t \mathcal{V}'_2]$$

where $\mathcal{V}'_2$ is a cycle cobordant to $\mathcal{V}_2$ and transverse to $\mathcal{V}_1$ (which exists by the transversality theorem) is a well defined group operation in $WH_k(\mathcal{X})$ for which the Goresky homology representation map $R_k : WH_k(\mathcal{X}) \to H_k(\mathcal{X})$ is a group isomorphism.

Theorem 4.2. *For every Whitney object $\mathcal{X}$ the sum operation defined in $WH^k(\mathcal{X})$ by:*

$$+ \quad : \quad WH^k(\mathcal{X}) \times WH^k(\mathcal{X}) \to WH^k(\mathcal{X}) \quad , \quad [\mathcal{Y}_1] + [\mathcal{Y}_2] = [\mathcal{Y}_1 \cup_t \mathcal{Y}'_2]$$

where $\mathcal{Y}'_2$ is a coycle cobordant to $\mathcal{Y}_2$ and transverse to $\mathcal{Y}_1$ (which exists by the transversality theorem) is a well defined group operation in $WH^k(\mathcal{X})$ and the Goresky cohomology representation map $R^k : WH^k(\mathcal{X}) \to H^k(\mathcal{X})$ is a group isomorphism.

Theorem 4.3. *For every stratified controlled map $f : \mathcal{X}_1 \to \mathcal{X}_2$ between Whitney stratifications there is an induced map f^* in Whitney cohomology defined by :*

$$f^* : WH^k(\mathcal{X}_2) \to WH^k(\mathcal{X}_1) \quad , \quad f^*([\mathcal{Y}]) = [f^{-1}(\mathcal{Y}')]$$

where $\mathcal{Y}'$ is cobordant to $\mathcal{Y}$ and transverse to f, and exists by the transversality theorem (f^* is given by the transverse preimage of a geometric cocycle). Moreover with respect to the transverse sum in $WH^k(\mathcal{X}_2)$ and $WH^k(\mathcal{X}_1)$, $f^* : WH^k(\mathcal{X}_2) \to WH^k(\mathcal{X}_1)$ is a group homomorphism.

Theorem 4.4. *In the geometric cohomology theory WH^* the cup product is defined by*

$$\cup \quad : \quad WH^*(\mathcal{X}) \times WH^*(\mathcal{X}) \to WH^*(\mathcal{X}) \quad , \quad [\mathcal{Y}_1] \cup [\mathcal{Y}_2] = [\mathcal{Y}_1 \cap_t \mathcal{Y}'_2]$$

where $\mathcal{Y}'_2$ is cobordant to $\mathcal{Y}_2$ and transverse to $\mathcal{Y}_1$ and exists by the transversality theorem. I.e. the cup product is given by the transverse intersection of two geometric cocycles.

Proposition 4.1. *In the geometric theory WH_*, WH^* the cap product is defined by*

$$\cap \quad : \quad WH_{p+k}(\mathcal{X}) \times WH^k(\mathcal{X}) \to WH_p(\mathcal{X}) \quad , \quad [\mathcal{V}] \cap [\mathcal{Y}] = [\mathcal{V} \cap_t \mathcal{Y}']$$

where $\mathcal{Y}'$ is cobordant to $\mathcal{Y}$ and transverse to $\mathcal{V}$ and exists by the transversality theorem. I.e. the cap product is the transverse intersection of a cocycle with a cycle.

Finally, although such a nice geometric interpretation of the cross product [15] and of Poincaré Duality [29] does not come from transversality we like to recall it :

Proposition 4.2. *In cohomology WH^* the cross product of cocycles is defined by*

$$\times \quad : \quad WH^k(\mathcal{X}) \times WH^h(\mathcal{Y}) \to WH^{k+h}(\mathcal{X} \times \mathcal{Y}) \quad , \quad [\mathcal{Y}_1] \times [\mathcal{Y}_2] = [\mathcal{Y}_1 \times \mathcal{Y}_2] .$$

I.e. the cross product is the cartesian product of two cocycles of $\mathcal{X}$ and $\mathcal{Y}$.

Proposition 4.3. *If $\mathcal{X}$ is a n-manifold the (inverse map of the) Poincaré Duality isomorphism $\mathcal{D}$ is the "identity" on the representative cocycles :*

$$\mathcal{D} : WH^k(\mathcal{X}) \to WH_{n-k}(\mathcal{X}) \quad , \quad \mathcal{D}([\mathcal{V}]) = [\mathcal{V}] .$$

I.e. it is obtained by interpreting each k-cocycle as a $(n-k)$-cycle of $\mathcal{X}$.

This nice geometric interpretation exists still when $\mathcal{X}$ is not a manifold but just a pseudomanifold (because in this case the the fundamental class $[\mathcal{X}]$ is a cycle of $\mathcal{X}$).

Finally, in Whitney cohomology there exists a geometric realisation of the main cohomology operations, the Steenrod Squares $\{Sq^\alpha\}_\alpha$ [27] and the Steenrod p-powers $\{P^\alpha\}_\alpha$ (p prime odd) [28] based on stratified transversality methods.

Theorem 4.5. *Let $A : \mathcal{X}^2 \times S^h \to \mathcal{X}^2 \times S^h$ be the $\mathbb{Z}_2$-action defined by $A(x,y,t) = (y,x,-t))$, φ the map $\varphi([W]) = [(W^2 \times S^h)/\mathbb{Z}_2]$, Δ the map $\Delta(x,[t]) = [((x,x),t)]$ and for every α, $pr_{k+\alpha}$ the Gysin homomorphism.*
   Then independently from $h \geq k$, the composition map:

$$Sq^\alpha : \quad WH^k(\mathcal{X}) \overset{\varphi}{\to} WH^{2k}((\mathcal{X}^2 \times S^h)/\mathbb{Z}_2) \overset{\Delta^*}{\to} WH^{2k}(\mathcal{X} \times \mathbb{P}^h) \cong$$

$$\cong \sum_{i+j=2k} WH^i(\mathcal{X}) \otimes WH^j(\mathbb{P}^h) \overset{pr_{k+\alpha}}{\to} WH^{k+\alpha}(\mathcal{X})$$

is a geometric construction of the Steenrod squares $\{Sq^\alpha : WH^k(\mathcal{X}) \to WH^{k+\alpha}(\mathcal{X})\}_{X,\alpha,k}$ in Whitney cohomology. I.e. we have :
   1) Sq^α is a group homomorphism ;
   2) $f : \mathcal{X} \to \mathcal{Y}$ is a controlled map $\quad \Rightarrow \quad Sq^\alpha f^* = f^* Sq^\alpha$;
   3) $k = \alpha \Rightarrow Sq^\alpha([W]) = [W] \cup [W]$ is the cup product ;
   4) $Sq^0 = 1_{WH^k(X)}$ is the identity map ;
   5) $Sq^i([W] \times [W']) = \sum_{\alpha+\beta=i} Sq^\alpha([W]) \times Sq^\beta([W'])$;
   6) $\alpha > k \Rightarrow Sq^\alpha([V]) = 0$.

Theorem 4.6. *Let $p \in \mathbb{N}$ be an odd prime, $h \in \mathbb{N}$, odd so $S^h \subseteq \mathbb{C}^r$, $r = (h+1)/2$ and $L_p = S^h/\mathbb{Z}_p$ be the Lens space quotient of S^h with respect to the multiplication for $e^{i\frac{2\pi}{p}}$ in $\mathbb{C}^r$.*

Let $A : \mathcal{X}^p \times S^h \to \mathcal{X}^p \times S^h$ be the $\mathbb{Z}_p$-action defined by $A((x_1,\ldots,x_p),z) = ((x_p,x_1\ldots,x_{p-1}),e^{i\frac{2\pi}{p}}z)$, φ the map defined by $\varphi([W]) = [(W^p \times S^h)/\mathbb{Z}_p]$, Δ the map $\Delta(x,[z]) = [((x,\ldots,x),z)]$ and for every $\alpha' = k + 2\alpha(p-1)$, $pr_{\alpha'}$ the Gysin homomorphism.

Then, independently from $h \geq (k - 2\alpha)(p-1)$, the composition map:

$$P^\alpha : \quad WH^k(\mathcal{X}) \quad \overset{\varphi}{\to} \quad WH^{kp}((\mathcal{X}^p \times S^h)/\mathbb{Z}_p) \quad \overset{\Delta^*}{\to} \quad WH^{kp}(\mathcal{X} \times L_p^h) \cong$$

$$\cong \sum_{i+j=kp} WH^i(\mathcal{X}) \otimes WH^j(L_p^h) \overset{pr_{\alpha'}}{\to} WH^{\alpha'}(\mathcal{X})$$

is a geometric construction of the Steenrod p-powers $\{P^\alpha : WH^k(\mathcal{X}) \to WH^{\alpha'}(\mathcal{X})\}_{X,\alpha,k}$ in Whitney cohomology. I.e. we have :

1) P^α is a group homomorphism ;

2) $f : \mathcal{X} \to \mathcal{Y}$ is a controlled map $\Rightarrow P^\alpha f^* = f^* P^\alpha$;

3) $k = 2\alpha \Rightarrow P^\alpha([W]) = [W] \cup [W] \cup \cdots \cup [W]$ is the cup product k times of $[W]$;

4) $P^0 = 1_{WH^k(X)}$ is the identity map ;

5) $P^i([W] \times [W']) = \sum_{\alpha+\beta=i} P^\alpha([W]) \times P^\beta([W'])$;

6) $2\alpha > k \Rightarrow P^\alpha([W]) = 0.$

5. Improvement of the Goresky Theorem. Applications. Open problems

Goresky's transversality theorem applies to those substratified objects W of $\mathcal{X}$ satisfying a π-*fibre condition* with respect to a fixed system of control data $\mathcal{F} = \{(\pi_X, \rho_X) : T_X \to X \times [0, \infty)\}_{X \in \Sigma}$ of $\mathcal{X}$, and to a stratified map $g : \mathcal{Y} \to \mathcal{X}$ which is controlled with respect to two systems of control data.

The π-fibre condition says that W is locally, near each point x of A, a union of fibres of the projection $\pi_S : T_S \to S$ where S is the stratum containing x, while the control condition on the map g imposes a similar property for the fibres of g (and of π_S). These conditions were used by Goresky to preserve transversality with respect to g of a deformation W' of W in his inductive proof. As explained above, Goresky's transversality theorem has been shown to be very useful in several important applications [15], [27, 28]; but the hypotheses of π-fibre on W and control on g prevent a wider use.

Definition 5.1. Let $\mathcal{X} = (A, \Sigma)$ be a stratified space. A *stratified isotopy* of $\mathcal{X}$ (or of A) $\Phi : A \times I \to A$ (denoted also $\{ \Phi_t : A \to A \}_{t \in I}$) is a stratified map such that for every $t \in I$, the map *at time t*, $\Phi_t : A \to A$ is a stratified homeomorphism.

Clearly, if $\{\Phi_t\}_{t \in I}$ and $\{\Psi_t\}_{t \in I}$ are stratified isotopies, so is $\{\Psi_t \circ \Phi_t\}_t$.

Definition 5.2. Let $\mathcal{W} = (W, \Sigma_{\mathcal{W}})$ and $\mathcal{W}' = (W', \Sigma_{\mathcal{W}'})$ be two S.S.O. of a stratified space $\mathcal{X} = (A, \Sigma)$. We say that $\mathcal{W}'$ *is a deformation by isotopy of $\mathcal{W}$ in A* if there exists a stratified isotopy $\Phi : A \times I \to A$ such that $\Phi_0 = 1_A$ and $\mathcal{W}' = \Phi_1(\mathcal{W})$.

If $\Phi : A \times I \to A$ is a stratified isotopy of $\mathcal{X} = (A, \Sigma)$ and $\mathcal{W}$ is a substratified object of $\mathcal{X}$, then for each $t \in I$ the image $\mathcal{W}' = \Phi_t(\mathcal{W})$ is a substratified object with stratification induced by Φ_t and $\mathcal{W}'$ is a deformation by isotopy of $\mathcal{W}$.

Let $h, h' : \mathcal{Y} \to \mathcal{X}$ be two stratified maps. We say that *h' is a deformation by isotopy of h in $\mathcal{X}$* if there exists a stratified isotopy $\Phi : \mathcal{X} \times I \to \mathcal{X}$ such that $\Phi_0 = 1_{\mathcal{X}}$ and $h' = \Phi_1 \circ h$, i.e. h' is the deformation via Φ and at time $t = 1$ of h.

"Deformation by isotopy" of S.S.O. of $\mathcal{X}$ and of maps $h, : \mathcal{Y} \to \mathcal{X}$ define clearly equivalence relations.

In two recent papers [31, 32], A. du Plessis, D. Trotman and myself, gave two different proofs of the stratified transversality theorem below, of which we recall here the ideas of the proofs.

Theorem 5.1. *Let $\mathcal{X} = (A, \Sigma)$ be an abstract stratified set, or a (w)-regular nice stratified subset of a manifold, and let $g : \mathcal{Y} \to \mathcal{X}$ be a stratified map.*

Then for each stratified map $h : \mathcal{Z} \to \mathcal{X}$ and each open neighbourhood U of $h(\mathcal{Z})$ in $\mathcal{X}$, there exists a deformation by isotopy h' of h in $\mathcal{X}$ which is transverse to g in $\mathcal{X}$ and such that $h'(\mathcal{Z}) \subseteq U$. If C is a closed subset of $\mathcal{X}$ on which h is transverse to g then one can obtain that $h' = h$ on C.

Proof. Both proofs are given by induction on the dimension $k \leq n = \dim \mathcal{X}$ of the skeleton $\mathcal{X}_k$ of $\mathcal{X}$ by constructing a stratified vector field $\zeta = \zeta_k$ of $\mathcal{X}$ having a time 1 flow $\Phi_1 = \Phi_1^k$ defined on the whole of $\mathcal{X}$ and such that the map $h'_k = \Phi_1 \circ h$ satisfies the inductives hypotheses.

To obtain this, ζ has to be 0 on $\mathcal{X}_{k-1}$ (so $\Phi_{1|\mathcal{X}_{k-1}} = id$) and the restriction $f = \Phi_{1\,S} : S \to S$ where $S = \mathcal{X}_k - \mathcal{X}_{k-1}$ has to be a diffeomorphism of S such that $f \circ h$ is transverse to g.

Outlines of proof in [32]. We first prove that the set of such diffeomorphisms f of S is open and dense in the connected component $Diff_0(S, S)$ of 1_S in $Diff(S, S)$. Then we apply the techniques used by Mather [20] to show that infinitesimal stability implies stability, and using the families of geodesics of S we prove that *"There exists a (sufficiently small) neighbourhood U' of $1_S \in Diff_0(S, S)$ such that every $f \in U'$ is the time 1 flow $f = \Phi_1$ of a <u>time-dependent</u> vector field $\zeta = \zeta(x, t)$ such that $\lim_{x \to X_{k-1}} \zeta(x, t) = 0$"* (we stress that this property is completely false [10] without the precision <u>time-dependent</u> as we said in §3).

This allows us to obtain a *"Stratified Extension Theorem"* [32] holding for the diffeomorphisms $f \in U'$ for which the inductive step follows easily by extending on the whole of $\mathcal{X}$ the time-dependent vector field $\zeta(x, t)$. This is possible by adapting the standard techniques of lifting of stratified vector fields [7, 21, 22, 34, 38]. It was a merit of Andrew du Plessis to discover this key idea of using Mather time-dependent vector fields when he was examiner for my Ph.D. thesis directed by D. Trotman in the summer of 1997. $\qquad\qquad\square$

Another way to prove the stratified transversality theorem could be the following. In [30], using a theorem of D. McDuff [26] on the classification of the distinguished subgroup of the diffeomorphisms of a compact manifold with boundary, I extend to non-compact manifolds a well known Epstein–Thurston theorem and show that :

Theorem 5.2. *If $S = intM$ is a manifold, diffeomorphic to the interior of a compact manifold with boundary, the image of the exponential map generates $Diff_0(S, S)$. In particular every $f \in Diff_0(S, S)$ can be written as a composition $f = \phi_1^1 \circ \cdots \circ \phi_1^s$ of diffeomorphisms ϕ_1^i which are the time 1 map of the flow ϕ^i of a vector field ζ^i on S.*

Question 5.1. *If $\lim_{x \to \partial S} f = 1_{\partial S}$ can we obtain in theorem 5.2, that $\lim_{x \to \partial S} \zeta^i = 0$, for every $i = 1, \dots, s$?*

If the answer was yes, then by the usual techniques of stratified lifting of vector fields one would deduce easily a new *Extension Theorem for stratified homeomorphisms,* and without using time-dependent vector fields. This seems to me to depend on an apparently difficult improvement of the McDuff theorem.

Proof. *Outlines of the proof in [31] of Theorem 5.1.* We construct $\Phi = \Phi^k$ by the following steps.

First we prove that :

i) "Every smooth manifold S, admits a finite family of smooth and complete vector fields $v_1(x), \ldots, v_r(x)$ which span $T_x S$ for every $x \in S$ and there exists a sufficiently small open ball $B = B(0, \epsilon) \subseteq \mathbb{R}^r$ such that for every $b = (b_1, \ldots, b_r) \in B$, the vector field $\zeta_b = \sum_i b_i v_i$ is again complete".

Then we prove that :

ii) "The smooth map $G : S \times B \to S$ defined by $G(x, b) = \psi_1^b(x)$, where $\psi^b : S \times \mathbb{R} \to S$ is the flow of ζ_b, satisfies the submersivity of all partial maps $G_x : B \to S$ defined by $G_x(b) = \psi_1^b(x)$".

Finally we use a very nice and ingenious way to apply Sard's Theorem, discovered by R. Abraham [1] (see also the 1987 book [17], p. 51 for comments) to prove that:

iii) "If $j : B \to C^\infty(S, S)$ denotes the map $j(b) = G_b$, then the subset $M = \{b \in B \mid j^0 G_b$ is not transverse to $h_{|h^{-1}(S)} \times g_{|g^{-1}(S)}\}$, where $j^0 G_b$ is the graph of the map $G_b = \psi_1^b$, has measure zero in $\mathbb{R}^r$.

So taking $b \in B - M$ the diffeomorphism $f = G_b = \psi_1^b : S \to S$ is transverse to $h_{|h^{-1}(S)} \times g_{|g^{-1}(S)}$. Thus $f \circ h_{|h^{-1}(S)}$ is transverse to $g_{|g^{-1}(S)}$".

Again, an extension of ζ_b on the whole of $\mathcal{X}$ (to obtain $\Phi_1 = \Phi_1^k$), and the conclusion of the inductive step follow by the usual techniques of lifting of vector fields. $\qquad\qquad\square$

In this proof (of [31]) we obtain as Goresky hoped in his transversality theorem ([15], 5.3.) a diffeomorphism f which is transversalizing and *simultaneously* the time 1 map ψ_1^b of the flow of a vector field ζ_b <u>not depending on time</u>.

I like also to make precise here that :

(1) : part of this idea *iii*) appears already, in Lemma 6.3.4, at the end of Goresky's thesis, where the author gave the following proposition and comment [14, p. 183] :

Proposition 5.1. *Suppose $\mathcal{W}_1$ and $\mathcal{W}_2$ are Whitney stratified objects in a manifold M and U an open subset of M such that $\mathcal{W}_1 \cap U$ is transverse to $\mathcal{W}_2 \cap U$. Let K be a closed subset of U. Then there is a smooth vector field η on M, $\eta_{|K} \equiv 0$ such that $\Phi_1(\mathcal{W}_1)$ is transverse to $\mathcal{W}_2$ where $\phi_1 : M \to M$ is the diffeomorphism generated at the time 1 by the flow of η.*

Comment. *"Note, however, that if η does not have a compact support then the one-parameter group of diffeomorphisms $\phi_t : M \to M$ do <u>not</u> describe a continuous path in $Diff(M)$ under the C^1-topology. This lemma is, therefore not a suitable substitute for the transversality theorem".*

So part *i)* and *ii)* of our proof in [31] fill this gap and allow one to obtain a complete proof of a more general stratified transversality theorem.

(2) : the idea to use the Abraham method to apply Sard's Thorem, appeared again in a better explained and formalized way in 1987 [17] (many years after the thesis) but also in this case is developed only in an example (*Example 1.3.7.* p. 39) and in the very particular case where the manifold S was the projective space $\mathbb{CP}^n$ which is a compact manifold.

On the other hand, the strata of a stratification are not in general compact manifolds, and moreover for compact manifolds, the properties *i)* and *ii)* become easy to prove.

Finally, using the definitions of Goresky–MacPherson in [17] the joint proofs of the *i)* and *ii)* may be stated in a more elegant way as follows :

Theorem 5.3. *Every smooth non compact manifold S admits a submersive family $G = \{G_b\}_{b \in B}$ of self maps.*

I like to recall here that (after we read and re-read Goresky's Ph.D. thesis), it was the merit of David Trotman in the summer 2000, to point-out that this was the *good property* to prove for a non compact manifold (such as the generic stratum of a regular stratification) in order to obtain diffeomorphisms which *at the same time* are transversalizing and lie in a one parameter group (this happened while we were working to answer L. Siebenmann, who asked us if we knew another way to prove the stratified transversality theorem which did not use time-dependent vector fields).

Given the technical and historical difference between our two proofs of Theorem 5.1, which extend (and also clarify) the Goresky transversality theorem, we now look at the main corollaries of this theorem.

Suppose now, as in the *Transversality Lemma* of Goresky, that $\mathcal{W}$ is a substratified object of $\mathcal{X}$, and that the map $h = i : \mathcal{W} \hookrightarrow \mathcal{X}$ is the stratified inclusion of $\mathcal{W}$ in $\mathcal{X}$, and consider the map $h' = \Phi_1 \circ h$.

Because the transversalizing deformation Φ_1 is a stratified homeomorphism, and hence is a diffeomorphism on each stratum, one can easily see that the condition "$h' = \Phi_1 \circ h$ *is transverse to* g" may be reread as "$\mathcal{W}' = \Phi_1(\mathcal{W})$ *is transverse to* g". Thus we have the following corollary which generalizes the *Transversality Lemma* of Goresky, without the π-fibre condition on the substratified object $\mathcal{W}$ to be deformed.

Corollary 5.1. *Let $\mathcal{X}$ be an abstract stratified set, or a (w)-regular nice stratified subspace of a manifold, and $g : \mathcal{Y} \to \mathcal{X}$ a stratified map defined* *on a stratified space $\mathcal{Y}$.*

Then for each substratified object $\mathcal{W}$ of $\mathcal{X}$ and each open neighbourhood U of W in $\mathcal{X}$, there exists a deformation by isotopy $\mathcal{W}'$ of $\mathcal{W}$ which is transverse to g and such that $W' \subseteq U$. Moreover if C is a closed subset of $\mathcal{X}$ on which $\mathcal{W}$ is transverse to g then we can obtain that $W' \cap C = W \cap C$.

Corollary 5.1 holds for stratifications and stratified maps that are more general than those of the *Transversality Lemma* of Goresky [15]. For we do not require either of the two conditions :

i) that g be controlled with respect to two fixed systems of control data $\mathcal{T}_1$ et $\mathcal{T}_2$ respectively of $\mathcal{Y}$ and $\mathcal{X}$ or that g be the restriction of a smooth map $\tilde{g} : M_1 \to M_2$ between two manifolds containing respectively $\mathcal{Y}$ and $\mathcal{X}$;

ii) that $\mathcal{W}$ satisfy the π-fibre condition.

The π-fibre condition (or to be more precise its version stratum by stratum redefined in [29, p. 160]) is a very strong restriction on the geometry of the substratified object $\mathcal{W}$ of $\mathcal{X}$ and ensures that (b)-regularity be preserved as was shown in [15]; possibly other regularity conditions are preserved. For example this is the case for (a)-regularity, but it could also be true for (w)-regularity or (c)-regularity.

In Corollary 5.1, as we do not consider any regularity condition for $\mathcal{W}$ other than being a substratified object of $\mathcal{X}$, the problem of the preservation of such a condition by deformation by isotopy does not arise. We will talk in §5.1. about this delicate problem.

Corollary 5.1 was also used by M. Grinberg, when $g : \mathcal{Y} \hookrightarrow \mathcal{X}$ is the inclusion map and $\dim(Y \cap S) + \dim(W \cap S) < \dim S$ for every stratum S of $\mathcal{X}$, to prove the existence of self-indexing stratified Morse functions on complex algebraic varieties ([18], 2005).

Corollary 5.2. *Let $\mathcal{X} = (A, \Sigma)$ be an abstract stratified or a (w)-regular nice stratified subspace of a manifold, and $\mathcal{V}$ a substratified object of $\mathcal{X}$.*

For each substratified object $\mathcal{W}$ of $\mathcal{X}$, and each open neighbourhood U of the support W of $\mathcal{W}$ in A there exists a deformation by isotopy $\mathcal{W}'$ of $\mathcal{W}$, transverse to $\mathcal{V}$ in $\mathcal{X}$, with support $W' \subseteq U$.

If Z is a closed subset of A at each point of which $\mathcal{W}$ is transverse to $\mathcal{V}$ one can obtain moreover that the transversalizing isotopy $\Phi : A \times I \to A$ satisfies $\Phi_{t|Z} = id$ for all $t \in I$ and so $W' \cap Z = W \cap Z$.

Remark 5.1. K. Bekka has shown [3] that (c)-regular stratifications admit a system of control data; so both (b)-regular and (c)-regular stratified sets are abstract stratified sets and hence Theorem 5.1 and its corollaries 5.1 and 5.2 hold for them.

Remark 5.2. Because (L)-regular nice stratifications are (w)-regular [38], Theorem 5.1 holds also for Mostowski's (L)-regular stratified spaces.

5.1. *Preservation of regularity after deformation. Open problems*

In general a deformation by isotopy $\mathcal{W}' = \Phi_1(\mathcal{W})$, without supposing the π-fibre condition on $\mathcal{W}$, does not preserve any regularity condition of $\mathcal{W}$ except *"to be an abstract stratified set"*.

In [31] we introduce then the following notion of differentiability :

Definition 5.3. We say that a stratified morphism $f : \mathcal{X} \to \mathcal{X}'$ is *semidifferentiable at x* of $X \in \Sigma$ if for each stratum $Y > X$ (i.e. $\overline{Y} \supseteq X$) and for each sequence $\{(y_n, v_n)\}_n$ in the tangent space TY we have that $\lim_{n\to\infty}(y_n, v_n) = (x, v) \in TX$ implies $\lim_{n\to\infty} f_{Y*y_n}(v_n) = f_{X*x}(v)$.

We say f is *semidifferentiable on a stratum X* iff it is semidifferentiable at every $x \in X$ and that f is *semidifferentiable* iff it is semidifferentiable on every stratum $X \in \Sigma$.

Semidifferentiability (at x) is weaker than C^1-differentiability of f (at x) and provides sufficient conditions for a stratified homeomorphism (C^1 diffeomorphism on each stratum) to preserve some regularity of stratified subspaces. In [31] we show :

Theorem 5.4. *Let $\mathcal{X} = (A, \Sigma)$ be a (c)-regular stratified space, with A a closed subset of a C^∞ manifold M and let $g : \mathcal{Y} \to \mathcal{X}$ be a stratified map defined on a stratification $\mathcal{Y}$.*

For each substratified object $\mathcal{W}$ of $\mathcal{X}$ and each open neighbourhood U of W in A there exists a stratified isotopy $\Phi_t : \mathcal{X} \to \mathcal{X}$ such that the deformation $\mathcal{W}' = \Phi_1(\mathcal{W})$ is transverse to g, and $W' \subseteq U$.

If moreover Φ_1 is semidifferentiable, then:
i) $\mathcal{W}$ (c)-regular $\Rightarrow$ $\mathcal{W}'$is (c)-regular;
ii) $\mathcal{W}$ (a)-regular $\Rightarrow$ $\mathcal{W}'$ is (a)-regular.

On the other hand, we do not know currently a sufficient condition for Φ_1 to be semidifferentiable. The use of *continuous* liftings of vector fields in the transversality theorem is a necessary condition [34]. In any case, without assuming the π-fibre condition on $\mathcal{W}$, the following problem remains open:

Problem 5.1. *Can we obtain a stratified transversality theorem in which the deformation by isotopy $W' = \Phi_1(W)$ transverse to V preserves some of the regularity conditions of W such as (a)- or (b)- or (c)- regularity ?*

This problem could be solved if one proves the smooth version of the Whitney fibering conjecture [35, 36] :

Conjecture 5.1. *Every Whitney stratification $\mathcal{X}$, is such that for each point x in a stratum X of $\mathcal{X}$, there exists a neighbourhood U of x in A which admits an l-dimensional stratified foliation $\mathcal{F} = \{F_y\}_y$ (where $l = \dim X$ and F_y denotes the leaf containing y) such that for every $x' \in U \cap X$, $\lim_{y \to x'} T_y F_y = T_{x'} X$.*

5.2. *Stratification of the transverse union and intersection. Open problem*

Let $\mathcal{W}' = (W', \Sigma_{\mathcal{W}'})$ be a deformation of $\mathcal{W} = (W, \Sigma_{\mathcal{W}})$ transverse to $\mathcal{V} = (V, \Sigma_{\mathcal{V}})$ and $\{V_\alpha\}_\alpha$, $\{W'_\beta\}_\beta$ the families of strata of $\mathcal{W}'$ and $\mathcal{V}$.

If $C(H)$ denotes the family of the connected components of a space H, then $V \cup W'$ and $V \cap W'$ have natural partitions in smooth manifolds defined by :

$$\Sigma_{V \cap W'} = \sqcup_{V_\beta \subseteq V, w'_\alpha \subseteq w'} \; C(V_\beta \cap W'_\alpha)$$

and respectively by

$$\Sigma_{V \cup W'} = \sqcup_{V_\beta \subseteq V, w'_\alpha \subseteq w'} C(V_\beta - W'_\alpha) \; \sqcup C(W'_\alpha - V_\beta) \; \sqcup \; C(V_\beta \cap W'_\alpha)$$

called *transverse intersection $\mathcal{V} \cap_t \mathcal{W}'$* and *transverse union $\mathcal{V} \cup_t \mathcal{W}'$* of $\mathcal{V}$ and $\mathcal{W}'$. Unfortunately, as we show in [31] (Examples 3.17 and 3.16), these partitions do not define in general natural stratifications for $V \cap W'$ and $V \cup W'$, for two reasons:

> *i) in general $\mathcal{V} \cap_t \mathcal{W}'$ and (thus) $\mathcal{V} \cup_t \mathcal{W}'$ are not locally finite ;*
> *ii) in general $\mathcal{V} \cup_t \mathcal{W}'$ does not satisfy the frontier condition.*

It follows that if $f : \mathcal{X}_1 \to \mathcal{X}_2$ is a stratified map, not necessarily controlled, and W a substratified space of $\mathcal{X}_2$, not necessarily π-fibre, in general it is not true that $f^{-1}(W)$ is a substratified space of $\mathcal{X}_1$. So the good morphisms for Whitney cohomology WH^* have to be controlled maps.

It also follows that we cannot define a *transverse sum operation* in the set $WH_k(\mathcal{X})$ when $\mathcal{X}$ is an arbitrary Whitney stratification (not a manifold). This is the main raison for which Whitney homology WH_* (whose cycles are not defined as π-fibres) is a theory less rich in geometric interpretations than Whitney cohomology WH^*. This obstruction could may be overcome by answering the following question :

Problem 5.2. *Is it possible to find <u>two</u> deformations by isotopy $\mathcal{V}' = \Phi_1(\mathcal{V})$ and $\mathcal{W}' = \Psi_1(\mathcal{W})$ such that $\mathcal{V}'$ is transverse to $\mathcal{W}'$ and moreover $\mathcal{V}' \cap_t \mathcal{W}'$ and $\mathcal{V}' \cup_t \mathcal{W}'$ are locally finite and satisfy the frontier condition, defining so two stratifications ?*

A positive answer to this question would give the possibility to structure the *homology* set $WH_k(\mathcal{X})$ as an abelian group with the *transverse sum operation* when $\mathcal{X}$ is an arbitrary Whitney stratification, and this group structure could be a powerful tool to approach in an algebraic way the Goresky conjecture (§3, Conjecture 3.1).

For a manifold transverse to all strata of an analytic stratification Σ such that for each stratum $S \in \Sigma$, $\overline{S}$ and $\overline{S} - S$ are analytic sets in 1972 D. Chéniot [6], without assuming any regularity condition for Σ, proved that :

Theorem 5.5. *Let V be a complex analytic set, in an open set U of $\mathbb{C}^n$, equipped with an analytic stratification Σ. Let M be a complex analytic submanifold of U, transverse to every stratum of Σ.*
Then the trace $\Sigma_{M \cap V} = \{S \cap M \mid S \in \Sigma\}$ of M over Σ is a stratification (so locally finite and satisfying the frontier condition).

Here the stratifications are intended by considering the strata to be not necessarily connected, with so some difference with respect to our definitions. In such a complex analytic context the essential property used was that *"for each stratum S of V, one has $\overline{S \cap M} = \overline{S} \cap M$"* ; D. Chéniot also proves that *this key property is a sufficient hypothesis to obtain the conclusion of theorem 5.5 also for V and M real analytic* [6]. This key property is always satisfied when Σ is a real or complex Whitney stratification.

6. Transverse intersections and other applications

The problem of knowing when a transverse intersection and a transverse union is a *good* stratification also arises for knowing when $\mathcal{W} \cap_t \mathcal{V}$ and $\mathcal{W} \cup_t \mathcal{V}$ preserve the regularity condition of $\mathcal{W}$ and $\mathcal{V}$. This becomes already very useful in the simpler case when the stratification of the ambient space $\mathcal{X}$ and/or one of the two subspaces (as an example $\mathcal{V}$) reduce to two manifolds.

In 1976 C. G. Gibson [12] proved that Whitney (b)-regularity is preserved by transverse intersection in a manifold :

Theorem 6.1. *Let $\mathcal{V}_1, \ldots, \mathcal{V}_n$ be Whitney stratifications in a smooth manifold M. If $\mathcal{V}_1, \ldots, \mathcal{V}_n$ are in general position then $\mathcal{V}_1 \cap \cdots \cap \mathcal{V}_n$ is a Whitney stratification.*

More recently in 2002, P. Orro and D. Trotman [37] gave a unique proof of the corresponding theorem holding for many regularity conditions (although for the $(a + \delta)$ regularity the result was first proved in 2000, [5]):

Theorem 6.2. *The regularity conditions (a), (b), (w), $(a + \delta)$, $(a + r^e)$ for every $e \in [0, 1[$, are invariant by transverse intersection in a manifold of two stratifications with C^2 strata.*

This result was essential in proving the main theorem of [37] :

Theorem 6.3. *Let $\mathcal{Z}$ be a closed set in a manifold, stratified by C^k strata, $k \geq 2$ and $(a + r^e)$-regular, with $e \in [0, 1[$, relatively to a stratum Y.*

For every $y \in Y$, the fibre $(C_Y \mathcal{Z})_y$ of the normal cone $C_Y \mathcal{Z}$, coincides with the fibre $C_y(\mathcal{Z}_y)$ of the tangent cone to the fibre $\mathcal{Z}_y = \mathcal{Z} \cap \pi^{-1}(y)$ of a C^1 retraction π over Y.

6.1. *Application to abstract stratified and (c)-regular homology*

For (c)-regular stratifications, in 1991 K. Bekka [3] proved the following very useful propositions :

Proposition 6.1. *Let $f : M \to N$ be a C^1 map between C^1-manifolds and let $\mathcal{W} \subseteq N$ a (c)-regular stratified space. If f is transverse to $\mathcal{W}$ then $f^{-1}(\mathcal{W})$ defines a (c)-regular stratified subspace of M.*

Proposition 6.2. *Let M be a C^1-manifold and let $\mathcal{V}, \mathcal{W} \subseteq M$ be two (c)-regular stratified spaces.*

If $\mathcal{V}$ and $\mathcal{W}$ are transverse in M then the transverse intersection $\mathcal{V} \cap_t \mathcal{W}$ and the transverse union $\mathcal{V} \cup_t \mathcal{W}$ are again two (c)-regular stratified spaces.

Proposition 6.3. *Let $f : M \to N$ be a C^1 map between C^1-manifolds and $\mathcal{V} \subseteq M$ and $\mathcal{W} \subseteq N$ two (c)-regular stratified spaces.*

If f sends $\mathcal{V}$ transversally on $\mathcal{W}$ then $f_{|\mathcal{V}}^{-1}(\mathcal{W}) = \mathcal{V} \cap_t f^{-1}(\mathcal{W})$ is a (c)-regular stratified space.

For abstract stratified sets, the proofs of the corresponding statements, already obtained by Goresky in his thesis [14], are again true with similar proofs. Thanks to these results (and following the ideas of Goresky in [15]) one can construct new homology and cohomology theories in which the ambient space $\mathcal{X}$ and its cycles $\mathcal{V}$ and cocycles $\mathcal{W}$ are Thom–Mather abstract stratified sets (see also [14]) and/or Bekka (c)-regular stratifications (instead of Whitney (b)-regular stratifications [15]).

One obtains so the new theories AH_*, AH^* and BH_*, BH^* [4, 29] (instead of WH_*, WH^* of [15]). In these new theories, Theorem 5.1 with its corollaries 5.1 and 5.2, play again the role of the fundamental transversality theorem, necessary to give a geometrical meaning to the homology and cohomology algebraic operations. In this way, *all* geometric results of §3 are again true for the abstract stratified homology AH, and most of them also hold for the (c)-regular homology BH [4], [29, Chapter IV].

6.2. *Some applications to homotopy of stratified spaces*

In 1999 C. Eyral used stratified transversality to study the homotopy of stratified spaces [8]. First he proved the following theorem in which the stratified transversality is a necessary hypothesis (see [8] for a counterexample):

Theorem 6.4. *Let M be a C^1 manifold of dimension n, Y a closed subset of M equipped with a Whitney stratification Σ of dimension d and let N be a submanifold of M transverse to each stratum of Σ.*

Then the pair $(N, N - (N \cap Y))$ is $(n - 1 - d)$-connected.

and by which one deduces immediately the familiar results :

Corollary 6.1. *Let M be a real analytic manifold of dimension n and Y a closed real analytic subspace of M of dimension d. Then the pair $(M, M - Y)$ is $(n - 1 - d)$-connected.*

Corollary 6.2. *Let M be a C^1 manifold and N a closed submanifold of M of codimension c. Then the pair $(M, M - N)$ is $(c - 1)$-connected.*

Then, for a compact and real analytic ambient manifold M, Eyral also improves Theorem 6.4 by proving that it remains true when one considers (instead of a submanifold N) a Whitney stratification Σ' of a closed subset X of M, transverse to Σ in M.

Theorem 6.5. *Let M be a compact real analytic manifold of dimension n, Y, X two closed real analytic subspaces of M equipped with two Whitney stratifications Σ and Σ' transverse in M. Then the pair $(X, X - (X \cap Y))$ is $(n - 1 - d)$-connected (where $d = \dim \Sigma$).*

Theorems 6.4 and 6.5 are useful in studying the *global rectified homotopical depth* (a notion introduced in 1997 by Eyral in his Ph.D. thesis by reconsidering the *Grothendieck rectified homotopical depth*, see [9]) to prove a theorem of Lefschetz type for quasi-projective singular varieties which extends previous theorems of Goresky–MacPherson ([17, Theorem II.5.2] and Hamm–Lê ([19 Theorem 2.1.4].

Eyral conjectures finally that Theorem 6.5 could also be true without the hypotheses of compactness and analyticity on M, X, Y.

7. More on Goresky's stratified transversality. Supertransversality

In chapter I of his thesis [14], M. Goresky gives the following definitions:

Definition 7.1. Let $f : M \to N$ be a C^∞ map, and $\mathcal{W}_1 \subseteq M$ and $\mathcal{W}_2 \subseteq N$ two closed Whitney substratified objects. One says that f *takes $\mathcal{W}_1$ transversally to $\mathcal{W}_2$ (on a closed subset $K \subseteq N$)* if the stratified map $f_{|\mathcal{W}_1} : \mathcal{W}_1 \to N$ is transverse to $\mathcal{W}_2$ (on K).

Of course if $f \in Diff(M, M)$ this means that $f(\mathcal{W}_1)$ is transverse to $\mathcal{W}_2$.

A map $f : M \to N$ is said to *take $\mathcal{W}_1$ supertransversally to $\mathcal{W}_2$* if for every pair of strata $A \subseteq \mathcal{W}_1$ and $B \subseteq \mathcal{W}_2$, f takes A transversally to B and moreover for every $p \in \overline{A} \subseteq M$ and τ_1 limit of tangent planes of A and whenever $q = f(p) \in \overline{B} \subseteq N$ and τ_2 is limit of tangent planes of B, then $f_{*p}(\tau_1) + \tau_2 = T_q N$.

Clearly if $\mathcal{W}_1 = A$ and $\mathcal{W}_2 = B$ are closed manifolds, f takes $\mathcal{W}_1$ supertransversally to $\mathcal{W}_2$ if and only if f takes $\mathcal{W}_1$ transversally to $\mathcal{W}_2$.

Moreover, if $\mathcal{W}_1 = A$ and $\mathcal{W}_2 = B$ are *two closed* Whitney objects, f takes $\mathcal{W}_1$ supertransversally to $\mathcal{W}_2$ if and only if f takes $\mathcal{W}_1$ transversally to $\mathcal{W}_2$. This follows by the (a)-regularity of $\mathcal{W}_1$ and $\mathcal{W}_2$ since if $A \subseteq M$ and $B \subseteq N$ are strata of $\mathcal{W}_1$ and $\mathcal{W}_2$, and (with the above notations), $p \in \overline{A}$, $q = f(p) \in \overline{B}$, by closedness there exist two strata $A' \leq A$ and $B' \leq B$ for which $p \in A'$, $q \in B'$ and such that by hypothesis $f_{*p}(T_p A') + T_q B' = T_q N$. So by the (a)-regularity of $A \leq A'$ and $B \leq B'$, $\tau_1 \supseteq T_p A'$ and $\tau_2 \supseteq T_q B'$ and one finds the supertransversality $f_{*p}(\tau_1) + \tau_2 = T_q N$ [14, 1.3.1].

Goresky introduced this new notion of supertransversality with the project to prove the following theorem [14, 1.2.2] :

Theorem 7.1. *Suppose M, N are manifolds, K is a closed subset of N, $\mathcal{W}_1 \subseteq M$ and $\mathcal{W}_2 \subseteq N$ are closed Whitney stratified objects. Suppose*

$f{:}M \to N$ is a smooth map, which takes $\mathcal{W}_1$ transversally to $\mathcal{W}_2$ on K. Then:

i) There is a neighborhood U of f in $C^\infty(M, N)$ so that if $g \in U$, then g takes $\mathcal{W}_1$ transversally to $\mathcal{W}_2$ on K.

ii) For any neighborhood U' of f in $C^\infty(M, N)$ there exists a map $f' \in U'$ which takes $\mathcal{W}_1$ transversally to $\mathcal{W}_2$ on N, and satisfies $f'_{|f^{-1}(K)} = f_{|f^{-1}(K)}$.

Furthermore f' may be chosen to be homotopic to f by a smooth homotopy which is constant on K.

The techniques used by Goresky in his thesis [14] to find a proof of Theorem 7.1 can be resumed in some utilisations of the *families of geodesics* of Mather [20] to construct isotopies sufficiently close to the identity (1.3.3 and 1.3.4), some lemmas (1.3.1, 1.3.2, 1.3.3) and discussions on supertransversality and the proposition (1.3.4) below :

Proposition 7.1. *Suppose $f \in C^\infty(M, N)$, A a closed submanifold of M and B a submanifold of N. Let $K \subseteq N$ be a closed subset and U be a C^1 neighbourhood of f in $C^\infty(M, N)$. Suppose f takes A transversally to B on K and let $K' = f^{-1}(K)$. Then there exists a $g \in U$ such that $g_{|K'} = f_{|K'}$ and g takes A transversally to B.*

I consider important to present it because :

i) this was historically the original project of a stratified transversality theorem by Goresky ;

ii) the fact that Theorem 7.1 above or (1.2.2) in [14] remained unpublished since 1976, induced probably Goresky to find a new formulation of it, in the completely revised version of 1981 [15].

Finally, Goresky ended Chapter I of his thesis [14], by deducing from it corollary 7.1 below and by adding the following proposition (already announced in §6.1) essential to give geometrical meaning to the operation involving the Whitney stratified cycles and cocycles of his geometric homology and cohomology theory :

Corollary 7.1. *If $\mathcal{W}_1$ and $\mathcal{W}_2$ are closed Whitney stratified objects in a manifold M which are transverse on a closed subset $K \subseteq M$, then there exists a diffeomorphism $\phi : M \to M$ arbitrarily close to 1_M in the C^1 topology, so that $\phi(\mathcal{W}_1)$ is transverse to $\mathcal{W}_2$ and $\phi_{|K} = 1_K$.*

Proposition 7.2. *Suppose $f : M \to N$ is a smooth map between smooth manifolds and suppose $\mathcal{W}_1 \subseteq M$ and $\mathcal{W}_2 \subseteq N$ are Whitney stratified objects.*

Suppose f takes $\mathcal{W}_1$ transversally to $\mathcal{W}_2$, then $\mathcal{W}_1 \cap f^{-1}(\mathcal{W}_2)$ is a Whitney stratified object.

784

References

1. R. Abraham, *Transversality in manifold of mapping*, Bull. Amer. Math. Soc. 143 (1969), 470-474.

2. E. Akin, *Manifold phenomena in the theory of polyedra*, Trans. Amer. Math. Soc. 143 (1969), 413-473.

3. K. Bekka, *C-régularité et trivialité topologique*, Singularity theory and its applications, Warwick 1989, Part I, Lecture Notes in Math. **1462**, Springer, Berlin (1991) 42-62.

4. K. Bekka et C. Murolo, *Homologie d'espaces stratifiés*, C. R. Acad. Sci. Paris, t. 331, Série I (2000) 703-708.

5. K. Bekka and D. Trotman, *Weakly Whitney stratified sets*, Real and complex singularities, Sao Carlos 1998 (edited by J.W. Bruce and F. Tari), Chapman and Hall/CRC Research Notes in Mathematics 412, Boca Raton, Florida (2000), 1-15.

6. D. Chéniot, *Sur les sections transversales d'un ensemble stratifié*, C. R. Acad. Sci., t. 275, série A, (1972), 915-916.

7. A. A. du Plessis, *Continuous controlled vector fields*, Singularity theory (Liverpool, 1996, edited by J. W. Bruce and D. M. Q. Mond), London Math. Soc. Lecture Notes **263**, Cambridge Univ. Press, Cambridge, (1999), 189-197.

8. C. Eyral, *Sur l'homotopie des espaces stratifiés*, International mathematics research Notices (1999), N. 13, 717-734.

9. C. Eyral, *Tomographie des variétés singulières et théorèmes de Lefschetz*, Proc. London Math. Soc. (3) 83 (2001) 141-175.

10. C. Freifeld, *One parameter subgroups do not fill a neighbourhood of the identity in an infinite dimensional Lie (pseudo-)group*, Battelle Rencontre, Lectures in Mathematics and Physics, Benjamin, New-York (1967), 538-543.

11. W. Fulton, *Intersection Theory*, Springer Verlag, Berlin-Heidelberg (1984).

12. C. G. Gibson, K. Wirthmüller, A. A. du Plessis and E. J. N. Looijenga, *Topological stability of smooth mappings*, Lecture Notes in Math. **552**, Springer–Verlag (1976).

13. M. Golubitsky and V. Guillemin, *Stable mappings and their singularities*, Graduate Texts in Mathematics 14, Springer–Verlag, Berlin and New York (1973).

14. M. Goresky, *Geometric Cohomology and homology of stratified objects*, Ph.D. thesis, Brown University (1976).

15. M. Goresky, *Whitney stratified chains and cochains*, Trans. Amer. Math. Soc. **267** (1981), 175-196.

16. M. Goresky and R. MacPherson, *Intersection homology*, Topology **19** (1980), 135-162.

17. M. Goresky and R. MacPherson, *Stratified Morse theory*, Springer–Verlag, Berlin (1987).

18. M. Grinberg, *Gradient-Like Flows and Self-Indexing in Stratified Morse Theory*, Topology 44 (2005), 175-202.

19. H.A. Hamm and D.T. Lê, *Lefschetz theorems on quasi-projective varieties*, Bull. Soc. Math. France 113 (1985), 123-142.

20. J. Mather, *Stability of C^∞ mappings: II. Infinitesimal stability implies stability*, Annals of Mathematics **89** (1969), 254-291.

21. J. Mather, *Notes on topological stability*, Mimeographed notes, Harvard University (1970).

22. J. Mather, *Stratifications and mappings*, Dynamical Systems (M. Peixoto, Editor), Academic Press, New York (1971), 195-223.

23. C. McCrory, *Poincaré duality in spaces with singularities*, Ph.D. thesis, Brandeis University (1972).

24. C. McCrory, *Stratified general position*, Algebraic and Geometric Topology, Santa Barbara 1977, Lecture Notes in Math. **664**, Springer–Verlag, Berlin and New York (1978), 142-146.

25. J. Milnor, *Remarks on Infinite-dimensional Lie groups*, Relativity, Groups and Topology II. Les Houches Session XL, 1983. B.S. de Witt & R. Stora Editors. NorthHolland. Amsterdam (1984).

26. D. McDuff, *The lattice of normal subgroups of the group of diffeomorphisms or homeomorphisms of an open manifold*, J. London Math. Soc. (2), 18 (1978), 353-364.

27. C. Murolo, *Whitney homology, cohomology and Steenrod squares*, Ricerche di Matematica **43** (1994), 175-204.

28. C. Murolo, *The Steenrod p-powers in Whitney cohomology*, Topology and its Applications **68**, (1996), 133-151.

29. C. Murolo, *Semidifférentiabilité, transversalité et homologie de stratifications régulières*, Ph.D. thesis, Université de Provence, Marseille (1997).

30. C. Murolo *Diffeomorphisms lying in one-parameter groups and extension of stratified homeomorphisms* Math. Proc. Cambridge Philos. Soc. 130 (2001), no. 2, 333–341.

31. C. Murolo, D. Trotman, A. du Plessis, *Stratified Transversality by Isotopy*, Trans. Amer. Math. Soc. 355 (2003), n. 12, 4881-4900.

32. C. Murolo, A. du Plessis, D. Trotman, *Stratified Transversality by Isotopy via time-dependent vector fields*, J. London Math. Soc. (2) 71 (2005), 516-530

33. C. Murolo and D. Trotman, *Semidifferentiable stratified morphisms*, C. R. Acad. Sci. Paris, t 329, Série I, 147-152 (1999).

34. C. Murolo and D. Trotman, *Relèvements continus de champs de vecteurs*, Bull. Sci. Math., 125, 4 (2001), 253-278.

35. C. Murolo and D. Trotman, *Horizontally-C^1 controlled stratified maps and Thom's first isotopy theorem*, C. R. Acad. Sci. Paris Sér. I Math. 330 (2000), n. 8, 707–712.

36. C. Murolo and D. Trotman, *Semidifférentiabilité de Morphismes Stratifiés et Version Lisse de la conjecture de fibration de Whitney*, Proceedings of 12th MSJ-IRI symposium "Singularity Theory and its Applications", to appear 2007.

37. P. Orro and D. Trotman, *Cône normal et régularité de Kuo-Verdier*, Bull. Soc. Math. France, **130** (1) (2002), 71-85.

38. A. Parusiński, *Lipschitz stratifications*, Global Analysis in Modern Mathematics (K. Uhlenbeck, ed.), Proceedings of a Symposium in Honor of Richard Palais' Sixtieth Birthday, Publish or Perish, Houston (1993), 73-91.

39. R. Thom, *Ensembles et morphismes stratifiés*, Bull.A.M.S. **75** (1969), 240-284.

40. D. J. A. Trotman, *Geometric versions of Whitney regularity*, Annales Scientifiques de l'Ecole Normale Supérieure, 4eme série, **t. 12**, (1979), 453-463.

41. J.-L. Verdier, *Stratifications de Whitney et théorème de Bertini-Sard*, Inventiones Math. **36** (1976), 295-312.

42. H. Whitney, *Local properties of analytic varieties*, Differential and Combinatorial Topology, Princeton Univ. Press, (1965), 205-244.

GRAPH 3-MANIFOLDS, SPLICE DIAGRAMS, SINGULARITIES

WALTER D. NEUMANN

Department of Mathematics, Barnard College, Columbia University
New York, NY 10027, USA
E-mail: neumann@math.columbia.edu

We describe how a coarse classification of graph manifolds can give clearer insight into their structure, and we relate this particularly to the manifolds that can occur as the links of points in normal complex surfaces. We relate this discussion to a special class of singularities; those of "splice type", which turn out to play a central role among singularities of complex surfaces.

An appendix gives a brief introduction to the relevant parts of classical 3-manifold theory.

Keywords: graph manifold, surface singularity, rational homology sphere, complete intersection singularity, abelian cover

1. Introduction

The early study of 3-manifolds and knots in 3-manifolds was motivated to a large extent by the theory of complex surfaces. For example, Poul Heergaard's 1898 thesis [7], in which he introduced the fundamental tool of 3-manifold theory now called a "Heegaard splitting," was on the topology of complex surfaces. For a thread from Heergaard's thesis through knot theory to the "splice diagrams" that will play a central role in this paper, see the survey [23] on topology of complex surface singularities.

The local topology of a normal complex surface ("normal" roughly means that any "inessential" singularities have been removed) at any point is the cone on a closed oriented 3-manifold. The manifold is called the "link" of the point. We call it a "singularity-link," even though we allow S^3, which can only be the link of a non-singular point (Mumford [16]).

Singularity links and other 3-manifolds that arise in the study of complex surfaces are of a special type, namely "graph manifolds." Graph manifolds were defined and classified by Waldhausen in his thesis [40]. The motivation was certainly that the set of graph manifolds includes all singularity-links, and Waldhausen's work together with Grauert's criterion effectively gave a description of exactly what 3-manifolds are singularity-links. This description was put in a more convenient algorithmic form in [17]. More elegant versions have emerged since, which depend on taking a coarser look at the classification of graph manifolds. These coarse classifications are a central theme of this paper. They will also lead us to a special class of singularities, the singularities of "splice type" which encompasses several important classes of singularities.

Sections 7 through 10 constitute an appendix designed to provide a convenient reference for some of the basic 3-manifold theory that we use.

Acknowledgements. This paper is based in part on lectures the author gave at CIRM (Luminy) in March 2005 and served as notes for a short course at ICTP Trieste in August 2005. The author thanks both institutions for their hospitality. The research was supported under NSF grants no. DMS-0083097 and DMS-0206464.

2. The place of graph manifolds in 3-manifold theory

Throughout this paper, 3-manifolds will be compact and oriented unless otherwise stated. They will also be *prime* — not decomposable as a non-trivial connected sum. One forms the *connected sum* of two 3-manifolds by removing the interior of a disk from each and then gluing the resulting punctured 3-manifolds along their S^2 boundaries. Kneser and Milnor [12,15] showed that any oriented 3-manifold has an essentially unique decomposition into prime 3-manifolds. Singularity links are always prime ([17]).

Definition 2.1. A *graph-manifold* is a 3-manifold M that can be cut along a family of disjoint embedded tori to decompose it into pieces $S_i \times S^1$, where each S_i is a compact surface (i.e., 2-manifold) with boundary.

The JSJ-decomposition is a natural decomposition of any prime 3-manifold into Seifert fibered and simple non-fibered pieces (see the appendix for relevant definitions and more detail). Its existence was proved in the mid-1970's independently by by Jaco and Shalen [9] and by Johannson [11], although it had been sketched earlier by Waldhausen [41]. From the point of view of JSJ-decomposition, a graph manifold is simply a 3-manifold which has no non-Seifert-fibered JSJ-pieces. There are various modifications of the JSJ decomposition, depending on the intended application, and they differ in essentially elementary ways (see e.g., [25]). One version is the "geometric decomposition" — a minimal decomposition along tori and Klein bottles into pieces that admit geometric structures in the sense of Thurston (finite volume locally homogeneous Riemannian metrics). The relevant geometry for simple non-Seifert-fibered pieces is hyperbolic geometry[a]. From this geometric point of view, graph manifolds are manifolds that have no hyperbolic pieces in their geometric decompositions.

In summary, a graph manifold is a 3-manifold that can be glued together from pieces of the form (surface)$\times S^1$, or more efficiently, from pieces which are Seifert fibered. Both points of view will be useful in the sequel.

3. Seifert manifolds

Let $M^3 \to F$ be a Seifert fibration of a closed 3-manifold. It is classified up to orientation preserving homeomorphism (or diffeomorphism) by the following data (see subsection 9.3 in the appendix for details):

- The homeomorphism type of the base surface F, which we can encode by its *genus* g. We use the convention that $g < 0$ refers to a non-orientable surface, so $g = -1, -2, \ldots$ means F is a projective place, Klein bottle, etc.
- A collection of rational numbers $0 < q_i/p_i < 1$, $i = 1, \ldots, n$, that encode the types of the singular fibers. Here p_i is the multiplicity of the i-th singular fiber and q_i encodes how nearby fibers twist around this singular fiber.
- A rational number $e = e(M \to F)$ called the *Euler number* of the Seifert fibration. Its only constraint is that $e + \sum_{i=1}^{n} \frac{q_i}{p_i}$ should be an integer.

It is most natural to think of the base surface F as an orbifold rather than a manifold, with orbifold points of degrees $p_1, \ldots, p_n$. As such, it has an *orbifold Euler characteristic*

$$\chi^{orb}(F) = \chi_g - \sum_i (1 - \frac{1}{p_i})$$

where χ_g is the Euler characteristic of the surface of genus g:

$$\chi_g = \begin{cases} 2 - 2g, & g \geq 0, \\ 2 + g, & g < 0. \end{cases}$$

---

[a]The existence of the hyperbolic structure when M is simple non-Siefert fibered and the JSJ decomposition is trivial was still conjectural until recently; although proved in many cases by Thurston, it is probably now proved in general by Perelman's work.

Note that an oriented 3-manifold M^3 may be Seifert fibered with non-orientable base. However, we do not need to consider this for links of singularities: a Seifert fibered 3-manifold is a singularity link if and only if it has a Seifert fibration over an orientable base and the Euler number $e(M \to F)$ is negative.

From the point of view of geometric structures and geometric decomposition, there are exactly six geometries that occur for Seifert fibered manifolds and the type of the geometry is determined by whether $\chi^{orb}(F)$ is $> 0, = 0, < 0$ and whether $e(M \to F)$ is $= 0$ or $\neq 0$ ([19,34]). These two invariants, which we will abbreviate simply as χ and e, are thus fundamental invariants for a Seifert fibered M^3. If $e \neq 0$ then M^3 has a unique orientation that makes $e < 0$, and we call this its "natural orientation," since it is the orientation that makes it (or a double cover of it if the base surface is non-orientable) into a singularity link.

The above discussion was for a closed 3-manifold M^3. If M^3 is allowed to have boundary (but is still compact) then the Euler number e is indeterminate unless one has extra data. The additional data consists of a choice of a simple closed curve in each boundary torus of M^3, transverse to the fibers of the Seifert fibration.

Definition 3.1. We call this collection of curves a *system of meridians for M*. Given this data, we form a closed Seifert fibered manifold $\bar{M}^3$ by gluing a solid torus onto each boundary component, matching a meridian of the solid torus with the chosen "meridian" on the boundary T^2. The Euler invariant $e(\bar{M})$ is called the *Euler invariant of M with its system of meridians*.

The classification of geometric structures on Seifert fibered manifolds among the six "Seifert geometries" mentioned above holds also in the non-compact case. The manifold is then interior of a Seifert fibered manifold with boundary with a system of meridians. "Geometric structure" means geometric structure in the usual sense (complete, locally homogeneous, and finite volume) with the additional constraint that the meridians should be represented by arbitrarily short curves in the manifold. To have a geometric structure, χ^{orb} must be negative, so there are just two geometries in question, determined by vanishing or non-vanishing of the Euler invariant e.

4. Decomposition graphs, decomposition matrices

We now return to a general graph manifold M, considering it from the point of view of JSJ-decomposition. So M can be cut along tori so that it breaks into pieces that are Seifert fibered 3-manifolds. "JSJ decomposition" means that no smaller collection of cutting tori will work (see section 8 in the Appendix for a proof of existence and uniqueness of JSJ decomposition).

If M fibers over the circle with torus fiber or is double covered by such a manifold then M admits a geometric structure, so the geometric version of JSJ decomposition would not decompose it, even though the standard JSJ usually cuts it along a torus. Such manifolds are completely understood (for a discussion close to the current point of view see [20]) so:

Assumption 4.1. *From now on we assume that M cannot be fibered over S^1 with T^2 fiber.*

Each piece M_i in the JSJ decomposition comes with a system of meridians (Definition 3.1) by choosing the meridian in each boundary torus of M_i to be a Seifert fiber of the piece across the torus from M_i. Thus the orbifold Euler characteristic and Euler number invariants are both defined for the i-th piece M_i, and we call them χ_i and e_i.

The *decomposition graph* is the graph with a vertex for each piece M_i and an edge for each gluing torus. The edge connects the vertices corresponding to the pieces of M that meet along the torus. We decorate this graph with weights as follows: At the vertex i corresponding to M_i we give the numbers χ_i and e_i, writing χ_i in square brackets to distinguish it. And for an edge E corresponding to a torus T^2 we record the absolute value of the intersection number $F.F'$, where F and F' are fibers in T^2 of the Seifert fibrations on the pieces M_i and M_j that meet along T^2. For example, if M is glued from two Seifert fibered pieces, each of which has one boundary component, then the decomposition graph has the form

$$\overset{\displaystyle e_1 \qquad p \qquad e_2}{\underset{\displaystyle [\chi_1] \qquad\quad [\chi_2]}{\bullet\!\!-\!\!-\!\!-\!\!-\!\!-\!\!-\!\!-\!\!\bullet}}.$$

There is one problem with the definition of the decomposition graph. If a piece M_i is the total space SMb of the unit tangent bundle of the Möbius band, then weights on adjacent edges of the decomposition graph are not well-defined. This is because SMb has two different Seifert fibrations, one as this circle bundle and another by orbits of the action of the circle on SMb induced by the non-trivial S^1 action on the Möbius band Mb. For this reason we always use the latter Seifert fibration if such a piece occurs. However, we will usually want to go further and avoid SMb pieces altogether. This can always be done by replacing M by a double cover. In fact:

Proposition 4.1 ([20]). *M always has a double cover whose JSJ decomposition satisfies:*

- *Every piece of M has a Seifert fibration over a orientable base surface.*
- *The fibers of each piece can be oriented so that the fiber intersection number in each torus is positive[b].*
- *No SMb pieces occur and no piece is glued to itself across a torus.*

If the first condition holds we say M is *good* and if the first two conditions hold M is *very good.*

Remark 4.1. The third of the above conditions is a condition on the decomposition graph: the absence of SMb pieces says that the χ–weight at each vertex is negative (unless the graph consists of a single vertex with $\chi = 0$), and the absence of "self-gluings" is absence of edges that have both ends at the same vertex.

Even though the decomposition graph carries much less information than is needed to reconstruct M, it determines M up to finite ambiguity:

Proposition 4.2. *There are only finitely many different manifolds for any given decomposition graph.*

The proof of this is an exercise, based on the fact that there are only finitely many 2-orbifolds with given orbifold Euler characteristic χ. But the number can grow quite rapidly with χ, so already for simple decomposition graphs the number of manifolds can be large. Nevertheless, the decomposition graph does determine M up to commensurability (recall that manifolds are *commensurable* if they have diffeomorphic finite covers):

Theorem 4.1 ([24]). *If M_1 and M_2 are graph manifolds with no SMb pieces and their decomposition graphs are isomorphic then there exist d-fold covers $\bar{M}_1$ and $\bar{M}_2$ of M_1 and M_2 for some $d \in \mathbb{N}$ such that $\bar{M}_1 \cong \bar{M}_2$.*

---

[b]To get a well defined intersection number we view the separating torus from one side and intersect the Seifert fibers in the torus in the order (fiber from the near side).(fiber from the far side). If we look from the other side we reverse the orientation of the torus and reverse the order of the fibers, so the intersection number stays the same.

For many properties of M even less information suffices. Namely, the *decomposition matrix* is the matrix $A = (a_{ij})$ with entries

$$a_{ij} = e_i + 2 \sum_{iEi} \frac{1}{|p(E)|} \quad \text{if } i = j$$
$$= \sum_{iEj} \frac{1}{|p(E)|} \quad \text{if } i \neq j \, ,$$

where iEj means E is an edge joining i and j, and $p(E)$ is the fiber intersection weight on this edge. So the decomposition matrix no longer retains the invariants χ_i nor the exact number of edges joining a vertex to another.

It turns out that a variety of questions about M are answered in the literature completely in terms of the decomposition matrix (in some cases variations of "good" or "very good" are needed, that are always achieved in some double cover):

- Is M a singularity link ([20])?
- Does M fiber over the circle ([20])?
- Does some cover of M fiber over the circle ([20])?
- Does M have an immersed incompressible surface of negative Euler number ([21])?
- Does some cover of M have an embedded incompressible surface of negative Euler number ([21])?
- Does M admit a metric of non-positive curvature ([1])?

For the first of these the answer is as follows:

Theorem 4.2. *M is a singularity link if and only if it is very good and the decomposition matrix is negative definite.*

This is proved in [20] by a combinatorial argument, but we can give a geometric reason why it might be expected. Grauert's criterion [5] characterizes singularity links among "plumbed manifolds" (another way of looking at graph manifolds) by the negative definiteness of the intersection matrix of a resolution of the singularity. Our decomposition matrix is the intersection matrix of a resolution, but not a full resolution. The so-called log-canonical resolution of a surface singularity resolves the singularity to the point where only cyclic quotient singularities remain. Although we then do not yet have a smooth manifold, it is a $\mathbb{Q}$-homology manifold, so intersection numbers are still defined (they are rational numbers rather than integers). The resulting intersection matrix is the decomposition matrix. The theorem can be interpreted to say that Grauert's criterion still holds in this situation.

5. Splice diagrams for rational homology spheres

This section and the next describe joint work of the author and J. Wahl. We will describe a different encoding of graph manifolds, that again brings focus to some information by throwing away other information. We now restrict to graph manifolds M which are rational homology spheres, that is $H_1(M; \mathbb{Z})$ is finite. We say, briefly, that M is a QHS. For the JSJ decomposition this implies that the decomposition graph must be a tree, and, moreover, that the base of each Seifert fibered piece is of genus zero. However, instead of using the JSJ decomposition we now use the Waldhausen decomposition — the minimal decomposition into pieces of the form (surface)$\times S^1$

We again form a graph for this decomposition. This graph, with weights on edges to be described, is called a *splice diagram.*

The Waldhausen decomposition differs from the JSJ decomposition in that for each singular fiber of a Seifert fibered piece we must cut out a $D^2 \times S^1$ neighborhood of that singular fiber. For example, the JSJ decomposition graph for a Seifert fibered manifold consists of a single vertex (decorated with two numerical weights e and χ), while the splice diagram is a star-shaped graph: a central node with an edge sticking out for each singular fiber of the Seifert fibration. We weight the edges by the degrees of the singular fibers. For example a Seifert fibered manifold with exactly three singular fibers of degrees $2, 3, 5$ would have splice diagram

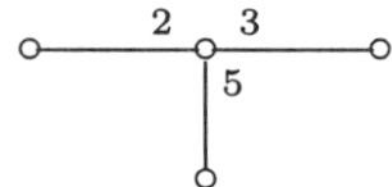

(There are infinitely many such Seifert manifolds, all with the same splice diagram; the corresponding decomposition graphs consist of a single vertex with weights $[\chi = 1/30]$, $e = q/30$, with q an arbitrary integer prime to 30.)

In general a splice diagram is a finite tree with vertices only of valence 1 ("leaves") or ≥ 3 ("nodes") and with non-negative integer weights decorating the edges around each node, and such that the weights on edges from nodes to leaves are ≥ 2. In addition, we decorate a node with an additional "−" sign if the linking number in M of two fibers of the corresponding Seifert piece is negative (this never occurs for splice diagrams of links of singularities).

Here is an example of a splice diagram for a certain singularity link M.

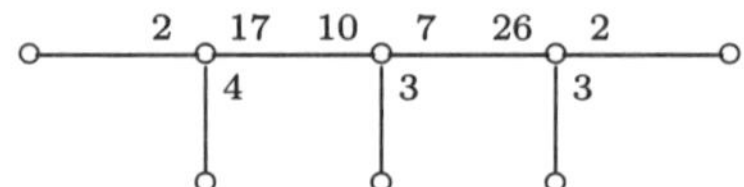

We describe the meaning of the weights by example of the weight 17. It is on an edge joining the two nodes, and this edge corresponds to a torus T^2 which cuts M into two pieces M_1 and M_2 (M_1 is Seifert fibered and M_2 is not). We look at the piece M_2 at the far side of T^2 and form a closed manifold $\bar{M}^2$ by gluing a solid torus into its boundary, matching – as in the previous section – meridian of the solid torus with the "meridian curve" on the boundary of M^2 (recall that this is a fiber in T^2 of the Seifert fibration across T^2 from M_2). The weight 17 is the order $|H_1(\bar{M}_2; \mathbb{Z})|$. This procedure weights an edge leading to a leaf with degree of the singular Seifert fiber corresponding to that leaf.

(It turns out that there is just one manifold with the above splice diagram. Its JSJ decomposition graph, obtained from the splice diagram by removing all the leaves and decorating with appropriate χ_i and e_i weights, is

$$
\underset{[\frac{-1}{4}]}{\overset{\frac{-5}{4}}{\circ}} \!\!-\!\!\!\!-\!\!\!\!-\!\! \underset{[\frac{-2}{3}]}{\overset{\frac{-5}{3}}{\circ}} \!\!-\!\!\!\!-\!\!\!\!-\!\! \underset{[\frac{-1}{6}]}{\overset{\frac{-7}{6}}{\circ}} \qquad)
$$

Definition 5.1. The *edge determinant* of an edge connecting two nodes in a splice diagram is defined to be the product of the two weights on that edge minus the product of the weights adjacent to that edge.

For example, both edge determinants in the above splice diagram are 2 since $10 \times 17 - 2 \times 4 \times 3 \times 7 = 2$ and $7 \times 26 - 10 \times 3 \times 2 \times 2 = 2$.

Theorem 5.1. *M is the link of a singularity if and only if no "$-$" decorations occur in the diagram and every edge determinant is positive.*

Again, although the splice diagram does not determine M uniquely in general, it does determine M up to commensurability. In fact, recall that the universal abelian cover of a space M is the Galois cover whose covering transformation group is $H_1(M; \mathbb{Z})$.

Theorem 5.2 (??). *If M_1 and M_2 are QHS graph manifolds with the same splice diagram then the universal abelian covers of M_1 and M_2 are diffeomorphic.*

The question marks are because we have not yet carefully written up a full proof of this theorem in the generality claimed. It is certainly correct when M_1 and M_2 are singularity links (the case that interests us most here).

In general the universal abelian cover of a QHS graph manifold M may be something quite horrible, with a complicated decomposition graph and lots of homology. But there is a case when we can describe it very nicely. For any splice diagram with pairwise coprime weights around each node there is a unique integral homology sphere ($\mathbb{Z}$HS) with the given splice diagram. It is its own universal abelian cover, so the theorem implies that this $\mathbb{Z}$HS is diffeomorphic to the universal abelian cover of any other graph manifold with the same splice diagram.

The splice diagram of a $\mathbb{Z}$HS graph manifold always has pairwise coprime weights around each node, so such diagrams classify $\mathbb{Z}$HS graph manifolds (see [3]).

We saw that the decomposition graph determines M up to finite ambiguity. The same is true for the splice diagram, except in the case of one-node splice diagrams (a one-node splice diagram always has infinitely many different manifolds associated with it). This is a consequence of Proposition 4.2 and the following:

Proposition 5.1 ([24]). *The splice diagram of M and the order of $H_1(M;\mathbb{Z})$ together determine the decomposition graph of M. The order of $H_1(M;\mathbb{Z})$ is a common divisor of the edge determinants of the splice diagram.*

6. Singularities of splice type

In general it has been very difficult to give explicit analytic realizations of singularities with given topology, but when the link is a QHS the recently discovered "singularities of splice type" [28] often do this.

Singularities of splice type have very strong properties: the universal abelian cover of a splice type singularity (by which we mean the maximal abelian cover that is ramified only at the singular point) is a complete intersection, defined by a quite elegant system of equations, and the covering transformation group acts diagonally in the coordinates. So the singularity is described by explicit equations and an explicit diagonal group action.

But, despite these strong properties, splice type singularities seem surprisingly common. For example, it has long been known that weighted homogeneous singularities with QHS link are of splice type ([18]), we (J. Wahl and the author) showed in [27] that Hirzebruch's quotient-cusp singularities are, and recently Okuma [32] has confirmed our conjecture that every rational singularity is of splice type and every minimally elliptic singularities with QHS link also is. Very recently we have proved a conjecture we had struggled with for some time, the "End Curve Conjecture", which postulated a characterization of this class of singularities in terms of curves through the singular point, and which has Okuma's theorem as a consequence.

To describe this result we need some terminology. Let (V, o) be a normal complex surface singularity and $\pi \colon (Y, E) \to (V, o)$ a good resolution. Recall that this means that $\pi^{-1}(o) = E$, π is biholomorphic between $Y - E$ and $V - o$, and E is a union of smooth curves E_j that intersect each other transversally, no three through a point. The link of the singularity is a QHS if and only if the resolution graph (the graph with a vertex for each E_j and an edge for each intersection of two E_j's) is a tree and each E_j is a rational curve. Let E_j correspond to a leaf j of the tree, so E_j intersects the rest of E in a single point x. An *end-curve* for j is a smooth curve germ cutting E_j transversally in a point other than x. An *end-curve function* for this leaf j is an analytic function germ $z_j \colon (V, o) \to (\mathbb{C}, 0)$ that "cuts out" an end-curve for j, in the sense that its zero set is the image in V of an end-curve for j (with some multiplicity).

Theorem 6.1 (End Curve Theorem, [30]). *Suppose (V, o) is a normal complex surface singularity with QHS link. It is of splice type if and only if an end-curve function exists for each leaf of the resolution graph. In this case appropriate roots of the end-curve functions can be used as coordinates on the universal abelian cover.*

The existence of end-curve functions is well known for rational singularities and for QHS-link minimally elliptic ones, so Okuma's theorem that these are of splice type follows.

To give the analytic description of splice type singularities we start with the weighted homogeneous case. Then there is a $\mathbb{C}^*$–action on the singularity which induces an S^1–action on the link M, so the the link is Seifert fibered. The splice diagram thus has the form

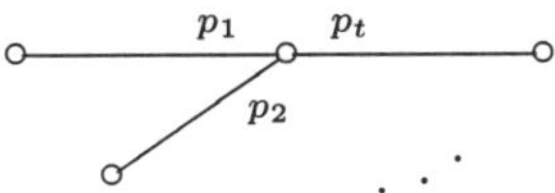

In this case it was shown in [18] that the universal abelian cover of the singularity is a Brieskorn complete intersection

$$\tilde{V}^{ab} \cong \{(z_1, \ldots, z_t) \in \mathbb{C}^k \mid a_{i1} z_1^{p_1} + \cdots + a_{ik} z_t^{p_t} = 0, \quad i = 1, \ldots, t - 2\},$$

for suitable coefficients a_{ij}. Moreover, an explicit action of $H_1(M; \mathbb{Z})$ on this Brieskorn complete intersection was given, with quotient the original singularity. Note that the Brieskorn equations are weighted homogeneous of total weight $p_1 \ldots p_t$ if we give the j-th variable weight $p_1 \ldots \hat{p}_j \ldots p_t$.

General splice type singularities generalize this situation. A variable z_i is associated to each leaf of the splice diagram, and for each node j of the diagram one associates a collection of $\delta - 2$ equations (δ the valence of the node) which are weighted homogeneous with respect to a system of weights associated to the node (one also allows higher weight perturbations of these equations). Doing this for all nodes gives a total of $t - 2$ equations, where t is the total number of leaves.

To describe these weights, fix the node v. The v-weight of the variable z_i corresponding to leaf i is the product of the weights directly adjacent to but not on the path from v to i in the splice diagram. We denote this number ℓ_{vi}. For example, if v is the left node in the splice diagram

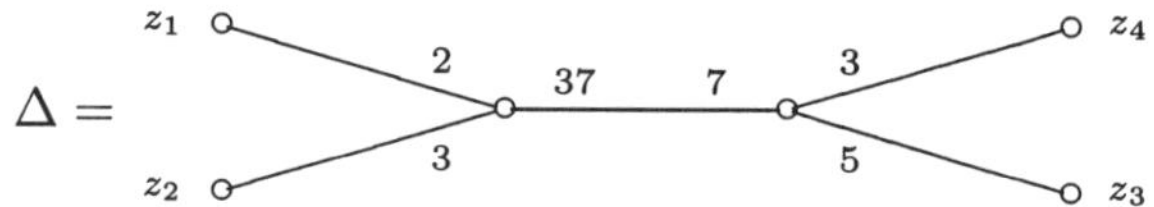

then the v-weights of the variables z_1, z_2, z_3, z_4 are:

$$\ell_{v1} = 3 \times 37 = 111, \quad \ell_{v2} = 74, \quad \ell_{v3} = 18, \quad \ell_{v4} = 30 .$$

The weight of the equations that we want to write down is the product of the weights at the node v, we denote this d_v; in our example $d_v = 222$. For each of the edges e departing v we choose a monomial M_e of total weight d_v in the variables corresponding to leaves beyond e from v. In this example the monomials z_1^2, z_2^3, and $z_3^4 z_4^5$ are suitable. Our equations will be equations which equate $\delta - 2$ generic linear combinations of these monomials to zero, where δ is the valence of v. So in this case there would be a single equation for the node v, of the form $a z_1^2 + b z_2^3 + c z_3^4 z_4^5 = 0$, for example

$$z_1^2 + z_2^3 + z_3^4 z_4^5 = 0 . \tag{1}$$

Note that a monomial M_e as above may not exist in general. The monomial $M_e = \prod_i z_i^{\alpha_i}$ has weight $\sum_i \alpha_i \ell_{vi}$, and the equation

$$d_v = \sum_i \alpha_i \ell_{vi}$$

may not have a solution in non-negative integers α_i as i runs through the leaves beyond e. The solubility of these equations gives a condition on the splice diagram that we call the *semigroup condition*. It is a fairly weak condition; for example the fact that rational and $\mathbb{Q}$HS-link minimally elliptic singularities are of splice type says that the semigroup condition is satisfied for the splice diagrams of the links of such singularities.

If the semigroup condition is satisfied, then we can write down equations as above for all nodes of the splice diagram, and we get a complete intersection singularity whose topology is the desired topology of a universal abelian cover. The other ingredient in defining splice type singularities is the group action that gives the covering transformations for the universal abelian cover. This group action in computed in a simple way from the desired topology of the singularity (as encoded by a resolution diagram; we describe this later), but the above equations will not necessarily be respected by it. Being able to choose the monomials so that the equations are respected is a further condition (on the resolution diagram for the singularity rather than just the splice diagram) which we call the *congruence condition*. Again, it is a condition that is satisfied for the classes of singularities that we mentioned above.

If both the semigroup condition and congruence condition are satisfied, so that the monomials can be chosen appropriately, then the complete intersection singularity we have described is the universal abelian cover of a singularity with the desired resolution diagram and the covering transformations are given by the group action in question.

We will carry this out for the explicit example of the splice diagram above. This is the splice diagram for a singularity with $\mathbb{Z}$HS link. The universal abelian cover is a trivial cover in this case, so the equations we construct will actually give such a singularity.

We have already seen that a possible equation for the left node is given by equation (1). In a similar way, we see that a possible equation for the right node is

$$z_1 z_2^2 + z_3^5 + z_4^3 = 0 \,. \tag{2}$$

The variety

$$V = \{(z_1, z_2, z_3, z_4) \mid z_1^2 + z_2^3 + z_3^4 z_4^5 = 0, \quad z_1 z_2^2 + z_3^5 + z_4^3 = 0\}$$

thus has an isolated singularity at 0 whose link is the $\mathbb{Z}$HS corresponding to the above splice diagram.

However, suppose the singularity we are really interested in is not the singularity with $\mathbb{Z}$HS link, which has resolution graph

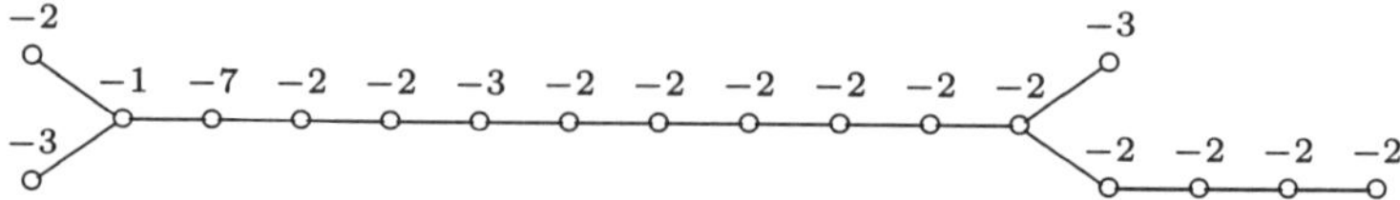

but instead the singularity with resolution graph

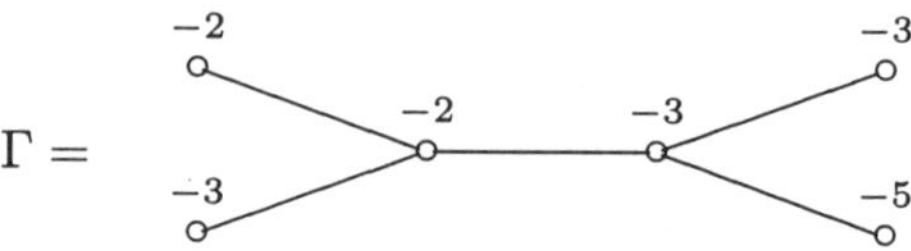

which has the same splice diagram, but its link M has first homology

$$H_1(M;\mathbb{Z}) = \mathbb{Z}/169\,.$$

By what we have already said, we expect the above variety V to be the universal abelian cover of what we want, so we want the $\mathbb{Z}/169$ action on V. As we describe in more detail below, the action of $\mathbb{Z}/169$ is generated by the map

$$(z_1, z_2, z_3, z_4) \mapsto (\xi^9 z_1, \xi^6 z_2, \xi^{38} z_3, \xi^7 z_4)\,,$$

where ξ is a primitive 169-th root of unity. This multiplies the first equation by ξ^{18} and the second by ξ^{21}, so it respects the equations and gives a free action of $\mathbb{Z}/169$ on the variety. The theory developed in [28] proves that $V' = V/(\mathbb{Z}/169)$ has the desired topology and that V is its universal abelian cover.

To describe the action of $H_1(M;\mathbb{Z})$ on $\mathbb{C}^t$ in general we first recall from [28] how to construct the splice diagram from the resolution graph Γ. Denote the incidence matrix of Γ by $A(\Gamma)$ — this is the intersection matrix of the resolution: the matrix whose diagonal entries are the self-intersection weights of Γ and which has has an entry 1 or 0 in the kl position with $k \neq l$ according as Γ does or does not have an edge connecting vertices k and l. The cokernel of $A(\Gamma)$ is isomorphic to $H_1(M\ \mathbb{Z})$, so $\det(-A(\Gamma)) = |H_1(M;\mathbb{Z})|$.

The splice diagram Δ has the same shape as Γ but with vertices of valence 2 suppressed. The splice diagram weights can be computed as follows. If one removes a node v of Γ and adjacent edges then Γ breaks into

δ subgraphs, where δ is the valence of v. The weights adjacent to v are the number $\det(-A(\Gamma'))$ as Γ' runs through these subgraphs.

This allows us to define a weight also adjacent to leaves of the splice diagram Δ, namely $\det(-A(\Gamma'))$ where Γ' is obtained by removing the leaf and adjacent edge. We now define ℓ_{ij} for any pair of leaves as the product of weights adjacent to the direct path from i to j (or just the weight adjacent to i if $i = j$).

With the leaves of Δ numbered $j = 1, \ldots, t$ we define for each leaf i a diagonal matrix

$$g_i = \mathrm{diag}(e^{2\pi i \ell_{ij}/d}; j = 1, \ldots, t)$$

where $d = |H_1(M; \mathbb{Z})|$. These matrices generate a diagonal subgroup of $\mathrm{GL}(\mathbb{C}^t)$ which is isomorphic to $H_1(M; \mathbb{Z})$. This gives the desired action of $H_1(M; \mathbb{Z})$ on $\mathbb{C}^t$ (see [28]).

In our particular example above, any one of $g_1, \ldots, g_4$ generates the cyclic group $H_1(M; \mathbb{Z}) = \mathbb{Z}/169$ and the actual element we gave above was g_3.

The congruence condition is the condition that for any node of the splice diagram we can choose the monomials M_e so that they all transform the same way under this group action. In this example the congruence condition turns out to be satisfied for any choice of monomials.

Appendix: Classical 3-manifold theory

The final four sections of this paper form an appendix which gives a quick survey of the "classical" 3-manifold theory underlying the paper. It is adapted from Chapter 2 of the notes [22]. Manifolds are always assumed to be smooth (or at least piecewise smooth).

7. Some basics

This section describes some fundamental classical tools of 3-manifold theory. The proofs of the results in this section can be found in several books on 3-manifolds, for example [8].

Theorem 7.1 (Dehn's Lemma). *If M^3 is a 3-manifold and $f \colon D^2 \to M^3$ a map of a disk such that for some neighborhood N of ∂D^2 the map $f|N$ is an embedding and $f^{-1}(f(N)) = N$. Then $f|\partial D^2$ extends to an embedding $g \colon D^2 \to M^3$.*

Dehn's proof of 1910 [4] had a serious gap which was pointed out in 1927 by Kneser. Dehn's Lemma was finally proved by Papakyriakopoulos in 1956, along with two other results, the loop and sphere theorems, which have been core tools ever since. These theorems have been refined by various authors since then. The following version of the loop theorem contains Dehn's lemma. It is due to Stallings [36].

Theorem 7.2 (Loop Theorem). *Let F^2 be a connected submanifold of ∂M^3, N a normal subgroup of $\pi_1(F^2)$ which does not contain $\ker(\pi_1(F^2) \to \pi_1(M^3))$. Then there is a proper embedding $g\colon (D^2, \partial D^2) \to (M^3, F^2)$ such that $[g|\partial D^2] \notin N$.*

Theorem 7.3 (Sphere Theorem). *If N is a $\pi_1(M^3)$-invariant proper subgroup of $\pi_2(M^3)$ then there is an embedding $S^2 \to M^3$ which represents an element of $\pi_2(M^3) - N$.*

(These theorems also hold if M^3 is non-orientable except that in the Sphere Theorem we must allow that the map $S^2 \to M^3$ may be a degree 2 covering map onto an embedded projective plane.)

Definition 7.1. An embedded 2-sphere $S^2 \subset M^3$ is *essential* or *incompressible* if it does not bound an embedded ball in M^3. M^3 is *irreducible* if it contains no essential 2-sphere.

Note that if M^3 has an essential 2-sphere that separates M^3 (i.e., M^3 falls into two pieces if you cut along S^2), then there is a resulting expression of M as a *connected sum* $M = M_1 \# M_2$ (to form connected sum of two manifolds, remove the interior of a ball from each and then glue along the resulting boundary components S^2). If M^3 has no essential separating S^2 we say M^3 is *prime*

Exercise 1. M^3 prime $\Leftrightarrow$ Either M^3 is irreducible or $M^3 \simeq S^1 \times S^2$. Hint[c].

Theorem 7.4 (Kneser and Milnor). *Any 3-manifold has a unique connected sum decomposition into prime 3-manifolds (the uniqueness is that the list of summands is unique up to order).*

We next discuss embedded surfaces other than S^2. Although we will mostly consider closed 3-manifolds (i.e., compact without boundary), it is sometimes necessary to consider manifolds with boundary. If M^3 has boundary, then there are two kinds of embeddings of surfaces that are of interest: embedding F^2 into ∂M^3 or embedding F^2 so that $\partial F^2 \subset \partial M^3$

---

[c]If M^3 is prime but not irreducible then there is an essential non-separating S^2. Consider a simple path γ that departs this S^2 from one side in M^3 and returns on the other. Let N be a closed regular neighborhood of $S^2 \cup \gamma$. What is ∂N? What is $M^3 - N$?

and $(F^2 - \partial F^2) \subset (M^3 - \partial M^3)$. The latter is usually called a "proper embedding." Note that ∂F^2 may be empty. In the following we assume without saying that embeddings of surfaces are of one of these types.

Definition 7.2. If M^3 has boundary, then a properly embedded disk $D^2 \subset M^3$ is *essential* or *incompressible* if it is not "boundary-parallel" (i.e., it cannot be isotoped to lie completely in ∂M^3, or equivalently, there is no ball in M^3 bounded by this disk and part of ∂M^3). M^3 is *boundary irreducible* if it contains no essential disk.

If F^2 is a connected surface $\neq S^2, D^2$, an embedding $F^2 \subset M^3$ is *incompressible* if $\pi_1(F^2) \to \pi_1(M^3)$ is injective. An embedding of a disconnected surface is incompressible if each component is incompressibly embedded.

It is easy to see that if you slit open a 3-manifold M^3 along an incompressible surface, then the resulting pieces of boundary are incompressible in the resulting 3-manifold. The loop theorem then implies:

Proposition 7.1. *If $F^2 \neq S^2, D^2$, then a two-sided embedding $F^2 \subset M^3$ is compressible (i.e., not incompressible) if and only if there is an embedding $D^2 \to M^3$ such that the interior of D^2 embeds in $M^3 - F^2$ and the boundary of D^2 maps to an essential simple closed curve on F^2.*

(For a one-sided embedding $F^2 \subset M^3$ one has a similar conclusion except that one must allow the map of D^2 to fail to be an embedding on its boundary: ∂D^2 may map 2-1 to an essential simple closed curve on F^2. Note that the boundary of a regular neighborhood of F^2 in M^3 is a two-sided incompressible surface in this case.)

Exercise 2. Show that if M^3 is irreducible then a torus $T^2 \subset M^3$ is compressible if and only if either

- it bounds an embedded solid torus in M^3, or
- it lies completely inside a ball of M^3 (and bounds a knot complement in this ball).

A 3-manifold is called *sufficiently large* if it contains an incompressible surface, and is called *Haken* if it is irreducible, boundary-irreducible, and sufficiently large. Fundamental work of Haken and Waldhausen analyzed Haken 3-manifolds by repeatedly cutting along incompressible surfaces until a collection of balls was reached (it is a theorem of Haken that this always happens). A main result is

Theorem 7.5 (Waldhausen). *If M^3 and N^3 are Haken 3-manifolds and we have an isomorphism $\pi_1(N^3) \to \pi_1(M^3)$ that "respects peripheral structure" (that is, it takes each subgroup represented by a boundary component of N^3 to a a conjugate of a subgroup represented by a boundary component of M^3, and similarly for the inverse homomorphism). Then this isomorphism is induced by a homeomorphism $N^3 \to M^3$ which is unique up to isotopy.*

The analogous theorem for surfaces is a classical result of Nielsen.

We mention one more "classical" result that is a key tool in Haken's approach.

Definition 7.3. Two disjoint surfaces $F_1^2, F_2^2 \subset M^3$ are *parallel* if they bound a subset isomorphic to $F_1 \times [0,1]$ between them in M^3.

Theorem 7.6 (Kneser-Haken finiteness theorem). *For any given 3-manifold M^3 there exists a bound on the number of disjoint pairwise non-parallel incompressible surfaces that can be embedded in M^3.*

8. JSJ Decomposition

We shall give a quick proof, originating in an idea of Swarup (see [25]), of the main "JSJ decomposition theorem" which describes a canonical decomposition of any irreducible boundary-irreducible 3-manifold along tori and annuli. The characterization of this decomposition that we actually use in these notes is here an exercise (Exercise 3 at the end of this section). F. Costantino gives a nice exposition in [2] of a proof, based on this proof and ideas of Matveev, that directly proves this characterization.

We shall just describe the decomposition in the case that the boundary of M^3 is empty or consists of tori, since that is what is relevant to these notes. Then only tori occur in the JSJ decomposition (see section 9.5). An analogous proof works in the general torus-annulus case (see [25]), but the general case can also be deduced from the case we prove here.

The theory of such decompositions for Haken manifolds with toral boundaries was first outlined by Waldhausen in [41]; see also [42] for his later account of the topic. The details were first fully worked out by Jaco and Shalen [9] and independently Johannson [11].

Definition 8.1. M is *simple* if every incompressible torus in M is boundary-parallel.

If M is simple we have nothing to do, so suppose M is not simple and let $S \subset M$ be an essential (incompressible and not boundary-parallel) torus.

Definition 8.2. S will be called *canonical* if any other properly embedded essential torus T can be isotoped to be disjoint from S.

Take a disjoint collection $\{S_1, \ldots, S_s\}$ of canonical tori in M such that

- no two of the S_i are parallel;
- the collection is maximal among disjoint collections of canonical tori with no two parallel.

A maximal system exists because of the Kneser-Haken finiteness theorem. The result of splitting M along such a system will be called a *JSJ decomposition* of M. The maximal system of pairwise non-parallel canonical tori will be called a *JSJ-system*.

The following lemma shows that the JSJ-system $\{S_1, \ldots, S_s\}$ is unique up to isotopy.

Lemma 8.1. *Let $S_1, \ldots, S_k$ be pairwise disjoint and non-parallel canonical tori in M. Then any incompressible torus T in M can be isotoped to be disjoint from $S_1 \cup \cdots \cup S_k$. Moreover, if T is not parallel to any S_i then the final position of T in $M - (S_1 \cup \cdots \cup S_k)$ is determined up to isotopy.*

By assumption we can isotop T off each S_i individually. Writing $T = S_0$, the lemma is thus a special case of the stronger:

Lemma 8.2. *Suppose $\{S_0, S_1, \ldots, S_k\}$ are incompressible surfaces in an irreducible manifold M such that each pair can be isotoped to be disjoint. Then they can be isotoped to be pairwise disjoint and the resulting embedded surface $S_0 \cup \ldots \cup S_k$ in M is determined up to isotopy.*

Proof. We just sketch the proof. We start with the uniqueness statement. Assume we have $S_1, \ldots, S_k$ disjointly embedded and then have two different embeddings of $S = S_0$ disjoint from $T = S_1 \cup \ldots \cup S_k$. Let $f\colon S \times I \to M$ be a homotopy between these two embeddings and make it transverse to T. The inverse image of T is either empty or a system of closed surfaces in the interior of $S \times I$. Now use Dehn's Lemma and Loop Theorem to make these incompressible and, of course, at the same time modify the homotopy (this procedure is described in Lemma 1.1 of [40] for example). We eliminate 2-spheres in the inverse image of T similarly. If we end up with nothing in the inverse image of T we are done. Otherwise each component T' in the inverse image is a parallel copy of S in $S \times I$ whose fundamental group maps injectively into that of some component S_i of T. This implies that S can be homotoped into S_i and its fundamental group $\pi_1(S)$ is conjugate into some $\pi_1(S_i)$. It is a standard fact (see, e.g., [37]) in this situation of two incompressible surfaces having comparable fundamental groups that, up to conjugation, either $\pi_1(S) = \pi_1(S_j)$ or S_j is one-sided and $\pi_1(S)$ is the fundamental group of the boundary of a regular neighborhood of T and thus of index 2 in $\pi_1(S_j)$. We thus see that either S is parallel to S_j and is being isotoped across S_j or it is a neighborhood boundary of a one-sided S_j and is being isotoped across S_j. The uniqueness statement thus follows.

A similar approach to proves the existence of the isotopy using Waldhausen's classification [39] of proper incompressible surfaces in $S \times I$ to show that S_0 can be isotoped off all of $S_1, \ldots, S_k$ if it can be isotoped off each of them. $\qquad \square$

The thing that makes decomposition along incompressible annuli and tori special is the fact that they have particularly simple intersection with other incompressible surfaces.

Lemma 8.3. *If a properly embedded incompressible torus T in an irreducible manifold M has been isotoped to intersect another properly embedded incompressible surface F with as few components in the intersection as possible, then the intersection consists of a family of parallel essential simple closed curves on T.*

Proof. Suppose the intersection is non-empty. If we cut T along the intersection curves then the conclusion to be proved is that T is cut into annuli. Since the Euler characteristics of the pieces of T must add to the Euler characteristic of T, which is zero, if not all the pieces are annuli then there must be at least one disk. The boundary curve of this disk bounds a disk in F by incompressibility of F, and these two disks bound a ball in M by irreducibility of M. We can isotop over this ball to reduce the number of intersection components, contradicting minimality. $\qquad \square$

Let $M_1, \ldots, M_m$ be the result of performing the JSJ-decomposition of M along the JSJ-system $\{S_1 \cup \cdots \cup S_s\}$.

Theorem 8.1. *Each M_i is either simple or Seifert fibered by circles (or maybe both).*

Proof. Suppose N is one of the M_i which is non-simple. We must show it is Seifert fibered by circles.

Since N is non-simple it contains essential tori. Consider a maximal disjoint collection of pairwise non-parallel essential tori $\{T_1, \ldots, T_r\}$ in N. Split N along this collection into pieces $N_1, \ldots, N_n$. We shall analyze these pieces and show that they are of one of nine basic types, each of which is evidently Seifert fibered. Moreover, we will see that the fibered structures match together along the T_i when we glue the pieces N_i together again to form N.

Consider N_1, say. It has at least one boundary component that is a T_j. Since T_j is not canonical, there exists an essential torus T' in N which essentially intersects T_1. We make the intersection of T' with the union $T = T_1 \cup \cdots \cup T_r$ minimal, and then by Lemma 8.3 the intersection consists of parallel essential curves on T'.

Let s be one of the curves of $T_j \cap T'$. Let P be the part of $T' \cap N_1$ that has s in its boundary. P is an annulus. Let s' be the other boundary component of P. It may lie on a T_k with $k \neq j$ or it may lie on T_j again. We first consider the case

Case 1: s' lies on a different T_k.

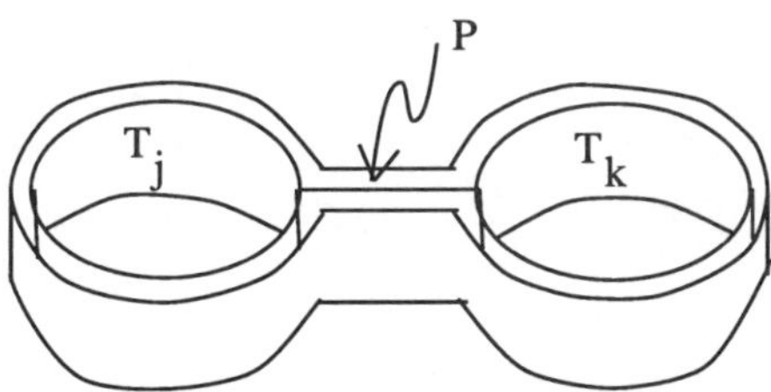

Fig. 1.

In Fig. 1 we have drawn the boundary of a regular neighborhood of the union $T_j \cup T_k \cup P$ in N_1. The top and the bottom of the picture should be identified, so that the whole picture is fibered by circles parallel to s and s'. The boundary torus T of the regular neighborhood is a new torus disjoint from the T_i's, so it must be parallel to a T_i or non-essential. If T is parallel to a T_i then N_1 is isomorphic to $X \times S^1$, where X is a the sphere with three disks removed. Moreover all three boundary tori are T_i's. If T is non-essential, then it is either parallel to a boundary component of N or it is compressible in N. In the former case N_1 is again isomorphic to $X \times S^1$, but with one of the three boundary tori belonging to ∂N. If T

808

is compressible then it must bound a solid torus in N_1 and the fibration by circles extends over this solid torus with a singular fiber in the middle (there must be a singular fiber there, since otherwise the two tori T_j and T_k are parallel).

We draw these three possible types for N_1 in items 1,2, and 3 of Fig. 2, suppressing the circle fibers, but noting by a dot the position of a possible singular fiber. Solid lines represent part of ∂N while dashed lines represent T_i's.

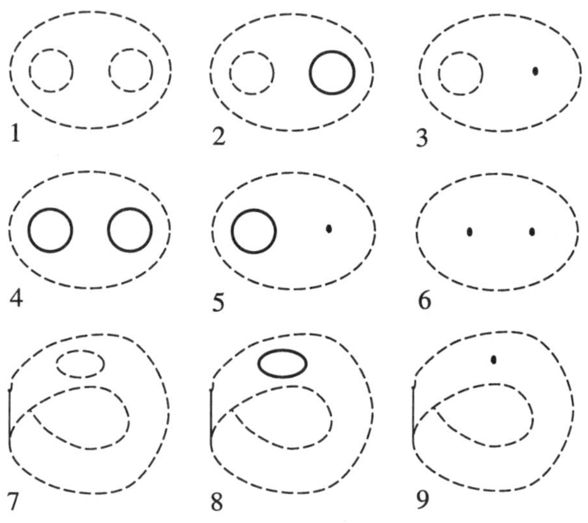

Fig. 2.

We next consider

Case 2. s' also lies on T_j, so both boundary components s and s' of P lie on T_j.

Now P may meet T_j along s and s' from the same side or from opposite sides, so we split Case 2 into the two subcases:

Case 2a. P meets T_j along s and s' both times from the same side;

Case 2b. P meets T_j along s and s' from opposite sides.

It is not hard to see that after splitting along T_j, Case 2b behaves just like Case 1 and leads to the same possibilities. Thus we just consider Case 2a. This case has two subcases 2a1 and 2a2 according to whether s and s' have the same or opposite orientations as parallel curves of T_j (we orient

s and s' parallel to each other in P). We have pictured these two cases in Fig. 3 with the boundary of a regular neighborhood of $T_j \cup P$ also pictured.

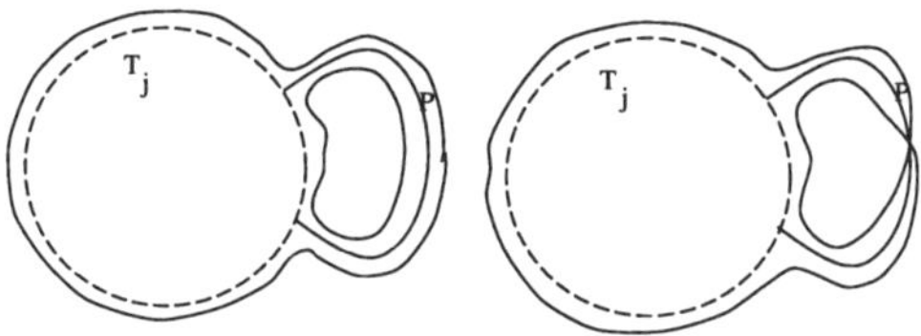

Fig. 3.

In Case 2a1 the regular neighborhood is isomorphic to $X \times S^1$ and there are two tori in its boundary, each of which may be parallel to a T_i, parallel to a boundary component of N, or bound a solid torus. This leads to items 1 through 6 of Fig. 2.

In Case 2b the regular neighborhood is a circle bundle over a möbius band with one puncture (the unique such circle bundle with orientable total space). The torus in its boundary may be parallel to a T_i, parallel to a component of ∂N, or bound a solid torus. This leads to cases 7, 8, and 9 of Fig. 2. In all cases but case 9 a dot signifies a singular fiber, but in case 9 it signifies a fiber which may or may not be singular.

We now know that N_1 is of one of the types of Fig. 2 and thus has a Seifert fibration by circles, and therefore similarly for each piece N_i. Moreover, on the boundary component T_j that we are considering, the fibers of N_1 are parallel to the intersection curves of T_j and T' and therefore match up with fibers of the Seifert fibration on the piece on the other side of T_j. We must rule out the possibility that, if we do the same argument using a different boundary component T_k of N_1, it would be a different Seifert fibration which we match across that boundary component. In fact, it is not hard to see that if N_1 is as in Fig. 2 with more than one boundary component, then its Seifert fibration is unique. To see this up to homotopy, which is all we really need, one can use the fact that the fiber generates a normal cyclic subgroup of $\pi_1(N_1)$, and verify by direct calculation that $\pi_1(N_1)$ has a unique such subgroup in the cases in question.

(In fact, the only manifold of a type listed in Fig. 2 that does not have a unique Seifert fibration is case 6 when the two singular fibers are both degree 2 singular fibers and case 9 when the possible singular fiber is in fact not singular. These are in fact two Seifert fibrations of the same manifold $T^1 Mb$, the unit tangent bundle of the Möbius band Mb. This manifold can also be fibered by lifting the fibration of the Möbius band by circles to a fibration of the total space of the tangent bundle of Mb by circles.) $\square$

An alternative characterization of the JSJ decomposition is as a minimal decomposition of M along incompressible tori into Seifert fibered and simple pieces. In particular, if some torus of the JSJ-system has Seifert fibered pieces on both sides of it, the fibrations do not match up along the torus.

Exercise 3. Verify the last statement.

9. Seifert fibered manifolds

In this section we describe all three-manifolds that can be Seifert fibered with circle or torus fibers.

Seifert's original concept of what is now called "Seifert fibration" referred to 3-manifolds fibered with circle fibers, allowing certain types of "singular fibers." For orientable 3-manifolds this gives exactly fibrations over 2-orbifolds, so it is reasonable to use the term "Seifert fibration" more generally to mean "fibration of a manifold over an orbifold." So we start by recalling what we need about orbifolds.

9.1. *Orbifolds*

An n-orbifold is a space that looks locally like $\mathbb{R}^n/G$ where G is a finite subgroup of $GL(n, \mathbb{R})$. Note that G varies from point to point, for example, a neighborhood of $[x] \in \mathbb{R}^n/G$ looks like $\mathbb{R}^n/G_x$ where $G_x = \{g \in G \mid gx = x\}$.

We will restrict, for simplicity, to locally orientable 2-orbifolds (i.e., the above G preserves orientation). Then the only possible local structures are $\mathbb{R}^2/C_p$, $p = 1, 2, 3, \ldots$, where C_p is the cyclic group of order p acting by rotations. the local structure is then a "cone point" with "cone angle $2\pi/p$" (Fig. 4).

Topologically, a 2-orbifold is thus simply a 2-dimensional manifold, in which certain points are singled out as being "orbifold points" where the total angle around the point is considered to be $2\pi/p$ instead of 2π. The underlying 2-manifold is classified by its genus g (we use negative numbers to refer to non-orientable surfaces, so genus $-1, -2, \ldots$ mean projective plane, Klein bottle, etc.). We can thus characterize a 2–orbifold by a tuple

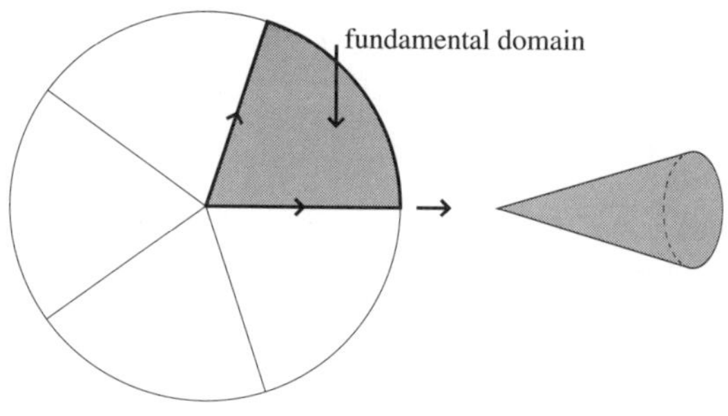

Fig. 4.

of numbers $(g; p_1, \ldots, p_k)$ where g is the genus and $p_1, \ldots, p_k$ describe the orbifold points.

9.2. *General concept of Seifert fibrations via orbifolds*

A map $M \to N$ is a Seifert fibration if it is locally isomorphic to maps of the form $(U \times F)/G \to U/G$, with U/G an orbifold chart in N (so U is isomorphic to an open subset of $\mathbb{R}^n$ with an action of the finite group G) and F a manifold with G-action such that the diagonal action of G on $U \times F$ is a free action. The freeness of the action is to make M a manifold rather than just an orbifold.

9.3. *Seifert circle fibrations*

We start with "classical" Seifert fibrations, that is, fibrations with circle fibers, but with some possibly "singular fibers." We first describe what the local structure of the singular fibers is. This has already been suggested by the proof of JSJ above.

We have a manifold M^3 with a map $\pi \colon M^3 \to F^2$ to a surface such that all fibers of the map are circles. Pick one fiber f_0 and consider a regular neighborhood N of it. We can choose N to be a solid torus fibered by fibers of π. To have a reference, we will choose a longitudingal curve l and a meridian curve m on the boundary torus $T = \partial N$. The typical fiber f on T is a simple closed curve, so it is homologous to $pl + rm$ for some coprime pair of integers p, r. We can visualize the solid torus N like an onion, made up of toral layers parallel to T (boundaries of thinner and thinner regular neighborhoods) plus the central curve f_0. Each toral shell is fibered just like the boundary T, so the typical fibers converge on pf_0 as one moves to the center of N.

Exercise 4. Let s be a closed curve on T that is a section to the boundary there. Then (with curves appropriately oriented) one has the homology relation $m = ps + qf$ with $qr \equiv 1 \pmod{p}$.

The pair $(p.q)$ is called the *Seifert pair* for the fiber f_0. It is important to note that the section s is only well defined up to multiples of f, so by changing the section s we can alter q by multiples of p. If we have chosen things so $0 \leq q < p$ we call the Seifert pair *normalized*.

By changing orientation of f_0 if necessary, we may assume $p \geq 0$. In fact:

Exercise 5. If M^3 contains a fiber with $p = 0$ then M^3 is a connected sum of lens spaces. (A lens space is a 3-manifold obtained by gluing two solid tori along their boundaries; it is classified by a pair of coprime integers (p, q) with $0 \leq q < p$ or $(p, q) = (0, 1)$. One usually writes it as $L(p, q)$. Special cases are $L(0, 1) = S^2 \times S^1$, $L(1, 0) = S^3$. For $p \geq 0$ $L(p, q)$ can also be described as the quotient of $S^3 = \{(z, w) \in \mathbb{C}^2 : |z|^2 + |w|^2 = 1\}$ by the action of $\mathbb{Z}/p$ generated by $(z, w) \mapsto (e^{2\pi i/p}z, e^{2\pi iq/p}w)$.)

We therefore rule out $p = 0$ and assume from now on that every fiber has $p > 0$. Note that $p = 1$ means that the fiber f_0 is a non-singular fiber, i.e., the whole neighborhood N of f_0 is fibered as the product $D^2 \times S^1$. If $p > 1$ then f_0 is a singular fiber, but the rest of N consists only of non-singular fibers. In particular, singular fibers are isolated, so there are only finitely many of them in M^3.

Now let $f_0, \ldots, f_r$ be a collection of fibers which includes all singular fibers. For each one we choose a fibered neighborhood N_i and a section s_i on ∂N_i as above, giving a Seifert pair (p_i, q_i) with $p_i \geq 1$ for each fiber. Now on $M_0 := M^3 - \bigcup \int(N_i)$ we have a genuine fibration by circles over a surface with boundary. Such a fibration always has a section, so we can assume that our sections s_i on ∂M_0 have come from a global section on M_0. This section on M_0 is not unique. If we change it, then each s_i is replaced by $s_i + n_i f$ for some integers n_i, and a homological calculation shows that $\sum n_i$ must equal 0. The effect on the Seifert pairs (p_i, q_i) is to replace each by $(p_i, q_i - n_i p_i)$. In summary, we see that changing the choice of global section on M_0 changes the Seifert pairs (p_i, q_i) by changing each q_i, keeping fixed:

- the congruence class $q_i \pmod{p_i}$
- $e := \sum \frac{q_i}{p_i}$

The above number e is called the *Euler number* of the Seifert fibration. We have not been careful about describing our orientation conventions here. With a standard choice of orientation conventions that is often used in the literature, e is more usually defined as $e := -\sum \frac{q_i}{p_i}$.

Note that we can also change the collection of Seifert pairs by adding or deleting pairs of the form $(1,0)$, since they correspond to non-singular fibers with choice of local section that extends across this fiber. Up to these changes the topology of the base surface F and the collection of Seifert pairs is a complete invariant of M^3. A convenient normalization is to take f_0 to be a non-singular fiber and $f_1, \ldots, f_s$ to be all the singular fibers and normalize so that $0 < q_i < p_i$ for $i \geq 1$. This gives a complete invariant:

$$(g; (1, q_0), (p_1, q_1), \ldots, (p_r, q_r)) \quad \text{with } g = \text{genus}(F)$$

which is unique up to permuting the indices $i = 1, \ldots, r$. A common convention is to use negative g for the genus of non-orientable surfaces (even though we are assuming M^3 is oriented, the base surface F need not be orientable).

Exercise 6. Explain why the base surface F most naturally has the structure of an orbifold of type $(g; p_1, \ldots, p_r)$.

Seifert manifolds can be given locally homogeneous Riemannian metrics (briefly "geometric structures"). There are six underlying types for the geometric structure. The orbifold Euler characteristic of this base orbifold and the Euler number e of the Seifert fibration together determine the type of natural geometric structure that can be put on M^3.

There exist a few manifolds M^3 that have more than one Seifert fibration. For example, the lens space $L(p, q)$ has infinitely many, all of them with base surface S^2 and at most two singular fibers (but if one requires the base to be a "good orbifold"– one that is globally the quotient of a group action on a manifold), then $L(p, q)$ has only one Seifert fibration up to isomorphism).

9.4. *"Seifert fibrations" with torus fiber*

There are two basic ways a 3-manifold M^3 can fiber with torus fibers. The base must be 1-dimensional so it is either the circle, or the 1-orbifold that one obtains by factoring the circle by the involution $z \mapsto \bar{z}$. The latter is the unit interval $[0, 1]$ considered as an orbifold.

In the case M^3 fibers over the circle, we can obtain it by taking $T^2 \times [0, 1]$ and then pasting $T^2 \times \{0\}$ to $T^2 \times \{1\}$ by an automorphism of the torus.

Thinking of the torus as $\mathbb{R}^2/\mathbb{Z}^2$, it is clear that an automorphism is given by a 2×2 integer matrix of determinant 1 (it is orientation preserving since we want M^3 orientable), that is, by an element $A \in \mathrm{SL}(2, \mathbb{Z})$.

Exercise 7. Show the resulting M^3 is Seifert fibered by circles if $|\operatorname{tr}(A)| \leq 2$. Work out the Seifert invariants.

If $|\operatorname{tr}(A)| > 2$ then the natural geometry for a geometric structure on M is the $\mathbb{S}ol$ geometry.

In case M^3 fibers over the orbifold $[0, 1]$ we can construct it as follows. The manifold SMb mentioned in Section 4 of this paper can also be described as the total space of the unique interval bundle over the Klein bottle with oriented total space. From this point of view, SMb is fibered by tori that are the boundaries of thinner versions of SMb obtained by shrinking the interval I, with the Klein bottle zero-section as special fiber. Gluing two copies of SMb by some identification of their torus boundaries gives M^3. This M^3 has a double cover that fibers over the circle, and it is Seifert fibered by circles if and only this double cover is Seifert fibered by circles, otherwise it again belongs to the $\mathbb{S}ol$ geometry.

9.5. *Simple Seifert fibered manifolds*

We said earlier that if M^3 is irreducible and all its boundary components are tori then only tori occur in the JSJ decomposition. This is essentially because of the following:

Exercise 8. Let M^3 be an orientable manifold, all of whose boundary components are tori, which is simple (no essential tori) and suppose M^3 contains an essential embedded annulus (i.e., incompressible and not boundary parallel). Then M^3 is Seifert fibered over D^2 with two singular fibers, or over the annulus or the Möbius band with at most one singular fiber.

For manifolds with boundary, "simple" is often defined by the absence of essential annuli and tori, rather than just tori. The difference between these definitions is just the manifolds of the above exercise. $D^2 \times S^1$ is simple by either definition. The only other simple Seifert fibered manifolds are those that are Seifert fibered over S^2 with at most three singular fibers or over $\mathbb{P}^2$ with at most one singular fiber and which moreover satisfy $e(M^3 \to F) \neq 0$.

10. Geometric versus JSJ decomposition

The JSJ decomposition does not give exactly the decomposition of M^3 into pieces with geometric structure. This is because of the fact that the manifold SMb (that caused us problems in Section 4 of this paper) may occur as a Seifert fibered piece in the decomposition, but it does not admit a geometric structure.

Recall (subsection 9.4) that SMb has an embedded Klein bottle, and splitting it along this Klein bottle gives $T^2 \times I$. Thus, whenever SMb occurs as a piece in the JSJ decomposition, instead of including the boundary of this piece as one of the surfaces to split M^3 along, we include its core Klein bottle. The effect of this is simply to eliminate all such pieces without affecting the topology of any other piece. The modified version of JSJ-decomposition that one gets this way is called *geometric decomposition*.

References

1. S.V. Buyalo and V.L. Kobel′skiĭ, Geometrization of graphmanifolds. II isometric geometrization, St. Petersburg Math. J. **7** (1996), 387–404.
2. F. Costantino, On a proof of the JSJ theorem, Rend. Sem. Mat. Univ. Pol. Torino **60** (2002), 129–146.
3. D. Eisenbud and W.D. Neumann, *Three-dimensional link theory and invariants of plane curve singularities.* Ann. Math. Stud. **110**, Princeton. Princeton Univ. Press (1985).
4. M. Dehn, Über die Topologie des dreidimensionalen Raumes. Math. Ann. **69** (1910), 137–168.
5. H. Grauert, Über Modifikation und exceptionelle analytische Mengen, Math. Ann. **146** (1962), 331–368.
6. R. Hartshorne, *Algebraic Geometry*, Graduate Texts in Mathematics 52 (Springer Verlag 1977).
7. P. Heergaard, *Forstudier til en topologisk teori for de algebraiske fladers sammenhæng*, Dissertation (Det Nordiske Forlag, Kopenhagen 1898). French translation: Sur l'Analysis situs, Bulletin de la Société Mathématique de France **44** (1916), 161–242.
8. J. Hempel, *3-Manifolds*, Ann. of Math. Studies **86** (Princton Univ. Press, 1976).
9. W.H. Jaco and P.B. Shalen, *Seifert fibered spaces in 3-manifolds*, Memoirs of American Math.Soc., Vol. 21, Number 220 (1979).
10. W.H. Jaco, *Lectures on Three-Manifold Topology*, Amer. Math. Soc. Regional Conference Series in Mathematics, **43** (1980).
11. K. Johannson, *Homotopy equivalences of 3-manifolds with boundary*, Lecture Notes in Mathematics **761** (Springer-Verlag, 1979).
12. H. Kneser, Geschlossenen Flächen in dreidimensionalen Mannigfaltigkeiten, Jahresbericht der DMV **38** (1929), 248–260.
13. H.B. Laufer, On rational singularities. Amer. J. Math. **94** (1972), 597–608.

14. H.B. Laufer, On minimally elliptic singularities. Amer. J. Math. **99** (1977), 1257–1295.

15. J. Milnor, A unique factorization theorem for 3-manifolds, Amer. J. Math. 84 (1962), 1–7.

16. D. Mumford, The topology of normal singularities of an algebraic surface and a criterion for simplicity. Inst. Hautes tudes Sci. Publ. Math. **9** (1961), 5–22.

17. W.D. Neumann, A calculus for plumbing applied to the topology of complex surface singularities and degenerating complex curves, Trans. Amer. Math. Soc. **268** (1981), 299–343.

18. W.D. Neumann, Abelian covers of quasihomogeneous surface singularities, *Singularities, Arcata 1981*, Proc. Symp. Pure Math. **40** (Amer. Math. Soc. 1983), 233–243.

19. W.D. Neumann, Geometry of quasihomogeneous surface singularities, *Singularities, Arcata 1981*, Proc. Symp. Pure Math. **40** (Amer. Math. Soc. 1983), 244–258.

20. W.D. Neumann, Commensurability and virtual fibration for graph manifolds, Topology **36** (1997), 355–378.

21. W.D. Neumann, Immersed and virtually embedded π_1-injective surfaces in graph manifolds, Algebraic and Geometric Topology **1** (2001), 411–426.

22. W.D. Neumann, Notes on Geometry and 3-Manifolds (with appendix by Paul Norbury), *Low Dimensional Topology*, Böröczky, Neumann, Stipsicz, Eds., Bolyai Society Mathematical Studies **8** (1999), 191–267.

23. W.D. Neumann, Topology of Hypersurface Singularities, in *Erich Kähler – Mathematische Werke, Mathematical works*, R. Berndt and O. Riemenschneider eds., Berlin, New York: Walter de Gruyter Verlag 2003, pp. 727–736.

24. W.D. Neumann, paper in preparation.

25. W.D. Neumann, G.A. Swarup, Canonical decompositions of 3-manifolds, Geometry and Topology **1** (1997), 21–40.

26. W.D. Neumann, J. Wahl, Universal abelian covers of surface singularities, *Trends on Singularities*, A. Libgober and M. Tibar, eds. (Birkhäuser Verlag, 2002), 181–190.

27. W.D. Neumann, J. Wahl, Universal abelian covers of quotient-cusps. Math. Ann. **326** (2003), 75–93.

28. W.D. Neumann, J. Wahl, Complete intersection singularities of splice type as universal abelian covers, Geom. Topol. **9** (2005), 699–755.

29. W.D. Neumann, J. Wahl, Complex surface singularities with homology sphere links, Geom. Topol. **9** (2005), 757–811.

30. W.D. Neumann, J. Wahl, paper in preparation.

31. T. Okuma, Universal abelian covers of rational surface singularities. J. London Math. Soc. **70** (2004), 307–324.

32. T. Okuma, Universal abelian covers of certain surface singularities, `http://arxiv.org/math.AG/0503733`.

33. M. Reid, Chapters on Algebraic Surfaces, *Complex algebraic geometry (Park City, UT, 1993)*, IAS/Park City Math. Ser. 3, Amer. Math. Soc., (Providence, RI, 1997), 3–159.

34. G.P. Scott, The geometries of 3-manifolds, Bull. London Math. Soc. **15**

(1983), 401–478.

35. G.P. Scott, Strong annulus and torus theorems and the enclosing property of characteristic submanifolds of 3-manifolds, Quarterly J. Math. Oxford (2), **35** (1984), 485–506,

36. J. Stallings, On the loop theorem, Ann. of Math. **72** (1960), 12–19.

37. G.A. Swarup, On incompressible surfaces in the complements of knots. J. Indian Math. Soc. (N.S.) **37** (1973), 9–24. Addendum: J. Indian Math. Soc. (N.S.) **38** (1974) 411–413.

38. W.P. Thurston, The geometry and topology of 3-manifolds. Mimeographed lecture notes, Princeton Univ., 1977.

39. F. Waldhausen, On irreducible 3-manifolds which are sufficiently large. Ann. of Math. (2) **87** (1968), 56–88.

40. F. Waldhausen, Gruppen mit Zentrum und 3-dimensionale Mannigfaltigkeiten, Topology **6** (1967), 505–517.

41. F. Waldhausen, On the determination of some 3-manifolds by their fundamental groups alone, Proc. of Inter. Sympo. Topology, Hercy-Novi, Yugoslavia, 1968, Beograd (1969), 331–332.

42. F. Waldhausen, On some recent results in 3-dimensional topology, Proc. Symp. in Pure Math. Amer. Math. Soc. **32** (1977), 21–38.

RIGID GEOMETRY AND THE MONODROMY CONJECTURE

JOHANNES NICAISE*

*Laboratoire Painlevé, CNRS - UMR 8524, Université Lille 1,
Cité Scientifique, 59655 Villeneuve d'Ascq Cédex (France),
E-mail: johannes.nicaise@math.univ-lille1.fr*

JULIEN SEBAG

*Institut Mathématique de Bordeaux, Laboratoire A2X, Université Bordeaux I,
351 Cours de la Libération, 33405 Talence (France),
E-mail: julien.sebag@math.u-bordeaux1.fr*

Dedicated to Jean-Paul Brasselet on the occasion of his 60th birthday

1. Introduction

The results in this survey will appear in the articles [20] (the curve case) and [21], and were announced in [23].

Let L be a number field, with ring of integers $\mathcal{O}$, and fix an embedding $L \subset \mathbb{C}$. Let f be a polynomial in $\mathcal{O}[x_1, \ldots, x_m]$. The Monodromy Conjecture predicts an intriguing connection between the arithmetic properties of f, and its complex topology. More precisely, it relates the poles of the Igusa zeta function associated to f, to the eigenvalues of the monodromy transformations at the points of the complex hypersurface defined by f.

The most famous instance of the interaction between arithmetic properties and complex topology, is probably given by the Weil Conjectures. We will show in this survey that there exists a strong formal connection between the Weil Conjectures and the Monodromy Conjecture, using the theory of motivic integration, and étale cohomology for non-archimedean analytic spaces.

In Section 2, we recall the statement of the Monodromy Conjecture. One

---

* During the preparation of this article, the first author was a Research Assistant of the Fund for Scientific Research – Flanders (Belgium)(F.W.O.)

of the aspects of the Monodromy Conjecture that make it hard to handle, is the fact that it postulates a bridge between two worlds: the arithmetic universe, and the universe of complex geometry. These worlds can be unified by means of the theory of motivic integration. In Section 3, we recall the definitions of Denef and Loeser's motivic zeta function and naïve motivic zeta function, as well as their motivic Monodromy Conjecture.

We propose a new point of view in Section 4. We use non-archimedean analytic geometry to construct an analytic Milnor fiber. This object is the bridge between the jet spaces used to define the motivic zeta function, and the monodromy action on the cohomology of the topological Milnor fiber. In Section 5, we introduce the motivic Serre invariant as a means to count rational points on rigid spaces. Using this invariant, we establish the motivic zeta function as a Weil generating series, counting rational points on the nearby fiber of f. We establish a corresponding Grothendieck trace formula, and state some applications to the theory of motivic zeta functions.

2. The Monodromy Conjecture

2.1. *Igusa's zeta function*

We fix a number field L, and an embedding $L \subset \mathbb{C}$. We denote by $\mathcal{O}$ its ring of integers. Let f be a non-constant polynomial in $\mathcal{O}[x_1, \ldots, x_m]$, and let $\mathfrak{P}$ be a maximal ideal in $\mathcal{O}$. We denote by $\mathcal{O}_{\mathfrak{P}}$ the $\mathfrak{P}$-adic completion of $\mathcal{O}$.

For $s \in \mathbb{C}$, $\Re(s) > 0$, one defines

$$Z_{\mathfrak{P}}(f; s) = \int_{\mathcal{O}_{\mathfrak{P}}^m} |f(x)|_{\mathfrak{P}}^s |dx|$$

where $|.|_{\mathfrak{P}}$ is the $\mathfrak{P}$-adic norm, and $|dx|$ is the Haar measure on $\mathcal{O}_{\mathfrak{P}}^m$. By a simple transformation formula, $Z_{\mathfrak{P}}(f; s)$ is related to the generating series

$$\sum_{n=0}^{\infty} |\{z \in (\mathcal{O}/\mathfrak{P}^{n+1}\mathcal{O})^m \mid f(z) \equiv 0 \mod \mathfrak{P}^{n+1}\}| T^n$$

where $|A|$ denotes the cardinality of the set A.

Taking an embedded resolution of singularities for f, Igusa proved in [15,16] that $Z_{\mathfrak{P}}(f; s)$ is a rational function in q^{-s}, where q is the cardinality of the residue field of $\mathcal{O}_{\mathfrak{P}}$. In particular, $Z_{\mathfrak{P}}(f; s)$ allows a meromorphic continuation to $\mathbb{C}$. This meromorphic continuation is Igusa's zeta function, associated to f and $\mathfrak{P}$. For a general survey on Igusa's zeta function, we refer the reader to Denef's Bourbaki report [8].

2.2. *The Milnor fibration*

Let X be a complex manifold, and let $g : X \to \mathbb{C}$ be a non-constant analytic function. We denote by X_0 the analytic subvariety of X defined by g. If x is a point of the hypersurface X_0, the topology of X_0 near x can be studied by means of the Milnor fibration [19]. Choosing an analytic chart, we may assume $X = \mathbb{C}^m$. Let $B(0,\eta)$ be an open disc in $\mathbb{C}$, with center 0, and radius η, and let $B(x,\varepsilon)$ be an open disc in X, with center x, and radius ε. We denote the punctured disc $B(0,\eta) \setminus \{0\}$ by $B(0,\eta)^\times$. For $0 < \eta \ll \varepsilon \ll 1$, the map

$$g : B(x,\varepsilon) \cap g^{-1}(B(0,\eta)^\times) \to B(0,\eta)^\times$$

is a locally trivial fibration [12], called the Milnor fibration of g at x. The fiber F_x over $t \in B(0,\eta)^\times$ is the topological Milnor fiber of g at x. Its singular cohomology is concentrated in degrees 0 and $m - s - 1, \ldots, m - 1$, where s is the dimension at x, of the locus of critical points of g.

The Milnor fibration induces a canonical representation of $\pi_1(B(0,\eta)^\times, t) \cong \mathbb{Z}$ on the singular cohomology spaces $H^*_{sing}(F_x, \mathbb{C})$. The image of the canonical generator of $\pi_1(B(0,\eta)^\times, t)$ (i.e. the class of a loop encircling the origin once counterclockwise) is called the monodromy transformation of f at the point x. The monodromy zeta function of f at x, denoted by $\zeta_x(T)$, is the alternating product $\prod_{i=0}^{m-1} P_i(T)^{((-1)^{i+1})}$, where $P_i(T)$ is the characteristic polynomial of the monodromy transformation on $H^i_{sing}(F_x, \mathbb{C})$.

2.3. *The Monodromy Conjecture*

We fix a number field L, and an embedding $L \subset \mathbb{C}$. We denote by $\mathcal{O}$ its ring of integers. Let f be a non-constant polynomial in $\mathcal{O}[x_1, \ldots, x_m]$. Via the embedding $L \subset \mathbb{C}$, we can also consider f as a complex polynomial.

The Monodromy Conjecture predicts an intriguing relationship between the arithmetic properties of f over $\mathcal{O}$, and the topology of the complex hypersurface X_0 defined by f.

Conjecture 2.1 (Monodromy conjecture). *For almost all maximal ideals $\mathfrak{P}$ in $\mathcal{O}$ (i.e. all but a finite number), the following holds: if α is a pole of $Z_{\mathfrak{P}}(f; s)$, then there exists a complex point x on X_0, such that $e^{2\pi i \Re(\alpha)}$ is an eigenvalue of the monodromy transformation of f at x.*

Remark 2.1. To be precise, this is only the most basic set-up in which the Monodromy Conjecture can be stated. See [8, 2.3.2] for a more complete statement.

822

3. A first geometric reduction: the motivic Monodromy Conjecture

One of the aspects of the Monodromy Conjecture that make it hard to handle, is the fact that it postulates a bridge between two worlds: the arithmetic universe, and the universe of complex geometry. These worlds can be unified by means of the theory of motivic integration.

The key idea of the construction, is to replace the ring of $\mathfrak{P}$-adic integers $\mathcal{O}_{\mathfrak{P}}$, by the ring of formal power series $L[[t]]$. On the one hand, Denef and Loeser show that the behaviour of f over $L[[t]]$ reflects the essential properties of f over $\mathcal{O}_{\mathfrak{P}}$, for almost all $\mathfrak{P}$. On the other hand, it captures important information on the structure of the singularities of the complex hypersurface defined by f.

3.1. *The Grothendieck ring of varieties*

Let k be a field. By a variety over a scheme S, we mean a separated reduced scheme, of finite type over S. For any scheme S, we denote by S_{red} the underlying reduced scheme.

Let Z be a variety over k. Consider the free Abelian group, generated by the isomorphism classes $[X]$ of Z-varieties X. We take the quotient of this group w.r.t. the following relations: whenever X is a Z-variety, and Y is a closed subvariety of X, we have $[X] = [X \setminus Y] + [Y]$. This quotient is called the Grothendieck group of varieties over Z, and is denoted by $K_0(Var_Z)$. We denote the class $[\mathbb{A}^1_Z]$ of the affine line over Z by $\mathbb{L}_Z$.

A constructible subset C of a Z-variety X can be written as a disjoint union of locally closed subsets, and defines unambiguously an element $[C]$ of $K_0(Var_Z)$. When Z is a separated scheme of finite type over k, we will write $K_0(Var_Z)$ instead of $K_0(Var_{Z_{\mathrm{red}}})$. For any separated scheme X of finite type over Z, we will write $[X]$ instead of $[X_{\mathrm{red}}]$.

We can define a product on $K_0(Var_Z)$ as follows: for any pair of Z-varieties X, Y, we put $[X].[Y] = [X \times_Z Y]$. This definition extends bilinearly to a product on $K_0(Var_Z)$, and makes it into a ring, the Grothendieck ring of varieties over Z. The localized Grothendieck ring $\mathcal{M}_Z$ is obtained by inverting $\mathbb{L}_Z$ in $K_0(Var_Z)$.

A morphism of k-varieties $f : W \to Z$ induces base-change ring morphisms $K_0(Var_Z) \to K_0(Var_W)$ and $\mathcal{M}_Z \to \mathcal{M}_W$, as well as forgetful morphisms of Abelian groups $K_0(Var_W) \to K_0(Var_Z)$ and $\mathcal{M}_W \to \mathcal{M}_Z$. The definition of the latter morphism deserves some care: an element $[X]\mathbb{L}_W^{-m}$ of $\mathcal{M}_W$, for some W-variety X and some positive integer m, is mapped to $[X]\mathbb{L}_Z^{-m}$ in $\mathcal{M}_Z$. One checks that this yields a well-defined morphism of Abelian groups $\mathcal{M}_W \to \mathcal{M}_Z$.

If $Z = \operatorname{Spec} k$, we write $K_0(Var_k)$, $\mathcal{M}_k$, and $\mathbb{L}$, rather than $K_0(Var_{\operatorname{Spec} k})$, $\mathcal{M}_{\operatorname{Spec} k}$, and $\mathbb{L}_{\operatorname{Spec} k}$.

The Grothendieck group $K_0(Var_Z)$ is a universal additive invariant for Z-varieties: if A is an Abelian group, and χ is an invariant of Z-varieties taking values in A, such that, for any Z-variety X and any closed subvariety $Y \subset X$, $\chi(X) = \chi(X \setminus Y) + \chi(Y)$, then χ factors uniquely through a group morphism $\chi : K_0(Var_Z) \to A$, defined by $\chi([X]) = \chi(X)$. If A is a ring, and χ is multiplicative, i.e. $\chi((X \times_Z Y)_{\mathrm{red}}) = \chi(X).\chi(Y)$ for any pair of Z-varieties X, Y, then $\chi : K_0(Var_Z) \to A$ is a morphism of rings.

For instance, for any k-variety X, we can consider its topological Euler characteristic $\chi_{top}(X)$. Fix a prime ℓ, invertible in k. Then $\chi_{top}(X)$ is defined as

$$\chi_{top}(X) := \sum_{i \geq 0} (-1)^i \dim H_c^i(X \times_k k^s, \mathbb{Q}_\ell)$$

where $H_c^i(\,.\,, \mathbb{Q}_\ell)$ is ℓ-adic étale cohomology with proper support, and k^s is a separable closure of k. This is an additive invariant, hence defines a morphism of groups

$$\chi_{top} : K_0(Var_Z) \to \mathbb{Z}$$

for any base variety Z. It is multiplicative for $Z = \operatorname{Spec} k$, so we get a morphism of rings

$$\chi_{top} : K_0(Var_k) \to \mathbb{Z}$$

3.2. *Arc spaces*

Let k be a field, and let X be a variety over k. For any integer $n \geq 0$, we consider the functor

$$F_n : (k - schemes) \to (Sets) : Y \mapsto Hom_{(k-schemes)}(Y \times_k k[t]/(t^{n+1}), X)$$

from the category of k-schemes to the category of sets.

This functor is representable by a separated k-scheme of finite type $\mathcal{L}_n(X)$, the n-th jet scheme of X. A K-valued n-jet on X, for some field $K \supset k$, is a K-valued point on $\mathcal{L}_n(X)$, or, equivalently, a morphism $\operatorname{Spec} K[t]/(t^{n+1}) \to X$. Note that $\mathcal{L}_0(X) \cong X$. The origin of the n-jet is the image in X of the unique point of $\operatorname{Spec} K[t]/(t^{n+1})$. For $m \geq n \geq 0$, the truncation morphism $k[t]/(t^{m+1}) \to k[t]/(t^{n+1})$ induces a natural transformation of functors $F_m \to F_n$, and, by Yoneda's Lemma, a morphism of k-schemes

$$\pi_n^m : \mathcal{L}_m(X) \to \mathcal{L}_n(X)$$

The natural projection $\pi_0^n : \mathcal{L}_n(X) \to X$ maps an n-jet to its origin.

The construction of the jet schemes shows that the truncation maps π_n^m are affine. Hence, we can take the projective limit

$$\mathcal{L}(X) := \varprojlim_{n \in \mathbb{N}} \mathcal{L}_n(X)$$

in the category of k-schemes, and we obtain the arc scheme $\mathcal{L}(X)$ of X. This scheme is not Noetherian, in general. It comes with natural projections $\pi_n : \mathcal{L}(X) \to \mathcal{L}_n(X)$.

If X is affine, this arc scheme satisfies $\mathcal{L}(X)(A) = X(A[[t]])$ for any k-algebra A. For any k-variety X, and any field $K \supset k$, we have $\mathcal{L}(X)(K) = X(K[[t]])$. A K-valued arc on X, for some field $K \supset k$, is a K-valued point on $\mathcal{L}(X)$, or, equivalently, a morphism $\operatorname{Spec} K[[t]] \to X$. The origin of the arc is the image in X of the closed point of $\operatorname{Spec} K[[t]]$. The natural projection $\pi_0 : \mathcal{L}(X) \to X$ maps an arc to its origin.

In what follows, we will always endow $\mathcal{L}_n(X)$ and $\mathcal{L}(X)$ with their reduced structure.

3.3. *Motivic zeta functions*

Let k be a field of characteristic zero, let X be a smooth irreducible k-variety of dimension d, and let $f : X \to \mathbb{A}_k^1 = \operatorname{Spec} k[x]$ be a dominant morphism. We denote by X_0 the (reduced) hypersurface in X defined by f.

Fix an integer $n > 0$. We can compose any n-jet $\psi : \operatorname{Spec} K[t]/(t^{n+1}) \to X$ on X with the morphism f; the pullback of the coordinate function x on $\mathbb{A}_k^1$ via $f \circ \psi$ yields an element of $K[t]/(t^{n+1})$, which we'll denote by $f(\psi)$. If ψ is a K-valued arc on X, $f(\psi) \in K[[t]]$ is defined analogously.

For $\alpha \in K[t]/(t^{n+1})$, the order $ord_t \, \alpha$ is by definition equal to ∞ when $\alpha = 0$, and equal to $i \in \{0, \dots, n\}$, if α can be written as $a_i t^i + a_{i-1} t^{i-1} + \dots + a_0$, with $a_i \neq 0$.

For any integer $n > 0$, we consider the following locally closed subschemes of the jet scheme $\mathcal{L}_n(X)$:

- $\mathfrak{X}_n := \{\psi \in \mathcal{L}_n(X) \,|\, ord_t \, f(\psi) = n\}$
- $\mathfrak{X}_{n,1} := \{\psi \in \mathcal{L}_n(X) \,|\, f(\psi) = t^n \mod t^{n+1}\}$.

We endow $\mathfrak{X}_n$ and $\mathfrak{X}_{n,1}$ with their reduced structure. The truncation map $\pi_0^n : \mathcal{L}_n(X) \to X$ induces morphisms $\mathfrak{X}_n \to X$ and $\mathfrak{X}_{n,1} \to X$. Since $n > 0$, the origins of n-jets in $\mathfrak{X}_n$ and $\mathfrak{X}_{n,1}$ are contained in X_0. Hence, the morphisms $\mathfrak{X}_n \to X$ and $\mathfrak{X}_{n,1} \to X$ factor through X_0, and $\mathfrak{X}_n$ and $\mathfrak{X}_{n,1}$ get the structure of a X_0-variety.

Definition 3.1 (Denef-Loeser). *The naïve motivic zeta function associated to f, is the generating series*

$$Z^{\text{naïve}}(f; T) := \sum_{n>0} [\mathfrak{X}_n] \mathbb{L}^{-dn} T^n \in \mathcal{M}_{X_0}[[T]]$$

The motivic zeta function associated to f, is the generating series

$$Z(f; T) := \sum_{n>0} [\mathfrak{X}_{n,1}] \mathbb{L}^{-dn} T^n \in \mathcal{M}_{X_0}[[T]]$$

Remark 3.1. Actually, Denef and Loeser define the motivic zeta function over a more refined, equivariant localized Grothendieck ring.

An embedded resolution of singularities for f, and the change of variables formula for motivic integrals, yield explicit expressions for both motivic zeta functions, in terms of the geometry of the resolution. In particular, they are rational functions over $\mathcal{M}_{X_0}$, and the limit

$$-\lim_{T \to \infty} Z(f; T) =: \mathcal{S}_f \in \mathcal{M}_{X_0}$$

is well-defined. Denef and Loeser call $\mathcal{S}_f$ the "motivic nearby cycles". For any closed point x on X_0, the image of $\mathcal{S}_f$ under the base-change morphism $\mathcal{M}_{X_0} \to \mathcal{M}_x$ is denoted by $\mathcal{S}_{f,x}$, and is called the motivic Milnor fiber of f at x. Denef and Loeser show that, if $k = \mathbb{C}$, the Hodge characteristic of the Milnor fiber of f at x can be recovered from $\mathcal{S}_{f,x}$ [9].

3.4. *The motivic Monodromy Conjecture*

We keep the notations of Section 3.3. Denef and Loeser state the following conjecture [9, Section 2.4].

Conjecture 3.1 (Motivic Monodromy Conjecture). *Suppose $k \subset \mathbb{C}$. There exists a finite set*

$$S = \{(a, b) \,:\, a, b \in \mathbb{N}, b > 0\}$$

such that the image of $Z^{\text{naïve}}(f; T)$ under the forgetful morphism $\mathcal{M}_{X_0}[[T]] \to \mathcal{M}_k[[T]]$ belongs to the subring

$$\mathcal{M}_k[T][\frac{1}{1 - \mathbb{L}^{-a} T^b}]_{(a,b) \in S}$$

and such that, for each $(a,b) \in S$, $e^{2\pi ia/b}$ is an eigenvalue of the monodromy transformation of f at some complex point of X_0.

Suppose that $k = L$ is a number field, with ring of integers $\mathcal{O}$, and fix an embedding $L \subset \mathbb{C}$. Let f be a polynomial in $\mathcal{O}[x_1,\dots,x_m]$. Denef and Loeser show that the naïve motivic zeta function associated to $f : \mathbb{A}_L^m \to \mathbb{A}_L^1$ specializes to Igusa's zeta function $Z_{\mathfrak{P}}(f;s)$, for almost all maximal ideals $\mathfrak{P}$ in $\mathcal{O}$, in some appropriate sense [11, 4.2.1.(2)]. As a consequence, the motivic Monodromy Conjecture for $f : \mathbb{A}_L^m \to \mathbb{A}_L^1$, implies the Monodromy Conjecture for the polynomial $f \in \mathcal{O}[x_1,\dots,x_m]$.

4. A new point of view: rigid geometry

We saw that the motivic Monodromy Conjecture relates the structure of the jet spaces, to the monodromy action on the cohomology of the Milnor fibers. In this section, we will try to show how the framework of rigid geometry can explain the relation between arc spaces and monodromy.

4.1. *Construction of the analytic Milnor fiber*

Let k be an algebraically closed field of characteristic zero, let X be a smooth irreducible variety over k, and let $f : X \to \mathbb{A}_k^1 = \operatorname{Spec} k[t]$ be a dominant morphism. We'll denote the ring of formal power series $k[[t]]$ by R, and its field of fractions $k((t))$ by K.

We can mimic the topological construction of the Milnor fibration in Section 2.2, in the setting of formal geometry. Consider the t-adic completion of the morphism $f : X \to \mathbb{A}_k^1 = \operatorname{Spec} k[t]$. This is a separated, flat, generically smooth formal R-scheme X_∞, topologically of finite type over R. To this formal scheme, one can associate the following data [5, 0.2.3]:

- its special fiber X_0, which is simply the subscheme of X defined by f,
- its generic fiber X_η, which is a separated, quasi-compact, smooth rigid analytic space over K,
- a specialization morphism of ringed sites $sp : X_\eta \to X_\infty$, so, in particular, a map between the underlying sets

$$|sp| : |X_\eta| \to |X_\infty| = |X_0|$$

For any closed subscheme Z of X_0, we denote $|sp|^{-1}(|Z|)$ by $]Z[$. It is an open analytic subspace of X_η in a natural way, and can be identified with the generic fiber of the formal completion of X along Z [5, 0.2.7].

For any closed point x on X_0, we define the analytic Milnor fiber $\mathcal{F}_x$ of f at x by $\mathcal{F}_x :=]x[$.

Remark 4.1. The topological intuition behind the construction is the following: the formal neighborhood $\operatorname{Spf} R$ of the origin in $\mathbb{A}_k^1 = \operatorname{Spec} k[t]$, corresponds to a sufficiently small disc around the origin in $\mathbb{C}$. Its inverse image under f is realized as the t-adic completion of the morphism f; the formal scheme X_∞ should be seen as a tubular neighborhood of the special fiber X_0 defined by f on X. The inverse image of the punctured disc becomes the complement of X_0 in X_∞; this complement makes sense in the category of rigid spaces, and we obtain the generic fiber X_η of X_∞. The specialization map sp can be seen as a canonical retraction of X_η on X_0.

4.2. *Points of the analytic Milnor fiber*

We keep the notations from Section 4.1. Finite extensions of the field K are of the form $K' = k((\sqrt[d]{t}))$, with $d > 0$ a strictly positive integer. We denote by R' the normalization of R in K', i.e. R' is the ring of formal power series $k[[\sqrt[d]{t}]]$.

The analytic space $X_\eta \times_K K'$ is canonically isomorphic to the generic fiber of the formal R'-scheme $X_\infty \times_R R'$, and the special fiber of this formal scheme is canonically isomorphic to X_0.

The K'-rational points on $X_\eta \times_K K'$ are in natural bijective correspondence with R'-rational points on the formal scheme $X_\infty \times_R R'$. These are, at their turn, in natural bijective correspondence with morphisms $\psi : \operatorname{Spec} R' \to X$, with $f(\psi) = t$, or, modulo a reparametrization $t \mapsto t^d$, with the set of k-rational arcs

$$\psi : \operatorname{Spec} k[[t]] \to X$$

on X, with $f(\psi) = t^d$. The specialization map

$$sp : X_\eta \times_K K' \to X_0$$

maps an arc ψ to its origin. Hence, if x is a closed point on X_0, the K'-valued points on the analytic Milnor fiber $\mathcal{F}_x$ of f at x, are exactly the k-rational arcs $\psi : \operatorname{Spec} k[[t]] \to X$ on X, with $f(\psi) = t^d$, and with origin x. We can think of the analytic Milnor fiber $\mathcal{F}_x$ as an object consisting of the generic points of arcs ψ on X with origin x, and with $f(\psi) \neq 0$.

The extension K'/K is Galois, with Galois group $\mu_d(k)$. This Galois group acts on $X_\eta(K')$ as follows: if $\xi \in \mu_d(k)$, and ψ is a k-rational arc on X with $f(\psi) = t^d$, then $\xi.\psi$ is the composition of ψ with the k-automorphism of $\operatorname{Spec} k[[t]]$ induced by $t \mapsto \xi t$.

828

4.3. *Cohomology of the analytic Milnor fiber*

We keep the notations from Section 4.1. We'll denote by $\widehat{K^s}$ the completion of the algebraic closure of the non-Archimedean field K. We fix a prime ℓ. For any integer $n > 0$, we consider the complex of étale nearby cycles $R\psi_\eta(\mathbb{Z}/\ell^n)$ with coefficients in $\mathbb{Z}/\ell^n$, associated to f. It is an object of $D_c^b(X_0, \mathbb{Z}/\ell^n)$.

In [3], Berkovich developed a theory of étale cohomology for non-Archimedean spaces. If $\mathfrak{X}$ is a $\widehat{K^s}$-analytic space, we put

$$H^i(\mathfrak{X}, \mathbb{Z}_\ell) := \varprojlim_n H^i(\mathfrak{X}, \mathbb{Z}/\ell^n)$$

$$H^i(\mathfrak{X}, \mathbb{Q}_\ell) := H^i(\mathfrak{X}, \mathbb{Z}_\ell) \otimes_{\mathbb{Z}_\ell} \mathbb{Q}_\ell$$

$$H(\mathfrak{X}, \mathbb{Q}_\ell) := \oplus_i H^i(\mathfrak{X}, \mathbb{Q}_\ell) \text{ as a graded } \mathbb{Q}_\ell\text{-vector space.}$$

From the comparison theorem [4, 3.5], we obtain the following result:

Prop 4.1. Let x be a closed point of X_0. For any $n > 0$, and any $i \geq 0$, there is a canonical isomorphism

$$H^i(\mathcal{F}_x \widehat{\times}_K \widehat{K^s}, \mathbb{Z}/\ell^n) \cong R^i\psi_\eta(\mathbb{Z}/\ell^n)_x$$

and the Galois action of $G(K^s/K)$ on $H^i(\mathcal{F}_x \widehat{\times}_K \widehat{K^s}, \mathbb{Z}/\ell^n)$ corresponds to the monodromy action of $G(K^s/K)$ on $R^i\psi_\eta(\mathbb{Z}/\ell^n)_x$.

If $k = \mathbb{C}$, we can combine this result with the comparison results for étale and transcendent nearby cycles [1].

Prop 4.2. Fix an embedding of $\mathbb{Q}_\ell$ in $\mathbb{C}$. Let x be a closed point of X_0. For any $i \geq 0$, there is a (non-canonical) isomorphism

$$H^i(\mathcal{F}_x \widehat{\times}_K \widehat{K^s}, \mathbb{Q}_\ell) \otimes_{\mathbb{Q}_\ell} \mathbb{C} \cong H^i_{sing}(F_x, \mathbb{C})$$

such that the Galois action of the canonical generator of $G(K^s/K)$ on the left hand side corresponds to the monodromy transformation on $H^i_{sing}(F_x, \mathbb{C})$.

Remark 4.2. If we replace F_x by the canonical topological Milnor fiber of f at x, the isomorphisms in Proposition 4.2 become canonical [17, (8.11.7)].

Hence, the analytic Milnor fiber provides us with an object whose points are closely related to the arcs on X, and the order of f on these arcs, and whose cohomology spaces, endowed with the natural Galois action, correspond to the singular cohomology of the topological Milnor fiber, endowed with he monodromy action.

5. The motivic zeta function as a Weil generating series

The results in the previous section led us to consider the motivic zeta function as a Weil generating series, counting rational points on analytic spaces. In the classical setting, the Weil generating series of a variety V over a finite field $\mathbb{F}_q$ is given by $\sum_{n>0} |V(\mathbb{F}_{q^n})| T^n$. The Weil conjectures probably constitute the best-known instance of the remarkable interaction between the arithmetic properties and the complex topology of a variety. Philosophically, but also formally, the Monodromy Conjecture inscribes itself in this framework in a natural way.

5.1. *Counting points on rigid spaces: the motivic Serre invariant*

In fact, we can give a precise meaning to the expression "counting rational points on analytic spaces". In [18], François Loeser and the second author have developed a theory of motivic integration on quasi-compact smooth separated rigid analytic varieties X_η, over the quotient field K of a complete discrete valuation ring R, with perfect residue field k. In particular, using the theory of Néron smoothening, they associate in a canonical way to a gauge form on X_η, its volume, which is an element of $\mathcal{M}_k$.

Taking the quotient of $\mathcal{M}_k$ by the ideal generated by $\mathbb{L} - 1$, this volume only depends on X_η (and no longer on the gauge form). Loeser and the second author called it the motivic Serre invariant of X_η, and denote it by $S(X_\eta)$. It is a motivic generalization of the p-adic Serre invariant of a compact smooth p-adic variety [25].

We can refine this construction in the following way. Let X_∞ be a separated, flat formal R-scheme, generically smooth, topologically of finite type over R. A weak Néron model for X_∞ consists of a composition of admissible formal blow-ups $Y_\infty \to X_\infty$, such that the formal R-scheme Y_∞ has the following property: for any finite unramified extension R'/R, all R'-rational points on Y_∞ are contained in the smooth locus $Sm(Y_\infty)$ of Y_∞/R.

By [6], weak Néron models always exist. If $Y_\infty \to X_\infty$ is any weak Néron model for X_∞, we define the motivic Serre invariant of X_∞ as

$$S(X_\infty) := [Sm(Y_\infty)_0] \in K_0(Var_{X_0})/(\mathbb{L}_{X_0} - [X_0])$$

Using the theory of motivic integration on formal schemes [24], one shows that this definition is independent of the choice of the weak Néron model Y_∞ [20]. The forgetful morphism

$$K_0(Var_{X_0})/(\mathbb{L}_{X_0} - [X_0]) \to K_0(Var_k)/(\mathbb{L} - 1) = \mathcal{M}_k/(\mathbb{L} - 1)$$

830

maps $S(X_\infty)$ to $S(X_\eta)$.

Since the affine line over k has Euler characteristic 1, the group morphism $\chi_{top} : K_0(Var_Z) \to \mathbb{Z}$ from Section 3.1 factors through

$$\chi_{top} : K_0(Var_Z)/(\mathbb{L}_Z - [Z]) \to \mathbb{Z}$$

for any base k-variety Z.

Remark 5.1. If X_η is a separated quasi-compact smooth rigid space over K, we can use the above constructions to define its Serre invariant $S(X_\eta)$ in

$$\varprojlim_{X_\infty} K_0(Var_{X_0})/(\mathbb{L}_{X_0} - [X_0])$$

where X_∞ runs through the projective system of admissible formal R-models of X_η, ordered by admissible formal blow-ups. The projection of $S(X_\eta)$ on

$$K_0(Var_{X_0})/(\mathbb{L}_{X_0} - [X_0])$$

is simply $S(X_\infty)$. Compare this to the construction of the Zariski-Riemann space associated to X_η in [13].

If Z is any closed subscheme of X_0, we define the motivic Serre invariant of X_∞ with support in Z, denoted by $S_Z(X_\infty)$, as the image of $S(X_\infty)$ under the base change morphism

$$K_0(Var_{X_0})/(\mathbb{L}_{X_0} - [X_0]) \to K_0(Var_Z)/(\mathbb{L}_Z - [Z])$$

In other words, if $h : Y_\infty \to X_\infty$ is any weak Néron model for X_∞, then

$$S_Z(X_\infty) = [Sm(Y_\infty)_0 \cap h^{-1}(Z)] \in K_0(Var_Z)/(\mathbb{L}_Z - [Z])$$

We would like to consider $S_Z(X_\infty)$ as a measure for the "number" of K'-rational points in the tube $]Z[$, where K' runs through the finite unramified extensions of K. The idea is that, since all such points on Y_η are contained in the smooth locus of Y_∞, the special fiber of this smooth locus is a good indicator for the number of rational points on the generic fiber.

5.2. *A Trace formula*

If the motivic Serre invariant can be interpreted as a measure for the number of rational points on a rigid space, we can look for a corresponding Grothendieck trace formula.

Let k be an algebraically closed field of characteristic zero, denote $k[[t]]$ by R, and $k((t))$ by K. For any integer $d > 0$, we denote by K_d the unique extension $k((\sqrt[d]{t}))$ of degree d over K, and by R_d the normalization $k[[\sqrt[d]{t}]]$ of R in K_d. Let φ be a topological generator of the absolute Galois group $G(K^s/K)$.

Theorem 5.1 (Trace formula). *Let X be a separated reduced scheme, flat and of finite type over R, with smooth generic fiber $X \times_R K$, and let X_∞/R be the t-adic completion of X. Let Z be a closed subvariety of X_0, proper over k. Then, for any integer $d > 0$,*

$$\chi_{top}(S_Z(X_\infty \times_R R_d)) = Trace(\varphi^d \mid H(]Z[\widehat{\times}_K \widehat{K^s}, \mathbb{Q}_\ell))$$

Remark 5.2. It is possible to prove a trace formula under more general conditions, assuming tame resolution of singularities. See [21].

5.3. *Computation of the Serre Poincaré series*

Let k be a field of characteristic zero, denote $k[[t]]$ by R, and $k((t))$ by K. For any integer $d > 0$, we denote by K_d the unique totally ramified extension $k((\sqrt[d]{t}))$ of degree d over K, and by R_d the normalization $k[[\sqrt[d]{t}]]$ of R in K_d.

Let X be a separated reduced scheme, flat and of finite type over R, with smooth generic fiber $X \times_R K$, and let X_∞/R be the t-adic completion of X.

Definition 5.1. The Serre Poincaré series of X_∞ is the generating series

$$S(X_\infty; T) := \sum_{d>0} S(X_\infty \times_R R_d)T^d \in (K_0(Var_{X_0})/(\mathbb{L}_{X_0} - [X_0]))[[T]]$$

Under the forgetful morphism

$$K_0(Var_{X_0})/(\mathbb{L}_{X_0} - [X_0]) \to K_0(Var_k)/(\mathbb{L} - 1)$$

it specializes to the Serre Poincaré series of the generic fiber X_η of X_∞, given by

$$S(X_\eta; T) := \sum_{d>0} S(X_\eta \times_K K_d)T^d \in (K_0(Var_k)/(\mathbb{L} - 1))[[T]]$$

Remark 5.3. If we consider the motivic Serre invariant of X_∞ as a measure for the number of rational points on X_η, the Serre Poincaré series is a Weil generating series of X_η.

We can give an explicit formula for $S(X_\infty; T)$, in terms of a resolution of singularities. Since R is an excellent ring, Hironaka's resolution of singularities [14] tells us that we can find a finite composition of blow-ups

$Y \to X$ with centers in the special fiber, such that Y is regular, and its special fiber Y_0 is a strict normal crossing divisor $\sum_{i \in I} N_i E_i$. We'll call any such $Y \to X$ an embedded resolution for the pair (X, X_0). The morphism $Y \to X$ induces an isomorphism between the generic fiber Y_η of the t-adic completion Y_∞ of Y, and X_η.

For each $i \in I$, we denote $E_i \setminus (\cup_{j \in I, j \neq i} E_j)$ by E_i^o. We construct a finite étale cover $\tilde{E}_i^o$ of E_i^o as follows: for each point x on E_i^o, we can find an affine open neighborhood $U = \operatorname{Spec} A$ of x in X, such that $t = u.v^{N_i}$ with $u, v \in A$, and u a unit. Consider the finite étale cover $\tilde{U} := \operatorname{Spec} A[z]/(uz^{N_i} - 1)$ of U, and restrict it over $E_i^o \cap U$. These covers glue together to a finite étale cover $\tilde{E}_i^o$ of E_i^o.

Remark 5.4. If X is obtained by base change from a morphism $f : Z \to \mathbb{A}_k^1 = \operatorname{Spec} k[t]$, with Z a smooth irreducible k-variety, this is just the construction from [10, 2.3].

Theorem 5.2. *Let X be a separated reduced scheme, flat and of finite type over R, with smooth generic fiber $X \times_R K$, and let X_∞/R be the t-adic completion of X. Let $Y \to X$ be an embedded resolution for (X, X_0), such that Y_0 is a strict normal crossing divisor $\sum_{i \in I} N_i E_i$. Then, for any integer $d > 0$,*

$$S(X_\infty \times_R R_d) = \sum_{N_i \mid d} [\tilde{E}_i^o] \in K_0(Var_{X_0})/(\mathbb{L}_{X_0} - [X_0])$$

and

$$S(X_\infty; T) = \sum_{i \in I} [\tilde{E}_i^o] \frac{T^{N_i}}{1 - T^{N_i}} \in (K_0(Var_{X_0})/(\mathbb{L}_{X_0} - [X_0]))[[T]]$$

The proof is based on an explicit construction of weak Néron models for the formal schemes $X_\infty \times_R R_d$, if X_∞ is a regular *stft* formal R-scheme whose special fiber is a strict normal crossing divisor.

5.4. *Motivic Serre invariant over the separable closure*

We keep the assumptions and notation of Section 5.3.

If V is a variety over a finite field $\mathbb{F}_q$, and if we denote by $Z(T)$ its Weil generating series $\sum_{n>0} |V(\mathbb{F}_{q^n})| T^n$, then Grothendieck's trace formula, combined with the identity [7, 1.5.3], shows that

$$\chi_{top}(V) = - \lim_{T \to \infty} Z(T)$$

Inspired by this result, we can state the following definitions.

We see from Theorem 5.2, that the limit

$$- \lim_{T \to \infty} S(X_\infty; T) \in K_0(Var_{X_0})/(\mathbb{L}_{X_0} - [X_0])$$

is well-defined, and equal to $\sum_{i \in I}[\tilde{E}_i^o]$. We call this limit the motivic Serre invariant of X_∞ over $\widehat{K^s}$, and denote it by $S(X_\infty; \widehat{K^s})$. For any subvariety Z of X_0, the image of $S(X_\infty; \widehat{K^s})$ under the base-change morphism

$$K_0(Var_{X_0})/(\mathbb{L}_{X_0} - [X_0]) \to K_0(Var_Z)/(\mathbb{L}_Z - [Z])$$

is called the motivic Serre invariant of X_∞ over $\widehat{K^s}$ with support in Z, and denoted by $S_Z(X_\infty; \widehat{K^s})$.

Remark 5.5. The invariants $S(X_\infty; \widehat{K^s})$ and $S_Z(X_\infty; \widehat{K^s})$ can already be defined over $\mathcal{M}_{X_0}$, resp. $\mathcal{M}_Z$, using motivic integrals of a gauge form on X_η. See [21].

A priori, $S_Z(X_\infty; \widehat{K^s})$ depends on the embedding of $]Z[$ in X_η, and not only on the space $]Z[\widehat{\times}_K \widehat{K^s}$. Taking limits $T \to \infty$ in our trace formula in Theorem 5.1, we see that, at least, $\chi_{top}(S_Z(X_\infty; \widehat{K^s}))$ is intrinsic, if Z is proper over k.

Prop 5.1. If Z is proper over k, then

$$\chi_{top}(S_Z(X_\infty; \widehat{K^s})) = \chi_{\acute{e}t}(]Z[\widehat{\times}_K \widehat{K^s})$$

where $\chi_{\acute{e}t}$ is the Euler characteristic with respect to Berkovich' ℓ-adic cohomology $H(]Z[\widehat{\times}_K \widehat{K^s}, \mathbb{Q}_\ell)$.

5.5. *Applications to the motivic zeta function*

Let k be an algebraically closed field of characteristic zero, denote $k[[t]]$ by R, and $k((t))$ by K. Let X be a smooth irreducible k-variety, and let $f : X \to \mathbb{A}^1_k = \mathrm{Spec}\, k[t]$ be a dominant morphism of k-varieties. We denote by X_∞/R the t-adic completion of f. The following result compares Denef and Loeser's motivic zeta function $Z(f; T)$ associated to f (Definition 3.1), to the Serre Poincaré series $S(X_\infty; T)$ associated to X_∞ (Definition 5.1).

Theorem 5.3.

$$Z(f; T) = S(X_\infty; T) \in (K_0(Var_{X_0})/(\mathbb{L}_{X_0} - [X_0]))[[T]]$$

This result immediately follows from our computation in Theorem 5.2, and the expression for $Z(f; T)$ in [10, 2.4.1]. Interpreting the motivic Serre invariant as a measure for the number of rational points on a rigid variety, we see that $Z(f;T)$ appears as a Weil generating series, counting rational points on finite extensions of the nearby fiber X_η.

Remark 5.6. It is possible to formulate a stronger result: one can recover $Z(f;T) \in \mathcal{M}_{X_0}[[T]]$ as a "Mellin transform" of a local singular series, via the motivic integral of the Gelfand Leray-form ω/df on the extensions $X_\eta \times_K K_d$ of the generic fiber X_η of X_∞. See [21].

In fact, Denef and Loeser endow their motivic zeta function with an additional action of the profinite group $\hat{\mu}(k)$ of roots of unity in k (see, for instance, [10, 2.9]). For each integer $d > 0$, $\mu_d(k)$ acts on the space $\mathfrak{X}_{d,1}$ as follows: if ψ is a point on $\mathfrak{X}_{d,1}$, and ξ is an element of $\mu_d(k)$, then $\xi.\psi(t) = \psi(\xi.t)$. This definition is remarkably close to our description of the Galois action of $\mu_d(k) = G(K_d/K)$ on $X_\eta \times_K K_d$ in Section 4.2. Any "decent" equivariant definition of the motivic Serre invariant, keeping track of this Galois action, should yield an equality between the equivariant Serre Poincaré series, and the motivic zeta function with $\hat{\mu}$-action. See [22].

Using Theorem 5.3, we can compare the motivic nearby cycles $\mathcal{S}_f$ associated to f (Section 3.3) to the Serre invariant of X_∞ over $\widehat{K^s}$. We can also compare the motivic Milnor fiber $\mathcal{S}_{f,x}$ of f at a closed point x of X_0 to the Serre invariant of X_η over $\widehat{K^s}$ with support in x.

Corollary 5.1. *We have*

$$\mathcal{S}_f = S(X_\infty; \widehat{K^s}) \in K_0(Var_{X_0})/(\mathbb{L}_{X_0} - [X_0])$$

and, for any closed point x on X_0,

$$\mathcal{S}_{f,x} = S_x(X_\infty; \widehat{K^s}) \in K_0(Var_k)/(\mathbb{L} - 1)$$

Remark 5.7. By means of Remark 5.5 and Remark 5.6, these equalities can be strengthened to equalities in $\mathcal{M}_{X_0}$, resp. $\mathcal{M}_k$.

To put it imprecisely, we recover Denef and Loeser's motivic Milnor fiber as the motivic Serre invariant of the analytic Milnor fiber $\mathcal{F}_x \widehat{\times}_K \widehat{K^s}$. Again, any "decent" equivariant definition of the motivic Serre invariant, should yield an equality between the motivic Serre invariant of $\mathcal{F}_x \widehat{\times}_K \widehat{K^s}$ (with $\hat{\mu}(k) = G(K^s/K)$-action), and the motivic Milnor fiber with $\hat{\mu}(k)$-action. See [22].

From Proposition 4.2, our trace formula Theorem 5.1, and Theorem 5.3, we recover the following important result of Denef and Loeser [10, Theorem 1.1].

Corollary 5.2 (Denef-Loeser). *Suppose $k = \mathbb{C}$. For any closed point x on X_0, and any integer $n > 0$, we denote by $\mathfrak{X}_{n,1,x}$ the reduced fiber of $\mathfrak{X}_{n,1}$ over x. Then*

$$\chi_{top}(\mathfrak{X}_{n,1,x}) = Trace(\, M^n \mid H_{sing}(F_x, \mathbb{C}))$$

where F_x is the topological Milnor fiber of f at x, M is the monodromy transformation, and H_{sing} is graded singular cohomology $\oplus_{i>0} H^i_{sing}$.

Combining Proposition 4.2, our trace formula Theorem 5.1, and the computation in Theorem 5.2, we recover A'Campo's formula [2].

Corollary 5.3 (A'Campo's formula). *Let X be a smooth irreducible variety over $\mathbb{C}$, and let $f : X \to \mathbb{A}^1_{\mathbb{C}}$ be a dominant morphism. Let x be a complex point of the hypersurface X_0 defined by f. Let $h : X' \to X$ be an embedded resolution of singularities for f, with $(f \circ h) = \sum_i N_i E_i$. Denote by $\zeta_x(T)$ the monodromy zeta function of f at x. We have*

$$Trace(M^d \mid H_{sing}(F_x, \mathbb{C})) = \sum_{N_i \mid d} N_i \chi_{top}(E_i^o \cap h^{-1}(x)),$$

$$\zeta_x(T) = \prod_i (T^{N_i} - 1)^{-\chi_{top}(E_i^o \cap h^{-1}(x))}.$$

6. Conclusion

Rigid geometry provides a natural framework to study the Monodromy Conjecture, and other problems related to birational geometry and the geometry of arc spaces. In general, this point of view makes it possible to apply cohomological methods to the study of the arc spaces.

We established the motivic zeta function as a Weil generating series of the nearby fiber X_η. However, in the motivic Monodromy Conjecture, it is the *naïve* motivic zeta function that appears. We believe it has an interpretation as a Hasse–Weil zeta function, counting orbits of rational points under the Galois action. This will be worked out in a future project.

Bibliography

1. *Groupes de monodromie en géométrie algébrique. II*, Séminaire de Géométrie Algébrique du Bois-Marie 1967–1969 (SGA 7 II), Dirigé par P. Deligne et N. Katz, Lecture Notes in Mathematics 340 (Springer-Verlag, 1973).
2. A'Campo, N., *La Fonction Zêta d'une Monodromie*, Comment. Math. Helvetici 50 (1975), 233–248.
3. Berkovich, V. G., *Étale cohomology for non-Archimedean analytic spaces*, Publ. Math., Inst. Hautes Étud. Sci. 78 (1993), 5–171.

4. Berkovich, V. G., *Vanishing cycles for formal schemes II*, Invent. Math. 125:2 (1996), 367–390.

5. Berthelot, P., *Cohomologie rigide et cohomologie rigide à supports propres*, Prépublication, Inst. Math. de Rennes (1996).

6. Bosch, S. and Schlöter, K., *Néron models in the setting of formal and rigid geometry*, Math. Ann. 301:2 (1995), 339–362.

7. Deligne, P., *La conjecture de Weil. I*, Publ. Math., Inst. Hautes Étud. Sci. 43 (1973), 273–307.

8. Denef, J., *Report on Igusa's local zeta function*, Séminaire Bourbaki, Vol. 1990/91, Exp. No.730-744, Astérisque 201-203 (1991), 359–386.

9. Denef, J., and Loeser, F., *Motivic Igusa zeta functions*, J. Algebraic Geom. 7 (1998), 505–537.

10. Denef, J. and Loeser, F., *Lefschetz numbers of iterates of the monodromy and truncated arcs*, Topology 41:5 (2002), 1031–1040.

11. Denef, J., and Loeser, F., *On some rational generating series occuring in arithmetic geometry*, in: A. Adolphson, F. Baldassarri, P. Berthelot, N. Katz and F. Loeser (ed.), *Geometric aspects of Dwork theory I* (de Gruyter, 2004), 509–526.

12. Dimca, A., *Singularities and Topology of Hypersurfaces* (Springer-Verlag, 1992).

13. Fujiwara, K., *Theory of tubular neighborhood in étale topology*, Duke Math. J. 80:1 (1995), 15–57.

14. Hironaka, H., *Resolution of singularities of an algebraic variety over a field of characteristic zero. I, II*, Ann. of Math. 79:2 (1964), 109–326.

15. Igusa, J., *Complex powers and asymptotic expansions I*, J. Reine Angew. Math. 268/269 (1974), 110–130.

16. Igusa, J., *Complex powers and asymptotic expansions II*, J. Reine Angew. Math. 278/279 (1975), 307–321.

17. Kulikov, V., *Mixed Hodge Structures and Singularities*, Cambridge Tracts in Mathematics 132 (Cambridge University Press, 1998).

18. Loeser, F. and Sebag, J., *Motivic integration on smooth rigid varieties and invariants of degenerations*, Duke Math. J. 119 (2003), 315–344.

19. Milnor, J., *Singular points of complex hypersurfaces*, Annals of Math. Studies 61 (Princeton University Press, 1968).

20. Nicaise, J. and Sebag, J., *Motivic Serre invariants of curves*, preprint (2006).

21. Nicaise, J. and Sebag, J., *The motivic Serre invariant, ramification, and the analytic Milnor fiber*, preprint (2006).

22. Nicaise, J. and Sebag, J., *Galois action on motivic Serre invariants, and motivic zeta functions*, preprint (2006).

23. Nicaise, J. and Sebag, J., *Invariant de Serre et fibre de Milnor analytique*, C.R., Math., Ac. Sci. Paris, 341:1 (2005), 21–24.

24. Sebag, J., *Intégration motivique sur les schémas formels*, Bull. Soc. Math. France 132:1 (2004), 1–54.

25. Serre, J.-P., *Classification des variétés analytiques p-adiques compactes*, Topology 3 (1965), 409–412.

Zariski pairs on sextics II

Dedicated to Professor Jean Paul Brasselet for his 60th birthday

Mutsuo Oka

Department of Mathematics,
Tokyo University of Science,
1-3 Kagurazaka,Shinjuku-ku,
Tokyo 162-8601
E-mail: oka@rs.kagu.tus.ac.jp

Keywords: torus type, Zariski pair, flex points, conical flex points

1. Introduction

We continue to study Zariski pairs in sextics. In this paper, we study Zariski pairs of sextics which are not irreducible. The idea of the construction of Zariski partner sextic for reducible cases is quit different from the irreducible case. It is crucial to take the geometry of the components and their mutual intersection data into account. When there is a line component, flex geometry (i.e., linear geometry) is concerned to the geometry of sextics of torus type and non-torus type. When there is no linear components, the geometry is more difficult to distinguish sextics of torus type. For this reason, we introduce the notion of conical flexes.

We have observed in [9] that the case $\rho(C,5) = 6$ is critical in the sense that the Alexander polynomial $\Delta_C(t)$ can be either trivial or non-trivial for sextics. If $\rho(C,5) > 6$ (resp. $\rho(C,5) < 6$), the Alexander polynomial is not trivial (resp. trivial) ([9]). For the definition of $\rho(C,5)$-invariant, see [9]. Thus we concentrate ourselves in this paper the case $\rho(C,5) = 6$. In [10], we have classified the possible configurations for reducible sextics of torus type. In particular, the configurations with $\rho(C,5) = 6$ are given as in Theorem 1 below. Hereafter we use the same notations as [9] for denoting component types. For example, $C = B_1 + B_5$ implies that C has a linear component B_1 and a quintic component B_5. We denote the configuration

838

of the singularities of C by $\Sigma(C)$.

Theorem 1. *([10]) Assume that C is a reducible sextic of torus type with $\rho(C,5) = 6$ and only simple singularities. Let Σ_{in} be the inner singularities. Then the possible configurations of simple singularities are as follows.*

(1) $\Sigma_{in} = [A_5, 4A_2]$: $C = B_5 + B_1$ *and* $\Sigma(C) = [A_5, 4A_2, 2A_1]$, $[A_5, 4A_2, 3A_1]$, $[A_5, 4A_2, 4A_1]$.

(2) $\Sigma_{in} = [2A_5, 2A_2]$:

    (a) $C = B_1 + B_5$: $\Sigma(C) = [2A_5, 2A_2, 2A_1]$, $[2A_5, 2A_2, 3A_1]$.

    (b) $C = B_1 + B_1' + B_4$: $\Sigma(C) = [2A_5, 2A_2, 3A_1]$, $[2A_5, 2A_2, 4A_1]$.

    (c) $C = B_2 + B_4$: $\Sigma(C) = [2A_5, 2A_2, 2A_1]$, $[2A_5, 2A_2, 3A_1]$.

    (d) $C = B_3 + B_3'$: $\Sigma(C) = [2A_5, 2A_2, 3A_1]$.

(3) $\Sigma_{in} = [E_6, A_5, 2A_2]$: $C = B_1 + B_5$, $\Sigma(C) = [E_6, A_5, 2A_2, 2A_1]$, $[E_6, A_5, 2A_2, 3A_1]$.

(4) $\Sigma_{in} = [3A_5]$:

    (a) $C = B_1 + B_5$: $\Sigma(C) = [3A_5, 2A_1]$.

    (b) $C = B_2 + B_4$: $\Sigma(C) = [3A_5, 2A_1]$.

    (c) $C = B_1 + B_1' + B_4$: $\Sigma(C) = [3A_5, 3A_1]$.

    (d) $C = B_3 + B_3'$: $\Sigma(C) = [3A_5]$, $[3A_5, A_1]$, $[3A_5, 2A_1]$.

    (e) $C = B_1 + B_2 + B_3$: $\Sigma(C) = [3A_5, 2A_1]$, $[3A_5, 3A_1]$.

    (f) $C = B_1 + B_1' + B_1'' + B_3$: $\Sigma(C) = [3A_5, 3A_1]$, $[3A_5, 4A_1]$,

    (g) $C = B_2 + B_2' + B_2''$: $\Sigma(C) = [3A_5, 3A_1]$.

(5) $\Sigma_{in} = [2A_5, E_6]$:

    (a) $C = B_1 + B_5$: $\Sigma(C) = [E_6, 2A_5, 2A_1]$.

    (b) $C = B_2 + B_4$: $\Sigma(C) = [E_6, 2A_5, 2A_1]$.

    (c) $C = B_1 + B_1' + B_4$: $\Sigma(C) = [E_6, 2A_5, 3A_1]$.

(6) $\Sigma_{in} = [A_8, A_5, A_2]$: $C = B_1 + B_5$, $\Sigma(C) = [A_8, A_5, A_2, 2A_1]$, $[A_8, A_5, A_2, 3A_1]$.

(7) $\Sigma_{in} = [A_{11}, 2A_2]$:

    (a) $C = B_2 + B_4$: $\Sigma(C) = [A_{11}, 2A_2, 2A_1]$, $[A_{11}, 2A_2, 3A_1]$.

    (b) $C = B_3 + B_3'$: $\Sigma(C) = [A_{11}, 2A_2, 3A_1]$.

(8) $\Sigma_{in} = [A_{11}, A_5]$:

    (a) $C = B_1 + B_5$: $\Sigma(C) = [A_{11}, A_5, 2A_1]$.

    (b) $C = B_2 + B_4$: $\Sigma(C) = [A_{11}, A_5, 2A_1]$.

    (c) $C = B_3 + B_3'$: $\Sigma(C) = [A_{11}, A_5]$, $[A_{11}, A_5, A_1]$, $[A_{11}, A_5, 2A_1]$,

    (d) $C = B_1 + B_2 + B_3$: $\Sigma(C) = [A_{11}, A_5, 2A_1]$, $[A_{11}, A_5, 3A_1]$,

(9) $\Sigma_{in} = [A_{17}]$: $C = B_3 + B_3'$, $\Sigma(C) = [A_{17}]$, $[A_{17}, A_1]$, $[A_{17}, 2A_1]$.

Our main result in this paper is:

Theorem 2. *There are Zariski partner sextics with the above configurations with the following exceptions:*

(1) $\Sigma(C) = [A_5, 4A_2, 4A_1]$ *with* $C = B_5 + B_1$.
(2) $\Sigma(C) = [2A_5, 2A_2, 4A_1]$ *with* $C = B_1 + B_1' + B_4$.
(3) $\Sigma(C) = [E_6, A_5, 2A_2, 3A_1]$ *with* $C = B_5 + B_1$.
(4) $\Sigma(C) = [3A_5, 4A_1]$ *with* $C = B_3 + B_1 + B_1' + B_1''$.
(5) $\Sigma(C) = [E_6, 2A_5, 3A_1]$ *with* $C = B_4 + B_1 + B_1'$.

The non-existence of sextics of non-torus type with the above exceptional configurations will be explained by flex geometry. The existence will be also explained by the flex geometry for those which has a line components and by conical flex geometry for the component type $B_4 + B_2$, $B_2 + B_2' + B_2''$.

Remark 3. (1) In the list of Theorem 2, there are certainly several cases which are already known. For example, the configuration $C = B_3 + B_3'$ with one singularity A_{17} is given by Artal [1].
(2) In this paper, we only studied possible Zariski pairs of reducible sextics (C, C') where C is of torus type and C' is not of torus type. On the other hand, the possibility of Zariski pairs among reducible sextics of the same class is not discussed here. Several examples are known among reducible sextics of non-torus type. For such cases, Alexander polynomials can not distinguish the differnece. See papers [2–5]

2. Reducible sextics of non-torus type

To compute explicit polynomials defining reducible sextics, it is not usually easy to look for special degenerations into several irreducible components starting from the generic sextics $\sum_{i+j\leq 6} a_{ij} x^i y^j$. Recall that we have classified all possible reducible simple configurations in [10] and it is easier to start from a fixed reducible decomposition. In fact, the geometry of the configuration of a reducible sextic depends very much on the geometry of each components. A smooth point $P \in C$ is called a *flex* point if the intersection multiplicity of the tangent line and C at P is strictly greater than 2. First we recall the following fact for flex points ([6,8]).

Lemma 4. *Let* $C : F(X, Y, Z) = 0$ *be an irreducible plane curve of degree* n *with singularities* $\{P_1, \ldots, P_k\}$. *Then the number of flexes* $\iota(C)$ *is given* *by*

$$\iota(C) = 3\,n\,(n-2) - \sum_{i=1}^{k} \varepsilon(P_i; C)$$

where the second term $\varepsilon(P_i; C)$ is the flex defect and given by the local intersection number of C and the hessian curve of C at P_i.

Generic flex defect of simple singularities we use are

$$\varepsilon(A_1) = 6,\ \varepsilon(A_2) = 8,\ \varepsilon(A_{3\iota-1}) = 9\iota,\ (\iota \geq 2),\ \varepsilon(E_6) = 22 \qquad (1)$$

Recall that flex points of a curve are described by the hessian of the defining homogeneous equation. When we have an affine equation $C : f(x,y) = 0$, flex points in $\mathbf{C}^2$ are described by $f(x,y) = flex_f(x,y) = 0$ ([8]) where

$$flex_f(x,y) := f_{xx}\,f_y^2 - 2\,f_{xy}\,f_x\,f_y + f_{yy}\,f_x^2$$

This is an easy way to check flex points from the affine equation.

A sextic C is of (2,3)-torus type if we can take a defining polynomial of the form $f_2(x,y)^3 + f_3(x,y)^2 = 0$ where degree $f_j = j$. The intersections $f_2 = f_3 = 0$ are singular points of C and we call them *inner singularities*. For a given sextic C of torus type whose singularities are simple, the possible inner singularities are

$$(\sharp):\quad \{A_2, A_5, A_8, A_{11}, A_{14}, A_{17}, E_6\}.$$

A convenient criterion for C to be of torus type is the existence a certain conic C_2 such that $C_2 \cap C \subset \Sigma(C)$ (Tokunaga's criterion [11], Lemma 3, [7]).

A sextic of torus type C is called *of linear torus type* if the conic polynomial f_2 can be written as $f_2(x,y) = \ell(x,y)^2$ for some linear form $\ell(x,y)$ ([9]). A sextic of linear torus type can have only A_5, A_{11}, A_{17} as inner singularties and the location of these singularities are colinear.

The proof of Theorem 2 is done by giving explicit examples. For the better understanding of the existence or non-existence of the Zariski pairs, we divide the above configurations into the following classes.

(1) C has a quintic component. The corresponding component type is $B_5 + B_1$ and B_1 is a flex tangent line.

(2) C has a quartic component. There are two subcases.

    (a) $C = B_4 + B_1 + B_1'$. In this case, two line components are flex tangent lines.

    (b) $C = B_4 + B_2$.

(3) C has a cubic component. There are two subcases.

(a) Sextics of linear torus type.

(b) Sextics, not of linear torus type.

(4) $C = B_2 + B_2' + B_2''$.

3. Configuration coming from quintic flex geometry

Let B_5 be an irreducible quintic and let P be a flex point of B_5. We denote the tangent line at P by L_P. We say that P is *a flex of torus type* (respectively *a flex of non-torus type*) if $B_5 \cup L_P$ is a sextic of torus type (resp. of non-torus type). The following configurations are mainly related to the flex geometry of certain quintics. (By 'flex geometry', we mean the geometry of the tangent lines at the flex points and the curve.) Recall that $\Sigma(B_5)$ is the configuration of the singularities of B_5. Let ι be the number of flex points on B_5.

(1) $C = B_5 + B_1$ with $\Sigma(C) = [A_5, 4A_2, kA_1]$, $k = 2$, 3, 4. Then $\Sigma(B_5) = [4A_2, (k-2)A_1]$ for $k = 2$, 3, 4 and $\iota = 13, 7, 1$ respectively.

(2) $C = B_5 + B_1$ with $\Sigma(C) = [2A_5, 2A_2, kA_1]$, $k = 2, 3$. Then $\Sigma(B_5) = [A_5, 2A_2, (k-2)A_1]$ and $\iota = 11, 5$ respectively.

(3) $C = B_5 + B_1$ with $\Sigma(C) = [E_6, A_5, 2A_2, kA_1]$, $k = 2$, 3. Then $\Sigma(B_5) = [E_6, 2A_2, (k-2)A_1]$ and $\iota = 7, 1$ respectively for $k = 2, 3$. The case $k = 3$ corresponds to sextics of torus type.

(4) $C = B_5 + B_1$ with $\Sigma(C) = [E_6, 2A_5, 2A_1]$. The quintic B_5 has $\Sigma(C) = [E_6, A_5]$ and $\iota = 5$.

(5) $C = B_5 + B_1$ with $\Sigma(C) = [3A_5, 2A_1]$. Then $\Sigma(B_5) = [2A_5]$ and $\iota = 9$.

(6) $C = B_5 + B_1$ with $\Sigma(C) = [A_8, A_5, A_2, kA_1]$, $k = 2, 3$. Then $\Sigma(B_5) = [A_8, A_2, (k-2)A_1]$ and $\iota = 10, 4$ for $k = 2, 3$.

(7) $C = B_5 + B_1$ with $\Sigma(C) = [A_{11}, A_5, 2A_1]$. Then $\Sigma(B_5) = [A_{11}]$ and $\iota = 9$.

We are going to show the stronger assertion for the above configurations: *the Zariski partner sextic of non-torus type are simply given by replacing the flex line components B_1 for the above cases, if B_5 has at least two flex points.*

Let Ξ be a configuration of singularities on B_5, which is one of the above list. Let $\mathcal{M}(\Xi; 5)$ be *the configuration space* of quintics B_5 such that $\Sigma(B_5) = \Xi$. We considere it as a topological subspace of the space of quintics. For our purpose, it is enough to consider the marked configuration subspace $\mathcal{M}(\Xi; 5)'$ which consists of the pair (B_5, P), where $B_5 \in \mathcal{M}(\Xi; 5)$ and P is a flex point of torus type. The following describes the existence of sextics of non-torus type with the above configurations.

Theorem 5. *Let Ξ be a configuration of singularities on B_5, which is one of the above list. 1. The configuration subspace $\mathcal{M}(\Xi;5)'$ is connected for each Ξ.*

2. For each $\Xi \neq [4A_2, 2A_1]$, $[E_6, 2A_2, A_1]$ and $B_5 \in \mathcal{M}(\Xi;5)'$, a Zariski pair sextics are given as $\{B_5 \cup L_P, B_5 \cup L_Q\}$ where P and Q are flex points of torus-type and of non-torus type respectively.

3. For these two exceptional cases, we have the equality $\mathcal{M}(\Xi;5)' = \mathcal{M}(\Xi;5)$ and a quintic $B_5 \in \mathcal{M}(\Xi;5)$ does not contain any flexes of non-torus type.

Remark 6. Let ι_t, ι_{nt} be the respective number of flex points of torus type and of non-torus type on a generic $B_5 \in \mathcal{M}(\Xi;5)'$. We do not need the precise number ι_t, ι_{nt} for our purpose. The sum $\iota = \iota_t + \iota_{nt}$ is described by Lemma 3. By an explicit computation, we have the next table which describes the distributions of number of flex points. The second line is the configuration of singularity and the last line is the pair of flex numbers (ι_t, ι_{nt}).

1	2	3	4	5
$4A_2$ $4A_2 + A_1$ $4A_2 + 2A_1$	$A_5 + 2A_2$ $A_5 + 2A_2 + A_1$	$E_6 + 2A_2$ $E_6 + 2A_2 + A_1$	$E_6 + A_5$	$2A_5$
(1,12) (1,6) (1,0)	(1,10) (1,4)	(1,6) (1,0)	(1,4)	(1,8)

6	7
$A_8 + A_2$ $A_8 + A_2 + A_1$	A_{11}
(1,9) (1,3)	(1,8)

Proof. First recall that the topology of the complement of the sextics $B_5 \cup L_P$ for a flex point P of torus type and non-torus type are different. They can be distinguished by Alexander polynomial ([9]). Therefore to show the assertion about the positivity $\iota_{nt} > 0$, it is enough to check the assertion by some quintic B_5. Examples will be given in the next subsection. Secondly, the irreducibility of the configuration space $\mathcal{M}(\Xi;5)$ of quintics $f_5(x,y) = 0$ with singularities $\Xi = [4A_2, 2A_1]$, $[E_6, 2A_2, A_1]$ are easily proves as follows. For $\Xi = [4A_2, 2A_1]$, the dual curves of quintics in this configuration space are quartics with configuration $[A_2, 2A_1]$. As the irreducibility of the configuration space $\mathcal{M}([A_2, 2A_1]; 4)$ is easy to be checked, the irreducibility of $\mathcal{M}([4A_2, 2A_1]; 5)$ follows. Take $\Xi = [E_6, 2A_2, A_1]$. For a quintic B_5 with

$\Sigma(B_5) = \Xi$, the dual curve B_5^* is again a quartic with $[A_2, 2A_1]$ (thus mapped into the same configuration space with the dual of quintics with $[4A_2, 2A_1]$). Hoever we can not apply the same argument. The reason is that the dual curve B_5^* is not generic in the configuration space $\mathcal{M}([A_2, 2A_1]; 4)$: the quartic B_5^* has not 4 flexes but three flexes, one flex of flex order 4 (=dual of E_6) and 2 flexes of flex order 3 (i.e., dual of $2A_2$). Thus we need another argument. Note that any three singular points can not be colinear on B_5 by Bézout theorem. We can consider the slice condition:

$(\star)$: E_6 is at $(-1, 0)$ and two A_2 are at $(0, 1)$, $(0, -1)$ and one A_1 at $(1, 0)$.

It is easy to compute that a Zariski open subset of this slice has the normal form:

$$h := e_1{}^3 - 4\,y^3\,x^2 - 2\,e_1{}^3\,x^3 + x^4\,e_1{}^3 + x\,e_1{}^3 + 3\,y^5\,e_1{}^2 + e_1{}^3\,y^4 + 3\,y\,e_1{}^2$$
$$+ 10\,e_1{}^3\,y^2\,x^3 + 18\,e_1{}^3\,y^2\,x^2 - 12\,e_1\,y^2\,x^2 + 6\,e_1{}^3\,y^2\,x - 12\,e_1\,y^2\,x - 9\,y\,x^4\,e_1{}^2$$
$$- 12\,y\,e_1{}^2\,x^3 + 12\,y\,x\,e_1{}^2 + 6\,y\,x^2\,e_1{}^2 - 6\,y^3\,x^2\,e_1{}^2 - 6\,y^3\,e_1{}^2 - 2\,e_1{}^3\,y^2$$
$$+ e_1{}^3\,x^5 - 7\,e_1{}^3\,y^4\,x + 12\,e_1\,y^4\,x - 12\,y^3\,x\,e_1{}^2 - 2\,x^2\,e_1{}^3 + 8\,y^3 - 4\,y$$
$$- 4\,y^5 - 8\,y\,x + 8\,y\,x^3 + 4\,y\,x^4 + 8\,y^3\,x, \quad e_1 \neq 0, \pm 2/\sqrt{3}$$

Thus the irreducibility of $\mathcal{M}([E_6, 2A_2, A_1]; 5)$ follows. For each of them we know that the number of flex points is 1 and $\mathcal{M}(\Xi; 5)' \neq \emptyset$. On the other hand, the topology of sextics $B_5 \cup L_P$, of torus type and non-torus type, are distinguished by the Alexander polynomials $(t^2 - t + 1)(t - 1)$ and $(t - 1)$ respectively. This implies $\mathcal{M}(\Xi; 5)' = \mathcal{M}(\Xi; 5)$. qed.

3.1. *Example for sextics with quintic components*

We gives examples of sextics with a quintic components.

1. $C = B_5 + B_1$, $\Sigma = [A_5, 4A_2, k\,A_1]$, $k = 2, 3$: First we consider the case $k = 2$. The quintic has $4A_2$ and 13 flex points.

$$C : \left(\left(\frac{28}{153}x + \frac{8}{153}\right)y^4 + \left(\frac{52}{51}x^3 - \frac{20}{51}x^2 - \frac{152}{153}x - \frac{16}{153}\right)y^2 + x^5\right.$$
$$\left. - \frac{2}{17}x^4 - \frac{193}{51}x^3 + \frac{116}{51}x^2 + \frac{124}{153}x + \frac{8}{153}\right)\left(\frac{80}{17}x + \frac{880}{51} - \frac{320}{51}y\right)$$

The sextics of torus type is obtained by replacing the line component by $\frac{28}{153}x + \frac{8}{153}$.

Next, we consider the case $k = 3$. Th equintic B_5 has $4A_2 + A_1$. The

844

flex which gives a sextic of torus type is $(1,0)$.

$$f_5 := -\frac{385}{16}\,x^4\,y + \frac{3885}{16}\,x\,y^2 + \frac{1897}{128}\,x\,y^4 + \frac{345}{16}\,y\,x^3 - \frac{441}{4}\,y^3\,x + \frac{529}{4}\,y\,x^2$$
$$+\ 73\,y + 72\,y^3 + \frac{403}{128}\,x^3\,y^2 - \frac{16783}{128}\,x^2\,y^2 - \frac{811}{4}\,y\,x - \frac{869}{64}\,y^4 + \frac{7087}{128}\,x$$
$$-\frac{3675}{32}\,y^2 + \frac{3201}{512}\,x^5 + \frac{601}{32}\,x^4 - y^5 + \frac{313}{8}\,x^2\,y^3 - \frac{11511}{256}\,x^2 - \frac{997}{64} - \frac{10167}{512}\,x^3$$

B_5 has two obvious flex points: $P := (1,0)$ and $Q := (-\frac{1520}{293}, -\frac{287}{293})$, where P is a flex of torus type and Q is a flex of non-torus type. There are 5 other flex points whose x-coordinates are the solution of

$$R_1 := 5926214587003\,x^5 - 32698277751050\,x^4 + 69779834665700\,x^3$$
$$-\ 72918583611000\,x^2 + 37638730560000\,x - 7728486400000 = 0$$

We can check that the roots of $R_1 = 0$ corresponds to flexes of non-torus type as follows. (The same argument applies to other cases.) Note that any conics which is passing through 4 A_2 of B_5 are given by

$$h_2 := y^2 - \frac{1}{2}\,d_{01}\,y\,x + d_{01}\,y - \frac{1}{2}\,x^2 + \frac{1}{19}\,x^2\,d_{01} - \frac{5}{2}\,x - \frac{10}{19}\,d_{01}\,x + 3 + \frac{16}{19}\,d_{01}$$

Thus if there is a flex $P(a,b)$ of torus type (so $R_1(a) = 0$), there is a cubic form $h_3(x,y)$ such that the sextic $C = B_5 \cup L_P$ is described as $C := \{h_3^2 + h_2^3 = 0\}$. On the other hand, put $S_2(x, d_{01})$ be the polynomial of degree 2 in x defined by $S_2(x, d_{01}) = R(h_2, f_5, y)/P(x)^2$ where $R(h_2, f_5, y)$ is the resultant of h_2 and f_5 in y and $P(x) = 0$ is the defining polynomial for the x-coordinates of 4 A_2. Then S_2 must be $c\,(x - a)^2$ for some $c \neq 0$. Let $b_1(d_{01})$ be the discriminant polynomial of S_2 in x and let $b_2(d_{01})$ be the resultant of $S_2(x, d_{01})$ and $R_1(x)$ in x. Thus we obtain two polynomials $b_1(d_{01})$, $b_2(d_{01})$ of the parameter d_{01} which must have a common root: We can check that $b_1(d_{01}) = b_2(d_{01}) = 0$ has no common root in d_{01}.

2. $C = B_5 + B_1$, $\Sigma = [2A_5, 2A_2, j\,A_1]$, $j = 2, 3$. The quintic B_5 has $A_5 + 2A_2$.

$j = 2, [2A_5, 2A_2, 2A_1]$:

$$\left(-\frac{16145}{1024}\,y^3 - \frac{93}{64}\,y^5 - \frac{877727}{8192}\,y^2 - \frac{110055}{1024}\,y - \frac{329525}{16384}\,x + \frac{11025}{16384}\,x^3 - \frac{543975}{16384}\,x^2\right.$$
$$+\ \frac{1751733}{16384}\,y^2\,x^2 + \frac{79625}{4096}\,x^5 - \frac{235529}{16384}\,y^4 + \frac{1999101}{16384}\,y^2\,x^3 - \frac{100809}{8192}\,y^3\,x$$
$$+\ \frac{1199495}{8192}\,y\,x^3 + \frac{289995}{4096}\,y\,x^2 - \frac{1199495}{8192}\,y\,x + \frac{150225}{4096}\,y\,x^4 - \frac{498845}{4096}\,y^2\,x$$
$$\left. -\ \frac{275703}{16384}\,y^4\,x + \frac{23889}{8192}\,y^3\,x^2 + \frac{18625}{512}\,x^4 - \frac{52025}{16384}\right)(y + 1)$$

A sextic of torus type is give by replacing the line component by $x - 1 = 0$.

$j = 3$, $[2A_5, 2A_2, 3A_1]$:

$$\left(\frac{2}{7} + x^5 - \frac{4}{7}\,x^2 - 2\,x^3 + \frac{2}{7}\,x^4 + x - \frac{2}{7}\,y^2\,x^3 + \frac{12}{7}\,x^2\,y^2 - \frac{1}{7}\,x\,y^4 - \frac{6}{7}\,y^2\,x - \frac{4}{7}\,y^2 + \frac{2}{7}\,y^4 \right)$$

$$\times \left(\frac{44064}{34157767}\,y\,\sqrt{-963 + 1182\,\sqrt{6}} + \frac{16521840}{34157767} - \frac{7198560}{34157767}\,\sqrt{6} \right.$$

$$\left. - \frac{28320}{34157767}\,y\,\sqrt{-963 + 1182\,\sqrt{6}}\,\sqrt{6} + \frac{7328592}{34157767}\,\sqrt{6}\,x - \frac{2704104}{4879681}\,x \right)$$

The quintic B_5 has $A_5 + 2A_2 + A_1$ and 5 flex points and among them, there exists a unique flex of torus type. The tangent line at this flex of torus type is given by $2 - x = 0$.

3. A sextic $C = B_5 + B_1$ with $\Sigma(C) = [E_6, A_5, 2A_2, 2A_1]$ is given by

$$f := \left(\frac{53}{141}\,x + \frac{3}{47}\,y + y^5 - \frac{50}{47}\,y^3 + \frac{4}{47}\,y^2\,x - \frac{769}{141}\,y^4\,x - \frac{614}{141}\,y^2\,x^3 + 2\,y^2 \right.$$

$$+ \frac{53}{141}\,x^5 + \frac{56}{141}\,y\,x^2 - \frac{10}{3}\,y\,x + \frac{1174}{141}\,y^3\,x + \frac{1256}{141}\,y^3\,x^2 + \frac{10}{3}\,x^3\,y - \frac{69}{47}\,y^4 - \frac{25}{47}\,x^4$$

$$\left. + \frac{50}{47}\,x^2 - \frac{106}{141}\,x^3 - \frac{1462}{141}\,y^2\,x^2 - \frac{65}{141}\,y\,x^4 - \frac{25}{47} \right) \left(y + 1 - \frac{8}{3}\,x \right)$$

The quintic has 7 flex points and there is a unique one among them which is of torus type at $\left(\frac{2400}{1357}, \frac{357}{1357} \right)$.

4. $C = B_5 + B_1$ with $[E_6, 2A_5, 2A_1]$. The quintic B_5 has $E_6 + A_5$ and it has 5 flexes. Among them, there is a unique flex of torus type. A sextic of non-torus type:

$$f := (4451 + 9742\,y^2\,x + 4639\,y^4\,x - 9501\,y - 14381\,x - 423\,y^5\,\sqrt{33} - 351\,\sqrt{33}$$

$$- 16343\,x^3 + 6546\,y^3 - 8005\,y^4 + 3554\,y^2 + 19373\,x^2 - 19373\,y^2\,x^2 + 9836\,x^4$$

$$+ 2955\,y^5 + 10266\,y\,x^3 - 2936\,x^5 - 19020\,y^3\,x + 19020\,y\,x - 14661\,y\,x^2$$

$$+ 14661\,y^3\,x^2 + 1521\,\sqrt{33}\,y - 1098\,y^3\,\sqrt{33} + 1593\,y^4\,\sqrt{33} + 756\,x^4\,\sqrt{33}$$

$$+ 1215\,x^2\,\sqrt{33} - 1242\,y^2\,\sqrt{33} - 1917\,x^3\,\sqrt{33} + 297\,x\,\sqrt{33} + 999\,y^2\,x^3\,\sqrt{33}$$

$$+ 36\,y\,x^4\,\sqrt{33} + 1458\,y^2\,x\,\sqrt{33} + 918\,y\,x^3\,\sqrt{33} + 2052\,y^3\,x\,\sqrt{33} - 423\,y\,x^2\,\sqrt{33}$$

$$- 2052\,y\,x\,\sqrt{33} + 423\,y^3\,x^2\,\sqrt{33} - 1215\,y^2\,x^2\,\sqrt{33} - 1755\,y^4\,x\,\sqrt{33} + 6077\,y^2\,x^3$$

$$- 5124\,y\,x^4)(y + 1)$$

A sextic torus type is given by replacing $y + 1$ by the flex tangent at (α, β)

846

where

$$\alpha := \frac{1476423}{6805087} + \frac{176748}{6805087}\sqrt{33}, \ \beta := \frac{1469468}{6805087} - \frac{931392}{6805087}\sqrt{33}$$

5. $C = B_5 + B_1$ with $[3A_5, 2A_1]$.

$$-\frac{12}{6279955}(-2516+27\sqrt{69})(-2516-10064\,x^4-45\,x\,y^4\,\sqrt{69}+90\,x\,y^2\,\sqrt{69}$$
$$+108\,\sqrt{69}\,y^3\,x^2-108\,\sqrt{69}\,y\,x^2+180\,x^5\,\sqrt{69}+54\,y^2\,\sqrt{69}-45\,x\,\sqrt{69}-27\,y^4\,\sqrt{69}$$
$$-108\,x^4\,\sqrt{69}-27\,\sqrt{69}+10060\,y\,x^4-5030\,y^3-10064\,y\,x^2-2516\,y^4$$
$$+10064\,y^3\,x^2+2515\,y^5+10060\,x^2-840\,x\,y^4+1680\,x\,y^2+5032\,y^2-10060\,y^2\,x^2$$
$$+3360\,x^5+2515\,y-840\,x)$$
$$\times\left(\left(\frac{31104}{2497}-\frac{1620}{2497}\sqrt{69}\right)y+\frac{108}{2497}\left(52\,\sqrt{69}-499\right)(x-1)\right)$$

The flex point which gives a sextic of torus type is (α, β),

$$\alpha = \frac{957138004}{22902646825} + \frac{2339358408}{22902646825}\sqrt{69}, \ \beta = \frac{540908244}{916105873} - \frac{52210443}{916105873}\sqrt{69}$$

6. a. A sextic of torus type with $[A_8, A_5, A_2, 2A_1]$ with line component is given by:

$$f := \left(-60\,y^2 + 60\,y - x^2\right)^3 +$$
$$\left(-\frac{81}{25}\,y^3+\left(-\frac{6849}{25}\,x-\frac{9639}{25}\right)y^2+\left(\frac{162}{25}\,x^2+\frac{6849}{25}\,x+\frac{1944}{5}\right)y+x^3-\frac{162}{25}\,x^2\right)^2$$

and the line component is defined by $y - 1 = 0$. It has 10 flexes and the flex at $(0, 1)$ gives a flex of torus type. Other flex tangent lines give a sextic of non-torus type. For example

$$f_6 := \left(\frac{324}{25}\,x^5 - \frac{26244}{625}\,x^4 - \frac{2223126}{625}\,y^2\,x^3 + \frac{40132557}{625}\,y^3\,x^2 - \frac{428706}{625}\,y\,x^4\right.$$
$$-\frac{43281837}{625}\,y^2\,x^2 - \frac{134993439}{625}\,y^5 + \frac{271568079}{625}\,y^4 - \frac{8419248}{125}\,y^3 - \frac{3779136}{25}\,y^2$$
$$+\frac{629856}{125}\,y\,x^2 + \frac{1733076}{625}\,y\,x^3 - \frac{26628912}{125}\,y^2\,x + \frac{132035022}{625}\,y^3\,x$$
$$\left.+\frac{1109538}{625}\,y^4\,x\right)\left(-\frac{64039734}{25}\,y+\frac{18974736}{5}+\frac{33205788}{25}\,x\right)$$

b. A sextic of torus type with $[A_8, A_5, A_2, 3A_1]$ and with component type $B_5 + B_1$ is given by

$$f := \left(\frac{15}{2}\, y^2 - \frac{15}{2}\, y - 16\, x^2\right)^3 +$$

$$\left(-\frac{455}{24}\, y^3 + \left(-\frac{80}{3}\, x + \frac{245}{6}\right) y^2 + \left(\frac{140}{3}\, x^2 + \frac{80}{3}\, x - \frac{175}{8}\right) y + 64\, x^3 - \frac{140}{3}\, x^2\right)^2$$

It has the line component $y - 1 = 0$ and the quintic has 4 flex points among which only the flex $(0,1)$ is of torus type. An example of sextic of non-torus type is given by

$$f_6 := \left(\frac{17920}{3}\, x^5 - \frac{19600}{9}\, x^4 + \frac{3500}{3}\, y^2\, x - \frac{6125}{3}\, y\, x^2 + \frac{47600}{9}\, x^3\, y - \frac{30625}{64}\, y^2 \right.$$

$$+ \frac{332125}{192}\, y^3 - \frac{11275}{3}\, y^3\, x^2 + 5800\, y^2\, x^2 - \frac{19600}{9}\, x\, y^3 + \frac{9100}{9}\, x\, y^4 - \frac{44240}{9}\, x^3\, y^2$$

$$\left. + \frac{40720}{9}\, y\, x^4 - \frac{1170775}{576}\, y^4 + \frac{450025}{576}\, y^5\right)$$

$$\times \left(-\frac{2222000000}{255584169}\, y + \frac{2156000000}{255584169} + \frac{492800000}{85194723}\, x\right)$$

7. A quintic with A_{11} has 9 flex points, among which there exists a unique flex of torus type. In the following example, our quintic has a flex of torus type at $(0, 1)$ so the the sextic of torus type is given by

$$f := \left(-\frac{28}{25}\, y^2 + \frac{28}{25}\, y - x^2\right)^3 +$$

$$\left(\frac{511}{100}\, y^3 + \left(-\frac{28}{25}\, x - \frac{7}{500}\right) y^2 + \left(-\frac{91}{20}\, x^2 + \frac{28}{25}\, x - \frac{637}{125}\right) y - x^3 + \frac{91}{20}\, x^2\right)^2$$

and the line component is $y - 1 = 0$. B_5 has 8 flex of non-torus type. We can take one at $(1, -1)$ so that a sextic of non-torus type is given by

$$f_6 := \left(\frac{6176793}{250000}\, y^5 - \frac{7154}{625}\, y^4\, x + \frac{7194719}{250000}\, y^4 - \frac{245049}{5000}\, y^3\, x^2 - \frac{859901}{31250}\, y^3 \right.$$

$$+ \frac{98}{3125}\, y^3\, x + \frac{35672}{3125}\, y^2\, x - \frac{405769}{15625}\, y^2 + \frac{13181}{5000}\, y^2\, x^2 - \frac{7}{250}\, y^2\, x^3 - \frac{2548}{125}\, y\, x^3$$

$$\left. + \frac{57967}{1250}\, y\, x^2 + \frac{7833}{400}\, y\, x^4 - \frac{8281}{400}\, x^4 + \frac{91}{10}\, x^5\right)\left(\frac{3306744}{15625}\, y + \frac{1928934}{15625} + \frac{275562}{3125}\, x\right)$$

4. Configuration coming from quartic geometry

4.1. *Configuration coming from quartic flex geometry*

We consider the sextics C with component type $B_4 + B_1 + B_1'$. The corresponding possible configurations are

(a) $\Sigma(C) = [3A_5 + 3A_1]$ and $\Sigma(B_4) = [A_5]$ or (b) $\Sigma(C) = [2A_5 + 2A_2 + kA_1]$, $k = 3, 4$ and $\Sigma(B_4) = [2A_2 + (k-3)A_1]$ or (c) $\Sigma(C) = [E_6 + 2A_5 + 3A_1]$ and $\Sigma(B_4) = [E_6]$.

Let P, Q be two flex points on B_4 and let L_P, L_Q be the flex tangents. We say that a pair of flex points $\{P, Q\}$ are a *flex pair of torus type* if the sextic $B_4 \cup L_P \cup L_Q$ is a sextic of torus type.

Theorem 7. *Case (a)* $\Sigma(C) = [3A_5 + 3A_1]$*. The quartic B_4 has one A_5 and 6 flex points and two line components are flex tangent lines. There exist two flex pairs of torus type. The other choices give sextics of non-torus type.*

Case (b) $\Sigma(C) = [2A_5 + 2A_2 + kA_1]$, $k = 3, 4$ *. The quartic B_4 has $2A_2$ or $2A_2 + A_1$ according to $k = 3$ or 4 and B_4 has 8 or 2 flex points respectively. For the case, $k = 3$, there are both flex pairs of torus type and of non-torus type. For $k = 4$, the choice of $\{P, Q\}$ is unique and it is a pair of torus type.*

Case (c) $\Sigma(C) = [E_6 + 2A_5 + 3A_1]$ *. B_4 has two flexes and they gives a pair of torus type. Thus there is no sextic of non-torus type with $E_6 + 2A_5 + 3A_2$ with component type $B_4 + B_1 + B_1'$.*

Proof. As the configuration spaces of quartics with one A_5, or $2A_2$ or $2A_2 + A_1$ or E_6 are connected, it is enough to check the assertion by an example.

For the non-existence, note that a quartic with $\Sigma(B_4) = 2A_2 + A_1$ or $\Sigma(B_4) = [E_6]$ has exactly 2 flexes. Thus the existence of sextic of torus type $B_4 + B_1 + B_1'$ with $[2A_5 + 2A_2 + 4A_1]$ or $[E_6 + 2A_5 + 3A_1]$ implies that there does not exist sextic of non-torus type with these two configurationsqed.

Example I. We consider the quartic $B_4 := \{g_4 = 0\}$ with one A_5 where g_4 is given by

$$639 - 1350\,x^2 + 351\,x^4 + 468\,x^3 - 108\,x + 288\,y + 1608\,I\,y^2\,x^2\,\sqrt{3} + 1452\,I\,y\,x\,\sqrt{3}$$
$$- 676\,I\,y\,x^3\,\sqrt{3} - 1032\,I\,y^2\,x\,\sqrt{3} - 1452\,I\,y^3\,x\,\sqrt{3} - 288\,y^3 - 918\,y^2 + 279\,y^4$$
$$- 648\,y^3\,x + 1350\,y^2\,x^2 + 108\,y^2\,x - 936\,y\,x^3 + 648\,y\,x - 432\,I\,y^2\,\sqrt{3} + 776\,I\,y^3\,\sqrt{3}$$
$$- 1608\,I\,x^2\,\sqrt{3} - 152\,I\,\sqrt{3} - 776\,I\,y\,\sqrt{3} + 1032\,I\,x\,\sqrt{3} + 584\,I\,\sqrt{3}\,y^4 + 728\,I\,x^3\,\sqrt{3}$$

B_4 has an A_5 singularity at $(1,0)$ and 6 flexes at

$$P_1 = (0,-1),\; P3 = (-\frac{130}{1069} + \frac{370}{1069}\,I\,\sqrt{3},\; \frac{263}{1069} + \frac{156}{1069}\,I\,\sqrt{3})$$

$$P_2 = (0,1),\; P4 := (\frac{2190}{13333} - \frac{4790}{39999}\,I\,\sqrt{3},\; -\frac{12671}{13333} - \frac{4492}{39999}\,I\,\sqrt{3}),$$

$$P_5 = (\frac{2116}{6841}\,I\,\sqrt{3} + \frac{632}{20160427}\,\sqrt{90988874 - 65462588\,I\,\sqrt{3}} + \frac{4086}{6841}$$

$$+ \frac{498}{20160427}\,I\,\sqrt{3}\,\sqrt{90988874 - 65462588\,I\,\sqrt{3}},$$

$$-\frac{9056}{47887} - \frac{6133}{47887}\,I\,\sqrt{3} + \frac{4}{47887}\,\sqrt{90988874 - 65462588\,I\,\sqrt{3}})$$

$$P_6 = (-\frac{632}{20160427}\,\sqrt{90988874 - 65462588\,I\,\sqrt{3}} + \frac{4086}{6841}$$

$$- \frac{498}{20160427}\,I\,\sqrt{3}\,\sqrt{90988874 - 65462588\,I\,\sqrt{3}} + \frac{2116}{6841}\,I\,\sqrt{3},$$

$$- \frac{9056}{47887} - \frac{6133}{47887}\,I\,\sqrt{3} - \frac{4}{47887}\,\sqrt{90988874 - 65462588\,I\,\sqrt{3}})$$

It is easy to check that $\{P_1, P_3\}$, $\{P_2, P_4\}$, $\{P_5, P_6\}$ give sextic of torus type. The other cases give sextics of non-torus type. For example, a nice sextic of non-torus type is given by taking the tangent lines at P_1 and P_2: $B_1 + B_1' : (y-1)(y+1) = 0$.

II. Now we consider the quartic with $2A_2$ (case (b) with $k=3$).

$$f_4(x,y) := \frac{254143}{4096}\,x^4 - \frac{251}{16}\,x^3 + \frac{11}{32}\,y\,x^3 - \frac{5893}{2048}\,y^2\,x^2 - \frac{2761}{1024}\,y\,x^2 - \frac{126093}{2048}\,x^2$$
$$- \frac{11}{32}\,y\,x + \frac{1}{32}\,y^3\,x + \frac{251}{16}\,x + \frac{5893}{2048}\,y^2 - \frac{1957}{4096} + \frac{211}{4096}\,y^4 + \frac{2761}{1024}\,y - \frac{251}{1024}\,y^3$$

It has 8 flex points and four flexes are explicitly written as

$$P_1 = (1,0),\; P_2 = (-1,0),\; P_3 = (0,-1),\; P_4 = \left(\frac{16064}{64025}, \frac{-61977}{64025}\right)$$

Sextics of torus type are given by taking tangent lines at $\{P_1, P_2\}$, $\{P_3, P_4\}$. As a sextic of non-torus type, we can take the tangent lines at P_1, P_3 so that the sextics is given by adding two lines $(x-1)(-4y-4+16x) = 0$. The configuration space of quartic with $2A_2 + A_1$ is connected. Each quartic has two flex points and with two tangent lines B_1, B_1', $B_4 \cup B_1 \cup B_1'$ gives a sextic of torus type with $[2A_5, 2A_2, 4A_1]$. Thus there is no sextic of non-torus type $C = B_4 + B_1 + B_1'$ with configuration $[2A_5, 2A_2, 4A_1]$.

4.2. *Conical geometry of quartic*

Now we consider the configuration with component type $B_4 + B_2$ will be considered here. The corresponding configurations are

(1) $\Sigma(C) = [2A_5, 2A_2, 2A_1]$, $[2A_5, 2A_2, 3A_1]$.
(2) $\Sigma(C) = [3A_5, 2A_1]$.
(3) $\Sigma(C) = [E_6, 2A_5, 2A_1]$.
(4) $\Sigma(C) = [A_{11}, 2A_2, 2A_1]$, $[A_{11}, 2A_2, 3A_1]$.
(5) $\Sigma(C) = [A_{11}, A_5, 2A_1]$.

Zariski pairs with the above configurations with fixed component type $B_4 + B_2$ can not be explained by the flex geometry.

We have to generalize the notion of flex points. Let B be a given irreducible plane curve of degree d. Let Φ be a linear system of conics and let α be the dimension of Φ. For a general smooth point $P \in B$, the maximal intersection number of $I(B, B_2; P)$ for $B_2 \in \Phi$ is α. We say P is a *conical flex point with respect to* Φ if the intersection number $I(B, B_2; P) \geq \alpha + 1$. If $\dim \Phi = 5$ (so Φ is the family of all conics), we say simply that P is a *conical flex point*.

1. Let us consider the case $\Sigma(C) = [2A_5, 2A_2, 2A_1]$, $[2A_5, 2A_2, 3A_1]$. We consider first a sextic of torus type $C = \{f_2^3 + f_3^2 = 0\}$ which decomposes into a quartic B_4 and a quartic B_2:

$$f(x, y) := \left(y^2 - 2 + 2\,x^2\right)^3 + \left(-25\,y^3 + (13\,x - 23)\,y^2 + (-26\,x^2 + 26)\,y + 13\,x^3\right.$$
$$\left. - 23\,x^2 - 13\,x + 23\right)^2 = (y^2 + x^2 - 1)(177\,x^4 - 598\,x^3 - 676\,y\,x^3 + 344\,x^2$$
$$+ 849\,x^2\,y^2 + 1196\,y\,x^2 - 650\,y^3\,x + 598\,x + 676\,y\,x - 598\,x\,y^2 - 521 - 151\,y^2$$
$$+ 1150\,y^3 - 1196\,y + 626\,y^4)$$

Our quartic is defined by

$$g_4(x, y) = (177\,x^4 - 598\,x^3 - 676\,y\,x^3 + 344\,x^2 + 849\,x^2\,y^2 + 1196\,y\,x^2 - 650\,y^3\,x$$
$$+ 598\,x + 676\,y\,x - 598\,x\,y^2 - 521 - 151\,y^2 + 1150\,y^3 - 1196\,y + 626\,y^4) = 0$$

Note that the singularities B_4 are $2\,A_2$. The intersection $B_2 \cap B_4$ makes two A_5 at $P := (1, 0)$ and $Q := (-1, 0)$. We consider the linear system Φ of conics of dimension 2 which are defined by the conics $C_2 := \{h_2(x, y) = 0\}$ such that $I(C_2, B_4; P) \geq 3$. Then we consider the conical flex points $R = (a, b) \in B_4$ with respect to Φ, which is described by the condition $\exists h_2 \in \Phi$

such that $I(h_2, B_4; R) \geq 3$. We found that there are 11 conical flex points. Two of them can be explicitly given as $S_1 = Q$ and $S_2 = (0, -1)$. The corresponding conics are given as

$$h_2(x, y) = x^2 + y^2 - 1, \quad \{h_2 = 0\} \cap B_4 \supset \{P, Q\}$$
$$g_2(x, y) = (y^2 + 2yx - 2y - x^2 + 4x - 3), \quad \{g_2 = 0\} \cap B_4 \supset \{P, S_2\}$$

We can easily check that the sextic

$$(y^2 + 2yx - 2y - x^2 + 4x - 3)(177\,x^4 - 598\,x^3 - 676\,yx^3 + 344\,x^2 + 849\,x^2y^2$$
$$+ 1196\,yx^2 - 650\,xy^3 + 598\,x + 676\,yx - 598\,xy^2 - 521 - 151\,y^2 + 1150\,y^3$$
$$- 1196\,y + 626\,y^4) = 0$$

is not of torus type. Thus $B_4 \cup \{h_2(x, y) = 0\}$ and $B_4 \cup \{g_2(x, y) = 0\}$ is a Zariski pair.

Similarly the case $[2A_5, 2A_2, 3A_1]$ can be treated in the same way. We start from a sextic of torus type:

$$f(x, y) = \left(-\frac{49}{64}y^2 - \frac{15}{64} + \frac{15}{64}x^2 \right)^3 + \left(-\frac{131}{256}y^3 + \left(\frac{729}{512}x - \frac{297}{256}\right)y^2 \right.$$
$$+ \left(-\frac{387}{256}x^2 + \frac{387}{256} \right)y + \frac{729}{512}x^3 - \frac{297}{256}x^2 - \frac{729}{512}x + \frac{297}{256} \Bigg)^2$$
$$= \frac{27}{262144}\left(y^2 + x^2 - 1 \right)\left(19808\,x^4 - 41796\,yx^3 - 32076\,x^3 + 40521\,y^2x^2 \right.$$
$$- 6865\,x^2 + 34056\,x^2y - 32076\,xy^2 + 32076\,x + 41796\,xy - 14148\,xy^3$$
$$- 7770\,y^2 - 34056\,y - 1815\,y^4 + 11528\,y^3 - 12943 \Bigg)$$

The quartic B_4 has two $2A_2 + A_1$ and we consider the linear system Φ of conics intersecting B_4 at $P = (1, 0)$ with intersection number 3. We find that there exist 5 conical flex points with respect to Φ, and among them we have two explicit ones: $Q = (-1, 0)$ and $(0, -1)$. We see that the conic corresponding to $(0, -1)$ gives a Zariski partner sextic $f_6 = 0$ to $C = \{f = 0\}$.

$$f_6 := (5y^2 - 64\,xy + 64\,y + 69\,x^2 - 128\,x + 59)(19808\,x^4 - 41796\,yx^3$$
$$- 32076\,x^3 + 40521\,y^2x^2 - 6865\,x^2 + 34056\,x^2y - 32076\,xy^2 + 32076\,x$$
$$+ 41796\,xy - 14148\,xy^3 - 7770\,y^2 - 34056\,y - 1815\,y^4 + 11528\,y^3 - 12943)$$

Remark 8. The calculation of conical flex points are usually very heavy. We used maple 7 to compute in the following recipe. a. First compute the normal form of $h_2 \in \Phi$. It contains two parameters. b. Assume $(u,v) \in B_4$. Put $gg_4(x,y) := (x+u, y+v)$ and $hh_2(x,y) := h_2(x+u, y+v)$. Consider the maximal contact coordinate at (u,v): $\Phi(x) = a_1\,x + a_2\,x^2 + a_3\,x^3$ and put $GG4(x) := gg_4(x, \Phi(x))$ and $HH_2(x) := hh_2(x, \Phi(x))$. Our assumption implies that $\mathrm{Coeff}(GG4, x, 1) = \mathrm{Coeff}(GG2, x, 2) = 0$ and $\mathrm{Coeff}(HH_2, x, 0) = \mathrm{Coeff}(HH_2, x, 1) = \mathrm{Coeff}(HH_2, x, 2) = 0$. Solve the equations $\mathrm{Coeff}(GG4, x, 1) = \mathrm{Coeff}(GG2, x, 2) = 0$ in a_1, a_2. Then solve the equations $\mathrm{Coeff}(HH_2, x, 0) = \mathrm{Coeff}(HH_2, x, 1) = 0$ in the remaining parameters of the linear system. Then we get two equations in u, v:

$$g_4(u,v) = \mathrm{Coeff}(HH_2, x, 2) = 0$$

c. Use the resultant computation to solve the above equations to obtain the possibility of conical flex points.

2. $\Sigma(C) = [3A_5, 2A_1]$ with $B_4 + B_2$:

$$B_4 : g_4(x,y) = (6\,y\,x + \frac{1710}{91}\,y\,x^2 - \frac{1466}{91}\,x^3\,y - \frac{1992}{91}\,y^2\,x - 6\,y^3\,x + \frac{790}{91}\,y^3 +$$
$$y^4 - \frac{4904}{91}\,x^3 + \frac{1992}{91}\,x - \frac{790}{91}\,y + \frac{939}{91}\,y^2\,x^2 + \frac{1161}{91}\,y^2 + \frac{1968}{91}\,x^2 - \frac{1252}{91} + \frac{2196}{91}\,x^4)$$

It has an A_5 at $P := (1,0)$. We consider the linear system Φ of conics of dimension 2 whose conic are intersecting with B_4 at $(0,1)$ with intersection number 3. We find 14 conical flex points with respect to Φ in which two are explicit: $R = (\frac{498727}{500817}, \frac{-266266}{500817})$ and $Q = (0,-1)$. The corresponding conics $f_2 = 0$, $k_2 = 0$ intersecting B_4 at P, Q oe P, R are given by the following and they gives sextics of non-torus type and of torus type respectively.

$$f_2(x,y) := y^2 - 1 + \frac{171}{79}\,x^2$$
$$k_2(x,y) = (y^2 + (\frac{268}{759}\,x + \frac{4424}{2277})\,y + \frac{85291}{19987}\,x^2 - \frac{268}{759}\,x - \frac{6701}{2277})$$

3. $\Sigma(C) = [E_6, 2A_5, 2A_1]$: We start the next quartic B_4:

$$y^4 + \left(-\frac{195}{64}\,x + \frac{169}{64}\right)y^3 + \left(\frac{105}{32}\,x^2 - \frac{33}{8}\,x + \frac{27}{32}\right)y^2 + \left(-\frac{143}{64}\,x^3 + \frac{117}{64}\,x^2\right.$$
$$\left. + \frac{195}{64}\,x - \frac{169}{64}\right)y + \frac{45}{32}\,x^4 - \frac{19}{8}\,x^3 - \frac{21}{16}\,x^2 + \frac{33}{8}\,x - \frac{59}{32} = 0$$

Sextics of non-torus type and of torus type are given by the conics B_2, B_2':

$$B_2 : y^2 - 1 + \frac{9}{13}\,x^2 = 0$$
$$B_2' : y^2 + (-\frac{770}{1147}\,x + \frac{156}{1147})\,y + \frac{11025}{14911}\,x^2 + \frac{770}{1147}\,x - \frac{1303}{1147} = 0$$

4. We consider the configurations $[A_{11}, 2A_2, 2A_1]$, $[A_{11}, 2A_2, 3A_1]$. First we consider two cuspidal quartics with $[2A_2]$:

$$f_4 := 5805\, x^4 - 2916\, I\, x^3\, \sqrt{2} + 3888\, I\, x^3\, y\, \sqrt{2} - 1269\, x^2\, y^2 - 729\, x^2$$
$$-3834\, x^2\, y - 3888\, I\, x\, y^2\, \sqrt{2} + 108\, I\, x\, \sqrt{2}\, y^3 + 2916\, I\, x\, y\, \sqrt{2} + 1323\, y^3$$
$$-1971\, y^2 - 81\, y^4 + 729\, y = 0$$

It can makes sextics of torus type and non-torus type with configuration $[A_{11}, 2A_2, 2A_1]$ with respective conics:

$$f_2(x, y) = y - x^2$$
$$h_2(x, y) = (\frac{1}{8}\, I\, \sqrt{2}\, y^2 + x\, y - \frac{3}{4}\, I\, y\, \sqrt{2} - \frac{5}{4}\, I\, \sqrt{2}\, x^2 - 3\, x + \frac{5}{8}\, I\, \sqrt{2})$$

They correspond to the conical flex points $(0, 0)$ and $(0,1)$. The other conical flexes are very heavy to be computed.

Next we consider the configuration $[A_{11}, 2A_2, 3A_1]$ which is produced by a quartic B_4 with $2A_2 + A_1$ and a conic B_2 with a single tangent at a conical flex.

$$B_4 : \frac{1}{16}\, y^4 + \frac{3}{4}\, x\, y^3 + \frac{59}{8}\, y^2\, x^2 - y^2\, x - \frac{1}{8}\, y^2 + \frac{27}{4}\, y\, x^3 - 6\, y\, x^2 - \frac{3}{4}\, y\, x + \frac{17}{16}\, x^4$$
$$- x^3 - \frac{9}{8}\, x^2 + x + \frac{1}{16} = 0$$

We find three conical flex points $P_1 = (-1/19, 12/19)$, $P_2 = (-4/13, 15/13)$ and $P_3 = (-1, 0)$. The corresponding conic which are tangent at the respective conical flex point P_i, $i = 1, 2, 3$ are given by

$$g_{21} := y^2 + (\frac{1270}{141}\, x + \frac{10}{141})\, y + \frac{38711}{423}\, x^2 - \frac{5486}{423}\, x - \frac{457}{423}$$
$$g_{22} := y^2 + (\frac{7462}{2517}\, x - \frac{5462}{2517})\, y + \frac{13007}{7551}\, x^2 - \frac{21902}{7551}\, x + \frac{8831}{7551}$$
$$g_{23} := -\frac{3}{22}\, y^2 - \frac{61}{11}\, y\, x + y - \frac{73}{66}\, x^2 - \frac{1}{33}\, x + \frac{71}{66}$$

The corresponding sextic $f_4(x, y)\, g_{2i}(x, y) = 0$ is of torus type for $i = 1$ and of non-torus type for i=2,3. The torus decomposition of $f_4\, g_{21}$ is given by $z_{21}^3 + z_{31}^2 = 0$ where

$$z_{21} := \frac{1}{423}\, 423^{(2/3)}\, (3\, y^2 + 24\, y\, x + 293\, x^2 - 34\, x - 3)$$

$$z_{31} := \frac{1}{6768}\, I(5 - 189\, x - 3477\, x^2 + 20045\, x^3 - 3\, y - 54\, y\, x$$
$$+ 2361\, y\, x^2 - 5\, y^2 + 247\, y^2\, x + 3\, y^3)\sqrt{6768}$$

5. Lastly, we consider the configuration $[A_{11}, A_5, 2A_1]$ which is associated to a quartic B_4 with an A_5 singularity and a conic B_2 tangent at a conical flex point with intersection number 6. As a quartic, we take:

$$f_4 := y^4 + x\,y^3 + \frac{7}{15}\,x^2\,y^2 - 2\,x\,y^2 - 3\,y^2 + \frac{2}{15}\,x^3\,y - \frac{2}{15}\,x^2\,y + x\,y + 2\,y + \frac{4}{75}\,x^4$$
$$-\frac{2}{15}\,x^3 - \frac{1}{3}\,x^2$$

B_4 has apparently 26 conical flex points. (The calculation is very heavy.) We take four explicit conical flex points: $P_1 = (0,0)$, $P_2 := (-\frac{540}{493}, \frac{250}{493})$, $P_3 := (\frac{-270}{301}, \frac{58}{301})$ and $P_4 := (\frac{-270}{193}, \frac{-50}{193})$. After an easy computation, the respective conics are given as

$$n_{21} := y^2 + (-\frac{5}{59}\,x - \frac{50}{59})\,y + \frac{25}{177}\,x^2$$

$$n_{22} := y^2 + (-\frac{10845}{262699}\,x - \frac{273650}{262699})\,y + \frac{135675}{262699}\,x^2 + \frac{153000}{262699}\,x + \frac{70000}{262699}$$

$$n_{23} := y^2 + (-\frac{16681}{32607}\,x - \frac{17318}{10869})\,y + \frac{30251}{163035}\,x^2 - \frac{1544}{32607}\,x - \frac{112}{10869}$$

$$n_{24} := y^2 + (\frac{5225}{1633}\,x + \frac{350}{71})\,y + \frac{4225}{1633}\,x^2 + \frac{13000}{1633}\,x + \frac{10000}{1633}$$

Put $f_{6j}(x,y) := f_4(x,y)\,n_{2j}(x,y)$ and $C^{(j)} = \{f_{6j} = 0\}$. It is also easy to see that $C^{(1)}$, $C^{(2)}$ are of non-torus type and $C^{(3)}$, $C^{(4)}$ are of torus type.

5. Flex geometry of cubic curves

5.1. *Configurations coming from cubic flex geometry: a cubic component and a line component*

Let us consider first configurations which occurs in sextics which have at least a cubic component B_3. We divides into the following cases.

(1) $C = B_3 + B_3'$.

    (a) $\Sigma(C) = [A_{17}]$, $[A_{17}, A_1]$, $[A_{17}, 2A_1]$.
    (b) $\Sigma(C) = [A_{11}, A_5]$, $[A_{11}, A_5, A_1]$, $[A_{11}, A_5, 2A_1]$, .
    (c) $\Sigma(C) = [A_{11}, 2A_2, 3A_1]$
    (d) $\Sigma(C) = [3A_5]$, $[3A_5 + A_1]$, $[3A_5, 2A_1]$.

(2) $C = B_3 + B_2 + B_1$.

    (a) $\Sigma(C) = [A_{11} + A_5 + 2A_1]$, $[A_{11} + A_5 + 3A_1]$.
    (b) $\Sigma(C) = [3A_5 + 2A_1]$, $[3A_5 + 3A_1]$.

(3) $C = B_3 + B_1 + B_1' + B_1''$ with configuration $[3A_5 + 3A_1]$, $[3A_5 + 4A_1]$.

We first consider the cases (2) and (3). In these cases, there are one cubic component B_3 and at least one line component B_1. Recall that the configurations in (2) and (3) occurs as sextics of linear torus type. For a reducible sextic C which is classified in either (2) or (3), the necessary and sufficient condition for C to be of torus type is there exists a line L containing inner singularities. In the case of $\Sigma(C) = [A_{11}, A_5]$, L is also tangent to the tangent cone of A_{11}. We first recall the following basic geometry for cubic curves.

Proposition 9. *1. A smooth cubic C has 9 flex points. Among 84 choices of three flex points, 12 colinear triples of flexes.*

2. A nodal cubic has 3 flex points, and they are colinear.

3. A cuspidal cubic has one flex point.

For the proof of the assertion 1, see Example below.

Corollary 10. *The configuration $[3A_5 + 4A_1]$ with components type $B_3 + B_1 + B_1' + B_1''$ does not exist as a sextic of non-torus type.*

Proof. The cubic has a node and three line components are flex tangent lines at three flex points. We know that such configuration exists as a sextics of linear torus type [10]. As the configuration space of one nodal cubics is connected, every sextics $B_3 + B_1 + B_1' + B_1''$ is of torus type.qed.

(2) $C = B_3 + B_2 + B_1$ with $\Sigma(C) = [3A_5, kA_1]$, $k = 2, 3$. In this case, B_3 is either smooth or nodal and two intersection points $B_3 \cap B_2$ generates $2\,A_5$. The third A_5 is generated by a flex tangent line B_1.

Proposition 11. *Assume that a cubic B_3 and a conic B_2 are intersecting at two points P, Q with respective intersection number 3, producing $2A_5$-singularities. Then the line passing through P, Q intersects B_3 at another point, say R, and R is a flex point of B_3.*

This Proposition describes sextics of torus type and non-torus type with configuration $[3A_5, j\,A_1]$, $j = 0, 1, 2$ ((2-b)).

Proof of Proposition 11. Assume that $P = (0, 1)$ and $Q = (0, -1)$ with the tanget lines $y = \pm 1$ respectively. Then by an easy computation, the cubic B_3 is defined by a polynomial

$$f_3 := y^3 + y^2\,a_{12}\,x - y^2\,a_{00} + y\,a_{21}\,x^2 - y + a_{30}\,x^3 - a_{00}\,a_{21}\,x^2 - a_{12}\,x + a_{00} = 0$$

and the conic is given as $y^2 + a_{21}\,x^2 - 1 = 0$. Then R is given as $(0, a_{00})$ and we can easily see that R is a flex of B_3.qed.

By the same calculation, we see that

Proposition 12. *Assume that a cubic B_3 and a conic B_2 are intersecting at one points P with intersection multiplicity 6 producing an A_{11}-singularity Then the tangent line passing at P intersects another point $R \in B_3$ and R is a flex point of B_3.*

Proof is similar. Putting $P = (0,0)$ and assuming $y = 0$ as the tangent line, the cubic is written as

$$f_3 := (y^3 {t_2}^4 - y^2 x\, t_3\, a_{01}\, t_4\, t_2 + y^2 x\, a_{21}\, t_3\, {t_2}^2 + 2\, y^2 x\, a_{11}\, {t_3}^2\, t_2 + 2\, y^2 x\, a_{01}\, {t_3}^3$$
$$- y^2 x\, a_{11}\, t_4\, {t_2}^2 - y^2 {t_2}^2\, a_{01}\, t_4 - y^2 {t_2}^3\, a_{21} - y^2 {t_2}^2\, a_{11}\, t_3 + y\, {t_2}^4\, a_{21}\, x^2$$
$$+ y\, {t_2}^4\, a_{11}\, x + y\, {t_2}^4\, a_{01} - x^3 {t_2}^4\, a_{01}\, t_3 - x^3 {t_2}^5\, a_{11} - a_{01}\, {t_2}^5\, x^2)/{t_2}^4$$

and $R = (0, x\, t_2\, a_{11} + a_{01}\, t_2 + x\, a_{01}\, t_3)$.

Lemma 13. *Assume that a conic B_2 is tangent to an irreducible curve C of degree $d \geq 3$ at a smooth point $P \in C$ so that $I(B_2, C; P) \geq 3$. Then P is not a flex point of C.*

Proof. Let $h_2(x, y) = 0$ be a conic equation which defines B_2. In fact, if $P = (a, b)$ is a flex point of C, $I(C, B_2; P) \geq 3$ implies that C is locally parametrized as $y_1(x) = t_1\, x_1 + t_3\, x_1^3 + (\text{higher terms})$ where $(x_1, y_1) = (x - a, y - b)$ assuming the tangent line is not $x - a = 0$. As B_2 does not have any flex, the equation $h_2(x, y) = 0$ is solved as $y_1 = s_1\, x_1 + s_2\, x_1^2 + (\text{higher terms})$ with $s_2 \neq 0$. Thus $I(C, B_2; P) = 1$ or 2 according to $s_1 \neq t_1$ or $s_1 = t_1$.qed.

First we consider the case (2-b). Then the cubic is either smooth or have a node. Thus it has at least 3 flexes. As the intersection $B_3 \cap B_2$ are not flex points, we can find another flex point $S \in B_3$, $S \neq R$. Taking flex tangent $L := T_S\, B_3$ as the line component, the corresponding sextic is not of torus type. As a cuspidal cubic has a unique flex points, we see that a sextic $C = B_3 + B_2 + B_1$ with $[3A_5 + A_2 + 2A_1]$ does not appear as a sextic of non-torus type.

Now we consider (3). Assume that the cubic is smooth. Then there are 9 flex points and B_1, B_1', B_1'' are flex tangents. The sextic is of torus type if and only if three flexes are colinear [9]. This case, the configuration is $[3A_5 + 3A_1]$.

Finally we consider the case (2-a). In this case, $B_3 \cap B_2$ is a single point P and $I(B_3, B_2; P) = 6$ and the intersection singularity is A_{11}. B_3 has at most a node and so it has at least 3 flex points. Taking a line component which is the flex tangent at S other than R, we get a sextic $C = B_3 + B_2 + B_1$ with $[A_{11} + A_5 + 2A_1]$ or $[A_{11} + A_5 + 3A_1]$. qed.

We omit explicit examples for (2-a) and (2-b) as they can be easily obtained from sextic of torus type with the same configuration ([10]) and replacing the flex line B_1. We only gives an example of (3).

Example $B_3 + B_1 + B_1' + B_1''$, with configuration $[3A_5 + 3A_1]$. In this case, the cubic B_3 is smooth and has 9 flexes and three lines B_1, B_1', B_1'' are the tangent lines at flexes P_1, P_2, P_3 (see below) of B_3. Three A_1 are the intersections of lines. We know that $B_3 \cup L_P \cup L_Q \cup L_R$ is of torus type if and only if P, Q, R are colinear, where L_P is the tangent line at the flex point P. Let $B_3 : f_3(x, y) = 0$. An example of such a sextic of non-torus type is given by

$$f_3(x, y) = y^3 + \left(- (-3 + \frac{1}{2} I \sqrt{3}) x - 2 \right) y^2 - y$$
$$+ (1 - \frac{1}{2} I \sqrt{3}) x^3 + (-3 + \frac{1}{2} I \sqrt{3}) x + 2$$

$$f(x, y) = (y^3 + (-(-3 + \frac{1}{2} I \sqrt{3}) x - 2) y^2 - y + (1 - \frac{1}{2} I \sqrt{3}) x^3$$
$$+ (-3 + \frac{1}{2} I \sqrt{3}) x + 2)(y - 1)(y + 1)(-I \sqrt{3} (x - 1) - y)$$

We can moreover explicitly compute 9 flex points $P_1, \ldots, P_9$ as follows. $P_1 = (1, 0, 1)$, $P_2 = (0, 1, 1)$, $P_3 = (0, -1, 1)$, $P_4 = (0, 2, 1)$, $P_5 = (\frac{3}{5}, \frac{4}{5}, 1)$,

$$P_6 = (\frac{1}{2} + \frac{1}{6} I \sqrt{3}, \frac{1}{2} - \frac{1}{6} I \sqrt{3}, 1), \ P_7 = (\frac{15}{14} + \frac{3}{14} I \sqrt{3}, \frac{1}{14} + \frac{3}{14} I \sqrt{3}, 1),$$
$$P_8 = (\frac{21}{38} + \frac{9}{38} I \sqrt{3}, \frac{31}{38} - \frac{3}{38} I \sqrt{3}, 1), \ P_9 = (\frac{33}{62} + \frac{3}{62} I \sqrt{3}, \frac{37}{62} + \frac{9}{62} I \sqrt{3}, 1)$$

Thus by a direct checking, we find the following 12 triples which are colinear.

$$C_1 = [P_1, P_2, P_6], C_2 = [P_1, P_3, P_7], C_3 = [P_1, P_4, P_5]$$
$$C_4 = [P_1, P_8, P_9], C_5 = [P_2, P_3, P_4], C_6 = [P_2, P_5, P_8]$$
$$C_7 = [P_3, P_5, P_9], C_8 = [P_3, P_6, P_8], C_9 = [P_4, P_6, P_9]$$
$$C_{10} = [P_4, P_7, P_8], C_{11} = [P_5, P_6, P_7], C_{12} = [P_2, P_7, P_9]$$

5.2. *Sextics with two cubic components: $C = B_3 + B_3'$*

Now we consider sextics with two cubic curves B_3, B_3'. The possible configurations are

(1) $C = B_3 + B_3'$.

    (a) $\Sigma(C) = [A_{17}], [A_{17}, A_1], [A_{17}, 2A_1]$.

(b) $\Sigma(C) = [A_{11}, A_5], [A_{11}, A_5, A_1], [A_{11}, A_5, 2A_1],$.
(c) $\Sigma(C) = [A_{11}, 2A_2, 3A_1]$
(d) $\Sigma(C) = [3A_5], [3A_5 + A_1], [3A_5, 2A_1]$.

First we consider two cubics B_3, B_3' which are tangent at the origin with intersection number 9. Let $f(x,y) = 0$ and $f'(x,y) = 0$ be the defining polynomials of B_3 and B_3' respectively and we may assume that the tangent line of B_3 is given by $y = 0$. Let $y = \sum_{i=2}^{\infty} t_i x^i$ be the solution of $f(x,y) = 0$ at O. Then by the assumption $I(B_3, B_3'; O) = 9$, we must have $\mathrm{ord}_x f'(x, \sum_{i=2}^{\infty} t_i x^i) = 9$.

Lemma 14. *The sextic $C := B_3 \cup B_3'$ is of torus type if and only if $t_2 = 0$. This implies that O is a flex of B_3.*

Proof. This assertion is given by Artal in [1]. Our proof is computational. In fact, if $t_2 = 0$, O is a flex for both B_3, B_3' and we see that $y = 0$ is a flex tangent line for B_3, B_3'. Thus by Tokunaga's criterion, $y^2 = 0$ is the conic which gives a linear torus decomposition. For the detail about linear torus decomposition, we refer [10].

Assume that $t_2 \neq 0$ and we prove that any such C is of non-torus type. In fact, supposing C to be a sextic of torus type, take a torus decomposition $f(x,y)f'(x,y) = f_2(x,y)^3 + f_3(x,y)^2$. Put $y_1 := y - \sum_{i=2}^{\infty} t_i x^i$. So the assumption implies that

$$f(x, y_1 + \phi(x)) \times f'(x, y_1 + \phi(x)) = y_1(y_1 + cx^9) + \text{(higher terms)}, \quad c \neq 0$$

and thus $y_1' := y - \sum_{i=2}^{8} t_i x^i$ is the maximal contact coordinate and it is also the solution of $f_3(x,y) = 0$ in $y \bmod x^9$ and $f_2(x, \sum_{i=2}^{8} t_i x^i) \equiv 0 \mod (x^6)$. For the existence of a non-trivial conic $f_2(x,y)$, we see that the coefficient must satisfy:

$$J_0 := -3t_4 t_3 t_2 + 2t_3^3 + t_5 t_2^2 = 0 \tag{2}$$

(conics are five dimensional but we have 6 equation $coeff(f_2(x, \sum_{i=2}^{8} t_i x^i), x, j) = 0$ for $j = 0, \dots, 5$). Then we examine the other equations

$$(\sharp) \quad \mathrm{Coeff}(g(x, \sum_{i=2}^{8} t_i x^i), x, j) = 0, \quad j = 0, \dots, 8$$

where $g(x,y)$ is a cubic which corresponds to either $f(x,y)$ or $f'(x,y)$. Write $g(x,y)$ as a generic cubic form with 10 coefficients (but by scalar multiplication, one coefficient can be normalized to be 1, say that of x^6 is 1, and we have 9 free coefficients), we solve the equations $(\sharp)$ from $j =$

0 to $j = 0$ to $j = 8$ to express the coefficients in rational functions of $t_2, \ldots, t_7$. At the last step, we have one free coefficient undetermined, say the coefficient c of $x^a y^b$ and a linear equation $\mathrm{Coeff}(g(x, \sum_{i=2}^{8} t_i x^i), x, 8) = 0$. This is written as $K_1 c + K_0 = 0$ where K_1, K_0 are rational functions of $t_2, \ldots, t_7$. Thus to have two non-trivial cubic solutions, we need to have $K_1 = K_0 = 0$. Now we can easily check that there are no solutions if we assume that $J_0(t_2, \ldots, t_5) = 0$. The other assertion will be checked in the explicit construction of examples.qed.

If $t_2 \neq 0$, there is no line L such that L intersects only at O. Thus the sextic can not be of torus type by the classification in [10]. However the above argument is useful for the explicit computation of non-torus sextics.

(1-a) Let us consider the case $[A_{17}]$, $[A_{17}, A_1]$, $[A_{17}, 2A_1]$. The sextics of non-torus type with above configurations are obtained using above computation ($t_2 \neq 0$). Their Zariski partners are cubics intersecting at flex points.

(a-1) $[A_{17}]$, $C = B_3 + B_3'$, $C := \{f_3(x, y)g_3(x, y) = 0\}$ where

$$f_3 := -y^3 - y^2 + (-x^2 + x)\, y - x^3 + x$$
$$g_3 := x - \frac{10}{9}\, y^3 - x^2\, y + \frac{10}{9}\, x\, y - y^2 - \frac{10}{9}\, x^3$$

(a-2) $[A_{17} + A_1]$, $C = B_3 + B_3'$ with B_3 has a node: Put $C := \{f_3(x, y)g_3(x, y) = 0\}$ where

$$f_3 := x - 3\, x\, y + 3\, x^2\, y - 2\, x^2 - y^2 + 2\, y^3 + 4\, y^2\, x + x^3$$
$$g_3 := x + x\, y + 11\, x^2\, y - 10\, x^2 - y^2 - 2\, y^3 + 8\, y^2\, x + 5\, x^3$$

(a-3) $[A_{17} + 2A_1]$, $C = B_3 + B_3'$ $C := \{f_3(x, y)g_3(x, y) = 0\}$, where cubics are nodal and

$$f_3 := \frac{1}{48}\, I(48\, x - 96\, x^2 - 264\, x\, y + 124\, y^3 - 48\, y^2 + 411\, y^2\, x + 264\, x^2\, y + 48\, x^3$$
$$- 104\, I\, x\, y\, \sqrt{3} + 181\, I\, y^2\, x\, \sqrt{3} + 104\, I\, x^2\, y\, \sqrt{3} + 16\, I\, x\, \sqrt{3} - 32\, I\, x^2\, \sqrt{3}$$
$$+ 16\, I\, x^3\, \sqrt{3} - 16\, I\, y^2\, \sqrt{3} + 52\, I\, y^3\, \sqrt{3})\sqrt{3}\, /(-1 + I\, \sqrt{3})$$

$$g_3 := -\frac{1}{8}(48\, x - 72\, x^2 + 10\, I\, x^2\, y\, \sqrt{3} - 25\, I\, y^2\, x\, \sqrt{3} + 56\, I\, x\, y\, \sqrt{3} - 216\, x\, y + 68\, y^3$$
$$- 48\, y^2 + 231\, y^2\, x + 138\, x^2\, y + 23\, x^3 + 7\, I\, x^3\, \sqrt{3} + 8\, I\, x^2\, \sqrt{3} - 16\, I\, x\, \sqrt{3}$$
$$- 12\, I\, y^3\, \sqrt{3} + 16\, I\, y^2\, \sqrt{3})\, /((3 + I\, \sqrt{3})\, (-1 + I\, \sqrt{3}))$$

860

5.3. Exceptional configuration: $[A_{11}, 2A_2, 3A_1]$ with two cubic components

In this case, we do the similar computation. We compute sextics $C = B_3 \cup B_3'$ such that B_3 and B_3' have two A_2 at $(0,1)$ and $(1,0)$ respectively and they intersect at $(0,-1)$ with intersection number 6 to make A_{11}. We have the following sextics of non-torus type.

$$f(x,y) := f_3(x,y)\, g_3(x,y)$$

$$f_3 := -\frac{1}{44652}(47\sqrt{3}+168+195\,I+74\,I\sqrt{3})(195\,y+168\,I\,y^3-168\,I\,y^2-168\,I\,y$$
$$-156\,I\,x^2+74\sqrt{3}\,y^3-74\sqrt{3}\,y^2-74\sqrt{3}\,y-60\sqrt{3}\,x^2+60\sqrt{3}\,y\,x^2-47\,I\sqrt{3}\,y^3$$
$$+47\,I\sqrt{3}\,y^2+156\,I\,y\,x^2+48\,I\,x^2\sqrt{3}+168\,x^2+47\,I\,y\sqrt{3}-48\,I\,y\,x^2\sqrt{3}$$
$$-168\,y\,x^2+195\,y^2-195\,y^3-195+168\,I+74\sqrt{3}-47\,I\sqrt{3}+48\,x^3)$$

$$g_3(x,y) = \frac{1}{35636460}(-3033-1989\,I+1313\,I\sqrt{3}+1361\sqrt{3})(-90720\,x+14898\,y$$
$$-15336\,y\,x+3021\sqrt{3}-52722\,x^2-11749\sqrt{3}\,y^3-8305\sqrt{3}\,y^2+6465\sqrt{3}\,y$$
$$+3021\sqrt{3}\,x^2-1785\sqrt{3}\,y\,x^2+438\,y\,x^2-87573\,y^2-24417\,y^3+75384\,y^2\,x$$
$$+65388\,x^3-6042\,x\sqrt{3}-7839\,I\,y^3+36333\,I\,y^2+60705\,I\,y+16533\,I\,x^2$$
$$+5736\,I\sqrt{3}-33066\,I\,x-4680\,y\,x\sqrt{3}+1362\,y^2\,x\sqrt{3}-6383\,I\sqrt{3}\,y^3$$
$$+22753\,I\sqrt{3}\,y^2+231\,I\,y\,x^2+5736\,I\,x^2\sqrt{3}+34872\,I\,y\sqrt{3}-60936\,I\,y\,x$$
$$-11472\,I\,x\sqrt{3}-27870\,I\,y^2\,x-5532\,I\,y\,x^2\sqrt{3}-29340\,I\,y\,x\sqrt{3}$$
$$-17868\,I\,y^2\,x\sqrt{3}+78054+16533\,I)$$

5.4. Examples of (b) and (d)

As the corresponding sextics of torus type are linear, we only need to check the singularities are not colinear.

(b) $\Sigma(C) \supset \{A_{11}, A_5\}$. We put A_{11} at $(0,0)$ with tangent line $x = 0$ and A_5 at $(1,0)$.

(b-1) $\Sigma(C) = [A_{11}, A_5]$:

$$f(x,y) = (-y^3 + (9x-1)y^2 + 7x^3 - 8x^2 + x) \times$$
$$(-2y^3 + (5x-1)y^2 + (-x^2 + x)y + 4x^3 - 5x^2 + x)$$

(b-2) $\Sigma(C) = [A_{11}, A_5, A_1]$:

$$f(x,y) = \frac{1}{55}(-16\,x\,y - 2\,y^2 + 16\,x^2\,y + 14\,y^2\,x - 8\,x^2 + 5\,x^3 + 3\,x)$$
$$(11\,x\,y + 4\,y^2 - 11\,x^2\,y + 98\,y^2\,x - 5\,x^2 + 11\,x^3 + 14\,y^3 - 6\,x)$$

(b-3) $\Sigma(C) = [A_{11}, A_5, 2A_1]$:

$$f(x,y) := (-175\,y^3 + 11\,x^2\,\sqrt{3} - 88\,x^2 - y^3\,\sqrt{3} - 30\,y^2\,x\,\sqrt{3} - 18\,y\,x^2\,\sqrt{3} + 27\,x$$
$$+ 61\,x^3 + 48\,y^2 - 94\,y\,x + 83\,y\,x^2 - 126\,y^2\,x + I\,x^2\,\sqrt{3} + 5\,I\,y^2\,\sqrt{3} - 17\,I\,y^2\,x$$
$$- I\,x\,\sqrt{3} - 3\,I\,y\,x + 11\,I\,\sqrt{3}\,y^3 - 6\,y^2\,\sqrt{3} - 11\,x\,\sqrt{3} - 8\,I\,x^2 + 21\,I\,y^2 + 8\,I\,x$$
$$- 27\,I\,y^3 + 25\,I\,y^2\,x\,\sqrt{3} + 8\,I\,y\,x\,\sqrt{3} + 15\,I\,y\,x^2\,\sqrt{3} + 2\,I\,y\,x^2 + 27\,y\,x\,\sqrt{3})(175\,y^3$$
$$+ 11\,x^2\,\sqrt{3} - 88\,x^2 + y^3\,\sqrt{3} - 30\,y^2\,x\,\sqrt{3} + 18\,y\,x^2\,\sqrt{3} + 27\,x + 61\,x^3 + 48\,y^2$$
$$+ 94\,y\,x - 83\,y\,x^2 - 126\,y^2\,x - 3\,I\,y\,x + 11\,I\,\sqrt{3}\,y^3 - 6\,y^2\,\sqrt{3} - 11\,x\,\sqrt{3} - 27\,I\,y^3$$
$$+ 8\,I\,y\,x\,\sqrt{3} + 15\,I\,y\,x^2\,\sqrt{3} + 2\,I\,y\,x^2 - 27\,y\,x\,\sqrt{3} - 25\,I\,y^2\,x\,\sqrt{3}$$
$$+ 8\,I\,x^2 - 21\,I\,y^2 - 8\,I\,x + I\,x\,\sqrt{3} - I\,x^2\,\sqrt{3} - 5\,I\,y^2\,\sqrt{3} + 17\,I\,y^2\,x)$$

(d) $\Sigma(C) \supset \{3A_5\}$, $C = B_3 + B_3'$. We put three A_5 at $(0,1)$, $(0,-1)$ and $(1,0)$.

$$[3A_5] : f(x,y) := (\frac{3}{7} + x - y + y^3 - \frac{3}{7}\,x^3 - x^2 - \frac{1}{7}\,y\,x^2 - y^2\,x - \frac{3}{7}\,y^2)$$
$$(-\frac{4}{5}\,x - y + 1 + \frac{4}{5}\,y^2\,x - \frac{3}{5}\,x^2 + \frac{3}{5}\,y\,x^2 + y^3 - y^2 + \frac{2}{5}\,x^3)$$

$$[3A_5, A_1] : f(x,y) := (y^3 + \frac{57}{4}\,y^2\,x + \frac{1}{4}\,y^2 - y - y\,x + \frac{1}{4}\,x^3 + \frac{1}{4}\,x^2 - \frac{1}{4}\,x - \frac{1}{4})$$
$$(y^3 + \frac{71}{5}\,y^2\,x - y^2 + \frac{49}{5}\,y\,x^2 - 20\,y\,x - y + \frac{67}{5}\,x^3 - \frac{101}{5}\,x^2 + \frac{29}{5}\,x + 1)$$

$$[3A_5 + 2A_1] : f(x,y) := (21\,y^2\,x - 12\,I\,\sqrt{3}\,y^2\,x + 12\,y\,x^2 - I\,\sqrt{3}\,y\,x^2 + 12\,y\,x$$
$$- 2\,I\,\sqrt{3}\,y\,x - I\,y\,\sqrt{3} + I\,y^3\,\sqrt{3} - 3 + 3\,y^2 + 3\,x^3 - 3\,x + 3\,x^2)(-1 + 49\,x^3 - 17\,y^2\,x$$
$$+ 84\,y\,x^2 - 63\,x^2 - 12\,y\,x + y^2 + 15\,x - 7\,I\,\sqrt{3}\,y\,x^2 - 6\,I\,\sqrt{3}\,y\,x - 4\,I\,\sqrt{3}\,y^2\,x$$
$$+ I\,y\,\sqrt{3} - I\,y^3\,\sqrt{3})$$

6. Three conics

In this section, we study the last case $C = B_2 + B_2' + B_2''$ with the configuration of the singularities $[3A_5, 3A_1]$. Such a sextic is given when each pair of conics are intersecting at two points: at one point, with intersection multiplicity 3 and at another point, transversely. We can understand Zariski pairs in this situation using conical flex points. Assume that the respective defining polynomials of B_2, B_2', B_2'' are $f_2(x,y)$, $g_2(x,y)$, $h_2(x,y)$ and the location of two A_5's are $P_1 = (0,1)$, $P_2 = (0,-1)$ with respective tangent cones are $y \mp 1 = 0$. We assume further $P_1 \in B_2 \cap B_2'$ and $P_2 \in B_2 \cap B_2''$. We fix B_2, B_2' generically and consider a linear system Φ of conics B_2'' of dimension 2 such that B_2'' and B_2 are tangent at P_2. Under this situation we assert that

Proposition 15. *There exist 5 conical flex points Q_i, $i = 1,\ldots,5$ on B_2' with respect to Φ so that Q_1 is a conical flex of torus type and the other are of non-torus type.*

Proof. To avoid the complexity of the equation, we choose a generic B_2, B_2' so that

$$f_2(x,y) = (y^2 - 1 + x^2)$$
$$g_2(x,y) = y^2 + \left(-\frac{2}{15}\sqrt{130}\,x - \frac{2}{3}\right)y + \frac{2}{3}x^2 + \frac{2}{15}\sqrt{130}\,x - \frac{1}{3}$$

We find 5 conical flex points on B_2':

$$Q_1 = \left(-\frac{1}{9}\sqrt{130}, \frac{-17}{9}\right)$$
$$Q_2 = \left(\frac{7}{3}I\sqrt{10}\sqrt{3} - \frac{2}{3}I\sqrt{13}\sqrt{10}\sqrt{3}, -4 + \sqrt{13} + \frac{5}{3}I\sqrt{13}\sqrt{3} - \frac{19}{3}I\sqrt{3}\right)$$
$$Q3 = \left(-\frac{7}{3}I\sqrt{10}\sqrt{3} + \frac{2}{3}I\sqrt{13}\sqrt{10}\sqrt{3}, -4 + \frac{19}{3}I\sqrt{3} + \sqrt{13} - \frac{5}{3}I\sqrt{13}\sqrt{3}\right)$$
$$Q4 = \left(-\frac{7}{3}I\sqrt{10}\sqrt{3} - \frac{2}{3}I\sqrt{13}\sqrt{10}\sqrt{3}, -4 - \frac{5}{3}I\sqrt{13}\sqrt{3} - \frac{19}{3}I\sqrt{3} - \sqrt{13}\right),$$
$$Q5 = \left(\frac{7}{3}I\sqrt{10}\sqrt{3} + \frac{2}{3}I\sqrt{13}\sqrt{10}\sqrt{3}, -4 + \frac{19}{3}I\sqrt{3} - \sqrt{13} + \frac{5}{3}I\sqrt{13}\sqrt{3}\right)$$

Q_1 gives a sextic of torus type so that B_2'' is given by

$$h_2(x,y) = \left(\frac{338}{201}y - \frac{104}{1005}y\sqrt{130}\,x - \frac{104}{1005}\sqrt{130}\,x + \frac{137}{201} + y^2 + \frac{32}{201}x^2\right)$$

The other conical flex points give sextics of non-torus type. For example, Q_2 gives B_2'' described as:

$$h_2(x,y) = 75 + 150\,I\,\sqrt{3} + 50\,\sqrt{13} - 45\,I\,\sqrt{13}\,\sqrt{3} - 104\,\sqrt{13}\,\sqrt{10}\,x - 80\,\sqrt{10}\,x$$
$$+\,72\,I\,\sqrt{10}\,x\,\sqrt{3} + 300\,I\,y\,\sqrt{3} + 790\,y - 90\,I\,y\,\sqrt{13}\,\sqrt{3} + 100\,y\,\sqrt{13}$$
$$-\,104\,y\,\sqrt{13}\,\sqrt{10}\,x - 80\,x\,y\,\sqrt{10} + 72\,I\,x\,y\,\sqrt{10}\,\sqrt{3} - 45\,I\,y^2\,\sqrt{13}\,\sqrt{3} + 50\,y^2\,\sqrt{13}$$
$$+\,150\,I\,y^2\,\sqrt{3} + 715\,y^2 + 320\,x^2$$

References

1. E. Artal Bartolo. Sur les couples des Zariski. *J. Algebraic Geometry*, 3:223–247, 1994.
2. E. Artal Bartolo, J. Carmona, J. I. Cogolludo, and H.-O. Tokunaga. Sextics with singular points in special position. *J. Knot Theory Ramifications*, 10(4):547–578, 2001.
3. E. Artal Bartolo, J. Carmona Ruber, and J. I. Cogolludo Agustín. Essential coordinate components of characteristic varieties. *Math. Proc. Cambridge Philos. Soc.*, 136(2):287–299, 2004.
4. E. Artal Bartolo and H.-o. Tokunaga. Zariski pairs of index 19 and Mordell-Weil groups of $K3$ surfaces. *Proc. London Math. Soc. (3)*, 80(1):127–144, 2000.
5. E. Artal Bartolo and H.-o. Tokunaga. Zariski k-plets of rational curve arrangements and dihedral covers. *Topology Appl.*, 142(1-3):227–233, 2004.
6. M. Namba. *Geometry of projective algebraic curves*. Decker, New York, 1984.
7. M. Oka. Zariski pairs on sextics 1. *Vietnam J. Math.*,33: SI 81-92, 2005.
8. M. Oka. Geometry of cuspidal sextics and their dual curves. In *Singularities—Sapporo 1998*, pages 245–277. Kinokuniya, Tokyo, 2000.
9. M. Oka. Alexander polynomial of sextics. *J. Knot Theory Ramifications*, 12(5):619–636, 2003.
10. M. Oka. Geometry of reduced sextics of torus type. *Tokyo J. Math.*, 26(2):301–327, 2003.
11. H.-o. Tokunaga. (2,3) torus sextics and the Albanese images of 6-fold cyclic multiple planes. *Kodai Math. J.*, 22(2):222–242, 1999.

A Survey of Characteristic Classes of Singular Spaces

Jörg Schürmann

Westf. Wilhelms-Universität,
SFB 478 "Geometrische Strukturen in der Mathematik",
Hittorfstr. 27, 48149 Münster, Germany
** E-mail: jschuerm@math.uni-muenster.de*

Shoji Yokura *

Department of Mathematics and Computer Science,
Faculty of Science, University of Kagoshima,
21-35 Korimoto 1-chome, Kagoshima 890-0065, Japan
** E-mail: yokura@sci.kagoshima-u.ac.jp*

The theory of characteristic classes of vector bundles and smooth manifolds plays an important role in the theory of smooth manifolds. An investigation of reasonable notions of characteristic classes of singular spaces started since a systematic study of singular spaces such as singular algebraic varieties. We give a quick survey of characteristic classes of singular varieties, mainly focusing on the functorial aspects of some important ones such as the singular versions of the Chern class, the Todd class and Thom–Hirzebruch's L-class. Further we explain our recent "motivic" characteristic classes, which in a sense unify these three different theories of characteristic classes. We also discuss bivariant versions of them and characteristic classes of proalgebraic varieties, which are related to the motivic measures/integrations. Finally we explain some recent work on "stringy" versions of these theories, together with some references for "equivariant" counterparts.

Dedicated to Jean-Paul Brasselet on the occasion of his 60th birthday

Keywords: Bivariant theory, Characteristic (co)homology class, Grothendieck rings of varieties, Grothendieck transformations, Motivic measure/integration, Pro-algebraic varieties, Riemann–Roch, Stringy characteristic class

1. Introduction

Characteristic classes are usually certain kinds of cohomology classes for vector bundles over spaces and characteristic classes of smooth manifolds are defined

---

*Partially supported by Grant-in-Aid for Scientific Research (No. 17540088), the Ministry of Education, Culture, Sports, Science and Technology (MEXT), Japan

via their tangent bundles. The most basic ones are *Stiefel–Whitney*, *Euler* and *Pontrjagin* classes in the real case, and *Chern* classes in the complex case. They were introduced in 1930's and 1940's and constructed in a topological manner, i.e., via the obstuction theory, and in a differential-geometrical manner, i.e., via the Chern–Weil theory. Various important characteristic classes of vector bundles and invariants of manifolds are expressed as polynomials of them. The theory of cohomological characteristic classes were used for classifying manifolds and the study of structures of manifolds.

In 1960's a systematic study of singular spaces was started by *R. Thom*, *H. Whitney*, *H. Hironaka*, *S. Łojasiewicz*, et al.; they studied triangulations, stratifications, resolution of singularities (in characteristic zero) and so on. Already in 1958 *R. Thom* introduced in [Thom2] *rational Pontrjagin and L-classes* for oriented rational PL-homology manifolds. In 1965 *M.-H. Schwartz* defined in [Schw1] certain characteristic classes using obstruction theory. of the so-called radial vector fields; the *Schwartz class* is defined for a singular complex variety embedded in a complex manifold as a cohomology class of the manifold supported on the singular variety. In 1969, *D. Sullivan* [Sull] proved that a real analytic space is mod 2 Euler space, i.e., the Euler–Poincaré characteristic of the link of any point is even, which implies that the sum of simplices in the first barycentric subdivision of any triangulation is mod 2 cycle. This enabled Sullivan to define the "singular" *Stiefel–Whitney class* as a mod 2 homology class, which is equal to the Poincaré dual of the above cohomological Stiefel–Whitney class for a smooth variety. Moreover, in his beautiful "MIT notes" of 1970 [Sull2, Chapter 6], Sullivan introduced for an oriented rational PL-homology manifold M an *orientation class* $\triangle(M) \in \mathbf{KO}_*(M)[\frac{1}{2}]$ in the KO-homology with 2 inverted, whose Pontrjagin-Chern character are the rational L-classes of Thom.

P. Deligne and *A.Grothendieck* (cf. [Sull]) conjectured the unique existence of the *Chern class* version of the Sullivan's Stiefel–Whitney class, and in 1974 *R. MacPherson* [Mac1] proved their conjecture affirmatively. Motivated by MacPherson's proof of the conjecture, *P. Baum*, *W. Fulton* and *R. MacPherson* [BFM1] proved a "singular Riemann–Roch theorem", which is nothing but the *Todd class* transformation in the case of singular varieties.

M. Goresky and *R. MacPherson* ([GM1], [GM2]) have introduced *Intersection Homology Theory*, by using the notion of "perversity". In [GM1] they extended the work of [Thom2] to stratified spaces with even (co)dimensional strata and introduced a *homology L-class* $L_*^{\mathrm{GM}}(X)$ such that if X is nonsingular it becomes the Poincaré dual of the original Thom–Hirzebruch L-class: $L^{\mathrm{GM}}_*(X) = L^*(TX) \cap [X]$. Independently, these were also discovered by *J. Cheeger* in his work [Che] on analysis on singular spaces. In particular he obtained under suitable assumptions a "local formula" for these L-classes in terms of η-invariants of links of simplices for a given triangulation of the singular space X. In [Si] the work of Goresky-MacPherson and Sullivan was further extended to so-called stratified "Witt-spaces", whose intersection (co)homology complex (for the middle perversity) becomes self-dual (compare also with [Ban] for a more recent extension). Later, *S. Cappell* and *J. Shaneson* [CS1](see also [CS2] and [Sh]) introduced a *homology L-class* transformation L_*, which turns out to be a natural transformation from the abelian group $\Omega(X)$ (see §7) of cobordism classes of self-dual constructible complexes to the rational homology group [BSY3] (cf. [Y2]).

In the case of singular varieties, the characteristic cohomology classes have been individually extended to the corresponding characteristic homology classes without any unifying theory of characteristic classes of singular varieties, unlike the case of smooth manifolds and vector bundles. Only very recently such a unifying theory of "motivic characteristic classes" for singular spaces appeared in our work [BSY3,BSY4]. The purpose of the present paper is to make a quick survey on the development of characteristic classes and the up date situation of characteristic classes of singular spaces. This includes our motivic characteristic classes, bivariant versions, characteristic classes of proalgebraic varieties and finally "stringy" versions of these theories, together with some references for "equivariant" counterparts.

The present survey is a kind of extended and up-dated version of MacPherson's survey article [Mac2] of more than 30 years ago. There are other surveys, e.g., [Alu1,Br2,Pa,Sea1,Sch4,Su2] on characteristic classes of singular varieties written from different viewpoints. Here we recommend also the monographs in preparation [Br3,BSS].

2. Euler–Poincaré characteristic

The simplest, but most fundamental and most important topological invariant of a compact topological space is the *Euler number* or *Euler–Poincaré characteristic*. Its definition is quite simple; for a compact triangulable space or more generally for a cellular decomposable space X, it is defined to be the alternating sum of the numbers of cells and denoted by $\chi(X)$:

$$\chi(X) = \sum_n (-1)^n \sharp(n - \text{cells}). \tag{2.1}$$

By the homology theory, the Euler–Poincaré characteristic turns out to be equal to the alternating sum of Betti numbers, i.e.,

$$\chi(X) = \sum_n (-1)^n \dim H_n(X;\mathbb{R}). \tag{2.2}$$

With this fact, the Euler–Poincaré characteristic is defined for any topological space as long as the right-hand-side of (2.2) is defined, e.g. for locally compact semialgebraic sets. Note that taking the alternating sum is essential in the definition (2.1), but it is not the case in the definition (2.2). The following general form is called the *Poincaré polynomial*:

$$P_t(X) := \sum_n \dim H_n(X;\mathbb{R})t^n,$$

which is also a topological invariant.

The Euler–Poincaré characteristic has the following properties:

(1) $\chi(X) = \chi(X')$ if $X \cong X'$,
(2) $\chi(X) = \chi(X,Y) + \chi(Y)$ for any closed subspace $Y \subset X$, where the relative Euler–Poincaré characteristic $\chi(X,Y)$ is defined by the relative homology groups $H_*(X,Y)$,
(3) $\chi(X \times Y) = \chi(X) \cdot \chi(Y)$.

For a fiber bundle $f : X \to Y$ we have $\chi(X) = \chi(F) \cdot \chi(Y)$, if the Euler characteristic $\chi(F)$ of all fibers F is constant, e.g. Y is connected. This generalizes the above property (3).

In most cases when one deals with non-compact spaces, we need to deal with *cohomology groups with compact supports*, for example, as they play a key role in Deligne's theory of mixed Hodge structures. One can define it as a direct limit over compact subspaces, but here we take a sheaf-theoretic approach, which is more effective (e.g., see [Dim, Chapter 2]).

Let $f : X \to Y$ be a continuous map of locally compact spaces and let $\mathcal{F}$ be a sheaf of vector spaces on X. The *the functor of direct image with compact supports under f*, denoted by $f_!$, is defined by

$$f_!\mathcal{F}(V) := \{s \in \Gamma(f^{-1}(V), \mathcal{F}) \mid f|_{\mathrm{supp}(s)} : \mathrm{supp}(s) \to V \text{ is proper}\}.$$

Note that if f is proper, then the usual functor f_* of direct image and the functor $f_!$ of direct image with compact supports are the same. For a map $a_X : X \to pt$ to a point, $a_{X!}\mathcal{F}$ is nothing but

$$\Gamma_c(X, \mathcal{F}) := \{s \in \Gamma(X, \mathcal{F}) \mid \mathrm{supp}(s) \text{ is compact}\},$$

which is the *functor of global sections with compact supports*. Namely, $\Gamma_c(X, \mathcal{F}) = (a_X)_!\mathcal{F}$. Then the higher derived functor of this direct image $(a_X)_!\mathcal{F}$

with compact support is called *the cohomology with compact supports*:

$$H_c^k(X; \mathcal{F}) := R^k(a_X)_! \mathcal{F}.$$

Let X be a locally compact space and Y be a closed subset of X. Let $i : Y \to X$ and $j : X \setminus Y \to X$ be the inclusions. For a sheaf $\mathcal{F}$ of modules on X we have the following exact sequence

$$0 \to j_! j^{-1}\mathcal{F} \to \mathcal{F} \to i_* i^{-1}\mathcal{F} \to 0.$$

Then by taking the higher direct image with compact support we get the following long exact sequence

$$\cdots \to H_c^i(X \setminus Y; \mathcal{F}) \to H_c^i(X; \mathcal{F}) \to H_c^i(Y; \mathcal{F}) \to H_c^{i+1}(X \setminus Y; \mathcal{F}) \cdots \to .$$

Here for a subspace $W \subset X$ with $\iota : W \to X$ the inclusion and $\mathcal{F}$ a sheaf over X, $H_c^k(W; \mathcal{F}) := H_c^k(W; \iota^{-1}\mathcal{F})$. This long exact sequence gives rise to

$$\chi_c(X, \mathcal{F}) = \chi_c(X \setminus Y, \mathcal{F}) + \chi_c(Y, \mathcal{F})$$

as long as the *Euler characteristic with compact support of $\mathcal{F}$*

$$\chi_c(W, \mathcal{F}) := \sum_n (-1)^n \dim H_c^n(W; \mathcal{F})$$

is well-defined for $W = X, Y$ and $X \setminus Y$.

Remark 2.3. It is worthwhile to mention that one can define the cohomology with compact support using a (in fact any) compactification; this description is useful for the theory of mixed Hodge structures (e.g., see [Sri]). Let X be a locally compact topological space and Y be a closed subspace of X. Let $j : X \setminus Y \to X$ be the inclusion as above. Then the relative cohomology group $H^k(X, Y; \mathcal{F})$ is defined by

$$H^k(X, Y; \mathcal{F}) := H^k(X, j_! j^{-1}\mathcal{F}).$$

The natural transformation $\Gamma_c(X, j_! j^{-1}\mathcal{F}) \to \Gamma(X, j_! j^{-1}\mathcal{F})$ induces the following commutative diagrams:

$$
\begin{array}{ccccccccc}
\to & H_c^i(X \setminus Y; \mathcal{F}) & \longrightarrow & H_c^i(X; \mathcal{F}) & \longrightarrow & H_c^i(Y; \mathcal{F}) & \longrightarrow & H_c^{i+1}(X \setminus Y; \mathcal{F}) & \to \\
& \downarrow & & \downarrow & & \downarrow & & \downarrow & \\
\to & H^i(X, Y; \mathcal{F}) & \longrightarrow & H^i(X; \mathcal{F}) & \longrightarrow & H^i(Y; \mathcal{F}) & \longrightarrow & H^{i+1}(X, Y; \mathcal{F}) & \to .
\end{array}
$$

If X is compact, then $H_c^i(X, \mathcal{F}) = H^i(X, \mathcal{F})$ and $H_c^i(Y, \mathcal{F}) = H^i(Y, \mathcal{F})$ and it follows from the 5-lemma that for any integer i we get the isomorphism

$$H_c^i(X \setminus Y; \mathcal{F}) \cong H^i(X, Y; \mathcal{F}).$$

In particular we get the following: Let X be a locally compact space and $\overline{X}$ a compactification of X such that X is open in $\overline{X}$. Then we have

$$H_c^k(X; \mathcal{F}) \cong H^k(\overline{X}, \partial X; \mathcal{F})$$

where $\partial X := \overline{X} \setminus X$ is called the boundary. This implies that the cohomology group with compact support can be defined using *any* such compactification.

If $\mathcal{F} = \mathbb{R}_X$ is the constant sheaf associated to the real numbers $\mathbb{R}$, then $\chi_c(X, \mathbb{R}_X)$ is simply denoted by $\chi_c(X)$:

$$\chi_c(X) := \sum_n (-1)^n \dim H_c^n(X; \mathbb{R}), \tag{2.4}$$

and called the *Euler characteristic with compact support*. Then the same properties as (1) and (3) above also hold for the Euler characteristic with compact support and (2) is simply replaced by

$$\chi_c(X) = \chi_c(X \setminus Y) + \chi_c(Y) \tag{2.5}$$

for any closed subspace $Y \subset X$.

Remark 2.6. For two topological spaces X, Y, let $X + Y$ denote the topological sum, which is the disjoint sum, we clearly have

$$\chi(X + Y) = \chi(X) + \chi(Y).$$

However, we should note that for a closed subspace $Y \subset X$ the following additivity property <u>does not hold in general</u>:

$$\chi(X) = \chi(X \setminus Y) + \chi(Y), \tag{2.7}$$

although $X = (X \setminus Y) + Y$ as a set, since the topological sum $Y + (X \setminus Y)$ is not equal to the original topological space X. In other words, $\chi(X, Y) \neq \chi(X \setminus Y)$ in general.

However, in the category of complex algebraic varieties, the above formula (2.7) holds, i.e., for any closed subvariety $Y \subset X$ we have that $\chi(X) = \chi(X \setminus Y) + \chi(Y)$. The key geometric reason for the equality $\chi(X) = \chi(X \setminus Y) + \chi(Y)$ is that a closed subvariety Y always has a neighborhood deformation retract N such that the Euler–Poincaré characteristic of the "link" $\chi(N \setminus Y)$ vanishes due to a result of Sullivan (see [Fu2, Exercise on p.95 and comments on p.141-142]). In other words $\chi(X \setminus Y) = \chi_c(X \setminus Y)$ in the complex algebraic context, which also can be extended and proved in the language of complex algebraically constructible functions (see [Sch3, §6.0.6]).

Remark 2.8. In the above we consider the cohomology with compact support. Here we remark that the dual $Hom_\kappa(H^n_c(X;\kappa),\kappa)$ of the cohomology with compact support for any field coefficient κ is isomorphic to the so-called Borel–Moore homology group $H^{BM}_n(X;\kappa)$. For the Borel–Moore homology groups, e.g., see [CG] and [Fu1].

3. Characteristic classes of vector bundles

Very nice references for this section are the books [MiSt,Hir2,Hus,Stong]. A characteristic class of vector bundles over a topological space X is defined to be a map from the set of isomorphism classes of vector bundles over X to the cohomology group (ring) $H^*(X;\Lambda)$ with a coefficient ring Λ, which is supposed to be compatible with the pullback of vector bundle and cohomology group for a continuous map. Namely, it is an assignment $cl : \mathrm{Vect}(X) \to H^*(X;\Lambda)$ such that the following diagram commutes for a continuous map $f : X \to Y$:

$$
\begin{array}{ccc}
\mathrm{Vect}(Y) & \xrightarrow{\;\;cl\;\;} & H^*(Y;\Lambda) \\
{\scriptstyle f^*}\downarrow & & \downarrow{\scriptstyle f^*} \\
\mathrm{Vect}(X) & \xrightarrow[\;\;cl\;\;]{} & H^*(X;\Lambda).
\end{array}
$$

Here $\mathrm{Vect}(W)$ is the set of isomorphism classes of vector bundles over W.

The theory of characteristic classes started in Stiefel's paper [Sti], in which he considered the problem of the existence of tangential frames, i.e., linearly independent vector fields on a differentiable manifold. And at the same year H. Whitney defined such characteristic classes for sphere bundles over a simplicial complex [Wh1], and some time later he invented cohomology and proved his important "sum formula" [Wh2]. Then Pontrjagin [Pontr] introduced other characteristic classes of real vector bundles, based on the study of the homology of real Grassmann manifolds. Finally Chern [Ch1,Ch2] defined similar characteristic classes of complex vector bundles.

The most fundamental characteristic classes of a real vector bundle E over X are the *Stiefel–Whitney classes* $w^i(E) \in H^i(X;\mathbb{Z}_2)$, *Pontrjagin classes* $p^i(E) \in H^{4i}(X;\mathbb{Z}[\frac{1}{2}])$, and for a complex vector bundle E the *Chern classes* $c^i(E) \in H^{2i}(X;\mathbb{Z})$. These characteristic classes $cl^i(E) \in H^*(X;\Lambda)$ are described axiomatically in a unified way (compare [MiSt, Chapter 4,8,14,15], [Hir2, Chapter 1.4] and [Hus, Chapter 17]):

Definition 3.1. The Stiefel Whitney classes and the Pontrjagin classes of real vector bundles, resp. the Chern classes of complex vector bundles, are operators assigning to each real (resp. complex) vector bundle $E \to X$ cohomology classes

$$c\ell^i(E) := \begin{cases} w^i(E) & \in H^i(X; \mathbb{Z}_2) \\ p^i(E) & \in H^{4i}(X; \mathbb{Z}[\frac{1}{2}]) \\ c^i(E) & \in H^{2i}(X; \mathbb{Z}) \end{cases}$$

of the base space X such that the following four axioms are satisfied:

Axiom-1: (finiteness) For each vector bundle E one has $cl^0(E) := 1$ and $cl^i(E) = 0$ for $i > \operatorname{rank} E$ (in fact $p^i(E) = 0$ for $i > [\frac{\operatorname{rank} E}{2}]$). $cl^*(E) := \sum_i cl^i(E)$ is called the corresponding *total characteristic class*. In particular $cl^*(0_X) = 1$ for the zero vector bundle 0_X of rank zero.

Axiom-2: (naturality) One has $cl^*(F) = cl^*(f^*E) = f^*cl^*(E)$ for any cartesian diagram

$$\begin{array}{ccc} F \simeq f^*E & \longrightarrow & E \\ \downarrow & & \downarrow \\ Y & \xrightarrow{\ f\ } & X\,. \end{array}$$

Axiom-3: (Whitney sum formula)

$$cl^*(E \oplus F) = cl^*(E)cl^*(F)\,,$$

or more generally

$$cl^*(E) = cl^*(E')cl^*(E'')$$

for any short exact sequence $0 \to E' \to E \to E'' \to 0$ of vector bundles.

Axiom-4: (normalization or the "projective space" condition) For the canonical (i.e., the dual of the tautological) line bundle $\gamma_n^1(\mathbb{K}) := \mathcal{O}_{\mathbf{P}^n(\mathbb{K})}(1)$ over the projective space $\mathbf{P}^n(\mathbb{K})$ (with $\mathbb{K} = \mathbb{R}, \mathbb{C}$) one has:

(w^1): $w^1(\gamma_n^1(\mathbb{R}))$ is non-zero.

(p^1): $p^1(\gamma_n^1(\mathbb{C})) = c^1(\gamma_n^1(\mathbb{C}))^2$.

(c^1): $c^1(\gamma_n^1(\mathbb{C})) = [\mathbf{P}^{n-1}(\mathbb{C})] \in H^2(\mathbf{P}^n(\mathbb{C}); \mathbb{Z})$ is the cohomology class represented by the hyperplane $\mathbf{P}^{n-1}(\mathbb{C})$, i.e., the Poincaré dual of the homology class $[\mathbf{P}^{n-1}(\mathbb{C})]$ of the hyperplane $[\mathbf{P}^{n-1}(\mathbb{C})]$.

Remark 3.2. We use the superscript notation cl^* for contravariant functorial characteristic classes of vector bundles in cohomology, to distinguish them from the subscript notation cl_* for covariant functorial characteristic classes of singular spaces in homology, which we consider later on. Also note that in topology any short exact sequence of vector bundles over a reasonable (i.e. paracompact) space splits (by using a metric on E). But this is not the case in the algebraic or complex analytic context, where one should ask the "Whitney sum formula" for short exact sequences.

The existence of such a class for vector bundles of rank n can be shown, for example, with the help of a classifying space, i.e., the infinite dimensional Grassmanian manifolds $\mathbf{G}_n(\mathbb{K}^\infty)$ (with $\mathbb{K} = \mathbb{R}, \mathbb{C}$), and the fact that the cohomology ring of this Grassmanian manifold is a polynomial ring

$$H^*(\mathbf{G}_n(\mathbb{K}^\infty); \Lambda) = \begin{cases} \mathbb{Z}_2[w^1, w^2, \cdots, w^n] & \text{for } \mathbb{K} = \mathbb{R} \text{ and } \Lambda = \mathbb{Z}_2, \\ \mathbb{Z}[\frac{1}{2}][p^1, p^2, \cdots, p^{[\frac{n}{2}]}] & \text{for } \mathbb{K} = \mathbb{R} \text{ and } \Lambda = \mathbb{Z}[\frac{1}{2}], \\ \mathbb{Z}[c^1, c^2, \cdots, c^n] & \text{for } \mathbb{K} = \mathbb{C} \text{ and } \Lambda = \mathbb{Z}. \end{cases}$$

The most important axiom is Axiom-2 and the uniqueness of such a class follows from Axiom-3 and Axiom-4. By the "splitting principle" one can assume (after pulling back to a suitable bundle so that the pullback on the cohomology level is injective) that a given non-zero vector bundle E splits into a sum of line (or 2-plane) bundles. These line (or 2-plane) bundles are then called the "Chern roots" of E. Then Axiom-3 reduces the calculation of characteristic classes to the case of line bundles (for $c\ell = w, c$) or real 2-plane bundles (for $c\ell = p$). By naturality these are uniquely determined by Axiom-4, since

$$\mathbf{G}_1(\mathbb{K}^\infty) = \lim_k \mathbf{P}^k(\mathbb{K}) \quad (\text{for } \mathbb{K} = \mathbb{R}, \mathbb{C}),$$

for the case $c\ell = w, c$, or from the fact that the canonical projection

$$\lim_k \mathbf{P}^k(\mathbb{C}) \to \mathbf{G}_2(\mathbb{R}^\infty)$$

is the orientation double cover for the case $c\ell = p$.

From the axioms one gets that in all cases w^1, p^1 and c^1 are *nilpotent* on finite dimensional spaces and that $c\ell^*(E) = 1$ for a trivial vector bundle E. Note that a real oriented line bundle is always trivial so that a real line bundle $L \to X$ has no interesting characteristic class $c\ell^j(L) = 0 \in H^j(X; \mathbb{Z}[\frac{1}{2}])$ for $j > 0$. Just pullback to an orientation double cover $\pi : \tilde{X} \to X$ so that π^*L is orientable with $\pi^* : H^j(X; \mathbb{Z}[\frac{1}{2}]) \to H^j(\tilde{X}; \mathbb{Z}[\frac{1}{2}])$ injective (since $2 \in \mathbb{Z}[\frac{1}{2}]$ is invertible). In particular a real vector bundle E of rank r is orientable if and only if $w^1(E) = w^1(\Lambda^r E) = 0$.

If a characteristic class $c\ell^* : \text{Vect}(X) \to H^*(X; \Lambda)$ satisfies the Whitney sum condition

$$c\ell^*(E \oplus F) = c\ell^*(E)c\ell^*(F) \quad \text{with} \quad c\ell^*(0_X) = 1 ,$$

then $c\ell^*$ is called a *multiplicative* characteristic class. Another important multiplicative characteristic class of an *oriented* real vector bundle $E \to X$ of rank r is the *Euler class* $e(E) \in H^r(X; \mathbb{Z})$, with $e(E)\ mod\ 2 = w^r(E)$, $e(E)^2 = p^{\frac{r}{2}}(E)$ for r even and $e(E) = c^r(E)$ in case E is given by a complex vector bundle E of rank r. But the *Euler class* is not a *normalized* characteristic class with $c\ell^0(L) = 1$.

The *Stiefel-Whitney, Pontrjagin and Chern classes* are essential in the sense that any *multiplicative* characteristic class $c\ell^*$ over finite dimensional base spaces is uniquely expressed as a polynomial (or power series) in these classes, i.e. the "splitting principle" implies (compare [Hus, chapter 20: thm.4.3, thm.5.5 and thm.7.1]):

Theorem 3.3. *Let Λ be a $\mathbb{Z}_2$-algebra (resp. a $\mathbb{Z}[\frac{1}{2}]$-algebra) for the case of real vector bundles, or a $\mathbb{Z}$-algebra for the case of complex vector bundles. Then there is a one-to-one correspondence between*

(1) multiplicative *characteristic classes $c\ell^*$ over finite dimensional base spaces,*
     and
(2) formal power series $f \in \Lambda[[z]]$

such that $c\ell^*(L) = f(w^1(L))$ or $c\ell^*(L) = f(c^1(L))$ for any real or complex line bundle L (resp. $c\ell^*(L) = f(p^1(L))$ for any real 2-plane bundle L). In this case f is called the characteristic power series *of the corresponding multiplicative characteristic class $c\ell_f^*$.*

Remark 3.4. For the result above it is important that characteristic classes of vector bundles live in cohomology so that one can build new classes by multiplication (i.e. by the cup-product) of the basic ones. This is not possible in the case of characteristic classes of singular spaces, which live in homology (except in the case of homology manifolds where Poincaré duality is available).

Moreover $c\ell_f^*$ is invertible with inverse $c\ell_{\frac{1}{f}}^*$, if $f \in \Lambda[[z]]$ is invertible, i.e. if $f(0) \in \Lambda$ is a unit (e.g. f is a normalized power series with $f(0) = 1$). Then the corresponding multiplicative characteristic class $c\ell^*$ extends over finite dimensional base spaces X to a natural transformation of groups

$$c\ell^* : (\mathbf{K}(X), \oplus) \to (H^*(X; \Lambda), \cup)$$

on the Grothendieck group $\mathbf{K}(X)$ of real or complex vector bundles over X (compare [Hus, loc.cit.]).

4. Characteristic classes of smooth manifolds

Let us now switch to smooth manifolds, which will be an important intermediate step on the way to characteristic classes of singular spaces. For a smooth (or almost complex) manifold M its real (or complex) tangent bundle TM is available and a characteristic class $cl^*(TM)$ of the tangent bundle TM is called a *characteristic cohomology class* $cl^*(M)$ of the manifold M. We also use the notation

$$cl_*(M) := cl^*(TM) \cap [M] \in H_*^{BM}(M; \Lambda)$$

for the corresponding *characteristic homology class* of the manifold M, with $[M] \in H_*^{BM}(M; \Lambda)$ the fundamental class in Borel-Moore homology (e.g., see [BoMo], [Bre], [CG], [Fu1]) of the (oriented) manifold M. Note that $H_*^{BM}(X; \Lambda) = H_*(X; \Lambda)$ for X compact.

Remark 4.1. Using a relation to suitable cohomology operations, i.e., Steenrod squares, Thom [Thom1] has shown that the Stiefel–Whitney classes $w^*(M)$ of a smooth manifold M are *topological* invariants. Later he introduced in [Thom2] *rational Pontrjagin and L-classes* for compact oriented rational PL-homology manifolds so that the *rational* Pontrjagin classes $p^*(M) \in H^*(M; \mathbb{Q})$ of a closed smooth manifold M are *combinatorial or piecewise linear* invariants. A deep result of Novikov [Nov] implies the *topological* invariance of these *rational* Pontrjagin classes $p^*(M) \in H^*(M; \mathbb{Q})$ of a smooth manifold M.

For a *closed oriented* manifold M one has the interesting formula (compare [MiSt, cor.11.12]):

$$deg(e(M)) = \int_M e(TM) \cap [M] = \chi(M) , \qquad (4.2)$$

which justifies the name "Euler class". For a closed complex manifold M this formula becomes

$$deg(c_*(M)) = \int_M c^*(TM) \cap [M] = \chi(M) ,$$

which is called the *Gauss–Bonnet–Chern Theorem* (see [Ch3]). In this sense, the Chern class is a higher cohomology class version of the Euler–Poincaré characteristic. Similarly

$$deg(w_*(M)) = \int_M w^*(TM) \cap [M] = \chi(M) \, mod \, 2$$

for any closed manifold M.

More generally let $\mathrm{Iso}(G)_n$ be the set of isomorphism classes of smooth closed (and oriented) pure n-dimensional manifolds M for $G = O$ (or $G = SO$), or of pure n-dimensional weakly ("= stably") almost complex manifolds M for $G = U$, i.e. $TM \oplus \mathbb{R}_M^N$ is a complex vector bundle (for suitable N, with $\mathbb{R}_M$ the trivial real line bundle over M). Then

$$\mathrm{Iso}(G)_* := \bigoplus_n \mathrm{Iso}(G)_n$$

becomes a commutative graded semiring with addition and multiplication given by disjoint union and exterior product, with 0 and 1 given by the classes of the empty set and one point space. Moreover any multiplicative characteristic class $c\ell_f$ coming from the power series f in the variable $z = w^1, p^1$ or c^1 induces by

$$M \mapsto deg(c\ell_{f*}(M)) := \int_M c\ell_f^*(TM) \cap [M]$$

a semiring homomorphism

$$\Phi_f : \mathrm{Iso}(G)_* \to \Lambda = \begin{cases} \text{a } \mathbb{Z}_2\text{-algebra for } G = O \text{ and } z = w^1, \\ \text{a } \mathbb{Z}[\tfrac{1}{2}]\text{-algebra for } G = SO \text{ and } z = p^1, \\ \text{a } \mathbb{Z}\text{-algebra for } G = U \text{ and } z = c^1. \end{cases}$$

Let $\Omega_*^G := \mathrm{Iso}(G)_*/ \sim$ be the corresponding *cobordism ring* of closed ($G = O$) and oriented ($G = SO$) or weakly ("= stably") almost complex manifolds ($G = U$) as dicussed for example in [Stong]. Here $M \sim 0$ for a closed pure n-dimensional G-manifold M if and only if there is a compact pure $n + 1$-dimensional G-manifold B with boundary $\partial B \simeq M$. Note that this is indeed a ring with $-[M] = [M]$ for $G = O$ or $-[M] = [-M]$ for $G = SO, U$, where $-M$ has the opposite orientation of M. Moreover, for B as above with $\partial B \simeq M$ one has

$$TB|\partial B \simeq TM \oplus \mathbb{R}_M$$

so that $c\ell_f^*(TM) = i^* c\ell_f^*(TB)$ for $i : M \simeq \partial B \to B$ the closed inclusion of the boundary. This also explains the use of the stable tangent bundle for the definition of a stably or weakly almost complex manifold. By a simple argument due to Pontrjagin one gets (compare [Stong, Theorem. on p.32]):

$$M \sim 0 \quad \Rightarrow \quad deg(c\ell_{f*}(TM)) = \int_M c\ell_f^*(TM) \cap [M] = 0.$$

Hence any multiplicative characteristic class $c\ell_f^*$ coming from the power series f in the variable $z = w^1, p^1$ or c^1 induces a ring homomorphism called *genus*

$$\Phi_f : \Omega_*^G \to \Lambda = \begin{cases} \text{a } \mathbb{Z}_2\text{-algebra for } G = O \text{ and } z = w^1, \\ \text{a } \mathbb{Z}[\tfrac{1}{2}]\text{-algebra for } G = SO \text{ and } z = p^1, \\ \text{a } \mathbb{Z}\text{-algebra for } G = U \text{ and } z = c^1. \end{cases} \tag{4.3}$$

In fact for Λ a $\mathbb{Q}$-algebra this induces a one-to-one correspondence (compare [Hir2, Theorem 6.3.1] and [HBJ, Chapter 1]) between

(1) normalized power series f in the variable $z = p^1$ (or c^1),
(2) *normalized and multiplicative* characteristic classes $c\ell_f^*$ over finite dimensional base spaces, and
(3) genera $\Phi : \Omega_*^G \to \Lambda$ for $G = SO$ (or $G = U$).

Here one uses the following structure theorem (compare [Stong, Theorems on p.177 and p.110]):

Theorem 4.4. *(1)* (Thom) $\Omega_*^{SO} \otimes \mathbb{Q} = \mathbb{Q}[[\mathbf{P}^{2n}(\mathbb{C})]|n \in \mathbb{N}]$ *is a polynomial algebra in the classes of the complex even dimensional projective spaces.*
(2) (Milnor) $\Omega_*^U \otimes \mathbb{Q} = \mathbb{Q}[[\mathbf{P}^n(\mathbb{C})]|n \in \mathbb{N}]$ *is a polynomial algebra in the classes of the complex projective spaces.*

In particular, the corresponding genus Φ_f with values in a $\mathbb{Q}$-algebra Λ, or the corresponding normalized and multiplicative characteristic class $c\ell_f^*$, is uniquely determined by the values $\Phi_f(M) = \int_M c\ell_f^*(TM) \cap [M]$ for all (complex even dimensional) complex projective spaces $M = P^n(\mathbb{C})$. These are best codified by the *logarithm* $g \in \Lambda[[t]]$ of Φ_f:

$$g(t) := \sum_{i=0}^{\infty} \Phi_f(P^i(\mathbb{C})) \cdot \frac{t^{i+1}}{i+1} . \tag{4.5}$$

Moreover, a genus $\Phi_f : \Omega_*^U \otimes \mathbb{Q} \to \Lambda$ factorizes over the canonical map

$$\Omega_*^U \otimes \mathbb{Q} \to \Omega_*^{SO} \otimes \mathbb{Q}$$

if and only if $f(z)$ is an even power series in $z = c^1$, $f(z) = h(z^2)$ with $z^2 = (c^1)^2 = p^1$ (compare [Stong, Proposition on p.177 and Theorem on p.180] and [HBJ, Chapter 1]).

Consider for example the *signature* $\sigma(M)$ of the cup-product pairing on the middle dimensional cohomology of the closed oriented manifold M of real dimension $4n$, with $\sigma(M) := 0$ in all other dimensions. This defines a genus $\sigma : \Omega_*^{SO} \otimes \mathbb{Q} \to \mathbb{Q}$, as observed by Thom, with $\sigma(P^{2n}(\mathbb{C})) = 1$ for all n (compare [Hir2, Chapter II.8] and [Stong, Theorem on p.220]). The signature genus comes from the normalized power series $h(z) = \frac{\sqrt{z}}{\tanh \sqrt{z}}$ in the variable $z = p^1$ (or $f(z) = \frac{z}{\tanh z}$ in the variable $z = c^1$), whose corresponding characteristic class $c\ell^* = L^*$ is by definition the Hirzebruch-Thom L-class. This is the content of the famous *Hirzebruch's Signature Theorem* (compare [Hir2, Theorem 8.2.2] and also also with [Hir3]):

$$\sigma(M) = \int_M L^*(TM) \cap [M].$$

Remark 4.6. The first structure theorem about cobordism rings due to Thom is the description of Ω_*^O as a polynomial algebra $\mathbb{Z}_2[[M^n]|n \in \mathbb{N}, n+1 \neq 2^k]$ in the classes of suitable closed manifolds M^n of dimension n, with one generator in each dimension n with $n+1$ not a power of 2 (compare [Stong, Theorem on p.96]). Then each genus $\Omega_*^O \to \Lambda$ to a $\mathbb{Z}_2$-algebra Λ is coming form a normalized and multiplicative characteristic class $c\ell_f^*$, but this correspondence is not injective.

The value $\Phi(M)$ of a genus Φ on the closed manifold M is also called a characteristic number of M. All these numbers can be used to classify closed manifolds up to cobordism.

Theorem 4.7. *(1)* (Pontrjagin–Thom) *Two closed C^∞-manifolds are cobordant (i.e., represent the same element in Ω_*^O) if and only if all their Stiefel–Whitney numbers are the same.*

(2) (Thom–Wall) *Two closed oriented C^∞-manifold are corbordant up to two-torsion (i.e., represent the same element in $\Omega_*^{SO} \otimes \mathbb{Z}[\frac{1}{2}]$) if and only if all their Pontrjagin numbers are the same.*

(3) (Milnor–Novikov) *Two closed stably or weakly almost complex manifold are cobordant (i.e., represent the same element in Ω_*^U) if and only if all their Chern numbers are the same.*

Compare for example with [Stong, Theorem on p.95] for (1), [Stong, Theorems on p.180 and 183] for (2), and [Stong, Theorem on p.117] for (3).

5. Hirzebruch–Riemann–Roch and Grothendieck–Riemann–Roch

Let X be a non-singular complex projective variety and E a holomorphic vector bundle over X. Note that in this context we do not need to distinguish between holomorphic and algebraic vector bundles, and similarly for coherent sheaves, by the so-called "GAGA-principle" [Serre]. Then the Euler–Poincaré characteristic of E is defined by

$$\chi(X, E) = \sum_{i \geq 0} (-1)^i \dim_{\mathbb{C}} H^i(X; \Omega(E)),$$

where $\Omega(E)$ is the coherent sheaf of germs of sections of E. *J.-P. Serre* conjectured in his letter to *Kodaira and Spencer* (dated September 29, 1953) that there exists a polynomial $P(X, E)$ of Chern classes of the base variety X and the vector bundle E such that

$$\chi(X, E) = \int_X P(X, E) \cap [X].$$

Within three months (December 9, 1953) *F. Hirzebruch* solved this conjecture affirmatively: the above looked for polynomial $P(X, E)$ can be expressed as

$$P(X, E) = ch^*(E)td^*(X)$$

where $ch^*(E)$ is the total *Chern character* of E and $td^*(TX)$ is the total *Todd class* of the tangent bundle TX of X . Let us recall that the cohomology classes $ch^*(V)$ and $td^*(V)$ are defined as follows:

$$ch^*(V) = \sum_{i=1}^{\text{rank } V} e^{\alpha_i} \in H^{2*}(X; \mathbb{Q})$$

and

$$td^*(V) = \prod_{i=1}^{\text{rank } V} \frac{\alpha_i}{1 - e^{-\alpha_i}} \in H^{2*}(X; \mathbb{Q})$$

where α_i's are the Chern roots of V . So td^* is just the normalized and multiplicative characteristic class corrsponding to the normalized power series $f(z) = \frac{z}{1-e^{-z}}$ in $z = c^1$. Similarly the Chern character defines a contravariant natural transformation of rings

$$ch^* : (\mathbf{K}(X), \oplus, \otimes) \to (H^{2*}(X; \mathbb{Q}), +, \cup)$$

on the Grothendieck group $\mathbf{K}(X)$ of complex vector bundles over X. Then we have the following celebrated theorem of Hirzebruch (compare [Hir2, Theorem 21.1.1]):

Theorem 5.1. *(Hirzebruch–Riemann–Roch)*

$$\chi(X, E) = T(X, E) := \int_X (ch^*(E)td^*(X)) \cap [X] . \qquad \textbf{(HRR)}$$

$T(X, E)$ is called the T-characteristic ([Hir2]). For a more detailed historical aspect of **HRR**, see [Hir3].

Remark 5.2. The T-characteristic $T(X, E)$ is *a priori* a rational number by the definitions of the Todd class and Chern character, but it has to be *an integer* as a consequence of **HRR**. The T-characteristic $T(X, E)$ of a complex vector bundle E can be defined for any almost complex manifold and Hirzebruch [Hir1] asked if the T-genus $T(X) := T(X, \mathbb{1})$ with $\mathbb{1}$ being a trivial line bundle is always an integer. Of course this follows from **HRR** and the later result of Quillen that $\Omega^U_* \otimes \mathbb{Q}$ is generated by complex projective algebraic manifolds. The identity

$$\frac{z}{1-e^{-z}} = \frac{z \cdot e^{\frac{z}{2}}}{2\sinh\frac{z}{2}}$$

allows one to introduce the Todd class

$$Td^*(X) := e^{\frac{c^1(TX)}{2}} \cdot \hat{A}^*(TX)\,,$$

and therefore also the T-characteristic $T(X,E)$, more generally for a so-called $Spin^c$-manifold X. Here $\hat{A}$ is the so-called A *hat genus or characteristic class* corresponding to the even normalized power series $f(z) = \frac{z}{2\sinh\frac{z}{2}}$ in the variable $z = c^1$ or to $f(z) = \frac{\sqrt{z}}{2\sinh\frac{\sqrt{z}}{2}}$ in the variable $z = p^1$. The T-characteristic $T(X,E)$ of a complex vector bundle E is then an *integer* by an application of the *Atiyah-Singer Index theorem* [AS] for a suitable *Dirac operator* (compare [Hir1, p.197, Theorem 26.1.1]).

A. Grothendieck (cf. [BoSe]) generalized **HRR** for non-singular quasi-projective algebraic varieties over any field and proper morphisms with Chow cohomology ring theory instead of ordinary cohomology theory (compare also with [Fu1, chapter 15]). For the complex case we can still take the ordinary cohomology theory (or the homology theory by the Poincaré duality). Here we stick ourselves to *complex projective algebraic varieties* for the sake of simplicity. For a variety X, let $\mathbf{G}_0(X)$ denote the Grothendieck group of algebraic coherent sheaves on X and for a morphism $f : X \to Y$ the pushforward $f_! : \mathbf{G}_0(X) \to \mathbf{G}_0(Y)$ is defined by

$$f_!(\mathcal{F}) := \sum_{i \geq 0}(-1)^i \mathbf{R}^i f_* \mathcal{F},$$

where $\mathbf{R}^i f_* \mathcal{F}$ is (the class of) the higher direct image sheaf of $\mathcal{F}$. Then $\mathbf{G}_0$ is a covariant functor with the above pushforward (see [Grot1] and [Man]). Similarly let $\mathbf{K}^0(X)$ be the Grothendieck group of complex algebraic vector bundles over X so that one has a canonical contravariant transformation of rings $\mathbf{K}^0(\ ) \to \mathbf{K}(\ )$ to the Grothendieck group of complex vector bundles. Note that on a smooth algebraic manifold the canonical map $\mathbf{K}^0(\ ) \to \mathbf{G}_0(\ )$ taking the sheaf of sections is an isomorphism. With this isomorphism one can define characteristic classes of any algebraic coherent sheaf. Then Grothendieck showed the existence of a natural transformation from the covariant functor $\mathbf{G}_0$ to the $\mathbb{Q}$-homology covariant functor $H_{2*}(\ ; \mathbb{Q})$ (see [BoSe]):

Theorem 5.3. *(Grothendieck–Riemann–Roch) Let the transformation* $\tau_* :$ $\mathbf{G}_0(\ ) \to H_{2*}(\ ; \mathbb{Q})$ *be defined by* $\tau_*(\mathcal{F}) = td^*(X)ch^*(\mathcal{F}) \cap [X]$ *for any* *smooth variety X. Then τ_* is actually natural, i.e., for any morphism $f : X \to Y$ the following diagram commutes:*

$$
\begin{array}{ccc}
\mathbf{G}_0(X) & \xrightarrow{\ \tau_*\ } & H_{2*}(X;\mathbb{Q}) \\
{\scriptstyle f_!}\downarrow & & \downarrow{\scriptstyle f_*} \\
\mathbf{G}_0(Y) & \xrightarrow[\ \tau\ ]{} & H_{2*}(Y;\mathbb{Q})
\end{array}
$$

i.e.,

$$
td^*(T_Y)ch^*(f_!\mathcal{F}) \cap [Y] = f_*(td^*(TX)ch^*(\mathcal{F}) \cap [X]). \tag{GRR}
$$

Clearly **HRR** is induced from **GRR** by considering a map from X to a point. Note that the target of the transformation of the original **GRR** is the cohomology $H^{2*}(\ ;\mathbb{Q})$ with the Gysin homomorphism instead of the homology $H_{2*}(\ ;\mathbb{Q})$, but, by the definition of the Gysin homomorphism the original **GRR** can be put in as above. For a far reaching generalization of **GRR** in the context of "oriented cohomology theories", which also explains why the Todd class appears as a "correction factor" for the the pushforward of the Chern character, we recommend the paper [Pan].

6. The Generalized Hirzebruch–Riemann–Roch

In Hirzebruch's book [Hir2, §12.1 and §15.5] he has generalized the characteristics $\chi(X, E)$ and $T(X, E)$ to the so-called χ_y-*characteristic* $\chi_y(X, E)$ and T_y-*characteristic* $T_y(X, E)$ as follows, using a parameter y (see also [HBJ, Chapter 5]).

Definition 6.1.

$$
\chi_y(X, E) := \sum_{p \geq 0}\left(\sum_{q \geq 0}(-1)^q \dim_{\mathbf{C}} H^q(X, \Omega(E) \otimes \Lambda^p T^*X)\right) y^p
$$

$$
= \sum_{p \geq 0}\chi(X, E \otimes \Lambda^p T^*X))y^p
$$

where T^*X is the dual of the tangent bundle TX, i.e., the cotangent bundle of X.

$$
T_y(X, E) := \int_X \widetilde{td}_{(y)}(TX)ch_{(1+y)}(E) \cap [X],
$$

$$
\widetilde{td}_{(y)}(TX) := \prod_{i=1}^{\dim X}\left(\frac{\alpha_i(1 + y)}{1 - e^{-\alpha_i(1+y)}} - \alpha_i y\right),
$$

$$
ch_{(1+y)}(E) := \sum_{j=1}^{\operatorname{rank} E} e^{\beta_j(1+y)},
$$

882

where $\alpha_i's$ are the Chern roots of TX and $\beta_j's$ are the Chern roots of E .

F. Hirzebruch [Hir2, §21.3] showed the following theorem:

Theorem 6.2. *(The generalized Hirzebruch–Riemann–Roch)*

$$\chi_y(X, E) = T_y(X, E). \qquad \textbf{(g-HRR)}$$

The **g-HRR** can be shown as follows, using **HRR**:

$$\chi_y(X, E) = \int_X \sum_{p \geq 0} \chi(X, E \otimes \Lambda^p T^* X)) y^p \quad \text{(by definition)}$$

$$= \int_X \sum_{p \geq 0} (ch^*(E \otimes \Lambda^p T^* X) td^*(X) \cap [X]) \, y^p \quad \text{(by \textbf{HRR})}$$

$$= \int_X \left(\sum_{p \geq 0} ch^*(E \otimes \Lambda^p T^* X) td^*(X) y^p \right) \cap [X]$$

$$= \int_X \left(ch^*(E) td^*(X) \sum_{p \geq 0} ch^*(\Lambda^p T^* X) y^p \right) \cap [X]$$

$$= \int_X \left(ch^*(E) td^*(X) \prod_{i=1}^{\dim X} \left(1 + ye^{-\alpha_i}\right) \right) \cap [X]$$

$$= \int_X \left(\sum_{j=1}^{\operatorname{rank} E} e^{\beta_j} \prod_{i=1}^{\dim X} \left(1 + ye^{-\alpha_i}\right) \frac{\alpha_i}{1 - e^{-\alpha_i}} \right) \cap [X].$$

However, the power series $\left(1 + ye^{-\alpha_i}\right) \dfrac{\alpha_i}{1 - e^{-\alpha_i}}$ is not a normalized power series because the 0-degree part is $1 + y$, not 1. So, by dividing this non-normalized power series by $1 + y$ and furthermore by changing β_j to $\beta_j(1 + y)$ and α_i to $\alpha_i(1 + y)$, which does not change the value of $\chi_y(X, E)$ at all, and by noticing that

$$\frac{1 + ye^{-\alpha_i(1+y)}}{1 + y} \frac{\alpha_i(1 + y)}{1 - e^{-\alpha_i(1+y)}} = \frac{\alpha_i(1 + y)}{1 - e^{-\alpha_i(1+y)}} - \alpha_i y,$$

we can see that the right hand side of the last equation is $T_y(X, E)$ (compare [HBJ, p.61-62]). In general, letting $g(z)$ be a normalized power series and $f(z)$ be a non-normalized power series with $a := f(0)$ a unit, we have

$$\left(\sum_{j=1}^{\operatorname{rank} E} g(\beta_j) \prod_{i=1}^{\dim X} f(\alpha_i) \right) \cap [X] = \left(\sum_{j=1}^{\operatorname{rank} E} g(a\beta_j) \prod_{i=1}^{\dim X} \frac{f(a\alpha_i)}{a} \right) \cap [X].$$

In particular, a non-normalized power series $f(z)$ with $a := f(0) \in \Lambda$ a unit induces the same genus as the normalized power series $\frac{f(az)}{a}$ does.

Remark 6.3. The generalized Hirzebruch Riemann-Roch theorem is also true for a holomorphic vector bundle E over a compact complex manifold X, by an application of the *Atiyah-Singer Index theorem* [AS].

The above *modified Todd class* $\widetilde{td}_{(y)}$ is the normalized and multiplicative characteristic class corresponding to the normalized power series (in $z = c^1$):

$$f(z) = f_y(z) = \frac{z(1+y)}{1 - e^{-z(1+y)}} - zy \in \mathbb{Q}[y][[z]].$$

The associated genus $\chi_y : \Omega_*^U \to \mathbb{Q}[y]$ is called the Hirzebruch χ_y-*genus*. A simple residue calculation in [Hir2, Lemma 1.8.1] implies that for all $n \in \mathbb{N}$:

$$\chi_y(P^n(\mathbb{C})) = \sum_{i=0}^{n} (-y)^i \in \mathbb{Z}[y] \subset \mathbb{Q}[y] . \tag{6.4}$$

So these values on $P^n(\mathbb{C})$ determine the χ_y-genus and the modified Todd class $\widetilde{td}_{(y)}$. Moreover, the normalized power series $f_y(z)$ specializes to

$$f_y(z) = \begin{cases} 1 + z & \text{for } y = -1, \\ \frac{z}{1 - e^{-z}} & \text{for } y = 0, \\ \frac{z}{\tanh z} & \text{for } y = 1. \end{cases}$$

So the modified Todd class $\widetilde{td}_{(y)}$ defined above unifies the following three important characteristic cohomology classes:

$(y = -1)$ the total Chern class

$$\widetilde{td}_{(-1)}(TX) = c^*(TX),$$

$(y = 0)$ the total Todd class

$$\widetilde{td}_{(0)}(TX) = td^*(TX),$$

$(y = 1)$ the total Thom–Hirzebruch L-class

$$\widetilde{td}_{(1)}(TX) = L^*(TX).$$

Therefore, when $E =$ the trivial line bundle, for these special values $y = -1, 0, 1$ the **g-HRR** reads as follows:

$(y = -1)$ *Gauss–Bonnet–Chern Theorem*:

$$\chi(X) = \int_X c^*(TX) \cap [X],$$

(y = 0)   *Riemann–Roch Theorem*: denoting $\chi_a(X) := \chi(X, \mathcal{O}_X)$, called the arithmetic genus of X, to avoid a possible confusion with the above topological Euler–Poincaré characteristic $\chi(X)$,

$$\chi_a(X) = \int_X td^*(TX) \cap [X],$$

(y = 1)   *Hirzebruch's Signature Theorem*:

$$\sigma(X) = \int_X L^*(TX) \cap [X].$$

Remark 6.5. (Poincaré–Hopf Theorem) The above Gauss–Bonnet–Chern Theorem due to Chern [Ch3] is a generalization of the original Gauss–Bonnet theorem saying that the integration of the Guassian curvature is equal to 2π times the topological Euler–Poincaré characteristic. There is another well-known differential-topological formula concerning the topological Euler–Poincaré characteristic. That is the so-called *Poincaré –Hopf theorem*, saying that the index of a smooth vector field V with only isolated singularites on a smooth compact manifold M is equal to the topological Euler–Poincaré characteristic of the manifold M;

$$\text{Index}(V) = \chi(M),$$

where the index $\text{Index}(V)$ is defined to be the sum of the indices of the vector field at the isolated singularities. See [Mi1] for a beautiful introduction to the Poincaré –Hopf theorem. Note that the Gauss–Bonnet–Chern Theorem follows from the Poincaré–Hopf theorem (cf. [Wi] and [Zh]).

7. Characteristic classes of singular varieties

In the following we consider for simplicity only *compact* spaces. For a singular algebraic or analytic variety X its tangent bundle is not available any longer because of the existence of singularities, thus one cannot define its characteristic class $c\ell_*(X)$ as in the previous case of manifolds, although a "tangent-like" bundle such as Zariski tangents is available. A main theme for defining reasonable characteristic classes for singular varieties is that reasonable ones should be interesting enough; for example, they should be geometrically or topologically interesting and quite well related to other well-known invariants of varieties and singularities (e.g., see [Mac2]).

The theory of characteristic classes of vector bundles is *a natural transformation* from the contravariant functor Vect to the contravariant cohomology functor $H^*(\ ;\Lambda)$. This *naturality* is an important guide for developing various theories of characteristic classes for singular varieties. The known *functorial characteristic classes* for singular spaces are *covariant* functorial maps

$$cl_* : A(X) \to H_*(X; \Lambda)$$

from a suitable covariant theory A depending on the choice of cl_*. Moreover, there is always a *distinguished element* $\mathbb{1}_X \in A(X)$ such that the corresponding *characteristic class of the singular space* X is defined as $cl_*(X) := cl_*(\mathbb{1}_X)$. Finally one has the *normalization*

$$cl_*(\mathbb{1}_M) = c\ell^*(TM) \cap [M] \in H_*(M; \Lambda)$$

for M a smooth manifold, with $c\ell^*(TM)$ the corresponding characteristic cohomology class of M. This justifies the notation cl_* for this homology class transformation, which should be seen as a relative homology class version of the following *characteristic number* of the singular space X:

$$\sharp(X) := cl_*((a_X)_* \mathbb{1}_X) = (a_X)_*(cl_*(\mathbb{1}_X)) \in H_*(\{pt\}; \Lambda) = \Lambda,$$

with $a_X : X \to \{pt\}$ a constant map. Note that the *normalization* implies that for M smooth:

$$\sharp(M) = deg(cl_*(M)) = \int_M c\ell^*(TM) \cap [M]$$

so that this is consistent with the notion of characteristic number of the smooth manifold M as used before.

7.1. *Stiefel–Whitney classes w_**

The first example of functorial characteristic classes is the theory of singular Stiefel–Whitney homology classes due to *Dennis Sullivan* [Sull] (also see [FM]). A crucial fact about the original Stiefel–Whitney class is the following fact: if T is any triangulation of a manifold X, then the sum of all the simplices of the first barycentric subdivision is a *mod 2 cycle* and its homology class is equal to the Poincaré dual of the Stiefel–Whitney class. In [Sull] D. Sullivan observed that also a *singular real algebraic variety X is a mod 2 Euler space*, i.e. the link of any point of X has even Euler characteristic. And this condition implies that the sum of all the simplices of the first barycentric subdivision of any triangulation of X is always a *mod 2 cycle* and he defined its homology class to be the singular Stiefel–Whitney class of the variety X. Then, with an insight of *Deligne*, Sullivan's Stiefel–Whitney homology classes where enhanced as a natural tansformation from a certain covariant functor to the mod 2 homology theory.

Let X be a complex (or real) algebraic set and let $F(X)$ (or $F^{mod2}(X)$) be the abelian group of $\mathbb{Z}$- (or $\mathbb{Z}_2$-)valued complex (or real) algebraically constructible functions on a variety X. Then the assignment F (or F^{mod2}) : $\mathcal{V} \to \mathcal{A}$ is a *contravariant* functor (from the category of algebraic varieties to the category of abelian groups) by the usual functional pullback for a morphism $f : X \to Y$: $f^*(\alpha) := \alpha \circ f$. For a constructible set $Z \subset X$, we define

$$\chi(Z;\alpha) := \sum_{n\in\mathbb{Z}} n \cdot \chi_c(Z \cap \alpha^{-1}(n)) \quad (mod\ 2).$$

Then it turns out that the assignment F (or F^{mod2}): $\mathcal{V} \to \mathcal{A}$ also becomes a *covariant* functor by the following pushforward defined by

$$f_*(\alpha)(y) := \chi(f^{-1}(y);\alpha) \quad \text{for } y \in Y.$$

To show that this is well-defined (i.e., $f_*(\alpha)$ is again constructible) and functorial requires, for example, stratification theory (see [Mac1]) or a suitable theory of constructible sheaves (see [Sch3]). For later use we also point out, that here in the (semi-)algebraic context we do *not* need the assumption that our spaces are compact or the morphism f is proper for the defintion of f_*. This properness of f for the definition of f_* is only needed in the corresponding (sub-)analytic context.

The above Sullivan's Stiefel–Whitney class is now the special case of the following *Stiefel–Whitney class transformation* (compare also with [FuMC]):

Theorem 7.1. *On the category of compact real algebraic varieties there exists a unique natural transformation*

$$w_* : F^{mod2}(\ ) \to H_*(\ ;\mathbb{Z}_2)$$

satisfying the normalization condition that for a nonsingular variety X

$$w_*(\mathbb{1}_X) = w^*(TX) \cap [X]\,.$$

Here $\mathbb{1}_X := 1_X$ is the characteristic function of X.

Note that $\sharp(X) = deg(w_*(\mathbb{1}_X)) = \chi(X)\,mod\,2$ is just the Euler characteristic $mod\,2$ of the singular space X.

7.2. Chern classes c_*

Based on *Grothendieck's* ideas or modifying Grothendieck's conjecture on a *Riemann–Roch type formula* concerning the constructible étale sheaves and Chow rings (see [Grot2, Part II, note (87_1), p.361 ff.]), *Deligne* made the following conjecture — this is usually simply phrased "Deligne and Grothendieck made the following conjecture" — and *R. MacPherson* [Mac1] proved it affirmatively:

Theorem 7.2. *There exists a unique natural transformation*

$$c_* : F(\ ) \to H_{2*}(\ ; \mathbb{Z})$$

from the constructible function covariant functor F to the integral homology covariant functor (in even degrees) H_{2*}, satisfying the "normalization" that the value of the characteristic function $\mathbb{1}_X := 1_X$ of a smooth complex algebraic variety X is the Poincaré dual of the total Chern cohomology class:

$$c_*(\mathbb{1}_X) = c^*(TX) \cap [X].$$

The main ingredients are *Chern–Mather classes, local Euler obstruction and "graph construction"*. The uniqueness follows from the above normalization condition and resolution of singularities. For an algebraic version of MacPherson's Chern class transformation c_* over a base field of characteristic zero (taking values in Chow groups), compare with [Ken]. MacPherson's approach [Mac1] also works in the complex analytic context, since the analyticity of the "graph construction" was solved by Kwieciński in his thesis [Kw2].

Remark 7.3. (see [KMY]) The individual component $c_i : F(\ ) \to H_{2i}(\ )$ of the transformation $c_* : F(\ ) \to H_{2*}(\ )$ is also a natural transformation and also any linear combination of these components is a natural transfomation. Let us consider *projective* varieties. Then, *modulo torsion*, these linear combinations are the *only* natural tansformations from the covariant functor F to the homology functor. In particular, the *rationalized* MacPherson's Chern class transformation $c_* \otimes \mathbb{Q}$ is the only such natural tansformation satisfying the *weaker normalization condition* that for each complex projective space $\mathbf{P}$ the top dimensional component of $c_*(\mathbf{P})$ is the fundamental class $[\mathbf{P}]$. A noteworthy feature of the proof of these statements is that one does *not* need to appeal to resolution of singularities.

J.-P. Brasselet and M.-H. Schwartz [BrSc] showed that the distinguished value $c_*(\mathbb{1}_X)$ of the characteristic function of a complex variety embedded into a complex manifold is isomorphic to the *Schwartz class* [Schw1,Schw2] via the Alexander duality. Thus for a complex algebraic variety X, singular or nonsingular, $c_*(X) := c_*(\mathbb{1}_X)$ is called the total *Chern–Schwartz–MacPherson class of X*. By considering mapping X to a point, one gets

$$\chi(X) = deg(c_*(\mathbb{1}_X)) = \sharp(X),$$

which is a singular version of the Gauss–Bonnet–Chern theorem.

Remark 7.4. For a singular version of the Poincaré–Hopf theorem for radial vector fields, see [Schw3] and for the Poincaré–Hopf theorem for general stratified vector fields compare with [BLSS] and the survey paper [Sea1]. For a version in terms of 1-forms and characteristic cycles of constructible functions, for example see [Sch3, §5.0.3] and [Sch5].

There are also other notions of Chern classes of a singular complex algebraic variety X: Chern–Mather classes $c_*^{Ma}(X)$ ([Mac1]), Fulton's Chern classes and Fulton–Johnson Chern classes $c_*^F(X), c_*^{FJ}(X)$ ([FJ] and [Fu1, Ex. 4.2.6]), and for "stringy and arc Chern classes" $c_*^{str}(X), c_*^{arc}(X)$ see subsection 11.4. In many interesting cases these can be described as $c_*(\alpha_X)$ for a suitable constructible function α_X related to some geometric properties of the singular space X (compare [Alu1,Br2,Pa,PP1,PP2,Sch1,Sch4,Sch5,Su2]). Of course $\alpha_X = 1_X$ for X smooth, but in general $\alpha_X \neq 1_X$ so that the MacPherson Chern class transformation c_* is the basic one, but in general $\mathbb{1}_X = 1_X$ is not the only possible choice of a distinguished element $\mathbb{1}_X$. In particular for a local complete intersection X the difference between $c_*^F(X)$ and $c_*(X)$ is called the *Milnor class* of X (compare loc.cit.), since in the case of isolated singularities its information reduces to the *local Milnor number* of an isolated complete intersection singularity [SeSu,Su3].

7.3. *Todd classes td_**

Motivated by the formulation of MacPherson's Chern class transformation c_* : $F \to H_*$, *P. Baum, W. Fulton and R. MacPherson* [BFM1] have extended **GRR** to singular varieties, by introducing the so-called *localized Chern character* $ch_X^M(\mathcal{F})$ of a coherent sheaf $\mathcal{F}$ with X embedded into a non-singular quasi-projective variety M, as a substitute of $ch^*(F) \cap [X]$ in the above **GRR**. Note that if X is smooth $ch_X^X(\mathcal{F}) = ch^*(F) \cap [X]$. For other constructions of localized Chern characters, see [Kw2], [Schw2] and [Su1].

In [BFM] they showed the following theorem:

Theorem 7.5. *(Baum–Fulton–MacPherson's Riemann–Roch)*
(i) $td_*(\mathcal{F}) := td^*(i_M^* T_M) \cap ch_X^M(\mathcal{F})$ *is independent of the embedding* $i_M : X \to M$.
(ii) Let the transformation $td_* : \mathbf{G}_0(\ \ ) \to H_{2*}(\ \ ; \mathbb{Q})$ *be defined by*

$$td_*(\mathcal{F}) = td^*(i_M^* T_M) \cap ch_X^M(\mathcal{F})$$

for any variety X. Then td_* is actually natural, i.e., for any morphism $f : X \to Y$ the following diagram commutes:

$$
\begin{array}{ccc}
\mathbf{G}_0(X) & \xrightarrow{\ td_*\ } & H_{2*}(X;\mathbb{Q}) \\
{\scriptstyle f_!}\downarrow & & \downarrow{\scriptstyle f_*} \\
\mathbf{G}_0(Y) & \xrightarrow[\ td_*\ ]{} & H_{2*}(Y;\mathbb{Q})
\end{array}
$$

i.e., for any embeddings $i_M : X \to M$ and $i_N : Y \to N$

$$td^*(i_N^* T_N) \cap ch_Y^N(f_! \mathcal{F}) = f_*(td^*(i_M^* T_M) \cap ch_X^M(\mathcal{F})) . \qquad \textbf{(BFM-RR)}$$

For a complex algebraic variety X, singular or nonsingular, $td_*(X) := td_*(\mathcal{O}_X)$ is called the Baum–Fulton–MacPherson's Todd homology class of X, i.e. the class of the structure sheaf is the distingiuished element $\mathbb{1}_X := [\mathcal{O}_X]$. And we get

$$\chi_a(X) = \int_X td_*(X) = \sharp(X) ,$$

which is a singular version of the Riemann–Roch theorem. And in [BFM2] this Todd class transformation is moreover factorized through complex K-homology, which maybe is the most natural formulation of this transformation. For the algebraic version of the Todd class transformation td_* over any base field compare with [Fu1, chapter 18].

Remark 7.6 (Euler homology class e_0). Even though the formulation of the BFM–RR was motivated by that of MacPherson's Chern class transformation, it was proved in a completely different way. And now there is available a similar proof of MacPherson's theorem for the embedded context based on the theory of characteristic cycles CC of constructible functions, with the Segre class $s_* CC$ of these conic characteristic cycles playing the role of the localized Chern character in the proof of Baum–Fulton–MacPherson. Here these characteristic cycles are conic Lagrangian cycles in $T^* M | X$, and the pullback

$$e_0 := k^* CC : F(X) \to H_0(X; \mathbb{Z})$$

by the zero section $k : X \to T^* M | X$ can be seen as a functorial *Euler homology class transformation* even in the context of real geometry. In particular

$$\chi(X) = deg(e_0(\mathbb{1}_X)) = \sharp(X)$$

also in this context. For more details of this, see [Sch4,Sch5]. Finally, this approach by characteristic cycles also gives a new approach to the Stiefel–Whitney class transformation w_* of Sullivan as observed and explained in [FuMC].

7.4. *L-classes L_**

Using the notion of "perversity", *M. Goresky and R. MacPherson* ([GM1], [GM2]) have introduced *Intersection Homology Theory*. In [GM1] they introduced a homology L-class $L_*^{\mathrm{GM}}(X)$ for stratified spaces X with even (co)dimensional strata such that if X is nonsingular it becomes the Poincaré dual of the original Thom–Hirzebruch L-class: $L_*^{GM}(X) = L^*(TX) \cap [X]$. Another approach to these classes is due to J. Cheeger [Che]. And for *rational PL-homology manifolds*, these L-classes agree with the classes introduced by Thom long ago in [Thom2] as one of the first characteristic classes of suitable singular spaces.

Later, *S. Cappell and J. Shaneson* [CS1] (see also [CS2] and [Sh]) introduced a homology L-class transformation L_*, which turns out to be a natural transformation from the abelian group $\Omega(X)$ of cobordism classes of selfdual constructible complexes, whose definition we now explain, to the rational homology group [BSY3] (cf. [Y2]).

Let X be a compact complex analytic (algebraic) space with $D_c^b(X)$ the bounded derived category of complex analytically (algebraically) constructible complexes of sheaves of $\mathbb{Q}$-vector spaces (compare [KS] and [Sch3]). So we consider bounded sheaf complexes $\mathcal{F}$, which have locally constant cohomology sheaves with finite dimensional stalks along the strata of a complex analytic (algebraic) Whitney stratification of X. This is a triangulated category with translation functor $T = [1]$ given by shifting a complex one step to the left. It also has a duality in the sense of Youssin [You] induced by the *Verdier duality functor* (compare [Sch3, Chap.4] and [KS, Chap.VIII]):

$$D_X := Rhom(\cdot, k^!\mathbb{Q}_{pt}) : \ D_c^b(X) \to D_c^b(X) \, ,$$

with $k : X \to \{pt\}$ a constant map, together with its *biduality isomorphism* $can : id \xrightarrow{\sim} D_X \circ D_X$. A constructible complex $\mathcal{F} \in ob(D_c^b(X))$ is called *selfdual*, if there is an isomorphism

$$d : \mathcal{F} \xrightarrow{\sim} D_X(\mathcal{F}) \, .$$

The pair $(\mathcal{F}, d)$ is called *symmetric* or *skew-symmetric*, if

$$D_X(d) \circ can = d \quad \text{or} \quad D_X(d) \circ can = -d \, .$$

Finally an isomorphism or *isometry* of selfdual objects $(\mathcal{F}, d)$ and $(\mathcal{F}', d')$ is an isomorphism u such that the following diagram commutes:

$$
\begin{array}{ccc}
\mathcal{F} & \xrightarrow{\ \ u\ \ } & \mathcal{F}' \\
\ \downarrow{\scriptstyle d} & & \ \downarrow{\scriptstyle d'} \\
D_X(\mathcal{F}) & \xleftarrow[D_X(u)]{\ \sim\ } & D_X(\mathcal{F}') \, .
\end{array}
$$

The isomorphism classes of such (skew-)symmetric selfdual complexes form a set, which becomes a *monoid* with addition induced by the direct sum. Using a definition of Youssin [You], the *cobordism groups* $\Omega_{\pm}(X)$ of (skew-)symmetric selfdual constructible complexes on X are defined by introducing a suitable *cobordism relation* in terms of an *octahedral diagram*, i.e. a diagram (Oct) of the following form:

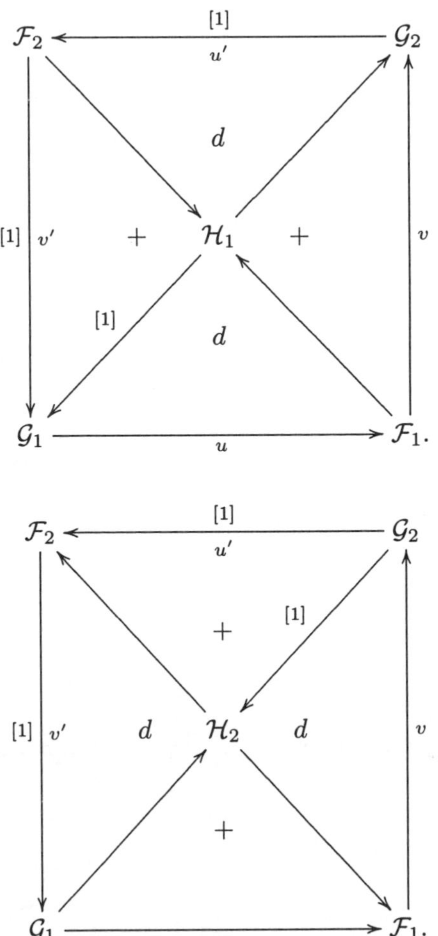

Here the morphism marked by [1] are of degree one, the triangles marked $+$ are commutative, and the ones marked d are distinguished. Finally the two composite morphisms from $\mathcal{H}_1$ to $\mathcal{H}_2$ (via $\mathcal{G}_1$ and $\mathcal{G}_2$) have to be the same, and similarly for the two composite morphisms from $\mathcal{H}_2$ to $\mathcal{H}_1$ (via $\mathcal{F}_1$ and $\mathcal{F}_2$).

Application of the duality functor $D := D_X$ and a rotation by $180°$ about the axis connecting upper-left and lower-right corner induces another octahedral diagram $(RD \cdot Oct)$ such that RD applied to $(RD \cdot Oct)$ gives the octahedral diagram $(D^2 \cdot Oct)$ which one gets from (Oct) by application of D^2 (compare with [You, p.387/388]). Then the octahedral diagram (Oct) is called *symmetric* or *skew-symmetric*, if there is an isomorphism $d : (Oct) \to (RD \cdot Oct)$ of octahedral diagrams such that

$$RD(d) \circ can = d \quad \text{or} \quad RD(d) \circ can = -d$$

as maps of octahedral diagrams $(Oct) \to (RD \cdot Oct)$. Note that this induces in particular (skew-)symmetric dualities d_1 and d_2 of the corners $\mathcal{F}_1$ and $\mathcal{F}_2$, and (Oct, d) is called an *elementary cobordism* between $(\mathcal{F}_1, d_1)$ and $(\mathcal{F}_2, d_2)$. This notion is a symmetric and reflexive relation. $(\mathcal{F}, d)$ and $(\mathcal{F}', d')$ are called *cobordant* if there is a sequence

$$(\mathcal{F}, d) = (\mathcal{F}_0, d_0),\ (\mathcal{F}_1, d_1),\ \ldots,\ (\mathcal{F}_m, d_m) = (\mathcal{F}', d')$$

with $(\mathcal{F}_i, d_i)$ elementary cobordant to $(\mathcal{F}_{i+1}, d_{i+1})$ for $i = 0, \ldots, m - 1$. This *cobordism relation* is then an equivalence relation.

The *cobordism group* $\Omega_{\pm}(X)$ of selfdual constructible complexes on X is the quotient of the monoid of isomorphism classes of (skew-)symmetric selfdual complexes by this cobordism relation. These are indeed abelian groups and not just monoids.

Consider now an algebraic (or holomorphic) map $f : X \to Y$, with X, Y compact so that f is proper. Then $Rf_* \simeq Rf_!$ maps $D_c^b(X)$ to $D_c^b(X)$. Moreover, the *adjunction isomorphism*

$$Rf_* Rhom(\mathcal{F}, f^! k^! \mathbb{Q}_{pt}) \simeq Rhom(Rf_! \mathcal{F}, k^! \mathbb{Q}_{pt})$$

induces the isomorphism

$$Rf_* D_X \xrightarrow{\sim} D_Y Rf_! \simeq D_Y Rf_* \tag{7.7}$$

so that Rf_* *commutes with Verdier-duality*. In particular Rf_* maps selfdual constructible complexes on X to selfdual constructible complexes on Y inducing group homomorphisms

$$f_* : \Omega_{\pm}(X) \to \Omega_{\pm}(Y);\ [(\mathcal{F}, d)] \mapsto [(Rf_* \mathcal{F}, Rf_*(d))]\,.$$

A simple example of a selfdual constructible complex is the shifted constant sheaf $\mathbb{Q}_Z[n]$ on a complex manifold Z of pure dimension n, with the duality isomorphism induced from the *complex orientation* of Z by Poincaré–Verdier duality:

$$k^! \mathbb{Q}_{pt} \simeq \mathbb{Q}_Z[2n] \quad , \text{with } k : X \to \{pt\} \text{ a constant map.}$$

This is (skew-)symmetric for n even (or odd).

Then the results of Cappell–Shaneson [CS1, §5] can be reformulated as in [BSY3] (cf. [Y2, Corollary 2.3]):

Theorem 7.8 (Cappell–Shaneson). *For a compact complex analytic (or algebraic) space X there is a* homology L-class transformation

$$L_* : \Omega(X) := \Omega_+(X) \oplus \Omega_-(X) \to H_*(X, \mathbb{Q}) \, ,$$

which is a group homomorphism functorial for the pushdown f_* induced by a holomorphic (or algebraic) map. The degree of $L_0((\mathcal{F}, d))$ is the signature *of the induced pairing*

$$H^0(X, \mathcal{F}) \otimes_{\mathbb{Q}} \mathbb{R} \times H^0(X, \mathcal{F}) \otimes_{\mathbb{Q}} \mathbb{R} \to \mathbb{R}$$

(by definition this is 0 for a skew-symmetric pairing). Moreover, for X smooth of pure dimension n one has the normalization

$$L_*((\mathbb{Q}_X[n], d)) = L^*(TX) \cap [X] \, .$$

There is also a *uniqueness* statement in [CS1, §5] for such an L-class transformation, but for this one has to go outside the complex algebraic or analytic context.

For X pure dimensional (otherwise one should only look at the top dimensional irreducible components of X) one has the distinguished self-dual constructible intersection cohomology complex $\mathbb{1}_X := \mathcal{IC}_X$, whose global cohomology calculates the intersection (co)homology of Goresky–MacPherson. By definition one gets $L_*(X) := L_*(\mathcal{IC}_X) = L_*^{\mathrm{GM}}(X)$ so that

$$\int_X L_*(X) = \sharp(X)$$

is the signature of the global intersection (co)homology.

Remark 7.9. Thom used in [Thom2] his combinatorial L-classes for the definition of *combinatorial Pontrjagin classes* of rational PL-homology manifolds. Note that in the context of rational homology manifolds, *rational L- and Pontrjagin classes* carry the same information (i.e. can be deduced from each other).

But this is not the case for more singular spaces, and only a corresponding L-class transformation exists for suitable singular spaces, but not a Pontrjagin class transformation.

So all these theories of characteristic homology class transformations for singular spaces have the same formalism, but their existence and construction is due to completely different underlying ideas: *mod 2 Euler spaces* for w_*, *local Euler obstruction* for c_*, *localized Chern character* for td_* and *duality* for L_*. Nevertheless it is natural to ask for another theory of characteristic homology classes of singular spaces, which unifies these theories for complex algebraic varieties:

Problem 7.10. *(cf. [Mac2] and [Y3]) Is there a "unifying and singular version"* $\boxed{?}_y$ *of the generalized Hirzebruch–Riemann–Roch* **g-HRR** *such that*

$(y = -1)$ $\boxed{?}_{-1}$ *gives rise to the rationalized MacPherson's Chern class* $c_* \otimes \mathbb{Q}$,

$(y = 0)$ $\boxed{?}_0$ *gives rise to the Baum–Fulton–MacPherson's Todd class* td_*, *and*

$(y = 1)$ $\boxed{?}_1$ *gives rise to the Cappell–Shaneson's homology* L-class L_*.

An obvious obstacle for this problem is that the source covariant functors of these three natural transformations are all different. And even if such a theory is not known, its *normalization condition* for a smooth complex algebraic manifold M has to be

$$cl_*(\mathbb{1}_M) = \widetilde{td}_{(y)}(TM) \cap [M]$$

by **g-HRR** so that this transformation has to be called a *Hirzebruch $\widetilde{td}_{(y*)}$-* or T_{y*}-class transformation*.

8. Relative Grothendieck rings of varieties and motivic characteristic classes

A "reasonable" answer for the above Problem 7.10 has been obtained in [BSY3, BSY4] via the so-called *relative Grothendieck ring of complex algebraic varieties over X*, denoted by $K_0(\mathcal{V}/X)$. This ring was introduced by E. Looijenga in [Lo] and further studied by F. Bittner in [Bit]. The relative Grothendieck group $K_0(\mathcal{V}/X)$ (of morphisms over a variety X) is the quotient of the free abelian group of isomorphism classes of morphisms to X (denoted by $[Y \to X]$ or $[Y \xrightarrow{h} X]$), modulo the following *additivity* relation:

$$[Y \xrightarrow{h} X] = [Z \hookrightarrow Y \xrightarrow{h} X] + [Y \setminus Z \hookrightarrow Y \xrightarrow{h} X]$$

for $Z \subset Y$ a closed subvariety of Y. The ring structure is given by the fiber square: for $[Y \xrightarrow{f} X], [W \xrightarrow{g} X] \in K_0(\mathcal{V}/X)$

$$[Y \xrightarrow{f} X] \cdot [W \xrightarrow{g} X] := [Y \times_X W \xrightarrow{f \times_X g} X].$$

Here $Y \times_X W \xrightarrow{f \times_X g} X$ is $g \circ f' = f \circ g'$ where f' and g' are as in the following diagram

$$
\begin{array}{ccc}
Y \times_X W & \xrightarrow{\ f'\ } & W' \\
{\scriptstyle g'}\downarrow & & \downarrow{\scriptstyle g} \\
Y & \xrightarrow{\ f\ } & X.
\end{array}
$$

The relative Grothendieck ring $K_0(\mathcal{V}/X)$ has the unit $1_X := [X \xrightarrow{id_X} X]$, which later becomes the distinguished element $\mathbb{1}_X := [id_X]$. Similarly one gets an exterior product

$$\times : K_0(\mathcal{V}/X) \times K_0(\mathcal{V}/Y) \to K_0(\mathcal{V}/X \times Y) .$$

Note that when $X = \{pt\}$ is a point, then the relative Grothendieck ring $K_0(\mathcal{V}/\{pt\})$ is nothing but the usual Grothendieck ring $K_0(\mathcal{V})$ of $\mathcal{V}$, which is the free abelian group generated by the isomorphism classes of varieties modulo the subgroup generated by elements of the form $[V] - [V'] - [V \setminus V']$ for a subvariety $V' \subset V$, and the ring structure is given by the Cartesian product of varieties.

Remark 8.1. In some sense the Grothendieck ring $K_0(\mathcal{V})$ can be seen as an algebraic substitute for cobordism rings Ω_* of smooth manifolds, based on the *additivity* instead of a cobordism relation.

For a morphism $f : X' \to X$, the pushforward

$$f_* : K_0(\mathcal{V}/X') \to K_0(\mathcal{V}/X)$$

is defined by

$$f_*[Y \xrightarrow{h} X'] := [Y \xrightarrow{f \circ h} X].$$

With this pushforward, the assignment $X \longmapsto K_0(\mathcal{V}/X)$ is a covariant functor. The pullback

$$f^* : K_0(\mathcal{V}/X) \to K_0(\mathcal{V}/X')$$

is defined as follows: for a fiber square

$$
\begin{array}{ccc}
Y' & \xrightarrow{\ g'\ } & X' \\
{\scriptstyle f'}\downarrow & & \downarrow{\scriptstyle f} \\
Y & \xrightarrow{\ g\ } & X
\end{array}
$$

the pullback $f^*[Y \xrightarrow{g} X] := [Y' \xrightarrow{g'} X']$. With this pullback, the assignment $X \longmapsto K_0(\mathcal{V}/X)$ is a contravariant functor. Let $\mathrm{Iso}^{\mathrm{pr}}(\mathcal{SV}/X)$ be the free abelain groups on isomorphism classes of proper morphisms from smooth varieties to a given variety X. Then we get the canonical quotient homomorphism

$$quo : \mathrm{Iso}^{\mathrm{pr}}(\mathcal{SV}/X) \to K_0(\mathcal{V}/X)$$

which is surjective by the above additivity relation and Hironaka's resolution of singularities [Hi]. And it turns out that the kernel of this surjective map is generated by the "blow-up relation", more precisely we have the following theorem, which is due to F. Bittner [Bi, Theorem 5.1], based on the very deep "weak factorization theorem" ([AKMW] and [W]):

Theorem 8.2. *The relative Grothendieck group $K_0(\mathcal{V}/X)$ is isomorphic to the quotient of the free abelian group $\mathrm{Iso}^{\mathrm{pr}}(\mathcal{SV}/X)$ modulo the following "blow-up relation"*

$$[\emptyset \to X] := 0 \quad and \quad [Bl_Y X' \to X] - [E \to X] = [X' \to X] - [Y \to X]$$

for any Cartesian "blow-up" diagram

$$
\begin{array}{ccc}
E & \xrightarrow{\ i'\ } & Bl_Y X' \\
\pi' \downarrow & & \downarrow \pi \\
Y & \xrightarrow{\ i\ } & X' \xrightarrow{\ f\ } X,
\end{array}
$$

with i being a closed embedding of smooth (pure dimensional) varieties and f : $X' \to X$ proper. Here $\pi : Bl_Y X' \to X'$ is the blow-up of X' along Y with E denoting the exceptional divisor.

From this theorem we can get the following corollary:

Theorem 8.3. *Let $B_* : \mathcal{V}/k \to \mathcal{A}$ be a functor from the category of reduced separated schemes of finite type over $\mathbb{C}$ to the category of abelian groups such that*
(i) $B_*(\emptyset) := 0$,
(ii) it is covariantly functorial for proper morphisms, and
(iii) for any smooth variety X there exists a distinguished element $d_X \in B_*(X)$ such that
(iii-1) for any isomorphism $h : X' \to X$, $h_*(d_{X'}) = d_X$ and
(iii-2) for any Cartesian "blow-up" diagram as in the above Theorem 8.2 with $f = \mathrm{id}_X$,

$$\pi_*(d_{Bl_Y X}) - i_* \pi'_*(d_E) = d_X - i_*(d_Y) \in B_*(X).$$

Then we have by (iii-1) that there exists a unique natural transformation of covariant functors

$$\Phi : \mathrm{Iso}^{\mathrm{pr}}(\mathcal{SV}/\ ) \to B_*(\ )$$

satisfying the normalization condition that for smooth X

$$\Phi([X \xrightarrow{\mathrm{id}} X]) = d_X,$$

and furthermore by (iii-2) there exists a unique natural transformation of covariant functors

$$\widetilde{\Phi} : K_0(\mathcal{V}/\ ) \to B_*(\ )$$

satisfying the normalization condition that for smooth X

$$\widetilde{\Phi}([X \xrightarrow{\mathrm{id}} X]) = d_X.$$

Then, using results of [Gros, IV.1.2.1] or [GNA, Proposition 3.3], we can get the following corollary about a *motivic Chern class transformation mC_**.

Corollary 8.4. *There exisits a unique natural transformation (with respect to proper maps)*

$$mC_* : K_0(\mathcal{V}/\ ) \to \mathbf{G}_0(\ ) \otimes \mathbb{Z}[y]$$

satisfying the normalization condition that for X smooth

$$mC_*([X \xrightarrow{\mathrm{id}} X]) = \sum_{i=0}^{\dim X} [\Lambda^i T^* X] y^i =: \Lambda_y([T^* X]) \cap [\mathcal{O}_X].$$

Here $\Lambda_y(\ )$ is the so-called total Λ-class.

If we compose $mC_*|_{y=-1,0,1}$ with the natural transformation $\mathbf{G}_0(\ ) \to \mathbf{K}_0^{top}(\ )$ to topological K-homology constructed in [BFM2], then $mC_*(X)$ unifies for X smooth the following K-theoretical homology classes:

(y=-1)   the top-dimensional Chern class $c_K^{top}(TX) \cap [X]_K$ in K-theory:

$$mC_*|_{y=-1}([id_X]) = \Lambda_{-1}([T^* X]) \cap [X]_K,$$

(y=0)   the fundamental class in K-homology of the complex manifold X:

$$mC_*|_{y=0}([id_X]) = [X]_K,$$

(y=1)   the class of the signature operator of the underlying spin^c manifold of X (compare with [RW]):

$$mC_*|_{y=1}([id_X]) = \Lambda_1([T^* X]) \cap [X]_K.$$

Its image in $\mathbf{KO}(M)[\frac{1}{2}] \subset \mathbf{K}(M)[\frac{1}{2}]$ is exactly Sullivan's orientation class $\triangle(X)$ (up to an identification of a suitable Bott periodicity factor, compare [Sull2, p.201-203]).

Consider the twisted BFM–RR transformation

$$td_{(1+y)} : \mathbf{G}_0(X) \otimes \mathbb{Z}[y] \to H_{2*}(X) \otimes \mathbb{Q}[y, (1+y)^{-1}]$$

defined by

$$td_{(1+y)}([\mathcal{F}]) := \sum_{i \geq 0} td_i([\mathcal{F}])(1+y)^{-i}$$

and extending it linearly with respect to $\mathbb{Z}[y]$ ([Y3]). Using this twisted BFM–RR transformation $td_{(1+y)}$ and the above transformation mC_*, we define the *Hirzebruch class transformation* T_{y*} as the composite $T_{y*} := td_{(1+y)} \circ mC_*$. Then we get the following theorem:

Theorem 8.5. *Let $K_0(\mathcal{V}/X)$ be the Grothendieck group of complex algebraic varieties over X. Then there exists a unique natural transformation (with respect to proper maps)*

$$T_{y_*} : K_0(\mathcal{V}/\ \ ) \to H_{2*}^{BM}(\ \ ) \otimes \mathbb{Q}[y] \subset H_{2*}^{BM}(\ \ ) \otimes \mathbb{Q}[y, (1+y)^{-1}]$$

such that for X nonsingular

$$T_{y_*}([X \xrightarrow{\mathrm{id}} X]) = \widetilde{td}_{(y)}(TX) \cap [X].$$

Remark 8.6. The transformations mC_* and T_{y*} can also be defined in the same way in the *algebraic context* over a base field of characteristic zero, using the algebraic version of the Todd tranformation td_* as in [Fu1, chapter 18], and in the *compactifiable complex analytic context*, using the analytic version of the Todd tranformation td_* constructed in [Levy] (compare with [BSY3] for more details).

For a later use, we observe that T_{y_*} commutes with the exterior product (and similarly for mC_*), i.e., the following diagram commutes:

$$
\begin{array}{ccc}
K_0(\mathcal{V}/X) \times K_0(\mathcal{V}/Y) & \xrightarrow{\ \times\ } & K_0(\mathcal{V}/X \times Y) \\
{\scriptstyle T_{y_*} \times T_{y_*}} \downarrow & & \downarrow {\scriptstyle T_{y_*}} \\
H_{2*}^{BM}(X) \otimes \mathbb{Q}[y] \times H_{2*}^{BM}(Y) \otimes \mathbb{Q}[y] & \xrightarrow{\ \times\ } & H_{2*}^{BM}(X \times Y) \otimes \mathbb{Q}[y].
\end{array}
$$

And we have the following theorem for a compact complex algebraic variety X:

Theorem 8.7. *($y = -1$) There exists a unique natural transformation $\epsilon :$ $K_0(\mathcal{V}/\ ) \to F(\ )$ such that for X nonsingular $\epsilon([X \xrightarrow{\mathrm{id}} X]) = 1_X$. And the following diagram commutes*

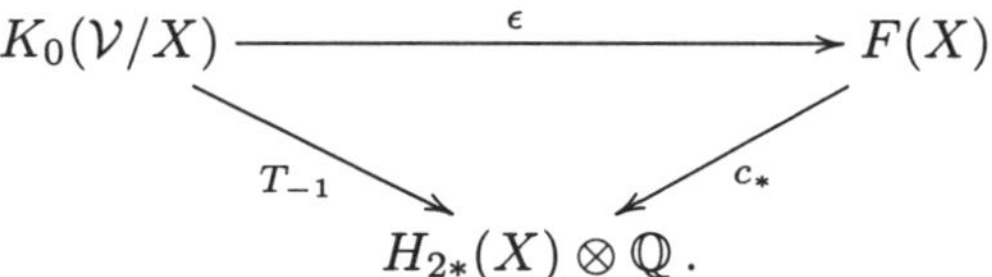

($y = 0$) There exists a unique natural transformation $\gamma : K_0(\mathcal{V}/\ ) \to \mathbf{G}_0(\ )$ such that for X nonsingular $\gamma([X \xrightarrow{\mathrm{id}} X]) = [\mathcal{O}_X]$. And the following diagram commutes

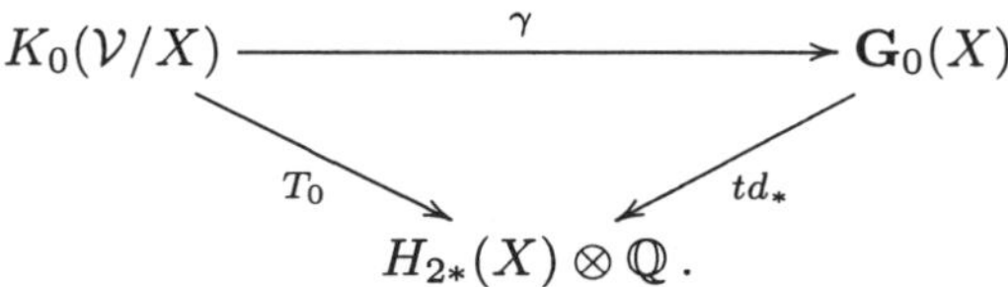

($y = 1$) There exists a unique natural transformation $\omega : K_0(\mathcal{V}/\ ) \to \Omega(\ )$ such that for X nonsingular $\omega([X \xrightarrow{\mathrm{id}} X]) = [\mathbb{Q}_X[\dim X]]$. And the following diagram commutes

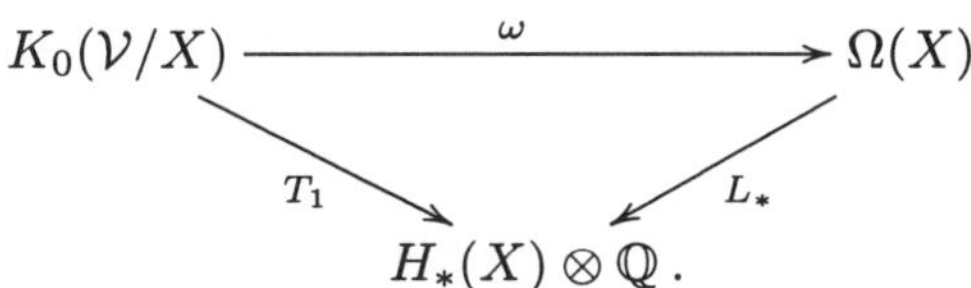

An original proof of the above Theorem 8.5 uses Saito's theory of mixed Hodge modules [Sai] instead of the above Theorem 8.2. In this way one can also study such characteristic classes of mixed Hodge modules, especially those associated to the intersection (co)homolgy complex (compare [To,CMS]). And an even more elementary proof can be given based on some classical results of [DuBo] about the so-called DuBois complex of a singular complex algebraic variety. Only the proof of the case ($y = 1$) of the above Theorem 8.7 depends, up to now, on Bittner's theorem, i.e., the above Theorem 8.2, in other words, on the "weak factorization theorem" ([AKMW] and [W]). Also note that the transformation ϵ is defined for *any* algebraic map of not necessarily compact algebraic varieties, and it also commutes with pullback and (exterior) products. For more details, see [BSY3].

Remark 8.8. The reader should be warned that the transformations γ and ω above do *not* preserve the distinguished elements in general. For any compact singular

900

complex algebraic variety X one has $\epsilon([id_X]) = 1_X$, so that the *Hirzebruch class* $T_{y*}(X) := T_{y*}([id_X])$ specializes to $T_{-1*}(X) = c_*(X) \in H_{2*}(X; \mathbb{Q})$. But in general

$$\gamma([id_X]) \neq [\mathcal{O}_X] \in G_0(X) \quad \text{and} \quad T_{0*}(X) \neq td_*(X).$$

But $T_{0*}(X) = td_*(X)$ if X has at most "Du Bois singularities", e.g. "rational singularities" like, for example, toric varieties. Similarly

$$\omega([id_X]) \neq [\mathcal{IC}_X] \in \Omega(X) \quad \text{and} \quad T_{1*}(X) \neq L_*(X)$$

in general, but we *conjecture* that $T_{1*}(X) = L_*(X)$ for X a *rational homology manifold*.

Moreover, the Hirzebruch characteristic class $\widetilde{td}_{(y)} = T_y^*$ is the *most general* normalized and multiplicative characteristic class of complex vector bundles

$$cl_f^* : \mathrm{Vect}(X) \to H^{2*}(X; \Lambda),$$

with Λ a $\mathbb{Q}$-algebra, which satisfies the condition of Theorem 8.3 with

$$d_X := cl_f^*(TX) \cap [X] \in H_{2*}^{BM}(X; \Lambda)$$

for X smooth. In fact, the correspondig *genus* Φ_f factorizes as

$$\begin{array}{ccc}
\mathrm{Iso}^{\mathrm{pr}}(\mathcal{SV}/\{pt\}) & \longrightarrow & \Omega_*^U \otimes \mathbb{Q} \\
\downarrow & & \downarrow \Phi_f \\
K_0(\mathcal{V}) & \xrightarrow{\ \Phi_f\ } & \Lambda = H_{2*}(\{pt\}; \Lambda).
\end{array} \qquad (8.9)$$

Moreover, the characteristic class cl_f^* or its genus Φ_f is uniquely determined by

$$\Phi_f([P^n(\mathbb{C})]) = \int_{P^n(\mathbb{C})} (cl_f^*(TP^n(\mathbb{C}))) \cap [P^n(\mathbb{C})]$$

for all n. But if Φ_f also factorizes over $K_0(\mathcal{V})$ then we get from the decomposition

$$P^n(\mathbb{C}) = \{pt\} \cup \mathbb{C} \cup \cdots \cup \mathbb{C}^n$$

by "additivity" and "multiplicativity" (and compare with equation (6.4)):

$$\Phi_f([P^n(\mathbb{C})]) = 1 + (-y) + \cdots + (-y)^n \quad \text{with} \quad y := 1 - \Phi_f([P^1(\mathbb{C})]). \quad (8.10)$$

So Φ_f is a specialization of the *Hirzebruch χ_y-genus* corresponding to the *Hirzebruch characteristic class T_y^**. Of course here we use a decomposition into the *non-compact* manifolds $\mathbb{C}^n$, which "is classically forbidden for a genus", with $y = -\Phi_f([\mathbb{C}])$.

Remark 8.11. So *additivity* is the underlying principle which "singles out" those normalized and multiplicative characteristic classes $c\ell_f^*$, which have (so far) a functorial extension to singular spaces. Also note that the specialization $y = 1$ corresponding to the *signature genus $sign = \chi_1$* and the *characteristic L-class transformation $L^* = T_1^*$* is the only one that factorizes by the canonical map $\Omega_*^U \otimes \mathbb{Q} \to \Omega_*^{SO} \otimes \mathbb{Q}$ over the *cobordism ring Ω_*^{SO} of oriented manifolds*, since $[P^1(\mathbb{C})] = 0 \in \Omega_*^{SO}$. In particular this "explains" why there is no functorial *Pontrjagin class transformation* for singular spaces.

For X a compact complex algebraic variety one can also deduce from Theorem 8.3 the Chern class transformation

$$c_* : K_0(\mathcal{V}/X) \to H_{2*}(X; \mathbb{Z}) ,$$

on the relative Grothendieck group $K_0(\mathcal{V}/X)$ without appealing to MacPherson's theorem, since the distinguished element

$$d_X := c^*(TX) \cap [X] \in H_{2*}(X; \mathbb{Z})$$

of a smooth space X satisfies the corresponding conditions. Condition (iii-1) follows from the projection formula, and condition (iii-2) is an easy application (by pushing down to X) of the classical "blowing up formula for Chern classes" [Fu1, Theorem 15.4] . And recent work of Aluffi [Alu3] can be interpreted as showing that this transformation c_* factorizes over $\epsilon : K_0(\mathcal{V}/\ ) \to F(\ )$.

9. Bivariant Characteristic classes

In [FM] (also, see [Fu1]) *W. Fulton and R. MacPherson* introduced the notion of *Bivariant Theory*, which is a simultaneous generalization of a pair of covariant and contravariant functors. Most pairs of covariant and contravariant theories, e.g., such as homology theory, K-theory, etc., extend to bivariant theories. A bivariant theory $\mathbb{B}$ on a suitable category $\mathcal{C}$ (with a distinguished class of so-called "proper" or "confined" maps) with values in the category of abelian groups is an assignment to each morphism $X \xrightarrow{f} Y$ in the category $\mathcal{C}$ an abelian group $\mathbb{B}(X \xrightarrow{f} Y)$, which is equipped with the following three basic operations:

(Product operations): For morphisms $f : X \to Y$ and $g : Y \to Z$, the product operation

$$\bullet : \mathbb{B}(X \xrightarrow{f} Y) \otimes \mathbb{B}(Y \xrightarrow{g} Z) \to \mathbb{B}(X \xrightarrow{gf} Z)$$

is defined.

(Pushforward operations): For morphisms $f : X \to Y$ and $g : Y \to Z$ with f

902

proper, the pushforward operation

$$f_* : \mathbb{B}(X \xrightarrow{gf} Z) \to \mathbb{B}(Y \xrightarrow{g} Z)$$

is defined.

(Pullback operations): For a fiber (or more generally a so-called independent) square

$$
\begin{array}{ccc}
X' & \xrightarrow{g'} & X \\
{\scriptstyle f'}\downarrow & & \downarrow{\scriptstyle f} \\
Y' & \xrightarrow{\quad g \quad} & Y,
\end{array}
$$

the pullback operation

$$g^\star : \mathbb{B}(X \xrightarrow{f} Y) \to \mathbb{B}(X' \xrightarrow{f'} Y')$$

is defined. And these three operations are required to satisfy *seven compatibility axioms* (see [FM, Part I, §2.2] for details). In particular, the class of "proper" maps has to be stable under composition and base change, and should contain all identity maps. Let $\mathbb{B}, \mathbb{B}'$ be two bivariant theories on such a category $\mathcal{C}$. Then a *Grothendieck transformation* from $\mathbb{B}$ to $\mathbb{B}'$

$$\gamma : \mathbb{B} \to \mathbb{B}'$$

is a collection of homomorphisms

$$\mathbb{B}(X \to Y) \to \mathbb{B}'(X \to Y)$$

for a morphism $X \to Y$ in the category $\mathcal{C}$, which preserves the above three basic operations:

(i) $\gamma(\alpha \bullet_{\mathbb{B}} \beta) = \gamma(\alpha) \bullet_{\mathbb{B}'} \gamma(\beta)$,

(ii) $\gamma(f_*\alpha) = f_*\gamma(\alpha)$, and

(iii) $\gamma(g^\star\alpha) = g^\star\gamma(\alpha)$.

$\mathbb{B}_*(X) := \mathbb{B}(X \to pt)$ and $\mathbb{B}^*(X) := \mathbb{B}(X \xrightarrow{\mathrm{id}} X)$ become a covariant functor for proper maps and a contravariant functor, respectively. And a Grothendieck transformation $\gamma : \mathbb{B} \to \mathbb{B}'$ induces natural transformations $\gamma_* : \mathbb{B}_* \to \mathbb{B}'_*$ and $\gamma^* : \mathbb{B}^* \to \mathbb{B}'^*$ such that γ_* commutes with the (bivariant) exterior product, i.e. the following diagram commutes:

$$
\begin{array}{ccc}
\mathbb{B}_*(X) \times \mathbb{B}_*(Y) & \xrightarrow{\quad \times \quad} & \mathbb{B}_*(X \times Y) \\
{\scriptstyle \gamma_* \times \gamma_*}\downarrow & & \downarrow{\scriptstyle \gamma_*} \\
\mathbb{B}'_*(X) \times \mathbb{B}'_*(Y) & \xrightarrow{\quad \times \quad} & \mathbb{B}'_*(X \times Y) \,.
\end{array}
$$

If we have a Grothendieck transformation $\gamma : \mathbb{B} \to \mathbb{B}'$, then via a bivariant class $b \in \mathbb{B}(X \xrightarrow{f} Y)$ we get the commutative diagram

$$\begin{array}{ccc}
\mathbb{B}_*(Y) & \xrightarrow{\;\gamma_*\;} & \mathbb{B}'_*(Y) \\
{\scriptstyle b\bullet}\downarrow & & \downarrow{\scriptstyle \gamma(b)\bullet} \\
\mathbb{B}_*(X) & \xrightarrow[\;\gamma_*\;]{} & \mathbb{B}'_*(X).
\end{array} \qquad (9.1)$$

This is called *the Verdier-type Riemann–Roch formula associated to the bivariant class b*.

Bivariant Todd class transformation τ. The most important (and motivating) example of such a Grothendieck transformation of bivariant theories is the *bivariant Riemann–Roch transformation τ* from the *bivariant algebraic K-theory* $\mathbb{K}_{\mathrm{alg}}$ *of perfect complexes* to *rational bivariant homology* $\mathbb{H}_{\mathbb{Q}}$

$$\tau : \mathbb{K}_{\mathrm{alg}} \to \mathbb{H}_{\mathbb{Q}}$$

constructed in [FM, Part II] in the complex quasi-projective context. Here $\mathbb{H}_{\mathbb{Q}}$ is the bivariant homology theory corresponding to usual cohomology with rational coefficients constructed in [FM, §3.1] for more general cohomology theories. Then the associated contravariant theory $\mathbb{H}_{\mathbb{Q}}^*(X) = H^*(X; \mathbb{Q})$ is the cohomology, and the associated covariant theory $\mathbb{H}_{\mathbb{Q}*}(X) = H_*^{BM}(X; \mathbb{Q})$ is the Borel–Moore homology. Similarly $\mathbb{K}_{\mathrm{alg}}^* \simeq \mathbf{K}^0$ is the Grothendieck group of algebraic vector bundles, and $\mathbb{K}_{\mathrm{alg}*} \simeq \mathbf{G}_0$ is the Grothendieck group of algebraic coherent sheaves. Then the associated contravariant transformation τ^* is the *Chern character*

$$ch^* : \mathbb{K}_{\mathrm{alg}}^*(\;) \simeq \mathbf{K}^0(\;) \to H^*(\;;\mathbb{Q}) \simeq \mathbb{H}_{\mathbb{Q}}^*(\;),$$

and the associated covariant transformation

$$\tau_* : \mathbb{K}_{\mathrm{alg}*}(\;) \simeq \mathbf{G}_0(\;) \to H_*^{BM}(\;;\mathbb{Q}) \simeq \mathbb{H}_{\mathbb{Q}*}(\;)$$

is just Baum–Fulton–MacPherson's Todd class transformation td_* constructed in [BFM1]. And the bivariant transformation τ unifies many different known Riemann–Roch type theorems. In particular for a *smooth* morphism $f : X \to Y$ of possible singular varieties one has

$$\mathbb{1}_f := [\mathcal{O}_X] \in \mathbb{K}_{\mathrm{alg}}(X \xrightarrow{f} Y),$$

with $\tau(\mathbb{1}_f) = td^*(T_f) \bullet [f]$. Here T_f is the vector bundle of tangent spaces of fibers of f, and $[f] \in \mathbb{H}_{\mathbb{Q}}(X \xrightarrow{f} Y)$ is the *canonical orientation* of the smooth morphism f. Then the Verdier-type Riemann–Roch formula (9.1) associated to

$\mathbb{1}_f$ becomes the usual *Verdier–Riemann–Roch theorem* for the Todd class transformation td_*:

$$td_*(f^*\beta) = td^*(T_f) \cap f^! td_*(\beta) \quad \text{for } \beta \in \mathbf{G}_0(Y). \tag{9.2}$$

Here $f^! = [f]\bullet : H_*^{BM}(Y;\mathbb{Q}) \simeq \mathbb{H}_{\mathbb{Q}*}(Y) \to \mathbb{H}_{\mathbb{Q}*}(X) \simeq H_*^{BM}(X;\mathbb{Q})$ is the *smooth pullback* in Borel–Moore homology. And for an *algebraic version* of this bivariant Riemann-Roch transformation τ compare with [Fu1, Ex. 18.3.16].

Bivariant Stiefel–Whitney class transformation ω. In the context of real geometry (e.g. the piecewise linear, (semi-)algebraic or subanalytic context) one has the following interesting example of a bivariant theory (with "proper" the usual meaning). Here *Fulton–MacPherson's bivariant group* $\mathbb{F}^{mod2}(X \xrightarrow{f} Y)$ *of $\mathbb{Z}_2$-valued constructible functions* consists of all the constructible functions on X which satisfy the local Euler condition with respect to f. Here a $\mathbb{Z}_2$-valued constructible function $\alpha \in F^{mod2}(X)$ is said to satisfy the *local Euler condition with respect to f*, if for any point $x \in X$ and for any local embedding $(X, x) \to (\mathbf{R}^N, 0)$ the equality

$$\alpha(x) = \chi\left(B_\epsilon \cap f^{-1}(z); \alpha\right) \ mod \ 2$$

holds, where B_ϵ is a sufficiently small *open* ball of the origin 0 with radius ϵ and z is any point close to $f(x)$ (cf. [Br1], [Sa]). In particular, if $\mathbb{1}_f := 1_X$ belongs to the bivariant group $\mathbb{F}^{mod2}(X \xrightarrow{f} Y)$, then the morphism $f : X \to Y$ is called an *Euler morphism*. For $f : X \to \{pt\}$ a constant map this just means (by the "local conic structure" of X) that X is a *mod 2 Euler space*, i.e. the link $\partial B_\epsilon \cap X$ of any point $x \in X$ has vanishing Euler characteristic modulo 2:

$$\begin{aligned}
\chi(\partial B_\epsilon \cap X) &= \chi_c(\partial B_\epsilon \cap X) \\
&= 1 - \chi_c(B_\epsilon \cap X) \\
&= 1 - \chi(B_\epsilon \cap X; 1_X) = 0 \quad mod \ 2
\end{aligned}$$

Also a *smooth* morphism, or a locally trivial fibration with fiber a mod 2 Euler space, is always an Euler morphism.

The three operations on $\mathbb{F}^{mod2}(X \xrightarrow{f} Y)$ are defined as follows:
(i) the product operation

$$\bullet : \mathbb{F}^{mod2}(X \xrightarrow{f} Y) \otimes \mathbb{F}^{mod2}(Y \xrightarrow{g} Z) \to \mathbb{F}^{mod2}(X \xrightarrow{gf} Z)$$

is defined by $\alpha \bullet \beta := \alpha \cdot f^*\beta$.
(ii) the pushforward operation $f_\star : \mathbb{F}^{mod2}(X \xrightarrow{gf} Z) \to \mathbb{F}^{mod2}(Y \xrightarrow{g} Z)$ is the

usual pushforward f_*, i.e.,

$$f_*(\alpha)(y) := \chi(f^{-1}(\{y\}); \alpha) \, mod \, 2.$$

(iii) for a fiber square

$$
\begin{array}{ccc}
X' & \xrightarrow{g'} & X \\
f' \downarrow & & \downarrow f \\
Y' & \xrightarrow{g} & Y,
\end{array}
$$

the pullback operation $g^\star : \mathbb{F}^{mod2}(X \xrightarrow{f} Y) \to \mathbb{F}^{mod2}(X' \xrightarrow{f'} Y')$ is the functional pullback g'^*, i.e.,

$$g^\star(\alpha)(x') := \alpha(g'(x')).$$

Note that for f proper and any *bivariant* constructible function $\alpha \in \mathbb{F}^{mod2}(X \xrightarrow{f} Y)$, the Euler–Poincaré characteristic $\chi(f^{-1}(y); \alpha)$ of α restricted to each fiber $f^{-1}(y)$ is *locally constant* on Y mod 2 (by the local Euler condition for $f_*(\alpha)$).

The correspondence $s\mathbb{F}^{mod2}(X \to Y) := F^{mod2}(X)$ assigning to a morphism $f : X \to Y$ the abelian group $F^{mod2}(X)$ of the source variety X, whatever the morphism f is, becomes a bivariant theory with the same operations above. This bivariant theory is called the *simple* bivariant theory of constructible functions (see [Sch2] and [Y6]). In passing, what we then need to do for showing that the Fulton–MacPherson's group of $\mathbb{Z}_2$-valued constructible functions satisfying the local Euler condition with respect to a morphism is a bivariant theory, is to show that the local Euler condition with respect to a morphism is preserved by each of these three operations.

For later use let us point out the abstract properties needed for the definition of a *simple bivariant theory* [Sch2, Definition, p.25-26]:

(SB1) We have a contravariant functor $G : \mathcal{C} \to Rings$ with values in the category of rings with unit.

(SB2) G is also covariantly functorial with respect to proper maps (as a functor to the category of Abelian groups).

(SB3) G satisfies the *two-sided projection-formula*, i.e. for $f : X \to Y$ proper and $\alpha \in G(Y)$ and $\beta \in G(X)$,

$$f_*((f^*\alpha) \cup \beta) = \alpha \cup (f_*\beta),$$

i.e., f_* is a left $G(Y)$-module and

$$f_*(\beta \cup (f^*\alpha)) = (f_*\beta) \cup \alpha \,,$$

i.e., f_* is a right $G(Y)$-module. (Note that we do not assume $(G, \cup)$ is (graded) commutative so that both versions of the usual projection formula are needed.)
(SB4) F has the *base-change property* $g^* f_* = f'_* g'^* : G(X) \to G(Y')$ for any fiber (or independent) square

$$
\begin{array}{ccc}
X' & \xrightarrow{\;g'\;} & X \\
{\scriptstyle f'}\downarrow & & \downarrow{\scriptstyle f} \\
Y' & \xrightarrow{\;g\;} & Y \,,
\end{array}
$$

with f, f' proper.

Then one gets a (simple) bivariant theory $s\mathbf{G}$ by $s\mathbf{G}(X \xrightarrow{f} Y) := G(X)$, with the obvious pushforward and pullback transformations as above. Finally the bivariant product

$$\bullet : s\mathbf{G}(X \xrightarrow{f} Y) \times s\mathbf{G}(Y \xrightarrow{g} Z) \to s\mathbf{G}(X \xrightarrow{gf} Z)$$

is just given by $\alpha \bullet \beta := \alpha \cup f^*(\beta)$, with $\cup$ the given product of the ring-structure. Note that this construction does apply not only to constructible functions $G(\ ) = F^{mod2}(\ )$ but also to the relative Grothendieck group of complex algebraic varieties $G(\ ) = K_0(\mathcal{V}/\ )$, even if we allow all algebraic morphisms as "proper" morphisms.

Let $\mathbb{H}^{mod2}(X \xrightarrow{f} Y)$ be Fulton–MacPherson's *bivariant homology theory* with $\mathbb{Z}_2$ coefficients, constructed from the corresponding cohomology theory in [FM, §3.1] so that $\mathbb{H}^{mod2,*}(X) = H^*(X; \mathbb{Z}_2)$ and $\mathbb{H}^{mod2}_*(X) = H^{BM}_*(X; \mathbb{Z}_2)$. Then in the *piecewise linear context* Fulton and MacPherson [FM, Theorem 6A] showed the following theorem, which is a bivariant version of the singular Stiefel–Whitney class transformation $w_* : F^{mod2}(\ ) \to H^{BM}_*(\ : \mathbb{Z}_2)$:

Theorem 9.3. *There existis a unique Grothendieck transformation*

$$\omega : \mathbb{F}^{mod2} \to \mathbb{H}^{mod2}$$

satisfying the normalization condition that for a morphism from a smooth variety X to a point

$$\omega(1_X) = w^*(TX) \cap [X] \in \mathbb{H}^{mod2}_*(X) = H^{BM}_*(X; \mathbb{Z}_2) \,.$$

Remark 9.4. As to the bivariant mod 2 constructible functions, in the context of real geometry, the definition and the theory of them can be given in any of the following categories: the PL-category, the (semi-)algebraic category and the subanalytic category. Note that the above bivariant Stiefel–Whitney class transformation is only proved and known in the PL-category.

Bivariant Chern class transformation γ. Instead of mod 2 constructible functions, in the complex analytic or algebraic context we certainly have similarly the bivariant group $\mathbb{F}(X \to Y)$ of $\mathbb{Z}$-valued constructible functions satisfying the local Euler condition with values in $\mathbb{Z}$ and the bivariant homology theory $\mathbb{H}(X \to Y)$ with integer coefficients, and W. Fulton and R. MacPherson conjectured or posed as a question the existence of a so-called *bivariant Chern class transformation* and *J.-P. Brasselet* [Br1] solved it:

Theorem 9.5. *For the category of embeddable complex analytic varieties with* cellular morphisms, *there exists a Grothendieck transformation*

$$\gamma : \mathbb{F} \to \mathbb{H}$$

such that for a morphism $f : X \to \{pt\}$ from a nonsingular variety X to a point $\{pt\}$ and the bivariant constructible function $\mathbb{1}_f := 1_X$ the following normalization condition holds:

$$\gamma(\mathbb{1}_f) = c^*(TX) \cap [X] \in \mathbb{H}_*(X) = H_*^{BM}(X;\mathbb{Z}) \,.$$

Since then, the *uniqueness* of the Brasselet bivariant Chern class and the problem of whether " cellularness" of morphisms (which is not so easy to check) can be dropped or not have been unresolved. In [Sa] *C. Sabbah* constructed a bivariant Chern class transformation "micro-local analytically" in some cases. In [Z1], [Z2] *J. Zhou* showed that the bivariant Chern classes constructed by J.-P. Brasselet [Br1] and the ones constructed by C. Sabbah [Sa] in some cases are identical in the case when the target variety is a *nonsingular curve*. And in [Y5, Theorem (3.7)] we showed the following more general *uniqueness theorem* of bivariant Chern classes for morphisms whose target varieties are *nonsingular of any dimension*:

Theorem 9.6. *If there exists a bivariant Chern class transformation $\gamma : \mathbb{F} \to \mathbb{H}$, then it is unique when restricted to morphisms whose target varieties are nonsingular; explicitly, for a morphism $f : X \to Y$ with Y nonsingular and for any bivariant constructible function $\alpha \in \mathbb{F}(X \xrightarrow{f} Y)$ the bivariant Chern class $\gamma(\alpha)$ is expressed by*

$$\gamma(\alpha) = f^* s(TY) \cap c_*(\alpha)$$

where $s(TY) := c^*(TY)^{-1}$ is the Segre class of the tangent bundle.

The twisted class $f^* s(TY) \cap c_*(\alpha)$ shall be called the *Ginzburg–Chern class* of α ([Gi1,Gi2] and [Y7,Y8]). Here, the above equality needs a bit of explanation. The left-hand-side $\gamma(\alpha)$ belongs to the bivariant homology group $\mathbb{H}(X \xrightarrow{f} Y)$ and the right-hand-side $f^* s(TY) \cap c_*(\alpha)$ belongs to the homology group $H_*^{BM}(X)$, and this equality is up to the isomorphism

$$\mathbb{H}(X \xrightarrow{f} Y) \xrightarrow[\cong]{\bullet [Y]} \mathbb{H}(X \to pt) \xrightarrow[\cong]{\mathcal{A}} H_*^{BM}(X) ,$$

where the first isomorphism is the bivariant product with the fundamental class $[Y]$ and the second isomorphism $\mathcal{A}$ is the Alexander duality map. Since we usually identify $\mathbb{H}(X \to pt)$ as $H_*^{BM}(X)$ via this Alexander duality, we ignore this Alexander duality isomorphism, unless we have to mention it. Hence we have

$$\gamma(\alpha) \bullet [Y] = f^* s(TY) \cap c_*(\alpha).$$

We remark that this formula follows from the *simple but crucial* observation that

$$\gamma_f(\alpha) \bullet \gamma_{Y \to pt}(\mathbb{1}_Y) = \gamma_{X \to pt}(\alpha)$$

and the fact that $\gamma_{Y \to pt}$ is nothing but MacPherson's Chern class transformation c_*. And in [BSY1] the above theorem is furthermore generalized to the case when the target variety can be singular but is "like a manifold".

Definition 9.7. (cf. [BM]) Let A be a Noetherian ring. A complex variety X is called an *A-homology manifold (of dimension $2n$)* or is said to be *A-smooth* if for all $x \in X$

$$H_i(X, X \setminus x; A) = \begin{cases} A & i = 2n \\ 0 & \text{otherwise.} \end{cases}$$

In this case X has to be locally pure n-dimensional, where we consider n as a locally constant function on X. Just look at the regular part of X, because a pure n-dimensional complex manifold is a homology manifold of dimension $2n$. Moreover the local orientation system or_X with stalk $or_{X,x} = H_{2n}(X, X \setminus x; A) \simeq A_X$ is then already trivial (on each connected component of X) so that X becomes an *oriented A-homology manifold.*

Example 9.8. If $A = \mathbb{Z}$, a $\mathbb{Z}$-homology manifold is called simply a *homology manifold* (cf. [MiSt]). There are singular complex varieties which are homology manifolds. Such examples are (products of) suitable singular hypersurfaces with isolated singualrities (see [Mi2]). If $A = \mathbb{Q}$, a $\mathbb{Q}$-manifold is called a *rational homology manifold.* As remarked in [BM, §1.4 Rational homology manifolds], examples of rational homology manifolds include surfaces with Kleinian singularities, the moduli space for curves of a given genus, and more generally *Satake's V-manifolds* or *orbifolds*. In particular, the quotient of a nonsingular variety by a finite group is a rational homology manifold.

Theorem 9.9. *Let Y be a complex analytic variety which is an* oriented A-homology manifold *for some commutative Noetherian ring A. If there exists a bivariant Chern class transformation $\gamma : \mathbb{F} \otimes A \to \mathbb{H} \otimes A$, then for any morphism $f : X \to Y$ the bivariant Chern class $\gamma_f : \mathbb{F}(X \xrightarrow{f} Y) \otimes A \to \mathbb{H}(X \xrightarrow{f} Y) \otimes A$ is uniquely determined and it is described by*

$$\gamma_f(\alpha) = f^* c^*(Y)^{-1} \cap c_*(\alpha) .$$

Here $c^*(Y)$ is the unique cohomology class *such that $c_*(1_Y) = c^*(Y) \cap [Y]$. (Note that $c^*(Y)$ is invertible.)*

When Y is nonsingular, we see that the cohomolgy class $c^*(Y)$ is nothing but the total Chern class $c^*(TY)$ of the tangent bundle TY, hence the inverse $c^*(Y)^{-1}$ is the total Segre class $s(TY)$. Therefore the twisted class $f^* c^*(Y)^{-1} \cap c_*(\alpha)$ shall also be called the *Ginzburg–Chern class* of α and still denoted by $\gamma^{\mathrm{Gin}}(\alpha)$. Note that we also have in this more general context the isomorphism

$$\mathbb{H}(X \xrightarrow{f} Y) \otimes A \xrightarrow[\cong]{\bullet [Y]} \mathbb{H}(X \to pt) \otimes A \xrightarrow[\cong]{A} H_*^{BM}(X) \otimes A ,$$

since for an oriented A-homology manifold Y the fundamental class $[Y] \in H_*^{BM}(X) \otimes A \simeq \mathbb{H}(X \to pt) \otimes A$ is a *strong orientation* in the sense of bivariant theories (compare [BSY1]).

Existence and uniqueness of bivariant characteristic classes. Note that the proof of Theorem 9.9 also applies in the real (semi-)algebraic or subanalytic context to a bivariant *Stiefel–Whitney class transformation* $\gamma : \mathbb{F}^{mod2} \to \mathbb{H}^{mod2}$ (with the obvious modification of the notations from c^*, c_* to w^*, w_*). In a similar manner, we can show the following theorem, which is an extended version of [Y5, Theorem (3.7)]:

Theorem 9.10. *The Grothendieck transformation from the bivariant algebraic K-theory $\mathbb{K}_{\mathrm{alg}}$ of perfect complexes*

$$\tau : \mathbb{K}_{\mathrm{alg}} \to \mathbb{H}_{\mathbb{Q}}$$

*constructed in [FM, Part II] is unique on morphisms whose target varieties are rational homology manifolds. Explicitly, for a bivariant element $\alpha \in \mathbb{K}_{\mathrm{alg}}(X \xrightarrow{f}$

910

Y) with Y being a rational homology manifold

$$\tau(\alpha) = f^*td^*(Y)^{-1} \cap td_*(\alpha \bullet [\mathcal{O}_Y]).$$

Here $[\mathcal{O}_Y] \in \mathbb{K}_{\mathrm{alg}*}(Y) \simeq \mathbf{G}_0(Y)$ is the class of the structure sheaf and the associated covariant transformation $\tau_* : \mathbb{K}_{\mathrm{alg}*}(\ ) \simeq \mathbf{G}_0(\ ) \to H_*^{BM}(\ ;\mathbb{Q})$ is Baum–Fulton–MacPherson's Todd class transformation td_* constructed in [BFM1]. Moreover $td^*(Y) \in H^*(Y;\mathbb{Q})$ is the Poincaré dual of the Todd class $td_*(Y) := td_*([\mathcal{O}_Y])$, which is invertible.

Conversely we ask ourselves whether the above Ginzburg–Chern class becomes a Grothendieck transformation for morphisms whose target varieties are oriented A - homology manifolds.

Theorem 9.11. *For a morphism of complex analytic varieties $f : X \to Y$ with Y an oriented A-homology manifold, we define $\overline{\mathbb{F}}(X \xrightarrow{f} Y)$ to be the set of all constructible functions $\alpha \in F(X)$ satisfying the following two conditions ($\sharp$) and ($\flat$) : for any fiber square*

$$
\begin{array}{ccc}
X' & \xrightarrow{\ g'\ } & X \\
{\scriptstyle f'}\downarrow & & \downarrow{\scriptstyle f} \\
Y' & \xrightarrow{\ g\ } & Y,
\end{array}
$$

with Y' an oriented A-homology manifold the following equalities hold:
($\sharp$) for any constructible function $\beta' \in F(Y')$:

$$\gamma^{\mathrm{Gin}}(g^\star\alpha \bullet \beta') = \gamma^{\mathrm{Gin}}(g^\star\alpha) \bullet \gamma^{\mathrm{Gin}}(\beta'),$$

($\flat$)

$$\gamma^{\mathrm{Gin}}(g^\star\alpha) = g^\star\gamma^{\mathrm{Gin}}(\alpha).$$

Then $\overline{\mathbb{F}}$ becomes a bivariant theory with the same operations as in $s\mathbb{F}$ and furthermore the transformation

$$\gamma^{\mathrm{Gin}} : \overline{\mathbb{F}} \to \mathbb{H}$$

is well-defined and becomes the unique Grothendieck transformation satisfying that γ^{Gin} for morphisms to a point is MacPherson's Chern class transformation $c_* : F \to H_*$. And also $\overline{\mathbb{F}}(X \to pt) = F(X)$.

The proof of the theorem is the same as in [Y9], in which the case when the target variety Y is nonsingular is treated. Note that to prove $\overline{\mathbb{F}}(X \to pt) = F(X)$ we need the *cross product formula* or *multiplicativity* of MacPherson's Chern class transformation c_* due to *Kwieciński* [Kw1] (cf. [KY]), i.e. the commutativity of the following diagram:

$$
\begin{array}{ccc}
F(X) \times F(Y) & \xrightarrow{\times} & F(X \times Y) \\
{\scriptstyle c_* \times c_*}\downarrow & & \downarrow{\scriptstyle c_*} \\
H_*^{BM}(X;\mathbb{Z}) \times H_*^{BM}(Y;\mathbb{Z}) & \xrightarrow{\times} & H_*^{BM}(X \times Y;\mathbb{Z}) \, .
\end{array}
$$

The cross product formula for Stiefel–Whitney classes in the *real algebraic context* can be shown similarly by using "resolution of singularities", or the corresponding product formula for "characteristic cycles" of constructible functions so that a variant of this theorem also works in the real algebraic context.

And for a *much more general version* of Theorem 9.11, see [Sch2].

The above theorem led us to another *uniqueness theorem*, which in a sense gives a positive solution to the general uniqueness problem concerning Grothendieck transformations posed in [FM, §10 Open Problems]. For more details, see [BSY2].

Theorem 9.12. *We define*

$$
\widetilde{\mathbb{F}}(X \xrightarrow{f} Y)
$$

to be the set consisting of all $\alpha \in s\mathbb{F}(X \xrightarrow{f} Y)$ satisfying the following condition: there exists a bivariant class $B_\alpha \in \mathbb{H}(X \xrightarrow{f} Y)$ such that for any base change $g : Y' \to Y$ (without any requirement) of an independent square

$$
\begin{array}{ccc}
X' & \xrightarrow{g'} & X \\
{\scriptstyle f'}\downarrow & & \downarrow{\scriptstyle f} \\
Y' & \xrightarrow{g} & Y,
\end{array}
$$

and for any $\beta' \in F(Y')$ the following equality holds:

$$
c_*(g^*\alpha \bullet \beta') = g^* B_\alpha \bullet c_*(\beta').
$$

Then $\widetilde{\mathbb{F}}$ is a bivariant theory. Furthermore $\widetilde{\mathbb{F}}(X \to pt) = F(X)$.

The above bivariant class B_α should ideally be the unique bivariant Chern class of α. However, so far we still do not know if it is the case or not. So, provisionally we call B_α *a pseudo-bivariant c_*-class of α*.

Example 9.13 (VRR for smooth morphisms). Let $f : X \to Y$ be a *smooth* morphism of possibly singular varieties. Then we have

$$
\mathbb{1}_f := 1_X \in \widetilde{\mathbb{F}}(X \xrightarrow{f} Y)
$$

912

with $c^*(T_f) \bullet [f]$ being a pseudo-bivariant c_*-class of $\mathbb{1}_f$. Here T_f is the vector bundle of tangent spaces of fibers of f, and $[f] \in \mathbb{H}(X \xrightarrow{f} Y)$ is the *canonical orientation* of the smooth morphism f. Then as in Theorem 9.12 we have for $\beta' \in F(Y')$:

$$\begin{aligned}
c_*(g^*\mathbb{1}_f \bullet \beta') &= c_*(f'^*\beta') \\
&= c^*(T_{f'}) \cap f'^! c_*(\beta') \\
&= c^*(T_{f'}) \bullet [f'] \bullet c_*(\beta') \\
&= g^* c^*(T_f) \bullet g^*[f] \bullet c_*(\beta') \\
&= g^*(c^*(T_f) \bullet [f]) \bullet c_*(\beta').
\end{aligned}$$

Here $f'^! = [f']\bullet : H_*^{BM}(Y') \simeq \mathbb{H}_*(Y') \to \mathbb{H}_*(X') \simeq H_*^{BM}(X')$ is the *smooth pullback* in Borel–Moore homology, and the equality

$$c_*(f'^*\beta') = c^*(T_{f'}) \cap f'^! c_*(\beta') \tag{9.14}$$

is the so-called *Verdier–Riemann–Roch theorem* for the smooth morphism f' and the Chern class transformation c_* (compare [FM,Sch1,Y4]).

In order to remedy this unpleasant possible non-uniqueness of the bivariant class B_α above, we set

$$\mathbb{PH}(X \xrightarrow{f} Y) :=$$

$$\left\{ B \in \mathbb{H}(X \xrightarrow{f} Y) | B \text{ is a pseudo-bivariant } c_*\text{-class of some } \alpha \in \widetilde{\mathbb{F}}(X \xrightarrow{f} Y) \right\}$$

to be the set of all pseudo-bivariant c_*-classes for the morphism $f : X \to Y$. It is clear that $\mathbb{PH}$ is a *bivariant subtheory* of $\mathbb{H}$, i.e, it is a subgroup stable under the three bivariant operations. Then we define

$$\widetilde{\mathbb{H}}(X \xrightarrow{f} Y) := \mathbb{PH}(X \xrightarrow{f} Y)/\sim$$

where the relation $\sim$ is defined by

$$B \sim B' \iff g^*B \bullet c_*(\beta') = g^*B' \bullet c_*(\beta')$$

for all independent squares with $g : Y' \to Y$ and all $\beta' \in F(Y')$. Certainly the relation $\sim$ is an equivalence relation. In other words, with this identification we want to make possibly many pseudo-bivariant c_*-classes into one unique bivariant c_*-class. Indeed we have

Theorem 9.15. $\widetilde{\mathbb{H}}(X \xrightarrow{f} Y)$ *is an Abelian group and* $\widetilde{\mathbb{H}}$ *is a bivariant theory with the canonical operations induced from those of* $\mathbb{H}$. *Furthermore we have*

$$\widetilde{\mathbb{H}}(X \to pt) = \text{Image}\Big(c_* : F(X) \to H_*^{BM}(X)\Big).$$

And we have the following theorem

Theorem 9.16. *There exists a unique Grothendieck transformation*

$$\widetilde{\gamma} : \widetilde{\mathbb{F}} \to \widetilde{\mathbb{H}}$$

whose associated covariant transformation is $c_* : F \to \mathrm{Im}(c_*)$, where

$$\mathrm{Im}(c_*)(X) := \mathrm{Image}\Big(c_* : F(X) \to H_*^{BM}(X) \Big).$$

Remark 9.17. As mentioned above, a key for the above argument is the fact that $c_*(\alpha) = \gamma(\alpha) \bullet c_*(\mathbb{1}_Y)$. So, putting it very vaguely, the bivariant class $\gamma(\alpha)$ could be said to be a kind of "$c_*(\alpha)$ divided by $c_*(\mathbb{1}_Y)$", whatever it is meant to be. In our previous paper [Y5] we posed the problem of whether or not there is a reasonable bivariant homology theory so that such a "quotient"

$$\frac{c_*(\alpha)}{c_*(\mathbb{1}_Y)}$$

is well-defined. The above theory $\widetilde{\mathbb{H}}$ is in a sense a positive answer to this problem.

The above construction works for a more general situation such as
(1) there exists a natural transformation $\tau_* : F_*(X) \to H_*(X)$ between two covariant functors F_* and H_* (covariant with respect to proper maps) such that $F_*(pt)$ and $H_*(pt)$ are commutative rings with unit and such that τ_* maps the unit to the unit,
(2) there are two bivariant theories $\mathbb{F}$ and $\mathbb{H}$ such that the associated covariant theories are

$$\mathbb{F}(X \to pt) = F_*(X) \quad \text{and} \quad \mathbb{H}(X \to pt) = H_*^{BM}(X),$$

(3) τ_* commutes with the bivariant exterior products, i.e., the following diagram commutes

$$
\begin{array}{ccc}
F_*(X) \times F_*(Y) & \xrightarrow{\;\;\times\;\;} & F_*(X \times Y) \\
\Big\downarrow{\scriptstyle \tau_* \times \tau_*} & & \Big\downarrow{\scriptstyle \tau_*} \\
H_*^{BM}(X) \times H_*^{BM}(Y) & \xrightarrow{\;\;\times\;\;} & H_*^{BM}(X \times Y).
\end{array}
$$

Here we assume that for $X = Y = \{pt\}$ a point this exterior product agrees with the given ring structure.

Certainly this construction works for the previous *motivic Chern class transformation*

$$mC_* : K_0(\mathcal{V}/\ \ ) \to \mathbf{G}_0(\ \ ) \otimes \mathbb{Z}[y]$$

and the *motivic Hirzebruch class transformation*

$$T_{y*} : K_0(\mathcal{V}/\ \ ) \to H_*^{BM}(\ \ ) \otimes \mathbb{Q}[y].$$

Indeed, the bivariant theory for $K_0(\mathcal{V}/\ \ )$ is the simple bivariant theory

$$s\mathbf{K}_0(X \to Y) := K_0(\mathcal{V}/X),$$

the bivariant theory for $\mathbf{G}_0(\ \ ) \otimes \mathbb{Z}[y]$ is Fulton–MacPherson's bivariant algebraic K-theory $\mathbb{K}_{\mathrm{alg}}$ tensored with $\mathbb{Z}[y]$, and the bivariant theory for $H_*(\ \ ) \otimes \mathbb{Q}[y]$ is of course Fulton–MacPherson's bivariant homology theory $\mathbb{H}$ tensored with $\mathbb{Q}[y]$. It also applies in the real algebraic context to the *Stiefel–Whitney class transformation*

$$w_* : F^{mod2}(\ \ ) \to H_*^{BM}(\ \ ; \mathbb{Z}_2)$$

by using the simple bivariant theory $s\mathbb{F}^{mod2}$ of $\mathbb{Z}_2$-valued real algebraically constructible functions.

Remark 9.18. Let $f : X \to Y$ be a *smooth* morphism of possible singular varieties. Then also Example 9.13 works in this context, with

$$\mathbb{1}_f := \mathbb{1}_X = [id_X] \in s\mathbf{K}_0(X \xrightarrow{f} Y) \quad \text{or} \quad \mathbb{1}_f := 1_X \in s\mathbb{F}^{mod2}(X \xrightarrow{f} Y),$$

and $c\ell^*(T_f) \bullet [f]$ being a pseudo-bivariant class of $\mathbb{1}_f$ for $c\ell^*(T_f) = \lambda_y(T_f^*), \widetilde{td}_{(y)}(T_f)$ or $w^*(T_f)$. Here the corresponding *Verdier–Riemann–Roch theorem* for the smooth morphism f follows for the motivic characteristic classes mC_* and T_{y*} from [BSY3, Corollary 2.1 and Corollary 3.1]. For the Stiefel–Whitney class transformation w_* it can be shown as for Chern classes by using "resolution of singularities" or "characteristic cycles of constructible functions".

This *Verdier–Riemann–Roch theorem for smooth morphisms* is also very important for the definition of *G-equivariant characteristic class transformations* in the equivariant algebraic context with G a reductive linear algebraic group. Here we refer to [EG1,EG2,BZ] for the *equivariant Todd class transformation* td_*^G, and to [Oh] for the *equivariant Chern class transformation* c_*^G. In fact, in future work we will construct in this equivariant algebraic context equivariant versions mC_*^G and T_{y*}^G of our motivic characteristic classes, together with the equivariant version of Theorem 8.5, relating T_{-1*}^G with c_*^G and T_{0*}^G with td_*^G.

Bivariant L-classes. At the moment we have no bivariant version of the L-class transformation L_* with values in bivariant homology

$$L_* : \Omega(X) \to H_*(X, \mathbb{Q}),$$

since we do not know a suitable bivariant theory, whose associated covariant theory is the cobordism group $\Omega(\ )$ of selfdual constructible sheaf complexes. Note that in this case we *cannot* define a *simple bivariant theory $s\Omega$*. Of course the Grothendieck group of constructible sheaf complexes $K_c(\ )$ satisfies the properties (SB1-4) with respect to the induced proper push down f_*, pullback f^* and tensor product $\otimes$ so that one gets a simple bivariant theory $s\mathbf{K}_c$. But the problem is that f^* and $\otimes$ do *not* commute with duality in general so that this approach doesn't apply to $\Omega(\ )$.

A similar problem appears in the context of real semialgebraic and subanalytic geometry for the group $F_{Eu}^{mod2}(\ )$ of $\mathbb{Z}_2$-valued constructible functions satisfying the *mod 2 local Euler condition* (for a constant map), which also can be interpreted as a "duality" condition (compare [Sch3, p.135 and Remark 5.4.4, p.367]). This group (or condition) is also not stable under general pullback or product so that one *cannot* define a simple bivariant theory $s\mathbb{F}_{Eu}^{mod2}$ in this context (compareable to $s\mathbb{F}^{mod2}$ in the real algebraic context). Nevertheless one can define a *Stiefel–Whitney class transformation*

$$w_* : F_{Eu}^{mod2}(\ ) \to H_*^{BM}(\ ;\mathbb{Z}_2)$$

with the help of "characteristic cycles of constructible functions" (compare [FuMC]), which is *multiplicative for exterior products* and satisfies the *Verdier–Riemann–Roch theorem for smooth morphisms*.

Similarly one can define in the complex algebraic or analytic context an *exterior product and smooth pullback* for the cobordism group $\Omega(\ )$ of selfdual constructible sheaf complexes (compare [BSY3]), and the L-class transformation L_* is also *multiplicative* by an argument similarly as in the recent paper [Wo, p.26, Proposition 5.16]. Also the corresponding Verdier–Riemann–Roch theorem for smooth morphisms is true, as will be explained in a forthcoming paper. Of course on the image of the transformation $\omega : K_0(\mathcal{V}/\ ) \to \Omega(\ )$ this VRR theorem also follows from Theorem 8.5 (compare [BSY3]).

Then in both these cases, L-class and Stiefel–Whitney class transformations, we can apply the results of [Y6] to get at least bivariant versions of these theories for the corresponding *operational bivariant theories*.

10. Characteristic classes of proalgebraic varieties

A *pro-algebraic variety* is defined to be a projective system of complex algebraic varieties and a *proalgebraic variety* is defined to be the projective limit of a pro-algebraic variety. Proalgebraic varieties are the main objects in [Grom]. A pro-category was first introduced by A. Grothendieck [Grot1] and it was used to develope the Etale Homotopy Theory [AM] and Shape Theory (e.g., see [Bor], [MaSe], etc.) and so on. In [Grom] M. Gromov investigated the *surjunctivity*, i.e. being either surjective or non-injective, in the category of proalgebraic varieties. The original or classical surjunctivity theorem is the so-called *Ax' Theorem* [Ax], saying that every regular selfmapping of a complex algebraic variety is surjunctive; thus if it is injective then it has to be surjective.

A very simple example of a proalgebraic variety is the Cartesian product $X^{\mathbb{N}}$ of countable infinitely many copies of a complex algebraic variety X, which is one of the main objects treated in [Grom]. Then, what would be the *"Chern–Schwartz–MacPherson class"* of $X^{\mathbb{N}}$? In particular, what would be the *"Euler–Poincaré characteristic"* of $X^{\mathbb{N}}$? This simple question led us to a study of characteristic classes of proalgebraic varieties and it naturally led us to the so-called *motivic measures* (see [Y10,Y11]). The motivic measures/integrations have been actively studied by many people (e.g., see [Cr], [DL1], [DL2], [Kon], [Lo], [Ve2] etc.).

In a general set-up one can deal with the so-called *bifunctors*. The bifunctors which we consider are bifunctors $\mathcal{F} : \mathcal{C} \to \mathcal{A}$ from a category $\mathcal{C}$ to the category $\mathcal{A}$ of abelian groups, i.e., $\mathcal{F}$ is a pair $(\mathcal{F}_*, \mathcal{F}^*)$ of a *covariant functor* $\mathcal{F}_*$ and a *contravariant functor* $\mathcal{F}^*$ such that $\mathcal{F}_*(X) = \mathcal{F}^*(X)$ for any object X. Unless some confusion occurs, we just denote $\mathcal{F}(X)$ for $\mathcal{F}_*(X) = \mathcal{F}^*(X)$. A typical example is the constructible function functor $F(X)$. Furthermore we assume that for a final object $pt \in Obj(\mathcal{C})$, $\mathcal{F}(pt)$ is a commutative ring $\mathcal{R}$ with a unit. The morphism from an object X to a final object pt shall be denoted by $\pi_X : X \to pt$. Then the covariance of the bifunctor $\mathcal{F}$ induces the homomorphism $\pi_{X*} := \mathcal{F}(\pi_X) : \mathcal{F}(X) \to \mathcal{F}(pt) = \mathcal{R}$, which shall be denoted by

$$\chi_{\mathcal{F}} : \mathcal{F}(X) \to \mathcal{R}$$

and called the $\mathcal{F}$-*characteristic*, just mimicking the Euler–Poincaré characteristic (with compact support) $\chi : F(X) \to \mathbb{Z}$ in the case when $\mathcal{F} = F$.

Let $X_\infty = \varprojlim_{\lambda \in \Lambda} \left\{ X_\lambda, \pi_{\lambda\mu} : X_\mu \to X_\lambda \right\}$ be a proalgebraic variety. Then we define

$$\mathcal{F}^{\mathrm{ind}}(X_\infty) := \varinjlim_{\lambda \in \Lambda} \left\{ \mathcal{F}(X_\lambda), \pi_{\lambda\mu}{}^* : \mathcal{F}(X_\lambda) \to \mathcal{F}(X_\mu)(\lambda < \mu) \right\},$$

which *may not belong to* the category $\mathcal{A}$. Another finer one can be defined as follows. Let $P = \{p_{\lambda\mu}\}$ be a *projective system* of elements of $\mathcal{R}$ by the directed set Λ, i.e., a set such that $p_{\lambda\lambda} = 1$ (the unit) and $p_{\lambda\mu} \cdot p_{\mu\nu} = p_{\lambda\nu}$ $(\lambda < \mu < \nu)$. For each $\lambda \in \Lambda$ the *subobject $\mathcal{F}_P^{\mathrm{st}}(X_\lambda)$ of $\chi_{\mathcal{F}}$-stable elements* in $\mathcal{F}(X_\lambda)$ is defined to be

$$\mathcal{F}_P^{\mathrm{st}}(X_\lambda)$$
$$:= \left\{ \alpha_\lambda \in \mathcal{F}(X_\lambda) \mid \chi_{\mathcal{F}}\left(\pi_{\lambda\mu}{}^* \alpha_\lambda\right) = p_{\lambda\mu} \cdot \chi_{\mathcal{F}}(\alpha_\lambda) \text{ for any } \mu \text{ such that } \lambda < \mu \right\}.$$

The *inductive limit*

$$\varinjlim_{\Lambda} \left\{ \mathcal{F}_P^{\mathrm{st}}(X_\lambda), \quad \pi_{\lambda\mu}{}^* : \mathcal{F}_P^{\mathrm{st}}(X_\lambda) \to \mathcal{F}_P^{\mathrm{st}}(X_\mu) \quad (\lambda < \mu) \right\}$$

considered for a proalgebraic variety $X_\infty = \varprojlim_{\lambda \in \Lambda} X_\lambda$ is denoted by

$$\mathcal{F}_P^{\mathrm{st.ind}}(X_\infty).$$

Of course this definition is not intrinsic to the proalgebraic variety X_∞, but depends on the given projective system $\left\{ X_\lambda, \pi_{\lambda\mu} : X_\mu \to X_\lambda \right\}$. But for simplicity we use this notation. Our key observation, which is an application of standard facts on indcutive systems and limits, is the following:

Theorem 10.1. *(i) For a proalgebraic variety $X_\infty = \varprojlim_{\lambda \in \Lambda} \left\{ X_\lambda, \pi_{\lambda\mu} : X_\mu \to X_\lambda \right\}$ and a projective system $P = \{p_{\lambda\mu}\}$ of elements of $\mathcal{R}$, we have the homomorphism*

$$\chi_{\mathcal{F}}^{\mathrm{ind}} : \mathcal{F}_P^{\mathrm{st.ind}}(X_\infty) \to \varinjlim_{\lambda \in \Lambda} \left\{ \times p_{\lambda\mu} : \mathcal{R} \to \mathcal{R} \right\},$$

which is called the proalgebraic $\mathcal{F}$-characteristic homomorphism.

(ii) Assume $\Lambda = \mathbb{N}$. For a proalgebraic variety $X_\infty = \varprojlim_{n \in \mathbb{N}} \left\{ X_n, \pi_{nm} : X_m \to X_n \right\}$ and a projective system $P = \{p_{nm}\}$ of elements of $\mathcal{R}$, the proalgebraic $\mathcal{F}$-characteristic homomorphism $\chi_{\mathcal{F}}^{\mathrm{ind}} : \mathcal{F}_P^{\mathrm{st.ind}}(X_\infty) \to \varinjlim_n \left\{ \times p_{nm} : \mathcal{R} \to \mathcal{R} \right\}$ is realized as the homomorphism

$$\widetilde{\chi_{\mathcal{F}}^{\mathrm{ind}}} : \mathcal{F}_P^{\mathrm{st.ind}}(X_\infty) \to \mathcal{R}_P$$

defined by

$$\widetilde{\chi_{\mathcal{F}}^{\mathrm{ind}}}\left([\alpha_n]\right) := \frac{\chi_{\mathcal{F}}(\alpha_n)}{p_{01} \cdot p_{12} \cdot p_{23} \cdots p_{(n-1)n}}.$$

918

Here $p_{01} := 1$ and $\mathcal{R}_P$ is the ring $\mathcal{R}_S$ of fractions of $\mathcal{R}$ with respect to the multiplicatively closed set S consisting of all the finite products of powers of elements in P.

(iii) In particular, in the case when the above projective system $P = \{p^s\}$ consists of powers of an element p, we get the homomorphism

$$\widetilde{\chi_{\mathcal{F}}^{\mathrm{ind}}} : \mathcal{F}_P^{\mathrm{st.ind}}(X_\infty) \to \mathcal{R}\left[\frac{1}{p}\right]$$

defined by

$$\widetilde{\chi_{\mathcal{F}}^{\mathrm{ind}}}\left([\alpha_n]\right) := \frac{\chi_{\mathcal{F}}(\alpha_n)}{p^{n-1}}.$$

Here $\mathcal{R}\left[\frac{1}{p}\right]$ is the localization by the multiplicatively closed set $S := \{p^s | s \in \mathbb{N}_0\}$.

Note that $\mathcal{R}_S$ or $\mathcal{R}\left[\frac{1}{p}\right]$ is the zero ring in the case when $0 \in S$ for the corresponding muliplicatively closed set S. A typical example for the above theorem is the following.

Example 10.2. Let $X_\infty = \varprojlim_{n \in \mathbb{N}} \left\{X_n, \pi_{nm} : X_m \to X_n\right\}$ be a proalgebraic variety such that for each n the structure morphism $\pi_{n,n+1} : X_{n+1} \to X_n$ satisfies the condition that the Euler–Poincaré characteristics of the fibers of $\pi_{n,n+1}$ are non-zero (which implies the surjectivity of the morphism $\pi_{n,n+1}$) and constant; for example, $\pi_{n,n+1} : X_{n+1} \to X_n$ is a locally trivial fiber bundle with fiber variety being F_n and $\chi(F_n) \neq 0$ Let us denote the constant Euler–Poincaré characteristic of the fibers of the morphism $\pi_{n,n+1} : X_{n+1} \to X_n$ by e_n and we set $e_0 := 1$. Then we get the canonical proalgebraic Euler–Poincaré characteristic homomorphism

$$\chi^{\mathrm{ind}} : F^{\mathrm{ind}}(X_\infty) \to \mathbb{Q}$$

described by

$$\chi^{\mathrm{ind}}\left([\alpha_n]\right) = \frac{\chi(\alpha_n)}{e_0 \cdot e_1 \cdot e_2 \cdots e_{n-1}}.$$

In particular, if the Euler–Poincaré characteristics e_n are all the same, say $e_n = e$ for any n, then the canonical proalgebraic Euler–Poincaré characteristic homomorphism $\chi^{\mathrm{ind}} : F^{\mathrm{ind}}(X_\infty) \to \mathbb{Q}$ is described by $\chi^{\mathrm{ind}}\left([\alpha_n]\right) = \frac{\chi(\alpha_n)}{e^{n-1}}$, and furthermore the target ring $\mathbb{Q}$ can be replaced by the ring $\mathbb{Z}\left[\frac{1}{e}\right]$.

Note that this example applies especially to the Cartesian product $X^{\mathbb{N}}$ of countable infinitely many copies of a complex algebraic variety X with $\chi(X) \neq 0$. In fact this example of Cartesian products is a special case of the following more general example:

Example 10.3. We make the following additional assumptions for our bifunctor:

(1) The contravariant functor $\mathcal{F}^*$ takes values in the category of *commutative rings with unit*. The corresponding unit in $\mathcal{F}(X)$ is denoted by $\mathbb{1}_X$, and $\mathcal{F}(X)$ becomes an $\mathcal{R} := \mathcal{F}(pt)$-algebra by the pullback for $\pi_X : X \to pt$.

(2) $\mathcal{F}^*$ and $\mathcal{F}_*$ are related for a morphism $f : X \to Y$ by the *projection formula*

$$f_*(\alpha \cdot f^*\beta) = f_*(\alpha) \cdot \beta \qquad \text{for all } \alpha \in \mathcal{F}(X) \text{ and } \beta \in \mathcal{F}(Y)$$

so that $f_* : \mathcal{F}(X) \to \mathcal{F}(Y)$ is $\mathcal{F}(Y)$- and $\mathcal{R}$-linear. (This is just a special case of our simple bivariant theories, where all morphisms are "proper" and only the "trivial fiber squares" are "independent".)

Consider a proalgebraic variety $X_\infty = \varprojlim_{n \in \mathbb{N}} \{X_n, \pi_{nm} : X_m \to X_n\}$ such that for each n the structure morphism $\pi_{n,n+1} : X_{n+1} \to X_n$ satisfies the condition

$$\pi_{n,n+1*}(\mathbb{1}_{X_{n+1}}) = e_n \cdot \mathbb{1}_{X_n} \in \mathcal{F}(X_n) \quad \text{for some } e_n \in \mathcal{R}, \text{ with } e_0 := \mathbb{1}_{pt}.$$

Then we get the canonical proalgebraic $\mathcal{F}$-characteristic homomorphisms

$$\chi^{\mathrm{ind}}_{\mathcal{F},X_1} : \mathcal{F}^{\mathrm{ind}}(X_\infty) \to \mathcal{F}(X_1)_E \quad \text{and} \quad \chi^{\mathrm{ind}}_{\mathcal{F}} : \mathcal{F}^{\mathrm{ind}}(X_\infty) \to \mathcal{R}_E$$

described by

$$\chi^{\mathrm{ind}}_{\mathcal{F},X_1}([\alpha_n]) = \frac{\pi_{1,n*}(\alpha_n)}{e_0 \cdot e_1 \cdot e_2 \cdots e_{n-1}} \quad \text{and} \quad \chi^{\mathrm{ind}}_{\mathcal{F}}([\alpha_n]) = \frac{\chi(\alpha_n)}{e_0 \cdot e_1 \cdot e_2 \cdots e_{n-1}}.$$

Here $\mathcal{R}_E$ (or $\mathcal{F}(X_1)_E$) is the ring of fractions of $\mathcal{R}$ with respect to the multiplicatively closed set consisting of all the finite products of powers of the elements e_i (or their pullbacks to X_1).

Consider a bifunctor as in example 10.3, with $f : X \to Y$ being a morphism such $f_*(\mathbb{1}_X) = e_f \cdot \mathbb{1}_Y$ for some $e_f \in \mathcal{R}$. Then one gets that for any $\alpha \in \mathcal{F}(Y)$:

$$f_* f^* \alpha = f_*(\mathbb{1}_X \cdot f^*\alpha) = e_f \cdot \alpha,$$

so that for any morphism $g : Y \to Z$ (e.g., $g = \pi_Y : Y \to pt$):

$$(g \circ f)_*(f^*\alpha) = g_*(f_* f^*\alpha)$$
$$= g_*(e_f \cdot \alpha)$$
$$= e_f \cdot g_*(\alpha).$$

Hence, if we set in the context of the example

$$p_{nm} = \begin{cases} 1 & n = m \\ e_n \cdot e_{n+1} \cdots e_{m-1} & n < m, \end{cases}$$

then $P := \{p_{nm}\}$ is a projective system and $\mathcal{F}_P^{\mathrm{st.ind}}(X_\infty) = \mathcal{F}^{\mathrm{ind}}(X_\infty)$ for both notions of Euler characteristics working over the base space X_1 or over pt. Thus the above description of $\chi_{\mathcal{F},X_1}^{\mathrm{ind}}$ and $\chi_{\mathcal{F}}^{\mathrm{ind}}$ follows from Theorem 10.1.

A "motivic" version of the Euler–Poincaré characteristic $\chi : F(X) \to \mathbb{Z}$ is the homomorphism $\Gamma_X : F(X) \to K_0(\mathcal{V}/X)$ "tautologically" defined by

$$\Gamma_X\left(\sum_W a_W \mathbb{1}_W\right) := \sum_W a_W [W \hookrightarrow X],$$

or better is the composite $\Gamma := \pi_{X*} \circ \Gamma_X : F(X) \to K_0(\mathcal{V})$. Note that Γ_X commutes with the pullback f^* (but not with the pushforward f_*). Then we get the following theorem, which is a generalization of the (naïve) motivic measure:

Theorem 10.4. *(i) For a proalgebraic variety* $X_\infty = \varprojlim_{\lambda \in \Lambda}\left\{X_\lambda, \pi_{\lambda\mu} : X_\mu \to X_\lambda\right\}$ *and a projective system* $G = \{\gamma_{\lambda\mu}\}$ *of Grothendieck classes, we get the* proalgebraic Grothendieck class homomorphism

$$\Gamma^{\mathrm{ind}} : F_G^{\mathrm{st.ind}}(X_\infty) \to \varinjlim_{\lambda \in \Lambda}\left\{\times \gamma_{\lambda\mu} : K_0(\mathcal{V}) \to K_0(\mathcal{V})\right\}.$$

(ii) Assume $\Lambda = \mathbb{N}$. *For a proalgebraic variety* $X_\infty = \varprojlim_{n \in \mathbb{N}}\left\{X_n, \pi_{nm} : X_m \to X_n\right\}$ *and a projective system* $G = \{\gamma_{n,m}\}$ *of Grothendieck classes, we have the following canonical proalgebraic Grothendieck class homomorphism*

$$\widetilde{\Gamma^{\mathrm{ind}}} : F_G^{\mathrm{st.ind}}(X_\infty) \to K_0(\mathcal{V})_G$$

which is defined by

$$\widetilde{\Gamma^{\mathrm{ind}}}\left([\alpha_n]\right) := \frac{\Gamma(\alpha_n)}{\gamma_{01} \cdot \gamma_{12} \cdot \gamma_{23} \cdots \gamma_{(n-1)n}}.$$

Here we set $\gamma_{01} := \mathbb{1}$ *and* $K_0(\mathcal{V})_G$ *is the ring of fractions of* $K_0(\mathcal{V})$ *with respect to the multiplicatively closed set consisting of finite products of powers of elements of* G.

(iii) Let $X_\infty = \varprojlim_{n \in \mathbb{N}}\left\{X_n, \pi_{nm} : X_m \to X_n\right\}$ *be a proalgebraic variety such that each structure morphism* $\pi_{n,n+1} : X_{n+1} \to X_n$ *satisfies the condition:*

$$\pi_{n,n+1*}([id_{X_{n+1}}]) = \gamma_n \cdot [id_{X_n}] \in K_0(\mathcal{V}/X_n) \quad \textit{for some } \gamma_n \in K_0(\mathcal{V});$$

for example $\pi_{n,n+1} : X_{n+1} \to X_n$ is a Zariski locally trivial fiber bundle with fiber variety being F_n (in which case one can take $\gamma_n := [F_n] \in K_0(\mathcal{V})$). Then the canonical proalgebraic Grothendieck class homomorphisms

$$\Gamma_{X_1}^{\mathrm{ind}} : F^{\mathrm{ind}}(X_\infty) \to K_0(\mathcal{V}/X_1)_G \quad \text{and} \quad \Gamma^{\mathrm{ind}} : F^{\mathrm{ind}}(X_\infty) \to K_0(\mathcal{V})_G$$

are described by

$$\Gamma_{X_1}^{\mathrm{ind}}([\alpha_n]) = \frac{\pi_{1,n*}(\Gamma_{X_n}(\alpha_n))}{\gamma_0 \cdot \gamma_1 \cdot \gamma_2 \cdots \gamma_{n-1}} \quad \text{and} \quad \Gamma^{\mathrm{ind}}([\alpha_n]) = \frac{\Gamma(\alpha_n)}{\gamma_0 \cdot \gamma_1 \cdot \gamma_2 \cdots \gamma_{n-1}}.$$

Here $\gamma_0 := \mathbb{1}$ and $K_0(\mathcal{V})_G$ (or $K_0(\mathcal{V}/X_1)_G$) is the ring of fractions of $K_0(\mathcal{V})$ with respect to the multiplicatively closed set consisting of finite products of powers of γ_m ($m = 1, 2, 3 \cdots$) (or their pullbacks to X_1).

(iv) In particular, if $\gamma_n = \gamma$ for all n, then the canonical proalgebraic Grothendieck class homomorphisms

$$\Gamma_{X_1}^{\mathrm{ind}} : F^{\mathrm{ind}}(X_\infty) \to K_0(\mathcal{V}/X_1)_G \quad \text{and} \quad \Gamma^{\mathrm{ind}} : F^{\mathrm{ind}}(X_\infty) \to K_0(\mathcal{V})_G$$

are described by

$$\Gamma_{X_1}^{\mathrm{ind}}([\alpha_n]) = \frac{\pi_{1,n*}(\Gamma_{X_n}(\alpha_n))}{\gamma^{n-1}} \quad \text{and} \quad \Gamma^{\mathrm{ind}}([\alpha_n]) = \frac{\Gamma(\alpha_n)}{\gamma^{n-1}}.$$

In this special case the quotient ring $K_0(\mathcal{V})_G$ (or $K_0(\mathcal{V}/X_1)_G$) shall be simply denoted by $K_0(\mathcal{V})_\gamma$ (or $K_0(\mathcal{V}/X_1)_\gamma$).

Example 10.5. The arc space $\mathcal{L}(X)$ of an algebraic variety X is defined to be the projective limit of the projective system consisting of the truncated arc varieties $\mathcal{L}_n(X)$ of jets of order n together with the canonical projections $\pi_{n,n+1} : \mathcal{L}_{n+1}(X) \to \mathcal{L}_n(X)$. Note that $\mathcal{L}_0(X) = X$ so that this time we use $\Lambda = \mathbb{N}_0$. Thus the arc space is a nontrivial example of a proalgebraic variety. If X is *nonsingular* and of complex dimension d, then the projection $\pi_{n,n+1} : \mathcal{L}_{n+1}(X) \to \mathcal{L}_n(X)$ is a Zariski locally trivial fiber bundle with fiber being $\mathbb{C}^d$. Thus in this case, in (iv) of Theorem 10.4 the Grothendieck class γ is $\mathbf{L}^d$, with $\mathbf{L} := [\mathbb{C}]$.

An element of $F^{\mathrm{ind}}(X_\infty) = \varinjlim_{\lambda \in \Lambda} F(X_\lambda)$ is called an *indconstructible function* and up to now we have not discussed the role of functions, even though it is called "function". In fact, the indconstructible function can be considered in a natural way as a function on the proalgebraic variety simply as follows: for $[\alpha_\lambda] \in F^{\mathrm{ind}}(X_\infty) = \varinjlim_{\lambda \in \Lambda} F(X_\lambda)$ the value of $[\alpha_\lambda]$ at a point $(x_\mu) \in X_\infty = \varprojlim_{\lambda \in \Lambda} X_\lambda$ is defined by

$$[\alpha_\lambda]\big((x_\mu)\big) := \alpha_\lambda(x_\lambda)$$

which is well-defined. So, if we let $Fun(X_\infty, \mathbb{Z})$ be the abelian group of $\mathbb{Z}$-valued functions on X_∞, then the homomorphism

$$\Psi : \varinjlim_{\lambda \in \Lambda} F(X_\lambda) \to Fun(X_\infty, \mathbb{Z}) \quad \text{defined by} \quad \Psi\left([\alpha_\lambda]\right)\left((x_\mu)\right) := \alpha_\lambda(x_\lambda)$$

shall be called the "functionization" homomorphism.

One can describe this in a fancier way as follows. Let $\pi_\lambda : X_\infty \to X_\lambda$ denote the canonical projection. Consider the following commutative diagram (which follows from $\pi_\lambda = \pi_{\lambda\mu} \circ \pi_\mu (\lambda < \mu)$):

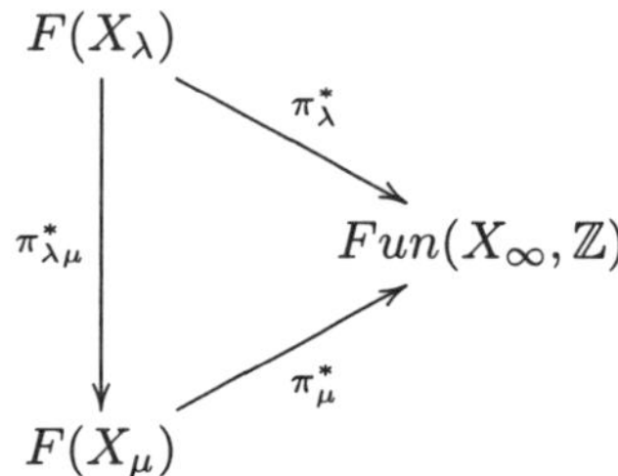

Then the "functionization" homomorphism $\Psi : \varinjlim_{\lambda \in \Lambda} F(X_\lambda) \to Fun(X_\infty, \mathbb{Z})$ is the unique homomorphism such that the following diagram commutes:

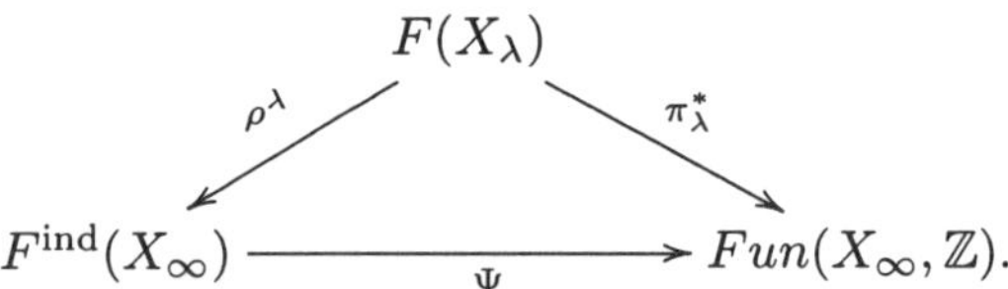

To avoid some possible confusion, the image $\Psi\left([\alpha_\lambda]\right) = \pi_\lambda^* \alpha_\lambda$ shall be denoted by $[\alpha_\lambda]_\infty$. For a constructible set $W_\lambda \in X_\lambda$, by definition we have

$$[\mathbb{1}_{W_\lambda}]_\infty = \mathbb{1}_{\pi_\lambda^{-1}(W_\lambda)}.$$

$\pi_\lambda^{-1}(W_\lambda)$ is called a *proconstructible or a cylinder set*, mimicking [Cr]. The characteristic function supported on a proconstructible set is called a *procharacteristic function* and a finite linear combination of procharacteristic functions is called a *proconstructible function*. Let $F^{\mathrm{pro}}(X_\infty)$ denote the abelian group of all proconstructible functions on the proalgebraic variety $X_\infty = \varprojlim_{\lambda \in \Lambda}\left\{X_\lambda, \pi_{\lambda\mu} : X_\mu \to X_\lambda\right\}$. Thus we have the following

Proposition 10.6. *For a proalgebraic variety* $X_\infty = \varprojlim_{\lambda \in \Lambda} \left\{ X_\lambda, \pi_{\lambda\mu} : X_\mu \to X_\lambda \right\}$

$$F^{\mathrm{pro}}(X_\infty) = \mathrm{Image}\left(\Psi : F^{\mathrm{ind}}(X_\infty) \to Fun(X_\infty, \mathbb{Z}) \right) = \bigcup_\mu \pi_\mu^*(F(X_\mu)).$$

If the structure morphisms $\pi_{\lambda\mu} : X_\mu \to X_\lambda$ $(\lambda < \mu)$ *are all surjective, then we have*

$$F^{\mathrm{ind}}(X_\infty) \cong F^{\mathrm{pro}}(X_\infty).$$

In the case of the arc space $\mathcal{L}(X)$ of a nonsingular variety X, since each structure morphism $\pi_{n,n+1} : \mathcal{L}_{n+1}(X) \to \mathcal{L}_n(X)$ is always surjective, we get the following

Corollary 10.7. *Assume* X *is a* nonsingular *variety of dimension d. Then we have for the arc space* $\mathcal{L}(X)$ *the canonical isomorphism*

$$F^{\mathrm{ind}}\big(\mathcal{L}(X)\big) \cong F^{\mathrm{pro}}\big(\mathcal{L}(X)\big),$$

together with the following canonical Grothendieck class homomorphisms

$$\Gamma_X^{\mathrm{ind}} : F^{\mathrm{pro}}(\mathcal{L}(X)) \to K_0(\mathcal{V}/X)_{[\mathbf{L}^d]} \quad and \quad \Gamma^{\mathrm{ind}} : F^{\mathrm{pro}}(\mathcal{L}(X)) \to K_0(\mathcal{V})_{[\mathbf{L}^d]}$$

described by

$$\Gamma_X^{\mathrm{ind}}\left([\alpha_n]_\infty\right) = \frac{\pi_{0,n*}(\Gamma_{\mathcal{L}_n(X)}(\alpha_n))}{[\mathbf{L}]^{nd}} \quad and \quad \Gamma^{\mathrm{ind}}\left([\alpha_n]_\infty\right) = \frac{\Gamma(\alpha_n)}{[\mathbf{L}]^{nd}}.$$

In particular, we get that $\Gamma_X^{\mathrm{ind}}\left(\mathbb{1}_{\mathcal{L}(X)}\right) = [id_X]$ *and* $\Gamma^{\mathrm{ind}}\left(\mathbb{1}_{\mathcal{L}(X)}\right) = [X]$.

So Γ_X^{ind} and Γ^{ind} define finitely additive measures μ_X and μ on the algebra of cylinder sets in the arc space $\mathcal{L}(X)$ of a *nonsingular* variety X, which are called *naïve motivic measures*. So we can rewrite $\Gamma_X^{\mathrm{ind}}(\alpha)$ and $\Gamma^{\mathrm{ind}}(\alpha)$ for $\alpha \in F^{\mathrm{pro}}(\mathcal{L}(X))$ as motivic integrals

$$\Gamma_X^{\mathrm{ind}}(\alpha) = \int_{\mathcal{L}(X)} \alpha \, d\mu_X \quad and \quad \Gamma^{\mathrm{ind}}(\alpha) = \int_{\mathcal{L}(X)} \alpha \, d\mu.$$

Therefore we see that our proalgebraic Grothendieck class homomorphisms of Theorem 10.4 are a generalization of these naïve motivic measures. Here for "naïve" we point out that for the applications of a good motivic integration theory (e.g., as described in the next section) one needs to consider a suitable *completion* of $K_0(\mathcal{V}/X)_{[\mathbf{L}^d]}$ or $K_0(\mathcal{V})_{[\mathbf{L}^d]}$ so that more general sets than just cylinder sets become "measurable". Also the use of the "relative measure" Γ_X^{ind} over the base space X due to Looijenga [Lo] is more recent, and will become important in the next section.

When we extend *MacPherson's Chern class transformation* [Mac1] to a category of proalgebraic varieties, we appeal to the *Bivariant Theory*. To fit it in with the notion of *bifunctors* used before, we assume for simplicity that *all* morphisms in the underlying category are "proper", e.g. in the topological context we work only with *compact* spaces. More generally, applying *bivariant characteristic classes, namely Grothendieck transformations* (as in Theorem **??**), given in the previous section, we can get a general theory of *characteristic classes of proalgebraic varieties* as follows:

For a morphism $f : X \to Y$ and a bivariant class $b \in \mathbb{B}(X \xrightarrow{f} Y)$, the pair $(f; b)$ is called a *bivariant-class-equipped morphism* and we just express $(f; b) : X \to Y$. Let $\mathbb{B}$ be a bivariant theory having units. If a system $\{b_{\lambda\mu}\}$ of bivariant classes satisfies that

$$b_{\lambda\lambda} = 1_{X_\lambda} \quad \text{and} \quad b_{\mu\nu} \bullet b_{\lambda\mu} = b_{\lambda\nu} \quad (\lambda < \mu < \nu),$$

then we call the system *a projective system of bivariant classes*. If $\{\pi_{\lambda\mu} : X_\mu \to X_\lambda\}$ and $\{b_{\lambda\mu}\}$ are projective systems, then the system $\{(\pi_{\lambda\mu}; b_{\lambda\mu}) : X_\mu \to X_\lambda\}$ shall be called *a projective system of bivariant-class-equipped morphisms*.

For a bivariant theroy $\mathbb{B}$ having units on the category $\mathcal{C}$ and for a projective system $\{(\pi_{\lambda\mu}; b_{\lambda\mu}) : X_\mu \to X_\lambda\}$ of bivariant-class-equipped morphisms, the inductive limit

$$\varinjlim_\Lambda \left\{ \mathbb{B}_*(X_\lambda), b_{\lambda\mu}\bullet : \mathbb{B}_*(X_\lambda) \to \mathbb{B}_*(X_\mu) \right\}$$

shall be denoted by

$$\mathbb{B}_*^{\mathrm{ind}}\left(X_\infty; \{b_{\lambda\mu}\} \right)$$

emphasizing the projective system $\{b_{\lambda\mu}\}$ of bivariant classes, because the above inductive limit surely depends on the choice of it. So we make the covariant functor $\mathbb{B}_*$ into a bifunctor using the functorial "Gysin homomorphisms" $b_{\lambda\mu}\bullet : \mathbb{B}_*(X_\lambda) \to \mathbb{B}_*(X_\mu)$ induced by the projective system $\{b_{\lambda\mu}\}$. For example, in the above Example 10.2 we have that

$$F^{\mathrm{ind}}(X_\infty) = \mathbb{F}_*^{\mathrm{ind}}\left(X_\infty; \{1_{\pi_{\lambda\mu}}\} \right).$$

Definition 10.8. Let $\{f_\lambda : X_\lambda \to Y_\lambda\}_{\lambda \in \Lambda}$ be a pro-morphism of pro-algebraic varieties $\left\{ X_\lambda, \pi_{\lambda\mu} : X_\mu \to X_\lambda \right\}$ and $\left\{ Y_\lambda, \rho_{\lambda\mu} : Y_\mu \to Y_\lambda \right\}$. If the following commutative diagram for $\lambda < \mu$

$$\begin{array}{ccc} X_\mu & \xrightarrow{\ f_\mu\ } & Y_\mu \\ {\scriptstyle \pi_{\lambda\mu}}\Big\downarrow & & \Big\downarrow{\scriptstyle \rho_{\lambda\mu}} \\ X_\lambda & \xrightarrow[\ f_\lambda\ ]{} & Y_\lambda \end{array}$$

is a fiber square, then we call the pro-morphism $\{f_\lambda : X_\lambda \to Y_\lambda\}_{\lambda\in\Lambda}$ a *fiber-square pro-morphism*, abusing words.

With these definitions we have the following theorem:

Theorem 10.9. *(i) Let $\gamma : \mathbb{B} \to \mathbb{B}'$ be a Grothendieck transformation between two bivariant theories $\mathbb{B}, \mathbb{B}' : \mathcal{C} \to \mathcal{A}$ and let $\{(\pi_{\lambda\mu}; b_{\lambda\mu}) : X_\mu \to X_\lambda\}$ be a projective system of bivariant-class-equipped morphisms. Then we get the following pro-version of the natural transformation $\gamma_* : \mathbb{B}_* \to \mathbb{B}'_*$:*

$$\gamma_*^{\mathrm{ind}} : \mathbb{B}_*^{\mathrm{ind}}\Big(X_\infty; \{b_{\lambda\mu}\}\Big) \to \mathbb{B}'^{\,\mathrm{ind}}_*\Big(X_\infty; \{\gamma(b_{\lambda\mu})\}\Big).$$

(ii) Let $\{f_\lambda : Y_\lambda \to X_\lambda\}$ be a fiber-square pro-morphism between two projective systems $\{(\rho_{\lambda\mu}; d_{\lambda\mu}) : Y_\mu \to Y_\lambda\}$ and $\{(\pi_{\lambda\mu}; b_{\lambda\mu}) : X_\mu \to X_\lambda\}$ of bivariant-class-equipped morphisms such that $d_{\lambda\mu} = f_\lambda^* b_{\lambda\mu}$. Then we have the following commutative diagram:

$$\begin{array}{ccc} \mathbb{B}_*^{\mathrm{ind}}(Y_\infty; \{d_{\lambda\mu}\}) & \xrightarrow{\ \gamma_*^{\mathrm{ind}}\ } & \mathbb{B}'^{\,\mathrm{ind}}_*(Y_\infty; \{\gamma(d_{\lambda\mu})\}) \\ {\scriptstyle f_{\infty*}}\Big\downarrow & & \Big\downarrow{\scriptstyle f_{\infty*}} \\ \mathbb{B}_*^{\mathrm{ind}}(X_\infty; \{b_{\lambda\mu}\}) & \xrightarrow[\ \gamma_*^{\mathrm{ind}}\ ]{} & \mathbb{B}'^{\,\mathrm{ind}}_*(X_\infty; \{\gamma(b_{\lambda\mu})\}). \end{array}$$

(iii) Let $\mathbb{B}_*(pt) = \mathbb{B}'_*(pt)$ be a commutative ring $\mathcal{R}$ with a unit and we assume that the homomorphism $\gamma : \mathbb{B}_*(pt) \to \mathbb{B}'_*(pt)$ is the identity. Let $P = \{p_{\lambda\mu}\}$ be a projective system of elements $p_{\lambda\mu} \in \mathcal{R}$. Then we get the commutative diagram

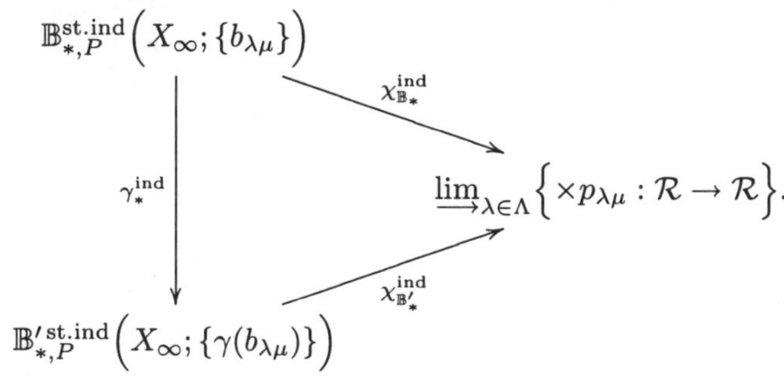

$$\begin{array}{ccc} \mathbb{B}_{*,P}^{\mathrm{st.ind}}\Big(X_\infty; \{b_{\lambda\mu}\}\Big) & & \\ & \searrow{\scriptstyle \chi_{\mathbb{B}_*}^{\mathrm{ind}}} & \\ {\scriptstyle \gamma_*^{\mathrm{ind}}}\Big\downarrow & & \varinjlim_{\lambda\in\Lambda}\Big\{\times p_{\lambda\mu} : \mathcal{R} \to \mathcal{R}\Big\}. \\ & \nearrow{\scriptstyle \chi_{\mathbb{B}'_*}^{\mathrm{ind}}} & \\ \mathbb{B}'^{\,\mathrm{st.ind}}_{*,P}\Big(X_\infty; \{\gamma(b_{\lambda\mu})\}\Big) & & \end{array}$$

If we apply this theorem to Brasselet's bivariant Chern class [Br1] or to the one of [BSY1], we get a proalgebraic version c_*^{ind} of MacPherson's Chern class transformation $c_* : F \to H_*$. But of course we also can apply it to the bivariant versions of our motivic characteristic class transformations mC_* and T_{y*}.

As a very simple example, consider a proalgebraic variety $X_\infty = \varprojlim_{\lambda \in \Lambda} \{X_\lambda, \pi_{\lambda\mu} : X_\mu \to X_\lambda\}$, whose structure maps $\pi_{\lambda\mu}$ are smooth (and therefore "Euler morphisms") and proper. Then we can apply the proalgebraic MacPherson's Chern class transformation c_*^{ind} to

$$F^{\mathrm{ind}}(X_\infty) = \mathbb{F}_*^{\mathrm{ind}}\Big(X_\infty; \{\mathbb{1}_{\pi_{\lambda\mu}}\}\Big).$$

Note that in this case $\gamma(\mathbb{1}_{\pi_{\lambda\mu}}) = c^*(T_{\pi_{\lambda\mu}}) \bullet [\pi_{\lambda\mu}]$ by the *Verdier Riemann-Roch theorem for a smooth morphism*, so that $\mathbb{H}_*{}^{\mathrm{ind}}\Big(X_\infty; \{\gamma(\mathbb{1}_{\lambda\mu})\}\Big)$ is just the inductive limit of the following system of "twisted" smooth pullbacks in homology:

$$\pi_{\lambda\mu}^{!!} := c^*(T_{\pi_{\lambda\mu}}) \cap \pi_{\lambda\mu}^! : H_*(X_\lambda; \mathbb{Z}) \to H_*(X_\mu; \mathbb{Z}).$$

Suitable modifications of such "inductive limits of twisted smooth pullback morphisms" are closely related to the construction of *equivariant characteristic classes* (e.g., see [Oh, §3.3, p.12-13]).

11. Stringy and arc characteristic classes of singular spaces

In this last section we explain another and more recent extension of characteristic classes to singular spaces. These are *not* functorial theories as before, but have a better "birational invariance", in particular for *K-equivalent* manifolds, i.e. M_i ($i = 1, 2$) are irreducible (or pure dimensional) complex algebraic manifolds dominated by a third such manifold M, with $\pi_i : M \to M_i$ proper birational ($i = 1, 2$) such that the pullbacks of their canonical bundles (or divisors) $\pi_1^* K_{M_1} \simeq \pi_2^* K_{M_2}$ are isomorphic (or linearly equivalent). For example M_1 and M_2 are both *Calabi-Yau manifolds* in the sense that their canonical bundle is trivial. In fact the origin of these classes and invariants goes back to two different generalizations of Hirzebruch's χ_y-genus (which was related to our motivic characteristic classes mC_* and T_{y*}).

The first one is the *E-polynomial or Hodge characteristic* $E(X)(u, v) \in \mathbb{Z}[u, v]$ defined in terms of Deligne's *mixed Hodge structure* [De1,De2] for the cohomology with compact support $H_c^*(X, \mathbb{Q})$ of a complex algebraic variety (e.g., see [Sri]). We have that $E(X)(1, 1) = \chi(X)$ for any variety X and $E(X)(-y, 1) = \chi_y(X)$ for X smooth and compact. In the 90's *V. Batyrev* [Bat1] extended this E-polynomial to a *stringy E-function* E_{str} *and stringy Euler numbers* χ^{str} of "log-terminal pairs" (X, D) relating them in some cases known as the "McKay correspondence" to orbifold invariants of suitable quotient varieties. He also used in [Bat2] methods from p-adic integration theory to prove that different "crepant resolutions" of a given singular space, and also *birationally equivalent Calabi–Yau manifolds*, have equal Betti numbers. Later on *M. Kontsevich* [Kon] invented "motivic integration" (with some analogy to p-adic integration) for extending these results from Betti numbers to Hodge numbers.

The other generalization of the χ_y-genus is the (complex) *elliptic genus ell_k* studied by I. Krichever [Krich] and G. Höhn [Höhn]. As observed by *B. Totaro* [To] (also see [BF]), this is the most general genus on the complex cobordism ring $\Omega_*^U \otimes \mathbb{Q}$, which can be invariant under a suitable notion of "flops". Later on this was extended by *L. Borisov and A. Libgober* [BL1] and *C.-L. Wang* [Wang] for showing the invariance of this elliptic genus ell_k for K-equivalent complex algebraic manifolds, a notion coming from "minimal model theory". Both works use the very deep "weak factorization theorem" ([AKMW] and [W]) for the comparison of different resolution spaces. They also introduced in this way the *elliptic homology class $\mathcal{E}ll_*(X)$* of a $\mathbb{Q}$-*Gorenstein log-terminal* singular complex algebraic variety X [BL2,Wang]. Here $\mathbb{Q}$-*Gorenstein* for a normal irreducible (or pure dimensional) variety X just means that some multiple $r \cdot K_X$ ($r \in \mathbb{N}$) of the canonical Weil divisor K_X is already a Cartier divisor, with $r = 1$ corresponding to a *Gorenstein variety* (e.g. X is smooth). Here K_X is just the closure of a canonical divisor on the regular part. In fact, Borisov–Libgober proved in [BL2] a very general version of the "McKay correspondence" for this elliptic homology class.

More recently simpler *stringy Chern classes $c_*^{str}(X)$* were introduced by *Aluffi* [Alu4], based on the "weak factorization theorem", and independently by *de Fernex, Lupercio, Nevins and Uribe* [FLNU], based on "motivic integration" and MacPherson's functorial Chern class transformation c_*. In fact Aluffi pointed out that there are two possible notions of such classes, depending on two different choices of a system of "relative canonical divisors" K_π for suitable resolution of singularities $\pi : M \to X$ (i.e. π is proper and M smooth), which he calls the "Ω-flavor" and "ω-flavor".

The "ω-flavor" is related to "stringy invariants and stringy characteristic classes" (like $E_{str}, \mathcal{E}ll_*$ and c_*^{str}). Here one assumes X is irreducible and $\mathbb{Q}$-Gorenstein so that the relative canonical divisor $K_\pi := K_M - \pi^* K_X$ is at least a $\mathbb{Q}$- Cartier divisor (class). Moreover it is supported on the exceptional locus E of the resolution, which is supposed to be (contained in) a normal crossing divisor with smooth irreducible components E_i. Then $K_\pi \simeq \sum_i a_i \cdot E_i$ for some fixed $a_i \in \mathbb{Q}$ (depending on the resolution). And for the definition of all these "stringy invariants" one needs the condition $a_i > -1$ for all i, which exactly means that X has only *log-terminal singularities*. If this condition holds for one such resolution, then it is true for any resolutions of this type. A resolution π is called *crepant*, if $K_M \simeq \pi^* K_X$, e.g., all $a_i = 0$ for E a normal crossing divisor as before.

The "Ω-flavor" is related to what we call "arc invariants and arc characteristic classes", because these generalize corresponding "arc invariants" of *Denef and Loeser* ([DL1, §6] and [DL2, §4.4.1]), which they introduced already before by their work on "motivic integration". In this case X is only assumed to be pure d-dimensional and K_π is defined for all resolutions π such that the canonical map $\pi^* \Omega^d_X \to \Omega^d_M$ of Kähler differentials has an image $\mathcal{I} \otimes \Omega^d_M$ with $\mathcal{I}$ a principal ideal in $\mathcal{O}_M$ (this can always be achieved by Hironaka [Hi]). Then K_π is defined by $\mathcal{I} = \mathcal{O}_M(-K_\pi)$. The effective Cartier divisor K_π is again supported on the exceptional locus E of the resolution, which can also be supposed to be (contained in) a normal crossing divisor with smooth irreducible components E_i. Then one can introduce the $a_i \in \mathbb{N}_0$ as before.

For X already smooth, both notions of a relative canonical divisor K_π agree with the divisor of the Jacobian of π defined by the section s of $K_M \otimes \pi^* K_X^*$ corresponding to the canonical map $\pi^* \Omega^d_X \to \Omega^d_M$. Note that in both cases the corresponding resolutions $\pi : M \to X$ as above form a directed set, i.e. two of them can be dominated by a third one of this type (and taking suitable limits over this directed set corresponds to the view point of Aluffi [Alu4]). If $\pi' : M' \to M$ is a proper birational map with π and $\pi \circ \pi'$ as above, then the relative canonical divisors have (in both cases) the following crucial transitivity property:

$$K_{\pi \circ \pi'} \simeq K_{\pi'} + \pi'^* K_\pi . \tag{11.1}$$

Then all these new invariants $I(X)$ for a singular space X as above are described as

$$I(X) := \pi_* \left(I(M) \cdot J(\{E_i, a_i\}) \right) \in \mathcal{B}_*(X)$$

for such a special resolution $\pi : M \to X$, with E a normal crossing divisor with smooth irreducible components E_i, where $I(M) \in \mathcal{B}_*(M)$ is the corresponding invariant of the smooth space M, together with some "correction term" $J(\{E_i, a_i\}) \in \mathcal{B}^*(M)$ depending on the exceptional divisor E and the multiplicities a_i defined by the relative canonical divisor K_π. Here $\mathcal{B}_*$ and $\mathcal{B}^*$ are suitable covariant and contravariant theories taking values in the category of Abelian groups and commutative rings with unit, related by the projection formula as in Example 10.3. Typical examples are

(1) $\mathcal{B}_*(X) = \mathcal{B}^*(X) = \Lambda$ is a commutative ring with unit (with all pullbacks and pushforwards being the identity transformation id_Λ), so that $I(M) \in \Lambda$ corresponds to a suitable generalized "Euler characteristic type invariant".

(2) $\mathcal{B}_*$ and $\mathcal{B}^*$ correspond to suitable homology and cohomology theories like $(\mathcal{B}_*(X), \mathcal{B}^*(X)) = (H_*^{BM}(X) \otimes \Lambda, H^*(X) \otimes \Lambda)$ or $(\mathcal{B}_*(X), \mathcal{B}^*(X)) = (\mathbf{G}_0(X) \otimes \Lambda, \mathbf{K}^0(X) \otimes \Lambda)$, so that $I(M) \in \mathcal{B}_*(M)$ is a suitable characteristic class of M.

(3) $\mathbb{B}(X) := \mathcal{B}_*(X) = \mathcal{B}^*(X)$ is a bifunctor as in Example 10.3, e.g. like constructible functions $\mathcal{B}(X) = F(X) \otimes \Lambda$ or relative Grothendieck rings of varieties $K_0(\mathcal{V}/X) \otimes \Lambda$ coming up from "motivic integrals".

If $I(X) \in \mathcal{B}_*(X)$ is such an invariant not depending on the choice of the resolution π, then the same is true for $\gamma_*(I(X)) \in \mathcal{B}'(X)$ for any natural transformation of covariant theories $\gamma_* : \mathcal{B}_* \to \mathcal{B}'_*$. For example $I(X) \in H_*(X) \otimes \Lambda$ is a characteristic homology class with X compact, and $deg := \gamma_* : H_*(X) \otimes \Lambda \to H_*(\{pt\}) \otimes \Lambda = \Lambda$ is just its degree (or push down to a point). Or we apply suitable "completions" of our motivic characteristic class transformations mC_* and T_{y*} to invariants $I(X)$ coming from motivic integration!

There are two ways to show that the final result $I(X)$ does not depend on the choice of the resolution. One is to use "motivic integration with its transformation rule" related to the "Jacobian factor" $J(\{E_i, a_i\})$:

$$\int_{\mathcal{L}(M)} \mathbf{L}^{-\alpha} \, d\tilde{\mu}_M = \pi'_* \int_{\mathcal{L}(M')} \mathbf{L}^{-(\pi'^*\alpha + K_{\pi'})} \, d\tilde{\mu}_{M'} \tag{11.2}$$

for $\pi' : M' \to M$ a proper birational map of manifolds and $\mathbf{L} := [\mathbb{C}] \in K_0(\mathcal{V})$. This suggests to think of $I(X)$ as the pushforward of an "integral with respect to the invariant $I(M)$":

$$I(X) = \pi_* \int_M \mathbf{L}^{-K_\pi} \, dI(M) \,.$$

The other one is to use the "weak factorization theorem", in which case only the invariance under suitable "blowing ups" has to be checked.

Moreover $J(\{E_i, a_i\}) = 1$ in case all $a_i = 0$, so that $I(X) = \pi_*(I(M))$ in the case of a *crepant resolution*. In particular $\pi_*(I(M))$ does *not* depend on the choice of this crepant resolution. Suppose two possibly *singular* spaces X_i $(i = 1, 2)$ are K-equivalent in the sense that they are dominated by a manifold M, with $\pi_i : M \to X_i$ a resolution of singularities such that the relative canonical divisors K_{π_i} are defined $(i = 1, 2)$ and equal. After taking another resolution of M, we can even assume that the exceptional locus of both maps is contained in a normal crossing divisor E with smooth irreducible components E_i (here we use the transitivity property of the relative canonical divisors). But then the correction factor $J(\{E_i, a_i\})$ for both maps is the same, so that

$$I(X_1) = \pi_{1*}(I(M) \cdot J(\{E_i, a_i\})) \quad \text{and} \quad I(X_2) = \pi_{2*}(I(M) \cdot J(\{E_i, a_i\})),$$

i.e. both invariants $I(X_1)$ and $I(X_2)$ are "dominated" by the same element coming from M. In particular

$$I(X_1) = I(X_2)$$

in the case of "Euler characteristic type invariants", and

$$deg(I(X_1)) = deg(I(X_2))$$

in the case of "characteristic homology classes" for compact spaces X_i. If we are working in the "ω-flavor" of stringy homology classes $I(X_i) \in H_*^{BM}(X_i) \otimes \Lambda$ for $\mathbb{Q}$-Gorenstein varieties X_i, we can use the first Chern class $c^1(K_{X_i}) := \frac{c^1(r \cdot K_{X_i})}{r} \in H^2(X_i; \mathbb{Q})$ (for Λ a $\mathbb{Q}$-algebra) to modify $I(X_i)$ into

$$I'(X_i) := f(c^1(K_{X_i})) \cdot I(X_i) \in H_*^{BM}(X_i) \otimes \Lambda.$$

By the *projection formula* these new invariants $I'(X_1)$ and $I'(X_2)$ are also "dominated" by the same element coming from M, where $f \in \Lambda[[z]]$ can be any power series. If X_i are both Gorenstein, we can do the same thing for corresponding invariants $I(X_i) \in \mathbf{G}_0(X) \otimes \Lambda$ by using polynomials in the (inverse) classes $[K_{X_i}^{\pm 1}] \in \mathbf{K}^0(X_i)$ of the canonical Cartier divisors (instead of their first Chern classes).

Note that the approach by resolution of singularities is different from our approach to functorial "motivic characteristic classes" based on "additivity" (i.e. decomposing a singular space into smooth pieces), but nevertheless they nicely fit together as we now explain.

11.1. *Elliptic classes*

Let us start with the definition of the *(complex) elliptic class* $\mathcal{E}LL(E)$ of a complex vector bundle $E \to X$. Consider the formal power series

$$\Lambda_t(E) := \sum_{n \geq 0} t^n \Lambda^n E \quad \text{and} \quad S_t(E) := \sum_{n \geq 0} t^n S^n E,$$

with $\Lambda^n E$ and $S^n E$ the corresponding exterior and symmetric power of E (so $\Lambda^n E = 0$ for $n > \text{rank } E$, with Λ_t the total Λ class, which was also used in our definition of the motivic Chern class transformation mC_* in Corollary 8.4). Then one has

$$\Lambda_t(E \oplus F) = \Lambda_t(E)\Lambda_t(F),\, S_t(E \oplus F) = S_t(E)S_t(F),\, \text{and}\, \Lambda_t(E)S_{-t}(E) = 1\,.$$

So these operations extend to the Grothendieck group of complex vector bundles (and similarly in the algebraic context):

$$\Lambda_t, S_t : (\mathbf{K}(X), \oplus) \to (1 + \mathbf{K}(X)[[t]], \otimes) \subset (\mathbf{K}(X)[[t]], \otimes)\,.$$

Then we define the *complex elliptic class*

$$\mathcal{E}LL(E) = \mathcal{E}LL(y, q)(E) \in \mathbf{K}(X)[[q]][y^{\pm 1}]$$

of a complex vector bundle $E \to X$ as $\mathcal{E}LL(y, q)(E) := \Lambda_y(E^*) \otimes \mathcal{W}(E)$, with

$$\mathcal{W}(E) := \bigotimes_{n \geq 1} \left(\Lambda_{yq^n}(E^*) \otimes \Lambda_{y^{-1}q^n}(E) \otimes S_{q^n}(E^*) \otimes S_{q^n}(E) \right)\,. \tag{11.3}$$

More generally the *elliptic class of order* k

$$\mathcal{E}LL_k(E) = \mathcal{E}LL_k(y, q)(E) \in \mathbf{K}(X)[[q]][y^{\pm 1}] \quad \text{with } k \in \mathbb{Z}$$

of a complex vector bundle $E \to X$ is defined as the twisted class

$$\mathcal{E}LL_k(E) := det(E)^{\otimes -k} \otimes \mathcal{E}LL(E)\,, \tag{11.4}$$

with $det(E) := \Lambda^{\mathrm{rank}\ E}(E)$ being the determinant line bundle of E. So $\mathcal{E}LL(E)$ (or $\mathcal{E}LL_k(E)$) is a one (or two) parameter deformation of the total Lambda class $\Lambda_y(E^*)$, with

$$\mathcal{E}LL_0(E) = \mathcal{E}LL(E) \quad \text{and} \quad \mathcal{E}LL(E)|_{q=0} = \Lambda_y(E^*)\,.$$

For M a complex projective algebraic manifold (or a compact almost complex manifold) one can introduce as in §5 the χ-characteristic

$$\chi(M, \mathcal{E}LL_k(E)) \in \mathbb{Q}[[k, q]][y^{\pm 1}]$$

of $\mathcal{E}LL_k(E)$ as

$$\chi(M, \mathcal{E}LL_k(E)) := \int_M ch^*(\mathcal{E}LL_k(E)) \cdot td^*(TM) \cap [M]$$

$$= \int_M e^{-k \cdot c^1(E)} \cdot ch^*(\mathcal{E}LL(E)) \cdot td^*(TM) \cap [M]\,.$$

Note that in the last term one can introduce k as a formal parameter. $ch^*(\mathcal{E}LL_k(E))$ and $ch^*(\mathcal{E}LL(E)))$ are *multiplicative* (but not normalized) characteristic classes so that we get the induced Krichever–Höhn *elliptic genus*

$$ell_k : \Omega_*^U \otimes \mathbb{Q} \to \mathbb{Q}[[k, q]][y^{\pm 1}]\,,$$

932

with

$$ell_k(M) := \chi(M, \mathcal{E}LL_k(TM))$$
$$= \int_M e^{-k \cdot c^1(TM)} \cdot ch^*(\mathcal{E}LL(TM)) \cdot td^*(TM) \cap [M].$$
(11.5)

The corresponding *complex elliptic genus* $ell := ell_0 : \Omega^U_* \otimes \mathbb{Q} \to \mathbb{Q}[[q]][y^{\pm 1}]$ given by

$$ell_0(M)$$
$$= \int_M ch^*(\mathcal{W}(TM)) \cdot ch^*(\Lambda_y T^* M) \cdot td^*(TM) \cap [M]$$
$$= \chi_y(M, \mathcal{W}(TM)) \quad \text{(by \textbf{g-HRR})}$$
$$= \chi_y \left(M, \bigotimes_{n \geq 1} \left(\Lambda_{yq^n}(TM^*) \otimes \Lambda_{y^{-1}q^n}(TM) \otimes S_{q^n}(TM^*) \otimes S_{q^n}(TM) \right) \right)$$

was formally interpreted by E. Witten as the S^1-equivariant χ_y-genus $\chi_y(S^1, \mathcal{L}M)$ of the free loop space $\mathcal{L}M = \{f : S^1 \to M | f \text{ smooth}\}$ of M (see [HBJ, Appendix III] and [BF]).

$$\chi_{k,y}(M) := ell_k(M)|_{q=0} \in \mathbb{Q}[y][[k]]$$

is called the *twisted χ_y-genus* of M:

$$\chi_{k,y}(M) = \int_M e^{-k \cdot c^1(TM)} \cdot ch^*(\Lambda_y(T^*M)) \cdot td^*(TM) \cap [M].$$
(11.6)

Another specialization is the *real elliptic genus $ell|_{y=1}$*, which factorizes over the oriented cobordism ring

$$ell|_{y=1} : \Omega^{SO}_* \otimes \mathbb{Q} \to \mathbb{Q}[[q]].$$

This one parameter genus interpolates between the signature genus (for $q \to 0$) and the $\hat{A}$-genus (for $q \to \infty$), and was formally interpreted by Witten as the S^1-equivariant signature $\sigma(S^1, LM)$ of the free loop space LM of the oriented manifold M (compare [HBJ, §6] and [BF]).

Remark 11.7. We point out that there are many different normalizations of the elliptic genus and classes in the literature. First of all many authors (like [BL1,BL2, To,Wang]) use $-y$ instead of y so that their elliptic genus is related to the χ_{-y}-genus. But what is maybe more important, we do not work with "normalized characteristic classes", i.e. the power series $f(z) \in \mathbb{Q}[[k,q]][y^{\pm 1}][[z]]$ in the variable $z = c^1$ corresponding to the multiplicative characteristic class $ch^*(\mathcal{E}LL_k(\ ))$ has a constant coefficient $a := f(0) \neq 1$, since $ch^*(\mathcal{E}LL(E))|_{q=0} = ch^*(\Lambda_y(E^*))$ implies $a = 1 + y \in \mathbb{Q}[y^{\pm 1}](k = 0, q = 0)$. So twisting $f(z)$ to a normalized power series $\frac{f(z)}{a}$ (as used in [BF,To,Wang]) would change the elliptic genus only to $\frac{ell_k(M)}{a^n}$ for M an (almost) complex manifold of complex dimension n, and similarly a characteristic homology class $cl_i(\ ) \in H_{2i}^{BM}(\ ) \otimes \Lambda$ would just be multiplied by a^{-i}. For example, in Theorem 8.5 we could have started with the natural transformation (with respect to proper maps):

$$\tilde{T}_{y*} := td_* \circ mC_* : K_0(\mathcal{V}/\ ) \to H_{2*}^{BM}(\ ) \otimes \mathbb{Q}[y] ,$$

satisfying the normalization that for M nonsingular

$$\tilde{T}_{y*}([M \xrightarrow{\mathrm{id}} M]) = ch^*(\Lambda_y T^*M) \cdot td^*(TM) \cap [M].$$

And "twisting" by $1 + y$ would then give our motivic characteristic class transformation T_{y*} with

$$T_{y,i}(\ ) = (1 + y)^{-i} \cdot \tilde{T}_{y,i}(\ ) \in H_{2i}^{BM}(\ ) \otimes \mathbb{Q}[y, (1 + y)^{-1}] . \tag{11.8}$$

But since we work in this section only with pure dimensional spaces, this "twisting" does not matter for the question of getting invariants of pure dimensional singular complex algebraic varieties. Similarly it will be enough to consider only the complex elliptic genus and classes corresponding to $k = 0$ (as in [BL1,BL2]), since the case of general k follows from the projection formula (as already explained before). So the elliptic classes $\mathcal{E}ll^*(z, \tau)$ used in [BL1,BL2] correspond in our notation to

$$\mathcal{E}ll^*(z, \tau)(TM) := y^{-\frac{dim(M)}{2}} \cdot td^*(TM) \cdot ch^*(\mathcal{E}LL(TM))(-y, q) ,$$

with $y = e^{2\pi i z}$ and $q = e^{2\pi i \tau}$.

With these notations, we can now explain the definition of Borisov and Libgober ([BL2, Definition 3.2] with $G := \{id\}$) for their *elliptic class* $\mathcal{E}ll_*((X, D))$ of a "Kawamata log-terminal pair (X, D)", i.e. X is a normal irreducible complex algebraic variety, with D a $\mathbb{Q}$-Weil divisor on X such that $K_X + D$ is a $\mathbb{Q}$-Cartier divisor satisfying the following condition: There is a resolution of singularities $\pi : M \to X$ with the exceptional locus E and the support of $K_\pi(D) := K_M - \pi^*(K_X + D)$ contained in a normal crossing divisor with smooth irreducible components E_i ($i \in I$) such that $K_\pi(D) \simeq \sum_i a_i \cdot E_i$, with all $a_i \in \mathbb{Q}$ satisfying the inequality $a_i > -1$. Note that the last condition is then independent of the choice of such a resolution (compare [KM, Definition 2.34, Corollary 2.31]), with the case $D = 0$ corresponding to the case "X is $\mathbb{Q}$-Gorenstein with only log-terminal singularities". Moreover, the "relative canonical divisor $K_\pi(D)$ of D" also satisfies the transitivity property

$$K_{\pi \circ \pi'}(D) \simeq K_{\pi'}(D) + \pi'^* K_\pi(D) \tag{11.9}$$

for $\pi' : M' \to M$ a proper birational map with π and $\pi \circ \pi'$ as before. Then the Borisov–Libgober elliptic class is:

$$\mathcal{E}ll_*((X,D))(z,\tau) :=$$
$$\pi_* \left((\mathcal{E}ll^*(z,\tau)(TM) \cap [M]) \cap \prod_i J(E_i, a_i)(z,\tau) \right) , \tag{11.10}$$

with

$$J(E_i, a_i)(z,\tau) := \frac{\theta(\frac{e_i}{2\pi i} - (a_i + 1)z, \tau)\theta(-z, \tau)}{\theta(\frac{e_i}{2\pi i} - z, \tau)\theta(-(a_i + 1)z, \tau)} \in H^*(M; \mathbb{Q})[[y, q]] .$$

Here $\theta(z, \tau)$ is the Jacobi theta function in $y = e^{2\pi i z}$ and $q = e^{2\pi i \tau}$, with $e_i = c^1(E_i) \in H^2(M, \mathbb{Z})$ the first Chern class of the smooth divisor E_i.

The proof of the independence of the resolution π uses the "weak factorization theorem" for reducing it to the comparison with a suitable blowing up along a smooth center. Using some modularity properties of the θ-function, this is finally reduced to the vanishing of a suitable residue (of an elliptic function with exactly one pole, compare [BL2, p.11] and [Wang, §4]). If X is compact, then

$$ell((X,D)) := deg\big(\mathcal{E}ll_*((X,D))\big) \tag{11.11}$$

is just the *singular elliptic genus* of the Kawamata log-terminal pair (X, D) as defined in [BL1, Definition 3.1] (up to a normalization factor).

Later on we only need the following limit formula (with $y = e^{2\pi i z}$):

$$\lim_{\tau \to i\infty} J(E_i, a_i)(z, \tau) = \frac{(y-1)(1 - y^{a_i+1}e^{-e_i})}{(y^{a_i+1} - 1)(1 - ye^{-e_i})} \tag{11.12}$$
$$= 1 + \frac{(y - y^{a_i+1})(1 - e^{-e_i})}{(y^{a_i+1} - 1)(1 - ye^{-e_i})} .$$

Note that the multiplicative characteristic class

$$\tilde{T}_y^*(E) := ch^*(\mathcal{E}LL(E))|_{q=0} \cdot td^*(E) = ch^*(\Lambda_y(E^*)) \cdot td^*(E)$$

exactly corresponds to the non-normalized power series $f(z) = \frac{z(1+ye^{-z})}{1-e^{-z}}$ in the variable $z = c^1$ (see §6). If we denote for $J \subset I$ the closed embedding $i_J : E_J := \bigcap_{i \in J} E_i \to M$ of the submanifold E_J (with $E_\emptyset := M$), then one has by the "adjunction formula"

$$i_{J*}i_J^* = \prod_{i \in J} e_i \cap ,$$

with $TE_J = i_J^*(TM - \sum_{i \in J} \mathcal{O}(E_i))$ (compare [HBJ, p.36]):

$$i_{J*}(\tilde{T}_y^*(TE_J) \cap [E_J]) = (\tilde{T}_y^*(TM) \cap [M]) \cap \prod_{i \in J} \frac{1 - e^{-e_i}}{1 + ye^{-e_i}} \, .$$

So altogether we get the following "limit formula" (with $y = e^{2\pi i z}$):

$$\lim_{\tau \to i\infty} y^{dim(X)/2} \cdot \mathcal{E}ll_*((X, D))$$

$$= \pi_*((\tilde{T}_{-y}^*(TM) \cap [M]) \cap \prod_{i \in I} (1 + \frac{(y - y^{a_i+1})(1 - e^{-e_i})}{(y^{a_i+1} - 1)(1 - ye^{-e_i})}))$$

$$= \pi_*(\sum_{J \subset I} i_{J*}(\tilde{T}_{-y*}(E_J))) \cdot \prod_{i \in J} \frac{y - y^{a_i+1}}{y^{a_i+1} - 1}) \tag{11.13}$$

$$= \sum_{J \subset I} \pi_* i_{J*}(\tilde{T}_{-y*}(E_J)) \cdot \prod_{i \in J} \frac{y - y^{a_i+1}}{y^{a_i+1} - 1} \, .$$

Recall that we use the notation $cl_*(E_J) = cl^*(TE_J) \cap [E_J]$ for the characteristic homology class of a manifold (corresponding to a characteristic class cl^* of vector bundles).

11.2. *Motivic integration*

Motivic integration was invented by Kontsevich [Kon] for showing that birational equivalent Calabi–Yau manifolds have equal Hodge numbers. In all details with many different applications it was developed by Denef and Loeser (e.g. [DL1,DL2,DL3]), with some improvements by Looijenga [Lo], who in particular introduced the calculus of relative Grothendieck rings $K_0(\mathcal{V}/X)$ of algebraic varieties. For a nice introduction to "stringy invariants of singular spaces" we recommend [Ve1,Ve2]. Even though motivic integration can be directly studied on singular spaces, we restrict ourselves to the simpler case of smooth spaces, which will be enough for our applications. Also in this way it can easily be compared to results coming from the use of the "weak factorization theorem". For a quick introduction to "motivic integration on smooth spaces" compare with [Cr] (where by Corollary 10.7 all arguments of [Cr] extend to the framework of "relative motivic measures").

Let M be a pure d-dimensional complex algebraic manifold and $E = \sum_{i=1}^k a_i E_i$ be an effective normal crossing divisor (e.g. $a_i \in \mathbb{N}_0$) on M, with smooth irreducible components E_i. Then one can introduce on the arc space $\mathcal{L}(M) = \{\gamma_u | u \in M\}$ the order function along E:

$$ord(E) := \sum_i a_i \cdot ord(E_i) : \mathcal{L}(M) \to \mathbb{N}_0 \cup \infty \, ,$$

with $ord(E_i)(\gamma_u) := ord_0\ f_i \circ \gamma_u(t)$ the zero order of $f_i \circ \gamma_u(t) \in \mathbb{C}[[t]]$ at the origin, if f_i is a local defining equation of E_i near the point $u \in M$. In particular

$$ord(D_i)(\gamma_u) = 0 \Leftrightarrow u \notin D_i \quad \text{and} \quad ord(D_i)(\gamma_u) = \infty \Leftrightarrow \gamma_u \subset D_i\,.$$

Then $\{ord(E) = n\} \subset \mathcal{L}(M)$ is for all $n \in \mathbb{N}_0$ a *proconstructible or cylinder set* in the sense of §10. Then one would like to introduce the following motivic integral:

$$\int_{\mathcal{L}(M)} \mathbf{L}^{-ord(E)} d\mu_M := \sum_{p \in \mathbb{N}_0} \mu_M(\{ord(E) = p\}) \cdot \mathbf{L}^{-p} \tag{11.14}$$

with values in the localized ring $K_0(\mathcal{V}/M)_{[\mathbf{L}^d]}$ as in Corollary 10.7. Recall that we normalized the (naive) motivic measure μ_M in such a way that for $E = 0$ we get :

$$\int_{\mathcal{L}(M)} 1\, d\mu_M = [M] \in K_0(\mathcal{V}/M)_{[\mathbf{L}^d]}\,.$$

But the problems with the definition (11.14) are that this is not a finite series and that $\{ord(E) = \infty\}$ is *not* a cylinder set in $\mathcal{L}(M)$. Both problems are solved by taking a suitable completion of $K_0(\mathcal{V}/M)_{[\mathbf{L}^d]}$. More precisely for X a complex algebraic variety let $\widehat{M}(\mathcal{V}/X)$ be the completion of $K_0(\mathcal{V}/X)[\mathbf{L}^{-1}]$ with respect to the following dimension filtration (for $k \to -\infty$):

$$F_k(K_0(\mathcal{V}/X)[\mathbf{L}^{-1}])\ \text{is generated by}\ [X' \to X]\mathbf{L}^{-n}\ \text{with}\ dim(X') - n \le k.$$

Remark 11.15. Here we consider $K_0(\mathcal{V}/X)$ as an algebra over $K_0(\mathcal{V}) := K_0(\mathcal{V}/\{pt\})$ by the pullback a_X^* for $a_X : X \to \{pt\} = Spec(\mathbb{C})$ the constant structure map. If $S \subset K_0(\mathcal{V})$ is a multiplicatively closed subset, then we can localize the commutative ring $K_0(\mathcal{V}/X)$ with respect to the induced multiplicatively closed subset $a_X^*(S) \subset K_0(\mathcal{V}/X)$, or we can localize $K_0(\mathcal{V}/X)$ as an $K_0(\mathcal{V})$-module with respect to S. Both localizations can be identified, since a_X^* is injective (compose with any map $\{pt\} \to X$), and are denoted by $K_0(\mathcal{V}/X)_S$. In case $S = \{\mathbf{L}^n | n \in \mathbb{N}_0\}$, with $\mathbf{L} := [\mathbb{C}] \in K_0(\mathcal{V})$, we also use the notation $K_0(\mathcal{V}/X)[\mathbf{L}^{-1}]$ above.

Also note that the filtration and completion as above are compatible with push down f_* and exterior product $\boxtimes$ so that in particular $\widehat{M}(\mathcal{V}/X)$ is a $\widehat{M}(\mathcal{V}) := \widehat{M}(\mathcal{V}/\{pt\})$-module, with an induced $\widehat{M}(\mathcal{V})$-linear push down $f_* : \widehat{M}(\mathcal{V}/X) \to \widehat{M}(\mathcal{V}/Y)$ for $f : X \to Y$ an algebraic morphism.

Let us come back to our motivic integral (11.14) on the manifold M. The composed *relative motivic measure*

$$\tilde{\mu}_M : F^{pro}(\mathcal{L}(M)) \to \widehat{M}(\mathcal{V}/M)$$

can now be extended from cylinder sets to a more general class of "measurable subsets" of the arc space $\mathcal{L}(M)$ in such a way that $\{ord(E) = \infty\}$ becomes measurable with measure 0, and the series (11.14) above converges in $\widehat{M}(\mathcal{V}/M)$. So now one can define

$$\int_{\mathcal{L}(M)} \mathbf{L}^{-ord(E)} d\tilde{\mu}_M := \sum_{p \in \mathbb{N}_0} \tilde{\mu}_M(\{ord(E) = p\}) \cdot \mathbf{L}^{-p} \in \widehat{M}(\mathcal{V}/M). \quad (11.16)$$

Moreover it can easily be computed with $b_i := \frac{\mathbf{L}-1}{\mathbf{L}^{a_i+1}-1} \in \widehat{M}(\mathcal{V})$:

$$\begin{aligned}
\int_{\mathcal{L}(M)} \mathbf{L}^{-ord(E)} d\tilde{\mu}_M &= \sum_{I \subset \{1,\dots,k\}} [E_I^o \to M] \cdot \prod_{i \in I} \frac{\mathbf{L}-1}{\mathbf{L}^{a_i+1}-1} \\
&= \prod_{i=1}^{k} \left(b_i \cdot [E_i \to M] + [M\backslash E_i \to M] \right) \\
&= \prod_{i=1}^{k} \left((b_i - 1) \cdot [E_i \to M] + [id_M] \right) \\
&= \sum_{I \subset \{1,\dots,k\}} [E_I \to M] \cdot \prod_{i \in I} \left(\frac{\mathbf{L}-1}{\mathbf{L}^{a_i+1}-1} - 1 \right).
\end{aligned} \quad (11.17)$$

Here we use the notation:

$$E_I := \bigcap_{i \in I} E_i \quad (\text{with } E_\emptyset := M), \text{ and} \quad E_I^o := E_I \backslash \bigcup_{i \in \{1,\dots,k\}\backslash I} E_i,$$

and the factor $\frac{1}{\mathbf{L}^{a_i+1}-1} = \frac{\mathbf{L}^{-(a_i+1)}}{1-\mathbf{L}^{-(a_i+1)}}$ has to be developed as the corresponding geometric series in $\widehat{M}(\mathcal{V})$. Recall that multiplication in $\widehat{M}(\mathcal{V}/M)$ is induced from taking the fiber product over M, with $[id_M]$ the corresponding unit element. Also note that the second equality above follows from

$$[E_i \to M] \cdot [M\backslash E_i \to M] = [\emptyset \to M] = 0.$$

The other piece of information that we need is the *transformation rule*

$$\int_{\mathcal{L}(M)} \mathbf{L}^{-ord(E)} d\tilde{\mu}_M = \pi'_* \int_{\mathcal{L}(M')} \mathbf{L}^{-ord(\pi'^*E+K_{\pi'})} d\tilde{\mu}_{M'} \quad (11.18)$$

for $\pi' : M' \to M$ a proper birational map of pure dimensional complex algebraic manifolds such that $\pi'^*E + K_{\pi'}$ is a normal crossing divisor with smooth irreducible components.

Assume now that we have a proper birational map $\pi : M \to X$, with X pure dimensional but possibly singular, together with a Cartier divisor D on M such

938

that D and the exceptional locus of π are contained in (the support of) E. Finally we assume

$$K_\pi(D) := K_\pi - D \simeq \sum_i a_i \cdot E_i \,,$$

with all $a_i \in \mathbb{Z}$ satisfying the inequality $a_i > -1$ (i.e. $a_i \in \mathbb{N}_0$). Here we of course use the relative canonical divisor K_π in the "Ω-flavor". Then we define the following *motivic arc invariant*

$$\mathcal{E}^{arc}((X,D)) \in \widehat{\mathbf{M}}(\mathcal{V}/X)$$

of the pair (X,D):

$$\mathcal{E}^{arc}((X,D)) := \pi_*\Big(\int_{\mathcal{L}(M)} \mathbf{L}^{-ord(K_\pi(D))} d\tilde{\mu}_M \Big) \,, \qquad (11.19)$$

which more explicitly can be calculated as in (11.17). This invariant is "independent" of the choice of π in the following sense. Let $\pi' : M' \to M$ be a proper birational map of pure dimensional complex algebraic manifolds such that $\pi'^* D$ and the exceptional locus of $\pi \circ \pi' : M' \to X$ is contained in a normal crossing divisor with smooth irreducible components. Then

$$K_{\pi\circ\pi'}(\pi'^* D) = K_{\pi\circ\pi'} - \pi'^* D = \pi'^* K_\pi(D) + K_{\pi'}$$

is also an effective Cartier divisor with

$$\mathcal{E}^{arc}((X,D)) = \mathcal{E}^{arc}((X,\pi'^* D))$$

by the transformation rule. So this is an invariant of the pair (X,D), if we consider D as a Cartier divisor (in the sense of Aluffi [Alu4]) on the directed set of all such resolutions $\pi : M \to X$. In particular $\mathcal{E}^{arc}(X) := \mathcal{E}^{arc}((X,0))$ is an invariant of the singular space X. In fact in the language of [DL1, sec.6] and [DL2, sec.4.4] it is just the "motivic volume of the arc space $\mathcal{L}(X)$" of the singular space X:

$$\mathcal{E}^{arc}(X) = \int_{\mathcal{L}(X)} 1 \, d\tilde{\mu}_X \,.$$

And this fits with our general description in the introduction of this section, if we set

$$I(M) := [id_M] \in \widehat{\mathbf{M}}(\mathcal{V}/M) \,, \quad \text{with} \quad J(\{E_i, a_i\}) := \int_{\mathcal{L}(M)} \mathbf{L}^{-ord(K_\pi)} d\tilde{\mu}_M \,.$$

For the corresponding "stringy invariant" in the "ω-flavor", first one has to extend these motivic integrals to $\mathbb{Q}$-Cartier divisors supported on a normal crossing divisor with smooth irreducible components E_i, i.e. we start with a strict normal crossing divisor $E = \sum_{i=1}^k a_i E_i$ on the smooth manifold M, with $a_i \in \mathbb{Q}$ such that $r \cdot E$ is a Cartier divisor for some $r \in \mathbb{N}$, i.e. $r \cdot a_i \in \mathbb{Z}$ for all i. Add a formal variable $\mathbf{L}^{\frac{1}{r}}$ to $\widehat{\mathbf{M}}(\mathcal{V})$ (and $a_X^* \mathbf{L}^{\frac{1}{r}}$ to $\widehat{\mathbf{M}}(\mathcal{V}/X)$), with $(\mathbf{L}^{\frac{1}{r}})^r = \mathbf{L}$. Then one can introduce and evaluate the integral

$$\int_{\mathcal{L}(M)} \mathbf{L}^{-ord(E)} d\tilde{\mu}_M := \sum_{p \in \mathbb{Z}} \tilde{\mu}_M(\{ord(rE) = p\}) \cdot (\mathbf{L}^{\frac{1}{r}})^{-p} , \qquad (11.20)$$

with value in $\widehat{\mathbf{M}}(\mathcal{V}/M)[\mathbf{L}^{\frac{1}{r}}]$, if $a_i > -1$ for all i. Moreover the corresponding formula (11.17) with $\mathbf{L}^{a_i+1} := (\mathbf{L}^{\frac{1}{r}})^{r \cdot (a_i+1)}$, and transformation rule (11.18) are also true in this more general context (compare with [Ve1, Appendix] for more details).

With these improvements, one can introduce for a "Kawamata log-terminal pair (X, D)" the corresponding *motivic stringy invariant* (for a suitable $r \in \mathbb{N}$):

$$\mathcal{E}^{str}((X, D)) \in \widehat{\mathbf{M}}(\mathcal{V}/X)[\mathbf{L}^{\frac{1}{r}}] .$$

Let D be a $\mathbb{Q}$-Weil divisor on the normal and irreducible complex variety X such that $K_X + D$ is a $\mathbb{Q}$-Cartier divisor (with $r \cdot (K_X + D)$ a Cartier divisor) satisfying the following condition: There is a resolution of singularities $\pi : M \to X$ with the exceptional locus E and the support of $K_\pi(D) := K_M - \pi^*(K_X + D)$ contained in a normal crossing divisor with smooth irreducible components E_i $(i \in I)$ such that $K_\pi(D) \simeq \sum_i a_i \cdot E_i$, with all $a_i \in \mathbb{Q}$ satisfying the inequality $a_i > -1$. Then we set

$$\mathcal{E}^{str}((X, D)) := \pi_* \left(\int_{\mathcal{L}(M)} \mathbf{L}^{-ord(K_\pi(D))} d\tilde{\mu}_M \right) , \qquad (11.21)$$

which can be more explicitly calculated as in (11.17). Once more this is an invariant of the pair (X, D), not depending on the resolution π by the transformation rule. In the language of [DL1,DL2,DL3] it is for $D = 0$ just the "motivic Gorenstein volume of the arc space $\mathcal{L}(X)$" of the singular space X, i.e. the following "motivic integral" on the singular space X:

$$\mathcal{E}^{str}((X)) = \int_{\mathcal{L}(X)} \mathbf{L}^{-ord(K_X)} d\tilde{\mu}_X .$$

Note that by our conventions $\mathcal{E}^{str}((X, D)) = \mathcal{E}^{arc}((X, D))$ in case D a Cartier divisor (with strict normal crossing) on a smooth manifold $X = M$.

11.3. *Stringy/arc E-function and Euler characteristic*

By application of suitable transformations, one can build from the motivic invariants $\mathcal{E}^{str}((X, D))$ and $\mathcal{E}^{arc}((X, D))$ other invariants. For example by pushing

940

down by a constant map:

$$const_* : \widehat{\mathbf{M}}(\mathcal{V}/X)[\mathbf{L}^{\frac{1}{r}}] \to \widehat{\mathbf{M}}(\mathcal{V})[\mathbf{L}^{\frac{1}{r}}],$$

one can transform these "relative invariants over X" to "absolute invariants" (with $r = 1$ in the case of "arc invariants"). And then one can apply for example the "E-function characteristic"

$$E : \widehat{\mathbf{M}}(\mathcal{V})[\mathbf{L}^{\frac{1}{r}}] \to \mathbb{Z}[u,v][[(uv)^{-1}]][(uv)^{\frac{1}{r}}],$$

which is defined with the help of Deligne's mixed Hodge theory. Then

$$E_{str}((X,D)) := E\big(\mathcal{E}^{str}((X,D))\big)$$

becomes Batyrev's *stringy E-function* of the Kawamata log terminal pair (X,D) (as in [Bat1]). Similarly

$$E_{arc}(X) := E\big(\mathcal{E}^{arc}(X)\big)$$

is the "Hodge-arc invariant" of X in the sense of [DL1, §6] and [DL2, §4.4.1] (up to a normalization factor $(uv)^{dim(X)}$ coming from a different normalization of the motivic measure).

Here $E : K_0(\mathcal{V}) \to \mathbb{Z}[u,v]$ is induced from

$$X \mapsto E(X) := \sum_{i,p,q \geq 0} (-1)^i \cdot dim_C \big(gr_F^p gr_{p+q}^W H_c^i(X^{an}, \mathbb{C})\big) u^p v^q, \quad (11.22)$$

with F the decreasing Hodge filtration and W the increasing weight filtration of Deligne's canonical and functorial *mixed Hodge structure* on $H_c^i(X^{an}, \mathbb{Q})$ [De1, De2]. Here X^{an} means the complex algebraic variety X with its classical (and not the Zariski) topology. This E-polynomial satisfies the defining "additivity" relation of $K_0(\mathcal{V})$, because the corresponding long exact cohomology sequence is strictly compatible with the filtrations F and W (i.e. the sequence remains exact after application of $gr_F^p gr_{p+q}^W$).

In particular, $E(1,1)(X) = \chi(X)$ is the topological Euler characteristic of X. Finally classical Hodge theory implies, for X *smooth and compact*, the "purity result" $gr_{p+q}^W H^i(X^{an}, \mathbb{C}) = 0$ for $p + q \neq i$, together with

$$h^{p,q}(X) := \sum_{i \geq 0} (-1)^i (-1)^{p+q} \cdot dim_C \big(gr_F^p gr_{p+q}^W H_c^i(X^{an}, \mathbb{C})\big)$$

$$= dim_C \big(gr_F^p H^{p+q}(X^{an}, \mathbb{C})\big) = dim_C H^q(X^{an}, \Lambda^p T^* X^{an})$$

$$= dim_C H^q(X, \Lambda^p T^* X).$$

Remark 11.23. One can get the transformation $E : K_0(\mathcal{V}) \to \mathbb{Z}[u, v]$ also as an application of Theorem 8.3 (but in a less explicit way), since the invariant

$$d_X := E(X) = \sum_{p,q \geq 0} (-1)^{p+q} \cdot dim_{\mathbb{C}} H^q(X, \Lambda^p T^* X) u^p v^q$$

for X compact and smooth satisfies the corresponding properties (iii-1) and (iii-2).

In particular, $\chi_y(X) = E(-y, 1)(X)$ for X smooth and compact by (**g-HRR**), so that this E-function is another generalization of the χ_y-genus. But the classes $[X]$ for X smooth and compact generate $K_0(\mathcal{V})$ so that we get the following Hodge theoretic description for any X (with T_{y*} our Hirzebruch class transformation of Theorem 8.5):

$$T_{y*}([X]) = \sum_{i,p \geq 0} (-1)^i dim_{\mathbb{C}} \left(gr_F^p H_c^i(X^{an}, \mathbb{C}) \right) (-y)^p = E(-y, 1)(X) .$$

(11.24)

Moreover, for $X \neq \emptyset$ of dimension d, $\chi_y(X) := E(-y, 1)(X)$ is a polynomial of degree d, with $E(\mathbf{L}) = E(\mathbb{C}) = uv \in \mathbb{Z}[u, v]$ so that one gets an induced map

$$E : \widehat{\mathbf{M}}(\mathcal{V})[\mathbf{L}^{\frac{1}{r}}] \to \mathbb{Z}[u, v][[(uv)^{-1}]][(uv)^{\frac{1}{r}}] .$$

By (11.17) we get the following explicit description of $E_{str}((X, D))$, with $\pi : M \to X$ a resolution of singularities such that $K_\pi(D) \simeq \sum_i a_i \cdot E_i$ is a strict normal crossing divisor with $a_i > -1$ for all i as before (and similarly for $E_{arc}((X))$):

$$\begin{aligned} E_{str}((X, D)) &= \sum_{I \subset \{1,\ldots,k\}} E(E_I^o) \cdot \prod_{i \in I} \frac{uv - 1}{(uv)^{a_i+1} - 1} \\ &= \sum_{I \subset \{1,\ldots,k\}} E(E_I) \cdot \prod_{i \in I} \left(\frac{uv - 1}{(uv)^{a_i+1} - 1} - 1 \right) . \end{aligned}$$

(11.25)

Putting $(u, v) = (-y, 1)$ gives a similar formula for (or defines) the "stringy χ_y-characteristic" $\chi_y^{str}((X, D))$ (or the "arc χ_y-characteristic" $\chi_y^{arc}((X))$), and also the limit $u, v \to 1$ exists with

$$\begin{aligned} \chi^{str}((X, D)) &:= \lim_{u,v \to 1} E_{str}((X; D)) \\ &= \sum_{I \subset \{1,\ldots,k\}} \chi(E_I^o) \cdot \prod_{i \in I} \frac{1}{a_i + 1} \\ &= \sum_{I \subset \{1,\ldots,k\}} \chi(E_I) \cdot (-1)^{|I|} \cdot \prod_{i \in I} \frac{a_i}{a_i + 1} . \end{aligned}$$

(11.26)

This $\chi^{str}((X, D))$ is just Batyrev's *stringy Euler number* of the log-terminal pair (X, D) (as defined in [Bat1]). Similarly $\chi^{arc}(X)$ is just the *arc Euler characteristic* of X in the sense of [DL1, §6] and [DL2, §4.4.1]. Finally note that (11.25) and the "limit formula" (11.13) for the elliptic class $\mathcal{E}ll((X, D))$ of the pair (X, D) imply for X compact (with $y = e^{2\pi i z}$)):

$$\lim_{\tau \to i\infty} y^{dim(X)/2} \cdot ell((X, D)) = \chi^{str}_{-y}((X, D)) = E_{str}((X, D))(y, 1) . \quad (11.27)$$

11.4. *Stringy and arc characteristic classes*

Recall our motivic characteristic class transformations mC_* form Corollary 8.4, T_{y*} from Theorem 8.5 and $\tilde{T}_{y*}$ from Remark 11.7. Here $T_{y,i}(\ ) = (1 + y)^{-i} \cdot \tilde{T}_{y,i}(\ )$ for all i, so that both classes carry the same information. These classes all satisfy $cl_*([\mathbb{C}]) = -y$, so that they induce similar transformations on $K_0(\mathcal{V}/X)[\mathbf{L}^{-1}]$:

$$mC_* : K_0(\mathcal{V}/X)[\mathbf{L}^{-1}] \to \mathbf{G}_0(X) \otimes \mathbb{Z}[y, y^{-1}] ,$$
$$T_{y*}, \tilde{T}_{y*} : K_0(\mathcal{V}/X)[\mathbf{L}^{-1}] \to H_*^{BM}(X) \otimes \mathbb{Q}[y, y^{-1}] .$$

And these extend by [BSY3, Corollary 2.1.1, Corollary 3.1.1] to the completions

$$mC_*^\wedge : \widehat{M}(\mathcal{V}/X)[\mathbf{L}^{\frac{1}{r}}] \to \mathbf{G}_0(X) \otimes \mathbb{Z}[y][[y^{-1}]][(-y)^{\frac{1}{r}}] ,$$
$$T_{y*}^\wedge, \tilde{T}_{y*}^\wedge : \widehat{M}(\mathcal{V}/X)[\mathbf{L}^{\frac{1}{r}}] \to H_*^{BM}(X) \otimes \mathbb{Q}[y][[y^{-1}]][(-y)^{\frac{1}{r}}] . \quad (11.28)$$

So we can introduce for $cl_* = mC_*, T_{y*}, \tilde{T}_{y*}$ the corresponding *stringy characteristic homology class* $cl_*^{str}((X, D))$ of the Kawamata log-terminal pair (X, D) by

$$cl_*^{str}((X, D)) := cl_*^\wedge\left(\mathcal{E}^{str}((X, D))\right) . \quad (11.29)$$

Moreover these transformations $cl_*^\wedge$ commute with proper push down and exterior products, in particular they are a ring homomorphisms for $X = \{pt\}$. Therefore one gets from the commutative diagram

$$
\begin{array}{ccccc}
X \times \{pt\} & \xrightarrow[\sim]{p_X} & X & \xrightarrow{a_X} & \{pt\} \\
{\scriptstyle f \times id_{pt}}\downarrow & & \downarrow{\scriptstyle f} & & \| \\
Y \times \{pt\} & \xrightarrow[\sim]{p_Y} & Y & \xrightarrow{a_Y} & \{pt\} ,
\end{array}
$$

with $f : X \to Y$ proper, $\alpha \in \widehat{M}(\mathcal{V}/X)$ and $\beta \in \widehat{M}(\mathcal{V})$, the following important equality:

$$
\begin{aligned}
cl_*^\wedge\big(f_*(\alpha \cdot (a_X)^*\beta)\big) &= cl_*^\wedge\big(f_* p_{X*}(\alpha \boxtimes \beta)\big) \\
&= cl_*^\wedge\big(p_{Y*}(f \times id_{pt})_*(\alpha \boxtimes \beta)\big) \\
&= p_{Y*}\big(cl_*^\wedge(f_*(\alpha)) \boxtimes cl_*^\wedge(\beta)\big) \\
&= \big(f_*(cl_*^\wedge(\alpha))\big) \cdot (a_Y)^* cl_*^\wedge(\beta) \, .
\end{aligned}
\tag{11.30}
$$

By (11.17) we get the following explicit description of $cl_*^{str}((X, D))$, with $\pi : M \to X$ a resolution of singularities such that $K_\pi(D) \simeq \sum_i a_i \cdot E_i$ is a strict normal crossing divisor with $a_i > -1$ for all i as before:

$$
\begin{aligned}
cl_*^{str}((X, D)) &= \sum_{I \subset \{1,\ldots,k\}} cl_*([E_I^o \to X]) \cdot \prod_{i \in I} \frac{(-y) - 1}{(-y)^{a_i+1} - 1} \\
&= \sum_{I \subset \{1,\ldots,k\}} cl_*([E_I \to X]) \cdot \prod_{i \in I} \frac{(-y) - (-y)^{a_i+1}}{(-y)^{a_i+1} - 1} \, .
\end{aligned}
\tag{11.31}
$$

But E_I is a closed smooth submanifold of M so that $cl_*([E_I \to X])$ is just the proper pushforward to X of the corresponding characteristic (homology) class

$$
cl_*(E_I) = cl^*(TE_I) \cap [E_I] \quad \text{for} \quad cl_* = mC_*, T_{y*}, \tilde{T}_{y*} \, .
$$

The *stringy Hirzebruch classes* $T_{y*}^{str}((X, D))$ and $\tilde{T}_{y*}^{str}((X, D))$ interpolate by (11.13) and (11.31) in the following sense between the *elliptic class* $\mathcal{E}ll_*((X, D))$ of Borisov-Libgober defined in (11.10):

$$
\lim_{\tau \to i\infty} y^{dim(X)/2} \cdot \mathcal{E}ll((X, D))(z, \tau) = \tilde{T}_{-y*}^{str}((X, D)) \quad \text{for } y = e^{2\pi i z} \, , \tag{11.32}
$$

and for compact X the *stringy E-function* $E_{str}((X, D))$ of Batyrev as in (11.25):

$$
\begin{aligned}
\chi_{-y}^{str}((X, D)) :&= deg\big(T_{-y*}^{str}((X, D))\big) \\
&= deg\big(\tilde{T}_{-y*}^{str}((X, D))\big) = E_{str}((X, D))(y, 1) \, .
\end{aligned}
\tag{11.33}
$$

So these stringy Hirzebruch classes are "in between" the elliptic class and the stringy E-function, and as suitable limits they are "weaker" than these more general invariants. But they have the following good properties of both of them:

- The stringy Hirzebruch classes come from a functorial "additive" characteristic homology class.
- The stringy E-function comes from the "additive" *E-polynomial* defined by Hodge theory, which does not have a homology class version (compare with [BSY3, §5]).

- The elliptic class is a homology class, which does not come from an "additive" characteristic class (of vector bundles), since the corresponding *elliptic genus* is more general than the *Hirzebruch χ_y-genus*, which is the most general "additive" genus of such a class.

Finally the stringy Hirzebruch class $T_{y*}^{str}((X, D))$ specializes for $y = -1$ in the following way to the *stringy Chern class $c_*^{str}((X, D))$* of (X, D) as introduced in [Alu4,FLNU]:

$$\lim_{y \to -1} T_{y*}^{str}((X, D)) = c_*^{str}((X, D)) \in H_*^{BM}(X) \otimes \mathbb{Q}. \tag{11.34}$$

In fact

$$
\begin{aligned}
&\lim_{y \to -1} T_{y*}^{str}((X, D)) \\[2mm]
&= \sum_{I \subset \{1,\dots,k\}} T_{-1*}([E_I^o \to X]) \cdot \prod_{i \in I} \frac{1}{a_i + 1} \\[2mm]
&= \sum_{I \subset \{1,\dots,k\}} T_{-1*}([E_I \to X]) \cdot (-1)^{|I|} \cdot \prod_{i \in I} \frac{a_i}{a_i + 1} \, .
\end{aligned}
\tag{11.35}
$$

So by Theorem 8.7 (for $y = -1$) we get:

$$
\begin{aligned}
&\lim_{y \to -1} T_{y*}^{str}((X, D)) \\[2mm]
&= c_* \Big(\sum_{I \subset \{1,\dots,k\}} \pi_*(1_{E_I^o}) \cdot \prod_{i \in I} \frac{1}{a_i + 1} \Big) \\[2mm]
&= \sum_{I \subset \{1,\dots,k\}} (-1)^{|I|} \cdot \prod_{i \in I} \frac{a_i}{a_i + 1} \cdot \pi_*(c_*(E_I)) \, .
\end{aligned}
\tag{11.36}
$$

And the right hand side is just $c_*^{str}((X, D))$ by [Alu4, §§3.4,5.5,6.5] and [FLNU, Corollary 2.5, §4]. In a similar way one gets for $cl_* = mC_*, T_{y*}, \tilde{T}_{y*}$ the *arc characteristic classes*

$$cl_*^{arc}((X, D)) := cl_*^{\wedge}(\mathcal{E}^{arc}((X, D))) \, , \tag{11.37}$$

with

$$\lim_{y \to -1} T_{y*}^{arc}((X, D)) = c_*^{arc}((X, D)) \in H_*^{BM}(X) \otimes \mathbb{Q} \tag{11.38}$$

the Chern class $\int_X \mathbb{1}(-D) \, dc_X$ of the pair $(X, -D)$ as introduced and studied in [Alu4, §§3.3,5.5], with "$\mathbf{L}^{-ord(K_\pi(D))}$ corresponding to $\mathbb{1}(-D)$ for $\mathbf{L} \to -y \to 1$".

Of course it is also natural to look at the other specializations $y \to 0$ and $y \to 1$ of the stringy and arc characteristic classes $cl_*^{str/arc}((X, D))$ for $cl_* =$

$mC_*, T_{y*}, \tilde{T}_{y*}$. But the limit $y \to 1$ doesn't exist in general so that one *cannot* introduce "stringy or arc L-classes and signature" in this generality. But if we specialize in (11.31) for $D = 0$ to $y = 0$, then we get by "additivity":

$$\lim_{y \to 0} mC_*^{str}(X) = \pi_*([\mathcal{O}_M]) = \lim_{y \to 0} mC_*^{arc}(X)$$

and

$$\lim_{y \to 0} T_{y*}^{str}(X) = \pi_*(Td^*(TM) \cap [M]) = \lim_{y \to 0} T_{y*}^{arc}(X) \,.$$

In particular the middle terms are independent of a resolution $\pi : M \to X$, whose exceptional locus is contained in a strictly normal crossing divisor. And by the "weak factorization theorem" one can even conclude (compare [BSY3, Corollary 3.2]):

Proposition 11.39. *Let* $\pi : M \to X$ *be a resolution of singularities of the pure dimensional complex algebraic variety* X. *Then the classes*

$$\pi_*([\mathcal{O}_M]) \in \mathbf{G}_0(X) \quad \text{and} \quad \pi_*(Td^*(TM) \cap [M]) \in H_*^{BM}(X) \otimes \mathbb{Q}$$

are independent of π.

Note that this result implies by the projection formula a conjecture of Rosenberg [Ro] about "an analogue of the Novikov Conjecture in complex algebraic geometry" (compare also with [BW]). Similarly one can use our stringy characteristic classes in the context of "higher genera" in the spirit of the "higher elliptic genera" of [BL3], even in the context of K-homology. This will be explained in a future work.

Acknowledgments

It is a pleasure to give our thanks to P. Aluffi, J.-P. Brasselet, A. Libgober, P. Pragacz, J. Seade, T. Suwa, W. Veys and A. Weber for valuable conversations on different aspects of this subject.

This survey is a combined, modified and extended version of the authors' two talks at "Singularities in Geometry and Topology" (the 5th week of Ecole de la Formation Permanente du CNRS - Session résidentielle de la FRUMAM) held at Luminy, Marseille, during the period of 21 February – 25 February 2005. The authors would like to thank the organizers of the conference for inviting us to give these talks. The second named author (S.Y.) also would like to thank the staff of ESI (Erwin Schrödinger International Institute for Mathematical Physics, Vienna, Austria), where a part of the paper was written in August 2005, for providing a nice atmosphere in which to work.

References

AKMW. D. Abramovich, K. Karu, K. Matsuki and J. Włodarczyk, *Torification and factorization of birational maps*, J. Amer. Math. Soc., **15** (2002), 531–572.

Alu1. P. Aluffi, *Characteristic classes of singular varieties*, Topics in Cohomological Studies of Algebraic Varieties (Ed. P. Pragacz), Trends in Mathematics, Birkhäuser (2005), 1–32.

Alu2. P. Aluffi, *Chern classes of birational varieties*, Int. Math. Res. Notices **63**(2004), 3367–3377.

Alu3. P. Aluffi, *Limits of Chow groups, and a new construction of Chern–Schwartz–MacPherson classes*, math.AG/0507029.

Alu4. P. Aluffi, *Modification systems and integration in their Chow groups*, Selecta Math. **11** (2005), 155–202.

AM. M. Artin and B. Mazur, *Etale Homotopy*, Springer Lecture Notes in Math. No. **100**, Springer-Verlag, Berlin, 1969.

AS. M. F. Atiyah, I. M. Singer, *The index of elliptic operators on compact manifolds*, Bull. Amer. Math. Soc. **69** (1963), 422-433.

Ax. J. Ax, *Injective endomorphisms of varieties and schemes*, Pacific J. Math. **31** (1969), 1–17.

Ban. M. Banagl, *Extending intersection homology type invariants to non-Witt spaces*, Mem. Amer. Math. Soc. 160, 2002.

Bat1. V. Batyrev, *Non-Archimedean integrals and stringy Euler numbers of log-terminal pairs*, J. Eur. Math. Soc. **1** (1999, 5-33.

Bat2. V. Batyrev, *Birational Calabi–Yau n-folds have equal Betti numbers*, In *New trends in algebraic geometry*, London Math. Soc. Lecture Note Ser. **264** (1999), 1–11.

BFM1. P. Baum, W. Fulton and R. MacPherson, *Riemann–Roch for singular varieties*, Publ. Math. I.H.E.S. **45** (1975), 101–145.

BFM2. P. Baum, W. Fulton and R. MacPherson, *Riemann–Roch and topological K-theory for singular varieties*, Acta Math. **143** (1979), 155-192.

Bit. F. Bittner, *The universal Euler characteristic for varieties of characteristic zero*, Compositio Math. **140** (2004), 1011–1032.

BW. J. Block and S. Weinberger, *Higher Todd classes and holomorphic group actions*, math.AG/0511305.

BoMo. A. Borel and J. Moore, Homology theory for locally compact spaces, Michigan J. Math., **7** (1960), 137–159.

BoSe. A. Borel and J.-P. Serre, *Le théorème de Riemann–Roch (d'apres Grothendieck)*, Bull. Soc. Math. France, **86** (1958), 97–136.

BM. W. Borho and R. MacPherson, *Partial Resolutions of Nilpotent Varieties*, Astérisque **101–102** (1983), 23–74.

BL1. L. Borisov and A. Libgober, *Elliptic genera for singular varieties*, Duke Math. J. **116** (2003), 319-351.

BL2. L. Borisov and A. Libgober, *McKay correspondence for elliptic genera*, Ann. of Math. **161** (2005), 1521–1569.

BL3. L. Borisov and A. Libgober, *Higher elliptic genera*, math.AG/0602387.

Bor. K. Borusk, *Theory of Shape*, Lecture Notes No. **28**, Matematisk Inst. Aarhus Univ. (1971), 1–145.

BF.   H.W. Braden and K.E. Feldman, *Functional equations and the generalized elliptic genus*, math-ph/0501011.

Br1.   J.-P. Brasselet, *Existence des classes de Chern en théorie bivariante*, Astérisque,**101–102** (1981), 7–22.

Br2.   J.-P. Brasselet, *From Chern classes to Milnor classes – a history of characteristic classes for singular varieties*, Singularities–Sapporo 1998, Adv. Stud. Pure Math. **29** (2000), 31–52

Br3.   J.-P. Brasselet, *Characteristic classes of Singular Varieties*, monograph in preparation.

BLSS.   J.-P. Brasselet, D. Lehman, J. Seade and T. Suwa, *Milnor classes of local complete intersection*, Tran. Amer. Math. Soc. **354** (2001), 1351–1371.

BSS.   J.-P. Brasselet, J. Seade and T. Suwa, *Indices of vector fields and characteristic classes of singular varieties*, monograph in preparation.

BSY1.   J.-P. Brasselet, J. Schürmann and S. Yokura, *On the uniqueness of bivariant Chern classes and bivariant Riemann–Roch transformations*, to appear in Adv. in Math.

BSY2.   J.-P. Brasselet, J. Schürmann and S. Yokura, *On Grothendieck transformation in Fulton–MacPherson's bivariant theory*, preprint (2006).

BSY3.   J.-P. Brasselet, J. Schürmann and S. Yokura, *Hirzebruch classes and motivic Chern classes for singular spaces*, math. AG/0503492

BSY4.   J.-P. Brasselet, J. Schürmann and S. Yokura, *Classes de Hirzebruch et classes de Chern motivique*, C. R. Acad. Sci. Paris Ser. I **342** (2006), 325–328.

BrSc.   J.-P. Brasselet and M.-H. Schwartz, *Sur les classes de Chern d'une ensemble analytique complexe*, Astérisque **82–83**(1981), 93–148.

Bre.   G. E. Bredon, *Sheaf Theory (2nd Edition)*, Graduate Texts in Math., Springer, 1997.

BZ.   J.-L. Brylinski and B. Zhang, *Equivariant Todd Classes for Toric Varieties*, math. AG/0311318.

CS1.   S. Cappell and J. L. Shaneson, *Stratifiable maps and topological invariants*, J. Amer. Math. Soc. **4** (1991), 521–551.

CS2.   S. Cappell and J. L. Shaneson, *Genera of algebraic varieties and counting lattice points*, Bull. Amer. Math. Soc. **30** (1994), 62–69.

CMS.   S. Cappell, L. Maxim and J. L. Shaneson, *Hodge genera of algebraic varieties, I*, math.AG/0606655.

Che.   J. Cheeger, *Spectral geometry of singular Riemannian spaces*, J. Diff. Geom. **18** (1983), 575–657.

Ch1.   S. S. Chern, *Characteristic classes of Hermitian manifolds*, Ann. Math. **47** (1946), 85–121.

Ch2.   S. S. Chern, *On the multiplication in the characteristic ring of a sphere bundle*, Ann. Math. **49** (1948), 362–372.

Ch3.   S. S. Chern, *A simple intrinsic proof of the Gauss–Bonnet formula for closed Riemann manifolds*, Ann. of Math. **45** (1944), 747–752.

CG.   N. Chriss and V. Ginzburg, *Representation theory and complex geometry*, Birkhäuser, 1997.

Cr.   A. Craw, *An introduction to motivic integration*, in *Strings and Geometry*, Clay Math. Proc., **3** (2004), Amer. Math. Soc., 203–225.

DL1.   J. Denef and F. Loeser, *Germs of arcs on singular algebraic varieties and motivic integration*, Invent. Math. **135** (1999), 201–232

DL2.   J. Denef and F. Loeser, *Geometry on arc spaces of algebraic varieties*, European Congress of Mathematicians (Barcelona, 2000), **1** (2001) Birkhäuser, 327–348

DL3.   J. Denef and F. Loeser, *Motivic integration, quotient singularitis and the McKay correspondence*, Comp. Math. **131** (2002), 267–290.

EG1.   D. Edidin and W. Graham, *Riemann–Roch for equivariant Chow groups*, Duke Math. J. **102** (2000), 567–594.

EG2.   D. Edidin and W. Graham, *Riemann–Roch for quotients and Todd classes of simplicial toric varieties*, Commun. Algebra **31**, No.8 (2003), 3735–3752.

De1.   P. Deligne, *Théorie des Hodge II*, Publ. Math. IHES **40** (1971), 5-58.

De2.   P. Deligne, *Théorie des Hodge III*, Publ. Math. IHES **44** (1974), 5-78.

Dim.   A. Dimca, *Sheaves in Topology*, Springer Universitext, 2004

DuBo.   Ph. DuBois, *Complexe de De Rham filtré d'une variété singulière*, Bull. Soc. Math. France, **109** (1981), 41–81.

Er.   L. Ernström, *Topological Radon transforms and the local Euler obstruction*, Duke Math. J. **76** (1994), 1–21

FLNU.   T. de Fernex, E. Lupercio, T. Nevins and B. Uribe, *Stringy Chern classes of singular varieties*, arXiv:math.AG/0407314, to appear in Adv. in Math.

Fu1.   W. Fulton, *Intersection Theory*, Springer Verlag, 1981.

Fu2.   W. Fulton, *Introduction to Toric Varieties*, Ann. of Math. Studies, No. **131**, Princeton Univ. Press, 1993.

FJ.   W. Fulton and K. Johnson, *Canonical classes on singular varieties*, Manus. Math. **32** (1980), 381–389.

FM.   W. Fulton and R. MacPherson, *Categorical frameworks for the study of singular spaces*, Memoirs of Amer. Math. Soc. **243**, 1981.

FuMC.   J. Fu and C. McCrory, *Stiefel–Whitney classes and the conormal cycle of a singular variety*, Trans. Amer. Math. Soc. **349** (1997), 809–835

Gi1.   V. Ginzburg, *𝔊–Modules, Springer's Representations and Bivariant Chern Classes*, Adv. in Math., **61** (1986), 1–48.

Gi2.   V. Ginzburg, *Geometric methods in the representation theory of Hecke algebras and quantumgroups*, In: A. Broer and A. Daigneault (Eds.): *Representation theories and algebraic geometry (Montreal, PQ, 1997)*, Kluwer Acad. Publ., Dordrecht, 1998, 127–183.

GM1.   M. Goresky and R. MacPherson, *Intersection homology theory*, Topology **149** (1980), 155–162.

GM2.   M. Goresky and R. MacPherson, *Intersection homology, II*, Inventiones Math. **149** (1980), 77–129.

Grom.   M. Gromov, *Endomorphisms of symbolic algebraic varieties*, J. Eur. Math. Soc. **1** (1999), 109–197.

Gros.   M. Gros, *Classes de Chern et classes de cycles en cohomologie de Hodge-Witt logarithmique*, Bull. Soc. Math. France Mem. 21, 1985.

Grot1. A. Grothendieck, *Technique de descente et théorèmes d'existence en géométrie algébrique, II*, Séminaire Bourbaki, 12 ème année, exposé 190-195 (1959-60).

Grot2. A. Grothendieck, *Récoltes er Semailles– Réflexions et Témoignages sur un passé de mathématicien*, Reprint, 1985.

GNA.   F. Guillén and V. Navarro Aznar, *Un critère d'extension des foncteurs définis sur les schémas lisses*, Publ. Math. I.H.E.S. **95** (2002), 1–91.

Höhn.   G. Höhn, *Komplexe elliptische Geschlechter und S^1-äquivariante Kobordismustheorie*, Diplomarbeit, Bonn 1991.

Hi.   H. Hironaka, *Resolution of singularities of an algebraic variety over a field of characteristic zero*, Ann. of Math. **79** (1964), 109–326.

Hir1.   F. Hirzebruch, *Some problems on differentiable and complex manifolds*, Ann. of Math. **60** (1954), 213–236.

Hir2.   F. Hirzebruch, *Topological Methods in Algebraic Geometry, 3rd ed. (1st German ed. 1956)*, Springer-Verlag, 1966.

Hir3.   F. Hirzebruch, *The Signature Theorem: Reminiscences and Recreation*, in "Prospects in Mathematics", Ann. of Math. Studies, No.**70**, Princeton Univ. Press, 1971.

HBJ.   F. Hirzebruch, T. Berger and R. Jung, *Manifolds and Modular Forms*, Vieweg, 1992.

Hus.   D. Husemoller, *Fibre Bundles, 3rd ed.*, Springer-Verlag, 1994.

KS.   M. Kashiwara and P. Schapira, *Sheaves on Manifolds*, Springer-Verlag, Berlin, Heidelberg, 1990.

Ken.   G. Kennedy, *MacPherson's Chern classes of singular varieties*, Com. Algebra. **9** (1990), 2821–2839.

KMY.   G. Kennedy, C. McCrory and S. Yokura, *Natural transformations from constructible functions to homology*, C. R. Acad. Sci. Paris, **t.319** (1994), 969–973.

KM.   J. Kollár and S. Mori, *Birational geometry of algebraic varieties*, Cambridge Tracts in Math. 124, Cambridge Univ. Press, 1998.

Kon.   M. Kontsevich, *Lecture at Orsay*, 1995.

Krich. I. Krichever, *Generalized elliptic genera and Baker-Akhiezer functions*, Math. Notes **47** (1990), 132–142.

Kw1.   M. Kwieciński, *Formule du produit pour les classes caractéristiques de Chern-Schwartz-MacPherson et homologie d'intersection*, C. R. Acad. Sci. Paris, **314** (1992), 625–628.

Kw2.   M. Kwieciński, *Sur le transformé de Nash et la construction du graph de MacPherson*, In Thèse, Université de Provence, 1994.

KY.   M. Kwieciński and S. Yokura, *Product formula of the twisted MacPherson class*, Proc. Japan Acad., **68** (1992), 167–171.

Levy.   R. Levy, *Riemann-Roch theorem for complex spaces*, Acta Math. **158** (1987), 149-188.

Lo.   E. Looijenga, *Motivic measures*, Séminaire Bourbaki, Astérisque **276** (2002), 267–297.

Mac1.   R. MacPherson, *Chern classes for singular algebraic varieties*, Ann. of Math. **100** (1974), 423–432.

Mac2.   R. MacPherson, *Characteristic classes for singular varieties*, Proceedings of the 9-th Brazilian Mathematical Colloquium (Poços de Caldas, 1973) Vol.II, Instituto de Matemática Pura e Aplicada, São Paulo, (1977), 321–327.

Man.   Yu. I. Manin, *Lectures on the K-functor in algebraic geometry*, Russ. Math. Survey, **24** (1969), 1–89.

MaSe.   S. Mardesić and J. Segal, *Shape Theory*, North-Holland, 1982.

Mi1.   J. W. Milnor , *Topology from the differentiable Viewpoint*, Univ. Virginia Press 1965.

Mi2.   J. W. Milnor , *Singular points of complex hypersurfaces*, Ann. of Math. Studies, No. **61**, Princeton Univ. Press, 1968.

MiSt.   J. W. Milnor and J. D. Stasheff, *Characteristic classes*, Ann. of Math. Studies **76**, Princeton Univ. Press 1974.

Nov.   S. P. Novikov, *Topological invariance of rational classes of Pontrjagin*, Dokl. Akad. Nauk SSSR **163** (1965), 298–300, English translation, Soviet Math. Dokl. **6** (1965), 921–923.

Oh.   T. Ohmoto, *Equivariant Chern classes of singular algebraic varieties with group actions*, Math. Proc. Cambridge Phil. Soc. **140** (2006), 115–134.

Pan.   I. Panin, *Riemann-Roch theorems for oriented cohomology*, Axiomatic, Enriched and Motivic Homotopy Theory, (Ed. J.P.C. Greenless), NATO Science Series, Kluwer Academic Publ. (2004), 261–333.

Pa.   A. Parusiński, *Characteristic classes of singular varieties*, Singularity Theory and Its Applications, Sapporo, September 16–25, 2003.

PP1.   A. Parusiński and P. Pragacz, *Chern–Schwartz–MacPherson classes and the Euler characteristic of degeneracy loci and special divisors*, J. Amer. Math. Soc., **8** (1995), 793–817.

PP2.   A. Parusiński and P. Pragacz, *Characteristic classes of hypersurfaces and characteristic cycles*, J. Algebraic Geometry, **10** (2001), 63–79.

Pontr.   L. Pontrjagin, *Characteristic cycles on differentiable manifolds*, Matematicheskii Sbornik (NS), 21 (63) , 233–284 (1947); AMS (1950), Translation n. 32.

Ro.   J. Rosenberg, *An analogue of the Novikov Conjecture in complex algebraic geometry*, math.AG/0509526.

RW.   J. Rosenberg and S. Weinberger, *The signature operator at 2*, Topology **45** (2006), 47–63.

Sa.   C. Sabbah, *Espaces conormaux bivariants*, Thèse, l'Université Paris, 1986.

Sai.   M. Saito, *Mixed Hodge Modules*, Publ. RIMS., Kyoto Univ. **26** (1990), 221–333.

Sch1.   J. Schürmann, *A generalized Verdier-type Riemann–Roch theorem for Chern–Schwartz– MacPherson classes*, math.AG/0202175.

Sch2.   J. Schürmann, *A general construction of partial Grothendieck transformations*, math. AG/0209299.

Sch3.   J. Schürmann, *Topology of singular spaces and constructible sheaves*, Monografie Matematyczne **63** (New Series), Birkhäuser, Basel, 2003.

Sch4.   J. Schürmann, *Lecture on characteristic classes of constructible functions*, Topics in Cohomological Studies of Algebraic Varieties (Ed. P. Pragacz), Trends in Mathematics, Birkhäuser (2005), 175–201.

Sch5.   J. Schürmann, *Characteristic cycles and indices of 1-forms on singular spaces*, in preparation.

Schw1.   M.-H. Schwartz, *Classes caractéristiques définies par une stratiification d'une variétié analytique complex*, C. R. Acad. Sci. Paris t. **260** (1965), 3262-3264, 3535–3537.

Schw2.   M.-H. Schwartz, *Classes et caractères de Chern des espaces linéaires*, C. R. Acad. Sci. Paris Sér. I. Math. **295** (1982), 399–402.

Schw3.   M.-H. Schwartz, *Champs radiaux sur une stratification analytique*, Travaux en cours, **39** Hermann, Paris, 1991.

Sea1.   J. Seade, *Indices of vector fields and Chern classes for singular varieties*, Invitations to geometry and topology (Ed. M.R. Bridson), Oxf. Grad. Texts Math. **7** (2002), 292–320.

Sea2.   J. Seade, *Indices of vector fields on singular varieties: an overview*, math. AG/0603582.

SeSu.   J. Seade and T. Suwa *An adjunction formula for local complete intersections*, Intern. J. Math. **9** (1998), 759–768.

Serre.   J.-P. Serre, *Géométrie algébrique et géométrie analytique*, Ann. Inst. Fourier **6** (1956), 1–42.

Sh.   J. L. Shaneson, *Characteristic classes, lattice points and Euler–Maclaurin formulae*, Proceedings of the ICM'94 (Zürich, Switerland) Birkhäuser Verlag (1995), 612–624.

Si.   P. H. Siegel, *Witt spaces: A geometric cycle theory for KO-homology at odd primes*, Amer. J. of Math. **105**, (1983) 1067-1105.

Sti.   E. Stiefel, *Richtungsfelder und Fernparallelismus in Mannigfaltigkeiten*, Comm. Math. Helv. **8** (1936), 3–51.

Stong.   R. E. Stong, *Notes on Cobordism Theory*, Princeton Math. Notes, Princeton Univ. Press 1968.

Sri.   V. Srinivas, *The Hodge Characteristic*, Lecture at Jet Scheme Seminar, MSRI, December 2002.

Sull.   D. Sullivan, *Combinatorial invariants of analytic spaces*, Springer Lecture Notes in Math., **192** (1970), 165–168.

Sull2.   D. Sullivan, *Geometric Topology: Localization, Periodicity and Galois Symmetry*, K-Monographs in Mathematics **8**, Springer, 2005.

Su1.   T. Suwa, *Characteristic classes of coherent sheaves on singular varieties*, in *Singularities–Sapporo 1998*, Adv. Stud. Pure Math. **29** (2000), 279–297.

Su2.   T. Suwa, *Characteristic classes of singular varieties*, Sugaku Expositions, Amer. Math. Soc. **16** (2003), 153–175.

Su3.   T. Suwa, *Classes de Chern des intersections complétes locales*, C. R. Acad. Sci. Paris Sér. **324** (1996), 76–70.

Thom1.   R. Thom, *Espaces fibrés en spheres et carrés de Steenrod*, Ann. Sci. Ecole Norm. Sup. **69** (1952), 109–181.

Thom2.   R. Thom, *Les Classes Caractéristiques de Pontrjagin des Variétés Triangulées*, Symp. Intern. de Topologia Algebraica. Unesco (1958).

To.   B. Totaro, *Chern numbers for singular varieties and elliptic homology*, Ann. Math. **151** (2000), 757-791.

Ve1.   W. Veys, *Zeta functions and 'Kontsevich invariants' on singular varieties*, Can. J. Math. **53** (2001), 834–865.

Ve2.   W. Veys, *Arc spaces, motivic integration and stringy invariants*, Singularity Theory and Its Applications, Sapporo, September 16–25, 2003, math.AG/0401374.

Wang.   C.-L. Wang, *K-equivalence in birational geometry and characterizations of complex elliptic genera*, J. Alg. Geometry **12** (2003), 285–306.

Wh1.   H. Whitney, *Sphere spaces*, Proc. Nat. Acad. Sci. **21** (1935), 462–468.

Wh2.   H. Whitney, *On the theory of sphere bundles*, Proc. Nat. Acad. Sci. **26** (1940), 148–153.

Wi.   E. Witten, *Supersymmetry and Morse theory*, J. Diff. Geom. **17** (1982), 661–692.

W.   J. Włodarczyk, *Toroidal varieties and the weak factorization theorem*, Invent. Math., **154** (2003), 223–331.

Wo.   J. Woolf, *Witt groups of sheaves on topological spaces*, math. AT/0510196.

Y1.   S. Yokura, *A generalized Grothendieck–Riemann–Roch theorem for Hirzebruch's χ_y-characteristic and T_y-characteristic*, Publ. Res. Inst. Math. Sci., **30** (1994), 603–610.

Y2.   S. Yokura, *On Cappell–Shaneson's homology L-class of singular algebraic varieties*, Trans. Amer. Math. Soc., **347** (1995), 1005–1012.

Y3.   S. Yokura, *A singular Riemann–Roch theorem for Hirzebruch characteristics*, Banach Center Publ., **44** (1998), 257–268.

Y4.   S. Yokura, *On a Verdier-type Riemann–Roch for Chern–Schwartz–MacPherson class*, Topology and Its Applications., **94** (1999), 315–327.

Y5.   S. Yokura, *On the uniqueness problem of the bivariant Chern classes*, Documenta Mathematica, **7** (2002), 133–142.

Y6.   S. Yokura, *Bivariant theories of constructible functions and Grothendieck transformations*, Topology and Its Applications, **123** (2002), 283–296.

Y7.   S. Yokura, *On Ginzburg's bivariant Chern classes*, Trans. Amer. Math. Soc., **355** (2003), 2501–2521.

Y8.   S. Yokura, *On Ginzburg's bivariant Chern classes,II*, Geometriae Dedicata, **101** (2003), 185–201.

Y9.   S. Yokura, *Bivariant Chern classes for morphisms with nonsingular target varieties*, Central European J. Math., **3** (2005), 614–626.

Y10.   S. Yokura, *Chern classes of proalgebraic varieties and motivic measures*, math.AG/0407237.

Y11.   S. Yokura, *Characteristic classes of proalgebraic varieties and motivic measures*, math. AG/0606352.

You.   B. Youssin, *Witt Groups of Derived Categories*, K-Theory, **11** (1997), 373-395.

Zh.   W. Zhang, *Lecture on Chern–Weil Theory and Witten Deformations*, Nankai Tracts in Mathematics Vol.4, World Scientific, 2001.

Z1.   J. Zhou, *Classes de Chern en théorie bivariante*, Thèse, Université Aix-Marseille, 1995.

Z2.   J. Zhou, *Morphisme cellulaire et classes de Chern bivariantes*, Ann. Fac. Sci. Toulouse Math., **9** (2000), 161–192.

INDICES OF VECTOR FIELDS ON SINGULAR VARIETIES: AN OVERVIEW

José Seade *

Instituto de Matemáticas, UNAM,
Unidad Cuernavaca,
A. P. 273-3, Cuernavaca, Morelos,
México.
** E-mail: jseade@matcuer.unam.mx*

Dedicado a Jean-Paul,
con gran respeto y afecto.

The Poincaré-Hopf local index is the most basic invariant of a vector field on a smooth manifold at an isolated zero; and the theorem of Poincaré-Hopf about the total index of vector fields on compact manifolds establishes a deep connection between the local indices and the Euler characteristic of the manifold in question. All of this generalizes to frames (sets of vector fields) and leads toward the Chern classes of complex manifolds. In this article we overview how these concepts and results generalize to singular complex analytic varieties.

1. Introduction

The Poincaré-Hopf total index of a vector field with isolated singularities on a smooth, closed manifold M can be regarded as the obstruction to constructing a non-zero section of the tangent bundle TM. In this way it extends naturally to complex vector bundles in general and leads to the notion of Chern classes. When working with singular analytic varieties, it is thus natural to ask what should be the notion of "the index" of a vector field.

Indices of vector fields on singular varieties were first considered by M. H. Schwartz in [52,53] in her study of Chern classes for singular varieties. For her purpose there was no point in considering vector fields in general, but only a special class of vector fields (and frames) that she called "ra-

---

*Supported by CNRS (France), CONACYT and DGAPA-UNAM (Mexico)

dial", which are obtained by the important process of *radial extension*. The generalisation of this index to other vector fields was defined independently in [16,39,57] (see also [2]), and its extension for frames in general was done in [7]. This index, that we call *Schwartz index*, is sometimes called "radial index" because it measures how far the vector field is from being radial. In [2,16] this index is defined also for vector fields on real analytic varieties.

MacPherson in [47] introduced the local Euler obstruction, also for constructing Chern classes of singular complex algebraic varieties. In [9] this invariant was defined via vector fields, interpretation that was essential to prove (also in [9]) that the Schwartz classes of a singular variety coincide with MacPherson's classes. This viewpoint brings the local Euler obstruction into the frame-work of "indices of vector fields on singular varieties" and yields to another index, that we may call *the local Euler obstruction* of the vector field at each isolated singularity; the Euler obstruction of the singular variety corresponding to the case of the radial vector field. This index relates to the previously mentioned Schwartz index by a formula known as the "Proportionality Theorem" of [9]. When the vector field is determined by the gradient of a function on the singular variety, this local Euler obstruction is the defect studied in [8].

On the other hand, one of the basic properties of the local index of Poincaré-Hopf is that it is stable under perturbations. In other words, if v is a vector field on an open set U in $\mathbb{R}^n$ and $x \in U$ is an isolated singularity of v, and if we perturb v slightly, then its singularity at x may split into several singular points of the new vector field $\hat{v}$, but the sum of the indices of $\hat{v}$ at these singular points equals the index of v at x. If we now consider an analytic variety V defined, say, by a holomorphic function $f : (\mathbb{C}^n, 0) \to (\mathbb{C}, 0)$ with an isolated critical point at 0, and if v is a vector field on V, non-singular away from 0, then one would like "the index" of v at 0 to be stable under small perturbations of both, the function f and the vector field v. The extension of this index to the case of vector fields on isolated complete intersection singularity germs (ICIS for short) is immediate. This leads naturally to another concept of index, now called the *GSV-index*, introduced in [27,54,57]. There is also the analogous index for continuous vector fields on real analytic varieties (see [2,32,33]).

One also has the virtual index, introduced in [43] for holomorphic vector fields; the extension to continuous vector fields is immediate and was done in [7,57]. This index is defined via Chern-Weil theory. The idea is that the usual Poincaré-hopf index can be regarded as a localisation, at the singular points of a vector field, of the nth-Chern class of a manifold. Similarly, for an ICIS $(V, 0)$ in $\mathbb{C}^{n+k}$, defined by functions $f = (f_1, \cdots, f_k)$, one has a localisation at 0 of the top Chern class of the ambient space, defined by the gradient vector fields of the f_i and the given vector field, tangent to V. This localisation defines *the virtual index* of the vector field; this definition extends to a rather general setting, providing a topological way for looking at the top Chern class of the so-called virtual tangent bundle of singular varieties which are local complete intersections. In the case envisaged above, when $(V, 0)$ is an ICIS, this index coincides with the GSV-index.

Another remarkable property of the local index of Poincaré-Hopf is that in the case of germs of holomorphic vector fields in $\mathbb{C}^n$ with an isolated singularity at 0, the local index equals the integer:

$$\dim \mathcal{O}_{\mathbb{C}^n,0}/(X_1, \cdots, X_n), \qquad (*)$$

where $(X_1, \cdots, X_n)$ is the ideal generated by the components of the vector field. This and other facts motivated the search for algebraic formulae for the index of vector fields on singular varieties. The *homological index* of Gomez-Mont [26] is a beautiful answer to that search. It considers an isolated singularity germ $(V, 0)$ of any dimension, and a holomorphic vector field on V, singular only at 0. One has the Kähler differentials on V, and a Koszul complex $(\Omega^\bullet_{V,0}, v)$:

$$0 \to \Omega^n_{V,0} \to \Omega^{n-1}_{V,0} \to \ldots \to \mathcal{O}_{V,0} \to 0,$$

where the arrows are given by contracting forms by the vector field v. The homological index of v is defined to be the Euler characteristic of this complex. When the ambient space V is smooth at 0, the complex is exact in all dimensions, except in degree 0 where the corresponding homology group has dimension equal to the local index of Poincaré-Hopf of v. When $(V, 0)$ is a hypersurface germ, this index coincides with the GSV-index, but for more general singularities the homological index is still waiting to be understood!

In fact, in [22] there is given the corresponding notion of *homological index* for holomorphic 1-forms on singular varieties, and recent work of Schürmann throws light into this, yet mysterious, invariant.

When considering smooth (real) manifolds, the tangent and cotangent bundles are canonically isomorphic and it does not make much difference to consider either vector fields or 1-forms in order to define their indices and their relations with characteristic classes. When the ambient space is a complex manifold, this is no longer the case, but there are still ways for comparing indices of vector fields and 1-forms, and to use these to study

Chern classes of manifolds. To some extent this is also true for singular varieties, but there are however important differences and each of the two settings has its own advantages.

The first time that indices of 1-forms on singular varieties appeared in the literature was in MacPherson's work [47], where he defined the local Euler obstruction in this way. But the systematic study of these indices was begun by W. Ebeling and S. Gusein-Zade in a series of articles (see for instance [17,19–21]). This has been, to some extent, a study parallel to the one for vector fields, outlined in this article. Also along this lines is [11], which adapts to 1-forms the radial extension technique of M. H. Schwartz and proves the corresponding Proportionality Theorem.

Also, J. Schürmann in his book [50] introduces powerful methods to studying singular varieties via micro-local analysis and Lagrangian cycles, and much of the theory of indices of 1-forms can also be seen in that way. Furthermore, he has recently found a remarkable method for assigning an index of 1-forms to each constructible function on a Whitney stratified complex analytic space, in such a way that each of the known indices corresponds to a particular choice of a constructible function. This is closely related to MacPherson work in [47] for defining characteristic classes of singular varieties.

In this article we briefly review the various indices of vector fields on singular varieties. I am presently working with Jean-Paul Brasselet and Tatsuo Suwa writing [12], a monograph with a detailed account of all these indices, through the viewpoints of algebraic topology (obstruction theory) and differential geometry (Chern-Weil theory), together with their relations with Chern classes of singular varieties. This will include some applications of these indices to other fields of singularity theory.

This article grew from my talk in the singularities meeting at the CIRM in Luminy in celebration of the 60th anniversary of Jean-Paul Brasselet, and I want to thank the organizers for the invitations to participate in that meeting and to write these notes, particularly to Anne Pichon. I am also grateful to Tatsuo Suwa, Jean-Paul Brasselet and Jörg Schürmann for many helpful conversations.

2. The Schwartz index

Consider first the case when the ambient space is an affine irreducible complex analytic variety $V \subset \mathbb{C}^N$ of dimension $n > 1$ with an isolated singularity at 0. Let U be an open ball around $0 \in \mathbb{C}^N$, small enough so that every sphere in U centered at 0 meets V transversally (see [48]). For simplicity we restrict the discussion to U and set $V = V \cap U$. Let v_{rad} be a continuous vector field on $V \setminus \{0\}$ which is transversal (outwards-pointing) to all spheres around 0, and scale it so that it extends to a continuous section of $T\mathbb{C}^N|_V$ with an isolated zero at 0. We call v_{rad} a *radial vector field* at $0 \in V$. Notice v_{rad} can be further extended to a *radial* vector field $v^{\#}_{rad}$ on all of U, *i.e.* transversal to all spheres centered at 0. By definition *the Schwartz index* of v_{rad} is the Poincaré-Hopf index at 0 of the radial extension $v^{\#}_{rad}$, so it is 1. Of course we could have started with the zero-vector at 0, then extend this to v_{rad} on V as above, and then extend it further to all of U being transversal to all the spheres, getting the same answer; this is the viewpoint that generalises when the singular set of V has dimension more than 0.

Let us continue with the case when V has an isolated singularity at 0, and assume now that v is a continuous vector field on V with an isolated singularity at 0. By this we mean a continuous section v of $T\mathbb{C}^N|_V$ which is tangent to $V^* = V \setminus \{0\}$. We want to define *the Schwartz index* of v; this index somehow measures the "radiality" of the vector field. It has various names in the literature (c.f. [2,18,39,57]), one of them being *radial index*.

Let v_{rad} be a radial vector field at 0, *i.e.* v_{rad} is transversal, outwards-pointing, to the intersection of V with every sufficiently small sphere $\mathbb{S}_\varepsilon$ centered at 0. We may now define the difference between v and v_{rad} at 0: consider small spheres $\mathbb{S}_\varepsilon$, $\mathbb{S}_{\varepsilon'}$; $\varepsilon > \varepsilon' > 0$, and let w be a vector field on the cylinder X in V bounded by the links $K_\varepsilon = \mathbb{S}_\varepsilon \cap V$ and $K_{\varepsilon'} = \mathbb{S}_{\varepsilon'} \cap V$, such that w has finitely many singularities in the interior of X, it restricts to v on K_ε and to v_{rad} on $K_{\varepsilon'}$. The *difference* $d(v, v_{rad}) = d(v, v_{rad}; V)$ of v and v_{rad} is:

$$d(v, v_{rad}) = \mathrm{Ind}_{PH}(w; X) \,,$$

the Poincaré-Hopf index of w on X. Then define *the Schwartz* (or radial) *index* of v at $0 \in V$ to be:

$$\mathrm{Ind}_{Sch}(v, 0; V) = 1 + d(v, v_{rad}) \,.$$

The following result is well known (see for instance [2,18,39,57]). For vector fields with radial singularities, this is a special case of the work of M. H. Schwartz; the general case follows easily from this.

Theorem 2.1. *Let V be a compact complex analytic variety with isolated singularities $q_1, \cdots, q_r$ in a complex manifold M, and let v be a continuous vector field on V, singular at the q_i's and possibly at some other isolated* *points in V. Let $\mathrm{Ind}_{Sch}(v; V)$ be the sum of the Schwartz indices of v at the q_i plus its Poincaré-Hopf index at the singularities of v in the regular part of V. Then:*

$$\mathrm{Ind}_{Sch}(v; V) = \chi(V) \,.$$

The proof is fairly simple and we refer to the literature for details.

The idea for defining the Schwartz index in general, when the singular set has dimension more than 0, is similar in spirit to the case above, but it presents some technical difficulties. Consider a compact, complex analytic variety V of dimension n embedded in a complex manifold M, equipped with a Whitney stratification $\{V_\alpha\}_{\alpha \in A}$ adapted to V. The starting point to define the Schwartz index of a vector field is the *radial extension* introduced by M. H. Schwartz. To explain this briefly, let v be a vector field defined on a neighbourhood of 0 in the stratum V_α of V that contains 0. The fact that the stratification is Whitney implies (see [6,52,53] for details) that one can make a *parallel extension* of v to a stratified vector field v' on a neighbourhood of 0 in M. Now, if 0 is an isolated singularity of v on V_α, then v' will be singular in a disc of dimension $(\dim_{\mathbb{R}} M - \dim_{\mathbb{R}} V_\alpha)$, transversal to V_α in M at 0. So this extension is not good enough by itself. We must add to it another vector field v'': the gradient of the square of the function "distance to" V_α, defined near 0. This vector field is transversal to the boundaries of all tubular neighbourhoods of V_α in M; using the Whitney conditions we can make v'' be a continuous, stratified vector field near 0. The zeroes of v'' are the points in V_α. Adding v' and v'' at each point near 0 we get a stratified, continuous vector field $v^{\#}$ defined on a neighbourhood of 0 in M, which restricts to the given vector field v on V_α. This vector field has the additional property of being radial in all directions which are normal to the stratum V_α. In other words, if we take a small smooth disc Σ in M transversal to V_α at 0 of dimension complementary to that of V_α. Then the restriction of $v^{\#}$ to Σ can be projected into a vector field tangent to Σ with Poincaré-Hopf index 1 at 0. Hence the Poincaré-Hopf index of v on the stratum V_α equals the Poincaré-Hopf index of $v^{\#}$ in the ambient space M: this is a basic property of the vector fields obtained by radial extension.

Definition 2.1. The *Schwartz index* of v at $0 \in V_\alpha \subset V$ is defined to be the Poincaré-Hopf index at 0 of its radial extension $v^{\#}$ to a neighbourhood of 0 in M.

From the previous discussion we deduce:

Proposition 2.1. *If the stratum V_α has dimension > 0, the Schwartz index of v equals the Poincaré-Hopf index of v at 0 regarded as a vector field on the stratum V_α.*

Now, more generally, let v be a stratified vector field on V with an isolated singularity at $0 \in V \subset M$. Let v_{rad} be a stratified radial vector field at 0, *i.e.* v_{rad} is transversal (outwards-pointing) to the intersection of V with every sufficiently small sphere $\mathbb{S}_\varepsilon$ in M centered at 0, and it is tangent to each stratum. We define the difference between v and v_{rad} at 0 as follows. Consider sufficiently small spheres $\mathbb{S}_\varepsilon$, $\mathbb{S}_{\varepsilon'}$ in M, $\varepsilon > \varepsilon' > 0$, and put the vector field v on $K_\varepsilon = \mathbb{S}_\varepsilon \cap V$ and v_{rad} on $K_{\varepsilon'} = \mathbb{S}_{\varepsilon'} \cap V$. We now use the Schwartz's technique of radial extension explained before, to get a stratified vector field w on the cylinder X in V bounded by the links K_ε and $K_{\varepsilon'}$, such that w extends v and v_{rad}, it has finitely many singularities in the interior of X and at each of these singular points its index in the stratum equals its index in the ambient space M (see [6] for details). The *difference* of v and v_{rad} is defined as:

$$d(v, v_{rad}) = \sum \mathrm{Ind}_{PH}(w; X) \,,$$

where the sum on the right runs over the singular points of w in X and each singularity is being counted with the local Poincaré-Hopf index of w in the corresponding stratum. As in the work of M. H. Schwartz, we can check that this integer does not depend on the choice of w.

Definition 2.2. *The Schwartz* (or radial) *index of v at $0 \in V$ is:*

$$\mathrm{Ind}_{Sch}(v, 0; V) = 1 + d(v, v_{rad}) \,.$$

It is clear that if V is smooth at 0 then this index coincides with the usual Poincaré-Hopf index; it also coincides with the index defined above when 0 is an isolated singularity of V and with the usual index of M. H. Schwartz for vector fields obtained by radial extension. In order to give a unified picture of what this index measures in the various cases, it is useful to introduce a concept that picks up one of the essential properties of the vector fields obtained by radial extension:

Definition 2.3. A stratified vector field on V is *normally radial* at $0 \in V_\alpha$ if it is radial in the direction of each stratum $V_\beta \neq V_\alpha$ containing 0 in its closure.

In other words, v is normally radial if its projection to each small disc Σ around 0, which is transversal to V_α at 0 and has dimension $(\dim_{\mathbb{R}} M -$

$\dim_{\mathbb{R}} V_{\alpha}$), is a radial vector field in Σ, *i.e.* it is transversal to each sphere in Σ centered at 0. The vector fields obtained by radial extension satisfy this condition at all points.

The proof of the following proposition is immediate from the definitions.

Proposition 2.2. *Let v be a stratified vector field on V with an isolated singularity at 0, and let V_{α} be the Whitney stratum that contains 0. If v is normally radial at 0, then its Schwartz index $\mathrm{Ind}_{Sch}(v, 0; V)$ equals its Poincaré-Hopf index $\mathrm{Ind}_{PH}(v, 0; V_{\alpha})$ in V_{α}. Otherwise, its Schwartz index $\mathrm{Ind}_{Sch}(v, 0; V)$ is the sum:*

$$\mathrm{Ind}_{Sch}(v, 0; V) = \mathrm{Ind}_{PH}(v, 0; V_{\alpha}) + \sum_{\beta \neq \alpha} d(v, v_{rad}; V_{\beta}),$$

where $\mathrm{Ind}_{PH}(v, 0; V_{\alpha})$ is defined to be 1 if the stratum V_{α} has dimension 0, and the sum in the right runs over all strata that contain V_{α} in their closures; $d(v, v_{rad}; V_{\beta})$ is the difference in each stratum V_{β} between v and a stratified radial vector field v_{rad} at 0 .

3. The local Euler obstruction

Let $(V, 0)$ be a reduced, pure-dimensional complex analytic singularity germ of dimension n in an open set $U \subset \mathbb{C}^N$. Let $G(n, N)$ denote the Grassmanian of complex n-planes in $\mathbb{C}^N$. On the regular part V_{reg} of V there is a map $\sigma : V_{reg} \to U \times G(n, N)$ defined by $\sigma(x) = (x, T_x(V_{reg}))$. The *Nash transformation* (or *Nash blow up*) $\widetilde{V}$ of V is the closure of $\mathrm{Im}(\sigma)$ in $U \times G(n, N)$. It is a (usually singular) complex analytic space endowed with an analytic projection map

$$\nu : \widetilde{V} \to V$$

which is a biholomorphism away from $\nu^{-1}(Sing(V))$, where $Sing(V) := V - V_{reg}$. Notice each point $y \in Sing(V)$ is being replaced by all limits of planes $T_{x_i} V_{reg}$ for sequences $\{x_i\}$ in V_{reg} converging to x.

Let us denote by $U(n, N)$ the tautological bundle over $G(n, N)$ and denote by $\mathbb{U}$ the corresponding trivial extension bundle over $U \times G(n, N)$. We denote by π the projection map of this bundle. Let $\widetilde{T}$ be the restriction of $\mathbb{U}$ to $\widetilde{V}$, with projection map π. The bundle $\widetilde{T}$ on $\widetilde{V}$ is called *the Nash bundle* of V. An element of $\widetilde{T}$ is written (x, T, v) where $x \in U$, T is a d-plane in $\mathbb{C}^N$ and v is a vector in T. We have maps:

$$\widetilde{T} \xrightarrow{\ \pi\ } \widetilde{V} \xrightarrow{\ \nu\ } V,$$

where π is the projection map of the Nash bundle over the Nash blow up $\widetilde{V}$.

Let us consider a complex analytic stratification $(V_\alpha)_{\alpha \in A}$ of V satisfying the Whitney conditions. Adding the stratum $U \setminus V$ we obtain a Whitney stratification of U. Let us denote by $TU|_V$ the restriction to V of the tangent bundle of U. We know that a stratified vector field v on V means a continuous section of $TU|_V$ such that if $x \in V_\alpha \cap V$ then $v(x) \in T_x(V_\alpha)$. The Whitney condition (a) implies that given $x \in Sing(V)$, any limit T of tangent spaces of points in $V_{reg} = V - Sing(V)$ converging to x contains the tangent space $T_x V_\alpha$ where V_α is the stratum that contains x. Hence one has the following lemma of [9]:

Lemma 3.1. *Every stratified vector field v on a set $A \subset V$ has a canonical lifting to a section $\widetilde{v}$ of the Nash bundle $\widetilde{T}$ over $\nu^{-1}(A) \subset \widetilde{V}$.*

Now consider a stratified radial vector field $v(x)$ in a neighborhood of $\{0\}$ in V; *i.e.* there is ε_0 such that for every $0 < \varepsilon \leq \varepsilon_0$, $v(x)$ is pointing outwards the ball $\mathbb{B}_\varepsilon$ over the boundary $V \cap \mathbb{S}_\varepsilon$ with $\mathbb{S}_\varepsilon := \partial \mathbb{B}_\varepsilon$. Recall that, essentially by the Theorem of Bertini-Sard (see [48]), for ε small enough the spheres $\mathbb{S}_\varepsilon$ are transverse to the strata $(V_\alpha)_{\alpha \in A}$.

One has the following interpretation of the local Euler obstruction [9]. We refer to [47] for the original definition which uses 1-forms instead of vector fields.

Definition 3.1. Let v be a stratified radial vector field on $V \cap \mathbb{S}_\varepsilon$ and $\widetilde{v}$ the lifting of v on $\nu^{-1}(V \cap \mathbb{S}_\varepsilon)$ to a section of the Nash bundle. The *local Euler obstruction* (or simply the Euler obstruction) $\mathrm{Eu}_V(0)$ is defined to be the obstruction to extending $\widetilde{v}$ as a nowhere zero section of $\widetilde{T}$ over $\nu^{-1}(V \cap \mathbb{B}_\varepsilon)$.

More precisely, let $\mathcal{O}(\widetilde{v}) \in H^{2d}\big(\mathbb{B}_\varepsilon, \nu^{-1}(V \cap \mathbb{S}_\varepsilon)\big)$ be the obstruction cocycle for extending $\widetilde{v}$ as a nowhere zero section of $\widetilde{T}$ inside $\nu^{-1}(V \cap \mathbb{B}_\varepsilon)$, where $\mathbb{B}_\varepsilon$ is a small ball around 0 and $\mathbb{S}_\varepsilon$ is its boundary. The local Euler obstruction $Eu_V(0)$ is the evaluation of $\mathcal{O}(\widetilde{v})$ on the fundamental class of the pair $\big(\nu^{-1}(V \cap \mathbb{B}_\varepsilon), \nu^{-1}(V \cap \mathbb{S}_\varepsilon)\big)$. The Euler obstruction is an integer.

The following result summarises some basic properties of the Euler obstruction:

Theorem 3.1. *The Euler obstruction satisfies:*

(i) $Eu_V(0) = 1$ if 0 is a regular point of V;
(ii) $Eu_{V \times V'}(0 \times 0') = Eu_V(0) \cdot Eu_{V'}(0')$;

(iii) If V is locally reducible at 0 and V_i are its irreducible components, then $Eu_V(0) = \sum Eu_{V_i}(0)$;

(iv) $Eu_V(0)$ is a constructible function on V, in fact it is constant on Whitney strata.

These statements are all contained in [47], except for the second part of (iv) which is not explicitly stated there and we refer to [9,45] for a detailed proof.

More generally, for every point $x \in V$, we will denote by $V_\alpha(x)$ the stratum containing x. Now suppose v is a stratified vector field on a small disc $\mathbb{B}_x$ around $x \in V$, and v has an isolated singularity at x. By 3.1 we have that v can be lifted to a section $\tilde{v}$ of the Nash bundle $\tilde{T}$ of V over $\nu^{-1}(\mathbb{B}_x \cap V)$ and $\tilde{v}$ is never-zero on $\nu^{-1}(\partial\mathbb{B}_x \cap V)$. The obstruction for extending $\tilde{v}$ without singularity to the interior of $\nu^{-1}(\mathbb{B}_x \cap V)$ is a cohomology class in $H^{2n}(\nu^{-1}(\mathbb{B}_x \cap V), \nu^{-1}(\partial\mathbb{B}_x \cap V))$; evaluating this class in the fundamental cycle $[\mathbb{B}_x, \partial\mathbb{B}_x]$ one gets *an index $Eu(v, x; V) \in \mathbb{Z}$ of v at x*. If v is radial at x then $Eu(v, x; V)$ is by definition the local Euler obstruction of V at x, $Eu_V(x)$.

Definition 3.2. The integer $Eu(v, x; V)$ is *the (local) Euler obstruction of the stratified vector field v at $x \in V$.*

As mentioned in the introduction, this index is related to the Schwartz index by the Proportionality Theorem of [9]. To state this result, recall that we introduced in section 1 the concept of normally radial vector fields, which essentially characterises the vector fields obtained by radial extension.

Theorem 3.2. *(Proportionality Theorem [9]) Let v be a stratified vector field on V which is normally radial at a singularity $0 \in V_\alpha$. Then one has:*

$$Eu(v, 0; V) = \mathrm{Ind}_{Sch}(v, 0) \cdot Eu_V(0)$$

where $Eu_V(0)$ is the Euler obstruction of V at 0 and $\mathrm{Ind}_{Sch}(v, 0)$ is the Schwartz index of v at 0.

In short, this theorem says that the obstruction $Eu(v, 0; V)$ to extend the lifting $\tilde{v}$ as a section of the Nash bundle inside $\nu^{-1}(V \cap \mathbb{B}_\varepsilon(p)$ is proportional to the Schwartz index of v at 0, the proportionality factor being precisely the local Euler obstruction. We refer to [13] for a short proof of this theorem.

The invariant $Eu(v, 0; V)$ was studied in [8] when v is the "gradient vector field" ∇_f of a function f on V. More precisely, if V has an isolated singularity at 0 and the (real or complex valued) differentiable function f has an isolated critical point at 0, then v is truely the (complex conjugate if f is complex valued) gradient of the restriction of f to $V \setminus \{0\}$. In general, if V has a non-isolated singularity at 0 but f has an isolated critical point at 0 (in the stratified sense [34,44]), then v is obtained essentially by projecting the gradient vector field of f to the tangent space of the strata in V, and then using the Whitney conditions to put these together in a continuous, stratified vector field. One may define this invariant even if f has non-isolated critical points, using intersections of characteristic cycles (see [8]), and it is a measure of how far the germ $(V, 0)$ is from satisfying the local Euler condition (in bivariant theory) with respect to the function f. Thus it was called in [8] *the Euler defect* of f at $(V, 0)$. The Euler obstruction of MacPherson corresponds to the case when f is the function distance to 0. As noticed in [21], this invariant can be also defined using the 1-form df instead of the gradient vector field. This avoids several technical difficulties and is closer to MacPherson's original definition of the local Euler obstruction.

In [55] it is proved that if f has an isolated critical point at $0 \in V$ (in the stratified sense), then its "defect" equals the number of critical points in the regular part of V of a morsification of f. This fact can also be deduced easily from [49].

4. The GSV-index

Let us denote by $(V, 0)$ the germ of a complex analytic n-dimensional, isolated complete intersection singularity, defined by a function

$$f = (f_1, ..., f_k) : \ (\mathbb{C}^{n+k}, 0) \to (\mathbb{C}^k, 0) \,,$$

and let v be a continuous vector field on V singular only at 0. If $n = 1$ we further assume (for the moment) that V is irreducible. We use the notation of [46]: an ICIS means an isolated complete intersection singularity.

Since 0 is an isolated singularity of V, it follows that the (complex conjugate) gradient vector fields $\{\overline{\nabla} f_1, ..., \overline{\nabla} f_k\}$ are linearly independent everywhere away from 0 and they are normal to V. Hence the set $\{v, \overline{\nabla} f_1, ..., \overline{\nabla} f_k\}$ is a $(k+1)$-frame on $V \setminus \{0\}$. Let $K = V \cap \mathbb{S}_\varepsilon$ be the link of 0 in V. It is an oriented, real manifold of dimension $(2n - 1)$ and the above frame defines a map

$$\phi_v = (v, \overline{\nabla} f_1, ..., \overline{\nabla} f_k) : \ K \to W_{k+1}(n + k) \,,$$

into the Stiefel manifold of complex $(k+1)$-frames in $\mathbb{C}^{n+k}$. Since $W_{k+1}(n+k)$ is simply connected, its first non-zero homology group is in dimension

$(2n - 1)$ and it is isomorphic to $\mathbb{Z}$. Hence the map ϕ_v has a well defined degree $\deg(\phi_v) \in \mathbb{Z}$. To define it we notice that $W_{k+1}(n + k)$ is a fibre bundle over $W_k(n + k)$ with fibre the sphere $\mathbb{S}^{2n-1}$; if $(e_1, \cdots, e_{n+k})$ is the canonical basis of $\mathbb{C}^{n+k}$, then the fiber γ over the k-frame $(e_1, \cdots, e_k)$ determines the canonical generator $[\gamma]$ of $H_{(2n-1)}(W_k(n + k)) \cong \mathbb{Z}$. If $[K]$ is the fundamental class of K, then $(\phi_v)_*[K] = \lambda \cdot [\gamma]$ for some integer λ. Then the degree of ϕ_v is defined by:

$$\deg(\phi_v) = \lambda.$$

Alternatively one can prove that every map from a closed oriented $(2n-1)$-manifold into $W_{k+1}(n+k)$ factors by a map into the fibre $\gamma \cong \mathbb{S}^{2n-1}$, essentially by transversality. Hence ϕ_v represents an element in $\pi_{2n-1}W_{k+1}(n + k) \cong \mathbb{Z}$, so ϕ_v is classified by its degree.

Definition 4.1. The GSV-index of v at $0 \in V$, $\mathrm{Ind}_{GSV}(v, 0; V)$, is the degree of the above map ϕ_v.

This index depends not only on the topology of V near 0, but also on the way V is embedded in the ambient space. For instance the singularities in $\mathbb{C}^3$ defined by

$$\{x^2 + y^7 + z^{14} = 0\} \ \text{ and } \ \{x^3 + y^4 + z^{12} = 0\},$$

are orientation preserving homeomorphic, but one can prove that the GSV-index of the radial vector field is 79 in the first case and 67 in the latter; this follows from the fact (see 4.1 below) that for radial vector fields the GSV-index is $1 + (-1)^{\dim V}\mu$, where μ is the Milnor number, which in the examples above is known to be 78 and 66 respectively.

We recall that one has a Milnor fibration associated to the function f, see [36,46,48] and the Milnor fibre F can be regarded as a compact $2n$-manifold with boundary $\partial F = K$. Moreover, by the Transversal Isotopy Lemma (see for instance [1]) there is an ambient isotopy of the sphere $\mathbb{S}_\varepsilon$ taking K into ∂F, which can be extended to a collar of K, which goes into a collar of ∂F in F. Hence v can be regarded as a non-singular vector field on ∂F.

Theorem 4.1. *This index has the following properties:*
(i) The GSV-index of v at 0 equals the Poincaré-Hopf index of v in the Milnor fibre:

$$\mathrm{Ind}_{GSV}(v, 0; V) = \mathrm{Ind}_{PH}(v, F).$$

(ii) If v is everywhere transversal to K, then

$$\mathrm{Ind}_{GSV}(v, 0; V) = 1 + (-1)^n \mu \,.$$

(iii) One has:

$$\mathrm{Ind}_{GSV}(v, 0; V) = \mathrm{Ind}_{Sch}(v, 0; V) + (-1)^n \mu \,,$$

where μ is the Milnor number of 0 and Ind_{Sch} is the Schwartz index.

Notice that the last statement says that the Milnor number of f equals (up to sign) the difference of the Schwartz and GSV indices of every vector field on V with an isolated singularity (cf. [22]).

In [10] there is a generalisation of this index to the case when the variety V has non-isolated singularities, but the vector field is stratified and it has an isolated singularity. In [2] is studied the real analytic setting and relations with other invariants of real singularities are given.

If V has dimension 1 and is not irreducible, then the GSV-index of vector fields on V was actually introduced by M. Brunella in [14,15] and by Khanedani-Suwa [37], in relation with the geometry of holomorphic 1-dimensional foliations on complex surfaces. In this case one has two possible definitions of the index: as the Poincaré-Hopf index of an extension of the vector field to a Milnor fibre, or as the sum of the degrees in 4.1 corresponding to the various branches of V. One can prove [2,60] that for plane curves these integers differ by the intersection numbers of the branches of V.

5. The Virtual Index

We now let V be a compact local complete intersection of dimension n in a manifold M of dimension $m = n+k$, defined as the zero set of a holomorphic section s of a holomorphic vector bundle E of rank k over M. The singular set of V, $Sing(V)$, may have dimension ≥ 0. Let v be a C^∞ vector field on V. We denote by Σ the singular set of v, which is assumed to consist of $Sing(V)$ and possibly some other connected components in the regular part of V, disjoint from $Sing(V)$.

The virtual index is an invariant that assigns an integer to each connected component S of Σ. When S consists of one point, this index coincides with the GSV index, and for a component $S \subset V_{reg}$ this is just the sum of the local indices of the singularities into which S splits under a morsification of v.

Given a connected component S of Σ, the idea to define the virtual index $\mathrm{Ind}_{Vir}(v; S)$ of v at S is to localize at S a certain characteristic class.

We know that if V is smooth, then the usual Poincaré-Hopf local index can be regarded as the top Chern class $c_n(TD, v)$ of the tangent bundle of a small disc around the singular point of the vector field, relative to the section v given by the vector field on the boundary of D. In other words, the Poincaré-Hopf local index is obtained by localizing at 0 the top Chern class $c_n(TD)$ using the vector field v. We can of course replace the point 0 by a component S of Σ contained in V_{reg}; in this case we replace D by a compact tubular neighbourhood $\mathcal{T}$ of S in V and we localize $c_n(TV)|_{\mathcal{T}}$ at S using the vector field v, which is assumed to be non-singular on $\mathcal{T} \setminus S$. This means that we consider the Chern class $c_n(TV|_{\mathcal{T}})$ relative to v on $\mathcal{T} \setminus S$. The class we get lives in $H^{2n}(\mathcal{T}, \mathcal{T} \setminus S) \cong H_0(S) \cong \mathbb{Z}$. The integer that we get in this way is the Poincaré-Hopf index of v at S, $\mathrm{Ind}_{PH}(v; S)$, *i.e.* the number of singularities of a generic perturbation of v inside $\mathcal{T}$, counted with signs.

The question now is what to do when S is contained in the singular set of V, so there is not a tangent bundle. The idea to define the virtual index is to make a similar "localisation" using the vector field and the *virtual tangent bundle of V*, defined below.

To define this bundle we notice that the restriction $E|_{V_{reg}}$ coincides with the (holomorphic) normal bundle $N(V_{reg})$ of the regular part $V_{reg} = V - Sing(V)$. We denote by TM the holomorphic tangent bundle of M and we set $N = E|_V$.

Definition 5.1. (c.f. [25]) The *virtual tangent bundle* of V is

$$\tau(V) = TM|_V - N,$$

regarded as an element in the complex K-theory $KU(V)$.

It is known that the equivalence class of this virtual bundle does not depend on the choice of the embedding of V in M.

We denote by

$$c_*(TM|_V) = 1 + c_1(TM|_V).... + c_m(TM|_V),$$

and

$$c_*(N) = 1 + c_1(N).... + c_k(N),$$

the total Chern classes of these bundles. These are elements in the cohomology ring of V and can be inverted, *i.e.* there is a unique class $c_*(N)^{-1} \in H^*(V)$ such that

$$c_*(N) \cdot c_*(N)^{-1} = c_*(N)^{-1} \cdot c_*(N) = 1.$$

Using this one has the total Chern class of the virtual tangent bundle defined in the usual way:

$$c_*(\tau(V)) = c_*(TM|_V) \cdot c_*(N)^{-1} \in H^*(V).$$

The ith Chern class of $TM|_V - N$ is by definition the component of $c_*(\tau(V))$ in dimension $2i$, for $i = 1, ..., n$.

It is clear that if V is smooth, then its virtual tangent bundle is equivalent in $KU(V)$ to its usual tangent bundle, and the Chern classes of the virtual tangent bundle are the usual Chern classes.

Consider the component $c_n(\tau(V))$ of $c_*(\tau(V))$ in dimension $2n$. This is the top Chern class of the virtual tangent bundle. As we said before, the idea to define the local index of the vector field v at a component S of $Sing(V)$ is to localize $c_n(\tau(V))$ at S using v. For this one needs to explain how to localize the Chern classes of the virtual tangent bundle. This is carefully done in [60], and we refer to that text for a detailed account on the subject, particularly in relation with indices of vector fields.

In the particular case when the component S has dimension 0, so that we can assume we have a local ICIS germ $(V, 0)$ of dimension n in $\mathbb{C}^{n+k}$, defined by functions

$$f = (f_1, \cdots, f_k) : U \subset \mathbb{C}^{n+k} \to \mathbb{C}^k,$$

with U an open set in $\mathbb{C}^{n+k}$, one has that the virtual tangent bundle of V is:

$$\tau(V) = T\mathbb{C}^{n+k}|_V - (V \times T\mathbb{C}^k).$$

If $\mathbb{B}$ denotes a small ball in U around 0, then one has the Chern class $c_{n+k}(T\mathbb{B}|_V)$ relative to the $(k+1)$-frame $(v, \overline{\nabla}f_1, \cdots, \overline{\nabla}f_k)$ on $\partial\mathbb{B} \cap V$. This is a cohomology class in $H^{2n+2k}(\mathbb{B} \cap V, \partial\mathbb{B} \cap V)$, and one can prove that its image in $\cong H_0(\mathbb{B}) \cong \mathbb{Z}$ under the Alexander homomorphism is the virtual index of v at 0 (see [43,57,60]); which in this case coincides with the GSV-index.

6. The Homological Index

The basic references for this section are the articles by Gomez-Mont and various co-authors, see [26] and also [3,28–31]. There are also important algebraic formulas for the index of holomorphic vector fields (and 1-forms) given by various authors, as for instance in [43] (see also [27,40–42]). In the real analytic case, interesting algebraic formulas for the index are given

968

in [16,32,33], which generalize to singular hypersurfaces the remarkable formula of Eisenbud-Levine and Khimshiashvili [23,38], that expresses the index of an analytic vector field in $\mathbb{R}^m$ as the signature of an appropriate bilinear form. Here we only describe (briefly) the homological index of holomorphic vector fields.

Let $(V,0) \subset (\mathbb{C}^N, 0)$ be the germ of a complex analytic (reduced) variety of pure dimension n with an isolated singular point at the origin. A vector field v on $(V,0)$ can always be defined as the restriction to V of a vector field $\tilde{v}$ in the ambient space which is tangent to $V \setminus \{0\}$; v is holomorphic if $\tilde{v}$ can be chosen to be holomorphic. So we may write v as $v = (v_1, \cdots, v_N)$ where the v_i are restriction to V of holomorphic functions on a neighbourhood of 0 in $(\mathbb{C}^N, 0)$.

It is worth noting that given every space V as above, there are always holomorphic vector fields on V with an isolated singularity at 0. This (non-trivial) fact is indeed a weak form of stating a stronger result ([3, 2.1, p. 19]): in the space $\Theta(V,0)$ of germs of holomorphic vector fields on V at 0, those having an isolated singularity form a connected, dense open subset $\Theta_0(V,0)$. Essentially the same result implies also that every $v \in \Theta_0(V,0)$ can be extended to a germ of holomorphic vector field in $\mathbb{C}^N$ with an isolated singularity, though it can possibly be also extended with a singular locus of dimension more that 0, a fact that may be useful for explicit computations (c.f. [26]).

A (germ of) holomorphic j-form on V at 0 means the restriction to V of a holomorphic j-form on a neighbourhood of 0 in $\mathbb{C}^N$; two such forms in $\mathbb{C}^N$ are equivalent if their restrictions to V coincide on a neighbourhood of $0 \in V$. We denote by $\Omega^j_{V,0}$ the space of all such forms (germs); these are the Kähler differential forms on V at 0. So, $\Omega^0_{V,0}$ is the local structure ring $\mathcal{O}_{(V,0)}$ of holomorphic functions on V at 0, and each $\Omega^j_{V,0}$ is a $\Omega^0_{V,0}$ module. Notice that if the germ of V at 0 is determined by $(f_1, \cdots, f_k)$ then one has:

$$\Omega^j_{V,0} := \frac{\Omega^j_{\mathbb{C}^N,0}}{\left(f_1\Omega^j_{\mathbb{C}^N,0} + df_1 \wedge \Omega^{j-1}_{\mathbb{C}^N,0}, \cdots, f_k\Omega^j_{\mathbb{C}^N,0} + df_k \wedge \Omega^{j-1}_{\mathbb{C}^N,0}\right)},$$

where d is the exterior derivative.

Now, given a holomorphic vector field $\tilde{v}$ at $0 \in \mathbb{C}^N$ with an isolated singularity at the origin, and a Kähler form $\omega \in \Omega^j_{\mathbb{C}^N,0}$, we can always contract ω by v in the usual way, thus getting a Kähler form $i_v(\omega) \in \Omega^{j-1}_{\mathbb{C}^N,0}$. If $v = \tilde{v}|_V$ is tangent to V, then contraction is well defined at the level of

Khäler forms on V at 0 and one gets a complex $(\Omega^{\bullet}_{V,0}, v)$:

$$0 \to \Omega^n_{V,0} \to \Omega^{n-1}_{V,0} \to \ldots \to \mathcal{O}_{V,0} \to 0\,,$$

where the arrows are contraction by v and n is the dimension of V; of course one also has Khäler forms of degree $> n$, but those forms do not play a significant role here. We consider the homology groups of this complex:

$$H_j(\Omega^{\bullet}_{V,0}, v) = \frac{ker\,(\Omega^j_{V,0} \to \Omega^{j-1}_{V,0})}{Im\,(\Omega^{j+1}_{V,0} \to \Omega^j_{V,0})}$$

An important observation in [26] is that if V is regular at 0, so that its germ at 0 is that of $\mathbb{C}^n$ at the origin, and if $v = (v_1, \cdots, v_n)$ has an isolated singularity at 0, then this is the usual Koszul complex (see for instance [35, p. 688]), so that all its homology groups vanish for $j > 0$, while

$$H_0(\Omega^{\bullet}_{V,0}, v) \cong \mathcal{O}_{\mathbb{C}^n,0)}/(v_1, \cdots, v_n)\,.$$

In particular the complex is exact when $v(0) \neq 0$. Since the contraction maps are $\mathcal{O}_{V,0}$-modules maps, this implies that if V has an isolated singularity at the origin, then the homology groups of this complex are concentrated at 0, and they are finite dimensional because the sheaves of Khäler forms on V are coherent. Hence, for V a complex analytic affine germ with an isolated singularity at 0 and v a holomorphic vector field on V with an isolated singularity at 0, it makes sense to define:

Definition 6.1. The *homological index* $\mathrm{Ind}_{\mathrm{hom}}(v, 0; V)$ of the holomorphic vector field v on $(V, 0)$ is the Euler characteristic of the above complex:

$$\mathrm{Ind}_{\mathrm{hom}}\,(v, 0; V) = \sum_{i=0}^{n}(-1)^i h_i(\Omega^{\bullet}_{V,0}, v)\,,$$

where $h_i(\Omega^{\bullet}_{V,0}, v)$ is the dimension of the corresponding homology group as a vector space over $\mathbb{C}$.

We recall that an important property of the Poincaré-Hopf local index is its stability under perturbations. This means that if we perturb v slightly in a neighbourhood of an isolated singularity, then this zero of v may split into a number of isolated singularities of the new vector field v', whose total number (counted with their local indices) is the index of v. When the ambient space V has an isolated singularity at 0, then every vector field on V necessarily vanishes at 0, since in the ambient space the vector field defines a local 1-parameter family of diffeomorphisms. Hence every perturbation of v producing a vector field tangent to V must also vanish at 0, but new singularities may arise with this perturbation. The homological index also satisfies the stability under such perturbations. This is called the "Law of Conservation of Number" in [26,29]:

Theorem 6.1. *(Gomez-Mont [26, Theorem 1.2]) For every holomorphic vector field v' on V sufficiently close to v one has:*

$$\mathrm{Ind}_{\mathrm{hom}}\,(v,0;V) = \mathrm{Ind}_{\mathrm{hom}}\,(v',0;V) + \sum \mathrm{Ind}_{PH}(v')\,,$$

where Ind_{PH} is the local Poincaré-Hopf index and the sum on the right runs over the singularities of v' at regular points of V near 0.

This result is a special case of a more general theorem in [29].

This theorem is a key property of the homological index. In particular this allows us to identify this index with the GSV-index when $(V,0)$ is a hypersurface germ [26]. In fact, it is easy to see that the GSV-index also satisfies the above "Law of Conservation of Number" for vector fields on complete intersection germs. This implies that if both indices coincide for a given vector field on $(V,0)$, then they coincide for every vector field on $(V,0)$, since the space $\Theta_0(V,0)$ is connected. Hence, in order to prove that both indices coincide for all vector fields on hypersurface (or complete intersection) germs, it is enough to show that given every such germ, there exists a holomorphic vector field v for which the GSV and homological indices coincide. This is what Gomez-Mont does in [26]. For that, he first gives a very nice algebraic formula to compute the homological index of vector fields on hypersurface singularities, which he then uses to perform explicit computations and prove that, for holomorphic vector fields on hypersurface singularities, the homological index coincides with the GSV index.

It is not known whether or not these indices coincide on complete intersection germs in general (c.f. [22]), but recently, Bothmer, Ebeling and Gómez-Mont [4] found a remarkable formula for the homological index of vector fields on complete intersection germs, and they obtain as corollary that the GSV and the homological indices coincide on quasi-homogeneous complete intersection germs.

7. Relations with Chern classes of singular varieties

The local index of Poincaré-Hopf is the most basic invariant of a vector field at an isolated singularity, and the theorem of Poincaré-Hopf about the total index of a vector field on a manifold is a fundamental result, giving rise, in particular, to obstruction theory and the theory of characteristic classes, such as the Chern classes of complex manifolds.

In the case of singular varieties, there are several definitions of characteristic classes, given by various authors. Somehow they correspond to the various extensions one has of the concept of "tangent bundle" as we go from manifolds to singular varieties, and they are closely related to the indices of vector fields discussed above.

The first one is due to M.H. Schwartz in [52,53], considering a singular complex analytic variety V embedded in a smooth complex manifold M which is equipped with a Whitney stratification adapted to V; she then replaces the tangent bundle by the union of tangent bundles of all the strata in V, and considers a class of stratified frames to define characteristic classes of V, which do not depend on M nor on the various choices. These classes live in the cohomology groups of M with support in V, $i.e.$ $H^*(M, M \setminus V; \mathbb{Z})$, and they are equivalent to the usual Chern classes when V is non-singular. The top degree Schwartz class is defined precisely using the Schwartz index presented in Section 1: consider a Whitney stratification of M adapted to V, a triangulation (K) compatible with the stratification, and the dual cell decomposition (D) (c.f. [5,6] for details). By construction, the cells of (D) are transverse to the strata. Consider now a stratified vector field v on V obtained by radial extension. Then (see [6,9,52,53]) the radial extension technique allows us to construct a vector field on a regular neighborhood U of V in M, union of (D)-cells, which is normally radial (in the sense of section 1), and has at most a singular point at the barycenter of each (D)-cell of top dimension $2m$, m being the complex dimension of M. This defines a cochain in the usual way, by assigning to each such cell the Schwartz index of this vector field ($i.e.$ its Poincaré-Hopf index in the ambient space). This cochain is a cocycle that represents a cohomology class in $H^*(U, U \setminus V) \cong H^*(M, M \setminus V)$, and this class is by definiton the top Schwartz class of V. Its image in $H_0(V)$ under Alexander duality gives the Euler-Poincaré characteristic of V. The Schwartz classes of lower degrees are defined similarly, considering stratified r-frames, $r = 1, \cdots, n = \dim V$, defined by radial extension on the $(2m-2r+2)$-skeleton of (D), and the corresponding Schwartz index of such frames: just as the concept of Schwartz index can be extended to stratified vector fields in general (section 1), so too one can define the Schwartz index of stratified frames in general, and use any such frame to define the corresponding Schwartz class (see [7,12]).

The second extension of the concept of tangent bundle is given by the Nash bundle $\widetilde{T} \to \widetilde{V}$ over the Nash Transform $\widetilde{V}$, which is biholomorphic to V_{reg} away from the divisor $\nu^{-1}(Sing(V))$, where $\nu : \widetilde{V} \to V$ is the projection. Thus $\widetilde{T}$ can be regarded as a bundle that extends $T(V_{reg})$ to a bundle over $\widetilde{V}$. The Chern classes of $\widetilde{T}$ lie in $H^*(\widetilde{V})$, which is mapped into $H_*(\widetilde{V})$ by the Alexander homomorphism (see [5,6]); the Mather classes of V, introduced in [47], are by definition the image of these classes under the morphism $\nu : H_*(\widetilde{V}) \to H_*(V)$. MacPherson's Chern classes for singular varieties [47], lie in the homology of V and can be thought of as being the Mather classes of V weighted (in a sense that is made precise below) by the local Euler obstruction (§2 above). In fact, it is easy to show that the local Euler obstruction satisfies that there exists unique integers $\{n_i\}$ for which the equation

$$\sum n_i \, Eu_{\overline{V}_\alpha}(x) = 1 \tag{1}$$

is satisfied for all points x in V, where the sum runs over all strata V_α containing x in their closure. Then the *MacPherson class* of degree r is defined by:

$$c_r(V) = c_r^M\Big(\sum n_i \overline{V}_\alpha\Big) = \sum n_i \, \iota_* c_r^M(\overline{V}_\alpha),$$

where $c_r^M(\overline{V}_\alpha)$ is the Mather class of degree r of the analytic variety $\overline{V}_\alpha$.

MacPherson's classes satisfy important axioms and functoriality properties conjectured by Deligne and Grothendieck in the early 1970's.

Later, Brasselet and Schwartz [9] proved that the Alexander isomorphism $H^*(M, M \setminus V) \cong H_*(V)$, carries the Schwartz classes into MacPherson's classes, so they are now called the *Schwartz-MacPherson classes* of V. As we briefly explained before, Schwartz classes are defined via the Schwartz indices of vector fields and frames; the MacPherson classes are defined from the Chern classes of the Nash bundle (which determine the Mather classes) and the local Euler obstructions.

A third way of extending the concept of tangent bundle to singular varieties was introduced by Fulton and Johnson [25]. The starting point is that if a variety $V \subset M$ is defined by a regular section s of a holomorphic bundle E over M, then one has the virtual tangent bundle $\tau V = [TM|_V - E|_V]$, introduced in §4 above. The Chern classes of the virtual tangent bundle τV (cap product the fundamental cycle $[V]$) are the *Fulton-Johnson classes* of V. One has ([43,57]) that the 0-degree Fulton-Johnson class of such varieties equals the total virtual index of every continuous vector field with isolated singularities on V_{reg}. Similar considerations hold for the higher degree Fulton-Johnson classes, using frames and the corresponding virtual classes (see [7,12]).

Summarizing, the various indices we presented in sections 1-4 are closely related to various known characteristic classes of singular varieties that generalize the concept of Chern classes of complex manifolds. There is left the homological index of section 5: this ought to be related with a fourth way for extending the concept of tangent bundle to singular varieties (with its corresponding generalisation of Chern classes), introduced and studied by Suwa in [59]. This is by considering the tangent sheaf Θ_V, which is by definition the dual of Ω_V, the sheaf of Khäler differentials on V introduced in §5. The latter is defined by the exact sequence:

$$I_V/I_V^2 \xrightarrow{f \mapsto df \otimes 1} \Omega_M \otimes \mathcal{O}_V \to \Omega_V \to 0 \,,$$

and $\Theta_V := \mathrm{Hom}(\Omega_V, \mathbb{C})$. Both sheaves Ω_V and Θ_V are coherent sheaves and one can use them to define characteristic classes of V that coincide with the usual Chern classes when V is non-singular. In particular, if V is a local complete intersection in M, then one has a canonical locally free resolution of Ω_V and the corresponding Chern classes essentially coincide with the Fulton-Johnson classes, though the corresponding classes for Θ_V differ from these. Recent work of J. Schürmann points out in this direction, at least if one considers the homological index of 1-forms.

We refer to [6] for a rather complete presentation of characteristic classes of singular varieties, including the constructions of Schwartz and MacPherson that we sketched above, and to [12] for a discussion of indices of vector fields and their relation with characteristic classes of singular varieties, much deeper than the one we presented here.

References

1. R. Abraham and J. Robin, *Transversal mappings and flows* W.A. Benjamin, Inc., (1967).
2. M. Aguilar, J. Seade and A. Verjovsky, *Indices of vector fields and topological invariants of real analytic singularities* Crelle's, J. Reine u. Ange. Math. 504 (1998), 159-176.
3. Ch. Bonatti and X. Gómez-Mont, *The index of a holomorphic vector field on a singular variety* Astérisque 222 (1994), 9-35.
4. H.-Chr. Graf von Bothmer, W. Ebeling, X. Gómez-Mont, *The homological index of a vector field on an isolated complete intersection singularity.* Preprint 2006, math.AG/0601640.
5. J. P. Brasselet, *Définition combinatoire des homomorphismes de Poincaré, Alexander et Thom pour une pseudo-variété* in "Caractéristique d'Euler-Poincaré", Astérisque 82-83, Société Mathématique de France (1981), 71-91.

974

6. J.-P. Brasselet, *Characteristic Classes of Singular Varieties*, book in preparation.

7. J. -P. Brasselet, D. Lehmann, J. Seade and T. Suwa, *Milnor classes of local complete intersections*, Transactions, Amer. Math. Soc. *354* (2001), 1351-1371.

8. J.-P. Brasselet, D.B. Massey, A.J. Parameswaran and J. Seade, *Euler obstruction and defects of functions on singular varieties*, J. London Math. Soc. *70* (2004), 59-76.

9. J.-P. Brasselet et M.-H. Schwartz, *Sur les classes de Chern d'un ensemble analytique complexe*, Astérisque *82-83* (1981), 93-147.

10. J.-P. Brasselet, J. Seade and T. Suwa, *An explicit cycle representing the Fulton-Johnson class, I*, Singularités Franco-Japonaises, Séminaires et Congrès *10*, Soc. Math. France (2005), 21-38.

11. J.P. Brasselet, J. Seade, T. Suwa, *Proportionality of Indices of 1-Forms on Singular Varieties*, Preprint 2005, math.AG/0503428.

12. J.P. Brasselet, J. Seade, T. Suwa, *Indices of vector fields and Chern classes of singular varieties*, monograph in preparation.

13. J.P. Brasselet, J. Seade, T. Suwa, *A proof of the proportionality theorem*, preprint 2006, http://arxiv.org/abs/math.AG/0511601.

14. M. Brunella, *Feuilletages holomorphes sur les surfaces complexes compactes*, Ann. Sci. E.N.S **30** (1997), 569-594.

15. M. Brunella, *Some remarks on indices of holomorphic vector fields*, Publicacions Matematiques **41** (1997), 527-544.

16. W. Ebeling and S. Gusein-Zade, *On the index of a vector field at an isolated singularity*. The Arnoldfest (Toronto, ON, 1997), 141–152, Fields Inst. Commun., 24, Amer. Math. Soc., Providence, RI, 1999.

17. W. Ebeling and S. Gusein-Zade, *On the index of a holomorphic 1-form on an isolated complete intersection singularity*. (Russian) Dokl. Akad. Nauk 380 (2001), no. 4, 458-461.

18. W. Ebeling and S. Gusein-Zade, *Indices of 1-forms on an isolated complete intersection singularity*, Moscow. Math. J. *3*, no. 2 (2003), 439-455.

19. W. Ebeling and S. Gusein-Zade, *Indices of 1-forms on an isolated complete intersection singularity*. Dedicated to Vladimir I. Arnold on the occasion of his 65th birthday. Mosc. Math. J. 3 (2003), no. 2, 439–455, 742–743.

20. W. Ebeling and S. Gusein-Zade, *Indices of 1-forms on an isolated complete intersection singularity*, Moscow. Math. J. **3**, no. 2 (2003).

21. W. Ebeling and S. Gusein-Zade, *Radial index and Euler obstruction of a 1-form on a singular variety*. Geom. Dedicata 113 (2005), 231-241.

22. W. Ebeling, S. Gusein-Zade, J. Seade, *Homological index for 1-forms and a Milnor number for isolated singularities*, International J. Maths. 15 (2004), 895-905.

23. D. Eisenbud, H. Levine, *An algebraic formula for the degree of a C^∞ map germ*, Ann. Math. (2) 106 (1977), no. 1, 19-44.

24. W. Fulton, *Intersection Theory*, Springer-Verlag (1984).

25. W. Fulton, K. Johnson, *Canonical classes on singular varieties*,

Manuscripta Math. **32** (1980), 381-389.

26. X. Gómez-Mont, *An algebraic formula for the index of a vector field on a hypersurface with an isolated singularity* J. Algebraic Geom. **7** (1998), 731-752.

27. X. Gómez-Mont, J. Seade and A. Verjovsky, *The index of a holomorphic flow with an isolated singularity*, Math. Ann. *291* (1991), 737-751.

28. L. Giraldo, X. Gómez-Mont, *On the complex formed by contracting differential forms with a vector field on a hypersurface singularity*. Bol. Soc. Mat. Mexicana (3) 7 (2001), no. 2, 211–221.

29. L. Giraldo, X. Gómez-Mont, *A law of conservation of number for local Euler characteristics*. Complex manifolds and hyperbolic geometry (Guanajuato, 2001), 251–259, Contemp. Math., 311, Amer. Math. Soc., Providence, RI, 2002.

30. L. Giraldo, X. Gómez-Mont, P. Mardesić, *Computation of topological numbers via linear algebra: hypersurfaces, vector fields and vector fields on hypersurfaces*. Complex geometry of groups (Olmu, 1998), 175-182, Contemp. Math., 240, Amer. Math. Soc., Providence, RI, 1999.

31. L. Giraldo, X. Gómez-Mont, P. Mardesić, *On the index of vector fields tangent to hypersurfaces with non-isolated singularities*. J. London Math. Soc. (2) 65 (2002)

32. X. Gómez-Mont and P. Mardesic, *The index of a vector field tangent to a hypersurface and signature of the relative Jacobian determinant*, Ann. Inst. Fourier (Grenoble) **47** (1997), 1523-1539.

33. X. Gómez-Mont and P. Mardesic, *The index of a vector field tangent to an odd-dimensional hypersurface, and the signature of the relative Hessian*. Funct. Anal. Appl. 33 (1999), no. 1, 1-10.

34. M. Goresky and R. MacPherson, *Stratified Morse theory*, Ergebnisse der Mathematik und

35. Ph. Griffiths, J. Harris, *Principles of algebraic geometry*, John Wiley and Sons, 1978.

36. H. Hamm, *Lokale topologische Eigenschaften komplexer Räume*, Math. Ann. **191** (1971), 235-252.

37. B. Khanedani and T. Suwa, *First variation of holomorphic forms and some applications*, Hokkaido Math. J. **26** (1997), 323-335.

38. G.N. Khimshiashvili, *On the local degree of a smooth mapping*, Comm. Acad. Sci. Georgian SSR 85, No. 2 (1977), 309-312 (in Russian).

39. H. King and D. Trotman, *Poincaré-Hopf theorems on stratified sets*, preprint 1996.

40. O. Klehn, *Local residues of holomorphic 1-forms on an isolated surface singularity*, Manuscripta Math. **109** (2002), 93-108.

41. O. Klehn, *On the index of a vector field tangent to a hypersurface with non-isolated zero in the embedding space*. Math. Nachr. 260 (2003), 48-57.

42. O. Klehn, *Real and complex indices of vector fields on complete intersection curves with isolated singularity*, to appear in Compositio Mathematica.

43. D. Lehmann, M. Soares and T. Suwa, *On the index of a holomorphic*

vector field tangent to a singular variety, Bull. Braz. Math. Soc. (N.S.) 26 (1995), pp 183–199.

44. D. T. Lê, *Le concept de singularité isolée de fonction analytique*, Adv. Stud. Pure Math. **8** (1986), 215-227, North Holland.

45. D. T. Lê, B. Teissier, *Variétés polaires locales et classes de Chern des variétés singulieres*, Ann. of Math **114** (1981), 457-491.

46. E.J.N. Looijenga, *Isolated Singular Points on Complete Intersections*, LMS Lecture Notes **77**, Cambridge Univ. Press 1984.

47. R. MacPherson, *Chern classes for singular varieties*, Ann. of Math. *100* (1974), 423-432.

48. J. Milnor, *Singular points of complex hypersurfaces*, Ann. of Math. Studies 61, Princeton 1968.

49. J. Schürmann, *Topology of Singular Spaces and Constructible Sheaves*, Monografie Matematyczne, Intytut Matematyczny PAN, *63*, New Series, Birkhäuser, 2003.

50. J. Schürmann, *A general intersection formula for Lagrangian cycles.* Compos. Math. *140* (2004), 1037–1052.

51. J. Schürmann and M. Tibăr, *Characteristic cycles and indices of 1-forms on singular spaces*, preprint in preperation.

52. M.-H. Schwartz, *Classes caractéristiques définies par une stratification d'une variété analytique complexe*, C.R. Acad. Sci. Paris *260* (1965), 3262-3264, 3535-3537.

53. M.-H. Schwartz, *Champs radiaux sur une stratification analytique complexe*, Travaux en Cours *39*, Hermann, Paris, 1991.

54. J. Seade, *The index of a vector field on a complex surface with singularities*, in "The Lefschetz Centennial Conf.", ed. A. Verjovsky, Contemp. Math. 58, Part III, Amer. Math. Soc. (1987), 225-232.

55. J. Seade, M. Tibăr and A. Verjovsky, *Milnor numbers and Euler obstruction*, Bol. Soc. Bras. Mat. Bull. Braz. Math. Soc. (N.S.) 36 (2005), 275-283.

56. J. Seade and T. Suwa, *A residue formula for the index of a holomorphic flow*, Math. Ann. *304* (1996), 621–634.

57. J. Seade, T. Suwa, *An adjunction formula for local complete intersections*, Internat. J. Math. *9* (1998), 759-768.

58. N. Steenrod, *The Topology of Fiber Bundles*, Princeton Univ. Press, 1951.

59. T. Suwa, *Characteristic classes of coherent sheaves on singular varieties.* (Singularities-Sapporo 1998) Edit. J.-P. Brasselet and T. Suwa, 279-297, Adv. Stud. Pure Math., 29, Kinokuniya, Tokyo, 2000.

60. T. Suwa, *Indices of vector fields and residues of singular holomorphic foliations*, Actualités Mathématiques, Hermann, Paris, 1998.

Direct Connections and Chern Character

Nicolae Teleman

*Dipartimento di Scienze Matematiche, Universita' Politecnica delle Marche,
60131-Ancona, Italia * E-mail: teleman@dipmat.univpm.it*

This paper is dedicated to Jean-Paul Brasselet's 60th birthday.

We show how the Chern character of the tangent bundle of a smooth manifold may be extracted from the geodesic distance function by means of cyclic homology. Such considerations lead us to define in Sect. 5 the notion of *linear direct connection* in vector bundles, notion which generalizes the notion of linear connection.

The basic difference between a linear direct connection and a linear connection consists of the fact that while a linear connection provides a transport of fibers along *curves*, a linear direct connection provides a *direct* transport of fibres *from point to point*.

For this reason, linear direct connections could be defined in contexts where differentiability is not available.

We show next that the algebraic procedure for constructing the Chern character, discussed in Sect. 4 applies also in the case of linear direct connections.

This paper provides a geometric interpretation of the non commutative Chern character due to A. Connes [1,2]. We show that this interpretation goes along the same lines as those presented by N. Teleman [11] and C. Teleman [10].

The arguments discussed here may be extended to the language of groupoids. In a subsequent paper we are going to improve some of the considerations presented here and extend their field of application to more singular situations.

Remark 0.1. The notion of linear direct connection, introduced in this paper, replaces the notion of *linear quasi connection*, introduced by the author in [12].

Keywords: Direct connection; Linear connection; Chern-Weil theory; Periodic cyclic homology

1. Introduction

This paper is motivated by the intent to make intrinsic the main constructions of global analysis, as much as possible. The purpose of this is not only to find elegant formulas, but also to allow further generalizations, mostly in the case of singular spaces.

In Sect.4 we present a procedure which allows one to extract, from the geodesic distance function, by means of cyclic homology, the Chern character of the tangent bundle of a smooth manifold. In Sect. 5 we introduce the notion of linear direct connection in vector bundles, notion which generalizes the notion of linear connection. Linear direct connections have the advantage that they may be defined in contexts where any kind of differentiability could be absent. We show next that the algebraic procedure discussed in Sect.4 for constructing the Chern character of tangent bundles applies also to the case of arbitrary linear direct connections in vector bundles.

This paper provides a geometric interpretation of the non commutative Chern character due to A. Connes [1,2]. We show that this interpretation goes along the same lines as those presented by N. Teleman [11] and C. Teleman [10].

The arguments discussed here may be extended to the language of groupoids. In a subsequent paper we intend to improve some of the considerations presented here and to extend their field of application to more singular situations.

Parts of this paper were written while visiting I.H.E.S.; the author thanks the Institut des Hautes Etudes Scientifiques for hospitality.

2. Geometrical Construction of the Chern Character

2.1. *The Cube Construction*

In this section we recall an old geometrical contruction of the Chern character due to the author [11].

We recall first the algebraic construction of the Chern character, see e.g. [7]. Let ξ be a smooth complex vector bundle ξ over the manifold M. Let ∇ be a hermitian linear connection in ξ and let $R = \nabla^2$ be its curvature. The Chern character of ξ, $Ch_*(\xi) \in H_{dR}^{even}(M)$, is by definition the de Rham cohomology class of the non homogeneous differential form

$$\Psi_*(\xi) = Tre^{-\frac{R}{2\pi i}}. \tag{1}$$

The homogeneous component of degree $2q$ of the above differential form is

$$\Psi_q(\xi) = \frac{(i)^q}{(2\pi)^q(q)!} TrR^q. \tag{2}$$

The differential form $\Psi_q(\xi)$ has the following two basic properties: -i) it is closed, -ii) its de Rham homology class is independent of the connection

∇; it depends only on the underlying topological structures.

In local co-ordinates, this differential form is given by

$$\Psi_q(\xi) = \frac{(i)^q}{(2\pi)^q(q)!} R^{j_0}_{j_1 i_1 i_2} R^{j_1}_{j_2 i_3 i_4} \cdots\cdots R^{j_q}_{j_0 i_{2q-1} i_{2q}}.$$

$$dx^{i_1} \wedge dx^{i_2} \wedge dx^{i_3} \wedge dx^{i_4} \wedge \ldots \wedge dx^{i_{2q-1}} \wedge dx^{i_{2q}},$$

where summation is supposed with respect to all repeated indices.

We present here a geometrical interpretation of the differential form $\Psi_q(\xi)$.

Let x_0 be a point in M and let $\{X_1, X_2, ..., X_{2q-1}, X_{2q}\}$ be tangent vectors to M at the point x_0. We intend to interpret geometrically the value of the differential form $\Psi_q(\xi)$ on these tangent vectors.

For, we choose a local system of co-ordinates $(x^1, x^2, ..., x^{2q}, ..., x^n)$ about the point x_0, $(n = dim(M).)$ Let $I^{2q} = [0,1]^{2q}$ be the unit cube in R^{2q}.

Let $Q : I^{2q} \longrightarrow M$ be a smooth mapping which carries the origin of the cube onto the point x_0 and the unit tangent vectors to the edges of the cube, $\{e_1, e_2, ..., e_{2q-1}, e_{2q}\}$, respectively, onto the tangent vectors $\{X_1, X_2, ..., X_{2q-1}, X_{2q}\}$.

Let $\iota = (i_1, i_2, ..., i_{2q}) \in \Pi_{2q}$ be an arbitrary permutation of the numbers $\{1, 2, ..., 2q\}$. For any $t \in [0,1]$ and for any such permutation ι, let $\lambda_\iota(t)$ be the unique polygonal path traced on the cube I^{2q}, which connects the origin with the point $(t, t, ..., t)$, subject to the requirements:

-i) each of the $2q$ co-ordinates of the points of the path varies just once on the interval $[0,t]$,

-ii) the order in which such co-ordinates vary is prescribed by the permutation ι.

Next, let $\tilde{\lambda}_\iota(t)$ be the loop of the cube consisting of the path $\lambda_\iota(t)$, followed by the segment of the diagonal connecting the point $(t, t, ..., t)$ with the origin. Let $\tau_\iota(t)$ be the parallel transport defined by the linear connection ∇ along the path $Q \circ \tilde{\lambda}_\iota(t)$.

The same cube construction may be performed on an odd number of tangent vectors $\{X_1, X_2, ..., X_{2q}, X_{2q+1}\}$.

Theorem 2.1 (N. Teleman, 1966). *-i.) For an even number of tangent vectors, the cube construction gives*

$$TrLim_{t\searrow 0}\frac{1}{t^{2q}} \sum_{\iota \in \Pi_{2q}} Sign(\iota).\tau_\iota(t) = \frac{q!(2q)!(-2\pi i)^q}{2^{2q}}\Psi_q(\xi). \tag{3}$$

-ii) For an odd number of tangent vectors $\{X_1, X_2, ..., X_{2q}, X_{2q+1}\}$, the

980

cube construction gives

$$TrLim_{t\searrow 0}\frac{1}{t^{2q}}\sum_{\iota\in\Pi_{2q}}Sign(\iota).\tau_\iota(t) =$$

$$\frac{q!(2q)!(-2\pi i)^q}{2^{2q}}\sum_{i=1}^{i=2q+1}(-1)^i\Psi_q(\xi)(X_1,X_2,...,\hat{X}^i,...,X_{2q},X_{2q+1}).$$

Remark 2.1. The part -ii) of the Theorem shows that the cube construction applied on an odd number of tangent vectors leads to an object which is no longer a differential form, as it is not a homogeneous function in each of the tangent vector arguments.

Proof. We treat first the case of an even number of tangent vectors.

We define an equivalence relation $\sim$ into the set of permutations Π_{2q}. To define the equivalence relation, think of any permutation $\iota = (i_1, i_2, ..., i_{2q})$ as being decomposed into consecutive pairs of indices $((i_1, i_2), (i_3, i_4), ..., (i_{2q-1}, i_{2q}))$; two permutations will be called equivalent if one is obtained from the other by an arbitrary number of transpositions of some of the pairs of the decomposition. The equivalence class $[\iota]$ of the permutation ι will have, obviously, 2^q elements.

Let $\iota_0 = (i_1, i_2, ..., i_{2q})$ be a permutation; we intend to estimate

$$S_{[\iota_0]} := \sum_{\iota\in[\iota_0]} Sign(\iota).\tau_\iota(t). \tag{4}$$

Denoting by $\tau(t)$ the parallel transport along the diagonal, from the point $(t, t, ..., t)$ to the origin (composed with the mapping Q), the main point is that $S_{[\iota_0]}$ decomposes into a product of linear homomorphisms

$$S_{[\iota_0]} = Sign(\iota_0)\tau(t)\circ\tau_{(i_{2q-1},i_{2q})}(t)\circ\tau_{(i_{2q-3},i_{2q-2})}(t)\circ...\circ\tau_{(i_3,i_4)}(t)\circ\tau_{(i_1,i_2)}(t),$$

where $\tau_{(i_{2k-1},i_{2k})}(t)$, (k = 1,2,..,q) is defined below.

Given an arbitrary permutation $\iota = (i_1, i_2, ..., i_{2q})$ and an index r, $1 \leq r \leq 2q$, denote by $P_{\iota,r}(t) \in I^{2q}$ that point which has all co-ordinates of orders $i_1, i_2, ..., i_r$ equal to t while all others equal to 0.

Then, $\tau_{(i_{2k-1},i_{2k})}(t)$ is by definition the parallel transport on the two-segments path

$$\overline{P_{\iota_0,2k-2}(t)P_{\iota_0,2k-1}(t)}, \overline{P_{\iota_0,2k-1}(t)P_{\iota_0,2k}(t)}$$

minus the parallel transport on the two-segments path

$$\overline{P_{\tilde{\iota}_0,2k-2}(t)P_{\tilde{\iota}_0,2k-1}(t)}, \overline{P_{\tilde{\iota}_0,2k-1}(t)P_{\tilde{\iota}_0,2k}(t)},$$

where $\overline{\iota_0}$ is the permutation ι_0 in which the pair (i_{2k-1}, i_{2k}) was transposed.

An elementary calculation, using the Taylor formula and the parallel transport equations, applied on the components of the vectors involved, shows that

$$\tau_{(i_{2k-1}, i_{2k})}(t) = t^2 R_{Q(P_{\iota_0, 2k-2}(t))}(X_{i_{2k-1}}, X_{i_{2k}}) + O(t^3)$$
$$= t^2 R_{x_0}(X_{i_{2k-1}}, X_{i_{2k}}) + O(t^3).$$

Note that this formula is essentially the classical definition of the curvature as the principal part of the asymmetry of the parallel transport on the sides of an infinitesimal rectangle, which in the present situation consists of the above pair of two-segments paths.

Substituting this estimate into the formula for $S_{[\iota_0]}$, we get

$$S_{[\iota_0]} =$$
$$t^{2q} Sign(\iota_0).\{\tau(t) \circ R_{x_0}(X_{i_{2q-1}}, X_{i_{2q}}) \circ \ldots \circ R_{x_0}(X_{i_3}, X_{i_4}) \circ R_{x_0}(X_{i_1}, X_{i_2}) + O(t)\}$$

and given that $\tau(t) = I + 0(t)$, one has further

$$S_{[\iota_0]} = (-1)^q t^{2q} Sign(\iota_0).$$
$$R_{x_0}(X_{i_{2q}}, X_{i_{2q-1}}) \circ \ldots \circ R_{x_0}(X_{i_4}, X_{i_3}) \circ R_{x_0}(X_{i_2}, X_{i_1}) + O(t^{2q+1}) =$$
$$t^{2q} Sign(\iota_0) R_{x_0}(X_{i_1}, X_{i_2}) \circ R_{x_0}(X_{i_3}, X_{i_4}) \circ \ldots \circ R_{x_0}(X_{i_{2q-1}}, X_{i_{2q}}) + O(t^{2q+1}) =$$
$$\frac{t^{2q}}{2^q} \sum_{j \in [\iota_0]} Sign(j) R_{x_0}(X_{j_1}, X_{j_2}) \circ R_{x_0}(X_{j_3}, X_{j_4}) \circ \ldots \circ R_{x_0}(X_{j_{2q-1}}, X_{j_{2q}}) + O(t^{2q+1}),$$

where $j = (j_1, j_2, \ldots, j_{2q})$.

From here we get

$$\sum_{\iota \in \Pi_{2q}} Sign(\iota).\tau_\iota(t) = \sum_{\gamma \in (\Pi_{2q}/\sim)} \sum_{j \in \gamma} Sign(j).\tau_j(t) = \sum_{\gamma \in (\Pi_{2q}/\sim)} S_\gamma =$$

$$\sum_{\gamma \in (\Pi_{2q}/\sim)} \frac{t^{2q}}{2^q} \sum_{j \in [\iota_0]} Sign(j) R_{x_0}(X_{j_1}, X_{j_2}) \circ R_{x_0}(X_{j_3}, X_{j_4}) \circ \ldots \circ R_{x_0}(X_{j_{2q-1}}, X_{j_{2q}})$$
$$+ O(t^{2q+1}) =$$
(where $j = (j_1, j_2, \ldots, j_{2q}) \in \gamma$)
$$\frac{t^{2q}}{2^q} \sum_{j \in \Pi_{2q}} Sign(j) R_{x_0}(X_{j_1}, X_{j_2}) \circ R_{x_0}(X_{j_3}, X_{j_4}) \circ \ldots \circ R_{x_0}(X_{j_{2q-1}}, X_{j_{2q}}) + O(t^{2q+1})$$

$$= \frac{t^{2q}}{2^q} \frac{(2q)!}{2^q} \sum_{j \in \Pi_{2q}} Sign(j)(R_{x_0} \wedge R_{x_0} \wedge \ldots \wedge R_{x_0})(X_1, X_2, \ldots, X_{2q}) + O(t^{2q+1}) =$$

$$\frac{(2q)! t^{2q}}{2^{2q}} (R_{x_0})^q (X_1, X_2, \ldots, X_{2q}) + O(t^{2q+1}).$$

Therefore,

$$TrLim_{t\searrow 0}\frac{1}{t^{2q}}\sum_{\iota\in\Pi_{2q}}Sign(\iota).\tau_\iota(t) =$$

$$= TrLim_{t\searrow 0}\frac{1}{t^{2q}}\{\frac{(2q)!t^{2q}}{2^{2q}}(R_{x_0})^q(X_1,X_2,...,X_{2q})+O(t^{2q+1})\}$$

$$= \frac{(2q)!}{2^{2q}}Tr(R_{x_0})^q(X_1,X_2,...,X_{2q})$$

$$= \frac{q!(2q)!(-2\pi i)^q}{2^{2q}}\Psi_q(\xi)(X_1,X_2,...,X_{2q}),$$

which proves the theorem for an even number of tangent vectors. The same proof may be easily adapted to the case of an odd number of tangent vectors.

2.2. *The Simplex Construction*

An analogous construction in which cubes are replaced by simplices and the parallel transport is performed along geodesics on the base space of the bundle is due to C. Teleman [10]. The simplex construction provides the important information that the Chern character of a vector bundle can be derived from the representation of the groupoid Ω of loops based at one point x_0 into the holonomy group of a linear connection.

3. Recall of Periodic Cyclic Homology

In this section we recall some basic notions and results due to A. Connes which lay to the foundations of the non commutative geometry.

Given a locally convex associative algebra $\mathcal{A}$, the space of k-chains $C_k(\mathcal{A})$ over the algebra A, $C_*(\mathcal{A})$ is, by definition, a topological completion (usually, projective completion) of the algebraic tensor product $\otimes^{k+1}A$. Two boundary operators, b' and b are introduced by the formulas

$$b'(f_0\otimes f_1\otimes....f_{k-1}\otimes f_k) = \sum_{r=0}^{r=k-1}(-1)^r f_0\otimes f_1\otimes....\otimes(f_r.f_{r+1})\otimes...\otimes f_{k-1}\otimes f_k$$

$$(5)$$

and

$$b(f_0\otimes f_1\otimes....f_{k-1}\otimes f_k) =$$

$$\sum_{r=0}^{r=k-1}(-1)^r f_0\otimes f_1\otimes....\otimes(f_r.f_{r+1})\otimes...\otimes f_{k-1}\otimes f_k$$

$$+(-1)^k f_0.f_1\otimes....\otimes f_{k-1}.$$

The boundary operator b' defines the bar complex; if the algebra $\mathcal{A}$ is unitary than the bar complex is acyclic.

The complex based on the boundary operator b is the Hochschild complex of the algebra $\mathcal{A}$; its homology is the Hochschild homology of the algebra.

The graded cyclic permutation $T : C_k(\mathcal{A}) \longrightarrow C_k(\mathcal{A})$ is defined on generators by

$$T(f_0 \otimes f_1 \otimes ....f_{k-1} \otimes f_k) = (-1)^k f_1 \otimes ....f_{k-1} \otimes f_k \otimes f_0. \tag{6}$$

The operator $N : C_k(\mathcal{A}) \longrightarrow C_k(\mathcal{A})$ is given by

$$N = 1 + T + T^2 + ... + T^k.$$

The cyclic complex $C_*^\lambda(\mathcal{A})$ of the algebra $\mathcal{A}$ is defined by

$$C_*^\lambda(\mathcal{A}) = \{Ker(1-T), b'\} \cong \{Coker(1-T), b\}. \tag{7}$$

By definition, its homology, $H_*^\lambda(\mathcal{A})$, is the cyclic homology of the algebra $\mathcal{A}$, due to A. Connes [1,2]; see also J. L. Loday [6], Sect. 2.1.

The cyclic homology could be defined also as the homology of the total complex associated to a first quadrant bicomplex made with the homorphisms $b', b, 1 - T, N$, see [6], Sect. 2.1.

Theorem 3.1 (A. Connes, [1,2]). *-i)*

$$H_k^\lambda(C^\infty(M)) = \Omega^k(M)/d\Omega^{k-1}(M) \bigoplus H_{dR}^{k-2}(M) \bigoplus H_{dR}^{k-4}(M) \bigoplus .... \bigoplus H_{dR}^\epsilon(M), \tag{8}$$

where $\Omega^k(M)$ denotes the vector space of smooth k-differential forms on M, $H_{dR}^*(M)$ denotes de Rham cohomology, and $\epsilon = 0$, or 1, depending on the parity of k.

-ii) Given the cyclic cycle $f \in C_*^\lambda(C^\infty)$, its top degree component $[f]_k$ belonging to $\Omega^k(M)/d\Omega^{k-1}(M)$, is the equivalence class (modulo exact forms) of the differential form (independent of local coordinates)

$$[f(x_0, x_1, ..., x_k)]_k(x) =$$

$$\frac{1}{k!}\frac{\partial}{\partial x_1^{i_1}}\frac{\partial}{\partial x_2^{i_2}}.....\frac{\partial}{\partial x_k^{i_k}}f(x_0, x_1, ..., x_k)_{|x_0=x_1=...=x_k=x}dx_1^{i_1} \wedge dx_2^{i_2} \wedge ... \wedge dx_k^{i_k}$$

$$(mod.d\Omega^{k-1}(M)).$$

A more natural homology is the periodic cyclic homology, $H_*^{per}(\mathcal{A})$, also due to A. Connes. The periodic cyclic homology eliminates the first unatural term from the above direct sum decomposition. It is defined (see [6], Sect. 5.1) as the homology of the total complex associated to a first and second quadrant direct product (rather than direct sum) bicomplex $\{C_{p,q}\}_{p\in Z, q\geq 0}$ defined by:

-i) $C_{p,q} = C_q(\mathcal{A})$; the boundary operators are considered of degree -1,

-ii)the columns consist of alternating bar and Hochschild complexes - the Hochschild complex on each even order column and the bar complex (with b' replaced by $-b'$) on each odd order column,

-iii) the boundary homomorphisms of the horizontal complexes are given by the alternating homorphisms N and $1 - T$

$$.... \xleftarrow{1-T} C_{-1,q} \xleftarrow{N} C_{0,q} \xleftarrow{1-T} C_{1,q} \xleftarrow{N} .... \tag{9}$$

This complex is called *periodic cyclic bicomplex*. The bicomplex mentioned at the beginning of this section (providing the cyclic homology, rather than the periodic cyclic homology) consists precisely of the first quadrant of the periodic cyclic bicomplex.

Given the periodicity of the periodic cyclic bicomplex, there are essentialy only two periodic cyclic homologies: $H_{even}^{\lambda,per}(\mathcal{A})$, and $H_{odd}^{\lambda,per}(\mathcal{A})$.

Theorem 3.2 (A. Connes [1,2]). *-i)*

$$H_{even}^{\lambda,per}(C^\infty(M)) = \overset{k=even}{\bigoplus} H_{dR}^k(M), \tag{10}$$

-ii)

$$H_{odd}^{\lambda,per}(C^\infty(M)) = \overset{k=odd}{\bigoplus} H_{dR}^k(M), \tag{11}$$

-iii)
Given the periodic cyclic cycle $f = \prod_{p=0}^{p=\infty} f_{\epsilon-p,p} \in \prod_{p=0}^{p=\infty} C_{\epsilon-p,p}(C^\infty(M))$, ($\epsilon = 0, 1$), its $(\epsilon + 2p)$-degree component is the de Rham cohomology class of the (closed) differential form (which is independent of local coordinates)

$$\frac{1}{(\epsilon + 2p)!} \frac{\partial^{\epsilon+2p}}{\partial x_1^{i_1}.....\partial x_{\epsilon+2p}^{i_{\epsilon+2p}}} f_{\epsilon+2p}(x_0, x_1, ..., x_{\epsilon+2p})_{|\Delta} dx_1^{i_1} \wedge ..... \wedge dx_{\epsilon+2p}^{i_{\epsilon+2p}}, \tag{12}$$

where Δ is the diagonal.

4. Distance Function and Chern Character

Combining the geometric construction of the Chern character presented in Sect. 3 with the tools of non commutative geometry recalled in the previous section, we intend to extract the Chern character of the tangent bundle of a smooth manifold from a Riemannian geodesic distance defined on the manifold.

4.1. *Riemannian Geometry Preliminaries*

Let g be a smooth Riemannian metric on the manifold M of dimension n.

Let $g_{i,j}(x) = <\frac{\partial}{\partial x^i}, \frac{\partial}{\partial x^j}>$ be the components of the metric tensor g with respect to the local coordinate system (x) and let $g^{ij}(x)$ be the components of the inverse matrix. Customary notation is used for the components of the curvature tensor $R_{rklh} = g_{rs}R_{klh}^s$, where $R_{klh}^s = \frac{\partial \Gamma_{kh}^s}{\partial x^l} - \frac{\partial \Gamma_{kl}^s}{\partial x^h} - \sum_{i=1}^{i=n}(\Gamma_{kh}^i \Gamma_{il}^s - \Gamma_{kl}^i \Gamma_{ih}^s)$.

The curvature tensor satisfies the identities, see e.g. [9]

Proposition 4.1. *-i)* $R_{rklh} = -R_{rkhl}, \qquad R_{rklh} = -R_{krlh}$

$\quad$ *-ii)* $R_{rklh} = R_{lhrk}$

$\quad$ *-iii)* $R_{rklh} + R_{rlhk} + R_{rhkl} = 0.$

The following identity is a corollary of this proposition; it will be used below.

Proposition 4.2. *-i)*

$$R_{i\alpha j\beta} - R_{i\beta j\alpha} = R_{ij\alpha\beta} \tag{13}$$

$\quad$ *-ii)*

$$\sum_{\alpha\beta}(R_{ij\alpha\beta} - R_{i\beta j\alpha})dx^\alpha \wedge dx^\beta = \frac{3}{2}\sum_{\alpha\beta}R_{ij\alpha\beta}dx^\alpha \wedge dx^\beta. \tag{14}$$

Proof. -i) The identity

$$R_{i\alpha j\beta} + R_{ij\beta\alpha} + R_{i\beta\alpha j} = 0 \tag{15}$$

given by Proposition 2.1 -iii) may be rewritten

$$R_{i\alpha j\beta} - R_{ij\alpha\beta} - R_{i\beta j\alpha} = 0, \tag{16}$$

which proves the desired relation.

$\quad$ -ii) We have

$$\sum_{\alpha\beta}(R_{ij\alpha\beta} - R_{i\beta j\alpha})dx^\alpha \wedge dx^\beta = \tag{17}$$

(skew-symmetry of the wedge product)

$$\sum_{\alpha\beta}R_{ij\alpha\beta}dx^\alpha \wedge dx^\beta - \frac{1}{2}\sum_{\alpha\beta}(R_{i\beta j\alpha} - R_{i\alpha j\beta})dx^\alpha \wedge dx^\beta = \tag{18}$$

(identity -i) above)

$$\sum_{\alpha\beta}R_{ij\alpha\beta}dx^\alpha \wedge dx^\beta + \frac{1}{2}\sum_{\alpha\beta}R_{ij\alpha\beta}dx^\alpha \wedge dx^\beta = \frac{3}{2}\sum_{\alpha\beta}R_{ij\alpha\beta}dx^\alpha \wedge dx^\beta. \tag{19}$$

Let $r : M \times M \longrightarrow [0, \infty)$ be the induced geodesic distance function. The function r^2 is smooth on a neighborhood of the diagonal, see e.g. [5]. $\square$

Proposition 4.3 (H. Donnelly, 1976).
Let $x_0 \in M$ and let $(x^1, x^2, ..., x^n)$ be a system of normal coordinates at the point x_0, corresponding to an ortho-normal tangent frame at x_0. Then
-i)

$$g_{rl}(x) = \delta_{rl} - \frac{1}{3} R_{rklh}(0) x^k x^h + O(x^3) \tag{20}$$

-i')

$$g^{rl}(x) = \delta^{rl} + O(x^2) \tag{21}$$

-ii) the first terms of the Taylor expansion of the function r^2 about the origin of the normal coordinates are given by the formula

$$r^2(x, y) = \sum_{i=1}^{n} (x^i - y^i)^2 - \frac{1}{3} R_{rklh}(0) x^k x^h (y^r - x^r)(y^l - x^l) + O((x, y)^5). \tag{22}$$

This result was also used by A. Connes and H. Moscovici in [3].

4.2. *The Characteristic Cyclic Functions* Φ_k

Introduce on $M \times M$ the double form

$$\varphi(x, y) := d_x d_y (\chi . r^2)(x, y) = \sum_{ij} \frac{\partial^2 (\chi . r^2)(x, y)}{\partial x^i \partial y^j} dx^i \otimes dy^j, \tag{23}$$

where χ is a real valued smooth cut-off function, identically 1 on a neighborhood of the diagonal, with sufficiently small support, so that r^2 be well defined and smooth on such support.

Let T_x denote the tangent space to M at x. We use the Riemannian metric to change the double form φ into a tensor $A(x, y)$, which remains covariant in the first argument and becomes contra-variant in the second. Let $A(x, y)$ be the tensor defined locally by the formula

$$A(x, y)(\sum_i \xi^i \frac{\partial}{\partial x^i}) = \sum_{i,j,k} \xi^i \frac{\partial^2 (\chi \circ r^2)(x, y)}{\partial x^i \partial y^j} g^{jk}(y) \frac{\partial}{\partial y^k}. \tag{24}$$

Then, $A(x, y) : T_x \to T_y$ is an well defined linear mapping, independent of the local co-ordinates. Let

$$A_i^k(x, y) = \sum_j \frac{\partial^2 (\chi \circ r^2)(x, y)}{\partial x^i \partial y^j} g^{jk}(y) \tag{25}$$

denote the components of the matrix associated to A.

Taken an arbitrary chain of points $x_0, x_1, ..., x_k$ together with corresponding local coordinates about them, one defines

$$\Phi_k(x_0, x_1, ..., x_k) := Tr A(x_0, x_1) A(x_1, x_2) ..... A(x_{k-1}, x_k) A(x_k, x_0). \qquad (26)$$

This function is reminiscent of the idea of composed parallel transports used in the geometrical construction of the Chern character.

The explicit formula for Φ_k is

$$\Phi_k(x_0, x_1, ..., x_k) =$$

$$\sum_{i_0, i_1, ..., i_k, j_0, j_1, ..., j_k} \frac{\partial^2 r^2(x_0, x_1)}{\partial x_0^{i_0} \partial x_1^{j_1}} g^{j_1 i_1}(x_1) \frac{\partial^2 r^2(x_1, x_2)}{\partial x_1^{i_1} \partial x_2^{j_2}} g^{j_2 i_2}(x_2) \cdots$$

$$\cdots \frac{\partial^2 r^2(x_{k-1}, x_k)}{\partial x_{k-1}^{i_{k-1}} \partial x_k^{j_k}} g^{j_k i_k}(x_k) \frac{\partial^2 r^2(x_k, x_0)}{\partial x_k^{i_k} \partial x_0^{j_0}} g^{j_0 i_0}(x_0).$$

The function Φ_k is a well defined smooth real valued function on a neighborhood of the diagonal in M^{k+1}.

We intend to study the function Φ_k within the context of cyclic homology.

4.3. *The Characteristic Cyclic Functions Φ_k and Chern Character*

Theorem 4.1. *For any smooth Riemannian metric on the smooth manifold M,*

-i) Φ_k, $(k = even)$, is a cyclic cycle over the algebra $\mathcal{A} = C^\infty(\mathcal{M})$,

-ii) the top degree component of the cyclic homology class of Φ_k is

$$[\Phi_k]_k = -\frac{2^{\frac{k}{2}+1}}{k!}(\frac{k}{2})!(2\pi i)^{\frac{k}{2}} Ch_k(M), \qquad (27)$$

where $Ch_k(M)$ is the k-component of the Chern character of the tangent bundle of M.

Proof. The proof will use the estimates from Proposition 4.3.

-i) We use the first definition of the cyclic complex. The cyclicity of the function Φ_k follows from the parity of k and the cyclicity of the trace.

It remains to show that $b'\Phi_k = 0$. To begin with, let's evaluate the first boundary face $\Phi_k(x_0, x_0, x_1, ..., x_{k-1})$.

We need to evaluate $\frac{\partial^2 r^2(x_0, x_1)}{\partial x_0^{i_0} \partial x_1^{j_1}} g^{j_1 i_1}(x_1)$ at $x_1 = x_0$. To do this we choose a system of normal coordinates centered at the point x_0 and we apply the

Taylor formula for the distance function. We have

$$\frac{\partial^2 r^2(x_0, x_1)}{\partial x_0^i \partial x_1^j}\bigg|_{x_0 = x_1 = 0} = \delta_{ij}, \quad g^{ij}(0) = \delta^{ij}. \tag{28}$$

From here we get that for sufficiently close points x, y, $A(x, y)$ is an isomorphism.

Given that $A(x, x)$ is the identity, one obtains that

$$\Phi_k(x_0, x_0, x_1, ..., x_{k-1}) = \Phi_{k-1}(x_0, x_1, ..., x_{k-1}). \tag{29}$$

Analogously, one gets that

$$\Phi_k(x_0, x_1, .., x_i, x_i, .., x_{k-1}) = \Phi_{k-1}(x_0, x_1, ..., x_{k-1}). \tag{30}$$

The needed property follows from the fact that $b'(\Phi_k)$ consists of a sum of an even number of equal terms multiplied by alternating signs.

-ii) We use Theorem 3.1.(ii) to compute the top degree component $[\Phi_k]_k \in \Omega^k(M)/d\Omega^{k-1}(M)$. The envisioned formula produces a differential form. We intend to determine this form at an arbitrary point $\mathbf{x} \in M$.

We choose a common normal coordinate system with the center at $\mathbf{x}$ for all variable points $x_0, x_1, ..., x_k$. We have to consider first order partial derivatives of the function Φ_k with respect to each of the variables $x_1^{\alpha_1}, x_2^{\alpha_2}, ..., x_k^{\alpha_k}$ and then to evaluate these derivatives at $x_0 = x_1 = ... = x_k = 0$.

Let us see how such derivatives $\frac{\partial}{\partial x^\alpha}, \frac{\partial}{\partial y^\beta}$ operate on a typical factor $A_i^k(x, y) = \sum_j \frac{\partial^2 r^2(x,y)}{\partial x^i \partial y^j} g^{jk}(y)$ of Φ_k.

If one of these partial derivatives operates on a factor $g^{jk}(y) = \delta^{jk} + O(y^2)$, then this factor is $O(y^1)$ and hence vanishes at 0. Such a factor could not be differentiated twice, or more, because it depends only on one point, y. Therefore, each such factor contributes in the formula for $[\Phi_k]_k$ by its value $g^{jk}(0) = \delta^{jk}$.

Let us analize the effect of the first and second order partial derivatives onto the factors $\frac{\partial^2 r^2(x,y)}{\partial x^i . \partial y^j}$. One has

$$\frac{\partial^2 r^2(x, y)}{\partial x^i \partial y^j} = -2\delta_{ij} - \frac{1}{3}\frac{\partial^2}{\partial x^i \partial y^j} R_{rklh}(0) x^k x^h (y^r - x^r)(y^l - x^l) + O((x, y)^3).$$

If such a term is differentiated once, it is $O((x, y)^1)$ at 0 and hence it vanishes at the origin. Therefore, only those factors which are differentiated either no time or two times could give a non zero contribution into the formula for $[\Phi_k]_k$.

If this factor is differentiated twice, its value at 0 is

$$-\frac{1}{3}R_{rklh}(0)\frac{\partial^2}{\partial x^\alpha \partial y^\beta}\frac{\partial^2}{\partial x^i \partial y^j}\{x^k x^h (y^r - x^r)(y^l - x^l)\} =$$

$$= -\frac{1}{3}R_{rklh}(0)\frac{\partial^2}{\partial x^\alpha \partial x^i}\frac{\partial^2}{\partial y^\beta \partial y^j}\{x^k x^h (y^r - x^r)(y^l - x^l)\}$$

$$= -\frac{1}{3}R_{rklh}(0)\frac{\partial^2}{\partial x^\alpha \partial x^i}\{x^k x^h\}\frac{\partial^2}{\partial y^\beta \partial y^j}\{(y^r - x^r)(y^l - x^l)\}$$

$$= -\frac{1}{3}R_{rklh}(0)(\delta_\alpha^k \delta_i^h + \delta_i^k \delta_\alpha^h)(\delta_\beta^r \delta_j^l + \delta_j^r \delta_\beta^l)$$

$$= -\frac{1}{3}(R_{\beta\alpha ji}(0) + R_{j\alpha\beta i}(0) + R_{\beta ij\alpha}(0) + R_{ji\beta\alpha}(0))$$

$$= -\frac{2}{3}(R_{ij\alpha\beta}(0) - R_{i\beta j\alpha}(0).)$$

This factor contributes further into the formula for $[\Phi_k]_k$ by juxtaposing the external product factor $dx^\alpha \wedge dx^\beta$, i.e. it contributes through the factor

$$-\frac{2}{3}(R_{ij\alpha\beta}(0) - R_{i\beta j\alpha}(0))dx^\alpha \wedge dx^\beta = -R_{ij\alpha\beta}(0)dx^\alpha \wedge dx^\beta, \qquad (31)$$

the last equality being given by Proposition 4.2. (ii).

We notice that if a factor

$$\frac{\partial^2 r^2(x_r, x_{r+1})}{\partial x_r^{i_r} \partial x_{r+1}^{j_{r+1}}}g^{j_{r+1}i_{r+1}}(x_{r+1}) \qquad (32)$$

of the formula for Φ_k is differentiated twice, the consecutive factor may not be differentiated because the two factors have one variable in common. Additionally, notice that the first and the last factors do not contribute into the final formula for $[\Phi_k]_k$ if they are differentiated because they contain the variable x_0, and no partial derivative with respect to such variable is contempleted.

We conclude that

$$[\Phi_k]_k =$$

$$\frac{-2^{\frac{k}{2}+1}}{k!}R_{i_1 i_2 \alpha_1 \alpha_2}R_{i_2 i_3 \alpha_3 \alpha_4}\cdots\cdots R_{i_{\frac{k}{2}} i_1 \alpha_{k-1}\alpha_k}dx^{\alpha_1}\wedge dx^{\alpha_2}\wedge\ldots\wedge dx^{\alpha_k} =$$

$$= \frac{-2^{\frac{k}{2}+1}}{k!}Tr(R^{\frac{k}{2}}),$$

where R is the curvature tensor, seen as an endomorphism valued 2-form (summation with respect to all repeating indices assumed).

Recalling that the k-component of the Chern character of the tangent bundle of M is $Ch_k(M) = 1/(\frac{k}{2})!.Tr(\frac{R}{2\pi i})^{\frac{k}{2}}$, we get

$$[\Phi_k]_k = -\frac{2^{\frac{k}{2}+1}}{k!}(\frac{k}{2})!(2\pi i)^{\frac{k}{2}}Ch_k(M). \tag{33}$$

This completes the proof of the theorem. $\qquad\square$

Let $\mathcal{M}(\mathcal{A})$ denote the matrix algebra over the associative algebra $\mathcal{A}$). The Morita isomorphism, see e.g. [1] or [6], states that

$$H_*^\lambda(\mathcal{M}(\mathcal{A})) \cong \mathcal{H}_*^\lambda(\mathcal{A}), \tag{34}$$

which, combined with Theorem 3.2 -ii), gives

$$H_*^\lambda(\mathcal{M}(\mathcal{C}^\infty)) \stackrel{\|\equiv *}{\cong} \bigoplus \mathcal{H}_{\lceil\mathcal{R}}^\|(\mathcal{M}) \tag{35}$$

Theorem 4.2 (A. Connes, [1,2]). *For any idempotent $e \in \mathcal{M}(\mathcal{A})$ the periodic Chern character of e and denoted $Ch^{per}(e) \in H_{even}^\lambda(\mathcal{A})$, is defined by*

$$f_{-2p,2p} = (-1)^p \frac{(2p)!}{p!} e^{\otimes 2p+1} \in C_{-2p,2p}(\mathcal{M}(\mathcal{A}))$$

$$f_{-(2p-1),2p-1} = (-1)^{p-1} \frac{(2p)!}{p!} e^{\otimes 2p} \in C_{-(2p-1),2p-1}(\mathcal{M}(\mathcal{A})).$$

Then

-i) $Ch^{per}(e)$ is a periodic cyclic cycle.

-ii) If e is an idempotent in $\mathcal{M}(\mathcal{C}^\infty(\mathcal{M}))$ then the image of the periodic cyclic homology class of $Ch^{per}(e)$ in the de Rham cohomology, through the Morita-Connes isomorphism, is the total Chern character of e, up to multiplicative constant.

Proof. See J.-L. Loday [6], Lemma 8.3.3. $\qquad\square$

Combining Theorem 4.1 with Theorem 4.2, we get

Theorem 4.3. *The infinite chain*

$$f_{-2p,2p} := (-1)^p \frac{(2p)!}{p!} \Phi_{2p} \in C_{-2p,2p}(C^\infty(M))$$

$$f_{-(2p-1),2p-1} := (-1)^{p-1} \frac{(2p)!}{p!} \Phi_{2p-1} \in C_{-(2p-1),2p-1}(C^\infty(M))$$

defines (up to a multiplicative constant) the periodic Chern character of the complexified tangent bundle to M.

Proof. The theorem follows from Theorem 4.1 along with the same kind of considerations as those used in the proof of Theorem 4.2; the role of the property $e^2 = e$ is played here by the property $A(x, x) = 1$.

Based on this last observation, we are going to introduce the notion of linear direct connection, discussed in the next section. $\qquad\square$

5. Direct Connections and Chern Character.

A *linear direct connection* in the vector bundle ξ over the smooth manifold M is by definition a function τ which assigns to any pair of points $x, y \in M$, sufficiently close one to each other, an isomorphism $\tau(y, x) : \xi_x \to \xi_y$, where ξ_x is the fiber at x, such that $\tau(x, x) = identity$.

Remark. This definition replaces the notion of *linear quasi connection* introduced by the author in [12]. We remark that the notion of linear quasi connection was also used in the previous paper by A.S. Mishchenko, N. Teleman [8]. For the reason explained in the next sentence, we believe that the new term is definitely more appropriate.

The basic difference between a linear direct connection and a linear connection consists of the fact that while a linear connection provides a transport of fibers along *curves*, a linear direct connection provides a *direct* transport of fibres *from point to point*.

The connection τ is *smooth* provided the isomorphism $\tau(y, x)$ depends smoothly on the pair x, y.

A. Connes, H. Moscovici [3] use linear connection parallel transport functions τ along geodesics. In fact, the parallel transport defined by a linear connection in ξ along the small geodesics of an affine connection in M induces a linear direct connection in ξ.

Remark, however, that a linear direct connection does not necessarily has to satisfy the property $\tau(y, x)\tau(x, y) = id.$, and hence there exist linear direct connections which do not derive from linear connections.

As in the previous section we associate with τ the function $\Phi_k : M^{k+1} \to C$ by the formula

$$\Phi_k(x_0, x_1, ..., x_k) := Tr\ \tau(x_0, x_1)\tau(x_1, x_2).....\tau(x_{k-1}, x_k)\tau(x_k, x_0). \quad (36)$$

Proposition 5.1. *Any two smooth linear direct connections in a smooth vector bundle are smoothly homotopic.*

Proof. Let τ_0, τ_1 be two linear direct connections in ξ. The affine deformation $(1 - t)\tau_0 + t\tau_1$ provides the desired homotopy. $\qquad\square$

992

Theorem 5.1. *Let ξ be a complex vector bundle over the paracompact manifold M and let τ be a smooth linear direct connection in ξ.*

For any natural number k define $\Phi_k \in C_k(C^\infty(M))$ by the formula

$$\Phi_k(x_0, x_1, ..., x_k) := Tr \; \tau(x_0, x_1) \circ \tau(x_1, x_2) \circ ..... \circ \tau(x_k, x_0) \qquad (37)$$

-i) The infinite chain

$$f_{-2p,2p} := (-1)^p \frac{(2p)!}{p!} \Phi_{2p} \in C_{-2p,2p}(C^\infty(M))$$

$$f_{-(2p-1),2p-1} := (-1)^{p-1} \frac{(2p)!}{p!} \Phi_{2p-1} \in C_{-(2p-1),2p-1}(C^\infty(M))$$

is an even periodic cyclic cycle over the algebra $C^\infty(M)$;

-ii) its homology class is (up to a multiplicative constant) the total Chern character of ξ.

Proof. Embedd the bundle ξ into a trivial bundle and associate an idempotent e to it. The idempotent e defines a linear direct connection

$$\tau(y, x) = e(y).e(x) \qquad (38)$$

The corresponding function Φ_k is

$$\Phi_k(x_0, x_1, ..., x_k) := Tr[e(x_0)e(x_1)][e(x_1)e(x_2)]....[e(x_k)e(x_0)] =$$
$$Tre(x_0)e(x_1)e(x_2)....e(x_k) = Tr(e \otimes e \otimes .... \otimes e)(x_0, x_1, ..., x_k).$$

The Theorem 4.2 by A. Connes shows that the statement of the theorem is true for the particular linear direct connection manufactured with the idempotent e. On the other hand, the periodic cyclic homology is invariant under homotopy. Proposition 5.1 states that any two linear direct connections are homotopic. This completes the proof of the theorem. $\square$

Remark 5.1. For linear connections, the fact that the differential form representing $[\Phi_k]_k$ is closed under the exterior derivative is a consequence of the Bianchi identity.

The curvature of a linear direct connection τ may be defined as a quantized object. By definition, the quantized curvature associated to the linear direct connection τ is the function $R : M^3 \longrightarrow End(E)$ given by

$$R(x_0, x_1, x_2) = \tau(x_0, x_1)\tau(x_1, x_2)\tau(x_2, x_0) \in EndE_{x_0}, \qquad (39)$$

where E_{x_0} is the fibre over x_0, see A.S.Mishchenko, N. Teleman [8].

Remark 5.2. The Levy-Civita connection involved in the construction of the Chern character of an idempotent uses the embedding of the bundle defined by the idempotent into a trivial bundle. The extraction of the Chern character from a direct connection does not require such an embedding.

Remark 5.3. We may compare the geometrical construction of the classical Chern character due to N. Teleman [11], C. Teleman [10], discussed here and in [12] in the non commutative context, on the one hand, and the construction of the Chern character due to A. Connes [1,2], on the other hand. Both constructions use the trace of iterated transports. Both constructions *involve* anti-symmetrizations and limits to the diagonal; while the first construction uses explicitly anti-symmetrizations and limits to the diagonal, the Connes' construction involves implicitly such operations, as these are built-in in the computation of the cyclic homology (for the algebra of smooth functions).

6. Further Extensions.

The notion of linear direct connection could be used in contexts where differentiability structure is not available.

The notion of linear direct connection may be extended to groupoids. The Chern character of spectral triples as well as the local index theorem due to A. Connes and H. Moscovici [3] could be reinterpreted along the same lines.

Acknowledgments

This paper was written with the partial support of the MIUR Contract N 2005010942_002/2005.

References

1. A. Connes A, *Non-commutative differential Geometry* (Publ. Math. IHES 62, pp.257 - 360, 1985).
2. A. Connes, *Non Commutative Geometry*, (Academic Press, 1994).
3. A. Connes, H. Moscovici, *Cyclic cohomology, the Novikov conjecture and hyperbolic groups* (Topology Vol. 29, Nr. 3 pp. 345-388, 1990).
4. H. Donnelly, *Spectrum and the Fixed Point Sets of Isometries*, (Math. Ann. 224, pp. 161-170, 1976).
5. G. de Rham, *Varietés Différentiables*, (Hermann Paris, 1960).
6. J.-L.Loday, *Cyclic Homology*, (Springer Verlag, 1992).
7. J. Milnor, *Lectures on Caracteristic Classes*, (Princeton Mathematical Studies Nr. 76, 1974).

8. A. S. Mishchenko, N. Teleman, *Almost flat bundles and almost flat structures in* (Topological Methods in Non Linear Analysis. To appear).

9. M. Spivak, *A Comprehensive Introduction to Differential Geometry, Vol. II*, (Publish or Perish, 1979).

10. C. Teleman, *Sur le charactère de Chern d'un fibré complexe différentiable*, (Rev. Roumaine Math. Pures Applic. 12, pp. 725-731, 1967) .

11. N. Teleman, *A geometrical definition of some André Weil's forms which can be associated with an infinitesimal connection*, (St. Cerc. Math. Tom. 18, No. 5, pp. 753-762, Bucarest, 1966).

12. N. Teleman, *Distance Function, Linear quasi connections and Chern Character*, (IHES Prepublications M/04/27, June 2004).

LOGARITHMIC COMPARISON THEOREM AND $\mathcal{D}$-MODULES: AN OVERVIEW

Tristan TORRELLI

Laboratoire J.A. Dieudonné, UMR du CNRS 6621
Université de Nice Sophia-Antipolis
Parc Valrose, 06108 Nice Cedex 2, France
E-mail: tristan_torrelli@yahoo.fr

Let $D \subset X$ be a divisor in a complex analytic manifold. A natural problem is to determine when the de Rham complex of meromorphic forms on X with poles along D is quasi-isomorphic to its subcomplex of logarithmic forms. In this mostly expository note, we recall the main results about this problem. In particular, we point out the relevance of the theory of $\mathcal{D}$-modules to this topic.

Keywords: de Rham complexes, $\mathcal{D}$-modules, free divisors, hyperplane arrangements, logarithmic comparison theorem, logarithmic vector fields, Bernstein polynomial.

2000 Mathematics Subject Classification: 32C38, 32S25, 14F10, 14F40.

Introduction

Let X be a complex analytic manifold of dimension $n \geq 2$. Given a divisor $D \subset X$, we denote j the natural inclusion $X \backslash D \hookrightarrow X$. Let $\Omega_X^\bullet(\star D)$ denote the complex of meromorphic forms on X with poles along D. From the Grothendieck Comparison Theorem [17], the de Rham morphism

$$\Omega_X^\bullet(\star D) \longrightarrow \mathbf{R}j_* \mathbf{C}_{X \backslash D}$$

is a quasi-isomorphism. In particular, if $X = \mathbf{C}^n$, then for each cohomology class $c \in H^p(\mathbf{C}^n \backslash D, \mathbf{C})$, there exists a differential form $w \in \Omega_X^p(\star D)$ such that for any p-cycle σ on $\mathbf{C}^n \backslash D$, one has $c(\sigma) = \int_\sigma w$.

It is natural to ask what one can say about the form w. For example, if D is a complex submanifold then the order of the pole of w can be

996

taken to be 1. The question of the order of the pole goes back to P.A. Griffiths [16]. We recall that a meromorphic form $w \in \Omega_X^p(\star D)$ is *logarithmic* if w and dw have at most a simple pole along D; let $\Omega_X^\bullet(\log D) \subset \Omega_X^\bullet(\star D)$ denote the subcomplex of logarithmic forms with pole along D, introduced in full generality by K. Saito in [27]. In the initial case of normal crossing divisors, P. Deligne [14] proved that the filtered morphism $(\Omega_X^\bullet(\log D), \sigma) \hookrightarrow (\Omega_X^\bullet(\star D), P)$ where P is the pole order filtration and σ is induced by P, is a quasi-isomorphism compatible with filtrations. This fact was crucial in order to defined a mixed Hodge structure on the cohomology of a quasi-projective algebraic variety. Hence, one says that D satisfies the *logarithmic comparison theorem* if

LCT(D) : The inclusion $\Omega_X^\bullet(\log D) \hookrightarrow \Omega_X^\bullet(\star D)$ is a quasi-isomorphism.

A natural problem is therefore to find classes of divisors satisfying this condition, and also to understand its meaning. Initiated by F.J. Castro-Jiménez, D. Mond and L. Narváez-Macarro [9], this problem has been intensively studied these last years. In this note, we gather together the main open questions* and the main results. Essentially, they were obtained for hypersurfaces with isolated singularities, hyperplane arrangements and free divisors (see §1). In this last case, we recall the characterization in terms of $\mathcal{D}$-modules due to the Sevillian group around F.J. Castro-Jiménez and L. Narváez-Macarro (Theorem 2.2). Finally, we explain how enlightening this viewpoint is for the general study of the condition **LCT**(D) (see §3).

1. Main results about LCT(D)

There are few families of divisors for which this condition **LCT**(D) has been studied. Indeed, it is difficult to work directly with the complex $\Omega_X^\bullet(\log D)$ since we do not have in general a description of the logarithmic forms.

1.1. *The case of weighted homogeneous hypersurfaces with an isolated singularity*

We recall that a polynomial $h \in \mathbf{C}[x] = \mathbf{C}[x_1, \ldots, x_n]$ is *weighted homogeneous* of weight $d \in \mathbf{Q}^+$ for a system $\alpha = (\alpha_1, \ldots, \alpha_n) \in (\mathbf{Q}^{*+})^n$ if h is a (nontrivial) $\mathbf{C}$-linear combination of monomials $x_1^{\gamma_1} \cdots x_n^{\gamma_n}$ with $\sum_{i=1}^n \alpha_i \gamma_i = d$. In other words, we have the relation $\chi(h) = dh$ where χ is the Euler-vector field $\alpha_1 x_1 \partial_1 + \cdots + \alpha_n x_n \partial_n$ associated with α.

---

*See the appendix.

As usual, the case of weighted homogeneous polynomials defining an isolated singularity at the origin provides combinatorial formulas in terms of the weights associated with the Jacobian algebra $A_h = \mathbf{C}[x]/(h'_{x_1}, \ldots, h'_{x_n})$. D. Mond and M. Holland [18] have obtained the following characterization:

Theorem 1.1. *Assume that $n \geq 3$. Let $h \in \mathbf{C}[x]$ be a weighted homogeneous polynomial of degree d for a system $\alpha \in (\mathbf{Q}^{*+})^n$. Assume that h defines an isolated singularity at the origin. Let $D \subset \mathbf{C}^n$ be the hypersurface defined by h. The following conditions are equivalent:*

(1) The logarithmic comparison theorem holds for D.
(2) The link of 0 in D is a $\mathbf{Q}$-homology sphere.
(3) There is no weighted homogeneous element in A_h whose weight belongs to the set $\{k \times d - \sum_{i=1}^{n} \alpha_i \ ; \ 1 \leq k \leq n - 2\} \subset \mathbf{Q}$.

In particular, the logarithmic comparison theorem does not hold in general (see also Proposition 3.2). For example, if $h = x_1^2 + \cdots + x_n^2$ then we can take $d = 2$, $\alpha_1 = \cdots = \alpha_n = 1$ and $A_h = \mathbf{C} \cdot \bar{1}$. Thus $\mathbf{LCT}(D)$ is satisfied if and only if $n = 2$ or n is odd.

1.2. *The case of hyperplane arrangements*

Let D be a finite union of affine hyperplanes H in $X = \mathbf{C}^n$, *i.e.* $H = \{\alpha_H = 0\}$ where $\alpha_H \in \mathbf{C}[x_1, \ldots, x_n]$ are polynomials of degree one. We can associate with D the $\mathbf{C}$-subalgebra of $\Omega_X^{\bullet}(\star D)$ generated by 1 and the 1-forms $d\alpha_H/\alpha_H$. Let $R^{\bullet}(D)$ denote this algebra of differential forms. It is well known that $R^{\bullet}(D)$ is isomorphic to the so-called Orlik-Solomon algebra. Moreover:

Theorem 1.2. [2] *For all $k \geq 0$, we have $R^k(D) \cong H^k(X \backslash D, \mathbf{C})$.*

On the other hand, we can consider the following complex of $\mathbf{C}$-vector spaces: $0 \to R^0(D) \xrightarrow{0} \cdots \xrightarrow{0} R^n(D) \to 0$ as a subcomplex of $\Omega_X^{\bullet}(\log D)$. Thus, a natural question is: does the logarithmic comparison theorem hold for any hyperplane arrangement? This was conjectured by H. Terao in [30]. This is true for tame arrangements (such as free arrangements, generic arrangements or complex reflection arrangements) and when $n \leq 4$ (see [35]). But in general, the question is still open.

1.3. *The case of free divisors*

Let $\mathcal{O}_X$ be the sheaf of holomorphic functions on X. Given a divisor $D \subset X$, we will denote by $h_D \in \mathcal{O}_{X,m} \cong \mathcal{O} = \mathbf{C}\{x_1, \ldots, x_n\}$ a defining equation of

D at $m \in D$.

A holomorphic vector field v is *logarithmic* along D if for any point $m \in D$, $v(h_D)$ belongs to $h_D \mathcal{O}_{X,m}$. Let $\mathrm{Der}(-\log D)$ denote the (coherent) $\mathcal{O}_X$-module of logarithmic vector fields. We recall a property studied by K. Saito in [27].

Definition 1.1. A divisor $D \subset X$ is *free at the point* $m \in D$ if $\mathrm{Der}(-\log D)_m$ is $\mathcal{O}_{X,m}$-free. It is a *free divisor* if $\mathrm{Der}(-\log D)$ is locally free.

From the inclusions $h_D \mathrm{Der}(\mathcal{O}_X)_m \subset \mathrm{Der}(-\log D)_m \subset \mathrm{Der}(\mathcal{O}_X)_m$, the rank of $\mathrm{Der}(-\log D)$ is also equal to n.

Example 1.1. Free divisors appear in many different contexts.

(i) Normal crossing divisors are free. Indeed, in local coordinates such that $h_D = x_1 \cdots x_p$, then

$$\mathrm{Der}(-\log D) = \mathcal{O}x_1\partial_1 \oplus \cdots \oplus \mathcal{O}x_p\partial_p \oplus \mathcal{O}\partial_{p+1} \oplus \cdots \oplus \mathcal{O}\partial_n.$$

(ii) Plane curves are free (K. Saito [27]).
(iii) Complex reflection arrangements are free (H. Terao [31]). For example, the braid arrangement, defined by $\prod_{1 \leq i < j \leq n}(x_i - x_j)$ in $\mathbf{C}^n$, is free.
(iv) The discriminant of a versal deformation of an isolated complete intersection singularity is a free divisor (see [28]; [22], Corollary 6.13; [1]).

Most of the known results about the logarithmic comparison theorem have been obtained for free divisors. Indeed, the logarithmic de Rham complex $\Omega_X^{\bullet}(\log D)$ is also explicit. More precisely, because of the duality between $\Omega_X^1(\log D)$ and $\mathrm{Der}(-\log D)$, $\Omega_X^1(\log D)$ is also a free $\mathcal{O}_X$-module and we have $\Omega_X^q(\log D) = \bigwedge^q \Omega_X^1(\log D)$ for $1 \leq q \leq n$ (see [27]). Firstly, we have the following characterization for plane curves:

Theorem 1.3. [5] *If $D \subset X = \mathbf{C}^2$ is a plane curve, then the logarithmic comparison theorem holds if and only if D is locally weighted homogeneous.*

This last condition means that for all $m \in D$, there exists an analytic change of coordinates ϕ such that $h_D \circ \phi$ is a weighted homogeneous polynomial; for example, weighted homogeneous hypersurfaces with an isolated singularity and hyperplane arrangements are locally weighted homogeneous. This unusual condition is the suitable one in this context for doing inductions on the dimension of D (see the proof of Proposition 3.1 for example). More generally, we have

Theorem 1.4. [9] *Let $D \subset X$ be a locally weighted homogeneous free divisor. Then the logarithmic comparison theorem holds for D.*

Among the free divisors in Example 1.1, the one given in (i), (iii) and some[†] of (iv) are locally weighted homogeneous.

The converse is false in general. For example, the polynomial $h = x_1 x_2 (x_1 + x_2)(x_1 + x_2 x_3)$ defines a free divisor $D \subset \mathbf{C}^3$ such that $\mathbf{LCT}(D)$ is true and h is not weighted homogeneous (see [5], §4). Meanwhile, h belongs to the ideal of its partial derivatives. In other words, there exists locally a vector field v such that $v(h_D) = h_D$; one says sometimes that h is *Euler-homogeneous*. In fact, we have no example of a free divisor $D = V(h)$ verifying $\mathbf{LCT}(D)$ which is not Euler-homogeneous. This is true for a Koszul-free divisor (see Definition 2.1, Theorem 3.1); moreover, M. Granger and M. Schulze [15] have obtained the following result:

Theorem 1.5. *Let $D = V(h) \subset X = \mathbf{C}^3$ be a free divisor. If the logarithmic comparison theorem holds for D, then h is Euler-homogeneous.*

For $n \geq 4$, this question is still open (see [5], Conjecture 1.4) and it can be extended for a general divisor.

2. A differential viewpoint for free divisors

Here we recall how the condition $\mathbf{LCT}(D)$ may be interpreted in terms of $\mathcal{D}_X$-modules for free divisors $D \subset X$, as it was initiated by F.J. Calderón-Moreno in [4].

2.1. *Preliminaries*

Given a complex analytic manifold X of dimension $n \geq 2$, we denote $\Omega_X^\bullet$ the complex of holomorphic differential forms on X and $(\mathcal{D}_X, F_\bullet)$ the sheaf of linear differential operators with holomorphic coefficients filtered by order. Locally at a point $m \in X$, we have $\mathcal{O}_{X,m} \cong \mathcal{O} = \mathbf{C}\{x_1, \ldots, x_n\}$ and $\mathcal{D}_{X,m} \cong \mathcal{D} = \mathcal{O}\langle \partial_1, \ldots, \partial_n \rangle$; moreover we identify $\mathrm{gr}^F \mathcal{D}$ with $\mathcal{O}[\xi] = \mathcal{O}[\xi_1, \ldots, \xi_n]$.

The so-called Riemann-Hilbert correspondence of Z. Mebkhout and M. Kashiwara [21,24,26] asserts that there is an equivalence of categories between the category $hr(\mathcal{D}_X)$ of (left) regular holonomic $\mathcal{D}_X$-modules and the

---

[†] For more details, see [9].

one of perverse sheaves $Perv_X(\mathbf{C})$ on X *via* the de Rham functor

$$hr(\mathcal{D}_X) \longrightarrow Perv_X(\mathbf{C})$$
$$\mathcal{M} \longmapsto \mathrm{DR}(\mathcal{M}) = \Omega_X^\bullet \otimes_{\mathcal{O}_X} \mathcal{M}.$$

Roughly speaking, a perverse sheaf on X is a special type of complex of sheaves on X whose cohomology groups are constructible in $\mathbf{C}$-vector spaces of finite dimension on a stratification of X. For example, $\mathcal{O}_X$ is regular holonomic [17] and $\mathrm{DR}(\mathcal{O}_X) = \Omega_X^\bullet$ is quasi-isomorphic to the constant sheaf $\mathbf{C}_X$ by the Poincaré lemma.

2.2. *On the perversity of* $\Omega_X^\bullet(\log D)$

Given a divisor $D \subset X$, we consider the sheaf $\mathcal{O}_X(\star D)$ of meromorphic functions with poles along D. As $\mathcal{O}_X(\star D)$ is regular holonomic [17,20,25], the meromorphic de Rham complex $\mathrm{DR}(\mathcal{O}_X(\star D)) = \Omega_X^\bullet(\star D)$ is a perverse sheaf too. Thus it is natural to investigate conditions on D in order to get the perversity of $\Omega_X^\bullet(\log D)$. In the case of free divisors, this question was studied by F.J. Calderón-Moreno and L. Narváez-Macarro in [4,8]. They obtained the following characterization:

Theorem 2.1. *Let $D \subset X$ be a free divisor. Then the logarithmic complex $\Omega_X^\bullet(\log D)$ is perverse if and only if the following conditions are satisfied:*

(1) the complex $\mathcal{D}_X \otimes_{V_0^D(\mathcal{D}_X)}^L \mathcal{O}_X$ is concentrated in degree 0;
(2) the $\mathcal{D}_X$-module $\mathcal{D}_X \otimes_{V_0^D(\mathcal{D}_X)} \mathcal{O}_X$ is holonomic.

From [11], we say also that D is of *Spencer type*. Here $V_0^D(\mathcal{D}_X) \subset \mathcal{D}_X$ is the coherent sheaf of rings[‡] of logarithmic operators [4] (that is, $P \in \mathcal{D}_X$ such that locally $P \cdot (h_D^k) \subset h_D^k \mathcal{O}$ for any integer k). Let us notice that this condition 1 has no clear meaning. Thus, the problem is now to find (geometrical) criteria on a free divisor to be of Spencer type (see [8], §5). The only known condition is to be a Koszul-free divisor (see [4], Theorem 4.2.1).

Definition 2.1. A free divisor $D \subset X$ is *Koszul-free* if there exists locally a basis[§] $\{\delta_1, \ldots, \delta_n\}$ of $\mathrm{Der}(-\log D)$ such that the sequence of principal symbols $(\sigma(\delta_1), \ldots, \sigma(\delta_n))$ is $\mathrm{gr}^F\mathcal{D}$-regular.

[‡]For some results about this sheaf for a general divisor, see [29].
[§]In fact, every basis of $\mathrm{Der}(-\log D)$ satisfies this property when D is Koszul-free.

For example, plane curves and locally weighted homogeneous free divisors are Koszul-free [6,7,27]. This notion means also that the free divisor D has a holonomic stratification in the sense of K. Saito (see [27], §3; [3], Proposition 6.3; [4], Corollary 1.9).

Finally, we make the remark that being Koszul-free is not necessary for the perversity of $\Omega_X^\bullet(\log D)$; for example, the free divisor $D \subset \mathbf{C}^3$ defined by $x_1 x_2(x_1 + x_2)(x_1 + x_2 x_3)$ is not Koszul-free but $\Omega_X^\bullet(\log D)$ is perverse [4,5]. On the other hand, the complex $\Omega_X^\bullet(\log D)$ is not perverse for any free divisor. For example, the divisor $(x_1^5 + x_2^4 + x_1^4 x_2)(x_1 + x_2 x_3) = 0$ is free but not of Spencer type (see [8], §5).

2.3. *A differential characterization of* LCT(D)

Let us now give a differential analogue of condition **LCT**(D). F.J. Castro-Jiménez and J.M. Ucha-Enríquez began work on this problem in [10,11], and they obtained a characterization for free divisors of Spencer type [12]. For a general free divisor, we have the following generalization[¶] due to F.J. Calderón-Moreno and L. Narváez-Macarro [8]:

Theorem 2.2. *Let $D \subset X$ be a free divisor. Then the inclusion*

$$\Omega_X^\bullet(\log D) \hookrightarrow \Omega_X^\bullet(\star D)$$

is a quasi-isomorphism if and only if the following conditions are satisfied:

(1) the complex $\mathcal{D}_X \otimes_{\mathcal{V}_0^D(\mathcal{D}_X)}^L \mathcal{O}_X(D)$ is concentrated in degree 0;
(2) the natural morphism

$$\varphi_D : \mathcal{D}_X \otimes_{\mathcal{V}_0^D(\mathcal{D}_X)} \mathcal{O}_X(D) \longrightarrow \mathcal{O}_X(\star D)$$

is an isomorphism.

Here $\mathcal{O}_X(D)$ denotes the $\mathcal{V}_0^D(\mathcal{D}_X)$-module of meromorphic functions with at most a simple pole along D. Unfortunately, this characterization is no more explicit than condition **LCT**(D); meanwhile, condition 1 is verified by Koszul-free divisors.

A key point in the proofs of Theorems 2.2 and 2.1 is the relationship between the duals of any integral logarithmic connection over the base ring $\mathcal{D}_X$ and $\mathcal{V}_0^D(\mathcal{D}_X)$ (see [11]; [8], §3). In particular, the $\mathcal{D}_X$-modules

---

[¶]In fact, they obtained this result for any integral logarithmic connection (see [8], Theorem 4.1).

$\mathcal{D}_X \otimes_{\mathcal{V}_0^P(\mathcal{D}_X)} \mathcal{O}_X(D)$ and $\mathcal{D}_X \otimes_{\mathcal{V}_0^P(\mathcal{D}_X)} \mathcal{O}_X$ are holonomic and dual when D is a free divisor of Spencer type, and we have $\mathrm{DR}(\mathcal{D}_X \otimes_{\mathcal{V}_0^P(\mathcal{D}_X)} \mathcal{O}_X(D)) \cong \Omega_X^\bullet(\log D)$. A generalization of this duality has been obtained in [13].

3. Towards an algebraic analogue of LCT(D)?

In this part, we explain how useful the $\mathcal{D}_X$-modules are in the general study of the condition **LCT**(D).

3.1. *Preliminaries*

What can one say about the free divisor D when the morphism φ_D is an isomorphism? First, we have locally $\mathcal{V}_0^P(\mathcal{D}) = \mathcal{O}[\delta_1, \ldots, \delta_n]$ where $\{\delta_1, \ldots, \delta_n\}$ is a basis of $\mathrm{Der}(-\log D)$ [4] and $\mathcal{O}(D) \cong \mathcal{V}_0^P(\mathcal{D})/\mathcal{V}_0^P(\mathcal{D})(\delta_1 + a_1, \ldots, \delta_n + a_n)$ with $\delta_i(h_D) = a_i h_D$, $1 \leq i \leq n$. Hence the morphism φ_D is given locally by

$$\varphi_D : \mathcal{D}/\tilde{I}^{log} \longrightarrow \mathcal{O}[1/h_D]$$
$$P + \tilde{I}^{log} \longmapsto P \cdot \frac{1}{h_D}$$

where $\tilde{I}^{log} \subset \mathcal{D}$ is the left ideal generated by $\mathrm{Ann}_\mathcal{D}\, 1/h_D \cap F_1\mathcal{D}$. In particular, it is bijective if and only if the two conditions:

A($1/h_D$)**:** the left ideal $\mathrm{Ann}_\mathcal{D}\, 1/h_D$ of operators annihilating $1/h_D$ is generated by operators of order 1,

B(h_D)**:** the $\mathcal{D}$-module $\mathcal{O}[1/h_D]$ is generated by $1/h_D$,

are satisfied. From a well known result of M. Kashiwara (see [19], Proposition 6.2), this last condition means that -1 is the only integral root of the Bernstein polynomial of h_D. We recall that the *Bernstein polynomial $b_h(s)$ of $h \in \mathcal{O}$ at the origin* is the (nonzero) unitary polynomial $b(s) \in \mathbf{C}[s]$ of smallest degree which satisfies a functional identity of the form:

$$b(s)h^s = P(s) \cdot h^{s+1}$$

with $P(s) \in \mathcal{D}[s] = \mathcal{D} \otimes \mathbf{C}[s]$ (see [19]). This is an analytic invariant of the ideal $h\mathcal{O}$. When $h \in \mathcal{O}$ is not a unit, it is easy to check that -1 is also a root of $b_h(s)$. For example, if $h = x_1^2 + \cdots + x_n^2$ then $b_h(s) = (s+1)(s+n/2)$ with the functional identity:

$$(s+1)(s+\frac{n}{2})h^s = \frac{1}{4}\left[\frac{\partial}{\partial x_1}^2 + \cdots + \frac{\partial}{\partial x_n}^2\right] \cdot h^{s+1}.$$

3.2. *The conditions* $\mathbf{B}(h_D)$ *and* $\mathbf{LCT}(D)$

The differential viewpoint above is relevant since it was not at all clear that for a free divisor, $\mathbf{LCT}(D)$ needs the condition $\mathbf{B}(h_D)$; in particular, every locally weighted homogeneous free divisor D satisfies the condition $\mathbf{B}(h_D)$ at any point (Theorem 1.4 with Theorem 2.2 or Proposition 3.1, or more directly [11], Theorem 5.2). Moreover, from the inclusions

$$\Omega_X^\bullet(\log D) \subset \Omega_X^\bullet \otimes (\mathcal{D} \cdot 1/h_D) \subset \Omega_X^\bullet \otimes \mathcal{O}[1/h_D] = \Omega_X^\bullet(\star D) \qquad (1)$$

it appears natural that conditions $\mathbf{LCT}(D)$ and $\mathbf{B}(h_D)$ are linked for any divisor D. As an illustration, we have the following result:

Proposition 3.1. *Let* $D \subset X$ *be a divisor which satisfies the condition* $\mathbf{LCT}(D)$. *Assume that one of the following conditions is satisfied:*

(1) The divisor D is free except at isolated points.
(2) The divisor D is locally weighted homogeneous.

Then $\mathbf{B}(h_D)$ *is satisfied at any point of D.*

Proof. Firstly we prove the assertion when the condition 1 is satisfied. Let $U \subset X$ be a neighborhood of a point $m \in D$ such that $D \cap U$ is free at any point different from m. Let h_D be a defining equation of D on U. From Theorem 2.2, the condition $\mathbf{B}(h_D)$ is satisfied at any point in $D \cap U - \{m\}$ - since D is free at such a point. In particular, the $\mathcal{D}_U$-module $\mathcal{C}$ in the short exact sequence

$$0 \to \mathcal{D}_U \cdot \frac{1}{h_D} \xrightarrow{\ j\ } \mathcal{O}_U(\star D) \longrightarrow \mathcal{C} \to 0$$

is supported at m. We just have to prove that $\mathcal{C}$ is zero. The associated long exact sequence of de Rham cohomology provides the short exact sequence of $\mathbf{C}$-vector spaces

$$0 \to H_{DR}^n(\mathcal{D}_U \cdot \frac{1}{h_D}) \longrightarrow H_{DR}^n(\mathcal{O}_U(\star D)) \longrightarrow H_{DR}^n(\mathcal{C}) \to 0.$$

In particular, $H^n(j)$ is injective. On the other hand, $\mathbf{LCT}(D)$ is satisfied; hence, we deduce from the inclusions (1) of de Rham complexes that the morphisms $H^i(j) : H_{DR}^i(\mathcal{D}_U(1/h_D)) \longrightarrow H_{DR}^i(\mathcal{O}_U(\star D))$, $0 \leq i \leq n$, are surjective. Therefore $H^n(j)$ is an isomorphism, that is, $H_{DR}^n(\mathcal{C}) = 0$. From classic results about $\mathcal{D}$-modules supported at a point (see [23] for example), the $\mathcal{D}_U$-module $\mathcal{C}$ is necessarily zero and the condition $\mathbf{B}(h_D)$ is satisfied at any point of D.

Now, we assume that D is locally weighted homogeneous. Let us prove the assertion by induction on dimension. If $n = 2$, then D is locally defined by a (reduced) weighted homogeneous polynomial in two variables; thus we can conclude as in the proof of Proposition 3.2 below. Let us assume that $n \geq 3$ and let m denote a point in D. From [9], Proposition 2.4, there exists a neighborhood U of m such that, for each point $w \in U \cap D$, $w \neq m$, the germ of pair (X, D, w) is isomorphic to a product $(\mathbf{C}^{n-1} \times \mathbf{C}, D' \times \mathbf{C}, (0, 0))$ where D' is a locally weighted homogeneous divisor of dimension $n - 2$. Moreover, the condition $\mathbf{LCT}(D)$ implies that $\mathbf{LCT}(D')$ is satisfied on a neighborhood of the origin (see [9], Lemma 2.2). Let $h_{D'} \in \mathcal{O}_{\mathbf{C}^{n-1},0}$ be a local equation of D'. By using the induction hypothesis, the $\mathcal{D}_{\mathbf{C}^{n-1},0}$-module $\mathcal{O}_{\mathbf{C}^{n-1},0}[1/h_{D'}]$ is generated by $1/h_{D'}$, thus so is $\mathcal{O}_{\mathbf{C}^{n-1} \times \mathbf{C},0}[1/h_{D'}]$. In particular, the condition $\mathbf{B}(h_D)$ is satisfied at any point in $D \cap U - \{m\}$. We conclude with the first part of the proof. $\qquad\square$

In particular, $\mathbf{LCT}(D)$ implies the condition $\mathbf{B}(h_D)$ for any divisor D with isolated singularities. For a general divisor, this question is open.

3.3. *The conditions* $\mathbf{A}(1/h_D)$ *and* $\mathbf{LCT}(D)$

From [33], Proposition 1.3, the condition $\mathbf{A}(1/h)$ implies $\mathbf{B}(h)$ for any nonzero germ $h \in \mathcal{O}$; in particular, $\mathbf{A}(1/h_D)$ is a local analogue of $\mathbf{LCT}(D)$ for any Koszul-free divisor D, and more generally, for any free divisor of Spencer type (from Theorem 2.2; see also [12]). Is $\mathbf{LCT}(D)$ locally equivalent to $\mathbf{A}(1/h_D)$ in general? This question is motivated by the following significant results. Firstly, this is true for weighted homogeneous hypersurfaces with an isolated singularity.

Proposition 3.2. *Let $h \in \mathbf{C}[x]$ be a weighted homogeneous polynomial. Assume that h defines an isolated singularity at the origin. Let $D \subset \mathbf{C}^n$ be the hypersurface defined by h. Then the logarithmic comparison theorem holds for D if and only if the condition $\mathbf{A}(1/h)$ is satisfied.*

Proof. Under our assumptions, the condition $\mathbf{A}(1/h)$ is in fact equivalent to $\mathbf{B}(h)$ by Theorem 3.2. On the other hand, the polynomial $b_h(s)$ is given by the formula $b_h(s) = (s + 1) \prod_{q \in \Pi} (s + |\alpha| + q)$ where $\alpha \in (\mathbf{Q}^{*+})^n$ is the system of weights such that the degree of h is equal to 1, the expression $|\alpha|$ denotes the sum $\sum_{i=1}^{n} \alpha_i \in \mathbf{Q}^{*+}$, and $\Pi \subset \mathbf{Q}^+$ is the set of the degrees of the weighted homogeneous elements in $A_h = \mathbf{C}[x]/(h'_{x_1}, \ldots, h'_{x_n})$ (see [36], §11). We recall that $n - 2|\alpha|$ is the maximal element of Π; in particular,

$\mathbf{A}(1/h)$ is satisfied if $n = 2$ and so is $\mathbf{LCT}(D)$ (Theorem 1.3). Moreover, the set Π is symmetric about $(n/2) - |\alpha|$; hence, we deduce easily that $\mathbf{B}(h)$ is equivalent to the last condition of Theorem 1.1 when $n \geq 3$. This completes the proof. $\qquad\square$

Moreover, the condition $\mathbf{B}(h_D)$ is satisfied by any hyperplane arrangement (A. Leykin [34], Theorem 5.1), and the condition $\mathbf{A}(1/h_D)$ is true for the union of a generic arrangement with a hyperbolic arrangement [33]; this agrees with Terao's conjecture (see §1.2). The general problem is still open, and condition $\mathbf{A}(1/h)$ may only be necessary. A difficulty is the lack of families of divisors which satisfy the condition $\mathbf{LCT}(D)$.

3.4. *The condition* $\mathbf{A}(1/h)$

Let $h \in \mathcal{O}$ be a nonzero germ such that $h(0) = 0$. We give here some results about the meaning of the condition $\mathbf{A}(1/h)$ (see [33]).

First, we have the following easy criterion:

Lemma 3.1. *Let $h \in \mathcal{O}$ be a nonzero germ such that $h(0) = 0$. Assume that the following conditions are satisfied:*

  $\mathbf{H}(h)$: *h belongs to the ideal of its partial derivatives;*
  $\mathbf{B}(h)$: *-1 is the smallest integral root of $b_h(s)$;*
  $\mathbf{A}(h)$: *the ideal $\mathrm{Ann}_{\mathcal{D}}\, h^s$ is generated by operators of order 1.*

Then the ideal $\mathrm{Ann}_{\mathcal{D}}\, 1/h$ is generated by operators of order 1.

Proof. By Euclidean division, we have also a decomposition

$$\mathrm{Ann}_{\mathcal{D}[s]}\, h^s = \mathcal{D}[s](s - v) + \mathcal{D}[s]\mathrm{Ann}_{\mathcal{D}}\, h^s$$

where v is a vector field such that $v(h) = h$ (condition $\mathbf{H}(h)$). Moreover, under the condition $\mathbf{B}(h)$, the ideal $\mathrm{Ann}_{\mathcal{D}}\, 1/h$ is obtained by fixing $s = -1$ in a system of generators of $\mathrm{Ann}_{\mathcal{D}[s]}\, h^s$. $\qquad\square$

Reciprocally, what does remain true? We recall that the condition $\mathbf{A}(1/h)$ always implies $\mathbf{B}(h)$. On the other hand, does $\mathbf{A}(1/h)$ imply $\mathbf{H}(h)$? This is true for isolated singularities [32], Koszul-free germs, and suspensions of unreduced plane curve $z^N + g(x_1, x_2)$ (see [33]); this question is still open. Finally, the condition $\mathbf{A}(1/h)$ does not imply $\mathbf{A}(h)$ in general. Indeed, Calderón's example $h = x_1 x_2 (x_1 + x_2)(x_1 + x_2 x_3)$ satisfies $\mathbf{LCT}(D)$,

$\mathbf{A}(1/h)$, $\mathbf{B}(h)$, $\mathbf{H}(h)$ and not $\mathbf{A}(h)$ (see [4–6,10,33]). Meanwhile, condition $\mathbf{A}(h)$ is not unrealistic, since we have the following characterization of $\mathbf{A}(1/h)$ for Koszul-free germs:

Theorem 3.1. [33] *Let $h \in \mathcal{O}$ be a Koszul-free germ. Then the left ideal $\mathrm{Ann}_{\mathcal{D}}\, 1/h$ is generated by operators of order one if and only if the conditions $\mathbf{H}(h)$, $\mathbf{B}(h)$ and $\mathbf{A}(h)$ are satisfied.*

Moreover, condition $\mathbf{A}(h)$ is satisfied when h defines an isolated singularity (see below). Thus, we have

Theorem 3.2. [32] *Let $h \in \mathcal{O}$ be a germ of holomorphic function defining an isolated singularity. Then the ideal $\mathrm{Ann}_{\mathcal{D}}\, 1/h$ is generated by operators of order one if and only if the germ h is weighted homogeneous and the condition $\mathbf{B}(h)$ is satisfied.*

In fact, the condition $\mathbf{A}(h)$ may be considered almost as a geometric condition. Indeed, the following condition implies $\mathbf{A}(h)$:

$\mathbf{W}(h)$ **:** the relative conormal space W_h is defined by linear equations in ξ

since $W_h = \overline{\{(x, \lambda\, \mathrm{d}h) \,:\, \lambda \in \mathbf{C})\}} \subset T^*\mathbf{C}^n$ is the characteristic variety of $\mathcal{D}h^s$ [19]. For example, $\mathbf{W}(h)$ is true for hypersurfaces with an isolated singularity [36] and for locally weighted homogeneous free divisors [6].

Acknowledgements

This research has been supported by a Marie Curie Fellowship of the European Community (programme FP5, contract HPMD-CT-2001-00097). The author is very grateful to the Departamento de Álgebra, Geometría y Topología (Universidad de Valladolid) for hospitality during the fellowship. Moreover, he is happy to thank Francisco Jesús Castro-Jiménez, Antoine Douai, Luis Narváez-Macarro and José María Ucha-Enríquez for judicious comments.

Appendix A.

The following diagram summarizes the expected relations between the conditions studied in this note:

Here D denotes a general divisor, and the solid headed arrows represent the open questions.

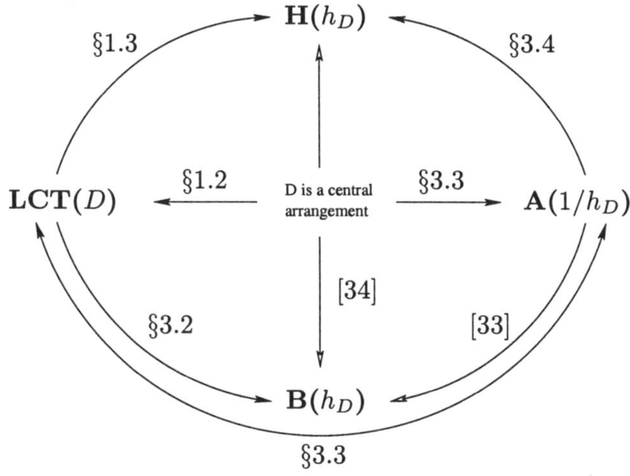

References

1. Aleksandrov A. G., *Euler-homogeneous singularities and logarithmic differential forms*, Ann. Global Anal. Geom. 4 (1986) 225–242.

2. Brieskorn E., *Sur les groupes de tresses [d'après V. I. Arnold]*, séminaire Bourbaki, 24ème année (1971/1972), Exp. No. 401, pp. 21–44. Lect. Notes in Math., Vol. 317, Springer, Berlin, 1973.

3. Bruce F.W., Roberts R.M., *Critical points of functions on analytic varieties*, Topology 27 (1988) 57–90.

4. Calderón-Moreno F.J., *Logarithmic differential operators and logarithmic de Rham complexes relative to a free divisor*, Ann. Sci. École Norm. Sup. 32 (1999) 577–595.

5. Calderón-Moreno F.J., Castro-Jiménez F.J., Mond D., Narváez-Macarro L., *Logarithmic cohomology of the complement of a plane curve*, Comment. Math. Helv. 77 (2002) 24–38.

6. Calderón-Moreno F.J., Narváez-Macarro L., *The module $\mathcal{D}f^s$ for locally quasi-homogeneous free divisors*, Compos. Math. 134 (2002) 59–74.

7. Calderón-Moreno F.J., Narváez-Macarro, L., *Locally quasi-homogeneous free divisors are Koszul free*, Tr. Mat. Inst. Steklova 238 (2002) 81–85

8. Calderón-Moreno F.J., Narváez-Macarro L., *Dualité et comparaison sur les complexes de de Rham logarithmiques par rapport aux diviseurs libres*, Ann. Inst. Fourier 55 (2005) 47–75.

9. Castro-Jiménez F.J., Mond D., Narváez-Macarro L., *Cohomology of the complement of a free divisor*, Trans. Amer. Math. Soc. 348 (1996) 3037–3049.

10. Castro-Jiménez F.J., Ucha-Enríquez J.M., *Explicit comparison theorems for $\mathcal{D}$-modules*, J. Symbolic Comput. 32 (2001) 677–685.

11. Castro-Jiménez F.J., Ucha-Enríquez J.M., *Free divisors and duality for $\mathcal{D}$-modules*, Tr. Mat. Inst. Steklova 238 (2002) 97–105.

12. Castro-Jiménez F.J., Ucha-Enríquez J.M., *Testing the logarithmic compari-*

son theorem for free divisors, Experiment. Math. 13 (2004) 441–449.

13. Castro-Jiménez F.J., Ucha-Enríquez J.M., *Quasi-free divisors and duality*, C. R. Acad. Sci. Paris 338 (2004) 461–466.

14. Deligne P., *Équations différentielles à points singuliers réguliers*, Lect. Notes in Math. 163 (1970).

15. Granger M., Schulze M., *On the formal structure of logarithmic vector fields*, Compos. Math. 142 (2006) 765–778.

16. Griffiths P.A., *On the periods of certain rational integrals I*, Ann. of Math. 90 (1969) 460–495.

17. Grothendieck A., *On the de Rham cohomology of algebraic varieties*, Pub. Math. I.H.E.S. 29 (1966) 95–105.

18. Holland M., Mond D., *Logarithmic differential forms and the cohomology of the complement of a divisor*, Math. Scand. 83 (1998) 235–254.

19. Kashiwara M., *B-functions and holonomic systems*, Invent. Math. 38 (1976) 33–53.

20. Kashiwara M., *On the holonomic systems of linear differential equations II*, Invent. Math. 49 (1978) 121–135.

21. Kashiwara M., *The Riemann-Hilbert problem for holonomic systems*, Publ. RIMS, Kyoto Univ. 20 (1984) 319–365.

22. Looijenga E.J.N., *Isolated singular points on complete intersections*, London Mathematical Society, Lecture Note Series 77 (1984).

23. Malgrange B., *Le polynôme de Bernstein d'une singularité isolée*, Lect. Notes in Math. 459 (1975) 98–119.

24. Mebkhout Z., *Une équivalence de catégories. Une autre équivalence de catégories*, Compos. Math. 51 (1984) 51–88.

25. Mebkhout Z., *Le formalisme des six opérations de Grothendieck pour les $\mathcal{D}_X$-modules cohérents*, Travaux en Cours 35, Hermann, Paris, 1989.

26. Mebkhout Z., *Le théorème de positivité, le théorème de comparaison et le théorème d'existence de Riemann*, Séminaires et Congrès 8, Soc. Math. France, Paris (2004) 165–310.

27. Saito K., *Theory of logarithmic differential forms and logarithmic vector fields*, J. Fac. Sci. Univ. Tokyo 27 (1980) 265–291.

28. Saito K., *Primitive forms for a universal unfolding of a function with an isolated critical point*, J. Fac. Sci. Univ. Tokyo Sect. IA Math. 28 (1982) 775–792.

29. Schulze M., *A criterion for the logarithmic differential operators to be generated by vector fields*, arXiv.org math.CV/0406023.

30. Terao H., *Forms with logarithmic pole and the filtration by the order of the pole*, Proceedings of the International Symposium on Algebraic Geometry (Kyoto University, Kyoto, 1977) (Kinokuniya Book Store, Tokyo, 1978) 673–685.

31. Terao H., *Free arrangements of hyperplanes and unitary reflection groups*, Proc. Japan Acad. Ser. A Math. Sci. 56 (1980) 389–392.

32. Torrelli T., *Polynômes de Bernstein associés à une fonction sur une intersection complète à singularité isolée*, Ann. Inst. Fourier 52 (2002) 221-244.

33. Torrelli T., *On meromorphic functions defined by a differential system of*

order 1, Bull. Soc. Math. France 132 (2004) 591–612.

34. Walther U., *Bernstein-Sato polynomial versus cohomology of the Milnor fiber for generic hyperplane arrangements*, Compos. Math. 141 (2005) 121–145.
35. Wiens J., Yuzwinsky S., *De Rham cohomology of logarithmic forms on arrangements of hyperplanes*, Trans. Amer. Math. Soc. 349 (1997) 1653–1662.
36. Yano T., *On the theory of b-functions*, Publ. R.I.M.S. Kyoto Univ. 14 (1978) 111–202.

ON TORSION IN HOMOLOGY
OF SINGULAR TORIC VARIETIES

ANDRZEJ WEBER[*]

Instytut Matematyki
Uniwersytet Warszawski
ul, Banacha 2
02-097 Warszawa, Poland
E-mail: aweber@mimuw.edu.pl

Dedicated to Jean-Paul Brasselet on the occasion of his 60th birthday

Let X be a toric variety. It follows from [FW] that the rational Borel-Moore homology of X is isomorphic to the homology of the Koszul complex $A_*^T(X) \otimes \Lambda^* M$. Here $A_*^T(X)$ is the equivariant Chow group and M is the character group of T. We give a simple geometric proof of that fact and we show that the same holds for coefficients which are the integers with certain primes inverted.

Keywords: Toric varieties, integral equivariant homology, weight filtration

1. Introduction

Let G be a connected Lie group which acts on a topological space X. The equivariant cohomology of X is defined to be the cohomology of the space obtained from the Borel construction

$$H_G^*(X) = H^*(EG \times_G X).$$

It often happens that this group is easily computable. This is so for example when X is a complex algebraic manifold with an action of a complex algebraic group with only finitely many orbits. Then the equivariant cohomology with rational coefficients is the direct sum

$$H_G^*(X; \mathbf{Q}) = \bigoplus_{\text{orbit}} H^{*-2c}(BG_x; \mathbf{Q}). \tag{1}$$

---

[*]Supported by KBN grant 1 P03A 005 26

Here G_x is the stabilizer of a point from the orbit and c is the complex codimension of the orbit.

There is an other invariant which even in the singular case is easy to compute. That is the equivariant Borel-Moore homology, [EG, §2.8]. It can be interpreted as the equivariant cohomology with coefficients in the dualizing sheaf. The equivariant Borel-Moore homology usually is nontrivial in the negative degrees. Again we have

$$H_*^{BM,G}(X;\mathbf{Q}) = \bigoplus_{\text{orbit}} H^{2d-*}(BG_x;\mathbf{Q}). \tag{2}$$

Here d is the complex dimension of the orbit. This formula, as well as the previous one, follows from the fact that the rational cohomology of the classifying space is concentrated in even degrees.

A passage from the equivariant cohomology to the usual one is possible due to the Eilenberg-Moore spectral sequence. The second table is of the form

$$E_2^{-p,q} = \mathrm{Tor}_p^{q,H^*(BG;\mathbf{Q})}(H_G^*(X;\mathbf{Q}),\mathbf{Q}) \Rightarrow H^{q-p}(X;\mathbf{Q}).$$

(The torsion functor has two gradings: p is the usual grading of the left derived functor and q is the internal grading.) A generalization of this spectral sequence for cohomology with sheaf coefficients was described in [FW]. If G is a complex algebraic group, X is an algebraic variety and the action is algebraic then all the cohomological invariants are equipped with the weight filtration. This often forces the spectral sequence to degenerate. In particular according to [FW, Th. 1.6] we have:

Theorem 1.1. *If X is smooth and the action has only finitely many orbits then the rational cohomology of X is given additively by:*

$$H^i(X;\mathbf{Q}) = \bigoplus_{q-p=i} \mathrm{Tor}_p^{q,H^*(BG;\mathbf{Q})}(H_G^*(X;\mathbf{Q}),\mathbf{Q}).$$

Having the decomposition Eq. (2) and using the fact that the equivariant cohomology of an orbit is pure we apply [FW, Th. 1.3]. We obtain

Theorem 1.2. *If the action has only finitely many orbits then the rational Borel-Moore homology of X is given by:*

$$H_i^{BM}(X;\mathbf{Q}) = \bigoplus_{p-q=i} \mathrm{Tor}_p^{q,H^*(BG;\mathbf{Q})}(H_*^{BM,G}(X;\mathbf{Q}),\mathbf{Q}).$$

In both theorems above we assume that spaces and actions are algebraic.

In this note we want to specialize our results to toric varieties. The intersection cohomology was already described in [W]. Here we study ordinary homology but, in addition, we care about integral coefficients. First we note that for a torus T the Eilenberg-Moore spectral sequence can be replaced by an easier one (which is in fact isomorphic after a renumbering of entries). This is just the spectral sequence of the fibration

$$T \subset ET \times X \longrightarrow ET \times_T X.$$

The second table

$$E_2^{p,q} = H_T^p(X) \otimes H^q(T)$$

with its differential is exactly the Koszul complex. Therefore

$$E_3^{p,q} = \operatorname{Tor}_q^{p+2q, H^*(BT)}(H_T^*(X), \mathbf{Z}).$$

The exact degrees are slightly surprising, but they agree with the weight filtration when X is smooth. For possibly singular varieties we will describe the homological variant of the spectral sequence in an elementary way.

Now we apply the Frobenius endomorphism of the toric variety. This allows to show that the spectral sequence degenerates not only over $\mathbf{Q}$ but also with small primes inverted. We prove two theorems.

Theorem 1.3. *The above spectral sequence degenerates on E^3 for rational coefficients and for coefficients in $\mathbf{F}_q$ if $q > \lceil \frac{\dim X}{2} \rceil$.*

Theorem 1.4. *If $q > \lceil \frac{\dim X + 1}{2} \rceil$ then the q-torsion of the integral homology is the direct sum of the q-torsions in E^3.*

As a consequence we obtain:

Theorem 1.5. *Let X be a toric variety and let R be the ring of integers with inverted primes which are smaller or equal to $\lceil \frac{\dim X + 1}{2} \rceil$. Then*

$$H_i^{BM}(X; R) = \bigoplus_{p-q=i} \operatorname{Tor}_p^{q, H^*(BT;R)}(H_*^{BM,T}(X; R), R).$$

It remains to remark that $H_{2i}^{BM}(X; R) \simeq A_i^T(X) \otimes R$ is the equivariant Chow group.

We suspect that the assumption about q is redundant. The authors of [BFMH] were studying the cohomology of toric varieties with coefficients in $\mathbf{F}_2$. They also conjecture that (a complex equivalent to) the Koszul complex computes the homology of X. It is shown that this is so if the dimension is smaller or equal to 3. Many cases are verified by a computer. Moreover the same complex (with another grading) computes the homology of the corresponding real toric variety.

1014

2. Equivariant Borel-Moore homology

From now on we omit coefficients in the notation. Let X be an algebraic variety acted by the torus $T = (\mathbf{C}^*)^n$. Let

$$ET_d = (\mathbf{C}^{d+1} \setminus \{0\})^n$$

be an approximation of ET. The equivariant Borel-Moore homology is defined by the formula:

$$H_i^{BM,T}(X) = \varprojlim_d H_{i+2nd}^{BM}(ET_d \times_T X).$$

The limit is taken with respect to the inverse system

$$\iota_d^! : H_{i+2n(d+1)}^{BM}(ET_{d+1} \times_T X) \to H_{i+2nd}^{BM}(ET_d \times_T X).$$

The Gysin map $\iota_d^!$ is defined since the inclusion (of the real codimension $2n$)

$$\iota_d : ET_d \times_T X \hookrightarrow ET_{d+1} \times_T X$$

is normally nonsingular. In fact the limit stabilizes: $H_i^{BM,T}(X) = H_{i+2d}^{BM}(E_{d+1} \times_T X)$ for $i > (1 - 2d) + 2 \dim X$.

Let X be a toric variety. By Eq. (2) or [BZ] the equivariant homology is the sum of the homologies of the orbits:

$$H_i^{BM,T}(X) = \bigoplus_{\sigma \in \Sigma} H_i^{BM,T}(O_\sigma). \tag{3}$$

The orbits σ are labelled by the fan Σ. The equivariant Borel-Moore homology is isomorphic to the equivariant Chow group $A_*^T(X)$ considered e.g. by Brion [Br]. Each $H_*^{BM,T}(O_\sigma)$ is isomorphic to $Sym(\langle\sigma\rangle)[-2\mathrm{codim}\,\sigma]$ (the symmetric power $Sym^i(\langle\sigma\rangle)$ is placed in the degree $2(\mathrm{codim}\,\sigma - i)$). In particular it is a free abelian group. The odd part of the equivariant Borel-Moore homology vanishes. The module structure over $H^*(BT)$ is described in [Br].

3. Frobenius endomorphism

Let $p > 1$ be a natural number. Complex toric varieties are equipped with Frobenius endomorphism (power map) $\phi_p : X \to X$, see [T] . The power map of T is denoted by ψ_p. If we embedd T into X as the open orbit, then ϕ_p is the unique extension of ψ_p. Both maps induce a map at each step of the approximation of the Borel construction $ET_d \times_T X$. We denote this map by ϕ_p^T. (We note that ϕ_p^T is not the same as $1 \otimes \phi_p$ considered in [BZ].)

We would like to encode the action of ϕ_p^T in equivariant homology. The map $\iota_d^!$ does not commute with $\phi_{p,*}^T$, but $\phi_{p,*}^T \iota_d^! = p\,\iota_d^! \phi_{p,*}^T$. Then we set $H_i^{BM,T}(X) = H_{i+2d}^{BM}(ET_d \times_T X)(d)$ for d sufficiently large. If we study homology with coefficients in the field $\mathbf{F} = \mathbf{Q}$ or $\mathbf{F}_q$, provided that $(p,q) = 1$, the symbol (d) denotes tensoring with $\mathbf{F}$ acted by ϕ_p via the multiplication by p^{-d}. If we want to study integral homology we just analyze homology of a sufficiently large approximation of the Borel construction. Now we can state

Proposition 3.1. $\phi_{p,*}^T$ *induces the multiplication by p^i on $H_{2i}^{BM,T}(X)$.*

Proof. The homology of an orbit is generated over $H^*(BT)$ by its fundamental class. The map $\phi_{p,*}^T$ restricted to the orbit O_σ is a covering of the degree p^i, where $i = \dim O_\sigma = \operatorname{codim} \sigma$. Thus $\phi_{p,*}^T[\sigma] = p^i[\sigma] \in H_{2i}^{BM,T}(O_\sigma)$. Now we apply the additivity Eq. (3). $\qquad\square$

We consider the system of T-fibrations

$$ET_d \times X \to ET_d \times_T X\,.$$

We obtain a system of spectral sequences with

$$_d E_{k,l}^2 = H_k^{BM}(ET_d \times_T X) \otimes H_l^{BM}(T)\,.$$

The map $\iota_d^!$ passes to a map of spectral sequences

$$_{d+1}E_{k+2,l}^j(1) \to {}_d E_{k,l}^j\,.$$

We set $_\infty E_{k,l}^j = {}_d E_{k+2nd,l}^j(d)$ for d large enough. The map ϕ_p^* of $H_q^{BM}(T)$ is the multiplication by p^{q-n} (for $q < n$ this group is trivial). Therefore the resulting map of the spectral sequences is the multiplication by $p^{\frac{k}{2}+l-n}$ on $_\infty E_{k,l}^2$.

Theorem 3.1. *The spectral sequence $_\infty E_{k,l}^j$ converges to $H_{k+l-2n}^{BM}(X)$. For rational coefficients it degenerates on E^3. The resulting filtration coincides with the weight filtration of homology.*

Remark 3.1. There is a shift $-2n$ in the degree which is repaired when we move the generators of $H_*^{BM}(T)$ to the negative degrees. They should be placed there since we compare E_2 with the Koszul complex, see below.

Remark 3.2. The homology $H_*^{BM,T}(X)$ is a module over $H^*(BT) = Sym\,M$, where $M = Hom(T, \mathbf{C}^*)$. The differential $d_2 : {}_\infty E_{k,l}^2 \to {}_\infty E_{k-2,l+1}^2$

after the identification $H_*^{BM}(T) = H^{2n-*}(T)$ with $\Lambda^* M$ becomes the Koszul differential

$$H_k^{BM,T}(X) \otimes \Lambda^{2n-l} M \to H_{k-2}^{BM,T}(X) \otimes \Lambda^{2n-l-1} M .$$

Remark 3.3. The Koszul complex contains a complex constructed in [T]. Totaro considered rational coefficients, but he remarked that some information about the torsion can be obtained, see also [J, Remark 2.4.8].

Remark 3.4. From another point of view the Koszul duality and Frobenius endomorphism appears in [B] and [BL] for dual pairs of affine toric varieties.

Proof of Theorem 3.1. At each step $_dE_{k,l}^*$ converges to $H_{k+l}^{BM}(X \times (\mathbf{C}^{d+1} \setminus \{0\})^n)$, which is equal to $H_{k+l}^{BM}(X \times (\mathbf{C}^{(d+1)n}) = H_{k+l-2n(d+1)}^{BM}(X)$ for d sufficiently large. Therefore $_\infty E_{k,l}^*$, which is equal to $_dE_{k+2nd,l}^*$ (for d sufficiently large) converges to $H_{k+l-2n}^{BM}(X)$. The eigenvalue of ϕ_p acting on $_\infty E_{k-r,l+r-1}^*$ is equal to $p^{\frac{k}{2}+l-n+\frac{r}{2}-1}$. There is no obstruction for the differential d_2, but the higher differentials have to vanish. $\square$

4. Torsion

It is not possible to detect the q-torsion of X for small prime q, but we prove Theorems 1.3 and 1.4 announced in the introduction:

Proof of Theorem 1.3. To show that the higher differentials vanish we consider the eigenvalues of ϕ_p acting on $_dE_{k-r,l+r-1}^2$ for $r = 2, \ldots n + 1$ and $k - r$ even. These are at most $\lceil \frac{n}{2} \rceil$ subsequent powers of p, as the reader may easily check (see the picture below). We will choose p such that all these powers are different modulo q. It is enough to take p which generate the group $\mathbf{F}_q^* \simeq \mathbf{Z}/(q-1)$. $\square$

$$2n \quad \ldots \circ \quad \cdot \quad \bullet \quad \cdot \quad \circ \quad \cdot \quad \circ \quad \cdot \quad \circ \quad \cdot \quad \circ \ldots$$
$$\ldots \circ \quad \cdot \quad \circ \quad \cdot \quad \circ \quad \cdot \quad \circ \quad \cdot \quad \circ \quad \cdot \quad \circ \ldots$$
$$\ldots \circ \quad \cdot \quad \circ \quad \cdot \quad \bullet \quad \cdot \quad \circ \quad \cdot \quad \circ \quad \cdot \quad \circ \ldots$$
$$\ldots \circ \quad \cdot \quad \circ \quad \cdot \quad \circ \quad \cdot \quad \circ \quad \cdot \quad \circ \quad \cdot \quad \circ \ldots$$
$$\ldots \circ \quad \cdot \quad \circ \quad \cdot \quad \circ \quad \cdot \quad \star \quad \cdot \quad \circ \quad \cdot \quad \circ \ldots$$
$$\ldots \circ \quad \cdot \quad \circ \quad \cdot \quad \circ \quad \cdot \quad \circ \quad \cdot \quad \star \quad \cdot \quad \circ \ldots$$
$$n \quad \ldots \circ \quad \cdot \quad \circ \quad \cdot \quad \circ \quad \cdot \quad \circ \quad \cdot \quad \circ \quad \cdot \quad \circ \ldots$$

A picture of the spectral sequence for $n = 6$.
$\star$ denotes the source and the target of d_2 (which always have the same weight),
$\bullet$ denotes the remaining possibly nonzero entries of the spectral sequence which are hit by the higher differentials.

If we want to determine the q-torsion for integral homology we may meet problems with extensions.

Proof of Theorem 1.4. The part of the integral spectral sequence computing the q-torsion degenerates as in the previous proof. To avoid problems with extensions we have to know that the eigenvalues of ϕ_p on $E^3_{k,l}$ are different along the lines $k + l = const$. There are exactly $\lceil \frac{n+1}{2} \rceil$ subsequent powers of p. We proceed as before, that is we find p with different powers modulo q. $\qquad\square$

$$2n \quad \ldots \circ \quad \cdot \quad \bullet \quad \cdot \quad \circ \quad \cdot \quad \circ \quad \cdot \quad \circ \quad \cdot \quad \circ \ldots$$
$$\ldots \circ \quad \cdot \quad \circ \quad \cdot \quad \circ \quad \cdot \quad \circ \quad \cdot \quad \circ \quad \cdot \quad \circ \ldots$$
$$\ldots \circ \quad \cdot \quad \circ \quad \cdot \quad \bullet \quad \cdot \quad \circ \quad \cdot \quad \circ \quad \cdot \quad \circ \ldots$$
$$\ldots \circ \quad \cdot \quad \circ \quad \cdot \quad \circ \quad \cdot \quad \circ \quad \cdot \quad \circ \quad \cdot \quad \circ \ldots$$
$$\ldots \circ \quad \cdot \quad \circ \quad \cdot \quad \circ \quad \cdot \quad \bullet \quad \cdot \quad \circ \quad \cdot \quad \circ \ldots$$
$$\ldots \circ \quad \cdot \quad \circ \quad \cdot \quad \circ \quad \cdot \quad \circ \quad \cdot \quad \circ \quad \cdot \quad \circ \ldots$$
$$n \quad \ldots \circ \quad \cdot \quad \circ \quad \cdot \quad \circ \quad \cdot \quad \circ \quad \cdot \quad \bullet \quad \cdot \quad \circ \ldots$$

A picture of the spectral sequence for $n = 6$.
The entries $\bullet$ should have different weights in $\mathbf{F}_q$.

We conjecture that the theorems above are true without assumptions on q. The conjecture holds if X is smooth by the work of M. Franz [F].

5. Weight filtration and gradation

Let V_j be the eigensubspace of ϕ_p acting on $H^{BM}_*(X; \mathbf{Q})$ for the eigenvalue p^j. It does not depend on p. The weight filtration in homology usually is denoted by $W^i = W_{-i}$ and it is decreasing. Our gradation is related to the weight filtration:

$$W^i = \bigoplus_{j \geq \frac{i}{2}} V_j \, .$$

In particular $W^{2j} = W^{2j-1}$.

CONJUGATION. Toric varieties are defined over real numbers. The complex conjugation acts on the complex points of X. We can also determine the action on the homology: it acts by $(-1)^{\frac{i}{2}}$ on the i-th term of the weight gradation.

References

BFMH.   F. Bihan, M. Franz, C. McCrory, J. van Hamel *Is every toric variety an M-variety?* Manuscripta Math. 120, 217-232 (2006)

B.   T. Braden *Koszul duality for toric varieties*, arXiv.org:math/0308216

BL.   T. Braden, V. A. Lunts *Equivariant-constructible Koszul duality for dual toric varieties*, Adv. Math. 201 (2006), no. 2, 408–453.

Br.   M. Brion, *Equivariant Chow groups for torus actions*, Transformation groups, Vol 2, No 3 (1997), pp. 1-43

BZ.   J-L. Brylinski, B. Zhang: *Equivariant Todd classes for toric varieties*, arXiv.org:math.AG/0311318

EG.   D. Edidin, W. Graham *Equivariant intersection theory* Invent. Math. vol 131 (1998) no. 3, pp. 595–634

F.   M. Franz *On the integral cohomology of smooth toric varieties*, Proc. Steklov Inst. Math. 252, 53-62 (2006)

FW.   M. Franz, A. Weber *Weights in cohomology and the Eilenberg-Moore spectral sequence*, Ann. Inst Fourier (Grenoble) 55 (2005), no. 2, 673–691

J.   A. Jordan *Homology and cohomology of toric varieties*, thesis, Konstanz 1998.

T.   B. Totaro *Chow groups, Chow cohomology, and linear varieties*. Journal of Algebraic Geometry, to appear. http://www.dpmms.cam.ac.uk/~bt219/papers.html

W.   A. Weber *Weights in the cohomology of toric varieties*, Cent. Eur. J. Math. 2 (2004), no. 3, 478–492, arXiv.org:math.AG/0301314

CENTRE SYMMETRY SETS AND OTHER INVARIANTS OF ALGEBRAIC SETS

MARIUSZ ZAJĄC

Faculty of Mathematics and Information Science,
Warsaw University of Technology,
Pl. Politechniki 1, 00-661 Warszawa, Poland
E-mail: zajac@mini.pw.edu.pl

The present note aims to reformulate some classical concepts of the theory of planar curves in the language of algebraic geometry. Section 1 presents the definitions and several basic properties of the centre symmetry sets of a closed planar curve. Section 2 recalls the classical notions of the centre of curvature and the evolute, i.e. the locus of the centres of curvature, and explains how they can be expressed in algebro-geometric terms. Finally, in Section 3 we combine the above approaches in order to obtain algebraic definitions of the centre symmetry sets. In particular we study the case of a cubic curve. We also suggest some possibilities for future research and mention some difficulties that are likely to appear.

The author wishes to thank the CIRM at Luminy for the hospitality and Peter Giblin (Liverpool) for discussions and clarifying some points.

Keywords: Plane algebraic curves; Centre symmetry set

1. The centre symmetry sets

Let us consider a smooth curve $C \in \mathbf{R}^2$. If C is an *oval*, i.e. a closed curve without inflection points, we can say that a point O is the *centre of symmetry* of C if it is the midpoint of all chords passing through O. This is, obviously, a very restrictive property and a generic oval has no centre of symmetry. One can, however, generalize this notion and replace the classical centre of symmetry by a certain set that can be defined for any oval C and reduces to a single point if C is centrally symmetric. The main idea is to draw the segments joining those pairs of points on C at which the tangents are parallel. This one-parameter family of chords is depicted in Fig. 1 for the bounded connected component of the cubic curve $x^2 = y^3 - y$ (this is the first interesting case, as obviously all closed conics, i.e. ellipses, are centrally symmetric).

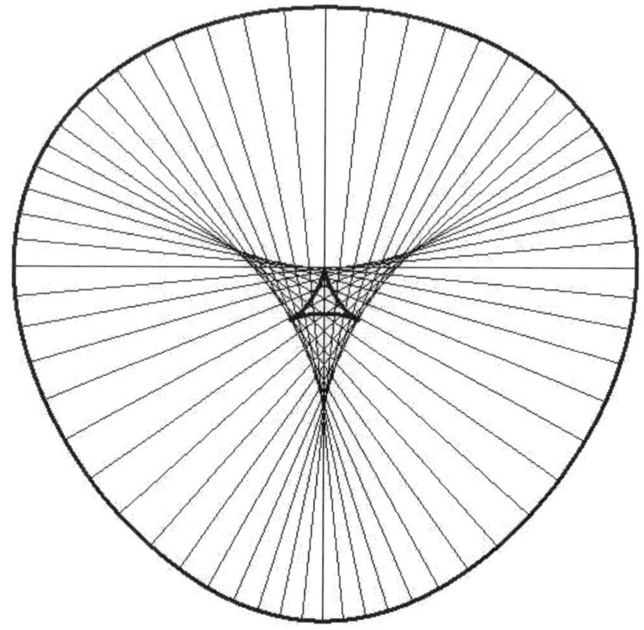

Fig. 1. Symmetry sets of an oval

We see that the following two definitions seem natural:

- the envelope of the above-mentioned family of chords (the larger curve with three cusps in Fig. 1) will be called the *centre symmetry set* (CSS);
- the set of midpoints of the chords considered (the smaller curve with three cusps in Fig. 1) will be called the *anti-centre symmetry set* (ACSS).

Of course if C has a centre of symmetry O then both the CSS and the ACSS reduce to the point O. These sets are also invariant under the affine transformations of $\mathbf{R}^2$ because we only use the notions of tangency, parallel lines and the midpoint of a segment.

The differential properties of these sets were studied in [4], [1] and [2], and in the case of a generic oval they can be summarized as follows:

- apart from a finite number of cusps both sets are smooth curves with no inflection points;
- the CSS and the ACSS have the same odd number of cusps;
- the cusps of the ACSS are the midpoints of the chords joining those pairs of points at which the curvatures of C are equal;
- the cusps of the CSS lie on the chords joining a point of large curvature with a point of small curvature (more precisely the ratio of curvatures must have a critical point).

2. Focal loci

In this section we shall recall the concept of the *focal locus* of a curve (traditionally referred to as the *evolute* of the curve), which is the locus of the centres of curvature of this curve. We shall generally follow the exposition of [5], restricting ourselves, however, to the case of algebraic curves in the (affine or projective) plane.

There are several equivalent ways of defining the centre of curvature of a curve in elementary differential geometry. We choose one of them because it can easily be rewritten in purely algebraic terms.

Let $\gamma : (a, b) \to \mathbf{R}^2$ be a smooth curve parametrized by the arclength, and let $T(t)$ and $N(t)$ be the unit vectors tangent and normal to γ at $\gamma(t)$, respectively. If we now define the following function:

$$e : (a, b) \times \mathbf{R} \to \mathbf{R}^2; \; (t, r) \mapsto \gamma(t) + rN(t),$$

i.e. the *endpoint map* associating to every normal vector of length r beginning at $\gamma(t)$ its end, then using the well-known Frenet formulae:

$$\gamma'(t) = T(t), \; T'(t) = k(t)N(t), \; N'(t) = -k(t)T(t),$$

where $k(t)$ is the curvature, we can easily prove the following

Proposition 2.1. *The evolute of a planar curve is equal to the set of critical values of the endpoint mapping.*

It should be mentioned here that the notions of critical and regular values are invariant under any diffeomorphic change of coordinates in the domain. Therefore the above proposition will remain equally valid if we take any smooth parametrization t instead of the arclength or any smooth normal vector field $N(t)$ instead of the vectors of length 1. However, the evolute is invariant under isometries only, and not all affine transformations, because we always require that $N(t)$ should be orthogonal to the tangent.

Focal loci of algebraic curves

As usual, the most natural setting for dealing with algebraic curves is the (complex) projective plane. If we consider a polynomial $f(x,y)$ of degree d and its homogeneous counterpart $F(X,Y,Z) = Z^d f(X/Z, Y/Z)$ then instead of the affine normal line to $C = \{F(x,y) = 0\}$ at a regular point $P = (x_0, y_0) \in C$ with the equation

$$F_x(x_0, y_0) \cdot (y - y_0) - F_y(x_0, y_0) \cdot (x - x_0) = 0$$

we can talk of its projectivization: the line in $\mathbf{P}^2$ joining $P = (x_0 : y_0 : 1)$ with the point at infinity $P_\infty = (F_X(P) : F_Y(P) : 0)$.

The projectivization of the endpoint mapping e is now

$$E : C \times \mathbf{P}^1 \to \mathbf{P}^2;$$

$$((X : Y : Z), (\lambda : \mu)) \mapsto (\lambda X + \mu F_X(X, Y, Z) : \lambda Y + \mu F_Y(X, Y, Z) : \lambda Z).$$

In this setting the focal locus is the critical value set of an explicitly defined rational mapping defined on a smooth algebraic variety. Therefore the evolute of an algebraic curve is also an algebraic set. More detailed analysis of this set, which can be found in [5], leads for instance to the following

Theorem 2.1. *Let $X \subset \mathbf{P}^2_{\mathbf{C}}$ be a general algebraic curve of degree $d > 1$ and Z its focal locus. Then $\deg Z = 3d(d - 1)$.*

Note that the degree of the real affine view of the focal locus can be equal to $3d(d - 1)$ or lower, e.g. for an ellipse the algebraic degree of the evolute equals $3d(d - 1) = 6$ but for a parabola it is 3.

3. Algebraic approach to centre symmetry sets

Let $C \subset \mathbf{R}^2$ be a smooth curve. As we see from the very definition of CSS and ACSS, the most important objects are pairs of points on C with parallel tangents. Let us define

$$S = \{(P, Q) \in C \times C : \text{ the tangents to } C \text{ at } P \text{ and } Q \text{ are parallel}\}.$$

If the projectivized equation of C is $F = 0$, this amounts to saying that these tangents meet at a point with $Z = 0$, or equivalently

$$S = \{(P, Q) \in C \times C : F_X(P)F_Y(Q) = F_X(Q)F_Y(P)\}.$$

The set S can be rather complicated in the general (nonconvex) case, but if C is an oval then S is the sum of the diagonal $\{(P,P)\}$ and the set of pairs of opposite points $\{(P,P')\}$. Thus topologically S has two components isomorphic to C itself, but they cannot be distinguished algebraically, which is a serious drawback (see below).

Once we have defined S we can define both CSS and ACSS in algebraic terms.

Proposition 3.1. *The ACSS of a convex algebraic curve is (a connected component of) the image of S under the 'midpoint mapping'*

$$m((X_1 : Y_1 : 1), (X_2 : Y_2 : 1)) = (X_1 + X_2 : Y_1 + Y_2 : 2),$$

whereas the CSS is (a connected component of) the set of critical values of the restriction $j|_{S \times \mathbf{P}^1}$, where j is the 'chord mapping'

$$j((X_1 : Y_1 : 1), (X_2 : Y_2 : 1), (\lambda : \mu)) = (\lambda X_1 + \mu X_2 : \lambda Y_1 + \mu Y_2 : \lambda + \mu).$$

This proposition can be proved analogously to Prop. 2.1.

3.1. *ACSS of an algebraic curve – cubic case*

In the affine setting we are tempted to say that the point (x, y) belongs to the ACSS of the curve $C = \{f(x, y) = 0\}$ if the following equations hold for some x_1, y_1, x_2, y_2 :

$$\begin{cases} f(x_1, y_1) = 0 \\ f(x_2, y_2) = 0 \\ f_x(x_1, y_1) f_y(x_2, y_2) = f_x(x_2, y_2) f_y(x_1, y_1) \\ x = \frac{x_1 + x_2}{2} \\ y = \frac{y_1 + y_2}{2} \end{cases} \tag{1}$$

If we eliminate x_1, y_1, x_2, y_2 from the above system, we should in principle obtain the equation defining the ACSS. However, as mentioned before, if $(x, y) = (x_1, y_1) = (x_2, y_2) \in C$ then the parallel tangents condition is tautologically fulfilled, so in fact the system (1) defines the union of the ACSS and the curve C itself. There are also other interesting phenomena, which will be visible in the example.

Let $f(x, y) = x^3 - 3x - y^2$. The elimination of x_1, y_1, x_2 and y_2 from (1), performed with the Singular package, gives the following 12-th degree equation for x and y :

$$(x^3 - 3x - y^2)(8x^9 - 12x^6 y^2 - 30x^7 + 6x^3 y^4 + 3x^4 y^2 - y^6 +$$
$$+ 42x^5 + 6xy^4 + 6x^2 y^2 - 26x^3 - y^2 + 6x) = 0$$

1024

The divisibility of this equation by $f(x, y)$ agrees with our previous analysis, and the zero set of the second factor is shown in Fig. 2 as a thin line together with the original (thick) curve consisting of an oval and an unbounded branch with two inflection points.

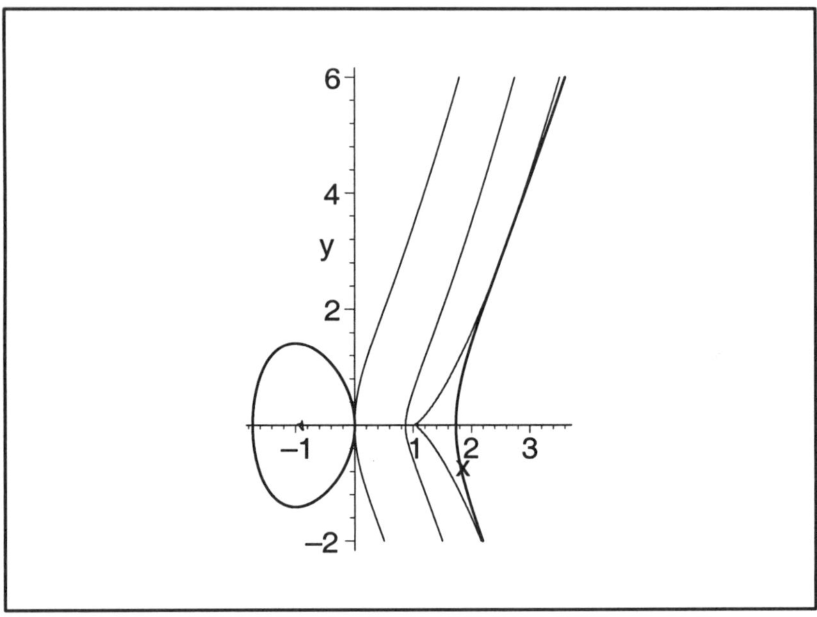

Fig. 2. ACSS of $x^3 - 3x - y^2 = 0$

Let us make some observations:

- First and foremost, what we see is the real picture of the complex ACSS, i.e. x and y are real, but x_1, y_1, x_2 and y_2 need not be.
- Beside the ACSS of the oval we can see three unbounded branches. Their appearance is easy to understand: for a fixed point $(x, y) \in C$ with large coordinates x, y and almost vertical tangent line there are three other points where the tangents are parallel, and they are approximately $(-\sqrt{3}, 0), (0, 0), (\sqrt{3}, 0)$. Therefore as (x, y) goes to infinity, the three respective midpoints tend asymptotically to three 'parallel' cubic curves.
- From the equation of the ACSS one can directly compute its singular points. It turns out that in the complex domain there are ten of them, four of which have real coordinates, namely the three cusps of

the ACSS of the oval: $(-1, 0)$ and $(-0.8475, \pm 0.153036)$ (hardly visible in Fig. 2, but cf. Fig. 1) plus the point $(1, 0)$, whose appearance is confusing at first sight, as it does not seem to be the midpoint of any chord of C. However, in precisely the same way as $(-1, 0)$ is the midpoint of the chord joining the points $(-1, \pm\sqrt{2})$, the cusp $(1, 0)$ is the midpoint of the chord between $(1, \sqrt{2}i)$ and $(1, -\sqrt{2}i)$. Indeed, apart from the obvious symmetry $(x, y) \mapsto (x, -y)$ there is also another one: $(x, y) \mapsto (-x, iy)$.

- In fact, the segments of the rightmost branch between its cusp $(1, 0)$ and the inflection points of C consist of real points that are midpoints of pairs of conjugate complex points. Let us discuss it in detail.

  In order to find a point on C with tangent parallel to a given line $y = ax + b$ we have to solve the system of equations

$$\begin{cases} x^3 - 3x - y^2 = 0 \\ \frac{3(x^2-1)}{2y} = a \end{cases} \tag{2}$$

  Substituting y from the second equation to the first one we get an equation of order 4, which can have 2, 3 or 4 real solutions, depending on a. Geometrically speaking, the two limiting values of a are just the slopes of the tangents at the inflection points of C: there are four parallel tangents going in 'more vertical' directions but only two in 'more horizontal' ones. Nevertheless, even if there are only two real solutions to the system (2), there are also two such points (x, y) with complex conjugate coordinates, and their midpoint is real.

For a general cubic curve, however, the degree of the ACSS can be higher.

Theorem 3.1. *Let $X \subset \mathbf{R}^2$ be a general algebraic curve of degree 3 and Z its ACSS. Then* $\deg Z = 12$.

This means that performing the procedure of eliminating x_1, y_1, x_2, y_2 from the system (1) for a general cubic polynomial $f(x, y)$ gives $G(x, y) = 0$, where G is a divisible by f polynomial of degree 15, so the degree of G/f is 12.

Before sketching a proof of this result, let us show why the curve $x^3 - 3x - y^2 = 0$ is not general for this problem. One can easily see that the system (2) for a general cubic leads to an equation of degree 6, not 4, and indeed, a cubic can have as much as six parallel tangents. This is not the case for $x^3 - 3x - y^2 = 0$ because its projectivization $X^3 - 3XZ^2 - Y^2Z = 0$ has at infinity an inflection point $(0 : 1 : 0)$ with the tangent $Z = 0$. Then $(0 : 1 : 0)$ is a double solution of the projectivization of (2) for any a. One could also say informally that the line at infinity $Z = 0$ is parallel to any line $Y = aX + bZ$ because two lines are called parallel whenever their intersection lies at infinity.

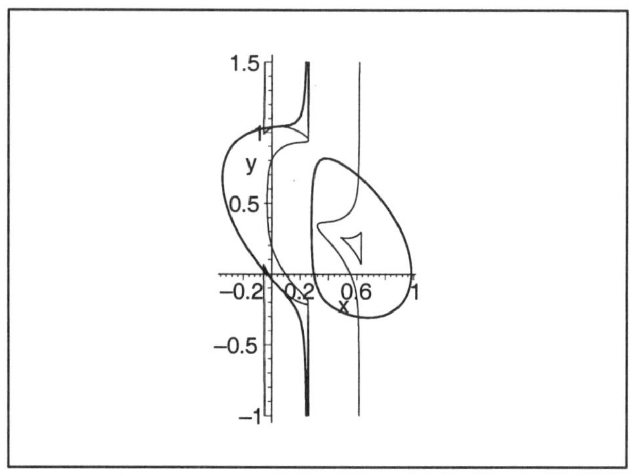

Fig. 3. ACSS of $(x^2 + xy + y^2)(x - \frac{1}{4}) + \frac{1}{100} = 0$

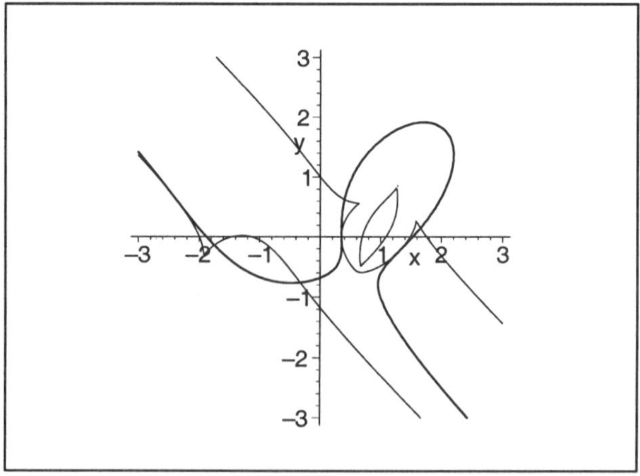

Fig. 4. ACSS of $x^3 + y^3 - 3xy - 3x + y + 1 = 0$

In Figs. 3 and 4 we see more examples of the ACSS of cubic curves. Please pay attention to the neighbourhoods of the inflection points. As a rule only one branch consists of midpoints of visible (real) chords.

3.2. *ACSS of an algebraic curve – another approach*

We intend to derive an algebraic condition for an arbitrary point $P = (x_0, y_0)$ to belong to the ACSS of the curve $C = \{f(x, y) = 0\}$. By translation invariance this happens whenever $(0, 0)$ belongs to the ACSS of the shifted curve $f_P(x, y) = f(x + x_0, y + y_0) = 0$. The coefficients of f_P are polynomials in x_0 and y_0, therefore it suffices to find a condition for $(0, 0)$ to belong to the ACSS of the curve $C = \{f(x, y) = 0\}$ algebraic in the coefficients of f.

However, the origin $(0, 0)$ belongs to the ACSS of C if and only if:

for some x and y both points (x, y) and $(-x, -y)$ belong to C and the respective tangents are parallel, or equivalently:

for some x and y the point (x, y) belongs both to C and to the symmetric curve $-C = \{f(-x, -y) = 0\}$ and the respective tangents coincide, which means exactly:

the curves C and $-C$ have at least one nontransversal intersection point.

We can also observe that the system

$$\begin{cases} f(x, y) = 0 \\ f(-x, -y) = 0 \end{cases} \tag{3}$$

is equivalent to

$$\begin{cases} f_e(x, y) = \frac{f(x,y)+f(-x,-y)}{2} = 0 \\ f_o(x, y) = \frac{f(x,y)-f(-x,-y)}{2} = 0 \end{cases} \tag{4}$$

where f_e and f_o denote the sums of monomials of even (respectively odd) degrees included in f.

In particular for a cubic equation $f = \sum_{i+j \leq 3} a_{ij} x^i y^j$ we obtain

$$\begin{cases} a_{20}x^2 + a_{11}xy + a_{02}y^2 + a_{00} = 0 \\ a_{30}x^3 + a_{21}x^2y + a_{12}xy^2 + a_{03}y^3 + a_{10}x + a_{01}y = 0 \end{cases}$$

We solve this system in the following steps:

We complete the square in the first equation, i.e. perform a linear change of coordinates in order to have $a_{11} = 0$. By abuse of notation the new x and y will have unchanged names.

We solve the first equation for y^2 obtaining $y^2 = Ax^2 + B$ and substitute $Ax^2 + B$ for y^2 and $(Ax^2 + B)y$ for y^3 in the second equation, obtaining

$$\begin{cases} y^2 = Ax^2 + B \\ Cx^3 + Dx^2y + Ex + Fy = 0 \end{cases}$$

(here $A, \ldots, F$ are some rational functions of the a_{ij}'s).

The second equation gives now

$$y = -\frac{Cx^2 + E}{Dx^2 + F}x$$

and from the first equation we have

$$\left(\frac{Cx^2 + E}{Dx^2 + F}\right)^2 x^2 = Ax^2 + B \tag{5}$$

which, when multiplied by the denominator, becomes a cubic equation in x^2. This equation has a multiple root if and only if its discriminant is 0, and that imposes a polynomial condition on $A, \ldots, F$, hence also on the a_{ij}'s.

The preceding sketch omits some details, but it shows in principle that we can derive not only the algebraic equation of the ACSS of any fixed cubic curve, but also the general formula transforming any set of 10 coefficients a_{ij} into the equation of the ACSS. Computations using Maple showed that this equation has in fact degree 12.

3.3. *Critical points of the ACSS*

From the above considerations it is clear that the ACSS has a critical point if and only if the system (3), or equivalently (4), has a solution of multiplicity at least 3, which gives an extra algebraic condition that the cubic equation in x^2, obtained from (5) should have a triple root rather than only a double one.

It is worth pointing out that the ACSS appeared in [3] under the name MPTL (mid-parallel-tangents locus) as an auxiliary set for studying one of generalizations of the axis of symmetry, namely the AESS (affine envelope symmetry set), which is the set of the centres of conics having two points of (at least) triple contact with the given curve (so called 3+3 conics). According to Propositions 2.4.7 and 2.4.9 of [3] the critical points of the ACSS are simultaneously critical points of the AESS and they are the centres of 3+3 conics with parallel tangents.

3.4. *CSS of an algebraic curve – preliminary remarks*

In this part we shall skip the details of computation and show only the general idea that one can treat the CSS in the same way as the ACSS in

3.2 and try to derive a condition for $(0,0)$ to belong to the CSS of the curve $C = \{f(x,y) = 0\}$.

According to [4] and [1], any point on the CSS of C divides its respective chord in the ratio equal to the ratio of the curvatures of C at the ends of this chord. In other words the origin $(0,0)$ belongs to the CSS of C if and only if there exists such a negative real k that the curves C and $kC = \{f(kx, ky) = 0\}$ have an intersection point with common tangents and curvatures, i.e. a triple intersection point (side remark: in the oval case there can be no such positive real $k \neq 1$).

This amounts to considering the system of equations:

$$\begin{cases} f(x,y) = 0 \\ f(kx, ky) = 0, \end{cases} \tag{6}$$

which unfortunately cannot be transformed to such a special form as (4) because of the existence of k. However if $\deg f = n$, then the following equivalent system

$$\begin{cases} k^n f(x,y) - f(kx, ky) = 0 \\ f(x,y) - f(kx, ky) = 0 \end{cases} \tag{7}$$

is simpler than (6) because the first equation has now degree $n-1$ and the second one has no constant term.

Generally speaking, there are several ways of checking whether two given plane curves have a triple intersection point. One of them requires equating the derivatives $y'(x)$ and $y''(x)$ for both curves computed by the Implicit Function Theorem; another method can be applied when one of the curves has a rational parametrization (this is the case for (7) when $n = 3$, since then the first equation has degree 2) – it involves substituting this parametrization to the equation to the other curve and writing the discriminant conditions in order to check if the resulting polynomial has a triple root.

Quite obviously, although we could in principle write the corresponding algebraic equations explicitly, they are much more complicated than in the ACSS case. Last but not least, we still have to eliminate k from these equations, and remember that there may exist nonreal numbers k for which the system (6) has a triple real solution (x,y).

It should be hoped, however, that for some special classes of curves the computations could simplify considerably – e.g. for the curves of constant width the ACSS coincides with the focal locus. These aspects will be dealt with in the further research.

1030

3.5. *Final remarks*

- A single point on the evolute corresponds to (a neighbourhood of) a single point on the original curve C, while one point on the CSS or ACSS reflects properties of two points on C.
- One algebraic equation usually defines a sum of several disjoint ovals, some of which may be convex and some nonconvex. The maximal number of ovals grows proportionally to the square of the degree of C, and therefore the maximal number of components of CSS and ACSS is asymptotically proportional to the fourth power of this degree.
- Finally, though the classical methods of algebraic geometry were used successfully in [5], one must be aware that some extra structure on $\mathbf{P}^2$ is necessary. For instance, in order to define parallel tangents and the midpoint of a segment we must fix a projective line $Z = 0$, and the evolute requires some notion of distance or orthogonality (in fact there exist also so called affine normals and affine evolutes, but we do not address them here).

The above remarks suggest that, although various types of symmetry sets arise quite naturally and find various applications e.g. in image recognition and processing, their detailed algebraic study encounters serious obstacles.

References

1. P. Giblin, P. Holtom, *The centre symmetry set*, in: Janeczko, Stanislaw (ed.) et al., Geometry and topology of caustics – CAUSTICS '98, Banach Cent. Publ. **50** (1999), 91–105
2. P.Giblin, V. Zakalyukin, *Singularities of centre symmetry sets*, Proc. London Math. Soc. (3) **90** (2005), 132–166
3. P. Holtom, *Affine-invariant symmetry sets*, PhD thesis, available at http://www.liv.ac.uk/~pjgiblin
4. S. Janeczko, *Bifurcations of the center of symmetry*, Geom. Dedicata **60** No.1 (1996), 9–16
5. F. Trifogli, *Focal loci of algebraic hypersurfaces: A general theory*, Geom. Dedicata **70** No.1 (1998), 1–26

Week 1 Participants

Introduction to Singularity Theory: 24–28th January 2005

Week 2 Participants

Singularities: 31th January –4th February 2005

Week 3 Participants
Applications of Singularities: 7–11th February 2005

Week 4 Participants

Young Researchers in Singularities: 14–18th February 2005

Week 5 Participants

Topology and Geometry of Singularities: 21–25th February 2005

PROGRAMS

Introduction to Singularity Theory
24-28th January 2005

5 courses of 5 hours each :

Jean-Paul Brasselet, Characteristic classes of singular varieties
Herwig Hauser Basic techniques for resolution of singularities
Anatoly Libgober Topology of the complements to hypersurfaces in projective space
David Trotman Subanalytic and semialgebraic sets: stratifications, equisingularity and metric properties

Vladimir Zakalyukin Lagrangian and Legendrian singularities

1042

Singularities
31th January - 4th February 2005

Monday 31th January

9:40-10:30 **Anatoly Libgober**, Topology of the complements and geometry of non normal crossings

11:00-11:50 **David Mond**, Linear free divisors and Lie groups

16:00-16:50 **Wim Veys**, Are principal value integrals birational invariants ?

17:00-17:50 **Grigory Mikhalkin**, Complex, real and tropical curves

Tuesday 1st February

8:30-9:20 **András Némethi**, On the classification of cuspidal rational projective plane curves

9:40-10:30 **Philippe Du Bois**, Dévissage de la forme de Seifert entière des germes de courbe plane à deux branches

11:00-11:50 **Paolo Aluffi**, Celestial integration and applications

16:00-16:50 **Laurentiu Maxim**, Alexander invariants of hypersurface complements

17:00-17:50 **Anne Frühbis-Krüger**, Algorithmic resolution of singularities from a practical point of view

18:00-18:50 **Andrzej Weber**, Pure homology of singular varietes

Wednesday 2nd February

8:30-9:20 **Richard Rimanyi**, Thom polynomials for quivers 9:40-10:30 **Lazlo Feher**, Relations among Thom Polynomials

11:00-11:50 **Mutsuo Oka**, Zariski pairs in sextics

Thursday 3rd February

8:30-9:20 **Alexandru Dimca**, Idaux d'adjonction et idaux des multiplicateurs

9:40-10:30 **Michel Vaquié**, Extension de valuation et polygone de Newton

11:00-11:50 **Adam Parusinski**, Recent results on blow-analytic equivalence of analytic function germs

16:00-16:50 **Inna Scherback**, Intersections of Schubert Cells, Tensor Products of sl_n-module and Critical Points of The Generating Function

17:00-17:50 **Piotr Mormul**, Small growth vectors: a Gödel-like encoding of the geometric classes of Goursat flags

Friday 4th February

8:30-9:20 **Lê Dung Trang** 9:40-10:30 **Fuensenta Aroca**, An extension of the Newton polyhedron construction to compute solutions of a Partial Differential Equation

11:00-11:50 **Jawad Snoussi**, The Nash modification and point blow-up on surfaces

14:00-14:50 **Andrew Du Plessis**, Projective hypersurfaces admitting a unipotent group action

Applications of Singularities
7-11th February 2005

Monday 7th February

10:00-10:50 **Shyuichi Izumiya**, Pseudo-spherical geometry in Minkowski space and Legendrian dualities

11:30-12:20 **Stanislaw Janeczko**, Integrable singularities of Dirac's generalized Hamiltonian dynamics

16:00-16:50 **Maria Carmen Romero-Fuster**, Generic contacts of surfaces in $I\!R^5$

17:00-17:50 **Hiroaki Terao**, Arrangements of hyperplanes and hypergeometric integrals

18:00-19:20 **Fernand Pelletier**, Stratifications en dimension infinie et applications

Tuesday 8th February

9:00-9:50 **Gert Vegter**, The Geometry of Resonance Tongues: A Singularity Theory Approach

10:00-10:50 **David Chillingworth**, Bifurcation from a 2-manifold: wavefronts and line congruences

11:30-12:20 **Mike Field**, Singularity and stratification theory applied to dynamical systems

16:00-16:50 **Jacques Furter** Bifurcations in elasticity : gradient, symmetry breaking structures and path formulation

17:00-17:50 **Isabel Labouriau**, Synchronization and desynchronization of coupled nerve cells

Wednesday 9th February

9:00-9:50 **Peter Giblin**, Recent work on symmetry sets and medial axes in 2 and 3 dimensions

10:00-10:50 **André Diatta**, Vanishing inflexions and vertices in 1-parameter families of plane curves

11:30-12:20 **James Damon**, Understanding the geometry of shapes in $I\!R^n$ via medial and skeletal structures

Thursday 10th February

9:00-9:50 **Ole Fogh Olsen**, Deep Structure Analysis of Images

10:00-10:50 **Xavier Pennec**, Statistical Computing on Riemannian Manifolds with Applications in Medical image Analysis
11:30-12:20 **Yannick Kergosien**, Singularities and genericity in Medical Imaging and Vision
16:00-16:50 **Alain Joets**, Optical singularities (caustics) as visualization technics. Application to liquid cristals
17:00-17:50 **Arlie Petters**, Singularities in gravitational lensing
18:00-19:20 **Dirk Siersma**, Shape and Morse Theory of the distance function

Friday 11th February

9:00-9:50 **Peter Donelan**, Singularities in Robotics
10:00-10:50 **Bernard Mourrain**, Meshing implicit algebraic surfaces
11:30-12:20 **Yosef Yomdin**, What is "Computational Singularity Theory" and what is it good for?

Young researchers in Singularities
14-18th February 2005

Monday 14th February

9:00-9:40 **Patrick Popescu-Pampu**, On the contact boundary of an isolated singularity

9:55-10:35 **José Ignacio Cogolludo**, Topology of real and complex line arrangements in $\mathbb{CP}^2$

11:00-12:20 **Dirk Siersma**, The topology of non-isolated singularities I

15:50-16:30 **Gil Guibert**, Motivic Milnor fiber and applications

17:00-17:40 **Johannes Nicaise**, Rigid geometry and the Monodrony conjecture

17:55-18:35 **Goulwen Fichou**, The corank is a blow-Nash invariant

Tuesday 15th February

9:00-9:40 **Didier D'Acunto**, Singularités de fonctions modérées : trajectoires de gradient et fonds de vallées

9:55-10:35 **Vincent Grandjean**, Gradient trajectories and singularity at infinity

11:00-12:20 **Walter Neumann**, Splice diagrams I

15:00-15:40 **Mariusz Zajac**, Centre symmetry sets and other affine invariants of algebraic sets

15:50-16:30 **Guillaume Valette**, Hardt's theorem : a bilipschitz version

17:00-17:40 **Rémi Soufflet**, Arc analycity of sub-analytuc functions and related topics

Wednesday 16th February

9:00-9:40 **Maria Alberich-Carramiñana**, Contractibility of exceptional curves

9:55-10:35 **Guillaume Rond**, Artin function and a counter-example to a conjecture of Spivakovsky

11:00-11:40 **Dmitry Kerner**, On enumeration of singular curves/hypersurfaces

11:50-12:30 **Wojciech Domitrz**, Volume-preserving diffeomorphims on germs of singular quasi-homogeneous varieties and $\mathcal{A}_\Omega$-equivalence of maps-germs

Thursday 17th February

9:00-9:40 **Pedro Gonzalez-Perez**, Harnack's deformations of real plane branch singularities and their amoebas

9:55-10:35 **Romain Bondil**, Multiplicities and vertical components

11:00-12:20 **Dirk Siersma**, The topology of non-isolated singularities II

15:00-15:40 **Christian Sevenheck**, Geometry and deformation theory of lagrangian singularities

15:50-16:30 **Nero Budur**, Applications of D-modules

17:00-17:40 **Tristan Torrelli**, Logarithmic comparison theorem and D-modules

17:55-18:35 **Céline Roucairol**, Irregularity of a differential system associated to two polynomials

Friday 18th February

9:00-9:40 **Ricardo Uribe-Vargas**, A new projective invariant associated to special parabolic points of surfaces and to swallowtails

9:55-10:35 **Arnaud Bodin**, Topology of families of complex polynomials

11:00-12:20 **Walter Neumann**, Splice diagrams II

15:00-15:40 **Evelia Garcia-Barroso**, The Lojasiewicz numbers and plane curve singularities

15:50-16:30 **Etienne Mann**, Correspondence between the orbifold cohomology of weighted projective spaces and singularity of a Laurent polynomial

Topology and Geometry of Singularities
21-25th February 2005

Monday 21th February

9:40-10:30 **Alejandro Melle**, Links and analytic invariants of superisolated singularities

11:00-11:50 **Stanislaw Janeczko**, Local symplectic invariants and relative Darboux Theorem

14h30-15:20 **Sabir Gusein-Zade**, Curves on rational surface singularities and Poincaré series

16:00-16:50 **Françoise Michel**, On the boundary of the Milnor Fiber for non isolated singularities of germs of surfaces

17:10-18:00 **Anne Pichon**, Real analytic germ from a singularity of normal surface

18:20-19:10 **Terry Gaffney**, The multiplicity polar theorem, and an application to the indices of vector fields on singular spaces

Tuesday 22th February

8:30-9:20 **Ignacio Luengo**, Open problems on plane rational curves

9:40-10:30 **Georges Comte**, Courbes intégrales de champs lipschitz

11:00-11:50 **Nicolas Dutertre**, Curvature integrals of the real Milnor fibre

14:30-15:20 **Marcio Soares**, Singularities of logarithmic foliations

16:00-16:50 **Patrice Orro**, Some properties of the SR geodesic distance

17:10-18:00 **Clément Caubel**, Contact structure and isolated singularities : some recent results

18:20-19:10 **Lev Birbrair**, Metric Theory of Real Surfaces

Wednesday 23th February

8:30-9:20 **Walter Neumann**, The Casson Invariant Conjecture

9:40-10:30 **Victor Goryunov**, Simple symmetric matrix singularities and the subgroups of Weyl groups A_μ, D_μ, E_μ

11:00-11:50 **Andrew du Plessis**, Fibration, stratification and resolution

Thursday 24th February

Special day for Jean-Paul Brasselet on his 60th Birthday

8:30-9:20 **Shoji Yokura**, Characteristic classes of symbolic algebraic varieties

9:40-10:30 **Jörg Schürman**, Motivic characteristic classes for singular spaces,

11:00-11:50 **Nicolae Teleman**, Sur le caractère de Chern

14:30-15:20 **André Legrand**, Singularités coniques isolées et caractère de Chern

16:00-16:50 **Markus Pflaum**, Towards an algebraic index theorem for symplectic orbifolds

17:15 **Concert : Anne Méhat** (violon) and **Ludger Kaup** (piano)

18:00 Apéritif

Birthday dinner

Friday 25th February

8:30-9:20 **José Seade**, Indices of vector fields on singular varieties

9:40-10:30 **Tatsuo Suwa**, Localization of characteristic classes and applications

11:00-11:50 **Lê Dũng Tráng**, Equisingularité sur les surfaces

14h30-15:20 **Gottfried Barthel**, Homologie des variétés toriques

16:00-16:50 **Bernard Teissier**, Valuations and toric geometry

LIST OF PARTICIPANTS

Haydée AGUILAR

Universidad Nacional Autónoma
de México
Cuernavaca, MEXICO
haydee@matcuer.unam.mx

Eric Dago AKEKE

Université de Provence
Marseille, FRANCE
akeke@cmi.univ-mrs.fr

Maria ALBERICH-CARRAMIÑANA

Universitat de Barcelona
Barcelona, SPAIN
maria.alberich@upc.edu

Lionel ALBERTI

École Normale Supérieure
de Cachan
Cachan, FRANCE
alberti@dptmaths.ens-cachan.fr

Paolo ALUFFI

Florida State University
Tallahassee, U.S.A.
aluffi@math.fsu.edu

Fuensanta AROCA

Universidad Nacional Autónoma
de México
Cuernavaca, MEXICO
fuen@matcuer.unam.mx

Gilles BAILLY-MAITRE

Université de La Rochelle
La Rochelle, FRANCE
gilles.baillymaitre@gmail.com

Gottfried BARTHEL

Universität Konstanz
Konstanz, GERMANY
Gottfried.Barthel@Uni-Konstanz.de

Lev BIRBRAIR

Universidade Federal do Ceará
Fortaleza, BRAZIL
lev@agt.uva.es

Rocio BLANCO

Universidad de Valladolid
Valladolid, SPAIN
rblanco@modulor.arq.uva.es

Vincent BLANLŒIL

Université de Strasbourg
Strasbourg, FRANCE
blanloeil@math.u-strasbg.fr

1052

Arnaud BODIN
Université Lille 1
Villeneuve d'Ascq, FRANCE
Arnaud.Bodin@math.univ-lille1.fr

Romain BONDIL
Université de Montpellier
Montpellier, FRANCE
bondil.romain@neuf.fr

Jean-Paul BRASSELET
Institut de Mathématiques de Luminy, CNRS
Marseille, FRANCE
jpb@iml.univ-mrs.fr

Gábor BRAUN
Eötvös Loránd Tudományegyetem
Budapest, HUNGARY
braung@renyi.hu

Thomas BRÉLIVET
Universidad de Valladolid
Valladolid, SPAIN
brelivet@agt.uva.es

Shirley BROMBERG
Universidad Autónoma Metropolitana
Mexico, MEXICO
stbs@xanum.uam.mx

Erwan BRUGALLE
Université Toulouse 3
Toulouse, FRANCE
brugalle@picard.ups-tlse.fr

Clemens BRUSCHEK
Universität Innsbruck
Innsbruck, AUSTRIA
csac8572@uibk.ac.at

Nero BUDUR
Johns Hopkins University
Baltimore, U.S.A.
nbudur@math.jhu.edu

Igor BURBAN
Universität Bonn
Bonn, GERMANY
burban@mpim-bonn.mpg.de

Pierrette CASSOU-NOGUÈS
Université Bordeaux I
Talence, FRANCE
cassou@math.u-bordeaux1.fr

Clément CAUBEL
Université Paris 7
Paris, FRANCE
caubel@math.jussieu.fr

Denis CHÉNIOT
Université de Provence
Marseille, FRANCE
cheniot@cmi.univ-mrs.fr

David CHILLINGWORTH	University of Southampton Southampton, UNITED KINGDOM drjc@maths.soton.ac.uk
José Luis CISNEROS	Universidad Nacional Autónoma de México Cuernavaca, MEXICO jlcm@matcuer.unam.mx
Helena COBO PABLOS	Universidad de Madrid Madrid, SPAIN hcobopab@mat.ucm.es
José Ignacio COGOLLUDO	Universidad de Zaragoza Zaragoza, SPAIN jicogo@unizar.es
Georges COMTE	Université de Nice Nice, FRANCE comte@math.unice.fr
Nuria CORRAL	Universidad de Vigo Pontevedra, SPAIN ncorral@uvigo.es
Didier D'ACUNTO	Università di Pisa Pisa, ITALY ddacu@univ-savoie.fr
James DAMON	University of North Carolina Chapel Hill, U.S.A. jndamon@math.unc.edu
Declan DAVIS	University of Liverpool Liverpool, UNITED KINGDOM d.davis@liverpool.ac.uk
André DIATTA	University of Liverpool Liverpool, UNITED KINGDOM adiatta@liv.ac.uk
Alexandru DIMCA	Université de Nice Nice, FRANCE dimca@math.unice.fr
DINH Si Tiep	Université de Savoie Le Bourget du Lac, FRANCE dinhsitiep@voila.fr
Vladimir DOBRYNSKIY	University of Kiev Kiev, UKRAINE dobry@imp.kiev.ua

Wojciech DOMITRZ

Politechnika Warszawska
Warszawa, POLAND
domitrz@mini.pw.edu.pl

Peter DONELAN

University of Wellington
Wellington, NEW ZEALAND
peter.donelan@vuw.ac.nz

Philippe DU BOIS

Université d'Angers
Angers, FRANCE
pdubois@univ-angers.fr

Andrew DU PLESSIS

Aarhus Universitet
Aarhus, DENMARK
matadp@imf.au.dk

Delphine DUPONT

Université de Nice
Nice, FRANCE
ddupont@math.unice.fr

Nicolas DUTERTRE

Université de Provence
Marseille, FRANCE
dutertre@cmi.univ-mrs.fr

Wolfgang EBELING

Universität Hannover
Hannover, GERMANY
ebeling@math.uni-hannover.de

Mohammed EL AMRANI

Université d'Angers
Angers, FRANCE
mohammed.elamrani@univ-angers.fr

László FEHÉR

Eötvös Loránd Tudományegyetem
Budapest, HUNGARY
lfeher@renyi.hu

Javier FERNÁNDEZ DE BOBADILLA

Universiteit Utrecht
Utrecht, NETHERLANDS
bobadilla@math.uu.nl

Goulwen FICHOU

Université de Rennes
Rennes, FRANCE
goulwen.fichou@univ-rennes1.fr

Michael FIELD

University of Houston
Houston, U.S.A.
mf@uh.edu

Karl Heinz FIESELER

Uppsala Universitet
Uppsala, SWEDEN
khf@math.uu.se

Sergey FINASHIN

Orta Doğu Teknik Üniversitesi
Ankara, TURKEY
serge@metu.edu.tr

Anne FRÜHBIS-KRÜGER

Universität Kaiserslautern
Kaiserslautern, GERMANY
anne@mathematik.uni-kl.de

Jacques-Élie FURTER

Brunel University
Uxbridge, UNITED KINGDOM
mastjef@brunel.ac.uk

Terence GAFFNEY

Northeastern University
Boston, U.S.A.
gaff@neu.edu

Mario GARCÍA FERNÁNDEZ

Universitat de València
Burjassot, SPAIN
garferma@alumni.uv.es

Evelia GARCÍA BARROSO

Universidad de La Laguna
La Laguna, SPAIN
ergarcia@ull.es

Peter GIBLIN

University of Liverpool
Liverpool, UNITED KINGDOM
pjgiblin@liv.ac.uk

Arturo GILES

Universidad Nacional Autónoma
de México
Cuernavaca, MEXICO
arturo@matcuer.unam.mx

Paweł GOLDSTEIN

Universytet Warszawski
Warszawa, POLAND
goldie@mimuw.edu.pl

Pedro Daniel GONZÁLEZ PÉREZ

Universidad de Madrid
Madrid, SPAIN
pgonzalez@mat.ucm.es

Manuel GONZÁLEZ VILLA

Universidad de Madrid
Madrid, SPAIN
mgv@mat.ucm.es

Victor GORYUNOV

University of Liverpool
Liverpool, UNITED KINGDOM
goryunov@liv.ac.uk

Vincent GRANDJEAN

University of Bath
Bath, UNITED KINGDOM
cssvg@bath.ac.uk

Jean-Michel GRANGER — Université d'Angers, Angers, FRANCE — granger@univ-angers.fr

Gil GUIBERT — Lyce Chaptal, Paris, FRANCE — guibert9@wanadoo.fr

Sabir GUSEIN-ZADE — Moscow State University, Moscow, RUSSIA — sabir@mccme.ru

Helmut HAMM — Universität Münster, Münster, GERMANY — hamm@math.uni-muenster.de

Herwig HAUSER — Universität Innsbruck, Innsbruck, AUSTRIA — herwig.hauser@uibk.ac.at

Fernando HERNANDO — Universidad de Valladolid, Valladolid, SPAIN — hernando@agt.uva.es

Michel HILSUM — Université Paris 7, Paris, FRANCE — hilsum@math.jussieu.fr

Kevin HOUSTON — Leeds University, Leeds, UNITED KINGDOM — k.houston@leeds.ac.uk

Dimce IVANOVSKI — Université Toulouse 3, Toulouse, FRANCE — divanovski@yahoo.fr

Shuichi IZUMIYA — Hokkaido University, Sapporo, JAPAN — izumiya@math.sci.hokudai.ac.jp

Stanislaw JANECZKO — Instytut Matematyczny Polskiej Akademii Nauk, Warszawa, POLAND — janeczko@impan.gov.pl

Alain JOETS — Université Paris-Sud, Orsay, FRANCE — joets@lps.u-psud.fr

Ludger KAUP — Universität Konstanz
Konstanz, GERMANY
Ludger.Kaup@Uni-Konstanz.de

Yannick KERGOSIEN — Université de Cergy-Pontoise
Cergy-Pontoise, FRANCE
yannick.kergosien@libertysurf.fr

Dmitry KERNER — Tel Aviv University
Tel Aviv, ISRAEL
kernerdm@post.tau.ac.il

Balázs KÔMÛVES — Közép-Európai Egyetem
Budapest, HUNGARY
komuves@renyi.hu

Nikolai KRYLOV — Universität Bremen
Bremen, GERMANY
n.krylov@iu-bremen.de

Isabel LABOURIAU — Universidade do Porto
Porto, PORTUGAL
islabour@fc.up.pt

Mustapha LAHYANE — Universidad de Valladolid
Valladolid, SPAIN
lahyane@agt.uva.es

Peter LAMBERSON — Columbia University
New York, U.S.A.
pjl2002@columbia.edu

Radu LAZA — Columbia University
New York, U.S.A.
laza@math.columbia.edu

LÊ Dũng Tráng — International Centre for Theoretical Physics
Trieste, ITALY
ledt@ictp.trieste.it

André LEGRAND — Université Toulouse 3
Toulouse, FRANCE
legrand@picard.ups-tlse.fr

Ann LEMAHIEU — Katholieke Universiteit Leuven
Leuven, BELGIUM
ann.lemahieu@wis.kuleuven.ac.be

Andrzej LENARCIK — Politechnika Świętokrzyska w Kielcach
Kielce, POLAND
ztpal@tu.kielce.pl

Anatoly LIBGOBER

University of Illinois
Chicago, U.S.A.
libgober@math.uic.edu

Gábor LIPPNER

Eötvös Loránd Tudományegyetem
Budapest, HUNGARY
lipi@cs.elte.hu

Michael LÖNNE

Universität Hannover
Hannover, GERMANY
loenne@math.uni-hannover.de

Ursula LUDWIG

Universität Freiburg
Freiburg, GERMANY
ursula.ludwig@math.uni-freiburg.de

Ignacio LUENGO

Universidad de Madrid
Madrid, SPAIN
iluengo@mat.ucm.es

Daniela Anca MĂCINIC

Institutul de Matematică "Simion Stoilow"
Bucureşti, ROUMANIA
Anca.Macinic@imar.ro

Etienne MANN

Université de Strasbourg
Strasbourg, FRANCE
mann@math.u-strasbg.fr

Matilde MARCOLLI

Max-Planck-Institut fr Mathematik
Bonn, GERMANY
marcolli@mpim-bonn.mpg.de

Thomas MARKWIG

Universität Kaiserslautern
Kaiserslautern, GERMANY
keilen@mathematik.uni-kl.de

Sergio MARTÍNEZ

Universidad de Zaragoza
Zaragoza, SPAIN
sergiomj@unizar.es

Rodrigo MARTINS

Universidad de São Paulo
São Carlos, BRAZIL
rmartins@icmc.usp.br

Laurentiu MAXIM

University of Pennsylvania
Philadelphia, U.S.A.
lmaxim@math.upenn.edu

Alejandro MELLE HERNÁNDEZ

Universidad de Madrid
Madrid, SPAIN
amelle@mat.ucm.es

Joël MERKER — Laboratoire d'Analyse, Topologie et Probabilités, CNRS, Marseille, FRANCE — merker@cmi.univ-mrs.fr

Françoise MICHEL — Université Toulouse 3, Toulouse, FRANCE — fmichel@picard.ups-tlse.fr

Grigory MIKHALKIN — University of Toronto, Toronto, CANADA — mikha@math.toronto.edu

Małgorzata MIKOSZ — Politechnika Warszawska, Warszawa, POLAND — emmikosz@adam.mech.pw.edu.pl

David MOND — University of Warwick, Coventry, UNITED KINGDOM — mond@maths.warwick.ac.uk

Simone MORAES — Universidade Federal de Viçosa, Viçosa, BRAZIL — smoraes@ufv.br

Heidi Camilla MORK — Universitetet i Oslo, Oslo, NORWAY — heidisu@math.uio.no

Piotr MORMUL — Universytet Warszawski, Warszawa, POLAND — mormul@mimuw.edu.pl

Bernard MOURRAIN — Institut National de Recherche en Informatique et en Automatique, Sophia Antipolis, FRANCE — mourrain@sophia.inria.fr

Julio-José MOYANO-FERNÁNDEZ — Universidad de Valladolid, Valladolid, SPAIN — moyano@agt.uva.es

Claudio MUROLO — Université de Provence, Marseille, FRANCE — murolo@cmi.univ-mrs.fr

András NÉMETHI — Rényi Alfréd Matematikai Kutatóintézet, Budapest, HUNGARY — nemethi@renyi.hu

Orlando NETO — Universidade de Lisboa, Lisboa, PORTUGAL — orlando@ptmat.fc.ul.pt

Walter NEUMANN — Columbia University, New York, U.S.A. — neumann@math.columbia.edu

NGUYÊN Viêt Anh — Max-Planck-Institut fr Mathematik, Bonn, GERMANY — nguyen@mpim-bonn.mpg.de

Johannes NICAISE — Katholieke Universiteit Leuven, Leuven, BELGIUM — johannes.nicaise@wis.kuleuven.ac.be

Alicia NIETO REYES — University of Warwick, Coventry, UNITED KINGDOM — alicia@maths.warwick.ac.uk

Mounir NISSE — Université Paris 6, Paris, FRANCE — nisse@math.jussieu.fr

Aleksandra NOWEL — Uniwersytet Gdański, Gdańsk, POLAND — olanowel@math.univ.gda.pl

Oliver OBLANCA — Universidad de Valladolid, Valladolid, SPAIN — oblanca@agt.uva.es

Mutsuo OKA — Tokyo Metropolitan University, Tokyo, JAPAN — oka@comp.metro-u.ac.jp

Ole Fogh OLSEN — IT-Universitetet i København, København, DENMARK — fogh@itu.dk

Patrice ORRO — Université de Savoie, Le Bourget du Lac, FRANCE — orro@univ-savoie.fr

Raúl OSET SINHA — Universitat de València, Burjassot, SPAIN — soyperrymason@hotmail.com

Adam PARUSIŃSKI — Université d'Angers, Angers, FRANCE — adam.parusinski@univ-angers.fr

Helge Møller PEDERSEN — Aarhus Universitet, Aarhus, DENMARK — helge@imf.au.dk

Fernand PELLETIER — Université de Savoie, Le Bourget du Lac, FRANCE, pelletier@univ-savoie.fr

Xavier PENNEC — Institut National de Recherche en Informatique et en Automatique, Sophia Antipolis, FRANCE, xavier.pennec@sophia.inria.fr

Arlie PETTERS — Duke University, Durham, U.S.A., petters@math.duke.edu

Markus PFLAUM — Johann Wolfgang Goethe Universität, Frankfurt am Main, GERMANY, pflaum@math.uni-frankfurt.de

Anne PICHON — Université de la Méditerranée, Marseille, FRANCE, pichon@iml.univ-mrs.fr

Rafał PIERZCHAŁA — Uniwersytet Jagielloński w Krakowie, Kraków, POLAND, Rafal.Pierzchala@im.uj.edu.pl

Camille PLÉNAT — Université du Mans, Le Mans, FRANCE, camille.plenat@univ-lemans.fr

Dorin POPESCU — Universitatea din Bucureşti, Bucureşti, ROUMANIA, dorin.popescu@uni-due.de

Patrick POPESCU-PAMPU — Université Paris 7, Paris, FRANCE, ppopescu@math.jussieu.fr

Anna PRATOUSSEVITCH — Universität Bonn, Bonn, GERMANY, anna@math.uni-bonn.de

Serge RANDRIAMBOLOLONA — Université de Savoie, Le Bourget du Lac, FRANCE, srand@univ-savoie.fr

Carine REYDY — Université Bordeaux 1, Talence, FRANCE, Carine.Reydy@math.u-bordeaux1.fr

Richárd RIMÁNYI — University of North Carolina, Chapel Hill, U.S.A., rimanyi@email.unc.edu

Nicholas ROBBINS

Duke University
Durham, U.S.A.
robbins@math.duke.edu

Ana RODRIGUES

Universidade do Porto
Porto, PORTUGAL
ana.rodrigues@fc.up.pt

Beatriz RODRÍGUEZ GONZÁLEZ

Universidad de Sevilla
Sevilla, SPAIN
rgbea@algebra.us.es

María Carmen ROMERO FUSTER

Universitat de València
Burjassot, SPAIN
carmen.romero@uv.es

Guillaume ROND

Université Toulouse 3
Toulouse, FRANCE
rond@picard.ups-tlse.fr

Céline ROUCAIROL

Université de Nice
Nice, FRANCE
roucair@math.unice.fr

Nermin SALEPÇİ

Orta Doğu Teknik Üniversitesi
Ankara, TURKEY
nermins@metu.edu.tr

Esther SANABRIA-CODESAL

Universitat Politècnica de València
València, SPAIN
esanabri@mat.upv.es

Luis SANHERMELANDO

Universitat de València
Burjassot, SPAIN
dep.geometria.i.topologia@uv.es

Jan SCHEPERS

Katholieke Universiteit Leuven
Leuven, BELGIUM
jan.schepers@wis.kuleuven.ac.be

Inna SCHERBAK

Tel Aviv University
Tel Aviv, ISRAEL
scherbak@post.tau.ac.il

Jörg SCHÜRMANN

Universität Münster
Münster, GERMANY
jschuerm@math.uni-muenster.de

José SEADE

Universidad Nacional Autónoma de México
Cuernavaca, MEXICO
jseade@matcuer.unam.mx

Christian SEVENHECK — Universität Mannheim, Mannheim, GERMANY — Christian.Sevenheck@uni-mannheim.de

Dirk SIERSMA — Universiteit Utrecht, Utrecht, NETHERLANDS — siersma@math.uu.nl

Alexandre SINE — Université d'Angers, Angers, FRANCE — Adam.Parusinski@univ-angers.fr

Dana SKRBO — Universitetet i Oslo, Oslo, NORWAY — danaskr@math.uio.no

Jawad SNOUSSI — Universidad Nacional Autónoma de México, Cuernavaca, MEXICO — jsnoussi@matcuer.unam.mx

Márcio SOARES — Universidade Federal de Minas Gerais, Belo Horizonte, BRAZIL — msoares@mat.ufmg.br

Rémi SOUFFLET — Université Lyon 1, Villeurbanne, FRANCE — soufflet@igd.univ-lyon1.fr

Anna STASICA — Uniwersytet Jagielloński w Krakowie, Kraków, POLAND — anna.stasica@im.uj.edu.pl

Tatsuo SUWA — Niigata University, Niigata, JAPON — suwa@ie.niigata-u.ac.jp

Aviva SZPIRGLAS — Université de Poitiers, Chasseneuil, FRANCE — aviva@math.univ-poitiers.fr

Jean-Pierre TECOURT — Institut National de Recherche en Informatique et en Automatique, Sophia Antipolis, FRANCE — Jean-Pierre.Tecourt@sophia.inria.fr

Bernard TEISSIER — Institut de Mathématiques de Jussieu, CNRS, Paris, FRANCE — teissier@math.jussieu.fr

Nicolae TELEMAN — Università Politecnica delle Marche, Ancona, ITALY — teleman@dipmat.univpm.it

Hiroaki TERAO
Tokyo Metropolitan University
Tokyo, JAPAN
hterao@comp.metro-u.ac.jp

Tristan TORRELLI
Universidad de Valladolid
Valladolid, SPAIN
torrelli@math.unice.fr

David TROTMAN
Université de Provence
Marseille, FRANCE
trotman@cmi.univ-mrs.fr

Ricardo URIBE-VARGAS
Collège de France
Paris, FRANCE
uribe@math.jussieu.fr

Guillaume VALETTE
Uniwersytet Jagielloński w Krakowie
Kraków, POLAND
guillaume.valette@im.uj.edu.pl

Michel VAQUIÉ
Université Toulouse 3
Toulouse, FRANCE
vaquie@picard.ups-tlse.fr

Gert VEGTER
Rijksuniversiteit Groningen
Groningen, NETHERLANDS
G.Vegter@cs.rug.nl

Willem VEYS
Katholieke Universiteit Leuven
Leuven, BELGIUM
wim.veys@wis.kuleuven.ac.be

Dominique WAGNER
Universität Innsbruck
Innsbruck, AUSTRIA
csac8698@uibk.ac.at

Andrzej WEBER
Uniwersytet Warszawski
Warszawa, POLAND
aweber@mimuw.edu.pl

Claude WEBER
Université de Genève
Genève, SWITZERLAND
Claude.Weber@math.unige.ch

Shoji YOKURA
University of Kagoshima
Kagoshima, JAPAN
yokura@sci.kagoshima-u.ac.jp

Yosef YOMDIN
Weizmann Institute
Rehovot, ISRAEL
yosef.yomdin@weizmann.ac.il

Mariusz ZAJĄC   Politechnika Warszawska
                       Warszawa, POLAND
                       zajac@mini.pw.edu.pl

Vladimir ZAKALYUKIN   Moscow University
                       Moscou, RUSSIA
                       zakalyu@mail.ru